Quantum Neural Computation

T0134740

International Series on
INTELLIGENT SYSTEMS, CONTROL, AND AUTOMATION:
SCIENCE AND ENGINEERING

VOLUME 40

For other titles published in this series, go to
www.springer.com/series/6259

Vladimir G. Ivancevic · Tijana T. Ivancevic

Quantum Neural Computation

Dr. Vladimir G. Ivancevic
Department of Defence
Defence Science &
Technology Organisation
(DSTO)
75 Labs.
Edinburgh SA 5111
Australia
vladimir.ivancevic@dsto.defence.gov.au

Dr. Tijana T. Ivancevic
School of Electrical &
Information Engineering
University of South Australia
Mawson Lakes Campus
Adelaide SA 5095
Australia
tijana.ivancevic@unisa.edu.au

ISBN 978-90-481-3349-9   e-ISBN 978-90-481-3350-5
DOI 10.1007/978-90-481-3350-5
Springer Dordrecht Heidelberg London New York

Library of Congress Control Number: 2009939613

Printed on acid-free paper

Springer is part of Springer Science+Business Media (www.springer.com)

Dedicated to Nick-Nitya, Atma and Kali

Preface

Quantum Neural Computation is a graduate-level monographic textbook. It presents a comprehensive introduction, both non-technical and technical, into modern quantum neural computation. Classical computing systems perform classical computations (i.e., Boolean operations, such as AND, OR, NOT gates) using devices that can be described classically (e.g., MOSFETs). On the other hand, quantum computing systems perform classical computations using quantum devices (quantum dots), that is devices that can be described only using quantum mechanics. Any information transfer between such computing systems involves a state measurement. This book describes this information transfer at the edge of classical and quantum chaos and turbulence, where mysterious quantum-mechanical linearity meets even more mysterious brain's nonlinear complexity, in order to perform a super-high-speed and error-free computations. This monograph describes a crossroad between quantum field theory, brain science and computational intelligence.

Quantum Neural Computation has six Chapters and Appendix. The Introduction gives a glimpse of what is to come later in the book, mostly various forms of quantum computation, quantum neural networks and quantum brain, together with modern adaptive path-integral methods. Chapter 2 gives a modern review of classical neurodynamics, including brain physiology, biological and artificial neural networks, synchronization, spike neural nets and wavelet resonance, motor control and learning. Chapter 3 presents a modern review of quantum physics, including quantum mechanics and quantum field theory (mostly Feynman pat-integral-based), as well as both Abelian and non-Abelian gauge theories (with their pat-integral quantizations). Chapter 4 presents several fields from nonlinear dynamics that are related to quantum neural computation, including classical and quantum chaos, turbulence and solitons, with the special treatment to *nonlinear Schrödinger equation* (NLS, the core model of quantum neural networks, in which quantum superposition meets neural nonlinearity). Chapter 5 gives a personalized review (mostly based on the authors' own papers) of the current research in quantum-brain and quantum-mind. Chapter 6 presents a review of quantum information, quantum game theory, quantum computation and classical electronic for quantum computing. The Appendix gives a brief review of some mathematical and physiological concepts (necessary for comprehensive reading of the book), followed by classical computational tools used in quantum neural computation.

The objective of *Quantum Neural Computation* is to provide a serious reader with a serious scientific tool that will enable them to actually *perform* a competitive research in the rapidly-growing field of quantum neural computation. The monograph includes a very comprehensive bibliography on the subject and a detailed index.

Target readership for *Quantum Neural Computation* includes all researchers and students of complex, classical + quantum nonlinear systems (in computer science, physics, mathematics, engineering, medicine, chemistry, biology, psychology, sociology, economics, etc.), working in industry, clinics and academia.

Adelaide   V. Ivancevic
May 2009   T. Ivancevic

Acknowledgments

The authors wish to thank Land Operations Division, Defence Science & Technology Organisation, Australia, for the support in developing the *Human Biodynamics Engine* (HBE) and crowd modelling simulator. In particular, we thank Dr. Darryn Reid, LOD–LRR Task Leader, for his support.

We also express our gratitude to *Springer* book series *Intelligent Systems, Control and Automation: Science and Engineering* and especially to the Editor, Professor Spyros Tzafestas.

Contents

1 Introduction 1
  1.1 Neurodynamics 2
  1.2 Quantum Computation 3
  1.3 Discrete Quantum Computers 4
  1.4 Topological Quantum Computers 14
  1.5 Computation at the Edge of Chaos and Quantum Neural
      Networks 18
  1.6 Adaptive Path Integral: An ∞-Dimensional QNN 22
      1.6.1 Computational Partition Function 22
      1.6.2 From Thermodynamics to Quantum Field Theory 24
      1.6.3 ∞-Dimensional QNNs 24
  1.7 Brain Topology vs. Small-World Topology 26
  1.8 Quantum Brain and Mind 30
      1.8.1 Connectionism, Control Theory and Brain Theory 30
      1.8.2 Neocortical Biophysics 31
      1.8.3 Quantum Neurodynamics 35
      1.8.4 Bi-Stable Perception and Consciousness 37
  1.9 Notational Conventions 40

2 Brain and Classical Neural Networks 43
  2.1 Human Brain 43
      2.1.1 Basics of Brain Physiology 50
      2.1.2 Modern 3D Brain Imaging 70
  2.2 Biological versus Artificial Neural Networks 78
      2.2.1 Common Discrete ANNs 80
      2.2.2 Common Continuous ANNs 98
  2.3 Synchronization in Neurodynamics 106
      2.3.1 Phase Synchronization in Coupled Chaotic Oscillators .106
      2.3.2 Oscillatory Phase Neurodynamics 109
      2.3.3 Kuramoto Synchronization Model 112

        2.3.4 Lyapunov Chaotic Synchronization 113
    2.4 Spike Neural Networks and Wavelet Resonance 114
        2.4.1 Ensemble Neuron Model 117
        2.4.2 Wavelet Neurodynamics 119
        2.4.3 Wavelets of Epileptic Spikes 129
    2.5 Human Motor Control and Learning 133
        2.5.1 Motor Control 134
        2.5.2 Human Memory 138
        2.5.3 Human Learning 140
        2.5.4 Spinal Musculo-Skeletal Control 142
        2.5.5 Cerebellum and Muscular Synergy 145

3 **Quantum Theory Basics** 151
    3.1 Basics of Non-Relativistic Quantum Mechanics 151
        3.1.1 Soft Introduction to Quantum Mechanics 152
        3.1.2 Quantum States and Operators 157
        3.1.3 The Tree Standard Quantum Pictures 163
        3.1.4 Dirac's Probability Amplitude and Perturbation 165
        3.1.5 State-Space for n Non-Relativistic Quantum Particles . . 168
        3.1.6 Quantum Fourier Transform 170
    3.2 Introduction to Quantum Fields 171
        3.2.1 Amplitude, Relativistic Invariance and Causality 171
        3.2.2 Gauge Theories 173
        3.2.3 Free and Interacting Field Theories 176
        3.2.4 Dirac's Quantum Electrodynamics (QED) 177
        3.2.5 Abelian Higgs Model 179
        3.2.6 Topological Quantum Computation 181
    3.3 The Feynman Path Integral 188
        3.3.1 The Action-Amplitude Formalism 188
        3.3.2 Correlation Functions and Generating Functional 191
        3.3.3 Quantization of the Electromagnetic Field 193
        3.3.4 Wavelet-Based QFT 194
    3.4 The Path-Integral TQFT 199
        3.4.1 Schwarz-Type and Witten-Type Theories 199
        3.4.2 Hodge Decomposition Theorem 201
        3.4.3 Hodge Decomposition and Chern–Simons Theory 204
    3.5 Non-Abelian Gauge Theories 207
        3.5.1 Introduction to Non-Abelian Theories 207
        3.5.2 Yang–Mills Theory 207
        3.5.3 Quantization of Yang–Mills Theory 212
        3.5.4 Basics of Conformal Field Theory (CFT) 214

4 **Spatio-Temporal Chaos, Solitons and NLS** 219
    4.1 Reaction–Diffusion Processes and Ricci Flow 219
        4.1.1 Bio-Reaction–Diffusion Systems 223

4.1.2   Reactive Neurodynamics 236
4.1.3   Dissipative Evolution Under the Ricci Flow 246
4.2   Turbulence and Chaos in PDEs 258
4.3   Quantum Chaos and Its Control......................... 266
4.3.1   Quantum Chaos vs. Classical Chaos 271
4.3.2   Optimal Control of Quantum Chaos 280
4.4   Solitons 288
4.4.1   Short History of Solitons 288
4.4.2   Lie–Poisson Bracket 309
4.4.3   Solitons and Muscular Contraction 310
4.5   Dispersive Wave Equations and Stability of Solitons 312
4.5.1   KdV Solitons 321
4.5.2   The Inverse Scattering Approach 323
4.6   Nonlinear Schrödinger Equation (NLS) 327
4.6.1   Cubic NLS 327
4.6.2   Nonlinear Wave and Schrödinger Equations 331
4.6.3   Physical NLS-Derivation 335
4.6.4   A Compact Attractor for High-Dimensional NLS 337
4.6.5   Finite-Difference Scheme for NLS................... 345
4.6.6   Method of Lines for NLS 345

5   Quantum Brain and Cognition............................. 349
5.1   Biochemistry of Microtubules 349
5.2   Kink Soliton Model of MT-Dynamics 351
5.3   Fractal Neurodynamics................................. 353
5.3.1   Open Liouville Equation 355
5.4   Dissipative Quantum Brain Model 360
5.5   QED Brain Model 366
5.6   Stochastic NLS-Filtering and Robotic Eye Tracking 373
5.7   QNN-Based Neural Motor Control....................... 376
5.7.1   Spinal Reflex Control of Biodynamics 378
5.7.2   Local Muscle-Joint Mechanics...................... 379
5.7.3   Cerebellum: Adaptive Path-Integral Comparator 383
5.8   Quantum Cognition in the Life Space Foam................ 388
5.8.1   Classical versus Quantum Probability 390
5.8.2   The Life Space Foam 395
5.8.3   Geometric Chaos and Topological Phase Transitions ... 402
5.8.4   Joint Action of Several Agents 408
5.8.5   Chaos and Bernstein–Brooks Adaptation 411
5.9   Quantum Cognition in Crowd Dynamics................... 413
5.9.1   Cognition and Crowd Behavior..................... 414
5.9.2   Generic Three-Step Crowd Behavioral Dynamics 416
5.9.3   Formal Individual, Aggregate and Crowd dynamics 417
5.9.4   Crowd Entropy, Chaos and Phase Transitions 430
5.9.5   Crowd Ricci Flow and Perelman Entropy............. 430

      5.9.6   Chaotic Inter-Phase in Crowd Dynamics 433

      5.9.7   Crowd Phase Transitions 434

6   Quantum Information, Games and Computation 437

   6.1   Quantum Information and Computing 437

      6.1.1   Entanglement, Teleportation and Information 437

      6.1.2   The Circuit Model for Quantum Computers 440

      6.1.3   Elementary Quantum Algorithms.................... 441

   6.2   Quantum Games 442

      6.2.1   Quantum Strategies 444

      6.2.2   Quantum Games 447

      6.2.3   Two-Qubit Quantum Games 449

      6.2.4   Quantum Cryptography and Quantum Gambling 460

      6.2.5   Formal Quantum Games 467

   6.3   Hardware for Quantum Computers 478

      6.3.1   Josephson Effect and Pendulum Analog 482

      6.3.2   Dissipative Josephson Junction..................... 484

      6.3.3   Josephson Junction Ladder (JJL) 489

      6.3.4   Synchronization in Arrays of Josephson Junctions 498

   6.4   Topological Machinery for Quantum Computation 511

      6.4.1   Non-Abelian Anyons............................. 512

      6.4.2   Emergent Anyons 519

   6.5   Option Price Modeling Using Quantum Neural Computation .. 551

      6.5.1   Bidirectional, Spatio-Temporal, Complex-Valued

            Associative Memory Machine 554

7   Appendix: Mathematical and Computational Tools 559

   7.1   Meta-Language of Categories and Functors 559

      7.1.1   Maps 559

      7.1.2   Categories 567

      7.1.3   Functors 570

      7.1.4   Natural Transformations 573

      7.1.5   Limits and Colimits 575

      7.1.6   Adjunction 576

      7.1.7 n-Categories and n-Functors 578

      7.1.8   Topological Structure of n-Categories 583

   7.2   Frequently Used Mathematical Concepts 586

      7.2.1   Groups and Related Algebraic Structures.............. 586

      7.2.2   Manifolds, Bundles and Lie Groups 590

      7.2.3   Unitary Matrix and Group 594

      7.2.4   Differential Forms and Stokes Theorem............... 600

      7.2.5   Symmetry Breaking and Partition Function 604

      7.2.6   Basics of Kalman Filtering 607

      7.2.7   Basics of Wavelet Transforms 619

      7.2.8   Basic of Nonlinear Dynamics and Chaos Theory 639

        7.2.9   Basics of Nash's Game Theory 713
    7.3   Frequently Used Computational Tools 729
        7.3.1   Basic Numerical Algorithms 729
        7.3.2   Numerical Integration of Functions 734
        7.3.3   Numerical Integration of ODEs 738
        7.3.4   Vector-Field and Lyapunov Function 755
        7.3.5   Basics of Qualitative Dynamics 759
        7.3.6   Kuramoto's Neural Model 761
        7.3.7   Boundary Value Problems and PDEs 767
        7.3.8   Evolution PDEs in $Mathematica^{TM}$ 788
        7.3.9   Fourier Transforms 791
        7.3.10 Sophisticated Random Algorithms 800
        7.3.11 Sophisticated Integration Algorithms 810

References 841

Index 913

1

Introduction

In this Introductory Chapter we give a glimpse of what is to come in later the book, mostly various forms of quantum computation, quantum neural networks[1] and quantum brain, together with modern adaptive path-integral methods.

---

[1] Roughly speaking, *artificial neural network* (ANN) is a system loosely modeled on the human brain (see, e.g. [II07b]). The field goes by many names, such as *connectionism, parallel distributed processing, neuro-computing*, natural intelligent systems, *machine learning* algorithms and ANNs. It is an attempt to simulate within specialized hardware or sophisticated software, the multiple layers of simple processing elements called neurons. Each neuron is linked to certain of its neighbors with varying coefficients of connectivity that represent the strengths of these connections. Learning is accomplished by adjusting these strengths to cause the overall network to output appropriate results.

Although currently there is a large variety of models for ANNs, they all share eight major aspects (for technical details, see Sect. 2.2 below): (i) A set of processing units, or 'neurons', represented by a set of integers; (ii) An activation for each unit, represented by a vector of time-dependent functions; (iii) An output function for each unit, represented by a vector of functions on the activations; (iv) A pattern of connectivity among units, represented by a matrix of real numbers indicating connection strength; (v) A propagation rule spreading the activations via the connections, represented by a function on the output of the units; (vi) An activation rule for combining inputs to a unit to determine its new activation, represented by a function on the current activation and propagation; (vii) A learning rule, which can be either unsupervised such as Hebbian [Heb49], or supervised, such as backpropriation [Hay98, II07b], for modifying connections based on experience, represented by a change in the 'synaptic weights' based on any number of variables; (viii) An environment which provides the system with experience, represented by sets of activation vectors for some subset of the units.

V.G. Ivancevic, T.T. Ivancevic, *Quantum Neural Computation*,
Intelligent Systems, Control and Automation: Science and Engineering 40,
DOI 10.1007/978-90-481-3350-5_1, © Springer Science+Business Media B.V. 2010

1.1 Neurodynamics

To give a brief introduction to classical neurodynamics, we start from the fully recurrent, N-dimensional, RC transient circuit, given by a nonlinear vector differential equation [Hay98, Kos92, II07b]:

$$C_j \dot{v}_j = I_j - \frac{v_j}{R_j} + w_{ij} f_i(v_i), \quad (i, j = 1, \ldots, N), \tag{1.1}$$

where $v_j = v_j(t)$ represent the activation potentials in the jth neuron, C_j and R_j denote input capacitances and leakage resistances, synaptic weights w_{ij} represent conductances, I_j represent the total currents flowing toward the input nodes, and the functions f_i are sigmoidal.

Geometrically, (1.1) defines a smooth autonomous vector-field $X(t)$ in ND neurodynamical phase-space manifold M, and its (numerical) solution for the given initial potentials $v_j(0)$ defines the autonomous neurodynamical phase-flow $\Phi(t) : v_j(0) \to v_j(t)$ on M.

In AI parlance, (1.1) represents a generalization of three well-known recurrent NN models (see [Hay98, Kos92, II07b]):

  (i)   continuous Hopfield model [Hop84],
 (ii)   Grossberg ART-family cognitive system [CG83b], and
(iii)   Hecht–Nielsen counter-propagation network [Hec87, Hec90].

Physiologically, (1.1) is based on the Nobel-awarded *Hodgkin–Huxley equation* of the neural action potential (for the single squid giant axon membrane) as a function of the conductances g of sodium, potassium and leakage [HH52a, Hod64]:

$$C\dot{v} = I(t) - g_{Na}(v - v_{Na}) - g_K(v - v_K) - g_L(v - v_L),$$

where bracket terms represent the electromotive forces acting on the ions.

The *continuous Hopfield circuit* model [Hop84]:

$$C_j \dot{v}_j = I_j - \frac{v_j}{R_j} + T_{ij} u_i, \quad (i, j = 1, \ldots, N), \tag{1.2}$$

where u_i are output functions from processing elements, and T_{ij} is the inverse of the resistors connection-matrix becomes (1.1) if we put $T_{ij} = w_{ij}$ and $u_i = f_i[v_j(t)]$.

The Grossberg *analogous ART2 system* is governed by activation equation:

$$\varepsilon \dot{v}_j = -A v_j + (1 - B v_j) I_j^+ - (C + D v_j) I_j^-, \quad (j = 1, \ldots, N),$$

where A, B, C, D are positive constants (A is dimensionally conductance), $0 \le \varepsilon \ll 1$ is the fast-variable factor (dimensionally capacitance), and I_j^+, I_j^- are excitatory and inhibitory inputs to the jth processing unit, respectively.

General *Cohen–Grossberg activation equations* [CG83b] have the form:

$$\dot{v}_j = -a_j(v_j)[b_j(v_j) - f_k(v_k)m_{jk}], \quad (j = 1, \ldots, N), \tag{1.3}$$

and the *Cohen–Grossberg theorem* ensures the global stability of the system (1.3). If

$$a_j = 1/C_j, \qquad b_j = v_j/R_j - I_j, \qquad f_j(v_j) = u_j,$$

and constant $m_{ij} = m_{ji} = T_{ij}$, the system (1.3) reduces to the Hopfield circuit model (1.2).

The Hecht–Nielsen *counter-propagation network* is governed by the activation equation [Hec87, Hec90]:

$$\dot{v}_j = -Av_j + (B - v_j)I_j - v_j \sum_{k \neq j} I_k, \quad (j = 1, \ldots, N),$$

where A, B are positive constants and I_j are input values for each processing unit.

Provided some simple conditions are satisfied, namely, say symmetry of weights $w_{ij} = w_{ij}$, non-negativity of activations v_j and monotonicity of transfer functions f_j, the system (1.1) is globally asymptotically stable (in the sense of Lyapunov energy functions). The fixed-points (stable states) of the system correspond to the fundamental memories to be stored, so it works as content-addressable memory (AM). The initial state of the system (1.1) lies inside the basin of attraction of its fixed-points, so that its initial state is related to appropriate memory vector. Various variations on this basic model are reported in the literature [Hay98, Kos92], and more general form of the vector-field can be given, preserving the above stability conditions.

1.2 Quantum Computation

Quantum computers promise to perform calculations believed to be impossible for ordinary computers. Some of those calculations are of great real-world importance. For example, certain widely used encryption methods could be cracked given a computer capable of breaking a large number into its component factors within a reasonable length of time. Virtually all encryption methods used for highly sensitive data are vulnerable to one quantum algorithm or another. The extra power of a quantum computer comes about because it operates on information represented as *qubits* (or, quantum bits, see Fig. 1.1) instead of bits of the conventional, or so-called *Von Neumann computer*.[2] Recall that an ordinary classical bit can be either a 0 or a 1, and standard microchip architectures enforce that dichotomy rigorously. A qubit,

---

[2] The so-called *Von Neumann architecture* [Neu58] is a design model for a stored-program digital computer that uses a *central processing unit* (CPU) and a single separate storage structure to hold both instructions and data. It is named after mathematician and early computer scientist John Von Neumann. Such a computer implements a *universal Turing machine*, and the common 'referential model' of spec-

in contrast, can be in a superposition state, which entails proportions of 0 and 1 coexisting together. One can think of the possible qubit states as *points on a sphere:* the north pole is a classical 1, the south pole a 0, and all the points in between are all the possible superpositions of 0 and 1. The freedom that qubits have to roam across the entire sphere helps to give quantum computers their unique capabilities (see [Col06]).

Quantum computers can be roughly divided into two classes: (i) discrete ones, pioneered by Richard P. Feynman several decades ago; and (ii) recently proposed topological ones. In this section we will give a brief, nontechnical description of both types. In later Chapters we will elaborate on the most interesting aspects (using the necessary technical machinery).

1.3 Discrete Quantum Computers

The concept of *quantum computing*, or *quantum computation*, was first stated by Feynman [Fey82] and Benioff [Ben82], and formalized by Deutsch [Deu85], Bernstein and Vazirani [BV93], and Yao [Yao93]. For review, see papers [Aha98, Ber97, RP98, Ste98], books [Gru99, WC98], and courses available on the web [Pre98b, Ven07].

---

ifying sequential architectures, in contrast with more recent parallel architectures. In particularly, a stored-program digital computer is one that keeps its program instructions as well as its data in read-write, random access memory. Stored-program computers were an advancement over the program-controlled computers of the 1940s, such as Colossus and ENIAC, which were programmed by setting switches and inserting patch leads to route data and control signals between various functional units. In the majority of modern computers, the same memory is used for both data and program instructions.

The separation between the CPU and memory leads to the *von Neumann bottleneck*, the limited throughput (data transfer rate) between the CPU and memory compared to the amount of memory. In modern machines, throughput is much smaller than the rate at which the CPU can work. This seriously limits the effective processing speed when the CPU is required to perform minimal processing on large amounts of data. The CPU is continuously forced to wait for vital data to be transferred to or from memory. As CPU speed and memory size have increased much faster than the throughput between them, the bottleneck has become more of a problem.

The performance problem is reduced by a *cache* between CPU and main memory, and by the development of branch prediction algorithms. Note that a cache is a collection of data duplicating original values stored elsewhere or computed earlier, where the original data is expensive to fetch (owing to longer access time) or to compute, compared to the cost of reading the cache. In other words, a cache is a temporary storage area where frequently accessed data can be stored for rapid access. Once the data is stored in the cache, future use can be made by accessing the cached copy rather than re-fetching or recomputing the original data, so that the average access time is shorter. A cache, therefore, helps expedite data access that the CPU would otherwise need to fetch from main memory.

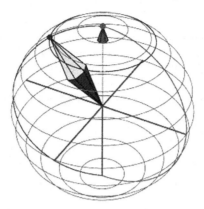

Fig. 1.1. A qubit rotation can be visualized by a rotation of the unit vector in the so-called *Bloch sphere*. For example, here we have a spin-1/2 state of a qubit in a magnetic field (modified and adapted from [Tha99]).

Roughly, *quantum computer* is a computation device that makes direct use of distinctively quantum-mechanical phenomena, such as *superposition* and *entanglement*,[3] to perform operations on data. Whilst in a conventional computer information is stored as bits, in a quantum computer it is stored as quantum binary digits, or *qubits*. The basic principle of quantum computation is that the quantum properties can be used to represent and structure data, and that quantum mechanisms can be devised and built to perform operations with these data [GC98].

A classical computer uses strings of 0s and 1s. It can do calculations on only one set of numbers at once. A quantum computer uses quantum states which can be in a superposition of many different numbers at once. A classical computer is made up of bits while a quantum computer is made up of quantum

---

[3] *Quantum entanglement*, a phenomenon referred to by E. Schrödinger as 'the essence of quantum physics', is a property of quantum superpositions involving more than one system. Just as two classical bits can be in any of four states $(00, 01, 10, 11)$, the general quantum state of two qubits is a superposition of the form $c_{00}|00\rangle + c_{01}|01\rangle + c_{10}|10\rangle + c_{11}|11\rangle$ and the quantum state of N qubits can be represented by a complex-valued vector with $2N$ components. This is the basis of the exponential superiority of quantum computation: instead of N Boolean registers, one has $2N$ complex variables, even though there are only N physical switches. But to be computationally useful, the joint quantum state must be 'non-separable'. A separable state can be expressed as an abstract product of individual states:

$$|00\rangle = |0\rangle_A|0\rangle_B, \qquad |00\rangle + |01\rangle = |0\rangle_A(|0\rangle + |1\rangle)_B.$$

However, the so-called *Bell state*, $|00\rangle + |11\rangle$, cannot be factorized in this way, and is therefore non-separable. The entanglement of a state is a measure of its non-separability, and arguably represents the fundamental resource used in quantum computation [CT07].

bits, or qubits. A quantum computer can do an arbitrary reversible classical computation on all the numbers simultaneously. A quantum computer can be treated as an interacting multi-spin system. Technically, in quantum computation, the traditional Ising-like bits of a classical computer are promoted to Heisenberg-like spin-1/2 systems. Computation in an N-bit system then takes place via unitary transformations of a $2N$D Hilbert space generated by pairwise spin–spin interactions.

There are many routes to build a quantum computer. The 0 and 1 of a qubit might be the ground and excited states of an atom in a linear *ion trap* or of a *quantum dot*; they might be polarizations of photons that interact in an *optical cavity*; they might be the excess of one nuclear spin state over another in a liquid sample in an *nuclear magnetic resonance* (NMR) machine. As long as one can put the system in a *quantum superposition* and there is a way to interact multiple qubits, a system can potentially be used as a quantum computer. In order for a system to be a good choice, it should fulfill five criteria [Bar08]:

  (i) be a scalable physical system with well-defined qubits;
 (ii) be initializable to a simple fiducial state such as $|000\ldots\rangle$;
(iii) have much longer decoherence times (i.e., one can do many operations before losing quantum coherence);
 (iv) have a universal set of quantum gates; and
  (v) permit high quantum efficiency, qubit-specific measurements.

In particular, NMR is a very promising approach. The computers are molecules in a liquid, and information is encoded in atomic nuclei in the molecules. Instead of trying to coax results out of a few fragile qubits, the technique is based on manipulating, or, in effect, programming, enormous numbers of nuclei with radio-frequency pulses and then harnessing statistics to filter the right answers (about one result in a million) out of the background of noise.

If large-scale quantum computers can be built, they will be able to solve certain problems much faster than any of conventional computers, e.g., famous *Shor's algorithm*, which is a quantum algorithm for integer factorization, first introduced by mathematician Peter Shor in 1994. On a quantum computer, to factor an integer N, Shor's algorithm takes polynomial time in $\log N$, specifically $O((\log N)^3)$, demonstrating that integer factorization is in the *complexity class BQP*. This is exponentially faster than the best-known classical factoring algorithm, the general number field sieve, which works in sub-exponential time, about $O(2^{(\log N)^{1/3}})$. Shor's algorithm is important because it can, in theory, be used to 'break' the widely used public-key cryptography scheme known as RSA, which is based on the assumption that factoring large numbers is computationally infeasible. So far as is known, this assumption is valid for conventional computers; no classical algorithm is known that can factor in polynomial time in $\log N$. However, Shor's algorithm shows that factoring is efficient on a quantum computer, so an appropriately large quantum computer can 'break' RSA. It was also a powerful motivator for the design and construction of quantum computers and for the study of new *quantum computer algorithms*.[4]

In particular, Shor's algorithm is based on the *quantum Fourier transform*, which is the discrete Fourier transform (DFT) with a particular decomposition into a product of simpler *unitary matrices* (see Appendix). Using this decomposition, the discrete Fourier transform can be implemented as a *quantum circuit*[5] consisting of *Hadamard transform gates*[6] and *controlled phase shifter gates*.[7] The quantum Fourier transform has many applications in quantum al-

---

[4] In 2001, Shor's algorithm was demonstrated by a group at IBM, who factored 15 into 3×5, using an NMR implementation of a quantum computer with 7 qubits [VSB01]. However, some doubts have been raised as to whether IBM's experiment was a true demonstration of quantum computation, since no entanglement was observed. Since IBM's implementation, several other groups have implemented Shor's algorithm using photonic qubits, emphasizing that entanglement was observed [LBY07].

[5] In quantum information theory, a quantum circuit is a model for quantum computation in which a computation is a sequence of reversible transformations on a quantum mechanical analog of an n bit register. This analogous structure is referred to as an n-qubit register.

To consider *quantum gates*, we need to specify the quantum replacement of an n-bit datum. The quantized version of classical n-bit space $\{0, 1\}^n$ is given by $H_{QB(n)} = \ell^2(\{0, 1\}^n)$. This is by definition the space of complex-valued functions on $\{0, 1\}^n$ and is naturally an *inner-product space*. This space can also be regarded as consisting of linear superpositions of classical bit strings. Using *Dirac bra-ket notation* (see Chap. 3), if $x_1, x_2, \ldots, x_n$ is a classical bit string, then

$$|x_1, x_2, \ldots, x_n\rangle$$

is an n-qubit; these special n-qubits (of which there are 2^n) are called *computational basis states*. All n-qubits are complex linear combinations of computational basis states. Note that HQB(n) has complex dimension 2^n.

[6] The *Hadamard transform* (also known as the Walsh–Hadamard transform) is an example of a generalized class of Fourier transforms. It performs an orthogonal, symmetric, involutary, linear operation on 2^n real numbers (or complex numbers, although the Hadamard matrices themselves are purely real). The Hadamard transform can be regarded as being built out of size-2 DFTs and is in fact equivalent to a multidimensional DFT of size $2 \times 2 \times \cdots \times 2 \times 2$. It decomposes an arbitrary input vector into a superposition of *Walsh functions*. The Hadamard transform can be computed in $n \log n$ operations, using the *fast Hadamard transform* algorithm.

Many quantum algorithms use the Hadamard transform as an initial step, since it maps n qubits initialized with $|0\rangle$ to a superposition of all 2^n orthogonal states in the $|0\rangle, |1\rangle$ basis with equal weight.

[7] Phase shifter gates operate on a single qubit. They are represented by 2×2 matrices of the form

$$R(\theta) = \begin{bmatrix} 1 & 0 \\ 0 & e^{2\pi i \theta} \end{bmatrix},$$

where θ is the phase shift.

gorithms as it provides the theoretical basis to the phase estimation procedure. This procedure is the key to quantum algorithms such as Shor's algorithm, the order finding algorithm and the hidden subgroup problem.

Quantum computers are different from other computers such as *DNA computers*[8] and traditional computers based on transistors. Some computing architectures such as *optical neural networks*[9] may use classical superposition

---

[8] DNA computing is a form of computing which uses DNA, biochemistry and molecular biology, instead of the traditional silicon-based computer technologies. DNA computing, or, more generally, molecular computing, is a fast developing interdisciplinary area. Research and development in this area concerns theory, experiments and applications of DNA computing. This field was initially developed by L. Adleman of the University of Southern California, in 1994 [Adl94]. Adleman demonstrated a proof-of-concept use of DNA as a form of computation which solved the seven-point *Hamiltonian path problem*. Since the initial Adleman experiments, advances have been made and various Turing machines have been proven to be constructible. DNA computing is fundamentally similar to parallel computing in that it takes advantage of the many different molecules of DNA to try many different possibilities at once. For certain specialized problems, DNA computers are faster and smaller than any other computer built so far. But DNA computing does not provide any new capabilities from the standpoint of computability theory, the study of which problems are computationally solvable using different models of computation.

[9] An optical neural network is a physical implementation of an artificial neural network (ANN, see Sect. 2.2 below) with optical components. Some ANNs that have been implemented as optical neural networks include the Hopfield net [YLS04] and the Kohonen self-organizing map with liquid crystals [LYG05]. While biological neural networks function on an electrochemical basis, optical neural networks use electromagnetic waves. Optical interfaces to biological neural networks can be created with optogenetics, but is not the same as an optical neural networks. In biological neural networks there exist a lot of different mechanisms for dynamically changing the state of the neurons, these include short-term and long-term synaptic plasticity. Synaptic plasticity is among the electrophysiological phenomena used to control the efficiency of synaptic transmission, long-term for learning and memory, and short-term for short transient changes in synaptic transmission efficiency. Implementing this with optical components is difficult, and ideally requires advanced photonic materials. Properties that might be desirable in photonic materials for optical ANNs include the ability to change their efficiency of transmitting light, based on the intensity of incoming light. There is one recent (2007) model of Optical Neural Network: the Programmable Optical Array/Analogic Computer (POAC). It had been implemented in the year 2000 and reported based on modified Joint Fourier Transform Correlator (JTC) and Bacteriorhodopsin (BR) as a holographic optical memory. Full parallelism, large array size and the speed of light are three promises offered by POAC to implement an optical CNN. They had been investigated during the last years with their practical limitations and considerations yielding the design of the first portable POAC version. POAC is a general purpose and programmable array computer that has a wide range of applications including: image processing; pattern recognition; target tracking; real-time video processing; document security; and optical switching.

of electromagnetic waves. However, it is conjectured that an exponential advantage over conventional computers is not possible without some specifically quantum mechanical resources such as *entanglement* [BCJ99].

In particular, a so-called *trapped ion quantum computer* is a type of quantum computer, in which ions (charged atomic particles) can be confined and suspended in free space using electromagnetic fields. Qubits are stored in stable electronic states of each ion, and quantum information can be processed and transferred through the collective quantized motion of the ions in the trap (interacting through the Coulomb force). Lasers are applied to induce coupling between the qubit states (for single qubit operations) or coupling between the internal qubit states and the external motional states (for entanglement between qubits). The fundamental operations of a quantum computer have been demonstrated experimentally with high accuracy (or 'high fidelity' in quantum computing language) in trapped ion systems and a strategy has been developed for scaling the system to arbitrarily large numbers of qubits by shuttling ions in an array of ion traps. This makes the trapped ion quantum computer system one of the most promising architectures for a scalable, universal quantum computer. As of June 2008, the largest number of entangled particles ever achieved in any quantum computer is eight calcium ions by way of the trapped ion method first achieved in 2005.

Generic Components of a Quantum Computer

A quantum computer has the following generic components:

1. Qubits: Any two-level quantum system can form a qubit, and there are two ways to form a qubit using the electronic states of an ion: (i) two ground state hyperfine levels (these are called 'hyperfine qubits'); and (ii) a ground state level and an excited level (these are called the 'optical qubits'). Hyperfine qubits are extremely long-lived (decay time of the order of thousands to millions of years) and phase/frequency stable (traditionally used for atomic frequency standards). Optical qubits are also relatively long-lived (with a decay time of the order of a second), compared to the logic gate operation time (which is of the order of microseconds). The use of each type of qubit poses its own distinct challenges in the laboratory.

2. Initialization: Ions can be prepared in a specific qubit state using a process called optical pumping. In this process, a laser couples the ion to some excited states which eventually decay to one state which is not coupled to by the laser. Once the ion reaches that state, it has no excited levels to couple to in the presence of that laser and, therefore, remains in that state. If the ion decays to one of the other states, the laser will continue to excite the ion until it decays to the state that does not interact with the laser. This initialization process is standard in many physics experiments and can be performed with extremely high fidelity ($> 99.9\%$).

3. Measurement: Measuring the state of the qubit stored in an ion is quite simple. Typically, a laser is applied to the ion that couples only one of the qubit states. When the ion collapses into this state during the measurement process, the laser will excite it, resulting in a photon being released when the ion decays from the excited state. After decay, the ion is continually excited by the laser and repeatedly emits photons. These photons can be collected by a photomultiplier tube (PMT) or a charge-coupled device (CCD) camera. If the ion collapses into the other qubit state, then it does not interact with the laser and no photon is emitted. By counting the number of collected photons, the state of the ion may be determined with a very high accuracy ($> 99.9\%$).

4. Arbitrary rotation of single qubit: One of the requirements of universal quantum computing is to coherently change the state of a single qubit. For example, this can transform a qubit starting out in 0 into any arbitrary superposition of 0 and 1 defined by the user. In a trapped ion system, this is often done using magnetic dipole transitions or stimulated Raman transitions for hyperfine qubits and electric quadrupole transitions for optical qubits. Gate fidelity can be greater than 99%.

5. Two-qubit entangling gates: Besides the controlled-NOT gate proposed by Cirac and Zoller in 1995, many equivalent, but more robust, schemes have been proposed and implemented experimentally since. Recent theoretical work has shown that there are no fundamental limitations to the speed of entangling gates, but gates in this impulsive regime (faster than 1 microsecond) have not yet been demonstrated experimentally (current gate operation time is of the order of microseconds). The fidelity of these implementations has been greater than 97%.

6. Scalable trap designs: Several groups have successfully fabricated ion traps with multiple trap regions and have shuttled ions between different trap zones. Ions can be separated from the same interaction region to individual storage regions and brought back together without losing the quantum information stored in their internal states. Ions can also be made to turn corners at a 'T' junction, allowing a two dimensional trap array design. Semiconductor fabrication techniques have also been employed to manufacture the new generation of traps, making the 'ion trap on a chip' a reality. These developments bring great promise to making a 'quantum charged-coupled device' (QCCD) for quantum computation using a large number of qubits.

Considerable interest has been generated in quantum computing since Shor [Sho97] showed that numbers can be factored in polynomial time on a quantum computer. From a practical viewpoint, Shor's result shows that a working quantum computer can violate the security of transactions that use the RSA protocol, a standard for secure transactions on the Internet. From a theoretical viewpoint, the result seemingly violates the polynomial version of the Church–Turing thesis; it is generally believed that factoring cannot be done in polynomial time on a deterministic or probabilistic Turing machine. What makes Shor's breakthrough result possible on a quantum Turing machine is that exponentially many computations can be performed in parallel in one step and certain quantum steps enable one to extract the desired information.

Even though simple quantum computers have been built, enormous practical issues remain for larger-scale machines. The problems seem to be exacerbated with more qubits and more computation steps. In this section, we initiate the study of quantum computing within the constraints of using a polylogarithmic ($O(\log^k n)$, $k \geq 1$) number of qubits and a polylogarithmic number of computation steps. The current research in the literature has focused on using a polynomial number of qubits. Recently, researchers have initiated the study of quantum computing using a polynomial number of qubits and a polylogarithmic number of steps [MN98, Moo99, GHP00, CW00].

The concept of *quantum neural networks* (QNNs) was initially built in [GZ01] upon Deutsch's model of quantum computational network [Deu89]. The QNN model introduces a nonlinear, irreversible, and dissipative operator, called D gate, similar to the speculative operator introduced by [AL98]. We also define the precise dynamics of this operator and while giving examples in which nonlinear Schrödinger's equations are applied, we speculate on the possible implementation of the D gate.

Within a general framework of size, depth, and precision complexity, we study the computational power of QNNs. We show that QNNs of logarithmic size and constant depth have the same computational power as threshold circuits, which are used for modeling ANNs. QNNs of polylogarithmic size and polylogarithmic depth can solve the problems in NC, the class of problems that have theoretically fast parallel solutions. Thus, the new model subsumes the computation power of various theoretical models of parallel computation.

We believe that the true advantage of quantum computation lies in overcoming the communication bottleneck that has plagued the implementation of various theoretical models of parallel computation. For example, NC circuits elegantly capture the class of problems that can be theoretically solved fast in parallel using simple gates. While fast implementations of individual gates have been achieved with semiconductors and millions of gates have been put on a single chip, we do not have the implementation of full NC circuits because of the communication and synchronization costs involved in wiring a polynomial number of gates. We believe that this hurdle can be overcome using the nonlocal interactions present in quantum systems—there is no need to explicitly wire the entangled units and the synchronization is instantaneous. This advantage is manifest in the standard unitary operator, where operations on one qubit can affect probability amplitudes on all the qubits, without requiring explicit physical connections and a global clock. Thus, the new model has the potential to overcome the practical problems associated with both quantum computing as well as classical parallel computing.

There are two equivalent models for quantum computing, quantum Turing machines [Deu85, BV93] based on reversible Turing machines [Ben73, Ben89] and quantum computational network [Deu89]. We briefly review the latter here. The basic unit in quantum computation is a *qubit*, a superposition of two independent states $|0\rangle$ and $|1\rangle$, denoted $\alpha_0|0\rangle + \alpha_1|1\rangle$, where α_0, α_1 are complex numbers such that $|\alpha_0|^2 + |\alpha_1|^2 = 1$. A system with n qubits is described using 2^n independent states $|i\rangle, 0 \leq i \leq 2^n - 1$, each associated with *probability amplitude* α_i, a complex number, as follows: $\sum_{i=0}^{2^n-1} \alpha_i|i\rangle$, where $\sum_{i=0}^{2^n-1} |\alpha_i|^2 = 1$. The direction of α_i on the complex plane is called the *phase* of state $|i\rangle$ and the absolute value $|\alpha_i|$ is called the *intensity* of state $|i\rangle$ [GZ01].

The computation unit in Deutsch's model consists of *quantum gates* whose inputs and outputs are qubits. A gate can perform any local unitary operation on the inputs. It has been shown that one-qubit gates together with two-qubit controlled NOT gates are universal [BBC95].

The quantum gates are interconnected by *wires*. A quantum computational network is a computing machine consisting of quantum gates with synchronized steps. By convention, the computation proceeds from left to right. The outputs of some of the gates are connected to the inputs of others. Some of the inputs are used as the input to the network. Other inputs are connected to *source* gates for 0 and 1 qubits. Some of the outputs are connected to *sink* gates, where the arriving qubits are discarded. An output qubit can be measured along state $|0\rangle$ or $|1\rangle$, and is observed based on the probability amplitudes associated with the qubit [GZ01].

Even though simple quantum computers have been built, enormous practical issues remain for larger-scale machines. Landauer [Lan95] exposes three main problems: *decoherence*, *localization*, and *manufacturing defects*. Decoherence is the process by which a quantum system decays to a classical state through interaction with the environment. In the best case, coherence is maintained for some 10^4 seconds, and, in the worst case, for about 10^{-10} seconds for single qubits. Some decoherence models show the coherence time declining exponentially as the number of qubits increases [Unr95]. Furthermore, the physical media that allow fast operations are also the ones with short coherence times.

The computation may also suffer from localization, that is, from reflection of the computational trajectory, causing the computation to turn around. Landauer points out that this problem is largely ignored by the research community [Lan95]. The combination of decoherence and localization makes the physical realization of quantum computation particularly difficult. On the one hand, we need to isolate a quantum computing system from the environment to avoid decoherence, and on the other hand, we need to control it externally to compel it to run forward to avoid reflection. Finally, minor manufacturing defects can engender major errors in the computations.

Introduction of the techniques of *error-correcting codes* and *fault-tolerant computation* to quantum computation has generated considerable optimism for building quantum computers, because these techniques can alleviate the problems of decoherence and manufacturing defects [GZ01]. Though this line of research is elegant and exciting, the codes correct only local errors. For example, one qubit can be encoded into the nonlocal interactions among three qubits to correct one qubit errors. However, in principle, any (nonlocal) unitary operator can be applied to all the qubits. These nonlocal errors easily subvert the error-correcting codes. Also, nonlocal interactions provide the exponential speed-ups in quantum computing. The hope that nature might allow computational speed-ups via nonlocal interactions, while errors are constrained to occur only locally, seems unavailing. For an excellent exposition on error-correcting codes and fault-tolerant quantum computing, the reader is referred to [Pre97].

It has been shown that one-qubit gates together with two-qubit controlled NOT gates are universal [BBC95]; that is, any $2^n \times 2^n$ unitary operator can be decomposed into a polynomial number of one and two qubit operators. However, in general, any error operator can be applied in one step that cannot even be detected without observing all the involved qubits. Having the ability to operate on many qubits does not solve the problem, for error-correcting codes for k qubit errors can be subverted by a $(k + 1)$-qubit error operator. Eventually, construction of a $2^n \times 2^n$ operator will itself be more time consuming than the actual computation.

There are some additional difficulties with computing using a polynomial number of qubits for a polynomial number of steps that are not discussed in the literature. For example, if $n = 1000$ and an $O(n^2)$ quantum algorithm is used, we need one million *uniquely identifiable* but *identical* carriers of quantum information. Clearly, the carriers need to be uniquely identifiable because we are not using their statistical properties, but encoding $2^{O(n^2)}$ computations in their interactions. However, the carriers need to be absolutely identical for the following reason. In describing the Hamiltonian for the whole system, there is a phase oscillation associated with each carrier. If all carriers have the same frequency, it does not affect the computation, which essentially changes the state relative to the global oscillation. But each qubit is likely to be encoded in carriers with a much larger state space, and even slight frequency differences can result in substantial errors over a polynomial number of steps. The task of preparing one million absolutely identical carriers, while exploiting the 2^{10^6} interactions, most of which are nonlocal, for speeding-up computation appears insurmountable. Controlling a polynomial number of entangled qubits for a polynomial number of steps, while compelling the computation forward, seems hard even with the help of error-correcting codes. To address the above problems, we initiate the study of quantum computation under the constraints of a poly-logarithmic number of qubits and a poly-logarithmic number of steps [GZ01].

1.4 Topological Quantum Computers

Unfortunately, quantum computers seem to be extremely difficult to build. The qubits are typically expressed as certain quantum properties of *trapped particles*, such as individual atomic ions or electrons. But their superposition states are exceedingly fragile and can be spoiled by the tiniest stray interactions with the ambient environment, which includes all the material making up the computer itself. If qubits are not carefully isolated from their surroundings, such disturbances will introduce errors into the computation. Most schemes to design a quantum computer therefore focus on finding ways to minimize the interactions of the qubits with the environment. Researchers know that if the error rate can be reduced to around one error in every 10,000 steps, then *error-correction procedures* can be implemented to compensate for decay of individual qubits. Constructing a functional machine that has a large number of qubits isolated well enough to have such a low error rate is a daunting task that physicists are far from achieving [Col06].

For this reason, a few researchers are pursuing a very different, topological way to build a quantum computer. In their approach the delicate quantum states depend on *topological properties* of a quantum system.[10] The so-called *topological quantum computer* is a theoretical quantum computer that employs 2D quasi-particles called *anyons*, whose *world lines* cross over one another to form *braids*[11] (see Fig. 1.2) in a $(1 + 2)$-space-time.

These braids form the *logic gates* that make up the quantum computer. The advantage of a quantum computer based on quantum braids over using *trapped quantum particles* is that the former is much more stable. While the smallest perturbations can cause a quantum particle to decohere and introduce errors in the computation, such small perturbations do not change the topological properties of the quantum braids (see Fig. 1.3). This is like the effort required to cut a string and reattach the ends to form a different braid, as opposed to a ball (representing an ordinary quantum particle in

---

[10] Recall that topology is called a rubber-sheet geometry, i.e., a geometrical study of properties that are unchanged when an object is smoothly deformed, by actions such as stretching, squashing and bending but not by cutting or joining. It embraces such subjects as *knot theory*, in which small perturbations do not change a topological property. For example, a closed loop of string with a knot tied in it is topologically different from a closed loop with no knot. The only way to change the closed loop into a closed loop plus knot is to cut the string, tie the knot and then reseal the ends of the string together. Similarly, the only way to convert a topological qubit to a different state is to subject it to some such violence.

[11] In topology, *braid theory* is an abstract geometric theory studying the everyday braid concept, and some generalizations. The idea is that braids can be organized into groups, in which the group operation is 'do the first braid on a set of strings, and then follow it with a second on the twisted strings'. Such groups may be described by explicit presentations, as was shown by E. Artin. Braid groups may also be given a deeper mathematical interpretation: as the fundamental group of certain configuration spaces (see Appendix, Sect. 7.2.1).

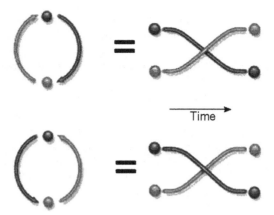

Fig. 1.2. Braiding: two anyons can encounter a clockwise swap (*top*) and a counterclockwise swap (*bottom*). These two moves in a plane generate all the possible braidings of the world lines of a pair of anyons (modified and adapted from [Col06]).

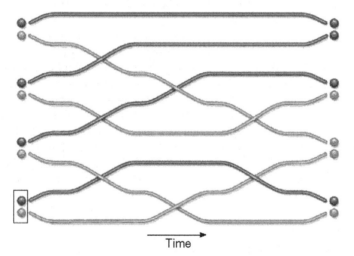

Fig. 1.3. Computing with braids of anyons. First, pairs of anyons are created and lined up in a row to represent the qubits, or quantum bits, of the computation. The anyons are moved around by swapping the positions of adjacent anyons in a particular sequence. These moves correspond to operations performed on the qubits. Finally, pairs of adjacent anyons are brought together and measured to produce the output of the computation. The output depends on the topology of the particular braiding produced by those manipulations. Small disturbances of the anyons do not change that topology, which makes the computation impervious to normal sources of errors (modified and adapted from [Col06]).

4D space-time) simply bumping into a wall. While the elements of a topo-logical quantum computer originate in a purely mathematical realm, recent experiments indicate these elements can be created in the real world using semiconductors made of gallium arsenide near absolute zero and subjected to strong magnetic fields.

Anyons are quasi-particles in a 2D space. Anyons are not strictly fermions or bosons, but do share the characteristic of fermions in that they cannot occupy the same state. Thus, the world lines of two anyons cannot cross or merge. This allows braids to be made that make up a particular circuit. In the real world, anyons form from the excitations in an electron gas in a very strong magnetic field, and carry fractional units of magnetic flux in a particle-like manner. This phenomenon is called the *fractional quantum Hall effect*.[12] The electron 'gas' is sandwiched between two flat plates of gallium arsenide, which create the 2D space required for anyons, and is cooled and subjected to intense transverse magnetic fields.

When anyons are braided, the transformation of the quantum state of the system depends only on the topological class of the anyons' trajectories (which are classified according to the *braid group*, see Fig. 1.4, as well as Appendix, Sect. 7.2.1). Therefore, the quantum information which is stored in the state of the system is impervious to small errors in the trajectories.

In 2005, Fields Medalist Michael Freedman and collaborators from the Microsoft Station Q proposed a quantum Hall device which would realize a *topological qubit*. The original proposal for topological quantum computation is due to Alexei Kitaev from CalTex in 1997. The problem of finding specific braids for doing specific computations was tackled in 2005 by N.E. Bonesteel of Florida State University, along with colleagues from the Bell Laboratories. The team showed explicitly how to construct a so-called controlled NOT (or CNOT) gate to an accuracy of two parts in 10^3 by braiding six anyons (see Fig. 1.5). A CNOT gate takes two input qubits and produces two output qubits. Those qubits are represented by triplets (green and blue) of so-called

---

[12] The fractional quantum Hall effect (FQHE) is a physical phenomenon in which a certain system behaves as if it were composed of particles with charge smaller than the elementary charge. Its discovery and explanation were recognized by the 1998 Nobel Prize in Physics. The FQHE is a manifestation of simple collective behav-ior in a 2D system of strongly-interacting electrons. At particular magnetic fields, the electron gas condenses into a remarkable state with liquid-like properties. This state is very delicate, requiring high quality material with a low carrier concen-tration, and extremely low temperatures. As in the integer quantum Hall effect, a series of plateaus forms in the Hall resistance. Each particular value of the magnetic field corresponds to a filling factor (the ratio of electrons to magnetic flux quanta) $\nu = p/q$, where p and q are integers with no common factors. In particular, *frac-tionally charged quasi-particles* are neither bosons nor fermions and exhibit *anyonic statistics*. The FQHE continues to be influential in theories about topological or-der. Certain fractional quantum Hall phases appear to have the right properties for building a topological quantum computer.

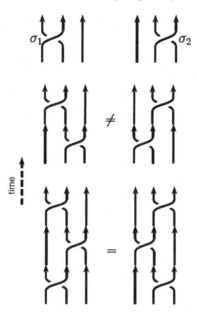

Fig. 1.4. Graphical representation of elements of the braid group. *Top*: the two elementary braid operations σ_1 and σ_2 on three anyons. *Middle*: non-commutativity is shown here as $\sigma_2\sigma_1 \neq \sigma_1\sigma_2$; hence the braid group is non-Abelian. *Bottom*: the *braid relation*: $\sigma_i\sigma_{i+1}\sigma_i = \sigma_{i+1}\sigma_i\sigma_{i+1}$ (modified and adapted from [NSS08]).

Fibonacci anyons. The particular style of braiding, leaving one triplet in place and moving two anyons of the other triplet around its anyons, simplified the calculations involved in designing the gate.

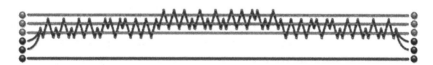

Fig. 1.5. Recently, a *quantum logic gate* known as a *CNOT-gate* has been produced by a complicated braiding of six anyons. This braiding produces a CNOT gate that is accurate to about 10^{-3} (modified and adapted from [Col06]).

Topological quantum computers are equivalent in computational power to other standard models of quantum computation, in particular to the *quantum circuit* model and to the *quantum Turing machine* model. That is, any of these models can efficiently simulate any of the others. Nonetheless, certain algorithms may be a more natural fit to the topological quantum computer model. For example, algorithms for evaluating the *Jones polynomial* were first developed in the topological model, and only later converted and extended in the discrete quantum circuit model.

Even though quantum braids are inherently more stable than trapped quantum particles, there is still a need to control for error inducing thermal fluctuations, which produce random stray pairs of anyons which interfere with adjoining braids. Controlling these errors is simply a matter of separating the anyons to a distance where the rate of interfering strays drops to near zero. It has been estimated that the error rate for a CNOT operation of a qubit state could be as low as 10^{-30} or less. Although this number has been criticized as being strongly overstated, there is nonetheless good reason to believe that topologically protected systems will be particularly immune to many sources of error that plague other schemes for quantum information processing. Simulating the dynamics of a topological quantum computer may be a promising method of implementing fault-tolerant quantum computation even with a standard quantum information processing scheme.

To build a topological quantum computer requires one additional complication: the anyons must be what is called non-Abelian, or, non-commutative, which means that the order in which particles are swapped is important.[13] If *non-Abelian anyons* actually exist,[14] topological quantum computers could well leapfrog discrete quantum computer designs in the race to scale up from individual qubits and logic gates to fully fledged machines more deserving of the name 'computer'. Carrying out calculations with quantum knots and braids, a scheme that began as an esoteric alternative, could become the standard way to implement practical, error-free quantum computation [Col06].

We will continue our expose on topological quantum computation in Sect. 3.2.6.

1.5 Computation at the Edge of Chaos and Quantum Neural Networks

Depending on the connectivity, the so-called *recurrent neural networks* (see Sect. 2.2 below), consisting of simple computational units, can show very different types of dynamics, ranging from totally ordered (linear-like behavior) to chaotic. Using the *mean-field theory* approach with *evolving Hamming distance*, Bertschinger and Natschläger analyzed how the type of dynamics (ordered or chaotic) exhibited by randomly connected networks of threshold gates driven by a time-varying input signal depended on the parameters describing the distribution of the connectivity matrix [BN04]. In particular, the authors calculated the critical boundary in *parameter space* where the transition from ordered to chaotic dynamics takes place. Employing a recently

---

[13] This is similar to the rotation group (in which the order of rotations matters) versus the translation group (in which the order of translations does not matter).

[14] Non-Abelian anyons probably exist in certain gapped 2D systems, including *fractional quantum Hall effect* and possibly also ruthenates, *topological insulators*, rapidly rotating *Bose–Einstein condensates*, *quantum loop gases*/string nets.

developed framework for analyzing real-time computations, they showed that only near the critical boundary could such networks perform complex computations on time series. This result strongly supports conjectures that dynamical systems that are capable of doing complex computational tasks should operate near the *edge of chaos*, that is, the transition from ordered to chaotic dynamics.

In particular, the authors pointed out the following: (i) Dynamics near the critical line are a general property of input-driven dynamical systems that support complex real-time computations; and (ii) Such systems can be used by training a simple readout function to approximate any arbitrary complex filter. This suggests that to exhibit sophisticated information processing capabilities, an adaptive system should stay close to *critical dynamics*. Since the dynamics is also influenced by the statistics of the input signal, these results indicate a new role for plasticity rules: stabilize the dynamics at the critical line. Such rules would then implement what could be called *dynamical homeostasis*, which could be achieved in an unsupervised learning manner. We will expand on this view in Chap. 4.

Now, if critical dynamics at the edge of chaos of classical adaptive systems is essentially nonlinear, then what can we say about modern *adaptive quantum systems*? While quantum mechanics is based on superposition principle (see Chap. 3), and thus is essentially linear, adaptation introduces critical nonlinearity. In this way, we come to the nonlinear deformation/extension of the linear *Schrödinger equation*.

An impetus to hypothesize a quantum brain model comes from the brain's necessity to unify the neuronal response into a single percept. Anatomical, neurophysiological and neuropsychological evidence, as well as brain imaging using fMRI and PET scans, show that separate functional maps exist in the biological brain to code separate features such as direction of motion, location, color and orientation. How does the brain compute all this data to have a coherent perception?

To provide a partial answer to the above question, a *quantum neural network* (QNN) (see Fig. 1.6) has been proposed [BKE05, BKE06], in which a collective response of a neuronal lattice is modeled using the Schrödinger equation:

$$i\partial_t \psi(x,t) = -\frac{1}{2}\Delta\psi(x,t) + V(x)\psi(x,t), \qquad (1.4)$$

where $\psi(x,t)$ is the wave function, or probability amplitude, associated with the quantum-mechanical system at a space-time point (x,t) and Δ is the standard *Laplacian operator*. It is shown that an external stimulus reaches each neuron in a lattice with a probability amplitude function ψ_i. Such a hypothesis would suggest that the carrier of the stimulus performs quantum computation. The collective response of all the neurons is given by the *quantum superposition equation*,

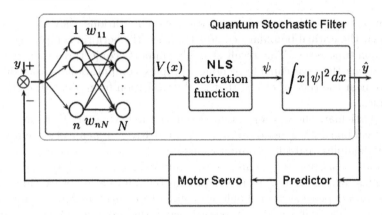

Fig. 1.6. Quantum neural network has three blocks: quantum stochastic filter based on the *nonlinear Schrödinger equation*, Kalman-like predictor and neural motor servo, corresponding to the Kalman-like corrector (modified and adapted from [BKE05, BK05, BKE06])

$$\psi = \sum_{i=0}^{N} c_i \psi_i = c_i \psi_i.$$

The QNN hypothesis suggests that the time evolution of the collective response $\psi = \psi(x,t)$ is described by (1.4). A neuronal lattice sets up a spatial potential field $V(x)$. A quantum process described by a quantum state ψ which mediates the collective response of a neuronal lattice evolves in the spatial potential field according to (1.4). Thus the *'classical brain'* sets up a spatio-temporal potential field $V(x,t)$ and the *'quantum brain'* is excited by this potential field to provide a collective response $\psi = \psi(x,t)$ [BKE05, BKE06].

Mathematical basis for the QNN presented in Fig. 1.6 is the *nonlinear Schrödinger equation* (NLS, see Sect. 4.6 below), which gives a closed-loop feedback dynamics for the plant defined by the linear Schrödinger equation (1.4). Informally, QNN is represented by *adaptive NLS*, that is NLS with 'synaptic weights' w_i replacing quantum superposition constants c_i. Formally, it is the Schrödinger (1.4) with added cubic nonlinearity and synaptic weights. We will analyze this QNN in Sect. 5.6 below.

Also, it was shown in [KLM01] that quantum computation on optical modes using only beam splitters, phase shifters, photon sources and photo detectors is possible. Following this concept, [Alt07] assumed the existence of a qubit

$$|x\rangle = \alpha|0\rangle + \beta|1\rangle, \tag{1.5}$$

where $|\alpha|^2 + |\beta|^2 = 1$, where the states $|0\rangle$ and $|1\rangle$ are understood as different polarization states of light.

Consider a *classical perceptron* [MP69], i.e., the system with n input channels $x_1, \ldots, x_n$ and one output channel y, given by

$$y = f\left(\sum_{j=1}^{n} w_j x_j\right), \tag{1.6}$$

where $f(\cdot)$ is the perceptron activation function and w_j are the weights tuning during learning process (see Sect. 2.2 below).

The *perceptron learning algorithm* works as follows:

1. The weights w_j are initialized to small random numbers.
2. A *pattern vector* $(x_1, \ldots, x_n)$ is presented to the perceptron and the output y generated according to the rule (1.6).
3. The weights are updated according to the rule

$$w_j(t+1) = w_j(t) + \eta(d - y)x_j, \tag{1.7}$$

where t is discreet time, d is the desired output provided for training and $0 < \eta < 1$ is the step size.

It will be hardly possible to construct an exact analog of the nonlinear activation function f, like sigmoid and other functions of common use in neural networks, but we will show that the leaning rule of the type (1.7) is possible for a quantum system too.

Consider a quantum system with n inputs $|x_1\rangle, \ldots, |x_n\rangle$ of the form (1.5), and the output $|y\rangle$ derived by the rule

$$|y\rangle = \hat{F} \sum_{j=1}^{n} \hat{w}_j |x_j\rangle, \tag{1.8}$$

where $\hat{w}_j$ become 2×2 matrices acting on the basis $(|0\rangle, |1\rangle)$, combined of phase shifters $e^{i\theta}$ and beam splitters, and possibly light attenuators, $\hat{F}$ is an unknown operator that can be implemented by the network of quantum gates.

Consider the simplistic case with $\hat{F} = 1$ being the identity operator. The output of the quantum perceptron at the time t will be [Alt07]

$$|y(t)\rangle = \sum_{j=1}^{n} \hat{w}_j(t)|x_j\rangle.$$

In analogy with classical case (1.7), let us provide a learning rule

$$\hat{w}_j(t+1) = \hat{w}_j(t) + \eta(|d\rangle - |y(t)\rangle)\langle x_j|, \tag{1.9}$$

where $|d\rangle$ is the desired output.

It can be shown that the learning rule (1.9) drives the quantum perceptron into desired state $|d\rangle$ used for learning. Using the rule (1.9) and taking the module-square difference of the real and desired outputs, we yield

$$\||d\rangle - |y(t+1)\rangle\|^2 = \left\||d\rangle - \sum_{j=1}^{n} \hat{w}_j(t+1)|x_j\rangle\right\|^2 = (1 - n\eta)^2 \||d\rangle - |y(t)\rangle\|^2.$$

For small η $(0 < \eta < 1/n)$ and normalized input states $\langle x_j|x_j\rangle = 1$ the result of iteration converges to the desired state $|d\rangle$. The whole network can be then composed from the primitive elements using the standard rules of ANN architecture. For more details, see [Alt07].

1.6 Adaptive Path Integral: An ∞-Dimensional QNN

1.6.1 Computational Partition Function

Recall that a thermodynamic *partition function* Z represents a quantity that encodes the statistical properties of a system in thermodynamic equilibrium (see, e.g. [II06b, II08a]). It is a function of temperature and other parameters, such as the volume enclosing a gas. Other thermodynamic variables of the system, such as the total energy, free energy, entropy, and pressure, can be expressed in terms of the partition function or its partial derivatives.

A canonical ensemble is a statistical ensemble representing a probability distribution of microscopic states of the system. Its probability distribution is characterized by the proportion p_i of members of the ensemble which exhibit a measurable macroscopic state i, where the proportion of microscopic states for each macroscopic state i is given by the Boltzmann distribution,

$$p_i = \frac{1}{Z} e^{-E_i/(kT)} = e^{-(E_i - A)/(kT)},$$

where E_i is the energy of state i. It can be shown that this is the distribution which is most likely, if each system in the ensemble can exchange energy with a heat bath, or alternatively with a large number of similar systems. In other words, it is the distribution which has *maximum entropy* for a given average energy $\langle E_i\rangle$.

The partition function of a *canonical ensemble* is defined as a sum

$$Z(\beta) = \sum_j e^{-\beta E_j}, \quad \text{where } \beta = 1/(k_B T)$$

is the 'inverse temperature', while T is an ordinary temperature and k_B is the *Boltzmann's constant*. However, as the position x^i and momentum p_i variables of an ith particle in a system can vary continuously, the set of microstates is actually uncountable. In this case, some form of *coarse-graining* procedure must be carried out, which essentially amounts to treating two mechanical states as the same microstate if the differences in their position and momentum variables are 'small enough'. The partition function then takes the form of an integral. For instance, the partition function of a gas consisting of N molecules is proportional to the $6N$-dimensional phase-space integral,

$$Z(\beta) \sim \int_{\mathbb{R}^{6N}} d^3 p_i \, d^3 x^i \exp[-\beta H(p_i, x^i)],$$

where $H = H(p_i, x^i)$, $(i = 1, \ldots, N)$ is the classical Hamiltonian (total energy) function.

More generally, the so-called *configuration integral*, as used in probability theory, information science and dynamical systems, is an abstraction of the above definition of a partition function in statistical mechanics. It is a special case of a normalizing constant in probability theory, for the Boltzmann distribution. The partition function occurs in many problems of probability theory because, in situations where there is a natural symmetry, its associated probability measure, the *Gibbs measure*, which generalizes the notion of the canonical ensemble, has the *Markov property* [IR08].

Given a set of random variables X_i taking on values x^i, and purely potential Hamiltonian function $H(x^i)$, $(i = 1, \ldots, N)$, the partition function is defined as

$$Z(\beta) = \sum_{x^i} \exp\left[-\beta H(x^i)\right]. \qquad (1.10)$$

The function H is understood to be a real-valued function on the space of states $\{X_1, X_2, \cdots\}$ while β is a real-valued free parameter (conventionally, the inverse temperature). The sum over the x^i is understood to be a sum over all possible values that the random variable X_i may take. Thus, the sum is to be replaced by an integral when the X_i are continuous, rather than discrete. Thus, one writes

$$Z(\beta) = \int dx^i \exp\left[-\beta H(x^i)\right],$$

for the case of continuously-varying random variables X_i.

The Gibbs measure of a random variable X_i having the value x^i is defined as the *probability density function* (PDF)

$$P(X_i = x^i) = \frac{1}{Z(\beta)} \exp\left[-\beta E(x^i)\right] = \frac{\exp[-\beta H(x^i)]}{\sum_{x^i} \exp[-\beta H(x^i)]},$$

where $E(x^i) = H(x^i)$ is the energy of the configuration x^i. This probability, which is now properly normalized so that $0 \leq P(x^i) \leq 1$, can be interpreted as a likelihood that a specific configuration of values x^i, $(i = 1, 2, \ldots, N)$ occurs in the system. $P(x^i)$ is also closely related to Ω, the probability of a *random partial recursive function halting*.

As such, the partition function $Z(\beta)$ can be understood to provide the Gibbs measure on the space of states, which is the unique statistical distribution that maximizes the entropy for a fixed expectation value of the energy,

$$\langle H \rangle = -\frac{\partial \log(Z(\beta))}{\partial \beta}.$$

The associated entropy is given by

$$S = -\sum_{x^i} P(x^i) \ln P(x^i) = \beta \langle H \rangle + \log Z(\beta),$$

representing 'ignorance' + 'randomness' [IR08].

1.6.2 From Thermodynamics to Quantum Field Theory

In addition, the number of variables X_i in the standard partition Z need not be countable, in which case the set of coordinates $\{x^i\}$ becomes a field $\phi = \phi(x)$, so the sum is to be replaced by the *Euclidean path integral* (that is a Wick-rotated Feynman transition amplitude in imaginary time, see Sect. 3.3 below), as

$$Z_{\text{Ham}}(\phi) = \int \mathcal{D}[\phi] \exp[-H(\phi)]. \qquad (1.11)$$

More generally, in quantum field theory, instead of the field Hamiltonian $H(\phi)$ we have the action $S(\phi)$ of the theory. Both Euclidean path integral,

$$Z_{\text{Euc}}(\phi) = \int \mathcal{D}[\phi] \exp[-S(\phi)], \quad \text{real path integral in imaginary time}$$
$$(1.12)$$

and Lorentzian one,

$$Z_{\text{Lor}}(\phi) = \int \mathcal{D}[\phi] \exp[iS(\phi)], \quad \text{complex path integral in real time} \quad (1.13)$$

as well as their common Lebesgue-type measure $\mathcal{D}[\phi]$, given by

$$\mathcal{D}[\phi] = \lim_{N \to \infty} \prod_{s=1}^{N} \phi_s^i, \quad (i = 1, \dots, n = \text{number of fields}),$$

so that we can 'safely integrate over a continuous field-spectrum and sum over a discrete field spectrum'—*represent quantum field theory* (QFT) partition functions.

We will give more formal definitions of the above path integrals in Sect. 3.3 below (see [II08a, II08b] for technical details). For the moment, we only remark that the Lorentzian path integral (1.13) represents a QFT generalization of the (nonlinear) Schrödinger equation, while the Euclidean path integral (1.12) in the rectified real time represents a statistical field theory (SFT) generalization of the Fokker–Planck equation (to be used in Sect. 5.9 below).

1.6.3 ∞-Dimensional QNNs

Now, a countably-infinite set of abstract synaptic weights, $\{w_s, s \in \mathbb{N}\}$, can be inserted into each of the path integrals (1.11), (1.12) and (1.13), transforming each of them into an ∞-dimensional quantum neural network,[15] respectively given by:

$$Z_{\text{Ham}}^{\text{QNN}}(\phi) = \int \mathcal{D}[w, \phi] \exp\left[-H(\phi)\right],$$

$$Z_{\text{Euc}}^{\text{QNN}}(\phi) = \int \mathcal{D}[w, \phi] \exp\left[-S(\phi)\right],$$

$$Z_{\text{Lor}}^{\text{QNN}}(\phi) = \int \mathcal{D}[w, \phi] \exp\left[iS(\phi)\right],$$

such that their *adaptive functional measure* $\mathcal{D}[w, \phi]$, given by

$$\mathcal{D}[\phi] = \lim_{N \to \infty} \prod_{s=1}^{N} w_s \phi_s^i, \quad (i = 1, \ldots, n = \text{ number of fields}),$$

is trained in a neural-networks fashion, that is, by the general rule

$$new\ value(t + 1) = old\ value(t) + innovation(t),$$

or more formally, by one of the standard learning rules in which the micro-time level is traversed in discrete steps, i.e., if $t = t_0, t_1, \ldots, t_s$ then $t + 1 = t_1, t_2, \ldots, t_{s+1}$:[16]

1. A *self-organized*, *unsupervised* (e.g., Hebbian-like [Heb49]) learning rule:

$$w_s(t + 1) = w_s(t) + \frac{\sigma}{\eta}(w_s^d(t) - w_s^a(t)), \tag{1.14}$$

   where $\sigma = \sigma(t)$, $\eta = \eta(t)$ denote *signal* and *noise*, respectively, while superscripts d and a denote *desired* and *achieved* micro-states, respectively; or

2. A certain form of a *supervised gradient descent learning*:

$$w_s(t + 1) = w_s(t) - \eta \nabla J(t), \tag{1.15}$$

   where η is a small constant, called the *step size*, or the *learning rate*, and $\nabla J(n)$ denotes the gradient of the 'performance hyper-surface' at the t-th iteration; or

---

[15] ∞ is that; ∞ is this; ∞ has come into existence from ∞. From ∞, when ∞ is taken away, ∞ remains—*Upanishads*.

[16] The traditional neural networks approaches are known for their classes of functions they can represent. Here we are talking about functions in an *extensional* rather than merely *intensional* sense; that is, function can be read as input/output behavior [Bar84, Ben91, For03, Han04]. This limitation has been attributed to their low-dimensionality (the largest neural networks are limited to the order of 10^5 dimensions [IE08]). The proposed path integral approach represents a new family of function-representation methods, which potentially offers a basis for a fundamentally more expansive solution.

3. A certain form of a reward-based, *reinforcement learning* rule [SB98], in
which system learns its optimal policy:

$$innovation(t) = |reward(t) - penalty(t)|. \qquad (1.16)$$

We will use various versions of these ∞-dimensional QNNs in the following
chapters.

1.7 Brain Topology vs. Small-World Topology

Now, from a physiological point of view, a *neural-networks complexity* has
been studied as an interplay between *dynamics and topology of brain networks* [LBM05a]. This approach is of interest from several points of view:
the brain and its structural features can be seen as a prototype of a physical system capable of highly complex and adaptable patterns in connectivity,
selectively improved through evolution; architectural organization of brain
cortex is one of the key features of how brain system evolves, adapts itself
to the experience, and to possible injuries. Brain activity can be modeled as
a dynamical process acting on a network; each vertex of the structure represents an elementary component, such as brain areas, groups of neurons or
individual cells. Recall that a *complexity measure*, has been introduced (see
[LBM05b] and reference therein) with the purpose to get a sensible measure of
two important features of the brain activity: segregation and integration. The
former is a measure of the relative statistical independence of small subsets;
the latter is the measure of statistical deviation from independence of large
subsets. Complexity is based on the values of the *Shannon entropy* calculated
over the dynamics of the different sized subgraphs of the whole network. It
is sensitive both to the statistical properties of the dynamics and to the connectivity. It has been shown by means of genetic algorithms that the graphs
showing high values of complexity are characterized by being both segregated
and integrated; the complexity is low when the system is either completely
independent (segregated), or completely dependent (integrated) [LBM05a].

In general this approach has been developed within the framework of *small-
world networks*, comprising World Wide Web structure etc. which have received much attention from researchers in various disciplines, since they were
introduced by Watts and Strogatz [WS98] as models of real networks that lie
somewhere between being random and being regular.

Watts and Strogatz introduced a simple model for tuning collections of
coupled dynamical systems between the two extremes of random and regular
networks. In this model, connections between nodes in a regular array are
randomly rewired with a probability p, such that $p = 0$ means the network is
regularly connected, while $p = 1$ results in a random connection of nodes. For
a range of intermediate values of p between these two extremes, the network
retains a property of regular networks (a large clustering coefficient) and also

acquires a property of random networks (a short characteristic path length between nodes).

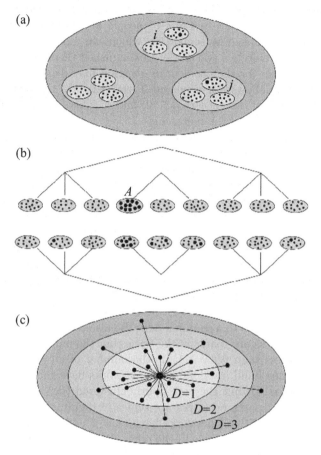

Fig. 1.7. Model of a *social network* (modified and adapted from [MNL03]—see text for explanation.

Many examples of such small worlds, both natural and human-made, have been discussed [Str, New00]. For example, a model of a *social network* is given in Fig. 1.7, where (a) denotes people (dots) belonging to groups (ellipses), which in turn belong to groups of groups and so on. The largest group corresponds to the entire community. As we go down in this hierarchical organization, each group represents a set of people with increasing social affinity. In the example, there are $l = 3$ hierarchical levels, each representing a subdivision in $b = 3$ smaller groups, and the lowest groups are composed of $g = 11$ people, on average. This defines a *social hierarchy*. The distance between the highlighted individuals i and j in this hierarchy is 3. (b) Each hierarchy can be represented as a tree-like structure. Different hierarchies are correlated, in the sense that distances that are short along one of them are more likely to be short along the others as well. The figure shows an example with $H = 2$ hierarchies, where highlighted in the second hierarchy are those people belonging to group A in the first one. (c) Pairs of people at shorter social distances are more likely to be linked by social ties, which can represent either friendship or acquaintanceship ties (we do not distinguished them here because the ones that are relevant for the problem in question may depend on the social context). The figure shows, for a person in the network, the distribution of acquaintances at social distance $D = 1, 2$, and 3, where D is the minimum over the distances along all the hierarchies.

Not surprisingly, there has been much interest in the synchronization of dynamical systems connected in a *small-world geometry* [BP02, NML03]. Generically, such studies have shown that the presence of small-world connections make it easier for a network to synchronize, an effect generally attributed to the reduced path length between the linked systems. This has also been found to be true for the special case in which the dynamics of each oscillator is described by a *Kuramoto model* [HCK02a, HCK02b].

Small-world networks are characterized by two numbers: the average path length L and the clustering coefficient C. L, which measures efficiency of communication or passage time between nodes, is defined as being the average number of links in the shortest path between a pair of nodes in the network. C represents the degree of local order, and is defined as being the probability that two nodes connected to a common node are also connected to each other.

Many real networks are sparse in the sense that the number of links in the network is much less than $N(N-1)/2$, the number of all possible (bidirectional) links. On one hand, random sparse networks have short average path length (i.e., $L \sim \log N$), but they are poorly clustered (i.e., $C \ll 1$). On the other hand, regular sparse networks are typically highly clustered, but L is comparable to N. (All-to-all networks have $C = 1$ and $L = 1$, so they are most efficient, but most expensive in the sense that they have all $N(N-1)/2$ possible connections and so they are dense rather than sparse.) The small-world network models have advantages of both random and regular sparse networks: they have small L for fast communication between nodes, and they have large C, ensuring sufficient redundancy for high fault tolerance. Many networks in the real world, such as the world-wide web (WWW) [AJB99], the neural network of *C. elegans* [WS98, Wat99], collaboration networks of actors [WS98, Wat99], and networks of scientific collaboration [New00], have been shown to have this property. The models of small-world networks are constructed from a regular lattice by adding a relatively small number of shortcuts at random, where a link between two nodes u and v is called a *shortcut* if the shortest path length between u and v in the absence of the link is more than two [Wat99]. The regularity of the underlying lattice ensures high clustering, while the shortcuts reduce the size of L.

Most work has focused on average properties of such models over different realizations of *random* shortcut configurations. However, a different point of view is necessary when a network is to be designed to optimize its performance with a restricted number of long-range connections. For example, a transportation network should be designed to have the smallest L possible, so as to maximize the ability of the network to transport people efficiently, while keeping a reasonable cost of building the network. The same can be said about communication networks for efficient exchange of information between nodes.

Most random choices of shortcuts result in a suboptimal configuration, since they do not have any special structures or organizations. On the contrary, many real networks have highly structured configurations of shortcuts. For example, in long-range transportation networks, the airline connections between major cities which can be regarded as shortcuts, are far from being random, but they are organized around hubs. Efficient travel involves ground transportation to a nearest airport, then flights through a hub to an airport closest to the destination, and ground transportation again at the end.

It has been shown in [NML02] that the average path length L of a small-world network with a fixed number of shortcuts attains its minimum value when there exists a 'center' node, from which all shortcuts are connected to uniformly distributed nodes in the network (see Fig. 1.8(a)). If a small-world network has several 'centers' and its subnetwork of shortcuts is *connected*, then L is almost as small as the minimum value (see Fig. 1.8(b)).

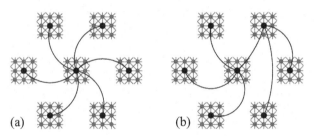

Fig. 1.8. Examples of shortcut configuration with (**a**) a single center and (**b**) two centers.

For example, a small-world geometry has been explored in a family of 1D networks of N integrate-and-fire neurons arranged on a ring [AT06]. To simplify the formalism, the geometry of a neuron is reduced to a point, and the time evolution of the membrane potential V_i of cell i, $i = 1, \ldots, N$ during an inter-spike interval follows the equation

$$\dot{V}_i(t) = -[V_i(t) - V^{res}]/\tau_m + \frac{R_m}{\tau_m}[I^{syn}(t) + I^{inh}(t) + I_i(t)], \quad (1.17)$$

where τ_m is the time constant of the neural-cell membrane, R_m its passive resistance, and V^{res} the resting potential. The total current entering cell i consists of a synaptic current $I_i^{syn}(t)$ due to the spiking of all other cells in the network that connect to cell i, an inhibitory current $I^{inh}(t)$, which depends on the network activity, and a small external current $I_i(t)$, which represents all other inputs into cell i.

In the subsequent Chapters we will explore various aspects of the *quantum neural computation*, as briefly outlined so far.

1.8 Quantum Brain and Mind

In this section we will introduce modern concepts of quantum brain and mind, mostly following a modern trend set-up in the journal *NeuroQuantology*. For example, according to some of the recent papers from this journal (see [Per03a]), 'brain is classical' (i.e., governed by classical biophysics), while 'mind is quantum',[17] and human consciousness is generated by 'neuro-quantization', which takes place inside brain's microtubules (see, e.g. [Pen89, Pen94]).

1.8.1 Connectionism, Control Theory and Brain Theory

Let us start with the classical *connectionist brain theory*[18] (see [Arb98] for a general overview). A recent paper [Roy08] proposes a engineering new theory for the supposed internal mechanisms of the brain function. It postulates that there are controllers in the brain and that there are parts of the brain that control other parts. Thus, the new theory refutes the standard connectionist theory that there are no separate controllers in the brain for higher level functions and that all control is 'local and distributed' at the level of the cells. Connectionist algorithms themselves are used to prove this theory. The author claims that there is evidence in the neuroscience literature to support this theory. Thus, this paper proposes a control theoretic approach for understanding how the brain works and learns. That means that control theoretic principles should be applicable to developing systems similar to the brain.

Having established the connection between *connectionism* and the control theory, it is a small step to argue that if connectionist models are brain-like, then, given that they use executive controllers, the internal mechanisms of the brain must also use executive controllers. Hence, it can be generalized

---

[17] From Delight we all came into existence; in Delight we grow and play our respective roles; at the end of our journey's close we return to Delight—*Upanishads*.

[18] As already mentioned, *connectionism* is an approach in the fields of artificial intelligence, cognitive psychology/cognitive science, neuroscience and philosophy of mind, that models mental or behavioral phenomena as the emergent processes of interconnected networks of simple units. There are many forms of connectionism, but the most common forms use ANNs.

and claimed that there are parts of the brain that control other parts. A control theoretic approach to understanding how the brain works and learns could overcome many of the limitations of connectionism and lead to the development of more powerful models that can properly replicate the functions of the brain, such as autonomous learning without external intervention. None of the supervised connectionist algorithms can learn without human intervention that requires setting various learning parameters by a trial-and-error process. Thus, it is impossible to build autonomous robots (software or hardware), ones that can learn on their own, with these connectionist algorithms. The new proposed control theoretic paradigm, however, should allow the field to freely explore other means of learning in neural networks [Roy08].

1.8.2 Neocortical Biophysics

Basic brain physiology will be presented in Chap. 2. In this section we focus on its latest and most important part, called *neocortex*[19] performs a large variety of complex computations in real time. It is conjectured that these

---

[19] The *neocortex* (Latin for 'new bark' or 'new rind') is a part of the brain of mammals. It is the outer layer of the cerebral hemispheres, and made up of six layers, labeled I to VI (with VI being the innermost and I being the outermost). The neocortex is part of the *cerebral cortex* (along with the archicortex and paleocortex, which are cortical parts of the limbic system). It is involved in higher functions such as sensory perception, generation of motor commands, spatial reasoning, conscious thought and, in humans, language.

The neocortex consists of the grey matter, or neuronal cell bodies and unmyelinated fibers, surrounding the deeper white matter (myelinated axons) in the cerebrum. Whereas the neocortex is smooth in rodents and other small mammals, it has deep grooves (sulci) and wrinkles (gyri) in primates and other larger mammals. These folds increase the surface area of the neocortex considerably without taking up too much more volume. This has allowed primates and especially humans to evolve new functional areas of neocortex that are responsible for enhanced cognitive skills such as working memory, speech, and language. The neocortex contains two primary types of neurons, excitatory pyramidal neurons (80% of neocortical neurons) and inhibitory interneurons (20%). The neurons of the neocortex are also arranged in vertical structures called neocortical columns.

The neocortex is divided into frontal, parietal, temporal, and occipital lobes, which perform different functions. For example, the occipital lobe contains the primary visual cortex, and the temporal lobe contains the primary auditory cortex. Further subdivisions or areas of neocortex are responsible for more specific cognitive processes. In humans, the frontal lobe contains areas devoted to abilities that are enhanced in or unique to our species, such as complex language processing localized to the ventrolateral prefrontal cortex (Broca's area) and social and emotional processing localized to the orbito-frontal cortex.

The female human neocortex contains approximately 19 billion neurons while the male human neocortex has 23 billion on average [PG97]. Additionally, the female neocortex has more white matter, while the male neocortex contains more grey matter. The implications of such differences are not fully known.

computations are carried out by a network of cortical micro-circuits, where each micro-circuit is a rather stereotypical circuit of neurons within a cortical column. A characteristic property of these circuits and networks is an abundance of feedback connections. But the computational function of these feedback connections is largely unknown. Two lines of research have been engaged to solve this problem. In one approach, which one might call the constructive approach, one builds hypothetical circuits of neurons and shows that (under some conditions on the response behavior of its neurons and synapses) such circuits can perform specific computations. In another research strategy, which one might call the analytical approach, one starts with data-based models for actual cortical micro-circuits, and analyzes which computational operations such given circuits can perform under the assumption that a learning process assigns suitable values to some of their parameters (e.g., synaptic efficacies of readout neurons). An underlying assumption of the analytical approach is that complex recurrent circuits, such as cortical micro-circuits, cannot be fully understood in terms of the usually considered properties of their components. Rather, system-level approaches that directly address the dynamics of the resulting recurrent neural circuits are needed to complement the bottom-up analysis [MJS07]. This line of research started with the identification and investigation of so-called *canonical micro-circuits* [DKM95]. Several issues related to cortical micro-circuits have also been addressed in the work of S. Grossberg and collaborators from Boston (see [Gro03] and the references therein). Subsequently it was shown that quite complex real-time computations on spike trains can be carried out by such 'given' models for cortical micro-circuits (see [DM04] for a review). A fundamental limitation of this approach was that only those computations could be modeled that can be carried out with a fading memory, more precisely only those computations that require integration of information over a time-span of 200 ms to 300 ms (its maximal length depends on the amount of noise in the circuit and the complexity of the input spike trains). In particular, computational tasks that require a representation of elapsed time between salient sensory events or motor actions, or an internal representation of expected rewards, working memory, accumulation of sensory evidence for decision making, the updating and holding of analog variables such as for example the desired eye position, and differential processing of sensory input streams according to attentional or other internal states of the neural system could not be modeled in this way. Previous work on concrete examples of artificial neural networks and cortical micro-circuit models had already indicated that these shortcomings of the model might arise only if one assumes that learning affects exclusively the synapses of readout neurons that project the results of computations to other circuits or areas, without giving feedback into the circuit from which they extract information. This scenario is unrealistic from a biological perspective, since pyramidal neurons in the cortex typically have in addition to their long projecting axon a large number of axon collaterals that provide feedback to the local circuit [MJS07]. Also, abundant feedback connections exist on the network level between different brain areas. For statistical mechanics of neocortical interactions (using path integral-related methods), see [Ing82, Ing97, Ing98].

Schematically, *neocortical dynamics* is depicted in Fig. 1.9(a), which illustrates the universally accepted six laminae of the neocortex [Sze78]. The pyramidal apical dendrites finish in a tuft-like branching in lamina I. It has been accepted (see, e.g. [Bec08]) that the apical bundles of dendrites are the basic anatomical units of the neocortex. They are observed in all areas of the cortex that have been investigated in all mammals, including humans. It has been proposed that these bundles are the cortical units for reception, which would give them a preeminent role. Since they are composed essentially of dendrites, the name *dendron* was adopted by John Eccles[20] [Ecc90, Ecc94]. Fig. 1.9(b) illustrates a typical *spine synapse* that makes an intimate contact with an apical dendrite of a pyramidal cell. The inner surface of a bouton confronting the synaptic cleft d forms the presynaptic vesicular grid (PVG). A nerve impulse propagating into a bouton causes a process called exocytosis. A nerve impulse evokes at most a single exocytosis from a PVG [Bec08]. Exocytosis is the basic unitary activity of the cerebral cortex. Each all-or-nothing *exocytosis* of synaptic transmitter substance results in a brief excitatory postsynaptic depolarization (EPSP). Summation by electrotonic transmission of many hundreds of these milli-EPSPs is required for an EPSP large enough (10–20 mV) to generate the discharge of an impulse by a pyramidal cell. This impulse will travel along its axon to make effective excitation at its many synapses. This is the conventional macro-operation of a pyramidal cell of the neocortex, and it can be satisfactorily described by conventional neuroscience, even in the most complex design of neural network theory and neural group selection (see, e.g. [Sze78]).

Experimental analysis of transmitter release by spine synapses of hippocampal pyramidal cells has revealed a remarkably low exocytosis probability per excitatory impulse. This means that there must exist an activation barrier against opening of an ion channel in the PVG. Activation can either occur purely stochastically by thermal fluctuations, or by stimulation of a trigger process. Recently, F. Beck has proposed a two-state *quantum trigger* which is realized by *quasi-particle tunneling*.[21] This is motivated by the

---

[20] Sir John Carew Eccles (January 27, 1903 – May 2, 1997) was an Australian neurophysiologist who won the 1963 Nobel Prize in Physiology or Medicine for his work on the synapse. He shared the prize together with Alan Lloyd Hodgkin and Andrew Fielding Huxley, who formulated the celebrated *Hodgkin–Huxley neural model*.

[21] The so-called *quantum tunneling effect* is a nanoscopic phenomenon in which a quantum particle violates the principles of classical mechanics by penetrating a potential barrier or impedance higher than the kinetic energy of the particle. A barrier, in terms of quantum tunneling, may be a form of energy state analogous to a 'hill' or incline in classical mechanics, which classically suggests that passage through or over such a barrier would be impossible without sufficient energy. On the quantum

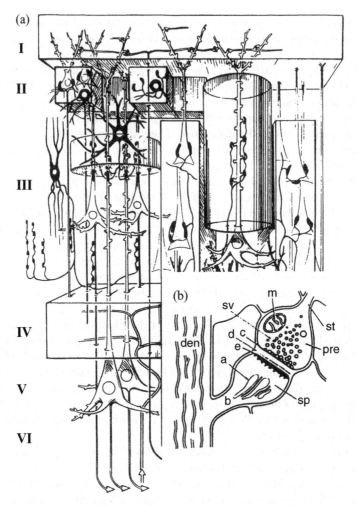

Fig. 1.9. Schematic of *neocortical biophysics* (adapted from [Bec08]): (**a**) 3D construct by [Sze78] showing cortical neurons of various types; there are two pyramidal cells in lamina V and three in lamina III, one being shown in detail in a column to the right, and two in lamina II. (**b**) Detailed structure of a spine (sp) synapse on a dendrite (den); st, axon terminating in the synaptic bouton or presynaptic terminal (Pre); sv, synaptic vesicles; c, presynaptic vesicular grid (PVG in text); d, synaptic cleft; e, postsynaptic membrane; a, spine apparatus; b, spine stalk; m, mitochondrion [Gra82].

predominant role of exocytosis as the synaptic regulator of cortical activity which is certainly not completely at random [Bec08]. The quasi-particle assumption allows the treatment of the complicated molecular transition as an effective one-body problem whose solution follows from the time dependent Schrödinger equation (in normal units, see Chap. 3)

$$i\partial_t \psi(q,t) = -\frac{1}{2}\Delta\psi(q,t) + V(q)\psi(q,t).$$

Here it is assumed that the activated state of the presynaptic cell lasts for a finite time period $[t_0, t_1]$ only before it recombines. Recombination of the activated state defines the measurement process, leading to von Neumann's *collapse of the tunneling state* [Bec08].

The total wave-function $\psi(q,t)$ can be separate at the final time t_1 into two components, representing left and right parts:

$$\psi(q,t_1) = \psi_{\text{left}}(q,t_1) + \psi_{\text{right}}(q,t_1),$$

which constitutes the two amplitudes for the alternative results of the same process: which, after collapse, determine: either exocytosis has happened (ψ_{right}), or exocytosis has not happened (ψ_{right}) (inhibition). State collapse transforms this into the probabilities [Bec08]:

$$\text{Exocytosis probability:} \quad P_{\text{exocyt}}(t_1) = \int |\psi_{\text{right}}(q,t_1)|^2 dq,$$

$$\text{Inhibition probability:} \quad P_{\text{inhibit}}(t_1) = \int |\psi_{\text{right}}(q,t_1)|^2 dq.$$

1.8.3 Quantum Neurodynamics

All theories that hold the neuron as the functional basis of consciousness must bridge a gap between a property of conscious experience and a fundamental tenet of neuronal processing. While evidence suggests that the brain subdivides perceptual processing into modality (e.g., the visual, the tactile) and submodality (e.g., color, temperature), our perceptions themselves are a unified experience [Mas04]. If the anatomic substrate of perceptual modality is functionally and spatially discrete neuronal subpopulations, how is information ultimately synthesized to create the manifest oneness of experience? The *cognitive binding problem*[22] is a central question in the study of consciousness:

---

scale, objects exhibit wave-like behavior; in quantum theory, quanta moving against a potential energy 'hill' can be described by their wave-function, which represents the probability amplitude of finding that particle in a certain location at either side of the 'hill'. If this function describes the particle as being on the other side of the 'hill', then there is the probability that it has moved through, rather than over it, and has thus 'tunneled'.

[22] The term *cognitive binding problem* is attributed to C. von der Mahlsburg [Mal81, Ma96].

how does the brain synthesize its modal and submodal processing systems to generate a unity of conscious experience? *Binding* is thought to occur at virtually all levels of perceptual (and motor) processing, and is thought to be a crucial event for consciousness itself [CK90, Cri94]. Although a relatively recent question in neuroscience, the *binding problem* may have made its first appearance in Immanuel Kant's *Critique of Pure Reason*. Kant's principle of a *transcendental unity of apperception*[23] describes the synthesis of the *knowledge of the manifold* [Kan65]:

> There can be in us no modes of knowledge, no connection or unity of consciousness of one mode of consciousness with another, without that unity of consciousness which precedes all data of intuitions, and by relation to which representation of objects is alone possible... For the mind could never think its identity in the manifoldness of its representations, and indeed think this identity *a priori*, if it did not have before its eyes the identity of its act, whereby it subordinates all synthesis of apprehension (which is empirical) to a transcendental unity, thereby rendering possible their interconnection according to a priori rules.

There have been various solutions proposed for the binding problem, which can be summarized as *binding by convergence* (information is bound by higher order neurons that collect various responses and fire as a *binding unit* when the full set of inputs converge), *binding by assembly* (binding occurs as a result of self-organizing Hebbian cell assemblies), and *binding by synchrony* (binding results from the synchronized firing of neuronal sub-populations that are spatially discrete) [Mas04].

The discussion of distributed domains of reality and consciousness leads us into a recent view of the brain that does not regard the neuron as a unit of cognitive perception, but rather regards the brain as one indivisible entity. This view is based on the emerging field of quantum neurodynamics [Mas04]. The brain, instead of being a Newtonian object obeying classical laws, is posited to be a macroscopic quantum object obeying the same laws found at the Planck scale [Pen94]. The large-scale quantum coherence would render the brain a *Bose–Einstein condensate*, where properties of quantum wave-functions hold at a macroscopic level. These properties would include

---

[23] *Apperception* (from Latin ad + percipere = to perceive) has the following meanings: (i) in epistemology, it is the introspective or reflective apprehension by the mind of its own inner states; (ii) in psychology, it is the process by which new experience is assimilated to and transformed by the residuum of past experience of an individual to form a new whole (in short, it is to perceive new experience in relation to past experience); (iii) in philosophy, Kant distinguished empirical apperception from transcendental apperception (the first is 'the consciousness of the concrete actual self with its changing states', the so-called 'inner sense'; the second is 'the pure, original, unchangeable consciousness which is the necessary condition of experience as such and the ultimate foundation of the synthetic unity of experience').

non-locality and quantum superposition. Such large-scale coherence can be found in superconductivity and superfluidity, where the environment is disentangled from the quantum events. Superconductivity and superfluidity occur at temperatures just above absolute zero, where it is difficult for the environment to interfere with the quantum coherence. H. Frölich, however, predicted that quantum coherence could also occur in the temperature of a biologic environment (e.g. the brain) with high metabolic energy and extreme dielectric properties of the material involved [FK83]. It has been posited by S. Hameroff that the cytoskeletal elements of *microtubules* could be just such a material [Ham87]. Microtubules are composed of tubulin dimers, which can exist in multiple conformations based on dielectric properties, leading to dipole oscillations that could be transmitted through the length of the tubule. These superconducting *Frölich waves* could be transmitted throughout the protein network and gap junctions: from cytoskeleton to membrane protein to extracellular matrix to adjacent membrane protein to cytoskeleton, and so on. It is of interest that the supporting glial cells, the cellular majority in the brain, would contribute to this network, recalling C. Golgi's conception of the *syncytium*. It is currently a point of theoretical contention whether quantum coherence in the brain would decohere from a waveform collapse of superposed states (termed objective reduction) [HP96a], or from the 'warm, wet, and noisy' environment of the brain [Teg00]. It has been suggested that the hollow core of microtubules, ordered water, and actin gelation may buffer the quantum events from the environment. Magnetic resonance imaging of the brain by quantum coherence (albeit an induced artifact of the procedure) has also been suggested as proof of principle [HHT02].

Hameroff has suggested that quantum computation within the microtubules could be involved in consciousness, and proposes the activity of general anesthetics as support for this claim [HW83]. He has argued that the similar activity of diverse chemical structures within the class of anesthetics can be explained by a common action of binding to hydrophobic domains and modulating dipole moments in the tubulin components of microtubules. This is supported by the fact that anesthetics inhibit prokaryotic motility, a function mediated by microtubules. It is of interest that recent studies using quantitative EEG suggest that anesthetics of various pharmacologic properties may act by interrupting cognitive binding.

1.8.4 Bi-Stable Perception and Consciousness

Recall from Gestalt psychology (see, e.g. [HHB01a]) that *bi-stable perception* arises whenever a stimulus can be thought in two different alternatives ways. The essential points are here on the terms 'be aware of ambiguity' and 'manifest state of only one percept at time'. For example consider the ambiguous Gestalt picture called *Rubin face*[24] given in Fig. 1.10. A quantum-mechanical

---

[24] Rubin's vase, also known as the *figure-ground vase*, is a famous set of cognitive optical illusions developed around 1915 by the Danish Gestalt psychologist Edgar

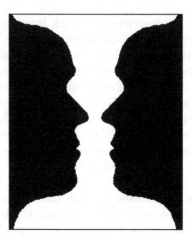

Fig. 1.10. Rubin face: ambiguous figure inducing two different representations, vase or faces. The subject is aware that different representations are possible, but he can perceive only one of the two possible percepts at any one time. The bi-stable perception is described as: $|\psi\rangle = a_{\text{vase}}|1\rangle + b_{\text{fase}}|2\rangle$. In manifest or actual state of consciousness the subject will select or the vase or the faces [Con08].

Rubin. His fundamental *principle of cognitive optical illusions* reads: When two fields have a common border, and one is seen as figure and the other as ground, the immediate perceptual experience is characterized by a shaping effect which emerges from the common border of the fields and which operates only on one field or operates more strongly on one than on the other (see [HHB01a]). The illusion generally presents the viewer with a mental choice of two interpretations, each of which is valid. Often, the viewer sees only one of them, and only realizes the second, valid, interpretation after some time or prompting. When they attempt to simultaneously see the second and first interpretations, they suddenly cannot see the first interpretation anymore, and no matter how they try, they simply cannot encompass both interpretations simultaneously—one occludes the other. The illusions are useful because they are an excellent and intuitive demonstration of the figure-ground distinction the brain makes during visual perception. Rubin's figure-ground distinction, since it involved higher-level cognitive pattern matching, in which the overall picture determines its mental interpretation, rather than the net effect of the individual pieces, influenced the Gestalt psychologists, who discovered many similar illusions themselves. Normally the brain classifies images by what surrounds what, establishing depth and relationships. If something surrounds another thing, the surrounded object is seen as figure, and the presumably further away (and hence background) object is the ground, and vice versa. This makes sense, since if a piece of fruit is lying on the ground, one would want to pay attention to the 'figure' and not the 'ground'. However, when the contours are not so unequal, ambiguity starts to creep into the previously simple inequality, and the brain must begin 'shaping' what it sees; it can be shown that this shaping overrides and is at a higher level than feature recognition processes that pull together the face and the vase images—one can think of the lower levels putting together distinct regions of the picture (each region of counterpart of such mind processes was proposed in [Con08]. In this quantum model of mental states, the state of the potential consciousness is represented by a quantum state-vector in Hilbert space. If we indicate e.g., a 2D case with potential states $|1\rangle$ and $|2\rangle$, the potential state $|\psi\rangle$ of bi-stable perception is given by

$$|\psi\rangle = a|1\rangle + b|2\rangle,$$

with the same *perception energy* E given by the *Schrödinger equation*,

$$i\partial_t|\psi\rangle = E|\psi\rangle.$$

Here a and b represent the respective probability amplitudes (see Sect. 3.1.4 below) of the two stable perception states, so that $|a|^2$ gives the probability that the state of consciousness, represented by the percept $|1\rangle$ will be finally actualized or manifested during perception; conversely, $|b|^2$ represents the probability that state (percept) $|2\rangle$ of consciousness will be actualized or manifested during perception. Closely-related is *binocular rivalry model* of [Man07], with a wave-function between perceptual events describing a state of potential consciousness and *wave-function collapse* when perceptual event is manifested.

On the other hand, *consciousness*, which has traditionally been based on Descartes' statement "Cogito ergo sum," refers to our capacity of asking questions about the so-called reality (see [Pop08]). It concerns the field of 'potential information' instead of the so-called 'actual information'.[25] In order to describe its contents, a measure of potential information was suggested in terms of the capacity of actualities in the possible contents. This definition includes a variety of relevant consequences, i.e., the appearance of a memory, the question of time down to relation of molecular *biophotons*.[26] Consciousness works for evolutionary purposes which is not a matter of fact but a process. The improvement and the optimization of consciousness is at the same time the most important healing power of life. In other words, the

---

which makes sense in isolation), but when the brain tries to make sense of it as a whole, contradictions ensue, and patterns must be discarded.

[25] The simplest system where actual and potential information are mutually transformed are *cavity resonators*. The potential information (in bit) is identical to the Q-value of the resonator. The actual information reflects the distribution of the coded sequence on the emitted wave. The actual information cannot exceed there the potential one. It corresponds to the maximum number of switch-on/off processes of a wave that is reflected N times within the resonator.

[26] Biophotons (BPHs) are weak photons within or emitted from living organisms. BPH emission originates from a de-localized coherent electromagnetic field within living organisms and is regulated by the field. Based on experimental results concerning Poisson and sub-Poisson distributions of photocount statistics, the coherent properties of BPHs and their functions in cell communication are described in [Cha08]. Functions of BPH roles are discussed in some important processes, including DNA replication, transcription, protein synthesis, and cell signaling, and in the processes of oxidative phosphorylation and photosynthesis.

communication system of biological systems is based on cavity resonator waves and light guides. This example shows the identity of mitotic figures with the force pattern of cavity resonator waves within the cells. These forces regulate the migration of the molecules in such a way that no error takes place. At the same time such a pattern represents in the average one coherent biophoton. They are the origin of consciousness.

Popp distinguishes four different 'global' forms of consciousness [Pop08]:

1. Self confidence according to the reference point of Descartes: "I am," resulting from the transformation of the physical existence of the body into 'doubts' and back to the control, and so on. The end of this process is the adjustment of actual and potential information.
2. Identification (awareness) according to the transformation of the actual information of the physical existence of matter or radiation into 'doubts' and back to the control, and so on. This leads after adjusting actual and potential information to the confirmation of "You are" or "It is."
3. Prediction: Repeated transformations of the actual information of the physical existence of different objects into the possibility field of the potential information under adjustment of the potential information of the memory which provide boundary conditions in the control processes. The transformations are again finished as soon as there is an equality of actual and potential information. The result is a statement: "It will be."
4. Memory and Inspiration. Repeated transformation and re-transformation of potential information of the past and the presence. After adjustment of these probability distributions the statement will follow: "It could have been," or "It could become."

1.9 Notational Conventions

Throughout the book we will use some of the following *conventions*:

(i) natural units, in which the Planck constant, particle mass and speed of light are all normalized to unity, i.e., $\hbar = m = c = 1$;
(ii) $i = \sqrt{-1}$, $\dot{z} = dz/dt$, $\partial_z = \partial/\partial z$;
(iii) Einstein's summation convention over repeated indices;
(iv) nD means n-dimensional;
(v) functorial meta-language (see Appendix) is often used; for example, a *generic system evolution* 2-functor $\mathcal{E}$, is given by

$$
\begin{array}{ccc}
A \xrightarrow{\;\;f\;\;} B & & \mathcal{E}(A) \xrightarrow{\;\;\mathcal{E}(f)\;\;} \mathcal{E}(B) \\
h\downarrow \;\;\text{CURRENT}\;\; \downarrow g & \xrightarrow{\;\;\mathcal{E}\;\;} & \mathcal{E}(h)\downarrow \;\;\text{DESIRED}\;\; \downarrow \mathcal{E}(g) \\
\;\;\text{SYSTEM}\;\; & & \;\;\text{SYSTEM}\;\; \\
\;\;\text{STATE}\;\; & & \;\;\text{STATE}\;\; \\
C \xrightarrow[\;\;k\;\;]{} D & & \mathcal{E}(C) \xrightarrow[\;\;\mathcal{E}(k)\;\;]{} \mathcal{E}(D)
\end{array}
$$

Here $\mathcal{E}$ represents an association/projection functor from the *source 2-category* of the current system state, defined as a commutative square of small system categories $A, B, C, D, \ldots$ of *current system components* and their causal interrelations $f, g, h, k, \ldots$, onto the *target 2-category* of the desired system state, defined a as a commutative square of small system categories $\mathcal{E}(A), \mathcal{E}(B), \mathcal{E}(C), \mathcal{E}(D), \ldots$ of *evolved system components* and their causal interrelations $\mathcal{E}(f), \mathcal{E}(g), \mathcal{E}(h), \mathcal{E}(k)$. As in the previous section, each causal arrow in above diagram, e.g., $f : A \to B$, stands for a *generic system dynamorphism*.

2

Brain and Classical Neural Networks

In this Chapter we give a modern review of classical neurodynamics, including brain physiology, biological and artificial neural networks, synchronization, spike neural nets and wavelet resonance, motor control and learning.

2.1 Human Brain

Recall that *human brain* is the most complicated mechanism in the Universe. Roughly, it has its own physics and physiology. A closer look at the brain points to a rather *recursively hierarchical structure* and a very elaborate organization (see [II06b] for details). An average brain weighs about 1.3 kg, and it is made of: $\sim 77\%$ water, $\sim 10\%$ protein, $\sim 10\%$ fat, $\sim 1\%$ carbohydrates, $\sim 0.01\%$ DNA/RNA. The largest part of the human brain, the cerebrum, is found on the top and is divided down the middle into left and right cerebral hemispheres, and front and back into *frontal lobe*, *parietal lobe*, *temporal lobe*, and *occipital lobe*. Further down, and at the back lies a rather smaller, spherical portion of the brain, the *cerebellum*, and deep inside lie a number of complicated structures like the *thalamus*, *hypothalamus*, *hippocampus*, etc.

Both the cerebrum and the cerebellum have comparatively thin outer surface layers of grey matter and larger inner regions of white matter. The grey regions constitute what is known as the *cerebral cortex* and the *cerebellar cortex*. It is in the *grey matter* where various kinds of *computational tasks* seem to be performed, while the *white matter* consists of long nerve fibers (axons) carrying signals from one part of the brain to another. However, despite of its amazing computational abilities, brain is not a computer, at least not a 'Von Neumann computer' [Neu58], but rather a huge, hierarchical, neural network. It is the cerebral cortex that is central to the higher brain functions, speech, thought, complex movement patterns, etc. On the other hand, the cerebellum seems to be more of an 'automaton'. It has to do more with precise coordination and control of the body, and with skills that have become our 'second nature'. Cerebellum actions seem almost to take place by themselves, without

V.G. Ivancevic, T.T. Ivancevic, *Quantum Neural Computation*,
Intelligent Systems, Control and Automation: Science and Engineering 40,
DOI 10.1007/978-90-481-3350-5_2, © Springer Science+Business Media B.V. 2010

thinking about them. They are very similar to the unconscious reflex actions, e.g., reaction to pinching, which may not be mediated by the brain, but by the upper part of the spinal column.

Various regions of the cerebral cortex are associated with very specific functions. The *visual cortex*, a region in the occipital lobe at the back of the brain, is responsible for the reception and interpretation of vision. The *auditory cortex*, in the temporal lobe, deals mainly with analysis of sound, while the *olfactory cortex*, in the frontal lobe, deals with smell. The *somatosensory cortex*, just *behind* the division between frontal and parietal lobes, has to do with the sensations of touch. There is a very specific mapping between the various parts of the surface of the body and the regions of the somatosensory cortex. In addition, just in *front* of the division between the frontal and parietal lobes, in the frontal lobe, there is the *motor cortex*. The *motor cortex* activates the movement of different parts of the body and, again here, there is a very specific mapping between the various muscles of the body and the regions of the motor cortex. All the above mentioned regions of the cerebral cortex are referred to as *primary*, since they are the one most directly concerned with the input and output of the brain. Near to these primary regions are the *secondary* sensory regions of the cerebral cortex, where information is processed, while in the *secondary motor* regions, conceived plans of motion get translated into specific directions for actual muscle movement by the primary motor cortex. The most abstract and sophisticated activity of the brain is carried out in the remaining regions of the cerebral cortex, the *association cortex*.

The basic building blocks of the brain are the nerve cells or neurons. Among about 200 types of different basic types of human cells, the neuron is one of the most specialized, exotic and remarkably versatile cell. The neuron is highly unusual in three respects: its *variation in shape*, its *electrochemical function*, and its *connectivity*, i.e., its ability to link up with other neurons in networks. Let us start with a few elements of neuron microanatomy (see, e.g. [KSJ91, BS91]). There is a central starlike bulb, called the *soma*, which contains the nucleus of the cell. A long nerve fibre, known as the *axon*, stretches out from one end of the soma. Its length, in humans, can reach up to few *cm*. The function of an axon is to transmit the neuron's output signal, in which it acts like a *coaxial cable*. The axon has the ability of multiple bifurcation, branching out into many smaller nerve fibers, and the very end of which there is always a *synaptic knob*. At the other end of the soma and often springing off in all directions from it, are the tree-like *dendrites*, along which input data are carried into the soma. The whole nerve cell, as basic unit, has a cell membrane surrounding soma, axon, synoptic knobs, and dendrites. Signals pass from one neuron to another at junctions known as *synapses*, where a synaptic knob of one neuron is attached to another neuron's soma or dendrites. There is very narrow gap, of a few *nm*, between the synaptic knob and the soma/dendrite to where the *synaptic cleft* is attached. The signal from one neuron to another has to propagate across this gap.

A nerve fibre is a cylindrical tube containing a mixed solution of NaCl and KCl, mainly the second, so there are Na^+, K^+, and Cl^- ions within the tube. Outside the tube the same type of ions are present but with more Na^+ than K^+. In the *resting* state there is an excess of Cl^- over Na^+ and K^+ inside the tube, giving it a negative charge, while it has positive charge outside. A nerve signal is a region of *charge reversal* traveling along the fibre. At its head, *sodium gates* open to allow the sodium to flow inwards and at its tail *potassium gates* open to allow potassium to flow outwards. Then, metabolic pumps act to restore order and establish the *resting state*, preparing the nerve fibre for another signal. There is no major material (ion) transport that produces the signal, just in and out local movements of ions, across the cell membranes, i.e., a *small* and *local* depolarization of the cell. Eventually, the nerve signal reaches the attached synaptic knob, at the very end of the nerve fibre, and triggers it to emit chemical substances, known as *neurotransmitters*. It is these substances that travel across the synaptic cleft to another neuron's soma or dendrite. It should be stressed that the signal here is not electrical, but a chemical one. What really is happening is that when the nerve signal reaches the synaptic knob, the local depolarization cause little bags immersed in the *vesicular grid*, the *vesicles* containing molecules of the neurotransmitter chemical (e.g., acetylcholine) to release their contents from the neuron into the synaptic cleft, the phenomenon of *exocytosis*. These molecules then diffuse across the cleft to interact with *receptor proteins* on receiving neurons. On receiving a neurotransmitter molecule, the receptor protein opens a gate that causes a local depolarization of the receiver neuron.

It depends on the nature of the synaptic knob and of the specific synaptic junction, if the next neuron would be encouraged to *fire*, i.e., to start a new signal along its own axon, or it would be discouraged to do so. In the former case we are talking about *excitatory synapses*, while in the latter case about *inhibitory synapses*. At any given moment, one has to add up the effect of all excitatory synapses and subtract the effect of all the inhibitory ones. If the net effect corresponds to a positive electrical potential difference between the inside and the outside of the neuron under consideration, *and* if it is bigger than a critical value, then the neuron *fires*, otherwise it stays mute.

The basic dynamical process of *neural communication* can be summarized in the following three steps: [Nan95]

1. The neural axon is an *all or none* state. In the *all state* a signal, called a *spike* or *action potential* (AP), propagates indicating that the summation performed in the soma produced an amplitude of the order of tens of mV. In the *none state* there is no signal traveling in the axon, only the resting potential (~ -70mV). It is essential to notice that the presence of a traveling signal in the axon, *blocks* the possibility of transmission of a second signal.

2. The nerve signal, upon arriving at the ending of the axon, triggers the emission of neurotransmitters in the synaptic cleft, which in turn cause

the receptors to open up and allow the penetration of ionic current into the *post synaptic* neuron. The *efficacy* of the synapse is a parameter specified by the amount of penetrating current per presynaptic spike.

3. The post synaptic potential (PSP) diffuses toward the soma, where all inputs in a short period, from all the presynaptic neurons connected to the postsynaptic are summed up. The amplitude of individual PSP's is about 1 mV, thus quite a number of inputs is required to reach the 'firing' threshold, of tens of mV. Otherwise the postsynaptic neuron remains in the *resting* or *none* state.

The cycle-time of a neuron, i.e., the time from the emission of a spike in the presynaptic neuron to the emission of a spike in the postsynaptic neuron is of the order of 1–2 ms. There is also some recovery time for the neuron, after it fired, of about 1–2 ms, independently of how large the amplitude of the depolarizing potential would be. This period is called the absolute refractory period of the neuron. Clearly, it sets an upper bound on the spike frequency of 500–1000/s. In the types of neurons that we will be interested in, the spike frequency is considerably lower than the above upper bound, typically in the range of 100/s, or even smaller in some areas, at about 50/s. It should be noticed that this rather exotic neural communication mechanism works very efficiently and it is employed universally, both by vertebrates and invertebrates. The vertebrates have gone even further in perfection, by protecting their nerve fibers by an insulating coating of myelin, a white fatty substance, which incidentally gives the white matter of the brain, discussed above, its color. Because of this insulation, the nerve signals may travel undisturbed at about 120 m/s [Nan95].

A very important anatomical fact is that each neuron receives some 10^4 synaptic inputs from the axons of other neurons, usually one input per presynaptic neuron, and that each branching neural axon forms about the same number ($\sim 10^4$) of synaptic contacts on other, postsynaptic neurons. A closer look at our cortex then would expose a mosaic-type structure of assemblies of a few thousand densely connected neurons. These assemblies are taken to be the basic cortical processing *modules*, and their size is about $1(\text{mm})^2$. The neural connectivity gets much sparser as we move to larger scales and with much less feedback, allowing thus for autonomous local collective, parallel processing and more serial and integrative processing of local collective outcomes. Taking into account that there are about 10^{11} nerve cells in the brain (about 7×10^{10} in the cerebrum and 3×10^{10} in the cerebellum), we are talking about 10^{15} synapses.

While the dynamical process of neural communication suggests that the brain action looks a lot like a computer action, there are some fundamental differences having to do with a basic brain property called *brain plasticity*. The interconnections between neurons are not fixed, as is the case in a computer-like model, but are changing all the time. These are *synaptic junctions* where the communication between different neurons actually takes place. The synaptic junctions occur at places where there are *dendritic spines* of suitable form such that contact with the synaptic knobs can be made. Under certain conditions these dendritic spines can shrink away and break contact, or they can grow and make new contact, thus determining the *efficacy* of the synaptic junction. Actually, it seems that it is through these dendritic spine changes, in synaptic connections, that long-term memories are laid down, by providing the means of storing the necessary information. A supporting indication of such a conjecture is the fact that such dendritic spine changes occur within *seconds*, which is also how long it takes for permanent memories to be laid down [Pen89, Pen94, Pen97, Sta93].

Furthermore, a very useful set of phenomenological rules has been put forward by Hebb [Heb49], the *Hebb rules*, concerning the underlying mechanism of brain plasticity. According to Hebb, a synapse between neuron 1 and neuron 2 would be strengthened whenever the firing of neuron 1 is followed by the firing of neuron 2, and weakened whenever it is not. It seems that *brain plasticity* is a *fundamental property* of the activity of the brain.

Many mathematical models have been proposed to try to simulate *learning process*, based upon the close resemblance of the dynamics of neural communication to computers and implementing, one way or another, the essence of the Hebb rules. These models are known as *neural networks*. They are closely related to *adaptive Kalman filtering* (see Sect. 7.2.6) as well as *adaptive control* (see [II06c]).

Let us try to construct a *neural network* model for a set of N interconnected neurons (see e.g., [Ama77]). The activity of the neurons is usually parameterized by N functions $\sigma_i(t)$, $(i = 1, 2, \ldots, N)$, and the synaptic strength, representing the synaptic efficacy, by $N \times N$ functions $j_{ik}(t)$. The total stimulus of the network on a given neuron (i) is assumed to be given simply by the sum of the stimuli coming from each neuron

$$S_i(t) = j_{ik}(t)\sigma_k(t), \quad \text{(summation over } k)$$

where we have identified the individual stimuli with the product of the synaptic strength j_{ik} with the activity σ_k of the neuron producing the individual stimulus. The dynamic equations for the neuron are supposed to be, in the simplest case

$$\dot{\sigma}_i = F(\sigma_i, S_i), \tag{2.1}$$

with F a nonlinear function of its arguments. The dynamic equations controlling the time evolution of the synaptic strengths $j_{ik}(t)$ are much more involved and only partially understood, and usually it is assumed that the j-dynamics is such that it produces the synaptic couplings. The simplest version of a neural network model is the *Hopfield model* [Hop82]. In this model the neuron activities are conveniently and conventionally taken to be *switch*-like, namely ± 1, and the time t is also an integer-valued quantity. This *all*$(+1)$ or *none*(-1) neural activity σ_i is based on the neurophysiology. The choice ± 1 is more natural the usual 'binary' one ($b_i = 1$ or 0), from a physicist's point of view corresponding to a two-state system, like the fundamental elements of the ferromagnet, i.e., the electrons with their spins up $(+)$ or $(-)$.

The increase of time t by one unit corresponds to one step for the dynamics of the neuron activities obtainable by applying (for all i) the rule

$$\sigma_i\left(t + \frac{i+1}{N}\right) = \text{sign}(S_i(t + i/N)), \tag{2.2}$$

which provides a rather explicit form for (2.1). If, as suggested by the Hebb rules, the j matrix is *symmetric* $(j_{ik} = j_{ki})$, the Hopfield dynamics [Hop82] corresponds to a sequential algorithm for looking for the minimum of the Hamiltonian

$$H = -S_i(t)\sigma_i(t) = -j_{ik}\sigma_i(t)\sigma_k(t).$$

The Hopfield model, at this stage, is very similar to the dynamics of a statistical mechanics *Ising-type* [Gol92], or more generally a *spin-glass*, model [Ste92]. This *mapping* of the Hopfield model to a spin-glass model is highly advantageous because we have now a justification for using the statistical mechanics language of phase transitions, like critical points or attractors, etc, to describe neural dynamics and thus brain dynamics. This simplified Hopfield model has many *attractors*, corresponding to many different *equilibrium* or *ordered* states, endemic in spin-glass models, and an unavoidable prerequisite for successful storage, in the brain, of many different patterns of activities. In the neural network framework, it is believed that an internal representation (i.e., a pattern of neural activities) is associated with each object or category that we are capable of recognizing and remembering. According to neurophysiology, it is also believed that an object is memorized by suitably changing the synaptic strengths. The so-called *associative memory* (see next section) is generated produced in this scheme as follows [Nan95]: An external stimulus, suitably involved, produces synaptic strengths such that a specific learned pattern $\sigma_i(0) = P_i$ is 'printed' in such a way that the neuron activities $\sigma_i(t) \sim P_i$ (II *learning*), meaning that the σ_i will remain for all times close to P_i, corresponding to a stable attractor point (III *coded brain*). Furthermore, if a *replication signal* is applied, pushing the neurons to σ_i values *partially* different from P_i, the neurons should evolve toward the P_i. In other words, the memory is able to retrieve the information on the whole object, from the knowledge of a part of it, or even in the presence of wrong information (IV *recall process*). Clearly, if the external stimulus is very different from any preexisting $\sigma_i = P_i$ pattern, it may either create a new pattern, i.e., create a new attractor point, or it may reach a chaotic, random behavior (I *uncoded brain*).

Despite the remarkable progress that has been made during the last few years in understanding brain function using the neural network paradigm, it is fair to say that neural networks are rather artificial and a very long way from providing a realistic model of brain function. It seems likely that the mechanisms controlling the changes in synaptic connections are much more complicated and involved than the ones considered in NN, as utilizing cytoskeletal restructuring of the sub-synaptic regions. *Brain plasticity* seems to play an essential, central role in the workings of the brain! Furthermore, the 'binding problem, i.e., how to *bind* together all the neurons firing to different features of the same object or category, especially when more than one object is perceived during a *single* conscious perceptual moment, seems to remain unanswered [Nan95]. In this way, we have come a long way since the times of the 'grandmother neuron', where a *single* brain location was invoked for self observation and control, identified with the pineal glands by Descartes [Ami89b].

It has been long suggested that different groups of neurons, responding to a common object/category, fire *synchronously*, implying *temporal correlations* [SG95]. If true, such correlated firing of neurons may help us in resolving the binding problem [Cri94]. Actually, brain waves recorded from the scalp, i.e., the EEGs, suggest the existence of some sort of *rhythms*, e.g., the 'α-rhythms' of a frequency of 10 Hz. More recently, oscillations were clearly observed in the visual cortex. Rapid oscillations, above EEG frequencies in the range of 35 to 75 Hz, called the 'γ-oscillations' or the '40 Hz oscillations', have been detected in the cat's visual cortex [SG95]. Furthermore, it has been shown that these oscillatory responses can become *synchronized* in a stimulus-dependent manner. Studies of auditory-evoked responses in humans have shown inhibition of the 40 Hz coherence with *loss of consciousness* due to the induction of general anesthesia [SB74]. These striking results have prompted Crick and Koch to suggest that this *synchronized firing* on, or near, the beat of a 'γ-oscillation' (in the 35–75 Hz range) might be the *neural correlate* of *visual awareness* [Cri94]. Such a behavior would be, of course, a very special case of a much more general framework where coherent firing of *widely-distributed, non-local* groups of neurons, in the 'beats' of x-oscillation (of specific frequency ranges), *bind* them together in a mental representation, expressing the *oneness of consciousness* or *unitary sense of self*. While this is a bold suggestion [Cri94], it is should be stressed that in a physicist's language it corresponds to a phenomenological explanation, not providing the underlying physical mechanism, based on neuron dynamics, that triggers the synchronized neuron firing [Nan95]. On the other hand, the *Crick–Koch binding hypothesis* [CK90, Cri94] is very suggestive (see Fig. 2.1) and in compliance with the *central biodynamic adjunction* [II06b] (see Appendix, Section on categories and functors)

$$coordination = sensory \dashv motor : brain \leftrightarrows body.$$

On the other hand, E.M. Izhikevich, Editor-in-Chief of the new Encyclopedia of Computational Neuroscience, considers brain as a *weakly-connected neural network* [Izh99b], consisting of n quasi-periodic cortical oscillators $X_1, \ldots, X_n$ forced by the thalamic input X_0 (see Fig. 2.2)

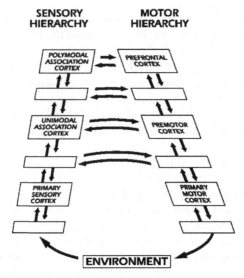

Fig. 2.1. Fiber connections between cortical regions participating in the *perception-action cycle*, reflecting again our sensory-motor adjunction. Empty rhomboids stand for intermediate areas or subareas of the labeled regions. Notice that there are connections between the two hierarchies at several levels, not just at the top level.

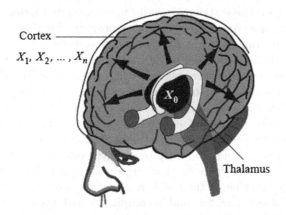

Fig. 2.2. A 1-to-many relation: *Thalamus* $\Rightarrow$ *Cortex* in the human brain (with permission from E. Izhikevich).

2.1.1 Basics of Brain Physiology

The nervous system consists basically of two types of cells: neurons and glia. *Neurons* (also called nerve cells, see Fig. 2.3) are the primary cells, morphologic and functional units of the nervous system. They are found in the brain, the spinal cord and in the peripheral nerves and ganglia. Neurons consist of four major parts, including the *dendrites* (shorter projections), which are responsible for receiving stimuli; the *axon* (longer projection), which sends the nerve impulse away from the cell; the *cell body*, which is the site of metabolic activity in the cell; and the *axon terminals*, which connect neurons to other neurons, or neurons to other body structures. Each neuron can have several hundred axon terminals that attach to another neuron multiple times, attach to multiple neurons, or both. Some types of neurons, such as Purkinje cells, have over 1000 dendrites. The body of a neuron, from which the axon and dendrites project, is called the soma and holds the nucleus of the cell. The nucleus typically occupies most of the volume of the soma and is much larger in diameter than the axon and dendrites, which typically are only about a micrometer thick or less. Neurons join to one another and to other cells through synapses.

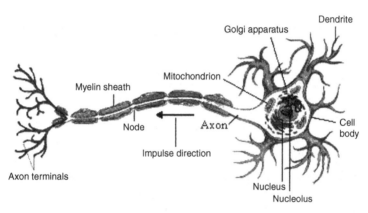

Fig. 2.3. A typical neuron, containing all of the usual cell organelles. However, it is highly specialized for the conductance of nerve impulse.

A defining feature of neurons is their ability to become 'electrically excited', that is, to undergo an action potential—and to convey this excitation rapidly along their axons as an impulse. The narrow cross section of axons and dendrites lessens the metabolic expense of conducting action potentials, although fatter axons convey the impulses more rapidly, generally speaking.

Many neurons have insulating sheaths of myelin around their axons, which enable their action potentials to travel faster than in unmyelinated axons of the same diameter. Formed by glial cells, the myelin sheathing normally runs along the axon in sections about 1 mm long, punctuated by unsheathed nodes of Ranvier. Neurons and glia make up the two chief cell types of the nervous system.

An action potential that arrives at its terminus in one neuron may provoke an action potential in another through release of neurotransmitter molecules across the synaptic gap.

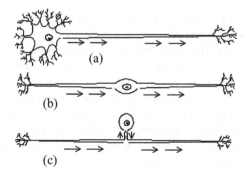

Fig. 2.4. Three structural classes of human neurons: (a) multipolar, (b) bipolar, and (c) unipolar.

There are three structural classes of neurons in the human body (see Fig. 2.4):

1. The *multipolar neurons*, the majority of neurons in the body, in particular in the central nervous system.
2. The *bipolar neurons*, sensory neurons found in the special senses.
3. The *unipolar neurons*, sensory neurons located in dorsal root ganglia.

Neuronal Circuits

Figure 2.5 depicts a general model of a convergent circuit, showing two neurons converging on one neuron. This allows one neuron or *neuronal pool* to receive input from multiple sources. For example, the neuronal pool in the brain that regulates rhythm of breathing receives input from other areas of the brain, baroreceptors, chemoreceptors, and stretch receptors in the lungs.

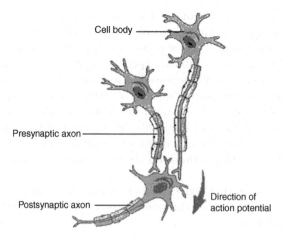

Fig. 2.5. A convergent neural circuit: nerve impulses arriving at the same neuron.

Glia are specialized cells of the nervous system whose main function is to 'glue' neurons together. Specialized glia called *Schwann cells* secrete myelin sheaths around particularly long axons. Glia of the various types greatly outnumber the actual neurons.

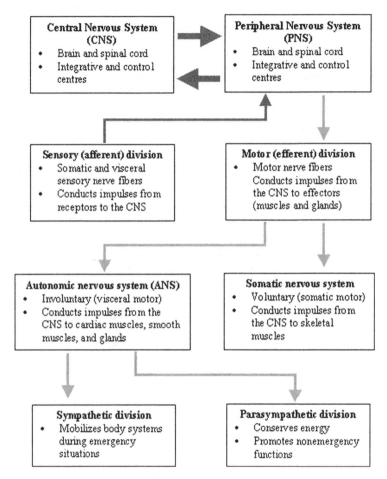

Fig. 2.6. Organization of the human nervous system.

The human nervous system consists of the central and peripheral parts (see Fig. 2.6). The *central nervous system* (CNS) refers to the core nervous system, which consists of the *brain* and *spinal cord* (as well as *spinal nerves*). The *peripheral nervous system* (PNS) consists of the nerves and neurons that reside or extend outside the central nervous system—to serve the limbs and organs, for example. The peripheral nervous system is further divided into the somato-motoric nervous system and the autonomic nervous system (see Fig. 2.7).

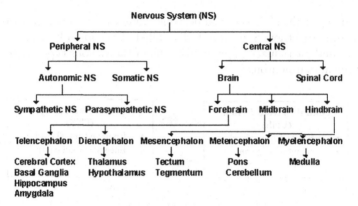

Fig. 2.7. Basic divisions of the human nervous system.

The CNS is further divided into two parts: the *brain* and the *spinal cord*. The average adult human brain weighs 1.3 to 1.4 kg (approximately 3 pounds). The brain contains about 100 billion nerve cells (neurons) and trillions of 'support cells' called glia. Further divisions of the human brain are depicted in Fig. 2.8. The spinal cord is about 43 cm long in adult women and 45 cm long in adult men and weighs about 35–40 grams. The vertebral column, the collection of bones (back bone) that houses the spinal cord, is about 70 cm long. Therefore, the spinal cord is much shorter than the vertebral column.

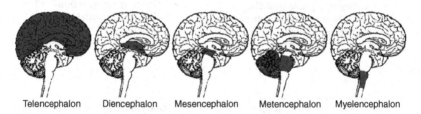

Telencephalon   Diencephalon   Mesencephalon   Metencephalon   Myelencephalon

Fig. 2.8. Basic divisions of the human brain.

The PNS is further divided into two major parts: the *somatic nervous system* and the *autonomic nervous system*.

The somatic nervous system consists of *peripheral nerve fibers* that send sensory information to the central nervous system and *motor nerve fibers* that project to skeletal muscle.

The autonomic nervous system (ANS) controls smooth muscles of the viscera (internal organs) and glands. In most situations, we are unaware of the workings of the ANS because it functions in an involuntary, reflexive manner. For example, we do not notice when blood vessels change size or when our heart beats faster. The ANS is most important in two situations:

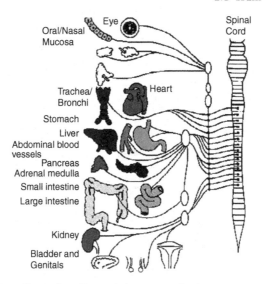

Fig. 2.9. Basic functions of the sympathetic nervous system.

1. In *emergencies* that cause stress and require us to 'fight' or take 'flight', and
2. In *non-emergencies* that allow us to 'rest' and 'digest'.

The ANS is divided into three parts:

1. The *sympathetic nervous system* (see Fig. 2.9),
2. The *parasympathetic nervous system* (see Fig. 2.10), and
3. The *enteric nervous system*, which is a meshwork of nerve fibers that innervate the viscera (gastrointestinal tract, pancreas, gall bladder).

In the PNS, neurons can be functionally divided in 3 ways:

1. • *sensory* (*afferent*) neurons—carry information into the CNS from sense organs, and
   • *motor* (*efferent*) neurons—carry information away from the CNS (for muscle control).
2. • *cranial* neurons—connect the brain with the periphery, and
   • *spinal* neurons—connect the spinal cord with the periphery.
3. • *somatic* neurons—connect the skin or muscle with the central nervous system, and
   • *visceral* neurons—connect the internal organs with the central nervous system.

Some differences between the PNS and the CNS are:

1. • In the CNS, collections of neurons are called *nuclei*.
   • In the PNS, collections of neurons are called *ganglia*.

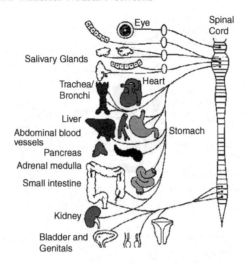

Fig. 2.10. Basic functions of the parasympathetic nervous system.

2. • In the CNS, collections of axons are called *tracts*.
   • In the PNS, collections of axons are called *nerves*.

Basic Brain Partitions and Their Functions

Cerebral Cortex. The word 'cortex' comes from the Latin word for 'bark' (of a tree). This is because the cortex is a sheet of tissue that makes up the outer layer of the brain. The thickness of the cerebral cortex varies from 2 to 6 mm. The right and left sides of the cerebral cortex are connected by a thick band of nerve fibers called the 'corpus callosum'. In higher mammals such as humans, the cerebral cortex looks like it has many bumps and grooves. A bump or bulge on the cortex is called a gyrus (the plural of the word gyrus is 'gyri') and a groove is called a sulcus (the plural of the word sulcus is 'sulci'). Lower mammals like rats and mice have very few gyri and sulci. The main cortical functions are: *thought, voluntary movement, language, reasoning*, and *perception*.

Cerebellum. The word 'cerebellum' comes from the Latin word for 'little brain'. The cerebellum is located behind the brain stem. In some ways, the cerebellum is similar to the cerebral cortex: the cerebellum is divided into hemispheres and has a cortex that surrounds these hemispheres. Its main functions are: *movement, balance*, and *posture*.

Brain Stem. The brain stem is a general term for the area of the brain between the thalamus and spinal cord. Structures within the brain stem include the medulla, pons, tectum, reticular formation and tegmentum. Some of these areas are responsible for the most basic functions of life such as breathing, heart rate and blood pressure. Its main functions are: *breathing, heart rate*, and *blood pressure*.

Hypothalamus. The *hypothalamus* is composed of several different areas and is located at the base of the brain. Although it is the size of only a pea (about 1/300 of the total brain weight), the hypothalamus is responsible for some very important functions. One important function of the hypothalamus is the control of body temperature. The hypothalamus acts like a 'thermostat' by sensing changes in body temperature and then sending signals to adjust the temperature. For example, if we are too hot, the hypothalamus detects this and then sends a signal to expand the capillaries in your skin. This causes blood to be cooled faster. The hypothalamus also controls the pituitary. Its main functions are: *body temperature, emotions, hunger, thirst, sexual instinct,* and *circadian rhythms*. The *hypothalamus* is 'the boss' of the ANS.

Thalamus. The thalamus receives sensory information and relays this information to the cerebral cortex. The cerebral cortex also sends information to the thalamus which then transmits this information to other areas of the brain and spinal cord. Its main functions are: *sensory processing* and *movement*.

Limbic System. The limbic system (or the limbic areas) is a group of structures that includes the amygdala, the hippocampus, mammillary bodies and cingulate gyrus. These areas are important for controlling the emotional response to a given situation. The hippocampus is also important for memory. Its main function is *emotions*.

Hippocampus. The hippocampus is one part of the limbic system that is important for memory and learning. Its main functions are: *learning* and *memory*.

Basal Ganglia. The basal ganglia are a group of structures, including the globus pallidus, caudate nucleus, subthalamic nucleus, putamen and substantia nigra, that are important in coordinating movement. Its main function is *movement*.

Midbrain. The midbrain includes structures such as the superior and inferior colliculi and red nucleus. There are several other areas also in the midbrain. Its main functions are: *vision, audition, eye movement* (see Fig. 2.11), and *body movement*.

Nerves

A *nerve* is an enclosed, cable-like bundle of *nerve fibers* or *axons*, which includes the glia that ensheathe the axons in myelin (see [Mar98, II06b]).

Nerves are part of the peripheral nervous system. *Afferent nerves* convey sensory signals to the brain and spinal cord, for example from skin or organs, while *efferent* nerves conduct stimulatory signals from the *motor neurons* of the brain and spinal cord to the muscles and glands.

These signals, sometimes called nerve impulses, are also known as action potentials: Rapidly traveling electrical waves, which begin typically in the cell body of a neuron and propagate rapidly down the axon to its tip or terminus'.

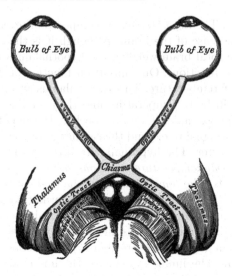

Fig. 2.11. Optical chiasma: the point of cross-over for optical nerves. By means of it, information presented to either left or right visual half-field is projected to the contralateral occipital areas in the visual cortex. For more details, see e.g. [II06b].

Nerves may contain fibers that all serve the same purpose; for example *motor nerves*, the axons of which all terminate on muscle fibers and stimulate contraction. Or they be *mixed nerves*.

An *axon*, or 'nerve fibre', is a long slender projection of a nerve cell or neuron, which conducts electrical impulses away from the neuron's cell body or soma. Axons are in effect the primary transmission lines of the nervous system, and as bundles they help make up nerves. The axons of many neurons are sheathed in myelin.

On the other hand, a *dendrite* is a slender, typically branched projection of a nerve cell or neuron, which conducts the electrical stimulation received from other cells through synapses to the body or soma of the cell from which it projects.

Many dendrites convey this stimulation passively, meaning without action potentials and without activation of voltage-gated ion channels. In such dendrites the voltage change that results from stimulation at a synapse may extend both towards and away from the soma. In other dendrites, though an action potential may not arise, nevertheless voltage-gated channels help to propagate excitatory synaptic stimulation. This propagation is efficient only toward the soma due to an uneven distribution of channels along such dendrites.

The structure and branching of a neuron's dendrites strongly influences how it integrates the input from many others, particularly those that input only weakly (more at synapse). This integration is in aspects 'temporal'—involving the summation of stimuli that arrive in rapid succession—as well as 'spatial'—entailing the aggregation of excitatory and inhibitory inputs from separate branches or 'arbors'.

Spinal nerves take their origins from the spinal cord. They control the functions of the rest of the body. In humans, there are 31 pairs of spinal nerves: 8 *cervical*, 12 *thoracic*, 5 *lumbar*, 5 *sacral* and 1 *coccygeal*.

Neural Action Potential

As the traveling signals of nerves and as the localized changes that contract muscle cells, *action potentials* are an essential feature of animal life. They set the pace of thought and action, constrain the sizes of evolving anatomies and enable centralized control and coordination of organs and tissues (see [Mar98]).

Basic Features

When a biological cell or patch of membrane undergoes an action potential, the polarity of the transmembrane voltage swings rapidly from negative to positive and back. Within any one cell, consecutive action potentials typically are indistinguishable. Also between different cells the amplitudes of the voltage swings tend to be roughly the same. But the speed and simplicity of action potentials vary significantly between cells, in particular between different cell types.

Minimally, an action potential involves a *depolarization*, a *re-polarization* and finally a *hyperpolarization* (or 'undershoot'). In specialized muscle cells of the heart, such as the pacemaker cells, a 'plateau phase' of intermediate voltage may precede re-polarization.

Underlying Mechanism

The transmembrane voltage changes that take place during an action potential result from changes in the permeability of the membrane to specific ions, the internal and external concentrations of which are in imbalance. In the axon fibers of nerves, depolarization results from the inward rush of sodium ions, while re-polarization and hyperpolarization arise from an outward rush of potassium ions. Calcium ions make up most or all of the depolarizing currents at an axon's pre-synaptic terminus, in muscle cells and in some dendrites.

The imbalance of ions that makes possible not only action potentials but the resting cell potential arises through the work of pumps, in particular the sodium-potassium exchanger.

Changes in membrane permeability and the onset and cessation of ionic currents reflect the opening and closing of 'voltage-gated' ion channels, which provide portals through the membrane for ions. Residing in and spanning the membrane, these enzymes sense and respond to changes in transmembrane potential.

Initiation

Action potentials are triggered by an initial depolarization to the point of *threshold*. This threshold potential varies but generally is about 15 millivolts above the resting potential of the cell. Typically action potential initiation occurs at a synapse, but may occur anywhere along the axon. In his discovery of 'animal electricity', L. Galvani elicited an action potential through contact of his scalpel with the motor nerve of a frog he was dissecting, causing one of its legs to kick as in life.

Wave Propagation

In the fine fibers of simple (or 'unmyelinated') axons, action potentials propagate as waves, which travel at speeds up to 120 meters per second.

The propagation speed of these 'impulses' is faster in fatter fibers than in thin ones, other things being equal. In their *Nobel Prize* winning work uncovering the wave nature and ionic mechanism of action potentials, *Alan L. Hodgkin* and *Andrew F. Huxley* performed their celebrated experiments on the 'giant fibre' of Atlantic squid [HH52a]. Responsible for initiating flight, this axon is fat enough to be seen without a microscope (100 to 1000 times larger than is typical). This is assumed to reflect an adaptation for speed. Indeed, the velocity of nerve impulses in these fibers is among the fastest in nature.

Saltatory propagation

Many neurons have insulating sheaths of myelin surrounding their axons, which enable action potentials to travel faster than in unmyelinated axons of the same diameter. The myelin sheathing normally runs along the axon in sections about 1 mm long, punctuated by unsheathed 'nodes of Ranvier'.

Because the salty cytoplasm of the axon is electrically conductive, and because the myelin inhibits charge leakage through the membrane, depolarization at one node is sufficient to elevate the voltage at a neighboring node to the threshold for action potential initiation. Thus in myelinated axons, action potentials do not propagate as waves, but recur at successive nodes and in effect hop along the axon. This mode of propagation is known as *saltatory conduction*. Saltatory conduction is faster than smooth conduction. Some typical action potential velocities are as follows:

Fiber	Diameter	AP Velocity
Unmyelinated	0.2–1.0 micron	0.2–2 m/s
Myelinated	2–20 microns	12–120 m/s

The disease called *multiple sclerosis* (MS) is due to a breakdown of myelin sheathing, and degrades muscle control by destroying axons' ability to conduct action potentials.

Detailed Features

Depolarization and re-polarization together are complete in about two milliseconds, while undershoots can last hundreds of milliseconds, depending on the cell. In neurons, the exact length of the roughly two-millisecond delay in re-polarization can have a strong effect on the amount of neurotransmitter released at a synapse. The duration of the hyperpolarization determines a nerve's 'refractory period' (how long until it may conduct another action potential) and hence the frequency at which it will fire under continuous stimulation. Both of these properties are subject to biological regulation, primarily (among the mechanisms discovered so far) acting on ion channels selective for potassium.

A cell capable of undergoing an action potential is said to be *excitable*.

Synapses

Synapses are specialized junctions through which cells of the nervous system signal to one another and to non-neuronal cells such as muscles or glands (see Fig. 2.12). Synapses define the circuits in which the neurons of the central nervous system interconnect. They are thus crucial to the biological computations that underlie perception and thought. They also provide the means through which the nervous system connects to and controls the other systems of the body (see [II06b]).

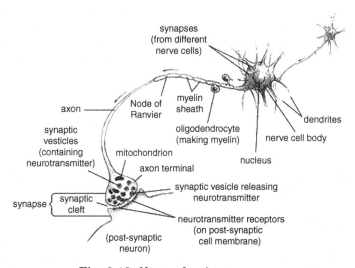

Fig. 2.12. Neuron forming a synapse.

Anatomy and Structure. At a classical synapse, a mushroom-shaped bud projects from each of two cells and the caps of these buds press flat against one another (see Fig. 2.13). At this interface, the membranes of the two cells flank each other across a slender gap, the narrowness of which enables signaling molecules known as neurotransmitters to pass rapidly from one cell to the other by diffusion. This gap is sometimes called the synaptic cleft.

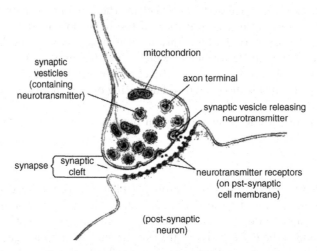

Fig. 2.13. Structure of a chemical synapse.

Synapses are asymmetric both in structure and in how they operate. Only the so-called pre-synaptic neuron secretes the neurotransmitter, which binds to receptors facing into the synapse from the post-synaptic cell. The pre-synaptic nerve terminal generally buds from the tip of an axon, while the post-synaptic target surface typically appears on a dendrite, a cell body or another part of a cell.

Signaling across the Synapse

The release of neurotransmitter is triggered by the arrival of a nerve impulse (or action potential) and occurs through an unusually rapid process of cellular secretion: Within the pre-synaptic nerve terminal, vesicles containing neuro-transmitter sit 'docked' and ready at the synaptic membrane. The arriving action potential produces an influx of calcium ions through voltage-dependent, calcium-selective ion channels, at which point the vesicles fuse with the membrane and release their contents to the outside. Receptors on the opposite side of the synaptic gap bind neurotransmitter molecules and respond by opening nearby ion channels in the post-synaptic cell membrane, causing ions to rush in or out and changing the local transmembrane potential of the cell. The result is *excitatory*, in the case of depolarizing currents, or *inhibitory* in the case of hyperpolarizing currents. Whether a synapse is excitatory or inhibitory depends on what type(s) of ion channel conduct the post-synaptic current, which in turn is a function of the type of receptors and neurotransmitter employed at the synapse.

Excitatory synapses in the brain show several forms of synaptic plasticity, including long-term potentiation (LTP) and long-term depression (LTD), which are initiated by increases in intracellular Ca^{2+} that are generated through NMDA (N-methyl-D-aspartate) receptors or voltage-sensitive Ca^{2+} channels. LTP depends on the coordinated regulation of an ensemble of enzymes, including Ca^{2+}/calmodulin-dependent protein kinase II, adenylyl cyclase 1 and 8, and calcineurin, all of which are stimulated by calmodulin, a Ca^{2+}-binding protein.

Synaptic Strength

The amount of current, or more strictly the change in transmembrane potential, depends on the 'strength' of the synapse, which is subject to biological regulation. One regulatory mechanism involves the simple coincidence of action potentials in the synaptically linked cells. Because the coincidence of sensory stimuli (the sound of a bell and the smell of meat, for example, in the experiments by *Nobel Laureate Ivan P. Pavlov*) can give rise to associative learning or conditioning, neuroscientists have hypothesized that synaptic strengthening through coincident activity in two neurons might underlie learning and memory. This is known as the *Hebbian theory* [Heb49]. It is related to *Pavlov's conditional-reflex learning*: it is learning that takes place when we come to associate two stimuli in the environment. One of these stimuli triggers a reflexive response. The second stimulus is originally neutral with respect to that response, but after it has been paired with the first stimulus, it comes to trigger the response in its own right.

Biophysics of Synaptic Transmission

Technically, synaptic transmission happens in transmitter-activated ion channels. Activation of a presynaptic neuron results in a release of neurotransmitters into the synaptic cleft. The transmitter molecules diffuse to the other side of the cleft and activate receptors that are located in the postsynaptic membrane. So-called ionotropic receptors have a direct influence on the state of an associated ion channel whereas metabotropic receptors control the state of the ion channel by means of a biochemical cascade of g-proteins and second messengers. In any case the activation of the receptor results in the opening of certain ion channels and, thus, in an excitatory or inhibitory postsynaptic current (EPSC or IPSC, respectively). The transmitter-activated ion channels can be described as an explicitly time-dependent conductivity $g_{syn}(t)$ that will open whenever a presynaptic spike arrives. The current that passes through these channels depends, as usual, on the difference of its reversal potential E_{syn} and the actual value of the membrane potential,

$$I_{syn}(t) = g_{syn}(t)(u - E_{syn}).$$

The parameter E_{syn} and the function $g_{syn}(t)$ can be used to characterize different types of synapse. Typically, a superposition of exponentials is used for $g_{syn}(t)$. For inhibitory synapses E_{syn} equals the reversal potential of potassium ions (about -75 mV), whereas for excitatory synapses $E_{syn} \approx 0$.

The effect of fast *inhibitory* neurons in the central nervous system of higher vertebrates is almost exclusively conveyed by a neuro-transmitter called γ-aminobutyric acid, or GABA for short. In addition to many different types of inhibitory interneurons, cerebellar Purkinje cells form a prominent example of projecting neurons that use GABA as their neuro-transmitter. These neurons synapse onto neurons in the deep cerebellar nuclei (DCN) and are particularly important for an understanding of cerebellar function.

The parameters that describe the conductivity of transmitter-activated ion channels at a certain synapse are chosen so as to mimic the time course and the amplitude of experimentally observed spontaneous postsynaptic currents. For example, the conductance $\bar{g}_{syn}(t)$ of inhibitory synapses in DCN neurons can be described by a simple exponential decay with a time constant of $\tau = 5$ ms and an amplitude of $\bar{g}_{syn} = 40$ pS,

$$
g_{syn}(t) = \sum_f \bar{g}_{syn} \exp\left(-\frac{t - t^{(f)}}{\tau}\right) \Theta(t - t^{(f)}),
$$

where $t^{(f)}$ denotes the arrival time of a presynaptic action potential. The reversal potential is given by that of potassium ions, viz. $E_{syn} = -75$ mV (see [GMK94]).

Clearly, more attention can be payed to account for the details of synaptic transmission. In cerebellar granule cells, for example, inhibitory synapses are also GABAergic, but their postsynaptic current is made up of two different components. There is a fast component, that decays with a time constant of about 5 ms, and there is a component that is ten times slower. The underlying postsynaptic conductance is thus of the form

$$
g_{syn}(t) = \sum_f \left(\bar{g}_{fast} \exp\left(-\frac{t - t^{(f)}}{\tau_{fast}}\right) + \bar{g}_{slow} \exp\left(-\frac{t - t^{(f)}}{\tau_{slow}}\right) \right) \Theta(t - t^{(f)}).
$$

Now, most of *excitatory* synapses in the vertebrate central nervous system rely on glutamate as their neurotransmitter. The postsynaptic receptors, however, can have very different pharmacological properties and often different types of glutamate receptors are present in a single synapse. These receptors can be classified by certain amino acids that may be selective agonists. Usually, NMDA (N-methyl-D-aspartate) and non-NMDA receptors are distinguished. The most prominent among the non-NMDA receptors are AMPA-receptors. Ion channels controlled by AMPA-receptors are characterized by a fast response to presynaptic spikes and a quickly decaying postsynaptic current. NMDA-receptor controlled channels are significantly slower and have additional interesting properties that are due to a voltage-dependent blocking by magnesium ions (see [GMK94]).

Excitatory synapses in cerebellar granule cells, for example, contain two different types of glutamate receptors, viz. AMPA- and NMDA-receptors. The time course of the postsynaptic conductivity caused by an activation of AMPA-receptors at time $t = t^{(f)}$ can be described as follows,

$$g_{AMPA}(t) = \bar{g}_{AMPA} \cdot N \cdot \left(\exp\left(-\frac{t - t^{(f)}}{\tau_{decay}} \right) - \exp\left(-\frac{t - t^{(f)}}{\tau_{rise}} \right) \right) \Theta(t - t^{(f)}),$$

with rise time $\tau_{rise} = 0.09$ ms, decay time $\tau_{decay} = 1.5$ ms, and maximum conductance $\bar{g}_{AMPA} = 720$ pS. The numerical constant $N = 1.273$ normalizes the maximum of the braced term to unity (see [GMK94]).

NMDA-receptor controlled channels exhibit a significantly richer repertoire of dynamic behavior because their state is not only controlled by the presence or absence of their agonist, but also by the membrane potential. The voltage dependence itself arises from the blocking of the channel by a common extracellular ion, Mg^{2+}. Unless Mg^{2+} is removed from the extracellular medium, the channels remain closed at the resting potential even in the presence of NMDA. If the membrane is depolarized beyond -50 mV, then the Mg^{2+}-block is removed, the channel opens, and, in contrast to AMPA-controlled channels, stays open for 10–100 milliseconds. A simple ansatz that accounts for this additional voltage dependence of NMDA-controlled channels in cerebellar granule cells is

$$g_{NMDA}(t) = \bar{g}_{NMDA} \cdot N \cdot \left(\exp\left(-\frac{t - t^{(f)}}{\tau_{decay}} \right) - \exp\left(-\frac{t - t^{(f)}}{\tau_{rise}} \right) \right) g_{\infty} \Theta(t - t^{(f)}),$$

where $g_{\infty} = (1 + e^{\alpha u}[Mg^{2+}]_o / \beta)$, $\tau_{rise} = 3$ ms, $\tau_{decay} = 40$ ms, $N = 1.358$, $\bar{g}_{NMDA} = 1.2$ nS, $\alpha = 0.062$ mV^{-1}, $\beta = 3.57$ mM, and the extracellular magnesium concentration $[Mg^{2+}]_o = 1.2$ mM (see [GMK94]).

Finally, Though NMDA-controlled ion channels are permeable to sodium and potassium ions, their permeability to Ca^{2+} is even five or ten times larger. Calcium ions are known to play an important role in intracellular signaling and are probably also involved in long-term modifications of synaptic efficacy. Calcium influx through NMDA-controlled ion channels, however, is bound to the coincidence of presynaptic (NMDA release from presynaptic sites) and postsynaptic (removal of the Mg^{2+}-block) activity. Hence, NMDA-receptors operate as a kind of a molecular coincidence detectors as they are required for a biochemical implementation of Hebb's learning rule [Heb49].

Reflex Action: the Basis of CNS Activity

The basis of all CNS activity, as well as the simplest example of our sensory-motor adjunction, is the *reflex (sensory-motor) action, RA*. It occurs at all neural organizational levels. We are aware of some reflex acts, while others occur without our knowledge.

In particular, the spinal reflex action is defined as a composition of neural pathways, $RA = EN \circ CN \circ AN$, where EN is the efferent neuron, AN is the afferent neuron and $CN = CN_1, \ldots, CN_n$ is the chain of n connector neurons ($n = 0$ for the simplest, stretch, reflex, $n \geq 1$ for all other reflexes). In other words, the following diagram commutes:

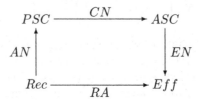

in which Rec is the receptor (for a complex-type receptor as eye, see Fig. 2.11), Eff is the effector (e.g., muscle), PSC is the posterior (or, dorsal) horn of the spinal cord, and ASC is the anterior (or, ventral) horn of the spinal cord. In this way defined map $RA : Rec \rightarrow Eff$ is the simplest, *one-to-one* relation between one receptor neuron and one effector neuron (e.g., patellar reflex, see Fig. 2.14).

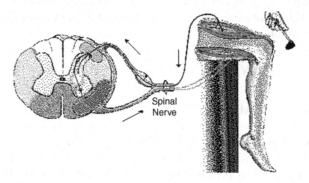

Fig. 2.14. Schematic of a simple knee-jerk reflex. Hammer strikes knee, generating sensory impulse to spinal cord. Primary neuron makes (monosynaptic, excitatory) synapse with anterior horn (motor) cell, whose axon travels via ventral root to quadriceps muscle, which contracts, raising foot. Hamstring (lower) muscle is simultaneously inhibited, via an internuncial neuron.

Now, in the majority of human reflex arcs a chain CN of many connector neurons is found. There may be link-ups with various levels of the brain and spinal cord. Every receptor neuron is potentially linked in the CNS with a large number of effector organs all over the body, i.e., the map $RA : Rec \rightarrow Eff$ is *one-to-many*. Similarly, every effector neuron is potentially in communication with receptors all over the body, i.e., the map $RA : Rec \rightarrow Eff$ is *many-to-one*.

However, the most frequent form of the map $RA : Rec \rightarrow Eff$ is *many-to-many*. Other neurons synapsing with the effector neurons may give a complex link-up with centers at higher and lower levels of the CNS. In this way, higher centers in the brain can modify reflex acts which occur through the spinal cord. These centers can send suppressing or facilitating impulses along their pathways to the cells in the spinal cord [II06b].

Through such 'functional' link-ups, neurons in different parts of the CNS, when active, can influence each other. This makes it possible for Pavlov's conditioned reflexes to be established. Such reflexes form the basis of all training, so that it becomes difficult to say where reflex (or involuntary) behavior ends and purely voluntary behavior begins.

In particular, the control of voluntary movements is extremely complex. Many different systems across numerous brain areas need to work together to ensure proper motor control. Our understanding of the nervous system decreases as we move up to higher CNS structures.

A Bird's Look at the Brain

The *brain* is the supervisory center of the nervous system consisting of *grey matter* (superficial parts called *cortex* and deep brain nuclei) and *white matter* (deep parts except the brain nuclei). It controls and coordinates behavior, *homeostasis*[1] (i.e., negative feedback control of the body functions such as heartbeat, blood pressure, fluid balance, and body temperature) and mental functions (such as cognition, emotion, memory and learning) (see [Mar98, II06b]).

---

[1] *Homeostasis* is the property of an open system to regulation its internal environment so as to maintain a stable state of structure and functions, by means of multiple dynamic equilibrium controlled by interrelated regulation mechanisms. The term was coined in 1932 by W. Cannon from two Greek words [*homeo-man*] and [*stasis-stationary*]. Homeostasis is one of the fundamental characteristics of living things. It is the maintenance of the internal environment within tolerable limits. All sorts of factors affect the suitability of our body fluids to sustain life; these include properties like temperature, salinity, acidity (carbon dioxide), and the concentrations of nutrients and wastes (urea, glucose, various ion, oxygen). Since these properties affect the chemical reactions that keep bodies alive, there are built-in physiological mechanisms to maintain them at desirable levels. This control is achieved with various organs and glands in the body. For example [Mar98, II06b]: The hypothalamus monitors water content, carbon dioxide concentration, and blood temperature, sending nerve impulses to the pituitary gland and skin. The pituitary gland synthesizes ADH (anti-diuretic hormone) to control water content in the body. The muscles can shiver to produce heat if the body temperature is too low. Warm-blooded animals (homeotherms) have additional mechanisms of maintaining their internal temperature through homeostasis. The pancreas produces insulin to control blood-sugar concentration. The lungs take in oxygen and give out carbon dioxide. The kidneys remove urea and adjust ion and water concentrations. More realistic is dynamical homeostasis, or *homeokinesis*, which forms the basis of the *Anochin's theory of functional systems*.

The *vertebrate brain* can be subdivided into: (i) *medulla oblongata* (or, *brain stem*); (ii) *myelencephalon*, divided into: *pons* and *cerebellum*; (iii) *mesencephalon* (or, *midbrain*); (iv) *diencephalon*; and (v) *telencephalon* (*cerebrum*).

Sometimes a gross division into three major parts is used: *hindbrain* (including medulla oblongata and myelencephalon), *midbrain* (mesencephalon) and *forebrain* (including diencephalon and telencephalon). The cerebrum and the cerebellum consist each of two *hemispheres*. The *corpus callosum* connects the two hemispheres of the cerebrum.

The cerebrum and the cerebellum consist each of two hemispheres. The corpus callosum connects the two hemispheres of the cerebrum. The cerebellum is a cauliflower-shaped section of the brain (see Fig. 2.15). It is located in the hindbrain, at the bottom rear of the head, directly behind the pons. The cerebellum is a complex computer mostly dedicated to the intricacies of managing walking and balance. Damage to the cerebellum leaves the sufferer with a gait that appears drunken and is difficult to control.

The spinal cord is the extension of the central nervous system that is enclosed in and protected by the vertebral column. It consists of nerve cells and their connections (axons and dendrites), with both gray matter and white matter, with the former surrounded by the latter.

Cranial nerves are nerves which start directly from the brainstem instead of the spinal cord, and mainly control the functions of the anatomic structures of the head. In human anatomy, there are exactly *12 pairs* of them: (I) *olfactory nerve*, (II) *optic nerve*, (III) oculomotor nerve, (IV) Trochlear nerve, (V) Trigeminal nerve, (VI) Abducens nerve, (VII) Facial nerve, (VIII) Vestibulocochlear nerve (sometimes called the auditory nerve), (IX) Glossopharyngeal nerve, (X) Vagus nerve, (XI) Accessory nerve (sometimes called the spinal accessory nerve), and (XII) Hypoglossal nerve.

The optic nerve consists mainly of axons extending from the ganglionic cells of the eye's retina. The axons terminate in the lateral geniculate nucleus, pulvinar, and superior colliculus, all of which belong to the primary visual center. From the lateral geniculate body and the pulvinar fibers pass to the visual cortex.

In particular, the *optic nerve* contains roughly one million nerve fibers. This number is low compared to the roughly 130 million receptors in the retina, and implies that substantial pre-processing takes place in the retina before the signals are sent to the brain through the optic nerve.

In most vertebrates the mesencephalon is the highest integration center in the brain, whereas in mammals this role has been adopted by the telencephalon. Therefore the cerebrum is the largest section of the mammalian brain and its surface has many deep fissures (sulci) and grooves (gyri), giving an excessively wrinkled appearance to the brain.

The *human brain* can be subdivided into several distinct regions:

The *cerebral hemispheres* form the largest part of the brain, occupying the anterior and middle cranial fossae in the skull and extending backwards over the tentorium cerebelli. They are made up of the cerebral cortex, the basal ganglia, tracts of synaptic connections, and the ventricles containing CSF.

The *diencephalon* includes the thalamus, hypothalamus, epithalamus and subthalamus, and forms the central core of the brain. It is surrounded by the cerebral hemispheres.

The *midbrain* is located at the junction of the middle and posterior cranial fossae.

The *pons* sits in the anterior part of the posterior cranial fossa; the fibres within the structure connect one cerebral hemisphere with its opposite cerebellar hemisphere.

The *medulla oblongata* is continuous with the spinal cord, and is responsible for automatic control of the respiratory and cardiovascular systems.

The *cerebellum* overlies the pons and medulla, extending beneath the tentorium cerebelli and occupying most of the posterior cranial fossa. It is mainly concerned with motor functions that regulate muscle tone, coordination, and posture.

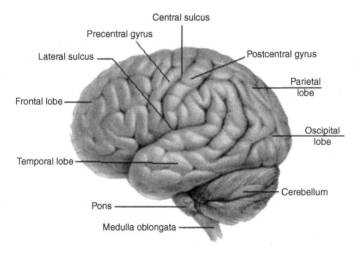

Fig. 2.15. The human cerebral hemispheres.

Now, the two *cerebral hemispheres* (see Fig. 2.15) can be further divided into *four lobes*:

The *frontal lobe* is concerned with higher intellectual functions, such as abstract thought and reason, speech (Broca's area in the left hemisphere only), olfaction, and emotion. Voluntary movement is controlled in the precentral gyrus (the primary motor area, see Fig. 2.16).

The *parietal lobe* is dedicated to sensory awareness, particularly in the postcentral gyrus (the primary sensory area, see Fig. 2.16). It is also associated with abstract reasoning, language interpretation and formation of a mental egocentric map of the surrounding area.

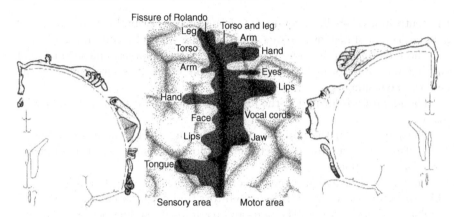

Fig. 2.16. Penfield's 'Homunculus', showing the primary somatosensory and motor areas of the human brain.

The *occipital lobe* is responsible for interpretation and processing of visual stimuli from the optic nerves, and association of these stimuli with other nervous inputs and memories.

The *temporal lobe* is concerned with emotional development and formation, and also contains the auditory area responsible for processing and discrimination of sound. It is also the area thought to be responsible for the formation and processing of memories.

2.1.2 Modern 3D Brain Imaging

Nuclear Magnetic Resonance and 2D Brain Imaging

The *Nobel Prize in Physiology or Medicine in 2003* was jointly awarded to *Paul C. Lauterbur* and *Peter Mansfield* for their discoveries concerning *magnetic resonance imaging* (MRI), a technique for using strong magnetic fields to produce images of the inside of the human body.

Atomic nuclei in a strong magnetic field rotate with a frequency that is dependent on the strength of the magnetic field. Their energy can be increased if they absorb radio waves with the same resonant frequency. When the atomic nuclei return to their previous energy level, radio waves are emitted. These discoveries were awarded the *Nobel Prize in Physics in 1952*, jointly to *Felix Bloch* and *Edward M. Purcell*. During the following decades, magnetic resonance was used mainly for studies of the chemical structure of substances. In the beginning of the 1970s, Lauterbur and Mansfield made pioneering contributions, which later led to the applications of nuclear magnetic resonance (NMR) in medical imaging.

Paul Lauterbur discovered the possibility to create a 2D picture by introducing gradients in the magnetic field. By analysis of the characteristics of the emitted radio waves, he could determine their origin. This made it possible to build up 2D pictures of structures that could not be visualized with other methods.

Peter Mansfield further developed the utilization of gradients in the magnetic field. He showed how the signals could be mathematically analyzed, which made it possible to develop a useful imaging technique. Mansfield also showed how extremely fast imaging could be achievable. This became technically possible within medicine a decade later.

Magnetic resonance imaging (MRI), is now a routine method within medical diagnostics. Worldwide, more than 60 million investigations with MRI are performed each year, and the method is still in rapid development. MRI is often superior to other imaging techniques and has significantly improved diagnostics in many diseases. MRI has replaced several invasive modes of examination and thereby reduced the risk and discomfort for many patients.

3D MRI of Human Brain

Modern technology of human brain imaging emphasizes 3D investigation of brain structure and function, using three variations of MRI. Brain *structure* is commonly imaged using *anatomical MRI*, or aMRI, while brain *physiology* is usually imaged using *functional MRI*, or fMRI. For bridging the gap between brain anatomy and function, as well as exploring natural brain connectivity, a *diffusion MRI*, or dMRI is used, based on the *diffusion tensor* (DT) technique (see [BJH95, Bih96, Bih00, Bih03]).

The ability to visualize anatomical connections between different parts of the brain, non-invasively and on an individual basis, has opened a new era in the field of functional neuro-imaging. This major breakthrough for neuroscience and related clinical fields has developed over the past ten years through the advance of *diffusion magnetic resonance imaging* (dMRI). The concept of dMRI is to produce MRI quantitative maps of microscopic, natural displacements of water molecules that occur in brain tissues as part of the physical diffusion process. Water molecules are thus used as a probe that can reveal microscopic details about tissue architecture, either normal or in a diseased state.

Molecular Diffusion in a 3D Brain Volume

Molecular diffusion refers to the *Brownian motion* of molecules (see Sect. 5.8.1), which results from the thermal energy carried by these molecules. Molecules travel randomly in space over a distance that is statistically well described by a *diffusion coefficient*, D. This coefficient depends only on the size (mass) of the molecules, the temperature and the nature (viscosity) of the medium (see Fig. 2.17).

dMRI is, thus, deeply rooted in the concept that, during their diffusion-driven displacements, molecules probe tissue structure at a *microscopic scale* well beyond the usual *millimetric image resolution*. During typical diffusion

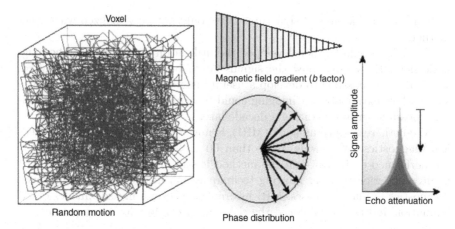

Voxel

Magnetic field gradient (*b* factor)

Signal amplitude

Random motion

Phase distribution

Echo attenuation

Fig. 2.17. Principles of dMRI: In the spatially varying magnetic field, induced through a magnetic field gradient, the amplitude and timing of which a characterized by a *b*-factor, moving molecules emit radiofrequency signals with slightly different phases. In a small 3D-volume (voxel) containing a large number of diffusing molecules, these phases become randomly distributed, directly reflecting the diffusion process, i.e., the trajectory of individual molecules. This diffusive phase distribution of the signal results in an *attenuation* A of the MRI signal, which quantitatively depends on the gradient characteristics of the *b*-factor and the diffusion coefficient D (adapted from [BMP01]).

times of about 50–100 ms, water molecules move in brain tissues on average over distances around 1–15 m, bouncing, crossing or interacting with many tissue components, such as cell membranes, fibres or macromolecules. Because of the tortuous movement of water molecules around those obstacles, the actual diffusion distance is reduced compared to free water. Hence, the non-invasive observation of the water diffusion-driven displacement distributions *in vivo* provides unique clues to the fine structural features and geometric organization of neural tissues, and to changes in those features with physiological or pathological states.

Imaging Brain Diffusion with MRI

While early water diffusion measurements were made in biological tissues using Nuclear Magnetic Resonance in the 1960s and 70s, it is not until the mid 1980s that the basic principles of dMRI were laid out. MRI signals can be made sensitive to diffusion through the use of a pair of sharp magnetic field gradient pulses, the duration and the separation of which can be adjusted. The result is a signal (echo) attenuation which is precisely and quantitatively linked to the amplitude of the molecular displacement distribution: Fast (slow) diffusion results in a large (small) distribution and a large (small) signal attenuation. Naturally, the effect also depends on the intensity of the magnetic field gradient pulses.

In practice, any MRI imaging technique can be sensitized to diffusion by inserting the adequate magnetic field gradient pulses. By acquiring data with various gradient pulse amplitudes one gets images with different degrees of diffusion sensitivity (see Fig. 2.18). Contrast in these images depends on diffusion, but also on other MRI parameters, such as the water relaxation times. Hence, these images are often numerically combined to determine, using a global diffusion model, an estimate of the diffusion coefficient in each image location. The resulting images are maps of the diffusion process and can be visualized using a quantitative scale.

Because the overall signal observed in a MRI image voxel, at a millimetric resolution, results from the integration, on a statistical basis, of all the microscopic displacement distributions of the water molecules present in this voxel it was suggested 6 to portray the complex diffusion processes that occur in a biological tissue on a voxel scale using a global, statistical parameter, the *apparent diffusion coefficient* (ADC). The ADC concept has been largely used since then in the literature. The ADC now depends not only on the actual diffusion coefficients of the water molecular populations present in the voxel, but also on experimental, technical parameters, such as the voxel size and the diffusion time.

3D Diffusion Brain Tensor

Now, as diffusion is really a 3D process, water molecular mobility in tissues is not necessarily the same in all directions. This *diffusion anisotropy* may result from the presence of obstacles that limit molecular movement in some directions. It is not until the advent of diffusion MRI that anisotropy was detected for the first time *in vivo*, at the end of the 1980s, in spinal cord and brain white matter (see [BJH95, Bih96, Bih00, Bih03]). Diffusion anisotropy in white matter grossly originates from its specific organization in bundles of more or less myelinated axonal fibres running in parallel: Diffusion in the direction of the fibres (whatever the species or the fiber type) is about 3–6 times faster than in the perpendicular direction. However the relative contributions of the intra-axonal and extracellular spaces, as well as the presence of the myelin sheath, to the ADC, and the exact mechanism for the anisotropy is still not completely understood, and remains the object of active research. It quickly became apparent, however, that this anisotropy effect could be exploited to map out the orientation in space of the white matter tracks in the brain, assuming that the direction of the fastest diffusion would indicate the overall orientation of the fibres. The work on diffusion anisotropy really took off with the introduction in the field of diffusion MRI of the more rigorous formalism of the diffusion tensor.

More precisely, with plain diffusion MRI, diffusion is fully described using a single (scalar) parameter, the diffusion coefficient, D. The effect of diffusion

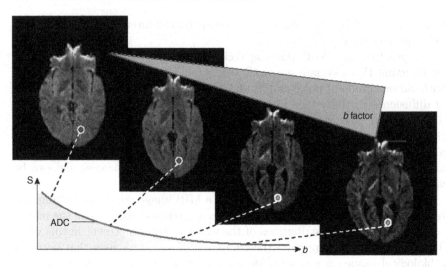

Fig. 2.18. Different degrees of *diffusion-weighted images* can be obtained using different values of the *b*-factor. The larger the *b*-factor the more the signal intensity becomes attenuated in the image. This attenuation, though, is modulated by the diffusion coefficient: signal in structures with fast diffusion (e.g., water filled ventricular cavities) decays very fast with *b*, while signal in tissues with low diffusion (e.g., gray and white matter) decreases more slowly. By fitting the signal decay as a function of *b*, one obtains the apparent diffusion coefficient (ADC) for each elementary volume (voxel) of the image. Calculated diffusion images (ADC maps), depending solely on the diffusion coefficient, can then be generated and displayed using a gray (or color) scale: High diffusion, as in the ventricular cavities, appears bright, while low diffusion is dark (adapted from [BMP01]).

on the MRI signal (most often a spin-echo signal) is an attenuation, A, which depends on D and on the b factor, which characterizes the gradient pulses (timing, amplitude, shape) used in the MRI sequence:

$$A = \exp(-bD).$$

However, in the presence of anisotropy, diffusion can no longer be characterized by a single scalar coefficient, but requires a 3D tensor-field $\mathbf{D} = \mathbf{D}(t)$ (see Chap. 3), give by the matrix of 'moments' (on the main diagonal) and 'product' (off-diagonal elements) [Bih03]:

$$\mathbf{D}(t) = \begin{pmatrix} D_{xx}(t) & D_{xy}(t) & D_{xz}(t) \\ D_{yx}(t) & D_{yy}(t) & D_{yz}(t) \\ D_{zx}(t) & D_{zy}(t) & D_{zz}(t) \end{pmatrix},$$

which fully describes molecular mobility along each direction and correlation between these directions. This tensor is symmetric ($D_{ij} = D_{ji}$, with $i, j = x, y, z$).

Now, in a *reference frame* $[x', y', z']$ that coincides with the principal or self directions of diffusivity, the off-diagonal terms do not exist, and the tensor is reduced only to its diagonal terms, $\{D_{x'x'}, D_{x'x'}, D_{x'x'}\}$, which represent molecular mobility along axes x', y', and z', respectively. The *echo attenuation* then becomes:

$$A = \exp\left(-b_{x'x'} D_{x'x'} - b_{y'y'} D_{y'y'} - b_{z'z'} D_{z'z'}\right),$$

where b_{ii} are the elements of the **b**-*tensor*, which now replaces the scalar b-factor, expressed in the coordinates of this reference frame.

In practice, unfortunately, measurements are made in the reference frame $[x, y, z]$ of the MRI scanner gradients, which usually does not coincide with the diffusion frame of the tissue [Bih03]. Therefore, one must also consider the coupling of the nondiagonal elements, b_{ij}, of the **b**-tensor with the nondiagonal terms, D_{ji}, $(i \neq j)$, of the diffusion tensor (now expressed in the scanner frame), which reflect correlation between molecular displacements in perpendicular directions:

$$A = \exp\left(-b_{xx} D_{xx} - b_{yy} D_{yy} - b_{zz} D_{zz} - 2b_{xy} D_{xy} - 2b_{xz} D_{xz} - 2b_{yz} D_{yz}\right).$$

Hence, it is important to note that by using diffusion-encoding gradient pulses along one direction only, signal attenuation not only depends on the diffusion effects along this direction but may also include contribution from other directions.

Now, calculation of the **b**-tensor may quickly become complicated when many gradient pulses are used, but the full determination of the diffusion tensor **D** is necessary if one wants to assess properly and fully all anisotropic diffusion effects.

To determine the diffusion tensor **D** fully, one must first collect diffusion-weighted images along several gradient directions, using diffusion-sensitized MRI pulse sequences such as *echoplanar imaging* (EPI) [BJH95]. As the diffusion tensor is symmetric, measurements along only 6 directions are mandatory (instead of 9), along with an image acquired without diffusion weighting ($b = 0$).

In the case of axial symmetry, only four directions are necessary (tetrahedral encoding), as suggested in the *spinal cord* [CWM00]. The acquisition time and the number of images to process are then reduced.

In this way, with *diffusion tensor imaging* (DTI), diffusion is no longer described by a single diffusion coefficient, but by an array of 9 coefficients (dependent on the sampling discrete time) that fully characterize how diffusion in space varies according to direction. Hence, diffusion anisotropy effects can be fully extracted and exploited, providing even more exquisite details on tissue microstructure.

With DTI, diffusion data can be analyzed in three ways to provide information on tissue microstructure and architecture for each voxel: (i) the *mean diffusivity*, which characterizes the overall mean-squared displacement of molecules and the overall presence of obstacles to diffusion; (ii) the *degree of anisotropy*, which describes how much molecular displacements vary in space and is related to the presence and coherence of oriented structures; (iii) the *main direction of diffusivities* (main ellipsoid axes), which is linked to the orientation in space of the structures.

Mean Diffusivity

To get an overall evaluation of the diffusion in a voxel or 3D-region, one must avoid anisotropic diffusion effects and limit the result to an invariant, i.e., a quantity that is independent of the orientation of the reference frame [BMB94]. Among several combinations of the tensor elements, the trace of the diffusion tensor, $\mathrm{Tr}(\mathbf{D}) = D_{xx} + D_{yy} + D_{zz}$, is such an invariant. The mean diffusivity is then given by $\mathrm{Tr}(\mathbf{D})/3$.

Diffusion Anisotropy Indices

Several scalar indices have been proposed to characterize diffusion anisotropy. Initially, simple indices calculated from diffusion-weighted images, or ADCs, obtained in perpendicular directions were used, such as $ADCx/ADCy$ and displayed using a color scale [DTP91]. Other groups have devised indices mixing measurements along x, y, and z directions, such as $\frac{\max[ADCx,ADCy,ADCz]}{\min[ADCx,ADCy,ADCz]}$, or the standard deviation of $ADCx$, $ADCy$, and $ADCz$ divided by their mean value [GVP94]. Unfortunately, none of these indices are really quantitative, as they do not correspond to a single meaningful physical parameter and, more importantly, are clearly dependent on the choice of directions made for the measurements. The degree of anisotropy would then vary according to the respective orientation of the gradient hardware and the tissue frames of reference and would generally be underestimated. Here again, invariant indices must be found to avoid such biases and provide an objective, intrinsic structural information [BP96].

Invariant indices are thus made of combinations of the eigen-values λ_1, λ_2, and λ_3 of the diffusion tensor $\mathbf{D}$ (see Fig. 2.19). The most commonly used invariant indices are the *relative anisotropy* (RA), the *fractional anisotropy* (FA), and the *volume ratio* (VR).

Fiber Orientation Mapping

The last family of parameters that can be extracted from the DTI concept relates to the mapping of the orientation in space of tissue structure. The assumption is that the direction of the fibers is collinear with the direction of the eigen-vector associated with the *largest eigen-diffusivity*. This approach opens a completely new way to gain direct and in vivo information on the organization in space of oriented tissues, such as muscle, myocardium, and brain or spine white matter, which is of considerable interest, both clinically and functionally. Direction orientation can be derived from DTI directly from

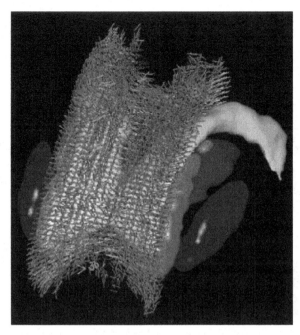

Fig. 2.19. 3D display of the diffusion tensor **D**. Main eigen-vectors are shown as cylinders, the length of which is scaled with the degree of anisotropy. Corpus callosum fibers are displayed around ventricles, thalami, putamen, and caudate nuclei (adapted from [BMP01]).

diffusion/orientation-weighted images or through the calculation of the diffusion tensor **D**. Here, a first issue is to display fiber orientation on a voxel-by-voxel basis. The use of color maps has first been suggested, followed by representation of ellipsoids, octahedra, or vectors pointing in the fiber direction [BMP01].

Brain Connectivity Studies

Studies of neuronal connectivity are important to interpret functional MRI data and establish the networks underlying cognitive processes. Basic DTI provides a means to determine the overall orientation of white matter bundles in each voxel, assuming that only one direction is present or predominant in each voxel, and that diffusivity is the highest along this direction. 3D vector-field maps representing fiber orientation in each voxel can then be obtained back from the image date through the diagonalization (a mathematical operation which provides orthogonal directions coinciding with the main diffusion directions) of the diffusion tensor determined in each voxel. A second step after this *inverse problem* is solved consists in *connecting* subsequent voxels on the basis of their respective fibre orientation to infer some continuity in the fibers (see Fig. 2.20). Several algorithms have been proposed. *Line propagation algorithms* reconstruct tracts from voxel to voxel from a seed point. Another approach is based on *regional energy minimization* (minimal bending) to select the most likely trajectory among several possible [BMP01, Bih03].

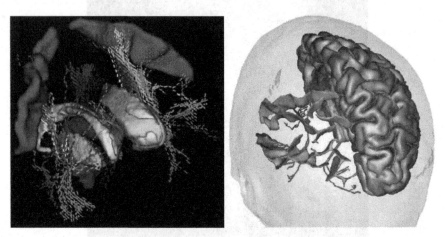

Fig. 2.20. Several approaches have been developed to 'connect' voxels after white matter fibers have been identified and their orientation determined. *Left*: 3D display of the motor cortex, central structures and connections. *Right*: 3D display from MRI of a brain hemisphere showing sulci and connections (adapted and modified from [BMP01]).

2.2 Biological versus Artificial Neural Networks

In biological neural networks, signals are transmitted between neurons by electrical pulses (action potentials or spike trains) traveling along the axon. These pulses impinge on the afferent neuron at terminals called synapses. These are found principally on a set of branching processes emerging from the cell body (soma) known as dendrites. Each pulse occurring at a synapse initiates the release of a small amount of chemical substance or neurotransmitter which travels across the synaptic cleft and which is then received at postsynaptic receptor sites on the dendritic side of the synapse. The neurotransmitter becomes bound to molecular sites here which, in turn, initiates a change in the dendritic membrane potential. This postsynaptic potential (PSP) change may serve to increase (hyperpolarize) or decrease (depolarize) the polarization of the postsynaptic membrane. In the former case, the PSP tends to inhibit generation of pulses in the afferent neuron, while in the latter, it tends to excite the generation of pulses. The size and type of PSP produced will depend on factors such as the geometry of the synapse and the type of neurotransmitter.

Each PSP will travel along its dendrite and spread over the soma, eventually reaching the base of the axon (axonhillock). The afferent neuron sums or integrates the effects of thousands of such PSPs over its dendritic tree and over time. If the integrated potential at the axonhillock exceeds a threshold, the cell fires and generates an action potential or spike which starts to travel along its axon. This then initiates the whole sequence of events again in neurons contained in the efferent pathway.

ANNs are very loosely based on these ideas. In the most general terms, a ANN consists of large numbers of simple processors linked by weighted connections. By analogy, the processing nodes may be called artificial neurons. Each node output depends only on information that is locally available at the node, either stored internally or arriving via the weighted connections. Each unit receives inputs from many other nodes and transmits its output to yet other nodes. By itself, a single processing element is not very powerful; it generates a scalar output, a single numerical value, which is a simple nonlinear function of its inputs. The power of the system emerges from the combination of many units in an appropriate way [FS92].

ANN is specialized to implement different functions by varying the connection topology and the values of the connecting weights. Complex functions can be implemented by connecting units together with appropriate weights. In fact, it has been shown that a sufficiently large network with an appropriate structure and property chosen weights can approximate with arbitrary accuracy any function satisfying certain broad constraints. In ANNs, the design motivation is what distinguishes them from other mathematical techniques: an ANN is a processing device, either an algorithm, or actual hardware, whose design was motivated by the design and functioning of animal brains and components thereof.

There are many different types of ANNs, each of which has different strengths particular to their applications. The abilities of different networks can be related to their structure, dynamics and learning methods.

Formally, here we are dealing with the *ANN evolution* 2-functor $\mathcal{E}$, given by

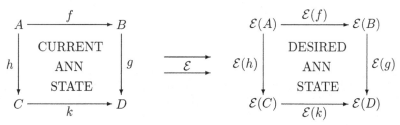

Here $\mathcal{E}$ represents ANN functor from the *source* *2-category* of the current ANN state, defined as a commutative square of small ANN categories $A, B, C, D, \ldots$ of *current ANN components* and their causal interrelations $f, g, h, k, \ldots$, onto the *target* *2-category* of the desired ANN state, defined as a commutative square of small ANN categories $\mathcal{E}(A), \mathcal{E}(B), \mathcal{E}(C), \mathcal{E}(D), \ldots$ of *evolved ANN*

components and their causal interrelations $\mathcal{E}(f), \mathcal{E}(g), \mathcal{E}(h), \mathcal{E}(k)$. As in the previous section, each causal arrow in above diagram, e.g., $f : A \rightarrow B$, stands for a *generic ANN dynamorphism*.

2.2.1 Common Discrete ANNs

Multilayer Perceptrons

The most common ANN model is the *feedforward neural network* with one input layer, one output layer, and one or more hidden layers, called *multilayer perceptron* (MLP). This type of neural network is known as a *supervised network* because it requires a desired output in order to learn. The goal of this type of network is to *create a model* $f : x \rightarrow y$ that correctly maps the input x to the output y using historical data so that the model can then be used to produce the output when the desired output is unknown [Kos92].

In MLP the inputs are fed into the input layer and get multiplied by interconnection weights as they are passed from the input layer to the first hidden layer. Within the first hidden layer, they get summed then processed by a nonlinear function (usually the hyperbolic tangent). As the processed data leaves the first hidden layer, again it gets multiplied by interconnection weights, then summed and processed by the second hidden layer. Finally the data is multiplied by interconnection weights then processed one last time within the output layer to produce the neural network output.

MLPs are typically trained with *static backpropagation*. These networks have found their way into countless applications requiring static pattern classification. Their main advantage is that they are easy to use, and that they can approximate any input/output map. The key disadvantages are that they train slowly, and require lots of training data (typically three times more training samples than the number of network weights).

McCulloch–Pitts Processing Element

MLPs are typically composed of *McCulloch–Pitts neurons* (see [MP43]). This processing element (PE) is simply a sum-of-products followed by a threshold nonlinearity. Its input–output equation is

$$y = f(\text{net}) = f\left(w_i x^i + b\right), \quad (i = 1, \ldots, D),$$

where D is the number of inputs, x^i are the inputs to the PE, w_i are the weights and b is a bias term (see e.g., [MP69]). The activation function is a *hard threshold* defined by *signum* function,

$$f(\text{net}) = \begin{cases} 1, & \text{for net} \geq 0, \\ -1, & \text{for net} < 0. \end{cases}$$

Therefore, McCulloch–Pitts PE is composed of an adaptive linear element (*Adaline*, the weighted sum of inputs), followed by a signum nonlinearity [PEL00].

Sigmoidal Nonlinearities

Besides the hard threshold defined by signum function, other nonlinearities can be utilized in conjunction with the McCulloch–Pitts PE. Let us now smooth out the threshold, yielding a sigmoid shape for the nonlinearity. The most common nonlinearities are the *logistic* and the *hyperbolic tangent threshold activation functions*,

$$\text{hyperbolic:} \quad f(\text{net}) = \tanh(\alpha \, \text{net}),$$

$$\text{logistic:} \quad f(\text{net}) = \frac{1}{1 + \exp(-\alpha \, \text{net})},$$

where α is a *slope parameter* and normally is set to 1. The major difference between the two sigmoidal nonlinearities is the range of their output values. The logistic function produces values in the interval $[0, 1]$, while the hyperbolic tangent produces values in the interval $[-1, 1]$. An alternate interpretation of this PE substitution is to think that the discriminant function has been generalized to

$$g(x) = f(w_i x^i + b), \quad (i = 1, \dots, D),$$

which is sometimes called a *ridge* function. The combination of the synapse and the tanh axon (or the sigmoid axon) is usually referred to as the modified McCulloch–Pitts PE, because they all respond to the full input space in basically the same functional form (a sum of products followed by a global nonlinearity). The output of the logistic function varies from 0 to 1. Under some conditions, the logistic function allows a very powerful interpretation of the output of the PE as a posteriori probabilities for Gaussian-distributed input classes. The tanh is closely related to the logistic function by a linear transformation in the input and output spaces, so neural networks that use either of these can be made equivalent by changing weights and biases [PEL00].

Gradient Descent on the Net's Performance Surface

The *search* for the weights to meet a *desired response* or internal constraint is the essence of any *connectionist* computation. The central problem to be solved on the road to machine-based classifiers is how to automate the process of *minimizing the error* so that the machine can independently make these weight changes, without need for hidden agents, or external observers. The optimality criterion to be minimized is usually the *mean square error* (MSE)

$$J = \frac{1}{2N} \sum_{i=1}^{N} \varepsilon_i^2,$$

where ε_i is the instantaneous error that is added to the output y_i (the linearly fitted value), and N is the number of observations. The function $J(w)$ is called the *performance surface* (the total error surface plotted in the space of weights w).

The search for the minimum of a function can be done efficiently using a broad class of methods based on *gradient information*. The gradient has two main advantages for the search:

1. It can be computed locally, and
2. It always points in the direction of maximum change.

The *gradient of the performance surface*, $\nabla J = \nabla_w J$, is a vector (with the dimension of w) that always points toward the direction of maximum J-change and with a magnitude equal to the slope of the tangent of the performance surface. The minimum value of the error J_{min} depends on both the input signal x^i and the desired signal d_i,

$$J_{min} = \frac{1}{2N} \left[\sum_i d_i^2 - \frac{(d_i x^i)}{\sum_i x^i} \right], \quad (i = 1, \ldots, D).$$

The location in coefficient space where the minimum w^* occurs also depends on both x^i and d_i. The performance surface shape depends only on the input signal x^i [PEL00].

Now, if the goal is to reach the minimum, the search must be in the direction opposite to the gradient. The overall method of gradient searching can be stated in the following way: Start the search with an arbitrary initial weight $w(0)$, where the iteration number is denoted by the index in parentheses. Then compute the gradient of the performance surface at $w(0)$, and modify the initial weight proportionally to the negative of the gradient at $w(0)$. This changes the operating point to $w(1)$. Then compute the gradient at the new position $w(1)$, and apply the same procedure again, that is,

$$w(n + 1) = w(n) - \eta \nabla J(n),$$

where η is a small constant and $\nabla J(n)$ denotes the gradient of the performance surface at the nth iteration. The constant η is used to maintain stability in the search by ensuring that the operating point does not move too far along the performance surface. This search procedure is called the *steepest descent method*.

In the late 1960s, Widrow proposed an extremely elegant algorithm to estimate the gradient that revolutionized the application of gradient descent procedures. His idea is very simple: Use the instantaneous value as the estimator for the true quantity:

$$\nabla J(n) = \frac{\partial}{\partial w(n)} J \approx \frac{1}{2} \frac{\partial}{\partial w(n)} \left(\varepsilon^2(n) \right) = -\varepsilon(n) x(n),$$

i.e., instantaneous estimate of the gradient at iteration n is simply the product of the current input $x(n)$ to the weight $w(n)$ times the current error $\varepsilon(n)$. The amazing thing is that the gradient can be estimated with one multiplication per weight. This is the gradient estimate that led to the celebrated *least means square algorithm* (LMS):

$$w(n + 1) = w(n) + \eta \varepsilon(n) x(n), \tag{2.3}$$

where the small constant η is called the *step size*, or the *learning rate*. The estimate will be noisy, however, since the algorithm uses the error from a single sample instead of summing the error for each point in the data set (e.g., the MSE is estimated by the error for the current sample).

Now, for fast convergence to the neighborhood of the minimum a large step size is desired. However, the solution with a large step size suffers from rattling. One attractive solution is to use a large learning rate in the beginning of training to move quickly toward the location of the optimal weights, but then the learning rate should be decreased to get good accuracy on the final weight values. This is called *learning rate scheduling*. This simple idea can be implemented with a variable step size controlled by

$$\eta(n + 1) = \eta(n) - \beta,$$

where $\eta(0) = \eta_0$ is the initial step size, and β is a small constant [PEL00].

Perceptron and Its Learning Algorithm

Rosenblatt perceptron (see [Ros58b, MP69]) is a *pattern-recognition* machine that was invented in the 1950s for optical character recognition. The perceptron has an input layer fully connected to an output layer with multiple McCulloch–Pitts PEs,

$$y_i = f(\underset{i}{\text{net}}) = f(w_i x^i + b_i), \quad (i = 1, \ldots, D),$$

where b_i is the bias for each PE. The number of outputs y_i is normally determined by the number of classes in the data. These PEs add the individual scaled contributions and respond to the entire input space.

F. Rosenblatt proposed the following procedure to directly minimize the error by changing the weights of the McCulloch–Pitts PE: Apply an input example to the network. If the output is correct do nothing. If the response is incorrect, tweak the weights and bias until the response becomes correct. Get the next example and repeat the procedure, until all the patterns are correctly classified. This procedure is called the *perceptron learning algorithm*, which can be put into the following form:

$$w(n + 1) = w(n) + \eta(d(n) - y(n)) x(n),$$

where η is the step size, y is the network output, and d is the desired response.

Clearly, the functional form is the same as in the LMS algorithm (2.3), that is, the old weights are incrementally modified proportionally to the product of the error and the input, but there is a significant difference. We cannot say that this corresponds to gradient descent since the system has a discontinuous nonlinearity. In the perceptron learning algorithm, $y(n)$ is the output of the nonlinear system. The algorithm is directly minimizing the difference between the response of the McCulloch–Pitts PE and the desired response, instead of minimizing the difference between the Adaline output and the desired response [PEL00].

This subtle modification has tremendous impact on the performance of the system. For one thing, the McCulloch–Pitts PE learns only when its output is wrong. In fact, when $y(n) = d(n)$, the weights remain the same. The net effect is that the final values of the weights are no longer equal to the linear regression result, because the nonlinearity is brought into the weight update rule. Another way of phrasing this is to say that the weight update became much more selective, effectively gated by the system performance. Notice that the LMS update is also a function of the error to a certain degree. Larger errors have more effect on the weight update than small errors, but all patterns affect the final weights implementing a 'smooth gate'. In the perceptron the net effect is that the placement of the discriminant function is no longer controlled smoothly by all the input samples as in the Adaline, only by the ones that are important for placing the discriminant function in a way that explicitly minimizes the output error.

The Delta Learning Rule

One can show that the LMS rule is equivalent to the chain rule in the computation of the *sensitivity* of the cost function J with respect to the unknowns. Interpreting the LMS equation (2.3) with respect to the sensitivity concept, we see that the gradient measures the sensitivity. LMS is therefore updating the weights proportionally to how much they affect the performance, i.e., proportionally to their sensitivity.

The LMS concept can be extended to the McCulloch–Pitts PE, which is a nonlinear system. The main question here is how can we compute the sensitivity through a nonlinearity? [PEL00] The so-called δ-*rule* represents a direct extension of the LMS rule to nonlinear systems with smooth nonlinearities. In case of the McCulloch–Pitts PE, *delta-rule* reads:

$$w_i(n+1) = w_i(n) + \eta \varepsilon_p(n) x_p^i(n) f'(\underset{p}{\mathrm{net}}(n)),$$

where $f'(\mathrm{net})$ is the partial derivative of the static nonlinearity, such that the *chain rule* is applied to the network topology, i.e.,

$$f'(\mathrm{net})x^i = \frac{\partial y}{\partial w_i} = \frac{\partial y}{\partial \mathrm{net}} \frac{\partial}{\partial w_i}. \tag{2.4}$$

As long as the PE nonlinearity is smooth we can compute how much a change in the weight δw_i affects the output y, or from the point of view of the sensitivity, how sensitive the output y is to a change in a particular weight δw_i.

Note that we compute this output sensitivity by a product of partial derivatives through intermediate points in the topology. For the nonlinear PE there is only one intermediate point, net, but we really do not care how many of these intermediate points there are. The chain rule can be applied as many times as necessary. In practice, we have an error at the output (the difference between the desired response and the actual output), and we want to adjust all the PE weights so that the error is minimized in a statistical sense. The obvious idea is to distribute the adjustments according to the sensitivity of the output to each weight.

To modify the weight, we actually *propagate back the output error* to intermediate points in the network topology and scale it along the way as prescribed by (2.4) according to the element transfer functions:

$$\text{forward path:} \quad x^i \longmapsto w_i \longmapsto \text{net} \longmapsto y$$

$$\text{backward path 1:} \quad w_i \overset{\partial \text{net}/\partial w}{\longleftarrow} \text{net} \overset{\partial y/\partial \text{net}}{\longleftarrow} y$$

$$\text{backward path 2:} \quad w_i \overset{\partial y/\partial w}{\longleftarrow} y.$$

This methodology is very powerful, because we do not need to know explicitly the error at intermediate places, such as net. The chain rule automatically derives the error contribution for us. This observation is going to be crucial for adapting more complicated topologies and will result in the *backpropagation* algorithm, discovered in 1988 by Werbos [Wer89].

Now, several key aspects have changed in the performance surface (which describes how the cost changes with the weights) with the introduction of the nonlinearity. The nice, parabolic performance surface of the linear least squares problem is lost. The performance depends on the topology of the network through the output error, so when nonlinear processing elements are used to solve a given problem the 'performance–weights' relationship becomes nonlinear, and there is no guarantee of a single minimum. The performance surface may have several minima. The minimum that produces the smallest error in the search space is called the *global minimum*. The others are called *local* minima. Alternatively, we say that the performance surface is *nonconvex*. This affects the search scheme because gradient descent uses local information to search the performance surface. In the immediate neighborhood, local minima are indistinguishable from the global minimum, so the gradient search algorithm may be caught in these suboptimal performance points, 'thinking' it has reached the global minimum [PEL00].

δ-rule extended to perceptron reads:

$$w_{ij}(n + 1) = w_{ij}(n) - \eta \frac{\partial J}{\partial w_{ij}} = w_{ij}(n) + \eta \delta_{ip} x_p^j,$$

which are local quantities available at the weight, that is, the activation x_p^j that reaches the weight w_{ij} from the input and the local error δ_{ip} propagated from the cost function J. This algorithm is local to the weight. Only the local error δ_i and the local activation x^j are needed to update a particular weight. This means that it is immaterial how many PEs the net has and how complex their interconnection is. The training algorithm can concentrate on each PE individually and work only with the local error and local activation [PEL00].

Backpropagation

The multilayer perceptron constructs input–output mappings that are a nested composition of nonlinearities, that is, they are of the form

$$y = f\left(\sum f\left(\sum(\cdot)\right)\right),$$

where the number of function compositions is given by the number of network layers. The resulting map is very flexible and powerful, but it is also hard to analyze [PEL00].

MLPs are usually trained by generalized δ-rule, the so-called *backpropagation* (BP). The weight update using backpropagation is

$$w_{ij}(n+1) = w_{ij}(n) + \eta f'(\underset{i}{\text{net}}(n))\left(\varepsilon^k(n)f'(\underset{k}{\text{net}}(n))w_{ki}(n)\right)y_j(n). \quad (2.5)$$

The summation in (2.5) is a sum of local errors δ_k at each network output PE, scaled by the weights connecting the output PEs to the ith PE. Thus the term in parenthesis in (2.5) effectively computes the total error reaching the ith PE from the output layer (which can be thought of as the ith PE's contribution to the output error). When we pass it through the ith PE nonlinearity, we have its local error, which can be written as

$$\delta_i(n) = f'(\underset{i}{\text{net}}(n))\delta^k w_{ki}(n).$$

Thus there is a unifying link in all the gradient-descent algorithms. All the weights in gradient descent learning are updated by multiplying the local error $\delta_i(n)$ by the local activation $x^j(n)$ according to Widrow's estimation of the instantaneous gradient first shown in the LMS rule:

$$\Delta w_{ij}(n) = \eta \delta_i(n)y_j(n).$$

What differs is the calculation of the local error, depending on whether the PE is linear or nonlinear and if the weight is attached to an output PE or a hidden-layer PE [PEL00].

Momentum Learning

Momentum learning is an improvement to the straight gradient-descent search in the sense that a memory term (the past increment to the weight) is used to speed up and stabilize convergence. In *momentum learning* the equation to update the weights becomes

$$w_{ij}(n+1) = w_{ij}(n) + \eta\delta_i(n)x_j(n) + \alpha\left(w_{ij}(n) - w_{ij}(n-1)\right),$$

where α is the momentum constant, usually set between 0.5 and 0.9. This is called momentum learning due to the form of the last term, which resembles the momentum in mechanics. Note that the weights are changed proportionally to how much they were updated in the last iteration. Thus if the search is going down the hill and finds a flat region, the weights are still changed, not because of the gradient (which is practically zero in a flat spot), but because of the rate of change in the weights. Likewise, in a narrow valley, where the gradient tends to bounce back and forth between hillsides, the momentum stabilizes the search because it tends to make the weights follow a smoother path. Imagine a ball (weight vector position) rolling down a hill (performance surface). If the ball reaches a small flat part of the hill, it will continue past this local minimum because of its momentum. A ball without momentum, however, will get stuck in this valley. Momentum learning is a robust method to speed up learning, and is usually recommended as the default search rule for networks with nonlinearities.

Advanced Search Methods

The popularity of *gradient descent method* is based more on its simplicity (it can be computed locally with two multiplications and one addition per weight) than on its search power. There are many other search procedures more powerful than backpropagation. For example, *Newtonian method* is a second-order method because it uses the information on the curvature to adapt the weights. However Newtonian method is computationally much more costly to implement and requires information not available at the PE, so it has been used little in neurocomputing. Although more powerful, Newtonian method is still a local search method and so may be caught in local minima or diverge due to the difficult neural network performance landscapes. Other techniques such as *simulated annealing*[2] and *genetic algorithms* (GA)[3] are global search procedures, that is, they can avoid local minima. The issue is that they are more costly to implement in a distributed system like a neural network, either because they are inherently slow or because they require nonlocal quantities [PEL00].

---

[2] Simulated annealing is a global search criterion by which the space is searched with a random rule. In the beginning the variance of the random jumps is very large. Every so often the variance is decreased, and a more local search is undertaken. It has been shown that if the decrease of the variance is set appropriately, the global optimum can be found with probability one. The method is called simulated annealing because it is similar to the annealing process of creating crystals from a hot liquid.

[3] Genetic algorithms are global search procedures proposed by J. Holland that search the performance surface, concentrating on the areas that provide better solutions. They use 'generations' of search points computed from the previous search points using the operators of crossover and mutation (hence the name).

The problem of search with local information can be formulated as an approximation to the functional form of the *matrix cost function* $J(\mathbf{w})$ at the operating point $\mathbf{w}_0$. This immediately points to the Taylor series expansion of J around $\mathbf{w}_0$,

$$J(\mathbf{w} - \mathbf{w}_0) = J_0 + (\mathbf{w} - \mathbf{w}_0)\nabla J_0 + \frac{1}{2}(\mathbf{w} - \mathbf{w}_0)\mathbf{H}_0(\mathbf{w} - \mathbf{w}_0)^T + \cdots,$$

where ∇J is our familiar gradient, and $\mathbf{H}$ is the Hessian matrix, that is, the matrix of second derivatives with entries

$$H_{ij}(\mathbf{w}_0) = \left. \frac{\partial^2 J(w)}{\partial w_i \partial w_j} \right|_{w=w_0},$$

evaluated at the operating point. We can immediately see that the Hessian cannot be computed with the information available at a given PE, since it uses information from two different weights. If we differentiate J with respect to the weights, we get

$$\nabla J(\mathbf{w}) = \nabla J_0 + \mathbf{H}_0(\mathbf{w} - \mathbf{w}_0) + \cdots \tag{2.6}$$

so we can see that to compute the full gradient at $\mathbf{w}$ we need all the higher terms of the derivatives of J. This is impossible. Since the performance surface tends to be bowl shaped (quadratic) near the minimum, we are normally interested only in the first and second terms of the expansion [PEL00].

If the expansion of (2.6) is restricted to the first term, we get the gradient-search methods (hence they are called *first-order-search methods*), where the gradient is estimated with its value at $\mathbf{w}_0$. If we expand to use the second-order term, we get *Newton method* (hence the name second-order method). If we equate the truncated relation (2.6) to 0 we immediately get

$$w = w_0 - \mathbf{H}_0^{-1}\nabla J_0,$$

which is the equation for the Newton method, which has the nice property of quadratic termination (it is guaranteed to find the exact minimum in a finite number of steps for quadratic performance surfaces). For most quadratic performance surfaces it can converge in one iteration.

The real difficulty is the memory and the computational cost (and precision) to estimate the Hessian. Neural networks can have thousands of weights, which means that the Hessian will have millions of entries. This is why methods of approximating the Hessian have been extensively researched. There are two basic classes of approximations [PEL00]:

1. Line search methods, and
2. Pseudo-Newton methods.

The information in the first type is restricted to the gradient, together with line searches along certain directions, while the second seeks approximations to the Hessian matrix. Among the line search methods probably the most effective is the *conjugate gradient method*. For quadratic performance surfaces the conjugate gradient algorithm preserves quadratic termination and can reach the minimum in D steps, where D is the dimension of the weight space. Among the Pseudo-Newton methods probably the most effective is the *Levenberg–Marquardt algorithm* (LM), which uses the Gauss–Newton method to approximate the Hessian. LM is the most interesting for neural networks, since it is formulated as a sum of quadratic terms just like the cost functions in neural networks.

The *extended Kalman filter* (EKF, see Sect. 7.2.6 in Appendix) forms the basis of a second-order neural network training method that is a practical and effective alternative to the batch-oriented, second-order methods mentioned above. The essence of the recursive EKF procedure is that, during training, in addition to evolving the weights of a network architecture in a sequential (as opposed to batch) fashion, an approximate error covariance matrix that encodes second-order information about the training problem is also maintained and evolved.

Homotopy Methods

The most popular method for solving nonlinear equations in general is the *Newton–Raphson method*. Unfortunately, this method sometimes fails, especially in cases when nonlinear equations possess multiple solutions (zeros). An emerging family of methods that can be used in such cases are homotopy (continuation) methods. These methods are robust and have good convergence properties.

Homotopy methods or *continuation methods* have increasingly been used for solving variety of nonlinear problems in fluid dynamics, structural mechanics, systems identifications, and integrated circuits (see [Wat90]). These methods, popular in mathematical programming, are globally convergent provided that certain coercivity and continuity conditions are satisfied by the equations that need to be solved [Wat90]. Moreover, they often yield all the solutions to the nonlinear system of equations.

The idea behind a homotopy or continuation method is to embed a parameter λ in the nonlinear equations to be solved. This is why they are sometimes referred to as *embedding methods*. Initially, parameter λ is set to zero, in which case the problem is reduced to an easy problem with a known or easily-found solution. The set of equations is then gradually deformed into the originally posed difficult problem by varying the parameter λ. The original problem is obtained for $\lambda = 1$. Homotopies are a class of continuation methods, in which parameter λ is a function of a path arc length and may actually increase or decrease as the path is traversed. Provided that certain coercivity conditions imposed on the nonlinear function to be solved are satisfied, the homotopy path does not branch (bifurcate) and passes through all the solutions of the nonlinear equations to be solved.

The zero curve of the homotopy map can be tracked by various techniques: an *ODE-algorithm*, a *normal flow algorithm*, and an *augmented Jacobian matrix algorithm*, among others [Wat90].

As a typical example, homotopy techniques can be applied to find the zeros of the gradient function $F : \mathbb{R}^N \to \mathbb{R}^N$, such that

$$F(\theta) = \frac{\partial E(\theta)}{\partial \theta_k}, \quad 1 \leq k \leq N,$$

where $E = (\theta)$ is the certain error function dependent on N parameters θ_k. In other words, we need to solve a system of nonlinear equations

$$F(\theta) = 0. \tag{2.7}$$

In order to solve (2.7), we can create a *linear homotopy* function

$$H(\theta, \lambda) = (1 - \lambda)(\theta - a) + \lambda F(\theta),$$

where a is an arbitrary starting point. Function $H(\theta, \lambda)$ has properties that equation $H(\theta, 0) = 0$ is easy to solve, and that $H(\theta, 1) \equiv F(\theta)$.

ANNs as Functional Approximators

The *universal approximation theorem* of Kolmogorov states [Hay98]:
Let $\phi(\cdot)$ be a nonconstant, bounded, and monotone-increasing continuous (C^0) function. Let I^N denote ND unit hypercube $[0, 1]^N$. The space of C^0-functions on I^N is denoted by $C(I^N)$. Then, given any function $f \in C(I^N)$ and $\epsilon > 0$, there exist an integer M and sets of real constants $\alpha_i, \theta_i, \omega_{ij}, i = 1, \ldots, M;$ $j = 1, \ldots, N$ such that we may define

$$F(x_1, \ldots, x_N) = \alpha_i \phi(\omega_{ij} x_j - \theta_i),$$

as an approximate realization of the function $f(\cdot)$; that is

$$|F(x_1, \ldots, x_N) - f(x_1, \ldots, x_N)| < \epsilon \quad \text{for all } \{x_1, \ldots, x_N\} \in I^N.$$

This theorem is directly applicable to *multilayer perceptrons*. First, the logistic function $1/[1 + \exp(-v)]$ used as the sigmoidal nonlinearity in a neuron model for the construction of a multilayer perceptron is indeed a nonconstant, bounded, and monotone-increasing function; it therefore satisfies the conditions imposed on the function $\phi(\cdot)$. Second, the upper equation represents the output of a multilayer perceptron described as follows:

1. The network has n input nodes and a single hidden layer consisting of M neurons; the inputs are denoted by $x_1, \ldots, x_N$.
2. ith hidden neuron has synaptic weights $\omega_{i1}, \ldots, \omega_{iN}$ and threshold θ_i.
3. The network output y_j is a linear combination of the outputs of the hidden neurons, with $\alpha_i, \ldots, \alpha_M$ defining the coefficients of this combination.

The theorem actually states that a single hidden layer is sufficient for a multilayer perceptron to compute a uniform ϵ approximation to a given training set represented by the set of inputs $x_1, \ldots, x_N$ and desired (target) output $f(x_1, \ldots, x_N)$. However, the theorem does not say that a single layer is *optimum* in the sense of learning time or ease of implementation.

Recall that training of multilayer perceptrons is usually performed using a certain clone of the BP algorithm (Sect. 2.2.1). In this forward-pass/backward-pass gradient-descending algorithm, the adjusting of synaptic weights is defined by the extended δ-*rule*, given by equation

$$\Delta \omega_{ji}(N) = \eta \cdot \delta_j(N) \cdot y_i(N), \tag{2.8}$$

where $\Delta \omega_{ji}(N)$ corresponds to the *weight correction*, η is the *learning-rate parameter*, $\delta_j(N)$ denotes the *local gradient* and $y_i(N)$—the *input signal of neuron j*; while the *cost function* E is defined as the instantaneous sum of squared errors e_j^2

$$E(n) = \frac{1}{2} \sum_j e_j^2(N) = \frac{1}{2} \sum_j [d_j(N) - y_j(N)]^2, \tag{2.9}$$

where $y_j(N)$ is the output of jth neuron, and $d_j(N)$ is the desired (target) response for that neuron. The slow BP convergence rate (2.8)–(2.9) can be accelerated using the faster LM algorithm (see Sect. 2.2.1 above), while its robustness can be achieved using an appropriate fuzzy controller (see [II07b]).

Summary of Supervised Learning Methods

Gradient Descent Method

Given the $(D + 1)$D weights vector $\mathbf{w}(n) = [w_0(n), \ldots, w_D(n)]^T$ (with $w_0 = $ bias), and the correspondent MSE-gradient (including partials of MSE w.r.t. weights)

$$\nabla \mathbf{e} = \left[\frac{\partial e}{\partial w_0}, \ldots, \frac{\partial e}{\partial w_D} \right]^T,$$

and the learning rate (step size) η, we have the vector learning equation

$$\mathbf{w}(n + 1) = \mathbf{w}(n) - \eta \nabla \mathbf{e}(n),$$

which in index form reads

$$w_i(n + 1) = w_i(n) - \eta \nabla e_i(n).$$

LMS Algorithm

$$\mathbf{w}(n + 1) = \mathbf{w}(n) + \eta \varepsilon(n) x(n),$$

where x is an input (measurement) vector, and ε is a zero-mean Gaussian noise vector uncorrelated with input, or

$$w_i(n + 1) = w_i(n) + \eta \varepsilon(n) x^i(n).$$

Newton's Method

$$\mathbf{w}(n+1) = \mathbf{w}(n) - \eta \mathbf{R}^{-1} \mathbf{e}(n),$$

where $\mathbf{R}$ is input (auto)correlation matrix, or

$$\mathbf{w}(n+1) = \mathbf{w}(n) + \eta \mathbf{R}^{-1} \varepsilon(n) x(n),$$

Conjugate Gradient Method

$$\mathbf{w}(n+1) = \mathbf{w}(n) + \eta \mathbf{p}(n),$$
$$\mathbf{p}(n) = -\nabla \mathbf{e}(n) + \beta(n) \mathbf{p}(n-1),$$
$$\beta(n) = \frac{\nabla \mathbf{e}(n)^T \nabla \mathbf{e}(n)}{\nabla \mathbf{e}(n-1)^T \nabla \mathbf{e}(n-1)}.$$

Levenberg–Marquardt Algorithm

Putting
$$\nabla \mathbf{e} = \mathbf{J}^T \mathbf{e},$$

where $\mathbf{J}$ is the Jacobian matrix, which contains first derivatives of the network errors with respect to the weights and biases, and $\mathbf{e}$ is a vector of network errors, LM algorithm reads

$$\mathbf{w}(n+1) = \mathbf{w}(n) - [\mathbf{J}^T \mathbf{J} + \mu \mathbf{I}]^{-1} \mathbf{J}^T \mathbf{e}. \tag{2.10}$$

Other Standard ANNs

Generalized Feedforward Nets

The *generalized feedforward network* (GFN) is a generalization of MLP, such that connections can jump over one or more layers, which in practice, often solves the problem much more efficiently than standard MLPs. A classic example of this is the two-spiral problem, for which standard MLP requires hundreds of times more training epochs than the generalized feedforward network containing the same number of processing elements. Both MLPs and GFNs are usually trained using a variety of backpropagation techniques and their enhancements like the nonlinear LM algorithm (2.10). During training in the spatial processing, the weights of the GFN converge iteratively to the analytical solution of the 2D Laplace equation.

Modular Feedforward Nets

The *modular feedforward networks* are a special class of MLP. These networks process their input using several parallel MLPs, and then recombine the results. This tends to create some structure within the topology, which will foster specialization of function in each submodule. In contrast to the MLP, modular networks do not have full inter-connectivity between their layers. Therefore, a smaller number of weights are required for the same size network (i.e., the same number of PEs). This tends to speed up training times and reduce the number of required training exemplars. There are many ways to segment a MLP into modules. It is unclear how to best design the modular topology based on the data. There are no guarantees that each module is specializing its training on a unique portion of the data.

Jordan and Elman Nets

Jordan and Elman networks (see [Elm90]) extend the multilayer perceptron with context units, which are processing elements (PEs) that remember past activity. Context units provide the network with the ability to extract temporal information from the data. In the Elman network, the activity of the first hidden PEs are copied to the context units, while the Jordan network copies the output of the network. Networks which feed the input and the last hidden layer to the context units are also available.

Kohonen Self-Organizing Map

Kohonen self-organizing map (SOM) is widely used for image pre-processing as well as a pre-processing unit for various hybrid architectures. SOM is a winner-take-all neural architecture that quantizes the input space, using a distance metric, into a discrete feature output space, where neighboring regions in the input space are neighbors in the discrete output space. SOM is usually applied to neighborhood clustering of random points along a circle using a variety of distance metrics: Euclidean, L^1, L^2, and L^n, Machalanobis, etc. The basic SOM architecture consists of a layer of Kohonen synapses of three basic forms: line, diamond and box, followed by a layer of winner-take-all axons. It usually uses added Gaussian and uniform noise, with control of both the mean and variance. Also, SOM usually requires choosing the proper initial neighborhood width as well as annealing of the neighborhood width during training to ensure that the map globally represents the input space.

The Kohonen SOM algorithm is defined as follows: Every stimulus $\mathbf{v}$ of an Euclidean input space V is mapped to the neuron with the position $\mathbf{s}$ in the neural layer R with the highest neural activity, the 'center of excitation' or 'winner', given by the condition

$$|\mathbf{w_s} - \mathbf{v}| = \min_{\mathbf{r} \in R} |\mathbf{w_r} - \mathbf{v}|,$$

where $|.|$ denotes the Euclidean distance in input space. In the Kohonen model the learning rule for each synaptic weight vector $\mathbf{w_r}$ is given by

$$\mathbf{w}_\mathbf{r}^{new} = \mathbf{w}_\mathbf{r}^{old} + \eta \cdot g_\mathbf{rs} \cdot (\mathbf{v} - \mathbf{w}_\mathbf{r}^{old}), \qquad (2.11)$$

with $g_\mathbf{rs}$ as a Gaussian function of Euclidean distance $|\mathbf{r} - \mathbf{s}|$ in the neural layer. Topology preservation is enforced by the common update of all weight vectors whose neuron $\mathbf{r}$ is adjacent to the center of excitation $\mathbf{s}$. The function $g_\mathbf{rs}$ describes the topology in the neural layer. The parameter η determines the speed of learning and can be adjusted during the learning process.

Radial Basis Function Nets

The *radial basis function network* (RBF) provides a powerful alternative to MLP for function approximation or classification. It differs from MLP in that the overall input–output map is constructed from local contributions of a layer of Gaussian axons. It trains faster and requires fewer training samples than MLP, using the hybrid supervised/unsupervised method. The unsupervised part of an RBF network consists of a competitive synapse followed by a layer of Gaussian axons. The means of the Gaussian axons are found through competitive clustering and are, in fact, the weights of the Conscience synapse. Once the means converge the variances are calculated based on the separation of the means and are associated with the Gaussian layer. Having trained the unsupervised part, we now add the supervised part, which consists of a single-layer MLP with a soft-max output.

Principal Component Analysis Nets

The *principal component analysis networks* (PCAs) combine unsupervised and supervised learning in the same topology. Principal component analysis is an unsupervised linear procedure that finds a set of uncorrelated features, principal components, from the input. A MLP is supervised to perform the nonlinear classification from these components. More sophisticated are the *independent component analysis networks* (ICAs).

Co-active Neuro-Fuzzy Inference Systems

The *co-active neuro-fuzzy inference system* (CANFIS), which integrates adaptable fuzzy inputs with a modular neural network to rapidly and accurately approximate complex functions. Fuzzy-logic inference systems (see next section) are also valuable as they combine the explanatory nature of rules (membership functions) with the power of 'black box' neural networks.

Support Vector Machines

The *support vector machine* (SVM), implementing the statistical learning theory, is used as the most powerful classification and decision-making system.

SVMs are a radically different type of classifier that has attracted a great deal of attention lately due to the novelty of the concepts that they bring to pattern recognition, their strong mathematical foundation, and their excellent results in practical problems. SVM represents the coupling of the following two concepts: the idea that transforming the data into a high-dimensional space makes linear discriminant functions practical, and the idea of large margin classifiers to train the MLP or RBF. It is another type of a kernel classifier: it places Gaussian kernels over the data and linearly weights their outputs to create the system output. To implement the SVM-methodology, we can use the Adatron-kernel algorithm, a sophisticated nonlinear generalization of the RBF networks, which maps inputs to a high-dimensional feature space, and then optimally separates data into their respective classes by isolating those inputs, which fall close to the data boundaries. Therefore, the Adatron-kernel is especially effective in separating sets of data, which share complex boundaries, as well as for the training for nonlinearly separable patterns. The support vectors allow the network to rapidly converge on the data boundaries and consequently classify the inputs.

The main advantage of SVMs over MLPs is that the learning task is a *convex optimization problem* which can be reliably solved even when the example data require the fitting of a very complicated function [Vap95, Vap98]. A common argument in computational learning theory suggests that it is dangerous to utilize the full flexibility of the SVM to learn the training data perfectly when these contain an amount of noise. By fitting more and more noisy data, the machine may implement a rapidly oscillating function rather than the smooth mapping which characterizes most practical learning tasks. Its prediction ability could be no better than random guessing in that case. Hence, modifications of SVM training [CS00] that allow for training errors were suggested to be necessary for realistic noisy scenarios. This has the drawback of introducing extra model parameters and spoils much of the original elegance of SVMs.

Mathematics of SVMs is based on real *Hilbert space* methods.

Genetic ANN-Optimization

Genetic optimization, added to ensure and speed-up the convergence of all other ANN-components, is a powerful tool for enhancing the efficiency and effectiveness of a neural network. Genetic optimization can fine-tune network parameters so that network performance is greatly enhanced. Genetic control applies a *genetic algorithm* (GA, see next section), a part of broader *evolutionary computation*, see MIT journal with the same name) to any network parameters that are specified. Also through the *genetic control*, GA parameters such as mutation probability, crossover type and probability, and selection type can be modified.

Time-Lagged Recurrent Nets

The *time-lagged recurrent networks* (TLRNs) are MLPs extended with short
term memory structures [Wer90]. Most real-world data contains information
in its time structure, i.e., how the data changes with time. Yet, most neural networks are purely static classifiers. TLRNs are the state of the art in
nonlinear time series prediction, system identification and temporal pattern
classification. Time-lagged recurrent nets usually use memory Axons, consisting of IIR filters with local adaptable feedback that act as a variable memory
depth. The time-delay neural network (TDNN) can be considered a special
case of these networks, examples of which include the Gamma and Laguerre
structures. The Laguerre axon uses locally recurrent all-pass IIR filters to
store the recent past. They have a single adaptable parameter that controls
the memory depth. Notice that in addition to providing memory for the input, we have also used a Laguerre axon after the hidden Tanh axon. This
further increases the overall memory depth by providing memory for that
layer's recent activations.

Fully Recurrent ANNs

The *fully recurrent networks* feed back the hidden layer to itself. Partially
recurrent networks start with a fully recurrent net and add a feedforward connection that bypasses the recurrency, effectively treating the recurrent part
as a state memory. These recurrent networks can have an infinite memory
depth and thus find relationships through time as well as through the instantaneous input space. Most real-world data contains information in its time
structure. Recurrent networks are the state of the art in nonlinear time series
prediction, system identification, and temporal pattern classification. In case
of large number of neurons, here the firing states of the neurons or their membrane potentials are the microscopic stochastic dynamical variables, and one
is mostly interested in quantities such as average state correlations and global
information processing quality, which are indeed measured by macroscopic
observables. In contrast to layered networks, one cannot simply write down
the values of successive neuron states for models of recurrent ANNs; here
they must be solved from (mostly stochastic) coupled dynamic equations. For
nonsymmetric networks, where the asymptotic (stationary) statistics are not
known, dynamical techniques from non-equilibrium statistical mechanics are
the only tools available for analysis. The natural set of macroscopic quantities (or order parameters) to be calculated can be defined in practice as the
smallest set which will obey closed deterministic equations in the limit of an
infinitely large network.

Being high-dimensional nonlinear systems with extensive feedback, the dynamics of recurrent ANNs are generally dominated by a wealth of attractors
(fixed-point attractors, limit-cycles, or even more exotic types), and the practical use of recurrent ANNs (in both biology and engineering) lies in the potential for creation and manipulation of these attractors through adaptation of

the network parameters (synapses and thresholds) (see [Hop82, Hop84]). Input fed into a recurrent ANN usually serves to induce a specific initial configuration (or firing pattern) of the neurons, which serves as a cue, and the output is given by the (static or dynamic) attractor which has been triggered by this cue. The most familiar types of recurrent ANN models, where the idea of creating and manipulating attractors has been worked out and applied explicitly, are the so-called attractor, associative memory ANNs, designed to store and retrieve information in the form of neuronal firing patterns and/or sequences of neuronal firing patterns. Each pattern to be stored is represented as a microscopic state vector. One then constructs synapses and thresholds such that the dominant attractors of the network are precisely the pattern vectors (in the case of static recall), or where, alternatively, they are trajectories in which the patterns are successively generated microscopic system states. From an initial configuration (the cue, or input pattern to be recognized) the system is allowed to evolve in time autonomously, and the final state (or trajectory) reached can be interpreted as the pattern (or pattern sequence) recognized by network from the input. For such programmes to work one clearly needs recurrent ANNs with extensive ergodicity breaking: the state vector will during the course of the dynamics (at least on finite time-scales) have to be confined to a restricted region of state-space (an ergodic component), the location of which is to depend strongly on the initial conditions. Hence our interest will mainly be in systems with many attractors. This, in turn, has implications at a theoretical/mathematical level: solving models of recurrent ANNs with extensively many attractors requires advanced tools from disordered systems theory, such as replica theory (statics) and generating functional analysis (dynamics).

Complex-Valued ANNs

It is expected that *complex-valued ANNs*, whose parameters (weights and threshold values) are all complex numbers, will have applications in all the fields dealing with complex numbers (e.g., telecommunications, quantum physics). A complex-valued, feedforward, multi-layered, back-propagation neural network model was proposed independently by [NF91, Nit97, Nit00, Nit04, GK92b, BP92], and demonstrated its characteristics:

(a) the properties greatly different from those of the real-valued back-propagation network, including 2D motion structure of weights and the orthogonality of the decision boundary of a complex-valued neuron;
(b) the learning property superior to the real-valued back-propagation;
(c) the inherent 2D motion learning ability (an ability to transform geometric figures); and
(d) the ability to solve the XOR problem and detection of symmetry problem with a single complex-valued neuron.

Following [NF91, Nit97, Nit00, Nit04], we consider here the complex-valued neuron. Its input signals, weights, thresholds and output signals are all complex numbers. The net input U_n to a complex-valued neuron n is defined as

$$U_n = W_{mn} X_m + V_n,$$

where W_{mn} is the (complex-valued) weight connecting the complex-valued neurons m and n, V_n is the (complex-valued) threshold value of the complex-valued neuron n, and X_m is the (complex-valued) input signal from the complex-valued neuron m. To get the (complex-valued) output signal, convert the net input U_n into its real and imaginary parts as follows: $U_n = x + iy = z$, where $i = \sqrt{-1}$. The (complex-valued) output signal is defined to be

$$\sigma(z) = \tanh(x) + i \tanh(y),$$

where $\tanh(u) = (\exp(u) - \exp(-u)) = (\exp(u) + \exp(-u))$, $u \in \mathbb{R}$. Note that $-1 < \text{Re}[\sigma], \text{Im}[\sigma] < 1$. Note also that σ is not regular as a complex function, because the Cauchy–Riemann equations do not hold.

A complex-valued ANN consists of such complex-valued neurons described above. A typical network has 3 layers: $m \to n \to 1$, with $w_{ij} \in \mathbb{C}$—the weight between the input neuron i and the hidden neuron j, $w_{0j} \in \mathbb{C}$—the threshold of the hidden neuron j, $c_j \in \mathbb{C}$—the weight between the hidden neuron j and the output neuron $(1 \le i \le m; 1 \le j \le n)$, and $c_0 \in \mathbb{C}$—the threshold of the output neuron. Let $y_j(z), h(z)$ denote the output values of the hidden neuron j, and the output neuron for the input pattern $z = [z_1, \ldots, z_m]^t \in \mathbb{C}^m$, respectively. Let also $\nu_j(z)$ and $\mu(z)$ denote the net inputs to the hidden neuron j and the output neuron for the input pattern $z \in \mathbb{C}^m$, respectively. That is,

$$\nu_j(z) = w_{ij} z_i + w_{0j}, \qquad \mu(z) = c_j y_j(z) + c_0,$$
$$y_j(z) = \sigma(\nu_j(z)), \qquad h(z) = \sigma(\mu(z)).$$

The set of all $m \to n \to 1$ complex-valued ANNs described above is usually denoted by $N_{m,n}$. The Complex-BP learning rule [NF91, Nit97, Nit00, Nit04] has been obtained by using a steepest-descent method for such (multilayered) complex-valued ANNs.

2.2.2 Common Continuous ANNs

Virtually all computer-implemented ANNs (mainly listed above) are discrete dynamical systems, mainly using supervised training (except Kohonen SOM) in one of gradient-descent searching forms. They are good as problem-solving tools, but they fail as models of animal nervous system. The other category of ANNs are continuous neural systems that can be considered as models of animal nervous system. However, *as models of the human brain, all current ANNs are simply trivial.*

Neurons as Functions

According to B. Kosko, neurons behave as functions [Kos92]; they transduce an unbounded input *activation* $x(t)$ into output *signal* $S(x(t))$. Usually a sigmoidal (S-shaped, bounded, monotone-nondecreasing: $S' \geq 0$) function describes the transduction, as well as the input-output behavior of many operational amplifiers. For example, the *logistic signal* (or, the *maximum-entropy*) function

$$S(x) = \frac{1}{1 + e^{-cx}}$$

is sigmoidal and strictly increases for positive scaling constant $c > 0$. Strict monotonicity implies that the *activation derivative* of S is positive:

$$S' = \frac{dS}{dx} = cS(1 - S) > 0.$$

An infinitely steep logistic signal function gives rise to a threshold signal function

$$S(x^{n+1}) = \begin{cases} 1, & \text{if } x^{n+1} > T, \\ S(x^n), & \text{if } x^{n+1} = T, \\ 0, & \text{if } x^{n+1} < T, \end{cases}$$

for an arbitrary real-valued threshold T. The index n indicates the discrete time step.

In practice signal values are usually binary or bipolar. *Binary signals*, like logistic, take values in the unit interval $[0, 1]$. *Bipolar signals* are signed; they take values in the bipolar interval $[-1, 1]$. Binary and bipolar signals transform into each other by simple scaling and translation. For example, the bipolar logistic signal function takes the form

$$S(x) = \frac{2}{1 + e^{-cx}} - 1.$$

Neurons with bipolar threshold signal functions are called *McCulloch–Pits neurons*.

A naturally occurring bipolar signal function is the *hyperbolic-tangent* signal function

$$S(x) = \tanh(cx) = \frac{e^{cx} - e^{-cx}}{e^{cx} + e^{-cx}},$$

with activation derivative

$$S' = c(1 - S^2) > 0.$$

The *threshold linear* function is a binary signal function often used to approximate neuronal firing behavior:

$$S(x) = \begin{cases} 1, & \text{if } cx \geq 1, \\ 0, & \text{if } cx < 0, \\ cx, & \text{else,} \end{cases}$$

which we can rewrite as

$$S(x) = \min(1, \max(0, cx)).$$

Between its upper and lower bounds the threshold linear signal function is trivially monotone increasing, since $S' = c > 0$.

Gaussian, or bell-shaped, signal function of the form $S(x) = e^{-cx^2}$, for $c > 0$, represents an important exception to signal monotonicity. Its activation derivative $S' = -2cxe^{-cx^2}$ has the sign opposite the sign of the activation x.

Generalized Gaussian signal functions define potential or radial basis functions $S_i(x^i)$ given by

$$S_i(x) = \exp\left[-\frac{1}{2\sigma_i^2} \sum_{j=1}^{n} (x_j - \mu_j^i)^2 \right],$$

for input activation vector $x = (x^i) \in \mathbb{R}^n$, variance σ_i^2, and mean vector $\boldsymbol{\mu}_i = (\mu_j^i)$. Each radial basis function S_i defines a spherical *receptive field* in $\mathbb{R}^n$. The ith neuron emits unity, or near-unity, signals for sample activation vectors x that fall in its receptive field. The mean vector $\boldsymbol{\mu}$ centers the receptive field in $\mathbb{R}^n$. The variance σ_i^2 localizes it. The radius of the Gaussian spherical receptive field shrinks as the variance σ_i^2 decreases. The receptive field approaches $\mathbb{R}^n$ as σ_i^2 approaches ∞.

The *signal velocity* $\dot{S} \equiv dS/dt$ is the *signal time derivative*, related to the activation derivative by

$$\dot{S} = S'\dot{x},$$

so it depends explicitly on *activation velocity*. This is used in unsupervised learning laws that adapt with *locally available information*.

The signal $S(x)$ induced by the activation x represents the neuron's firing frequency of action potentials, or pulses, in a sampling interval. The firing frequency equals the average number of pulses emitted in a sampling interval.

Short-term memory is modeled by *activation dynamics*, and *long-term memory* is modeled by *learning dynamics*. The overall neural network behaves as an *adaptive filter* (see [Hay91]).

In the simplest and most common case, neurons are not topologically ordered. They are related only by the synaptic connections between them. Kohonen calls this *lack of topological structure* in a *field of neurons* the *zeroth-order topology*. This suggests that ANN-models are *abstractions*, not *descriptions* of the brain neural networks, in which order does matter.

Basic Activation and Learning Dynamics

One of the oldest continuous training methods, based on Hebb's biological synaptic learning [Heb49], is *Oja–Hebb learning rule* [Oja82], which calculates the weight update according to the ODE

$$\dot{\omega}_i(t) = O(t)[I_i(t) - O(t)\omega_i(t)],$$

where $O(t)$ is the output of a simple, linear processing element; $I_i(t)$ are the inputs; and $\omega_i(t)$ are the synaptic weights.

Related to the Oja–Hebb rule is a special matrix of synaptic weights called *Karhunen–Loeve covariance matrix* **W** (KL), with entries

$$W_{ij} = \frac{1}{N}\omega_i^\mu \omega_j^\mu, \quad \text{(summing over } \mu\text{)}$$

where N is the number of vectors, and ω_i^μ is the ith component of the μth vector. The KL matrix extracts the principal components, or directions of maximum information (correlation) from a dataset.

In general, continuous ANNs are *temporal dynamical systems*. They have two coupled dynamics: activation and learning. First, a general system of coupled ODEs for the output of the ith *processing element* (PE) x^i, called the *activation dynamics*, can be written as

$$\dot{x}^i = g_i(x^i, \underset{i}{\mathrm{net}}), \tag{2.12}$$

with the *net input* to the ith PE x^i given by $\mathrm{net}_i = \omega_{ij}x^j$.

For example,

$$\dot{x}^i = -x^i + f_i(\underset{i}{\mathrm{net}}),$$

where f_i is called *output*, or *activation, function*. We apply some input values to the PE so that $\mathrm{net}_i > 0$. If the inputs remain for a sufficiently long time, the output value will reach an equilibrium value, when $\dot{x}^i = 0$, given by $x^i = f_i(\mathrm{net}_i)$. Once the unit has a nonzero output value, removal of the inputs will cause the output to return to zero. If $\mathrm{net}_i = 0$, then $\dot{x}^i = -x^i$, which means that $x \to 0$.

Second, a general system of coupled ODEs for the *update* of the synaptic weights ω_{ij}, i.e, *learning dynamics*, can be written as a generalization of the Oja–Hebb rule, i.e..

$$\dot{\omega}_{ij} = G_i(\omega_{ij}, x^i, x^i),$$

where G_i represents the *learning law*; the learning process consists of finding weights that encode the knowledge that we want the system to learn. For most realistic systems, it is not easy to determine a closed-form solution for this system of equations, so the approximative solutions are usually enough.

Standard Models of Continuous Nets

Hopfield Continuous Net

One of the first physically-based ANNs was developed by J. Hopfield. He first made a discrete, Ising-spin based network in [Hop82], and later generalized it to the continuous, graded-response network in [Hop84], which we briefly describe here. Later we will give full description of Hopfield models. Let $\text{net}_i = u_i$—the net input to the ith PE, biologically representing the summed action potentials at the axon hillock of a neuron. The PE *output function* is

$$v_i = g_i(\lambda u_i) = \frac{1}{2}(1 + \tanh(\lambda u_i)),$$

where λ is a constant called the *gain parameter*. The network is described as a transient RC circuit

$$C_i \dot{u}_i = T_{ij} v_j - \frac{u_i}{R_i} + I_i, \tag{2.13}$$

where I_i, R_i and C_i are inputs (currents), resistances and capacitances, and T_{ij} are synaptic weights.

The Hamiltonian energy function corresponding to (2.13) is given as

$$H = -\frac{1}{2}T_{ij}v_i v_j + \frac{1}{\lambda}\frac{1}{R_i}\int_0^{v_i} g_i^{-1}(v)\, dv - I_i v_i, \quad (j \neq i) \tag{2.14}$$

which is a generalization of a discrete, *Ising-spin Hopfield network* with energy function

$$E = -\frac{1}{2}\omega_{ij}x^i x^j, \quad (j \neq i),$$

where $g_i^{-1}(v) = u$ is the inverse of the function $v = g(u)$. To show that (2.14) is an appropriate *Lyapunov function* for the system, we shall take its time derivative assuming T_{ij} are symmetric:

$$\dot{H} = -\dot{v}_i\left(T_{ij}v_j - \frac{u_i}{R_i} + I_i\right) = -C_i\dot{v}_i\dot{u}_i = -C_i\dot{v}_i^2\frac{\partial g_i^{-1}(v_i)}{\partial v_i}. \tag{2.15}$$

All the factors in the summation (2.15) are positive, so $\dot{H}$ must decrease as the system evolves, until it eventually reaches the stable configuration, where $\dot{H} = \dot{v}_i = 0$.

Hecht–Nielsen Counterpropagation Net

Hecht–Nielsen counterpropagation network (CPN) is a full-connectivity, graded-response generalization of the standard BP algorithm (see [Hec87, Hec90]). The outputs of the PEs in CPN are governed by the set of ODEs

$$\dot{x}^i = -Ax_i + (B - x^i)I_i - x^i\sum_{j \neq i} I_j,$$

where $0 < x^i(0) < B$, and $A, B > 0$. Each PE receives a net excitation (on-center) of $(B - x^i)I_i$ from its corresponding input value, I. The addition of inhibitory connections (off-surround), $-x^i I_j$, from other units is responsible for preventing the activity of the processing element from rising in proportion to the absolute pattern intensity, I_i. Once an input pattern is applied, the PEs quickly reach an equilibrium state ($\dot{x}^i = 0$) with

$$x^i = \Theta_i \frac{BI_i}{A + I_i},$$

with the normalized *reflectance pattern* $\Theta_i = I_i(\sum_i I_i)^{-1}$, such that $\sum_i \Theta_i = 1$.

Competitive Net

Activation dynamics is governed by the ODEs

$$\dot{x}^i = -Ax_i + (B - x^i)[f(x^i) + \underset{i}{\text{net}}] - x^i \left[\sum_{j \neq i} f(x_j) + \sum_{j \neq i} \text{net}_j \right],$$

where $A, B > 0$ and $f(x^i)$ is an output function.

Kohonen's Continuous SOM and Adaptive Robotics Control

Kohonen continuous self organizing map (SOM) is actually the original Kohonen model of the biological neural process (see [Koh88]). SOM activation dynamics is governed by

$$\dot{x}^i = -r_i(x^i) + \underset{i}{\text{net}} + z_{ij}x_j, \tag{2.16}$$

where the function $r_i(x^i)$ is a general form of a loss term, while the final term models the lateral interactions between units (the sum extends over all units in the system). If z_{ij} takes the form of the Mexican-hat function, then the network will exhibit a bubble of activity around the unit with the largest value of net input.

   SOM learning dynamics is governed by

$$\dot{\omega}_{ij} = \alpha(t)(I_i - \omega_{ij})U(x^i),$$

where $\alpha(t)$ is the learning momentum, while the function $U(x^i) = 0$ unless $x^i > 0$ in which case $U(x^i) = 1$, ensuring that only those units with positive activity participate in the learning process.

   Kohonen's continuous SOM (2.16) is widely used in adaptive robotics control. Having an n-segment robot arm with n chained $SO(2)$-joints, for a particular initial position x and desired velocity $\dot{x}^j_{desir}$ of the end-effector, the required torques T_i in the joints can be found as

$$T_i = a_{ij}\dot{x}^j_{desir},$$

where the inertia matrix $a_{ij} = a_{ij}(x)$ is learned using SOM.

Adaptive Resonance Theory

Principles derived from an analysis of experimental literatures in vision, speech, cortical development, and reinforcement learning, including attentional blocking and cognitive-emotional interactions, led to the introduction of S. Grossberg's *adaptive resonance theory* (ART) as a theory of human *cognitive information processing* (see [CG03]). The theory has evolved as a series of real-time neural network models that perform unsupervised and supervised learning, pattern recognition, and prediction. Models of unsupervised learning include ART1, for binary input patterns, and fuzzy-ART and ART2, for analog input patterns [Gro82, CG03]. ARTMAP models combine two unsupervised modules to carry out supervised learning. Many variations of the basic supervised and unsupervised networks have since been adapted for technological applications and biological analyzes.

A central feature of all ART systems is a *pattern matching process* that compares an external input with the internal memory of an active code. ART matching leads either to a resonant state, which persists long enough to permit learning, or to a parallel memory search. If the search ends at an established code, the memory representation may either remain the same or incorporate new information from matched portions of the current input. If the search ends at a new code, the memory representation learns the current input. This match-based learning process is the foundation of ART *code stability*. Match-based learning allows memories to change only when input from the external world is close enough to internal expectations, or when something completely new occurs. This feature makes ART systems well suited to problems that require on-line learning of large and evolving databases (see [CG03]).

Many ART applications use fast learning, whereby adaptive weights converge to equilibrium in response to each input pattern. Fast learning enables a system to adapt quickly to inputs that occur rarely but that may require immediate accurate recall. Remembering details of an exciting movie is a typical example of learning on one trial. Fast learning creates memories that depend upon the order of input presentation. Many ART applications exploit this feature to improve accuracy by voting across several trained networks, with voters providing a measure of confidence in each prediction.

Match-based learning is complementary to *error-based learning*, which responds to a mismatch by changing memories so as to reduce the difference between a target output and an actual output, rather than by searching for a better match. Error-based learning is naturally suited to problems such as adaptive control and the learning of *sensory-motor maps*, which require ongoing adaptation to present statistics. Neural networks that employ error-based learning include backpropagation and other multilayer perceptrons (MLPs).

Activation dynamics of ART2 is governed by the ODEs [Gro82, CG03]

$$\epsilon \dot{x}_i = -Ax_i + (1 - Bx_i)I_i^+ - (C + Dx_i)I_i^-,$$

where ϵ is the 'small parameter', I_i^+ and I_i^- are excitatory and inhibitory inputs to the ith unit, respectively, and $A, B, C, D > 0$ are parameters.

General *Cohen–Grossberg activation equations* have the form:

$$\dot{v}_j = -a_j(v_j)[b_j(v_j) - f_k(v_k)m_{jk}], \quad (j = 1, \ldots, N), \tag{2.17}$$

and the *Cohen–Grossberg theorem* ensures the global stability of the system (2.17). If

$$a_j = 1/C_j, \qquad b_j = v_j/R_j - I_j, \qquad f_j(v_j) = u_j,$$

and constant $m_{ij} = m_{ji} = T_{ji}$, the system (2.17) reduces to the Hopfield circuit model (2.13).

ART and distributed ART (dART) systems are part of a growing family of self-organizing network models that feature attentional feedback and stable code learning. Areas of technological application include industrial design and manufacturing, the control of mobile robots, face recognition, remote sensing land cover classification, target recognition, medical diagnosis, electrocardiogram analysis, signature verification, tool failure monitoring, chemical analysis, circuit design, protein/DNA analysis, 3D visual object recognition, musical analysis, and seismic, sonar, and radar recognition. ART principles have further helped explain parametric behavioral and brain data in the areas of visual perception, object recognition, auditory source identification, variable-rate speech and word recognition, and *adaptive sensory-motor control* (see [CG03]).

Spatiotemporal Networks

In *spatiotemporal networks*, activation dynamics is governed by the ODEs

$$\dot{x}^i = A(-ax_i + b[I_i - \Gamma]^+),$$
$$\dot{\Gamma} = \alpha(S - T) + \beta\dot{S}, \quad \text{with}$$
$$[u]^+ = \begin{cases} u & \text{if } u > 0, \\ 0 & \text{if } u \leq 0, \end{cases}$$
$$A(u) = \begin{cases} u & \text{if } u > 0, \\ cu & \text{if } u \leq 0, \end{cases}$$

where $a, b, \alpha, \beta > 0$ are parameters, $T > 0$ is the *power-level target*, $S = \sum_i x^i$, and $A(u)$ is called the *attack function*.

Learning dynamics is given by *differential Hebbian law*

$$\dot{\omega}_{ij} = (-c\omega_{ij} + dx_i x_j)U(\dot{x}^i)U(-\dot{x}_j), \quad \text{with}$$
$$U(s) = \begin{cases} 1 & \text{if } s > 0, \\ 0 & \text{if } s \leq 0 \end{cases} \quad \text{where } c, d > 0 \text{ are constants.}$$

For review of *reactive neurodynamics*, see Sect. 4.1.2 below.

2.3 Synchronization in Neurodynamics

2.3.1 Phase Synchronization in Coupled Chaotic Oscillators

Over past two decades, *synchronization in chaotic oscillators* [FY83, PC90] has received much attention because of its fundamental importance in nonlinear dynamics and potential applications to laser dynamics [DBO01], electronic circuits [KYR98], chemical and biological systems [ESH98], and secure communications [KP95]. Synchronization in chaotic oscillators is characterized by the loss of exponential instability in the transverse direction through interaction. In coupled chaotic oscillators, it is known, various types of synchronization are possible to observe, among which are *complete synchronization* (CS) [FY83, PC90], *phase synchronization* (PS) [RPK96, ROH98], *lag synchronization* (LS) [RPK97] and *generalized synchronization* (GS) [KP96].

One of the noteworthy synchronization phenomena in this regard is PS which is defined by the phase locking between nonidentical chaotic oscillators whose amplitudes remain chaotic and uncorrelated with each other: $|\theta_1 - \theta_2| \leq$ const. Since the first observation of PS in mutually coupled chaotic oscillators [RPK96], there have been extensive studies in theory [ROH98] and experiments [DBO01]. The most interesting recent development in this regard is the report that the inter-dependence between physiological systems is represented by PS and *temporary phase-locking* (TPL) states, e.g., (a) *human heart beat and respiration* [SRK98], (b) a certain brain area and the tremor activity [TRW98, RGL99]. Application of the concept of PS in these areas sheds light on the analysis of non-stationary bivariate data coming from biological systems which was thought to be impossible in the conventional statistical approach. And this calls new attention to the PS phenomenon [KK00, KLR03].

Accordingly, it is quite important to elucidate a detailed transition route to PS in consideration of the recent observation of a TPL state in biological systems. What is known at present is that TPL[ROH98] transits to PS and then transits to LS as the coupling strength increases. On the other hand, it is noticeable that the phenomenon from non-synchronization to PS have hardly been studied, in contrast to the wide observations of the TPL states in the biological systems.

In this section, following [KK00, KLR03], we study the characteristics of TPL states observed in the regime from non-synchronization to PS in coupled chaotic oscillators. We report that there exists a special locking regime in which a TPL state shows maximal periodicity, which phenomenon we would call *periodic phase synchronization* (PPS). We show this PPS state leads to local negativeness in one of the vanishing Lyapunov exponents, taking the measure by which we can identify the maximal periodicity in a TPL state. We present a qualitative explanation of the phenomenon with a nonuniform oscillator model in the presence of noise.

We consider here the unidirectionally coupled non-identical Rössler oscillators for first example:

$$\dot{x}_1 = -\omega_1 y_1 - z_1, \qquad \dot{y}_1 = \omega_1 x_1 + 0.15 y_1, \qquad \dot{z}_1 = 0.2 + z_1(x_1 - 10.0),$$
$$\dot{x}_2 = -\omega_2 y_2 - z_2, \qquad \dot{y}_2 = \omega_2 x_2 + 0.165 y_2 + \epsilon(y_1 - y_2), \qquad (2.18)$$
$$\dot{z}_2 = 0.2 + z_2(x_2 - 10.0),$$

where the subscripts imply the oscillators 1 and 2, respectively, $\omega_{1,2}$ ($= 1.0 \pm 0.015$) is the overall frequency of each oscillator, and ϵ is the coupling strength. It is known that PS appears in the regime $\epsilon \geq \epsilon_c$ and that 2π phase jumps arise when $\epsilon < \epsilon_c$. Lyapunov exponents play an essential role in the investigation of the transition phenomenon with coupled chaotic oscillators and as generally understood that PS transition is closely related to the transition to the negative value in one of the vanishing Lyapunov exponents [PC90].

A vanishing Lyapunov exponent corresponds to a phase variable of an oscillator and it exhibits the neutrality of an oscillator in the phase direction. Accordingly, the local negativeness of an exponent indicates this neutrality is locally broken [RPK96]. It is important to define an appropriate phase variable in order to study the TPL state more thoroughly. In this regard, several methods have been proposed methods of using linear interpolation at a Poincaré section [RPK96], phase-space projection [RPK96, ROH98], tracing of the center of rotation in phase-space [YL97], Hilbert transformation [RPK96], or wavelet transformation [KK00, KLR03]. Among these we take the method of phase-space projection onto the $x_1 - y_1$ and $x_2 - y_2$ planes with the geometrical relation

$$\theta_{1,2} = \arctan(y_{1,2}/x_{1,2}),$$

and get *phase difference* $\varphi = \theta_1 - \theta_2$.

The system of coupled oscillators is said to be in a TPL state (or laminar state) when $\langle \varphi \rangle < \Lambda_c$ where $\langle \cdots \rangle$ is the running average over appropriate short time scale and Λ_c is the cutoff value to define a TPL state. The locking length of the TPL state, τ, is defined by time interval between two adjacent peaks of $\langle \varphi \rangle$.

In order to study the characteristics of the locking length τ, we introduce a measure [KK00, KLR03]: $P(\epsilon) = \sqrt{\mathrm{var}(\tau)}/\langle \tau \rangle$, which is the ratio between the average value of time lengths of TPL states and their standard deviation. In terminology of stochastic resonance, it can be interpreted as noise-to-signal ratio [PK97]. The measure would be minimized where the periodicity is maximized in TPL states.

To validate the argument, we explain the phenomenon in simplified dynamics. From (2.18), we get the equation of motion in terms of phase difference:

$$\dot{\varphi} = \Delta\omega + A(\theta_1, \theta_2, \epsilon)\sin\varphi + \xi(\theta_1, \theta_2, \epsilon), \qquad (2.19)$$

where

$$A(\theta_1, \theta_2, \epsilon) = (\epsilon + 0.15) \cos(\theta_1 + \theta_2) - \frac{\epsilon}{2}\left(\frac{R_1}{R_2}\right),$$

$$\xi(\theta_1, \theta_2, \epsilon) = \frac{\epsilon}{2}\frac{R_1}{R_2} \sin(\theta_1 + \theta_2) + \frac{z_1}{R_1} \sin(\theta_1) - \frac{z_2}{R_2} \sin(\theta_2)$$

$$+ (\epsilon + 0.015) \cos(\theta_2) \sin(\theta_2).$$

Here, $\Delta\omega = \omega_1 - \omega_2,$ $R_{1,2} = \sqrt{x_{1,2}^2 + y_{1,2}^2}.$

And from (2.19) we get the simplified equation to describe the phase dynamics:

$$\dot{\varphi} = \Delta\omega + \langle A \rangle \sin(\varphi) + \xi,$$

where $\langle A \rangle$ is the time average of $A(\theta_1, \theta_2, \epsilon)$. This is a nonuniform oscillator in the presence of noise where ξ plays a role of effective noise [Str94] and the value of $\langle A \rangle$ controls the width of bottleneck (i.e, non-uniformity of the flow). If the bottleneck is wide enough, (i.e., faraway from the saddle-node bifurcation point: $\Delta\omega \gg -\langle A \rangle$), the effective noise hardly contributes to the phase dynamics of the system. So the passage time is wholly governed by the width of the bottleneck as follows:

$$\langle \tau \rangle \sim 1/\sqrt{\Delta\omega^2 - \langle A \rangle^2} \sim 1/\sqrt{\Delta\omega^2 - \epsilon^2/4},$$

which is a slowly increasing function of ϵ. In this region while the standard deviation of TPL states is nearly constant (because the widely opened bottlenecks periodically appears and those lead to small standard deviation), the average value of locking length of TPL states is relatively short and the ratio between them is still large.

On the contrary as the bottleneck becomes narrower (i.e., near the saddle-node bifurcation point: $\Delta\omega \geq -\langle A \rangle$) the effective noise begins to perturb the process of bottleneck passage and regular TPL states develop into intermittent ones [ROH98, KK00]. It makes the standard deviation increase very rapidly and this trend overpowers that of the average value of locking lengths of the TPL states. Thus we understand that the competition between width of bottleneck and amplitude of effective noise produces the crossover at the minimum point of $P(\epsilon)$ which shows the maximal periodicity of TPL states.

Rosenblum et al. firstly observed the dip in mutually coupled chaotic oscillators [RPK96]. However the origin and the dynamical characteristics of the dip have been left unclarified. We argue that the dip observed in mutually coupled chaotic oscillators has the same origin as observed above in unidirectionally coupled systems.

Common apprehension is that near the border of synchronization the phase difference in coupled regular oscillators is periodic [RPK96] whereas in coupled chaotic oscillators it is irregular [ROH98]. On the contrary, we report that the special locking regime exhibiting the maximal periodicity of a TPL state also exists in the case of coupled chaotic oscillators. In general, the phase difference of coupled chaotic oscillators is described by the 1D Langevin equation,

$$\dot{\varphi} = F(\varphi) + \xi,$$

where ξ is the effective noise with finite amplitude. The investigation with regard to PS transition is the study of scaling of the laminar length around the virtual fixed-point φ^* where $F(\varphi^*) = 0$ [KK00, KT01] and PS transition is established when

$$\left| \int_\varphi^{\varphi^*} F(\varphi)d\phi \right| > \max |\xi|.$$

Consequently, the crossover region, from which the value of P grows exponentially, exists because intermittent series of TPL states with longer locking length τ appears as PS transition is nearer. Eventually it leads to an exponential growth of the standard deviation of the locking length. Thus we argue that PPS is the generic phenomenon mostly observed in coupled chaotic oscillators prior to PS transition.

In conclusion, analyzing the dynamic behaviors in coupled chaotic oscillators with slight parameter mismatch we have completed the whole transition route to PS. We find that there exists a special locking regime called PPS in which a TPL state shows maximal periodicity and that the periodicity leads to local negativeness in one of the vanishing Lyapunov exponents. We have also made a qualitative description of this phenomenon with the nonuniform oscillator model in the presence of noise. Investigating the characteristics of TPL states between non-synchronization and PS, we have clarified the transition route before PS. Since PPS appears in the intermediate regime between non-synchronization and PS, we expect that the concept of PPS can be used as a tool for analyzing weak inter-dependences, i.e., those not strong enough to develop to PS, between non-stationary bivariate data coming from biological systems, for instance [KK00, KLR03]. Moreover PPS could be a possible mechanism of the chaos regularization phenomenon [Har92, Rul01] observed in neurobiological experiments.

2.3.2 Oscillatory Phase Neurodynamics

In coupled oscillatory neuronal systems, under suitable conditions, the original dynamics can be reduced theoretically to a simpler phase dynamics. The state of the ith neuronal oscillatory system can be then characterized by a single phase variable φ_i representing the timing of the neuronal firings. The typical dynamics of *oscillator neural networks* are described by the *Kuramoto model* [Kur84, HI97, Str00], consisting of N equally weighted, all-to-all, phase-coupled limit-cycle oscillators, where each oscillator has its own natural frequency ω_i drawn from a prescribed distribution function:

$$\dot{\varphi}_i = \omega_i + \frac{K}{N} J_{ij} \sin(\varphi_j - \varphi_i + \beta_{ij}). \tag{2.20}$$

Here, J_{ij} and β_{ij} are parameters representing the effect of the interaction, while $K \geq 0$ is the coupling strength. For simplicity, we assume that all natural frequencies ω_i are equal to some fixed value ω_0. We can then eliminate ω_0 by applying the transformation $\varphi_i \to \varphi_i + \omega_0 t$. Using the complex representation $W_i = \exp(i\varphi_i)$ and $C_{ij} = J_{ij} \exp(i\beta_{ij})$ in (2.20), it is easily found that all neurons relax toward their stable equilibrium states, in which the relation $W_i = h_i/|h_i|$ ($h_i = C_{ij}W_j$) is satisfied. Following this line of reasoning, as a synchronous update version of the oscillator neural network we can consider the alternative discrete form [AN99],

$$W_i(t+1) = \frac{h_i(t)}{|h_i(t)|}, \quad h_i(t) = C_{ij}W_j(t). \tag{2.21}$$

Now we will attempt to construct an extended model of the oscillator neural networks to retrieve sparsely coded phase patterns. In (2.21), the complex quantity h_i can be regarded as the local field produced by all other neurons. We should remark that the phase of this field, h_i, determines the timing of the ith neuron at the next time step, while the amplitude $|h_i|$ has no effect on the retrieval dynamics (2.21). It seems that the amplitude can be thought of as the strength of the local field with regard to emitting spikes. Pursuing this idea, as a natural extension of the original model we stipulate that the system does not fire and stays in the resting state if the amplitude is smaller than a certain value. Therefore, we consider a network of N oscillators whose dynamics are governed by

$$W_i(t+1) = f(|h_i(t)|)\frac{h_i(t)}{|h_i(t)|}, \quad h_i(t) = C_{ij}W_j(t). \tag{2.22}$$

We assume that $f(x) = \Theta(x - H)$, where the real variable H is a threshold parameter and $\Theta(x)$ is the unit step function; $\Theta(x) = 1$ for $x \geq 0$ and 0 otherwise. Therefore, the amplitude $|W_i^t|$ assumes a value of either 1 or 0, representing the state of the ith neuron as firing or non-firing. Consequently, the neuron can emit spikes when the amplitude of the local field $h_i(t)$ is greater than the threshold parameter H.

Now, let us define a set of P patterns to be memorized as $\xi_i^\mu = A_i^\mu \exp(i\theta_i^\mu)$ ($\mu = 1, 2, \ldots, P$), where θ_i^μ and A_i^μ represent the phase and the amplitude of the ith neuron in the μth pattern, respectively. For simplicity, we assume that the θ_i^μ are chosen at random from a uniform distribution between 0 and 2π. The amplitudes A_i^μ are chosen independently with the probability distribution

$$P(A_i^\mu) = a\delta(A_i^\mu - 1) + (1 - a)\delta(A_i^\mu),$$

where a is the mean activity level in the patterns. Note that, if $H = 0$ and $a = 1$, this model reduces to (2.21).

For the synaptic efficacies, to realize the function of the associative memory, we adopt the *generalized Hebbian rule* in the form

$$C_{ij} = \frac{1}{aN}\xi_i^\mu \tilde{\xi}_j^\mu, \tag{2.23}$$

where $\tilde{\xi}_j^\mu$ denotes the complex conjugate of ξ_j^μ. The overlap $M_\mu(t)$ between the state of the system and the pattern μ at time t is given by

$$M_\mu(t) = m_\mu(t)e^{i\varphi_\mu(t)} = \frac{1}{aN}\tilde{\xi}_j^\mu W_j(t), \qquad (2.24)$$

In practice, the rotational symmetry forces us to measure the correlation of the system with the pattern μ in terms of the amplitude component $m_\mu(t) = |M_\mu(t)|$.

Let us consider the situation in which the network is recalling the pattern ξ_i^1; that is, $m_1(t) = m(t) \sim O(1)$ and $m_\mu(t) \sim O(1/\sqrt{N})(\mu \neq 1)$. The local field $h_i(t)$ in (2.22) can then be separated as

$$h_i(t) = C_{ij}W_j(t) = m_t e^{i\varphi_1(t)}\xi_i^1 + z_i(t), \qquad (2.25)$$

where $z_i(t)$ is defined by

$$z_i(t) = \frac{1}{aN}\xi_i^\mu \tilde{\xi}_j^\mu W_j(t). \qquad (2.26)$$

The first term in (2.25) acts to recall the pattern, while the second term can be regarded as the noise arising from the other learned patterns. The essential point in this analysis is the treatment of the second term as *complex Gaussian noise* characterized by

$$\langle z_i(t)\rangle = 0, \qquad \langle |z_i(t)|^2\rangle = 2\sigma(t)^2. \qquad (2.27)$$

We also assume that $\varphi_1(t)$ remains a constant, that is, $\varphi_1(t) = \varphi_0$. By applying the method of statistical neurodynamics to this model under the above assumptions [AN99], we can study the retrieval properties analytically. As a result of such analysis we have found that the retrieval process can be characterized by some macroscopic order parameters, such as $m(t)$ and $\sigma(t)$.

From (2.24), we find that the overlap at time $t + 1$ is given by

$$m(t+1) = \left\langle\!\!\!\left\langle f(|m(t) + z(t)|)\frac{m(t) + z(t)}{|m(t) + z(t)|}\right\rangle\!\!\!\right\rangle, \qquad (2.28)$$

where $\langle\!\langle \cdots \rangle\!\rangle$ represents an average over the *complex-valued Gaussian* $z(t)$ with mean 0 and variance $2\sigma(t)^2$. For the noise $z(t+1)$, in the limit $N \to \infty$ we get [AN99]

$$\begin{aligned}
z_i(t+1) \sim{} & \frac{1}{aN}\xi_i^\mu \tilde{\xi}_j^\mu f(|h_{j,\mu}(t)|)\frac{h_{j,\mu}(t)}{|h_{j,\mu}(t)|} \\
& + z_i(t)\left(\frac{f'(|h_{j,\mu}(t)|)}{2} + \frac{f(|h_{j,\mu}(t)|)}{2|h_{j,\mu}(t)|}\right),
\end{aligned} \qquad (2.29)$$

where $h_{j,\mu}(t) = 1/aN\xi_j^\nu \tilde{\xi}_k^\nu W_k(t)$.

2.3.3 Kuramoto Synchronization Model

The microscopic individual level dynamics of the Kuramoto model (2.20) is easily visualized by imagining oscillators as points running around on the unit circle. Due to rotational symmetry, the average frequency $\Omega = \sum_{i=1}^{N} \omega_i / N$ can be set to 0 without loss of generality; this corresponds to observing dynamics in the co-rotating frame at frequency Ω.

The governing equation (2.20) for the ith oscillator phase angle φ_i can be simplified to

$$\dot{\varphi}_i = \omega_i + \frac{K}{N} \sum_{i=1}^{N} \sin(\varphi_j - \varphi_i), \quad 1 \leq i \leq N. \tag{2.30}$$

It is known that as K is increased from 0 above some critical value K_c, more and more oscillators start to get synchronized (or phase-locked) until all the oscillators get fully synchronized at another critical value of K_{tp}. In the choice of $\Omega = 0$, the fully synchronized state corresponds to an exact steady state of the 'detailed', fine-scale problem in the co-rotating frame.

Such synchronization dynamics can be conveniently summarized by considering the fraction of the synchronized (phase-locked) oscillators, and conventionally described by a *complex-valued order parameter* [Kur84, Str00], $re^{i\psi} = \frac{1}{N} e^{i\varphi_j}$, where the radius r measures the phase coherence, and ψ is the average phase angle.

Transition from Full to Partial Synchronization

Following [MK05], here we restate certain facts about the nature of the second transition mentioned above, a transition between the full and the partial synchronization regime at $K = K_{tp}$, in the direction of decreasing K.

A fully synchronized state in the continuum limit corresponds to the solution to the mean-field type alternate form of (2.30),

$$\dot{\varphi}_i = \Omega = \omega_i + rK \sum_{i=1}^{N} \sin(\psi - \varphi_i), \tag{2.31}$$

where Ω is the common angular velocity of the fully synchronized oscillators (which is set to 0 in our case). Equation (2.31) can be further rewritten as

$$\frac{\Omega - \omega_i}{rK} = \sum_{i=1}^{N} \sin(\psi - \varphi_i), \tag{2.32}$$

where the absolute value of the r.h.s is bounded by unity.

As K approaches K_{tp} from above, the l.h.s for the 'extreme' oscillator (the oscillator in a particular family that has the maximum value of $|\Omega - \omega_i|$) first exceeds unity, and a real-valued solution to (2.32) ceases to exist. Different random draws of ω_i's from $g(\omega)$ for a finite number of oscillators result in slightly different values of K_{tp}. K_{tp} appears to follow the Gumbel type extreme distribution function [KN00], just as the maximum values of $|\Omega - \omega_i|$ do:

$$p(K_{tp}) = \sigma^{-1} e^{-(K_{tp}-\mu)/\sigma} \exp[-e^{-(K_{tp}-\mu)/\sigma}],$$

where σ and μ are parameters.

2.3.4 Lyapunov Chaotic Synchronization

The notion of *conditional Lyapunov exponents* was introduced by Pecora and Carroll in their study of synchronization of chaotic systems. First, in [PC91], they generalized the idea of driving a stable system to the situation when the drive signal is chaotic. This leaded to the concept of conditional Lyapunov exponents and also generalized the usual criteria of the linear stability theorem. They showed that driving with chaotic signals can be done in a robust fashion, rather insensitive to changes in system parameters. The calculation of the stability criteria leaded naturally to an estimate for the convergence of the driven system to its stable state. The authors focused on a homogeneous driving situation that leaded to the construction of synchronized chaotic subsystems. They applied these ideas to the Lorenz and Rössler systems, as well as to an electronic circuit and its numerical model. Later, in [PC98], they showed that many coupled oscillator array configurations considered in the literature could be put into a simple form so that determining the stability of the synchronous state could be done by a master stability function, which could be tailored to one's choice of stability requirement. This solved, once and for all, the problem of synchronous stability for any linear coupling of that oscillator.

It turns out, that, like the full Lyapunov exponent, the conditional exponents are well defined ergodic invariants, which are reliable quantities to quantify the relation of a global dynamical system to its constituent parts and to characterize dynamical self-organization [Men98].

Given a dynamical system defined by a map $f : M \rightarrow M$, with $M \subset R^m$ the conditional exponents associated to the splitting $R^k \times R^{m-k}$ are the eigenvalues of the limit

$$\lim_{n \rightarrow \infty} (D_k f^{n*}(x) D_k f^n(x))^{\frac{1}{2n}},$$

where $D_k f^n$ is the $k \times k$ diagonal block of the full Jacobian.

Mendes [Men98] proved that existence of the conditional Lyapunov exponents as well-defined ergodic invariants was guaranteed under the same conditions that established the existence of the Lyapunov exponents.

Recall that for measures μ that are absolutely continuous with respect to the Lebesgue measure of M or, more generally, for measures that are smooth along unstable directions (SBR measures) Pesin's [Pes77] identity holds

$$h(\mu) = \sum_{\lambda_i > 0} \lambda_i,$$

relating *Kolmogorov–Sinai entropy* $h(\mu)$ to the sum of the Lyapunov exponents. By analogy we may define the *conditional exponent entropies* [Men98] associated to the splitting $R^k \times R^{m-k}$ as the sum of the positive conditional exponents counted with their multiplicity

$$h_k(\mu) = \sum_{\xi_i^{(k)} > 0} \xi_i^{(k)}, \qquad h_{m-k}(\mu) = \sum_{\xi_i^{(m-k)} > 0} \xi_i^{(m-k)}.$$

The Kolmogorov–Sinai entropy of a dynamical system measures the rate of information production per unit time. That is, it gives the amount of randomness in the system that is not explained by the defining equations (or the minimal model [CY89]). Hence, the conditional exponent entropies may be interpreted as a measure of the randomness that would be present if the two parts $S^{(k)}$ and $S^{(m-k)}$ were uncoupled. The difference $h_k(\mu) + h_{m-k}(\mu) - h(\mu)$ represents the effect of the coupling.

Given a dynamical system S composed of N parts $\{S_k\}$ with a total of m degrees of freedom and invariant measure μ, one defines a *measure of dynamical self-organization* $I(S, \Sigma, \mu)$ as

$$I(S, \Sigma, \mu) = \sum_{k=1}^{N} \{h_k(\mu) + h_{m-k}(\mu) - h(\mu)\}.$$

For each system S, this quantity will depend on the partition Σ into N parts that one considers. $h_{m-k}(\mu)$ always denotes the conditional exponent entropy of the complement of the subsystem S_k. Being constructed out of ergodic invariants, $I(S, \Sigma, \mu)$ is also a well-defined ergodic invariant for the measure μ. $I(S, \Sigma, \mu)$ is formally similar to a mutual information. However, not being strictly a mutual information, in the information theory sense, $I(S, \Sigma, \mu)$ may take negative values.

2.4 Spike Neural Networks and Wavelet Resonance

In recent years much studies have been made for the stochastic resonance (SR), in which weak input signals are enhanced by background noises (see [GHJ98]). This paradoxical SR phenomenon was first discovered in the context of climate dynamics, and it is now reported in many nonlinear systems such as electric circuits, ring lasers, semiconductor devices and neurons.

For single neurons, SR has been studied by using various theoretical models such as the *integrate-and-fire model* (IF) [BED96, SPS99], the *FitzHough–Nagumo model* (FN) [Lon93, LC94] and the *Hodgkin–Huxley model* (HH) [LK99]. In these studies, a weak periodic (sinusoidal) signal is applied to the neuron, and it has been reported that the peak height of the *inter-spike-interval* (ISI) distribution [BED96, Lon93] or the *signal-to-noise ratio* (SNR) of output signals [WPP94, LK99] shows the maximum when the noise intensity is changed.

SR in coupled or ensemble neurons has been also investigated by using the IF model [SRP99, LS01], FN model [CCI95a, SM01] and HH model [PWM96, LHW00]. The transmission fidelity is examined by calculating various quantities: the peak-to-width ratio of output signals [CCI95a, SRP99, PWM96, TSP99], the cross-correlation between input and output signals [SRP99, LHW00], SNR [SRP99, KHO00][LHW00], and the mutual information [SM01]. One or some of these quantities has been shown to take a maximum as functions of the noise intensity and the coupling strength. [CCI95a] have pointed out that SR of ensemble neurons is improved as the size of an ensemble is increased. Some physiological experiments support SR in real, biological systems of crayfish [DWP93, PMW96], cricket [LM96] and rat [GNN96, NMG99].

SR studies mentioned above are motivated from the fact that peripheral sensory neurons play a role of transducers, receiving analog stimula and emitting spikes. In central neural systems, however, cortical neurons play a role of data-processors, receiving and transmitting spike trains. The possibility of SR in the spike transmission has been reported [CGC96, Mat98]. The response to periodic coherent spike-train inputs has been shown to be enhanced by an addition of weak spike-train noises whose inter-spike intervals (ISIs) have the Poison or gamma distribution.

It should be stressed that these theoretical studies on SR in neural systems have been performed mostly for stationary analog (or spike-train) signals although they are periodic or aperiodic [CCI95b]. There has been few theoretical studies on SR for non-stationary signals. Fakir [Fak98] has discussed SR for non-stationary analog inputs with finite duration by calculating the cross-correlation. By applying a single impulse, [PWM96] have demonstrated that the spike-timing precision is improved by noises in an ensemble of 1000 HH neurons. One of the reason why SR study for stationary signals is dominant, is mainly due to the fact that the stationary signals can be easily analyzed by the Fourier transform (FT) with which, for example, the SNR is evaluated from FT spectra of output signals. The FT requires that a signal to be examined is stationary, not giving the time evolution of the frequency pattern. Actual biological signals are, however, not necessarily stationary. It has been reported that neurons in different regions have different firing activities. In thalamus, which is the major gateway for the flow of information toward the cerebral cortex, *gamma oscillations* (30–70 Hz), mainly 40 Hz, are reported in arousal, whereas spindle oscillations (7–14 Hz) and slow oscillations (1–7 Hz) are found in early sleeping and deepen sleeping states, respectively [BM93]. In hippocampus, gamma oscillation occurs in vivo, following sharp waves [FB96]. In neo-cortex, gamma oscillation is observed under conditions of sensory signal as well as during sleep [KBE93]. It has been reported that spike signals in cortical neurons are generally not stationary, rather they are transient signals or bursts [TJW99], whereas periodic spikes are found in systems such as auditory systems of owl and the electro-sensory system of electric fish [RH85].

The limitation of the FT analysis can be partly resolved by using the *short-time Fourier transform* (STFT). Assuming that the signal is quasi-stationary in the narrow time period, the FT is applied with time-evolving narrow windows. Then STFT yields the time evolution of the frequency spectrum. The STFT, however, has a critical limitation violating the *uncertainty principle*, which asserts that if the window is too narrow, the frequency resolution will be poor whereas if the window is too wide, the time resolution will be less precise. This limitation becomes serious for signals with much transient components, like spike signals.

The disadvantage of the STFT is overcome in the wavelet transform (WT). In contrast to the FT, the WT offers the 2D expansion for a time-dependent signal with the scale and translation parameters which are interpreted physically as the inverse of frequency and time, respectively. As a basis of the WT, we employ the *mother wavelet* which is localized in both frequency and time domains. The WT expansion is carried out in terms of a family of wavelets which is made by dilation and translation of the mother wavelet. The time evolution of frequency pattern can be followed with an optimal time-frequency resolution.

The WT appears to be an ideal tool for analyzing signals of a non-stationary nature. In recent years the WT has been applied to an analysis of biological signals [SBR99], such as electroencephalographic (EEG) waves [BAI96, RBY01], and spikes [HSS00, Has01]. EEG is a reflection of the activity of ensembles of neurons producing oscillations. By using the WT, we can get the time-dependent decomposition of EEG signals to δ (0.3–3.5 Hz), θ (3.5–7.5 Hz), α (7.5–12.5 Hz), β (12.5–30.0 Hz) and γ (30–70 Hz) components [BAI96]–[RBY01]. It has been shown that the WT is a powerful tool to the spike sorting in which coherent signals of a single target neuron are extracted from mixture of response signals [HSS00, ZT97]. The WT has been probed [HSS00] to be superior than the conventional analysis methods like the *principal component analysis* (PCA) [Oja98]. Quite recently [Has01] has made an analysis of transient spike-train signals of a HH neuron with the use of WT, calculating the energy distribution and Shanon entropy.

It is interesting to analyze the response of ensemble neurons to *transient spike inputs* in noisy environment by using the WT. There are several sources of noises: (i) cells in sensory neurons are exposed to noises arising from the outer world, (ii) ion channels of the membrane of neurons are known to be stochastic [DMS], (iii) the synaptic transmission yields noises originating from random fluctuations of the synaptic vesicle release rate [Smi98], and (iv) synaptic inputs include leaked currents from neighboring neurons [SN94]. Most of existing studies on SR adopt the Gaussian noises, taking account of the items (i)–(iii). Simulating the noise of the item (iv), [CGC96, Mat98] include spike-train noises whose ISIs have the Poisson or gamma distribution. In this section, following [Has02] we take into account Gaussian noises.

We assume an ensemble of Hodgkin–Huxley (HH) neurons to receive transient spike trains with independent Gaussian noises. The HH neurons model is adopted because it is known to be the most realistic among theoretical models [HH52b]. The signal transmission is assessed by the cross-correlation between input and output signals and SNR, which are expressed in terms of WT expansion coefficients. In calculating the SNR, we adopt the de-noising technique within the WT method [BBD92]–[Qui00], by which the noise contribution is extracted from output signals. We study the cases of both sub-threshold and supra-threshold inputs. Usually SR is realized only for the sub-threshold input. Quite recently, [SM01, Sto00] has pointed out that SR can be occurred also for the supra-threshold signals in systems when they have multilevel threshold. It is the case also in our calculations when the synaptic coupling strength is distributed around the critical threshold value.

2.4.1 Ensemble Neuron Model

In order to study the propagation of spike trains in neural networks, we adopt a simple model of two layers which are referred to as the layer A and B, and which include N_A and N_B HH neurons, respectively. The HH neurons in the layer A receive a common input consisting of M impulses and independent Gaussian noises. Output impulses of neurons in the layer A are feed-forwarded to neurons in the layer A through synaptic couplings with time delays. Our model mimics a part of the *Sinfire neural network* with feedforward couplings but no intra-layer ones.

We assume a network consisting of N-unit HH neurons which receive the same spike trains but independent Gaussian white noises through excitatory synapses. Spikes emitted by the ensemble neurons are collected by a summing neuron. A similar model was previously adopted by several authors studying SR for analog signals [CCI95a, PWM96, TSP99]. An input signal in this section is a transient spike train consisting of M impulses ($M = 1$–3). Dynamics of the membrane potential V_i of the HH neuron i is described by the nonlinear differential equations given by [Has02]

$$\bar{C}\dot{V}_i = -I_i^{\mathrm{ion}} + I_i^{\mathrm{ps}} + I_i^{\mathrm{n}}, \quad (\text{for } 1 \le i \le N), \tag{2.33}$$

where $\bar{C} = 1\ \mu\mathrm{F/cm}^2$ is the capacity of the membrane. The first term I_i^{ion} of (2.33) denotes the ion current given by

$$I_i^{\mathrm{ion}} = g_{\mathrm{Na}}m_i^3 h_i(V_i - V_{\mathrm{Na}}) + g_{\mathrm{K}}n_i^4(V_i - V_{\mathrm{K}}) + g_{\mathrm{L}}(V_i - V_{\mathrm{L}}),$$

where the maximum values of conductivities of Na and K channels and leakage are $g_{\mathrm{Na}} = 120\ \mathrm{mS/cm}^2$, $g_{\mathrm{K}} = 36\ \mathrm{mS/cm}^2$ and $g_{\mathrm{L}} = 0.3\ \mathrm{mS/cm}^2$, respectively; the respective reversal potentials are $V_{\mathrm{Na}} = 50$ mV, $V_{\mathrm{K}} = -77$ mV and $V_{\mathrm{L}} = -54.5$ mV. Dynamics of the gating variables of Na and K channels, m_i, h_i and n_i, are described by the ordinary differential equations, whose details have been given elsewhere [HH52b, Has00].

The second term I_i^{ps} in (2.33) denotes the postsynaptic current given by

$$I_i^{\text{ps}} = \sum_{m=1}^{M} g_s (V_a - V_s) \alpha(t - t_{im}), \qquad (2.34)$$

which is induced by an input spike with the magnitude V_a given by

$$U_i(t) = V_a \sum_{m=1}^{M} \delta(t - t_{im}), \qquad (2.35)$$

with the alpha function

$$\alpha(t) = (t/\tau_s) e^{-t/\tau_s} \Theta(t). \qquad (2.36)$$

In (2.34)–(2.36) t_{im} is the mth firing time of the input, the Heaviside function is defined by $\Theta(t) = 1$ for $x \geq 0$ and 0 for $x < 0$, given by $t_{im} = (m - 1) : T_i$ for $m = 1 - M$, and g_s, V_s and τ_s stand for the conductance, reversal potential and time constant, respectively, of the synapse.

The third term I_i^{n} in (2.33) denotes the Gaussian noise given by

$$\tau_n \dot{I}_j^{\text{n}} = -I_j^{\text{n}} + \xi_j(t),$$

with the independent *Gaussian white noise* $\xi_j(t)$, given by

$$\langle \overline{I_i^{\text{n}}(t)} \rangle = 0, \qquad \langle \overline{I_i^{\text{n}}(t) : I_\ell^{\text{n}}(t')} \rangle = 2D : \delta_{i\ell} : \delta(t - t'),$$

where the overline $\overline{X}$ and the bracket $\langle X \rangle$ denote the temporal and spatial averages, respectively, and D the intensity of white noises.

The output spike of the neuron i in an ensemble is given by

$$U_{oi}(t) = V_a \sum_{n} \delta(t - t_{oin}), \qquad (2.37)$$

in a similar form as an input spike (2.35), where t_{oin} is the nth firing time when $V_i(t)$ crosses $V_z = 0$ mV from below.[4]

The above ODEs are solved by the forth-order Runge–Kutta method by the integration time step of 0.01 ms with double precision. Some results are examined by using the exponential method for a confirmation of the accuracy. The initial conditions for the variables are given by

$$V_i(t) = -65 \text{ mV}, \qquad m_i(t) = 0.0526, \qquad h_i(t) = 0.600, \qquad n_i(t) = 0.313,$$

---

[4] We should remark that the model given above does not include couplings among ensemble neurons. This is in contrast with some works on ensemble neurons [KHO00, SM01, WCW00, LHW00] where introduced couplings among neurons play an important role in SR besides noises.

at $t = 0$, which are the rest-state solution of a single HH neuron. The initial function for $V_i(t)$, whose setting is indispensable for the delay-ODE, is given by

$$V_j(t) = -65 \text{ mV}, \quad \text{for } j = 1, 2, \text{ at: } t \in [-\tau_d, 0).$$

Hereafter time, voltage, conductance, current, and D are expressed in units of ms, mV, mS/cm^2, μA/cm^2, and μA^2/cm^4, respectively. We have adopted parameters of $V_a = 30$, $V_c = -50$, and $\tau_s = \tau_n = 2$, and $\tau_{i\ell} = 10$. Adopted values of g_s, D, M and N will be described shortly.

2.4.2 Wavelet Neurodynamics

Recall (see Appendix) that there are two types of WTs: one is the continuous wavelet transform (CWT) and the other the discrete wavelet transform (DWT). In the former the parameters denoting the scale and translation are continuous variables while in the latter they are discrete variables.

CWT

The CWT for a given regular function $f(t)$ is defined by [Has02]

$$c(a, b) = \int \psi_{ab}^*(t) f(t) dt \equiv \langle \psi_{ab}(t), : f(t) \rangle,$$

with a family of wavelets $\psi_{ab}(t)$ generated by

$$\psi_{ab}(t) = |a|^{-1/2} \psi\left(\frac{t - b}{a}\right), \tag{2.38}$$

where $\psi(t)$ is the *mother wavelet*, the star denotes the complex conjugate, and a and b express the scale change and translation, respectively, and they physically stand for the inverse of the frequency and the time. Then the CWT transforms the time-dependent function $f(t)$ into the frequency- and time-dependent function $c(a, b)$. The mother wavelet is a smooth function with good localization in both frequency and time spaces. A wavelet family given by (2.38) plays a role of elementary function, representing the function $f(t)$ as a superposition of wavelets $\psi_{ab}(t)$.

The *inverse* of the wavelet transform may be given by

$$f(t) = C_\psi^{-1} \int \frac{da}{a^2} \int c(a, b) \psi_{ab}(t) db,$$

when the mother wavelet satisfies the following two conditions:

(i) the admissibility condition given by

$$0 < C_\psi < \infty, \quad \text{with } C_\psi = \int_{-\infty}^{\infty} |\hat{\Psi}(\omega)|^2 / |\omega| d\omega,$$

where $\hat{\Psi}(\omega)$ is the Fourier transform of $\psi(t)$, and

(ii) the zero mean of the mother wavelet:

$$\int_{-\infty}^{\infty} \psi(t)dt = \hat{\Psi}(0) = 0.$$

DWT

On the contrary, the DWT is defined for *discrete* values of $a = 2^j$ and $b = 2^j k$ (j, k: integers) as

$$c_{jk} \equiv c(2^j, 2^j k) = \langle \psi_{jk}(t), f(t) \rangle, \quad \text{with} \tag{2.39}$$
$$\psi_{jk}(t) = 2^{-j/2} \psi(2^{-j}t - k). \tag{2.40}$$

The ortho-normal condition for the wavelet functions is given by [Has02]

$$\langle \psi_{jk}(t), \psi_{j'k'}(t) \rangle = \delta_{jj'} \delta_{kk'},$$

which leads to the *inverse DWT*:

$$f(t) = \sum_j \sum_k c_{jk} \psi_{jk}(t). \tag{2.41}$$

In the multiresolution analysis (MRA) of the DWT, we introduce a scaling function $\phi(t)$, which satisfies the recurrent relation with $2K$ masking coefficients, h_k, given by

$$\phi(t) = \sqrt{2} \sum_{k=0}^{2K-1} h_k \phi(2t - k),$$

with the normalization condition for $\phi(t)$ given by

$$\int \phi(t)dt = 1.$$

A family of wavelet functions is generated by

$$\psi(t) = \sqrt{2} \sum_{k=0}^{2K-1} (-1)^k h_{2K-1-k} \phi(2t - k).$$

The scaling and wavelet functions satisfy the ortho-normal relations:

$$\langle \phi(t), \phi(t - m) \rangle = \delta_{m0}, \tag{2.42}$$
$$\langle \psi(t), \psi(t - m) \rangle = \delta_{m0}, \tag{2.43}$$
$$\langle \phi(t), \psi(t - m) \rangle = 0. \tag{2.44}$$

A set of masking coefficients h_j is chosen so as to satisfy the conditions shown above. Here $g_k = (-1)^k h_{2M-1-k}$ which is derived from the orthonormal relations among wavelet and scaling functions.

The simplest wavelet function for $K = 1$ is the *Harr wavelet* for which we get $h_0 = h_1 = 1/\sqrt{2}$, and

$$\psi_H(t) = \begin{cases} 1, & \text{for } 0 \leq t < 1/2 \\ -1, & \text{for } 1/2 \leq t < 1 \\ 0, & \text{otherwise.} \end{cases}$$

In the more sophisticated wavelets like the *Daubechies wavelet*, an additional condition given by

$$\int t^\ell \psi(t) dt = 0, \quad \text{for } \ell = 0, 1, 2, 3, \ldots, L - 1$$

is imposed for the smoothness of the wavelet function. Furthermore, in the *Coiflet wavelet*, for example, a similar smoothing condition is imposed also for the scaling function as

$$\int t^\ell \phi(t) dt = 0, \quad \text{for } \ell = 1, 2, 3, \ldots, L' - 1.$$

Once WT coefficients are obtained, we can calculate various quantities such as auto- and cross-correlations and SNR, as will be discussed shortly. In principle the expansion coefficients c_{jk} in DWT may be calculated by using (2.39) and (2.40) for a given function $f(t)$ and an adopted mother wavelet $\psi(t)$. This integration is, however, inconvenient, and in an actual fast wavelet transform, the expansion coefficients are obtained by a matrix multiplication with the use of the iterative formulae given by the masking coefficients and expansion coefficients of the neighboring levels of indices, j and k.

One of the advantages of the WT over FT is that we can choose a proper mother wavelet among many mother wavelets, depending on a signal to be analyzed. Among many candidates of mother wavelets, we have adopted the Coiflet with level 3, compromising the accuracy and the computational effort.

Simulation Results and Discussion

Input Pulses with $M = 1$

Firstly we discuss the case in which ensemble HH neurons receive a common single ($M = 1$) impulse. The input synaptic strength is assume to be the same for all neurons: $g_i = g_s$. When input synaptic strength is small: $g_s < g_{th}$, no neurons fire in the noise-free case, while it is sufficiently large ($g_s \geq g_{th}$) neurons fire, where $g_{th} = 0.091$ is the threshold value. For a while, we will discuss the sub-threshold case of $g_s = 0.06 < g_{th}$ with $N = 500$. The $M = 1$ input pulse is applied at $t = 100$ ms. We analyze the time dependence of the postsynaptic current $I_i = I_i^{ps} + I_i^n$ of the neuron $i = 1$ with added noises of $D = 0.5$. Because neurons receive noises for $0 \leq t < 100$ ms, the states of neurons when they receive input pulse are randomized [TSP99]. The neuron 1 fires with a delay of about 6 ms. This delay is much larger than the conventional value of 2–3 ms for the supra-threshold inputs because the integration of marginal inputs at synapses needs a fairly a long period before firing [Has00]. Spike of an ensemble neurons are forwarded to the summing neuron. Note that neurons fire not only by input pulses plus noises but also spuriously by noises only; when the noise intensity is increased to $D = 1.0$, spurious firings are much increased.

We assume that information is carried by firing times of spikes but not by details of their shapes. In order to study how information is transmitted through ensemble HH neurons with the use of the DWT, we divide the time scale by the width of time bin of T_b as $t = t_\ell = \ell T_b$ (ℓ: integer), and define the input and output signals within the each time bin by [Has02]

$$W_i(t) = \sum_{m=1}^{M} \Theta(|t - t_{im}| - T_b/2),$$

$$W_o(t) = \frac{1}{N} \sum_{i=1}^{N} \sum_{n} \Theta(|t - t_{oin}| - T_b/2),$$

$$(2.45)$$

where $\Theta(t)$ stands for the Heaviside function, $W_i(t)$ the external input signal (without noises), $W_o(t)$ the output signal averaged over the ensemble neurons, t_{im} the mth firing time of inputs, and t_{oin} the nth firing time of outputs of the neuron i (2.37). The time bin is chosen as $T_b = 5$ ms in our simulations. A single simulation has been performed for $320 (= 2^6 T_b)$ ms. The magnitude of $W_o(t)$ is much smaller than that of $W_i(t)$ because only a few neurons fire among 500 neurons. The peak position of $W_o(t)$ is slightly shifted compared with that of $W_i(t)$ because of a significant delay of neuron firings as mentioned above.

Wavelet Transformation

Now we apply the DWT to input and output signals. By substituting $f(t) = W_i(t)$ or $W_o(t)$ in (2.39), we get their WT coefficients given by [Has02]

$$c_{\lambda jk} = \int \psi_{jk}^*(t) W_\lambda(t) dt, \quad (\lambda = i, o)$$

where $\psi_{jk}(t)$ is a family of wavelets generated from the mother *Coiflet wavelet* (2.40).

$$\psi_{jk}(t) = 2^{-j/2} \psi(2^{-j}t - k).$$

Note that the lower and upper horizontal scales express b and $b T_b$ (in units of ms), respectively. We have calculated WT coefficients of $W_i(t)$ (as a function of $b(j) = (k - 0.5)2^j$ for various j values after convention) and performed the WT decomposition of the signal. The WT coefficients of $j = 1$ and 2 have large values near $b = 20$ ($bT_b = 100$ ms) where $W_i(t)$ has a peak. Contributions from $j = 1$ and $j = 2$ are predominant in $W_i(t)$. It is noted that the WT coefficient for $j = 4$ has a significant value at $b \sim 56$ far away from $b = 20$. As the noise intensity is increased, fine structures in the WT coefficients appear, in particular for small j.

Auto- and Cross-Correlations

The auto-correlation functions for input and output signals are defined by [Has02]

$$\Gamma_{\lambda\lambda} = M^{-1} \int W_\lambda(t)^* W_\lambda(t)dt = M^{-1} \sum_j \sum_k c^*_{\lambda jk} c_{\lambda jk}, \quad (\lambda = i, o)$$

where the ortho-normal relations of the wavelets given by (2.42)–(2.44) are employed. Similarly the cross-correlation between input and output signals is defined by

$$\Gamma_{io}(\beta) = M^{-1} \int W_i(t)^* W_o(t + \beta T_b)dt,$$
$$= M^{-1} \sum_j \sum_k c^*_{ijk} c_{ojk}(\beta),$$

where c_{ijk} and $c_{ojk}(\beta)$ are the expansion coefficients of $W_i(t)$ and $W_o(t+\beta T_b)$, respectively. The maxima in the cross-correlation and the normalized one are given by

$$\Gamma = \max_\beta[\Gamma_{io}(\beta)], \qquad \gamma = \max_\beta\left[\frac{\Gamma_{io}(\beta)}{\sqrt{\Gamma_{ii}}\sqrt{\Gamma_{oo}}}\right]. \tag{2.46}$$

It is noted that for the supra-threshold inputs in the noise-free case, we get $\Gamma_{ii} = \Gamma_{oo} = \Gamma_{io} = 1$ and then $\Gamma = \gamma = 1$. As increasing D from zero, Γ and Γ_{oo} are gradually increased. Because of the factor of $1/\sqrt{\Gamma_{oo}}$ in (2.46), the magnitude of γ is larger than those of Γ and Γ_{oo}. We note that γ is enhanced by weak noises and it is decreased at larger noises, which is a typical SR behavior.

Signal-to-Noise Ratio

We will evaluate the SNR by employing the de-noising method [BBD92]-[Qui00]. The key point in the de-noising is how to choose which wavelet coefficients are correlated with the signal and which ones with noises. The simple de-noising method is to neglect some DWT expansion coefficients when reproducing the signal by the inverse wavelet transform. We get the de-noising signal by the inverse WT (2.41):

$$W_\lambda^{dn}(t) = \sum_j \sum_k c_{\lambda jk}^{dn} \psi_{jk}(t),\qquad(2.47)$$

with the denoising WT coefficients c_{jk}^{dn} given by

$$c_{jk}^{dn}(t) = \sum_j \sum_k H_{jk} c_{\lambda jk} \psi_{jk}(t),$$

where H_{jk} is the *de-noising filter*. The simplest de-noising method, for example, is to assume that WT components for $a < a_c$ in the (a,b) plane arise from noises to set the de-noising WT coefficients as

$$c_{jk}^{dn} = \begin{cases} c_{jk}, & \text{for } j \geq j_c \ (a \geq a_c), \\ 0, & \text{otherwise}, \end{cases}\qquad(2.48)$$

where $j_c \ (= \log_2 a_c)$ is the critical j value [BBD92]. Alternatively, we may assume that the components for $b < b_L$ or $b > b_U$ in the (a,b)-plane are due to noises, and set $H(j,k)$ as

$$H(j,k) = \begin{cases} 1, & \text{for } k_L \leq k \leq k_U \ (b_L \leq b \leq b_U), \\ 0, & \text{otherwise}, \end{cases}$$

where $k_L(j) = b_L 2^{-j}$ and $k_U(j) = b_U 2^{-j}$ are j-dependent critical k values [Qui00].

In this study we adopt a more sophisticated method, assuming that the components for $b < b_L$ or $b > b_U$ at $a < a_c$ in the (a,b) plane are noises to set the de-noising WT coefficients as

$$c_{jk}^{dn} = \begin{cases} c_{jk}, & \text{for } j \geq j_c \text{ or } k_L \leq k \leq k_U, \\ 0, & \text{otherwise}, \end{cases}$$

where $j_c \ (= \log_2 a_c)$ denotes the critical j value, and $k_L \ (= b_L 2^{-j})$ and k_U $(= b_U 2^{-j})$ are the lower and upper critical k values, respectively. We get the inverse, de-noising signal by using (2.47) with the de-noising WT coefficients determined by (2.48).

From the above consideration, we may define the signal and noise components by [Has02]

$$A_s = \int W_o^{dn}(t)^* W_o^{dn}(t) dt = \sum_j \sum_k |c_{ojk}^{dn}|^2,$$

$$A_n = \int \left[W_o(t)^* W_o(t) - W_o^{dn}(t)^* W_o^{dn}(t) \right] dt = \sum_j \sum_k (|c_{ojk}|^2 - |c_{ojk}^{dn}|^2).$$

The SNR is defined by

$$\mathrm{SNR} = 10\log_{10}(A_s/A_n) \ \mathrm{dB}.$$

In the present case we can fortunately get the WT coefficients for *ideal* case of noise-free and supra-threshold inputs. We then properly determine the de-noising parameters of j_c, b_L and b_U. From the observation of the WT coefficients for the ideal case, we assume that the upper and lower bounds, may be chosen as

$$b_L = t_{o1}/T_b - \delta b, \qquad b_U = t_{oM}/T_b + \delta b, \qquad (2.49)$$

where t_{o1} (t_{oM}) are the first (Mth) firing times of output signals in the ideal case of noise-free and supra-threshold inputs, and δb denotes the marginal distance from the b values expected to be responsible to the signal transmission.

Note that the value of SNR is rather insensitive to a choice of the parameters of δb and j_c. Then we have decided to adopt $\delta b = 5$ and $j_c = 3$ for our simulations. Also, SNR shows a typical SR behavior: a rapid rise to a clear peak and a slow decrease for larger value of D.

The D dependence of the cross-correlation γ and SNR, respectively, ha s been calculated for $N = 1$, 10 100 and 500. SR effect for a single ($N = 1$) is marginally realized in SNR but not in γ. Large error bars for the results of $N = 1$ implies that the reliability of information transmission is very low in the sub-threshold condition [MS95a]. The signal can be transmitted only when the signal fortunately coincides with noises and the signal plus noise crosses the threshold level. Then only 13 among 100 trials are succeeded in the transmission of $M = 1$ inputs through a neuron. These calculations demonstrate that the ensemble of neurons play an *indispensable* role for information transmission of transient spike signals [Has02]. This is consistent with the results of [CCI95a] and [PWM96], who have pointed out the improvement of the information transmission by increasing the size of the network.

Next we change the value of g_s, the strength of input synapses. Note that in the noise-free case ($D = 0$), we get $\gamma = 1$ and SNR $= \infty$ for $g_s \geq g_{th}$ but $\gamma = 0$ and SNR $= -\infty$ for $g_s < g_{th}$. Moderate sub-threshold noises considerably improve the transition fidelity. We also note that the presence of such noises does not significantly degrade the transmission fidelity for supra-threshold cases in ensemble neurons.

Input Pulses with $M = 2$ and 3

Now we discuss the cases of $M = 2$ and 3. Input pulses are applied at $t = 100$ and 125 ms for the $M = 2$ case. The input ISI is assume to be $T_i = 25$ ms because spikes with this value of ISI are reported to be ubiquitous in cortical brains [TJW99]. The de-noising has been made by the procedure given by equations (2.47), (2.48) and (2.49).

The D-dependence of the cross-correlation γ [SNR] for $M = 1$, 2 and 3 is calculated. Both γ and SNR show typical SR behavior irrespective of the value of M, although a slight difference exists between the M dependence of

γ and SNR: for larger M, the former is larger but the latter is smaller at the moderate noise intensity of $D < 1.0$. When similar simulations are performed for different ISI values of $T_i = 15$ and 35 ms, we get results which are almost the same as that for $T_i = 25$ (not shown). This is because the output spikes for inputs with $M = 2$ and 3 are superposition of an output spike for a $M = 1$ input when the ISI is larger than the refractory period of neurons.

When we apply the $M = 2$ and 3 pulse trains to single neurons, we get the trial-dependent cross-correlations. As the value of M is increased, the number of trials with non-zero γ is increased while the maximum value of γ is decreased. The former is due to the fact that as M is increased, the opportunity that input pulses coincide with noises to fire increase. On the contrary, the latter arises from the fact that as M value is increased, the opportunity for all the multiple M pulses to coincide noises becomes small; note that in order to get $\gamma = 1$, it is necessary that all M pulses have to be fired. The γ value averaged over 100 trials is 0.127 ± 0.330, 0.149 ± 0.276 and 0.1645 ± 0.238 for $M = 1$, 2 and 3, respectively; the average γ is increased as the value of M is increased.

Discussion

It has been controversial how neurons communicate information by action potentials or spikes [RWS96, PZD00]. The one issue is whether information is encoded in the average firing rate of neurons (*rate code*) or in the precise firing times (*temporal code*). Since [And26] first noted the relationship between neural firing rate and stimulus intensity, the rate-code model has been supported in many experiments of motor and sensory neurons. In the last several years, however, experimental evidences have been accumulated, indicating a use of the temporal code in neural systems [CH86, TFM96]. Human visual systems, for example, have shown to classify patterns within 250 ms despite the fact that at least ten synaptic stages are involved from retina to the temporal brain [TFM96]. The transmission times between two successive stages of synaptic transmission are suggested to be no more than 10 ms on the average. This period is too short to allow rates to be determined accurately.

Although much of debates on the nature of the neural code has focused on the rate versus temporal codes, there is the other important issue to consider: information is encoded in the activity of single (or very few) neurons or that of a large number of neurons (*population code* or *ensemble code*). The population rate code model assumes that information is coded in the relative firing rates of ensemble neurons, and has been adopted in the most of the theoretical analysis. On the contrary, in the population temporal code model, it is assumed that relative timings between spikes in ensemble neurons may be used as an encoding mechanism for perceptional processing [Hop95, RT01]. A number of experimental data supporting this code have been reported in recent years [GS89, HOP98]. For example, data have demonstrated that temporally coordinated spikes can systematically signal sensory object feature, even in the absence of changes in firing rate of the spikes [DM96].

Our simulations based on the temporal-code model has shown that a population of neurons plays a very important role for the transmission of subthreshold transient spike signals. In particular for single neurons the transmission is quite unreliable and the appreciable SR effect is not realized. When the size of ensemble neurons is increased, the transmission fidelity is much improved in a fairly wide-range of parameters g_s including both the sub- and supra-threshold cases. We note that γ (or SNR) for $N = 100$ is different from and larger than that for $N = 1$ with 100 trials. This seems strange because a simulation for an ensemble of N neurons is expected to be equivalent to simulations for a single neuron with N trials, if there is no couplings among neurons as in our model. This is, however, not true, and it will be understood as follows. We consider a quantity of $X(N, N_r)$ which is γ (or SNR) averaged over N_r trials for an ensemble of N neurons. We implicitly express $X(N, N_r)$ as [Has02]

$$X(N, N_r) = \langle\!\langle F(\langle w_i^{(\mu)}\rangle)\rangle\!\rangle = \frac{1}{N_r} \sum_{\mu=1}^{N_r} F\left(\frac{1}{N} \sum_{i=1}^{N} w_i^{(\mu)}\right),$$

$$\text{with } w_i^{(\mu)} = w_i^{(\mu)}(t) = \sum_n \Theta(|t - t_{oin}^{(\mu)}| - T_b/2), \tag{2.50}$$

where $\langle\!\langle \cdot \rangle\!\rangle$ and $\langle \cdot \rangle$ stand for averages over trials and an ensemble neurons, respectively, defined by (2.50), $t_{oin}^{(\mu)}$ is the nth firing time of the neuron i in the μth trial, $w_i^{(\mu)}(t)$ is its output signal of the neuron i within a time bin of T_b (2.45), and $F(y(t))$ is a functional of a given function of $y(t)$ relevant to a calculation of γ (or SNR). Above calculations show that the relation: $X(100, 1) > X(1, 100)$, namely

$$F(\langle w_i^{(1)}\rangle) > \langle\!\langle F(w_1^{(\mu)})\rangle\!\rangle, \tag{2.51}$$

holds for our γ and SNR. Note that if $F(\cdot)$ is linear, we get $X(100, 1) = X(1, 100)$. This implies that the inequality given by (2.51) is expected to arise from a *nonlinear* character of $F(\cdot)$. This reminds us of the algebraic inequality: $f(\langle x \rangle) \geq \langle f(x) \rangle$ valid for a convex function $f(x)$, where the bracket $\langle \cdot \rangle$ stands for an average over a distribution of a variable x. It should be again noted that there is no couplings among our neurons in the adopted model. Then the enhancement of SNR with increasing N is only due to a population of neurons. This is quite different from the result of some papers [KHO00, SM01, WCW00, LHW00] in which the transmission fidelity is enhanced not only by noises but also by introduced couplings among neurons in an ensemble.

In the above simulations independent noises are applied to ensemble neurons. If instead we try to apply the same or *completely correlated* noise to them, it is equivalent to applying noises to a single neuron, and then appreciable SR effect is not realized as discussed above. Then SR for transient spikes requires independent noises to be applied to a large-scale ensemble of neurons. This is consistent with the result of Liu, Hu and Wang [LHW00] who discussed the effect of correlated noises on SR for stationary analog inputs.

Although spike trains with small values of $M = 1$–3 have been examined above, we can make an analysis of spikes with larger M or bursts, by using our method. In order to demonstrate its feasibility, we have presented simulations for transient spikes with larger M. We apply the WT to $W_o(t)$ to get its WT coefficients and its WT decomposition. The $j = 1$ and $j = 2$ components are dominant. After the de-noising, we get $\gamma = 0.523$ and $SNR = 18.6$ dB, which are comparable to those for $D = 1.0$ with $M = 1$–3. In our de-noising method given by (2.48) and (2.49), we extract noises outside the b region relevant to a cluster of spikes, but do not take account of noises between pulses. When a number of pulses M and/or the ISI T_i become larger, a contribution from noises between pulses become considerable, and it is necessary to modify the de-noising method such as to extract noises between pulses, for example, as given by [Has02]

$$c_{jk}^{dn} = \begin{cases} c_{jk}, & \text{for } j \geq j_c \text{ or } k_{Lm} \leq k \leq k_{Um} \quad (m = 1 - M), \\ 0 & \text{otherwise.} \end{cases} \qquad (2.52)$$

In (2.52) k_{Lm} and k_{Um} are m- and j-dependent lower and upper limits given by

$$k_{Lm} = 2^{-j} (t_{om}/T_b - \delta b), \qquad k_{Um} = 2^{-j} (t_{om}/T_b + \delta b),$$

where t_{om} is the mth firing time for the noise-free and supra-threshold input and δb the margin of b.

An ensemble of neurons adopted in our model can be regarded as the front layer (referred to as the layer I here) of a synfire chain [ABM93]: output spikes of the layer I are feed-forwarded to neurons on the next layer (referred to as the layer II) and spikes propagate through the synfire chain. The postsynaptic current of a neuron ℓ on the layer II is given by

$$I_\ell^{ps} = \frac{1}{N} \sum_{i=1}^{N} \sum_m w_{\ell i}(V_a - V_s)\alpha(t - \tau_{\ell i} - t_{oim}) + I_\ell^n, \qquad (2.53)$$

where $w_{\ell i}$ and $\tau_{\ell i}$ are the synaptic coupling and delay time, respectively, for spikes to propagate from neuron i on the layer I to neuron ℓ on layer II, t_{oim} the mth firing time of neuron i on the layer I, and I_ℓ^n noises. Transmission of spikes through the synfire chain depends on the couplings (excitatory or inhibitory) $w_{\ell i}$, the delay time $\tau_{\ell i}$ and noises I_ℓ^n. There are some discussions on the stability of spike transmission in a synfire chain. Quite recently [DGA99] has shown by simulations using IF models that the spike propagation with a precise timing is possible in fairly large-scale synfire chains with moderate noises.

By augmenting our neuron model including the coupling term given by (2.53) and tentatively assuming $w_{\ell i} = w = 1.5$ and $I_\ell^n = 0$, we have calculated the transmission fidelity by applying our WT analysis to output signals on the layer II as well as those on the layer I. We note that the transmission fidelity on the layer II is better that of the layer I because $\gamma_I < \gamma_{II}$ and $SNR_I < SNR_{II}$ at almost D values. We have chosen $w = 1.5$ such that neurons on the layer II can fire when more than 6% of neurons on the layer I fire. When we adopt smaller (larger) value of w, both γ_{II} and SNR_{II} abruptly increase at larger (smaller) D value. However, the general behavior of the D dependence of γ_{II} and SNR_{II} is not changed. This improvement of the transmission fidelity in the layer II than in the front layer I is expected to be partly due to the threshold-crossing behavior of HH neurons. It would be interesting to investigate the transmission of signals in a synfire chain by including SR of its front layer. For more details, see [Has02].

2.4.3 Wavelets of Epileptic Spikes

Recordings of human brain electrical activity (EEG) have been the fundamental tool for studying the dynamics of cortical neurons since 1929. Even though the gist of this technique has essentially remained the same, the methods of EEG data analysis have profoundly evolved during the last two decades. [BSN85] demonstrated that certain nonlinear measures, first introduced in the context of chaotic dynamical systems, changed during slow-wave sleep. The flurry of research work that followed this discovery focused on the application of nonlinear dynamics in quantifying brain electrical activity during different mental states, sleep stages, and under the influence of the epileptic process (for a review see for example [WNK95, PD92]). It must be emphasized that a straightforward interpretation of neural dynamics in terms of such nonlinear measures as the largest Lyapunov exponent or the correlation dimension is not possible since most biological time series, such as EEG, are non-stationary and consequently do not satisfy the assumptions of the underlying theory. On the other hand, traditional power spectral methods are also based on quite restrictive assumptions but nevertheless have turned out to be successful in some areas of EEG analysis. Despite these technical difficulties, the number of applications of nonlinear time series analysis has been growing steadily and now includes the characterization of encephalopaties [SLK99], monitoring of anesthesia depth [WSR00], characteristics of seizure activity [CIS97], and prediction of epileptic seizures [LE98]. Several other approaches are also used to elucidate the nature of electrical activity of the human brain ranging from coherence measures [NSW97, NSS99] and methods of non-equilibrium statistical mechanics [Ing82] to complexity measures [RR98, PS01].

One of the most important challenges of EEG analysis has been the quantification of the manifestations of epilepsy. The main goal is to establish a correlation between the EEG and clinical or pharmacological conditions. One of the possible approaches is based on the properties of the interictal EEG (electrical activity measured between seizures) which typically consists of linear stochastic background fluctuations interspersed with transient nonlinear spikes or sharp waves. These transient potentials originate as a result of a simultaneous pathological discharge of neurons within a volume of at least several mm^3.

The traditional definition of a spike is based on its amplitude, duration, sharpness and emergence from its background [GG76, GW82]. However, automatic epileptic spike detection systems based on this direct approach suffer from false detections in the presence of numerous types of artifacts and non-epileptic transients. This shortcoming is particularly acute for long-term EEG monitoring of epileptic patients which became common in 1980s. To reduce false detections [GW91] made the process of spike identification dependent upon the state of EEG (active wakefulness, quiet wakefulness, desynchronized EEG, phasic EEG, and slow EEG). This modification leads to significant overall improvement provided that state classification is correct.

The authors of [DM99] adopted nonlinear prediction for epileptic spike detection. They demonstrated that when the model's parameters are adjusted during the 'learning phase' to assure good predictive performance for stochastic background fluctuations, the appearance of an interictal spike is marked by a very large forecasting error. This novel approach is appealing because it makes use of changes in EEG dynamics. One expects good nonlinear predictive performance when the dynamics of the EEG interval used for building up the model is similar to the dynamics of the interval used for testing. However, it is uncertain at this point whether it is possible to develop a robust spike detection algorithm based solely on this idea.

As [CBE95] put it succinctly, automatic EEG analysis is a formidable task because of the lack of "... features that reflect the relevant information". Another difficulty is the nonstationary nature of the spikes and the background in which they are embedded. One technique developed for the treatment of such non-stationary time series is wavelet analysis [Mal99, UA96]. In this section, following [Lwk03], we characterize the epileptic spikes and sharp waves in terms of the properties of their wavelet transforms. In particular, we search for features which could be important in the detection of epileptic events.

Recall from introduction, that the wavelet transform is an integral transform for which the set of basis functions, known as wavelets, are well localized both in time and frequency. Moreover, the wavelet basis can be constructed from a single function $\psi(t)$ by means of translation and dilation:

$$\psi_{a;t_0} = \psi\left(\frac{t - t_0}{a}\right). \tag{2.54}$$

$\psi(t)$ is commonly referred to as the mother function or analyzing wavelet. The wavelet transform of function $h(t)$ is defined as

$$W(a, t_0) = \frac{1}{\sqrt{a}} \int_{-\infty}^{\infty} h(t)\psi_{a;t_0}^* dt, \tag{2.55}$$

where $\psi^*(t)$ denotes the complex conjugate of $\psi(t)$. The continuous wavelet transform of a discrete time series $\{h_i\}_{i=0}^{N-1}$ of length N and equal spacing δt

is defined as

$$W_n(a) = \sqrt{\frac{\delta t}{a}} \sum_{n'=0}^{N-1} h_{n'} \psi^* \left[\frac{(n'-n)\delta t}{a} \right]. \tag{2.56}$$

The above convolution can be evaluated for any of N values of the time index n. However, by choosing all N successive time index values, the convolution theorem allows us to calculate all N convolutions simultaneously in Fourier space using a discrete Fourier transform (DFT). The DFT of $\{h_i\}_{i=0}^{N-1}$ is

$$\hat{h}_k = \frac{1}{N} \sum_{n=0}^{N-1} h_n e^{-2\pi i k n/N}, \tag{2.57}$$

where $k = 0, \ldots, N-1$ is the frequency index. If one notes that the Fourier transform of a function $\psi(t/a)$ is $|a| \hat{\psi}(af)$ then by the convolution theorem

$$W_n(a) = \sqrt{a \delta t} \sum_{k=0}^{N-1} \hat{h}_n \psi^* (a f_k) e^{2\pi i f_k n \delta t}, \tag{2.58}$$

frequencies f_k are defined in the conventional way. Using (2.58) and a standard FFT routine it is possible to efficiently calculate the continuous wavelet transform (for a given scale a) at all n simultaneously [TC98]. It should be emphasized that formally (2.58) does not yield the discrete linear convolution corresponding to (2.56) but rather a discrete circular convolution in which the shift $n' - n$ is taken modulo N. However, in the context of this work, this problem does not give rise to any numerical difficulties. This is because, for purely practical reasons, the beginning and the end of the analyzed part of data stream are not taken into account during the EEG spike detection.

From a plethora of available mother wavelets, following [Lwk03] we employ the Mexican hat

$$\psi(t) = \frac{2}{\sqrt{3}} \pi^{-1/4} (1 - t^2) e^{-t^2/2}, \tag{2.59}$$

which is particularly suitable for studying epileptic events.

In the top panel of Fig. 2.21 two pieces of the EEG recording joined at approximately $t = 1s$ are presented. The digital 19 channel recording sampled at 240 Hz was obtained from a juvenile epileptic patient according to the international 10–20 standard with the reference average electrode. The epileptic spike in (marked by the arrow) is followed by two artifacts. For small scales, a, the values of the wavelet coefficients for the spike's ridge are much larger than those for the artifacts. The peak value along the spike ridge corresponds to $a = 7$. In sharp contrast, for the range of scales used in Fig. 2.21 the absolute value of coefficients $W(a, t_0)$ for the artifacts grow monotonically with a [Lwk03].

The question arises as to whether the behavior of the wavelet transform as a function of scale can be used to develop a reliable detection algorithm. The first step in this direction is to use the normalized wavelet power [Lwk03]

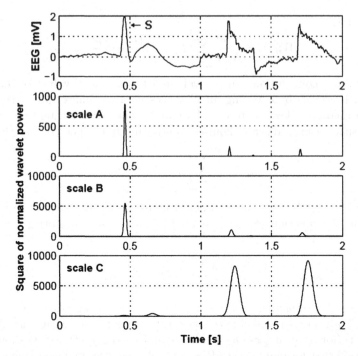

Fig. 2.21. Simple epileptic spike (marked by S) followed by two artifacts (top), together with the square of normalized wavelet power for three different scales $A < B < C$ (see text for explanation).

$$w(a, t_0) = W^2(a, t_0)/\sigma^2, \tag{2.60}$$

instead of the wavelet coefficients to reduce the dependence on the amplitude of the EEG recording. In the above formula σ^2 is the variance of the portion of the signal being analyzed (typically we use pieces of length 1024 for EEG tracings sampled at 240 Hz). In actual numerical calculations we prefer to use the square of $w(a, t_0)$ to merely increase the range of values analyzed during the spike detection process.

In the most straightforward approach, we identify an EEG transient potential as a simple or isolated epileptic spike iff [Lwk03]:

- the value of w^2 at $a = 7$ is greater than a predetermined threshold value T_1,
- the square of normalized wavelet power decreases from scale $a = 7$ to $a = 20$,
- the value of w^2 at $a = 3$ is greater than a predetermined threshold value T_2.

The threshold values T_1 and T_2 may be considered as the model's parameters which can be adjusted to achieve the desired *sensitivity* (the ratio of detected epileptic events to the total number of epileptic events present in the analyzed EEG tracing) and *selectivity* (the ratio of epileptic events to the total number of events marked by the algorithm as epileptic spikes).

While this simple algorithm is quite effective for simple spikes such as one shown in Fig. 2.21 (top) it fails for the common case of an epileptic spike accompanied by a slow wave with comparable amplitude. The normalized wavelet power decreases from scale B to C. Consequently, in the same vein as the argument we presented above, we can develop an algorithm which detects the epileptic spike in the vicinity of a slow wave by calculating the following linear combination of wavelet transforms:

$$\tilde{W}(a, t_0) = c_1 W(a, t_0) + c_2 W(a_s, t_0 + \tau), \tag{2.61}$$

and checking whether the square of corresponding normalized power $\tilde{w}(a, t_0) = \tilde{W}^2(a, t_0)/\sigma^2$ at scales $a = 3$ and $a = 7$ exceeds the threshold value $\tilde{T}_1$ and $\tilde{T}_2$, respectively. The second term in (2.61) allows us to detect the slow wave which follows the spike. The parameters a_s and τ are chosen to maximize the overlap of the wavelet with the slow wave. For the Mexican hat we use $a_s = 28$ and $\tau = 0.125s$. By varying the values of coefficients c_1 and c_2, it is possible to control the relative contribution of the spike and the slow wave to the linear combination (2.61).

The goal of wavelet analysis of the two types of spikes, presented in this section, was to elucidate the approach to epileptic events detection which explicitly hinges on the behavior of wavelet power spectrum of EEG signal *across* scales and not merely on its values. Thus, this approach is distinct not only from the detection algorithms based upon discrete multiresolution representations of EEG recordings [BDI96, DIS97, SE99, CER00, GAM01] but also from the method developed by [SW02] which employs continuous wavelet transform.

In summary, epilepsy is a common disease which affects 1–2% of the population and about 4% of children [Jal02]. In some epilepsy syndromes interictal paroxysmal discharges of cerebral neurons reflect the severity of the epileptic disorder and themselves are believed to contribute to the progressive disturbances in cerebral functions e.g., speech impairment, behavioral disturbances) [Eng01]. In such cases precise quantitative spike analysis would be extremely important. For more details, see [Lwk03].

2.5 Human Motor Control and Learning

Human motor control system can be formally represented by a 2-functor $\mathcal{E}$, is given by

$$A \xrightarrow{\quad f \quad} B \qquad \mathcal{E}(A) \xrightarrow{\quad \mathcal{E}(f) \quad} \mathcal{E}(B)$$

$$h \Big\downarrow \quad \text{CURRENT} \quad \Big\downarrow g \qquad \xrightarrow{\quad \mathcal{E} \quad} \qquad \mathcal{E}(h) \Big\downarrow \quad \text{DESIRED} \quad \Big\downarrow \mathcal{E}(g)$$

MOTOR   MOTOR
STATE   STATE

$$C \xrightarrow{\quad k \quad} D \qquad \mathcal{E}(C) \xrightarrow{\quad \mathcal{E}(k) \quad} \mathcal{E}(D)$$

Here $\mathcal{E}$ represents the *motor-learning* functor from the *source 2-category* of the current motor state, defined as a commutative square of small motor categories $A, B, C, D, \ldots$ of *current motor components* and their causal interrelations $f, g, h, k, \ldots$, onto the *target 2-category* of the desired motor state, defined as a commutative square of small motor categories $\mathcal{E}(A), \mathcal{E}(B), \mathcal{E}(C), \mathcal{E}(D), \ldots$ of *evolved motor components* and their causal interrelations $\mathcal{E}(f), \mathcal{E}(g), \mathcal{E}(h), \mathcal{E}(k)$. As in the previous section, each causal arrow in above diagram, e.g., $f : A \to B$, stands for a *generic motor dynamorphism*.

2.5.1 Motor Control

The so-called *basic biomechanical unit* consists of a pair of mutually antagonistic muscles producing a common muscular torque, T_{Mus}, in the same joint, around the same axes. The most obvious example is the biceps–triceps pair (see Fig. 2.22). Note that in the normal vertical position, the triceps downward action is supported by gravity, that is the torque due to the weight of the forearm and the hand (with the possible load in it).

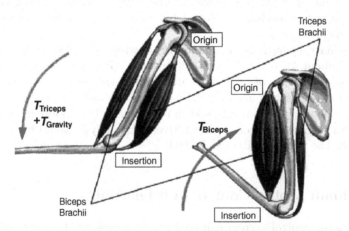

Fig. 2.22. Basic biomechanical unit: *left*—triceps torque T_{Triceps}; *right*—biceps torque T_{Biceps}.

Muscles have their structure and function (of generating muscular force), they can be exercised and trained in strength, speed, endurance and flexibility, but they are still only dumb effectors (just like excretory glands). They only respond to neural command impulses. To understand the process of training motor skills we need to know the basics of neural motor control. Within the motor control muscles are force generators, working in antagonistic pairs. They are controlled by neural reflex feedbacks and/or voluntary motor inputs.

Recall that a *reflex* is a spinal neural feedback, the simplest functional unit of a sensory-motor control. Its *reflex arc* involves five basic components (see Fig. 2.23): (1) sensory receptor, (2) sensory (afferent) neuron, (3) spinal interneuron, (4) motor (efferent) neuron, and (5) effector organ—skeletal muscle. Its purpose is the quickest possible reaction to the potential threat.

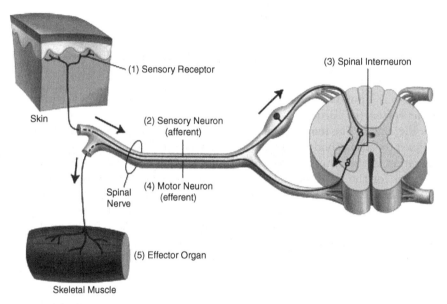

Fig. 2.23. Five main components of a reflex arc: (1) sensory receptor, (2) sensory (afferent) neuron, (3) spinal interneuron, (4) motor (efferent) neuron, and (5) effector organ—skeletal muscle.

All of the body's voluntary movements are controlled by the brain. One of the brain areas most involved in controlling these voluntary movements is the motor cortex (see Fig. 2.24).

In particular, to carry out goal-directed human movements, our motor cortex must first receive various kinds of information from the various lobes of the brain: information about the body's position in space, from the parietal lobe; about the goal to be attained and an appropriate strategy for attaining it, from the anterior portion of the frontal lobe; about memories of past strategies, from the temporal lobe; and so on.

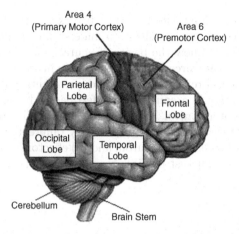

Fig. 2.24. Main lobes (parts) of the brain, including motor cortex.

There is a popular "motor map", a pictorial representation of the primary motor cortex (or, Brodman's area 4), called *Penfield's motor homunculus*[5] (see Fig. 2.25). The most striking aspect of this motor map is that the areas assigned to various body parts on the cortex are proportional not to their size, but rather to the complexity of the movements that they can perform. Hence, the areas for the hand and face are especially large compared with those for the rest of the body. This is no surprise, because the speed and dexterity of human hand and mouth movements are precisely what give us two of our most distinctly human faculties: the ability to use tools and the ability to speak. Also, stimulations applied to the precentral gyrus trigger highly localized muscle contractions on the contralateral side of the body. Because of this *crossed control*, this motor center normally controls the voluntary movements on the opposite side of the body.

Planning for any human movement is done mainly in the forward portion of the frontal lobe of the brain (see Fig. 2.26). This part of the cortex receives information about the player's current position from several other parts. Then, like the ship's captain, it issues its commands, to Brodman's area 6, the premotor cortex. Area 6 acts like the ship's lieutenants. It decides which set of muscles to contract to achieve the required movement, then issues the corresponding orders to the primary motor cortex (Area 4). This area in turn activates specific muscles or groups of muscles via the motor neurons in the spinal cord.

Between brain (that plans the movements) and muscles (that execute the movements), the most important link is the *cerebellum* (see Fig. 2.24). For you to perform even so simple a gesture as touching the tip of our nose, it is not enough for our brain to simply command our hand and arm muscles to

---

[5] Note that there is a similar "sensory map", with a similar complexity-related distribution, called *Penfield's sensory homunculus*.

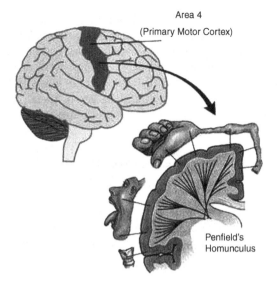

Fig. 2.25. Primary motor cortex and its motor map, the Penfield's homunculus.

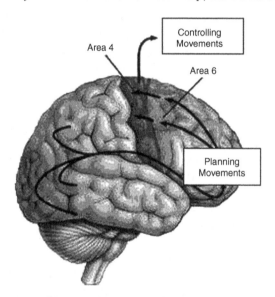

Fig. 2.26. Planning and control of movements by the brain.

contract. To make the various segments of our hand and arm deploy smoothly, you need an internal "clock" that can precisely regulate the sequence and duration of the elementary movements of each of these segments. That clock is the cerebellum.

The cerebellum performs this fine coordination of movement in the following way. First it receives information about the intended movement from the sensory and motor cortices. Then it sends information back to the motor cortex about the required direction, force, and duration of this movement (see Fig. 2.27). It acts like an air traffic controller who gathers an unbelievable amount of information at every moment, including (to return to our original example) the position of our hand, our arm, and our nose, the speed of their movements, and the effects of potential obstacles in their path, so that our finger can achieve a "soft landing" on the tip of our nose.

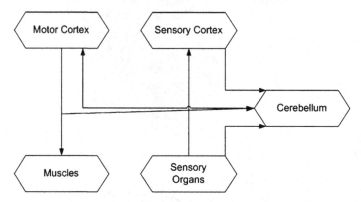

Fig. 2.27. The cerebellum loop for movement coordination.

2.5.2 Human Memory

The *human memory* is fundamentally associative. You can remember a new piece of information better if you can associate it with previously acquired knowledge that is already firmly anchored in our memory. And the more meaningful the association is to you personally, the more effectively it will help you to remember. Memory has three main types: sensory, short-term and long-term (see Fig. 2.28).

The *sensory memory* is the memory that results from our perceptions automatically and generally disappears in less than a second.

The *short-term memory* depends on the attention paid to the elements of sensory memory. Short-term memory lets you retain a piece of information for less than a minute and retrieve it during this time.

The *working memory* is a novel extension of the concept of short-term memory; it is used to perform cognitive processes (like reasoning) on the items that are temporarily stored in it. It has several components: a control system, a central processor, and a certain number of auxiliary "slave" systems.

The *long-term memory* includes both our memory of recent facts, which is often quite fragile, as well as our memory of older facts, which has become more consolidated (see Fig. 2.29). It consists of three main processes that

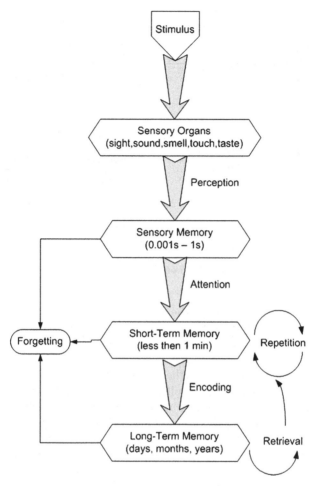

Fig. 2.28. Human cognitive memory.

take place consecutively: encoding, storage, and retrieval (recall) of information. The purpose of *encoding* is to assign a meaning to the information to be memorized. The so-called *storage* can be regarded as the active process of consolidation that makes memories less vulnerable to being forgotten. Lastly, retrieval (recall) of memories, whether voluntary or not, involves active mechanisms that make use of encoding indexes. In this process, information is temporarily copied from long-term memory into working memory, so that it can be used there.

Retrieval of information encoded in long-term memory is traditionally divided into two categories: recall and recognition. Recall involves actively reconstructing the information, whereas recognition only requires a decision as to whether one thing among others has been encountered before. Recall is

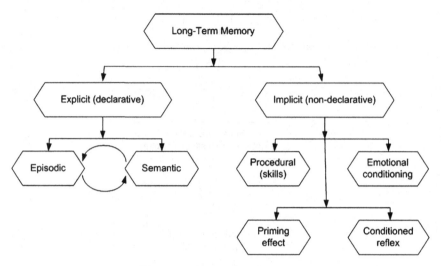

Fig. 2.29. Long-term memory.

more difficult, because it requires the activation of all the neurons involved in the memory in question. In contrast, in recognition, even if a part of an object initially activates only a part of the neural network concerned, that may then suffice to activate the entire network.

Long-term memory can be further divided into explicit memory (which involves the subjects' conscious recollection of things and facts) and implicit memory (from which things can be recalled automatically, without the conscious effort needed to recall things from explicit memory, see Fig. 2.29). Episodic (or, autobiographic) memory lets you remember events that you personally experienced at a specific time and place. Semantic memory is the system that you use to store our knowledge of the world; its content is thus abstract and relational and is associated with the meaning of verbal symbols.

Procedural memory, which is unconscious, enables people to acquire motor skills and gradually improve them. Implicit memory is also where many of our conditioned reflexes and conditioned emotional responses are stored. It can take place without the intervention of the conscious mind.

2.5.3 Human Learning

Learning is a relatively permanent change in behavior that marks an increase in knowledge, skills, or understanding thanks to recorded memories. A memory[6] is the fruit of this learning process, the concrete trace of it that is left in our neural networks.

---

[6] "The purpose of memory is not to let us recall the past, but to let us anticipate the future. Memory is a tool for prediction"—Alain Berthoz.

More precisely, learning is a process that lets us retain acquired information, affective states, and impressions that can influence our behavior. Learning is the main activity of the brain, in which this organ continuously modifies its own structure to better reflect the experiences that we have had. Learning can also be equated with encoding, the first step in the process of memorization. Its result, memory, is the persistence both of autobiographical data and of general knowledge.

But memory is not entirely faithful. When you perceive an object, groups of neurons in different parts of our brain process the information about its shape, color, smell, sound, and so on. Your brain then draws connections among these different groups of neurons, and these relationships constitute our perception of the object. Subsequently, whenever you want to remember the object, you must reconstruct these relationships. The parallel processing that our cortex does for this purpose, however, can alter our memory of the object.

Also, in our brain's memory systems, isolated pieces of information are memorized less effectively than those associated with existing knowledge. The more associations between the new information and things that you already know, the better you will learn it.

If you show a chess grand master a chessboard on which a game is in progress, he can memorize the exact positions of all the pieces in just a few seconds. But if you take the same number of pieces, distribute them at random positions on the chessboard, then ask him to memorize them, he will do no better than you or we. Why? Because in the first case, he uses his excellent knowledge of the rules of the game to quickly eliminate any positions that are impossible, and his numerous memories of past games to draw analogies with the current situation on the board.

Psychologists have identified a number of factors that can influence how effectively memory functions, including:

1. Degree of vigilance, alertness, attentiveness,[7] and concentration.
2. Interest, strength of motivation,[8] and need or necessity.
3. Affective values associated with the material to be memorized, and the individual's mood and intensity of emotion.[9]

---

[7] Attentiveness is often said to be the tool that engraves information into memory.

[8] It is easier to learn when the subject fascinates you. Thus, motivation is a factor that enhances memory.

[9] Your emotional state when an event occurs can greatly influence our memory of it. Thus, if an event is very upsetting, you will form an especially vivid memory of it. The processing of emotionally-charged events in memory involves norepinephrine, a neurotransmitter that is released in larger amounts when we are excited or tense. As Voltaire put it, "That which touches the heart is engraved in the memory".

4. Location, light, sounds, smells..., in short, the entire context in which the memorizing takes place is recorded along with the information being memorizes.[10]

Forgetting is another important aspect of memorization phenomena. Forgetting lets you get rid of the tremendous amount of information that you process every day but that our brain decides it will not need in future.

2.5.4 Spinal Musculo-Skeletal Control

About three decades ago, James Houk pointed out in [Hou67, Hou78, Hou79] that stretch and unloading reflexes were mediated by combined actions of several autogenetic neural pathways. In this context, "autogenetic" (or, autogenic) means that the stimulus excites receptors located in the same muscle that is the target of the reflex response. The most important of these muscle receptors are the primary and secondary endings in muscle spindles, sensitive to length change, and the Golgi tendon organs, sensitive to contractile force. The autogenetic circuits appear to function as servo-regulatory loops that convey continuously graded amounts of excitation and inhibition to the large (alpha) skeletomotor neurons. Small (gamma) fusimotor neurons innervate the contractile poles of muscle spindles and function to modulate spindle-receptor discharge. Houk's term "motor servo" [Hou78] has been used to refer to this entire control system, summarized by the block diagram in Fig. 2.30.

Prior to a study by Matthews [Mat69], it was widely assumed that secondary endings belong to the mixed population of "flexor reflex afferents", so called because their activation provokes the flexor reflex pattern—excitation of flexor motoneurons and inhibition of extensor motoneurons. Matthews' results indicated that some category of muscle stretch receptor other than the primary ending provides important excitation to extensor muscles, and he argued forcefully that it must be the secondary ending.

The primary and secondary muscle spindle afferent fibers both arise from a specialized structure within the muscle, the *muscle spindle*, a fusiform structure 4–7 mm long and 80–200 μm in diameter. The spindles are located deep within the muscle mass, scattered widely through the muscle body, and attached to the tendon, the endomysium or the perimysium, so as to be in parallel with the extrafusal or regular muscle fibers. Although spindles are scattered widely in muscles, they are not found throughout. Muscle spindle (see Fig. 2.31) contains two types of intrafusal muscle fibers (intrafusal means inside the fusiform spindle): the nuclear bag fibers and the nuclear chain fibers. The nuclear bag fibers are thicker and longer than the nuclear chain fibers, and they receive their name from the accumulation of their nuclei in the expanded bag-like equatorial region-the nuclear bag. The nuclear chain fibers

---

[10] Our memory systems are thus contextual. Consequently, when you have trouble remembering a particular fact, you may be able to retrieve it by recollecting where you learnt it or the book from which you learnt it.

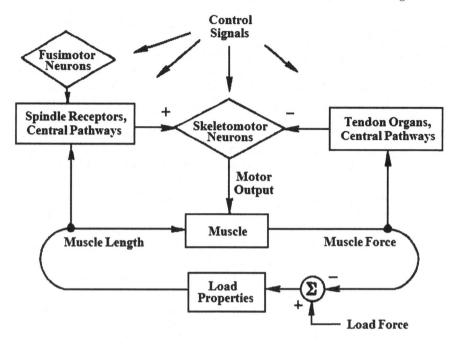

Fig. 2.30. Houk's autogenetic motor servo.

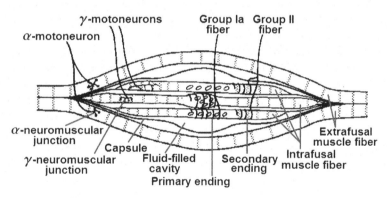

Fig. 2.31. Schematic of a muscular length-sensor, the muscle spindle.

have no equatorial bulge; rather their nuclei are lined up in the equatorial region-the nuclear chain. A typical spindle contains two nuclear bag fibers and 4–5 nuclear chain fibers.

The pathways from primary and secondary endings are treated commonly by Houk in Fig. 2.30, since both receptors are sensitive to muscle length and both provoke reflex excitation. However, primary endings show an additional sensitivity to the dynamic phase of length change, called dynamic responsiveness, and they also show a much-enhanced sensitivity to small changes in muscle length [Mat72].

The motor servo comprises three closed circuits (Fig. 2.30), two neural feedback pathways, and one circuit representing the mechanical interaction between a muscle and its load. One of the feedback pathways, that from spindle receptors, conveys information concerning muscle length, and it follows that this loop will act to keep muscle length constant. The other feedback pathway, that from tendon organs, conveys information concerning muscle force, and it acts to keep force constant.

In general, it is physically impossible to maintain both muscle length and force constant when external loads vary; in this situation the action of the two feedback loops will oppose each other. For example, an increased load force will lengthen the muscle and cause muscular force to increase as the muscle is stretched out on its length-tension curve. The increased length will lead to excitation of motoneurons, whereas the increased force will lead to inhibition. It follows that the net regulatory action conveyed by skeletomotor output will depend on some relationship between force change and length change and on the strength of the feedback from muscle spindles and tendon organs. A simple mathematical derivation [NH76] demonstrates that the change in skeletomotor output, the error signal of the motor servo, should be proportional to the difference between a regulated stiffness and the actual stiffness provided by the mechanical properties of the muscle, where stiffness has the units of force change divided by length change. The regulated stiffness is determined by the ratio of the gain of length to force feedback.

It follows that the combination of spindle receptor and tendon organ feedback will tend to maintain the stiffness of the neuromuscular apparatus at some regulated level. If this level is high, due to a high gain of length feedback and a low gain of force feedback, one could simply forget about force feedback and treat muscle length as the regulated variable of the system. However, if the regulated level of stiffness is intermediate in value, i.e. not appreciably different from the average stiffness arising from muscle mechanical properties in the absence of reflex actions, one would conclude that stiffness, or its inverse, compliance, is the regulated property of the motor servo.

In this way, the autogenetic reflex motor servo provides the local, reflex feedback loops for individual muscular contractions. A voluntary contraction force F of human skeletal muscle is reflexly excited (positive feedback $+F^{-1}$) by the responses of its *spindle receptors* to stretch and is reflexly inhibited (negative feedback $-F^{-1}$) by the responses of its *Golgi tendon organs* to contraction. Stretch and unloading reflexes are mediated by combined actions of several autogenetic neural pathways, forming the *motor servo* (see [II06a, II06b, II06c]).

In other words, branches of the afferent fibers also synapse with interneurons that inhibit motor neurons controlling the antagonistic muscles—*reciprocal inhibition*. Consequently, the stretch stimulus causes the antagonists to relax so that they cannot resists the shortening of the stretched muscle caused by the main reflex arc. Similarly, firing of the Golgi tendon receptors causes inhibition of the muscle contracting too strong and simultaneous *reciprocal activation* of its antagonist.

2.5.5 Cerebellum and Muscular Synergy

The cerebellum is a brain region anatomically located at the bottom rear of the head (the hindbrain), directly above the brainstem, which is important for a number of subconscious and automatic motor functions, including motor learning. It processes information received from the motor cortex, as well as from proprioceptors and visual and equilibrium pathways, and gives 'instructions' to the motor cortex and other subcortical motor centers (like the basal nuclei), which result in proper balance and posture, as well as smooth, coordinated skeletal movements, like walking, running, jumping, driving, typing, playing the piano, etc. Patients with cerebellar dysfunction have problems with precise movements, such as walking and balance, and hand and arm movements. The cerebellum looks *similar in all animals*, from fish to mice to humans. This has been taken as evidence that it performs a common function, such as regulating motor learning and the timing of movements, in all animals. Studies of simple forms of motor learning in the vestibulo-ocular reflex and eye-blink conditioning are demonstrating that timing and amplitude of learned movements are encoded by the cerebellum.

When someone compares learning a new skill to learning how to ride a bike they imply that once mastered, the task seems embedded in our brain forever. Well, embedded in the cerebellum to be exact. This brain structure is the commander of coordinated movement and possibly even some forms of cognitive learning. Damage to this area leads to motor or movement difficulties.

A part of a human brain that is devoted to the sensory-motor control of human movement, that is motor coordination and learning, as well as equilibrium and posture, is the cerebellum (which in Latin means "little brain"). It performs integration of sensory perception and motor output. Many neural pathways link the cerebellum with the motor cortex, which sends information to the muscles causing them to move, and the spino-cerebellar tract, which provides proprioception, or feedback on the position of the body in space. The cerebellum integrates these pathways, using the constant feedback on body position to fine-tune motor movements [Ito84].

The human cerebellum has 7–14 million Purkinje cells. Each receives about 200,000 synapses, most onto dendritic splines. Granule cell axons form the *parallel fibers*. They make excitatory synapses onto Purkinje cell dendrites. Each parallel fibre synapses on about 200 Purkinje cells. They create a strip of excitation along the cerebellar folia.

Mossy fibers are one of two main sources of input to the cerebellar cortex (see Fig. 2.32). A mossy fibre is an axon terminal that ends in a large, bulbous swelling. These mossy fibers enter the granule cell layer and synapse on the dendrites of granule cells; in fact the granule cells reach out with little 'claws'

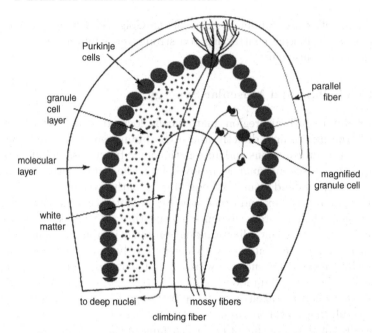

Fig. 2.32. Stereotypical ways throughout the cerebellum.

to grasp the terminals. The granule cells then send their axons up to the molecular layer, where they end in a T and run parallel to the surface. For this reason these axons are called *parallel fibers*. The parallel fibers synapse on the huge dendritic arrays of the Purkinje cells. However, the individual parallel fibers are not a strong drive to the Purkinje cells. The Purkinje cell dendrites fan out within a plane, like the splayed fingers of one hand. If we were to turn a Purkinje cell to the side, it would have almost no width at all. The parallel fibers run perpendicular to the Purkinje cells, so that they only make contact once as they pass through the dendrites.

Unless firing in bursts, parallel fibre EPSPs do not fire Purkinje cells. Parallel fibers provide excitation to all of the Purkinje cells they encounter. Thus, granule cell activity results in a strip of activated Purkinje cells.

Mossy fibers arise from the spinal cord and brainstem. They synapse onto granule cells and deep cerebellar nuclei. The Purkinje cell makes an inhibitory synapse (GABA) to the deep nuclei. Mossy fibre input goes to both cerebellar cortex and deep nuclei. When the Purkinje cell fires, it inhibits output from the deep nuclei.

The *climbing fibre* arises from the inferior olive. It makes about 300 excitatory synapses onto one Purkinje cell. This powerful input can fire the Purkinje cell.

The parallel fibre synapses are plastic—that is, they can be modified by experience. When parallel fibre activity and climbing fibre activity converge on the same Purkinje cell, the parallel fibre synapses become weaker (EP-SPs are smaller). This is called long-term depression. Weakened parallel fibre synapses result in less Purkinje cell activity and less inhibition to the deep nuclei, resulting in facilitated deep nuclei output. Consequently, the mossy fibre collaterals control the deep nuclei.

The *basket cell* is activated by parallel fibers afferents. It makes inhibitory synapses onto Purkinje cells. It provides lateral inhibition to Purkinje cells. Basket cells inhibit Purkinje cells lateral to the active beam.

Golgi cells receive input from parallel fibers, mossy fibers, and climbing fibers. They inhibit granule cells. Golgi cells provide feedback inhibition to granule cells as well as feedforward inhibition to granule cells. Golgi cells create a brief burst of granule cell activity.

Although each parallel fibre touches each Purkinje cell only once, the thousands of parallel fibers working together can drive the Purkinje cells to fire like mad.

The second main type of input to the folium is the *climbing fibre*. The climbing fibers go straight to the Purkinje cell layer and snake up the Purkinje dendrites, like ivy climbing a trellis. Each climbing fibre associates with only one Purkinje cell, but when the climbing fibre fires, it provokes a large response in the Purkinje cell.

The Purkinje cell compares and processes the varying inputs it gets, and finally sends its own axons out through the white matter and down to the *deep nuclei*. Although the inhibitory Purkinje cells are the main output of the cerebellar cortex, the output from the cerebellum as a whole comes from the deep nuclei. The three deep nuclei are responsible for sending excitatory output back to the thalamus, as well as to postural and vestibular centers.

There are a few other cell types in cerebellar cortex, which can all be lumped into the category of inhibitory interneuron. The *Golgi cell* is found among the granule cells. The *stellate* and *basket cells* live in the molecular layer. The basket cell (right) drops axon branches down into the Purkinje cell layer where the branches wrap around the cell bodies like baskets.

The cerebellum operates in 3's: there are 3 highways leading in and out of the cerebellum, there are 3 main inputs, and there are 3 main outputs from 3 deep nuclei. They are:

The 3 highways are the *peduncles*. There are 3 pairs (see [Mol97, Har97, Mar98]):

1. The *inferior cerebellar peduncle* (restiform body) contains the dorsal spinocerebellar tract (DSCT) fibers. These fibers arise from cells in the ipsilateral Clarke's column in the spinal cord (C8–L3). This peduncle contains the cuneo-cerebellar tract (CCT) fibers. These fibers arise from the ipsilateral accessory cuneate nucleus. The largest component of the inferior cerebellar peduncle consists of the olivo-cerebellar tract (OCT) fibers. These fibers arise from the contralateral inferior olive. Finally, vestibulo-cerebellar tract (VCT) fibers arise from cells in both the vestibular gan-

glion and the vestibular nuclei and pass in the inferior cerebellar peduncle
to reach the cerebellum.

2. The *middle cerebellar peduncle* (brachium pontis) contains the pontocere-
   bellar tract (PCT) fibers. These fibers arise from the contralateral pontine
   grey.
3. The *superior cerebellar peduncle* (brachium conjunctivum) is the primary
   efferent (out of the cerebellum) peduncle of the cerebellum. It contains
   fibers that arise from several deep cerebellar nuclei. These fibers pass
   ipsilaterally for a while and then cross at the level of the inferior colliculus
   to form the decussation of the superior cerebellar peduncle. These fibers
   then continue ipsilaterally to terminate in the red nucleus ('ruber-duber')
   and the motor nuclei of the thalamus (VA, VL).

The 3 inputs are: *mossy fibers* from the *spinocerebellar* pathways, climbing
fibers from the *inferior olive*, and more mossy fibers from the *pons*, which are
carrying information from *cerebral cortex* (see Fig. 2.33). The mossy fibers
from the spinal cord have come up ipsilaterally, so they do not need to cross.
The fibers coming down from cerebral cortex, however, do need to cross (as
the cerebrum is concerned with the opposite side of the body, unlike the cerebellum). These fibers synapse in the pons (hence the huge block of fibers in the
cerebral peduncles labeled 'cortico-pontine'), cross, and enter the cerebellum
as mossy fibers.

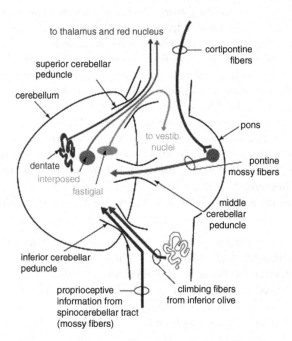

Fig. 2.33. Inputs and outputs of the cerebellum.

The 3 deep nuclei are the *fastigial, interposed,* and *dentate nuclei.* The fastigial nucleus is primarily concerned with balance, and sends information mainly to vestibular and reticular nuclei. The dentate and interposed nuclei are concerned more with voluntary movement, and send axons mainly to thalamus and the red nucleus.

The main function of the cerebellum as a motor controller is depicted in Fig. 5.3. A coordinated movement is easy to recognize, but we know little about how it is achieved. In search of the neural basis of coordination, a model of spinocerebellar interactions was recently presented in [AG05], in which the structure-functional organizing principle is a division of the cerebellum into discrete micro-complexes. Each micro-complex is the recipient of a specific motor error signal, that is, a signal that conveys information about an inappropriate movement. These signals are encoded by spinal reflex circuits and conveyed to the cerebellar cortex through climbing fibre afferents. This organization reveals salient features of cerebellar information processing, but also highlights the importance of systems level analysis for a fuller understanding of the neural mechanisms that underlie behavior.

The authors of [AG05] reviewed anatomical and physiological foundations of cerebellar information processing. The cerebellum is crucial for the coordination of movement. The authors presented a model of the cerebellar paravermis, a region concerned with the control of voluntary limb movements through its interconnections with the spinal cord. They particularly focused on the olivo-cerebellar climbing fibre system.

Climbing fibres are proposed to convey motor error signals (signals that convey information about inappropriate movements) related to elementary limb movements that result from the contraction of single muscles. The actual encoding of motor error signals is suggested to depend on sensorimotor transformations carried out by spinal modules that mediate nociceptive withdrawal reflexes.

The termination of the climbing fibre system in the cerebellar cortex subdivides the paravermis into distinct microzones. Functionally similar but spatially separate microzones converge onto a common group of cerebellar nuclear neurons. The processing units formed as a consequence are termed 'multizonal micro-complexes' (MZMCs), and are each related to a specific spinal reflex module.

The distributed nature of microzones that belong to a given MZMC is proposed to enable similar climbing fibre inputs to integrate with mossy fibre inputs that arise from different sources. Anatomical results consistent with this notion have been obtained.

Within an individual MZMC, the skin receptive fields of climbing fibres, mossy fibres and cerebellar cortical inhibitory interneurons appear to be similar. This indicates that the inhibitory receptive fields of Purkinje cells within a particular MZMC result from the activation of inhibitory interneurons by local granule cells.

On the other hand, the parallel fibre-mediated excitatory receptive fields of the Purkinje cells in the same MZMC differ from all of the other receptive fields, but are similar to those of mossy fibres in another MZMC. This indicates that the excitatory input to Purkinje cells in a given MZMC originates in non-local granule cells and is mediated over some distance by parallel fibres.

The output from individual MZMCs often involves two or three segments of the ipsilateral limb, indicative of control of multi-joint muscle synergies. The distal-most muscle in this synergy seems to have a roughly antagonistic action to the muscle associated with the climbing fibre input to the MZMC.

The model proposed in [AG05] indicates that the cerebellar paravermis system could provide the control of both single- and multi-joint movements. Agonist-antagonist activity associated with single-joint movements might be controlled within a particular MZMC, whereas coordination across multiple joints might be governed by interactions between MZMCs, mediated by parallel fibres.

Two main theories address the function of the cerebellum, both dealing with motor coordination. One claims that the cerebellum functions as a regulator of the "timing of movements". This has emerged from studies of patients whose timed movements are disrupted [IKD88].

The second, "Tensor Network Theory" provides a mathematical model of transformation of sensory (covariant) space-time coordinates into motor (contravariant) coordinates by cerebellar neuronal networks [PL80, PL82, PL85].

Studies of motor learning in the vestibulo-ocular reflex and eye-blink conditioning demonstrate that the timing and amplitude of learned movements are encoded by the cerebellum [BK04]. Many synaptic plasticity mechanisms have been found throughout the cerebellum. The *Marr–Albus model* mostly attributes motor learning to a single plasticity mechanism: the long-term depression of parallel fiber synapses. The Tensor Network Theory of sensory-motor transformations by the cerebellum has also been experimentally supported [GZ86].

3

Quantum Theory Basics

In this Chapter we give a modern review of quantum theory, including quantum mechanics and (mostly path-integral based) quantum field theory, as well as Abelian and non-Abelian gauge theories (with their quantization).

3.1 Basics of Non-Relativistic Quantum Mechanics

In this Chapter we give the necessary background in quantum theory. Main reference is [II08b], although we remark that there is a number of good textbooks in quantum field theory (QFT, see [BD65, IZ80, Ram90, PS95, Wei95, Del99, Zee03, Wit03, Nai05, Sre07], as well as the two most important papers: [Wit88] and [Ati88].

Recall that quantum theory was born with Max Planck's 1900 paper, in which he derived the correct shape of the black-body spectrum which now bears his name, eliminating the ultraviolet catastrophe—with the price of introducing a 'bizarre assumption' (today called *Planck's quantum hypothesis*) that energy was only emitted in certain finite chunks, or 'quanta'. In 1905, Albert Einstein took this bold idea one step further. Assuming that radiation could only transport energy in such chunks, 'photons', he was able to explain the so-called *photoelectric effect*. In 1913, Niels Bohr made a new breakthrough by postulating that the amount of angular momentum in an atom was quantized, so that the electrons were confined to a discrete set of orbits, each with a definite energy. If the electron jumped from one orbit to a lower one, the energy difference was sent off in the form of a photon. If the electron was in the innermost allowed orbit, there were no orbits with less energy to jump to, so the atom was stable. In addition, Bohr's theory successfully explained a slew of spectral lines that had been measured for Hydrogen. The famous *wave-particle duality of matter* was proposed by French prince Louis de Broglie in 1923 in his Ph.D. thesis: that electrons and other particles acted like standing waves. Such waves, like vibrations of a guitar string, can

V.G. Ivancevic, T.T. Ivancevic, *Quantum Neural Computation*,
Intelligent Systems, Control and Automation: Science and Engineering 40,
DOI 10.1007/978-90-481-3350-5_3, © Springer Science+Business Media B.V. 2010

only occur with certain discrete (quantized) frequencies.[1] In November 1925, Erwin Schrödinger gave a seminar on de Broglie's work in Zurich. When he was finished, P. Debye said in effect, "You speak about waves. But where is the wave equation?" Schrödinger went on to produce and publish his famous wave equation, the master key for so much of modern physics. An equivalent formulation involving infinite matrices was provided by Werner Heisenberg, Max Born and Pasquale Jordan around the same time. With this new powerful mathematical underpinning, quantum theory made explosive progress. Within a few years, a host of hitherto unexplained measurements had been successfully explained, including spectra of more complicated atoms and various numbers describing properties of chemical reactions. For more details, see, e.g. [TW01].

3.1.1 Soft Introduction to Quantum Mechanics

According to quantum mechanics, *light* consists of particles called *photons*, and Fig. 3.1 shows a photon source which we assume emits photons one at a time. There are two slits, A and B, and a screen behind them. The photons arrive at the screen as individual events, where they are detected separately, just as if they were ordinary particles. The curious quantum behavior arise in the following way [Pen94, Pen97, II08b]. If only slit A were open and the other closed, there would be many places on the screen which the photon could reach. If we now close the slit A and open the slit B, we may again find that the photon could reach the same spot on the screen. However, if we open *both slits*, and if we have chosen the point on the screen carefully, we may now find that the photon cannot reach that spot, even though it could have done so if either slit alone were open. Somehow, the two possible things which the photon *might* do cancel each other out. This type of behavior does not take place in classical physics. Either one thing happens or another thing happens—we do not get two possible things which might happen, somehow conspiring to cancel each other out.

The way we understand the outcome of this experiment in quantum mechanics is to say that when the photon is *en route* from the source to the screen, the state of the photon is not that of having gone through one slit or the other, but is some mysterious combination of the two, weighted by complex numbers. That is, we can write the state of the photon as a wave ψ-function,[2] which is the *linear superposition* of the two states, $|A\rangle$ and $|B\rangle$,[3]

---

[1] The idea was so new that the examining committee went outside its circle for advice on the acceptability of the thesis. Einstein gave a favorable opinion and the thesis was accepted.

[2] In the *Schrödinger picture*, the *unitary evolution* U of a quantum system is described by the *Schrödinger equation*, which provides the time rate of change of the quantum state or wave function $\psi = \psi(t)$.

[3] We are using here the standard Dirac 'bra-ket' notation for quantum states. Paul Dirac was one of the outstanding physicists of the 20th century. Among his achieve-

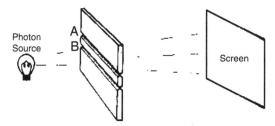

Fig. 3.1. The two-slit experiment, with individual photons of monochromatic light (see text for explanation).

corresponding to the A-slot and B-slot alternatives,

$$|\psi\rangle = z_1|A\rangle + z_2|B\rangle,$$

where z_1 and z_2 are complex numbers (not both zero), while $|\cdot\rangle$ denotes the *quantum state ket-vector*.

Now, in quantum mechanics, we are not so interested in the sizes of the complex numbers z_1 and z_2 themselves as we are in their ratio—it is only the ratio of these numbers which has direct physical meaning (as multiplying a quantum state with a nonzero complex number does not change the physical situation). Recall that the *Riemann sphere* is a way of representing complex numbers (plus ∞) and their ratios on a sphere on unit radius, whose equatorial plane is the complex-plane, whose center is the origin of that plane and the equator of this sphere is the unit circle in the complex-plane. We can project each point on the equatorial complex-plane onto the Riemann sphere, projecting from its south pole S, which corresponds to the *point at infinity* in the complex-plane. To represent a particular complex ratio, say $u = z/w$ (with $w \neq 0$), we take the stereographic projection from the sphere onto the plane.

The Riemann sphere plays a fundamental role in the quantum picture of two-state systems [Pen94, II08b]. If we have a spin-$\frac{1}{2}$ particle, such as an electron, a proton, or a neutron, then the various combinations of their spin states can be realized geometrically on the Riemann sphere. Spin $-\frac{1}{2}$ particles can have two spin states: (i) spin-up (with the rotation vector pointing upwards), and (ii) spin-down (with the rotation vector pointing downwards). The superposition of the two spin-states can be represented symbolically as

$$|\nearrow\rangle = w|\uparrow\rangle + z|\downarrow\rangle.$$

Different combinations of these spin states give us rotation about some other axis and, if we want to know where that axis is, we take the ratio of complex

---

ments was a general formulation of quantum mechanics (having Heisenberg matrix mechanics and Shrödinger wave mechanics as special cases) and also its relativistic generalization involving the 'Dirac equation', which he discovered, for the electron. He had an unusual ability to 'smell out' the truth, judging his equations, to a large degree, by their aesthetic qualities.

numbers $u = z/w$. We place this new complex number u on the Riemann sphere and the direction of u from the center is the direction of the spin axis (see Fig. 3.2).

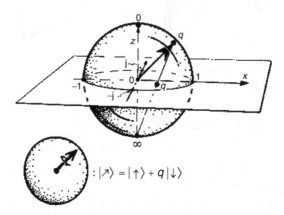

Fig. 3.2. The quantum Riemann sphere, represented as the space of physically distinct spin-states of a spin-$\frac{1}{2}$ particle (e.g., electron, proton, neutron): $|\nearrow\rangle = |\uparrow\rangle + q|\downarrow\rangle$. The sphere is projected stereographically from its south pole (∞) to the complex-plane through its equator (see text for explanation).

More general quantum state vectors might have a form such as [Pen94, II08b]:

$$|\psi\rangle = z_1|A_1\rangle + z_2|A_2\rangle + \cdots + z_n|A_n\rangle,$$

where $z_1, \ldots, z_n$ are complex numbers (not all zero) and the state vectors $|A_1\rangle, \ldots, |A_n\rangle$ might represent various possible locations for a particle (or perhaps some other property of a particle, such as its state of spin). Even more generally, infinite sums would be allowed for a wave ψ-function or quantum state vector.

Now, the most basic feature of *unitary quantum evolution* U[4] is that it is *linear*. This means that, if we have two states, say $|\psi\rangle$ and $|\phi\rangle$, and if the Schrödinger equation would tell us that, after some time t, the states $|\psi\rangle$ and

---

[4] Recall that unitary quantum evolution U is governed by the time-dependent *Schrödinger equation*,

$$i\hbar\partial_t|\psi(t)\rangle = H|\psi(t)\rangle,$$

where $\partial_t \equiv \partial/\partial t$, $\hbar$ is the *Planck's constant*, and H is the Hamiltonian (total energy) operator. Given the quantum state $|\psi(t)\rangle$ at some initial time ($t = 0$), we can integrate the Schrödinger equation to get the state at any subsequent time. In particular, if H is independent of time, then

$$|\psi(t)\rangle = \exp\left(-\frac{iHt}{\hbar}\right)|\psi(0)\rangle.$$

$|\phi\rangle$ would each individually evolve to new states $|\psi'\rangle$ and $|\phi'\rangle$, respectively then any linear superposition $z_1|\psi\rangle + z_2|\phi\rangle$, must evolve, after some time t, to the corresponding superposition $z_1|\psi'\rangle + z_2|\phi'\rangle$. Let us use the symbol $\rightsquigarrow$ to denote the evolution after time t, Then linearity asserts that if

$$|\psi\rangle \rightsquigarrow |\psi'\rangle \quad \text{and} \quad |\phi\rangle \rightsquigarrow |\phi'\rangle,$$

then the evolution

$$z_1|\psi\rangle + z_2|\phi\rangle \quad \rightsquigarrow \quad z_1|\psi'\rangle + z_2|\phi'\rangle$$

would also hold. This would consequently apply also to linear superpositions of more than two individual quantum states. For example, $z_1|\psi\rangle + z_2|\phi\rangle + z_3|\chi\rangle$ would evolve, after time t, to $z_1|\psi'\rangle + z_2|\phi'\rangle + z_3|\chi'\rangle$, if $|\psi\rangle, |\phi\rangle$, and $|\chi\rangle$ would each individually evolve to $|\psi'\rangle, |\phi'\rangle$, and $|\chi'\rangle$, respectively. Thus, the evolution always proceeds as though each different component of a superposition were oblivious to the presence of the others.

As a second experiment, consider a situation in which light impinges on a half-silvered mirror, that is a semi-transparent mirror that reflects just half the light (composed of a *stream of photons*) falling upon it and transmits the remaining half [Pen94, II08b]. We might well have imagined that for a stream of photons impinging on our half-silvered mirror, half the photons would be reflected and half would be transmitted. Not so! Quantum theory tells us that, instead, each individual photon, as it impinges on the minor, is separately put into a superposed state of reflection and transmission. If the photon before its encounter with the minor is in state $|A\rangle$, then afterwards it evolves according to U to become a state that can be written $|B\rangle + \mathrm{i}|C\rangle$, where $|B\rangle$ represents the state in which the photon is transmitted through the mirror and $|C\rangle$ the state where the photon is reflected from it (see Fig. 3.3). Let us write this as

$$|A\rangle \rightsquigarrow |B\rangle + \mathrm{i}|C\rangle.$$

The imaginary factor 'i' arises here because of a net phase shift by a quarter of a wavelength (see [II08b]), which occurs between the reflected and transmitted beams at such a mirror.

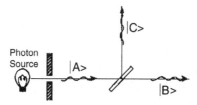

Fig. 3.3. A photon in state $|A\rangle$ impinges on a half-silvered mirror and its state evolves according to U into a superposition $|B\rangle + \mathrm{i}|C\rangle$ (see text for explanation).

Although, from the classical picture of a particle, we would have to imagine that $|B\rangle$ and $|C\rangle$ just represent alternative things that the photon *might* do, in quantum mechanics we have to try to believe that the photon is now actually doing *both things at once* in this strange, complex superposition. To see that it cannot just be a matter of classical probability-weighted alternatives, let us take this example a little further and try to bring the two parts of the photon state, i.e., the two photon beams, back together again [Pen94, II08b]. We can do this by first reflecting each beam with a fully silvered mirror. After reflection, the photon state $|B\rangle$ would evolve according to U, into another state $i|D\rangle$, whilst $|C\rangle$ would evolve into $i|E\rangle$,

$$|B\rangle \rightsquigarrow i|D\rangle \quad \text{and} \quad |C\rangle \rightsquigarrow i|E\rangle.$$

Thus the entire state $|B\rangle + i|C\rangle$ evolves by U into

$$|B\rangle + i|C\rangle \rightsquigarrow i|D\rangle + i(i|E\rangle)) = i|D\rangle - |E\rangle,$$

since $i^2 = -1$. Now, suppose that these two beams come together at a fourth mirror, which is now half silvered (see Fig. 3.4). The state $|D\rangle$ evolves into a combination $|G\rangle + i|F\rangle$, where $|G\rangle$ represents the transmitted state and $|F\rangle$ the reflected one. Similarly, $|E\rangle$ evolves into $|F\rangle + i|G\rangle$, since it is now the state $|F\rangle$ that is the transmitted state and $|G\rangle$ the reflected one,

$$|D\rangle \rightsquigarrow |G\rangle + i|F\rangle \quad \text{and} \quad |E\rangle \rightsquigarrow |F\rangle + i|G\rangle.$$

Our entire state $i|D\rangle - |E\rangle$ is now seen (because of the linearity of U) to evolve as:

$$i|D\rangle - |E\rangle \rightsquigarrow i(|G\rangle + i|F\rangle) - (|F\rangle + i|G\rangle)$$
$$= i|G\rangle - |F\rangle - |F\rangle - i|G\rangle = -2|F\rangle.$$

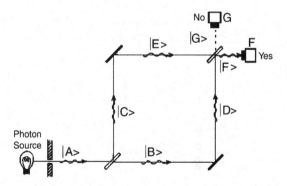

Fig. 3.4. *Mach–Zehnder interferometer*: the two parts of the photon state are brought together by two fully silvered mirrors (black), so as to encounter each other at a final half-silvered mirror (white). They interfere in such a way that the entire state emerges in state $|F\rangle$, and the detector at G cannot receive the photon (see text for explanation).

As mentioned above, the multiplying factor -2 appearing here plays no physical role, thus we see that the possibility $|G\rangle$ is *not* open to the photon; the two beams together combine to produce just a *single* possibility $|F\rangle$. This curious outcome arises because *both* beams are present *simultaneously* in the physical state of the photon, between its encounters with the first and last mirrors. We say that the two beams *interfere* with one another.[5]

3.1.2 Quantum States and Operators

More precisely, non-relativistic quantum-mechanical systems have two modes of evolution in time [Dir82, II08b]. The first, governed by standard, *time-dependent Schrödinger equation*:

$$i\hbar\partial_t|\psi\rangle = \hat{H}|\psi\rangle, \tag{3.1}$$

describes the time evolution of quantum systems when they are undisturbed by measurements. 'Measurements' are defined as *interactions* of the quantum system with its classical environment. As long as the system is sufficiently isolated from the environment, it follows Schrödinger equation. If an interaction with the environment takes place, i.e., a measurement is performed, the system abruptly *decoheres* i.e., collapses or reduces to one of its classically allowed states.

A *time-dependent state of a quantum system* is determined by a normalized, complex-valued, *wave psi-function* $\psi = \psi(t)$. In Dirac's words [Dir82], this is a unit *ket* vector $|\psi\rangle$, which is an element of the *Hilbert space* $L^2(\psi) \equiv \mathcal{H}$, with a coordinate basis (q^i).[6] The state ket-vector $|\psi(t)\rangle$ is subject to

---

[5] This is a property of single photons: each individual photon must be considered to feel out both routes that are open to it, but it remains one photon; it does not split into two photons in the intermediate stage, but its location undergoes the strange kind of complex-number-weighted *co-existence of alternatives* that is characteristic of quantum theory.

[6] The family of all possible states $(|\psi\rangle, |\phi\rangle$, etc.) of a quantum system configure what is known as a *Hilbert space*. It is a *complex vector space*, which means that can perform the complex-number-weighted combinations that we considered before for quantum states. If $|\psi\rangle$ and $|\phi\rangle$ are both elements of the Hilbert space, then so also is $w|\psi\rangle + z|\phi\rangle$, for any pair of complex numbers w and z. Here, we even allow $w = z = 0$, to give the element $\mathbf{0}$ of the Hilbert space, which does not represent a possible physical state. We have the normal algebraic rules for a vector space:

$$|\psi\rangle + |\phi\rangle = |\phi\rangle + |\psi\rangle,$$
$$|\psi\rangle + (|\phi\rangle + |\chi\rangle) = (|\psi\rangle + |\phi\rangle) + |\chi\rangle,$$
$$w(z|\psi\rangle) = (wz)|\psi\rangle,$$
$$(w + z)|\psi\rangle = w|\psi\rangle + z|\psi\rangle,$$
$$z(|\psi\rangle + |\phi\rangle) = z|\psi\rangle + z|\phi\rangle$$
$$0|\psi\rangle = \mathbf{0}, \qquad z\mathbf{0} = \mathbf{0}.$$

action of the Hermitian operators, obtained by the procedure of *quantization* of classical biodynamic quantities, and whose real eigenvalues are being measured.

Quantum superposition is a generalization of the algebraic principle of linear combination of vectors. The Hilbert space has a set of states $|\varphi_i\rangle$ (where the index i runs over the degrees-of-freedom of the system) that form a basis and the most general state of such a system can be written as $|\psi\rangle = \sum_i c_i |\varphi_i\rangle$. The system is said to be in a state $|\psi(t)\rangle$, describing the motion of the *de Broglie waves*, which is a linear superposition of the basis states $|\varphi_i\rangle$ with weighting coefficients c_i that can in general be complex. At the microscopic or quantum level, the state of the system is described by the wave function $|\psi\rangle$, which in general appears as a linear superposition of all basis states. This

A Hilbert space can sometimes have a finite number of dimensions, as in the case of the spin states of a particle. For spin $\frac{1}{2}$, the Hilbert space is just 2D, its elements being the complex linear combinations of the two states $|\uparrow\rangle$ and $|\downarrow\rangle$. For spin $\frac{1}{2}n$, the Hilbert space is $(n + 1)$D. However, sometimes the Hilbert space can have an infinite number of dimensions, as e.g., the states of position or momentum of a particle. Here, each alternative position (or momentum) that the particle might have counts as providing a separate dimension for the Hilbert space. The general state describing the quantum location (or momentum) of the particle is a complex-number superposition of all these different individual positions (or momenta), which is the wave ψ-function for the particle.

Another property of the Hilbert space, crucial for quantum mechanics, is the *Hermitian inner (scalar) product*, which can be applied to any pair of Hilbert-space vectors to produce a single complex number. To understand how important the Hermitian inner product is for quantum mechanics, recall that the Dirac's 'bra-ket' notation is formulated on the its basis. If we have the two quantum states (i.e., Hilbert-space vectors) are $|\psi\rangle$ and $|\phi\rangle$, then their Hermitian scalar product is denoted $\langle\psi|\phi\rangle$, and it satisfies a number of simple algebraic properties:

$$\overline{\langle\psi|\phi\rangle} = \langle\phi|\psi\rangle, \quad \text{(bar denotes complex-conjugate)}$$
$$\langle\psi|(|\phi\rangle + |\chi\rangle) = \langle\psi|\phi\rangle + \langle\psi|\chi\rangle,$$
$$(z\langle\psi|)|\phi\rangle = z\langle\psi|\phi\rangle,$$
$$\langle\psi|\phi\rangle \geq 0, \quad \langle\psi|\phi\rangle = 0 \quad \text{if} \quad |\psi\rangle = \mathbf{0}.$$

For example, probability of finding a quantum particle at a given location is a *squared length* $|\psi|^2$ of a Hilbert-space position vector $|\psi\rangle$, which is the scalar product $\langle\psi|\psi\rangle$ of the vector $|\psi\rangle$ with itself. A *normalized state* is given by a Hilbert-space vector whose squared length is *unity*.

The second important thing that the Hermitian scalar product gives us is the notion of *orthogonality* between Hilbert-space vectors, which occurs when the scalar product of the two vectors is *zero*. In ordinary terms, orthogonal states are things that are independent of one another. The importance of this concept for quantum physics is that the different alternative outcomes of any measurement are always orthogonal to each other. For example, states $|\uparrow\rangle$ and $|\downarrow\rangle$ are mutually orthogonal. Also, orthogonal are *all* different possible *positions* that a quantum particle might be located in.

can be interpreted as the system being in all these states at once. The coefficients c_i are called the *probability amplitudes* and $|c_i|^2$ gives the probability that $|\psi\rangle$ will collapse into state $|\varphi\rangle$ when it decoheres (interacts with the environment). By simple normalization we have the constraint that $\sum_i |c_i|^2 = 1$. This emphasizes the fact that the wave-function describes a *real, physical system*, which must be in one of its allowable classical states and therefore by summing over all the possibilities, weighted by their corresponding probabilities, one must get unity. In other words, we have the *normalization condition* for the psi-function, determining the unit length of the state ket-vector

$$\langle \psi(t)|\psi(t)\rangle = \int \psi^* \psi \, dV = \int |\psi|^2 \, dV = 1,$$

where $\psi^* = \langle\psi(t)|$ denotes the *bra* vector, the complex-conjugate to the ket $\psi = |\psi(t)\rangle$, and $\langle\psi(t)|\psi(t)\rangle$ is their scalar product, i.e., Dirac *bracket*. For this reason the scene of quantum mechanics is the functional space of square-integrable complex psi-functions, i.e., the Hilbert space $L^2(\psi)$.

When the system is in the state $|\psi(t)\rangle$, the average value $\langle f \rangle$ of any physical observable f is equal to

$$\langle f \rangle = \langle\psi(t)|\hat{f}|\psi(t)\rangle,$$

where $\hat{f}$ is the Hermitian operator corresponding to f.

A quantum system is *coherent* if it is in a linear superposition of its basis states. If a measurement is performed on the system and this means that the system must somehow interact with its environment, the superposition is destroyed and the system is observed to be in only one basis state, as required classically. This process is called *reduction* or *collapse* of the wave-function or simply *decoherence* and is governed by the form of the wave-function $|\psi\rangle$.

Entanglement, on the other hand, is a purely quantum phenomenon and has no classical analogue. It accounts for the ability of quantum systems to exhibit correlations in counterintuitive 'action-at-a-distance' ways. Entanglement is what makes all the difference in the operation of quantum computers versus classical ones. Entanglement gives 'special powers' to quantum computers because it gives quantum states the potential to exhibit and maintain correlations that cannot be accounted for classically. Correlations between bits are what make information encoding possible in classical computers. For instance, we can require two bits to have the same value thus encoding a relationship. If we are to subsequently change the encoded information, we must change the correlated bits in tandem by explicitly accessing each bit. Since quantum bits exist as superpositions, *correlations* between them also exist in superposition. When the superposition is destroyed (e.g., one qubit is measured), the correct correlations are *instantaneously* 'communicated' between the qubits and this communication allows *many qubits* to be accessed *at once*, preserving their correlations, something that is absolutely impossible classically.

More precisely, the *first quantization* is a *linear representation* of all classical dynamical variables (like coordinate, momentum, energy, or angular momentum) by linear *Hermitian* operators acting on the associated Hilbert state-space $\mathcal{H}$, which has the following properties [Dir82]:

1. *Linearity:* $\alpha f + \beta g \rightarrow \alpha \hat{f} + \beta \hat{g}$, (for all constants $\alpha, \beta \in \mathbb{C}$);
2. A 'dynamical' variable, equal to unity everywhere in the phase-space, corresponds to unit operator: $1 \rightarrow \hat{I}$; and
3. *Classical Poisson brackets*

$$\{f, g\} = \frac{\partial f}{\partial q^i} \frac{\partial g}{\partial p_i} - \frac{\partial f}{\partial p_i} \frac{\partial g}{\partial q^i}$$

quantize to the corresponding *commutators*

$$\{f, g\} \rightarrow -i\hbar[\hat{f}, \hat{g}], \quad [\hat{f}, \hat{g}] = \hat{f}\hat{g} - \hat{g}\hat{f}.$$

Like Poisson bracket, commutator is bilinear and skew-symmetric operation, satisfying Jacobi identity. For Hermitian operators $\hat{f}, \hat{g}$ their commutator $[\hat{f}, \hat{g}]$ is anti-Hermitian; for this reason i is required in $\{f, g\} \rightarrow -i\hbar[\hat{f}, \hat{g}]$.

Property (2) is introduced for the following reason. In Hamiltonian mechanics each dynamical variable f generates some transformations in the phase-space via Poisson brackets. In quantum mechanics it generates transformations in the state-space by direct application to a state, i.e.,

$$\dot{u} = \{u, f\}, \qquad \partial_t |\psi\rangle = \frac{i}{\hbar} \hat{f} |\psi\rangle. \tag{3.2}$$

Exponent of anti-Hermitian operator is unitary. Due to this fact, transformations, generated by Hermitian operators

$$\hat{U} = \exp \frac{i\hat{f}t}{\hbar},$$

are unitary. They are *motions*—scalar product preserving transformations in the Hilbert state-space $\mathcal{H}$. For this property i is needed in (3.2).

Due to property (2), the transformations, generated by classical variables and quantum operators, have the same algebra.

For example, the quantization of energy E gives:

$$E \rightarrow \hat{E} = i\hbar\partial_t.$$

The relations between operators must be similar to the relations between the relevant physical quantities observed in classical mechanics.

For example, the quantization of the classical equation $E = H$, where

$$H = H(p_i, q^i) = T + U$$

denotes the Hamilton's function of the total system energy (the sum of the kinetic energy T and potential energy U), gives the Schrödinger equation of motion of the state ket-vector $|\psi(t)\rangle$ in the Hilbert state-space $\mathcal{H}$

$$i\hbar\partial_t|\psi(t)\rangle = \hat{H}|\psi(t)\rangle.$$

In the simplest case of a single particle in the potential field U, the operator of the total system energy—Hamiltonian is given by:

$$\hat{H} = -\frac{\hbar^2}{2m}\nabla^2 + U,$$

where m denotes the mass of the particle and ∇ is the classical gradient operator. So the first term on the r.h.s. denotes the kinetic energy of the system, and therefore the momentum operator must be given by:

$$\hat{p} = -i\hbar\nabla.$$

Now, for each pair of states $|\varphi\rangle, |\psi\rangle$ their scalar product $\langle\varphi|\psi\rangle$ is introduced, which is [II08b]:

1. Linear (for right multiplier):

$$\langle\varphi|\alpha_1\psi_1 + \alpha_2\psi_2\rangle = \alpha_1\langle\varphi|\psi_1\rangle + \alpha_2\langle\varphi|\psi_2\rangle;$$

2. In transposition transforms to complex conjugated:

$$\langle\varphi|\psi\rangle = \overline{\langle\psi|\varphi\rangle};$$

this implies that it is 'anti-linear' for left multiplier:

$$\langle\alpha_1\varphi_1 + \alpha_2\varphi_2\rangle = \bar{\alpha}_1\langle\varphi_1|\psi\rangle + \bar{\alpha}_2\langle\varphi_2|\psi\rangle);$$

3. Additionally it is often required, that the scalar product should be positively defined:

$$\text{for all } |\psi\rangle, \quad \langle\psi|\psi\rangle \geq 0 \quad \text{and} \quad \langle\psi|\psi\rangle = 0 \quad \text{iff} \quad |\psi\rangle = 0.$$

Complex conjugation of classical variables is represented as Hermitian conjugation of operators.[7]

---

[7] Two operators $\hat{f}, \hat{f}^+$ are called Hermitian conjugated (or adjoint), if

$$\langle\varphi|\hat{f}\psi\rangle = \langle\hat{f}^+\varphi|\psi\rangle \quad \text{(for all } \varphi, \psi).$$

This scalar product is also denoted by $\langle\varphi|\hat{f}|\psi\rangle$ and called a matrix element of an operator.

– operator is Hermitian (self-adjoint) if $\hat{f}^+ = \hat{f}$ and anti-Hermitian if $\hat{f}^+ = -\hat{f}$;

If the two Hermitian operators $\hat{f}$ and $\hat{g}$ commute, i.e., $[\hat{f}, \hat{g}] = 0$ (see Heisenberg picture below), than the corresponding quantities can simultaneously have definite values. If the two operators do not commute, i.e., $[\hat{f}, \hat{g}] \neq 0$, the quantities corresponding to these operators cannot have definite values simultaneously, i.e., the general *Heisenberg uncertainty relation* is valid:

$$(\Delta \hat{f})^2 \cdot (\Delta \hat{g})^2 \geq \frac{\hbar}{4}[\hat{f}, \hat{g}]^2,$$

where Δ denotes the deviation of an individual measurement from the mean value of the distribution. The well-known particular cases are ordinary uncertainty relations for coordinate–momentum $(q - p)$, and energy–time $(E - t)$:

$$\Delta q \cdot \Delta p_q \geq \frac{\hbar}{2}, \quad \text{and} \quad \Delta E \cdot \Delta t \geq \frac{\hbar}{2}.$$

For example, the rules of commutation, analogous to the classical ones written by the Poisson's brackets, are postulated for canonically-conjugate coordinate and momentum operators:

$$[\hat{q}^i, \hat{q}^j] = 0, \qquad [\hat{p}_i, \hat{p}_j] = 0, \qquad [\hat{q}^i, \hat{p}_j] = i\hbar \delta^i_j \hat{I},$$

where δ^i_j is the Kronecker's symbol. By applying the commutation rules to the system Hamiltonian $\hat{H} = \hat{H}(\hat{p}_i, \hat{q}^i)$, the *quantum Hamilton's equations* are obtained:

$$\dot{\hat{p}}_i = -\partial_{\hat{q}^i} \hat{H} \quad \text{and} \quad \dot{\hat{q}}^i = \partial_{\hat{p}_i} \hat{H}.$$

A quantum state can be observed either in the *coordinate q-representation*, or in the *momentum p-representation*. In the q-representation, operators of coordinate and momentum have respective forms: $\hat{q} = q$, and $\hat{p}_q = -i\hbar\frac{\partial}{\partial q}$, while in the p-representation, they have respective forms: $\hat{q} = i\hbar\frac{\partial}{\partial p_q}$, and $\hat{p}_q = p_q$. The forms of the state vector $|\psi(t)\rangle$ in these two representations are mathematically related by a *Fourier-transform pair*.

– operator is unitary, if $\hat{U}^+ = \hat{U}^{-1}$; such operators preserve the scalar product:

$$\langle \hat{U}\varphi | \hat{U}\psi \rangle = \langle \varphi | \hat{U}^+ \hat{U} | \psi \rangle = \langle \varphi | \psi \rangle.$$

Real classical variables should be represented by Hermitian operators; complex conjugated classical variables $(a, \bar{a})$ correspond to Hermitian conjugated operators $(\hat{a}, \hat{a}^+)$.

Multiplication of a state by complex numbers does not change the state physically. Any Hermitian operator in Hilbert space has only real eigenvalues:

$$\hat{f}|\psi_i\rangle = f_i|\psi_i\rangle, \quad (\text{for all } f_i \in \mathbb{R}).$$

Eigenvectors $|\psi_i\rangle$ form complete orthonormal basis (eigenvectors with different eigenvalues are automatically orthogonal; in the case of multiple eigenvalues one can form orthogonal combinations; then they can be normalized).

3.1.3 The Tree Standard Quantum Pictures

In the q-representation, there are three main pictures (see e.g., [II08b]):

1. *Schrödinger picture*,
2. *Heisenberg picture*, and
3. *Dirac interaction picture*.

These three pictures mutually differ in the time-dependence, i.e., time-evolution of the state-vector wave-function $|\psi(t)\rangle$ and the Hilbert coordinate basis (q^i) together with the system operators.

1. In the *Schrödinger* (S) *picture*, under the action of the *evolution operator* $\hat{S}(t)$ the state-vector $|\psi(t)\rangle$ rotates:

$$|\psi(t)\rangle = \hat{S}(t)|\psi(0)\rangle,$$

and the coordinate basis (q^i) is fixed, so the operators are constant in time:

$$\hat{F}(t) = \hat{F}(0) = \hat{F},$$

and the system evolution is determined by the Schrödinger wave equation:

$$i\hbar\partial_t|\psi^S(t)\rangle = \hat{H}^S|\psi^S(t)\rangle.$$

If the Hamiltonian does not explicitly depend on time, $\hat{H}(t) = \hat{H}$, which is the case with the absence of variables of macroscopic fields, the state vector $|\psi(t)\rangle$ can be presented in the form:

$$|\psi(t)\rangle = \exp\left(-i\frac{E}{\hbar}t\right)|\psi\rangle,$$

satisfying the time-independent Schrödinger equation

$$\hat{H}|\psi\rangle = E|\psi\rangle,$$

which gives the eigenvalues E_m and eigenfunctions $|\psi_m\rangle$ of the Hamiltonian $\hat{H}$.

2. In the *Heisenberg* (H) *picture*, under the action of the evolution operator $\hat{S}(t)$, the coordinate basis (q^i) rotates, so the operators of physical variables evolve in time by the similarity transformation:

$$\hat{F}(t) = \hat{S}^{-1}(t)\hat{F}(0)\hat{S}(t),$$

while the state vector $|\psi(t)\rangle$ is constant in time:

$$|\psi(t)\rangle = |\psi(0)\rangle = |\psi\rangle,$$

and the system evolution is determined by the *Heisenberg equation of motion*:

$$i\hbar\partial_t\hat{F}^H(t) = [\hat{F}^H(t), \hat{H}^H(t)],$$

where $\hat{F}(t)$ denotes arbitrary Hermitian operator of the system, while the commutator, i.e., Poisson quantum bracket, is given by:

$$[\hat{F}(t), \hat{H}(t)] = \hat{F}(t)\hat{H}(t) - \hat{H}(t)\hat{F}(t) = \hat{\imath}K.$$

In both Schrödinger and Heisenberg picture the evolution operator $\hat{S}(t)$ itself is determined by the Schrödinger-like equation:

$$i\hbar\partial_t\hat{S}(t) = \hat{H}\hat{S}(t),$$

with the initial condition $\hat{S}(0) = \hat{I}$. It determines the Lie group of transformations of the Hilbert space $L^2(\psi)$ in itself, the Hamiltonian of the system being the generator of the group.

3. In the *Dirac interaction* (I) *picture* both the state vector $|\psi(t)\rangle$ and coordinate basis (q^i) rotate; therefore the system evolution is determined by both the Schrödinger wave equation and the Heisenberg equation of motion:

$$i\hbar\partial_t|\psi^I(t)\rangle = \hat{H}^I|\psi^I(t)\rangle, \quad \text{and} \quad i\hbar\partial_t\hat{F}^I(t) = [\hat{F}^I(t), \hat{H}^O(t)].$$

Here, $\hat{H} = \hat{H}^0 + \hat{H}^I$, where $\hat{H}^0$ corresponds to the Hamiltonian of the free fields and $\hat{H}^I$ corresponds to the Hamiltonian of the interaction.

In particular, the stationary (time-independent) Schrödinger equation,

$$\hat{H}\psi = \hat{E}\psi,$$

can be obtained from the condition for the minimum of the *quantum action*:

$$\delta S[\psi] = 0.$$

The quantum action is usually defined by the integral:

$$S[\psi] = \langle\psi(t)|\hat{H}|\psi(t)\rangle = \int \psi^*\hat{H}\psi \, dV,$$

with the additional normalization condition for the unit-probability of the psi-function:

$$\langle\psi(t)|\psi(t)\rangle = \int \psi^*\psi \, dV = 1.$$

When the functions ψ and ψ^* are considered to be formally independent and only one of them, say ψ^* is varied, we can write the condition for an extreme of the action:

$$\delta S[\psi] = \int \delta\psi^*\hat{H}\psi \, dV - E \int \delta\psi^*\psi \, dV = \int \delta\psi^*(\hat{H}\psi - E\psi) \, dV = 0,$$

where E is a Lagrangian multiplier. Owing to the arbitrariness of $\delta\psi^*$, the Schrödinger equation $\hat{H}\psi - \hat{E}\psi = 0$ must hold.

3.1.4 Dirac's Probability Amplitude and Perturbation

Most quantum-mechanical problems cannot be solved exactly. For such problems we can use *Dirac's perturbation method*, which consists in splitting up the time-dependent Hamiltonian $H = H(t)$ into two parts:

$$H(t) = H_0 + \epsilon H_1(t),$$

in which $H_0 = E$ must be simple, non-autonomous, energy function that can be dealt with exactly, while $\epsilon H_1(t) = V(t)$ is small time-dependent perturbation, which can be expanded as a power series in a small numerical factor ϵ. The first part, H_0, may then be considered as the Hamiltonian of a simplified, or unperturbed system that can be exactly solved, while the addition of the second part $\epsilon H_1(t)$ will require small corrections, of the nature of a power-series expanded perturbation in the solution for the unperturbed system. Provided the perturbation series in ϵ converges, the perturbation method will give the answer to our problem with any desired accuracy; even when the series does not converge, the first approximation obtained by means of it is usually fairly accurate [Dir82].

Therefore, we do not consider any modification to be made in the states of the unperturbed system $E = H_0$, but we suppose that the perturbed system $H(t)$, instead of remaining permanently in one of its states, is continually changing from one state to another (or, making transmissions), under the influence of the perturbation $V(t) = \epsilon H_1(t)$.

We will work in the *Heisenberg representation* for the unperturbed system E, assuming that we have a general set of linear Hermitian operators α's to label the representatives. Let us suppose that at initial time t_0 the system is in a state for which the α's certainly have the values α', so that the basic ket $|\alpha'\rangle$ would correspond to this state. This state would be stationary if there were no perturbation, i.e., if $H(t) = E$. The perturbation $V(t)$ cause the E to change. At time t the ket corresponding to the state $|\alpha'\rangle$ in the *Schrödinger's picture* will be $T|\alpha'\rangle$, according to equation (in natural units)

$$|At\rangle = T|At_0\rangle, \quad \text{as well as}$$
$$i\frac{dT}{dt} = H(t)T \text{ and } i\frac{d|At\rangle}{dt} = H(t)|At\rangle, \tag{3.3}$$

where T is a linear Hermitian operator independent of the ket $|At\rangle$ and depending only on time (t_0 and t). The probability of the α's having the values α'' is given by the absolute square of the *probability amplitude* (or, *transition amplitude*) $\langle\alpha''|T|\alpha'\rangle$ (for the system's transition from the state $|\alpha'\rangle$ to the state $|\alpha''\rangle$),

$$P(\alpha', \alpha'') = |\langle\alpha''|T|\alpha'\rangle|^2. \tag{3.4}$$

For $\alpha' \neq \alpha''$, $P(\alpha', \alpha'')$ is the probability of a transition taking place from state α' to state α'' during the time interval $[t_0, t]$; $P(\alpha', \alpha')$ is the probability of no transition taking place at all, while the sum of $P(\alpha', \alpha'')$ for all α'' is unity.

Let us now suppose more generally that initially the system is in one of the various states α' with the probability $P_{\alpha'}$ for each. To deal effectively with this problem, we introduce the *von Neumann's quantum density function* ρ, a quantum-mechanical analogue to the *Gibbs statistical density function* $\rho = \rho(t)$ of a *Gibbs ensemble* with the classical Hamiltonian $H(q, p)$, which evolves within the ensemble's n-dimensional phase-space $\mathcal{P} = \{(q^i, p_i) \mid i = 1, \ldots, n\}$ according to the *Poisson equation*

$$\partial_t \rho = -[\rho, H(q, p)],$$

with the normalizing condition: $\displaystyle\iint_{\mathcal{P}} \rho \, dq^i dp_i = 1.$

The von Neumann's quantum density function ρ corresponding to the initial probability distribution $P_{\alpha'}$ is given by

$$\rho_0 = \sum_{\alpha'} |\alpha'\rangle \, P_{\alpha'} \, \langle\alpha'| \, .$$

At time t, each ket $|\alpha'\rangle$ will have changed to $T|\alpha'\rangle$ and each bra $\langle\alpha'|$ will change to $\langle\alpha'|\bar{T}$ (where $\bar{T}$ is complex-conjugate to T), so ρ_0 will have changed to

$$\rho(t) = \sum_{\alpha'} T |\alpha'\rangle \, P_{\alpha'} \, \langle\alpha'| \bar{T}.$$

The *probability amplitude* of α's then having the values of α'' will be (using (3.4))

$$\langle\alpha''| \rho(t) |\alpha'\rangle = \sum_{\alpha'} \langle\alpha''| T |\alpha'\rangle \, P_{\alpha'} \, \langle\alpha'| \bar{T} |\alpha''\rangle = P_{\alpha'} P(\alpha', \alpha'').$$

This result expresses that the probability of the system being in the state α'' at time t equals the sum of the probabilities of the system being initially in any state $\alpha' \neq \alpha''$, and making a transition from state α' to the final state α''. Thus, the various transition probabilities act independently of one another, according to the ordinary laws of probability.

The whole problem of calculating transitions thus reduces to the determination of the probability amplitudes $\langle\alpha''|T|\alpha'\rangle$ [Dir82]. These can be worked out from (3.3), or

$$i\dot{T} = [H_0 + \epsilon H_1(t)]T = (E + V)T \quad (\text{where } \dot{T} = dT/dt). \tag{3.5}$$

This calculation can be simplified if instead of T and V operators, we are working with

$$T^* = \exp[iE(t - t_0)]T \quad \text{and} \quad V^* = \exp[iE(t - t_0)]V \exp[-iE(t - t_0)], \tag{3.6}$$

where V^* is the result of applying a unitary transformation to V. Using (3.6) we obtain

$$i\dot{T}^* = \exp[iE(t-t_0)]VT = V^*T^*, \tag{3.7}$$

which is more convenient then (3.5) as it makes the time evolution of T^* depend only on the (unitary transformed) perturbation V^* and not on the unperturbed state E. From (3.6) we get the probability amplitude

$$\langle \alpha''|T^*|\alpha'\rangle = \exp[iE(t-t_0)]\langle \alpha''|T|\alpha'\rangle, \quad \text{so that}$$
$$P(\alpha',\alpha'') = |\langle \alpha''|T^*|\alpha'\rangle|^2,$$

which shows that T and T^* are equally good for determining transition probabilities.

So far, our work in this subsection has been exact. Now we assume the perturbation $V(t) = \epsilon H_1(t)$ is a small quantity of the first order in ϵ and express T^* in the form

$$T^* = 1 + T_1^* + T_2^* + \cdots, \tag{3.8}$$

where $T_1^* = T_1^*(\epsilon)$, $T_2^* = T_2^*(\epsilon^2)$, etc. Substituting (3.8) into (3.7) we get the expansion

$$i\dot{T}_1^* = V^*,$$
$$i\dot{T}_2^* = V^*T_1^*,$$
$$i\dot{T}_3^* = V^*T_2^*,$$
$$\vdots$$

From the first of these equations we obtain

$$T_1^* = -i\int_{t_0}^t V^*(t')\,dt',$$

from the second we get

$$T_2^* = -\int_{t_0}^t V^*(t')\,dt' \int_{t_0}^t V^*(t'')\,dt'',$$

and so on.

Now, the perturbation form of the transition probability $P(\alpha',\alpha'') = |\langle \alpha''|T^*|\alpha'\rangle|^2$ is, if we retain only the first-order term T_1^* (which is sufficiently accurate for many practical problems), given by

$$P(\alpha',\alpha'') = \left| \int_{t_0}^t \langle \alpha''|V^*(t')|\alpha'\rangle \, dt' \right|^2.$$

If we retain the first two terms, T_1^* and T_2^*, the transition probability $P(\alpha', \alpha'')$ is given by

$$
P(\alpha', \alpha'') = \left| \int_{t_0}^t \langle \alpha'' | V^*(t') | \alpha' \rangle \, dt' \right.
$$

$$
\left. - \, \mathrm{i} \sum_{\alpha''' \neq \alpha'', \alpha'} \int_{t_0}^t \langle \alpha'' | V^*(t') | \alpha''' \rangle \, dt' \int_{t_0}^t \langle \alpha''' | V^*(t'') | \alpha' \rangle \, dt'' \right|^2,
$$

where α''' is the so-called intermediate state (between α' and α''). This shows the perturbative calculation of the transition probability $P(\alpha', \alpha'')$: we fist calculate the perturbative expansion of the transition amplitude $\langle \alpha'' | T^* | \alpha' \rangle$, and then take its absolute square to obtain the overall transition probability. For more technical details, see [Dir82].

Both Dirac's concepts introduced in this subsection, namely transition amplitude and time-dependent perturbation, will prove essential later in the development of the *Feynman path integral*, as well as the *Feynman diagrams* approach to *quantum field theory* (QFT).

3.1.5 State-Space for n Non-Relativistic Quantum Particles

Classical state-space for the system of n particles is its $6ND$ phase-space $\mathcal{P}$, including all position and momentum vectors, $\mathbf{r}_i = (x, y, z)_i$ and $\mathbf{p}_i = (p_x, p_y, p_z)_i$ respectively (for $i = 1, \dots, n$). The *quantization* is performed as a *linear representation* of the real Lie algebra $\mathcal{L}_P$ of the phase-space $\mathcal{P}$, defined by the Poisson bracket $\{A, B\}$ of classical variables A, B—into the corresponding real Lie algebra $\mathcal{L}_H$ of the Hilbert space $\mathcal{H}$, defined by the commutator $[\hat{A}, \hat{B}]$ of skew-Hermitian operators $\hat{A}, \hat{B}$ [II08b].

We start with the *Hilbert space* $\mathcal{H}_x$ for a single 1D quantum particle, which is composed of all vectors $|\psi_x\rangle$ of the form

$$
|\psi_x\rangle = \int_{-\infty}^{+\infty} \psi(x) |x\rangle \, dx,
$$

where $\psi(x) = \langle x | \psi \rangle$ are square integrable Fourier coefficients,

$$
\int_{-\infty}^{+\infty} |\psi(x)|^2 \, dx < +\infty.
$$

The position and momentum Hermitian operators, $\hat{x}$ and $\hat{p}$, respectively, act on the vectors $|\psi_x\rangle \in \mathcal{H}_x$ in the following way:

$$
\hat{x}|\psi_x\rangle = \int_{-\infty}^{+\infty} \hat{x}\psi(x) |x\rangle \, dx, \qquad \int_{-\infty}^{+\infty} |x\psi(x)|^2 \, dx < +\infty,
$$

$$
\hat{p}|\psi_x\rangle = \int_{-\infty}^{+\infty} -\mathrm{i}\hbar \partial_{\hat{x}} \psi(x) |x\rangle \, dx, \qquad \int_{-\infty}^{+\infty} |-\mathrm{i}\hbar \partial_x \psi(x)|^2 \, dx < +\infty.
$$

The *orbit Hilbert space* $\mathcal{H}_1^o$ for a single 3D quantum particle with the full set of compatible observable $\hat{\mathbf{r}} = (\hat{x}, \hat{y}, \hat{z}), \hat{\mathbf{p}} = (\hat{p}_x, \hat{p}_y, \hat{p}_z)$, is defined as

$$\mathcal{H}_1^o = \mathcal{H}_x \otimes \mathcal{H}_y \otimes \mathcal{H}_z,$$

where $\hat{\mathbf{r}}$ has the common generalized eigenvectors of the form

$$|\hat{\mathbf{r}}\rangle = |x\rangle \times |y\rangle \times |z\rangle.$$

$\mathcal{H}_1^o$ is composed of all vectors $|\psi_r\rangle$ of the form

$$|\psi_r\rangle = \int_{\mathcal{H}^o} \psi(\mathbf{r}) |\mathbf{r}\rangle \, d\mathbf{r} = \int_{-\infty}^{+\infty} \int_{-\infty}^{+\infty} \int_{-\infty}^{+\infty} \psi(x, y, z) |x\rangle \times |y\rangle \times |z\rangle \, dxdydz,$$

where $\psi(\mathbf{r}) = \langle \mathbf{r}|\psi_r\rangle$ are square integrable Fourier coefficients,

$$\int_{-\infty}^{+\infty} |\psi(\mathbf{r})|^2 \, d\mathbf{r} < +\infty.$$

The position and momentum operators, $\hat{\mathbf{r}}$ and $\hat{\mathbf{p}}$, respectively, act on the vectors $|\psi_r\rangle \in \mathcal{H}_1^o$ in the following way:

$$\hat{\mathbf{r}}|\psi_r\rangle = \int_{\mathcal{H}_1^o} \hat{\mathbf{r}}\psi(\mathbf{r}) |\mathbf{r}\rangle \, d\mathbf{r}, \qquad \int_{\mathcal{H}_1^o} |\mathbf{r}\psi(\mathbf{r})|^2 \, d\mathbf{r} < +\infty,$$

$$\hat{\mathbf{p}}|\psi_r\rangle = \int_{\mathcal{H}_1^o} -i\hbar\partial_{\hat{\mathbf{r}}}\psi(\mathbf{r}) |\mathbf{r}\rangle \, d\mathbf{r}, \qquad \int_{\mathcal{H}_1^o} |-i\hbar\partial_{\mathbf{r}}\psi(\mathbf{r})|^2 \, d\mathbf{r} < +\infty.$$

Now, if we have a system of n 3D particles, let $\mathcal{H}_i^o$ denote the orbit Hilbert space of the ith particle. Then the composite orbit state-space $\mathcal{H}_n^o$ of the whole system is defined as a direct product

$$\mathcal{H}_n^o = \mathcal{H}_1^o \otimes \mathcal{H}_2^o \otimes \cdots \otimes \mathcal{H}_n^o.$$

$\mathcal{H}_n^o$ is composed of all vectors

$$|\psi_r^n\rangle = \int_{\mathcal{H}_n^o} \psi(\mathbf{r}_1, \mathbf{r}_2, \ldots, \mathbf{r}_n) |\mathbf{r}_1\rangle \times |\mathbf{r}_2\rangle \times \cdots \times |\mathbf{r}_n\rangle \, d\mathbf{r}_1 d\mathbf{r}_2 \cdots d\mathbf{r}_n$$

where $\psi(\mathbf{r}_1, \mathbf{r}_2, \ldots, \mathbf{r}_n) = \langle \mathbf{r}_1, \mathbf{r}_2, \ldots, \mathbf{r}_n|\psi_r^n\rangle$ are square integrable Fourier coefficients

$$\int_{\mathcal{H}_n^o} |\psi(\mathbf{r}_1, \mathbf{r}_2, \ldots, \mathbf{r}_n)|^2 \, d\mathbf{r}_1 d\mathbf{r}_2 \cdots d\mathbf{r}_n < +\infty.$$

The position and momentum operators $\hat{\mathbf{r}}_i$ and $\hat{\mathbf{p}}_i$ act on the vectors $|\psi_r^n\rangle \in \mathcal{H}_n^o$ in the following way:

$$\hat{\mathbf{r}}_i|\psi_r^n\rangle = \int_{\mathcal{H}_n^o} \{\hat{\mathbf{r}}_i\} \psi(\mathbf{r}_1, \mathbf{r}_2, \ldots, \mathbf{r}_n) |\mathbf{r}_1\rangle \times |\mathbf{r}_2\rangle \times \cdots \times |\mathbf{r}_n\rangle \, d\mathbf{r}_1 d\mathbf{r}_2 \cdots d\mathbf{r}_n,$$

$$\hat{\mathbf{p}}_i|\psi_r^n\rangle = \int_{\mathcal{H}_n^o} \{-i\hbar\partial_{\hat{\mathbf{r}}_i}\} \psi(\mathbf{r}_1, \mathbf{r}_2, \ldots, \mathbf{r}_n) |\mathbf{r}_1\rangle \times |\mathbf{r}_2\rangle \times \cdots \times |\mathbf{r}_n\rangle \, d\mathbf{r}_1 d\mathbf{r}_2 \cdots d\mathbf{r}_n,$$

with the square integrable Fourier coefficients

$$\int_{\mathcal{H}_n^o} |\{\hat{\mathbf{r}}_i\}\, \psi\,(\mathbf{r}_1,\mathbf{r}_2,\ldots,\mathbf{r}_n)|^2 \; d\mathbf{r}_1 d\mathbf{r}_2 \cdots d\mathbf{r}_n < +\infty,$$

$$\int_{\mathcal{H}_n^o} |\{-i\hbar\partial_{\mathbf{r}_i}\}\, \psi\,(\mathbf{r}_1,\mathbf{r}_2,\ldots,\mathbf{r}_n)|^2 \; d\mathbf{r}_1 d\mathbf{r}_2 \cdots d\mathbf{r}_n < +\infty,$$

respectively. In general, any set of vector Hermitian operators $\{\hat{\mathbf{A}}_i\}$ corresponding to all the particles, act on the vectors $|\psi_r^n\rangle \in \mathcal{H}_n^o$ in the following way:

$$\hat{\mathbf{A}}_i|\psi_r^n\rangle = \int_{\mathcal{H}_n^o} \{\hat{\mathbf{A}}_i\}\psi\,(\mathbf{r}_1,\mathbf{r}_2,\ldots,\mathbf{r}_n)\,|\mathbf{r}_1\rangle\times|\mathbf{r}_2\rangle\times\cdots\times|\mathbf{r}_n\rangle\,d\mathbf{r}_1 d\mathbf{r}_2 \cdots d\mathbf{r}_n,$$

with the square integrable Fourier coefficients

$$\int_{\mathcal{H}_n^o} \left|\{\hat{\mathbf{A}}_i\}\psi\,(\mathbf{r}_1,\mathbf{r}_2,\ldots,\mathbf{r}_n)\right|^2 \; d\mathbf{r}_1 d\mathbf{r}_2 \cdots d\mathbf{r}_n < +\infty.$$

3.1.6 Quantum Fourier Transform

Recall that in *quantum computing*, the *quantum Fourier transform* (QFT) is a linear transformation on *quantum bits* (or, *qubits*), and is the quantum analogue of the *discrete Fourier transform* (DFT). The QFT is a part of many quantum algorithms, like the *Shor algorithm* for factoring. It acts not upon the data stored in the quantum system state, but upon the quantum state itself.

Formally, the QFT is the classical DFT applied to the vector of amplitudes of a quantum state. Recall that the classical (unitary) Fourier transform acts on a vector $(\alpha_0,\ldots,\alpha_{N-1}) \in \mathbb{C}^N$ and maps it to the vector $(\overline{\alpha}_0,\ldots,\overline{\alpha}_{N-1}) \in \mathbb{C}^N$ by

$$\overline{\alpha}_k = \frac{1}{\sqrt{N}} \sum_{j=0}^{N-1} \alpha_j \omega^{jk},$$

where $\omega = e^{\frac{2\pi i}{N}}$ is a primitive Nth root of unity.

In a similar way, the QFT acts on the amplitudes of a quantum state $\sum_{i=0}^{N-1}\alpha_i|i\rangle$ and maps it to a quantum state $\sum_{i=0}^{N-1}\overline{\alpha}_i|i\rangle$ by expression

$$\overline{\alpha}_k = \frac{1}{\sqrt{N}} \sum_{j=0}^{N-1} \alpha_j \omega^{jk},$$

which defines the following map:

$$|j\rangle \mapsto \frac{1}{\sqrt{N}} \sum_{k=0}^{N-1} \omega^{jk}|k\rangle.$$

In terms of unitary matrices (see Appendix), the QFT can be viewed as a unitary matrix F_N, given by

$$
F_N = \frac{1}{\sqrt{N}}
\begin{bmatrix}
1 & 1 & 1 & 1 & \cdots & 1 \\
1 & \omega & \omega^2 & \omega^3 & \cdots & \omega^{N-1} \\
1 & \omega^2 & \omega^4 & \omega^6 & \cdots & \omega^{2(N-1)} \\
1 & \omega^3 & \omega^6 & \omega^9 & \cdots & \omega^{3(N-1)} \\
\vdots & \vdots & \vdots & \vdots & & \vdots \\
1 & \omega^{N-1} & \omega^{2(N-1)} & \omega^{3(N-1)} & \cdots & \omega^{(N-1)(N-1)}
\end{bmatrix},
$$

—acting on quantum state vectors.

Most of the properties of the quantum Fourier transform follow from the fact that it is a unitary transformation (see Appendix). From the unitary property it follows that the inverse of the quantum Fourier transform is the Hermitian adjoint of the Fourier matrix, $F^{-1} = F^\dagger$. There is an efficient *quantum circuit* implementing the quantum Fourier transform, the circuit can be run in reverse to perform the inverse quantum Fourier transform. Thus both transforms can be efficiently performed on a *quantum computer*.

3.2 Introduction to Quantum Fields

3.2.1 Amplitude, Relativistic Invariance and Causality

We will see later that in quantum field theory (QFT), the fundamental quantity is not any more Schrödinger's *wave-function* but the rather the closely related, yet different, Dirac–Feynman's *amplitude*. To introduce the amplitude concept within the non-relativistic quantum mechanics, suppose that $\mathbf{x} = (x, y, z)$ and consider the amplitude for a free particle to propagate in time t from $\mathbf{x}_0$ to $\mathbf{x} = \mathbf{x}(t)$, which is given by

$$
U(t) = \left\langle \mathbf{x} \right| e^{-iHt} \left| \mathbf{x}_0 \right\rangle.
$$

As the kinetic energy of a free particle is $E = \mathbf{p}^2/2m$, we have

$$
U(t) = \left\langle \mathbf{x} \right| e^{-i(\mathbf{p}^2/2m)t} \left| \mathbf{x}_0 \right\rangle = \int d^3 p \left\langle \mathbf{x} \right| e^{-i(\mathbf{p}^2/2m)t} \left| \mathbf{p} \right\rangle \left\langle \mathbf{p} | \mathbf{x}_0 \right\rangle
$$

$$
= \frac{1}{(2\pi)^2} \int d^3 p \, e^{-i(\mathbf{p}^2/2m)t} \cdot e^{i\mathbf{p}\cdot(\mathbf{x}-\mathbf{x}_0)} = \left(\frac{m}{2\pi i t} \right)^{3/2} e^{im(\mathbf{x}-\mathbf{x}_0)^2/2t}.
$$

Later we will deal with the amplitude in the relativistic framework.

As we have seen in the previous section, in non-relativistic quantum mechanics observables are represented by self-adjoint operators that in the Heisenberg picture depend on time, so that for any *quantum observable* $\mathcal{O}$, the Heisenberg equation of motion holds (in natural units):

$$i\partial_t \mathcal{O} = [\mathcal{O}, H]. \tag{3.9}$$

Therefore measurements are localized in time but are global in space. The situation is radically different in the relativistic case. Because no signal can propagate faster than the speed of light, measurements have to be localized both in time and space. Causality demands then that two measurements carried out in causally-disconnected regions of space-time cannot interfere with each other. In mathematical terms this means that if $\mathcal{O}_{R_1}$ and $\mathcal{O}_{R_2}$ are the observables associated with two measurements localized in two causally-disconnected regions R_1, R_2, they satisfy the *commutator* relation [PS95]

$$[\mathcal{O}_{R_1}, \mathcal{O}_{R_2}] = 0, \quad \text{if } (x_1 - x_2)^2 < 0, \text{ for all } x_1 \in R_1, \; x_2 \in R_2. \tag{3.10}$$

Hence, in a relativistic theory, the basic operators in the Heisenberg picture must depend on the space-time position x^μ. Unlike the case in non-relativistic quantum mechanics, here the position x *is not* an observable, but just a label, similarly to the case of time in ordinary quantum mechanics. Causality is then imposed microscopically by requiring

$$[\mathcal{O}(x), \mathcal{O}(y)] = 0, \quad \text{if } (x - y)^2 < 0. \tag{3.11}$$

A smeared operator $\mathcal{O}_R$ over a space-time region R can then be defined as

$$\mathcal{O}_R = \int d^4 x \, \mathcal{O}(x) f_R(x),$$

where $f_R(x)$ is the *characteristic function* associated with R,

$$f_R(x) = \begin{cases} 1, & \text{for } x \in R, \\ 0, & \text{for } x \notin R. \end{cases}$$

Equation (3.10) follows now from the micro-causality condition (3.11).

Therefore, relativistic invariance forces the introduction of quantum fields. It is only when we insist in keeping a single-particle interpretation that we crash against causality violations. To illustrate the point, let us consider a single particle wave function $\psi(t, x)$ that initially is localized in the position $x = 0$

$$\psi(0, x) = \delta(x).$$

Evolving this wave function using the Hamiltonian $H = \sqrt{-\nabla^2 + m^2}$ we find that the wave function can be written as

$$\psi(t, x) = e^{-it\sqrt{-\nabla^2 + m^2}} \delta(x) = \int \frac{d^3 p}{(2\pi)^3} e^{ip \cdot x - it\sqrt{p^2 + m^2}}.$$

Integrating over the angular variables, the wave function can be recast in the form

$$\psi(t, x) = \frac{1}{2\pi^2 |x|} \int_{-\infty}^{\infty} p \, dk \, e^{ip|x|} e^{-it\sqrt{p^2 + m^2}}.$$

The resulting integral can be evaluated using the complex integration contour C. The result is that, for any $t > 0$, one finds that $\psi(t, x) \neq 0$ for any x. If we insist in interpreting the wave function $\psi(t, x)$ as the probability density of finding the particle at the location x in the time t we find that the probability leaks out of the light cone, thus violating causality.

In the Heisenberg picture, the amplitude for a particle to propagate from point y to point x in the *field* ϕ is defined as

$$D(x - y) = \langle 0|\phi(x)\phi(y)|0\rangle = \int \frac{d^3p}{(2\pi)^3} \frac{1}{2E} e^{-ip\cdot(x-y)}.$$

In this picture we can make time-dependent the *field operator* $\phi = \phi(x)$ and its canonically-conjugate *momentum operator* $\pi = \pi(x)$, as

$$\phi(x) = \phi(\mathbf{x}, t) = e^{iHt}\phi(\mathbf{x})e^{-iHt}, \qquad \pi(x) = \pi(\mathbf{x}, t) = e^{iHt}\pi(\mathbf{x})e^{-iHt}.$$

Using (3.9) we can compute the time dependence of ϕ and π as

$$i\dot{\phi}(\mathbf{x}, t) = i\pi(\mathbf{x}, t), \qquad i\dot{\pi}(\mathbf{x}, t) = -i(-\nabla^2 + m^2)\phi(\mathbf{x}, t).$$

Combining the two results we get the *Klein–Gordon equation*

$$\ddot{\phi} = (\nabla^2 - m^2)\phi.$$

3.2.2 Gauge Theories

Recall that a *gauge theory* is a theory that admits a symmetry with a local parameter. For example, in every quantum theory the global phase of the wave ψ-function is arbitrary and does not represent something physical. Consequently, the theory is invariant under a global change of phases (adding a constant to the phase of all wave functions, everywhere); this is a global symmetry. In quantum electrodynamics, the theory is also invariant under a local change of phase, that is, one may shift the phase of all wave functions so that the shift may be different at every point in space-time. This is a local symmetry. However, in order for a well-defined derivative operator to exist, one must introduce a new field, the *gauge field*, which also transforms in order for the local change of variables (the phase in our example) not to affect the derivative. In quantum electrodynamics this gauge field is the electromagnetic potential 1-form A (see Appendix), in components within the nD coframe $\{dx^\mu\}$ on a smooth manifold M (dual to the frame, i.e., basis of tangent vectors $\{\partial_\mu = \partial/\partial x^\mu\}$), given by

$$A = A_\mu dx^\mu, \quad \text{such that} \quad A_{\text{new}} = A_{\text{old}} + df, \quad (f \text{ is any scalar function})$$

—leaves the electromagnetic field 2-form $F = dA$ unchanged. This change df of local gauge of variable A is termed *gauge transformation*. In quantum field theory the excitations of fields represent particles. The particle associated with

excitations of the gauge field is the *gauge boson*. All the fundamental interactions in nature are described by gauge theories. In particular, in quantum
electrodynamics, whose gauge transformation is a local change of phase, the
gauge group is the circle group $U(1)$ (consisting of all complex numbers with
absolute value 1), and the gauge boson is the photon (see e.g., [Fra86]).

The *gauge field* of classical electrodynamics, given in local coframe $\{dx^\mu\}$
on M as an electromagnetic potential 1-form

$$A = A_\mu dx^\mu = A_\mu dx^\mu + df, \quad (f = \text{arbitrary scalar field}),$$

is globally a *connection* on a $U(1)$-bundle of M.[8] The corresponding electromagnetic field, locally the 2-form on M,

$$F = dA, \quad \text{in components given by}$$

$$F = \frac{1}{2}F_{\mu\nu}\, dx^\mu \wedge dx^\nu, \quad \text{with } F_{\mu\nu} = \partial_\nu A_\mu - \partial_\mu A_\nu$$

is globally the *curvature* of the connection A[9] under the gauge-covariant
derivative,

---

[8] Recall that in the 19th Century, Maxwell unified Faraday's electric and magnetic
fields. Maxwell's theory led to Einstein's special relativity where this unification
becomes a spin-off of the unification of space end time in the form of the *Faraday
tensor* [MTW73]

$$F = E \wedge dt + B,$$

where F is electromagnetic 2-form on space-time, E is electric 1-form on space,
and B is magnetic 2-form on space. Gauge theory considers F as secondary object
to a connection-potential 1-form A. This makes half of the Maxwell equations into
tautologies [BM94], i.e.,

$$F = dA \implies dF = 0: \text{ Bianchi identity},$$

but does not imply the second half of Maxwell's equations,

$$\delta F = -4\pi J: \text{ dual Bianchi identity}.$$

To understand the deeper meaning of the connection-potential 1-form A, we can
integrate it along a path γ in space-time, $x \xrightarrow{\gamma} y$. Classically, the integral $\int_\gamma A$
represents an *action* for a charged point particle to move along the path γ. Quantum-
mechanically, $\exp(i \int_\gamma A)$ represents a *phase* (within the unitary Lie group $U(1)$) by
which the particle's wave-function changes as it moves along the path γ, so A is a
$U(1)$-connection.

In other words, Maxwell's equations can be formulated using complex line bundles, or principal bundles with fibre $U(1)$. The connection ∇ on the line bundle has
a curvature $F = \nabla^2$ which is a 2-form that automatically satisfies $dF = 0$ and can
be interpreted as a field-strength. If the line bundle is trivial with flat reference
connection d, we can write $\nabla = d + A$ and $F = dA$ with A the 1-form composed of
the electric potential and the magnetic vector potential.

[9] The only thing that matters here is the *difference* α between two paths γ_1 and γ_2
in the action $\int_\gamma A$ [BM94], which is a 2-morphism (see e.g., [II06c, II07e])

$$D_\mu = \partial_\mu - ieA_\mu, \tag{3.12}$$

where e is the charge coupling constant.[10] In particular, in 4D space-time electrodynamics, the 1-form *electric current density* J has the components $J_\mu = (\rho, \mathbf{j}) = (\rho, j_x, j_y, j_z)$ (where ρ is the charge density), the 2-form *Faraday* F is given in components of electric field $\mathbf{E}$ and magnetic field $\mathbf{B}$ by

$$F_{\mu\nu} = \begin{pmatrix} 0 & E_x & E_y & E_z \\ -E_x & 0 & -B_z & B_y \\ -E_y & B_z & 0 & -B_x \\ -E_z & -B_y & B_x & 0 \end{pmatrix}, \quad \text{with } F_{\nu\mu} = -F_{\mu\nu},$$

while its dual 2-form *Maxwell* $\star F$ has the following components

$$\star F_{\mu\nu} = \begin{pmatrix} 0 & -B_x & -B_y & -B_z \\ B_x & 0 & -E_z & E_y \\ B_y & E_z & 0 & -E_x \\ B_z & -E_y & B_x & 0 \end{pmatrix}, \quad \text{with } \star F_{\nu\mu} = -\star F_{\mu\nu},$$

so that classical electrodynamics is governed by the *Maxwell equations*, which in modern exterior formulation read

$$dF = 0, \qquad \delta F = -4\pi J, \quad \text{or in components,}$$
$$F_{[\mu\nu,\eta]} = 0, \qquad F_{\mu\nu},^\mu = -4\pi J_\mu,$$

_____

[10] If a gauge transformation is given by

$$\psi \mapsto e^{i\Lambda}\psi$$

and for the gauge potential

$$A_\mu \mapsto A_\mu + \frac{1}{e}(\partial_\mu \Lambda),$$

then the gauge-covariant derivative,

$$D_\mu = \partial_\mu - ieA_\mu$$

transforms as

$$D_\mu \mapsto \partial_\mu - ieA_\mu - i(\partial_\mu \Lambda)$$

and $D_\mu \psi$ transforms as

$$D_\mu \mapsto \partial_\mu - ieA_\mu - i(\partial_\mu \Lambda).$$

where $\star$ is the Hodge star operator and δ is the Hodge codiferential (see Sect. 3.4.2 below), comma denotes the partial derivative and the 1-form of electric current $J = J_\mu dx^\mu$ is conserved, by the electrical *continuity equation*,

$$\delta J = 0, \quad \text{or in components,} \quad J_{\mu,}{}^\mu = 0.$$

The first, sourceless Maxwell equation, $dF = 0$, gives vector magnetostatics and magnetodynamics,

$$\text{Magnetic Gauss' law:} \quad \text{div } \mathbf{B} = 0,$$
$$\text{Faraday's law:} \quad \partial_t \mathbf{B} + \text{curl } \mathbf{E} = 0.$$

The second Maxwell equation with source, $\delta F = J$, gives vector electrostatics and electrodynamics,

$$\text{Electric Gauss' law:} \quad \text{div } \mathbf{E} = 4\pi\rho,$$
$$\text{Ampère's law:} \quad \partial_t \mathbf{E} - \text{curl } \mathbf{B} = -4\pi\mathbf{j}.$$

The standard *Lagrangian* for the free electromagnetic field, $F = dA$, is given by [II06c, II07e, II08b]

$$\mathcal{L}(A) = \frac{1}{2}(F \wedge \star F),$$

with the corresponding *action functional*

$$S(A) = \frac{1}{2} \int F \wedge \star F.$$

Maxwell's equations are generally applied to macroscopic averages of the fields, which vary wildly on a microscopic scale in the vicinity of individual atoms, where they undergo quantum effects as well (see below).

3.2.3 Free and Interacting Field Theories

A generic *gauge-covariant derivative* with Lorentz index μ is denoted by D_μ. For a Maxwell field, D_μ is given by (3.12).

$$\text{Dirac slash notation:} \quad \partial\!\!\!/ \overset{\text{def}}{=} \gamma^\mu \partial_\mu, \quad D\!\!\!\!/\, \overset{\text{def}}{=} \gamma^\mu D_\mu,$$
$$\text{Dirac algebra:} \quad \{\gamma^\mu, \gamma^\nu\} = \gamma^\mu\gamma^\nu + \gamma^\nu\gamma^\mu = 2g^{\mu\nu} \times \mathbf{1}_{n\times n}.$$

Standard free theories are Klein–Gordon and Dirac fields:

$$\text{Klein–Gordon equation:} \quad (\partial^2 + m^2)\psi = 0,$$
$$\text{Dirac equation:} \quad (i\partial\!\!\!/ - m)\psi = 0.$$

Two main examples of interacting theories are ϕ^4-theory and QED:

1. ϕ^4-theory:

$$\text{Lagrangian:} \quad \mathcal{L} = \frac{1}{2}(\partial_\mu \phi)^2 - \frac{1}{2}m^2\phi^2 - \frac{\lambda}{4!}\phi^4,$$

$$\text{Equation of motion:} \quad (\partial^2 + m^2)\phi = -\frac{\lambda}{3!}\phi^3.$$

2. QED:

$$\text{Lagrangian:} \quad \mathcal{L} = \mathcal{L}_{\text{Maxwell}} + \mathcal{L}_{\text{Dirac}} + \mathcal{L}_{\text{int}}$$

$$= -\frac{1}{4}(F_{\mu\nu})^2 + \bar{\psi}(i\slashed{\partial} - m)\psi.$$

$$\text{Gauge invariance:} \quad \psi(x) \to e^{i\alpha(x)}\psi(x) \implies A_\mu \to A_\mu - \frac{1}{e}\partial_\mu \alpha(x).$$

$$\text{Equation of motion:} \quad (i\slashed{\partial} - m)\psi = 0.$$

3.2.4 Dirac's Quantum Electrodynamics (QED)

The *Dirac equation* for a particle with mass m (in natural units) reads (see, e.g., [II08b])

$$(i\gamma^\mu \partial_\mu - m)\psi = 0, \quad (\mu = 0, 1, 2, 3) \tag{3.13}$$

where $\psi(x)$ is a 4-component spinor[11] wave-function, the so-called Dirac spinor, while γ^μ are 4×4 *Dirac γ-matrices*,

$$\gamma^0 = \begin{pmatrix} 1 & 0 & 0 & 0 \\ 0 & 1 & 0 & 0 \\ 0 & 0 & -1 & 0 \\ 0 & 0 & 0 & -1 \end{pmatrix}, \quad \gamma^1 = \begin{pmatrix} 0 & 0 & 0 & 1 \\ 0 & 0 & 1 & 0 \\ 0 & -1 & 0 & 0 \\ -1 & 0 & 0 & 0 \end{pmatrix},$$

$$\gamma^2 = \begin{pmatrix} 0 & 0 & 0 & -i \\ 0 & 0 & i & 0 \\ 0 & i & 0 & 0 \\ -i & 0 & 0 & 0 \end{pmatrix}, \quad \gamma^3 = \begin{pmatrix} 0 & 0 & 1 & 0 \\ 0 & 0 & 0 & -1 \\ -1 & 0 & 0 & 0 \\ 0 & 1 & 0 & 0 \end{pmatrix}.$$

They obey the *anticommutation relations*

$$\{\gamma^\mu, \gamma^\nu\} = \gamma^\mu \gamma^\nu + \gamma^\nu \gamma^\mu = 2g^{\mu\nu},$$

where $g_{\mu\nu}$ is the metric tensor.

Dirac's γ-matrices are conventionally derived as

---

[11] The most convenient definitions for the 2-spinors, like the Dirac spinor, are:

$$\phi^1 = \begin{bmatrix} 1 \\ 0 \end{bmatrix}, \quad \phi^2 = \begin{bmatrix} 0 \\ 1 \end{bmatrix} \quad \text{and} \quad \chi^1 = \begin{bmatrix} 0 \\ 1 \end{bmatrix}, \quad \chi^2 = \begin{bmatrix} 1 \\ 0 \end{bmatrix}.$$

$$\gamma^k = \begin{pmatrix} 0 & \sigma^k \\ -\sigma^k & 0 \end{pmatrix}, \quad (k = 1, 2, 3)$$

where σ^k are *Pauli sigma matrices*[12] (a set of 2×2 complex Hermitian and unitary matrices), defined as

$$\sigma_1 = \sigma_x = \begin{pmatrix} 0 & 1 \\ 1 & 0 \end{pmatrix}, \quad \sigma_2 = \sigma_y = \begin{pmatrix} 0 & -i \\ i & 0 \end{pmatrix}, \quad \sigma_3 = \sigma_z = \begin{pmatrix} 1 & 0 \\ 0 & -1 \end{pmatrix},$$

obeying both the commutation and anticommutation relations

$$[\sigma_i, \sigma_j] = 2i\varepsilon_{ijk}\sigma_k, \quad \{\sigma_i, \sigma_j\} = 2\delta_{ij} \cdot I,$$

where ε_{ijk} is the Levi-Civita symbol, δ_{ij} is the Kronecker delta, and I is the identity matrix.

Now, the Lorentz-invariant form of the Dirac equation (3.13) for an electron with a charge e and mass m_e, moving with a 4-momentum 1-form $p = p_\mu dx^\mu$ in a classical electromagnetic field defined by 1-form $A = A_\mu dx^\mu$, reads (see, e.g., [II08b]):

$$\{i\gamma^\mu [p_\mu - eA_\mu] - m_e\} \psi(x) = 0, \tag{3.14}$$

and is called the *covariant Dirac equation*.

The formal QED Lagrangian (density) includes three terms,

$$\mathcal{L}(x) = \mathcal{L}_{em}(x) + \mathcal{L}_{int}(x) + \mathcal{L}_{e-p}(x), \tag{3.15}$$

related respectively to the free electromagnetic field 2-form $F = F_{\mu\nu}dx^\mu \wedge dx^\nu$, the electron–positron field (in the presence of the external vector potential 1-form A_μ^{ext}), and the interaction field (dependent on the charge-current 1-form $J = J_\mu dx^\mu$). The free electromagnetic field Lagrangian in (3.15) has the standard electrodynamic form

$$\mathcal{L}_{em}(x) = -\frac{1}{4}F^{\mu\nu} F_{\mu\nu},$$

where the electromagnetic fields are expressible in terms of components of the potential 1-form $A = A_\mu dx^\mu$ by

$$F_{\mu\nu} = \partial_\mu A_\nu^{tot} - \partial_\nu A_\mu^{tot}, \quad \text{with } A_\mu^{tot} = A_\mu^{ext} + A_\mu.$$

The electron–positron field Lagrangian is given by Dirac's equation (3.14) as

---

[12] In quantum mechanics, each Pauli matrix represents an observable describing the spin of a spin $\frac{1}{2}$ particle in the three spatial directions. Also, $i\sigma_j$ are the generators of rotation acting on non-relativistic particles with spin $\frac{1}{2}$. The state of the particles are represented as two-component spinors.

In *quantum information*, single-qubit quantum gates are 2×2 unitary matrices. The Pauli matrices are some of the most important single-qubit operations.

$$\mathcal{L}_{e-p}(x) = \bar{\psi}(x) \left\{ i\gamma^{\mu} \left[p_{\mu} - eA_{\mu}^{\text{ext}} \right] - m_{e} \right\} \psi(x),$$

where $\bar{\psi}(x)$ is the Dirac adjoint spinor wave function.

The interaction field Lagrangian

$$\mathcal{L}_{\text{int}}(x) = -J^{\mu} A_{\mu},$$

accounts for the interaction between the uncoupled electrons and the radiation field.

The field equations deduced from (3.15) read

$$\left\{ i\gamma^{\mu} \left[p_{\mu} - eA_{\mu}^{\text{ext}} \right] - m_{e} \right\} \psi(x) = \gamma^{\mu} \psi(x) A_{\mu},$$
$$\partial^{\mu} F_{\mu\nu} = J_{\nu}. \tag{3.16}$$

The formal QED requires the solution of the system (3.16) when $A^{\mu}(x)$, $\psi(x)$ and $\bar{\psi}(x)$ are quantized fields.

3.2.5 Abelian Higgs Model

The Abelian[13] Higgs model is an example of gauge theory used in particle and condensed matter physics. Besides the electromagnetic field it contains a self-interacting scalar field, the so-called *Higgs field*, minimally coupled to electromagnetism. From the conceptual point of view, it is advantageous to consider this field theory in $(2+1)$D space-time and to extend it subsequently to $(3+1)$D for applications. The Abelian Higgs Lagrangian reads [Len04]

$$\mathcal{L} = -\frac{1}{4} F_{\mu\nu} F^{\mu\nu} + (D_{\mu}\phi)^{*}(D^{\mu}\phi) - V(\phi),$$

which contains the complex (charged), self-interacting scalar field ϕ. The Higgs potential is the *Mexican hat* function of the real and imaginary part of the Higgs field,

$$V(\phi) = \frac{1}{4} \lambda (|\phi|^{2} - a^{2})^{2}.$$

By construction, this Higgs potential is minimal along a circle $|\phi| = a$ in the complex ϕ plane. The constant λ controls the strength of the self-interaction of the Higgs field and, for stability reasons, is assumed to be positive, $\lambda \geq 0$. The Higgs field is minimally coupled to the radiation field A_{μ}, i.e., the partial derivative ∂_{μ} is replaced by the covariant derivative, $D_{\mu} = \partial_{\mu} + ieA_{\mu}$. Gauge fields and field strengths are related by

$$F_{\mu\nu} = \partial_{\mu} A_{\nu} - \partial_{\nu} A_{\mu} = \frac{1}{ie} [D_{\mu}, D_{\nu}].$$

---

[13] An *Abelian* (or, commutative) *group* (even better, Lie group, see Appendix), is such a group G that satisfies the condition: $a \cdot b = b \cdot a$ for all $a, b \in G$. In other words, its *commutator*, $[a, b] := a^{-1}b^{-1}ab$ equals the identity element.

The inhomogeneous Maxwell equations are obtained from the least action principle,

$$\delta S = 0, \quad \text{with } S = \int \mathcal{L} d^4 x = 0,$$

by variation of the action S with respect to the gauge fields A_μ (and their derivatives $\partial_\mu A_\mu$). With

$$\frac{\delta \mathcal{L}}{\delta \partial_\mu A_\nu} = -F^{\mu\nu}, \qquad \frac{\delta \mathcal{L}}{\delta A_\nu} = -j^\nu, \quad \text{we get}$$

$$\partial_\mu F^{\mu\nu} = j^\nu, \qquad j_\nu = ie(\phi^* \partial_\nu \phi - \phi \partial_\nu \phi^*) - 2e^2 \phi^* \phi A_\nu.$$

We remark here that the homogeneous Maxwell equations are not dynamical equations of motion. They are integrability conditions and guarantee that the field strength can be expressed in terms of the gauge fields. The homogeneous equations follow from the *Jacobi identity* of the covariant derivative

$$[D_\mu, [D_\nu, D_\sigma]] + [D_\sigma, [D_\mu, D_\nu]] + [D_\nu, [D_\sigma, D_\mu]] = 0.$$

Multiplication with the totally antisymmetric 4-tensor $\epsilon^{\mu\nu\rho\sigma}$, yields the homogeneous equations for the dual field strength $\tilde{F}^{\mu\nu}$

$$\left[D_\mu, \tilde{F}^{\mu\nu} \right] = 0, \qquad \tilde{F}^{\mu\nu} = \frac{1}{2} \epsilon^{\mu\nu\rho\sigma} F_{\rho\sigma}.$$

The transition: $F \longrightarrow \tilde{F}$ corresponds to the following duality relation of electric and magnetic fields, $\mathbf{E} \longrightarrow \mathbf{B}, \mathbf{B} \longrightarrow -\mathbf{E}$.

Variation with respect to the charged matter field yields the equation of motion:

$$D_\mu D^\mu \phi + \frac{\delta V}{\delta \phi^*} = 0.$$

Gauge theories contain redundant variables. This redundancy manifests itself in the presence of local symmetry transformations, or gauge transformations, $U(x) = e^{ie\alpha(x)}$, which rotate the phase of the matter field and shift the value of the gauge field in a space-time dependent manner

$$\phi \longrightarrow \phi^{[U]} = U(x)\phi(x), \qquad A_\mu \longrightarrow A_\mu^{[U]} = A_\mu + U(x)\frac{1}{ie}\partial_\mu U^\dagger(x). \quad (3.17)$$

The covariant derivative D_μ has been defined such that $D_\mu \phi$ transforms covariantly, i.e., like the matter field ϕ itself.

$$D_\mu \phi(x) \longrightarrow U(x) D_\mu \phi(x).$$

This transformation property together with the invariance of $F_{\mu\nu}$ guarantees invariance of $\mathcal{L}$ and of the equations of motion. A gauge field which is gauge equivalent to $A_\mu = 0$ is called a pure gauge. According to (3.17) a pure gauge satisfies

$$A_\mu^{pg}(x) = U(x)\frac{1}{\mathrm{i}e}\partial_\mu U^\dagger(x) = -\partial_\mu\alpha(x),$$

and the corresponding field strength vanishes.

Note that the non-Abelian Higgs model has the action:

$$S(\phi, A) = \frac{1}{4}\int \mathrm{Tr}(F^{\mu\nu}F_{\mu\nu}) + |D\phi|^2 + V(|\phi|),$$

where now the non-Abelian field A is contained both in the covariant derivative D and in the components $F^{\mu\nu}$ and $F_{\mu\nu}$ (see Yang–Mills theory below).

3.2.6 Topological Quantum Computation

Recall from Introduction that quantum computation encompasses several models of computation based on a theoretical ability to manufacture, manipulate and measure quantum states. In this subsection, following [FKL03b], we recall that there are three areas where remarkable algorithms have been found: searching a data base, Abelian groups (factoring and discrete logarithm), and simulating physical systems. To this list we may add a fourth class of algorithms which yield approximate, but rapid, evaluations of many quantum invariants of 3D manifolds, e.g., the absolute value of the Jones polynomial of a link L at certain roots of unity: $|V_L(\mathrm{e}^{\frac{2\pi i}{5}})|$. This seeming curiosity is actually the 'tip of an iceberg' which links quantum computation both to low-dimensional topology and the theory of anyons; the motion of anyons in a two dimensional system defines a braid in $(2+1)$D. This iceberg is a model of quantum computation based on topological, rather than local, degrees of freedom.

The class of functions, BQP (functions computable with bounded error, given quantum resources, in polynomial time), has been defined in three distinct but equivalent ways: via *quantum Turing machines*, *quantum circuits*, and *modular functors* (see section on categories and functors in Appendix). The last is the subject of this article. We may now propose a 'thesis' in the spirit of Alonzo Church: all 'reasonable' computational models which add the resources of quantum mechanics, or quantum field theory, to classical computation yield efficiently inter-simulable classes: there is one quantum theory of computation.

The case for quantum computation rests on three pillars: inevitability, *Moore's law* suggests we will soon be doing it whether we want to or not, desirability, the above mentioned algorithms, and finally feasibility, which in the past has been argued from combinatorial fault tolerant protocols [FKL03b].

Focusing on feasibility, we present a model of quantum computation in which information is stored and manipulated in 'topological degrees of freedom' rather than in localized degrees. The usual candidates for storing information are either localized in space (e.g., an electron or a nuclear spin) or localized in momentum (e.g., photon polarization.) Almost by definition (see 'code subspace' below) a topological degree of freedom is protected from local errors. In the presence of perturbation (of the system Hamiltonian) this protection will not be perfect, but physical arguments suggest undesirable tunneling amplitudes between orthogonal ground states will scale like $e^{-\alpha l}$, where l is a length scale for the system, e.g., the minimum separation maintained between point-like anyonic excitations. We will return to this crucial point.

But let us take a step backward and discuss the standard quantum circuit model and the presumptive path toward its physical implementation. To specify a quantum circuit Γ, we begin with a *tensor product* $\mathbb{C}_1^2 \otimes \cdots \otimes \mathbb{C}_n^2$ of n copies of $\mathbb{C}^2$, called *qubits*. Physically, this models a system of n non-interacting spin $= \frac{1}{2}$ particles. The circuit then consists of a sequence of K 'gates' $U_k, (1 \leq k \leq K)$, applied to individual or paired tensor factors. A gate is some coherent interaction; mathematically it is a unitary transformation on either $\mathbb{C}_i^2$ or $\mathbb{C}_i^2 \otimes \mathbb{C}_j^2, (1 \leq i, j \leq n)$, the identity on all remaining factors. The gates are taken from a fixed finite library of unitary 2×2 and 4×4 matrices (with respect to a fixed basis $\{|0\rangle, |1\rangle\}$ for each $\mathbb{C}^2$ factor) and must obey the surprisingly mild condition, called 'universality', that the set of possible gate applications generates the unitary group $U(2^n)$ densely (up to a physically irrelevant overall phase).

Popular choices include a *relative phase gate* $\begin{pmatrix} 1 & 0 \\ 0 & e^{\frac{2\pi i}{5}} \end{pmatrix}$ and CNOT $\begin{pmatrix} 1&0&0&0 \\ 0&1&0&0 \\ 0&0&0&1 \\ 0&0&1&0 \end{pmatrix}$ operating on one and two quasi-particles, respectively [FKL03b]. It is known that beyond the density requirement the particular choice of gates is not too important. Let $W_\Gamma = \prod_{i=1}^m U_i$ denote the operator effected by the circuit Γ. It is important for the fault tolerance theory that many gates can be applied simultaneously (to different qubits) without affecting the output of the circuit $W_\Gamma(|0\rangle \otimes \cdots \otimes |0\rangle)$.

Formally, information is extracted from the output by measuring the first qubit. The probability of observing $|1\rangle$ is given according to the usual axioms of quantum mechanics as [FKL03b]

$$p(\Gamma) = \langle 0|W_\Gamma^\dagger \Pi_1 W_\Gamma|0\rangle, \qquad (3.18)$$

where Π_1 is the projection to $|1\rangle$, $\begin{pmatrix} 0 & 0 \\ 0 & 1 \end{pmatrix}$, applied to the first qubit. Any *decision problem*, such as finding the kth binary digit of the largest prime factor of an integer x, can be modeled by a function F on binary strings, $F : \{0,1\}^* \rightarrow \{0,1\}$; in our example, the input string would encode (x,k). We say F belongs to *BQP* if there is a classical *polynomial-time algorithm* (in string length) for specifying a 'quantum circuit' $\Gamma(y)$ (in the example $y = (x,k)$) which satisfies [FKL03b]

$$p(\Gamma(y)) \geq \frac{2}{3} \quad \text{if } F(y) = 1 \quad \text{and}$$

$$p(\Gamma(y)) \leq \frac{1}{3} \quad \text{if } F(y) = 0.$$

This definition suggests that one needs to make an individual quantum circuit to solve each instance of a computational problem. However, it is possible to construct a single circuit to solve any instance of a given BQP problem of bounded size, e.g., factor integers with ≤ 1000 digits. Moreover, there is a universal circuit, univ.(n, k), which simulates all circuits of size k on n qubits:

$$W_{\text{univ.(n,k)}} \left(|0 \cdots 0\rangle \otimes |\Gamma\rangle \right) = W_\Gamma |0 \cdots 0\rangle \otimes |\Gamma\rangle.$$

Yet another definition allows one to do measurements in the middle of computation and choose the next unitary gate depending on the measurement outcome. This choice will generally involve some classical Boolean gates. This scheme is called *adaptive quantum computation*. In certain cases, it can squeeze general BQP computation out of a gate set which is *not* universal.

Implementation of a Quantum Computer

Recall from previous subsections that it is not possible to realize a unitary gate precisely, and even when we do not intend to apply a gate, the environment will interact with the qubits causing decoherence. Both imprecision and decoherence can be considered in a unified way as 'errors' which can be quantified by a fidelity distance or a super-operator norm. A crucial step in the theory of quantum computing has been the discovery of error-correcting quantum codes and fault-tolerant quantum computation. These techniques cope with sufficiently small errors. However, the error magnitude must be smaller than some constant (called *an accuracy threshold*) for these methods to work. According to rather optimistic estimates, this constant lies between 10^{-5} and 10^{-3}, beyond the reach of current technologies.

The presumptive path toward a quantum computer includes these steps [FKL03b]:

1. Build physical qubits and physical gates;
2. Minimize error level down below the threshold;
3. Implement decoherence-protected logical qubits and fault-tolerant logical gates, in which one logical qubit is realized by several qubits using an error-correcting code.

As a counterpoint, the theme of this article is that implementing physical qubits might be redundant. Indeed, one can 'encode' a logical qubit into a physical system which is not quite a set of qubits or even well separated into subsystems. Such a system must have macroscopic quantum degrees of freedom which would be decoherence-protected. A super-conducting phase or anything related to a local order parameter will not work for if a degree of freedom is accessible by local measurement, it is vulnerable to local error. However, there is another type of macroscopic quantum degree of freedom. It is related to topology and arises in collective electronic systems, e.g. the fractional quantum Hall effect and most recently in $2D$ cuprate superconductors above T_c.

Though much studied since the mid-1980's, the connection between *fractional quantum Hall effect* and quantum computation has only recently been realized. It was shown by that the ground state of the $\nu = \frac{1}{3}$ electron liquid on the torus is 3-fold degenerate. This follows from the fact that excitations in this system are Abelian anyons: moving one excitation around another multiplies the state vector by a phase factor $e^{i\phi}$ (in this case $\phi = \frac{2\pi}{3}$). The process of creating a particle-antiparticle pair, moving one of the particles around the torus, and annihilating it specifies a unitary operator on the ground state. By moving the particle in two different directions, one obtains two different unitary operators A_1 and A_2 with the commutation relation

$$A_1 A_2 A_1^{-1} A_2^{-1} = e^{i\phi},$$

implying a ground state degeneracy. This argument is very robust and only requires the existence of an energy gap or, equivalently, finite correlation length l_0. Indeed, the degeneracy is lifted only by spontaneous tunneling of virtual excitations around the torus. The resulting energy splitting scales as $e^{-\frac{l}{l_0}}$, where l is the size of the system. Interaction with the environment does not change this conclusion, although thermal noise can create actual excitation pairs rather than virtual ones.

Both the ground state degeneracy on the torus and the existence of anyons are manifestations of somewhat mysterious topological properties of the $\nu = \frac{1}{3}$ electron liquid itself. Anyons can be regarded as *topological defects* similar to Abrikosov vortices but without any local order parameter. The presence of a particle enclosed by a loop on the plane can be detected by *holonomy*, moving another particle around the loop.[14]

A more flexible and controllable way of storing quantum information is based on non-Abelian anyons. These are believed to exist in the $\nu = \frac{5}{2}$ fractional quantum Hall state. According to the theory, there should be charge $\frac{1}{4}$ anyonic particles and some other excitations. The quantum state of the system with $2n$ charge $\frac{1}{4}$ particles on the plane is 2^{n-1}-degenerate. The degeneracy is gradually lifted as two particles come close to each other. More precisely, the 2^{n-1}D Hilbert space $\mathbb{H}_n$ splits into two 2^{n-2}D subspaces. They correspond to two different types of charge $\frac{1}{2}$ particles which can result from fusion of charge $\frac{1}{4}$ particles. Thus observing the fusion outcome effects a measurement on the Hilbert space $\mathbb{H}_n$. This model supports adaptive quantum computation when surfaces of high genus are included in the theory and admits a combinatorial description apparently in the same universal class as the fractional quantum Hall fluid.

Beyond this, a discrete family of quantum Hall models exists based on $(k+1)$-fold hard-core interaction between electrons in a fixed Laudau level which

---

[14] At $\nu = \frac{1}{3}$, the electron liquid on the torus could be used as a *logical qutrit* (i.e., generalized qubit with 3 states). Unfortunately, this will hardly work in practice. Besides the obvious problem with implementing the torus topology, there is no known way to measure this logical qutrit or prepare it in a pure state.

appears to represent the same universality class as *Witten–Chern–Simons theory* (see Sect. 3.4.3) for $SU(2)$ at level $= k$. Anyons in these models behave as topological defects of a geometric construction and their braiding matrices have been shown to be universal for $k \geq 3$, $k \neq 4$.

Code Spaces and Quantum Media

Even after the particle types and positions of anyons are specified, there is an exponentially large (but finite-dimensional) Hilbert space describing topological degrees of freedom. In combinatorial models, this Hilbert space becomes a *code subspace* W of a larger 'quantum media' Y. Thus a fundamental concept of cryptography is transplanted into physics. Let V be a finite-dimensional complex vector space, perhaps $\mathbb{C}^2$, and $Y = V \otimes \cdots \otimes V$ an n-fold tensor product (where n is typically quite large). Let $W \subset Y$ be a linear subspace. We call $W \subset Y$ k-code iff $W \xrightarrow{\Pi_W} W$ is multiplication by a scalar whenever $\mathcal{O}$ is a k-local operator (an arbitrary linear map on any k-tensor factors of Y and the identity on the remaining $n - k$ factors) and Π_W is the orthogonal projection onto W. We think of such spaces as resisting local alteration and in the usual interpretation of Y as the Hilbert space of n particles, quantum information stored in W will be relatively secure. It is a theorem that the quantum information in W cannot be degraded by errors operating on fewer than $\frac{k}{2}$ of the n particles [FKL03b].

Let us define a (discrete) quantum medium to be a tensor product $Y = \bigotimes_i V_i$ as above, where now the set of indices $\{i\}$ consists of points distributed on a geometric surface T, together with a local Hamiltonian $H = \sum H_i$, where each H_i is a Hermitian operator defined only on those tensor factors whose index is within $\epsilon > 0$ of the ith point in the geometry of the surface. Local Hamiltonians H have been found with highly d-degenerate ground states corresponding to modular functors, (and thus *braid group* representation and link polynomials, see Appendix, Sect. 7.2.1). In these cases, the ground state G of H will be k-code for $k \sim$ injectivity radius of T $\sim \sqrt{\text{area T}}$. The topological degrees of freedom referred to above reside in G. But we do not attempt a precise mathematical definition of topological degrees of freedom since we would like it to extend beyond discrete system, e.g., to *fractional quantum Hall ground states*.

In the case when T is a disk D with *punctures*, physically realized as *anyonic excitations*, a sequence of local modifications to H effects a discrete 1-parameter family H_t of Hamiltonians, where the ground states G_{t_i} and $G_{t_{i+1}}$ at consecutive time steps differ by a $\lfloor \frac{k}{2} \rfloor$-local operator $\mathcal{O}_i$, such that $\mathcal{O}_i(G_{t_i}) = G_{t_{i+1}}$. Note that if G_{t_i} and $G_{t_{i+1}}$ are both k-code that for $\mathcal{O}_i$ as above, such that $\mathcal{O}_i|_{G_{t_i}}$ must be unique up to a scalar. This uniqueness property forces this discrete-cryptographic transport of code spaces to coincide (up to phase) with the differential geometric notion of *adiabatic transport*, that is, integration of the canonical connection in the 'tautological' bundle of d-planes in $Y = \mathbb{C}^{2^n}$. If the 1-parameter family is a closed loop, a projective representation of the braid group on (code) is obtained. By choosing H, these can be engineered to be precisely the Hecke algebra representations $\{\rho_\lambda\}$ associated to the *Jones polynomial*. From H one can build a concrete model of quantum computation; the model and its connection to the Jones polynomial are described below. Although the H found in is enormously cumbersome, it appears to share the universality class 'Witten–Chern–Simons theory of $SU(2)$ at level 3' with a simple 4-body Hamiltonian, which has been proposed as a model for the fractional quantum Hall plateaus at $\nu = \frac{13}{5}$ and $\frac{12}{5}$ [RR99].

An Anyonic Model for Quantum Computation

A family of unitary representations of all mapping class and braid groups with certain compatibility properties under fusion is known as a *unitary topological modular functor* [FKL03b]. To define our model, we take only the planar surface portion of the simplest universal modular functor, Witten–Chern–Simons $SU(2)$ modular functor at level 3, and this reduces to the Jones representations of braids $\{\rho_\lambda\}$. For an appropriate family of local Hamiltonians H_t, these representations describe the adiabatic transport of the lowest energy states with n anyonic excitation pairs, these states form a subspace W of dimension *Fibonacci(2n)*, as the $2n$ anyonic excitations are 'braided' around each other in $(2+1)$D space-time. In this theory, (we denote it $CS5$ because of its link to the Jones polynomial at a fifth root of unity), there are 4 label types 0, 1, 2, 3 corresponding to the complete list of irreducible representations of the quantum group $SU(2)_5$ of dimensions 1, 2, 3 and 4. We initialize our system on the disk D in a known state by pulling anyonic pairs out of the vacuum. This theory is self-dual so the two partners have identical types. Pairs are kept or returned to the vacuum according to the results of local holonomic measurement. Finally we have a known initial state in the disk with $2n$ punctures where each puncture has label $= 1$ and ∂D has label $= 0$. We assume n is even and group the punctures into $n/2$ batches of 4 punctures each. Similar to, each batch B is used to encode one qubit $\cong \mathbb{C}^2$: the basis $\{|0\rangle, |1\rangle\}$ is mapped into the type (0 or 2) of the 2-fold composite particle (round circle), which would result from fusing a (fixed) pair of the type 1 particles within B. Initially, both the double and 4-fold composites (ovals) have type 0. By maintaining, at least approximately, this condition on the ovals after the braiding is complete, we define a *computational summand* of the modular functor: it is spanned by n-bit strings of 0's and 2's residing on the 2-fold composites (round circles).

Now as in the quantum circuit model, a classical *poly-time algorithm* looks at the problem instance (F, y) and builds a sequence of 'gates', but this time the gates are braid generators (right half twist between adjacent anyons) σ_i, $(1 \leq i \leq 2n - 1)$, and a powerful approximation theorem is used to select the braid sequence which approximates the more traditional quantum circuit solving (F, y). So the topological model may be described as [FKL03b]:

1. Initialization of a known state in the modular functor.
2. Classical computation of braid b effecting a desired unitary transformation X of the computational subspace of the modular functor.
3. Adiabatic implementation of the braid by moving the anyons in D to draw b in space-time, where we keep the anyons separated by a scale l.
4. Application of a projection operator Π to measure the type (0 or 2) of the 'left most' composite particle.

The last step is the direct analogue of measuring the first qubit in the quantum circuit model and the same formula (1) applies: the probability of observing type 0 (the null particle) is:

$$\Pr(0) = \langle 0|X^{\dagger}\Pi X|0\rangle,$$

where the 0's on the right hand side represent our carefully prepared initial state with $2n$ type 1 excitations.

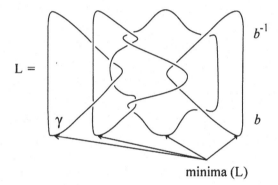

$$\text{minima (L)}$$

Fig. 3.5. Plat closure of a braid $L = \text{plat}(b^{-1}\gamma b)$, where γ is a small loop inserted to measure the left-most qubit (adapted from [FKL03b]).

To close the topological discussion, we note that the previous formula can be translated into a *plat closure* (see Fig. 3.5), so that the outcome of the quantum circuit calculation, prob(0), becomes a Jones evaluation:

$$\Pr(0) = \frac{1}{1 + [2]_5^2}\left(1 + \frac{(-1)^{c(L)+w(L)}(-a)^{3w(L)}V_L(e^{2\pi i/5})}{[2]_5^{m(L)-2}}\right),$$

where $[2]_5 = \frac{1+\sqrt{5}}{2}$; $c, m,$ and w are the number of components, number of local minima, and writhe of L respectively; and $a = e^{\pi i/10}$. For more details, see [FKL03b].

The braiding disturbs and reforms composite particle types with sufficient subtlety to effect universal computation. To reduce this model to engineering, very significant obstacles must be overcome: stable quantum media must be

maintained in a suitable phase, e.g., $CS5$; excitations must be readily manipulated, and electrical neutral particle types 0 and 2 must be distinguished,
presumably by holonomy experiments. Although these challenges are daunting, they are, perhaps, less difficult than a head-on assault on the accuracy
threshold in the quantum circuit model.

   We will give a technical review of topological quantum computation in
Sect. 6.4.

3.3 The Feynman Path Integral

3.3.1 The Action-Amplitude Formalism

The 'driving engine'[15] of quantum field theory is the Feynman path integral.
Very briefly, there are three basic forms of the path integral (see, e.g., [II08a,
II08b]):

   1. *Sum-over-histories*, developed in Feynman's version of quantum mechan-
      ics (QM)[16] [Fey48];

---

[15] AUM is the bow; Atma is the arrow; Brahman is the target—*Upanishads*.

[16] Feynman's *amplitude* is a space-time version of the Schrödinger's *wave-function*
ψ, which describes how the (non-relativistic) quantum state of a physical system
changes in space and time, i.e.,

$$\langle \text{Out}_{t_{fin}} | \text{In}_{t_{ini}} \rangle = \psi(\mathbf{x}, t), \quad (\text{for } \mathbf{x} \in [\text{In}, \text{Out}], \ t \in [t_{ini}, t_{fin}]).$$

In particular, quantum wave-function ψ is a complex-valued function of real space
variables $\mathbf{x} = (x_1, x_2, \ldots, x_n) \in \mathbb{R}^n$, which means that its domain is in $\mathbb{R}^n$ and
its range is in the complex plane, formally $\psi(\mathbf{x}) : \mathbb{R}^n \to \mathbb{C}$. For example, the one-
dimensional *stationary plane wave* with wave number k is defined as

$$\psi(x) = e^{ikx}, \quad (\text{for} x \in \mathbb{R}),$$

where the real number k describes the wavelength, $\lambda = 2\pi/k$. In n dimensions, this
becomes

$$\psi(x) = e^{i\mathbf{p} \cdot \mathbf{x}},$$

where the momentum vector $\mathbf{p} = \mathbf{k}$ is the vector of the wave numbers $\mathbf{k}$ (in natural
units).

   More generally, quantum wave-function is also time dependent, $\psi = \psi(\mathbf{x}, \mathbf{t})$. The
time-dependent plane wave is defined by

$$\psi(\mathbf{x}, \mathbf{t}) = e^{i\mathbf{p} \cdot \mathbf{x} - ip^2 t/2}. \tag{3.19}$$

   In general, $\psi(\mathbf{x}, \mathbf{t})$ is governed by the Schrödinger equation [Tha99, II08b]

$$i\frac{\partial}{\partial t}\psi(\mathbf{x}, t) = -\frac{1}{2}\Delta\psi(\mathbf{x}, t), \tag{3.20}$$

where Δ is the n-dimensional Laplacian. The solution of (3.20) is given by the
integral of the time-dependent plane wave (3.19),

2. *Sum-over-fields*, started in Feynman's version of quantum electrodynamics (QED) [Fey49] and later improved by Faddeev–Popov [FP67];
3. *Sum-over-geometries/topologies* in quantum gravity (QG), initiated by S. Hawking and properly developed in the form of causal dynamical triangulations (see [ALW08]; for a 'softer' review, see [Lol08]).

In all three versions, Feynman's *action–amplitude formalism* includes two components:

1. A real-valued, classical, *Hamilton's action functional*,

$$S[\Phi] := \int_{t_{ini}}^{t_{fin}} L[\Phi]\, dt, \qquad (3.21)$$

with the Lagrangian energy function defined over the Lagrangian density $\mathcal{L}$,

$$L[\Phi] = \int d^n x \mathcal{L}(\Phi, \partial_\mu \Phi), \quad (\partial_\mu \equiv \partial/\partial x^\mu),$$

while Φ is a common symbol denoting all three things to be summed upon (histories, fields and geometries). The action functional $S[\Phi]$ obeys the *Hamilton's least action principle*, $\delta S[\Phi] = 0$, and gives, using standard variational methods,[17] the Euler–Lagrangian equations, which define the shortest path, the extreme field, and the geometry of minimal curvature (and without holes).

$$\psi(\mathbf{x}, t) = \frac{1}{(2\pi)^{n/2}} \int_{\mathbb{R}^n} e^{i\mathbf{p}\cdot\mathbf{x} - ip^2 t/2} \hat{\psi}_0(\mathbf{p}) d^n p,$$

which means that $\psi(\mathbf{x}, t)$ is the inverse Fourier transform of the function

$$\hat{\psi}(\mathbf{p}, t) = e^{-ip^2 t/2} \hat{\psi}_0(\mathbf{p}),$$

where $\hat{\psi}_0(\mathbf{p})$ has to be calculated for each initial wave-function. For example, if initial wave-function is Gaussian,

$$f(x) = \exp\left(-a\frac{x^2}{2}\right), \quad \text{with the Fourier transform } \hat{f}(p) = \frac{1}{\sqrt{a}} \exp\left(-\frac{p^2}{2a}\right).$$

$$\text{then } \hat{\psi}_0(p) = \frac{1}{\sqrt{a}} \exp\left(-\frac{p^2}{2a}\right).$$

---

[17] In Lagrangian field theory, the fundamental quantity is the action

$$S[\Phi] = \int_{t_{in}}^{t_{out}} L\, dt = \int_{\mathbb{R}^4} d^n x \mathcal{L}(\Phi, \partial_\mu \Phi),$$

so that the least action principle, $\delta S[\Phi] = 0$, gives

$$0 = \int_{\mathbb{R}^4} d^n x \left\{ \frac{\partial \mathcal{L}}{\partial \Phi} \delta\Phi + \frac{\partial \mathcal{L}}{\partial(\partial_\mu \Phi)} \delta(\partial_\mu \Phi) \right\}$$

$$= \int_{\mathbb{R}^4} d^n x \left\{ \frac{\partial \mathcal{L}}{\partial \Phi} \delta\Phi - \partial_\mu \left(\frac{\partial \mathcal{L}}{\partial(\partial_\mu \Phi)} \right) \delta\Phi + \partial_\mu \left(\frac{\partial \mathcal{L}}{\partial(\partial_\mu \Phi)} \delta\Phi \right) \right\}.$$

2. A complex-valued, quantum *transition amplitude* (in natural units),

$$\langle \text{Out}_{t_{fin}} | \text{In}_{t_{ini}} \rangle := \int_{\Omega} \mathcal{D}[\Phi] \, e^{iS[\Phi]}, \tag{3.22}$$

where $\mathcal{D}[\Phi]$ is 'an appropriate' Lebesgue-type measure,

$$\mathcal{D}[\Phi] = \lim_{N \to \infty} \prod_{s=1}^{N} \Phi_s^i, \quad (i = 1, \ldots, n),$$

so that we can 'safely integrate over a continuous spectrum and sum over a discrete spectrum of our problem domain Ω', of which the absolute square is the real-valued probability density function,

$$P := |\langle \text{Out}_{t_{fin}} | \text{In}_{t_{ini}} \rangle \rangle|^2.$$

This procedure can be redefined in a mathematically cleaner way if we Wick-rotate the time variable t to imaginary values, $t \mapsto \tau = t$, thereby making all integrals real:

The last term can be turned into a surface integral over the boundary of the $\mathbb{R}^4$ (4D space-time region of integration). Since the initial and final field configurations are assumed given, $\delta\Phi = 0$ at the temporal beginning t_{in} and end t_{out} of this region, which implies that the surface term is zero. Factoring out the $\delta\Phi$ from the first two terms, and since the integral must vanish for arbitrary $\delta\Phi$, we arrive at the Euler–Lagrange equation of motion for a field,

$$\partial_\mu \left(\frac{\partial \mathcal{L}}{\partial(\partial_\mu \Phi)} \right) - \frac{\partial \mathcal{L}}{\partial \Phi} = 0.$$

If the Lagrangian (density) $\mathcal{L}$ contains more fields, there is one such equation for each. The momentum density $\pi(x)$ of a field, conjugate to $\Phi(x)$ is defined as: $\pi(x) = \frac{\partial \mathcal{L}}{\partial_\mu \Phi(x)}$.

For example, the standard electromagnetic action

$$S = -\frac{1}{4} \int_{\mathbb{R}^4} d^4x F_{\mu\nu} F^{\mu\nu}, \quad \text{where } F_{\mu\nu} = \partial_\mu A_\nu - \partial_\nu A_\mu,$$

gives the sourceless Maxwell's equations:

$$\partial_\mu F^{\mu\nu} = 0, \qquad \epsilon^{\mu\nu\sigma\eta} \partial_\nu F_{\sigma\eta} = 0,$$

where the field strength tensor $F_{\mu\nu}$ and the Maxwell equations are invariant under the *gauge transformations*,

$$A_\mu \longrightarrow A_\mu + \partial_\mu \epsilon.$$

The equations of motion of charged particles are given by the Lorentz-force equation,

$$m \frac{du^\mu}{d\tau} = e F^{\mu\nu} u_\nu,$$

where e is the charge of the particle and $u^\mu(\tau)$ its four-velocity as a function of the proper time.

$$\int \mathcal{D}[\Phi]\, e^{iS[\Phi]} \xrightarrow{\quad \text{Wick} \quad} \int \mathcal{D}[\Phi]\, e^{-S[\Phi]}. \tag{3.23}$$

For example, in non-relativistic quantum mechanics, the propagation amplitude from x_a to x_b is given by the *configuration path integral*

$$U(x_a, x_b; T) = \langle x_b | x_a \rangle = \langle x_b | e^{-iHT} | x_a \rangle = \int \mathcal{D}[x(t)]\, e^{iS[x(t)]},$$

which satisfies the Schrödinger equation (in natural units)

$$i\frac{\partial}{\partial T} U(x_a, x_b; T) = \hat{H} U(x_a, x_b; T), \quad \text{where } \hat{H} = -\frac{1}{2}\frac{\partial^2}{\partial x_b^2} + V(x_b).$$

The *phase-space path integral* (without peculiar constants in the functional measure) reads

$$U(q_a, q_b; T) = \left(\prod_i \int \mathcal{D}[q(t)]\mathcal{D}[p(t)] \right) \exp\left[i \int_0^T \left(p_i \dot{q}^i - H(q, p) \right) dt \right],$$

where the functions $q(t)$ (space coordinates) are constrained at the endpoints, but the functions $p(t)$ (canonically-conjugated momenta) are not. The functional measure is just the product of the standard integral over phase space at each point in time

$$\mathcal{D}[q(t)]\mathcal{D}[p(t)] = \prod_i \frac{1}{2\pi} \int dq^i\, dp_i.$$

Applied to a non-relativistic real scalar field $\phi(x, t)$, this path integral becomes

$$\langle \phi_b(x, t) | e^{-iHT} | \phi_a(x, t) \rangle = \int \mathcal{D}[\phi] \exp\left[i \int_0^T \mathcal{L}(\phi)\, d^4x \right],$$

$$\text{with } \mathcal{L}(\phi) = \frac{1}{2}(\partial_\mu \phi)^2 - V(\phi).$$

3.3.2 Correlation Functions and Generating Functional

If we have two fields in the interacting theory, the corresponding two-point correlation function, or two-point Green's function, is denoted by $\langle \Omega | T\{\phi(x)\phi(y)\} | \Omega \rangle$, where the notation $|\Omega\rangle$ is introduced to denote the ground state of the interacting theory, which is generally different from $|0\rangle$, the ground state of the free theory. The correlation function can be interpreted physically as the amplitude for propagation of a particle or excitation between y and x. In the free theory, it is simply the Feynman propagator

$$\langle 0 | T\{\phi(x)\phi(y)\} | 0 \rangle_{\text{free}} = D_F(x - y) = \int \frac{d^4p}{(2\pi)^4} \frac{i\, e^{-ip \cdot (x-y)}}{p^2 - m^2 + i\epsilon}.$$

We would like to know how this expression changes in the interacting theory. Once we have analyzed the two-point correlation functions, it will be easy to generalize our results to higher correlation functions in which more than two field operators appear.

In general we have:

$$\langle \Omega | T\{\phi(x)\phi(y)\}|\Omega\rangle = \lim_{T\to\infty(1-i\epsilon)} \frac{\langle 0|T\{\phi_I(x)\phi_I(y)\exp[-i\int_{-T}^{T} dt\, H_I(t)]\}|0\rangle}{\langle 0|T\{\exp[-i\int_{-T}^{T} dt\, H_I(t)]\}|0\rangle},$$

$$\langle 0|T\left\{\phi_I(x)\phi_I(y)\exp\left[-i\int_{-T}^{T} dt\, H_I(t)\right]\right\}|0\rangle$$
$$= \left(\begin{array}{c}\text{sum of all Feynman diagrams}\\ \text{with two external points}\end{array}\right),$$

where each diagram is built out of Feynman propagators, vertices and external points.

The virtue of considering the time-ordered product is clear: It allows us to put everything inside one large T-operator. A similar formula holds for higher correlation functions of arbitrarily many fields; for each extra factor of ϕ on the left, put an extra factor of ϕ_I on the right.

In the interacting theory, the corresponding two-point correlation function is given by

$$\langle \Omega | T\{\phi(x)\phi(y)\}|\Omega\rangle = \left(\begin{array}{c}\text{sum of all connected diagrams}\\ \text{with two external points}\end{array}\right).$$

This is generalized to higher correlation functions as

$$\langle \Omega | T\{\phi(x_1)\cdots\phi(x_n)\}|\Omega\rangle = \left(\begin{array}{c}\text{sum of all connected diagrams}\\ \text{with } n \text{ external points}\end{array}\right).$$

In a scalar field theory, the generating functional of correlation functions is defined as

$$Z[J] = \int \mathcal{D}[\phi]\exp\left[i\int d^4x\left[\mathcal{L} + J(x)\phi(x)\right]\right] = \langle \Omega | e^{-iHT}|\Omega\rangle = e^{-iE[J]};$$

this is a functional integral over $\phi(x)$ in which we have added a source term $J(x)\phi(x)$ to $\mathcal{L} = \mathcal{L}(\phi)$.

For example, the generating functional of the free Klein–Gordon theory is simply

$$Z[J] = Z_0 \exp\left[-\frac{1}{2}\int d^4x\, d^4y\, J(x)D_F(x-y)J(y)\right].$$

3.3.3 Quantization of the Electromagnetic Field

Consider the path integral

$$Z[A] = \int \mathcal{D}[A] \, e^{iS[A]},$$

where the action for the free e.-m. field is

$$S[A] = \int d^4x \left[-\frac{1}{4}(F_{\mu\nu})^2 \right] = \frac{1}{2} \int d^4x A_\mu(x) \left(\partial^2 g^{\mu\nu} - \partial^\mu \partial^\nu \right) A_\nu(x).$$

$Z[A]$ is the path integral over each of the four spacetime components:

$$\mathcal{D}[A] = \mathcal{D}[A]^0 \mathcal{D}[A]^1 \mathcal{D}[A]^2 \mathcal{D}[A]^3.$$

This functional integral is badly divergent, due to gauge invariance. Recall that $F_{\mu\nu}$, and hence L, is invariant under a general gauge transformation of the form

$$A_\mu(x) \rightarrow A_\mu(x) + \frac{1}{e} \partial_\mu \alpha(x).$$

The troublesome modes are those for which $A_\mu(x) = \partial_\mu \alpha(x)$, that is, those that are gauge-equivalent to $A_\mu(x) = 0$. The path integral is badly defined because we are redundantly integrating over a continuous infinity of physically equivalent field configurations. To fix this problem, we would like to isolate the interesting part of the path integral, which counts each physical configuration only once. This can be accomplished using the *Faddeev–Popov trick*, which effectively adds a term to the system Lagrangian and after which we get

$$Z[A] = \int \mathcal{D}[A] \exp \left[i \int_{-T}^{T} d^4x \left[\mathcal{L} - \frac{1}{2\xi}(\partial^\mu A_\mu)^2 \right] \right],$$

where ξ is any finite constant.

This procedure needs also to be applied to the formula for the two-point correlation function

$$\langle \Omega | T\mathcal{O}(A) | \Omega \rangle = \lim_{T \to \infty(1-i\epsilon)} \frac{\int \mathcal{D}[A]\mathcal{O}(A) \exp[i \int_{-T}^{T} d^4x \mathcal{L}]}{\int \mathcal{D}[A] \exp[i \int_{-T}^{T} d^4x \mathcal{L}]},$$

which after Faddeev–Popov procedure becomes

$$\langle \Omega | T\mathcal{O}(A) | \Omega \rangle = \lim_{T \to \infty(1-i\epsilon)} \frac{\int \mathcal{D}[A]\mathcal{O}(A) \exp[i \int_{-T}^{T} d^4x [\mathcal{L} - \frac{1}{2\xi}(\partial^\mu A_\mu)^2]]}{\int \mathcal{D}[A] \exp[i \int_{-T}^{T} d^4x [\mathcal{L} - \frac{1}{2\xi}(\partial^\mu A_\mu)^2]]}.$$

3.3.4 Wavelet-Based QFT

In this subsection, following [Alt07], we give a brief on the *wavelet-based QFT*.

The description of infinite-dimensional nonlinear systems in quantum field theory and statistical physics always faces the problem of divergent loop integrals emerging in the *Green functions*. Different *methods of regularization* have been applied to make the divergent integrals finite [ZJ89]. There are a few basic ideas connected with those regularizations. First, certain minimal scale $L = \frac{2\pi}{\Lambda}$, where Λ is the cut-off momentum, is introduced into the theory, with all the fields $\phi(x)$ being substituted by their Fourier transforms truncated at momentum Λ:

$$\phi(x) \rightarrow \phi_{(\frac{2\pi}{\Lambda})}(x) = \int_{|k| \leq \Lambda} e^{-ikx} \tilde{\phi}(k) \frac{n^n k}{(2\pi)^n}.$$

The physical quantities are than demanded to be independent on the rescaling of the parameter Λ. The second thing is the Kadanoff blocking procedure [Kad66], which averages the small-scale fluctuations up to a certain scale—this makes a kind of effective interaction.

These methods are related to the self-similarity assumption: blocks interact to each other similarly to the sub-blocks. Similarly, but not necessarily having the same interaction strength—the latter can be dependent on scale $g = g(a)$. It is the case for high energy physics, for the developed hydrodynamic turbulence, and for many other phenomena [Vas04]. However there is no place for such dependence if the fields are described solely in terms of their Fourier transform—except for the cut-off momentum. The latter representation of the scale-dependence is rather restrictive: it determines the effective interaction of all fluctuations up to a certain scale, but says nothing about the interaction of the fluctuations at a given scale [Alt07].

We have to admit that the origin of divergences is not the singular behavior of the interaction strength at small distance, but the inadequate choice of the functional space used to describe these interactions. Namely, the decomposition of the fields with respect to the representations of *translation group*, i.e., the *Fourier transform*

$$\phi(x) = \int e^{-ikx} \tilde{\phi}(k) \frac{n^n k}{(2\pi)^n}.$$

is physically sound only for the problems that clearly manifest translational invariance. For more general cases one can use decompositions with respect to other Lie groups, different from translation group ($x \rightarrow x+b$) (see, e.g. [IK91]). The problem is what groups are physically relevant for a field theory? In physical settings, along with translation invariance, the other symmetry is observed quite often—the symmetry with respect to scale transformations: $x \rightarrow \alpha x$. This suggests the affine group (3.27) given below may be more adequate for *self-similar phenomena* than the subgroup of translations. The discrete representation of the self-similarity idea can be found in the *Kadanoff spin-blocking procedure*, or in application of the discrete wavelet transform

$$\phi(x) = \sum n_k^j \psi_k^j(x)$$

in QFT models, considered in lattice settings by [Bat89, Fed81].

The decomposition with respect to the representations of affine group may have a natural probabilistic interpretation. In Euclidean quantum field theory the L^2-norm of the field $\phi(x)$ determines the *probability density* of registering that particle in a certain region $\Omega \subset \mathbb{R}^n$:

$$P(\Omega) = \int_{x \in \Omega} |\phi(x)|^2 dx, \qquad P(\mathbb{R}^n) = 1, \tag{3.24}$$

i.e., defines a measure. The unit normalization in (3.24) is understood as 'the probability of registering a particle anywhere in space is exactly one'. This tacitly assumes the existence of registration devices working at infinite coordinate resolution. There are no such devices in reality: even if particle is there, but its typical wavelength is much smaller or much bigger than the typical wavelength of the measuring device there is nonzero probability the particle will not be registered.

For this reason it seems beneficial for theoretical description to use wave-functions, or fields, that are explicitly labeled by resolution of the measuring equipment: $\phi_a(x)$. The incorporation of an observation parameter a is in excellent agreement with the Copenhagen interpretation of quantum mechanics: $\phi_a(x)$ describes our perception of the object ϕ at resolution a, rather than an 'object as it is', $\phi_{a \to 0}(x)$, the existence of which is at least questionable. Needless to say that infinitely small resolution ($a \to 0$) requires infinitely high energy ($E \to \infty$) and is therefore practically unreachable.

According to [Alt07], the normalization for the resolution-dependent functions $\phi_a(x)$ should be

$$\int_{-\infty}^{\infty} dx \int_0^{\infty} d\mu(a) |\phi_a(x)|^2 = 1, \tag{3.25}$$

where $\mu(a)$ is a measure of the resolution of the equipment. The normalization (3.25) will be read as 'the probability to register the object ϕ anywhere in space tuning the resolution of the equipment from zero to infinity is exactly one'.

In this section, following [Alt07], we show how the quantum field theory of scale-dependent fields $\phi_a(x)$ can be constructed using *continuous wavelet transform* (CWT, see Appendix). The integration over all scales a of course will drive us back to the standard theory. The advantage is that the Green functions $\langle \phi_{a_1}(x_1) \cdots \phi_{a_n}(x_n) \rangle$, i.e., those really observed in experiment, are finite—no further renormalization is required.

Let us now show how the field theory of scale-dependent fields $\phi_a(x)$ can be constructed using continuous wavelet transform. If $\mathcal{H}$ is the Hilbert space, with is a Lie group G acting transitively on that space, and there exists a vector $\psi \in \mathcal{H}$, called an *admissible vector*, such that [Alt07]

$$C_\psi = \frac{1}{\|\psi\|_2} \int_G |\langle\psi, U(g)\psi\rangle|^2 d\mu_L(g) < \infty,$$

where $U(g)$ is a representation of G in $\mathcal{H}$, and $d\mu_L(g)$ is the left-invariant measure, then for any $\phi \in \mathcal{H}$ the following decomposition holds [Car76, DM76]:

$$|\phi\rangle = C_\psi^{-1} \int_G |U(g)\psi\rangle\langle\psi|U(g)\phi\rangle d\mu_L(g), \quad \text{(for all } \phi \in \mathcal{H}\text{)}. \tag{3.26}$$

The Lie group that comprises two required operations (change of scale and translations) is the affine group

$$x \to ax + b, \qquad \psi(x) \to U(a,b)\psi[x] = a^{-\frac{n}{2}}\psi\left(\frac{x-b}{a}\right), \tag{3.27}$$

where $x, b \in \mathbb{R}^n$, $a \in \mathbb{R}_+$. The decomposition (3.26) with respect to affine group (3.27) is known as *continuous wavelet transform*.

To keep the scale-dependent fields $\phi_a(x)$ the same physical dimension as the ordinary fields $\phi(x)$ we write the coordinate representation of wavelet transform (3.26) in L^1-norm [Chu92, HM96]:

$$\phi(x) = \frac{1}{C_\psi} \int \frac{1}{a^n}\psi\left(\frac{x-b}{a}\right)\phi_a(b)\frac{na n^n b}{a}, \tag{3.28}$$

$$\phi_a(b) = \int \frac{1}{a^n}\overline{\psi\left(\frac{x-b}{a}\right)}\phi(x)d^n x. \tag{3.29}$$

In the latter equations the field $\phi_a(b)$ (the wavelet coefficient) has a physical meaning of the amplitude of the field ϕ measured at point b using a device with an aperture ψ and a tunable spatial resolution a. For isotropic wavelets, which we assume in this section, the normalization constant C_ψ is readily evaluated using Fourier transform:

$$C_\psi = \int_0^\infty |\tilde{\psi}(ak)|^2 \frac{da}{a} = \int |\tilde{\psi}(k)|^2 \frac{n^n k}{S_d|k|} < \infty,$$

where $S_n = \frac{2\pi^{n/2}}{\Gamma(d/2)}$ is the area of unit sphere in nD.

The idea of substituting CWT (3.29), (3.28) into quantum mechanics or field theory is not new (see, e.g. [HM96, Fed95]). However all attempts to substitute it into field theory models were aimed to take at the final end the inverse wavelet transform and calculate the Green functions for the 'true' fields $\langle\phi(x_1)\cdots\phi(x_n)\rangle$, i.e., for the case of infinite resolution. Our claim is that this last step should be avoided because the infinite resolution can not be achieved experimentally. Instead we suggest to calculate the functions, which correspond to experimentally observable finite resolution correlations. The integration over all scales a_i of course will drive us back to the standard divergent theory. The advantage of our approach is that the Green functions $\langle \phi_{a_1}(x_1) \cdots \phi_{a_n}(x_n) \rangle$ become finite under certain causality assumptions.

Consider the Euclidean field theory with the forth power interaction ϕ^4. The corresponding action functional can be written in the form

$$
S_E[\phi(x)] = \frac{1}{2} \int \phi(x_1) D(x_1 - x_2) \phi(x_2) dx_1 dx_2
$$
$$
+ \frac{\lambda}{4!} \int V(x_1, \ldots, x_4) \phi(x_1) \phi(x_2) \phi(x_3) \phi(x_4) dx_1 dx_2 dx_3 dx_4,
\tag{3.30}
$$

where D is the inverse propagator. To calculate the n-point Green functions of such a theory the generation functional is constructed

$$
\langle \phi(x_1) \cdots \phi(x_n) \rangle = \left. \frac{\delta^n \ln W[J]}{\delta J^n} \right|_{J=0},
$$
$$
W[J] = \int \mathcal{D}[\phi(x)] \, e^{-S_E[\phi] + \int J(x)\phi(x)dx}.
\tag{3.31}
$$

Similarly, to calculate the Green functions for scale-dependent fields $\langle \phi_{a_1}(x_1) \cdots \phi_{a_n}(x_n) \rangle$ we have to construct the generating functional for scale-dependent fields $\phi_a(x)$. This is readily done by substituting wavelet transform (3.28) into the action (3.30). This gives

$$
W_W[J_a] = \int \mathcal{D}[\phi_a(x)] \, e^{-S_W[\phi_a] + \int J_a(x)\phi_a(x)\frac{da\,dx}{a}},
$$
$$
S_W[\phi_a] = \frac{1}{2} \int \phi_{a_1}(x_1) D(a_1, a_2, x_1 - x_2) \phi_{a_2}(x_2) \frac{da_1 dx_1}{a_1} \frac{da_2 dx_2}{a_2}
$$
$$
+ \frac{\lambda}{4!} \int V_{x_1,\ldots,x_4}^{a_1,\ldots,a_4} \phi_{a_1}(x_1) \cdots \phi_{a_4}(x_4) \frac{da_1 dx_1}{a_1} \frac{da_2 dx_2}{a_2} \frac{da_3 dx_3}{a_3} \frac{da_4 dx_4}{a_4},
\tag{3.32}
$$

with $D(a_1, a_2, x_1 - x_2)$ and $V_{x_1,\ldots,x_4}^{a_1,\ldots,a_4}$ denoting the wavelet images of the inverse propagator and that of the interaction potential, respectively.

The functional (3.32) keeps the same form as its counterpart (3.31) with the difference that the functional integration over the two-argument fields $\phi_a(x)$ requires their ordering in both the position x and the scale a, in case the fields are operator-valued. It is important that if the interaction in the original theory (3.30) is local, $V \sim \prod_{i=2}^{4} \delta(x_1 - x_i)$, its wavelet image $V_{x_1,\ldots,x_4}^{a_1,\ldots,a_4}$ may be nonlocal, and vice versa. Here the dependence of interaction on scale is only due to wavelet transform:

$$
V(x_1, \ldots, x_n) \leftrightarrow V_{x_1,\ldots,x_n}^{a_1,\ldots,a_n}.
$$

Generally speaking the explicit scale dependence of the coupling constant $\lambda = \lambda(a)$ is also allowed. In the framework of modern field theory such dependence can not be tested: the running coupling constant $\lambda = \lambda(2\pi/\Lambda)$, obtained by renormalization group methods, accounts for the collective interaction of all modes up to the certain scale Λ, but says nothing about the interaction of modes precisely at the given scale.

The technical way to calculate the Green functions

$$\langle \phi_{a_1}(x_1) \cdots \phi_{a_n}(x_n) \rangle = \left. \frac{\delta^n \ln W_W[J_a]}{\delta J_a^n} \right|_{J_a=0}$$

is to apply the Fourier transform to the r.h.s. of wavelet transform (3.28) and then substitute the result

$$\phi(x) = \frac{1}{C_\psi} \int_0^\infty \frac{da}{a} \int \frac{n^n k}{(2\pi)^n} e^{-ikx} \tilde{\psi}(ak) \tilde{\phi}_a(k),$$

into the action (3.30). Doing so, we have the following modification of the Feynman diagram technique [Alt07]:

- each field $\tilde{\phi}(k)$ will be substituted by the scale component $\tilde{\phi}_a(k) = \overline{\tilde{\psi}(ak)} \tilde{\phi}(k)$.
- each integration in momentum variable will be accompanied by integration in corresponding scale variable:

$$\frac{n^n k}{(2\pi)^n} \rightarrow \frac{n^n k}{(2\pi)^n} \frac{da}{a}.$$

- each vertex is substituted by its wavelet transform.

For instance, for the massive scalar field propagator we have the correspondence

$$D(k) = \frac{1}{k^2 + m^2} \rightarrow D(a_1, a_2, k) = \frac{\tilde{\psi}(a_1 k) \tilde{\psi}(-a_2 k)}{k^2 + m^2}.$$

Surely the integration over all scale arguments in infinite limits drive us back to the usual theory in $\mathbb{R}^n$, since

$$\frac{1}{C_\psi} \int_0^\infty \frac{da}{a} |\tilde{\psi}(ak)|^2 = 1.$$

In physical settings the integration should not be performed over *all* scales $0 \le a < \infty$. In fact, if the system is affected at the point x with the resolution Δx and the response is measured at a point y with the resolution Δy, the modes that are essentially different from those two scales will hardly contribute to the result. In the simplest case of linear propagation the result will be proportional to the product of preparation and measuring filters

$$\int \frac{\tilde{\psi}(k\Delta x)\tilde{\psi}(-k\Delta y)}{k^2 + m^2} e^{-ik(x-y)} \frac{n^n k}{(2\pi)^n},$$

with the maximum achieved when Δx and Δy are of the same order.

A simplest assumption of this type formulated in the language of Feynman's diagrams is: there should be no scales in internal lines smaller than the minimal scale of all external lines. This means that there should be no virtual particles in internal lines unless there is sufficient energy in external lines to excite them. For more technical details, see [Alt07].

3.4 The Path-Integral TQFT

3.4.1 Schwarz-Type and Witten-Type Theories

Consider a set of fields $\{\phi_i\}$ on a Riemannian n-manifold M (with a metric $g_{\mu\nu}$) and real functional of these fields, $S[\phi_i]$, which is the action of the theory. Also consider a set of operators $\mathcal{O}_\alpha(\phi_i)$ (labeled by some set of indices α), which are arbitrary functionals of the fields $\{\phi_i\}$. The *vacuum expectation value* (VEV) of a product of these operators is defined as the path integral (see [Lab97])

$$\langle \mathcal{O}_{\alpha_1} \mathcal{O}_{\alpha_2} \cdots \mathcal{O}_{\alpha_p} \rangle = \int \mathcal{D}[\phi_i] \mathcal{O}_{\alpha_1}(\phi_i) \mathcal{O}_{\alpha_2}(\phi_i) \cdots \mathcal{O}_{\alpha_p}(\phi_i) \exp\left(-S[\phi_i]\right).$$

A quantum field theory is considered *topological* if it possesses the following property:

$$\frac{\delta}{\delta g^{\mu\nu}} \langle \mathcal{O}_{\alpha_1} \mathcal{O}_{\alpha_2} \cdots \mathcal{O}_{\alpha_p} \rangle = 0, \tag{3.33}$$

i.e., if the VEVs of some set of selected operators remain invariant under variations of the metric $g_{\mu\nu}$ on M. In this case, the operators $\mathcal{O}_\alpha(\phi_i)$ are called *observables*.

There are two ways to formally guarantee that condition (3.33) is satisfied. The first one corresponds to the situation in which both, the action, S, as well as the operators $\mathcal{O}_\alpha$, are metric independent. These TQFTs are called *Schwarz-type*. In the case of Schwarz-type theories one must first construct an action which is independent of the metric $g_{\mu\nu}$. The method is best illustrated by considering an example. Let us take into consideration the most interesting case of this type of theories: *Chern–Simons gauge theory*. The data in Chern–Simons gauge theory are the following: a differentiable compact 3-manifold M, a gauge group G, which will be taken simple and compact, and an integer parameter k. The action is the integral of the *Chern–Simons form* associated to a *gauge connection A* corresponding to the group G,

$$S_{\mathrm{CS}}[A] = \int_M \mathrm{Tr}\left(A \wedge dA + \frac{2}{3} A \wedge A \wedge A\right). \tag{3.34}$$

Observables are constructed out of operators which do not contain the metric $g_{\mu\nu}$. In gauge invariant theories, as it is the case, one must also demand for these operators invariance under gauge transformations. An important set of observables in Chern–Simons gauge theory is constituted by the trace of the holonomy of the gauge connection A in some representation R along a 1-cycle γ, that is the *Wilson loop*,[18]

$$\mathrm{Tr}_R\left(\mathrm{Hol}_\gamma(A)\right) = \mathrm{Tr}_R\, \mathrm{P}\exp\int_\gamma A. \tag{3.35}$$

The VEVs are labeled by representations R_i and embeddings γ_i of S^1 into M [Lab97]

$$\left\langle \mathrm{Tr}_{R_1}\, \mathrm{Pe}^{\int_{\gamma_1} A} \ldots \mathrm{Tr}_{R_n}\, \mathrm{Pe}^{\int_{\gamma_n} A} \right\rangle$$
$$= \int [DA]\, \mathrm{Tr}_{R_1}\, \mathrm{Pe}^{\int_{\gamma_1} A} \ldots \mathrm{Tr}_{R_n}\, \mathrm{Pe}^{\int_{\gamma_n} A}\, e^{\frac{ik}{4\pi} S_{CS}(A)}.$$

A non-perturbative analysis of Chern–Simons gauge theory shows that the invariants associated to the observables $\mathcal{O}_\alpha(\phi_i)$ are knot and link invariants of polynomial type as the Jones polynomial and its generalizations. The perturbative analysis has also led to this result and has shown to provide a very useful framework to study Vassiliev invariants.

The second way to guarantee (3.33) corresponds to the case in which there exists a symmetry, whose infinitesimal form is denoted by δ, satisfying the following properties:

$$\delta\mathcal{O}_\alpha(\phi_i) = 0, \qquad T_{\mu\nu}(\phi_i) = \delta G_{\mu\nu}(\phi_i), \tag{3.36}$$

where $T_{\mu\nu}(\phi_i)$ is the energy–momentum tensor of the theory, given by

$$T_{\mu\nu}(\phi_i) = \frac{\delta}{\delta g^{\mu\nu}} S[\phi_i], \tag{3.37}$$

while $G_{\mu\nu}(\phi_i)$ is some tensor.

The fact that δ in (3.36) is a symmetry of the theory means that the transformations $\delta\phi_i$ of the fields are such that $\delta S[\phi_i] = 0$ and $\delta\mathcal{O}_\alpha(\phi_i) = 0$. Conditions (3.36) lead formally to the following relation for VEVs:

---

[18] A *holonomy* on a smooth manifold is a general geometrical consequence of the curvature of the manifold connection, measuring the extent to which parallel transport around closed loops fails to preserve the geometrical data being transported. Related to holonomy is a *Wilson loop*, which is a gauge-invariant observable obtained from the holonomy of the gauge connection around a given loop. More precisely, a Wilson loop is a quantity defined by the trace of a path-ordered exponential of a gauge field A_μ transported along a closed curve (loop) γ, $W_\gamma = \mathrm{Tr}(P\exp[\mathrm{i}\oint_\gamma A_\mu dx^\mu])$, where P is the path-ordering operator.

$$\frac{\delta}{\delta g^{\mu\nu}} \langle \mathcal{O}_{\alpha_1} \mathcal{O}_{\alpha_2} \cdots \mathcal{O}_{\alpha_p} \rangle$$

$$= - \int \mathcal{D}[\phi_i] \mathcal{O}_{\alpha_1}(\phi_i) \mathcal{O}_{\alpha_2}(\phi_i) \cdots \mathcal{O}_{\alpha_p}(\phi_i) T_{\mu\nu} \exp\left(-S[\phi_i]\right)$$

$$= - \int \mathcal{D}[\phi_i] \delta \left(\mathcal{O}_{\alpha_1}(\phi_i) \mathcal{O}_{\alpha_2}(\phi_i) \cdots \mathcal{O}_{\alpha_p}(\phi_i) G_{\mu\nu} \exp\left(-S[\phi_i]\right)\right) = 0, \quad (3.38)$$

which implies that the quantum field theory can be regarded as topological. In (3.38) it has been assumed that the action and the measure $\mathcal{D}[\phi_i]$ are invariant under the symmetry δ. We have assumed also in (3.38) that the observables are metric-independent. This is a common situation in this type of theories, but it does not have to be necessarily so. In fact, in view of (3.38), it would be possible to consider a wider class of operators satisfying:

$$\frac{\delta}{\delta g_{\mu\nu}} \mathcal{O}_\alpha(\phi_i) = \delta O_\alpha^{\mu\nu}(\phi_i), \quad (3.39)$$

where $O_\alpha^{\mu\nu}(\phi_i)$ is a certain functional of the fields of the theory.

This second type of TQFTs are called *cohomological* of *Witten-type*. One of its main representatives is *Donaldson–Witten theory*, which can be regarded as a certain *twisted* version of $N = 2$ supersymmetric Yang–Mills theory. It is important to remark that the symmetry δ must be a scalar symmetry. The reason is that, being a global symmetry, the corresponding parameter must be covariantly constant and for arbitrary manifolds this property, if it is satisfied at all, implies strong restrictions unless the parameter is a scalar.

Most of the TQFTs of cohomological type satisfy the relation: $S[\phi_i] = \delta \Lambda(\phi_i)$, for some functional $\Lambda(\phi_i)$. This means that the topological observables of the theory (in particular the partition function itself) are independent of the value of the coupling constant. For example, consider the VEV [Lab97]

$$\langle \mathcal{O}_{\alpha_1} \mathcal{O}_{\alpha_2} \cdots \mathcal{O}_{\alpha_p} \rangle = \int \mathcal{D}[\phi_i] \mathcal{O}_{\alpha_1}(\phi_i) \mathcal{O}_{\alpha_2}(\phi_i) \cdots \mathcal{O}_{\alpha_p}(\phi_i) \exp\left(-\frac{1}{g^2} S[\phi_i]\right).$$
$$(3.40)$$

Under a change in the coupling constant, $1/g^2 \to 1/g^2 - \Delta$, one has (assuming that the observables do not depend on the coupling), up to first order in Δ:

$$\langle \mathcal{O}_{\alpha_1} \mathcal{O}_{\alpha_2} \cdots \mathcal{O}_{\alpha_p} \rangle \longrightarrow \langle \mathcal{O}_{\alpha_1} \mathcal{O}_{\alpha_2} \cdots \mathcal{O}_{\alpha_p} \rangle,$$

$$\Delta \int \mathcal{D}[\phi_i] \delta \left[\mathcal{O}_{\alpha_1}(\phi_i) \mathcal{O}_{\alpha_2}(\phi_i) \cdots \mathcal{O}_{\alpha_p}(\phi_i) \Lambda(\phi_i) \exp\left(-\frac{1}{g^2} S[\phi_i]\right) \right]$$

$$= \langle \mathcal{O}_{\alpha_1} \mathcal{O}_{\alpha_2} \cdots \mathcal{O}_{\alpha_p} \rangle.$$

Hence, observables can be computed either in the weak coupling limit, $g \to 0$, or in the strong coupling limit, $g \to \infty$.

3.4.2 Hodge Decomposition Theorem

The *Hodge star* operator $\star : \Omega^p(M) \to \Omega^{n-p}(M)$, which maps any exterior p-form $\alpha \in \Omega^p(M)$ into its *dual* $(n-p)$-form $\star\alpha \in \Omega^{n-p}(M)$ on a smooth

n-manifold M, is defined as (see, e.g. [Rha84, Voi02])

$$\alpha \wedge \star \beta = \beta \wedge \star \alpha = \langle \alpha, \beta \rangle \, \mu, \qquad \star \star \alpha = (-1)^{p(n-p)} \alpha, \quad (\text{for } \alpha, \beta \in \Omega^p(M)),$$

The $\star$ operator depends on the Riemannian metric $g = g_{ij}$ on M and also on the orientation (reversing orientation will change the sign) [II06c, II07e]. Using the star operator, for any two p-forms $\alpha, \beta \in \Omega^p(M)$ with compact support on M we define bilinear and positive-definite Hodge L^2-inner product as

$$\langle \alpha, \beta \rangle := \int_M \alpha \wedge \star \beta, \tag{3.41}$$

where $\alpha \wedge \star \beta$ is an n-form.

Given the exterior derivative $d : \Omega^p(M) \to \Omega^{p+1}(M)$ on a smooth manifold M (see Appendix), its Hodge dual (or, formal adjoint) is the *codifferential* δ, a linear map $\delta : \Omega^p(M) \to \Omega^{p-1}(M)$, which is a generalization of the divergence, defined by [Rha84, Voi02]

$$\delta = (-1)^{n(p+1)+1} \star d \star \quad \text{so that} \quad d = (-1)^{np} \star \delta \star.$$

That is, if the dimension n of the manifold M is even, then $\delta = - \star d \star$.

Applied to any p-form $\omega \in \Omega^p(M)$, the codifferential δ gives

$$\delta \omega = (-1)^{n(p+1)+1} \star d \star \omega, \qquad \delta d \omega = (-1)^{np+1} \star d \star d \omega.$$

If $\omega = f$ is a 0-form, or function, then $\delta f = 0$. If a p-form α is a codifferential of a $(p+1)$-form β, that is $\alpha = \delta \beta$, then β is called the *coexact* form. A p-form α is *coclosed* if $\delta \alpha = 0$; then $\star \alpha$ is closed (i.e., $d \star \alpha = 0$) and conversely.

The Hodge codifferential δ satisfies the following set of rules:

- $\delta \delta = \delta^2 = 0$, the same as $dd = d^2 = 0$;
- $\delta \star = (-1)^{p+1} \star d$; $\star \delta = (-1)^p \star d$;
- $d \delta \star = \star \delta d$; $\star d \delta = \delta d \star$.

The codifferential δ can be coupled with the exterior derivative d to construct the *Hodge Laplacian* $\Delta : \Omega^p(M) \to \Omega^p(M)$, a harmonic generalization of the Laplace–Beltrami differential operator, given by

$$\Delta = \delta d + d \delta = (d + \delta)^2.$$

Δ satisfies the following set of rules:

$$\delta \Delta = \Delta \delta = \delta d \delta; \qquad d \Delta = \Delta d = d \delta d; \qquad \star \Delta = \Delta \star.$$

A p-form α is called *harmonic* iff

$$\Delta \alpha = 0 \quad \Longleftrightarrow \quad d \alpha = \delta \alpha = 0.$$

Thus, α is harmonic in a compact domain $D \subset M$ iff it is both closed and coclosed in D. Informally, every harmonic form is both closed and coclosed. As a proof, we have:

$$0 = \langle \alpha, \Delta\alpha \rangle = \langle \alpha, d\delta\alpha \rangle + \langle \alpha, \delta d\alpha \rangle = \langle \delta\alpha, \delta\alpha \rangle + \langle d\alpha, d\alpha \rangle.$$

Since $\langle \beta, \beta \rangle \geq 0$ for any form β, $\langle \delta\alpha, \delta\alpha \rangle$ and $\langle d\alpha, d\alpha \rangle$ must vanish separately. Thus, $d\alpha = 0$ and $\delta\alpha = 0$. All harmonic p-forms on a smooth manifold M form the vector space $H_\Delta^p(M)$.

Now, the celebrated *Hodge decomposition theorem* (HDT) states that, on a compact orientable smooth n-manifold M (with $n \geq p$), any exterior p-form can be written as a unique sum of an *exact* form, a *coexact* form, and a *harmonic* form. More precisely, for any form $\omega \in \Omega^p(M)$ there are unique forms $\alpha \in \Omega^{p-1}(M)$, $\beta \in \Omega^{p+1}(M)$ and a harmonic form $\gamma \in \Omega^p(M)$, such that

$$\text{HDT:} \quad \overset{\text{anyform}}{\omega} = \overset{\text{exact}}{d\alpha} + \overset{\text{coexact}}{\delta\beta} + \overset{\text{harmonic}}{\gamma}.$$

For the proof, see [Rha84, Voi02].

In physics community, the exact form $d\alpha$ is called *longitudinal*, while the coexact form $\delta\beta$ is called *transversal*, so that they are mutually orthogonal. Thus any form can be orthogonally decomposed into a harmonic, a longitudinal and transversal form. For example, in fluid dynamics, any vector-field v can be decomposed into the sum of two vector-fields, one of which is divergence-free, and the other is curl-free.

Since γ is harmonic, $d\gamma = 0$. Also, by Poincaré lemma, $d(d\alpha) = 0$. In case ω is a closed p-form, $d\omega = 0$, then the term $\delta\beta$ in HDT is absent, so we have the *short Hodge decomposition*,

$$\omega = d\alpha + \gamma, \tag{3.42}$$

thus ω and γ differ by $d\alpha$. In topological terminology, ω and γ belong to the same *cohomology class* $[\omega] \in H^p(M)$. Now, by the de Rham theorems it follows that if C is any p-cycle, then

$$\int_C \omega = \int_C \gamma,$$

that is, γ and ω have the same periods. More precisely, if ω is any closed p-form, then there exists a unique harmonic p-form γ with the same periods as those of ω (see [Rha84, Fla63]).

The *Hodge–Weyl theorem* [Rha84, Voi02] states that every de Rham cohomology class has a unique harmonic representative. In other words, the space $H_\Delta^p(M)$ of harmonic p-forms on a smooth manifold M is isomorphic to the pth de Rham cohomology group,

$$H_{DR}^p(M) := \frac{Z^p(M)}{B^p M} = \frac{\text{Ker}(d : \Omega^p(M) \to \Omega^{p+1}(M))}{\text{Im}(d : \Omega^{p-1}(M) \to \Omega^p(M))}, \tag{3.43}$$

or, $H_\Delta^p(M) \cong H_{DR}^p(M)$. That is, the harmonic part γ of HDT depends only on the global structure, i.e., the topology of M.

For example, in $(2 + 1)$D electrodynamics, p-form Maxwell equations in the Fourier domain Σ are written as [TC99]

$$dE = i\omega B, \qquad dB = 0,$$
$$dH = -i\omega D + J, \qquad dD = Q,$$

where H and ω are 0-forms (magnetizing field and field frequency), D (electric displacement field), J (electric current density) and E (electric field) are 1-forms, while B (magnetic field) and Q (electric charge density) are 2-forms. From $d^2 = 0$ it follows that the J and the Q satisfy the *continuity equation*

$$dJ = i\omega Q.$$

Constitutive equations, which include all metric information in this framework, are written in terms of Hodge star operators (that fix an isomorphism between p forms and $(2 - p)$ forms in the $(2 + 1)$ case)

$$D = \star E, \qquad B = \star H.$$

Applying HDT to the electric field intensity 1-form E, we get [HT05]

$$E = d\phi + \delta A + \chi,$$

where ϕ is a 0-form (a scalar field) and A is a 2-form; $d\phi$ represents the static field and δA represents the dynamic field, and χ represents the harmonic field component. If domain Σ is contractible, χ is identically zero and we have the short Hodge decomposition,

$$E = d\phi + \delta A.$$

3.4.3 Hodge Decomposition and Chern–Simons Theory

Functional Measure on the Space of Differential Forms

The Hodge inner product (3.41) leads to a natural (metric-dependent) functional measure $\mathcal{D}\mu[\omega]$ on $\Omega^p(M)$, which normalizes the *Gaussian functional integral*

$$\int \mathcal{D}\mu[\omega]\, e^{i\langle\omega|\omega\rangle} = 1. \tag{3.44}$$

One can use the invariance of (3.44) to determine how the functional measure transforms under the Hodge decomposition. Using HDT and its orthogonality with respect to the inner product (3.41), it was shown in [GK94] that

$$\langle\omega, \omega\rangle = \langle\gamma, \gamma\rangle + \langle d\alpha, d\alpha\rangle + \langle\delta\beta, \delta\beta\rangle = \langle\gamma, \gamma\rangle + \langle\alpha, \delta d\alpha\rangle + \langle\beta, d\delta\beta\rangle, \tag{3.45}$$

where the following differential/conferential identities were used [CD82]

$$\langle d\alpha, d\alpha \rangle = \langle \alpha, \delta d\alpha \rangle \quad \text{and} \quad \langle \delta\beta, \delta\beta \rangle = \langle \beta, d\delta\beta \rangle.$$

Since, for any linear operator O, one has

$$\int \mathcal{D}\mu[\omega] \exp i\langle \omega|O\omega \rangle = \det{}^{-1/2}(O),$$

(3.44) and (3.45) imply that

$$\mathcal{D}\mu[\omega] = \mathcal{D}\mu[\gamma]\mathcal{D}\mu[\alpha]\mathcal{D}\mu[\beta]\det{}^{1/2}(\delta d)\det{}^{1/2}(d\delta).$$

Abelian Chern–Simons Theory

Recall that the classical action for an Abelian Chern–Simons theory,

$$S = \int_M A \wedge dA,$$

is invariant (up to a total divergence) under the gauge transformation:

$$A \longmapsto A + d\varphi. \tag{3.46}$$

We wish to compute the *partition function* for the theory

$$Z := \int \frac{1}{V_G} \mathcal{D}\mu[A] \, e^{iS[A]},$$

where V_G denotes the volume of the group of gauge transformations in (3.46), which must be factored out of the partition function in order to guarantee that the integration is performed only over physically distinct gauge fields. We can handle this by using the Hodge decomposition to parametrize the potential A in terms of its gauge invariant, and gauge dependent parts, so that the volume of the group of gauge transformations can be explicitly factored out, leaving a functional integral over gauge invariant modes only [GK94].

We now transform the integration variables:

$$A \longmapsto \alpha, \beta, \gamma,$$

where α, β, γ parameterize respectively the exact, coexact, and harmonic parts of the connection A. Using the Jacobian (3.45) as well as the following identity on 0-forms $\Delta = \delta d$, we get [GK94]

$$Z = \int \frac{1}{V_G} \mathcal{D}\mu[\alpha]\mathcal{D}\mu[\beta]\mathcal{D}\mu[\gamma]\det{}^{1/2}(\Delta)\det{}^{1/2}(d\delta) \, e^{iS},$$

from which it follows that

$$V_G = \int \mathcal{D}\mu[\alpha], \tag{3.47}$$

while the classical action functional becomes, after integrating by parts, using the harmonic properties of γ and the nilpotency of the exterior derivative operators, and dropping surface terms:

$$S = -\langle \beta, \star \delta d \delta \beta \rangle.$$

Note that S depends only the coexact (transverse) part of A. Using (3.47) and integrating over β yields:

$$Z = \int \mathcal{D}\mu[\gamma] \det{}^{-1/2}\left(\star \delta d \delta\right) \det{}^{1/2}\left(\Delta\right) \det{}^{1/2}\left(d\delta\right).$$

Also, it was proven in [GK94] that

$$\det(\star \delta d \delta) = \det{}^{1/2}((d\delta d)(\delta d\delta)) = \det^{\frac{3}{2}}(d\delta).$$

As a consequence of Hodge duality we have the identity

$$\det(\delta d) = \det(d\delta),$$

from which it follows that

$$Z = \int \mathcal{D}\mu[\gamma] \det{}^{-3/4}\left(\Delta_{(1)}^T\right) \det{}^{1/2}\left(\Delta\right) \det{}^{1/2}\left(\Delta_{(1)}^T\right).$$

The operator $\Delta_{(1)}^T$ is the transverse part of the Hodge Laplacian acting on 1-forms:

$$\Delta_{(1)}^T := (\delta d)_{(1)}.$$

Applying identity for the Hodge Laplacian $\Delta_{(p)}$ [GK94]

$$\det\left(\Delta_{(p)}\right) = \det\left((\delta d)_{(p)}\right) \det\left((\delta d)_{(p-1)}\right),$$

we get

$$\det\left(\Delta_{(1)}^T\right) = \det\left(\Delta_{(1)}\right)/\det\left(\Delta\right)$$

and hence

$$Z = \int \mathcal{D}\mu[\gamma] \det{}^{-1/4}\left(\Delta_{(1)}\right) \det{}^{3/4}\left(\Delta\right).$$

The space of harmonic forms γ (of any order) is a finite set. Hence, the integration over harmonic forms (Sect. 3.4.3) is a simple sum.

In the next section we will focus on the Yang–Mills theory.

3.5 Non-Abelian Gauge Theories

3.5.1 Introduction to Non-Abelian Theories

QED is the simplest example of a gauge theory coupled to matter based in the Abelian gauge symmetry of local $U(1)$ phase rotations. However, it is possible also to construct gauge theories based on non-Abelian groups. Actually, our knowledge of the strong and weak interactions is based on the use of such non-Abelian generalizations of QED.

Let us consider a gauge group G (see Appendix) with generators T^a, ($a = 1, \ldots, \dim G$) satisfying the Lie algebra

$$[T^a, T^b] = \mathrm{i} f^{abc} T^c.$$

A gauge field taking values on the Lie algebra of $\mathcal{G}$ can be introduced $A_\mu \equiv A_\mu^a T^a$, which transforms under a gauge transformations as

$$A_\mu \longrightarrow \frac{1}{\mathrm{i}g} U \partial_\mu U^{-1} + U A_\mu U^{-1}, \quad U = \mathrm{e}^{\mathrm{i}\chi^a(x)T^a},$$

where g is the coupling constant. The associated field strength is defined as

$$F_{\mu\nu}^a = \partial_\mu A_\nu^a - \partial_\nu A_\mu^a - g f^{abc} A_\mu^b A_\nu^c.$$

Notice that this definition of the $F_{\mu\nu}^a$ reduces to the one used in QED in the Abelian case when $f^{abc} = 0$. In general, however, unlike the case of QED the field strength is not gauge invariant. In terms of $F_{\mu\nu} = F_{\mu\nu}^a T^a$ it transforms as

$$F_{\mu\nu} \longrightarrow U F_{\mu\nu} U^{-1}.$$

The coupling of matter to a non-Abelian gauge field is done by introducing again the *covariant derivative*. For a field $\Phi \longrightarrow U\Phi$ in a representation of $\mathcal{G}$, the covariant derivative is given by

$$D_\mu \Phi = \partial_\mu \Phi - \mathrm{i}g A_\mu^a T^a \Phi.$$

With the help of this we can write a generic Lagrangian for a non-Abelian gauge field coupled to scalars ϕ and spinors ψ as

$$\mathcal{L} = -\frac{1}{4} F_{\mu\nu}^a F^{\mu\nu a} + \mathrm{i}\bar{\psi}\slashed{D}\psi + \overline{D_\mu\phi}D^\mu\phi - \bar{\psi}\left[M_1(\phi) + \mathrm{i}\gamma_5 M_2(\phi)\right]\psi - V(\phi).$$

3.5.2 Yang–Mills Theory

In non-Abelian gauge theories, gauge fields are matrix-valued functions of space-time. In $SU(N)$ gauge theories they can be represented by the generators of the corresponding Lie algebra, i.e., gauge fields and their color components are related by

$$A_\mu(x) = A_\mu^a(x)\frac{\lambda^a}{2}, \tag{3.48}$$

where the color sum runs over the $N^2 - 1$ generators. The generators are hermitian, traceless $N \times N$ matrices whose commutation relations are specified by the structure constants f^{abc} [Len04]

$$\left[\frac{\lambda^a}{2}, \frac{\lambda^b}{2}\right] = i f^{abc} \frac{\lambda^c}{2}.$$

The normalization is chosen as

$$\text{Tr}\left(\frac{\lambda^a}{2} \cdot \frac{\lambda^b}{2}\right) = \frac{1}{2}\delta_{ab}.$$

Most of our applications will be concerned with $SU(2)$ gauge theories; in this case the generators are the *Pauli matrices*,

$$\lambda^a = \tau^a, \quad \text{with structure constants } f^{abc} = \epsilon^{abc}.$$

Covariant derivative, field strength tensor, and its color components are respectively defined by

$$D_\mu = \partial_\mu + ig A_\mu, \tag{3.49}$$

$$F^{\mu\nu} = \frac{1}{ig}[D_\mu, D_\nu], \quad F_{\mu\nu}^a = \partial_\mu A_\nu^a - \partial_\nu A_\mu^a - g f^{abc} A_\mu^b A_\nu^c. \tag{3.50}$$

The definition of electric and magnetic fields in terms of the field strength tensor is the same as in electrodynamics

$$E^{ia}(x) = -F^{0ia}(x), \quad B^{ia}(x) = -\frac{1}{2}\epsilon^{ijk} F^{jka}(x). \tag{3.51}$$

The dimensions of gauge field and field strength in 4D space-time are

$$[A] = \ell^{-1}, \quad [F] = \ell^{-2},$$

and therefore in absence of a scale, $A_\mu^a \sim M_\mu^a \frac{x^\nu}{x^2}$, with arbitrary constants $M_{\mu\nu}^a$. In general, the action associated with these fields exhibits infrared and ultraviolet logarithmic divergencies. In the following we will discuss

- *Yang–Mills Theories:* Only gauge fields are present. The Yang–Mills Lagrangian is

$$\mathcal{L}_{YM} = -\frac{1}{4}F^{\mu\nu a}F_{\mu\nu}^a = -\frac{1}{2}\text{Tr}(F^{\mu\nu}F_{\mu\nu}) = \frac{1}{2}(\mathbf{E}^2 - \mathbf{B}^2). \tag{3.52}$$

- *Quantum Chromodynamics:* QCD contains besides the gauge fields (gluons), fermion fields (quarks). Quarks are in the fundamental representation, i.e., in $SU(2)$ they are represented by 2-component color spinors. The QCD Lagrangian is (flavor dependences suppressed)

$$\mathcal{L}_{QCD} = \mathcal{L}_{YM} + \mathcal{L}_m, \quad \mathcal{L}_m = \bar{\psi}\left(i\gamma^\mu D_\mu - m\right)\psi, \tag{3.53}$$

with the action of the covariant derivative on the quarks given by

$$(D_\mu\psi)^i = (\partial_\mu\delta^{ij} + igA_\mu^{ij})\psi^j, \quad (i,j = 1,\ldots,N).$$

- *Georgi–Glashow Model:* In the Georgi–Glashow model [GEG72] (non-Abelian Higgs model), the gluons are coupled to a scalar, self-interacting $(V(\phi))$ (Higgs) field in the adjoint representation. The Higgs field has the same representation in terms of the generators as the gauge field (3.48) and can be thought of as a 3-component color vector in $SU(2)$. Lagrangian and action of the covariant derivative are respectively

$$\mathcal{L}_{GG} = \mathcal{L}_{YM} + \mathcal{L}_m, \quad \mathcal{L}_m = \frac{1}{2}D_\mu\phi D^\mu\phi - V(\phi), \tag{3.54}$$

$$(D_\mu\phi)^a = [D_\mu, \phi]^a = (\partial_\mu\delta^{ac} - gf^{abc}A_\mu^b)\phi^c. \tag{3.55}$$

Yang–Mills Action

The general principle of least action,

$$\delta S = 0, \quad \text{with } S = \int \mathcal{L}d^4x,$$

applied to the gauge fields,

$$\delta S_{YM} = -\int d^4x\,\mathrm{Tr}\left(F_{\mu\nu}\delta F^{\mu\nu}\right) = -\int d^4x\,\mathrm{Tr}\left(F_{\mu\nu}\frac{2}{ig}\left[D^\mu, \delta A^\nu\right]\right)$$

$$= 2\int d^4x\,\mathrm{Tr}\left(\delta A^\nu\left[D^\mu, F_{\mu\nu}\right]\right)$$

gives the inhomogeneous field equations [Len04]

$$[D_\mu, F^{\mu\nu}] = j^\nu, \tag{3.56}$$

with j^ν the color current associated with the matter fields

$$j^{a\nu} = \frac{\delta\mathcal{L}_m}{\delta A_\nu^a}. \tag{3.57}$$

For QCD and the Georgi–Glashow model, these currents are given respectively by

$$j^{a\nu} = g\bar{\psi}\gamma^\nu \frac{\tau^a}{2}\psi, \qquad j^{a\nu} = gf^{abc}\phi^b(D^\nu\phi)^c. \tag{3.58}$$

As in electrodynamics, the homogeneous field equations for the Yang–Mills field strength

$$\left[D_\mu, \tilde{F}^{\mu\nu}\right] = 0,$$

with the dual field strength tensor

$$\tilde{F}^{\mu\nu} = \frac{1}{2}\epsilon^{\mu\nu\sigma\rho}F_{\sigma\rho},$$

are obtained as the *Jacobi identities* of the covariant derivative,

$$[D_\mu, [D_\nu, D_\rho]] + [D_\nu, [D_\rho, D_\mu]] + [D_\rho, [D_\nu, D_\mu]] = 0.$$

Gauge Transformations

Gauge transformations change the color orientation of the matter fields locally, i.e., in a space-time dependent manner, and are defined as

$$U(x) = \exp\left\{ig\alpha(x)\right\} = \exp\left\{ig\alpha^a(x)\frac{\tau^a}{2}\right\},$$

with the arbitrary gauge function $\alpha^a(x)$. Matter fields transform covariantly with U

$$\psi v U \psi, \qquad \phi \longrightarrow U\phi U^\dagger. \tag{3.59}$$

The transformation property of A is chosen such that the covariant derivatives of the matter fields $D_\mu\psi$ and $D_\mu\phi$ transform as the matter fields ψ and ϕ respectively. As in electrodynamics, this requirement makes the gauge fields transform inhomogeneously [Len04]

$$A_\mu(x) \longrightarrow U(x)\left(A_\mu(x) + \frac{1}{ig}\partial_\mu\right)U^\dagger(x) = A_\mu^{[U]}(x) \tag{3.60}$$

resulting in a covariant transformation law for the field strength

$$F_{\mu\nu} \longrightarrow UF_{\mu\nu}U^\dagger. \tag{3.61}$$

Under infinitesimal gauge transformations ($|g\alpha^a(x)| \ll 1$)

$$A_\mu^a(x) \longrightarrow A_\mu^a(x) - \partial_\mu\alpha^a(x) - gf^{abc}\alpha^b(x)A_\mu^c(x). \tag{3.62}$$

As in electrodynamics, gauge fields which are gauge transforms of $A_\mu = 0$ are called pure gauges and are, according to (3.60), given by

$$A_\mu^{pg}(x) = U(x)\frac{1}{ig}\partial_\mu U^\dagger(x). \tag{3.63}$$

Physical observables must be independent of the choice of gauge (coordinate system in color space). Local quantities such as the Yang–Mills action density $\text{Tr}(F^{\mu\nu}(x)F_{\mu\nu}(x))$ or matter field bilinears like $\bar{\psi}(x)\psi(x)$, $\phi^a(x)\phi^a(x)$ are gauge invariant, i.e., their value does not change under local gauge transformations. One also introduces non-local quantities which, in generalization of the transformation law (3.61) for the field strength, change homogeneously under gauge transformations. In this construction a basic building block is the path ordered integral

$$\Omega\left(x,y,\mathcal{C}\right) = P\exp\left\{-\mathrm{ig}\int_{s_0}^{s} d\sigma \frac{dx^\mu}{d\sigma} A_\mu\left(x(\sigma)\right)\right\} = P\exp\left\{-\mathrm{ig}\int_{\mathcal{C}} dx^\mu A_\mu\right\}. \tag{3.64}$$

It describes a gauge string between the space-time points $x = x(s_0)$ and $y = x(s)$. Ω satisfies the differential equation

$$\frac{d\Omega}{ds} = -\mathrm{ig}\frac{dx^\mu}{ds}A_\mu\Omega. \tag{3.65}$$

Gauge transforming this differential equation yields the transformation property of Ω

$$\Omega\left(x,y,\mathcal{C}\right) \longrightarrow U\left(x\right)\Omega\left(x,y,\mathcal{C}\right)U^\dagger\left(y\right). \tag{3.66}$$

With the help of Ω, non-local, gauge invariant quantities like

$$\text{Tr}\left(F^{\mu\nu}(x)\Omega\left(x,y,\mathcal{C}\right)F_{\mu\nu}(y)\right), \qquad \bar{\psi}(x)\Omega\left(x,y,\mathcal{C}\right)\psi(y),$$

or closed gauge strings, the following $SU(N)$-Wilson loops

$$W_{\mathcal{C}} = \frac{1}{N}\text{Tr}\left(\Omega\left(x,x,\mathcal{C}\right)\right) \tag{3.67}$$

can be constructed. For pure gauges (3.63), the differential equation (3.65) is solved by

$$\Omega^{pg}\left(x,y,\mathcal{C}\right) = U(x)U^\dagger(y). \tag{3.68}$$

While $\bar{\psi}(x)\Omega(x,y,\mathcal{C})\psi(y)$ is an operator which connects the vacuum with meson states for $SU(2)$ and $SU(3)$, fermionic baryons appear only in $SU(3)$ in which gauge invariant states containing an odd number of fermions can be constructed. In $SU(3)$ a point-like gauge invariant baryonic state is obtained by creating three quarks in a color antisymmetric state at the same space-time point

$$\psi(x) \sim \epsilon^{abc}\psi^a(x)\psi^b(x)\psi^c(x).$$

Under gauge transformations,

$$\psi(x) \longrightarrow \epsilon^{abc}U_{a\alpha}(x)\psi^\alpha(x)U_{b\beta}(x)\psi^\beta(x)U_{c\gamma}(x)\psi^\gamma(x)$$
$$= \det\left(U(x)\right)\epsilon^{abc}\psi^a(x)\psi^b(x)\psi^c(x).$$

Operators that create finite size baryonic states must contain appropriate gauge strings as given by the following expression

$$\psi(x,y,z) \sim \epsilon^{abc}[\Omega(u,x,\mathcal{C}_1)\psi(x)]^a [\Omega(u,y,\mathcal{C}_2)\psi(y)]^b [\Omega(u,z,\mathcal{C}_3)\psi(z)]^c.$$

The presence of these gauge strings makes ψ gauge invariant as is easily verified with the help of the transformation property (3.66). Thus, gauge invariance is enforced by color exchange processes taking place between the quarks.

3.5.3 Quantization of Yang–Mills Theory

Gauge theories are formulated in terms of redundant variables. Only in this way, a covariant, local representation of the dynamics of gauge degrees of freedom is possible. For quantization of the theory both canonically or in the path integral, redundant variables have to be eliminated. This procedure is called gauge fixing. It is not unique and the implications of a particular choice are generally not well understood. In the path integral one performs a sum over all field configurations. In gauge theories this procedure has to be modified by making use of the decomposition of the space of gauge fields into equivalence classes, the gauge orbits. Instead of summing in the path integral over formally different but physically equivalent fields, the integration is performed over the equivalence classes of such fields, i.e., over the corresponding gauge orbits. The value of the action is gauge invariant, i.e., the same for all members of a given gauge orbit. Therefore, the action is seen to be a functional defined on classes (gauge orbits) [Len04]. Also the integration measure

$$\mathcal{D}[A] = \prod_{x,\mu,a} dA_\mu^a(x)$$

is gauge invariant since shifts and rotations of an integration variable do not change the value of an integral. Therefore, in the naive path integral

$$Z[A] = \int \mathcal{D}[A]\, e^{iS[A]} \propto \int \prod_x dU(x).$$

A 'volume' associated with the gauge transformations $\prod_x dU(x)$ can be factorized and thereby the integration be performed over the gauge orbits. To turn this property into a working algorithm, redundant variables are eliminated by imposing a *gauge condition*, $f[A] = 0$, which is supposed to eliminate all gauge copies of a certain field configuration A. In other words, the functional f has to be chosen such that, for arbitrary field configurations, the equation, $f[A^{[U]}] = 0$, determines uniquely the gauge transformation U. If successful, the set of all gauge equivalent fields, the gauge orbit, is represented by exactly one representative. In order to write down an integral over gauge orbits, we insert into the integral the gauge-fixing δ-functional

$$\delta[f(A)] = \prod_x \prod_{a=1}^{N^2-1} \delta[f^a(A(x))].$$

This modification of the integral however changes the value depending on the representative chosen, as the following elementary identity shows

$$\delta\left(g\left(x\right)\right) = \frac{\delta(x-a)}{|g'(a)|}, \qquad g\left(a\right) = 0.$$

This difficulty is circumvented with the help of the Faddeev–Popov determinant $\Delta_f[A]$ defined implicitly by

$$\Delta_f\left[A\right] \int \mathcal{D}\left[U\right] \delta\left[f\left(A^{[U]}\right)\right] = 1.$$

Multiplication of the path integral $Z[A]$ with the above "1" and taking into account the gauge invariance of the various factors yields

$$Z[A] = \int \mathcal{D}\left[U\right] \int \mathcal{D}\left[A\right] e^{iS[A]} \Delta_f\left[A\right] \delta\left[f\left(A^{[U]}\right)\right]$$

$$= \int \mathcal{D}\left[U\right] \int \mathcal{D}\left[A\right] e^{iS[A^{[U]}]} \Delta_f\left[A^{[U]}\right] \delta\left[f\left(A^{[U]}\right)\right].$$

The gauge volume has been factorized and, being independent of the dynamics, can be dropped. In summary, the final definition of the generating functional for gauge theories is given in terms of a *sum over gauge orbits*,

$$Z\left[J\right] = \int \mathcal{D}\left[A\right] \Delta_f\left[A\right] \delta\left(f\left[A\right]\right) e^{iS[A]+i\int d^4x J^\mu A_\mu}.$$

Faddeev–Popov Determinant

For the calculation of $\Delta_f[A]$, we first consider the change of the gauge condition $f^a[A]$ under infinitesimal gauge transformations. Taylor expansion

$$f_x^a\left[A^{[U]}\right] \approx f_x^a\left[A\right] + \int d^4y \sum_{b,\mu} \frac{\delta f_x^a\left[A\right]}{\delta A_\mu^b\left(y\right)} \delta A_\mu^b\left(y\right)$$

$$= f_x^a\left[A\right] + \int d^4y \sum_b M\left(x,y;a,b\right) \alpha^b\left(y\right),$$

with δA_μ^a given by infinitesimal gauge transformations,

$$A_\mu^a\left(x\right) \rightarrow A_\mu^a\left(x\right) - \partial_\mu \alpha^a\left(x\right) - g f^{abc} \alpha^b\left(x\right) A_\mu^c\left(x\right), \quad \text{yields}$$

$$M\left(x,y;a,b\right) = \left(\partial_\mu \delta^{b,c} + g f^{bcd} A_\mu^d\left(y\right)\right) \frac{\delta f_x^a\left[A\right]}{\delta A_\mu^c\left(y\right)}.$$

In the second step, we compute the integral

$$\Delta_f^{-1}\left[A\right] = \int \mathcal{D}\left[U\right] \delta\left[f\left(A^{[U]}\right)\right],$$

by expressing the integration $\mathcal{D}[U]$ as an integration over the gauge functions α. We finally change to the variables $\beta = M\alpha$,

$$\Delta_f^{-1}[A] = |\det M|^{-1} \int \mathcal{D}[\beta]\, \delta\left[f(A) - \beta\right],$$

and arrive at the final expression for the Faddeev–Popov determinant [Len04]

$$\Delta_f[A] = |\det M|.$$

Examples:

• Lorentz gauge

$$f_x^a(A) = \partial^\mu A_\mu^a(x) - \chi^a(x),$$
$$M(x, y; a, b) = -\left(\delta^{ab}\Box - gf^{abc}A_\mu^c(y)\partial_y^\mu\right)\delta^{(4)}(x - y).$$

• Coulomb gauge

$$f_x^a(A) = \operatorname{div}\mathbf{A}^a(x) - \chi^a(x),$$
$$M(x, y; a, b) = \left(\delta^{ab}\Delta + gf^{abc}\mathbf{A}^c(y)\nabla_y\right)\delta^{(4)}(x - y).$$

• Axial gauge

$$f_x^a(A) = n^\mu A_\mu^a(x) - \chi^a(x),$$
$$M(x, y; a, b) = -\delta^{ab}n_\mu\partial_y^\mu\delta^{(4)}(x - y).$$

3.5.4 Basics of Conformal Field Theory (CFT)

A conformal field theory (CFT) is a quantum field theory (or, a statistical mechanics model at the critical point) that is invariant under *conformal transformations*. Conformal field theory is often studied in 2D where there is an infinite-dimensional group of local conformal transformations, described by the holomorphic functions. CFT has important applications in string theory, statistical mechanics, and condensed matter physics. For a good introduction to CFT see [BPZ84, DMS97]. We consider here only chiral CFTs in 2D (see [NSS08]), where 'chiral'[19] means that all of our fields will be functions of a complex number $z = x + iy$ only and not functions of its conjugate $\bar{z}$.

---

[19] In general, a chiral field is a holomorphic field $W(z)$ which transforms as

$$L_n W(z) = -z^{n+1}\frac{\partial}{\partial z}W(z) - (n+1)\Delta z^n W(z), \quad \text{with } \bar{L}_n W(z) = 0,$$

and similarly for an anti-chiral field. Here, Δ is the conformal weight of the chiral field W.

To formally describe a 2D CFT we give its 'conformal data', including a set of primary fields, each with a conformal dimension Δ, a table of fusion rules of these fields and a central charge c. Data for three CFTs are given in Table 3.1.

The *operator product expansion* (OPE) describes what happens to two fields when their positions approach each other. We write the OPE for two arbitrary fields ϕ_i and ϕ_j as

$$\lim_{z \to w} \phi_i(z)\phi_j(w) = \sum_k C_{ij}^k (z-w)^{\Delta_k - \Delta_i - \Delta_j} \phi_k(w),$$

where the *structure constants* C_{ij}^k are only nonzero as indicated by the fusion table. Note that the OPE works *inside* a correlator. For example, in the $\mathbb{Z}_3$ para-fermion CFT (see Table 3.1), since $\sigma_1 \times \psi_1 = \epsilon$, for arbitrary fields ϕ_i we have [NSS08]

$$\lim_{z \to w} \langle \phi_1(z_1) \cdots \phi_M(z_M) \sigma_1(z) \psi_1(w) \rangle$$
$$\sim (z-w)^{2/5 - 1/15 - 2/3} \langle \phi_1(z_1) \cdots \phi_M(z_M) \epsilon(w) \rangle.$$

In addition to the OPE, there is also an important 'neutrality' condition: a correlator is zero unless all of the fields can fuse together to form the identity field $\mathbf{1}$. For example, in the $\mathbb{Z}_3$ para-fermion field theory $\langle \psi_2 \psi_1 \rangle \neq 0$ since $\psi_2 \times \psi_1 = \mathbf{1}$, but $\langle \psi_1 \psi_1 \rangle = 0$ since $\psi_1 \times \psi_1 = \psi_2 \neq \mathbf{1}$.

Chiral Bose Vertex: ($c = 1$)

	Δ	$\times$	$e^{i\alpha\phi}$
$e^{i\alpha\phi}$	$\alpha^2/2$	$e^{i\beta\phi}$	$e^{i(\alpha+\beta)\phi}$

Ising CFT: ($c = 1/2$)

	Δ	$\times$	ψ	σ
ψ	$1/2$	ψ	$\mathbf{1}$	
σ	$1/16$	σ	σ	$\mathbf{1} + \psi$

$\mathbb{Z}_3$ Parafermion CFT: ($c = 4/5$)

	Δ	$\times$	ψ_1	ψ_2	σ_1	σ_2	ϵ
ψ_1	$2/3$	ψ_1	ψ_2				
ψ_2	$2/3$	ψ_2	$\mathbf{1}$	ψ_1			
σ_1	$1/15$	σ_1	ϵ	σ_2	$\sigma_2 + \psi_1$		
σ_2	$1/15$	σ_2	σ_1	ϵ	$\mathbf{1} + \epsilon$	$\sigma_1 + \psi_2$	
ϵ	$2/5$	ϵ	σ_2	σ_1	$\sigma_1 + \psi_2$	$\sigma_2 + \psi_1$	$\mathbf{1} + \epsilon$

Table 3.1. Conformal data for three CFTs. Given is the list of primary fields in the CFT with their conformal dimension Δ, as well as the fusion table. In addition, every CFT has an identity field $\mathbf{1}$ with dimension $\Delta = 0$ which fuses trivially with any field ($\mathbf{1} \times \phi_i = \phi_i$ for any ϕ_i). Note that fusion tables are symmetric so only the lower part is given. In the Ising CFT the field ψ is frequently notated as ϵ. This fusion table indicates the nonzero elements of the fusion matrix N_{ab}^c. For example in the $\mathbb{Z}_3$ CFT, since $\sigma_1 \times \sigma_2 = \mathbf{1} + \epsilon$, $N_{\sigma_1\sigma_2}^1 = N_{\sigma_1\sigma_2}^\epsilon = 1$ and $N_{\sigma_1\sigma_2}^c = 0$ for all c not equal to $\mathbf{1}$ or ϵ.

Let us look at what happens when a fusion has more than one possible result. For example, in the Ising CFT, $\sigma \times \sigma = 1 + \psi$. Using the OPE, we have

$$\lim_{w_1 \to w_2} \sigma(w_1)\sigma(w_2) \sim \frac{1}{(w_1 - w_2)^{1/8}} + (w_1 - w_2)^{3/8}\psi, \qquad (3.69)$$

where we have neglected the constants C_{ij}^k. If we consider $\langle \sigma\sigma \rangle$, the neutrality condition picks out only the first term in (3.69) where the two σ's fuse to form 1. Similarly, $\langle \sigma\sigma\psi \rangle$ results in the second term of (3.69) where the two σ's fuse to form ψ which then fuses with the additional ψ to make 1.

Fields may also fuse to form the identity in more than one way. For example, in the correlator $\langle \sigma(w_1)\sigma(w_2)\sigma(w_3)\sigma(w_4) \rangle$ of the Ising CFT, the identity is obtained via two possible fusion paths—resulting in two different so-called 'conformal blocks'. On the one hand, one can fuse $\sigma(w_1)$ and $\sigma(w_2)$ to form 1 and similarly fuse $\sigma(w_3)$ and $\sigma(w_4)$ to form 1. Alternately, one can fuse $\sigma(w_1)$ and $\sigma(w_2)$ to form ψ and fuse $\sigma(w_3)$ and $\sigma(w_4)$ to form ψ then fuse the two resulting ψ fields together to form 1. The correlator generally gives a linear combination of the possible resulting conformal blocks. We should thus think of such a correlator as living in a vector space rather than having a single value. (If we instead choose to fuse 1 with 3, and 2 with 4, we would obtain two blocks which are linear combinations of the ones found by fusing 1 with 2 and 3 with 4. The resulting vectors space, however, is independent of the order of fusion.) Crucially, transporting the coordinates w_i around each other makes a rotation within this vector space.

To be clear about the notion of conformal blocks, let us look at the explicit form of the Ising CFT correlator [NSS08]

$$\lim_{w \to \infty} \langle \sigma(0)\sigma(z)\sigma(1)\sigma(w) \rangle = a_+ F_+ + a_- F_-,$$

$$F_\pm(z) \sim (wz(1-z))^{-1/8}\sqrt{1 \pm \sqrt{1-z}},$$

where a_+ and a_- are arbitrary coefficients. When $z \to 0$ we have $F_+ \sim z^{-1/8}$ whereas $F_- \sim z^{3/8}$. Comparing to (3.69) we conclude that F_+ is the result of fusing $\sigma(0) \times \sigma(z) \to 1$ whereas F_- is the result of fusing $\sigma(0) \times \sigma(z) \to \psi$. As z is taken in a clockwise circle around the point $z = 1$, the inner square-root changes sign, switching F_+ and F_-. Thus, this 'braiding' (or 'monodromy') operation transforms

$$\begin{pmatrix} a_+ \\ a_- \end{pmatrix} \to e^{2\pi i/8} \begin{pmatrix} 0 & 1 \\ 1 & 0 \end{pmatrix} \begin{pmatrix} a_+ \\ a_- \end{pmatrix}.$$

Having a multiple valued correlator (i.e., multiple conformal blocks) is a result of having such branch cuts. Braiding the coordinates (w's) around each other results in the correlator changing values within its allowable vector space.

A useful technique for counting conformal blocks is the *Bratteli diagram*. In Fig. 3.6 we give the Bratteli diagram for the fusion of multiple σ fields in the Ising CFT. Starting with **1** at the lower left, at each step moving from the left to the right, we fuse with one more σ field. At the first step, the arrow points from **1** to $\boldsymbol{\sigma}$ since $\mathbf{1} \times \sigma = \sigma$. At the next step σ fuses with σ to produce either ψ or **1** and so forth. Each conformal block is associated with a path through the diagram. Thus to determine the number of blocks in $\langle \sigma\sigma\sigma\sigma \rangle$ we count the number of paths of four steps in the diagram starting at the lower left and ending at **1**.

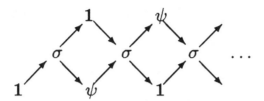

Fig. 3.6. Bratteli diagram for fusion of multiple σ_1 fields in the $\mathbb{Z}_3$ para-fermion CFT (modified and adapted from [NSS08]).

A particularly important CFT is obtained from a free Bose field theory in $(1+1)$D by keeping only the left moving modes. The free chiral Bose field $\phi(z)$, which is a sum of left moving creation and annihilation operators, has a correlator $\langle \phi(z)\phi(z') \rangle = -\log(z - z')$. We then define the normal ordered 'chiral vertex operator', $e^{i\alpha\phi(z)}$, which is a conformal field. Since ϕ is a free field, Wick's theorem can be used to obtain [DMS97]

$$\left\langle e^{i\alpha_1\phi}(z_1) \cdots e^{i\alpha_N\phi}(z_N) \right\rangle = e^{-\sum_{i<j} \alpha_i \alpha_j \langle \phi(z_i)\phi(z_j) \rangle} = \prod_{i<j}(z_i - z_j)^{\alpha_i \alpha_j}.$$

4

Spatio-Temporal Chaos, Solitons and NLS

In the introductory Sect. 1.5 we stated the conjecture [BN04] that dynamical systems that are capable of doing complex computational tasks should operate near the *edge of chaos*. Following this view, in this Chapter we present several fields from nonlinear dynamics that are related to quantum neural computation, including classical and quantum chaos, turbulence and solitons, with the special treatment to nonlinear Schrödinger equation (NLS, the basic model of quantum neural networks).

4.1 Reaction–Diffusion Processes and Ricci Flow

Parabolic *reaction–diffusion* systems are abundant in mathematical biology. They are mathematical models that describe how the concentration of one or more substances distributed in space changes under the influence of two processes: local chemical reactions in which the substances are converted into each other, and diffusion which causes the substances to spread out in space. More formally, they are expressed as semi-linear parabolic partial differential equations (PDEs, see e.g., [PBL05]). The evolution of the state vector $\mathbf{u}(\mathbf{x}, t)$ describing the concentration of the different reagents is determined by anisotropic diffusion as well as local reactions:

$$\partial_t \mathbf{u} = \mathbf{D}\Delta\mathbf{u} + \mathbf{R}(\mathbf{u}), \quad (\partial_t = \partial/\partial t), \tag{4.1}$$

where each component of the state vector $\mathbf{u}(\mathbf{x}, t)$ represents the concentration of one substance, Δ is the standard Laplacian operator, $\mathbf{D}$ is a symmetric positive-definite matrix of diffusion coefficients (which are proportional to the velocity of the diffusing particles) and $\mathbf{R}(\mathbf{u})$ accounts for all local reactions. The solutions of reaction–diffusion equations display a wide range of behaviors, including the formation of traveling waves and other self-organized patterns like *dissipative solitons* (DSs).

V.G. Ivancevic, T.T. Ivancevic, *Quantum Neural Computation*,
Intelligent Systems, Control and Automation: Science and Engineering 40,
DOI 10.1007/978-90-481-3350-5_4, © Springer Science+Business Media B.V. 2010

On the other hand, the *Ricci flow equation* (or, the parabolic Einstein equation), introduced by R. Hamilton in 1982 [Ham82], is the nonlinear heat-like evolution equation[1]

$$\partial_t g_{ij} = -2R_{ij}, \tag{4.2}$$

for a time-dependent Riemannian metric $g = g_{ij}(t)$ on a smooth real[2] n-manifold M with the Ricci curvature tensor R_{ij}.[3] This equation roughly says that we can deform any metric on a 2-surface or n-manifold by the negative of its curvature; after *normalization* (see Fig. 4.1), the final state of such deformation will be a metric with constant curvature. However, this is not true in general since, in addition to the presence of singularities, the limits could be Ricci solitons (see below). The factor of 2 in (4.2) is more or less arbitrary, but the negative sign is essential to insure a kind of global *volume exponential decay*,[4] since the Ricci flow equation (4.2) is a kind of nonlinear

---

[1] A current hot topic in geometric topology is the Ricci flow, a Riemannian evolution machinery that recently allowed G. Perelman to prove the celebrated *Poincaré Conjecture*, a century-old mathematics problem (and one of the seven Millennium Prize Problems of the Clay Mathematics Institute)—and won him the 2006 Fields Medal (which he declined in a public controversy) [Mac06]. The Poincaré Conjecture can roughly be put as a question: Is a closed 3-manifold M topologically a sphere if every closed curve in M can be shrunk continuously to a point? In other words, Poincaré conjectured: A simply-connected compact 3-manifold is diffeomorphic to the 3-sphere S^3 (see e.g., [Yau06]).

[2] For the related Kähler–Ricci flow on complex manifolds, see e.g., [II07e].

[3] This particular PDE (4.2) was chosen by Hamilton for much the same reason that A. Einstein introduced the Ricci tensor into his gravitation field equation,

$$R_{ij} - \frac{1}{2}g_{ij}R = 8\pi T_{ij},$$

where T_{ij} is the energy–momentum tensor. Einstein needed a symmetric 2-index tensor which arises naturally from the metric tensor g_{ij} and its first and second partial derivatives. The Ricci tensor R_{ij} is essentially the only possibility. In gravitation theory and cosmology, the Ricci tensor has the volume-decreasing effect (i.e., convergence of neighboring geodesics, see [HP96b]).

[4] This complex geometric process is globally similar to a generic exponential decay ODE:

$$\dot{x} = -\lambda f(x),$$

for a positive function $f(x)$. We can get some insight into its solution from the simple exponential decay ODE,

$$\dot{x} = -\lambda x \quad \text{with the solution } x(t) = x_0 e^{-\lambda t},$$

(where $x = x(t)$ is the observed quantity with its initial value x_0 and λ is a positive decay constant), as well as the corresponding nth order rate equation (where $n > 1$

geometric generalization of the standard linear *heat equation*[5]

$$\partial_t u = \Delta u. \tag{4.3}$$

Like the heat equation (4.3), the Ricci flow equation (4.2) is well behaved in forward time and acts as a kind of smoothing operator (but is usually impossible to solve in backward time). If some parts of a solid object are hot and others are cold, then, under the heat equation, heat will flow from hot to cold, so that the object gradually attains a uniform temperature. To some extent the Ricci flow behaves similarly, so that the Ricci curvature 'tries' to become more uniform [Mil03], thus resembling a monotonic *entropy growth*,[6] $\partial_t S \geq 0$, which is due to the positive definiteness of the metric $g_{ij} \geq 0$, and naturally implying the *arrow of time* [Pen79, II07e].

In a suitable local coordinate system, the Ricci flow equation (4.2) has a nonlinear heat-type form, as follows. At any time t, we can choose local harmonic coordinates so that the coordinate functions are locally defined harmonic functions in the metric $g(t)$. Then the Ricci flow takes the general form (see e.g., [And04])

$$\partial_t g_{ij} = \Delta_M g_{ij} + Q_{ij}(g, \partial g), \tag{4.4}$$

where Δ_M is the *Laplace–Beltrami operator* (4.5) and $Q = Q_{ij}(g, \partial g)$ is a lower-order term quadratic in g and its first order partial derivatives ∂g. From the analysis of nonlinear heat PDEs, one obtains existence and uniqueness of forward-time solutions to the Ricci flow on some time interval, starting at any smooth initial metric g_0.

The quadratic Ricci flow equation (4.4) is our geometric framework for general bio-reaction–diffusion systems, so that the spatio-temporal PDE (4.1) corresponds to the quadratic Ricci flow PDE

---

is an integer),

$$\dot{x} = -\lambda x^n \quad \text{with the solution} \quad \frac{1}{x^{n-1}} = \frac{1}{x_0{}^{n-1}} + (n-1)\lambda t.$$

[5] More precisely, the negative sign is to make the equation parabolic so that there is a theory of existence and uniqueness. Otherwise the equation would be backwards parabolic and not have any theory of existence, uniqueness, etc.

[6] Note that two different kinds of entropy functional have been introduced into the theory of the Ricci flow, both motivated by concepts of entropy in thermodynamics, statistical mechanics and information theory. One is Hamilton's entropy, the other is Perelman's entropy. While in Hamilton's entropy, the scalar curvature R of the metric g_{ij} is viewed as the leading quantity of the system and plays the role of a probability density, in Perelman's entropy the leading quantity describing the system is the metric g_{ij} itself. Hamilton established the monotonicity of his entropy along the volume-normalized Ricci flow on the 2-sphere S^2 [Ham88]. Perelman established the monotonicity of his entropy along the Ricci flow in all dimensions [Per02].

$$\partial_t \mathbf{u} = \mathbf{D}\Delta \mathbf{u} + \mathbf{R}(\mathbf{u})$$
$$\updownarrow \qquad \updownarrow \qquad \qquad \updownarrow$$
$$\partial_t g_{ij} = \Delta_M g_{ij} + Q_{ij}(g, \partial g)$$

with:

- the metric $g = g_{ij}$ on an n-manifold M corresponding to the n-dimensional (or n-component, or n-phase) concentration $\mathbf{u}(\mathbf{x}, t)$;
- the Laplace–Beltrami differential operator Δ_M, as defined on C^2-functions on an n-manifold M, with respect to the Riemannian metric g_{ij}, by

$$\Delta_M \equiv \frac{1}{\sqrt{\det(g)}} \frac{\partial}{\partial x^i} \left(\sqrt{\det(g)} g^{ij} \frac{\partial}{\partial x^j} \right) \qquad (4.5)$$

—corresponding to the n-dimensional bio-diffusion term $\mathbf{D}\Delta \mathbf{u}$; and

- the quadratic n-dimensional Ricci-term, $Q = Q_{ij}(g, \partial g)$, corresponding to the n-dimensional bio-reaction term, $\mathbf{R}(\mathbf{u})$.

As a simple example of the Ricci flow equations (4.2)–(4.4), consider a round spherical boundary S^2 of the 3-ball radius r. The metric tensor on S^2 takes the form

$$g_{ij} = r^2 \hat{g}_{ij},$$

where $\hat{g}_{ij}$ is the metric for a unit sphere, while the Ricci tensor

$$R_{ij} = (n-1)\hat{g}_{ij}$$

is independent of r. The Ricci flow equation on S^2 reduces to

$$\dot{r}^2 = -2(n-1), \quad \text{with the solution } r^2(t) = r^2(0) - 2(n-1)t.$$

Thus the boundary sphere S^2 collapses to a point in finite time (see [Mil03]).

More generally, the geometrization conjecture [Thu82] holds for any 3-manifold M (see below). Suppose that we start with a compact initial 3-manifold M_0 whose Ricci tensor R_{ij} is everywhere positive definite. Then, as M_0 shrinks to a point under the Ricci flow (4.2), it becomes rounder and rounder. If we rescale the metric g_{ij} on M_0 so that the volume of M_0 remains constant, then M_0 converges towards another compact 3-manifold M_1 of constant positive curvature (see [Ham82]).

In case of even more general 3-manifolds (outside the class of positive Ricci curvature metrics), the situation is much more complicated, as various singularities may arise. One way in which singularities may arise during the Ricci flow is that a spherical boundary $S^2 = \partial M$ of an 3-manifold M may collapse to a point in finite time. Such collapses can be eliminated by performing a kind of 'geometric surgery' on the 3-manifold M, that is a sophisticated sequence of cutting and pasting without accumulation of time errors[7] (see [Per03b]).

---

[7] Hamilton's idea was to perform surgery to cut off the singularities and continue his flow after the surgery. If the flow develops singularities again, one repeats the

After a finite number of such surgeries, each component either: (i) converges towards a 3-manifold of constant positive Ricci curvature which shrinks to a point in finite time, or possibly (ii) converges towards an $S^2 \times S^1$ which shrinks to a circle S^1 in finite time, or (iii) admits a 'thin-thick' decomposition of [Thu82]. Therefore, one can choose the surgery parameters so that there is a well defined Ricci flow with surgery, that exists for all time [Per03b].

In this section we use the evolving n-dimensional geometric machinery of the volume-decaying and entropy-growing Ricci flow $g(t)$, given by equations (4.2)–(4.4), for modeling various biological reaction–diffusion systems and dissipative solitons, defined by special cases of the general spatio-temporal model (4.1).

4.1.1 Bio-Reaction–Diffusion Systems

In case of ideal mixtures, the driving force for the general diffusion $\mathbf{D}\Delta\mathbf{u}$ (4.1) is the concentration gradient $-\nabla\mathbf{u}$, or the gradient of the chemical potential $-\nabla u_i$ of each species u_i, $(i = 1,\dots,n)$, giving the *diffusion flux* by the First Fick's law,

$$J = -\mathbf{D}\nabla\mathbf{u}. \qquad (4.6)$$

Assuming the diffusion coefficients $\mathbf{D}$ to be a constant, the Second Fick's law gives the linear parabolic heat equation,

$$\partial_t \mathbf{u} = \mathbf{D}\Delta\mathbf{u}, \qquad (4.7)$$

while, in case of variable diffusion coefficients $\mathbf{D}$, we get (slightly) more general parabolic *diffusion equation*,

$$\partial_t \mathbf{u} = \nabla \cdot (\mathbf{D}\nabla\mathbf{u}), \qquad (4.8)$$

which is still analogous to the 'linear' part of the quadratic Ricci flow equation (4.4),

$$\partial_t g_{ij} = \Delta_M g_{ij},$$

due to general 'diffusion properties' of the Laplace–Beltrami operator Δ_M.

The n-dimensional diffusion coefficient $\mathbf{D} = \mathbf{D}(T)$ at different temperatures T can be approximated by the Arrhenius exponential-decay relation,

$$\mathbf{D}(t) = \mathbf{D}_0 e^{-\frac{E_A}{rT}},$$

---

process of performing surgery and continuing the flow. If one can prove there are only a finite number of surgeries in any finite time interval, and if the long-time behavior of solutions of the Ricci flow (4.2) with surgery is well understood, then one would be able to recognize the topological structure of the initial manifold. Thus Hamilton's program, when carried out successfully, would lead to a proof of the Poincaré Conjecture and Thurston's Geometrization Conjecture [Yau06].

where $\mathbf{D}_0$ is the maximum possible diffusion coefficient (at infinite temperature T), E_A is the activation energy for diffusion (i.e., the energy that must be overcome in order for a chemical reaction to occur) and r is the gas constant.

Using the First Fick's first law (4.6), the diffusion equation (4.8) can be derived in a straightforward way from the *continuity equation*, which states that a change in density in any part of the system is due to inflow and outflow of material into and out of that part of the system (effectively, no material is created or destroyed),

$$\partial_t \mathbf{u} + \nabla \cdot \mathbf{j} = 0,$$

where $\mathbf{j}$ is the flux of the diffusing material.

The most important special case of (4.7) is at a steady state, when the concentrations $\mathbf{u}$ do not change in time, giving the *Laplace's equation*,

$$\Delta \mathbf{u} = 0, \quad \text{or} \quad \Delta u_i = 0, \tag{4.9}$$

for harmonic functions $\mathbf{u} = \{u_i\}$.

The stochastic version of the deterministic heat equation (4.7), connected with the study of Brownian motion,[8] is the *Fokker–Planck equation* (see e.g., [II08a]),

$$\partial_t f = -\partial_{x^i}\left[D_i^1(x^i)f\right] + \partial_{x^i x^j}\left[D_{ij}^2(x^i)f\right], \tag{4.10}$$

$(\partial_{x^i} = \frac{\partial}{\partial x_i},\ \partial_{x^i x^j} = \frac{\partial^2}{\partial x_i \partial x_j})$, where D_i^1 is the drift vector and D_{ij}^2 the diffusion tensor (which results from the presence of the stochastic force). The Fokker–Planck equation (4.10) is used for computing the probability densities of stochastic differential equations.[9]

Also, notice that the real-valued heat equation (4.7) is formally similar to the complex-valued *Schrödinger equation* (see Chap. 3).

---

[8] Brownian motion is the random movement of particles suspended in a liquid or gas or the mathematical model used to describe such random movements, often called a particle theory. The infinitesimal generator (and hence characteristic operator) of a Brownian motion on $\mathbb{R}^n$ is $\frac{1}{2}\Delta$, where Δ is the Laplacian on $\mathbb{R}^n$. More generally, a Brownian motion on an n-manifold M is given by one-half of the Laplace–Beltrami operator Δ_M (4.5).

[9] Consider the Itō stochastic differential equation,

$$d\mathbf{X}_t = \boldsymbol{\mu}(\mathbf{X}_t, t)\, dt + \boldsymbol{\sigma}(\mathbf{X}_t, t)\, d\mathbf{W}_t,$$

where $\mathbf{X}_t \in \mathbb{R}^n$ is the state of an n-dimensional stochastic system at time t and $\mathbf{W}_t \in \mathbb{R}^m$ is the standard mD Wiener process. If the initial distribution is $\mathbf{X}_0 \sim f(\mathbf{x}, 0)$, then the probability density of the state is given by the Fokker–Planck equation (4.10) with the drift and diffusion terms,

$$D_i^1(\mathbf{x}, t) = \mu_i(\mathbf{x}, t) \quad \text{and} \quad D_{ij}^2(\mathbf{x}, t) = \frac{1}{2}\sum_k \sigma_{ik}(\mathbf{x}, t)\sigma_{kj}^{\mathsf{T}}(\mathbf{x}, t).$$

In the remainder of this section, we will review a number of particular bio-reaction–diffusion systems, which can all be subsumed by the quadratic Ricci flow model (4.4).

1-Component Systems

Kolmogorov–Petrovsky–Piscounov Equation

The simplest bio-reaction–diffusion PDE concerning the concentration $u = u(x, t)$ of a single substance in one spatial dimension,

$$\partial_t u = D \partial_x^2 u + R(u), \tag{4.11}$$

is also referred to as the Kolmogorov–Petrovsky–Piscounov (KPP) equation. If the reaction term vanishes, then the equation represents a pure diffusion process described by the heat equation. In particular, the choice

$$R(u) = u(1 - u)$$

yields *Fisher's equation* that was originally used to describe the spreading of biological populations.[10]

The one-component KPP equation (4.11) can also be written in the variational (gradient) form

$$\partial_t u = -\frac{\delta F}{\delta u}, \tag{4.12}$$

and therefore describes a permanent decrease (a kind of exponential decay) of the system's *free energy* functional

$$F = \int_{-\infty}^{\infty} \left[\frac{D}{2} (\partial_x u)^2 + V(u) \right] dx,$$

where $V(u)$ is the potential such that

$$R(u) = -\frac{dV(u)}{du}. \tag{4.13}$$

---

[10] In addition, the effects of convection and quenched spatial disorder on the evolution of a population density are described by a generalization of the Fisher/KPP equation given by [NS98]

$$\partial_t u = D \nabla^2 u + Uu - qu^2,$$

where $u = (\mathbf{x}, t)$ represents the population density, D is a spatially homogeneous diffusion constant, $U = U(\mathbf{x})$ is a spatially inhomogeneous growth term, and $q = b\ell_0{}^d$ is a competition term (b is a competition rate and ℓ_0 is the microscopic length scale at which two particles will compete with one another). One simple form of inhomogeneity considered in these works is a 'square well' potential $U(\mathbf{x})$ which consists of a uniform space with negative growth rate (termed the 'desert'), in which a single region of positive growth rate (an 'oasis') is placed. This model has proven to be applicable to experiments with bacteria populations in adverse environments [MD07].

Swift–Hohenberg Equation

The Swift–Hohenberg (SH) equation, noted for its pattern-forming behavior, is the decaying reaction–diffusion PDE,

$$\partial_t u = -(1 + \Delta)^2 u + R(u), \tag{4.14}$$

given by the variational (gradient) equation (4.12) with the free energy functional

$$F = \int_\Omega \left[V(u) + \frac{1}{2}\left((1 + \Delta)u\right)^2 \right] dx dy,$$

where $R(u)$ is given by (4.13), while Ω is a 2-dimensional region in which (bio)chemical pattern formation occurs.

The time derivative of the free energy F is given by

$$\partial_t F = \int_\Omega \left[\frac{dV(u)}{du} + (1 + \Delta)u \right] \partial_t u \, dx dy,$$

and, since the expression in square brackets is equal to the negative right-hand side of (4.14), we have

$$\dot{F} = -\int_\Omega (\partial_t u)^2 \, dx dy \leq 0.$$

Therefore, the free energy F is the *Lyapunov functional* that may only decrease as it evolves along its trajectory in some phase space. If F has no minima, then when the horizontal scale of the liquid container is large compared to the instability wavelength, a propagating front will be observed (e.g., in chemically reacting flames). In this case, F will decrease continuously until the front approaches the boundary of the medium. An alternative possibility is realized when F has one or several minima, each corresponding to a local equilibrium state in time. In this case the so-called multi-stability is possible. Therefore, the limit behavior of gradient systems of the form of (4.13) is characterized by either a steady attractor or propagating fronts [REW00].

Ginzburg–Landau Equation

One of the most popular models in the pattern-formation theory is the complex Ginzburg–Landau equation (see e.g., [REW00]),

$$\partial_t A = \varepsilon A + (1 + i\alpha)\Delta A - (1 + i\beta)|A|^2 A, \tag{4.15}$$

where A is the complex wave amplitude, $i = \sqrt{-1}, \varepsilon$ is the super-criticality parameter, while α and β measure linear and nonlinear dispersion (the dependence of the frequency of the waves on the wave-number), respectively. Equation (4.15) describes a vast array of phenomena including nonlinear waves, second-order phase transitions, Rayleigh–Bénard convection and superconductivity. The equation describes the evolution of amplitudes of unstable modes for any process exhibiting a Hopf bifurcation, for which a continuous spectrum of unstable wave-numbers is taken into account. It can be viewed as a highly general normal form for a large class of bifurcations and nonlinear wave phenomena in spatially extended systems.[11]

In particular, if we put $\alpha = \beta = 0$ in (4.15), we get the real, or dissipative, Ginzburg–Landau equation,

$$\partial_t A = \varepsilon A + \Delta A - |A|^2 A, \tag{4.17}$$

which is a gradient equation: $\partial_t A = -\delta F/\delta A$, with the free energy functional

$$F = -\int_\Omega \left[\varepsilon|A|^2 - \frac{1}{2}|A|^4 + (\nabla A)^2 \right] dxdy.$$

Using the fact that

$$\dot{F} = -\int_\Omega |\partial_t A|^2 dxdy \le 0,$$

solutions of (4.17) at $t \to \infty$ are either stationary field-distributions satisfying, for $\varepsilon = 1$, the equation

$$\Delta A + A - |A|^2 A = 0, \tag{4.18}$$

of fronts whose propagation is accompanied by a decrease of the functional F. The functional must reach its minimum at stable stationary solutions of (4.18).

---

[11] The extension of the complex Ginzburg–Landau equation (4.15), which describes strongly resonant multi-frequency forcing of the form

$$F = f_1 e^{i\omega \bar{t}} + f_2 e^{2i\omega \bar{t}} + f_3 e^{3i\omega \bar{t}} + c.c.$$

was recently proposed in [CR08] by considering the analogous center-manifold reduction of the extended dynamical system in which the forcing amplitudes f_1, f_2, and f_3 are considered as dynamical variables that vary on the slow time scale t. Under time translations $T_\tau : A \to Ae^{i\omega \tau}$, they transform as $f_1 \to f_1 e^{i\omega \tau}, f_2 \to f_2 e^{2i\omega \tau}, f_3 \to f_3 e^{3i\omega \tau}$. To cubic order in A the most general equation that is equivariant under T_τ is then given by

$$\partial_t A = a_1 + a_2 A + a_3 \Delta A + a_4 A|A|^2 + a_5 \bar{A} + a_6 \bar{A}^2, \tag{4.16}$$

where $a_1 = b_{11} f_1 + b_{12} \bar{f}_2 f_3$, $a_2 = b_{21} + b_{22}|f_3|^2$, $a_5 = b_{51} f_2$, $a_6 = b_{61} f_3$. The forcing terms f_j satisfy decoupled evolution equations on their own. In the simplest case this evolution expresses a de-tuning ν_j of the forcing f_j from the respective resonance and the f_j satisfy

$$\dot{f}_j = i\nu_j f_j, \quad (j = 1 \ldots 3).$$

In general, the de-tuning introduces time dependence into (4.16).

Neural Field Theory

The dynamical system from which the temporal evolution of neural activation fields (see Appendix) is generated is constrained by the postulate that localized peaks of activation are stable objects, or formally, *fixed-point attractors*. Such a field dynamics has the generic form [Sch07]

$$\tau \partial_t u = -u + \text{resting level} + \text{input} + \text{interaction}, \tag{4.19}$$

where $u = u(x,t)$ is the activation field defined over the metric dimension x and time t. The first three terms define an input driven regime, in which attractor solutions have the form

$$u(x,t) = \text{resting level} + \text{input}.$$

The *rate of relaxation* is determined by the time scale parameter τ. The interaction stabilizes localized peaks of activation against decay by local excitatory interaction and against diffusion by global inhibitory interaction. In Amari's formulation [Ama77] the conceptual model (4.19) is specified as a *continuous model for neural activity in cortical structures*,

$$\tau \partial_t u(x,t) = -u(x,t) + h + S(x,t) + \int dx' w(x-x')\sigma(u(x',t)), \tag{4.20}$$

where $h < 0$ is a constant resting level, $S(x,t)$ is spatially and temporally variable input function, $w(x)$ is an interaction kernel and $\sigma(u)$ is a sigmoidal nonlinear threshold function. The interaction term collects input from all those field sites x' at which activation is sufficiently large. The interaction kernel determines if inputs from those sites are positive, driving up activation (excitatory), or negative, driving down activation (inhibitory). Excitatory input from nearby location and inhibitory input from all field locations generically stabilizes localized peaks of activation. For this class of dynamics, detailed analytical results provide a framework for the inverse dynamics task facing the modeler, determining a dynamical system that has the appropriate attractor solutions [II07a, IJP08].[12]

---

[12] Recently, a *neural attractor dynamics* (NAD) was designed (see [Sch07]) based on a discretization for single neurons of Amari's neural field equation (4.20). The so-called *discrete Amari equation* describes the temporal evolution of the activity of all single neurons considering positive and negative contributions from external input and internal neural interactions. Since only activated neurons can have an impact on other neurons, the neural attractor dynamics is nonlinear, and effects of bi-stability and hysteresis can be used for low-level memory and neural competition. The NAD describes the temporal rate of change of the dynamical variable u_i of neural activity for all behavioral neurons i. It is formulated as the following differential equation:

$$\tau \dot{u}_i = -u_i + h + s_i^{\text{beh}} + c_{\text{mot}} \cdot \sigma(m_i) + \alpha_{\text{selfexc},i}^{\text{beh}} + \alpha_{\text{exc},i}^{\text{beh}} - \alpha_{\text{inh},i}^{\text{beh}}, \tag{4.21}$$

where the system parameters have the following meaning:

2-Component Systems

Two-component systems allow for a much larger range of possible phenomena than their one-component counterparts. An important idea that was first proposed by A. Turing is that a state that is stable in the local system should become unstable in the presence of diffusion [Tur52]. This idea seems counterintuitive at first glance as diffusion is commonly associated with a stabilizing effect. However, the linear stability analysis shows that when linearizing the general two-component system

$$
\begin{pmatrix} \partial_t u \\ \partial_t v \end{pmatrix} = \begin{pmatrix} D_u & 0 \\ 0 & D_v \end{pmatrix} \begin{pmatrix} \partial_{xx} u \\ \partial_{xx} v \end{pmatrix} + \begin{pmatrix} F(u,v) \\ G(u,v) \end{pmatrix}
$$

and perturbing the system against plane waves

$$
\tilde{\mathbf{u}}_{\mathbf{k}}(\mathbf{x}, t) = \begin{pmatrix} \tilde{u}(t) \\ \tilde{v}(t) \end{pmatrix} e^{i\mathbf{k}\cdot\mathbf{x}}
$$

close to a stationary homogeneous solution, one finds [II07d]

$$
\begin{pmatrix} \partial_t \tilde{u}_{\mathbf{k}}(t) \\ \partial_t \tilde{v}_{\mathbf{k}}(t) \end{pmatrix} = -k^2 \begin{pmatrix} D_u \tilde{u}_{\mathbf{k}}(t) \\ D_v \tilde{v}_{\mathbf{k}}(t) \end{pmatrix} + \mathbf{R}' \begin{pmatrix} \tilde{u}_{\mathbf{k}}(t) \\ \tilde{v}_{\mathbf{k}}(t) \end{pmatrix}.
$$

τ, the constant relaxation rate, i.e., the time scale on which the dynamics reacts to changes;

h, the constant negative resting level of neural activation;

$\sigma(.)$, a sigmoidal function, which maps the value of neural activity onto $[0,1]$, given by

$\sigma(u) = \frac{1}{1+e^{-\beta u}}$, where β ($=100$) parameterizes the slope of the resulting function;

s_i^{beh}, the adequate stimulus provided by sensory input of a certain duration;

u_i, activity of behavioral neuron i, i.e., activity of behavior i;

c_{mot}, a constant for weighting the motivational contribution, $c_{\text{mot}} < |h|$;

$\alpha_{\text{selfexc},i}^{\text{beh}}$ excitatory contribution of neuron i's own activity u_i;

$\alpha_{\text{exc},i}^{\text{beh}}$, all excitatory contribution of active neurons connected to neuron i;

$\alpha_{\text{inh},i}^{\text{beh}}$, all inhibitory contribution of active neurons connected to neuron i;

m_i, activity of motivational neuron i, i.e., motivation of behavior i is in [Sch07] defined by the following NAD-equation, similar to (4.21):

$$
\tau \dot{m}_i = -m_i + h + s_i^{\text{mot}} + \alpha_{\text{selfexc},i}^{\text{mot}} + \alpha_{\text{exc},i}^{\text{mot}} - \alpha_{\text{inh},i}^{\text{mot}},
$$

where

$\alpha_{\text{selfexc},i}^{\text{mot}}$, excitatory contribution of neuron i's own motivation m_i;

$\alpha_{\text{exc},i}^{\text{mot}}$, all excitatory contribution of motivation neurons connected to neuron i;

$\alpha_{\text{inh},i}^{\text{mot}}$, all inhibitory contribution of motivation neurons connected to neuron i.

In this framework, a nonlinear neural dynamical and control system generates the temporal evolution of behavioral variables, such that desired behaviors are fixed-point attractor solutions while un-desired behaviors are repellers.

This kind of *attractor & repeller dynamics* [II07a] provides the basis for understanding *cognition*, both natural and artificial [IJP08, II07b, II07d].

Turing's idea can only be realized in four equivalence classes of systems characterized by the signs of the Jacobian $\mathbf{R}'$ of the reaction function. In particular, if a finite wave vector k is supposed to be the most unstable one, the Jacobian must have the signs

$$\begin{pmatrix} + & - \\ + & - \end{pmatrix}, \quad \begin{pmatrix} + & + \\ - & - \end{pmatrix}, \quad \begin{pmatrix} - & + \\ - & + \end{pmatrix}, \quad \begin{pmatrix} - & - \\ + & + \end{pmatrix}.$$

This class of systems is named activator-inhibitor system after its first representative: close to the ground state, one component stimulates the production of both components while the other one inhibits their growth. Its most prominent representative is the FitzHugh–Nagumo equation (4.24).

Brusselator

Classical model of an autocatalytic chemical reaction is Prigogine's Brusselator (see e.g., [Pri80])

$$\partial_t u = D_u^2 \Delta u + \alpha + u^2 v - (1 + \beta)u, \qquad \partial_t v = D_v^2 \Delta v - u^2 v + \beta u, \quad (4.22)$$

which describe the spatio-temporal evolution of the intermediate components u and v, with diffusion coefficients D_u and D_v, while reactions

$$\alpha \xrightarrow{r_1} u, \qquad 2u + v \xrightarrow{r_2} 3u, \qquad \beta + u \xrightarrow{r_3} v + d, \qquad u \xrightarrow{r_4} c$$

describe the concentration of the original substances α and β, for which the final products c and d are constant when all four reaction rates r_i equal unity.

A discretized (temporal only) version of the Brusselator PDE (4.22) reads

$$\dot{u} = \alpha + u^2 v - (1 + \beta)u, \qquad \dot{v} = \beta u - u^2 v.$$

The Brusselator displays oscillatory behavior in the species u and v when reverse reactions are neglected and the concentrations of α and β are kept constant.

2-Component Model of Excitable Media

Turbulence of scroll waves is a kind of spatio-temporal chaos that exists in 3-dimensional excitable media. Cardiac tissue and the Belousov–Zhabotinsky reaction are examples of such media. In cardiac tissue, chaotic behavior is believed to underlie fibrillation which, without intervention, precedes cardiac death. Fast computer-simulation of waves in excitable media have been often performed using the 2-component Barkley model of excitable media [Bar91],

$$\partial_t u = \frac{1}{\epsilon} u(1 - u) \left(u - \frac{v + b(t)}{a} \right) + \nabla^2 u + h(t), \qquad \partial_t v = u - v, \quad (4.23)$$

where ϵ is a small parameter $\epsilon \ll 1$ characterizing mutual time scales of the fast u and slow v variables, and a and b specify the kinetic properties of the system. Parameter b determines the excitation threshold and thus controls the excitability of the medium. The term $h(t)$ represents an 'extra transmembrane current'.

Suppression of the turbulence using stimulation of two different types, 'modulation of excitability' and 'extra transmembrane current' was performed in [MBB08], using the Barkley model (4.23). With cardiac defibrillation in mind, the authors used a single pulse as well as repetitive extra current with both constant and feedback controlled frequency. They show that turbulence can be terminated using either a resonant modulation of excitability or a resonant extra current. The turbulence is terminated with much higher probability using a resonant frequency perturbation than a non-resonant one. Suppression of the turbulence using a resonant frequency is up to fifty times faster than using a non-resonant frequency, in both the modulation of excitability and the extra current modes. They also demonstrate that resonant perturbation requires strength one order of magnitude lower than that of a single pulse, which is currently used in clinical practice to terminate cardiac fibrillation.

Gierer–Meinhardt Activator–Inhibitor System

Spontaneous pattern formation in initially almost homogeneous systems is common in both organic and inorganic systems. The Gierer–Meinhardt model [GM72] is a reaction–diffusion system of the activator–inhibitor type that appears to account for many important types of pattern formation and morphogenesis observed in biology, chemistry and physics. The model describes the concentration of a short-range autocatalytic substance, the activator, that regulates the production of its long-range antagonist, the inhibitor. It is given as a 2-component nonlinear PDE system,

$$\partial_t a = -\mu_a a + \rho a^2/h + D_a \partial_{x^2} a + \rho_a, \qquad \partial_t h = -\mu_h h + \rho a^2 + D_h \partial_{x^2} h + \rho_h,$$

where a is a short-range autocatalytic substance, i.e., *activator*, and h is its long-range antagonist, i.e., *inhibitor*. $\partial_t a$ and $\partial_t h$ describe respectively the changes of activator and inhibitor concentrations per second, μ_a and μ_h are the corresponding decay rates, while D_a and D_h are the corresponding diffusion coefficients. ρ is a positive constant. ρ_a is a small activator-independent production rate of the activator and is required to initiate the activator autocatalysis at very low activator concentration, e.g., in the case of regeneration. A low baseline production of the inhibitor, ρ_h, leads to a stable non-patterned steady state; the system can be asleep until an external trigger occurs by an elevation of the activator concentration above a threshold [GM06].

Fitzhugh–Nagumo Activator–Inhibitor System

An important example of bio-reaction–diffusion systems, frequently used in neurodynamics, is the 2-component Fitzhugh–Nagumo activator–inhibitor system [Fit61b, NAY60] (see also [II06b, II07b])

$$\tau_u \partial_t u = D_u^2 \Delta u + f(u) - \sigma v, \qquad \tau_v \partial_t v = D_v^2 \Delta v + u - v, \qquad (4.24)$$

with $f(u) = \lambda u - u^3 - \kappa$, which describes how an action potential travels through a nerve, D_u and D_v are diffusion coefficients, τ_u and τ_v are time characteristics, while κ, σ and λ are positive constants. In matrix form, system (4.24) reads

$$\begin{pmatrix} \tau_u \partial_t u \\ \tau_v \partial_t v \end{pmatrix} = \begin{pmatrix} D_u^2 & 0 \\ 0 & D_v^2 \end{pmatrix} \begin{pmatrix} \Delta u \\ \Delta v \end{pmatrix} + \begin{pmatrix} \lambda u - u^3 - \kappa - \sigma v \\ u - v \end{pmatrix}.$$

When an activator–inhibitor system undergoes a change of parameters, one may pass from conditions under which a homogeneous ground state is stable to conditions under which it is linearly unstable. The corresponding bifurcation may be either a *Hopf bifurcation* to a globally oscillating homogeneous state with a dominant wave number $k = 0$ or a *Turing bifurcation* to a globally patterned state with a dominant finite wave number. The latter in two spatial dimensions typically leads to stripe or hexagonal patterns.

In particular, for the Fitzhugh–Nagumo system (4.24), the neutral stability curves marking the boundary of the linearly stable region for the Turing and Hopf bifurcation are given by

$$q_n^H(k): \quad \frac{1}{\tau} + \left(d_u^2 + \frac{1}{\tau}d_v^2\right)k^2 = f'(u_h),$$

$$q_n^T(k): \quad \frac{\kappa_3}{1 + d_v^2 k^2} + d_u^2 k^2 = f'(u_h).$$

If the bifurcation is subcritical, often localized structures (i.e., dissipative solitons) can be observed in the hysteretic region where the pattern coexists with the ground state. Other frequently encountered structures comprise pulse trains, spiral waves and target patterns.

The reduced (temporal) non-dimensional Fitzhugh–Nagumo equations read:

$$\dot{v} = v(a - v)(v - 1) - w + I_a, \qquad (4.25)$$

$$\dot{w} = bv - \gamma w, \qquad (4.26)$$

where $0 < a < 1$ is essentially the threshold value, b and γ are positive constants and I_a is the applied current. The drift field for this model is given by

$$u_1(v, w) = v(a - v)(v - 1) - w, \qquad u_2(v, w) = bv - \gamma w.$$

As can be seen from (4.26) the null cline of the deterministic dynamics of this equations is the line $v = \frac{\gamma}{b}w$. By substitution on the r.h.s. of (4.25) we find the following equation for steady states: $v(a - v)(v - 1) - \frac{b}{\gamma}v = 0$.

When this system is in a noisy environment, in the limit of weak noise, we can approximate the dynamics of the fluctuations by the *Langevin equation* [II06b, II08a]

$$\dot{v} = v(a - v)(v - 1) - \frac{b}{\gamma}v + \xi(t),$$

that is, the fluctuations run along the line $v = \frac{\gamma}{b}w$.

In particular, parameters in the *FitzHugh–Nagumo neuron* model [IJP08, II07d]

$$\dot{v} = a + bv + cv^2 + dv^3 - u, \qquad \dot{u} = \varepsilon(ev - u),$$

can be tuned so that the model describes spiking dynamics of many resonator neurons. Since one needs to simulate the shape of each spike, the time step in the model must be relatively small, e.g., $\tau = 0.25$ ms. Since the model is a 2-dimensional system of ODEs, without a reset, it cannot exhibit autonomous chaotic dynamics or bursting. Adding noise to this, or some other 2-dimensional models, allows for stochastic bursting.

2-Component Belousov–Zhabotinsky Reaction

Classical Belousov–Zhabotinsky (BZ) reaction is a family of oscillating chemical reactions. During these reactions, transition-metal ions catalyze oxidation of various, usually organic, reductants by bromic acid in acidic water solution. Most BZ reactions are homogeneous. The BZ reaction makes it possible to observe development of complex patterns in time and space by naked eye on a very convenient human time scale of dozens of seconds and space scale of several millimeters. The BZ reaction can generate up to several thousand oscillatory cycles in a closed system, which permits studying chemical waves and patterns without constant replenishment of reactants [Zha07].

Consider the water-in-oil micro-emulsion BZ reaction [KVE06, KVE05]

$$\partial_t v = D_v \Delta v + \frac{1}{\varepsilon_0} \left[f_0 z + i_0 (1 - mz) \right] \frac{v - q_0}{v + q_0} + \frac{1}{\varepsilon_0} \left[\frac{1 - mz}{1 - mz + \varepsilon_1} \right] v - v^2,$$

$$\partial_t z = D_z \Delta z - z + v \left[\frac{1 - mz}{1 - mz + \varepsilon_1} \right], \tag{4.27}$$

where v, z are dimensionless concentrations of activator $HBrO_2$ and oxidized catalyst $[Ru(bpy)_3]^{3+}$ respectively; D_v and D_z are dimensionless diffusion coefficients of activator and catalyst; $f, \varepsilon_0, \varepsilon_1$ and q are parameters of the standard Tyson model [Tys85]; i_0 represents the photoinduced production of inhibitor, and m represents the strength of oxidized state of the catalyst with $0 < mz < 1$. This reaction was shown experimentally and numerically to admit localized spot patterns that persist for long time [KVE06, KVE05].

We can rescale the variables in (4.27) as [KT07]

$$z = 1/m - m^{-3/2} w \varepsilon_1, \qquad v = m^{-1/2} \hat{v}, \qquad t = \varepsilon_0 m^{1/2} \hat{t}.$$

In the new variables, after dropping the hats, we obtain the non-dimensional 2-component BZ reaction

$$\partial_t v = \varepsilon^2 \Delta v + f(v, w), \qquad \tau \partial_t w = D \Delta w + g(v, w),$$

where

$$f(v,z) = -\left[f_0 + f_1 w\right] \frac{v-q}{v+q} + \left[\frac{w}{1+\alpha w}\right] v - v^2, \qquad g(v,w) = 1 - \left[\frac{w}{1+\alpha w}\right] v,$$

with the non-dimensional constants given by

$$\alpha = m^{-1/2}, \qquad f_1 = \varepsilon_1 m^{1/2}\left(i_0 - \frac{f_0}{m}\right), \qquad q = q_0 m^{1/2},$$

$$\varepsilon^2 = \varepsilon_0 D_v m^{1/2}, \qquad D = D_z \varepsilon_1 m^{-1/2}, \qquad \tau = \frac{1}{m}\frac{\varepsilon_1}{\varepsilon_0}.$$

3-Component and Multi-Component Systems

Oregonator

The Oregonator model is based on the so-called FKN-mechanism [FKN72], which provided the first successful explanation of the chemical oscillations that occur in the experimental Belousov–Zhabotinsky reaction. It is composed of five coupled elementary chemical stoichiometries. During the last two decades, the Oregonator model has been modified in many ways by inclusion of additional chemical reaction steps or by changing the rate constants. If we denote the concentration of the species S by [S], then we define: $A = [\text{BrO}_3^-]$, $H = [\text{H}^+]$, $X = [\text{HBrO}_2]$, $Y = [\text{Br}^-]$, $Z = [\text{Ce}^{4+}]$. The original Oregonator model was described by the following three coupled nonlinear PDEs,

$$\partial_t X = k_1 A H^2 Y - k_2 H X Y - 2k_3 X^2 + k_4 A H X + D_X \nabla_r^2 X,$$
$$\partial_t Y = -k_1 A H^2 Y - k_2 H X Y + k_5 f Z + D_Y \nabla_r^2 Y, \qquad (4.28)$$
$$\partial_t Z = 2k_4 A H X - k_5 Z + D_Z \nabla_r^2 Z,$$

where f is a stoichiometric factor [HHS91], $k_i (i = 1, \ldots, 5)$ are rate constants, while D_X, D_Y, and D_Z are the diffusion constants of the species HBrO_2, Br^-, and Ce^{4+} respectively (for dilute solutions, the diffusion matrix is diagonal). For a thorough discussion of the chemistry on which the Oregonator is based, the reader is referred to [Tys85].

The Oregonator temporal mass-action dynamics is a well-stirred, homogeneous system of ODEs given by

$$\dot{X} = k_1 A Y - k_2 X Y + k_3 A X - 2k_4 X^2,$$
$$\dot{Y} = -k_1 A Y - k_2 X Y + 1/2 k_c f B Z,$$
$$\dot{Z} = 2k_3 A X - k_c B Z,$$

which are typically scaled as [Tys85]

$$\epsilon(dx/d\tau) = qy - xy + x(1-x), \qquad \epsilon'(dy/d\tau) = -qy - xy + fz, \qquad dz/d\tau = x - z.$$

The basic chemistry of the BZ-oscillations involves jumps between high and low HBrO2 (X) states, which is reflected in the relaxation oscillator nature of the Oregonator. This fundamental bistability may be stabilized in a flow reactor (CSTR) with reactants and Br$^-$ in the feed stream. Hysteresis between the two states is observed both experimentally and in the Oregonator. Quasiperiodicity and chaos also are observed in CSTR and can be modeled by the Oregonator [Fie07].

Multi-Phase Tumor Growth Equations

Our last reaction–diffusion system is a general model of multi-phase tumor growth, in the form of nonlinear parabolic PDE, as reviewed recently in [RCM07]

$$\partial_t \Phi_i = \nabla \cdot (D_i \Phi_i) - \nabla \cdot (\mathbf{v}_i \Phi_i) + \lambda_i(\Phi_i, C_i) - \mu_i(\Phi_i, C_i) \tag{4.29}$$

($\partial_t \equiv \partial/\partial t$), where for phase i, Φ_i is the volume fraction ($\sum_i \Phi_i = 1$), D_i is the random motility or diffusion, $\lambda_i(\Phi_i, C_i)$ is the chemical and phase dependent production, and $\mu_i(\Phi_i, C_i)$ is the chemical and phase dependent degradation/death, and $\mathbf{v}_i$ is the cell velocity defined by the constitutive equation

$$\mathbf{v}_i = -\mu \nabla p, \tag{4.30}$$

where μ is a positive constant describing the viscous-like properties of tumor cells and p is the spheroid internal pressure.

In particular, the multi-phase equation (4.29) splits into two heat-like mass-conservation PDEs [RCM07],

$$\partial_t \Phi^C = S^C - \nabla \cdot (\mathbf{v}^C \Phi^C), \qquad \partial_t \Phi^F = S^F - \nabla \cdot (\mathbf{v}^F \Phi^F), \tag{4.31}$$

where Φ^C and Φ^F are the tissue cell/matrix and fluid volume fractions, respectively, $\mathbf{v}^C$ and $\mathbf{v}^F$ are the cell/matrix and the fluid velocities (both defined by their constitutive equations of the form of (4.30)), S^C is the rate of production of solid phase tumor tissue and S^F is the creation/degradation of the fluid phase. Conservation of matter in the tissue, $\Phi^C + \Phi^F = 1$, implies that $\nabla \cdot (\mathbf{v}^C \Phi^C + \mathbf{v}^F \Phi^F) = \Phi^C + \Phi^F$. The assumption that the tumor may be described by two phases only implies that the new cell/matrix phase is formed from the fluid phase and vice versa, so that $S^C + S^F = 0$. The detailed biochemistry of tumor growth can be coupled into the model above through the growth term S^C, with equations added for nutrient diffusion, see [RCM07] and references therein.

The multi-phase tumor growth model (4.29) has been derived from the classical transport/mass conservation equations for different chemical species [RCM07],

$$\partial_t u_i = P_i - \nabla \cdot \mathbf{N}_i. \tag{4.32}$$

Here C_i are the concentrations of the chemical species, subindex a for oxygen, b for glucose, c for lactate ion, d for carbon dioxide, e for bicarbonate ion,

f for chloride ion, and g for hydrogen ion concentration; P_i is the net rate of consumption/production of the chemical species both by tumor cells and due to the chemical reactions with other species; and $\mathbf{N}_i$ is the flux of each of the chemical species inside the tumor spheroid, given (in the simplest case of uncharged molecules of glucose, O_2 and CO_2) by Fick's law,

$$\mathbf{N}_i = -D_i \nabla u_i,$$

where D_i are (positive) constant diffusion coefficients. In case of charged molecules of ionic species, the flux $\mathbf{N}_i$ contains also the (negative) gradient of the volume fractions Φ_i.

There are three distinct stages to cancer development: avascular, vascular, and metastatic—researchers often concentrate their efforts on answering specific OUPC-related questions on each of these stages [RCM07]. In particular, as some tumor cell lines grown in vitro form spherical aggregates, the relative cheapness and ease of in vitro experiments in comparison to animal experiments has made 3D *multicellular tumor spheroids* (MTS, see Fig. 6 in [RCM07]) very popular in vitro model system of avascular tumors[13] [Kun99]. They are used to study how local micro-environments affect cellular growth/decay, viability, and therapeutic response [Sut88b]. MTS provide, allowing strictly controlled nutritional and mechanical conditions, excellent experimental patterns to test the validity of the proposed mathematical models of tumor growth/decay [Pre03].

4.1.2 Reactive Neurodynamics

If we drop the diffusion term in the general reaction–diffusion equation (4.1), we get a purely reactive temporal dynamics, reducing a single nD parabolic PDE to an nD system of nonlinear ODEs. In this subsection we will explore nonlinear temporal dynamics of (Hopfield-type) neural networks.

---

[13] In vitro cultivation of tumor cells as multicellular tumor spheroids (MTS) has greatly contributed to the understanding of the role of the cellular micro-environment in tumor biology (for review see [Sut88b, Kun99]). These spherical cell aggregates mimic avascular tumor stages or micro-metastases in many aspects and have been studied intensively as an experimental model reflecting an in vivo-like micro-milieu with 3D metabolic gradients. With increasing size, most MCTS not only exhibit proliferation gradients from the periphery towards the center but they also develop a spheroid type-specific nutrient supply pattern, such as radial oxygen partial pressure gradients. Similarly, MCTS of a variety of tumor cell lines exhibit a concentric histo-morphology, with a necrotic core surrounded by a viable cell rim. The spherical symmetry is an important prerequisite for investigating the effect of environmental factors on cell proliferation and viability in a 3D environment on a quantitative basis.

Neurons as Functions

It is a common view in computational intelligence that neurons behave as functions (see [Kos92, II06b, II07b]): they transduce an unbounded input *activation* $x(t)$ into output *signal* $S(x(t))$. Usually a sigmoidal (S-shaped, bounded, monotone-nondecreasing: $S' \geq 0$) function describes the transduction, as well as the input–output behavior of many operational amplifiers. For example, the *logistic signal* (or, the *maximum-entropy*) function

$$S(x) = \frac{1}{1 + e^{-cx}}$$

is sigmoidal and strictly increases for positive scaling constant $c > 0$. Strict monotonicity implies that the *activation derivative* of S is positive:

$$S' = \frac{dS}{dx} = cS(1 - S) > 0.$$

Another frequent bipolar signal function is the *hyperbolic-tangent*,

$$S(x) = \tanh(cx) = \frac{e^{cx} - e^{-cx}}{e^{cx} + e^{-cx}},$$

with activation derivative

$$S' = c(1 - S^2) > 0.$$

Also, the *Gaussian*, or bell-shaped, signal function of the form $S(x) = e^{-cx^2}$, for $c > 0$, represents an important exception to signal monotonicity. Its activation derivative $S' = -2cxe^{-cx^2}$ has the sign opposite the sign of the activation x.

The *signal velocity* $\dot{S} \equiv dS/dt$ is the *signal time derivative*, related to the activation derivative by

$$\dot{S} = S'\dot{x},$$

so it depends explicitly on *activation velocity*. This is used in unsupervised learning laws that adapt with *locally available information*.

The signal $S(x)$ induced by the activation x represents the neuron's firing frequency of action potentials, or pulses, in a sampling interval. The firing frequency equals the average number of pulses emitted in a sampling interval.

Short-term memory is modeled by *activation dynamics*, and *long-term memory* is modeled by *learning dynamics*. The overall neural network behaves as an *adaptive filter* (see [Hay91]).

Continuous ANNs are *temporal dynamical systems* with two coupled dynamics: activation and learning. First, a general system of coupled ODEs for the output of the *ith processing element* (PE) x^i, called the *activation dynamics*, can be written as

$$\dot{x}^i = g_i(x^i, \underset{i}{\text{net}}), \tag{4.33}$$

with the *net input* to the *ith* PE x^i given by $\text{net}_i = \omega_{ij}x^j$.

For example,

$$\dot{x}^i = -x^i + f_i(\text{net}_i),$$

where f_i is called *output*, or *activation*, *function*. We apply some input values to the PE so that $\text{net}_i > 0$. If the inputs remain for a sufficiently long time, the output value will reach an equilibrium value, when $\dot{x}^i = 0$, given by $x^i = f_i(\text{net}_i)$. Once the unit has a nonzero output value, removal of the inputs will cause the output to return to zero. If $\text{net}_i = 0$, then $\dot{x}^i = -x^i$, which means that $x \to 0$.

Second, a general system of coupled ODEs for the *update* of the synaptic weights ω_{ij}, i.e., *learning dynamics*, can be written as a generalization of the Oja–Hebb rule, i.e.,

$$\dot{\omega}_{ij} = G_i(\omega_{ij}, x^i, x^i),$$

where G_i represents the *learning law*; the learning process consists of finding weights that encode the knowledge that we want the system to learn. For most realistic systems, it is not easy to determine a closed-form solution for this system of equations, so the approximative solutions are usually enough.

Activation Dynamics

Hopfield's graded-response neurons[14] have continuous input-output relation (like nonlinear operational amplifiers) of the form $V_i = g_i(\lambda u_i)$, where u_i

---

[14] The paradigm for the unsupervised, self-organizing, associative, and recurrent ANN is the discrete Hopfield network (see [Hop82]). Hopfield gives a collection of simple threshold automata, called *formal neurons* by McCulloch and Pitts (see [II06b, II07b]): two-state, 'all-or-none', firing or non-firing units that can be modeled by *Ising spins* (uniaxial magnets) $\{S_i\}$ such that $S_i = \pm 1$ (where $1 = |\uparrow\rangle = $ 'spin up' and $-1 = |\downarrow\rangle = $ 'spin down'; the label of the neuron is i and ranges between 1 and the size of the network N). The neurons are connected by synapses J_{ij}.

Firing *patterns* $\{\xi_i^\mu\}$ represent specific S_i-*spin configurations*, where the label of the pattern is μ and ranges between 1 and q. Using random patterns $\xi_i^\mu = \pm 1$ with equal probability $1/2$, we have the *synaptic efficacy* J_{ij} of jth neuron operating on ith neuron given by

$$J_{ij} = N^{-1}\xi_i^\mu\xi_j^\mu \equiv N^{-1}\boldsymbol{\xi_i} \cdot \boldsymbol{\xi_j}. \tag{4.34}$$

Postsynaptic potential (PSP) represents an *internal local field*

$$h_i(t) = J_{ij}S_j(t). \tag{4.35}$$

Now, the *sequential (threshold) dynamics* is defined in the form of discrete equation

$$S_i(t + \Delta t) = \text{sgn}[h_i(t)]. \tag{4.36}$$

Dynamics (4.36) is equivalent to the rule that the state of a neuron is changed, or a spin is flipped iff the total network *energy*, given by *Ising Hamiltonian*

$$H_N = -\frac{1}{2}J_{ij}S_iS_j, \tag{4.37}$$

denotes the input at i, a constant λ is called the gain parameter, and V_i is the output [Hop84]. Usually, g_i are taken to be sigmoid functions, odd, and monotonically increasing (e.g., $g(\cdot) = \frac{1}{2}(1+\tanh(\cdot))$), while discrete Ising spins have $g_i(u_i) = \operatorname{sgn}_i(u_i)$. The behavior of the *continuous Hopfield network* is usually described by a set of coupled RC-transient equations

$$C_i \dot{u}_i = I_i + J_{ij} V_j - \frac{u_i}{R_i}, \tag{4.38}$$

where $u_i = g^{-1}(V_i)$, R_i and C_i denote input capacitance and resistance, and I_i represents an external source.

Using random patterns $\xi_i^\mu = \pm 1$ with equal probability $1/2$, we have the *synaptic efficacy* J_{ij} of jth neuron operating on ith neuron given by[15]

$$J_{ij} = N^{-1} \xi_i^\mu \xi_j^\mu \equiv N^{-1} \boldsymbol{\xi_i} \cdot \boldsymbol{\xi_j}. \tag{4.39}$$

The Hamiltonian of the continuous system (4.38) is given by

$$H = -\frac{1}{2} J_{ij} V_i V_j + \sum_{i=1}^{N} R_i^{-1} \int_0^{V_i} dV \, g^{-1}(V) - I_i V_i. \tag{4.40}$$

However, according to Hopfield [Hop84] the synapses J_{ij} retain the form (4.39) with random patterns $\xi_i^\mu = \pm 1$ with equal probability $1/2$, and the synaptic symmetry $J_{ij} = J_{ji}$ implies that the continuous Hamiltonian (4.40) represents a *Lyapunov function* of the system (4.38), i.e., H decreases under the continual neuro-dynamics governed by (4.38) as time proceeds.

Hopfield's Overlaps

Assuming that the number q of stored patterns is small compared to the number of neurons, i.e., $q/N \to 0$, we find that the synapses (4.39) give rise to a local field of the form

$$h_i = \xi_i^\mu m_\mu, \quad \text{where} \tag{4.41}$$
$$m_\mu = N^{-1} \xi_i^\mu S_i \tag{4.42}$$

---

is lowered [Hop82]. Therefore, the Ising Hamiltonian H_N represents the monotonically decreasing *Lyapunov function* for the sequential dynamics (4.36), which converges to a local minimum or ground state of H_N. This holds for any *symmetric coupling* $J_{ij} = J_{ji}$ with $J_{ii} = 0$ and if spin-updating in (4.36) is asynchronous. In this case the patterns $\{\xi_i^\mu\}$ after convergence become identical, or very near to, ground states of H_N, each of them at the bottom of the valley.

[15] More general form of synapses is

$$J_{ij} = N^{-1} Q(\boldsymbol{\xi_i}; \boldsymbol{\xi_j}),$$

for some synaptic kernel Q on $\mathbb{R}^{\shortparallel} \times \mathbb{R}^{\shortparallel}$. The vector $\boldsymbol{\xi_i}$ varies as i travels from 1 to N, but remains on a corner of the *Hamming hypercube* $[-1, 1]^q$.

is the *auto-overlap* (or simply *overlap*)[16] of the network state $\{S_i\}$ with the pattern $\{\xi_i^\mu\}$, measuring the proximity between them. We can see that $m_\mu = 1$ (like peak-up in auto-correlation) if $\{S_i\}$ and $\{\xi_i^\mu\}$ are identical patterns, $m_\mu = -1$ (like peak-down in autocorrelation) if they are each other's complement, and $m_\mu = O(1/\sqrt{N})$ if they are uncorrelated (like no-peak in auto-correlation) with each other. Overlaps m_μ are related to the Hamming distance d_μ between the patterns (the fraction of spins which differ) by $d_\mu = \frac{1}{2}(1 - m_\mu)$.

As a pattern ξ_i^μ represents (in the simplest case) a specific Ising-spin S_i-configuration, then $(\xi_i^\mu)^2 = 1$. If $S_i = \xi_i^\mu$ for all i, then $m_\mu = 1$. Conversely, if $m_\mu = 1$, then $S_i = \xi_i^\mu$. In all other cases $m_\mu < 1$, by the Cauchy–Schwartz inequality. If ξ_i^μ and S_i are uncorrelated, we may expect m_μ to be of the order of $N^{-1/2}$, since the sum consists of N terms, each containing a ξ_i^μ. On the other hand, if the S_i are positively correlated with ξ_i^μ, then m_μ is of the order of unity. So the overlaps give the global information about the network and hence are good order parameters. Also, according to Hopfield [Hop84], the extension to the continual network is straightforward.

Using overlaps, the *Ising Hamiltonian* becomes

$$H_N = -\frac{1}{2} N \sum_{\mu=1}^{q} m_\mu^2. \tag{4.43}$$

The similarity between two different patterns ξ_i^μ and ξ_i^ν is measured by their *mutual overlap* or *cross-overlap* $m_{\mu\nu}$ (in other parlance it is called *Karhunen–Loeve covariance matrix* (see [II07b]), which extracts the principal components from a data set),[17] equal

$$m_{\mu\nu} = N^{-1} \xi_i^\mu \xi_i^\nu. \tag{4.44}$$

For similar patterns the cross-overlap is close to unity whereas for uncorrelated patterns it is random variable with zero mean and small $(1/\sqrt{N})$ variance.

The symmetric *Hopfield synaptic matrix* J_{ij} can be expressed in terms of the cross-overlaps $m_{\mu\nu}$ as

$$J_{ij} = N^{-1} \xi_i^\mu (m_{\mu\nu})^{-1} \xi_j^\nu = J_{ji}, \tag{4.45}$$

where $(m_{\mu\nu})^{-1}$ denotes the *Moore–Penrose pseudoinverse* of the cross-overlap matrix $m_{\mu\nu}$.

Besides the Hopfield model, the proposed pattern-overlap picture can be extended to cover some more sophisticated kinds of associative memory, among them [II07b]:

---

[16] Resembling the auto-correlation function of a time-series, where distinct peaks indicate that the series at the certain time t is similar to the series at time $t + \Delta t$

[17] Resembling the cross-correlation function of two time-series, with several distinct peaks, indicating that the two series are very similar at each point in time where the peaks occur.

1. Forgetful memories, characterized by iterative synaptic prescription

$$J_{ij}^{(\mu)} = \phi(\epsilon\xi_i^\mu\xi_j^\mu + J_{ij}^{(\mu-1)}),$$

   for some small parameter ϵ and some odd function ϕ. If $\phi(\cdot)$ saturates as $|\cdot| \to \infty$, the memory creates storage capacity for new patterns by forgetting the old ones.
2. Temporal associative memories, which can store and retrieve a sequence of patterns, through synapses

$$NJ_{ij} = \xi_i^\mu\xi_j^\mu + \epsilon\sum_{\mu=1}^{q}\xi_i^{(\mu+1)}\xi_j^\mu, \qquad (4.46)$$

   where the second term on the right is associated with a temporal delay, so that one can imagine that the second term 'pushes' the neural system through an energy landscape created by the first term.

Glauber vs. Overlap Dynamics

According to Hopfield [Hop84], the extension of the sequential dynamics $S_i = \text{sgn}(\sum_\mu m_\mu\xi_i^\mu)$ of the network made of the simplest Ising-spin-neurons to the network made of continual graded-response amplifier-neurons, is straightforward using the probabilistic *Glauber dynamics*

$$\text{Prob}\{S_i \mapsto -S_i\} = \frac{1}{2}[1 - \tanh(\beta h_i S_i)], \quad (i = 1,\ldots,N), \qquad (4.47)$$

where β represents the universal temperature ($\beta = \frac{1}{k_B T}$, k_B is the normalized Boltzman's constant and $k_B T$ has dimension of energy).

Under the Glauber's dynamics (4.47), and as $N \to \infty$ (transition from the single neurons to the neural field), for time-dependent patterns $\xi^\mu(t) = \xi_\mu(t)$, vector auto-overlaps $m_\mu(t)$, and tensor cross-overlaps $m_{\mu\nu}(t)$, we present the dynamics of overlaps governed by the following nonlinear differential equations [II07b]

$$\dot{m}_\mu(t) = -m_\mu(t) + \langle\xi_\mu(t)\tanh[\beta m_\mu(t)\xi^\mu(t)]\rangle, \qquad (4.48)$$

and in the tensor form

$$\dot{m}_{\mu\nu}(t) = -m_{\mu\nu}(t) + \langle\xi_\mu(t)\xi_\nu(t)\tanh[\beta m_{\mu\nu}(t)\xi^\mu(t)\xi^\nu(t)]\rangle, \qquad (4.49)$$

where the angular brackets denote an average over the q patterns $\xi^\mu(t)$.

The stationary solutions (for any fixed instant of time $t = \tau$) of (4.48) and (4.49) are given by corresponding fixed-point vector and tensor equations

$$m_\mu = \langle\xi_\mu\tanh[\beta m_\mu\xi^\mu]\rangle, \quad \text{and} \qquad (4.50)$$
$$m_{\mu\nu} = \langle\xi_\mu\xi_\nu\tanh[\beta m_{\mu\nu}\xi^\mu\xi^\nu]\rangle, \qquad (4.51)$$

respectively.

Generalized Hebbian Learning Dynamics

In terms of stochastic feed-forward multi-layer neural networks, the tensorial equation (4.49) corresponds to the average, general, self-organizing Hebbian neural learning scheme (see [Heb49, Kos92])

$$\dot{m}_{\mu\nu}(t) = -m_{\mu\nu}(t) + \langle \mathcal{I}_{\mu\nu} \rangle, \qquad (4.52)$$

with random signal *Hebbian innovation*

$$\mathcal{I}_{\mu\nu} = f_\mu[\xi^\mu(t)]f_\nu[\xi^\nu(t)] + \sigma_{\mu\nu}(t), \qquad (4.53)$$

where $\sigma_{\mu\nu}$, denotes the tensorial, additive, zero-mean, Gaussian white-noise, independent of the main innovation function $I_{\mu\nu}$, while $f_{\mu,\nu}[\cdot]$ represent the hyperbolic tangent (sigmoid) neural activation functions. A single-layer Hebbian learning scheme, corresponding to the tensor equation (4.52), gives

$$\dot{m}_\mu(t) = -m_\mu(t) + \langle \mathcal{I}_\mu \rangle, \qquad (4.54)$$

with the vector innovation

$$\mathcal{I}_\mu = f_\mu[\xi^\mu(t)] + \sigma_\mu(t),$$

where σ_μ, denotes the vector additive zero-mean Gaussian white-noise, also independent of the main innovation function I_μ, while $f_\mu[\cdot]$ represents the hyperbolic tangent (sigmoid) neural activation function.

If we assume the small absolute value of the average (stochastic) terms, the nonlinear overlap-dynamics equations (4.48) and (4.49) can be presented in the form of *weakly-connected neural networks* (see [HI97]), respectively, as a single-layer network

$$\dot{m}_\mu(t) = -m_\mu(t) + \varepsilon g_\mu(m_\mu, \varepsilon), \quad \varepsilon \ll 1, \qquad (4.55)$$

and a multi-layer network

$$\dot{m}_{\mu\nu}(t) = -m_{\mu\nu}(t) + \varepsilon g_{\mu\nu}(m_{\mu\nu}, \varepsilon), \quad \varepsilon \ll 1, \qquad (4.56)$$

where, g_μ and $g_{\mu\nu}$, corresponding to the average (bracket) terms in (4.48) and (4.49), describe (vector and tensor, respectively) synaptic connections and the 'small' parameter ε describes their (dimensionless) strength. These weakly-connected neural systems represent ε-perturbations of the corresponding linear systems

$$\dot{m}_\mu(t) = -m_\mu(t) \quad \text{and} \quad \dot{m}_{\mu\nu}(t) = -m_{\mu\nu}(t),$$

with exponential maps as solutions

$$m_\mu(t) = m_\mu \, e^{-t} \quad \text{and} \quad m_{\mu\nu}(t) = m_{\mu\nu} \, e^{-t},$$

using the stationary (fixed-point) solutions (4.50), (4.51) as initial conditions m_μ and $m_{\mu\nu}$. According to the *Hartman–Grobman theorem* from dynamical systems theory, the weakly-connected systems (4.55), (4.56) are topologically equivalent (homeomorphic) to the corresponding linear systems. Therefore the whole analysis for the linear vector and matrix flows can be applied here, with only difference that instead of increasing transients e^t here we have decreasing (i.e., asymptotically-stable) transients e^{-t}.

On the other hand, in terms of *synergetics* [Hak83], both nonlinear overlap-dynamics equations (4.48)–(4.49) and Hebbian learning equations (4.52)–(4.53), represent (covariant) *order parameter equations*.

By introducing the scalar quadratic potential fields, dependent on vector and tensor order parameters (overlaps), respectively

$$V(m_\mu) = -\frac{1}{2}\sum_{\mu=1}^{q} m_\mu^2 \quad \text{and} \quad V(m_{\mu\nu}) = -\frac{1}{2}m_{\mu\nu}^2,$$

we can generalize the overlap-dynamics equations (4.48)–(4.49) to the *stochastic-gradient order parameter equations*, in vector and tensor form, respectively

$$\dot{m}_\mu(t) = -\frac{\partial V(m_\mu)}{\partial m_\mu(t)} + F_\mu(t), \quad \text{and} \tag{4.57}$$

$$\dot{m}_{\mu\nu}(t) = -\frac{\partial V(m_{\mu\nu})}{\partial m_{\mu\nu}(t)} + F_{\mu\nu}(t). \tag{4.58}$$

$F_\mu(t)$ in (4.57) represents a vector fluctuating force, with average (over the stochastic process which produces the fluctuating force $F_\mu(t)$)

$$\langle F_\mu(t) \rangle = \langle \xi_\mu(t)\tanh[\beta m_\mu(t)\xi^\mu(t)] \rangle,$$

and variation

$$\langle F_\mu(t)F_\mu(t') \rangle = Q_\mu \delta(t - t'), \tag{4.59}$$

while $F_{\mu\nu}(t)$ in (4.58) represents a tensor fluctuating force, with average (over the stochastic process which produces the fluctuating force $F_{\mu\nu}(t)$)

$$\langle F_{\mu\nu}(t) \rangle = \langle \xi_\mu(t)\xi_\nu(t)\tanh[\beta m_{\mu\nu}(t)\xi^\mu(t)\xi^\nu(t)] \rangle,$$

and variation

$$\langle F_{\mu\nu}(t)F_{\mu\nu}(t') \rangle = Q_{\mu\nu}\delta(t - t'). \tag{4.60}$$

Coefficients Q_μ in (4.59) and $Q_{\mu\nu}$ in (4.60) represent strengths of the corresponding stochastic processes, while Dirac δ-functions $\delta(t - t')$ express their short-term memories.

Recall that standard interpretation of synergetics (see [Hak83]) describes the stochastic gradient systems (4.57)–(4.58) as the overdamped motion of (vector and tensor, respectively) representative particles in scalar potential fields $V(m_\mu)$ and $V(m_{\mu\nu})$, subject to fluctuating forces $F_\mu(t)$ and $F_{\mu\nu}(t)$. These particles undergo *non-equilibrium phase transitions* (in the similar way as the magnet undergoes transition from its unmagnetized state into a magnetized state, or a superconductor goes from its normal state into the superconducting state, only occurring now in systems far from thermal equilibrium), and associated phenomena, including a *symmetry breaking instability* and *critical slowing down* (see [Hak83]).

The non-equilibrium phase transitions of vector and tensor order parameters (overlaps) $m_\mu(t)$ and $m_{\mu\nu}(t)$, are in synergetics described in terms of probability distributions $p(m_\mu, t)$ and $p(m_{\mu\nu}, t)$, respectively, defined by corresponding *Fokker–Planck equations*

$$\dot{p}(m_\mu, t) = p(m_\mu, t) + \frac{1}{2} Q_\mu \frac{\partial^2 p(m_\mu, t)}{\partial m_\mu^2}, \quad \text{and}$$

$$\dot{p}(m_{\mu\nu}, t) = p(m_{\mu\nu}, t) + \frac{1}{2} Q_{\mu\nu} \frac{\partial^2 p(m_{\mu\nu}, t)}{\partial m_{\mu\nu}^2}.$$

Lotka–Volterra-based Associative Net

Hopfield recurrent associative memory network can be generalized to get a *bidirectional associative memory* network, the so-called BAM model of Kosko [Kos92]. Here we derive an alternative self-organizing neural net model with competitive *Lotka–Volterra ensemble dynamics*[18] (see [Vol31, II06b]).

We start from $(n + m)$D linear ODEs, describing two competitive neural ensembles participating in a two-party game,

$$\begin{aligned} \dot{R}^i &= -\alpha_B^j B_j, & R^i(0) &= R_0^i, \\ \dot{B}_j &= -\beta_i^R R^i, & B_j(0) &= B_j^0, & (i = 1, \dots, n; \ j = 1, \dots, m), \end{aligned} \tag{4.63}$$

where $R^i = R^i(t)$ and $B_j = B_j(t)$ respectively represent the numerical strengths of the two neural ensembles at time t, R_0^i, B_j^0 are their initial conditions, and α_B and β^R represent the effective spiking rates (which are either constant, or Poisson random process). In this way, we generate a $(n + m)$D smooth manifold M, a *neural state-space*, and two dynamical objects acting

---

[18] The competitive Lotka–Volterra equations are a simple model of the population dynamics of two species competing for some common resource. It is the pair of coupled logistic ODEs for two competing populations with sizes x_1, x_2,

$$\dot{x}_1 = r_1 x_1 \left(\frac{K_1 - x_1 - \alpha_{12} x_2}{K_1} \right), \tag{4.61}$$

$$\dot{x}_2 = r_2 x_2 \left(\frac{K_2 - x_2 - \alpha_{21} x_1}{K_2} \right), \tag{4.62}$$

where $r_{1,2}$ are their growth rates, $K_{1,2}$ are their respective carrying capacities, while $\alpha_{1,2}, \alpha_{2,1}$ are the mutual (coupling) effects of one species onto another.

on it: an nD smooth *vector-field* $\dot{R}^i$, and an mD differential 1-*form* $\dot{B}_j$. Their dot product $\dot{R}^i \cdot \dot{B}_j$, represents a hypothetical *neural outcome*. This is a linear system, with the passive-decay couplings $\alpha_B^j B_j$ and $\beta_i^R R^i$, fully predictable but giving only equilibrium solutions.

Secondly, to incorporate competitive dynamics of Volterra–Lotka style as commonly used in ecological modeling and known to produce a global chaotic attractor [II07a], we include to each of the neural ensembles a nonlinear competing term depending only on its own units,

$$
\begin{aligned}
\dot{R}^i &= a^i R^i (1 - b^i R^i) - \alpha_B^j B_j, \\
\dot{B}_j &= c_j B_j (1 - d_j B_j) - \beta_i^R R^i.
\end{aligned}
\tag{4.64}
$$

Now we have a *competition between the two chaotic attractors*, one for the R^i and one for the B_j ensemble, i.e., the two self-organization patterns emerging far-from-equilibrium.

Thirdly, to make this even more realistic, we include the ever-present *noise* in the form of Langevin-type random forces $F^i = F^i(t)$, and $G_j = G_j(t)$, thus adding the 'neural heating', i.e., noise induced entropy growth, to the competitive dynamics

$$
\begin{aligned}
\dot{R}^i &= a^i R^i (1 - b^i R^i) - \alpha_B^j B_j + F^i, \\
\dot{B}_j &= c_j B_j (1 - d_j B_j) - \beta_i^R R^i + G_j.
\end{aligned}
\tag{4.65}
$$

Finally, to overcome the deterministic chaos and stochastic noise with an adaptive *brain-like dynamics*, we introduce the *field competition potential V*, in the scalar form

$$
V = -\frac{1}{2}(\omega_i^j R^i B_j + \varepsilon_i^j B_j R^i),
\tag{4.66}
$$

where ω_i^j and ε_i^j represent *synaptic associative-memory* matrices for the R^i and B_j ensemble, respectively. From the negative potential V, we get a *Lyapunov-stable gradient system* $\dot{R}^i = -\frac{\partial V}{\partial B_j}$, $\dot{B}_j = -\frac{\partial V}{\partial R^i}$. This robust system, together with the *sigmoidal activation functions* $S(\cdot) = \tanh(\cdot)$, and *control inputs* $u_{OLN}^i = u_{OLN}^i(t)$ and $v_j^{OLN} = v_j^{OLN}(t)$, we incorporate into (4.65) to get the *full neural competitive-associative dynamics*

$$
\begin{aligned}
\dot{R}^i &= u_{OLN}^i - \alpha_B^j B_j + a^i R^i (1 - b^i R^i) + \omega_i^j S_j(B_j) + F^i, \\
\dot{B}_j &= v_j^{OLN} - \beta_i^R R^i + c_j B_j (1 - d_j B_j) + \varepsilon_i^j S^i(R^i) + G_j,
\end{aligned}
\tag{4.67}
$$

with initial conditions $R^i(0) = R_0^i$, $B_j(0) = B_j^0$.

Now, each ensemble learns by trial-and-error from the opposite side. In a standard ANN-fashion, we model this learning on the spot by initially setting the random values to the synaptic matrices ω_i^j and ε_i^j, and subsequently adjust these values using the standard *Hebbian learning scheme* [Heb49]:

$$
\text{New Value} = \text{Old Value} + \text{Innovation},
$$

which in our case reads:

$$\dot{\omega}_i^j = -\omega_i^j + \Phi_i^j(R^i, B_j),$$
$$\dot{\varepsilon}_i^j = -\varepsilon_i^j + \Psi_i^j(B_j, R^i),$$
(4.68)

with *innovation* given in tensor signal form (generalized from [Kos92])

$$\Phi_i^j = S_j(R^i)S_j(B_j) + \dot{S}_j(R^i)\dot{S}_j(B_j),$$
$$\Psi_i^j = S^i(R^i)S^i(B_j) + \dot{S}^i(R^i)\dot{S}^i(B_j),$$
(4.69)

where terms with overdots, equal $\dot{S}(\cdot) = 1 - \tanh(\cdot)$, denote the *signal velocities*.

4.1.3 Dissipative Evolution Under the Ricci Flow

In this section we will derive the geometric formalism associated with the quadratic Ricci-flow equation (4.4), as a general framework for all presented bio-reaction–diffusion systems. For the necessary mathematical background, see Appendix.

Geometrization Conjecture

Geometry and topology of smooth surfaces are related by the *Gauss–Bonnet formula* for a closed surface Σ (see, e.g., [II06c, II07e])

$$\frac{1}{2\pi} \iint_\Sigma K \, dA = \chi(\Sigma) = 2 - 2 \operatorname{gen}(\Sigma),$$
(4.70)

where dA is the area element of a metric g on Σ, K is the Gaussian curvature, $\chi(\Sigma)$ is the Euler characteristic of Σ and $\operatorname{gen}(\Sigma)$ is its *genus*, or number of handles, of Σ. Every closed surface Σ admits a metric of constant Gaussian curvature $K = +1, 0$, or -1 and so is uniformized by elliptic, Euclidean, or hyperbolic geometry, which respectively have $\operatorname{gen}(S^2) = 0$ (sphere), $\operatorname{gen}(T^2) = 1$ (torus) and $\operatorname{gen}(\Sigma) > 1$ (torus with several holes). The integral (4.70) is a *topological invariant* of the surface Σ, always equal to 2 for all topological spheres S^2 (that is, for all closed surfaces without holes that can be continuously deformed from the geometric sphere) and always equal to 0 for the topological torus T^2 (i.e., for all closed surfaces with one hole or handle).

Topological framework for the Ricci flow (4.2) is Thurston's *Geometrization Conjecture* [Thu82], which states that the interior of any compact 3-manifold can be split in an essentially unique way by disjoint embedded 2-spheres S^2 and tori T^2 into pieces and each piece admits one of 8 geometric structures (including (i) the 3-sphere S^3 with constant curvature $+1$;

(ii) the Euclidean 3-space $\mathbb{R}^3$ with constant curvature 0 and (iii) the hyperbolic 3-space $\mathbb{H}^3$ with constant curvature -1).[19] The geometrization conjecture (which has the Poincaré Conjecture as a special case) would give us a link between the geometry and topology of 3-manifolds, analogous in spirit to the case of 2-surfaces.

In higher dimensions, the Gaussian curvature K corresponds to the Riemann curvature tensor $\mathfrak{Rm}$ on a smooth n-manifold M, which is in local coordinates on M denoted by its $(4,0)$-components R_{ijkl}, or its $(3,1)$-components R^l_{ijk} (see Appendix, as well as e.g., [II06c, II07e]). The trace (or, contraction) of $\mathfrak{Rm}$, in $(4,0)$-case using the inverse metric tensor $g^{ij} = (g_{ij})^{-1}$, is the Ricci tensor $\mathfrak{Rc}$, the contracted curvature tensor, which is in a local coordinate system $\{x^i\}_{i=1}^n$ defined in an open set $U \subset M$, given by

$$R_{ij} = \text{tr}(\mathfrak{Rm}) = g^{kl} R_{ijkl}$$

(using Einstein's summation convention), while the *scalar curvature* is now given by the second contraction of $\mathfrak{Rm}$ as

$$R = \text{tr}(\mathfrak{Rc}) = g^{ij} R_{ij}.$$

In general, the Ricci flow $g_{ij}(t)$ is a one-parameter family of Riemannian metrics on a compact n-manifold M governed by (4.2), which has a unique solution for a short time for an arbitrary smooth metric g_{ij} on M [Ham82]. If $\mathfrak{Rc} > 0$ at any local point $x = \{x^i\}$ on M, then the Ricci flow (4.2) contracts the metric $g_{ij}(t)$ near x, to the future, while if $\mathfrak{Rc} < 0$, then the flow (4.2) expands $g_{ij}(t)$ near x. The solution metric $g_{ij}(t)$ of the Ricci flow equation (4.2) shrinks in positive Ricci curvature direction while it expands in the negative Ricci curvature direction, because of the minus sign in the front of the Ricci tensor R_{ij}. In particular, on a 2-sphere S^2, any metric of positive Gaussian curvature will shrink to a point in finite time. At a general point, there will be directions of positive and negative Ricci curvature along which the metric will locally contract or expand (see [And04]). Also, if a simply-connected compact 3-manifold M has a Riemannian metric g_{ij} with positive Ricci curvature then it is diffeomorphic to the 3-sphere S^3 [Ham82]. More generally speaking, the Ricci flow deforms manifolds with positive Ricci curvature to a point which can be renormalized to the 3-sphere.

---

[19] Another five allowed geometric structures are represented by the following examples: (iv) the product $S^2 \times S^1$; (v) the product $\mathbb{H}^2 \times S^1$ of hyperbolic plane and circle; (vi) a left invariant Riemannian metric on the special linear group $SL(2,\mathbb{R})$; (vii) a left invariant Riemannian metric on the solvable Poincaré-Lorentz group $E(1,1)$, which consists of rigid motions of a $(1+1)$-dimensional space-time provided with the flat metric $dt^2 - dx^2$; (viii) a left invariant metric on the nilpotent Heisenberg group, consisting of 3×3 matrices of the form $\left[\begin{smallmatrix} 1 & * & * \\ 0 & 1 & * \\ 0 & 0 & 1 \end{smallmatrix}\right]$. In each case, the universal covering of the indicated manifold provides a canonical model for the corresponding geometry [Mil03].

Reaction–diffusion-type evolution of curvatures and volumes

All three Riemannian curvatures (R, $\mathfrak{Rc}$ and $\mathfrak{Rm}$), as well as the associated volume forms, *evolve* during the Ricci flow (4.2). In general, the Ricci-flow evolution equation (4.2) for the metric tensor g_{ij} implies the reaction–diffusion-type evolution equation for the Riemann curvature tensor $\mathfrak{Rm}$ on an n-manifold M,

$$\partial_t \mathfrak{Rm} = \Delta \mathfrak{Rm} + Q_n, \tag{4.71}$$

where Q_n is a quadratic expression of the Riemann n-curvatures, corresponding to the n-component bio-chemical reaction, while the term $\Delta \mathfrak{Rm}$ corresponds to the n-component diffusion. From the general n-curvature evolution (4.71) we have two important particular cases:[20]

1. The evolution equation for the Ricci curvature tensor $\mathfrak{Rc}$ on a 3-manifold M,

$$\partial_t \mathfrak{Rc} = \Delta \mathfrak{Rc} + Q_3, \tag{4.72}$$

where Q_3 is a quadratic expression of the Ricci 3-curvatures, corresponding to the 3-component bio-chemical reaction, while the term $\Delta \mathfrak{Rc}$ corresponds to the 3-diffusion.

2. The evolution equation for the scalar curvature R,

$$\partial_t R = \Delta R + 2|\mathfrak{Rc}|^2, \tag{4.73}$$

in which the term $2|\mathfrak{Rc}|^2$ corresponds to the 2-component bio-chemical reaction, while the term ΔR corresponds to the 2-component diffusion. By the *maximum principle* (see Sect. 4.1.3), the minimum of the scalar curvature R is non-decreasing along the flow $g(t)$, both on M and on its boundary ∂M (see [Per02]).

---

[20] By expanding the maximum principle for tensors, Hamilton proved that Ricci flow $g(t)$ given by (4.2) preserves the positivity of the Ricci tensor $\mathfrak{Rc}$ on 3-manifolds (as well as of the Riemann curvature tensor $\mathfrak{Rm}$ in all dimensions); moreover, the eigenvalues of the Ricci tensor on 3-manifolds (and of the curvature operator $\mathfrak{Rm}$ on 4-manifolds) are getting pinched point-wisely as the curvature is getting large [Ham82, Ham86]. This observation allowed him to prove the convergence results: the evolving metrics (on a compact manifold) of positive Ricci curvature in dimension 3 (or positive Riemann curvature in dimension 4) converge, modulo scaling, to metrics of constant positive curvature.

However, without assumptions on curvature, the long time behavior of the metric evolving by Ricci flow may be more complicated [Per02]. In particular, as t approaches some finite time T, the curvatures may become arbitrarily large in some region while staying bounded in its complement. On the other hand, Hamilton [Ham93] discovered a remarkable property of solutions with nonnegative curvature tensor $\mathfrak{Rm}$ in arbitrary dimension, called the *differential Harnack inequality*, which allows, in particular, to compare the curvatures of the solution of (4.2) at different points and different times.

Let us now see in detail how various geometric quantities evolve given the *short-time solution* of the Ricci flow equation (4.2) on an arbitrary n-manifold M. For this, we need first to calculate the *variation formulas* for the Christoffel symbols and curvature tensors on M; then the corresponding evolution equations will naturally follow (see [Ham82, CC99, CK04]). If $g(s)$ is a 1-parameter family of metrics on M with $\partial_s g_{ij} = v_{ij}$, then the variation of the Christoffel symbols Γ_{ij}^k on M is given by

$$\partial_s \Gamma_{ij}^k = \frac{1}{2} g^{kl} \left(\nabla_i v_{jl} + \nabla_j v_{il} - \nabla_l v_{ij} \right), \qquad (4.74)$$

(where ∇ is the covariant derivative with respect to the Riemannian connection) from which follows the evolution of the Christoffel symbols Γ_{ij}^k under the Ricci flow $g(t)$ on M given by (4.2),

$$\partial_t \Gamma_{ij}^k = -g^{kl} \left(\nabla_i R_{jl} + \nabla_j R_{il} - \nabla_l R_{ij} \right).$$

From (4.74) we calculate the variation of the Ricci tensor R_{ij} on M as

$$\partial_s R_{ij} = \nabla_m \left(\partial_s \Gamma_{ij}^m \right) - \nabla_i \left(\partial_s \Gamma_{mj}^m \right), \qquad (4.75)$$

and the variation of the scalar curvature R on M by

$$\partial_s R = -\Delta V + \operatorname{div}(\operatorname{div} v) - \langle v, \mathfrak{Rc} \rangle, \qquad (4.76)$$

where $V = g^{ij} v_{ij} = \operatorname{tr}(v)$ is the trace of $v = (v_{ij})$.

If an n-manifold M is oriented, then the *volume n-form* on M is given, in a positively-oriented local coordinate system $\{x^i\} \in U \subset M$, by [II07e]

$$d\mu = \sqrt{\det(g_{ij})}\, dx^1 \wedge dx^2 \wedge \cdots \wedge dx^n. \qquad (4.77)$$

If $\partial_s g_{ij} = v_{ij}$, then $\partial_s d\mu = \frac{1}{2} V d\mu$. The evolution of the volume n-form $d\mu$ under the Ricci flow $g(t)$ on M is given by the exponential decay/growth relation with the scalar curvature R as the (variable) rate parameter,

$$\partial_t d\mu = -R d\mu, \qquad (4.78)$$

which gives an exponential decay for $R \geq a > 0$ (elliptic geometry) and exponential growth for $R \leq a < 0$ (hyperbolic geometry)—for any small constant a (scalar curvature must be bounded away from zero). The elementary volume evolution (4.78) implies the integral form of the exponential relation for the total n-volume

$$\operatorname{vol}(g) = \int_M d\mu, \quad \text{as } \partial_t \operatorname{vol}(g(t)) = -\int_M R d\mu,$$

which again gives an *exponential decay* for elliptic $R > 0$ and *exponential growth* for hyperbolic $R < 0$.

This is a crucial point for the *tumor suppression* by the body: the immune system needs to keep the elliptic geometry of the MTS, evolving by (4.29)—by all possible means.[21] In the healthy organism this normally happens, because the initial MTS started as a spherical shape with $R > 0$. The immune system just needs to keep the MTS in the spherical/elliptic shape and prevent any hyperbolic distortions of $R < 0$. Thus, it will naturally have an exponential decay and vanish.

Since the n-volume is not constant and sometimes we would like to prevent the solution from shrinking to an n-point on M (elliptic case) or expanding to an infinity (hyperbolic case), we can also consider the *normalized Ricci flow* on M (see Fig. 4.1 as well as ref. [CC99]):

$$\partial_t \hat{g}_{ij} = -2\hat{R}_{ij} + \frac{2}{n}\hat{r}\hat{g}_{ij}, \quad \text{where } \hat{r} = \text{vol}(\hat{g})^{-1} \int_M \hat{R}d\mu \qquad (4.79)$$

is the average scalar curvature on M. We then have the *n-volume conservation law:*

$$\partial_t \text{vol}(\hat{g}(t)) = 0.$$

To study the *long-time existence* of the normalized Ricci flow (4.79) on an arbitrary n-manifold M, it is important to know what kind of curvature conditions are preserved under the equation. In general, the Ricci flow $g(t)$ on M, as defined by the fundamental relation (4.2), tends to preserve some

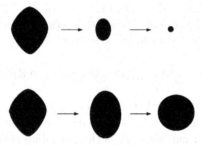

Fig. 4.1. An example of Ricci flow normalization: unnormalized flow (*up*) and normalized flow (*down*).

---

[21] As a tumor decay control tool, a monoclonal antibody therapy is usually proposed. Monoclonal antibodies (mAb) are mono-specific antibodies that are identical because they are produced by one type of immune cell that are all clones of a single parent cell. Given (almost) any substance, it is possible to create monoclonal antibodies that specifically bind to that substance; they can then serve to detect or purify that substance. The invention of monoclonal antibodies is generally accredited to Georges Köhler, César Milstein, and Niels Kaj Jerne in 1975 [KM75], who shared the Nobel Prize in Physiology or Medicine in 1984 for the discovery. The key idea was to use a line of myeloma cells that had lost their ability to secrete antibodies, come up with a technique to fuse these cells with healthy antibody producing B-cells, and be able to select for the successfully fused cells.

kind of positivity of curvatures. For example, positive scalar curvature R (i.e., elliptic geometry) is preserved both on M and on its boundary ∂M in any dimension. This follows from applying the maximum principle to the evolution equation (4.73) for scalar curvature R both on M and on ∂M. Similarly, positive Ricci curvature is preserved under the Ricci flow on a 3-manifold M. This is a special feature of dimension 3 and is related to the fact that the Riemann curvature tensor may be recovered algebraically from the Ricci tensor and the metric on 3-manifolds [CC99].

In particular, we have the following result for 2-surfaces (see [Ham88]): Let $S = \partial M$ be a closed 2-surface, which is a boundary of a compact 3-manifold M. Then for any initial 2-metric g_0 on ∂M, the solution to the normalized Ricci flow (4.79) on ∂M exists for all time. In other words, the normalized Ricci flow in 2D always converges. Moreover, (i) if the Euler characteristic of ∂M is non-positive, then the solution metric $g(t)$ on ∂M converges to a constant curvature metric as $t \to \infty$; and (ii) if the scalar curvature R of the initial metric g_0 is positive, then the solution metric $g(t)$ on ∂M converges to a positive constant curvature metric as $t \to \infty$. (For surfaces with non-positive Euler characteristic, the proof is based primarily on maximum principle estimates for the scalar curvature.)

Applying to the tumor evolution (4.29), the normalized Ricci flow (4.79) of the MTS will make it completely round with a geometric sphere shell, which is ideal for surgical removal. This is our second option for the MTS growth/decay control. If we cannot force it to exponential decay, then we must try to normalize into a round spherical shell—which is suitable for surgical removal.

The negative flow of the total n-volume $\mathrm{vol}(g(t))$ represents the *Einstein–Hilbert functional* (see [MTW73, CC99, And04])

$$E(g) = \int_M R d\mu = -\partial_t \mathrm{vol}(g(t)).$$

If we put $\partial_s g_{ij} = v_{ij}$, we have

$$\partial_s E(g) = \int_M \left(-\Delta V + \mathrm{div}(\mathrm{div}\, v) - \langle v, \mathfrak{Rc} \rangle + \frac{1}{2} RV \right) d\mu$$

$$= \int_M \left\langle v, \frac{1}{2} R g_{ij} - R_{ij} \right\rangle d\mu,$$

so, the critical points of $E(g)$ satisfy *Einstein's equation* $\frac{1}{2} R g_{ij} = R_{ij}$ in the vacuum. The gradient flow of $E(g)$ on an n-manifold M,

$$\partial_t g_{ij} = 2 \left(\nabla E(g) \right)_{ij} = R g_{ij} - 2 R_{ij},$$

is the Ricci flow (4.2) plus $R g_{ij}$. Thus, Einstein metrics are the fixed points of the normalized Ricci flow.[22]

---

[22] Einstein metrics on n-manifolds are metrics with constant Ricci curvature. However, along the way, the deformation will encounter singularities. The major question, resolved by Perelman, was how to find a way to describe all possible singularities.

Let Δ denote the Laplacian acting on functions on a closed n-manifold M, which is in local coordinates $\{x^i\} \in U \subset M$ given by

$$\Delta = g^{ij}\nabla_i\nabla_j = g^{ij}\left(\partial_{ij} - \Gamma^k_{ij}\partial_k\right).$$

For any smooth function f on M we have [Ham82, CK04]

$$\Delta\nabla_i f = \nabla_i\Delta f + R_{ij}\nabla_j f \quad \text{and}$$
$$\Delta|\nabla f|^2 = 2|\nabla_i\nabla_j f|^2 + 2R_{ij}\nabla_i f\nabla_j f + 2\nabla_i f\nabla_i\Delta f,$$

from which it follows that if we have

$$\mathfrak{Rc} \geq 0, \qquad \Delta f \equiv 0, \qquad |\nabla f| \equiv 1, \quad \text{then}$$
$$\nabla\nabla f \equiv 0 \quad \text{and} \quad \mathfrak{Rc}(\nabla f, \nabla f) \equiv 0.$$

Using this Laplacian Δ, we can write the linear heat equation on M as $\partial_t u = \Delta u$, where u is the temperature. In particular, the Laplacian acting on functions with respect to $g(t)$ will be denoted by $\Delta_{g(t)}$. If $(M, g(t))$ is a solution to the Ricci flow equation (4.2), then we have

$$\partial_t\Delta_{g(t)} = 2R_{ij}\nabla_i\nabla_j.$$

The evolution equation (4.73) for the scalar curvature R under the Ricci flow (4.2) follows from (4.76). Using (7.4) from Appendix, we have:

$$\text{div}(\mathfrak{Rc}) = \frac{1}{2}\nabla R, \quad \text{so that} \quad \text{div}(\text{div}(\mathfrak{Rc})) = \frac{1}{2}\Delta R,$$

showing again that the scalar curvature R satisfies a heat-type equation with a quadratic nonlinearity both on a 3-manifold M and on its boundary 2-surface ∂M.

Next we will find the exact form of the evolution equation (4.72) for the Ricci tensor $\mathfrak{Rc}$ under the Ricci flow $g(t)$ given by (4.2) on any 3-manifold M. (Note that in higher dimensions, the appropriate formula of huge complexity would involve the whole Riemann curvature tensor $\mathfrak{Rm}$.) Given a variation $\partial_s g_{ij} = v_{ij}$, from (4.75) we get

$$\partial_s R_{ij} = \frac{1}{2}\left(\Delta_L v_{ij} + \nabla_i\nabla_j V - \nabla_i(\text{div } v)_j - \nabla_j(\text{div } v)_i\right),$$

where Δ_L denotes the so-called Lichnerowicz Laplacian (which depends on $\mathfrak{Rm}$, see [Ham82, CK04]). Since

$$\nabla_i\nabla_j R - \nabla_i(\text{div}(\mathfrak{Rc}))_j - \nabla_j(\text{div}(\mathfrak{Rc}))_i = 0,$$

by (7.4) (after some algebra) we get that under the Ricci flow (4.2) the evolution equation for the Ricci tensor $\mathfrak{Rc}$ on a 3-manifold M is

$$\partial_t R_{ij} = \Delta R_{ij} + 3RR_{ij} - 6R_{im}R_{jm} + \left(2|\mathfrak{Rc}|^2 - R^2\right)g_{ij}.$$

So, just as in case of the evolution (4.73) of the scalar curvature $\partial_t R$ (both on a 3-manifold M and on its 2-boundary ∂M), we get a heat-type evolution equation with a quadratic nonlinearity for $\partial_t R_{ij}$, which means that positive Ricci curvature ($\Re c > 0$) of elliptic 3-geometry is preserved under the Ricci flow $g(t)$ on M.

More generally, we have the following result for 3-manifolds (see [Ham82]): Let (M, g_0) be a compact Riemannian 3-manifold with positive Ricci curvature $\Re c$. Then there exists a unique solution to the normalized Ricci flow $g(t)$ on M with $g(0) = g_0$ for all time and the metrics $g(t)$ converge exponentially fast to a constant positive sectional curvature metric g_∞ on M. In particular, M is diffeomorphic to a 3-sphere S^3. (As a consequence, such a 3-manifold M is necessarily diffeomorphic to a quotient of the 3-sphere by a finite group of isometries. It follows that given any homotopy 3-sphere, if one can show that it admits a metric with positive Ricci curvature, then the Poincaré Conjecture would follow [CC99].) In addition, compact and closed 3-manifolds which admit a non-singular solution can also be decomposed into geometric pieces [Ham99].

From the geometric evolution equations reviewed in this subsection, we see that both short-time and long-time geometric solution can always be found for 2-component bio-reaction–diffusion equations, as they correspond to evolution of the scalar 2-curvature R. Regarding the 3-component bio-reaction–diffusion equations, corresponding to evolution of the Ricci 3-curvature $\Re c$, we can always find the short-time geometric solution, while the long-time solution exists only under some additional (compactnes and/or closure) conditions. Finally, in case of n-component bio-reaction–diffusion equations, corresponding to evolution of the Riemann n-curvature $\Re m$, only short-time geometric solution is possible.

Dissipative Solitons and Ricci Breathers

An important class of bio-reaction–diffusion systems are *dissipative solitons* (DSs), which are stable solitary localized structures that arise in nonlinear spatially extended dissipative systems due to mechanisms of *self-organization*. They can be considered as an extension of the classical soliton concept in conservative systems. Apart from aspects similar to the behavior of classical particles like the formation of bound states, DSs exhibit entirely nonclassical behavior—e.g., scattering, generation and annihilation—all without the constraints of energy or momentum conservation. The excitation of internal degrees of freedom may result in a dynamically stabilized intrinsic speed, or periodic oscillations of the shape.

In particular, stationary DSs are generated by production of material in the center of the DSs, diffusive transport into the tails and depletion of material in the tails. A propagating pulse arises from production in the leading and depletion in the trailing end. Among other effects, one finds periodic oscillations of DSs, the so-called 'breathing' dissipative solitons [GAP06].

DSs in many different systems show universal particle-like properties. To understand and describe the latter, one may try to derive 'particle equations' for slowly varying order parameters like position, velocity or amplitude of the DSs by adiabatically eliminating all fast variables in the field description. This technique is known from linear systems, however mathematical problems arise from the nonlinear models due to a coupling of fast and slow modes [Fri04].

Similar to low-dimensional dynamic systems, for supercritical bifurcations of stationary DSs one finds characteristic normal forms essentially depending on the symmetries of the system; e.g., for a transition from a symmetric stationary to an intrinsically propagating DS one finds the Pitchfork normal form for the DS-velocity $\mathbf{v}$ [Bod97],

$$\dot{\mathbf{v}} = (\sigma - \sigma_0)\mathbf{v} - |\mathbf{v}|^2\mathbf{v}$$

where σ represents the bifurcation parameter and σ_0 the bifurcation point. For a bifurcation to a 'breathing' DS, one finds the Hopf normal form [II07a]

$$\dot{A} = (\sigma - \sigma_0)A - |A|^2A$$

for the amplitude A of the oscillation. Note that the above problems do not arise for classical solitons as inverse scattering theory yields complete analytical solutions [GAP06].

Ricci Breathers and Ricci Solitons

Closely related to dissipative solitons are the so-called *breathers*, solitonic structures given by localized periodic solutions of some nonlinear soliton PDEs, including the exactly-solvable sine-Gordon equation[23] and the focusing nonlinear Schrödinger equation.[24]

---

[23] An exact solution $u = u(x,t)$ of the $(1+1)$D sine-Gordon equation

$$\partial_{t^2}u = \partial_{x^2}u - \sin u, \quad \text{is [AK73]}$$

$$u = 4\arctan\left(\frac{\sqrt{1-\omega^2}\cos(\omega t)}{\omega\cosh(\sqrt{1-\omega^2}x)}\right),$$

which, for $\omega < 1$, is periodic in time t and decays exponentially when moving away from $x = 0$.

[24] The focusing nonlinear Schrödinger equation is the dispersive complex-valued $(1+1)$D PDE [AEK87],

$$i\partial_t u + \partial_{x^2}u + |u|^2u = 0,$$

with a breather solution of the form:

$$u = \left(\frac{2b^2\cosh(\theta) + 2ib\sqrt{2-b^2}\sinh(\theta)}{2\cosh(\theta) - \sqrt{2}\sqrt{2-b^2}\cos(abx)} - 1\right)a\exp(ia^2t)$$

$$\text{with } \theta = a^2b\sqrt{2-b^2}t,$$

which gives breathers periodic in space x and approaching the uniform value a when moving away from the focus time $t = 0$.

A metric $g_{ij}(t)$ evolving by the Ricci flow $g(t)$ given by (4.2) on any 3-manifold M is called a *Ricci breather*, if for some $t_1 < t_2$ and $\alpha > 0$ the metrics $\alpha g_{ij}(t_1)$ and $g_{ij}(t_2)$ differ only by a diffeomorphism; the cases $\alpha = 1, \alpha < 1, \alpha > 1$ correspond to steady, shrinking and expanding breathers, respectively. Trivial breathers on M, for which the metrics $g_{ij}(t_1)$ and $g_{ij}(t_2)$ differ only by diffeomorphism and scaling for each pair of t_1 and t_2, are called *Ricci solitons*. Thus, if one considers Ricci flow as a dynamical system on the space of Riemannian metrics modulo diffeomorphism and scaling, then breathers and solitons correspond to periodic orbits and fixed points respectively. At each time the Ricci soliton metric satisfies on M an equation of the form [Per02]

$$R_{ij} + cg_{ij} + \nabla_i b_j + \nabla_j b_i = 0,$$

where c is a number and b_i is a 1-form; in particular, when $b_i = \frac{1}{2}\nabla_i a$ for some function a on M, we get a gradient Ricci soliton. An important example of a gradient shrinking soliton is the *Gaussian soliton*, for which the metric g_{ij} is just the Euclidean metric on $\mathbb{R}^3$, $c = 1$ and $a = -|x|^2/2$.

Smoothing/Averaging Heat Equation and Ricci Entropy

Given a C^2 function $u : M \to \mathbb{R}$ on a Riemannian 3-manifold M, its Laplacian is defined in local coordinates $\{x^i\} \in U \subset M$ to be

$$\Delta u = \mathrm{tr}_g\left(\nabla^2 u\right) = g^{ij}\nabla_i\nabla_j u,$$

where ∇_i is the *covariant derivative* (Levi-Civita connection, see Appendix). We say that a C^2 function $u : M \times [0, T) \to \mathbb{R}$, where $T \in (0, \infty]$, is a solution to the heat equation if (4.3) holds. One of the most important properties satisfied by the heat equation is the *maximum principle*, which says that for any smooth solution to the heat equation, whatever point-wise bounds hold at $t = 0$ also hold for $t > 0$ [CC99]. More precisely, we can state: Let $u : M \times [0, T) \to \mathbb{R}$ be a C^2 solution to the heat equation (4.3) on a complete Riemannian 3-manifold M. If $C_1 \leq u(x, 0) \leq C_2$ for all $x \in M$, for some constants $C_1, C_2 \in \mathbb{R}$, then $C_1 \leq u(x, t) \leq C_2$ for all $x \in M$ and $t \in [0, T)$. This property exhibits the averaging behavior of the heat equation (4.3) on M.

Now, consider Perelman's *entropy functional* [Per02] on a 3-manifold M[25]

---

[25] Note that in the related context of Riemannian gravitation theory, the so-called *gravitational entropy* is embedded in the Weyl curvature $(4, 0)$-tensor $\mathfrak{W}$, which is the traceless component of the Riemann curvature tensor $\mathfrak{Rm}$ (i.e., $\mathfrak{Rm}$ with the Ricci tensor $\mathfrak{Rc}$ removed),

$$\mathfrak{W} = \mathfrak{Rm} - f(R_{ij}g_{ij}),$$

where $f(R_{ij}g_{ij})$ is a certain linear function of R_{ij} and g_{ij}. According to Penrose's *Weyl curvature hypothesis*, the entire *history of a closed universe* starts from a uniform low-entropy Big Bang with zero Weyl curvature tensor of the cosmologi-

$$\mathcal{F} = \int_M (R + |\nabla f|^2) e^{-f} d\mu \qquad (4.80)$$

for a Riemannian metric g_{ij} and a (temperature-like) scalar function f (which satisfies the backward heat equation) on a closed 3-manifold M, where $d\mu$ is the volume 3-form (4.77). During the Ricci flow (4.2), $\mathcal{F}$ evolves on M as

$$\partial_t \mathcal{F} = 2 \int |R_{ij} + \nabla_i \nabla_j f|^2 e^{-f} d\mu. \qquad (4.81)$$

Now, define $\lambda(g_{ij}) = \inf \mathcal{F}(g_{ij}, f)$, where infimum is taken over all smooth f, satisfying

$$\int_M e^{-f} d\mu = 1. \qquad (4.82)$$

$\lambda(g_{ij})$ is the lowest eigenvalue of the operator $-4\Delta + R$. Then the entropy evolution formula (4.81) implies that $\lambda(g_{ij}(t))$ is nondecreasing in t, and moreover, if $\lambda(t_1) = \lambda(t_2)$, then for $t \in [t_1, t_2]$ we have $R_{ij} + \nabla_i \nabla_j f = 0$ for f which minimizes $\mathcal{F}$ on M [Per02]. Thus a steady breather on M is necessarily a steady soliton.

If we define the conjugate *heat operator* on M as

$$\Box^* = -\partial/\partial t - \Delta + R$$

then we have the *conjugate heat equation*[26] [Per02]

---

cal gravitational field and ends with a high-entropy Big Crunch, representing the congealing of may black holes, with Weyl tensor approaching infinity (see [Pen79, HP96b]).

[26] In [Per02] Perelman stated a differential Li–Yau–Hamilton (LYH) type inequality [Hsu08] for the fundamental solution $u = u(x, t)$ of the conjugate heat equation (4.84) on a closed n-manifold M evolving by the Ricci flow (4.2). Let $p \in M$ and

$$u = (4\pi\tau)^{-\frac{n}{2}} e^{-f}$$

be the fundamental solution of the conjugate heat equation in $M \times (0, T)$,

$$\Box^* u = 0, \quad \text{or} \quad \partial_t u + \Delta u = Ru,$$

where $\tau = T - t$ and $R = R(\cdot, t)$ is the scalar curvature of M with respect to the metric $g(t)$ with $\lim_{t \nearrow T} u = \delta_p$ (in the distribution sense), where δ_p is the delta-mass at p. Let

$$v = [\tau(2\Delta f - |\nabla f|^2 + R) + f - n]u,$$

where $\tau = T - t$. Then we have a differential LYH-type inequality

$$v(x, t) \le 0 \quad \text{in } M \times (0, T). \qquad (4.83)$$

This result was used by Perelman to give a proof of the *pseudolocality theorem* [Per02] which roughly said that almost Euclidean regions of large curvature in closed manifold with metric evolving by Ricci flow $g(t)$ given by (4.2) remain localized.

$$\Box^* u = 0. \tag{4.84}$$

The entropy functional (4.80) is nondecreasing under the coupled *Ricci-diffusion flow* on M (see [Ye07, Li07])

$$\partial_t g_{ij} = -2R_{ij}, \qquad \partial_t u = -\Delta u + \frac{R}{2}u - \frac{|\nabla u|^2}{u}, \tag{4.85}$$

where the second equation ensures $\int_M u^2 d\mu = 1$, to be preserved by the Ricci flow $g(t)$ on M. If we define $u = e^{-\frac{f}{2}}$, then the right-hand equation in (4.85) is equivalent to the generic scalar-field f-evolution equation on M,

$$\partial_t f = -\Delta f - R + |\nabla f|^2,$$

which instead preserves (4.82).

The coupled Ricci-diffusion flow (4.85), or equivalently, the dual system

$$\partial_t g_{ij} = \Delta_M g_{ij} + Q_{ij}(g, \partial g), \qquad \partial_t f = -\Delta f - R + |\nabla f|^2, \tag{4.86}$$

is our *global decay* model for a general n-dimensional bio-reaction–diffusion process, including both geometric and bio-chemical multi-phase evolution.

The sole *hypothesis of this paper* is that any kind of reaction–diffusion processes in biology, chemistry and physics is subsumed by the geometric-diffusion system (4.85), or the dual system (4.86).

Thermodynamic Analogy

Perelman's functional $\mathcal{F}$ is analogous to negative entropy [Per02]. Recall that thermodynamic *partition function* for a generic canonical ensemble at temperature β^{-1} is given by

$$Z = \int e^{-\beta E} d\omega(E), \tag{4.87}$$

where $\omega(E)$ is a 'density measure', which does not depend on β. From it, the *average energy* is given by

$$\langle E \rangle = -\partial_\beta \ln Z,$$

the *entropy* is

In particular, let $(M, g(t))$, $0 \le t \le T$, $\partial M \ne \phi$, be a compact 3-manifold with metric $g(t)$ evolving by the Ricci flow $g(t)$ given by (4.2) such that the second fundamental form of the surface ∂M with respect to the unit outward normal $\partial/\partial\nu$ of ∂M is uniformly bounded below on $\partial M \times [0, T]$. A global Li–Yau gradient estimate [LY86] for the solution of the generalized conjugate heat equation was proved in [Hsu08] (using a variation of the method of P. Li and S.T. Yau, [LY86]) on such a manifold with Neumann boundary condition.

$$S = \beta \langle E \rangle + \ln Z,$$

and the *fluctuation* is

$$\sigma = \langle (E - \langle E \rangle)^2 \rangle = \partial_{\beta^2} \ln Z.$$

If we now fix a closed 3-manifold M with a probability measure m and a metric $g_{ij}(\tau)$ that depends on the temperature τ, then according to equation

$$\partial_\tau g_{ij} = 2(R_{ij} + \nabla_i \nabla_j f),$$

the partition function (4.87) is given by

$$\ln Z = \int \left(-f + \frac{n}{2}\right) dm. \tag{4.88}$$

From (4.88) we get (see [Per02])

$$\langle E \rangle = -\tau^2 \int_M \left(R + |\nabla f|^2 - \frac{n}{2\tau}\right) dm,$$

$$S = -\int_M (\tau(R + |\nabla f|^2) + f - n) dm,$$

$$\sigma = 2\tau^4 \int_M \left|R_{ij} + \nabla_i \nabla_j f - \frac{1}{2\tau} g_{ij}\right|^2 dm,$$

where

$$dm = u dV, \quad u = (4\pi\tau)^{-\frac{n}{2}} e^{-f}.$$

From the above formulas, we see that the fluctuation σ is nonnegative; it vanishes only on a gradient shrinking soliton. $\langle E \rangle$ is nonnegative as well, whenever the flow exists for all sufficiently small $\tau > 0$. Furthermore, if the heat function u: (a) tends to a δ-function as $\tau \to 0$, or (b) is a limit of a sequence of partial heat functions u_i, such that each u_i tends to a δ-function as $\tau \to \tau_i > 0$, and $\tau_i \to 0$, then the entropy S is also nonnegative. In case (a), all the quantities $\langle E \rangle, S, \sigma$ tend to zero as $\tau \to 0$, while in case (b), which may be interesting if $g_{ij}(\tau)$ becomes singular at $\tau = 0$, the entropy S may tend to a positive limit.

4.2 Turbulence and Chaos in PDEs

Recall that *chaos theory*, of which *turbulence* is the most extreme form, started in 1963, when Ed Lorenz from MIT took the *Navier–Stokes equations* from viscous fluid dynamics and reduced them into three first-order coupled nonlinear ODEs, to demonstrate the idea of sensitive dependence upon initial conditions and associated *chaotic behavior*.

It is well-known that the viscous fluid evolves according to the nonlinear Navier–Stokes PDEs[27]

$$\dot{\mathbf{u}} + \mathbf{u} \cdot \boldsymbol{\nabla}\mathbf{u} + \nabla p/\rho = \nu \varDelta \mathbf{u} + \mathbf{f}, \tag{4.89}$$

where $\mathbf{u} = \mathbf{u}(x^i, t)$, $(i = 1, 2, 3)$ is the fluid 3D velocity, $p = p(x^i, t)$ is the pressure field, ρ, ν are the fluid density and viscosity coefficient, while $\mathbf{f} = \mathbf{f}(x^i, t)$ is the nonlinear external energy source. To simplify the problem, we can impose to $\mathbf{f}$ the so-called *Reynolds condition*, $\langle \mathbf{f} \cdot \mathbf{u} \rangle = \varepsilon$, where ε is the average rate of energy injection.

Fluid dynamicists believe that *Navier–Stokes equations* (4.89) *accurately describe turbulence*. A mathematical proof of the global regularity of the solutions to the Navier–Stokes equations is a very challenging problem and yet such a proof or disproof does not solve the problem of turbulence. However, it may help understanding turbulence. Turbulence is more of a dynamical system problem. We will see below that studies on chaos in PDEs indicate that turbulence can have *Bernoulli shift dynamics* which results in the wandering of a turbulent solution in a fat domain in the phase space; thus, turbulence can not be averaged. The hope is that turbulence can be controlled [Li04].

Turbulent Flow

Recall that in fluid dynamics, *turbulent flow* is a flow regime characterized by low momentum diffusion, high momentum convection, and rapid variation of pressure and velocity in space and time. Flow that is not turbulent is called *laminar flow*. Also, recall that the *Reynolds number Re* characterizes whether flow conditions lead to laminar or turbulent flow. The structure of turbulent flow was first described by A. Kolmogorov. Consider the flow of water over a simple smooth object, such as a sphere. At very low speeds the flow is laminar, i.e., the flow is locally smooth (though it may involve vortices on a large scale). As the speed increases, at some point the transition is made to turbulent (or, chaotic) flow. In turbulent flow, unsteady vortices[28]

---

[27] Recall that the Navier–Stokes equations, named after C.L. Navier and G.G. Stokes, are a set of PDEs that describe the motion of liquids and gases, based on the fact that changes in momentum of the particles of a fluid are the product of changes in pressure and dissipative viscous forces acting inside the fluid. These viscous forces originate in molecular interactions and dictate how viscous a fluid is, so the Navier–Stokes PDEs represent a dynamical statement of the balance of forces acting at any given region of the fluid. They describe the physics of a large number of phenomena of academic and economic interest (they are useful to model weather, ocean currents, water flow in a pipe, motion of stars inside a galaxy, flow around an airfoil (wing); they are also used in the design of aircraft and cars, the study of blood flow, the design of power stations, the analysis of the effects of pollution, etc.).

[28] Recall that a *vortex* can be any circular or rotary flow that possesses vorticity. Vortex represents a spiral whirling motion (i.e., a spinning turbulent flow) with

appear on many scales and interact with each other. Drag due to boundary layer skin friction increases. The structure and location of boundary layer separation often changes, sometimes resulting in a reduction of overall drag. Because laminar-turbulent transition is governed by Reynolds number, the same transition occurs if the size of the object is gradually increased, or the viscosity of the fluid is decreased, or if the density of the fluid is increased.

Vorticity Dynamics

Vorticity $\omega = \omega(x^i, t)$, ($i = 1, 2, 3$) is a geometrical concept used in fluid dynamics, which is related to the amount of 'circulation' or 'rotation' in a fluid. More precisely, *vorticity* is the circulation per unit area at a point in the flow field, or formally, $\omega = \nabla \times \mathbf{u}$, where $\mathbf{u} = \mathbf{u}(x^i, t)$ is the fluid velocity. It is a vector quantity, whose direction is (roughly speaking) along the axis of the swirl. The movement of a fluid can be said to be vortical if the fluid moves around in a circle, or in a helix, or if it tends to spin around some axis. Such motion can also be called *solenoidal*. In the atmospheric sciences, vorticity is a property that characterizes large-scale rotation of air masses. Since the atmospheric circulation is nearly horizontal, the 3D vorticity is nearly vertical, and it is common to use the vertical component as a scalar vorticity.

A vortex can be seen in the spiraling motion of air or liquid around a center of rotation. Circular current of water of conflicting tides form vortex shapes. Turbulent flow makes many vortices. A good example of a vortex is the atmospheric phenomenon of a whirlwind or a *tornado*. This whirling air mass mostly takes the form of a helix, column, or spiral. Tornadoes develop from severe thunderstorms, usually spawned from squall lines and *supercell thunderstorms*, though they sometimes happen as a result of a *hurricane*.[29] Another example is a meso-vortex on the scale of a few miles (smaller than a hurricane but larger than a tornado). On a much smaller scale, a vortex is usually formed as water goes down a drain, as in a sink or a toilet. This occurs in water as the revolving mass forms a whirlpool.[30] This whirlpool is caused by water flowing out of a small opening in the bottom of a basin or reservoir. This swirling flow structure within a region of fluid flow opens downward from the water surface. In the hydrodynamic interpretation of the behavior of electromagnetic fields, the acceleration of electric fluid in a particular direction

---

closed streamlines. The shape of media or mass rotating rapidly around a center forms a vortex. It is a flow involving rotation about an arbitrary axis.

[29] Recall that a hurricane is a much larger, swirling body of clouds produced by evaporating warm ocean water and influenced by the Earth's rotation. In particular, polar vortex is a persistent, large-scale cyclone centered near the Earth's poles, in the middle and upper troposphere and the stratosphere. Similar, but far greater, vortices are also seen on other planets, such as the permanent Great Red Spot on Jupiter and the intermittent Great Dark Spot on Neptune.

[30] Recall that a whirlpool is a swirling body of water produced by ocean tides or by a hole underneath the vortex, where water drains out, as in a bathtub.

creates a positive vortex of magnetic fluid. This in turn creates around itself a corresponding negative vortex of electric fluid.

Dynamical Similarity and Eddies

In order for two flows to be similar they must have the same geometry and equal Reynolds numbers. When comparing fluid behavior at homologous points in a model and a full-scale flow, we have $Re^* = Re$, where quantities marked with * concern the flow around the model and the other the real flow. This allows us to perform experiments with reduced models in water channels or wind tunnels, and correlate the data to the real flows. Note that true dynamic similarity may require matching other dimensionless numbers as well, such as the Mach number used in compressible flows, or the Froude number that governs free-surface flows.

In a turbulent flow, there is a range of scales of the fluid motions, sometimes called *eddies*. A single packet of fluid moving with a bulk velocity is called an *eddy*. The size of the largest scales (eddies) are set by the overall geometry of the flow. For instance, in an industrial smoke-stack, the largest scales of fluid motion are as big as the diameter of the stack itself. The size of the smallest scales is set by Re. As Re increases, smaller and smaller scales of the flow are visible. In the smoke-stack, the smoke may appear to have many very small bumps or eddies, in addition to large bulky eddies. In this sense, Re is an indicator of the range of scales in the flow. The higher the Reynolds number, the greater the range of scales.

In their first edition of Fluid Mechanics, Landau and Lifschitz proposed a *route to turbulence* in spatio-temporal fluid systems. Since then, much work, in dynamical systems, experimental fluid dynamics, and many other fields has been done concerning the routes to turbulence. Ever since the discovery of chaos in low-dimensional systems, researchers have been trying to use the concept of chaos to understand turbulence [RT71]. recall that there are two types of fluid motions: laminar flows and turbulent flows. Laminar flows look regular, and turbulent flows are non-laminar and look irregular. Chaos is more precise, for example, in terms of the so-called *Bernoulli shift dynamics*. On the other hand, even in low-dimensional systems, there are solutions which look irregular for a while, and then look regular again. Such a dynamics is often called a *transient chaos*.

Low-dimensional chaos is the starting point of a long journey toward understanding turbulence. To have a better connection between chaos and turbulence, one has to study chaos in PDEs [Li04].

Sine-Gordon Equation

Consider the simple perturbed *sine-Gordon equation* [Li04c]

$$u_{tt} = c^2 u_{xx} + \sin u + \epsilon[-a u_t + \cos t \sin^3 u], \qquad (4.90)$$

subject to periodic boundary condition

$$u(t, x + 2\pi) = u(t, x),$$

as well as even or odd constraint,

$$u(t, -x) = u(t, x), \quad \text{or} \quad u(t, -x) = -u(t, x),$$

where u is a real-valued function of two real variables (t, x), c is a real constant, $\epsilon \geq 0$ is a small perturbation parameter, and $a > 0$ is an external parameter. One can view (4.90) as a *flow* (u, u_t) defined in the phase-space manifold $M \equiv H^1 \times L^2$, where H^1 and L^2 are the Sobolev spaces on $[0, 2\pi]$. A point in the phase-space manifold M corresponds to two profiles, $(u(x), u_t(x))$. [Li04c] has proved that there exists a homoclinic orbit $(u, u_t) = h(t, x)$ asymptotic to $(u, u_t) = (0, 0)$. Let us define two orbits segments

$$\eta_0 : \ (u, u_t) = (0, 0), \quad \text{and} \quad \eta_1 : (u, u_t) = h(t, x), \quad (t \in [-T, T]).$$

When T is large enough, η_1 is almost the entire homoclinic orbit (chopped off in a small neighborhood of $(u, u_t) = (0, 0)$). To any binary sequence

$$a = \{\cdots a_{-2} a_{-1} a_0, a_1 a_2 \cdots\}, \quad (a_k \in \{0, 1\}), \tag{4.91}$$

one can associate a pseudo-orbit

$$\eta_a = \{\cdots \eta_{a_{-2}} \eta_{a_{-1}} \eta_{a_0}, \eta_{a_1} \eta_{a_2} \cdots\}.$$

The pseudo-orbit η_a is not a true orbit but rather 'almost an orbit'. One can prove that for any such pseudo-orbit η_a, there is a unique true orbit in its neighborhood [Li04c]. Therefore, each binary sequence labels a true orbit. All these true orbits together form a chaos. In order to talk about sensitive dependence on initial data, one can introduce the *product topology* by defining the neighborhood basis of a binary sequence

$$a^* = \{\cdots a^*_{-2} a^*_{-1} a^*_0, a^*_1 a^*_2 \cdots\} \quad \text{as} \quad \Omega_N = \{a : \ a_n = a^*_n, \ |n| \leq N\}.$$

The Bernoulli shift on the binary sequence (4.91) moves the comma one step to the right. Two binary sequences in the neighborhood Ω_N will be of order Ω_1 away after N iterations of the Bernoulli shift. Since the binary sequences label the orbits, the orbits will exhibit the same feature. In fact, the Bernoulli shift is topologically conjugate to the perturbed sine-Gordon flow.

Replacing a homoclinic orbit by its fattened version—a homoclinic tube, or by a heteroclinic cycle, or by a heteroclinically tubular cycle; one can still obtain the same Bernoulli shift dynamics. Also, adding diffusive perturbation $\epsilon b u_{txx}$ to (4.90), one can still prove the existence of homoclinics or heteroclinics, but the Bernoulli shift result has not been established [Li04c].

Complex Ginzburg–Landau Equation

Consider the complex-valued *Ginzburg–Landau equation* [Li04a, Li04b],

$$iq_t = q_{xx} + 2[|q|^2 - \omega^2] + i\epsilon[q_{xx} - \alpha q + \beta], \qquad (4.92)$$

which is subject to periodic boundary condition and even constraint

$$q(t, x + 2\pi) = q(t, x), \qquad q(t, -x) = q(t, x),$$

where q is a complex-valued function of two real variables (t, x), (ω, α, β) are positive constants, and $\epsilon \geq 0$ is a small perturbation parameter. In this case, one can prove the existence of homoclinic orbits [Li04a]. But the Bernoulli shift dynamics was established under generic assumptions [Li04b].

A real fluid example is the amplitude equation of Faraday water wave, which is also a complex Ginzburg–Landau equation [Li04d],

$$iq_t = q_{xx} + 2[|q|^2 - \omega^2] + i\epsilon[q_{xx} - \alpha q + \beta \bar{q}], \qquad (4.93)$$

subject to the same boundary condition as (4.92). For the first time, one can prove the existence of homoclinic orbits for a water wave equation (4.93). The Bernoulli shift dynamics was also established under generic assumptions [Li04d]. That is, one can prove the existence of chaos in water waves under generic assumptions.

The nature of the complex Ginzburg–Landau equation is a parabolic equation which is near a hyperbolic equation. The same is true for the perturbed sine-Gordon equation with the diffusive term $\epsilon b u_{txx}$ added. They contain effects of diffusion, dispersion, and nonlinearity. The *Navier–Stokes equations* (4.89) are diffusion–advection equations. The advective term is missing from the perturbed sine-Gordon equation and the complex Ginzburg–Landau equation. Turbulence happens when the diffusion is weak, i.e., in the near hyperbolic regime. One should hope that turbulence should share some of the features of chaos in the perturbed sine-Gordon equation. There is a popular myth that turbulence is fundamentally different from chaos because turbulence contains many unstable modes. In both the perturbed sine-Gordon equation and the complex Ginzburg–Landau equation, one can incorporate as many unstable modes as one likes, the resulting Bernoulli shift dynamics is still the same. On a computer, the solution with more unstable modes may look rougher, but it is still chaos [Li04].

In a word, dynamics of strongly nonlinear classical fields is 'turbulent', not 'laminar'. On the other hand, field theories such as 4-dimensional QCD or gravity have many dimensions, symmetries, tensorial indices. They are far too complicated for exploratory forays into this forbidding terrain. Instead, we consider a simple spatio-temporally chaotic nonlinear system of physical interest [CCP96a, II08a].

Kuramoto–Sivashinsky System

One of the simplest and extensively studied spatially extended dynamical systems is the Kuramoto–Sivashinsky (KS) system [Kur76, Siv77]

$$u_t = (u^2)_x - u_{xx} - \nu u_{xxxx}, \qquad (4.94)$$

which arises as an amplitude equation for interfacial instabilities in a variety of contexts. The so-called *flame front* $u(x,t)$ has compact support, with $x \in [0, 2\pi]$ a periodic space coordinate. The u^2 term makes this a nonlinear system, $t \geq 0$ is the time, and ν is a 4-order 'viscosity' damping parameter that irons out any sharp features. Numerical simulations demonstrate that as the viscosity decreases (or the size of the system increases), the *flame front* becomes increasingly unstable and turbulent. The task of the theory is to describe this spatio-temporal turbulence and yield quantitative predictions for its measurable consequences.

For any finite spatial resolution, the KS system (4.94) follows approximately for a finite time a pattern belonging to a finite alphabet of admissible patterns, and the long term dynamics can be thought of as a walk through the space of such patterns, just as chaotic dynamics with a low dimensional attractor can be thought of as a succession of nearly periodic (but unstable) motions. The periodic orbit gives the machinery that converts this intuitive picture into precise calculation scheme that extracts asymptotic time predictions from the short time dynamics. For extended systems the theory gives a description of the asymptotics of partial differential equations in terms of recurrent spatio-temporal patterns.

The KS periodic orbit calculations of Lyapunov exponents and escape rates [CCP96a] demonstrate that the *periodic orbit theory* predicts observable averages for deterministic but classically chaotic spatio-temporal systems. The main problem today is not how to compute such averages—periodic orbit theory as well as direct numerical simulations can handle that—but rather that there is no consensus on what the sensible experimental observables worth are predicting [Cvi00].

Burgers Dynamical System

Consider the following *Burgers dynamical system* on a *functional manifold* $M \subset C^k(\mathbb{R}; \mathbb{R})$:

$$u_t = uu_x + u_{xx}, \qquad (4.95)$$

where $u \in M$, $t \in \mathbb{R}$ is an evolution parameter. The flow of (4.95) on M can be recast into a set of 2-forms $\{\alpha\} \subset \Lambda^2(J(\mathbb{R}^2; \mathbb{R}))$ upon the adjoint jet-manifold $J(\mathbb{R}^2; \mathbb{R})$ (see [II07e]) as follows [BPS98]:

$$\{\alpha\} = \big\{ du^{(0)} \wedge dt - u^{(1)} dx \wedge dt = \alpha^1,$$
$$du^{(0)} \wedge dx + u^{(0)} du^{(0)} \wedge dt + du^{(1)} \wedge dt = \alpha^2 :$$
$$\big(x, t; u^{(0)}, u^{(1)}\big)^\tau \in M^4 \subset J^1(\mathbb{R}^2; \mathbb{R})\big\}, \qquad (4.96)$$

where M^4 is some finite-dimensional submanifold in $J^1(\mathbb{R}^2; \mathbb{R}))$ with coordinates $(x, t, u^{(0)} = u, u^{(1)} = u_x)$. The set of 2-forms (4.96) generates the closed ideal $\mathfrak{I}(\alpha)$, since

$$d\alpha^1 = dx \wedge \alpha^2 - u^{(0)} dx \wedge \alpha^1, \qquad d\alpha^2 = 0, \qquad (4.97)$$

the integral submanifold $\bar{M} = \{x, t \in \mathbb{R}\} \subset M^4$ being defined by the condition $\mathfrak{I}(\alpha) = 0$. We now look for a reduced 'curvature' 1-form $\Gamma \in \Lambda^1(M^4) \otimes \mathcal{G}$, belonging to some not yet determined Lie algebra $\mathcal{G}$. This 1-form can be represented using (4.96), as follows:

$$\Gamma = b^{(x)}(u^{(0)}, u^{(1)}) dx + b^{(t)}(u^{(0)}, u^{(1)}) dt, \qquad (4.98)$$

where elements $b^{(x)}, b^{(t)} \in \mathcal{G}$ satisfy such determining equations [BPS98]

$$\frac{\partial b^{(x)}}{\partial u^{(0)}} = g_2, \qquad \frac{\partial b^{(x)}}{\partial u^{(1)}} = 0, \qquad \frac{\partial b^{(t)}}{\partial u^{(0)}} = g_1 + g_2 u^{(0)},$$

$$\frac{\partial b^{(t)}}{\partial u^{(1)}} = g_2, \qquad [b^{(x)}, b^{(t)}] = -u^{(1)} g_1. \qquad (4.99)$$

The set (4.99) has the following unique solution

$$b^{(x)} = A_0 + A_1 u^{(0)},$$

$$b^{(t)} = u^{(1)} A_1 + \frac{u^{(0)2}}{2} A_1 + [A_1, A_0] u^{(0)} + A_2, \qquad (4.100)$$

where $A_j \in \mathcal{G}, j = \overline{0, 2}$, are some constant elements on M of a Lie algebra $\mathcal{G}$ under search, obeying the *Lie structure equations* (see [II07e]):

$$[A_0, A_2] = 0,$$

$$[A_0, [A_1, A_0]] + [A_1, A_2] = 0,$$

$$[A_1, [A_1, A_0]] + \frac{1}{2}[A_0, A_1] = 0. \qquad (4.101)$$

From (4.99) one can see that the curvature 2-form $\Omega \in span_{\mathbb{R}}\{A_1, [A_0, A_1] : A_j \in \mathcal{G}, j = 0, 1\}$. Therefore, reducing via the *Ambrose–Singer theorem* the associated principal fibred frame space $P(M; G = GL(n))$ to the principal fibre bundle $P(M; G(h))$, where $G(h) \subset G$ is the corresponding holonomy Lie group of the connection Γ on P, we need to satisfy the following conditions for the set $\mathcal{G}(h) \subset \mathcal{G}$ to be a Lie subalgebra in $\mathcal{G} : \nabla_x^m \nabla_t^n \Omega \in \mathcal{G}(h)$ for all $m, n \in \mathbb{Z}_+$.

Let us try now to close the above transfinitive procedure requiring that [BPS98]

$$\mathcal{G}(h) = \mathcal{G}(h)_0 = span_{\mathbb{R}}\{\nabla_x^m \nabla_x^n \Omega \in \mathcal{G} : m + n = 0\}. \qquad (4.102)$$

This means that

$$\mathcal{G}(h)_0 = span_{\mathbb{R}}\{A_1, A_3 = [A_0, A_1]\}. \tag{4.103}$$

To enjoy the set of relations (4.101) we need to use expansions over the basis (4.103) of the external elements $A_0, A_2 \in \mathcal{G}(h)$:

$$A_0 = q_{01}A_1 + q_{13}A_3, \qquad A_2 = q_{21}A_1 + q_{23}A_3. \tag{4.104}$$

Substituting expansions (4.104) into (4.101), we get that $q_{01} = q_{23} = \lambda, q_{21} = -\lambda^2/2$ and $q_{03} = -2$ for some arbitrary real parameter $\lambda \in \mathbb{R}$, that is $\mathcal{G}(h) = span_{\mathbb{R}}\{A_1, A_3\}$, where

$$[A_1, A_3] = A_3/2; \qquad A_0 = \lambda A_1 - 2A_3, \qquad A_2 = -\lambda^2 A_1/2 + \lambda A_3. \tag{4.105}$$

As a result of (4.105) we can state that the holonomy Lie algebra $\mathcal{G}(h)$ is a real 2D one, assuming the following (2×2)-matrix representation [BPS98]:

$$A_1 = \begin{pmatrix} 1/4 & 0 \\ 0 & -1/4 \end{pmatrix}, \qquad A_3 = \begin{pmatrix} 0 & 1 \\ 0 & 0 \end{pmatrix},$$

$$A_0 = \begin{pmatrix} \lambda/4 & -2 \\ 0 & -\lambda/4 \end{pmatrix}, \qquad A_2 = \begin{pmatrix} -\lambda^2/8 & \lambda \\ 0 & \lambda^2/8 \end{pmatrix}. \tag{4.106}$$

Thereby from (4.98), (4.100) and (4.106) we obtain the *reduced curvature 1-form* $\Gamma \in \Lambda^1(M) \otimes \mathcal{G}$,

$$\Gamma = (A_0 + uA_1)dx + ((u_x + u^2/2)A_1 - uA_3 + A_2)dt, \tag{4.107}$$

generating *parallel transport* of vectors from the representation space Y of the holonomy Lie algebra $\mathcal{G}(h)$:

$$dy + \Gamma y = 0 \tag{4.108}$$

upon the integral submanifold $\bar{M} \subset M^4$ of the ideal $\mathcal{I}(\alpha)$, generated by the set of 2-forms (4.96). The result (4.108) means also that the Burgers dynamical system (4.95) is endowed with the standard Lax type representation, having the spectral parameter $\lambda \in \mathbb{R}$ necessary for its integrability in quadratures.

4.3 Quantum Chaos and Its Control

Recall that in 1917 Albert Einstein wrote a paper that was completely ignored for 40 years. In it he raised a question that physicists have only recently begun asking themselves: What would classical chaos, which lurks everywhere in our work, do to quantum mechanics, the theory describing the atomic and sub-atomic worlds? The effects of classical chaos have long been observed: even Kepler knew about the motion of the moon around the earth and Newton complained bitterly about the phenomenon. At the end of the 19th century the American astronomer William Hill demonstrated that the irregularity is the result entirely of the gravitational pull of the sun. So thereafter, the great French genius Henri Poincaré surmised that the moon's motion is only mild case of a congenital disease affecting nearly everything. In the long run Poincaré realized, most dynamic systems show no discernible regularity or repetitive pattern. The behavior of even a simple system can depend so sensitively on its initial conditions that the final outcome is uncertain. At about the time of Poincaré's seminal work on classical chaos, Max Planck started another revolution, which would lead to the modern theory of quantum mechanics. The simple systems that Newton had studied were investigated again, but this time on the atomic scale. The quantum analogue of the humble pendulum is the laser; the flying cannonballs of the atomic world consist of beams of protons or electrons, and the rotating wheel is the spinning electron (the basis of magnetic tapes). Even the solar system itself is mirrored in each of the atoms found in the periodic table of the elements. Perhaps the single most outstanding feature of the quantum world is its smooth and wavelike nature. This feature leads to the question of how chaos makes itself felt when moving from the classical world to the quantum world. How can the extremely irregular character of classical chaos be reconciled with the smooth and wavelike nature of phenomena on the atomic scale? Does chaos exist in the quantum world? Preliminary work seems to show that it does. Chaos is found in the distribution of energy levels of certain atomic systems; it even appears to sneak into the wave patterns associated with those levels. Chaos is also found when electrons scatter from small molecules. We must emphasize, however, that the term *quantum chaos* serves more to describe a conundrum than to define a well-posed problem [Gut90, Gut92].

Considering the following interpretation of the bigger picture may be helpful in coming to grips with quantum chaos. All our theoretical discussions of mechanics can be somewhat artificially divided into three compartments: **R** (for regular), **P** (for Poincaré) and **Q** (for quantum)—although nature recognizes none of these divisions. Elementary classical mechanics falls in the **R** compartment. This box contains all the nice, clean systems exhibiting simple and regular behavior. Also contained in **R** is an elaborate mathematical tool called *perturbation theory* (see Chap. 3 for Dirac's quantum version of this theory), which is used to calculate the effects of small interactions and extraneous disturbances, such as the influence of the sun on the moon's motion around the earth. With the help of perturbation theory, a large part of physics is understood nowadays as making relatively mild modifications of regular systems. Reality though, is much more complicated; chaotic systems lie outside the range of perturbation theory and they constitute the second, **P** compartment, initiated by Poincaré. It is stuffed with the chaotic dynamic systems that are the bread and butter of modern science. Among these systems are all the fundamental problems of mechanics, starting with *three*, rather than *only two* bodies interacting with one another, such as the Earth, moon and sun, or the three atoms in the water molecule, or the three quarks in the proton. Quantum mechanics, as it has been practiced for about 90 years, belongs in the third **Q** compartment. Recall that after the pioneering work of Planck, Einstein and Niels Bohr, quantum mechanics was given its definitive form in four short years, starting in 1924. The seminal work of Louis de Broglie, Werner Heisenberg, Erwin Schrödinger, Max Born, Wolfgang Pauli and Paul Dirac has stood the test of the laboratory without the slightest lapse. Miraculously, it provides physics with a mathematical framework that, according to Dirac, has yielded a deep understanding of "most of physics and all of chemistry." Nevertheless, even though most physicists and chemists have learned how to solve special problems in quantum mechanics, they have yet to come to terms with the incredible subtleties of the field. These subtleties are quite separate from the difficult, conceptual issues having to do with the interpretation of quantum mechanics. The three boxes **R** (classic, simple systems), **P** (classic chaotic systems) and **Q** (quantum systems) are linked by several connections. The connection between **R** and **Q** is known as *Bohr's correspondence principle*. The correspondence principle claims, quite reasonably, that classical mechanics must be contained in quantum mechanics in the limit where objects become much larger than the size of atoms. The main connection between **R** and **P** is the *Kolmogorov–Arnold–Moser theorem* (or, KAM theorem). The KAM theorem provides a powerful tool for calculating how much of the structure of a regular system survives when a small perturbation is introduced, and the theorem can thus identify perturbations that cause a regular system to undergo chaotic behavior. The new *quantum chaos* is concerned with establishing the relation between boxes **P** (chaotic systems) and **Q** (quantum systems). In establishing this relation, it is useful to introduce a concept of *phase space* [Gut92].[31]

The spectrum of a chaotic quantum system was first suggested by Eugene P. Wigner, another early master of quantum mechanics and inventor of the *quantum phase space*. Wigner observed, as had many others, that nuclear physics does not possess the safe underpinnings of atomic and molecular physics: the origin of the nuclear force is still not clearly understood. He therefore asked whether the statistical properties of nuclear spectra could be derived from the assumption that many parameters in the problem have definite, but unknown values. This rather vague starting point allowed him to find the most probable formula for the distribution. Today it is well-known

---

[31] Quite amazingly, this concept, which is now so widely exploited by experts in the field of dynamic systems, dates back to Newton. The notion of phase space can be found in *Newton's Principia* published in 1687. In the second definition of the first chapter, entitled "Definitions," Newton states (as translated from the original Latin in 1729): "The quantity of motion is the measure of the same, arising from the velocity and quantity of matter conjointly." In modern English this means that for every object there is a quantity, called momentum, which is the product of the mass and velocity of the object. Newton gives his laws of motion in the second chapter, entitled "Axioms, or Laws of motion." The second law says that the change of motion is proportional to the motive force impressed. Newton relates the force to the change of momentum (not to the acceleration as most textbooks do).

that the so-called *Wigner distribution* happens to be exactly what is found for the spectrum of a chaotic dynamic system.

Chaos does not seem to limit itself to the distribution of quantum energy levels, however, it even appears to work its way into the wavelike nature of the quantum world. The position of the electron in the hydrogen atom is described by a wave pattern. The electron cannot be pinpointed in space; it is a cloudlike smear hovering near the proton. Associated with each allowed energy level is a stationary state, which is a wave pattern that does not change with time. A stationary state corresponds quite closely to the vibrational pattern of a membrane that is stretched over a rigid frame, such as a drum. The stationary states of a chaotic system have surprisingly interesting structure, as demonstrated in the early 1980s by Eric Heller (the most recent reference is [WMH08]). Heller and his students calculated a series of stationary states for a 2D cavity in the shape of a stadium. The corresponding problem in classical mechanics was known to be chaotic, for a typical trajectory quickly covers most of the available ground quite evenly. Such behavior suggests that the stationary states might also look random, as if they had been designed without rhyme or reason. In contrast, Heller discovered that most stationary states are concentrated around narrow channels that form simple shapes inside the stadium, and he called these channels 'scars'. Similar structure can also be found in the stationary states of a hydrogen atom in a strong magnetic field. The smoothness of the quantum wave forms is preserved from point to point, but when one steps back to view the whole picture, the fingerprint of chaos emerges. It is possible to connect the chaotic signature of the energy spectrum to ordinary classical mechanics. A clue to the prescription is provided in Einstein's 1917 paper. He examined the phase space of a regular system from box **R** and described it geometrically as filled with surfaces in the shape of a donut; the motion of the system corresponds to the trajectory of a point over the surface of a particular donut. The trajectory winds its way around the surface of the donut in a regular manner, but it does not necessarily close on itself. In Einstein's picture, the application of Bohr's correspondence principle to find the energy levels of the analogous quantum mechanical system is simple. The only trajectories that can occur in nature are those in which the cross section of the donut encloses an area equal to an integral multiple of Planck's constant. It turns out that the integral multiple is precisely the number that specifies the corresponding energy level in the quantum system. Unfortunately as Einstein clearly saw, his method cannot be applied if the system is chaotic, for the trajectory does not lie on a donut and there is no natural area to enclose an integral multiple of Planck's constant. A new approach must be sought to explain the distribution of quantum mechanical energy levels in terms of the chaotic orbits of classical mechanics. Which features of the trajectory of classical mechanics help us to understand quantum chaos? Hill's discussion of the moon's irregular orbit because of the presence of the sun provides a clue. His work represented the first instance where a particular periodic orbit is found to be at the bottom of a difficult mechanical problem.[32] Inspiration can also be drawn from Poincaré, who emphasized the general importance of periodic orbits. In the beginning of his 3-volume work, "The New Methods of Celestial Mechanics" which appeared in 1892, he expresses the belief that periodic orbits "offer the only opening through which we might penetrate into the fortress that has the reputation of being impregnable." Phase space for a chaotic system can be organized, at least partially around periodic orbits, even though they are sometimes quite difficult to find [Gut90, Gut92].

In 1970 Martin Gutzwiller discovered a very general way to extract information about the quantum mechanical spectrum from a complete enumeration of the classical *periodic orbits*.[33] The mathematics of the approach is too difficult to delve into here, but the main result of the method is a relatively simple expression called a *trace formula*. The approach has now been used by a number of investigators, including Michael V. Berry, who has used the formula to derive the statistical properties of the spectrum. Gutzwiller has applied the trace formula to compute the lowest two dozen energy levels for an electron in a semiconductor lattice, near one of the carefully controlled impurities. The trajectory of the electron can be uniquely characterized by a string of symbols, which has a straightforward interpretation. The string is produced by defining an axis through the semiconductor and simply noting when the trajectory crosses the axis. A crossing to the 'positive' side of the axis gets the symbol $+$, and a crossing to the 'negative' side gets the symbol $-$. A trajectory then looks exactly like the record of a coin toss. Even if the past is known in all detail even if all the crossings have been recorded, the future is still wide open. The sequence of crossings can be chosen arbitrarily. Now, a periodic orbit consists of a binary sequence that repeats itself; the simplest such sequence is $(+-)$, the next is $(+-)$, and so on.[34] All periodic orbits are thereby enumerated, and it is possible to calculate an appropriate spectrum with the help of the trace formula. In other words, the quantum mechanical energy levels are obtained in an approximation that relies on quantities from classical mechanics only. The classical periodic orbits and the quantum me-

---

[32] A periodic orbit is like a closed track on which the system is made to run: there are many of them, although they are isolated and unstable.

[33] The periodic orbit theory has been developed for classical and quantum non-integrable systems, and for renormalization group flows. In particular, the cycle expansions have contributed to the recent progress in the semi-classical theory of chaotic systems and its applications: improved the convergence of periodic orbit expansions, unified the Ruelle's theory for classical flows and Gutzwiller's theory for semi-classical quantization, and set the stage for accurate tests of tunneling and other corrections to semi-classical approximations. Currently, the most promising directions lie in applying our expertise in 'wave chaos' to the new type of acoustics experiments initiated by our CATS experimental group, and developing periodic orbit theory for stochastic nonlinear flows [CAM05].

[34] Two crossings in a row having the same sign indicate that the electron has been trapped temporarily.

chanical spectrum are closely bound together through the Fourier analysis. The hidden regularities in one set, and the frequencies with which they show up, are exactly given by the other set. The energies at which the atoms absorb radiation appear to be quite random, but a *Fourier analysis* converts the jumble of peaks into a set of well-separated peaks. The important feature here is that each of the well-separated peaks corresponds precisely to one of several standard classical periodic orbits. Poincaré's insistence on the importance of periodic orbits now takes on a new meaning. Not only does the classical organization of phase space depend critically on the classical periodic orbits, but so too does the understanding of a chaotic quantum spectrum [Gut90, Gut92].

4.3.1 Quantum Chaos vs. Classical Chaos

The mathematical description of chaos has reached a mature state over the last decade. The tools being used are classical *phase space*, *Lyapunov exponents*, *Poincaré sections*, *Kolmogorov–Sinai entropy* and others. There is also chaos in the quantum systems [FW89, BTU93, GWW00]. People have investigated in quantum physics the analogues of classically chaotic systems. For example, a stadium-shaped billiard is classically chaotic. The corresponding quantum system is a system of ultra-cold atoms bouncing against walls of stadium shape, being created by interaction of the atoms with laser beams. Quantum chaos has been found to play an important role in dynamical tunneling [HHB01b, SOR01]. The fingerprints of chaos in quantum systems and the mathematical tools of its description are quite different from those used in classical chaos. The necessity for a different treatment is due to the nature of quantum mechanics: There is no proper phase space in quantum systems. Heisenberg's uncertainty principle forbids that a point in phase space (uncertainty zero in position and momentum) exists. Heisenberg's uncertainty relation is a direct consequence of quantum mechanical fluctuations. A common approach to describe quantum chaos is random matrix theory and the use of energy level spacing distributions [GWW00]. A fundamental conjecture by [BGS84] postulates that the energy level spacing distributions possesses a dominant part, which depends on the particular system, and a sub-leading universal part, independent of the particular system. The universal part gives a level spacing Wigner-type distribution, if the corresponding classical system is fully chaotic. Such level spacing distribution can be generated also by random matrices of a certain symmetry class. The conjecture has not been rigorously proven yet, but has been verified and found to be valid in almost all cases.

This approach is successful and very popular. However, it has some short comings: First, the conjecture by [BGS84] holds strictly only in the case of a fully chaotic system, while in nature most systems are only partly chaotic, i.e., so-called mixed systems where its classical counter part has coexistence of regular and chaotic phase space. In such cases the quantum system yields a level spacing distribution, which is neither Wignerian nor Poissonian (where the latter corresponds to a completely regular system). There is no mathematical prediction of the functional form of such distribution. However, a number of interpolations between the Poisson and Wigner distribution have been proposed (see, e.g. [BFF81, LH91]). Second, we may ask: How about the comparison of the classical with the quantum system? And what is the quantitative degree of chaos? How can we answer this when the instruments used to measure chaos are quite different for both systems?

Starting from this perspective and having in mind the goal to compare classical with quantum chaos, one may try the following strategy [KLM06]: Find a uniform description of chaotic phenomena, valid for both, classical and quantum systems. In more detail: Starting from non-linear dynamics and phase space in classical systems, one may look for a suitable quantum analogue phase space. Starting from random matrix theory and energy level spacing distributions of quantum systems, one may seek a random matrix description and a level spacing distribution of suitable dynamical objects in classical physics. In the following we will discuss some progress recently made in this direction. Using some of those results, we will compare for a particular system the chaotic behavior of the quantum system with the classical system. The numerical analysis shows that the quantum system is globally less chaotic than the classical system. We believe that such finding is not limited to the particular system. In particular, we want to understand the underlying reason for such behavior.

Weaker Quantum Chaos

First, [CGM98] considered the N-component Φ^4 theory in the presence of an external field and in the limit of large N. They used mean field theory and observed a strong suppression of chaos in the quantum system, due to quantum corrections causing the system the system to move away from a hyperbolic fixed point responsible for classical chaos. Second, [MM97] studied another field theoretic model which is classically chaotic, namely massless scalar electrodynamics. They investigated the corresponding quantum field theory using effective field theory and loop expansion. They noticed that quantum corrections increase the threshold for chaos due to a modification of the ground state of the system. A third example is the kicked rotor which is a classically chaotic system in 1D. [SF95] considered the corresponding quantum system using Bohm's interpretation of quantum mechanics to introduce trajectories and a quantum equivalent phase space. They found that the Kolmogorov–Sinai entropy goes to zero in the quantum system, i.e., it is non-chaotic. This approach has been applied also to study chaos in anisotropic harmonic oscillators, coupled anharmonic oscillators [Par02] and the hydrogen atom in an external electromagnetic field [IP96]. Finally, an anharmonic oscillator was considered in 2D, which is classically a mixed chaotic system [CHK04]. Using the concept of the *quantum action functional* (see [Krö02]), a quantum analogue phase space has been constructed. As a result, the phase space portrait of the quantum system was found to be slightly but globally less chaotic for all energies.

Because the quantum action has been constructed from the classical action by taking into account quantum fluctuations (see [JKL01, Krö02, JKL02, HKM03]), hence the softening of chaos in the quantum system must be due to quantum fluctuations. We suspect that such softening effect may not solely show up in chaos. Indeed, looking at a double-well potential $V(x) = \lambda(x^2 - \frac{1}{8\lambda})^2$ in the context of tunneling, we observe that quantum effects cause the quantum potential to be much 'weaker' than the classical potential, i.e. the potential wells are less pronounced and the potential barrier is much lower for the quantum potential (note that the quantum potential has a triple-well shape). The shape of the potential for tunneling translates into the shape of the instantons. Thus it comes as no surprise that the instanton of the quantum action (actually a double-instanton) is softer than the classical instanton.

Uniform Description of Quantum Chaos

Let us recall that the conjecture by [BGS84] is about random matrix theory and energy level spacing distributions to describe chaos in quantum systems, while chaos in classical systems is conventionally described in terms of phase space. That is, the tools to describe chaos in classical and quantum physics are different. For the purpose to better understand the physical content of this conjecture or eventually to find a mathematical proof, it is highly desirable to use the same language, respectively, tools in quantum physics as in classical physics. This can be viewed in two ways: Either one adopts the point of view that chaos should be analyzed in terms of phase space. Then a uniform description is achieved by use of classical phase space in classical physics and the quantum analogue phase space in quantum systems. How about the point of view that chaos should be analyzed in terms of random matrix theory and level spacing distributions? In the following we propose how to construct such a level spacing distribution, which will play the same role in classical physics as the energy level spacing distribution in the quantum system.

One may immediately object that the energy level spacing distribution in quantum physics is due to the fact that one considers a system of bound states (with a 'confining' potential), which due to the rules of quantum mechanics gives a discrete spectrum. The discreteness of the spectrum is a quantum effect, i.e., a physical effect. Having in mind to construct a counterpart in classical physics, one may object that classical physics is continuous. There is no discreteness inherent in classical physics. If any discreteness occurs it will be due to some (mathematical) approximation. Hence, how can a level spacing distribution derived from a classical function be discrete and physical? First we want to propose a function and show how to construct a meaningful level spacing distribution. Afterwards, we will try to answer the last question.

We would like to emphasize two properties: First, the level spacing distribution of the quantum system, in the case of a classically fully chaotic system, corresponds to a *Wigner distribution*. The type of Wigner distribution is determined by the symmetry of the Hamiltonian. E.g., there is the Gaussian orthogonal ensemble (GOE), the unitary ensemble and the symplectic ensemble. Thus, we need a function which carries the same symmetries. A function convenient for this purpose is the classical action. Another property to be emphasized is locality. Let us recall that the level spacing distribution, e.g. the GOE distribution, is invariant under orthogonal transformations. Because an orthogonal transformation maps any orthogonal basis of states onto another orthogonal basis, the level spacing distribution is essentially independent of the particular choice of the basis. That means, we may choose a basis that is quasi-local, i.e., built from square integrable functions which are identically zero everywhere except in a small interval where they are non zero and constant. Those box functions are almost local. The advantage of locality is the fact that symmetries of the Hamiltonian often have to do with transformations in position space, and locality facilitates the analysis of such symmetries. Why do we need to bother about symmetries? Because the energy level spacing distribution is meaningful only (and gives the Wigner distribution for a chaotic system, respectively a Poissonian distribution for an integrable system) if the energy levels have all the same quantum numbers (except for the quantum number of energy). For example, for the spectrum of the hydrogen atom, one should take the bound states all with the same angular momentum quantum numbers, e.g., $l = m = 0$ [KLM06].

Action Matrix

We consider a system defined by a Lagrangian function $L(q, \dot{q}, t)$. Let S denote the corresponding action,

$$S[q(t)] = \int_0^T L(q, \dot{q}, t)\, dt.$$

Let q_{traj} denote the trajectory, i.e. the solution of the *Euler–Lagrange equation* of motion. Such trajectory is a function, which makes the action functional stationary. Each trajectory is specified by indicating the initial and final boundary points, i.e., $q(t = 0) = q_{in}$ and $q(t = T) = q_{fi}$. We assume that those boundary points are taken from some finite set of nodes, $q_{in}, q_{fi} \in \{q_1, \ldots, q_N\}$. Thus with each pair of boundary points, (q_k, q_l) we associate a trajectory $q_{kl}^{traj}(t)$. Then we introduce an action matrix Σ, where the matrix element Σ_{kl} corresponds to the value of the action S evaluated along the trajectory $q_{kl}^{traj}(t)$, as: $\Sigma_{kl} = S[q_{kl}^{traj}(t)]$. All matrix elements Σ_{kl} are real. They are also symmetric, $\Sigma_{kl} = \Sigma_{lk}$. Thus the action matrix Σ is a Hermitian $N \times N$ matrix. Consequently, Σ has a discrete spectrum of action eigenvalues, which are all real, $\sigma(\Sigma) = \{\sigma_1, \ldots, \sigma_N\}$.

Symmetry

For the purpose to compute a level spacing distribution from the action eigenvalues, one must first address the issue of symmetry. Let us consider for example the harmonic oscillator in 2D. We may take the coordinates q_k to be located on the nodes of a regular grid (reaching from $-\Lambda$ to $+\Lambda$ on the x and y axis), with a spacing $\Delta x = \Delta y = a = \text{const}$. For example, the action of the harmonic oscillator, with the classical trajectory going from $\mathbf{x}_a$ to $\mathbf{x}_b$ in time T is given by [KLM06]

$$\Sigma_{a,b} = \frac{m\omega}{2\sin(\omega T)} \left[(x_a^2 + x_b^2)\cos(\omega T) - 2\mathbf{x}_a \cdot \mathbf{x}_b \right].$$

This function is invariant under rotations. When choosing the coordinates q_k to be located on the regular grid, the continuous symmetry of rotation will become a discrete symmetry of finite rotations (a group). We have to find the irreducible representations of such group and sort the action eigenvalues according to those irreducible representations (this can be done by inspecting the properties of the corresponding eigenvector). Then one has to select a particular representation and retain a subset of eigenvalues in that representation. The action level spacing distribution can then be obtained from the action eigenvalues in such subset.

This procedure is feasible. However, it has two disadvantages. First, finding the irreducible representations and classifying the eigenvectors accordingly is laborious. More importantly, the fact that one has to work with a subset of eigenvalues only and not the whole ensemble of eigenvalues means a drastic reduction of the size of the statistical ensemble. In other words, the statistics of the resulting level spacing distribution will deteriorate.

For those reasons it would be highly desirable to avoid the above strategy. This is indeed possible by using the following trick. One can camouflage the symmetry by choosing the coordinates off the nodes of the regular grid. That means, for example to define new coordinates as: $q_k^{deform} = q_k + \epsilon_k$, where ϵ_k denotes a randomly chosen small deformation (in angle and length). In this way the nodes are irregularly distributed. Consequently, the discrete symmetry of the action matrix Σ_{kl} disappears when replacing it by the 'deformed' action matrix [KLM06]

$$\Sigma_{kl}^{deform} = \text{action evaluated along the classical trajectory}$$
$$\text{from node } q_k^{deform} \text{ to node } q_l^{deform}.$$

In doing so we avoid a laborious symmetry analysis and secondly will have a better statistics!

Action Level Spacing Distribution

Now we want to construct a level spacing distribution of action levels. We proceed in analogy to random matrix theory and the method of constructing an energy level spacing distribution. For an overview on how to compute level spacing distributions see [Haa01]. One has to separate the dominant system dependent part from the sub-leading universal part which describes the properly normalized fluctuations. Because we are here interested only in the fluctuation part, we suppress the leading part. One should note that this means to discard all physical information which depends on the particular system. For example thermodynamical functions can not be computed from the sub-leading fluctuating part. The strategy to obtain the sub-leading part is called unfolding. One constructs a fit to the original spectrum and multiplies the spectrum such that on average the mean spacing distribution becomes unity. Also the integrated level spacing distribution will be normalized to unity.

Following [KLM06], we have applied this to simple integrable systems in 1D. For integrable systems one would expect a Poissonian distribution for the action level spacing distribution. Preliminary results are compatible with a Poissonian distribution. The following remarks are in order. The first numerical results for integrable systems have to be repeated with precision and analyzed carefully. Second, one wants to see what happens in chaotic systems. Possibly such strategy applied to a fully chaotic system will result in a Wignerian action level spacing distribution. Finding an answer will be computationally much more involved, simply because the action functions for the integrable systems considered above are analytically known, while for a chaotic system (i.e., non-integrable) this needs to be calculated numerically. Moreover, the numerical precision required needs to be sufficient to resolve small fluctuations. In the statistical sense, one is interested in a sample of large size. But that means that after unfolding the fluctuations will become small and hence require a high numerical precision for its resolution. Presently, numerical studies of such question are under way.

Let us get back to the question posed above: How can a level spacing distribution derived from a classical function be discrete and physical? In our opinion, the answer lies in the fact that the level spacing distribution is universal, that is, it does not depend, for example, on the parameters of the discrete grid. Different grids give the same result. This has been verified numerically. It also should not depend on the deformation ϵ_k (as long it is not too close to zero and the discrete symmetry is restored). Also this has been verified numerically and found to be satisfied. Thus one could in principle go with the volume of the lattice $V = (2\Lambda)^D$ to infinity and with the lattice spacing $\Delta x = \Delta y$ to zero. The result should not change, but one would have reached the continuum limit. The discreteness would then disappear.

Quantum Action Renormalization

Above we have seen examples for the observation that quantum chaos seems to be weaker than chaos in the corresponding classical system. Of course it would be interesting to explore a much wider class of systems in order to see if such observation holds more generally. Here we want to pick one of the above examples, namely the chaotic anharmonic oscillator in 2D and try to understand why quantum chaos is weaker than classical chaos. The classical action is given [KLM06]

$$S = \int_0^T \left[\frac{1}{2m}(\dot{x}^2 + \dot{y}^2) - V(x,y) \right] dt,$$

$$V = \frac{m\omega^2}{2}(x^2 + y^2) + \lambda x^2 y^2 = v_0 + v_2(x^2 + y^2) + v_{22}x^2y^2.$$

For $\lambda = 0$ the system is reduced to the standard harmonic oscillator, which is integrable. The chaoticity is introduced and controlled by the parameter λ. Thus small λ causes mild chaos, while large λ makes the system strongly chaotic. The quantum action has been postulated to be of the functional form like the classical action, i.e. the kinetic term of the quantum action may differ in the value of the mass, and the potential of the quantum action should also be local, depend only on coordinates, but may have a different functional form. Here let us consider an ansatz of the following form

$$\tilde{S} = \int_0^T \left[\frac{1}{2\tilde{m}}(\dot{x}^2 + \dot{y}^2) - \tilde{V}(x,y) \right] dt,$$

$$\tilde{V} = \tilde{v}_0 + \tilde{v}_2(x^2 + y^2) + \tilde{v}_{22}x^2y^2 + \text{higher order polynomials.}$$

As a quantitative measure of the strength of chaos we take the strength of the parameters of the action, in particular, the parameter λ. The parameters of the quantum action can be interpreted as a 'renormalization effect' of the parameters of the classical action. The calculation of those parameters has to be done numerically, following the definition of the quantum action to be a functional which fits the transition amplitudes [JKL01]. However, in a certain limit, the quantum action is known to be an exact representation of the transition amplitudes and moreover the action is related via differential equations to the classical action [Krö02]. This limiting case is using imaginary time and let time go to infinity (Feynman–Kac limit). Because we want to obtain an analytical result, we will use perturbation theory. This means that we consider the regime of small $\lambda = v_{22}$.

In order to simplify the matter, let us start by considering the system in 1D. Thus we have the potential [KLM06]

$$V(x) = \frac{1}{2}m\omega^2 x^2 + \lambda x^4 \equiv v_2 x^2 + v_4 x^4. \tag{4.109}$$

As we assume λ to be small, we have: $\frac{\lambda \Lambda_{sc}}{v_2} \ll 1$, where Λ_{sc} introduces a physical length scale, e.g. the analogue of the Bohr radius. According to [JKL02], the following relation between the classical and the quantum potential holds,

$$2m(V(x) - E_{gr}) = 2\tilde{m}(\tilde{V}(x) - \tilde{v}_0) - \frac{1}{2}\frac{\frac{d}{dx}2\tilde{m}(\tilde{V}(x) - \tilde{v}_0)}{\sqrt{2\tilde{m}(\tilde{V}(x) - \tilde{v}_0)}} \, \text{sgn}(x). \tag{4.110}$$

We define the functions

$$W(x) = 2m(V(x) - E_{gr}), \qquad U(x) = 2\tilde{m}(\tilde{V}(x) - \tilde{v}_0), \qquad (4.111)$$

where E_{gr} denotes the ground state energy of the lowest eigen-state of the quantum mechanical system (obtained from the Schrödinger equation). Due to (4.109), the function $W(x)$ must have the form

$$W(x) = w_0 + w_2 x^2 + w_4 x^4. \qquad (4.112)$$

This function is symmetric with respect to parity. Then the function $U(x)$ representing the quantum potential, will be parity symmetric also. We make an Ansatz of the form

$$U(x) = u_0 + u_2 x^2 + u_4 x^4 + u_6 x^6 + \cdots . \qquad (4.113)$$

The assumption that the expansion parameter λ is small is now expressed by

$$w_4 \equiv w_4^{(0)} \epsilon, \quad \epsilon \ll 1. \qquad (4.114)$$

Now using (4.112), (4.113) in combination with (4.110), we obtain

$$w_0 + w_2 x^2 + w_4 x^4 = u_2 x^2 + u_4 x^4 + u_6 x^6 + \cdots$$
$$- \frac{1}{2} \frac{2u_2 + 4u_4 x^2 + 6u_6 x^4 + \cdots}{\sqrt{u_2 + u_4 x^2 + u_6 x^4 + \cdots}} \quad \text{for } x > 0. \quad (4.115)$$

The smallness of w_4 implies that the terms of fourth order and higher in x occurring in the function $U(x)$ are small compared to the second order, i.e.

$$u_4 x^2 + u_6 x^4 + \cdots \ll u_2.$$

Now doing a Taylor expansion in the small terms $u_4 x^2 + u_6 x^4 + \cdots$ allows to express the r.h.s. of (4.115) as a polynomial in x. Then comparing terms in x order by order, we find the following relations [KLM06]

$$w_0 = -\frac{1}{4\sqrt{u_2}^3} 4u_2^2, \qquad w_2 = u_2 - \frac{1}{4\sqrt{u_2}^3} 6u_2 u_4,$$
$$w_4 = u_4 - \frac{1}{4\sqrt{u_2}^3} (10u_2 u_6 - 4u_4^2). \qquad (4.116)$$

Now we try to find the parameters u_2 and u_4 as solution of those equations. The first equation gives $u_2 = (w_0)^2$. On the other hand we have, due to (4.111),

$$w_0 = -2m E_{gr}. \qquad (4.117)$$

Due to the anharmonic perturbation the ground state energy is different from the ground state energy of the harmonic oscillator. However, because of the smallness of the perturbation, we can express the ground-state energy E_{gr} using perturbation theory as a power series in ϵ,

$$E_{gr} = E^{(0)} + \epsilon E^{(1)} + \epsilon^2 E^{(2)} + \cdots, \tag{4.118}$$

where $E^{(0)} = E_{gr}^{osc}$. Thus from (4.117), (4.118) we obtain

$$\sqrt{u_2} = m\omega + \epsilon 2mE^{(1)} + O(\epsilon^2), \quad \text{or}$$

$$u_2 = m^2\omega^2 + \epsilon 4m^2 E^{(1)} + O(\epsilon^2) = w_2 \left[1 + \frac{2E^{(1)}}{E^{(0)}}\epsilon + O(\epsilon^2)\right]. \tag{4.119}$$

Next let us consider (4.116b). We obtain

$$u_4 = -\frac{2}{3}\sqrt{u_2}(w_2 - u_2).$$

Recalling $w_2 = m^2\omega^2$ and (4.119), we find

$$u_4 = -\frac{2}{3}m^3\omega^3 \frac{E^{(1)}}{\omega}\epsilon + O(\epsilon^2). \tag{4.120}$$

In (4.119), (4.120) we have expressed the parameters of the quantum action in terms of the parameters of the classical action. However, it remains to compute the energy $E^{(1)}$. Again we use stationary perturbation theory. The Hamiltonian is given by, taking into account (4.112), (4.114),

$$H = H^{(0)} + \epsilon H^{(1)}, \quad \text{with}$$

$$H^{(0)} = \frac{p^2}{2m} + \frac{1}{2}m\omega^2 x^2, \qquad H^{(1)} = \frac{w_4^{(0)}}{2m}x^4.$$

To first order of perturbation theory in ϵ the energy $E^{(1)}$ is given by

$$E^{(1)} = \langle \psi_{gr}^{osc}|\frac{w_4^{(0)}}{2m}x^4|\psi_{gr}^{osc}\rangle,$$

which yields the result

$$E^{(1)} = \frac{3w_4^{(0)}}{8m^3\omega^2}. \tag{4.121}$$

Substituting this result into (4.119), (4.120) we finally obtain

$$u_2 = w_2 \left[1 + \left(\frac{3}{2m^3\omega^3}\right)w_4 + O(\epsilon^2)\right], \quad \text{and} \tag{4.122}$$

$$u_4 = -\frac{1}{4}w_4 + O(\epsilon^2). \tag{4.123}$$

Interpretation

Let us see what happens when we keep the classical parameters fixed, except for w_4, i.e., we keep $w_4^{(0)}$ fixed and vary ϵ. Note that $w_2 > 0$ and $w_4 > 0$. We also have $w_4^{(0)} > 0$ and $\epsilon > 0$. Now we want to study what happens when $\epsilon \to 0$. (4.122) yields [KLM06]

$$u_2 > w_2, \qquad u_2 \xrightarrow[\epsilon \to 0]{} w_2.$$

Likewise, (4.123) yields

$$u_4 < w_4, \qquad u_4 \xrightarrow[\epsilon \to 0]{} w_4 \xrightarrow[\epsilon \to 0]{} 0.$$

In other words, in the limit $\epsilon \to 0$ the classical potential approaches the potential of the harmonic oscillator. The potential of the quantum action asymptotically approaches the classical potential, hence also the harmonic oscillator potential. That is, the renormalization group flow of the parameters $u_2(\epsilon), u_4(\epsilon)$ goes to a Gaussian fixed point. Second, for any value of ϵ the value the quadratic term of the potential is larger for the quantum potential than for the classical potential. Third, for any value of ϵ the value the quartic term of the potential is smaller for the quantum potential than for the classical potential. Recall that the quadratic term is the term, which, if it would stand alone, would make the system integrable. On the other hand, the quartic term is the term which drives the system away from integrability (and introduces chaos in 2D). Thus we find that quantum fluctuations, which are the cause for the differences $\Delta_2 = w_2 - u_2$ and $\Delta_4 = w_4 - u_4$ to be non-zero, have the tendency to drive the quantum system closer to the regime of integrability.

The above perturbative calculations (see Chap. 3) were performed in 1D. Chaos in time-independent Hamilton systems exists only for $D \geq 2$. A similar, but more tedious calculation can be performed in 2D. It confirms the above result that quantum fluctuations drive the system closer to the regime of integrability and away from the regime of chaos. Clearly, such perturbative calculations are meaningful only in a neighborhood of the Gaussian fixed point. It would be desirable to extend the calculations to a larger regime. However going to higher order of perturbation theory would make those calculation much more tedious. Nevertheless the perturbative result gives some insight into the dynamical consequences of quantum fluctuations. For more technical details, see [KLM06].

4.3.2 Optimal Control of Quantum Chaos

Controlling quantum systems is one of hot topics in physics and chemistry as illustrated in the fields of quantum information processings [NC00, RR96, TV02] and laser control of atomic and molecular processes [RZ00]. As for the latter, there have been devised various control schemes: A π-pulse is a simple example to induce a transition between two eigen-states [AE87]. As a generalization of the π pulse or adiabatic rapid passage [MGH94], we can utilize the non-adiabatic transitions induced by laser fields [TN98]. For more than three level systems, STIRAP scheme uses a counterintuitive pulse sequence to achieve a perfect population transfer between two eigen-states [BTS98]. When more than two electronic states are involved in the controlled system, we can use a pulse-timing control (Tannor–Rice) scheme to selectively break a chemical bond on a desired potential surface by using a pump and dump pulses with an appropriate time interval [TR85]. When the controlled system has more than two pathways from an initial state to a target state, quantum mechanical interference between them can be utilized to modify the ratio of products, which is called coherent control (Shapiro–Brumer) scheme [SB03].

These control schemes are very effective for a certain class of processes but are not versatile and ineffective for, e.g., multi-level→multi-level transitions we shall consider in this section. There exist several mathematical studies which investigate controllability of general quantum-mechanical systems [HTC83, PDR88]. The theorem of controllability says that quantum mechanical systems with a discrete spectrum under certain conditions have complete controllability in the sense that an initial state can be guided to a chosen target state after some time. Although the theorem guarantees the existence of optimal fields, it does not tell us how to construct such a field for a given problem.

One of the method to practically design an optimal field is optimal control theory (OCT) [PDR88, ZBR98] or genetic algorithms [JR92, RZ00]. We focus on the former in this section as a theoretical vehicle. The equations derived from OCT are highly nonlinear (and coupled), so we must solve them using some iterative procedures. There are known some effective algorithms to carry out this procedure numerically, however, the field thus obtained is so complicated that it is difficult to analyze the results: What kinds of dynamical processes are involved in the controlled dynamics? In addition, the cost of the computation becomes larger if we want to apply OCT to realistic problems with many degrees of freedom. Several efforts have been paid to reduce computational costs; [ZR99] have introduced a non-iterative algorithm for the optimal field.

On the other hand, we know that some chemical reaction systems, especially when highly excited, exhibit quantum chaotic features [Gut90], i.e., statistical properties of eigen-energies and eigen-vectors are very similar to those of random matrix systems [Haa01]. We call such systems *quantum chaos systems* in short. It has been also studied how these quantum chaos systems behave under some external parameters [GRM90, TH92, ZD93]. These statistical properties of quantum chaos systems stem from multi-level→multi-level interactions of eigen-states, which are related to the existence of many avoided crossings [Tak92]. Hence it is necessary to consider the interaction between many eigen-states when we study dynamics in such a system. Furthermore, if our purpose is to control a Gaussian wave-packet in a quantum chaos system, the process also becomes a multi-level→multi-level transition because a Gaussian wave-packet in such a system contains many eigen-states. These are our motivations why we treat multi-level→multi-level transitions and want to control them.

In this section, following [TFM05], we study optimal control problems of quantum chaos systems. The goal of control is to obtain an optimal field $\varepsilon(t)$ which guides a quantum chaos system from an initial state $|\varphi_i\rangle$ at $t = 0$ to a given target state $|\varphi_f\rangle$ at some specific time $t = T$. One such method is optimal control theory (OCT), which has been successfully applied to atomic and molecular systems [RZ00].

OCT is usually formulated as a variational problem under constraints as follows: We start from the *Zhu–Botina–Rabitz functional* [ZBR98] (in normal units)

$$J = J_0 - \alpha \int_0^T [\varepsilon(t)]^2 dt - 2\,\mathrm{Re}\left[\langle\phi(T)|\varphi_f\rangle \int_0^T \langle\chi(t)|\partial_t - (H[\varepsilon(t)]/\mathrm{i})|\phi(t)\rangle dt\right].$$

The first term in the right-hand side is the squared absolute value of the final overlap,

$$J_0 = |\langle\phi(T)|\varphi_f\rangle|^2.$$

The second term is the penalty term with respect to an amplitude of the external field $\varepsilon(t)$. The factor $\langle\phi(T)|\varphi_f\rangle$ in the last term is introduced to decouple the conditions for the state $|\phi(t)\rangle$ and the inversely-evolving state $|\chi(t)\rangle$, which both evolve under the Hamiltonian $H[\varepsilon(t)]$ [RZ00, ZBR98]. The variation of J with respect to $|\phi(t)\rangle$ and $|\chi(t)\rangle$ gives Schrödinger's equations,

$$\mathrm{i}\partial_t|\phi(t)\rangle = H[\varepsilon(t)]|\phi(t)\rangle, \qquad \mathrm{i}\partial_t|\chi(t)\rangle = H[\varepsilon(t)]|\chi(t)\rangle.$$

Here we impose the following boundary conditions

$$|\phi(0)\rangle = |\varphi_i\rangle, \qquad |\chi(T)\rangle = |\varphi_f\rangle.$$

Another variation of J with respect to $\varepsilon(t)$ gives an expression for the external field

$$\varepsilon(t) = \frac{1}{\alpha}\,\mathrm{Im}\left[\langle\phi(t)|\chi(t)\rangle\langle\chi(t)|\partial_{\varepsilon(t)}H[\varepsilon(t)]|\phi(t)\rangle\right]. \tag{4.124}$$

In actual numerical calculations, we usually solve these equations with some iteration procedure [ZBR98] because they are nonlinear with respect to $|\phi(t)\rangle$ and $|\chi(t)\rangle$. The optimal field, (4.124), is finally given after a local maximum of the functional is reached.

In the following subsections, we numerically demonstrate to control multi-level→multi-level transition problems in quantum chaos systems: one is a random matrix system, and the other is a quantum kicked rotor [TFM05].

Controlled Random Matrix System

The *random matrix theory* was first introduced by E.P. Wigner as a model to mimick unknown interactions in nuclei, and has been studied to describe statistical natures of spectral fluctuations in quantum chaos systems [Haa01]. Here, we introduce a random matrix system driven by a time-dependent external field $\varepsilon(t)$, which is considered as a model of highly excited atoms or molecules under an electromagnetic field. We write the Hamiltonian

$$H[\varepsilon(t)] = H_0 + \varepsilon(t)V, \tag{4.125}$$

where H_0 and V are $N \times N$ random matrices subject to the Gaussian Orthogonal Ensemble (GOE), which represent generic quantum systems with time-reversal symmetry. The matrix elements of H_0 and V are scaled so that the nearest-neighbor spacing of eigenvalues of H_0 and the variance of the off-diagonal elements of V become both unity.

Once we fix the initial state $|\varphi_i\rangle$ and the final state $|\varphi_f\rangle$, the optimal field $\varepsilon(t)$ is obtained by some numerical procedures for appropriate values of the target time T and the penalty factor α. Though there should be many situations corresponding to the choice of $|\varphi_i\rangle$ and $|\varphi_f\rangle$, we only consider the case where they are *Gaussian random vectors*. It is defined by: $|\varphi\rangle = \sum_j c_j |\phi_j\rangle$, where c_j are complex numbers determined from the following *Gaussian distribution*, $P(c_j) \propto \exp(-|c_j|^2)$, and $|\phi_j\rangle$ is an orthonormal basis. We take this state because it is typical in a random matrix system [TFM05].

Controlled Quantum Kicked Rotor

The kicked rotor (or the standard map) is one of famous models in chaotic dynamical systems, and has been studied in various situations [Haa01]. One feature of its chaotic dynamics is the *deterministic diffusion* along the momentum direction. It is also well known that, if we quantize this system, this diffusion is suppressed by the effects of the wave-function localization in momentum space [Gut90].

Here we employ the quantum kicked rotor as a simple model of quantum chaos systems. The Hamiltonian of a kicked rotor is written as

$$H_{\mathrm{KR}}(t) = \frac{p^2}{2} + \frac{K}{\tau}\cos\theta \sum_{n=-\infty}^{\infty} \delta(t - n\tau),$$

where θ is an angle (mod 2π), p momentum, K a kick strength, and τ a period between kicks. An external field $\varepsilon(t)$ is applied through the coupling Hamiltonian: $H_{\mathrm{I}}[\varepsilon(t)] = -\mu(\theta)\varepsilon(t)$, where the dipole moment is assumed to be: $\mu(\theta) = -\cos(\theta + \delta\theta_0)$. The extra phase $\delta\theta_0$ is introduced to break symmetry of the system. We take $\delta\theta_0 = \pi/3$ in the numerical calculations throughout this section. The total Hamiltonian is given by: $H[\varepsilon(t)] = H_{\mathrm{KR}}(t) + H_{\mathrm{I}}[\varepsilon(t)]$.

For easiness of computation, we impose a periodic boundary condition for p as well as θ; the phase space of the corresponding classical system becomes a 2D torus [Izr86, CS86].

The kicked rotor is often described only at discrete time immediately after/before the periodic kicks. In our control problem, however, we must represent dynamics driven by $\varepsilon(t)$ between those kicks. Then, we can apply ZBR scheme as usual. According to (4.124), the optimal external field is given by

$$\varepsilon(t) = -\frac{1}{\alpha} \operatorname{Im}\left[\langle\phi(t)|\chi(t)\rangle\langle\chi(t)|\mu(\theta)|\phi(t)\rangle\right]. \tag{4.126}$$

Note that, because $\mu(\theta)$ commutes with the unitary operator $e^{-iK\cos\theta}$ of a kick, $\varepsilon(t)$ is obtained as a continuous function of time even at the moment of the delta kicks.

Coarse-Grained Picture

If we apply a resonant external field to a two-level system, we can observe a so-called *Rabi oscillation*. In such a case, the quantum state is well described by [TFM05]

$$|\phi(t)\rangle = e^{E_1 t/i}|\phi_1\rangle \cos|\Omega|t - ie^{-i\theta}e^{E_2 t/i}|\phi_2\rangle \sin|\Omega|t, \tag{4.127}$$

where $|\phi_1\rangle$ and $|\phi_2\rangle$ (E_1 and E_2) are two eigen-states (eigen-energies) of the system, $|\Omega| \equiv |\varepsilon_0\mu_{12}|$ the *Rabi frequency*, $\mu_{12} \equiv \langle\phi_1|\hat{\mu}|\phi_2\rangle$ matrix elements of a dipole operator $\hat{\mu}$, ε_0 an amplitude of the field, and θ a certain phase parameter.

In this section, we study the controlled dynamics from an initial state $|\varphi_i\rangle$ at $t = 0$ to a target state $|\varphi_f\rangle$ at $t = T$ in a multi-state quantum mechanical system described by (4.125). By introducing a 'coarse-grained' picture, which means neglecting highly oscillating terms as the case of *rotating-wave approximation* (RWA) [AE87] and assuming that $|\varphi_i\rangle$ and $|\varphi_f\rangle$ contain many eigen-states without any correlation between them, we show that the controlled dynamics can be represented as a transition between a pair of time-dependent states [TFM05].

Coarse-Grained Rabi State and Frequency

As shown above, the overlap in the controlled dynamics rapidly oscillates because the system contains many states. To analyze this complicated behavior more easily, we introduce the following two time-dependent states

$$|\phi_0(t)\rangle = \hat{U}_0(t,0)|\varphi_i\rangle, \qquad |\chi_0(t)\rangle = \hat{U}_0(t,T)|\varphi_f\rangle,$$
$$\text{where:} \quad \hat{U}_0(t_2,t_1) = e^{-iH_0(t_2-t_1)} \tag{4.128}$$

is a 'free' propagator with H_0 from $t = t_1$ to t_2, and T is a target time. These states are an analogue of eigen-states in the usual Rabi state (4.127), and we try to describe the controlled dynamics as a transition from $|\phi_0(t)\rangle$ to $|\chi_0(t)\rangle$.

We introduce another quantum state by a linear combination of the two time-dependent states,

$$|\phi(t)\rangle = |\phi_0(t)\rangle c(t) + |\chi_0(t)\rangle s(t),$$

where $c(t)$ and $s(t)$ are functions satisfying a normalization condition:

$$|c(t)|^2 + |s(t)|^2 = 1.$$

If we require $|\phi(t)\rangle$ to satisfy Schrödinger's equation, we obtain [TFM05]

$$\mathrm{i}\left[|\phi_0(t)\rangle \partial_t c(t) + |\chi_0(t)\rangle \partial_t s(t)\right] = \varepsilon(t) V \left[|\phi_0(t)\rangle c(t) + |\chi_0(t)\rangle s(t)\right].$$

Multiplying $\langle\phi_0(t)|$ and $\langle\chi_0(t)|$ from the left gives the following equations for $c(t)$ and $s(t)$

$$\mathrm{i}\partial_t \begin{pmatrix} c(t) \\ s(t) \end{pmatrix} = \begin{pmatrix} \langle\phi_0(t)|\varepsilon(t)V|\phi_0(t)\rangle & \langle\phi_0(t)|\varepsilon(t)V|\chi_0(t)\rangle \\ \langle\chi_0(t)|\varepsilon(t)V|\phi_0(t)\rangle & \langle\chi_0(t)|\varepsilon(t)V|\chi_0(t)\rangle \end{pmatrix} \begin{pmatrix} c(t) \\ s(t) \end{pmatrix}, \quad (4.129)$$

where we have used: $|\langle\phi_0(t)|\chi_0(t)\rangle| \ll 1$, which is satisfied when $|\varphi_i\rangle$ and $|\varphi_f\rangle$ are random vectors with a large number of elements.

Our aim is not to solve (4.129) exactly, but to find a coarse-grained (CG) solution by ignoring rapidly oscillating terms when the target time T is large enough. If we use the well-optimized field $\varepsilon(t)$, we expect that the following condition

$$|\langle\phi_0(t)|\varepsilon(t)V|\phi_0(t)\rangle|, |\langle\chi_0(t)|\varepsilon(t)V|\chi_0(t)\rangle| \ll |\langle\phi_0(t)|\varepsilon(t)V|\chi_0(t)\rangle|, \quad (4.130)$$

are satisfied for $T \to \infty$ under the coarse-grained picture.

Under this condition, we obtain the following simple equations [TFM05]

$$\mathrm{i}\partial_t \begin{pmatrix} c(t) \\ s(t) \end{pmatrix} = \begin{pmatrix} 0 & \Omega \\ \Omega^* & 0 \end{pmatrix} \begin{pmatrix} c(t) \\ s(t) \end{pmatrix},$$

$$\text{where } \Omega \equiv \langle\phi_0(t)|\varepsilon(t)V|\chi_0(t)\rangle_{\mathrm{CG}} \qquad (4.131)$$

is a frequency defined by ignoring rapidly oscillating terms. We also expect that Ω has a constant (time-independent) value, which will be justified below. Then, the boundary conditions $c(0) = 1$ and $s(0) = 0$ gives a solution

$$c(t) = \cos|\Omega|t, \qquad s(t) = -\mathrm{i}e^{-\mathrm{i}\theta}\sin|\Omega|t,$$

where $e^{\mathrm{i}\theta} = \Omega/|\Omega|$. The final expression of the controlled dynamics is

$$|\phi(t)\rangle = |\phi_0(t)\rangle \cos|\Omega|t - \mathrm{i}e^{-\mathrm{i}\theta}|\chi_0(t)\rangle \sin|\Omega|t. \qquad (4.132)$$

Note that this state is interpreted to represent a transition between $|\phi_0(t)\rangle$ and $|\chi_0(t)\rangle$ or that between $|\varphi_i\rangle$ and $|\varphi_f\rangle$. Since this is very similar to the usual Rabi state, (4.127), we call this state, (4.132), 'CG Rabi state', and the frequency, (4.131), 'CG Rabi frequency' [TFM05].

Actual Coarse-Graining Procedure

In the previous subsection, we have introduced the concept 'coarse-graining' (CG) to define the CG Rabi frequency Ω in (4.131). In the actual calculations, we carry out this procedure by averaging over a certain time interval,

$$\langle A(t) \rangle_{\text{CG}} \equiv \frac{1}{t_2 - t_1} \int_{t_1}^{t_2} A(t') dt'.$$

Although this result depends on the choice of t_1, t_2 in general, we consider that there exists a natural time scale where the time averaging is meaningful. In optimal control problems, if we choose the target time T large enough, we can substitute the range of the integration into above expression, i.e., $t_1 = 0$ to $t_2 = T$.

To check when the condition, (4.130), is fulfilled, and when the CG Rabi frequency Ω defined in (4.131) becomes constant, we introduce the following integrals [TFM05]

$$F(t) = \int_0^t \langle \phi_0(t') | \varepsilon(t') V | \chi_0(t') \rangle dt',$$

$$g_\phi(t) = \int_0^t \langle \phi_0(t') | \varepsilon(t') V | \phi_0(t') \rangle dt',$$

$$g_\chi(t) = \int_0^t \langle \chi_0(t') | \varepsilon(t') V | \chi_0(t') \rangle dt'.$$

Though the integrands are rapidly oscillating, a certain smoothness can be observed in those integrals, especially for $F(t)$. In such a case, we judge that 'coarse-graining' (CG) is appropriate. Note that $F(t)$ is a linear function of t when the CG Rabi frequency Ω is constant.

Analytic Expression for the Optimal Field

In this subsection, contrary to above, we assume that the dynamics is well approximated by the CG Rabi state, and try to derive an analytic optimal field by using OCT [TFM05]. We start from an assumption that optimally controlled quantum states are represented by the CG Rabi states, i.e., the forwardly evolving state $|\phi(t)\rangle$ and the inversely evolving state $|\chi(t)\rangle$ are assumed to be

$$|\phi(t)\rangle = |\phi_0(t)\rangle \cos|\Omega|t - \text{ie}^{-i\theta}|\chi_0(t)\rangle \sin|\Omega|t, \tag{4.133}$$

$$|\chi(t)\rangle = -\text{ie}^{-i\theta}|\phi_0(t)\rangle \sin|\Omega|(t-T) + |\chi_0(t)\rangle \cos|\Omega|(t-T). \tag{4.134}$$

The optimal field induces a smooth transition between $|\phi_0(t)\rangle$ and $|\chi_0(t)\rangle$. In this section, we employ OCT to study an analytic formulation of the optimal field. Substituting (4.133) and (4.134) into the expression of the optimal field, (4.124), and after some manipulations, we obtain

$$\varepsilon(t) = \frac{\sin 2|\Omega|T}{2\alpha} \, \mathrm{Re} \left[e^{-i\theta} \langle \phi_0(t)|V|\chi_0(t)\rangle \right], \tag{4.135}$$

where $|\langle\phi_0(t)|\chi_0(t)\rangle| \ll 1$ has been used as before. This is an analytic expression for the optimal field while the value of the CG Rabi frequency Ω and the phase parameter θ have not been determined yet.

The definition of the CG Rabi frequency, (4.131), is used to determine $|\Omega|$. Substituting (4.135) and using the relation $\Omega = e^{i\theta}|\Omega|$, we obtain

$$|\Omega| = \frac{\bar{V}^2 \sin 2|\Omega|T}{4\alpha}, \quad \text{where} \tag{4.136}$$

$$\bar{V}^2 \equiv \left\langle |\langle\phi_0(t)|V|\chi_0(t)\rangle|^2 + \left[e^{-i\theta}\langle\phi_0(t)|V|\chi_0(t)\rangle \right]^2 \right\rangle_{\mathrm{CG}} \tag{4.137}$$

is a CG transition element. This equation gives $|\Omega|$ when the penalty factor α and the target time T are fixed. For a large T, the second term in the right-hand side is considered small compared to the first term. In order to see this, we represent the initial and final state using the eigen-states $|\phi_k\rangle$ of H_0 as

$$|\varphi_i\rangle = \sum_j c_j |\phi_j\rangle, \qquad |\varphi_f\rangle = \sum_k d_k |\phi_k\rangle,$$

with the coefficients c_j and d_j. For a large T, we can ignore oscillating terms to obtain [TFM05]

$$|\langle\phi_0(t)|V|\chi_0(t)\rangle|^2 = \sum_{j,k} |c_j|^2 |V_{jk}|^2 |d_k|^2 + |R(T)|^2,$$

$$[\langle\phi_0(t)|V|\chi_0(t)\rangle]^2 = (R(T))^2,$$

$$\text{where } R(T) \equiv \sum_j c_j^* V_{jj} d_j e^{-E_j T/i} \tag{4.138}$$

becomes small for $N \to \infty$ when $|\varphi_i\rangle$ and $|\varphi_f\rangle$ are random vectors without any special correlation. Thus (4.136) is simplified as

$$\bar{V}^2 \approx \sum_{j,k} |c_j|^2 |V_{jk}|^2 |d_k|^2.$$

If the following condition is satisfied: $\frac{\bar{V}^2 T}{2\alpha} > 1$, at least one $|\Omega|$ ($\Omega \neq 0$) is obtained from (4.136). Using this $|\Omega|$, the final overlap J_0 is given by

$$J_0 = \sin^2 |\Omega|T, \tag{4.139}$$

and the averaged amplitude $\bar{\varepsilon}$ of the external field (4.135) is calculated as

$$\bar{\varepsilon} \equiv \sqrt{\frac{1}{T} \int_0^T |\varepsilon(t)|^2 dt} \approx \frac{\sqrt{2}|\Omega|}{\bar{V}}. \tag{4.140}$$

Analytic Solution for Perfect Control

In the ZBR scheme, we must choose a small penalty factor α to make the final overlap large enough. In our analytical results, if we take the limit $\alpha \to 0$, we find that

$$|\Omega| = \frac{(2k-1)\pi}{2T}, \quad (k = 1, 2, \ldots)$$

satisfies (4.136), and then $J_0 = 1$, i.e., a perfect control is achieved. Using (4.135) and (4.136), the optimal field for the perfect control in the small α limit is obtained as

$$\varepsilon(t) = \frac{(2k-1)\pi}{\bar{V}^2 T} \operatorname{Re}\left[e^{-i\theta}\langle\phi_0(t)|V|\chi_0(t)\rangle\right], \tag{4.141}$$

where θ can be determined by a normalization condition as

$$e^{2i\theta} = \frac{\langle\phi_0(T)|\varphi_f\rangle}{\langle\varphi_f|\phi_0(T)\rangle}.$$

This field is expected to be the optimal field which steers the quantum state $|\varphi_i\rangle$ at $t = 0$ to $|\varphi_f\rangle$ at $t = T$, as well as it induces a CG Rabi oscillation between $|\phi_0(t)\rangle$ and $|\chi_0(t)\rangle$. Note that the penalty factor α does not appear in (4.141), so this is different from other non-iterative optimal fields discussed in [ZR99]. For more technical details, see [TFM05].

4.4 Solitons

Recall that synergetics teaches us that *order parameters* (and their spatio-temporal evolution) are *patterns, emerging from chaos*. In our opinion, the most important of these order parameters, both natural and man made, are *solitons*, because of their self-organizing quality to create order out of chaos. From this perspective, *nonlinearity*—the essential characteristic of nature—is the *cause* of both *chaos* and *order*. Recall that the solitary particle-waves, also called the 'light bullets', are localized space–time excitations $\Psi(x, t)$, propagating through a certain medium Ω with constant velocities v_j. They describe a variety of nonlinear wave phenomena in one dimension and playing important roles in optical fibers, many branches of physics, chemistry and biology.

4.4.1 Short History of Solitons

In this subsection we will give a brief of soliton history, mainly following [Sco99, Sco04].

It all started in 1834, when young Scottish engineer named John Scott Russell was conducting experiments on the Union Canal (near Edinburgh) to measure the relationship between the speed of a boat and its propelling force, with the aim of finding design parameters for conversion from horse power to steam. One August day, a rope parted in his measurement apparatus and [Rus844] "The boat suddenly stopped—not so the mass of water in the channel which it had put in motion; it accumulated round the prow of the vessel in a state of violent agitation, then suddenly leaving it behind, rolled forward with great velocity, assuming the form of a large solitary elevation, a rounded, smooth and well defined heap of water, which continued its course along the channel without change of form or diminution of speed." Russell did not ignore this unexpected phenomenon, but "followed it on horseback, and overtook it still rolling on at a rate of some eight or nine miles an hour, preserving its original figure some thirty feet long and a foot to a foot and a half in height" until the wave became lost in the windings of the channel. He continued to study the solitary wave in tanks and canals over the following decade, finding it to be an independent dynamic entity moving with constant shape and speed.

Using a wave tank he demonstrated four facts [Rus844]:

(i) Solitary waves have the shape $h \operatorname{sech}^2[k(x - vt)]$;
(ii) A sufficiently large initial mass of water produces two or more independent solitary waves;
(iii) Solitary waves cross each other "without change of any kind";
(iv) A wave of height h and traveling in a channel of depth d has a velocity given by the expression

$$v = \sqrt{g(d + h)}, \qquad (4.142)$$

(where g is the acceleration of gravity), implying that a large amplitude solitary wave travels faster than one of low amplitude.

Although confirmed by observations on the Canal de Bourgogne, near Dijon, most subsequent discussions of the hydrodynamic solitary wave missed the physical significance of Russell's observations. Evidence that to the end of his life Russell maintained a much broader and deeper appreciation of the importance of his discovery is provided by a posthumous work, where he correctly estimated the height of the earth's atmosphere from (4.142) and the fact that "the sound of a cannon travels faster than the command to fire it" [Rus885].

In 1895, D. Korteweg and H. de Vries published a theory of shallow water waves that reduced Russell's problem to its essential features. One of their results was the nonlinear PDE

$$u_t + c u_x + \varepsilon u_{xxx} + \gamma u u_x = 0, \qquad (4.143)$$

which would play a key role in soliton theory [KdV895]. In this equation, $u(x, t)$ is the wave amplitude, $c = \sqrt{gd}$ is the speed of small amplitude waves, $\varepsilon \equiv c(d^2/6 - T/2\rho g)$ is a dispersive parameter, $\gamma \equiv 3c/2d$ is a nonlinear parameter, and T and ρ are respectively the surface tension and the density of water. The authors showed that (4.143) has a family of exact traveling wave solutions of the form $u(x,t) = \tilde{u}(x - vt)$, where $\tilde{u}(\cdot)$ is Russell's "rounded, smooth and well defined heap" and v is the wave speed. If the dispersive term (ε) and the nonlinear term (γ) in (4.143) are both zero, then the *Korteweg-de Vries equation* (KdV) becomes linear,

$$u_t + cu_x = 0,$$

with a traveling wave solution for any pulse shape at the fixed speed $v = c = \sqrt{gd}$. In general the KdV equation (4.143) is nonlinear with exact traveling wave solutions

$$u(x,t) = h\,\mathrm{sech}^2[k(x - vt)], \qquad (4.144)$$

where $k = \sqrt{\gamma h/12\varepsilon}$, implying that higher amplitude waves are more narrow. With this shape, the effects of dispersion balance those of nonlinearity at an adjustable value of the pulse speed. Thus the solitary wave is recognized as an independent dynamic entity, maintaining a dynamic balance between these two influences. Interestingly, solitary wave velocities are related to amplitudes by

$$v = c + \gamma h/3 = \sqrt{gd}(1 + h/2d), \qquad (4.145)$$

in accord with Russell's empirical results, given in (4.142), to $O(h)$.

Although unrecognized at the time, such an energy conserving solitary wave is related to the existence of a *Bäcklund transform* (BT), which was proposed by J.O. Bäcklund in 1885 [Lam76]. In such a transform, a known solution generates a new solution through a single integration, after which the new solution can be used to generate another new solution, and so on. It is straightforward to find a BT for any linear PDE, which introduces a new eigenfunction into the total solution with each application of the transform. Only special nonlinear PDEs are found to have BTs, but late nineteenth mathematicians knew that these include

$$u_{\xi\tau} = \sin u, \qquad (4.146)$$

which arose in research on the geometry of curved surfaces [Ste36].

In 1939, Y. Frenkel and T. Kontorova introduced a seemingly unrelated problem arising in solid state physics to model the relationship between dislocation dynamics and plastic deformation of a crystal [FK39]. From this study, a PDE describing dislocation motion is

$$u_{xx} - u_{tt} = \sin u, \qquad (4.147)$$

where $u(x,t)$ is atomic displacement in the x-direction and the 'sin' function represents periodicity of the crystal lattice. A *traveling wave solution* of (4.147), corresponding to the propagation of a dislocation, is

$$u(x,t) = 4\arctan\left[\exp\left(\frac{x - vt}{\sqrt{1 - v^2}}\right)\right], \qquad (4.148)$$

with velocity v in the range $(-1, +1)$. Since (4.147) is identical to (4.146) after an independent variable transformation $[\xi = (x-t)/2$ and $\tau = (x+t)/2]$, exact solutions of (4.147) involving arbitrary numbers of dislocation components as in (4.148) can be generated through a succession of Bäcklund transforms, but this was not known to Frenkel and Kontorova.

In the late 1940s, E. Fermi, J. Pasta and S. Ulam (the famous FPU-trio) suggested one of the first scientific problems to be assigned to the Los Alamos MANIAC computing machine: the dynamics of energy equipartition in a slightly nonlinear crystal lattice, which is related to thermal conductivity. The system they chose was a chain of 64 equal mass particles connected by slightly nonlinear springs, so from a linear perspective there were 64 normal modes of oscillation in the system. It was expected that if all the initial energy were put into a single vibrational mode, the small nonlinearity would cause a gradual progress toward equal distribution of the energy among all modes (thermalization), but the numerical results were surprising. If all the energy is originally in the mode of lowest frequency, it returns almost entirely to that mode after a period of interaction among a few other low frequency modes. In the course of several numerical refinements, no thermalization was observed [FPU55].

The original idea, proposed by the Nobel Laureate Enrico Fermi, was to simulate the 1D analogue of atoms in a crystal (see Fig. 4.2(a)): a long chain of particles linked by springs that obey *Hooke's law* (a linear elastic interaction), but with a weak nonlinear correction (quadratic for the FPU-α model or cubic for the FPU-β model). A purely linear law for the springs guarantees that energy given to a single *normal mode* always remains in that mode (see caption of Fig. 4.2(b) for the definition of normal modes in terms of atom displacements from their equilibrium positions).

Pursuit of an explanation for this *FPU-recurrence* led Zabusky and Kruskal to approximate the nonlinear spring-mass system by the KdV equation. In 1965, they reported numerical observations that KdV solitary waves pass through each other with no change in shape or speed, and coined the term "soliton" to suggest this property [ZK65a].

N.J. Zabusky and M.D. Kruskal were not the first to observe nondestructive interactions of energy conserving solitary waves. Apart from Russell's tank measurements [Rus844], Perring and Skyrme, had studied solutions of (4.147) comprising two solutions as in (4.148) undergoing a collision. In 1962, they published numerical results showing perfect recovery of shapes and speeds after a collision and went on to discover an exact analytical description of this phenomenon [PS62].

This result would not have surprised nineteenth century mathematicians; it is merely the second member of the hierarchy of solutions generated by a Bäcklund transform. Nor would it have been unexpected by Seeger and his colleagues, who had noted in 1953 the connections between the nineteenth century work [Ste36] and the studies of Frenkel and Kontorova [SDK53]. Since Perring and Skyrme were interested in (4.147) as a nonlinear model for ele-

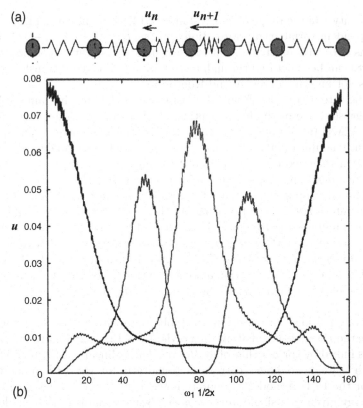

Fig. 4.2. (a) Schematic of the FPU model: masses that can move only in 1D are coupled by nonlinear springs. u_n is the relative displacement with respect to the equilibrium position of the nth mass. The two ends of the chain were assumed to be fixed, i.e., $u_0 = u_N = 0$. (b) FPU recurrence for a FPU$-\alpha$ model with $N = 32$ masses and fixed ends. The plot shows the time evolution of the total energy, $E_k = (\dot{A}_k^2 + \omega_k^2 A_k^2)/2$ of each of the three lowest normal modes, related to the displacements through $A_k = \sqrt{2/(N+1)} \sum_{n=1}^{N} u_n \sin(nk\pi/(N+1))$ with the frequencies $\omega_k^2 = 4\sin^2(k\pi/(2N+2))$. Initially, only mode $k = 1$ is excited. After flowing to other modes, $k = 2$, $k = 3$, etc., the energy almost fully returns to mode $k = 1$—this was a surprise! Therefore, the system behaved in a surprising way: contrary to the predictions of statistical mechanics when the number of particles is going to infinity, the energy equipartition state was not reached and energy was periodically returning to the initially excited 1 mode. This highly remarkable result, known as the *FPU-paradox*, shows that nonlinearity is not enough to guarantee the *equipartition of energy* (modified and adapted from [DR08]).

mentary particles of matter, however, the complete absence of scattering may have been disappointing.

Throughout the 1960s, (4.147) arose in a variety of problems including the propagation of ferromagnetic domain walls, self-induced transparency in nonlinear optics, and the propagation of magnetic flux quanta in long Josephson transmission lines. Eventually (4.147) became known as the *sine-Gordon (SG) equation*, which is a nonlinear version of the linear *Klein–Gordon equation*:

$$u_{xx} - u_{tt} = u.$$

Perhaps the most important contribution made by Zabusky and Kruskal in their 1965 paper was to recognize the relation between nondestructive soliton collisions and the riddle of FPU recurrence. Viewing KdV solitons as independent and localized dynamic entities, they explained the FPU observations as follows. The initial condition generates a family of solitons with different speeds, moving apart in the $x - t$ plane. Since the system studied was of finite length with perfect reflections at both ends, the solitons could not move infinitely far apart; instead they eventually reassembled in the $x - t$ plane, approximately recreating the initial condition after a surprisingly short *recurrence time*.

By 1967, this insight had led Kruskal and his colleagues to devise a nonlinear generalization of the Fourier transform method for constructing solutions of the KdV equation emerging from arbitrary initial conditions [GGK67]. Called the inverse scattering (or inverse spectral) method (ISM), this approach proceeds in three steps.

1. The nonlinear KdV dynamics are mapped onto an associated linear problem, where each eigenvalue of the linear problem corresponds to the speed of a particular KdV soliton.
2. Since the associated problem is linear, the time evolution of its solution is readily computed.
3. An inverse calculation then determines the time evolved KdV dynamics from the evolved solution of the linear associated problem. Thus the solution of a nonlinear problem is found from a series of linear computations.

Another development of the 1960s was M. Toda's discovery of exact two-soliton interactions on a nonlinear spring-mass system, called *Toda lattice* [Tod67]. As in the FPU system, equal masses were assumed to be interconnected with nonlinear springs, but Toda chose the potential

$$\left(\frac{a}{b}\right) [e^{-bu_j} - 1] + au_j, \tag{4.149}$$

where $u_j(t)$ is the longitudinal extension of the jth spring from its equilibrium value and both a and b are adjustable parameters. (In the limit $a \to \infty$ and $b \to 0$ with ab finite, this reduces to the quadratic potential of a linear spring. In the limit $a \to 0$ and $b \to \infty$ with ab finite, it describes the interaction between hard spheres.) Thus by the late 1960s, it was established (although not widely known) that solitons were not limited to PDEs (KdV and SG). Local solutions of difference-differential equations could also exhibit the unexpected properties of unchanging shapes and speeds after collisions.

These events are only the salient features of a growing panorama of nonlinear wave activities that became gradually less parochial during the 1960s. Solid state physicists began to see relationships between their solitary waves (magnetic domain walls, self-shaping pulses of light, quanta of magnetic flux, polarons, etc.), and those from classical hydrodynamics and oceanography, while applied mathematicians began to suspect that the ISM might be used for a broader class of nonlinear wave equations. It was amid this intellectual ferment that A.C. Newell and his colleagues organized the first soliton research workshop during the summer of 1972 [New74]. Interestingly, one of the most significant contributions to this conference came by post. From the Soviet Union arrived a paper by V.E. Zakharov and A.B. Shabat, formulating Kruskal's ISM for the nonlinear PDE [ZS72]

$$iu_t + u_{xx} + 2|u|^2 u = 0. \tag{4.150}$$

In contrast to KdV, SG and the Toda lattice, the dependent variable in this equation is complex rather than real, so the evolutions of two quantities (magnitude and phase of u) are governed by the equation. This reflects the fact that (4.150) is a nonlinear generalization of a linear PDE

$$iu_t + u_{xx} + u = 0,$$

solutions of which comprise both an envelope and a carrier wave. Since this linear equation is a *Schrödinger equation* for the quantum mechanical probability amplitude of a particle (like an electron) moving through a region of uniform potential, it is natural to call (4.150) the *nonlinear Schrödinger (NLS) equation*. When the NLS equation is used to model wave packets in such fields as hydrodynamics, nonlinear optics, nonlinear acoustics, plasma waves and biomolecular dynamics, however, its solutions are devoid of quantum character.

Upon appreciating the Zakharov and Shabat paper, many left the 1972 conference convinced that four nonlinear equations (KdV, SG, NLS, and the Toda lattice) display solitary wave behavior with the special properties that led Zabusky and Kruskal to coin the term soliton [New74]. Within two years, ISM formulations had been constructed for the SG equation and also for the Toda lattice.

Since the mid-1970s, the soliton concept has become established in several areas of applied science, and dozens of dynamic systems are now known to be integrable through the ISM. Even if a system is not exactly integrable, additionally, it may be close to an integrable system, allowing analytic insight to be gleaned from perturbation theory. Thus one is no longer surprised to find stable spatially localized regions of energy, balancing the opposing effects of nonlinearity and dispersion and displaying the essential properties of objects. In the last two decades, soliton studies have been focused on working out the details of such object-like behavior in a wide range of research areas [Sco99].

The Fermi–Pasta–Ulam Experiments

Perhaps the single most important event leading up to the explosive growth of soliton mathematics in the last decades was a seemingly innocuous computer computation, carried out by Enrico Fermi, John Pasta, and Stanislaw Ulam in 1954–55, on the Los Alamos MANIAC computer (originally published as Los Alamos Report LA1940 (1955) and reprinted in [FPU55]).

The following quotation is taken from Stanislaw Ulam's autobiography, *Adventures of a Mathematician* [Ula91].

> Computers were brand new; in fact the Los Alamos Maniac was barely finished.... As soon as the machines were finished, Fermi, with his great common sense and intuition, recognized immediately their importance for the study of problems in theoretical physics, astrophysics, and classical physics. We discussed this at length and decided to formulate a problem simple to state, but such that a solution would require a lengthy computation which could not be done with pencil and paper or with existing mechanical computers.... [W]e found a typical one... the consideration of an elastic string with two fixed ends, subject not only to the usual elastic force of stress proportional to strain, but having, in addition, a physically correct nonlinear term.... The question was to find out how... the entire motion would eventually thermalize....
>
> John Pasta, a recently arrived physicist, assisted us in the task of flow diagramming, programming, and running the problem on the Maniac....
>
> The problem turned out to be felicitously chosen. The results were entirely different qualitatively from what even Fermi, with his great knowledge of wave motion, had expected.

What Fermi, Pasta, and Ulam (FPU) were trying to do was to verify numerically a basic article of faith of statistical mechanics; namely the belief that if a mechanical system has many degrees of freedom and is close to a stable equilibrium, then a generic nonlinear interaction will "thermalize" the energy of the system, i.e., cause the energy to become equidistributed among the normal modes of the corresponding linearized system. In fact, Fermi believed he had demonstrated this fact in [Fer23]. Equipartition of energy among the normal modes is known to be closely related to the ergodic properties of such a system, and in fact FPU state their goal as follows: "The ergodic behavior of such systems was studied with the primary aim of establishing, experimentally, the rate of approach to the equipartition of energy among the various degrees of freedom of the system."

FPU make it clear that the problem that they want to simulate is the vibrations of a "1D continuum" or "string" with fixed end-points and non-linear elastic restoring forces, but that "for the purposes of numerical work this continuum is replaced by a finite number of points ... so that the PDE describing the motion of the string is replaced by a finite number of ODE." To rephrase this in the current jargon, FPU study a 1D lattice of N oscillators with nearest neighbor interactions and zero boundary conditions (for their computations, FPU take $N = 64$, see Fig. 4.2) [Pal97].

We imagine the original string to be stretched along the x-axis from 0 to its length ℓ. The N oscillators have equilibrium positions

$$p_i = ih, \quad (i = 0, \ldots, N-1), \text{ where } h = \ell/(N-1)$$

is the lattice spacing, so their positions at time t are $X_i(t) = p_i + x_i(t)$, (where the x_i represent the displacements of the oscillators from equilibrium). The force attracting any oscillator to one of its neighbors is taken as $k(\delta + \alpha\delta^2)$, δ denoting the "strain", i.e., the deviation of the distance separating these two oscillators from their equilibrium separation h. (Note that when $\alpha = 0$ this is just a linear Hooke's law force with spring constant k.) The force acting on the ith oscillator due to its right neighbor is

$$F(x)_i^+ = k[(x_{i+1} - x_i) + \alpha((x_{i+1} - x_i)^2],$$

while the force acting on it due to its left neighbor is

$$F(x)_i^- = k[(x_{i-1} - x_i) - \alpha((x_{i-1} - x_i)^2].$$

Thus the total force acting on the ith oscillator will be the sum of these two forces, namely:

$$F(x)_i = k(x_{i+1} + x_{i-1} - 2x_i)[1 + \alpha(x_{i+1} - x_{i-1})],$$

and assuming that all of the oscillators have the same mass, m, Newton's equations of motion read:

$$m\ddot{x}_i = k(x_{i+1} + x_{i-1} - 2x_i)[1 + \alpha(x_{i+1} - x_{i-1})],$$

with the boundary conditions

$$x_0(t) = x_{N-1}(t) = 0.$$

In addition, FPU looked at motions of the lattice that start from rest, i.e., they assumed that $\dot{x}_i(0) = 0$, so the motion of the lattice is completely specified by giving the $N-2$ initial displacements $x_i(0)$, $i = 1, \ldots, N-2$. We shall call this the FPU initial value problem (with initial condition $x_i(0)$).

It will be convenient to rewrite Newton's equations in terms of parameters that refer more directly to the original string that we are trying to model. Namely, if ρ denotes the density of the string, then $m = \rho h$, while if κ denotes the Young's modulus for the string (i.e., the spring constant for a piece of unit length), then $k = \kappa/h$ will be the spring constant for a piece of length h. Defining $c = \sqrt{\kappa/\rho}$ we can now rewrite Newton's equations as [Pal97]

$$\ddot{x}_i = c^2 \left(\frac{x_{i+1} + x_{i-1} - 2x_i}{h^2} \right) [1 + \alpha(x_{i+1} - x_{i-1})],$$

and in this form we shall refer to them as the FPU Lattice Equations. We can now 'pass to the continuum limit'; i.e., by letting N tend to infinity (so h tends to zero) we can attempt to derive a PDE for the function $u(x,t)$ that measures the displacement at time t of the particle of string with equilibrium position x. We shall leave the nonlinear case for later, and here restrict our attention to the linear case, $\alpha = 0$. If we take $x = p_i$, then by definition $u(x,t) = x_i(t)$, and since $p_i + h = p_{i+1}$ while $p_i - h = p_{i-1}$, with $\alpha = 0$ the latter form of Newton's equations gives:

$$u_{tt}(x,t) = c^2 \frac{u(x+h,t) + u(x-h,t) - 2u(x,t)}{h^2}.$$

By Taylor's formula:

$$f(x \pm h) = f(x) \pm hf'(x) + \frac{h^2}{2!}f''(x) \pm \frac{h^3}{3!}f'''(x) + \frac{h^4}{4!}f''''(x) + O(h^5),$$

and taking $f(x) = u(x,t)$ gives:

$$\frac{u(x+h,t) + u(x-h,t) - 2u(x,t)}{h^2} = u_{xx}(x,t) + \left(\frac{h^2}{12}\right) u_{xxxx}(x,t) + O(h^4);$$

so letting $h \to 0$, we find $u_{tt} = c^2 u_{xx}$, i.e., u satisfies the linear wave equation, with propagation speed c (and of course the boundary conditions $u(0,t) = u(\ell,t) = 0$, and initial conditions $u_t(x,0) = 0$, $u(x,0) = u_0(x)$).

This is surely one of the most famous initial value problems of mathematical physics, and nearly every mathematician sees a derivation of both the d'Alembert and Fourier version of its solution early in their careers. For each positive integer k there is a normal mode or 'standing wave' solution [Pal97]

$$u_k(x,t) = \cos\left(\frac{k\pi ct}{\ell}\right) \sin\left(\frac{k\pi x}{\ell}\right),$$

and the solution to the initial value problem is

$$u(x,t) = \sum_{k=1}^{\infty} a_k u_k(x,t),$$

where the a_k are the Fourier coefficients of u_0:

$$a_k = \frac{2}{l} \int_0^\ell u_0(x) \sin\left(\frac{k\pi x}{\ell}\right) dx.$$

Replacing x by $p_j = jh$ in $u_k(x,t)$ (and using $\ell = (N-1)h$) we get functions

$$\xi_j^{(k)}(t) = \cos\left(\frac{k\pi ct}{(N-1)h}\right)\sin\left(\frac{kj\pi}{N-1}\right),$$

and it is natural to conjecture that these will be the normal modes for the FPU initial value problem (with $\alpha = 0$ of course). This is easily checked using the addition formula for the sine function. It follows that, in the linearized case, the solution to the FPU initial value problem with initial conditions $x_i(0)$ is given explicitly by

$$x_j(t) = \sum_{k=1}^{N-2} a_k \xi_j^{(k)}(t),$$

where the Fourier coefficients a_k are determined from the formula:

$$a_k = \sum_{j=1}^{N-2} x_j(0)\sin\left(\frac{kj\pi}{N-1}\right).$$

Clearly, when α is zero and the interactions are linear, we are in effect dealing with $N-2$ un-coupled harmonic oscillators (the above normal modes) and there is no thermalization. On the contrary, the sum of the kinetic and potential energy of each of the normal modes is a constant of the motion.

But if α is small but non-zero, FPU expected (on the basis of then generally accepted statistical mechanics arguments) that the energy would gradually shift between modes so as to eventually roughly equalize the total of potential and kinetic energy in each of the $N-2$ normal modes $\xi^{(k)}$. To test this they started the lattice in the fundamental mode $\xi^{(1)}$, with various values of α, and integrated Newton's equations numerically for a long time interval, interrupting the evolution from time to time to compute the total of kinetic plus potential energy in each mode. What did they find? Here is a quotation from their report:

Let us say here that the results of our computations show features which were, from the beginning, surprising to us. Instead of a gradual, continuous flow of energy from the first mode to the higher modes, all of the problems showed an entirely different behavior. Starting in one problem with a quadratic force and a pure sine wave as the initial position of the string, we did indeed observe initially a gradual increase of energy in the higher modes as predicted (e.g., by Rayleigh in an infinitesimal analysis). Mode 2 starts increasing first, followed by mode 3, and so on. Later on, however, this gradual sharing of energy among the successive modes ceases. Instead, it is one or the other mode that predominates. For example, mode 2 decides, as it were, to increase rather rapidly at the cost of the others. At one time it has more energy than all the others put together. Then mode 3 undertakes this rôle. It is only the first few modes which exchange energy among themselves, and they do this in a rather regular fashion. Finally, at a later time, mode 1 comes back to within one percent of its initial value, so that the system seems to be almost periodic.

There is no question that Fermi, Pasta, and Ulam realized they had stumbled onto something big. In his autobiography [Ula91], Ulam devotes several pages to a discussion of this collaboration. Here is a little of what he says:

> I know that Fermi considered this to be, as he said, "a minor discovery." And when he was invited a year later to give the Gibbs Lecture (a great honorary event at the annual American Mathematical Society meeting), he intended to talk about it. He became ill before the meeting, and his lecture never took place. . . .
> The results were truly amazing. There were many attempts to find the reasons for this periodic and regular behavior, which was to be the starting point of what is now a large literature on nonlinear vibrations. Martin Kruskal, a physicist in Princeton, and Norman Zabusky, a mathematician at Bell Labs, wrote papers about it. Later, Peter Lax contributed significantly to the theory.

Unfortunately, Fermi died in 1955, even before the paper cited above was published. It was to have been the first in a series of papers, but with Fermi's passing it fell to others to follow up on the striking results of the Fermi–Pasta–Ulam experiments.

The MANIAC computer, on which FPU carried out their remarkable research, was designed to carry out some computations needed for the design of the first hydrogen bombs, and of course it was a marvel for its day. But it is worth noting that it was very weak by today's standards—not just when compared with current supercomputers, but even when compared with modest desktop machines. At a conference held in 1977 Pasta recalled, "The program was of course punched on cards. A DO loop was executed by the operator feeding in the deck of cards over and over again until the loop was completed!"

The Kruskal–Zabusky Experiments

Following the FPU experiments, there were many attempts to explain the surprising quasi-periodicity of solutions of the FPU Lattice Equations. However it was not until ten years later that Martin Kruskal and Norman Zabusky took the crucial steps that led to an eventual understanding of this behavior [ZK65a].

In fact, they made two significant advances. First they demonstrated that, in a continuum limit, certain solutions of the FPU Lattice Equations could be described in terms of solutions of the so-called Korteweg–de Vries (or KdV) equation. And second, by investigating the initial value problem for the KdV equation numerically on a computer, they discovered that its solutions had remarkable behavior that was related to, but if anything even more surprising and unexpected than the anomalous behavior of the FPU lattice that they had set out to understand.

Finding a good continuum limit for the nonlinear FPU lattice is a lot more sophisticated than one might at first expect after the easy time we had with the linear case. In fact the approach to the limit has to be handled with considerable skill to avoid inconsistent results, and it involves several non-obvious steps.

Let us return to the FPU Lattice Equations

$$\ddot{x}_i = c^2 \left(\frac{x_{i+1} + x_{i-1} - 2x_i}{h^2} \right) [1 + \alpha(x_{i+1} - x_{i-1})], \qquad (4.151)$$

and as before let $u(x,t)$ denote the function measuring the displacement at time t of the particle of string with equilibrium position x, so if $x = p_i$ then, by definition, $x_i(t) = u(x,t)$, $x_{i+1}(t) = u(x+h,t)$, and $x_{i-1}(t) = u(x-h,t)$. Clearly, $\ddot{x}_i = u_{tt}(x,t)$, and Taylor's Theorem with remainder gives

$$\frac{x_{i+1} + x_{i-1} - 2x_i}{h^2} = \frac{u(x+h,t) + u(x-h,t) - 2u(x,t)}{h^2}$$

$$= u_{xx}(x,t) + \left(\frac{h^2}{12} \right) u_{xxxx}(x,t) + O(h^4).$$

By a similar computation

$$\alpha(x_{i+1} - x_{i-1}) = (2\alpha h)u_x(x,t) + \left(\frac{\alpha h^3}{3} \right) u_{xxx}(x,t) + O(h^5),$$

so substitution in (FPU) gives

$$\left(\frac{1}{c^2} \right) u_{tt} - u_{xx} = (2\alpha h)u_x u_{xx} + \left(\frac{h^2}{12} \right) u_{xxxx} + O(h^4).$$

As a first attempt to derive a continuum description for the FPU lattice in the nonlinear case, it is tempting to just let h approach zero and assume that $2\alpha h$ converges to a limit ϵ. This would give the PDE [Pal97]

$$u_{tt} = c^2(1 + \epsilon u_x)u_{xx}$$

as our continuum limit for the FPU Lattice equations and the nonlinear generalization of the wave equation. But this leads to a serious problem. This equation is familiar in applied mathematics, it was studied by Rayleigh in the last century, and it is easy to see from examples that its solutions develop discontinuities (shocks) after a time on the order of $(\epsilon c)^{-1}$, which is considerably shorter than the time scale of the almost periods observed in the Fermi–Pasta–Ulam experiments. It was Zabusky who realized that the correct approach was to retain the term of order h^2 and study the equation

$$\left(\frac{1}{c^2} \right) u_{tt} - u_{xx} = (2\alpha h)u_x u_{xx} + \left(\frac{h^2}{12} \right) u_{xxxx}. \qquad (4.152)$$

If we differentiate this equation with respect to x and make the substitution $v = u_x$, we see that it reduces to the more familiar *Boussinesq equation*

$$\left(\frac{1}{c^2}\right) v_{tt} = v_{xx} + \alpha h \frac{\partial(v^2)}{\partial x^2} + \left(\frac{h^2}{12}\right) v_{xxxx},$$

where the effect of the fourth order term is to add dispersion to the equation, and this smoothes out incipient shocks before they can develop.

It is important to realize that, since $h \neq 0$, (4.152) cannot logically be considered a true continuum limit of the FPU lattice. It should rather be regarded as an asymptotic approximation to the lattice model that works for small lattice spacing h (and hence large N). Nevertheless, we shall now see how to pass from (4.152) to a true continuum description of the FPU lattice.

The next step is to notice that, with α and h small, the solutions of (4.152) should behave qualitatively like solutions of the linear wave equation

$$u_{tt} = c^2 u_{xx},$$

and increasingly so as α and h tend to zero. Now the general solution of the linear wave equation is obviously

$$u(x, t) = f(x + ct) + g(x - ct),$$

i.e., the sum of an arbitrary left moving traveling wave and an arbitrary right moving traveling wave, both moving with speed c. Recall that it is customary to simplify the analysis in the linear case by treating each kind of wave separately, and we would like to do the same here. That is, we would like to look for solutions $u(x, t)$ that behave more and more like (say) right moving traveling waves of velocity c—and for longer and longer periods of time—as α and h tend to zero.

It is not difficult to make precise sense out of this requirement. Suppose that $y(\xi, \tau)$ is a smooth function of two real variables such that the map $\tau \mapsto y(\cdot, \tau)$ is uniformly continuous from $\mathbb{R}$ into the bounded functions on $\mathbb{R}$ with the sup norm—i.e., given $\epsilon > 0$ there is a positive δ such that

$$|\tau - \tau_0| < \delta \quad \text{implies} \quad |y(\xi, \tau) - y(\xi, \tau_0)| < \epsilon.$$

Then for

$$|t - t_0| < T = \delta/(\alpha h c) \quad \text{we have}$$
$$|\alpha h c t - \alpha h c t_0| < \delta, \quad \text{so} \quad |y(x - ct, \alpha h c t) - y(x - ct, \alpha h c t_0)| < \epsilon.$$

In other words, the function $u(x, t) = y(x - ct, \alpha h c t)$ is uniformly approximated by the traveling wave $u^0(x, t) = y(x - ct, \alpha h c t_0)$ on the interval $|t - t_0| < T$ (and of course $T \to \infty$ as α and h tend to zero). To restate this a little more picturesquely, $u(x, t) = y(x - ct, \alpha h c t)$ is approximately a traveling wave whose shape gradually changes in time. Notice that if $y(\xi, \tau)$

is periodic or almost periodic in τ, the gradually changing shape of the approximate traveling wave will also be periodic or almost periodic.

To apply this observation, we define new variables $\xi = x - ct$ and $\tau = (\alpha h)ct$. Then by the chain rule, $\partial^k/\partial x^k = \partial^k/\partial \xi^k$, $\partial/\partial t = -c(\partial/\partial \xi - (\alpha h)\partial/\partial \tau)$, and $\partial^2/\partial t^2 = c^2(\partial^2/\partial \xi^2 - (2\alpha h)\partial^2/\partial \xi \partial \tau) + (\alpha h)^2 \partial^2/\partial \tau^2)$. Thus in these new coordinates the wave operator transforms to [Pal97]

$$\frac{1}{c^2}\frac{\partial^2}{\partial t^2} - \frac{\partial^2}{\partial x^2} = -2\alpha h \frac{\partial^2}{\partial \xi \partial \tau} + (\alpha h)^2 \frac{\partial^2}{\partial \tau^2},$$

so substituting $u(x,t) = y(\xi,\tau)$ in (4.152) (and dividing by $-2\alpha h$) gives:

$$y_{\xi\tau} - \left(\frac{\alpha h}{2}\right) y_{\tau\tau} = -y_\xi y_{\xi\xi} - \left(\frac{h}{24\alpha}\right) y_{\xi\xi\xi\xi},$$

and, at last, we are prepared to pass to the continuum limit. We assume that α and h tend to zero at the same rate, i.e., that as h tends to zero, the quotient h/α tends to a positive limit, and we define

$$\delta = \lim_{h \to 0} \sqrt{h/(24\alpha)} \quad \text{which implies} \quad \alpha h = O(h^2),$$

so letting h approach zero gives

$$y_{\xi\tau} + y_\xi y_{\xi\xi} + \delta^2 y_{\xi\xi\xi\xi} = 0.$$

Finally, making the substitution $v = y_\xi$ we arrive at the KdV equation:

$$v_\tau + v v_\xi + \delta^2 v_{\xi\xi\xi} = 0. \tag{4.153}$$

Note that if we re-scale the independent variables by $\tau \to \beta\tau$ and $\xi \to \gamma\xi$, then the KdV equation becomes:

$$v_\tau + \left(\frac{\beta}{\gamma}\right) v v_\xi + \left(\frac{\beta}{\gamma^3}\right) \delta^2 v_{\xi\xi\xi} = 0,$$

so by appropriate choice of β and γ we can get any equation of the form

$$v_\tau + \lambda v v_\xi + \mu v_{\xi\xi\xi} = 0,$$

and any such equation is referred to as "the KdV equation". A commonly used choice that is convenient for many purposes is

$$v_\tau + 6v v_\xi + v_{\xi\xi\xi} = 0,$$

although the form

$$v_\tau - 6v v_\xi + v_{\xi\xi\xi} = 0$$

(obtained by replacing v by $-v$) is equally common. We will use both these forms.

Let us recapitulate the relationship between the FPU Lattice and the KdV equation. Given a solution $x_i(t)$ of the FPU Lattice, we get a function $u(x,t)$ by interpolation, i.e.,

$$u(ih, t) = x_i(t), \quad (i = 0, \ldots, N).$$

For small lattice spacing h and nonlinearity parameter α there will be solutions $x_i(t)$ so that the corresponding $u(x,t)$ will be an approximate right moving traveling wave with slowly varying shape, i.e., it will be of the form

$$u(x,t) = y(x - ct, \alpha hct)$$

for some smooth function $y(\xi, \tau)$, and the function $v(\xi, \tau) = y_\xi(\xi, \tau)$ will satisfy the KdV equation

$$v_\tau + v v_\xi + \delta^2 v_{\xi\xi\xi} = 0, \quad \text{where } \delta^2 = h/(24\alpha).$$

Having found this relationship between the FPU Lattice and the KdV equation, Kruskal and Zabusky made some numerical experiments, solving the KdV initial value problem for various initial data. Before discussing the remarkable results that came out of these experiments, it will be helpful to recall some of the early history of this equation.

A First Look at the KdV Equation

Korteweg and de Vries derived their equation in 1895 to settle a debate that had been going on since 1844, when the naturalist and naval architect John Scott Russell, in an oft-quoted paper [Rus844], reported an experience a decade earlier in which he followed the bow wave of a barge that had suddenly stopped in a canal. This "solitary wave", some thirty feet long and a foot high, moved along the channel at about eight miles per hour, maintaining its shape and speed for over a mile as Russell raced after it on horseback. Russell became fascinated with this phenomenon and made extensive further experiments with such waves in a wave tank of his own devising, eventually deriving a (correct) formula for their speed as a function of height. The mathematicians Airy and Stokes made calculations which appeared to show that any such wave would be unstable and not persist for as long as Russell claimed. However, later work by Boussinesq (1872), Rayleigh (1876) and finally the Korteweg–de Vries paper in 1895 [KdV895] pointed out errors in the analysis of Airy and Stokes and vindicated Russell's conclusions.

The KdV equation is now accepted as controlling the dynamics of waves moving to the right in a shallow channel. Of course, Korteweg and de Vries did the obvious and looked for traveling-wave solutions for their equation by making the Ansatz [Pal97]

$$v(x,t) = f(x - ct).$$

When this is substituted in the standard form of the KdV equation, it gives

$$-cf' + 6ff' + f''' = 0.$$

If we add the boundary conditions that f should vanish at infinity, then a fairly routine analysis leads to the 1-parameter family of traveling-wave solutions

$$v(x,t) = 2a^2 \operatorname{sech}^2(a(x - 4a^2 t)),$$

now referred to as the one-soliton solutions of KdV (these are the solitary waves of Russell). Note that the amplitude $2a^2$ is exactly half the speed $4a^2$, so that taller waves move faster than their shorter brethren.

Now, back to Zabusky and Kruskal. For numerical reasons, they chose to deal with the case of periodic boundary conditions, in effect studying the KdV equation

$$u_t + uu_x + \delta^2 u_{xxx} = 0 \qquad (4.154)$$

on the circle instead of on the line. For their published report, they chose $\delta = 0.022$ and used the initial condition

$$u(x,0) = \cos(\pi x).$$

Here is an extract from their report (containing the first use of the term "soliton") in which they describe their observations:

(I) Initially the first two terms of (4.154) dominate and the classical overtaking phenomenon occurs; that is u steepens in regions where it has negative slope. (II) Second, after u has steepened sufficiently, the third term becomes important and serves to prevent the formation of a discontinuity. Instead, oscillations of small wavelength (of order δ) develop on the left of the front. The amplitudes of the oscillations grow, and finally *each* oscillation achieves an almost steady amplitude (that increases linearly from left to right) and has the shape of an individual solitary-wave of (4.154). (III) Finally, each "solitary wave pulse" or *soliton* begins to move uniformly at a rate (relative to the background value of u from which the pulse rises) which is linearly proportional to its amplitude. Thus, the solitons spread apart. Because of the periodicity, two or more solitons eventually overlap spatially and interact nonlinearly. Shortly after the interaction they reappear virtually unaffected in size or shape. In other words, solitons "pass through" one another without losing their identity. *Here we have a nonlinear physical process in which interacting localized pulses do not scatter irreversibly.*

Zabusky and Kruskal go on to describe a second interesting observation, a recurrence property of the solitons that goes a long way towards accounting for the surprising recurrence observed in the FPU Lattice. Let us explain again, but in somewhat different terms, the reason why the recurrence in the FPU Lattice is so surprising. The lattice is made up of a great many identical

oscillators. Initially the relative phases of these oscillators are highly correlated by the imposed cosine initial condition. If the interactions are linear
($\alpha = 0$), then the oscillators are harmonic and their relative phases remain
constant. But, when α is positive, the an-harmonic forces between the oscillators cause their phases to start drifting relative to each other in an apparently
uncorrelated manner. The expected time before the phases of all of the oscillators will be simultaneously close to their initial phases is enormous, and
increases rapidly with the total number N. But, from the point of view of the
KdV solitons, an entirely different picture appears. As mentioned in the above
paragraph, if δ is put equal to zero in the KdV equation, it reduces to the
so-called *inviscid Burgers' equation*, which exhibits steepening and breaking
of a negatively sloped wave front in a finite time T_B. (For the above initial
conditions, the breaking time, T_B, can be computed theoretically to be $1/\pi$.)
However, when $\delta > 0$, just before breaking would occur, a small number of
solitons emerge (eight in the case of the above initial wave shape, $\cos(\pi x)$)
*and this number depends only on the initial wave shape, not on the num-
ber of oscillators.* The expected time for their respective centers of gravity
to all eventually "focus" at approximately the same point of the circle is of
course much smaller than the expected time for the much larger number of
oscillators to all return approximately to their original phases. In fact, the
recurrence time T_R for the solitons turns out to be approximately equal to
$30.4T_B$, and at this time the wave shape $u(x, T_R)$ is uniformly very close to
the initial wave form $u(x, 0) = \cos(\pi x)$. There is a second (somewhat weaker)
focusing at time $t = 2T_R$, etc. (Note that these times are measured in units
of the "slow time", τ, at which the shape of the FPU traveling wave evolves,
not in the "fast time", t, at which the traveling wave moves.) In effect, the
KdV solitons are providing a hidden correlation between the relative phases
of the FPU oscillators.

Notice that, as Zabusky and Kruskal emphasize, it is the persistence or
shape conservation of the solitons that provides the explanation of recurrence.
If the shapes of the solitons were not preserved when they interacted, there
would be no way for them to all get back together and approximately recon-
stitute the initial condition at some later time. Here in their own words is
how they bring in solitons to account for the fact that thermalization was not
observed in the FPU experiment:

> Furthermore, because the solitons are remarkably stable entities, pre-
> serving their identities throughout numerous interactions, one would
> expect this system to exhibit thermalization (complete energy sharing
> among the corresponding linear normal modes) only after extremely
> long times, if ever.

But this explanation, elegant as it may be, only pushes the basic question back a step. A full understanding of FPU recurrence requires that we
comprehend the reasons behind the remarkable new phenomenon of solitonic
behavior, and in particular *why* solitons preserve their shape. In fact, it was

quickly recognized that the soliton was itself a vital new feature of nonlinear dynamics, so that understanding it better and discovering other nonlinear
wave equations that had soliton solutions became a primary focus for research
in both pure and applied mathematics. The mystery of the FPU Lattice recurrence soon came to be regarded as an important but fortuitous spark that
ignited this larger effort.

The next few short sections explain some elementary but important facts
about 1D wave equations. If you know about shock development, and how
dispersion smooths shocks, you can skip these sections without loss of continuity.

Split-Stepping KdV

In the KdV equation,

$$u_t = -6uu_x - u_{xxx},$$

if we drop the nonlinear term, we have a constant coefficient linear PDE
whose initial value problem can be solved explicitly by the Fourier Transform.
On the other hand, if we ignore the linear third-order term, then we are left
with the inviscid Burgers' equation, whose initial value problem can be solved
numerically by a variety of methods. Note that it can also be solved in implicit
form analytically, for short times, by the method of characteristics,

$$u = u_o(x - 6ut),$$

but the solution is not conveniently represented on a fixed numerical grid. So,
can we somehow combine the methods for solving each of the two parts into
an efficient numerical method for solving the full KdV initial value problem?

In fact we can, and indeed there is a very general technique that applies
to such situations. In the pure mathematics community it is usually referred
to as the Trotter Product Formula, while in the applied mathematics and
numerical analysis communities it is called *split-stepping*. Let me state it in
the context of ordinary differential equations. Suppose that Y and Z are two
smooth vector-fields on $\mathbb{R}^n$, and we know how to solve each of the ODEs

$$\dot{x} = Y(x) \quad \text{and} \quad \dot{x} = Z(x),$$

meaning that we know both of the flows ϕ_t and ψ_t on $\mathbb{R}^n$ generated by X and
Y respectively. The Trotter Product Formula is a method for constructing the
flow θ_t generated by $Y + Z$ out of ϕ and ψ; namely, letting [Pal97]

$$\Delta t = \frac{t}{n}, \qquad \theta_t = \lim_{n \to \infty} (\phi_{\Delta t} \psi_{\Delta t})^n.$$

The intuition behind the formula is simple. Think of approximating the solution of

$$\dot{x} = Y(x) + Z(x)$$

by Euler's Method. If we are currently at a point p_0, to propagate one more time step Δt we go to the point $p_0 + \Delta t(Y(p_0) + Z(p_0))$. Using the split-step approach on the other hand, we first take an Euler step in the $Y(p_0)$ direction, going to $p_1 = p_0 + \Delta t Y(p_0)$, then take a second Euler step, but now from p_1 and in the $Z(p_1)$ direction, going to $p_2 = p_1 + \Delta t Z(p_1)$. If Y and Z are constant vector-fields, then this gives exactly the same final result as the simple full Euler step with $Y + Z$, while for continuous Y and Z and small time step Δt it is a good enough approximation that the above limit is valid.

The situation is more delicate for flows on infinite dimensional manifolds. Nevertheless it was shown by [Tap74] that the *Cauchy Problem for KdV* can be solved numerically by using split-stepping to combine solution methods for

$$u_t = -6uu_x \quad \text{and} \quad u_t = -u_{xxx}.$$

In addition to providing a perspective on an evolution equation's relation to its component parts, split-stepping allows one to modify a code from solving KdV to the *Kuramoto–Sivashinsky equation*

$$u_t + uu_x = -u_{xx} - u_{xxxx},$$

or study the joint zero-diffusion–dispersion limits *KdV–Burgers' equation*

$$u_t + 6uu_x = \nu u_{xx} + \epsilon u_{xxxx},$$

by merely changing one line of code in the Fourier module.

Tappert uses an interesting variant, known as Strang splitting, which was first suggested in [Str68] to solve multi-dimensional hyperbolic problems by split-stepping 1D problems. The advantage of splitting comes from the greatly reduced effort required to solve the smaller bandwidth linear systems which arise when implicit schemes are necessary to maintain stability. In addition, Strang demonstrated that second-order accuracy of the component methods need not be compromised by the asymmetry of the splitting, as long as the pattern $\phi_{\frac{\Delta t}{2}} \psi_{\frac{\Delta t}{2}} \psi_{\frac{\Delta t}{2}} \phi_{\frac{\Delta t}{2}}$ is used, to account for possible non-commutativity of Y and Z. (This may be seen by multiplying the respective exponential series.) No higher-order analogue of Strang splitting is available. Serendipitously, when output is not required, several steps of Strang splitting require only marginal additional effort:

$$\left(\phi_{\frac{\Delta t}{2}} \psi_{\frac{\Delta t}{2}} \psi_{\frac{\Delta t}{2}} \phi_{\frac{\Delta t}{2}}\right)^n = \left(\phi_{\frac{\Delta t}{2}} \psi_{\Delta t} (\phi_{\Delta t} \psi_{\Delta t})^{n-1} \phi_{\frac{\Delta t}{2}}\right).$$

Now, the FPU Lattice is a classical finite dimensional mechanical system, and as such it has a natural Hamiltonian formulation. However its relation to KdV is rather complex, and KdV is a PDE rather than a finite dimensional system of ODE, so it is not clear that it too can be viewed as a Hamiltonian system. We shall now see how this can be done in a simple and natural way. Moreover, when interpreted as the infinite dimensional analogue of a Hamiltonian system, KdV turns out to have a key property one would expect from any

generalization to infinite dimensions of the concept of complete integrability
in the Liouville sense, namely the existence of infinitely many functionally
independent constants of the motion that are in involution.

In 1971, Gardiner and Zakharov independently showed how to interpret
KdV as a Hamiltonian system, starting from a Poisson bracket approach, and
from this beginning Poisson brackets have played a significantly more important rôle in the infinite dimensional theory of Hamiltonian systems than they
did in the more classical finite dimensional theory, and in recent years this has
led to a whole theory of *Poisson manifolds* and *Poisson Lie groups*. However,
we will start with the more classical approach to Hamiltonian systems, defining a symplectic structure for KdV first, and then get the Poisson bracket
structure as a derived concept (see [AMR88], as well as [II06c]). Thus, we can
exhibit a symplectic structure Ω for the phase-space P of the KdV equation
and a Hamiltonian function, $H : P \to \mathbb{R}$, such that the KdV equation takes
the form

$$\dot{u} = (\nabla_s H)_u.$$

Solitons from a Pendulum Chain

Consider the linear chain of N pendula, each coupled to the nearest neighbors
by the elastic bonds. Let us denote the angle of rotation of ith pendulum by
$\varphi_i = \varphi_i(t)$ and the angular velocity by $\omega_i = \omega_i(t)$. Each pendulum experiences
the angular momentum due to the gravity and the angular momenta from the
two elastic bonds, which are proportional to the difference in angles of rotation
of the coupled pendula with the coefficient κ. Under the assumption that all
pendula have the same inertia moment J, the set of equations of motion takes
the form

$$J\dot{\omega}_i = p_i^{el} + p_i^{gr}, \qquad (4.155)$$

where p_i^{el} is the angular momentum from the elastic bonds and p_i^{gr} is the
gravity angular momentum. The angular momenta from the left and right
elastic bonds are $\kappa(\varphi_{i-1} - \varphi_i)$ and $\kappa(\varphi_{i+1} - \varphi_i)$, respectively and hence, $p_i^{el} = \kappa(\varphi_{i+1} - 2\varphi_i + \varphi_{i-1})$. The gravity angular momentum can be expressed as
$p_i^{gr} = -Jg \sin \varphi_i$, where g is the gravity constant. In view of the expressions
for p_i^{el} and p_i^{gr}, the equations of motion (4.155) can be rewritten as

$$J\ddot{\varphi}_i = \kappa(\varphi_{i+1} - 2\varphi_i + \varphi_{i-1}) - Jg \sin \varphi_i. \qquad (4.156)$$

In the *continuous limit*, when κ/Jg tends to 0, the ODE system (4.156)
transforms into a single PDE

$$J\Phi_{tt} = \kappa\Phi_{xx} - K_G \sin \Phi, \qquad (4.157)$$

where $\Phi = \Phi(x, t)$, and $K_G = Jg$.

In the new spatial and time variables, $X = x\sqrt{\frac{K_G}{\kappa}}$, $T = t\sqrt{\frac{K_G}{J}}$, the PDE
(4.157) gets the standard form of the *sine-Gordon equation* for solitary waves,

$$\Psi_{tt} = \Psi_{xx} - \sin\Psi,$$

with $\Psi = \Psi(X,T)$.

In this way the system of coupled pendula represents the discrete analog to the continuous sine-Gordon equation. In the framework of the pendulum-chain the *kink* is the solitary wave of counter-clockwise rotation of pendula through the angle 2π. The *antikink* is the solitary wave of clockwise rotation of pendula through the angle 2π.

1D Crystal Soliton

Crystal is considered as a chain of the un-deformable molecules, which are connected with each other by the elastic hinges with elastic constant f.

The chain is compressed by the force p along its axis. The vertical springs make elastic support for the nodes of the chain. u_n denotes the transversal displacement of nth node.

The Hamiltonian of the crystal may be written in the dimensionless form:

$$H = \frac{1}{2}\sum_n \left[\dot{u}_n^2 + f(u_{n+1} - 2u_n + u_{n-1})^2 - p(u_n - u_{n-1})^2 + u_n^2 + \frac{1}{2}u_n^4\right], \quad (4.158)$$

where the first term is the kinetic energy, the second one is the potential energy of the elastic hinge, the third one is the work done by the external force and the last two terms give the energy of the elastic nonlinear support due to the vertical springs.

Equation of motion for nth hinge has the form

$$\ddot{u}_n + f(u_{n+2} - 4u_{n+1} + 6u_n - 4u_{n-1} + u_{n-2})$$
$$+ P(u_{n+1} - 2u_n + u_{n-1}) + u_n + u_n^3 = 0. \quad (4.159)$$

The role of the elastic hinges is to keep the chain as a straight line while the compression force plays a destructive role, it tends to destroy the horizontal arrangement of bars. The competition between these two factors, f and p, gives rise to modulation instability in the model.

4.4.2 Lie–Poisson Bracket

Instead of using symplectic structures arising in standard finite-dimensional Hamiltonian dynamics, we propose here the more general *Poisson manifold* $(\mathbf{g}^*, \{F,G\})$. Here $\mathbf{g}^*$ is a chosen Lie algebra with a $(\pm)$ *Lie–Poisson bracket* $\{F,G\}_{\pm}(\mu))$ and carries an abstract *Poisson evolution equation* $\dot{F} = \{F,H\}$. This approach is well-defined in both the finite- and the infinite-dimensional case. It is equivalent to the strong symplectic approach when this exists and offers a viable formulation for Poisson manifolds which are not symplectic (for technical details, see [II06c, II07e]).

For any two smooth functions $F, G : \mathbf{g}^* \to \mathbb{R}$, we define the ($\pm$) *Lie–Poisson bracket* by

$$\{F, G\}_\pm(\mu) = \pm \left\langle \mu, \left[\frac{\delta F}{\delta \mu}, \frac{\delta G}{\delta \mu} \right] \right\rangle. \tag{3.1}$$

Here $\mu \in \mathbf{g}^*$, $[\xi, \mu]$ is the Lie bracket in $\mathbf{g}$ and $\delta F/\delta \mu$, $\delta G/\delta \mu \in \mathbf{g}$ are the functional derivatives of F and G.

The ($\pm$) Lie–Poisson bracket (3.1) is clearly a bilinear and skew-symmetric operation. It also satisfies the Jacobi identity

$$\{\{F, G\}, H\}_\pm(\mu) + \{\{G, H\}, F\}_\pm(\mu) + \{\{H, F\}, G\}_\pm(\mu) = 0$$

thus confirming that $\mathbf{g}^*$ is a Lie algebra, as well as Leibniz' rule

$$\{FG, H\}_\pm(\mu) = F\{G, H\}_\pm(\mu) + G\{F, H\}_\pm(\mu). \tag{4.160}$$

If $\mathbf{g}$ is a finite-dimensional phase-space manifold with structure constants γ_{ij}^k, the ($\pm$) Lie–Poisson bracket (4.160) becomes

$$\{F, G\}_\pm(\mu) = \pm \mu_k \gamma_{ij}^k \frac{\delta F}{\delta \mu_i} \frac{\delta G}{\delta \mu_j}. \tag{4.161}$$

The ($\pm$) Lie–Poisson bracket represents a Lie-algebra generalization of the classical finite-dimensional Poisson bracket $[F, G] = \omega(X_f, X_g)$ on the symplectic phase-space manifold (P, ω) for any two real-valued smooth functions $F, G : P \to \mathbb{R}$.

As in the classical case, any two smooth functions $F, G : \mathbf{g}^* \to \mathbb{R}$ are *in involution* if $\{F, G\}_\pm(\mu) = 0$.

The Lie–Poisson theorem states that a Lie algebra $\mathbf{g}^*$ with a $\pm$ Lie–Poisson bracket $\{F, G\}_\pm(\mu)$ represents a Poisson manifold $(\mathbf{g}^*, \{F, G\}_\pm(\mu))$.

Given a smooth Hamiltonian function $H : \mathbf{g}^* \to \mathbb{R}$ on the Poisson manifold $(\mathbf{g}^*, \{F, G\}_\pm(\mu))$, the time evolution of any smooth function $F : \mathbf{g}^* \to \mathbb{R}$ is given by the abstract *Poisson evolution equation*

$$\dot{F} = \{F, H\}. \tag{4.162}$$

4.4.3 Solitons and Muscular Contraction

Recall that solitons are natural models of propagation of the neural action potential, as well as muscular contraction. In particular, the basis of the molecular model of muscular contraction is oscillations of Amid I peptide groups with associated dipole electric momentum inside a spiral structure of myosin filament molecules (see [Dav82, II06b, II06a]). There is a simultaneous resonant interaction and strain interaction generating a collective interaction directed along the axis of the spiral. The resonance excitation jumping from one peptide group to another can be represented as an exciton, the local molecule strain caused by the static effect of excitation as a phonon and the resultant collective interaction as a soliton.

The simplest model of Davydov's solitary particle-waves is given by the *nonlinear Schrödinger equation* [IP01b]

$$i\partial_t\psi = -\partial_{x^2}\psi + 2\chi|\psi|^2\psi, \tag{4.163}$$

for $-\infty < x < +\infty$. Here $\psi(x,t)$ is a smooth complex-valued wave function with initial condition $\psi(x,t)|_{t=0} = \psi(x)$ and χ is a nonlinear parameter. In the linear limit ($\chi = 0$) (4.163) becomes the ordinary Schrödinger equation for the wave function of the free 1D particle with mass $m = 1/2$.

We may define the infinite-dimensional phase-space manifold $\mathcal{P} = \{(\psi, \bar{\psi}) \in S(\mathbb{R}, \mathbf{C})\}$, where $S(\mathbb{R}, \mathbf{C})$ is the Schwartz space of rapidly-decreasing complex-valued functions defined on $\mathbb{R}$). We define also the algebra $\chi(\mathcal{P})$ of observables on $\mathcal{P}$ consisting of real-analytic functional derivatives $\delta F/\delta\psi$, $\delta F/\delta\bar{\psi} \in S(\mathbb{R}, \mathbf{C})$.

The Hamiltonian function $H : \mathcal{P} \to \mathbb{R}$ is given by

$$H(\psi) = \int_{-\infty}^{+\infty} \left(\left|\frac{\partial\psi}{\partial x}\right|^2 + \chi|\psi|^4 \right) dx$$

and is equal to the total energy of the soliton. It is a conserved quantity for (4.3) (see [II06c]).

The Poisson bracket on $\chi(\mathcal{P})$ represents a direct generalization of the classical finite-dimensional Poisson bracket

$$\{F, G\}_+(\psi) = i \int_{-\infty}^{+\infty} \left(\frac{\delta F}{\delta\psi}\frac{\delta G}{\delta\bar{\psi}} - \frac{\delta F}{\delta\bar{\psi}}\frac{\delta G}{\delta\psi} \right) dx. \tag{4.164}$$

It manifestly exhibits skew-symmetry and satisfies Jacobi identity. The functionals are given by $\delta F/\delta\psi = -i\{F, \bar{\psi}\}$ and $\delta F/\delta\bar{\psi} = i\{F, \psi\}$. Therefore the algebra of observables $\chi(\mathcal{P})$ represents the Lie algebra and the Poisson bracket is the (+) Lie–Poisson bracket $\{F, G\}_+(\psi)$.

The nonlinear Schrödinger equation (4.163) for the solitary particle-wave is a Hamiltonian system on the Lie algebra $\chi(\mathcal{P})$ relative to the (+) Lie–Poisson bracket $\{F, G\}_+(\psi)$ and Hamiltonian function $H(\psi)$. Therefore the Poisson manifold $(\chi(\mathcal{P}), \{F, G\}_+(\psi))$ is defined and the abstract Poisson evolution equation [II06c, II07e] (4.162), which holds for any smooth function $F : \chi(\mathcal{P}) \to \mathbb{R}$, is equivalent to (4.163).

A more subtle model of soliton dynamics is provided by the *Korteveg–de Vries equation* [IP01b]

$$f_t - 6ff_x + f_{xxx} = 0, \quad (f_x = \partial_x f), \tag{4.165}$$

where $x \in \mathbb{R}$ and f is a real-valued smooth function defined on $\mathbb{R}^{35}$. This equation is related to the ordinary Schrödinger equation by the inverse scattering method [II06c].

---

[35] The Korteweg–de Vries equation

We may define the infinite-dimensional phase-space manifold $\mathcal{V} = \{f \in S(\mathbb{R})\}$, where $S(\mathbb{R})$ is the Schwartz space of rapidly-decreasing real-valued functions $\mathbb{R}$. We define further $\chi(\mathcal{V})$ to be the algebra of observables consisting of functional derivatives $\delta F/\delta f \in S(\mathbb{R})$.

The Hamiltonian $H : \mathcal{V} \to \mathbb{R}$ is given by

$$H(f) = \int_{-\infty}^{+\infty} \left(f^3 + \frac{1}{2} f_x^2 \right) dx$$

and provides the total energy of the soliton. It is a conserved quantity for (4.165) (see [II06c]).

As a real-valued analogue to (4.164), the (+) Lie–Poisson bracket on $\chi(\mathcal{V})$ is given via (4.160) by

$$\{F, G\}_+(f) = \int_{-\infty}^{+\infty} \frac{\delta F}{\delta f} \frac{d}{dx} \frac{\delta G}{\delta f} dx.$$

Again it possesses skew-symmetry and satisfies Jacobi identity. The functionals are given by $\delta F/\delta f = \{F, f\}$.

The Korteveg–de Vries equation (4.165), describing the behavior of the molecular solitary particle-wave, is a Hamiltonian system on the Lie algebra $\chi(\mathcal{V})$ relative to the (+) Lie–Poisson bracket $\{F, G\}_+(f)$ and the Hamiltonian function $H(f)$. Therefore, the Poisson manifold $(\chi(\mathcal{V}), \{F, G\}_+(f))$ is defined and the abstract Poisson evolution equation (4.162), which holds for any smooth function $F : \chi(\mathcal{V}) \to \mathbb{R}$, is equivalent to (4.165).

4.5 Dispersive Wave Equations and Stability of Solitons

The theory of linear dispersive equations predicts that waves should spread out and disperse over time. However, it is a remarkable phenomenon, observed both in theory and practice, that once nonlinear effects are taken into account, *solitary wave* or *soliton* solutions can be created, which can be stable enough to persist indefinitely. The construction of such solutions can be relatively straightforward, but the fact that they are *stable* requires some significant amounts of analysis to establish, in part due to symmetries in the equation

$$u_t + u_{xxx} + uu_x = 0 \qquad (4.166)$$

arises in physical systems in which both nonlinear and dispersive effects are relevant. A vector-field

$$v = \xi(x, t, u)\partial_x + \tau(x, t, u)\partial_t + \phi(x, t, u)\partial_u$$

generates a one-parameter symmetry group iff

$$\phi_t + \phi_{xxx} + u\phi_x + u_x\phi = 0,$$

whenever u satisfies (4.166).

(such as translation invariance) which create degeneracy in the stability analysis [Tao08b, Tao06].

In physics, it has been realized for centuries that the behavior of idealized vibrating media (such as waves on a string, on the surface of a body of water, or in the air), in the absence of friction or other dissipative forces, can be modeled by a number of PDEs, known collectively as *dispersive wave equations* (which we right in a quantum-like form of the wave ψ-equations), including the following:

- The *free wave equation*
$$\psi_{tt} - c^2 \Delta \psi = 0,$$

  where $\psi : \mathbb{R} \times \mathbb{R}^n \to \mathbb{R}$ represents the amplitude $\psi(t, x)$ of a wave at a point in a spacetime with n spatial dimensions, the spatial Laplacian on $\mathbb{R}^n$ is defined by
$$\Delta := \sum_{i=1}^{n} \frac{\partial^2}{\partial x_i^2},$$

  ψ_{tt} is short for $\frac{\partial^2 \psi}{\partial t^2}$, and $c > 0$ is a fixed constant (which can be rescaled to equal 1 if one wishes). This equation models the evolution of waves in a medium which has a fixed speed c of propagation in all directions.

- The *linear Schrödinger equation* (1.4) which we rewrite here in physical units:
$$i\hbar\psi_t + \frac{\hbar^2}{2m}\Delta\psi = V\psi, \qquad (4.167)$$

  where $\psi : \mathbb{R} \times \mathbb{R}^n \to \mathbf{C}$ is the wave function of a quantum particle, $\hbar, m > 0$ are physical constants (which are in natural units rescaled to equal 1), and $V : \mathbb{R}^n \to \mathbb{R}$ is a potential function, which we assume to depend only on the spatial variable x. This equation models the evolution of a quantum particle in space in the presence of a classical potential well V.

- The *nonlinear Schrödinger equation (NLS)*
$$i\psi_t + \Delta\psi = \mu|\psi|^{p-1}\psi, \qquad (4.168)$$

  where $p > 1$ is an exponent and $\mu = \pm 1$ is a sign (the case $\mu = +1$ is known as the *defocussing* case, and $\mu = -1$ as the *focusing* case). This equation can be viewed as a variant of the linear Schrödinger equation (with the constants $\hbar$ and m normalized away), in which the potential V now depends in a nonlinear fashion on the solution itself. This equation no longer has a physical interpretation as the evolution of a quantum particle, but can be derived as a model for *quantum media* such as *Bose–Einstein condensates*.[36]

---

[36] A Bose–Einstein condensate (BEC, discovered by S.N. Bose and A. Einstein in 1924–1925) is a state of matter of bosons confined in an external potential and cooled to temperatures very near to absolute zero ($0K = -273.15°C$). Under such supercooled conditions, a large fraction of the atoms collapse into the lowest quantum

- The *Airy equation* (time-dependent)

$$\psi_t + \psi_{xxx} = 0 \tag{4.169}$$

where $\psi : \mathbb{R} \times \mathbb{R} \to \mathbb{R}$ is a scalar function. This equation can be derived as a very simplified model for propagation of low amplitude water waves in a shallow canal, by starting with the Euler equations, making a number of simplifying assumptions to discard nonlinear terms, and the normalizing all constants to equal 1.

- The *Korteveg–de Vries (KdV) equation* [KdV895]

$$\psi_t + \psi_{xxx} + 6\psi\psi_x = 0, \tag{4.170}$$

which is a more refined version of the Airy equation in which the first nonlinear term is retained. The constant 6 that appears here is not essential, but turns out to be convenient when connecting this equation to the theory of inverse scattering (of which more will be said later).

- The *generalized Korteveg–de Vries (gKdV) equation*

$$\psi_t + \psi_{xxx} + (\psi^p)_x = 0, \tag{4.171}$$

for $p > 1$ an integer; the case $p = 2$ is essentially the KdV equation, and the case $p = 3$ is known as the *modified Korteveg–de Vries (mKdV) equation*. The case $p = 5$ is particularly interesting due to its *mass-critical* nature, which we will discuss later.

For simplicity (to avoid some non-trivial topological phenomena, which are an interesting topic which we have no space to discuss here) we shall only

---

state of the external potential, at which point quantum effects become apparent on a macroscopic scale.

Condensates are extremely low-temperature fluids which contain properties and exhibit behaviors that are currently not completely understood, such as spontaneously flowing out of their containers. The effect is the consequence of quantum mechanics, which states that since continuous spectral regions can typically be neglected, systems can almost always acquire energy only in discrete steps. If a system is at such a low temperature that it is in the lowest energy state, it is no longer possible for it to reduce its energy, not even by friction. Without friction, the fluid will easily overcome gravity because of adhesion between the fluid and the container wall, and it will take up the most favorable position, all around the container (see e.g. [Spo80]). This transition occurs below a critical temperature, which for a uniform 3D gas consisting of non-interacting particles with no apparent internal degrees of freedom is given by:

$$T_c = \left(\frac{n}{\zeta(3/2)}\right)^{2/3} \frac{h^2}{2\pi m k_B},$$

where T_c is the critical temperature, n is the particle density, m is the mass per boson, h is the ordinary Planck's constant, k_b is the Boltzmann constant, and ζ is the Riemann zeta function.

consider equations on flat space-times $\mathbf{R}^n$, and only consider solutions which decay to zero at spatial infinity.

The above equations are all evolution equations; if we specify the initial position $\psi(0, x) = \psi_0(x)$ of a wave at time $t = 0$, we expect these equations to have a unique solution with that initial data[37] for all future times $t > 0$. Actually, all of the above equations are time-reversible (for instance, if $(t, x) \mapsto \psi(t, x)$ solves (4.170), then $(t, x) \mapsto \psi(-t, -x)$ also solves (4.170)) so we also expect the initial data at time $t = 0$ to determine the solution at all past times $t < 0$. (This is in sharp contrast to *dissipative* equations such as the heat equation $\psi_t = \Delta\psi$, which are solvable forward in time but are not solvable backwards in time, at least in the category of smooth functions, due to an irreversible loss of energy and information inherent in this equation as time moves forward.)

Solutions to the above equations have two properties, which may seem at first to contradict each other [Tao08b, Tao06]. The first property is that all of these equations are *conservative*; there exists a *Hamiltonian:* $v \mapsto H(v)$, which is a functional that assigns a real number $H(v)$ to any (sufficiently smooth and decaying[38]) function v on the spatial domain $\mathbf{R}^n$, such that the Hamiltonian[39] $H(\psi(t))$ of a (sufficiently smooth and decaying) solution $\psi(t)$ to the above equation is conserved in time:

$$H(\psi(t)) = H(\psi(0)), \quad \text{or equivalently} \quad \partial_t H(\psi(t)) = 0.$$

More specifically, the Hamiltonian is given by

$$H(\psi(t), \psi_t(t)) := \int_{\mathbf{R}^n} \frac{1}{2} |\psi_t(t, x)|^2 + \frac{1}{2} |\nabla_x \psi(t, x)|^2 \, dx$$

for the wave equation,

$$H(\psi(t)) := \int_{\mathbf{R}^n} \frac{\hbar^2}{2m} |\nabla_x \psi(t, x)|^2 + V(x) |\psi(t, x)|^2 \, dx$$

for the linear Schrödinger equation,

$$H(\psi(t)) := \int_{\mathbf{R}^n} \frac{1}{2} |\nabla_x \psi(t, x)|^2 + \frac{\mu}{p+1} |\psi(t, x)|^{p+1} \, dx$$

---

[37] For the wave equation, which is second-order in time, we also need to specify the initial velocity $\partial_t \psi(0, x) = \psi_1(x)$.

[38] To simplify the exposition, we shall largely ignore the important, but technical, analytic issues of exactly how much regularity and decay one needs in order to justify all the computations and assertions given here. In practice, one usually first works in the category of *classical solutions*, solutions that are smooth and rapidly decreasing, and then uses rigorous limiting arguments (and in particular, exploiting the *low-regularity well-posedness theory* of these equations) to extend all results to more general classes of solutions, such as solutions in the energy space $H^1(\mathbf{R}^n)$.

[39] Again, with the wave equation, the Hamiltonian depends on the instantaneous velocity $\psi_t(t)$ of the solution at time t as well as the instantaneous position $\psi(t)$.

for the nonlinear Schrödinger equation,

$$H(\psi(t)) := \int_R \psi_x(t,x)^2 \, dx$$

for the Airy equation,

$$H(\psi(t)) := \int_R \psi_x(t,x)^2 - 2\psi(t,x)^3 \, dx \tag{4.172}$$

for the Korteveg–de Vries equation, and

$$H(\psi(t)) := \int_R \psi_x(t,x)^2 - \frac{2}{p+1}\psi(t,x)^{p+1} \, dx \tag{4.173}$$

for the generalized Korteveg–de Vries equation. In all of these cases, the conservation of the Hamiltonian can be formally verified by computing $\partial_t H(\psi(t))$ via differentiation under the integral sign, substituting the evolution equation for ψ, and then integrating by parts.[40]

Actually, the Hamiltonian is not the only conserved quantity available for these equations; each of these equations also enjoy a number of symmetries (e.g. translation invariance), which (by the *Noether's theorem*) leads to a number of important additional conserved quantities, which we will discuss later. It is often helpful to interpret the Hamiltonian as describing the total *energy* of the wave.

The conservative nature of these equations means that even for very late times t, the state $\psi(t)$ of solution is still[41] 'similar' to the initial state $\psi(0)$, in the sense that they have the same energy. This may lead one to conclude that solutions to these evolution equations should evolve in a fairly static, or fairly periodic manner; after all, this is what happens to solutions to finite-dimensional systems of ordinary differential equations which have a conserved energy H which is *coercive* in the sense that the *energy surfaces:* $\{H = \text{const}\}$ are always bounded.

However, this intuition turns out to not be correct in the realm of dispersive equations, even though such equations can be thought of as infinite-dimensional systems of ODE with a conserved energy, and even though this energy usually exhibits coercive properties. This is ultimately because of a second property of all of these equations, namely *dispersion*. Informally, dispersion means that different components of a solution ψ to any of these equations travel at different velocities, with the velocity of each component determined by the frequency. As a consequence, even though the state of solution at late times has the same energy as the initial state, the different components of

---

[40] One can also formally establish conservation of the Hamiltonian by interpreting each of the above dispersive equations in turn as an infinite-dimensional Hamiltonian system.

[41] This is assuming that the solution exists all the way up to this time t, which can be a difficult task to establish rigorously, especially if the initial data was rough.

the solution are often so dispersed that the solution at late times tends to have much smaller amplitudes than at early times.[42] Thus, for instance, it is perfectly possible for the solution $\psi(t)$ to go to zero in $L^\infty(\mathbb{R}^n)$ norm as $t \to \pm\infty$, even as its energy stays constant (and non-zero). The ability to go to zero as measured in one norm, while staying bounded away from zero in another, is a feature of systems with infinitely many degrees of freedom, which is not present when considering systems of ODE with only boundedly many degrees of freedom [Tao08b, Tao06].

One can see this dispersive effect in a number of ways. One (somewhat informal) way is to analyze *plane wave* solutions

$$\psi(t, x) = A e^{it\tau + x \cdot \xi}, \tag{4.174}$$

for some non-zero amplitude A, some temporal frequency $\tau \in \mathbb{R}$, and some spatial frequency $\xi \in \mathbb{R}^n$. For instance, for the Airy equation (4.169), one easily verifies that (4.174) solves[43] (4.169) exactly when $\tau = \xi^3$; this equation is known as the *dispersion relation* for the Airy equation. If we rewrite the right-hand side of (4.174) in this case as $A e^{i\xi(x-(-\xi^2)t)}$, this asserts that a plane wave solution to (4.169) has a *phase velocity* $-\xi^2$ which is the negative square of its spatial frequency ξ. Thus we see that for this equation, higher frequency plane waves have a much faster phase velocity than lower frequency ones, and the velocity is always in a leftward direction. Similar analyses can be carried out for the other equations given above, though in those equations involving a nonlinearity or a potential, one has to restrict attention to small amplitude or high frequency solutions so that one can (non-rigorously) neglect the effect of these terms. For instance, for the Schrödinger equation (4.167) (at least with $V = 0$) one has the dispersion relation

$$\tau = -\frac{\hbar^2}{2m} |\xi|^2. \tag{4.175}$$

However, as is well known in physics, the phase velocity does not determine the speed of propagation of information in a system; that quality is instead controlled by the *group velocity*, which typically is slightly different from the phase velocity. To explain this quantity, let us modify the ansatz (4.174) by allowing the amplitude A to vary slowly in space, and propagate in time at some velocity $v \in \mathbb{R}^n$. More precisely, we consider solutions of the form

$$\psi(t, x) \approx A(\varepsilon(x - vt)) \, e^{it\tau + x \cdot \xi},$$

---

[42] This phenomenon may seem to be inconsistent with time reversal symmetry. However, this dispersive effect only occurs when the initial data is spatially localized; dispersion sends localized high-amplitude states to broadly dispersed, low-amplitude states, but (by time reversal) can also have the reverse effect.

[43] Strictly speaking, one needs to allow solutions ψ to (4.169) to be complex-valued here rather than real-valued, but this is of course a minor change.

where A is a smooth function and $\varepsilon > 0$ is a small parameter, and we shall be vague about what the symbol '$\approx$' means. If we have $\tau = \xi^3$ as before, then a short (and slightly non-rigorous) computation shows that

$$\psi_t + \psi_{xxx} \approx \varepsilon(v + 3\xi^2)A'(\varepsilon(x - vt))e^{it\tau + x \cdot \xi} + O(\varepsilon^2).$$

Thus we see that in order for ψ to (approximately) solve (4.169) up to errors of $O(\varepsilon^2)$, the group velocity v must be equal to $-3\xi^2$, which is three times the phase velocity $-\xi^2$. Thus, at a qualitative level at least, we still have the same predicted behavior as before; all frequencies propagate leftward, and higher frequencies propagate faster than lower ones. In particular we expect localized high-amplitude states, which can be viewed (via the Fourier inversion formula) as linear superpositions of plane waves of many different frequencies, to disperse leftwards over time into broader, lower-amplitude states (but still with the same energy as the original state, of course).

One can perform similar analyses for other equations. For instance, for the linear Schrödinger equation, and assuming either high frequencies or small potential, one expects waves to propagate at a velocity proportional to their frequency, according to *de Broglie's law*: $mv = \hbar\xi$; similarly for the nonlinear Schrödinger equation when one assumes either high frequencies or small amplitude. In contrast, for wave equation, this analysis suggests that waves of (non-zero) frequency ξ should propagate at velocities $\frac{\xi}{|\xi|}c$; thus the propagation speed c is constant but the propagation direction $\frac{\xi}{|\xi|}$ varies with frequency, leading to a weak dispersive effect. For more general dispersive equations, the group velocity can be read off of the dispersion relation $\tau = \tau(\xi)$ by the formula $v = -\nabla_\xi\tau$ (whereas in contrast, the phase velocity is $-\frac{\xi}{|\xi|^2}\tau$).

In the case of the *Schrödinger equation* with $V = 0$, one can see the dispersive effort more directly by using the explicit solution formula [Tao08b, Tao06]

$$\psi(t, x) = \frac{1}{(2\pi\hbar t/m)^{n/2}} \int_{\mathbb{R}^n} e^{i\frac{m|x-y|^2}{2\hbar^2 t}} \psi_0(y) \, dy,$$

for $t \neq 0$ and all sufficiently smooth and decaying initial data ψ_0. Indeed, we immediately conclude from this formula (formally, at least) that if ψ_0 is absolutely integrable, then $\|\psi(t)\|_{L^\infty(\mathbb{R}^n)}$ decays at the rate $O(|t|^{-n/2})$.

In the case of linear equations such as the Airy equation (4.169), there is a similar explicit formula (involving the Airy function $\mathrm{Ai}(x)$ instead of complex exponentials), but one can avoid the use of special functions by instead proceeding using the Fourier transform and the principle of stationary phase (see e.g. [Ste93]). Indeed, by starting with the Fourier inversion formula

$$\psi_0(x) = \int_R \hat{\psi}_0(\xi)e^{ix\xi} \, d\xi, \quad \text{where}$$

$$\hat{\psi}_0(\xi) := \frac{1}{2\pi} \int_R \psi_0(x)e^{-ix\xi} \, dx$$

is the *Fourier transform* of ψ_0, and noting as before that $e^{it\xi^3 + ix\xi}$ is the solution of the Airy equation with initial data $e^{ix\xi}$, we see from the principle of superposition (and ignoring issues of interchanging derivatives and integrals, etc.) that the solution ψ is given by the formula

$$\psi(t, x) = \int_R \hat{\psi}_0(x) e^{it\xi^3 + ix\xi} \, d\xi. \tag{4.176}$$

If ψ_0 is a Schwartz function (infinitely smooth, with all derivatives decreasing faster than any polynomial), then its Fourier transform is also Schwartz and thus slowly varying. On the other hand, as t increases, the phase $e^{it\xi^3 + i\xi}$ oscillates more and more rapidly (for non-zero ξ), and so we expect an increasing amount of cancellation in the integral in (4.176), leading to decay of ψ as $t \to \infty$. This intuition can be formalized using the methods of stationary phase (which can be viewed as advanced applications of the undergraduate calculus tools of integration by parts and changes of variable), and can for instance be used to show that $\|\psi(t)\|_{L^\infty(\mathbb{R})}$ decays at a rate $O(t^{-1/3})$ in general.

This technique of representing a solution as a superposition of plane waves also works (with a twist) for the linear Schrödinger equation (4.167) in the presence of a potential V, provided that the potential is sufficiently smooth and decaying. The basic idea is to replace the plane waves (4.174) by *distorted plane waves:* $\psi(\tau, x) e^{it\tau}$, where (in order to solve (4.167)) ψ has to solve the *time-independent Schrödinger equation*

$$-\tau \hbar \psi + \frac{\hbar^2}{2m} \Delta \psi = V \psi, \tag{4.177}$$

and then to try to represent solutions ψ to (4.167) as superpositions

$$\psi(t, x) = \int a(\tau) \psi(\tau, x) \, e^{it\tau} \, d\tau,$$

where we are being intentionally vague as to what the range of integration is. If we restrict attention to negative values of τ, then it turns out (by use of scattering theory) that we can construct distorted plane waves $\psi(\tau, x)$ which asymptotically resemble the standard plane waves $e^{ix \cdot \xi}$ as $|x| \to \infty$, where ξ is a frequency obeying the dispersion relation (4.175). If ψ is composed entirely of these waves, then one has a similar dispersive behavior to the free Schrödinger equation (for instance, under suitable regularity and decay hypotheses on V and ψ_0, $\|\psi(t)\|_{L^\infty(\mathbb{R}^n)}$ will continue to decay like $O(t^{-n/2})$). In such cases we say that ψ is in a *radiating state*. In many important cases (such as when the potential V is non-negative, or is small in certain function space norms), all states (with suitable regularity and decay hypotheses) are radiating states. However, when V is large and allowed to be negative, it is also possible[44] to contain *bound states*, in which τ is positive, and the distorted

---

[44] When V does not decay rapidly, then there can also be some intermediate states involving the singular continuous spectrum of the Schrödinger operator $-\frac{\hbar^2}{2m}\Delta + V$,

plane wave $\psi(\tau, x)$ is replaced by an *eigenfunction* ψ, which continues to solve the equation (4.177), but now ψ decays exponentially to zero as $|x| \to \infty$, instead of oscillating like a plane wave as before. (Informally, this is because once τ is positive, the dispersion relation (4.175) is forcing ξ to be imaginary rather than real.) In particular, ψ lies in $L^2(\mathbb{R}^n)$, and so $-\tau$ becomes an eigenvalue of the *Schrödinger operator*[45]

$$H := -\frac{\hbar^2}{2m}\Delta + V.$$

Because multiplication V is a compact operator relative to $-\frac{\hbar^2}{2m}\Delta$, standard spectral theory shows that the set of eigenvalues $-\tau$ is discrete (except possibly at the origin $-\tau = 0$). Note that it is necessary for V to take on negative values in order to obtain negative eigenvalues, since otherwise the operator H is positive semi-definite.

If ψ_0 consists of a superposition of one or more of these eigenfunctions, e.g.

$$\psi_0 = \sum_k c_k \psi(\tau_k, x),$$

where $-\tau_k$ ranges over finitely many of the eigenvalues of H, then we formally have

$$\psi(t) = \sum_k c_k e^{it\tau_k} \psi(\tau_k, x),$$

and so we see that $\psi(t)$ oscillates in time but does not disperse in space. In this case we say that ψ is a *bound state*. Indeed, the evolution is instead *almost periodic*, in the sense that

$$\liminf_{t \to \infty} \|\psi(t) - \psi_0\|_{L^2(\mathbb{R}^n)} = 0,$$

or equivalently, that the orbit $\{\psi(t) : t \in \mathbb{R}\}$ is a precompact subset of $L^2(\mathbb{R}^n)$.

By further application of spectral theory, one can show that an arbitrary state ψ_0 (in, say, $L^2(\mathbb{R}^n)$) can be decomposed as the orthogonal sum of a radiating state, which disperses as $t \to \infty$, and a bound state, which evolves in an almost periodic manner. Indeed this decomposition corresponds to the decomposition of the spectral measure of H into absolutely continuous and pure point components [Tao08b, Tao06].

---

which disperse over time slower than the radiating states but faster than the bound states. One can also occasionally have *resonances* corresponding to the boundary case $\tau = 0$, which exhibit somewhat similar behavior. For simplicity of exposition, we will not discuss these (important) phenomena.

[45] This operator H is related to the Hamiltonian $H(\psi)$ discussed earlier by the formula $H(\psi) = \langle H\psi, \psi \rangle$, where $\langle u, v \rangle := \int_{\mathbb{R}^n} u\bar{v}$ is the usual inner product on $L^2(\mathbb{R}^n)$.

4.5.1 KdV Solitons

We have seen how solutions to linear dispersive equations either disperse completely as $t \to \infty$, or else (in the presence of an external potential) decompose into a superposition of a radiative state that disperses to zero, plus a bound state that exhibits phase oscillation but is otherwise stationary.

In everyday physical experience with water waves, we of course see that such waves disperse to zero over time; once a rock is thrown into a pond, for instance, the amplitude of the resulting waves diminish over time. However, one does see in nature water waves which refuse to disperse for astonishingly long periods of time, instead moving at a constant speed without change in shape. Recall that the soliton phenomenon was first explained mathematically by Korteveg and de Vries [KdV895] in 1895, using (4.170) that now bears their name (although this equation was first proposed as a model for shallow wave propagation by Boussinesq a few decades earlier). Indeed, if one considers traveling wave solutions to (4.170) of the form

$$\psi(t, x) = f(x - ct),$$

for some velocity c, then this will be a solution to (4.170) as long as f solves the ODE

$$-cf' + f''' + 6ff' = 0.$$

If we assume that f decays at infinity, then we can integrate this third-order ODE to obtain a second-order ODE

$$-cf + f'' + 3f^2 = 0.$$

For $c > 0$, this ODE admits the localized explicit solutions

$$f(x) = cQ(c^{1/2}(x - x_0)),$$

for any $x_0 \in \mathbb{R}$, where Q is the explicit *Schwartz function*

$$Q(x) := \frac{1}{2} \operatorname{sech}^2\left(\frac{x}{2}\right).$$

For $c \leq 0$, one can show that there are no localized solutions other than the trivial solution $f \equiv 0$. Thus we obtain a family of explicit solitary wave solutions

$$\psi(t, x) = cQ(c^{1/2}(x - ct - x_0)) \tag{4.178}$$

to the KdV equation; the parameter c thus controls the speed, amplitude, and width of the wave, while x_0 determines the initial location.

Interestingly, all the solutions (4.178) move to the *right*, while radiating states move to the left. This phenomenon is somewhat analogous to the situation with the linear Schrödinger equation, in which the temporal frequency τ (which is somewhat like the propagation speed c in KdV) was negative for radiating states and positive for bound states. Similar solitary wave solutions can also be found for gKdV and NLS equations, though in higher dimensions $n > 1$ one cannot hope to obtain such explicit formulae for these solutions, and instead one needs to use more modern PDE tools, such as calculus of variations and other elliptic theory methods, in order to build such solutions [Tao08b, Tao06]. There are also larger and more oscillatory 'excited' solitary wave solutions which, unlike the 'ground state' solitary wave solutions described above, exhibit changes of sign, but we will not discuss them here.

Early numerical analyses of the KdV equation [ZK65b, FPU74] revealed that these soliton solutions (4.178) were remarkably stable. Firstly, if one perturbed a soliton by adding a small amount of noise, then the noise would soon radiate away from the soliton, leaving the soliton largely unchanged (other than some slight perturbation in the c and x_0 parameters); these phenomena are described mathematically by results on the *orbital stability* and *asymptotic stability* of solitons, of which more will be said later. This is perhaps unsurprising, given that solitons move rightwards and radiation moves leftwards, but one has to bear in mind that equations such as (4.170) are not linear, and in particular one cannot obviously superimpose a soliton and a radiative state to create a new solution to the KdV equation.

What was even more surprising was what happened if one considered collisions between two solitons, for instance imagining initial data of the form [Tao08b, Tao06]

$$\psi(0, x) = c_1 Q(c_1^{1/2}(x - x_1)) + c_2 Q(c_2^{1/2}(x - x_2)),$$

with $0 < c_2 < c_1$ and x_1 far to the left of x_2; thus initially we have a larger, fast-moving soliton to the left of a shallower, slow-moving soliton. If the KdV equation were linear, the solution would now take the form

$$\psi(t, x) = c_1 Q(c_1^{1/2}(x - c_1 t - x_1)) + c_2 Q(c_2^{1/2}(x - c_2 t - x_2))$$

and so the faster solitons would simply overtake the slower one, with no interaction between the two. At the other extreme, with a strongly nonlinear equation, one could imagine all sorts of scenarios when two solitons collide, for instance that they scatter into radiation or into many smaller solitons, combine into a large soliton, and so forth. However, the KdV equation exhibits an interesting intermediate behavior: the solitons do interact nonlinearly with each other during collision, but then emerge from that collision almost unchanged, except that the solitons have been shifted slightly by their collision. In other words, for very late times t, the solution approximately takes the form

$$\psi(t, x) \approx c_1 Q(c_1^{1/2}(x - c_1 t - x_1 - \theta_1)) + c_2 Q(c_2^{1/2}(x - c_2 t - x_2 - \theta_2)),$$

for some additional shift parameters $\theta_1, \theta_2 \in \mathbb{R}$.

More generally, if one starts with *arbitrary* (but smooth and decaying) initial data, what usually happens (numerically, at least) with evolutions of equations such as (4.170) is that some nonlinear (and chaotic-seeming) behavior happens for a while, but eventually most of the solution radiates away to infinity and a finite number of solitons emerge, moving away from each other at different rates. Quite remarkably, this behavior can in fact be justified rigorously for the KdV equation and a handful of other equations (such as the NLS equation in the cubic 1D case $n = 1, p = 3$) due to the inverse scattering method, which we shall discuss shortly, although even in those cases, there are some exotic solutions, such as 'breather' solutions, which occasionally arise and which do not evolve to a superposition of solitons and radiation, but instead exhibit periodic or almost periodic behavior in time. Nevertheless, it is widely believed (and supported by extensive numerics) that for many other dispersive equations (roughly speaking, those equations whose nonlinearity is not strong enough to cause finite time blowup, and more precisely for the *subcritical* equations), solutions with 'generic' initial data should eventually resolve into a finite number of solitons, moving at different speeds, plus a radiative term which goes to zero. This (rather vaguely defined) conjecture goes by the name of the *soliton resolution conjecture*. Except for those few equations which admit exact solutions (for instance, by inverse scattering methods), the conjecture remains unsolved in general, in part because we have very few tools available that can say anything meaningful about *generic* data in a certain class (e.g. with some function norm bounds) without also being applicable to *all* data in that class; thus the presence of a few exotic solutions that do not resolve into solitons and radiation seems to prevent us from tackling all the other cases. Nevertheless, there are certain important regimes in which we do have a good understanding. One of these is the perturbative regime near a single soliton, in which the initial state ψ_0 is close to that of a soliton such as (4.178); this case will be the main topic of our discussion. More recently, results have begun to emerge on *multisoliton states*, in which the solution is close to the superposition of many widely separated solitons, and even more recently still, there has been some results on the collision between a very fast narrow soliton and a very slow broad one. However, it seems that truly non-perturbative regimes, such as the collisions between two solitons of comparable size, remain beyond the reach of current tools (perhaps requiring a new advance in our understanding of dynamical systems in general) [Tao08b, Tao06].

4.5.2 The Inverse Scattering Approach

We now briefly mention the technique of *inverse scattering*, which is a non-perturbative approach which allows one to control the evolution of solutions to completely integrable equations such as (4.170). (Similar techniques apply to one-dimensional cubic NLS, see e.g. [AKN74, SA76, ZM76, Nov80].) This is a vast subject, which can be viewed from many different algebraic and geometric perspectives; we shall content ourselves with describing the approach based on *Lax pairs* [Lax68], which has the advantage of simplicity, provided that one is willing to accept a rather miraculous algebraic identity.

The identity in question is as follows [Tao08b, Tao06]. Suppose that ψ solves the KdV equation (4.170). As always we assume enough smoothness and decay to justify the computations that follow. For every time t, we consider the time-dependent differential operators $L(t)$, $P(t)$ acting on functions on the real line $\mathbb{R}$, defined by

$$L(t) := \frac{d^2}{dx^2} + \psi(t), \qquad P(t) := \frac{d^3}{dx^3} + \frac{3}{4}\left(\frac{d}{dx}\psi(t) + \psi(t)\frac{d}{dx}\right),$$

where we view $\psi(t)$ as a multiplication operator, $f \mapsto \psi(t)f$. One can view $P(t)$ as a truncated (non-commutative) Taylor expansion of $L(t)^{3/2}$. In view with this interpretation, it is perhaps not so surprising that $L(t)$ and $P(t)$ 'almost commute'; the commutator

$$[P(t), L(t)] := P(t)L(t) - L(t)P(t)$$

of the third order operator $P(t)$ and the second order operator $L(t)$ would normally be expected to be fourth order, but in fact things collapse to just be zeroth order. Indeed, after some computation, one eventually obtains

$$[P(t), L(t)] = \frac{1}{4}(\psi_{xxx}(t) + 6\psi(t)\psi_x(t)).$$

In particular, if we substitute in (4.170), we obtain the remarkable *Lax pair equation*

$$\frac{d}{dt}L(t) = 4[P(t), L(t)]. \qquad (4.179)$$

If we non-rigorously treat the operators $L(t)$, $P(t)$ as if they were matrices, we can interpret this equation as follows. Using the Newton approximation

$$L(t + dt) \approx L(t) + dt\frac{d}{dt}L(t); \qquad \exp(\pm P(t)dt) \approx (1 \pm P(t)dt)$$

for infinitesimal dt, we see from (4.179) that

$$L(t + dt) \approx \exp(4P(t)dt)L(t)\exp(-4P(t)dt). \qquad (4.180)$$

This informal analysis suggests that $L(t + dt)$ is a conjugate of $L(t)$, and so on iterating this we expect $L(t)$ to be a conjugate of $L(0)$. In particular, the *spectrum* of $L(t)$ should be time-invariant. Since $L(t)$ is determined by $\psi(t)$, this leads to a rich source of invariants for $\psi(t)$.

The above analysis can be made more rigorous. For instance, one can show that the traces[46] $\mathrm{tr}(e^{sL(t)})$ of heat kernels are independent of t for any fixed

---

[46] Actually, to avoid divergences we need to consider normalized traces, such as $\mathrm{tr}(e^{sL(t)} - e^{s\frac{d^2}{dx^2}})$.

$s > 0$; expanding those traces in powers of s one can recover an infinite number of conservation laws, which includes the conservation of the Hamiltonian (4.172) as a special case. We will not pursue this approach further here, and refer the reader to [HSW99]. Another way to proceed is to consider solutions to the generalized eigenfunction equation

$$L(t)\phi(t,x) = \tau\phi(t,x), \qquad (4.181)$$

for some $\tau \in \mathbb{R}$ and some smooth function $\phi(t,x) = \phi(t,x;\tau)$ (not necessarily decaying at infinity). If (4.181) holds for a single time t (e.g. $t = 0$), and if ϕ then evolves by the equation

$$\phi_t(t,x) = 4P(t)\phi(t,x), \qquad (4.182)$$

for all t, one can verify (formally, at least) that (4.181) persists for all t, by differentiating (4.181) in time and substituting in (4.179) and (4.182). This now suggests an strategy to solve the KdV equation exactly from an arbitrary choice of initial data $\psi(0) = \psi_0$.

1. Use the initial data ψ_0 to form the operator $L(0)$, and then locate the generalized eigenfunctions $\phi(0,x;\lambda)$ for each choice of spectral parameter τ.
2. Evolve each generalized eigenfunction ϕ in time by (4.182).
3. Use the generalized eigenfunctions $\phi(t,x;\tau)$ to recover $L(t)$ and $\psi(t)$.

This strategy looks very difficult to execute, because the operator $P(t)$ itself depends on $\psi(t)$, and so (4.182) cannot be solved exactly without knowing what $\psi(t)$ is, which is exactly what we are trying to find in the first place! But we can break this circularity by only seeking to solve (4.182) *at spatial infinity* $x = \pm\infty$. Indeed, if $\psi(t)$ is decaying, and $\tau = -\xi^2$ for some real number ξ, then we see that solutions $\phi(t,x)$ to (4.181) must take the form [Tao08b, Tao06]

$$\phi(t,x) \approx a_\pm(\xi;t)e^{i\xi x} + b_\pm(\xi;t)e^{-i\xi x},$$

as $x \to \pm\infty$, for some quantities $a_\pm(\xi;t), b_\pm(\xi;t) \in \mathbf{C}$, which we shall refer to as the *scattering data* of $L(t)$. (One can normalize, say, $a_-(0) = 1$ and $b_-(\xi;0) = 0$, and focus primarily on $a_+(\xi;t)$ and $b_+(\xi;t)$, if desired.) Applying (4.182) and using the decay of $\psi(t)$ once again, we are then led (formally, at least) to the asymptotic equations

$$\partial_t a_\pm(\xi;t) = 4(i\xi)^3 a_\pm(\xi;t); \qquad \partial_t b_\pm(\xi;t) = 4(-i\xi)^3 b_\pm(\xi;t),$$

which can be explicitly solved;[47]

$$a_\pm(\xi;t) = e^{-4i\xi^3 t}a_\pm(\xi;0); \qquad b_\pm(\xi;t) = e^{-4i\xi^3 t}b_\pm(\xi;0). \qquad (4.183)$$

---

[47] Note the resemblance of the phases here to those in (4.176). This is not a coincidence, and indeed the scattering and inverse scattering transforms can be viewed as nonlinear versions of the Fourier and inverse *Fourier transform*.

This only handles the case of negative energies $\lambda < 0$. For positive energies, say $\lambda = +\xi^2$ for some $\xi > 0$, the situation is somewhat similar; in this case, we have a discrete set of ξ for which we have a decaying solution $\phi(t, x)$, with $\phi(t, x) \approx c_\pm(\xi; t) e^{\mp \xi x}$ for $x \to \pm\infty$, where

$$c_\pm(\xi; t) = e^{\mp 4\xi^3 t} c_\pm(\xi; 0). \tag{4.184}$$

This suggests a revised strategy to solve the KdV equation exactly:

1. Use the initial data ψ_0 to form the operator $L(0)$, and then locate the scattering data $a_\pm(\xi; 0)$, $b_\pm(\xi; 0)$, $c_\pm(\xi; 0)$.
2. Evolve the scattering data by the equations (4.183), (4.184).
3. Use the scattering data at t to recover $L(t)$ and $\psi(t)$.

The main difficulty in this strategy is now the third step, in which one needs to solve the *inverse scattering problem* to recover $\psi(t)$ from the scattering data. This is a vast and interesting topic in its own right, and involves complex-analytic problems such as the *Riemann–Hilbert problem* [Tao08b, Tao06].

The relationship of all this to solitons is as follows. Recall from our discussion of the linear Schrödinger equation (4.167) that the operator

$$L(0) = \frac{d^2}{dx^2} + \psi_0$$

is going to have radiating states (or absolutely continuous spectrum) corresponding to negative energies $\tau = -\xi^2 < 0$, and a discrete set of positive eigenfunctions corresponding to positive energies $\tau = +\xi^2 > 0$. Generically, the eigenvalues are simple.[48] In that case, it turns out that the inverse scattering procedure relates each eigenvalue $+\xi^2$ of $L(0)$ to a soliton present inside ψ_0; the value of ξ determines the scaling parameter c of the soliton, and the scattering data $c_\pm(\xi; 0)$ determines (in a slightly complicated fashion, depending on the rest of the spectrum) the location of the solitons. The remaining scattering data $a_\pm(\xi; 0)$, $b_\pm(\xi; 0)$ determines the radiative portion of the solution. As the solution evolves, the spectrum stays constant, but the data $a_\pm, b_\pm, c_\pm$ changes in a controlled manner; this is what causes the solitons to move and the radiation to scatter. It turns out that the exact location of each soliton depends to some extent on the relative sizes of the constants $c_\pm$, which are growing or decaying exponentially at differing rates; it is because of this that as one soliton overtakes another, the location of each soliton gets shifted slightly.

For more technical details, see [Tao08b, Tao06].

---

[48] Repeated eigenvalues lead to more complicated behavior, including *breather solutions* and *logarithmically divergent solitons*.

4.6 Nonlinear Schrödinger Equation (NLS)

In this section we derive and explore the nonlinear Schrödinger equation (NLS), which is the basis for recurrent QNNs.

4.6.1 Cubic NLS

The most important case of the *nonlinear Schrödinger equation* is the *cubic NLS*,

$$i\psi_t + \Delta\psi = \pm|\psi|^2\psi, \tag{4.185}$$

with the cubic nonlinearity $|\psi|^2\psi$, which is described in some detail in this subsection. In (4.185), the sign + represents *defocussing NLS*, while the − sign represents *focusing NLS*. This equation is traditionally studied on Euclidean domains $\mathbb{R}^n$, but other domains, like circle, torus or hypersphere, are also studied. Note that in the theory of nonlinear PDEs, we use the terms *sub-critical*, *critical*, and *super-critical* to denote a significant transition in the behavior of a particular equation with respect to a specified regularity class (or conserved quantity). Typically, sub-critical equations behave in an approximately linear manner, supercritical equation behave in a highly nonlinear manner, and critical equations are very finely balanced between the two. Occasionally one also discusses the sub-criticality, criticality, or super-criticality of regularities with respect to other symmetries than scaling, such as Galilean invariance or Lorentz invariance. For survey of recent advances in nonlinear wave equations based on their criticality, see [Tao06].

Cubic NLS on $\mathbb{R}$

The $(1+1)$-NLS, defined on $\mathbb{R}$, is given by the following data:

$$
\begin{aligned}
\text{Equation:} \quad & i\psi_t + \psi_{xx} = \pm|\psi|^2\psi, \\
\text{Fields:} \quad & \psi : \mathbb{R} \times \mathbb{R} \to \mathbb{C}, \\
\text{Dataclass:} \quad & \psi(0) \in H^s(\mathbb{R}).
\end{aligned}
\tag{4.186}
$$

The $(1+1)$-NLS has a completely integrable Hamiltonian structure. Equation (4.186) is mass-subcritical, energy-subcritical and scattering-critical semilinear Schrödinger equation with critical regularity $\dot{H}^{-1/2}(\mathbb{R})$ and Galilean covariance.

The $(1+1)$-NLS (as well as the $(1+2)$-NLS) can be simulated in *Mathematica^{TM}*, using the efficient *method of lines* for evolution PDEs (see e.g., [HSG07]). For example, simulation of the following, more general, focusing NLS boundary value problem,

$$i\psi_t + \psi_{xx} = -2|\psi|^2\psi + 0.1\left(1 - \cos\frac{\pi x}{L}\right), \quad (\text{with } L = 50),$$

$$\psi(0, x) = \text{sech}(x)\exp(ix), \qquad \psi(t, -L) = \psi(t, L), \quad x \in [-L, L], \ t \in [0, 10],$$

is given in Fig. 4.3.

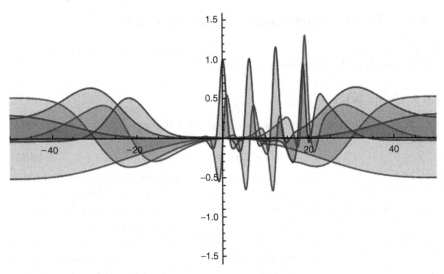

Fig. 4.3. Simulation of the focusing NLS boundary value problem in *Mathematica*[TM].

Cubic NLS on $\mathbb{R}^3$

The $(1+3)$-NLS, defined on $\mathbb{R}^3$, is given by the following data:

$$
\begin{aligned}
\text{Equation:} \quad & i\psi_t + \Delta\psi = \pm|\psi|^2\psi, \\
\text{Fields:} \quad & \psi : \mathbb{R} \times \mathbb{R}^3 \to \mathbb{C}, \\
\text{Dataclass:} \quad & \psi(0) \in H^s(\mathbb{R}^3).
\end{aligned}
\tag{4.187}
$$

The $(1+3)$-NLS has a Hamiltonian structure. Equation (4.187) is mass-subcritical, energy-subcritical and scattering-subcritical semi-linear Schrödinger equation with critical regularity $\dot{H}^{1/2}(\mathbb{R}^3)$ and Galilean covariance.

Cubic NLS on $\mathbb{R}^n$

The $(1+n)$-NLS, defined on $\mathbb{R}^n$, is given by the following data:

$$
\begin{aligned}
\text{Equation:} \quad & i\psi_t + \Delta\psi = \pm|\psi|^2\psi, \\
\text{Fields:} \quad & \psi : \mathbb{R} \times \mathbb{R}^n \to \mathbb{C}, \\
\text{Dataclass:} \quad & \psi(0) \in H^s(\mathbb{R}^n).
\end{aligned}
\tag{4.188}
$$

The $(1+n)$-NLS has a Hamiltonian structure. Equation (4.188) is semi-linear Schrödinger equation with varying criticality, $\dot{H}^{n/2-1}(\mathbb{R}^n)$ as critical regularity[49] and Galilean covariance.

---

[49] The cubic NLS on $\mathbb{R}^n$ obeys the scale invariance

In particular, in focusing equations with a critical conserved quantity (e.g. mass), the behavior for small values of this quantity is often quite different from that at large values. In many equations there is a critical value of this quantity, below which linear-type behavior is expected, and above which non-linear behavior can occur. When one is exactly at the critical value, soliton-type behavior is very typical. For example, in a mass-critical NLS, masses less than the mass $M(Q)$ of the *ground state*[50] are considered subcritical, masses greater than $M(Q)$ are supercritical, and the mass $M(Q)$ itself is critical.

When analyzing the long-time asymptotics of nonlinear equations with data which is small and highly localized, one can distinguish equations into scattering-subcritical (short-range), scattering-critical (critical-range), and scattering-supercritical (long-range) classes; this measures the relative asymptotic strength of the nonlinear and linear components of the equation, and is generally unrelated to the distinction arising from scale invariance or other symmetries. For example, the NLS on $\mathbb{R}^3$ is scattering-subcritical for $p > 5/3$, scattering-critical for $p = 5/3$, and scattering-supercritical for $p < 5/3$.

For equations that are mass-critical (resp. energy-critical), such as the mass critical NLS, *perturbation theory*[51] alone can often give global existence,

$$\psi(t, x) \mapsto \frac{1}{\lambda} \psi\left(\frac{t}{\lambda^2}, \frac{x}{\lambda}\right),$$

which gives the *critical regularity*

$$s_c = \frac{n}{2} - 1.$$

[50] A ground state is a *soliton solution* of minimal mass (or in some cases minimal energy). They are usually smooth, positive, rapidly decreasing, and (after translation) radially symmetric and unique. Ground states are important in the study of solitons.

[51] Recall that *perturbation theory* refers to any situation in which a solution to an equation is analyzed by using an existing nearby solution (possibly solving a nearby equation rather than the original equation) as a reference. In many cases the reference solution is trivial (the zero solution). In order for perturbation theory to work, one or more of the following should hold: (i) the desired initial data should be close to the reference initial data; (ii) the desired equation should be close to the reference equation; and (iii) the time interval on which the analysis is performed should be small.

A typical *perturbation problem* can be formulated with the *evolution PDE*,

$$\partial_t \psi = L(\psi) + \lambda V(\psi),$$

where λ is a small *order parameter*. A solution series with $\lambda \to 0$ can be computed as

$$\psi = \sum_{n=0}^{\infty} \lambda^n \psi_n,$$

which gives the following set of PDEs to be solved

regularity, and linear-type control when the mass (resp., energy) is sufficiently small. On the other hand, the existence of soliton solutions or blowup[52] solutions at high mass (resp., energy) in focusing settings indicates that nonlinear behavior can occur when the mass (resp., energy) is sufficiently large. This indicates that there must be a critical value of mass (or, energy), below

$$\partial_t \psi_0 = L(\psi_0), \qquad \partial_t \psi_1 = L'(\psi_0)\psi_1 + V(\psi_0) \cdots ,$$

where a derivative with respect to the order parameter is indicated by a prime. The choice of the order parameter is just a conventional matter and one can alternatively consider $L(\psi)$ as a perturbation with respect to the same parameter. Then one could formally write the alternative set of PDEs

$$\partial_t v_0 = V(v_0), \qquad \partial_t v_1 = V'(v_0)v_1 + L(v_0) \cdots ,$$

where $L(\psi)$ and $V(\psi)$ are interchanged with the new solution φ.

In order to understand the expansion parameter we rescale the time variable as $\tau = \lambda t$ into the equation to be solved obtaining

$$\lambda \partial_\tau \psi = L(\psi) + \lambda V(\psi),$$

and we introduce the small parameter $\epsilon = \frac{1}{\lambda}$. We see that applying again the small perturbation theory to the parameter $\epsilon \to 0$, we get the required set of equations but now the time is scaled as t/ϵ, that is, at the leading order the development parameter of the series will enter into the scale of the time evolution producing a proper slowing down ruled by the equation

$$\epsilon \partial_t \varphi_0 = V(\varphi_0),$$

which is an equation for *adiabatic evolution* that will in the proper limit $\epsilon \to 0$ give the static solution $V(\varphi_0) = 0$. So, the *dual series*

$$\varphi = \sum_{n=0}^{\infty} \frac{1}{\lambda^n} \varphi_n,$$

is obtained by simply interchanging the terms for doing perturbation theory. This is a *strong coupling expansion* holding in the limit $\lambda \to \infty$ dual to the small perturbation theory $\lambda \to 0$ that we started with and having an adiabatic equation at the leading order.

The main mathematical problem of this kind of methods is the existence of the solution series. For the most interesting cases these series are not converging and represent asymptotic approximations to the true solution. Finally, the success of this method relies on the possibility to obtain a proper analytical solution to the leading order equation.

[52] In the L^2-super-critical focussing NLS one has blowup whenever the Hamiltonian is negative, thanks to the *virial inequality* [Gla77, OT91]

$$\partial_t^2 \int x^2 |\psi|^2 dx \le H(\psi).$$

which linear behavior always occurs, and above which nonlinear behavior can sometimes occur. Determining the exact critical value is often an important challenge; in many cases it is suspected that this value is equal to the mass (or, energy) of the ground state.

4.6.2 Nonlinear Wave and Schrödinger Equations

Nonlinear Wave Equation

Let $n \geq 1$, and consider the *nonlinear wave equation* (NLW)

$$\partial^i \partial_i \psi = \mu |\psi|^{p-1} \psi, \tag{4.189}$$

where $\psi : \mathbb{R} \times \mathbb{R}^n \to \mathbb{C}$ is a complex scalar field, $p > 1$ is the power of the nonlinearity, and $\mu = \pm 1$ is the sign of the nonlinearity (the case $\mu = +1$ is *defocussing*, while the case $\mu = -1$ is *focusing*). The *wave d'Lambertian operator* (in natural units, with $c = 1$) is defined by

$$\partial^i \partial_i = \partial^0 \partial_0 + \cdots + \partial^n \partial_n = -\partial_t^2 + \Delta.$$

One often restricts attention to the case when ψ is real-valued, though most of the analysis extends without difficulty to the complex case also [Tao06, Tao08a]. This equation is also the *Euler–Lagrange equation* for the functional

$$\int_{\mathbb{R} \times \mathbb{R}^n} \frac{1}{2} \partial^i \overline{\psi} \partial_i \psi + \mu |\psi|^{p+1} \, dx dt$$

and is thus one of the simplest nonlinear Lagrangian perturbations of the free wave equation (which has the same Lagrangian but with $\mu = 0$).

Equation (4.189) has a *conserved energy*

$$E(\psi) = E(\psi(t)) := \int_{\mathbb{R}^n} \frac{1}{2} |\partial_t \psi(t,x)|^2 + \frac{1}{2} |\nabla \psi(t,x)|^2 + \mu \frac{1}{p+1} |\psi(t,x)|^{p+1} \, dx. \tag{4.190}$$

Here we adopt the useful convention that $\psi(t) := (\psi(t), \partial_t \psi(t))$ denotes the instantaneous state (both position and velocity) of the field ψ at time t. Indeed, one can easily verify from differentiating under the integral sign that $E(\psi(t))$ is independent of t if ψ is a sufficiently smooth and rapidly decreasing solution to (4.189); one can also observe that this energy is the Hamiltonian for (4.189) using the *symplectic structure* [Tao06, Tao08a]

$$\{(\psi, \psi_t), (v, v_t)\} := \Re \int_{\mathbb{R}^n} \overline{\psi} v_t - v \overline{\psi}_t \, dx,$$

where $\Re$ denotes the real part of the expression.

Note that in the defocussing case $\mu = +1$ the nonlinear component $\mu \frac{1}{p+1} |\psi|^{p+1}$ of the energy density has the same sign as the linear component $\frac{1}{2} |\psi_t|^2 + \frac{1}{2} |\nabla \psi|^2$, whereas in the focusing case these components have opposing signs. Thus in the defocussing case we heuristically expect the non-linearity to amplify the dispersive effects of the linear equation, while in the focusing case we expect the nonlinearity to oppose this dispersion.

Equation (4.189) also enjoys the scaling invariance

$$\psi(t,x) \mapsto \frac{1}{\lambda^{2/(p-1)}} \psi\left(\frac{t}{\lambda}, \frac{x}{\lambda}\right). \tag{4.191}$$

In the *energy-critical* case $n \geq 3$, $p = 1 + \frac{4}{n-2}$, the scaling (4.191) preserves the energy (4.190). Note also that in this case the exponent $\frac{2n}{n-2}$ appearing in the nonlinear component of the energy (4.190) is precisely the exponent appearing in the endpoint *Sobolev inequality* [Tao06, Tao08a]

$$\|f\|_{L_x^{2n/(n-2)}(\mathbb{R}^n)} \leq C_n \|\nabla f\|_{L_x^2(\mathbb{R}^n)}.$$

Historically, the energy-critical wave equation was one of the first critical nonlinear evolution equations to have a satisfactory global theory. This is due to a number of factors, including the finite speed of propagation property (which allows one to analyze blowup by localization in space), as well as the fact that the *conserved momentum*

$$\mathbf{p}(\psi) = \mathbf{p}(\psi[t]) := -\Re \int_{\mathbb{R}^n} \overline{\psi}_t(t,x) \nabla \psi(t,x) \, dx$$

(which will ultimately be the source for a key monotonicity formula in the defocussing case) has the same scaling as the conserved energy.

In the focusing case $\mu = -1$ we have the *stationary solutions* (or *solitons*)

$$\psi(t,x) = Q_\omega(x) e^{i\omega t},$$

where $\omega > 0$ is a *time-frequency* and Q_ω is the solution of the elliptic equation

$$\Delta Q_\omega + |Q_\omega|^{p-1} Q_\omega = \omega^2 Q_\omega.$$

One can also create traveling wave solutions by applying Lorentz transforms to the stationary solution. When Q_ω is a ground state (i.e. it is positive), then these solutions are believed to mark the transition between linear behavior (such as decay in time) and nonlinear behavior (such as blowup, or at least lack of decay in time); very recently there has been some progress in making this behavior rigorous. One also expects these stationary solutions to play a prominent role in analysis of singularities (blowup) of solutions, though the precise relationship here is presently rather poorly understood [Tao06, Tao08a].

When $n \leq 2$, or when $n \geq 3$ and $p < 1 + \frac{4}{n-2}$, (4.189) is *energy-subcritical*, because the scaling (4.191) for $\lambda > 1$ will decrease the energy rather than preserve it. Thus a bounded amount of energy at fine scales is equivalent (after scaling) to a small amount of energy at unit scales, and so we therefore expect the fine-scale behavior of bounded-energy solutions to be close to linear. Because of this, the local theory of subcritical equations is very well understood, though the global asymptotic behavior remains a mystery.

There are a number of other important exponents p, such as the *conformal power* $p = 1 + \frac{4}{n-1}$, which makes the equation (4.189) invariant under conformal transformations of spacetime, and in particular under the *Kelvin inversion*

$$\psi(t,x) \mapsto (t^2 - |x|^2)^{-(n-1)/2} \psi\left(\frac{t}{t^2 - |x|^2}, \frac{x}{t^2 - |x|^2}\right).$$

With this power the equation is energy-subcritical, though the symplectic structure is now critical.

Nonlinear Schrödinger Equation (NLS)

Take $n \geq 1$ and consider the energy-critical NLS [Tao06, Tao08a][53]

$$i\psi_t + \Delta\psi = \mu|\psi|^{p-1}\psi, \tag{4.192}$$

where $\psi : \mathbb{R} \times \mathbb{R}^n \to \mathbb{C}$ is a complex scalar field, and $\mu = \pm 1$ is the sign of the nonlinearity (again, $\mu = +1$ is *defocussing*, while the case $\mu = -1$ is *focusing*). These equations arise naturally as models describing various forms of weakly dispersive behavior [SS99]. The case $n = 1, p = 3$ happens to be completely integrable, but in general the equations are Hamiltonian.

The scaling symmetry is now given by

$$\psi(t,x) \mapsto \frac{1}{\lambda^{2/(p-1)}} \psi\left(\frac{t}{\lambda^2}, \frac{x}{\lambda}\right), \tag{4.193}$$

while the *conserved energy* is now [Tao06, Tao08a]

$$E(\psi) = E(\psi(t)) := \int_{\mathbb{R}^n} \frac{1}{2}|\nabla\psi(t,x)|^2 + \mu\frac{1}{p+1}|\psi|^{p+1}(t,x)\,dx. \tag{4.194}$$

Again, this energy can be interpreted as a Hamiltonian for (4.192), using the *symplectic form*

$$\{\psi, v\} = \int_{\mathbb{R}^n} \Im(\overline{\psi}v)\,dx,$$

where $\Im$ denotes the imaginary part of the expression. The NLS also has an additional phase rotation symmetry $\psi(t,x) \mapsto e^{i\theta}\psi(t,x)$, which leads (via

---

[53] It is sometimes convenient to replace the linear part $i\partial_t + \Delta$ of this operator with $-i\partial_t + \Delta$, $i\partial_t + \frac{1}{2}\Delta$, or $-i\partial_t + \frac{1}{2}\Delta$ to make certain formulae slightly prettier, however it is a trivial matter to transform one equation to the other (by conjugating, dilating, or stretching the solution u in space or time) and so all choices of operator here are essentially equivalent.

Noether's theorem) to a second important conserved quantity,[54] the *mass* (or *charge*)

$$M(\psi) = M(\psi(t)) = \int_{\mathbb{R}^n} |\psi(t,x)|^2 \, dx. \qquad (4.195)$$

The translation symmetry $\psi(t,x) \mapsto \psi(t-x_0)$ also leads to a third conserved quantity, the *vector momentum* [Tao06, Tao08a]

$$\mathbf{p}(\psi) = \mathbf{p}(\psi(t)) := 2 \int_{\mathbb{R}^n} \Im(\overline{\psi(t,x)} \nabla \psi(t,x)) \, dx. \qquad (4.196)$$

When $n \geq 3$ and $p = 1 + \frac{4}{n-2}$, (4.192) is energy-critical but mass-supercritical and momentum-supercritical; conversely, in the *pseudoconformal* case $p = \frac{4}{n}$ (4.192) is mass-critical but energy-subcritical and momentum-subcritical. Thus in both cases, the momentum (which supplies a crucial monotonicity formula in the large data theory) is not scale-invariant, which causes significant technical difficulties in the analysis.

Of the two critical equations, the mass-critical equation is considered harder to analyze. This is because in this case the NLS equation enjoys two less obvious symmetries, namely the *Galilean invariance*

$$\psi(t,x) \mapsto e^{-it|v|^2/4} e^{iv \cdot x/2} \psi(t, x - vt),$$

where $v \in \mathbb{R}^n$ is arbitrary,[55] as well as the *pseudoconformal symmetry* [Tao06, Tao08a]

$$\psi(t,x) \mapsto \frac{1}{|t|^{n/2}} e^{i|x|^2/4t} \psi\left(\frac{1}{t}, \frac{x}{t}\right), \qquad (4.197)$$

for $t \neq 0$. These two symmetries (as well as spatial translation symmetry) also preserve the mass (4.195), thus the mass is in fact critical with respect to quite a large group of symmetries. This wealth of symmetries complicates the analysis, because it implies quite a serious breakdown of compactness for the 'essential' part of the dynamics.

As with NLW, the focusing NLS ($\mu = -1$) also enjoys stationary solutions (or *solitons*)

$$\psi(t,x) = Q_\omega(x) e^{i\omega t},$$

where $\omega > 0$ is a time-frequency and Q_ω solves the elliptic PDE

$$\Delta Q_\omega + |Q_\omega|^{p-1} Q_\omega = \omega Q_\omega.$$

---

[54] The analogue of this quantity for NLW would be the charge $\int \Im(\overline{u} u_t) \, dx$, but this quantity vanishes for the most important case of real scalar fields u and so has not been of major importance in the analysis.

[55] Indeed, this invariance holds for all powers p, being the analogue of the Lorentz invariance for the NLW. The pseudoconformal symmetry however is restricted to the pseudoconformal exponent $p = 1 + \frac{4}{d}$.

One can apply Galilean invariance to also obtain traveling soliton solutions. As with NLW, the ground state solitons are expected to demarcate the transition between linear and nonlinear behavior, and to dominate the dynamics of blowup (at least in certain cases), and there are now several rigorous results that demonstrate this fact.

There is an algebraic embedding of NLS into NLW: if $\psi : \mathbb{R} \times \mathbb{R}^n \to \mathbb{C}$ solves (4.192) in n spatial dimensions, then the complex field $\tilde{u} : \mathbb{R}^{1+(n+1)} \to \mathbb{C}$ defined by

$$\tilde{u}(t, x_1, \ldots, x_{n+1}) := \mathrm{e}^{\mathrm{i}(t+x_{n+1})}\psi(t - x_{n+1}, x_1, \ldots, x_n)$$

solves (4.189) in $n + 1$ spatial dimensions. In the Fourier space, this fact becomes the geometric observation that a nD paraboloid can be viewed as a section of a $(n+1)$D cone. This allows one to deduce many algebraic identities for the nD NLS from the corresponding identities for the $(n+1)$D NLW—the so-called 'method of descent'.

However, this embedding of NLS into NLW, while exact, is not very useful analytically as it maps finite-energy solutions to infinite-energy ones. There is a more profitable *asymptotic* embedding from NLS to a variant of NLW, the *nonlinear Klein–Gordon equation* (NLKG)

$$\partial^i \partial_i \psi = \psi + \mu |\psi|^{p-1}\psi,$$

namely that if $\psi : \mathbb{R} \times \mathbb{R}^n$ solves NLS, then the complex field $\tilde{u} : \mathbb{R}^{1+n} \to \mathbb{C}$ defined by

$$\tilde{u}(t, x) := \mathrm{e}^{-\mathrm{i}^2 t}\psi(t/2, x)$$

solves NLKG. For more technical details, see [Tao06, Tao08a].

4.6.3 Physical NLS-Derivation

While it is generally believed that, at the fundamental level, the evolution of the entire universe is governed by a linear Schrödinger equation for "the wave-function of the universe," this view is not particularly helpful in practice. For limited physical systems, such as those found in our laboratories, the linear Schrödinger's equation (1.4) is only an approximation where certain degrees of freedom can be ignored or controlled. Thus, in the most celebrated example, an electron in a hydrogen atom (see, e.g. [Sha94]), we find (in natural units)

$$\mathrm{i}\partial_t \psi(\mathbf{x}, t) = \left(-\frac{1}{2}\Delta - \frac{\mathrm{e}^2}{r} \right) \psi(\mathbf{x}, t), \tag{4.198}$$

where $\psi(\mathbf{x}, t)$ is the wave function (probability amplitude) associated with the electron at space-time point $(\mathbf{x}, t)$, e the elementary charge, and r is the magnitude $|\mathbf{x}|$. Here, the nucleus (proton) has been placed at the origin of the coordinate system, providing only an 'external', Coulomb potential (e^2/r)

in which the electron moves. All the degrees of freedom associated with the proton are ignored in this equation. In fact, it is treated as a *classical* particle. Had we taken into account its quantum mechanical properties, or the fact that it is composed of three quarks, we would have to consider wave functions for these and to deal with the associated equations. The resultant would be far more complex than (4.198). Similarly, the Coulomb potential is known to be one aspect of a photon, so that its degree of freedom has also been *ignored* in this simple equation. If the quantum mechanics of the photon is also taken into account, we are necessarily faced with the full theory of relativistic quantum fields [Kak93]. The most venerable example of such kind of system is quantum electrodynamics [Kak93, Fey98], in which *interactions* between the electrons and photons are fully incorporated. Nevertheless, by ignoring all degrees of freedom except that associated with the electronic position, a simple linear equation such as (4.198) has been used to predict many properties of the hydrogen atom, to a high degree of accuracy. Parenthetically, notice that the spin degree of freedom, a favorite in quantum computing community, is also ignored in (4.198). When this is included into a more complex version of (4.198), the effects of 'spin–orbit' coupling can be discussed and a better approximation of the properties of hydrogen emerges [GZ01].

Between the simplest levels of approximation, such as (4.198), and the most complete/complex theories lies a vast set of approaches. Instead of simply ignoring some of the other degrees of freedom, these 'intermediate' theories incorporate some of their effects into 'effective interactions'. The results involve, typically, nonlinear equations. We give two examples here.

An early example, prior to Dirac's QED, takes into account the 'self interaction' of the electron in, say, the hydrogen atom. The idea is that, since quantum mechanical description of the electron is in terms of a probability distribution, $P(\mathbf{x}) = |\psi(\mathbf{x})|^2$ (where we have dropped the t-dependence for simplicity), we are faced with a charge distribution associated with this electron,

$$\rho(\mathbf{x}) = -eP(\mathbf{x}) = -e\,|\psi(\mathbf{x})|^2\,. \qquad (4.199)$$

By the known laws of electrodynamics [Jac98], this charged cloud would generate an electric potential, at every point in space, $\mathbf{x}$, of the form [GZ01]

$$V_\rho(\mathbf{x}) = \int \frac{\rho(\mathbf{x}')}{|\mathbf{x} - \mathbf{x}'|}\,d\mathbf{x}'. \qquad (4.200)$$

But then this potential should affect the electron in a way no different than the potential due to the proton (that is, the e^2/r term in (4.198)). Inserting this extra potential into (4.198), we have

$$\mathrm{i}\partial_t\psi(\mathbf{x}, t) = \left[-\frac{1}{2}\Delta - \frac{e^2}{r} - eV_\rho(\mathbf{x})\right]\psi(\mathbf{x}, t), \qquad (4.201)$$

Combining (4.199), (4.200), and (4.201) we arrive at the NLS with cubic nonlinearity [GZ01]

$$i\partial_t \psi(\mathbf{x}, t) = \left[-\frac{1}{2}\Delta - \frac{e^2}{r} \right] \psi(\mathbf{x}, t) + e^2 \int \frac{\psi(\mathbf{x}, t) |\psi(\mathbf{x}', t)|^2}{|\mathbf{x} - \mathbf{x}'|} \, d\mathbf{x}'. \qquad (4.202)$$

Notice that the last term describes the electron's interaction with *itself*, as the source of the potential is the electronic cloud. We should emphasize that this equation was eventually abandoned for a number of reasons. One of them is that, as an approximation which keeps only the electronic degree of freedom (ψ) while ignoring the photonic degree of freedom, it is too crude to describe self interaction adequately. As mentioned above, quantum electrodynamics was developed and the final picture is not simply embodied in (4.202).

Another example, which is similar in form, comes from the study of structure of atoms with more than one electron. Instead of self interactions, the interest here is the *mutual* interactions between the electrons. Known as the Hartree–Fock approximation, with the photonic degrees of freedom still ignored, an ND system of NLS equations reads (see [Sch68]):

$$i\partial_t \psi_k(\mathbf{x}_k, t) = \left[-\frac{1}{2}\Delta_k - \frac{Ze^2}{|\mathbf{x}_k|} + \sum_{j \neq k} \int |\psi_j(\mathbf{x}_j)|^2 \frac{e^2}{r_{jk}} \, d\mathbf{x}_j \right] \psi_k(\mathbf{x}_k, t). \quad (4.203)$$

Here, $\psi_k(\mathbf{x}_k, t)$ is the wave function of the kth electron in an atom with N electrons (and N protons in the nucleus) and $r_{jk} = |\mathbf{x}_j - \mathbf{x}_k|$ is the distance between it and the jth electron. Notice that this is a system of N nonlinear integro-differential equations for the N unknowns $\psi_k(\mathbf{x}_k, t)$.

Both of these examples illustrate the use of nonlinear Schrödinger equations (up to cubic terms) in the context of quantum mechanical systems. Very similar equations are also used in the context of nonlinear optics. In fact, many of these are even closer in form to the ones we proposed, for example [Boy91],

$$\partial_t \mathcal{A}(t) = c_1 \mathcal{A} + c_3 |\mathcal{A}|^2 \mathcal{A}, \qquad (4.204)$$

where c_1 and c_3 are constants. Similarly, nonlinear Schrödinger equations of this type appear frequently in the study of solitons [Jac91]. Finally, known as the time-dependent Landau–Ginzburg equation (4.92), nonlinear systems of the form of (4.204) are ubiquitous in many areas of condensed matter physics.

4.6.4 A Compact Attractor for High-Dimensional NLS

In this section, following [Tao07b] we describe some asymptotic properties of bounded-energy solutions of nonlinear Schrödinger (NLS) equations

$$i u_t + \Delta u = F(u), \qquad (4.205)$$

with moderate (but possibly focusing) nonlinearity $F : \mathbb{C} \to \mathbb{C}$ and high dimension n; we allow the nonlinearity F to be focusing in nature as long as the energy remains bounded. The main result is to describe a certain *compact attractor* for the *NLS-flow*.

Assumptions on the Equation

We will consider NLS equations (4.205) which obey the following hypotheses:

- High dimension: $n \geq 5$.
- Hamiltonian structure: there exists a C^1 function $G : \mathbb{R}^+ \to \mathbb{R}$ with $G(0) = 0$ such that $F(z) = G'(|z|^2)z$ for all $z \in \mathbb{C}$.
- Power-type nonlinearity: there exists an exponent $p > 1$, a constant $C_0 > 0$ and a *Hölder regularity index* $0 < \theta \leq \min(p-1, 1)$ for which, for all $z, w \in \mathbb{C}$, we have the estimates [Tao07b]

$$|F(z)| \leq C_0 |z|^p, \qquad |F'(z)| \leq C_0 |z|^{p-1},$$
$$|F'(z) - F'(w)| \leq C_0 |z - w|^\theta (|z| + |w|)^{p-1-\theta}.$$

Here we view the differential $F'(z)$ of F at w as a real-linear map from $\mathbb{C}$ to $\mathbb{C}$.

- Mass-supercriticality: $p > 1 + \frac{4}{n}$.
- Energy-subcriticality: $p < 1 + \frac{4}{n-2}$.

Throughout this section we fix n, p, θ, C_0, F and we shall always assume the above hypotheses to be in effect. Also, all quantities in this section are implicitly assumed to depend on the dimension n, the exponent p, the Hölder regularity θ, and the constant C_0. Also, note that we have made no assumptions about the sign of the nonlinearity F or the potential function G. Important examples of NLS of the above type to keep in mind are:

- Quadratic case: $n = 5$ and $p = 2$.
- Coercive case: $\liminf_{x \to +\infty} G(x)/x^{(n+2)/n} \geq 0$.
- Defocussing case: $F(z) = +|z|^{p-1}z$ (this is coercive).
- Focusing case: $F(z) = -|z|^{p-1}z$ (this is non-coercive).

We also note the *completely integrable case* when $n = 1, p = 3$, and the nonlinearity is either focusing or defocussing; this case is not, strictly speaking, covered by the above hypotheses (the dimension is too low), but is better understood than most other NLS equations and serves as motivation for the soliton resolution conjecture which we discuss later.

The NLS equation (4.205) is manifestly translation invariant. The assumption of Hamiltonian structure also gives us the symmetries of *phase invariance* $u \mapsto e^{i\alpha}u$ and *Galilean invariance*

$$u(t, x) \mapsto e^{iv \cdot x/2}e^{-i|v|^2 t/4}u(t, x - vt), \tag{4.206}$$

for any $v \in \mathbb{R}$. The Hamiltonian structure also gives several conserved quantities, including the *mass*

$$M(u) := \int_{\mathbb{R}^n} |u(t, x)|^2 \, dx \tag{4.207}$$

and the *Hamiltonian*

$$H(u) := \int_{\mathbb{R}^n} \frac{1}{2}|\nabla u(t,x)|^2 + \frac{1}{2}G(|u(t,x)|^2)\,dx. \tag{4.208}$$

Throughout this section we shall be working in the *energy space* $H :=$ $H_x^1(\mathbb{R}^n \to \mathbb{C})$, which is a *Hilbert space* with inner product

$$\langle u,v\rangle_H := \int_{\mathbb{R}^n} u(x)\overline{v(x)} + \nabla u(x) \cdot \overline{\nabla v(x)}\,dx.$$

Assumptions on the Solution

We shall only consider *bounded-energy solutions* to (4.205), although we allow this energy bound to be arbitrarily large. More precisely, we have

A *bounded-energy strong solution* to (4.205), or *solution* for short, will be any function $u \in C_t^0 H_x^1(I \times \mathbb{R}^n)$ on a non-empty time interval $I \subset \mathbb{R}$ taking values continuously in the energy space H such that [Tao07b]

$$u(t_1) = e^{i(t_1-t_0)\Delta}u(t_0) - i\int_{t_0}^{t_1} F(u(t))\,dt,$$

for all $t_0, t_1 \in I$, where we of course adopt the convention that $\int_{t_0}^{t_1} = -\int_{t_1}^{t_0}$ if $t_1 < t_0$, and such that the *energy*

$$E(u) := \sup_{t \in I} \|u(t)\|_H^2 \tag{4.209}$$

is finite. Here $e^{it\Delta}$ is the free Schrödinger propagator, defined via the Fourier transform

$$\hat{f}(\xi) := \int_{\mathbb{R}^n} e^{-ix\cdot\xi}f(x)\,dx \quad \text{by}$$

$$\widehat{e^{it\Delta}f}(\xi) := e^{-it|\xi|^2}\hat{f}(\xi), \tag{4.210}$$

or more directly as

$$e^{it\Delta}f(x) = \frac{1}{(4\pi it)^{n/2}} \int_{\mathbb{R}^n} e^{i|x-y|^2/4t}f(y)\,dy. \tag{4.211}$$

We say that a solution is *forward-global* if I contains $[0, +\infty)$, and *global* if $I = \mathbb{R}$.[56]

We make the trivial observation that the restriction of any solution to a sub-interval is still a solution. Also observe that the propagators $e^{it\Delta}$ are unitary on H.

We list some basic facts about solutions below. Define an exponent pair (q,r) to be *admissible* if $\frac{2}{q} + \frac{n}{r} = \frac{n}{2}$ and $2 \le q, r \le \infty$.

We have the following *theorem on local existence and uniqueness* [Tao07b]:

---

[56] One can of course talk about backward-global solutions, which contain $(-\infty, 0]$, but because of the time-reversal symmetry $u(t,x) \mapsto \overline{u(-t,x)}$ all the results here for forward-global solutions immediately have counterparts for backward-global solutions (and hence, by concatenation, for global solutions).

- Local existence: If $t_0 \in R$ and $B \subset H$ is bounded, then there exists an open time interval I containing t_0 such that for every $u_0 \in B$ there exists a solution $u : I \to H$ such that $u(t_0) = u_0$. Furthermore the map $u_0 \mapsto u(t)$ is Lipschitz continuous on B for all $t \in I$. If B is a sufficiently small neighborhood of the origin, one can take $I = R$.
- Uniqueness: If two solutions $u : I \to H$, $\tilde{u} : I \to H$ agree on at least one time, then they are equal for all time.
- Strichartz regularity: If $u : I \to H$ is a solution, J is a compact sub-interval of I, and (q, r) is an admissible pair, then $u, \nabla u \in L_t^q L_x^r (J \times \mathbb{R}^n)$.
- Finite time blowup condition: If a solution $u : I \to H$ with finite future endpoint $T_+ := \sup I < +\infty$ cannot be extended beyond T_+, then $\|u(t)\|_H$ goes to infinity as $t \to T_+$ from below. Similarly for solutions which cannot be extended beyond their finite past endpoint $T_- = \inf I > -\infty$.
- Conservation laws: The mass $M(u(t))$ and Hamiltonian $H(u(t))$ are constant for $t \in I$.

We define the nonlinear flow maps $S(t)$ on H for $t \in \mathbb{R}$ by setting $S(t)u(0) := u(t)$ whenever $u : [0, t] \times \mathbb{R}^n \to \mathbb{C}$ is a solution. These maps are not necessarily globally defined (except in the coercive case), but from the above theorem we see that they are continuous and obey the group law $S(t)S(t') = S(t + t')$ on their domain of definition, and for any bounded set in H the $S(t)$ are defined for all sufficiently small t.

The Soliton Resolution Conjecture

Suppose we have a forward-global solution $u : I \to H$ to the NLS (4.205). A natural question then arises as to what the asymptotic behavior of $u(t)$ is as $t \to +\infty$. In the case of small energy or defocussing nonlinearity, the answer is known as the Scattering theorem: Let $u_0 \in H$ and $t_0 \in \mathbb{R}$, and assume either that $\|u_0\|_H$ is sufficiently small, or that the nonlinearity is defocussing. Then there is a unique global solution $u : \mathbb{R} \to H$ with $u(t_0) = u_0$. Furthermore there is a unique *scattering state* $u_+ \in H$ such that

$$\lim_{t \to +\infty} \|u(t) - e^{it\Delta} u_+\|_H = 0.$$

For the proof, see [Tao07b].

Equivalently, we may write

$$u(t) = S(t - t_0)u_0 = e^{it\Delta} u_+ + o_H(1), \qquad (4.212)$$

where we use $o_H(1)$ to denote a time-dependent function which goes to zero in H norm as $t \to +\infty$.

In the focusing case and with large data u_0, the above theorem fails for at least two reasons. Firstly, as mentioned earlier, the solution can blow up in finite time, especially if the Hamiltonian is negative [Gla77, OT91]. Secondly, even if the solution remains global (or at least forward-global), it does not necessary scatter to a free solution $e^{it\Delta}u_+$. This can be seen by considering *stationary soliton solutions* of the form $u(t,x) = Q(x)e^{i\omega t}$, where $\omega > 0$ is a constant and $Q \in H$ solves the elliptic equation

$$\Delta Q + |Q|^{p-1}Q = \omega Q. \tag{4.213}$$

One can also apply Galilean symmetry (4.206) to create traveling soliton solutions, which at time t would be localized near $x_0 + vt$ for some $x_0, v \in \mathbb{R}^n$. Furthermore, it is possible to create *multi-soliton* solutions which as $t \to +\infty$ resemble superpositions of J divergent traveling solitons for any given $J \geq 1$ [Per97a, MMT06], for some constructions of such solutions for various choices of n, p, F. Finally, it is possible in some cases to superimpose a free solution $e^{it\Delta}u_+$ with a soliton or multi-soliton solution, at least for sufficiently late times.

It is tentatively conjectured that the above behavior is in fact generic. This leads to an (imprecise) *soliton resolution conjecture*, that for 'generic' large global solutions, the evolution asymptotically decouples into the superposition of divergent solitons, a free radiation term $e^{it\Delta}u_+$, and an error which goes to zero at infinity (see (4.212)). We leave questions such as the regularity and decay class of the solution, the sense in which the error goes to zero, and the definition of 'generic' as deliberately vague. Indeed, our understanding of this conjecture is still very poor (even with strong additional assumptions such as spherical symmetry and coercive nonlinearity), with the majority of results being concentrated either on the small data or defocussing cases (in which no solitons are present), or when the solution is very close to a soliton or multi-soliton solution, especially if the solitons are generated by a ground state. See [Sof06, Tao04] for some further discussion of this conjecture (referred to there as the *grand conjecture*).

As just mentioned, there is little direct progress on the soliton resolution conjecture for generic large data (not close to any soliton or multi-soliton). However one can consider weakening the conjecture by asking instead for an asymptotic resolution into a free solution $e^{it\Delta}u_+$, an error, and some sort of 'pseudo-multi-soliton' which exhibits behavior similar to that of a multi-soliton. This type of conjecture is easiest to formalize in the case of spherically symmetric solutions, in which traveling solitons are precluded and the only multi-soliton which is expected to be relevant is that of a single soliton placed at the origin. But in principle one could also imagine multiple solitons of different amplitudes and widths all superimposed on each other at the origin, or more generally some sort of exotic *breather solution* which is periodic or almost periodic, but which does not have the explicit form $Q(x)e^{i\omega t}$. While such solutions are expected to be very unstable, and in fact probably do not exist for most nonlinearities, we do not know how to rule them out with present technology. Thus we can try to weaken the conjecture in this case by allowing the pseudo-soliton component to merely be almost periodic in time, rather than be an actual soliton. As we shall see, this weakened statement is related to the *petite conjecture* in [Sof06].

Radial Case

The first set of results establishes the petite conjecture in the spherically symmetric case, by showing the existence of a compact attractor for the non-radiating component of the evolution. More precisely, we have the Compact attractor theorem (spherically symmetric case): Let $E > 0$. Then there exists a compact subset $\mathcal{K}_{E,\text{rad}} \subset H$ which is invariant under the NLS flow (thus $S(t)$ is well-defined and is a homeomorphism on $\mathcal{K}_{E,\text{rad}}$ for all $t \in \mathbb{R}$), and such that for every spherically symmetric forward-global solution u of energy at most E, there exists a unique *radiation state* $u_+ \in H$ such that [Tao07b]

$$\lim_{t \to +\infty} \text{dist}_H(u(t) - e^{it\Delta}u_+, \mathcal{K}_{E,\text{rad}}) = 0. \qquad (4.214)$$

We write

$$\text{dist}_H(f, K) := \inf\{\|f - g\|_H : g \in K\}$$

for the distance between f and K.

Thus $\mathcal{K}_{E,\text{rad}}$ is a compact attractor for spherically symmetric solutions of energy at most E, once the effect of the radiation term $e^{it\Delta}u_+$ is removed.[57] In other words, for spherically symmetric forward-global solutions u of energy at most E, we have a decomposition[58] of the form

$$u(t) = e^{it\Delta}u_+ + w(t) + o_H(1) \qquad (4.215)$$

where $w(t)$ ranges in the fixed compact set $\mathcal{K}_{E,\text{rad}}$ for all times t. Note that we do not assert that w itself evolves by NLS (which would make w an almost periodic solution); the problem is that the radiation terms may cause significant long-term drift in the 'secular modes' or 'modulation parameters' of the compact attractor $\mathcal{K}_{E,\text{rad}}$. In high dimension one expects that the strong dispersive properties of the equation will in fact rule out this scenario, but we were unable to do so here (it seems to require a linearized stability analysis of the almost periodic solutions, which we do not know how to do).[59]

---

[57] It is essential that we remove *radiation*, otherwise the concept of a compact attractor is incompatible with the time-reversibility of the NLS equation.

[58] This decomposition can be regarded as a nonlinear analogue of the spectral decomposition of a linear Schrödinger operator with potential into continuous (dispersive) and pure point (almost periodic) components. With this perspective, the point of the high dimension hypothesis $d \geq 5$ is to rule out *nonlinear resonances*.

[59] This theorem provides no information about the *rate* of convergence to the compact attractor. Indeed we expect this rate of convergence to be highly non-uniform, depending in a discontinuous way on the initial data $u(0)$. To give an example in the focusing case, suppose $u(0)$ was equal to the ground state Q (which is known to be orbitally unstable for the range of exponents p under consideration, see e.g. [SS85]). Then $u(t)$ will lie in the circle $\{e^{i\alpha}Q : \alpha \in \mathbb{R}\}$, which we have already observed to lie in $\mathcal{K}_{E,\text{rad}}$. If however we perturb the initial data $u(0)$ by a small amount (of size ε in the H norm, say), then a typical scenario would then be that after a relatively long

As one consequence of the above theorems we obtain the *petite conjecture* of [Sof06] in the radial high-dimensional case: A solution $u : I \times \mathbb{R}^n \to \mathbb{C}$ is *almost periodic* if the orbit $\{u(t) : t \in I\}$ is pre-compact in H. For example, the global soliton solution $u(t,x) = Q(x)e^{it}$ to a focusing NLS is almost periodic, as is any translate or rescaling of this soliton solution. Any other hypothetical periodic or quasi-periodic *breather solution* to an NLS would also qualify as being almost periodic. If one applies a Galilean transformation to give such solitons or breathers a non-zero velocity, then the solution is no longer almost periodic. Since the set $\mathcal{K}_{E,\mathrm{rad}}$ is compact and invariant, any initial data $u(0)$ in $\mathcal{K}_{E,\mathrm{rad}}$ gives rise to a global almost periodic solution. Thus if we could demonstrate that the only spherically symmetric almost periodic solutions were soliton solutions, the above theorem would yield the soliton resolution conjecture in the spherically symmetric case.

Let u be a spherically symmetric forward-global solution, and let u_+ be the radiation state. Then the following are equivalent [Tao07b]:

- u is almost periodic.
- $u_+ = 0$ (i.e. u is future non-radiating).
- u is spatially localized in the sense that

$$\lim_{R \to +\infty} \limsup_{t \to +\infty} \int_{|x|>R} |u(t,x)|^2 \, dx = 0.$$

- u is spatially localized in the sense that

$$\lim_{R \to +\infty} \limsup_{t \to +\infty} \int_{|x|>R} |u(t,x)|^2 + |\nabla u(t,x)|^2 \, dx = 0.$$

Our methods actually give an explicit rate of decay for the spatial localization, thus if u is an almost-periodic forward-global spherically symmetric solution of energy at most E then we have

$$\limsup_{t \to +\infty} \int_{|x|>R} |u(t,x)|^2 + |\nabla u(t,x)|^2 \, dx \le c_E(R)$$

for some explicit quantity $c_E(R)$ which goes to zero as $R \to \infty$. It would be of interest to obtain a good bound for this rate of decay, such as a polynomial decay $R^{-\varepsilon}$. Based on the observation that solitons are rapidly decreasing in space, one might even hope to get much more rapid decay, i.e. $O_N(R^{-N})$ for all $N > 0$. Unfortunately our methods here only give a much weaker decay, something like $1/\log^c R$. One important milestone might be to obtain a decay

---

time (e.g. of time $\log \frac{1}{\varepsilon}$, or perhaps ε^{-c} for some $c > 0$) the solution would eventually move away from this circle, and would most likely collapse entirely into radiation. Thus we see that the time required to reach the asymptotic state can be arbitrarily large as $\varepsilon \to 0$, leading to a discontinuity in the decay rates in the above theorem. In particular, we do *not* expect the compact attractor $\mathcal{K}_{E,\mathrm{rad}}$ to be orbitally stable.

better than $1/R^2$ for the mass density $|u(t,x)|^2$, as this would then place the weakly bound component of the solution in the *scattering space*

$$\Sigma = \{u : xu, u, \nabla u \in L^2_x(\mathbb{R}^n)\}$$

and allow for tools such as the pseudo-conformal identity to be applied.

General Case

We now turn to the general case, in which no spherical symmetry is assumed. The key difficulty here is that the class of solutions of energy E is now translation-invariant, and so the notion of almost periodicity needs to be replaced by a more general notion which is both translation-invariant and also closed under certain 'superposition' operations.

We now have the following *symmetry group* [Tao07b]: Given any $h \in \mathbb{R}^n$, we let $\tau_h : H \to H$ be the *unitary shift operator*

$$\tau_h f(x) := f(x - h),$$

and we let $G := \{\tau_h : h \in \mathbb{R}^n\}$ be the associated translation group. Note that this is a *non-compact Lie group* and so there is a well-defined notion of a sequence of group elements g_n going to infinity, indeed we have $\tau_{h_n} \to \infty$ iff $|h_n| \to \infty$. Given any set $K \subset H$, we let

$$GK := \{gf : g \in G, f \in K\}$$

be the *orbit* of K under G.

The G-invariance of the problem means in particular that the set K_E, being G-invariant, can no longer be compact (unless it consists only of $\{0\}$). One might still hope that K_E is an attractor in the sense of (4.214), but this can be easily seen to be false (at least in the focusing NLS) by taking a stationary soliton solution $Q(x)e^{it}$ and applying a Galilean transform to create a traveling soliton which is not almost periodic. The orbit of this traveling soliton is still almost periodic once one quotients out by the group G, so one might think to extend K_E to cover solutions which are 'almost periodic modulo G'. However, this is still not enough, as can be seen (at least heuristically) by considering *multi-soliton solutions*—the superposition of two or more diverging solitons. See [MMT06] for details of how to construct such solutions forward-globally in time. Observe that such solutions are not almost periodic even after quotienting out by G; in other words, there is no compact set $K \subset H$ such that the orbit $\{u(t) : t \in [0, +\infty)\}$ is contained in GK. Thus a 'concentration compactness' style definition of almost periodicity is needed, in order to account for the fact that solutions may be a superposition of components, each of which lives in a compact set after quotienting out by a different element of G. For more technical details, see [Tao07b].

4.6.5 Finite-Difference Scheme for NLS

The $(1 + 1)$-NLS equation (1.4) can be numerically integrated (for the given initial and boundary conditions) using the following finite-difference scheme (see Appendix for implementation). If we divide the x-axis into N mesh points so that x and t are represented as:

$$x_j = j\Delta x, \qquad t_n = n\Delta t,$$

(where j varies from $-N/2$ to $+N/2$), (1.4) can be rewritten in the finite-difference form [BK05]

$$i\frac{\psi(x, t + \Delta t) - \psi(x, t)}{\Delta t}$$
$$= -\frac{\psi(x + \Delta x, t) - 2\psi(x, t) + \psi(x - \Delta x, t)}{2\Delta x^2} + V(x)\psi(x, t). \quad (4.216)$$

If we now apply the following conventions:

$$\psi(x_j, t_n + \Delta t) := \psi_j^{n+1}, \qquad \psi(x_j, t_n) := \psi_j^n, \qquad \psi(x_j - \Delta x, t_n t) := \psi_{j-1}^n,$$

the finite-difference equation (4.216) becomes

$$\psi_j^{n+1} = \psi_j^n + i\Delta t \left(\frac{\psi_{j+1}^n - 2\psi_j^n + \psi_{j-1}^n}{2\Delta x^2} - V_j\psi_j^n \right). \quad (4.217)$$

Finally, (4.217) can be rewritten in a matrix form as:

$$F_{n+1} = F_n - i\Delta t\hat{H}F_n = UF_n \qquad \text{where } U = 1 - i\Delta t\hat{H}, \quad (4.218)$$

while the Hamiltonian operator $\hat{H}$ is defined (as usual) by: $\hat{H} = -\partial_{x^2} + V(x)$. Since it is required that the norm of F is $F^*F = 1$, U must be an orthonormal operator; however, as U in (4.218) does not have such a property, we need to impose the normalization after every step in the simulation (for more details, see [BK05]).

4.6.6 Method of Lines for NLS

Assuming that we already have a solid Runge–Kutta(–Fehlberg) integrator for systems of ODEs (see Appendix), we can implement the powerful *method of lines* as follows.

Backward Heat Equation

To start with, consider the *backward heat equation*:

$$\frac{\partial \psi}{\partial t} = -\frac{1}{2}\frac{\partial^2 \psi}{\partial x^2}, \quad (4.219)$$

where $\psi = \psi(x,t)$ is dependent real-valued variable. Solution of this PDE can be found if we have given one initial condition:

$$\text{IC:} \quad \psi(x, t = 0) = \psi_0$$

and two boundary conditions:

$$\text{BC:} \quad \psi(x = x_0, t) = \psi_b, \qquad \frac{\partial \psi(x = x_f, t)}{\partial x} = 0.$$

A commonly used second order, central finite difference approximation for $\frac{\partial^2 \psi}{\partial x^2}$ is

$$\frac{\partial^2 \psi}{\partial x^2} \approx \frac{\psi_{k+1} - 2u_k + \psi_{k-1}}{\Delta x^2},$$

where k is an index designating a position along a grid in x which has M (e.g. 10) points and Δx is the spacing in x along the grid. The initial PDE (4.219) can be approximated as a system of ODEs

$$\frac{d\psi_k}{dt} = -\frac{1}{2}\frac{\psi_{k+1} - 2\psi_k + \psi_{k-1}}{\Delta x^2}, \quad (k = 1, 2, \ldots, M = \# \text{ of ODEs}). \quad (4.220)$$

The set of ODEs (4.220) is then integrated, using the Runge–Kutta integrator, subject to IC and BC.

Introducing Potential $V(x)$

Once we can solve the set of ODEs (4.220), we will add some potential term:

$$\frac{\partial \psi}{\partial t} = -\frac{1}{2}\frac{\partial^2 \psi}{\partial x^2} + V(x)\psi,$$

to get the set of approximating ODEs:

$$\frac{d\psi_k}{dt} = -\frac{1}{2}\frac{\psi_{k+1} - 2\psi_k + \psi_{k-1}}{\Delta x^2} + V(k)\psi_k, \quad (k = 1, 2, \ldots, M = \# \text{ of ODEs}). \quad (4.221)$$

Linear Schrödinger Equation

Once we can solve the set of ODEs (4.221), we move into the complex plane, by introducing the imaginary unit:

$$\mathrm{i}\frac{\partial \psi}{\partial t} = -\frac{1}{2}\frac{\partial^2 \psi}{\partial x^2} + V(x)\psi,$$

to get the set of approximating complex-valued ODEs:

$$\mathrm{i}\frac{d\psi_k}{dt} = -\frac{1}{2}\frac{\psi_{k+1} - 2\psi_k + \psi_{k-1}}{\Delta x^2} + V(k)\psi_k, \quad (k = 1, 2, \ldots, M). \quad (4.222)$$

Computing NLS

Once we can solve the set of complex-valued ODEs (4.222), we replace the linear term with the cubic nonlinearity:

$$\mathrm{i}\frac{\partial\psi}{\partial t} = -\frac{1}{2}\frac{\partial^2\psi}{\partial x^2} + V(x)\psi|\psi|^2,$$

to get the set of approximating complex-valued ODEs:

$$\mathrm{i}\frac{d\psi_k}{dt} = -\frac{1}{2}\frac{\psi_{k+1} - 2\psi_k + \psi_{k-1}}{\Delta x^2} + V(k)\psi_k|\psi_k|^2, \quad (k = 1, 2, \dots, M). \quad (4.223)$$

Set of Weakly-Coupled NLS Equations

Once we can solve the set of complex-valued nonlinear ODEs (4.223), we can put them into another loop to get a set of decoupled NLS equations. Finally, to implement their weak coupling in a common adaptive potential $V(w, x)$, we will have to implement also a simple neural network to iteratively update the weights w_k. The obtained result is popularly called: "*network of networks, or QNN of QNNs.*"

5

Quantum Brain and Cognition

In this Chapter we present several aspects of the current research in quantum brain.

5.1 Biochemistry of Microtubules

Recent developments/efforts to understand aspects of the brain function at the *sub-neural* level are discussed in [Nan95]. Microtubules (MTs), protein polymers constructing the cytoskeleton of a neuron, participate in a wide variety of dynamical processes in the cell. Of special interest for this subsection is the MTs participation in bio-information processes such as *learning* and *memory*, by possessing a well-known binary error-correcting code $[K_1(13, 2^6, 5)]$ with 64 words. In fact, MTs and DNA/RNA are *unique* cell structures that possess a code system. It seems that the MTs' code system is strongly related to a kind of *mental code* in the following sense. The MTs' periodic paracrystalline structure make them able to support a *superposition* of coherent quantum states, as it has been recently conjectured by Hameroff and Penrose [HP96a], representing an *external* or *mental order*, for sufficient time needed for *efficient quantum computing* [II07b, II07d].

Living organisms are collective assemblies of cells which contain collective assemblies of organized material, including membranes, organelles, nuclei, and the *cytoplasm*, the bulk interior medium of living cells. Dynamic rearrangements of the cytoplasm within *eucaryotic cells*, the cells of all animals and almost all plants on Earth, account for their changing shape, movement, etc. This extremely important cytoplasmic structural and dynamical organization is due to the presence of networks of interconnected protein polymers, which are referred to as the *cytosceleton* due to their bone-like structure [HP96a, Dus84]. The cytoskeleton consists of MT's, actin micro-filaments, intermediate filaments and an *organizing complex*, the *centrosome* with its chief component the *centriole*, built from two bundles of microtubules in a separated **T** shape. Parallel-arrayed MTs are interconnected by cross-bridging proteins

V.G. Ivancevic, T.T. Ivancevic, *Quantum Neural Computation*,
Intelligent Systems, Control and Automation: Science and Engineering 40,
DOI 10.1007/978-90-481-3350-5_5, © Springer Science+Business Media B.V. 2010

(*MT-Associated Proteins*: MAPs) to other MTs, organelle filaments and membranes to form *dynamic networks* [HP96a, Dus84]. MAPs may be contractile, structural, or enzymatic. A very important role is played by contractile MAPs, like dynein and kinesin, through their participation in cell movements as well as in intra-neural, or axoplasmic transport which moves material and thus is of fundamental importance for the *maintenance* and *regulation* of *synapses* (see, e.g., [Ecc64]). The structural bridges formed by MAPs stabilize MTs and prevent their disassembly. The MT-MAP 'complexes' or *cyto-sceletal networks* determine the cell architecture and dynamic functions, such a *mitosis*, or *cell division, growth, differentiation, movement*, and for us here the very crucial, *synapse formation and function*, all essential to the living state. It is usually said that *microtubules* are ubiquitous through the entire biology [HP96a, Dus84].

MTs are hollow cylinders comprised of an exterior surface of cross-section diameter 25 nm (1 nm $= 10^{-9}$ meters) with 13 arrays (protofilaments) of protein dimers called tubulines [Dus84]. The interior of the cylinder, of cross-section diameter 14 nm, contains *ordered water* molecules, which implies the existence of an electric dipole moment and an electric field. The arrangement of the dimers is such that, if one ignores their size, they resemble triangular lattices on the MT surface. Each dimer consists of two hydrophobic protein pockets, and has an unpaired electron. There are two possible positions of the electron, called α and β *conformations*. When the electron is in the β-conformation there is a 29° distortion of the electric dipole moment as compared to the α conformation.

In standard models for the simulation of the MT dynamics [STZ93], the 'physical' DOF—relevant for the description of the energy transfer—is the projection of the electric dipole moment on the longitudinal symmetry axis (x-axis) of the MT cylinder. The 29° distortion of the β-conformation leads to a displacement u_n along the x-axis, which is thus the relevant physical DOF.

There has been speculation for quite some time that MTs are involved in information processing: it has been shown that the particular geometrical arrangement (packing) of the tubulin proto-filaments obeys an error-correcting mathematical code known as the $K_2(13, 2^6, 5)$-code [KHS93]. Error correcting codes are also used in classical computers to protect against errors while in quantum computers special error correcting algorithms are used to protect against errors by preserving quantum coherence among qubits.

Information processing occurs via interactions among the MT proto-filament chains. The system may be considered as similar to a model of *interacting Ising chains* on a triangular lattice, the latter being defined on the plane stemming from filleting open and flattening the cylindrical surface of MT. Classically, the various dimers can occur in either α or β conformations. Each dimer is influenced by the neighboring dimers resulting in the possibility of a transition. This is the basis for classical information processing, which constitutes the picture of a (classical) cellular automaton.

5.2 Kink Soliton Model of MT-Dynamics

The *quantum nature* of an MT network results from the *assumption* that each dimer finds itself in a *superposition* of α and β conformations. Viewed as a *two-state quantum mechanical system*, the MT tubulin dimers couple to conformational changes with 10^{-9}–10^{-11} s transitions, corresponding to an angular frequency $\omega \sim \mathcal{O}(10^{10})$–$\mathcal{O}(10^{12})$ Hz [Nan95].

The *quantum computer* character of the MT network [Pen89] results from the assumption that each dimer finds itself in a superposition of α and β conformations [Ham87]. There is a macroscopic coherent state among the various chains, which lasts for $\mathcal{O}(1$ sec) and constitutes the 'preconscious' state [Nan95]. The interaction of the chains with (non-critical stringy) quantum gravity, then, induces self-collapse of the wave function of the coherent MT network, resulting in quantum computation.

In [EMN92, EMN99a, MN95a, MN95b, Nan95] the authors assumed that the collapse occurs mainly due to the interaction of each chain with quantum gravity, the interaction from neighboring chains being taken into account by including mean-field interaction terms in the dynamics of the displacement field of each chain. This amounts to a modification of the effective potential by anharmonic oscillator terms. Thus, the effective system under study is 2D, possessing one space and one time coordinate.

Let u_n be the displacement field of the nth dimer in a MT chain. The continuous approximation proves sufficient for the study of phenomena associated with energy transfer in biological cells, and this implies that one can make the replacement

$$u_n \rightarrow u(x,t), \qquad (5.1)$$

with x a spatial coordinate along the longitudinal symmetry axis of the MT. There is a time variable t due to fluctuations of the displacements $u(x)$ as a result of the dipole oscillations in the dimers.

The effects of the neighboring dimers (including neighboring chains) can be phenomenologically accounted for by an effective potential $V(u)$. In the kink-soliton model[1] of [STZ93] a double-well potential was used, leading to a classical kink solution for the $u(x,t)$ field. More complicated interactions are allowed in the picture of Ellis *et al.*, where more generic polynomial potentials have been considered.

The effects of the surrounding water molecules can be summarized by a *viscous force* term that damps out the dimer oscillations,

$$F = -\gamma \partial_t u, \qquad (5.2)$$

with γ determined phenomenologically at this stage. This friction should be viewed as an environmental effect, which however does not lead to energy

---

[1] Recall that kinks are solitary (non-dispersive) waves arising in various 1D (bio)physical systems.

dissipation, as a result of the non-trivial solitonic structure of the ground-state and the non-zero constant force due to the electric field. This is a well known result, directly relevant to energy transfer in biological systems.

In mathematical terms, the effective equation of motion for the relevant field DOF $u(x,t)$ reads:

$$u''(\xi) + \rho u'(\xi) = P(u), \tag{5.3}$$

where $\xi = x - vt$, $u'(\xi) = du/d\xi$, v is the velocity of the soliton, $\rho \propto \gamma$ [STZ93], and $P(u)$ is a polynomial in u, of a certain degree, stemming from the variations of the potential $V(u)$ describing interactions among the MT chains. In the mathematical literature there has been a classification of solutions of equations of this form. For certain forms of the potential the solutions include *kink solitons* that may be responsible for dissipation-free energy transfer in biological cells:

$$u(x,t) \sim c_1 \left(\tanh[c_2(x - vt)] + c_3\right), \tag{5.4}$$

where c_1, c_2, c_3 are constants depending on the parameters of the dimer lattice model. For the form of the potential assumed in the model of [STZ93] there are solitons of the form $u(x,t) = c_1' + \frac{c_2' - c_1'}{1 + e^{c_3'(c_2' - c_1')(x - vt)}}$, where again c_i', $i = 1, \ldots, 3$ are appropriate constants.

A *semiclassical quantization* of such solitonic states has been considered by Ellis *et al.* The result of such a quantization yields a *modified soliton equation* for the (quantum corrected) field $u_q(x,t)$ [TF91]

$$\partial_t^2 u_q(x,t) - \partial_x^2 u_q(x,t) + \mathcal{M}^{(1)}[u_q(x,t)] = 0, \tag{5.5}$$

with the notation

$$M^{(n)} = e^{\frac{1}{2}(G(x,y,t) - G_0(x,y))\frac{\partial^2}{\partial z^2}} U^{(n)}(z)\big|_{z = u_q(x,t)}, \qquad U^{(n)} \equiv d^n U/dz^n.$$

The quantity U denotes the potential of the original soliton Hamiltonian, and $G(x,y,t)$ is a bilocal field that describes quantum corrections due to the modified boson field around the soliton. The quantities $M^{(n)}$ carry information about the quantum corrections. For the kink soliton (5.4) the quantum corrections (5.5) have been calculated explicitly in [TF91], thereby providing us with a concrete example of a large-scale quantum coherent state.

A typical propagation velocity of the kink solitons (e.g., in the model of [STZ93]) is $v \sim 2$ m/s, although, models with $v \sim 20$ m/s have also been considered. This implies that, for moderately long microtubules of length $L \sim 10^{-6}$ m, such kinks transport energy without dissipation in

$$t_F \sim 5 \times 10^{-7} \text{ s.} \tag{5.6}$$

Such time scales are comparable to, or smaller in magnitude than, the decoherence time scale of the above-described coherent (solitonic) states $u_q(x,t)$. This implies the possibility that fundamental quantum mechanical phenomena may then be responsible for frictionless energy (and signal) transfer across microtubular arrangements in the cell [Nan95].

5.3 Fractal Neurodynamics

Neuro- and psycho-dynamics have its physical behavior both on the macroscopic, classical, inter-neuronal level [IB05, II07d], and on the microscopic, quantum, intra-neuronal level [IA07, II07b, II08b]. On the macroscopic level, various models of neural networks (NNs, for short) have been proposed as goal-oriented models of the specific neural functions, like for instance, function-approximation, pattern-recognition, classification, or control. In the physically-based, Hopfield-type models of NNs [Hop82, Hop84] the information is stored as a content-addressable memory in which synaptic strengths are modified after the Hebbian rule (see [Heb49]). Its retrieval is made when the network with the symmetric couplings works as the point-attractor with the fixed points. Analysis of both activation and learning dynamics of Hopfield–Hebbian NNs using the techniques of statistical mechanics (see, e.g. [II08a]), provides us with the most important information of storage capacity, role of noise and recall performance [II06b, II07d].

Conversely, an indispensable role of quantum theory in the brain dynamics was emphasized in [Kur05]. On the general microscopic intra-cellular level, energy transfer across the cells, without dissipation, had been first conjectured to occur in biological matter by [FK83]. The phenomenon conjectured by them was based on their 1D superconductivity model: in one dimensional electron systems with holes, the formation of solitonic structures due to electron-hole pairing results in the transfer of electric current without dissipation. In a similar manner, Frölich and Kremer conjectured that energy in biological matter could be transferred without dissipation, if appropriate solitonic structures are formed inside the cells. This idea has lead theorists to construct various models for the energy transfer across the cell, based on the formation of kink classical solutions [STZ93, SZT98].

The interior of living cells is structurally and dynamically organized by cytoskeletons, i.e., networks of protein polymers. Of these structures, microtubules (MTs, for short) appear to be the most fundamental [Dus84]. Their dynamics has been studied by a number of authors in connection with the mechanism responsible for dissipation-free energy transfer. Hameroff and his colleagues [Ham94, HP96a, Ham98, HHT02] have conjectured another fundamental role for the MTs, namely being responsible for quantum computations in the human neurons. Penrose [Pen89, Pen94, Pen97, Pen98] further argued that the latter is associated with certain aspects of quantum theory that are believed to occur in the cytoskeleton MTs, in particular quantum superposition and subsequent collapse of the wave function of coherent MT networks. These ideas have been elaborated by [MN95b] and [Nan95], based on the quantum-gravity language of [EMN99b], where MTs have been physically modeled as non-critical (SUSY) bosonic strings. It has been suggested that the neural MTs are the microsites for the emergence of stable, macroscopic quantum coherent states, identifiable with the preconscious states; stringy-quantum space-time effects trigger an organized collapse of the coherent states down to a specific or conscious state. More recently, the evidence for biological self-organization and pattern formation during embryogenesis was presented in [TVP99].

In particular, MTs in the cytoskeletons of eukaryotic cells provide a wide range of micro-skeletal and micro-muscular functionalities. Some evidence has indicated that they can serve as a medium for intracellular signaling processing. For the inherent symmetry structures and the electric properties of tubulin dimers, the microtubule (MT) was treated as a 1D ferroelectric system in [CQD05]. The nonlinear dynamics of the dimer electric dipoles was described by virtue of the double-well potential and the physical problem was further mapped onto the pseudo-spin system, taking into account the effect of the external electric field on the MT.

More precisely, MTs are polymers of tubulin subunits (dimers) arranged on a hexagonal lattice. Each tubulin dimer comprises two monomers, the α-tubulin and β-tubulin, and can be found in two states. In the first state a mobile negative charge is located into the α-tubulin monomer and in the second into the β-tubulin monomer. Each tubulin dimer is modeled as an electrical dipole coupled to its neighbors by electrostatic forces. The location of the mobile charge in each dimer depends on the location of the charges in the dimer's neighborhood. Mechanical forces that act on the microtubule affect the distances between the dimers and alter the electrostatic potential. Changes in this potential affect the mobile negative charge location in each dimer and the charge distribution in the microtubule. The net effect is that mechanical forces affect the charge distribution in microtubules [KL07].

Various models of the mind have been based on the idea that neuron MTs can perform computation. From this point of view, information processing is the fundamental issue for understanding the brain mechanisms that produce consciousness. The cytoskeleton polymers could store and process information through their dynamic coupling mediated by mechanical energy. The problem of information transfer and storage in brain microtubules was analyzed in [FPR06], considering them as a communication channel.

Therefore, we have two space-time biophysical scales of neuro- and psycho-dynamics: classical and quantum. Naturally the question arises: are these two scales somehow inter-related, is there a space-time self-similarity between them?

The purpose of the present paper is to prove the formal positive answer to the self-similarity question. We try to describe neurodynamics on both physical levels by the unique form of a single equation, namely open Liouville equation: NN-dynamics using its classical form, and MT-dynamics using its quantum form in the Heisenberg picture. If this formulation is consistent, that would prove the existence of the formal neuro-biological space-time self-similarity. Even more, this would prove the existence of a Neurodynamical Law, which acts on different scales of brain's functioning.

5.3.1 Open Liouville Equation

Hamiltonian Framework

Suppose that on the macroscopic NN-level we have a conservative Hamiltonian system acting in a $2ND$ symplectic phase space $T^*Q = \{q^i(t), p_i(t)\}$, $i = 1, \ldots, N$ (which is the cotangent bundle of the NN-configuration manifold $Q = \{q^i\}$), with a Hamiltonian function $H = H(q^i, p_i, t) : T^*Q \times \mathbb{R} \to \mathbb{R}$ (see [Iva04, II06a, II06c]). The conservative dynamics is defined by classical Hamilton's canonical equations:

$$\begin{aligned} \dot{q}^i &= \partial_{p_i} H \quad \text{contravariant velocity equation,} \\ \dot{p}_i &= -\partial_{q^i} H \quad \text{covariant force equation,} \end{aligned} \tag{5.7}$$

(here and henceforth overdot denotes the total time derivative). Within the framework of the conservative Hamiltonian system (5.7) we can apply the formalism of classical Poisson brackets: for any two functions $A = A(q^i, p_i, t)$ and $B = B(q^i, p_i, t)$ their Poisson bracket is (using the summation convention) defined as [II06c, II07e]

$$[A, B] = (\partial_{q^i} A \partial_{p_i} B - \partial_{p_i} A \partial_{q^i} B).$$

Conservative Classical System

Any function $A(q^i, p_i, t)$ is called a constant (or integral) of motion of the conservative system (5.7) if [II06c, II07e]

$$\dot{A} \equiv \partial_t A + [A, H] = 0, \quad \text{which implies } \partial_t A = -[A, H]. \tag{5.8}$$

For example, if $A = \rho(q^i, p_i, t)$ is a density function of ensemble phase-points (or, a probability density to see a state $\mathbf{x}(t) = (q^i(t), p_i(t))$ of ensemble at a moment t), then equation

$$\partial_t \rho = -[\rho, H] \tag{5.9}$$

represents the *Liouville theorem*, which is usually derived from the continuity equation

$$\partial_t \rho + \operatorname{div}(\rho \dot{\mathbf{x}}) = 0.$$

Conserved quantity here is the Hamiltonian function $H = H(q^i, p_i, t)$, which the sum of kinetic and potential energy. For example, in case of an ND harmonic oscillator, we have the phase space $M = T^*\mathbb{R}^N \simeq \mathbb{R}^{2N}$, with the symplectic form $\omega = dp_i \wedge dq^i$ and Hamiltonian (total energy) function:

$$H = \frac{1}{2} \sum_{i=1}^{N} [p_i^2 + (q^i)^2]. \tag{5.10}$$

The corresponding Hamiltonian vector field X_H is given by

$$X_H = p_i \partial_{q^i} - q^i \partial_{p_i},$$

which gives canonical equations:

$$\dot{q}^i = p_i, \qquad \dot{p}_i = -\delta_{ij} q^j, \quad \text{(where } \delta_{ij} \text{ is the Kronecker symbol).} \qquad (5.11)$$

In addition, for any two smooth ND functions $f, g : \mathbb{R}^{2N} \to \mathbb{R}$, the Poisson bracket is given by

$$[f, g]_\omega = \frac{\partial f}{\partial q^i} \frac{\partial g}{\partial p_i} - \frac{\partial f}{\partial p_i} \frac{\partial g}{\partial q^i},$$

which implies that the particular functions $f = p_i p_j + q^i q^j$ and $g = p_i q^j + p_j q^i$ (for $i, j = 1, \ldots, N$)—are constants of motion. This system is integrable in an open set of $T^* \mathbb{R}^n$ with N integrability functions:

$$K_1 = H, \quad K_2 = p_2^2 + (q^2)^2, \quad \ldots, \quad K_N = p_N^2 + (q^N)^2.$$

Conservative Quantum System

We perform the formal quantization of the conservative equation (5.9) in the Heisenberg picture: all variables become Hermitian operators (denoted by '$\wedge$'), the symplectic phase space $T^*Q = \{q^i, p_i\}$ becomes the Hilbert state space $\mathcal{H} = \mathcal{H}_{\hat{q}^i} \otimes \mathcal{H}_{\hat{p}_i}$ (where $\mathcal{H}_{\hat{q}^i} = \mathcal{H}_{\hat{q}^1} \otimes \cdots \otimes \mathcal{H}_{\hat{q}^N}$ and $\mathcal{H}_{\hat{p}_i} = \mathcal{H}_{\hat{p}_1} \otimes \cdots \otimes \mathcal{H}_{\hat{p}_N}$), the classical Poisson bracket $[\, ,\,]$ becomes the quantum commutator $\{\, ,\,\}$ multiplied by $-i$ (in normal units) [II07c, II08b]

$$[\, ,\,] \longrightarrow -i\{\, ,\,\}. \qquad (5.12)$$

In this way, the classical Liouville equation (5.9) becomes the quantum Liouville equation [II07e, II08b]

$$\partial_t \hat{\rho} = i\{\hat{\rho}, \hat{H}\}, \qquad (5.13)$$

where $\hat{H} = \hat{H}(\hat{q}^i, \hat{p}_i, t)$ is the Hamiltonian evolution operator, while

$$\hat{\rho} = \sum_a P(a) |\Psi_a\rangle \langle \Psi_a|, \quad \text{where } \mathrm{Tr}(\hat{\rho}) = 1$$

denotes the von Neumann density matrix operator, where each quantum state $|\Psi_a >$ occurs with probability $P(a)$; $\hat{\rho} = \hat{\rho}(\hat{q}^i, \hat{p}_i, t)$ is closely related to another von Neumann concept: entropy

$$S = -\mathrm{Tr}(\hat{\rho}[\ln \hat{\rho}]).$$

Open Classical System

We now move to the open (nonconservative) system: on the macroscopic NN-level the opening operation equals to the adding of a covariant vector of external (dissipative and/or motor) forces $F_i = F_i(q^i, p_i, t)$ to (the right-hand-side of) the covariant Hamilton's force equation, so that Hamilton's equations obtain the open (dissipative and/or forced) form [Iva04, II06a, II06c]:

$$\dot{q}^i = \partial_{p_i} H, \qquad \dot{p}_i = -\partial_{q^i} H + F_i. \tag{5.14}$$

In the framework of the open Hamiltonian system (5.14) dynamics of any function $A(q^i, p_i, t)$ is defined by the open (dissipative and/or forced) evolution equation:

$$\partial_t A = -[A, H] + F_i[A, q^i], \quad ([A, q^i] = -\partial_{p_i} A). \tag{5.15}$$

In particular, if $A = \rho(q^i, p_i, t)$ represents the density function of ensemble phase-points then its dynamics is given by the open (dissipative and/or forced) Liouville equation [II06c, II07e]:

$$\partial_t \rho = -[\rho, H] + F_i[\rho, q^i]. \tag{5.16}$$

Equation (5.16) represents the open classical model of our microscopic NN-dynamics.

For example, in case of our ND oscillator, Hamiltonian function (5.10) is not conserved any more, the canonical equations (5.11) become

$$\dot{q}^i = p_i, \qquad \dot{p}_i = F_i - \delta_{ij} q^j,$$

and the system is not integrable any more.

Classical NN-Dynamics

The generalized NN-dynamics, including two special cases of graded response neurons (GRN) and coupled neural oscillators (CNO), can be presented in the form of a Langevin stochastic equation [II07b, II08a]

$$\dot{\sigma}_i = f_i + \eta_i(t), \tag{5.17}$$

where $\sigma_i = \sigma_i(t)$ are the continual neuronal variables of ith neurons (representing either membrane action potentials in case of GRN, or oscillator phases in case of CNO); J_{ij} are individual synaptic weights; $f_i = f_i(\sigma_i, J_{ij})$ are the deterministic forces (given, in GRN-case, by

$$f_i = \sum_j J_{ij} \tanh[\gamma \sigma_j] - \sigma_i + \theta_i, \quad \text{with } \gamma > 0$$

and with the θ_i representing injected currents, and in CNO-case, by

$$f_i = \sum_j J_{ij} \sin(\sigma_j - \sigma_i) + \omega_i,$$

with ω_i representing the natural frequencies of the individual oscillators); the noise variables are given as

$$\eta_i(t) = \lim_{\Delta \to 0} \zeta_i(t)\sqrt{2T/\Delta},$$

where $\zeta_i(t)$ denote uncorrelated Gaussian distributed random forces and the parameter T controls the amount of noise in the system, ranging from $T = 0$ (deterministic dynamics) to $T = \infty$ (completely random dynamics).

More convenient description of the neural random process (5.17) is provided by the Fokker–Planck equation describing the time evolution of the probability density $P(\sigma_i)$ [II07a, IA07, II08a]

$$\partial_t P(\sigma_i) = -\sum_i \partial_{\sigma_i}[f_i P(\sigma_i)] + T \sum_i \partial_{\sigma_i^2} P(\sigma_i). \tag{5.18}$$

Now, in the case of deterministic dynamics $T = 0$, (5.18) can be easily put into the form of the conservative Liouville equation (5.9), by making the substitutions:

$$P(\sigma_i) \to \rho, \qquad f_i = \dot{\sigma}_i, \quad \text{and} \quad [\rho, H] = \text{div}(\rho\dot{\sigma}_i) \equiv \sum_i \partial_{\sigma_i}(\rho\dot{\sigma}_i),$$

where $H = H(\sigma_i, J_{ij})$. Further, we can formally identify the stochastic forces, i.e., the second-order noise-term $T\sum_i \partial_{\sigma_i^2}\rho$ with $F^i[\rho, \sigma_i]$, to get the open Liouville equation (5.16).

Therefore, on the NN-level deterministic dynamics corresponds to the conservative system (5.9). Inclusion of stochastic forces corresponds to the system opening (5.16), implying the macroscopic arrow of time.

Open Quantum System

By formal quantization of (5.16), we obtain the quantum open Liouville equation [II07c, II08b]

$$\partial_t \hat{\rho} = \mathrm{i}\{\hat{\rho}, \hat{H}\} - \mathrm{i}\hat{F}_i\{\hat{\rho}, \hat{q}^i\}, \tag{5.19}$$

where $\hat{F}_i = \hat{F}_i(\hat{q}^i, \hat{p}_i, t)$ represents the covariant quantum operator of external friction forces in the Hilbert state space $\mathcal{H} = \mathcal{H}_{\hat{q}^i} \otimes \mathcal{H}_{\hat{p}_i}$.

Equation (5.19) represents the open quantum-decoherence model of our microscopic MT-dynamics.

Non-Critical Stringy MT-Dynamics

In EMN-language of non-critical (SUSY) bosonic strings, our MT-dynamics equation (5.19) reads [II07c, II07e, II08b]

$$\partial_t \hat{\rho} = \mathrm{i}\{\hat{\rho}, \hat{H}\} - \mathrm{i}\hat{g}_{ij}\{\hat{\rho}, \hat{q}^i\}\hat{\dot{q}}^j, \tag{5.20}$$

where the target-space density matrix $\hat{\rho}(\hat{q}^i, \hat{p}_i)$ is viewed as a function of coordinates $\hat{q}^i$ that parameterize the couplings of the generalized σ-models on the bosonic string world-sheet, and their conjugate momenta $\hat{p}_i$, while $\hat{g}_{ij} = \hat{g}_{ij}(\hat{q}^i)$ is the quantum operator of the positive definite metric in the space of couplings. Therefore, the covariant quantum operator of external friction forces is in EMN-formulation given as $\hat{F}_i(\hat{q}^i, \hat{\dot{q}}^i) = \hat{g}_{ij}\hat{\dot{q}}^j$.

Equation (5.20) establishes the conditions under which a large-scale coherent state appearing in the MT-network, which can be considered responsible for loss-free energy transfer along the tubulins.

The system-independent properties of (5.20), are:

(i) Conservation of probability P

$$\partial_t P = \partial_t[\mathrm{Tr}(\hat{\rho})] = 0. \tag{5.21}$$

(ii) Conservation of energy E, on the average

$$\partial_t \langle\!\langle E \rangle\!\rangle \equiv \partial_t[\mathrm{Tr}(\hat{\rho}E)] = 0. \tag{5.22}$$

(iii) Monotonic increase in entropy

$$\partial_t S = \partial_t[-\mathrm{Tr}(\hat{\rho}\ln\hat{\rho})] = (\hat{\dot{q}}^i \hat{g}_{ij}\hat{\dot{q}}^j)S \geq 0, \tag{5.23}$$

due to the positive definiteness of the metric $\hat{g}_{ij}$, and thus automatically and naturally implying a microscopic arrow of time [EMN99b].

Equivalence of Neurodynamic Forms

Both the macroscopic NN-equation (5.16) and the microscopic MT-equation (5.19) have the same open Liouville form, which implies the arrow of time [II07e, II08b]. These demonstrates the existence of a formal neuro-biological space-time self-similarity.

Therefore, we have described neuro- and psycho-dynamics of both NN and MT ensembles, belonging to completely different biophysical space-time scales, brain's neural networks and brain's microtubules, by the unique form of the *open Liouville equation*, which implies the arrow of time. In this way the existence of the formal neuro-biological space-time self-similarity has been proved.

This proof implies the *existence of a unique Neurodynamical Law*, which acts on different scales of brain's functioning. In both cases of macroscopic continuous neural networks and microscopic discrete microtubules we have a process which is expressible in the open Liouville equation form.

5.4 Dissipative Quantum Brain Model

The *conservative brain* model was originally formulated within the framework of the quantum field theory (QFT) by [RU67] and subsequently developed in [STU78, STU79, JY95, JPY96]. The conservative brain model has been recently extended to the *dissipative quantum dynamics* in the work of G. Vitiello and collaborators [Vit95, AV00, PV99, Vit01, PV03, PV04].

The motivations at the basis of the formulation of the quantum brain model by Umezawa and Ricciardi trace back to the laboratory observations leading Lashley to remark (in 1940) that "masses of excitations... within general fields of activity, without regard to particular nerve cells are involved in the determination of behavior" [Las42, Pri91]. In 1960's, K. Pribram, also motivated by experimental observations, started to formulate his *holographic hypothesis*. According to W. Freeman [Fre90, Fre96], "information appears indeed in such observations to be spatially uniform in much the way that the information density is uniform in a hologram". While the activity of the single neuron is experimentally observed in form of discrete and stochastic pulse trains and point processes, the 'macroscopic' activity of large assembly of neurons appears to be spatially coherent and highly structured in phase and amplitude.

Motivated by such an experimental situation, Umezawa and Ricciardi formulated in [RU67] the quantum brain model as a many-body physics problem, using the formalism of QFT with spontaneous breakdown of symmetry (which had been successfully tested in condensed matter experiments). Such a formalism provides the only available theoretical tool capable to describe long-range correlations such as the ones observed in the brain—presenting almost simultaneous responses in several regions to some external stimuli. The understanding of these long-range correlations in terms of modern biochemical and electrochemical processes is still lacking, which suggests that these responses could not be explained in terms of single neuron activity [Pri71, Pri91].

Lagrangian dynamics in QFT is, in general, invariant under some group G of continuous transformations, as proposed by the famous Noether theorem. Now, spontaneous symmetry breakdown, one of the corner-stones of Haken's synergetics [Hak83, Hak93], occurs when the minimum energy state (the ground, or vacuum, state) of the system is not invariant under the full group G, but under one of its subgroups. Then it can be shown [IZ80, Ume93] that collective modes, the so-called Nambu–Goldstone (NG) boson modes, are dynamically generated. Propagating over the whole system, these modes are the carriers of the *long-range correlation*, in which the order manifests itself as a global property dynamically generated. The long-range correlation modes are responsible for keeping the ordered pattern: they are coherently *condensed* in the ground state (similar to e.g., in the crystal case, where they keep the atoms trapped in their lattice sites). The long-range correlation thus forms a sort of net, extending over all the system volume, which traps the system components in the ordered pattern. This explains the "holistic" macroscopic collective behavior of the system components.

More precisely, according to the *Goldstone theorem* in QFT [IZ80, Ume93], the spontaneous breakdown of the symmetry implies the existence of long-range correlation NG-modes in the ground state of the system. These modes are massless modes in the infinite volume limit, but they may acquire a finite, non-zero mass due to boundary or impurity effects [ARV02]. In the quantum brain model these modes are called dipole-wave-quanta (DWQ). The density of their condensation in the ground states acts as a *code* classifying the state and the memory there recorded. States with different code values are unitarily inequivalent states, i.e., there is no unitary transformation relating states of different codes.[2]

Now, in formulating a proper mathematical model of brain, the conservative dynamics is not realistic: we cannot avoid to take into consideration the dissipative character of brain dynamics, since the brain is an intrinsically open system, continuously interacting with the environment. As Vitiello observed in [Vit01, PV03, PV04], the very same fact of "getting an information" introduces a partition in the time coordinate, so that one may distinguish between *before* "getting the information" (the past) and *after* "getting the information" (the future): the *arrow of time* is in this way introduced. ... "*Now* you know it!" is the familiar warning to mean that now, i.e. after having received a certain information, you are not the same person as before getting it. It has been shown that the psychological arrow of time (arising as an effect of memory recording) points in the same direction of the thermodynamical arrow of time (increasing entropy direction) and of the cosmological arrow of time (the expanding Universe direction) [AMV00].

The canonical quantization procedure of a dissipative system requires to include in the formalism also the system representing the environment (usually the heat bath) in which the system is embedded. One possible way to do that is to depict the environment as the time-reversal image of the system [CRV92]: the environment is thus described as the *double* of the system in the time-reversed dynamics (the system image in the mirror of time).

Within the framework of dissipative QFT, the brain system is described in terms of an *infinite collection of damped harmonic oscillators* A_κ (the simplest prototype of a dissipative system) representing the DWQ [Vit95]. Now, the collection of damped harmonic oscillators is ruled by the Hamiltonian [Vit95, CRV92]

$$H = H_0 + H_I,$$
$$\text{with } H_0 = \Omega_\kappa(A_\kappa^\dagger A_\kappa - \tilde{A}_\kappa^\dagger \tilde{A}_\kappa), \ H_I = \mathrm{i}\Gamma_\kappa(A_\kappa^\dagger \tilde{A}_\kappa^\dagger - A_\kappa \tilde{A}_\kappa),$$

---

[2] We remark that the spontaneous breakdown of symmetry is possible since in QFT there exist infinitely many ground states or vacua which are physically distinct (technically speaking, they are "unitarily inequivalent"). In quantum mechanics (QM), on the contrary, all the vacua are physically equivalent and thus there cannot be symmetry breakdown.

where Ω_κ is the frequency and Γ_κ is the damping constant. The $\tilde{A}_\kappa$ modes are the 'time-reversed mirror image' (i.e., the 'mirror modes') of the A_κ modes. They are the doubled modes, representing the environment modes, in such a way that κ generically labels their degrees-of-freedom. In particular, we consider the damped harmonic oscillator (DHO)

$$m\ddot{x} + \gamma\dot{x} + \kappa x = 0, \qquad (5.24)$$

as a simple prototype for dissipative systems (with intention that thus get results also apply to more general systems). The damped oscillator (5.24) is a non-Hamiltonian system and therefore the customary canonical quantization procedure cannot be followed. However, one can face the problem by resorting to well known tools such as the *density matrix* ρ and the *Wigner function* $W = W(x, p, t)$.

Let us start with the special case of a *conservative particle* in the absence of friction γ, with the standard Hamiltonian (in natural units),

$$H = -(\partial_x)^2/2 + V(x).$$

Recall (from the previous subsection) that the *density matrix equation of motion*, i.e., *quantum Liouville equation*, is given by

$$i\dot{\rho} = [H, \rho]. \qquad (5.25)$$

The density matrix function ρ is defined by

$$\left\langle x + \frac{1}{2}y \middle| \rho(t) \middle| x - \frac{1}{2}y \right\rangle = \psi^*\left(x + \frac{1}{2}y, t\right)\psi\left(x - \frac{1}{2}y, t\right) \equiv W(x, y, t),$$

with the associated standard expression for the *Wigner function* (see, e.g., [II07e]),

$$W(p, x, t) = \frac{1}{2\pi}\int W(x, y, t)\, e^{(-ipy)}\, dy.$$

Now, in the coordinate x-representation, by introducing the notation

$$x_\pm = x \pm \frac{1}{2}y, \qquad (5.26)$$

the Liouville equation (5.25) can be expanded as

$$i\partial_t \langle x_+ | \rho(t) | x_- \rangle$$
$$= \left\{ -\frac{1^2}{2}[\partial_{x_+}^2 - \partial_{x_-}^2] + [V(x_+) - V(x_-)] \right\} \langle x_+ | \rho(t) | x_- \rangle, \qquad (5.27)$$

while the Wigner function $W(p, x, t)$ is now given by

$$i\partial_t W(x, y, t) = H_o W(x, y, t),$$
$$\text{with } H_o = p_x p_y + V\left(x + \frac{1}{2}y\right) - V\left(x - \frac{1}{2}y\right),$$
$$\text{and } p_x = -i\partial_x, \; p_y = -i\partial_y. \qquad (5.28)$$

The new Hamiltonian H_o (5.28) may be obtained from the corresponding Lagrangian

$$L_o = m\dot{x}\dot{y} - V\left(x + \frac{1}{2}y\right) + V\left(x - \frac{1}{2}y\right). \tag{5.29}$$

In this way, Vitiello concluded that the density matrix and the Wigner function formalism *required*, even in the conservative case (with zero mechanical resistance γ), the introduction of a 'doubled' set of coordinates, $x_\pm$, or, alternatively, x and y. One may understand this as related to the introduction of the 'couple' of indices *necessary* to label the density matrix elements (5.27).

Let us now consider the case of the *particle interacting* with a *thermal bath* at temperature T. Let f denote the *random force* on the particle at the position x due to the bath. The interaction Hamiltonian between the bath and the particle is written as

$$H_{int} = -fx. \tag{5.30}$$

Now, in the *Feynman–Vernon formalism* (see [Fey72]), the *effective action* $A[x, y]$ for the particle is given by

$$A[x, y] = \int_{t_i}^{t_f} L_o(\dot{x}, \dot{y}, x, y)\, dt + I[x, y],$$

with L_o defined by (5.29) and

$$e^{iI[x,y]} = \langle(e^{-i\int_{t_i}^{t_f} f(t)x_-(t)dt})_-(e^{i\int_{t_i}^{t_f} f(t)x_+(t)dt})_+\rangle, \tag{5.31}$$

where the symbol $\langle . \rangle$ denotes the average with respect to the thermal bath; '$(.)_+$' and '$(.)_-$' denote time ordering and anti-time ordering, respectively; the coordinates $x_\pm$ are defined as in (5.26). If the interaction between the bath and the coordinate x (5.30) were turned off, then the operator f of the bath would develop in time according to

$$f(t) = e^{iH_\gamma t} f e^{-iH_\gamma t},$$

where H_γ is the Hamiltonian of the isolated bath (decoupled from the coordinate x). $f(t)$ is then the force operator of the bath to be used in (5.31).

The interaction $I[x, y]$ between the bath and the particle has been evaluated in [SVW95] for a linear passive damping due to thermal bath by following Feynman–Vernon and Schwinger. The final result from [SVW95] is:

$$I[x, y] = \frac{1}{2}\int_{t_i}^{t_f} dt[x(t)F_y^{ret}(t) + y(t)F_x^{adv}(t)]$$

$$+ \frac{i}{2}\int_{t_i}^{t_f}\int_{t_i}^{t_f} dtds\, N(t - s)y(t)y(s),$$

where the retarded force on y, F_y^{ret}, and the advanced force on x, F_x^{adv}, are given in terms of the retarded and advanced Green functions $G_{ret}(t-s)$ and $G_{adv}(t-s)$ by

$$F_y^{ret}(t) = \int_{t_i}^{t_f} ds\, G_{ret}(t-s)y(s), \qquad F_x^{adv}(t) = \int_{t_i}^{t_f} ds\, G_{adv}(t-s)x(s),$$

respectively. In (5.32), $N(t-s)$ is the *quantum noise* in the fluctuating random force given by

$$N(t-s) = \frac{1}{2}\langle f(t)f(s) + f(s)f(t)\rangle.$$

The real and the imaginary part of the action are given respectively by

$$\text{Re}\,(A[x,y]) = \int_{t_i}^{t_f} L\,dt, \tag{5.32}$$

$$L = m\dot{x}\dot{y} - \left[V\left(x + \frac{1}{2}y\right) - V\left(x - \frac{1}{2}y\right)\right] + \frac{1}{2}\left[xF_y^{ret} + yF_x^{adv}\right], \tag{5.33}$$

and

$$\text{Im}\,(A[x,y]) = \frac{1}{2}\int_{t_i}^{t_f}\int_{t_i}^{t_f} N(t-s)y(t)y(s)\,dtds. \tag{5.34}$$

Equations (5.32)–(5.34), are *exact* results for linear passive damping due to the bath. At quantum level nonzero y accounts for quantum noise effects in the fluctuating random force in the system-environment coupling arising from the imaginary part of the action (see [SVW95]). When in (5.33) we use

$$F_y^{ret} = \gamma\dot{y} \quad \text{and} \quad F_x^{adv} = -\gamma\dot{x}$$

we get,

$$L(\dot{x}, \dot{y}, x, y) = m\dot{x}\dot{y} - V\left(x + \frac{1}{2}y\right) + V\left(x - \frac{1}{2}y\right) + \frac{\gamma}{2}(x\dot{y} - y\dot{x}). \tag{5.35}$$

By using

$$V\left(x \pm \frac{1}{2}y\right) = \frac{1}{2}\kappa\left(x \pm \frac{1}{2}y\right)^2$$

in (5.35), the DHO equation (5.24) and its complementary equation for the y coordinate

$$m\ddot{y} - \gamma\dot{y} + \kappa y = 0. \tag{5.36}$$

are derived. The y-oscillator is the time-reversed image of the x-oscillator (5.24). From the manifolds of solutions to (5.24) and (5.36), we could choose those for which the y coordinate is constrained to be zero, they simplify to

$$m\ddot{x} + \gamma\dot{x} + \kappa x = 0, \qquad y = 0.$$

Thus we get the classical damped oscillator equation from a Lagrangian theory at the expense of introducing an 'extra' coordinate y, later constrained to vanish. Note that the constraint $y(t) = 0$ is *not* in violation of the equations of motion since it is a true solution to (5.24) and (5.36).

Therefore, the general scheme of the dissipative quantum brain model can be summarized as follows. The starting point is that the brain is permanently coupled to the environment. Clearly, the specific details of such a coupling may be very intricate and changeable so that they are difficult to be measured and known. One possible strategy is to average the effects of the coupling and represent them, at some degree of accuracy, by means of some 'effective' inter-action. Another possibility is to take into account the environmental influence on the brain by a suitable *choice* of the brain vacuum state. Such a choice is triggered by the external input (breakdown of the symmetry), and it actually is the end point of the internal (spontaneous) dynamical process of the brain (self-organization). The chosen vacuum thus carries the *signature* (memory) of the reciprocal brain-environment influence at a given time under given bound-ary conditions. A change in the brain-environment reciprocal influence then would correspond to a change in the choice of the brain vacuum: the brain state evolution or 'story' is thus the story of the trade of the brain with the surrounding world. The theory should then provide the equations describing the brain evolution 'through the vacua', each vacuum for each instant of time of its history.

The brain evolution is thus similar to a time-ordered sequence of pho-tograms: each photogram represents the 'picture' of the brain at a given in-stant of time. Putting together these photograms in 'temporal order' one gets a movie, i.e. the story (the evolution) of open brain, which includes the brain-environment interaction effects.

The evolution of a memory specified by a given code value, say $\mathcal{N}$, can be then represented as a trajectory of given initial condition running over time-dependent vacuum states, denoted by $|0(t)\rangle_{\mathcal{N}}$, each one minimizing the free energy functional. These trajectories are known to be *classical* trajectories in the infinite volume limit: transition from one representation to another inequivalent one would be strictly forbidden in a quantum dynamics.

Since we have now two-modes (i.e., non-tilde and tilde modes), the mem-ory state $|0(t)\rangle_{\mathcal{N}}$ turns out to be a two-mode coherent state. This is known to be an *entangled state*, i.e., it cannot be factorized into two single-mode states, the non-tilde and the tilde one. The physical meaning of such an entanglement between non-tilde and tilde modes is in the fact that the brain dynamics is permanently a dissipative dynamics. The entanglement, which is an unavoid-able mathematical result of dissipation, represents the impossibility of cutting the links between the brain and the external world.[3]

---

[3] We remark that the entanglement is permanent in the large volume limit. Due to boundary effects, however, a unitary transformation could disentangle the tilde and non-tilde sectors: this may result in a pathological state for the brain. It is known that forced isolation of a subject produces pathological states of various

In the dissipative brain model, noise and chaos turn out to be natural ingredients of the model. In particular, in the infinite volume limit the chaotic behavior of the trajectories in memory space may account for the high perceptive resolution in the recognition of the perceptual inputs. Indeed, small differences in the codes associated to external inputs may lead to diverging differences in the corresponding memory paths. On the other side, it also happens that codes differing only in a finite number of their components (in the momentum space) may easily be recognized as being the 'same' code, which makes possible that 'almost similar' inputs are recognized by the brain as 'equal' inputs (as in pattern recognition).

Therefore, the brain may be viewed as a complex system with (infinitely) many macroscopic configurations (the memory states). Dissipation is recognized to be the root of such a complexity.

5.5 QED Brain Model

In this subsection, mainly following [Sta95], we formulate a quantum electrodynamics brain model. Recall that quantum electrodynamics (extended to cover the magnetic properties of nuclei) is the theory that controls, as far as we know, the properties of the tissues and the aqueous (ionic) solutions that constitute our brains. This theory is our paradigm basic physical theory, and the one best understood by physicists. It describes accurately, as far as we know, the huge range of actual physical phenomena involving the materials encountered in daily life. It is also related to classical electrodynamics in a particularly beautiful and useful way.

In the low-energy regime of interest here it should be sufficient to consider just the classical part of the photon interaction defined in [Sta83]. Then the explicit expression for the unitary operator that describes the evolution from time t_1 to time t_2 of the quantum electromagnetic field in the presence of a set $L = \{L_i\}$ of specified classical charged-particle trajectories, with trajectory L_i specified by the function $x_i(t)$ and carrying charge e_i, is [Sta95]

$$U[L; t_2, t_1] = \exp\langle a^* \cdot J(L)\rangle \exp\langle -J^*(L) \cdot a\rangle \exp[-(J^*(L) \cdot J(L)/2)],$$

where, for any X and Y,

$$\langle X \cdot Y\rangle \equiv \int d^4k (2\pi)^{-4} 2\pi \delta^+(k^2) X(k) \cdot Y(k),$$

---

kinds. We also observe that the tilde mode is not just a mathematical fiction. It corresponds to a real excitation mode (quasi-particle) of the brain arising as an effect of its interaction with the environment: the couples of non-tilde/tilde dwq quanta represent the correlation modes dynamically created in the brain as a response to the brain-environment reciprocal influence. It is the interaction between tilde and non-tilde modes that controls the irreversible time evolution of the brain: these collective modes are confined to live *in* the brain. They vanish as soon as the links between the brain and the environment are cut.

$$(X \cdot Y) \equiv \int d^4k (2\pi)^{-4} \mathrm{i} (k^2 + \mathrm{i}\epsilon)^{-1} X(k) \cdot Y(k),$$

and $X \cdot Y = X_\mu Y^\mu = X^\mu Y_\mu$. Also,

$$J_\mu(L; k) \equiv \sum_i -\mathrm{i}e_i \int_{L_i} dx_\mu \exp(\mathrm{i}kx).$$

The integral along the trajectory L_i is

$$\int_{L_i} dx_\mu \exp(\mathrm{i}kx) \equiv \int_{t_1}^{t_2} dt (dx_{i\mu}(t)/dt) \exp(\mathrm{i}kx).$$

The $a^*(k)$ and $a(k)$ are the photon creation and annihilation operators:

$$[a(k), a^*(k')] = (2\pi)^3 \delta^3(k - k')2k_0.$$

The operator $U[L; t_2, t_1]$ acting on the photon vacuum state creates the coherent photon state that is the quantum-theoretic analog of the classical electromagnetic field generated by classical point particles moving on the set of trajectories $L = \{L_i\}$ between times t_1 and t_2.

The $U[L; t_2, t_1]$ can be decomposed into commuting contributions from the various values of k. The general coherent state can be written [Sta95]

$$|q, p\rangle \equiv \exp \mathrm{i}(\langle q \cdot P \rangle - \langle p \cdot Q \rangle)|0\rangle,$$

where $|0\rangle$ is the photon vacuum state and

$$Q(k) = (a_k + a_k^*)/\sqrt{2} \quad \text{and} \quad P(k) = \mathrm{i}(a_k - a_k^*)/\sqrt{2},$$

and $q(k)$ and $p(k)$ are two functions defined (and square integrable) on the mass shell $k^2 = 0$, $k_0 \geq 0$. The inner product of two coherent states is

$$\langle q, p | q', p' \rangle = \exp[-(\langle q - q' \cdot q - q' \rangle + \langle p - p' \cdot p - p' \rangle + 2\mathrm{i}\langle p - p' \cdot q + q' \rangle)]/4.$$

There is a decomposition of unity

$$I = \prod d^4k (2\pi)^{-4} 2\pi \delta^+(k^2) \int dq_k dp_k/\pi$$
$$\times \exp(\mathrm{i}q_k P_k - \mathrm{i}p_k Q_k)|0_k\rangle\langle 0_k| \exp[-(\mathrm{i}q_k P_k - \mathrm{i}p_k Q_k)].$$

Here meaning can be given by quantizing in a box, so that that the variable k is discretized. Equivalently,

$$I = \int d\mu(q, p)|q, p\rangle\langle q, p|,$$

where $\mu(q, p)$ is the appropriate measure on the functions $q(k)$ and $p(k)$. Then if the state $|\Psi\rangle\langle\Psi|$ were to jump to $|q, p\rangle\langle q, p|$ with probability density $\langle q, p|\Psi\rangle\langle\Psi|q, p\rangle$, the resulting mixture would be [Sta95]

$$\int d\mu(q,p)|q,p\rangle\langle q,p|\Psi\rangle\langle\Psi|q,p\rangle\langle q,p|,$$

whose trace is

$$\int d\mu(q,p)\langle q,p|\Psi\rangle\langle\Psi|q,p\rangle = \langle\Psi|\Psi\rangle.$$

To represent the limited capacity of consciousness let us assume, in this model, that the states of consciousness associated with a brain can be expressed in terms of a relatively small subset of the modes of the electromagnetic field in the brain cavity. Let us assume that events occurring outside the brain are keeping the state of the universe outside the brain cavity in a single state, so that the state of the brain can also be represented by a single state. The brain is represented, in the path-integral method of Feynman, by a superposition of the trajectories of the particles in it, with each element of the superposition accompanied by the coherent-state electromagnetic field that this set of trajectories generates. Let the state of the electromagnetic field restricted to the modes that represent consciousness be called $|\Psi(t)\rangle$. Using the decomposition of unity one can write

$$|\Psi(t)\rangle = \int d\mu(q,p)|q,p\rangle\langle q,p|\Psi(t)\rangle.$$

Hence the state at time t can be represented by the function $\langle q,p|\Psi(t)\rangle$, which is a complex-valued function over the set of arguments $\{q_1,p_1,q_2,p_2,\ldots,q_n,p_n\}$, where n is the number of modes associated with $|\Psi\rangle$. Thus in this model the contents of the consciousness associated with a brain is represented in terms of this function defined over a $2n$D space: the ith conscious event is represented by the transition

$$|\Psi_i(t_{i+1})\rangle \longrightarrow |\Psi_{i+1}(t_{i+1})\rangle = P_i|\Psi_i(t_{i+1})\rangle,$$

where P_i is a projection operator.

For each allowed value of k the pair of numbers (q_k,p_k) represents the state of motion of the kth mode of the electromagnetic field. Each of these modes is defined by a particular wave pattern that extends over the whole brain cavity. This pattern is an oscillating structure something like a sine wave or a cosine wave. Each mode is fed by the motions of all of the charged particles in the brain. Thus each mode is a representation of a certain integrated aspect of the activity of the brain, and the collection of values $q_1,p_1,\ldots,p_n$ is a compact representation of certain aspects the over-all activity of the brain.

The state $|q,p\rangle$ represents the conjunction, or collection over the set of all allowed values of k, of the various states $|q_k,p_k\rangle$. The function

$$V(q,p,t) = \langle q,p|\Psi(t)\rangle\langle\Psi(t)|q,p\rangle$$

satisfies $0 \leq V(q,p,t) \leq 1$, and it represents, according to orthodox thinking, the 'probability' that a system that is represented by a general state $|\Psi(t)\rangle$

just before the time t will be observed to be in the classically describable state $|q, p\rangle$ if the observation occurs at time t. The coherent states $|q, p\rangle$ can, for various mathematical and physical reasons, be regarded as the 'most classical' of the possible states of the electromagnetic quantum field.

To formulate a causal dynamics in which the state of consciousness itself controls the selection of the next state of consciousness one must specify a rule that determines, in terms of the evolving state $|\Psi_i(t)\rangle$ up to time t_{i+1}, both the time t_{i+1} when the next selection event occurs, and the state $|\Psi_{i+1}(t_{i+1})\rangle$ that is selected and actualized by that event.

In the absence of interactions, and under certain ideal conditions of confinement, the deterministic normal law of evolution entails that in each mode k there is an independent rotation in the (q_k, p_k) plane with a characteristic angular velocity $\omega_k = k_0$. Due to the effects of the motions of the particles there will be, added to this, a flow of probability that will tend to concentrate the probability in the neighborhoods of a certain set of 'optimal' classical states $|q, p\rangle$. The reason is that the function of brain dynamics is to produce some single template for action, and to be effective this template must be a 'classical' state, because, according to orthodox ideas, only these can be dynamically robust in the room temperature brain. According to the semi-classical description of the brain dynamics, only one of these classical-type states will be present, but according to quantum theory there must be a superposition of many such classical-type states, unless collapses occurs at lower (i.e., microscopic) levels. The assumption here is that no collapses occur at the lower brain levels: there is absolutely no empirical evidence, or theoretical reason, for the occurrence of such lower-level brain events.

So in this model the probability will begin to concentrate around various locally optimal coherent states, and hence around the various (generally) isolated points (q, p) in the $2n$D space at which the quantity [Sta95]

$$V(q, p, t) = \langle q, p | \Psi_i(t) \rangle \langle \Psi_i(t) | q, p \rangle$$

reaches a local maximum. Each of these points (q, p) represents a *locally-optimal solution* (at time t) to the search problem: as far as the myopic local mechanical process can see the state $|q, p\rangle$ specifies an analog-computed 'best' template for action in the circumstances in which the organism finds itself. This action can be either intentional (it tends to create in the future a certain state of the body/brain/environment complex) or attentional (it tends to gather information), and the latter action is a special case of the former. As discussed in [Sta93], the intentional and attentional character of these actions is a consequence of the fact that the template for action actualized by the quantum brain event is represented as a projected body-world schema, i.e., as the brains projected representation of the body that it is controlling and the environment in which it is situated.

Let a certain time $t_{i+1} > t_i$ be defined by an (urgency) energy factor $E(t) = \hbar(t_{i+1} - t_i)^{-1}$. Let the value of (q, p) at the largest of the local-maxima of $V(q, p, t_{i+1})$ be called $(q(t_{i+1}), p(t_{i+1}))_{max}$. Then the simplest possible reasonable selection rule would be given by the formula

$$P_i = |(q(t_{i+1}), p(t_{i+1}))_{max}\rangle\langle(q(t_{i+1}), p(t_{i+1}))_{max}|,$$

which entails that

$$\frac{|\Psi_{i+1}\rangle\langle\Psi_{i+1}|}{\langle\Psi_{i+1}|\Psi_{i+1}\rangle} = |(q(t_{i+1}), p(t_{i+1}))_{max}\rangle\langle(q(t_{i+1}), p(t_{i+1}))_{max}|.$$

This rule could produce a tremendous speed up of the search process. Instead of waiting until all the probability gets concentrated in one state $|q, p\rangle$, or into a set of isolated states $|q_i, p_i\rangle$ [or choosing the state randomly, in accordance with the probability function $V(q, p, t_{i+1})$, which could often lead to a disastrous result], this simplest selection process would pick the state $|q, p\rangle$ with the largest value of $V(q, p, t)$ at the time $t = t_{i+1}$. This process does not involve the complex notion of picking a random number, which is a physically impossible feat that is difficult even to define.

One important feature of this selection process is that it involves the state $\Psi(t)$ as a whole: the whole function $V(q, p, t_{i+1})$ must be known in order to determine where its maximum lies. This kind of selection process is not available in the semi-classical ontology, in which only one classically describable state exists at the macroscopic level. That is because this single classically describable macro-state state (e.g., some one actual state $|q, p, t_{i+1}\rangle$) contains no information about what the probabilities associated either with itself or with the other alternative possibilities would have been if the collapse had not occurred earlier, at some micro-level, and reduced the earlier state to some single classically describable state, in which, for example, the action potential along each nerve is specified by a well defined classically describable electromagnetic field. There is no rational reason in quantum mechanics for such a micro-level event to occur. Indeed, the only reason to postulate the occurrence of such premature reductions is to assuage the classical intuition that the action-potential pulse along each nerve 'ought to be classically describable even when it is not observed', instead of being controlled, when unobserved, by the local deterministic equations of quantum field theory. But the validity of this classical intuition is questionable if it severely curtails the ability of the brain to function optimally.

A second important feature of this selection process is that the actualized state Ψ_{i+1} is the state of the entire aspect of the brain that is connected to consciousness. So the feel of the conscious event will involve that aspect of the brain, taken as a whole. The 'I' part of the state $\Psi(t)$ is its slowly changing part. This part is being continually re-actualized by the sequence of events, and hence specifies the slowly changing background part of the felt experience. It is this persisting stable background part of the sequence of templates for action that is providing the over-all guidance for the entire sequence of selection events that is controlling the on-going brain process itself [Sta95].

A somewhat more sophisticated search procedure would be to find the state $|(q,p)_{max}\rangle$, as before, but to identify it as merely a candidate that is to be examined for its concordance with the objectives embedded in the current template. This is what a good search procedure ought to do: first pick out the top candidate by means of a mechanical process, but then evaluate this candidate by a more refined procedure that could block its acceptance if it does not meet specified criteria.

It may at first seem strange to imagine that nature could operate in such a sophisticated way. But it must be remembered that the generation of a truly random sequence is itself a very sophisticated (and indeed physically impossible) process, and that what the physical sciences have understood, so far, is only the mechanical part of nature's two-part process. Here it is the not-well-understood selection process that is under consideration. We have imposed on this attempt to understand the selection process the naturalistic requirement that the whole process be expressible in natural terms, i.e., that the universal process be a causal self-controlling evolution of the Hilbert-space state-vector in which all aspects of nature, including our conscious experiences, are efficacious.

It may be useful to describe the main features of this model in simple terms. If we imagine the brain to be, for example, a uniform rectangular box then each mode k would correspond to wave form that is periodic in all three directions: it would be formed as a combination of products of sine waves and cosine waves, and would cover the whole box-shaped brain. (More realistic conditions are needed, but this is a simple prototype.) Classically there would be an amplitude for this wave, and in the absence of interactions with the charged particles this amplitude would undergo a simple periodic motion in time. In analogy with the coordinate and momentum variables of an oscillating pendulum there are two variables, q_k and p_k, that describe the motion of the amplitude of the mode k. With a proper choice of scales for the variables q_k and p_k the motion of the amplitude of mode k if it were not coupled to the charges would be a circular motion in the (q_k, p_k)-plane. The classical theory would say that the physical system, mode k, would be represented by a point in q_k, p_k space. But quantum theory says that the physical system, mode k, must be represented by a wave (i.e., by a wave ψ-function) in (q_k, p_k) space. The reason is that interference effects between the values of this wave (function) at different points (q_k, p_k) can be exhibited, and therefore it is not possible to say the full reality is represented by any single value of (q_k, p_k): one must acknowledge the reality of the whole wave. It is possible to associate something like a 'probability density' with this wave, but the corresponding probability cannot be concentrated at a point: in units where Planck's constant is unity the bulk of the probability cannot be squeezed into a region of the (q_k, p_k) plane of area less that unity.

The mode k has certain natural states called 'coherent states', $|q_k, p_k\rangle$. Each of these is represented in (q_k, p_k)-space by a wave function that has a 'probability density' that falls off exponentially as one moves in any direction away from the center-point (q_k, p_k) at which the probability density is maximum. These coherent states are in many ways the 'most classical' wave functions allowed by quantum theory [Gla63a, Gla63b], and a central idea of the present model is to specify that it is to one of these 'most classical' states that the mode-k component of the electromagnetic field will jump, or collapse, when an observation occurs. This specification represents a certain 'maximal' principle: the second process, which is supposed to pick out and actualize some classically describable reality, is required to pick out and actualize one of these 'most classical' of the quantum states. If this selection/actualization process really exists in nature then the classically describable states that are actualized by this process should be 'natural classical states' from some point of view. The coherent states satisfy this requirement. This strong, specific postulate should be easier to disprove, if it is incorrect, than a vague or loosely defined one.

If we consider a system consisting of a collection of modes k, then the generalization of the single coherent state $|q_k, p_k\rangle$ is the product of these states, $|q, p\rangle$. Classically this system would be described by specifying the values all of the classical variables q_k and p_k as functions of time. But the 'best' that can be done quantum mechanically is to specify that at certain times t_i the system is in one of the coherent states $|q, p\rangle$. However, the equations of local quantum field theory (here quantum electrodynamics) entail that if the system starts in such a state then the system will, if no 'observation' occurs, soon evolve into a superposition (i.e., a linear combination) of many such states. But the next 'observation' will then reduce it again to some classically describable state. In the present model each a human observation is identified as a human conscious experience. Indeed, these are the same observations that the pragmatic Copenhagen interpretation of Bohr refers to, basically. The 'happening' in a human brain that corresponds to such an observation is, according to the present model, the selection and actualization of the corresponding coherent state $|q, p\rangle$.

The quantity $V(q, p, t_{i+1})$ defined above is, according to orthodox quantum theory, the predicted probability that a system that is in the state $\Psi(t_{i+1})$ at time t_{i+1} will be observed to be in state $|q, p\rangle$ if the observation occurs at time t_{i+1}. In the present model the function $V(q, p, t_{i+1})$ is used to specify not a fundamentally stochastic (i.e., random or chance-controlled) process but rather the causal process of the selection and actualization of some particular state $|q, p\rangle$. And this causal process is controlled by features of the quantum brain that are specified by the Hilbert space representation of the conscious process itself. This process is a nonlocal process that rides on the local brain process, and it is the nonlocal selection process that, according to the principles of quantum theory, is required to enter whenever an observation occurs.

5.6 Stochastic NLS-Filtering and Robotic Eye Tracking

According to R. Bucy [Buc70], every solution to a *stochastic filtering problem* involves the computation of time varying *probability density function* (PDF) on the state space of the observed system. Using the same concept, in this subsection, an architecture of Recurrent QNN is proposed following [BKE05, BK05, BKE06], where this PDF information of a stochastic variable can be transferred to $\psi(x,t)$, the wave amplitude function, of the NLS. Therefore, along the spatio-temporal evolution of ψ-function, the actual PDF is computed as the square of the modulus of the ψ-function that is a solution of the NLS (see Fig. 1.6 in Introduction).

The recurrent stochastic NLS-filter from Fig. 1.6 has been applied to model the mechanism of eye movements tracking a moving target consists of the following three stages:

(i) A *quantum stochastic filter* of noisy data that impact the eye sensors;
(ii) A *Kalman-like predictor* that predicts the next spatial position of the moving target; and
(iii) A *neural motor servo* [Hou79, HBB96] (acting as a Kalman-like corrector, see Appendix) that aligns the eye pupil along the moving targets trajectory.

The biological eye sensor fans out the input signal y to a specific neural lattice in the visual cortex. We have a 1D array of neurons whose receptive fields are excited by the signal input y reaching each neuron through a synaptic connection described by a nonlinear map. The neural lattice responds to the stimulus by setting up a spatial potential field, $V(x,t)$, which is a function of external stimulus y and estimated trajectory $\hat{y}$ of the moving target [BKE05, BKE06]

$$V(x,t) = \sum_{i=1}^{n} w_i(x,t)g_i[\nu(t)] \qquad (5.37)$$

where $g_i[.]$ is a Gaussian kernel, n represents the number of such Gaussian functions describing the nonlinear map that represents the synaptic connections, $\nu(t)$ represents the difference between y and $\hat{y}$ and w represents the synaptic weights. The Gaussian kernel is given by,

$$g_i(\nu(t)) = \exp(-(\nu(t) - m_i)^2)$$

where m_i is the center of the ith Gaussian function g_i. This center is chosen from input space described by the input signal, $\nu(t)$, through uniform random sampling.

Our quantum brain model proposes that a quantum process mediates the collective response of this neuronal lattice which sets up a spatial potential field $V(x,t)$. This happens when the quantum state associated with this quantum process evolves in this potential field. The spatio-temporal evolution follows as per (1.4). We hypothesize that this collective response is described by a wave packet, $f(x,t) = |\psi(x,t)|^2$, where the term $\psi(x,t)$ represents a quantum state. In a generic sense, we assume that a classical stimulus in a brain triggers a wave packet in the counterpart 'quantum brain'. This subjective response, $f(x,t)$, is quantified using the Maximum Likelihood Estimator (MLE)[4]

$$\hat{y}(t) = \int x(t)f(x,t)dx.$$

The estimate equation is motivated by the fact that the wave packet, $f(x,t) = |\psi(x,t)|^2$ is interpreted as the probability density function. Based on this estimate, $\hat{y}$, the predictor estimates the next spatial position of the moving target. To simplify our analysis, the predictor is made silent. Thus its output is the same as that of $\hat{y}$. The biological motor control is commanded to fixate the eye pupil to align with the target position, which is predicted to be at $\hat{y}$. Obviously, we have assumed that biological motor control is ideal.

After the above mentioned simplification, the closed form dynamics of the model becomes [BKE05, BKE06]

$$i\partial_t \psi(x,t) = -\frac{1}{2}\Delta\psi(x,t) + \zeta G\left[y(t) - \int x|\psi(x,t)|^2 dx\right]\psi(x,t) \qquad (5.38)$$

where $G[.]$ is a Gaussian kernel map, such that $V(x,t) = \zeta G[.]$, introduced to nonlinearly modulate the spatial potential field that excites the dynamics of the quantum object.

The NLS given by (5.38) is 1D with cubic nonlinearity. Interestingly, the closed form dynamics of the *recurrent QNN* (5.38) closely resembles a nonlinear Schrödinger wave equation with cubic nonlinearity studied in [GZ01]

$$i\partial_t \psi(x,t) = \left(-\frac{1}{2}\Delta - \frac{e^2}{r}\right)\psi(x,t) + e^2\int \frac{\psi(x,t)|\psi(x',t)|^2}{|x-x'|}dx', \qquad (5.39)$$

where e is the elementary charge and r is the magnitude of $|x|$. Also, nonlinear Schrödinger wave equations with cubic nonlinearity of the form

$$\partial_t \mathcal{A}(t) = c_1\mathcal{A} + c_3|\mathcal{A}|^2\mathcal{A},$$

where c_1 and c_3 are constants (see Sect. 4.6 above).

In (5.38), the unknown parameters are weights $w_i(x,t)$ associated with the Gaussian kernel, mass m, and ζ, the scaling factor to actuate the spatial potential field. The weights are updated using the *Hebbian learning* algorithm:

$$\partial_t w_i(x,t) = \beta\phi_i(\nu(t))f(x,t), \quad \text{with } \nu(t) = y(t) - \hat{y}(t). \qquad (5.40)$$

---

[4] In the stochastic filter model given by [Daw92], the author used a inverse filter in the feedback. Using this model we could not move the wave packet and the author agreed to this finding in our personal correspondence. We will see later in this subsection that the wave packet moves in the required direction in our new model.

The idea behind the proposed quantum computing model is as follows. As an individual observes a moving target, the uncertain spatial position of the moving target triggers a wave packet within the quantum brain. The quantum brain is so hypothesized that this wave packet turns out to be a collective response of a classical neural lattice. As we combine (5.38) and (5.40), it is desired that there exist some parameters m, ζ and β such that each specific spatial position $x(t)$ triggers a unique wave packet, $f(x,t) = |\psi(x,t)|^2$, in the quantum brain. This brings us to the question whether the closed form dynamics can exhibit soliton properties. As pointed out above, our equation has a form that is known to possess soliton properties for a certain range of parameters and we just have to find those parameters for each specific problem.

We would like to reiterate the importance of the soliton properties. According to our model, eye tracking means tracking of a wave packet in the domain of the quantum brain. The biological motor control aligns the eye pupil along the spatial position of the external target that the eye tracks. As the eye sensor receives data y from this position, the resulting error stimulates the quantum brain. In a noisy background, if the tracking is accurate, since when the estimate $\hat{y}(t)$ is the actual signal, then the signal that generates the potential field for the Schrödinger wave equation, $\hat{\nu}(t)$, is simply the noise that is embedded in the signal. If the statistical mean of the noise is zero, then this error correcting signal $\nu(t)$ has little effect on the movement of the wave packet. Precisely, it is the actual signal content in the input $y(t)$ that moves the wave packet along the desired direction which, in effect, achieves the goal of the stochastic filtering part of the eye movement for tracking purposes. It is expected that the synaptic weights evolve in such a manner so as to drive the ψ function to carry the exact information of the *pdf* of the observed stochastic variable $y(t)$.

This exhibits soliton property, i.e., the square of $|\psi(x,t)|$ is a wave packet which moves like a particle. The importance of this property is explained as follows [BKE05, BKE06]. Let the stochastic variable $y(t)$ be described by a Gaussian probability density function $f(x,t)$ with mean κ and standard deviation σ. Let the initial state of equation (1.4) correspond to zero mean Gaussian probability density function $f'(x,t)$ with standard deviation σ'. As the dynamics evolves with on-line update of the synaptic weights $K(x,t)$, the probability density function $f'(x,t)$ should ideally move toward the *pdf*, $f(x)$, of the signal $y(t)$. Thus the filtering problem in this new framework can be seen as the ability of the nonlinear Schrödinger wave equation to produce a wave packet solution that glides along with the time varying *pdf* corresponding to the signal $y(t)$.

Therefore, the nature of eye movement has been studied in this section using the proposed recurrent QNN model, where the predictor and motor control are assumed to be ideal. The most important finding is that our theoretical model of eye-tracking agrees with previously observed experimental results. The model predicts that eye movements will be of saccadic type while following a static trajectory. In the case of dynamic trajectory following, eye movement consists of saccades and smooth pursuits. In this sense, the proposed quantum brain concept in this section is very successful in explaining the nature of eye movements. Earlier explanation [BS79] for saccadic movement has been primarily attributed to motor control mechanism whereas the present model emphasizes that such eye movements are due to decision making process of the brain—albeit quantum brain. Thus the significant contribution of this section to explain biological eye-movement as a neural information processing event may inspire researchers to study quantum brain models from the biological perspective. The other significant contribution of this section is the prediction efficiency of the proposed model over the prevailing model. The stochastic filtering of a dc signal using recurrent QNN has been reported to be 1000 times more accurate compared to a Kalman filter [BKE05, BKE06].

5.7 QNN-Based Neural Motor Control

Recall (see [Iva04, IB05, II06a, II06b, II06c]) that realistic human biodynamics (RHB) is a science of human (and humanoid robot) motion in its full complexity. It is governed by both Newtonian dynamics and biological control laws.

There are over 200 bones in the human skeleton driven by about 640 muscular actuators (see, e.g. [Mar98, II06b]). While the muscles generate driving torques in the moving joints,[5] subcortical neural system performs both local and global (loco)motion control: first reflexly controlling contractions of individual muscles, and then orchestrating all the muscles into synergetic actions in order to produce efficient movements. While the local reflex control of individual muscles is performed on the *spinal control level*, the global integration of all the muscles into coordinated movements is performed within the *cerebellum* [II06a, II06b].

---

[5] Here we need to emphasize that human joints are significantly more flexible than humanoid robot joints. Namely, each humanoid joint consists of a pair of coupled segments with only Eulerian rotational degrees of freedom. On the other hand, in each human synovial joint, besides gross Eulerian rotational movements (roll, pitch and yaw), we also have some hidden and restricted translations along (X, Y, Z)-axes. For example, in the knee joint, patella (knee cap) moves for about 7–10 cm from maximal extension to maximal flexion). It is well-known that even greater are translational amplitudes in the shoulder joint. In other words, within the realm of rigid body mechanics, a segment of a human arm or leg is not properly represented as a rigid body fixed at a certain point, but rather as a rigid body hanging on rope-like ligaments. More generally, the whole skeleton mechanically represents a system of flexibly coupled rigid bodies. This implies the more complex kinematics, dynamics and control then in the case of humanoid robots.

All hierarchical subcortical neuro-muscular physiology, from the bottom level of a single muscle fiber, to the top level of cerebellar muscular synergy, acts as a *temporal* ⟨out|in⟩ *reaction*, in such a way that the higher level acts as a command/control space for the lower level, itself representing an abstract image of the lower one:

1. At the *muscular level*, we have *excitation-contraction dynamics* [Hat77a, Hat78, Hat77b], in which ⟨out|in⟩ is given by the following sequence of nonlinear diffusion processes [II06a, II06b]:

    neural action potential ⤳ synaptic potential ⤳ muscular action potential

    ⤳ excitation contraction coupling ⤳ muscle tension generating.

    Its purpose is the generation of muscular forces, to be transferred into driving torques within the joint anatomical geometry.

2. At the *spinal level*, ⟨out|in⟩ is given by *autogenetic-reflex stimulus-response control* [Hou79]. Here we have a neural image of all individual muscles. The main purpose of the spinal control level is to give both positive and negative feedbacks to stabilize generated muscular forces within the 'homeostatic' (or, more appropriately, 'homeokinetic') limits. The individual muscular actions are combined into flexor-extensor (or agonist-antagonist) pairs, mutually controlling each other. This is the mechanism of *reciprocal innervation of agonists and inhibition of antagonists*. It has a purely mechanical purpose to form the so-called *equivalent muscular actuators* (EMAs), which would generate driving torques $T_i(t)$ for all movable joints.

3. At the *cerebellar level*, ⟨out|in⟩ is given by *sensory-motor integration* [HBB96]. Here we have an abstracted image of all autogenetic reflexes. The main purpose of the cerebellar control level is integration and fine tuning of the action of all active EMAs into a synchronized movement, by *supervising* the individual autogenetic reflex circuits. At the same time, to be able to perform in new and unknown conditions, the cerebellum is continuously adapting its own neural circuitry by unsupervised (self-organizing) learning. Its action is subconscious and automatic, both in humans and in animals.

Naturally, we can ask the question: Can we assign a single ⟨out|in⟩ measure to all these neuro-muscular stimulus-response reactions? We think that we can do it; so in this Letter, we propose the concept of *adaptive sensory-motor transition amplitude* as a unique measure for this temporal ⟨out|in⟩ relation. Conceptually, this ⟨out|in⟩-*amplitude* can be formulated as the '*neural path integral*':

$$\langle \text{out}|\text{in}\rangle \equiv \underset{\text{amplitude}}{\langle \text{motor}|\text{sensory}\rangle} = \int \mathcal{D}[w,x]\, e^{iS[x]}. \tag{5.41}$$

Here, the integral is taken over all *activated* (or, 'fired') *neural pathways* $x^i = x^i(t)$ of the cerebellum, connecting its input *sensory*-state with its output *motor*-state, symbolically described by *adaptive neural measure* $\mathcal{D}[w,x]$, defined by the weighted product (of discrete time steps)

$$\mathcal{D}[w, x] = \lim_{n \to \infty} \prod_{t=1}^{n} w_i(t) \, dx^i(t), \qquad (5.42)$$

in which the *synaptic weights* $w_i = w_i(t)$, included in all active neural pathways $x^i = x^i(t)$, are updated by the standard learning rule

$$new \; value(t + 1) = old \; value(t) + innovation(t).$$

More precisely, the weights w_i in (5.42) are updated according to one of the two standard neural learning schemes, in which the micro-time level is traversed in discrete steps, i.e., if $t = t_0, t_1, \dots, t_n$ then $t + 1 = t_1, t_2, \dots, t_{n+1}$:[6]

1. A *self-organized, unsupervised* (e.g., Hebbian-like [Heb49]) learning rule:

$$w_i(t + 1) = w_i(t) + \frac{\sigma}{\eta}(w_i^d(t) - w_i^a(t)), \qquad (5.43)$$

   where $\sigma = \sigma(t)$, $\eta = \eta(t)$ denote *signal* and *noise*, respectively, while superscripts d and a denote *desired* and *achieved* micro-states, respectively; or

2. A certain form of a *supervised gradient descent learning*:

$$w_i(t + 1) = w_i(t) - \eta \nabla J(t), \qquad (5.44)$$

   where η is a small constant, called the *step size*, or the *learning rate*, and $\nabla J(n)$ denotes the gradient of the 'performance hyper-surface' at the t-th iteration.

Theoretically, (5.41)–(5.44) define an ∞-*dimensional neural network* (see [IA07, IAY08, II08c]). Practically, in a computer simulation we can use $10^7 \leq n \leq 10^8$, roughly corresponding to the number of neurons in the cerebellum [II07b, II07d].

   The exponent term $S[x]$ in (5.41) represents the *autogenetic-reflex action*, describing reflexly-induced motion of all active EMAs, from their initial *stimulus*-state to their final *response*-state, along the family of extremal (i.e., Euler–Lagrangian) paths $x^i_{\min}(t)$. ($S[x]$ is properly derived in (5.48)–(5.49) below.)

5.7.1 Spinal Reflex Control of Biodynamics

Sub-cerebellar biodynamics includes the following three components: (i) local muscle-joint mechanics, (ii) whole-body musculo-skeletal dynamics, and (iii) autogenetic reflex servo-control.

---

[6] Note that we could also use a reward-based, reinforcement learning rule [SB98], in which system learns its optimal policy:

$$innovation(t) = |reward(t) - penalty(t)|.$$

5.7.2 Local Muscle-Joint Mechanics

Local muscle-joint mechanics comprises of [Iva06, II06a, II06b]):

1. Synovial joint dynamics, giving the first stabilizing effect to the conservative skeleton dynamics, is described by the $(x, \dot{x})$-form of the Rayleigh—Van der Pol's dissipation function

$$R = \frac{1}{2} \sum_{i=1}^{n} (\dot{x}^i)^2 [\alpha_i + \beta_i (x^i)^2],$$

where α_i and β_i denote dissipation parameters. Its partial derivatives give rise to the viscous-damping torques and forces in the joints

$$\mathcal{F}_i^{joint} = \partial R / \partial \dot{x}^i,$$

which are linear in $\dot{x}^i$ and quadratic in x^i.

2. Muscular dynamics, giving the driving torques and forces $\mathcal{F}_i^{muscle} = \mathcal{F}_i^{muscle}(t, x, \dot{x})$ with $(i = 1, \ldots, n)$ for RHB, describes the internal excitation and contraction dynamics of equivalent muscular actuators [Hat78].

(a) Excitation dynamics can be described by an impulse force-time relation

$$F_i^{imp} = F_i^0 (1 - e^{-t/\tau_i}) \quad \text{if stimulation} > 0$$
$$F_i^{imp} = F_i^0 e^{-t/\tau_i} \quad \text{if stimulation} = 0,$$

where F_i^0 denote the maximal isometric muscular torques and forces, while τ_i denote the associated time characteristics of particular muscular actuators. This relation represents a solution of the Wilkie's muscular active-state element equation [Wil56]

$$\dot{\mu} + \gamma \mu = \gamma S A, \quad \mu(0) = 0, \ 0 < S < 1,$$

where $\mu = \mu(t)$ represents the active state of the muscle, γ denotes the element gain, A corresponds to the maximum tension the element can develop, and $S = S(r)$ is the 'desired' active state as a function of the motor unit stimulus rate r. This is the basis for the RHB force controller.

(b) Contraction dynamics has classically been described by the Hill's hyperbolic force-velocity relation [Hil38]

$$F_i^{Hill} = \frac{(F_i^0 b_i - \delta_{ij} a_i \dot{x}^j)}{(\delta_{ij} \dot{x}^j + b_i)},$$

where a_i and b_i denote the Hill's parameters, corresponding to the energy dissipated during the contraction and the phosphagenic energy conversion rate, respectively, while δ_{ij} is the Kronecker's δ-tensor.

In this way, RHB describes the excitation/contraction dynamics for the ith equivalent muscle-joint actuator, using the simple impulse-hyperbolic product relation

$$\mathcal{F}_i^{muscle}(t, x, \dot{x}) = F_i^{imp} \times F_i^{Hill}.$$

Now, for the purpose of biomedical engineering and rehabilitation, RHB has developed the so-called hybrid rotational actuator. It includes, along with muscular and viscous forces, the D.C. motor drives, as used in robotics (see [Iva06, II06a, II06c])

$$\mathcal{F}_k^{robo} = i_k(t) - J_k \ddot{x}_k(t) - B_k \dot{x}_k(t), \quad \text{with}$$
$$l_k i_k(t) + R_k i_k(t) + C_k \dot{x}_k(t) = u_k(t),$$

where $k = 1, \ldots, n$, $i_k(t)$ and $u_k(t)$ denote currents and voltages in the rotors of the drives, R_k, l_k and C_k are resistances, inductances and capacitances in the rotors, respectively, while J_k and B_k correspond to inertia moments and viscous dampings of the drives, respectively.

Finally, to make the model more realistic, we need to add some stochastic torques and forces [IS01, II07b]

$$\mathcal{F}_i^{stoch} = B_{ij}[x^i(t), t] \, dW^j(t),$$

where $B_{ij}[x(t), t]$ represents continuous stochastic diffusion fluctuations, and $W^j(t)$ is an N-variable Wiener process (i.e. generalized Brownian motion), with $dW^j(t) = W^j(t + dt) - W^j(t)$ for $j = 1, \ldots, N$.

Hamiltonian Biodynamics and Its Reflex Servo-Control

General form of Hamiltonian biodynamics on the configuration manifold of human motion is formulated in [IS01, Iva02, Iva04, IB05, II06a, II06c] using the concept of Euclidean group of motions $SE(3)$[7] (see Fig. 5.1).

Briefly, based on *affine Hamiltonian function of human motion*, formally $H_a : T^*Q \to \mathbb{R}$, in local canonical coordinates on the symplectic phase space (which is the cotangent bundle of the human configuration manifold Q) T^*Q given as

---

[7] Briefly, the Euclidean $SE(3)$-group is defined as a semidirect (noncommutative) product of 3D rotations and 3D translations, $SE(3) := SO(3) \triangleright \mathbb{R}^3$. Its most important subgroups are the following (for technical details see [II06c, PC05, II07e]):

Subgroup	Definition
$SO(3)$, group of rotations in 3D (a spherical joint)	Set of all proper orthogonal (3×3)-rotational matrices
$SE(2)$, special Euclidean group in 2D (all planar motions)	Set of all 3×3 - matrices: $\begin{bmatrix} \cos\theta & \sin\theta & r_x \\ -\sin\theta & \cos\theta & r_y \\ 0 & 0 & 1 \end{bmatrix}$
$SO(2)$, group of rotations in 2D subgroup of $SE(2)$-group (a revolute joint)	Set of all proper orthogonal (2×2)-rotational matrices included in $SE(2)$-group
$\mathbb{R}^3$, group of translations in 3D (all spatial displacements)	Euclidean 3D vector space

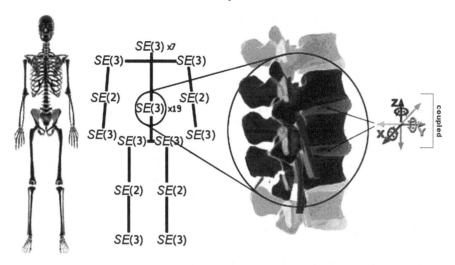

Fig. 5.1. The configuration manifold Q of the human musculoskeletal dynamics is defined as an anthropomorphic product of constrained Euclidean $SE(3)$-groups acting in all major (synovial) human joints. This concept has been used by the first author for development of the *Human Biodynamics Engine* (HBE), a world-class human motion simulator with 270 degrees of freedom.

$$H_a(x,p,u) = H_0(x,p) - H^j(x,p)u_j, \qquad (5.45)$$

where $H_0(x,p) = E_k(p) + E_p(x)$ is the physical Hamiltonian (kinetic + potential energy) dependent on joint coordinates x^i and their canonical momenta p_i, $H^j = H^j(x,p)$, $(j = 1,\ldots,m \le n$ are the coupling Hamiltonians corresponding to the system's active joints and $u_i = u_i(t,x,p)$ are (reflex) feedback-controls). Using (5.45) we come to the affine Hamiltonian control RHB-system, in deterministic form

$$\dot{x}^i = \partial_{p_i} H_0 - \partial_{p_i} H^j u_j + \partial_{p_i} R,$$
$$\dot{p}_i = \mathcal{F}_i - \partial_{x^i} H_0 + \partial_{x^i} H^j u_j + \partial_{x^i} R,$$
$$o^i = -\partial_{u_i} H_a = H^j, \qquad (5.46)$$
$$x^i(0) = x_0^i, \qquad p_i(0) = p_i^0,$$
$$(i = 1,\ldots,n;\ j = 1,\ldots,Q \le n),$$

(where $\partial_u \equiv \partial/\partial u$, $\mathcal{F}_i = \mathcal{F}_i(t,x,p)$, $H_0 = H_0(x,p)$, $H^j = H^j(x,p)$, $H_a = H_a(x,p,u)$, $R = R(x,p)$), as well as in the fuzzy-stochastic form [IS01, II07b]

$$dq^i = \left(\partial_{p_i} H_0(\sigma_\mu) - \partial_{p_i} H^j(\sigma_\mu) u_j + \partial_{p_i} R\right) dt,$$

$$dp_i = B_{ij}[x^i(t), t] dW^j(t)$$
$$+ \left(\bar{\mathcal{F}}_i - \partial_{x^i} H_0(\sigma_\mu) + \partial_{x^i} H^j(\sigma_\mu) u_j + \partial_{x^i} R\right) dt, \qquad (5.47)$$

$$d\bar{o}^i = -\partial_{u_i} H_a(\sigma_\mu) dt = H^j(\sigma_\mu) dt,$$

$$x^i(0) = \bar{x}_0^i, \qquad p_i(0) = \bar{p}_i^0.$$

In (5.46)–(5.47), $R = R(x, p)$ denotes the joint (nonlinear) dissipation function, o^i are affine system outputs (which can be different from joint coordinates); $\{\sigma\}_\mu$ (with $\mu \geq 1$) denote fuzzy sets of conservative parameters (segment lengths, masses and moments of inertia), dissipative joint dampings and actuator parameters (amplitudes and frequencies), while the bar $(\bar{.})$ over a variable denotes the corresponding fuzzified variable; $B_{ij}[q^i(t), t]$ denote diffusion fluctuations and $W^j(t)$ are discontinuous jumps as the n-dimensional Wiener process.

In this way, the force RHB servo-controller is formulated as affine control Hamiltonian-systems (5.46–5.47), which resemble the *autogenetic motor servo*, acting on the spinal-reflex level of the human locomotion control. A voluntary contraction force F of human skeletal muscle is reflexly excited (positive feedback $+F^{-1}$) by the responses of its spindle receptors to stretch and is reflexly inhibited (negative feedback $-F^{-1}$) by the responses of its Golgi tendon organs to contraction. Stretch and unloading reflexes are mediated by combined actions of several autogenetic neural pathways, forming the so-called 'motor servo'. The term 'autogenetic' means that the stimulus excites receptors located in the same muscle that is the target of the reflex response. The most important of these muscle receptors are the primary and secondary endings in the muscle-spindles, which are sensitive to length change—positive length feedback $+F^{-1}$, and the Golgi tendon organs, which are sensitive to contractile force—negative force feedback $-F^{-1}$.

The gain G of the length feedback $+F^{-1}$ can be expressed as the positional stiffness (the ratio $G \approx S = dF/dx$ of the force-F change to the length-x change) of the muscle system. The greater the stiffness S, the less the muscle will be disturbed by a change in load. The autogenetic circuits $+F^{-1}$ and $-F^{-1}$ appear to function as servoregulatory loops that convey continuously graded amounts of excitation and inhibition to the large (alpha) skeletomotor neurons. Small (gamma) fusimotor neurons innervate the contractile poles of muscle spindles and function to modulate spindle-receptor discharge.

This affine Hamiltonian control formalism has been used by the first author for development of the *Human Biodynamics Engine* (HBE), a world-class human motion simulator with 270 degrees of freedom, muscular excitation-contraction dynamics and hierarchical neural-like control. The simulator has been built and kinematically validated in the last five years at the Land Operations Division, Defence Science & Technology Organisation, Australia.

5.7.3 Cerebellum: Adaptive Path-Integral Comparator

Cerebellum as a Neural Controller

Having, thus, defined the spinal reflex control level, we proceed to model the top subcortical commander/controller, the *cerebellum*. The cerebellum is responsible for coordinating precisely timed ⟨out|in⟩ activity by integrating motor output with ongoing sensory feedback (see Fig. 5.2). It receives extensive projections from sensory-motor areas of the cortex and the periphery and directs it back to premotor and motor cortex [Ghe90, Ghe91]. This suggests a role in sensory-motor integration and the timing and execution of human movements. The cerebellum stores patterns of motor control for frequently performed movements, and therefore, its circuits are changed by experience and training. It was termed the *adjustable pattern generator* in the work of J. Houk and collaborators [HBB96]. Also, it has become the inspiring 'brain-model' in robotic research [SA98, Sch98, Sch99].

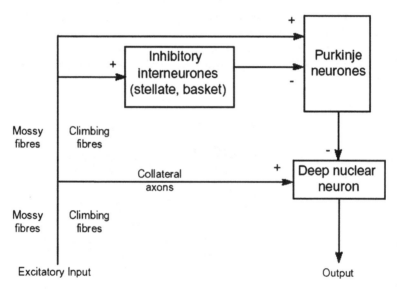

Fig. 5.2. Schematic ⟨out|in⟩ organization of the primary cerebellar circuit. In essence, excitatory inputs, conveyed by collateral axons of Mossy and Climbing fibers activate directly neurones in the Deep cerebellar nuclei. The activity of these latter is also modulated by the inhibitory action of the cerebellar cortex, mediated by the Purkinje cells.

The cerebellum is known to be involved in the production and learning of smooth coordinated movements [TGK92, FSB97]. Two classes of inputs carry information into the cerebellum: the mossy fibers (MFs) and the climbing fibers (CFs). The MFs provide both plant state and contextual information

[BC81]. The CFs, on the other hand, are thought to provide information that reflect errors in recently generated movements [Ito84, Ito90]. This information is used to adjust the programs encoded by the cerebellum. The MFs carry plant state, motor efference, and other contextual signals into the cerebellum. These fibers impinge on granule cells, whose axons give rise to parallel fibers (PFs). Through the combination of inputs from multiple classes of MFs and local inhibitory interneurons, the granule cells are thought to provide a sparse expansive encoding of the incoming state information [Alb71]. The large number of PFs converge on a much smaller set of Purkinje cells (PCs), while the PCs, in turn, provide inhibitory signals to a single cerebellar nuclear cell [FSB97]. Using this principle, the Cerebellar Model Arithmetic Computer, or CMAC-neural network has been built [Alb71, Mil56] and implemented in robotics [Sma98], using trial-and-error learning to produce bursts of muscular activity for controlling robot arms.

So, this 'cerebellar control' works for simple robotic problems, like nonredundant manipulation. However, comparing the number of its neurons (10^7–10^8), to the size of conventional neural networks (including CMAC), suggests that artificial neural nets *cannot* satisfactorily model the function of this sophisticated 'super-bio-computer', as its dimensionality is virtually infinite. Despite a lot of research dedicated to its structure and function (see [HBB96] and references there cited), the real nature of the cerebellum still remains a 'mystery'.

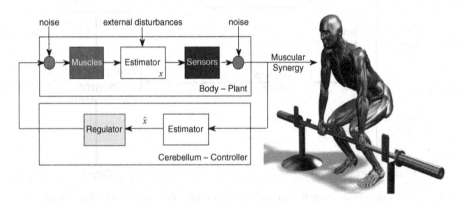

Fig. 5.3. The cerebellum as a motor controller.

The main function of the cerebellum as a motor controller is depicted in Fig. 5.3. A coordinated movement is easy to recognize, but we know little about how it is achieved. In search of the neural basis of coordination, a model of spinocerebellar interactions was recently presented in [AG05], in which the structural and functional organizing principle is a division of the cerebellum into discrete micro-complexes. Each micro-complex is the recipient of a specific motor error signal, that is, a signal that conveys information about an inappropriate movement. These signals are encoded by spinal reflex circuits and conveyed to the cerebellar cortex through climbing fibre afferents. This organization reveals salient features of cerebellar information processing, but also highlights the importance of systems level analysis for a fuller understanding of the neural mechanisms that underlie behavior.

Hamiltonian Action and Neural Path Integral

Here, we propose a *quantum-like adaptive control* approach to modeling the 'cerebellar mystery'. Corresponding to the affine Hamiltonian control function (5.45) we define the *affine Hamiltonian control action*,

$$S_{aff}[q,p] = \int_{t_{in}}^{t_{out}} d\tau \left[p_i \dot{q}^i - H_{aff}(q,p) \right] . \tag{5.48}$$

From the affine Hamiltonian action (5.48) we further derive the associated expression for the *neural phase-space path integral* (in normal units), representing the *cerebellar sensory-motor amplitude* $\langle \text{out} | \text{in} \rangle$,

$$
\begin{aligned}
&\langle q_{out}^i, p_i^{out} | q_{in}^i, p_i^{in} \rangle \\
&= \int \mathcal{D}[w,q,p] \, e^{iS_{aff}[q,p]} \\
&= \int \mathcal{D}[w,q,p] \exp \left\{ i \int_{t_{in}}^{t_{out}} d\tau \left[p_i \dot{q}^i - H_{aff}(q,p) \right] \right\}, \\
&\text{with} \int \mathcal{D}[w,q,p] = \int \prod_{\tau=1}^{n} \frac{w^i(\tau) dp_i(\tau) dq^i(\tau)}{2\pi},
\end{aligned}
\tag{5.49}
$$

where $w_i = w_i(t)$ denote the cerebellar synaptic weights positioned along its neural pathways, being continuously updated using the Hebbian-like self-organizing learning rule (5.43). Given the transition amplitude $out|in$ (5.49), the *cerebellar sensory-motor transition probability* is defined as its absolute square, $|\langle \text{out} | \text{in} \rangle|^2$.

In the phase-space path integral (5.49), $q_{in}^i = q_{in}^i(t)$, $q_{out}^i = q_{out}^i(t)$; $p_i^{in} = p_i^{in}(t)$, $p_i^{out} = p_i^{out}(t)$; $t_{in} \leq t \leq t_{out}$, for all discrete time steps, $t = 1, \ldots, n \to \infty$, and we are allowing for the affine Hamiltonian $H_{aff}(q,p)$ to depend upon all the $(M \leq N)$ EMA-angles and angular momenta collectively. Here, we actually systematically took a discretized differential time limit of the form $t_\sigma - t_{\sigma-1} \equiv d\tau$ (both σ and τ denote discrete time steps) and wrote $\frac{(q_\sigma^i - q_{\sigma-1}^i)}{(t_\sigma - t_{\sigma-1})} \equiv \dot{q}^i$. For technical details regarding the path integral calculations on Riemannian and symplectic manifolds (including the standard regularization procedures), see [Kla97, Kla00].

Now, motor learning occurring in the cerebellum can be observed using functional MR imaging, showing changes in the cerebellar action potential, related to the motor tasks (see, e.g., [MA02]). To account for these electro-physiological currents, we need to add the *source* term $J_i(t)q^i(t)$ to the affine Hamiltonian action (5.48), (the current $J_i = J_i(t)$ acts as a source $J_i A^i$ of the *cerebellar electrical potential* $A^i = A^i(t)$),

$$S_{aff}[q, p, J] = \int_{t_{in}}^{t_{out}} d\tau \left[p_i \dot{q}^i - H_{aff}(q, p) + J_i q^i \right],$$

which, subsequently gives the cerebellar path integral with the action potential source, coming either from the motor cortex or from other subcortical areas.

Note that the standard *Wick rotation*: $t \mapsto t$ (see [Kla97, Kla00]), makes our path integral real, i.e.,

$$\int \mathcal{D}[w, q, p] \, e^{iS_{aff}[q,p]} \xrightarrow{\text{Wick}} \int \mathcal{D}[w, q, p] \, e^{-S_{aff}[q,p]},$$

while their subsequent discretization gives the standard thermodynamic *partition function* (see Introduction),

$$Z = \sum_j {}^{-w_j E^j / T}, \tag{5.50}$$

where E^j is the energy eigenvalue corresponding to the affine Hamiltonian $H_{aff}(q, p)$, T is the temperature-like environmental control parameter, and the sum runs over all energy eigenstates (labeled by the index j). From (5.50), we can further calculate all statistical and thermodynamic system properties (see [Fey72]), as for example, *transition entropy* $S = k_B \ln Z$, etc.

Entropy and Motor Control

Our cerebellar path integral controller is closely related to *entropic motor control* [HN08a, HN08b], which deals with neuro-physiological feedback information and environmental uncertainty. The probabilistic nature of human motor action can be characterized by entropies at the level of the organism, task, and environment. Systematic changes in motor adaptation are characterized as task-organism and environment-organism tradeoffs in entropy. Such compensatory adaptations lead to a view of goal-directed motor control as the product of an underlying conservation of entropy across the task-organism-environment system. In particular, an experiment conducted in [HN08b] examined the changes in entropy of the coordination of isometric force output under different levels of task demands and feedback from the environment. The goal of the study was to examine the hypothesis that human motor adaptation can be characterized as a process of entropy conservation that is reflected in the compensation of entropy between the task, organism motor output, and environment. Information entropy of the coordination dynamics relative phase of the motor output was made conditional on the idealized situation of human movement, for which the goal was always achieved. Conditional entropy of the motor output decreased as the error tolerance and feedback frequency were decreased. Thus, as the likelihood of meeting the task demands was decreased increased task entropy and/or the amount of information from the environment is reduced increased environmental entropy, the subjects of this experiment employed fewer coordination patterns in the force output to achieve the goal. The conservation of entropy supports the view that context dependent adaptations in human goal-directed action are guided fundamentally by natural law and provides a novel means of examining human motor behavior. This is fundamentally related to the *Heisenberg uncertainty principle* [II08b] and further supports the argument for the primacy of a probabilistic approach toward the study of biodynamic cognition systems.

The action-amplitude formalism represents a kind of a generalization of the Haken–Kelso–Bunz (HKB) model of self-organization in the individual's motor system [HKB85, Kel95], including: multi-stability, phase transitions and hysteresis effects, presenting a contrary view to the purely feedback driven systems. HKB uses the concepts of synergetics (order parameters, control parameters, instability, etc.) and the mathematical tools of nonlinearly coupled (nonlinear) dynamical systems to account for self-organized behavior both at the cooperative, coordinative level and at the level of the individual coordinating elements. The HKB model stands as a building block upon which numerous extensions and elaborations have been constructed. In particular, it has been possible to derive it from a realistic model of the cortical sheet in which neural areas undergo a reorganization that is mediated by intra- and inter-cortical connections. Also, the HKB model describes phase transitions ('switches') in coordinated human movement as follows: (i) when the agent begins in the anti-phase mode and speed of movement is increased, a spontaneous switch to symmetrical, in-phase movement occurs; (ii) this transition happens swiftly at a certain critical frequency; (iii) after the switch has occurred and the movement rate is now decreased the subject remains in the symmetrical mode, i.e. she does not switch back; and (iv) no such transitions occur if the subject begins with symmetrical, in-phase movements. The HKB dynamics of the order parameter relative phase as is given by a nonlinear first-order ODE:

$$\dot{\phi} = (\alpha + 2\beta r^2)\sin\phi - \beta r^2 \sin 2\phi,$$

where ϕ is the phase relation (that characterizes the observed patterns of behavior, changes abruptly at the transition and is only weakly dependent on parameters outside the phase transition), r is the oscillator amplitude, while α, β are coupling parameters (from which the critical frequency where the phase transition occurs can be calculated).

5.8 Quantum Cognition in the Life Space Foam

Classical physics has provided a strong foundation for understanding brain function through measuring brain activity, modeling the functional connectivity of networks of neurons with algebraic matrices, and modeling the dynamics of neurons and neural populations with sets of coupled differential equations [Fre75, Fre00]. Various tools from classical physics enabled recognition and documentation of aspects of the physical states of the brain; the structures and dynamics of neurons, the operations of membranes and organelles that generate and channel electric currents; and the molecular and ionic carriers that implement the neural machineries of electrogenesis and learning. They support description of brain functions at several levels of complexity through measuring neural activity in the brains of animal and human subjects engaged in behavioral exchanges with their environments. One of the key properties of brain dynamics are the coordinated oscillations of populations of neurons that change rapidly in concert with changes in the environment [FV06, II06b, II07b]. Also, most experimental neurobiologists and neural theorists have focused on sensorimotor functions and their adaptations through various forms of learning and memory. Reliance has been placed on measurements of the rates and intervals of trains of action potentials of small numbers of neurons that are tuned to perceptual invariances and modeling neural interactions with discrete networks of simulated neurons. These and related studies have given a vivid picture of the cortex as a mosaic of modules, each of which performs a sensory or motor function; they have not given a picture of comparable clarity of the integration of modules.

According to [FV06], many-body quantum field theory appears to be the only existing theoretical tool capable to explain the dynamic origin of long-range correlations, their rapid and efficient formation and dissolution, their interim stability in ground states, the multiplicity of coexisting and possibly non-interfering ground states, their degree of ordering, and their rich textures relating to sensory and motor facets of behaviors. It is historical fact that many-body quantum field theory has been devised and constructed in past decades exactly to understand features like ordered pattern formation and phase transitions in condensed matter physics that could not be understood in classical physics, similar to those in the brain.

The domain of validity of the 'quantum' is not restricted to the microscopic world [Ume93]. There are macroscopic features of classically behaving systems, which cannot be explained without recourse to the quantum dynamics. This field theoretic model leads to the view of the phase transition as a condensation that is comparable to the formation of fog and rain drops from water vapor, and that might serve to model both the gamma and beta phase transitions. According to such a model, the production of activity with long-range correlation in the brain takes place through the mechanism of spontaneous breakdown of symmetry (SBS), which has for decades been shown to describe long-range correlation in condensed matter physics. The adoption of such a field theoretic approach enables modeling of the whole cerebral hemisphere and its hierarchy of components down to the atomic level as a fully integrated macroscopic quantum system, namely as a macroscopic system which is a quantum system not in the trivial sense that it is made, like all existing matter, by quantum components such as atoms and molecules, but in the sense that some of its macroscopic properties can best be described with recourse to quantum dynamics (see [FV06] and references therein).

It is well-known that non-equilibrium phase transitions [Hak83, Hak93, Hak96] are phenomena which bring about qualitative physical changes at the macroscopic level in presence of the same microscopic forces acting among the constituents of a system. Phase transitions can also be associated with autonomous robot competence levels, as informal specifications of desired classes of behaviors for robots over all environments they will encounter, as described by Brooks' subsumption architecture approach [Bro86a, Bro89, Bro90]. The distributed network of augmented finite-state machines can exist in different phases or modalities of their state-space variables, which determine the systems intrinsic behavior. The phase transition represented by this approach is triggered by either internal (a set-point) or external (a command) control stimuli, such as a command to transition from a sleep mode to awake mode, or walking to running.

On the other hand, it is well-known that humans possess more degrees of freedom than are needed to perform any defined motor task, but are required to co-ordinate them in order to reliably accomplish high-level goals, while faced with intense motor variability. In an attempt to explain how this takes place, Todorov and Jordan [TJ02] have formulated an alternative theory of human motor co-ordination based on the concept of stochastic optimal feedback control. They were able to conciliate the requirement of goal achievement (e.g., grasping an object) with that of motor variability (biomechanical degrees of freedom). Moreover, their theory accommodates the idea that the human motor control mechanism uses internal 'functional synergies' to regulate task-irrelevant (redundant) movement.

Until recently, research concerning sensory processing and research concerning motor control have followed parallel but independent paths. The partitioning of the two lines of research in practice partly derived from and partly fostered a bipartite view of sensorimotor processing in the brain—that a sensory/perceptual system creates a general purpose representation of the world which serves as the input to the motor systems (and other cognitive systems) that generate action/behavior as an output. Recent results from research on vision in natural tasks have seriously challenged this view, suggesting that the visual system does not generate a general-purpose representation of the world, but rather extracts information relevant to the task at hand [DHT05, LH01]. At the same time, researchers in motor control have developed an increasing understanding of how sensory limitations and sensory uncertainty can shape the motor strategies that humans employ to perform tasks. Moreover, many aspects of the problem of sensorimotor control are specific to the mapping from sensory signals to motor outputs and do not exist in either domain in isolation. Sensory feedback control of hand movements, coordinate transformations of spatial representations and the influence of processing speed and attention on sensory contributions to motor control are just a few of these. In short, to understand how human (and animal) actors use sensory information to guide motor behavior, we must study sensory and motor systems as an integrated whole rather than as decomposable modules in a sequence of discrete processing steps [KMT07].

Cognitive neuroscience investigations, including fMRI studies of human co-action, suggest that cognitive and neural processes supporting co-action include joint attention, action observation, task sharing, and action coordination [FFG05, KJ03, NNM07, SBK06]. For example, when two actors are given a joint control task (e.g., tracking a moving target on screen) and potentially conflicting controls (e.g., one person in charge of acceleration, the other—deceleration), their joint performance depends on how well they can anticipate each other's actions. In particular, better coordination is achieved when individuals receive real-time feedback about the timing of each other's actions [SBK06].

A developing field in coordination dynamics involves the theory of social coordination, which attempts to relate the DC to normal human development of complex social cues following certain patterns of interaction. This work is aimed at understanding how human social interaction is mediated by meta-stability of neural networks. fMRI and EEG are particularly useful in mapping thalamocortical response to social cues in experimental studies. In particular, a new theory called the *Phi complex* has been developed by S. Kelso and collaborators, to provide experimental results for the theory of social coordination dynamics (see the recent nonlinear dynamics paper discussing social coordination and EEG dynamics [TLD07]). According to this theory, a pair of phi rhythms, likely generated in the mirror neuron system, is the hallmark of human social coordination. Using a dual-EEG recording system, the authors monitored the interactions of eight pairs of subjects as they moved their fingers with and without a view of the other individual in the pair.

Recently developed Life Space Foam (LSF) model [IA07] is an integration of two modern approaches to cognition: (i) dynamical field theory (DFT) [Ama77, Sch07] and (ii) quantum-probabilistic dynamics (QP) [Gli05, BWT06]. In this section we expand the LSF-concept to model decision making process in human-robot joint action and related LSF-phase transitions.

5.8.1 Classical versus Quantum Probability

As *quantum probability* in human cognition and decision making has recently become popular, let us briefly describe this fundamental concept (see [II07e, II07d, II08b] for more details).

Classical Probability and Stochastic Dynamics

Recall that a *random variable* X is defined by its *distribution function* $f(x)$. Its *probabilistic description* is based on the following rules: (i) $P(X = x_i)$ is the probability that $X = x_i$; and (ii) $P(a \leq X \leq b)$ is the probability that X lies in a closed interval $[a, b]$. Its statistical description is based on: (i) μ_X or $E(X)$ is the mean or expectation of X; and (ii) σ_X is the standard deviation of X. There are two cases of random variables: discrete and continuous, each having its own probability (and statistics) theory.

A discrete random variable X has only a countable number of values $\{x_i\}$. Its distribution function $f(x_i)$ has the following properties:

$$P(X = x_i) = f(x_i), \quad f(x_i) \geq 0, \quad \sum_i f(x_i)\,dx = 1.$$

Statistical description of X is based on its discrete mean value μ_X and standard deviation σ_X, given respectively by

$$\mu_X = E(X) = \sum_i x_i f(x_i), \qquad \sigma_X = \sqrt{E(X^2) - \mu_X^2}.$$

Here $f(x)$ is a piecewise continuous function such that:

$$P(a \leq X \leq b) = \int_a^b f(x)\,dx, \quad f(x) \geq 0,$$

$$\int_{-\infty}^{\infty} f(x)\,dx = \int_{\mathbb{R}} f(x)\,dx = 1.$$

Statistical description of X is based on its continuous mean μ_X and standard deviation σ_X, given respectively by

$$\mu_X = E(X) = \int_{-\infty}^{\infty} x f(x)\,dx, \qquad \sigma_X = \sqrt{E(X^2) - \mu_X^2}.$$

Now, let us observe the similarity between the two descriptions. The same kind of similarity between discrete and continuous quantum spectrum stroke P. Dirac when he suggested the combined integral approach, that he denoted by ⨍—meaning 'both integral and sum at once': summing over a discrete spectrum and integration over a continuous spectrum.

To emphasize this similarity even further, as well as to set-up the stage for the path integral, recall the notion of a *cumulative distribution function* of a random variable X, that is a function $F : \mathbb{R} \longrightarrow \mathbb{R}$, defined by

$$F(a) = P(X) \leq a.$$

In particular, suppose that $f(x)$ is the distribution function of X. Then

$$F(x) = \sum_{x_i \leq x} f(x_i), \quad \text{or} \quad F(x) = \int_{-\infty}^{\infty} f(t)\, dt,$$

according to as x is a discrete or continuous random variable. In either case, $F(a) \leq F(b)$ whenever $a \leq b$. Also,

$$\lim_{x \longrightarrow -\infty} F(x) = 0 \quad \text{and} \quad \lim_{x \longrightarrow \infty} F(x) = 1,$$

that is, $F(x)$ is monotonic and its limit to the left is 0 and the limit to the right is 1. Furthermore, its cumulative probability is given by

$$P(a \leq X \leq b) = F(b) - F(a),$$

and the Fundamental Theorem of Calculus tells us that, in the continuum case,

$$f(x) = \partial_x F(x).$$

Now, recall that *Markov stochastic process* is a random process characterized by a *lack of memory*, i.e., the statistical properties of the immediate future are uniquely determined by the present, regardless of the past [Gar85, II06c].

For example, a *random walk* is an example of the *Markov chain*, i.e., a discrete-time Markov process, such that the motion of the system in consideration is viewed as a sequence of states, in which the transition from one state to another depends only on the preceding one, or the probability of the system being in state k depends only on the previous state $k-1$. The property of a Markov chain of prime importance in biomechanics is the existence of an *invariant distribution of states*: we start with an initial state x_0 whose absolute probability is 1. Ultimately the states should be distributed according to a specified distribution.

Between the pure deterministic dynamics, in which all DOF of the system in consideration are explicitly taken into account, leading to classical dynamical equations, for example in Hamiltonian form (using $\partial_x \equiv \partial/\partial x$),

$$\dot{q}^i = \partial_{p_i} H, \qquad \dot{p}_i = -\partial_{q^i} H, \tag{5.51}$$

(where q^i, p_i are coordinates and momenta, while $H = H(q,p)$ is the total system energy)—and pure stochastic dynamics (Markov process), there is so-called *hybrid dynamics*, particularly *Brownian dynamics*, in which some of DOF are represented only through their *stochastic influence* on others. As an example, suppose a system of particles interacts with a viscous medium. Instead of specifying a detailed interaction of each particle with the particles of the viscous medium, we represent the medium as a *stochastic force* acting on the particle. The stochastic force *reduces the dimensionally* of the dynamics.

Recall that the Brownian dynamics represents the phase-space trajectories of a collection of particles that individually obey *Langevin rate equations* in the field of force (i.e., the particles interact with each other via some deterministic force). For a free particle, the Langevin equation reads [Gar85]:

$$m\dot{v} = R(t) - \beta v,$$

where m denotes the mass of the particle and v its velocity. The right-hand side represent the coupling to a *heat bath*; the effect of the random force $R(t)$ is to heat the particle. To balance overheating (on the average), the particle is subjected to *friction* β. In humanoid dynamics this is performed with the Rayleigh–Van der Pol's *dissipation*. Formally, the solution to the Langevin equation can be written as

$$v(t) = v(0) \exp\left(-\frac{\beta}{m}t\right) + \frac{1}{m} \int_0^t \exp[-(t-\tau)\beta/m]R(\tau)\,d\tau,$$

where the integral on the right-hand side is a *stochastic integral* and the solution $v(t)$ is a random variable. The stochastic properties of the solution depend significantly on the stochastic properties of the random force $R(t)$. In the Brownian dynamics the random force $R(t)$ is Gaussian distributed. Then the problem boils down to finding the solution to the Langevin stochastic differential equation with the supplementary condition (zero and mean variance)

$$\langle R(t)\rangle = 0, \qquad \langle R(t)R(0)\rangle = 2\beta k_B T \delta(t),$$

where $\langle.\rangle$ denotes the mean value, T is temperature, k_B-*equipartition* (i.e., uniform distribution of energy) coefficient, Dirac $\delta(t)$-function.

Algorithm for computer simulation of the Brownian dynamics (for a single particle) can be written as [Hee90]:

1. Assign an initial position and velocity.
2. Draw a random number from a Gaussian distribution with mean zero and variance.
3. Integrate the velocity to get v^{n+1}.
4. Add the random component to the velocity.

Another approach to taking account the coupling of the system to a heat bath is to subject the particles to collisions with *virtual particles* [Hee90]. Such collisions are imagined to affect only momenta of the particles, hence they affect the kinetic energy and introduce fluctuations in the total energy. Each stochastic collision is assumed to be an instantaneous event affecting only one particle.

The collision-coupling idea is incorporated into the Hamiltonian model of dynamics (5.51) by adding a stochastic force $R_i = R_i(t)$ to the $\dot{p}$ equation

$$\dot{q}^i = \partial_{p_i} H, \qquad \dot{p}_i = -\partial_{q^i} H + R_i(t).$$

On the other hand, the so-called *Ito stochastic integral* represents a kind of classical Riemann–Stieltjes integral from linear functional analysis, which is (in 1D case) for an arbitrary time-function $G(t)$ defined as the *mean square limit*

$$\int_{t_0}^t G(t)dW(t) = \text{ms} \lim_{n \to \infty} \left\{ \sum_{i=1}^n G(t_{i-1})[W(t_i) - W(t_{i-1})] \right\}.$$

Now, the general ND Markov process can be defined by *Ito* stochastic differential equation (SDE),

$$dx_i(t) = A_i[x^i(t), t]dt + B_{ij}[x^i(t), t] \, dW^j(t),$$
$$x^i(0) = x_{i0}, \quad (i, j = 1, \ldots, N)$$

or corresponding *Ito stochastic integral equation*

$$x^i(t) = x^i(0) + \int_0^t ds \, A_i[x^i(s), s] + \int_0^t dW^j(s) B_{ij}[x^i(s), s],$$

in which $x^i(t)$ is the variable of interest, the vector $A_i[x(t), t]$ denotes deterministic *drift*, the matrix $B_{ij}[x(t), t]$ represents continuous stochastic *diffusion fluctuations*, and $W^j(t)$ is an N-variable *Wiener process* (i.e., generalized Brownian motion) [Wie61], and $dW^j(t) = W^j(t + dt) - W^j(t)$.

Now, there are three well-known special cases of the *Chapman–Kolmogorov equation* (see [Gar85]):

1. When both $B_{ij}[x(t), t]$ and $W(t)$ are zero, i.e., in the case of pure deterministic motion, it reduces to the *Liouville equation*

$$\partial_t P(x', t'|x'', t'') = -\sum_i \frac{\partial}{\partial x^i} \left\{ A_i[x(t), t] P(x', t'|x'', t'') \right\}.$$

2. When only $W(t)$ is zero, it reduces to the *Fokker–Planck equation*

$$\partial_t P(x', t'|x'', t'') = -\sum_i \frac{\partial}{\partial x^i} \left\{ A_i[x(t), t] P(x', t'|x'', t'') \right\}$$
$$+ \frac{1}{2} \sum_{ij} \frac{\partial^2}{\partial x^i \partial x^j} \left\{ B_{ij}[x(t), t] P(x', t'|x'', t'') \right\}.$$

3. When both $A_i[x(t), t]$ and $B_{ij}[x(t), t]$ are zero, i.e., the state-space consists of integers only, it reduces to the *Master equation* of discontinuous jumps

$$\partial_t P(x', t'|x'', t'') = \int dx \, W(x'|x'', t) P(x', t'|x'', t'')$$
$$- \int dx \, W(x''|x', t) P(x', t'|x'', t'').$$

The *Markov assumption* can now be formulated in terms of the conditional probabilities $P(x^i, t_i)$: if the times t_i increase from right to left, the conditional probability is determined entirely by the knowledge of the most recent condition. Markov process is generated by a set of conditional probabilities whose probability-density $P = P(x', t' | x'', t'')$ evolution obeys the general *Chapman–Kolmogorov integro-differential equation*

$$\partial_t P = -\sum_i \frac{\partial}{\partial x^i} \{A_i[x(t), t]P\} + \frac{1}{2} \sum_{ij} \frac{\partial^2}{\partial x^i \partial x^j} \{B_{ij}[x(t), t]P\}$$
$$+ \int dx \{W(x'|x'', t)P - W(x''|x', t)P\}$$

including *deterministic drift, diffusion fluctuations* and *discontinuous jumps* (given respectively in the first, second and third terms on the r.h.s.).

It is this general Chapman–Kolmogorov integro-differential equation, with its conditional probability density evolution, $P = P(x', t' | x'', t'')$, that we are going to model by the Feynman path integral $\int$, providing us with the physical insight behind the abstract (conditional) probability densities.

5.8.2 The Life Space Foam

General nonlinear attractor dynamics, both deterministic and stochastic, as well as possibly chaotic, developed in the framework of Feynman path integrals, have recently [IA07] been applied to Lewinian field-theoretic psychodynamics [Lew51, Lew97, Gol99b], resulting in the development of a new concept of life-space foam (LSF) as a natural medium for motivational and cognitive psychodynamics. According to the LSF-formalism, the classic Lewinian life space can be macroscopically represented as a smooth manifold with steady force-fields and behavioral paths, while at the microscopic level it is more realistically represented as a collection of wildly fluctuating force-fields, (loco)motion paths and local geometries (and topologies with holes).

We have used the new LSF concept to develop modeling framework for motivational dynamics (MD) and induced cognitive dynamics (CD) (see [IA07]). Motivation processes both precede and coincide with every goal-directed action. Usually these motivation processes include the sequence of the following four feedforward *phases* [IA07]: (*)

1. *Intention Formation $\mathcal{F}$*, including: decision making, commitment building, etc.
2. *Action Initiation $\mathcal{I}$*, including: handling conflict of motives, resistance to alternatives, etc.
3. *Maintaining the Action $\mathcal{M}$*, including: resistance to fatigue, distractions, etc.
4. *Termination $\mathcal{T}$*, including parking and avoiding addiction, i.e., staying in control.

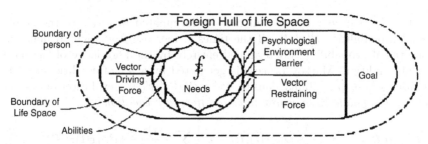

Fig. 5.4. Diagram of the *life space foam*: classical representation of Lewinian life space, with an adaptive path integral $\oint$ (denoting integration over continuous paths and summation over discrete Markov jumps) acting inside it and generating microscopic fluctuation dynamics.

With each of the phases $\{\mathcal{F}, \mathcal{I}, \mathcal{M}, \mathcal{T}\}$ in (*), we can associate a *transition propagator*—an ensemble of (possibly crossing) feedforward paths propagating through the 'wood of obstacles' (including topological holes in the LSF, see Fig. 5.5), so that the complete transition is a product of propagators (as well as sum over paths). All the phases-propagators are controlled by a unique *Monitor* feedback process.

A set of least-action principles is used to model the smoothness of global, macro-level LSF paths, fields and geometry, according to the following prescription. The action $S[\Phi]$, psycho-physical dimensions of *Energy* × *Time* = *Effort* and depending on macroscopic paths, fields and geometries (commonly denoted by an abstract field symbol Φ^i) is defined as a temporal integral from the initial time instant t_{ini} to the final time instant t_{fin},

$$S[\Phi] = \int_{t_{ini}}^{t_{fin}} \mathfrak{L}[\Phi]\, dt, \tag{5.52}$$

with Lagrangian density given by

$$\mathfrak{L}[\Phi] = \int d^n x \mathcal{L}(\Phi_i, \partial_{x^j}\Phi^i),$$

where the integral is taken over all n coordinates $x^j = x^j(t)$ of the LSF, and $\partial_{x^j}\Phi^i$ are time and space partial derivatives of the Φ^i-variables over coordinates. The standard least action principle

$$\delta S[\Phi] = 0, \tag{5.53}$$

gives, in the form of the so-called Euler–Lagrangian equations, a shortest (loco)motion path, an extreme force-field, and a life-space geometry of minimal curvature (and without holes). In this way, we effectively derive a unique globally smooth transition map

$$F : INTENTION_{t_{ini}} \longrightarrow ACTION_{t_{fin}}, \tag{5.54}$$

performed at a macroscopic (global) time-level from some initial time t_{ini} to the final time t_{fin}. In this way, we have obtained macro-objects in the global LSF: a single path described by Newtonian-like equation of motion, a single force-field described by Maxwellian-like field equations, and a single obstacle-free Riemannian geometry (with global topology without holes).

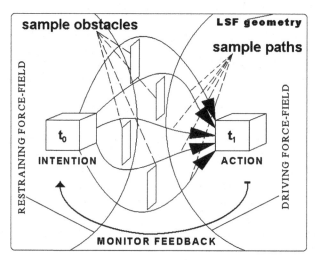

Fig. 5.5. *Transition-propagator* corresponding to each of the motivational phases $\{\mathcal{F}, \mathcal{I}, \mathcal{M}, \mathcal{T}\}$, consisting of an ensemble of feedforward paths propagating through the 'wood of obstacles'. The paths affected by driving and restraining force-fields, as well as by the local LSF-geometry. Transition goes from *Intention*, occurring at a sample time instant t_0, to *Action*, occurring at some later time t_1. Each propagator is controlled by its own *Monitor* feedback.

To model the corresponding local, micro-level LSF structures of rapidly fluctuating cognitive dynamics, an adaptive path integral is formulated, defining a multi-phase and multi-path (multi-field and multi-geometry) transition amplitude from the state of *Intention* to the state of *Action*,

$$\langle Action|Intention\rangle_{total} := \sum \int \mathcal{D}[w\Phi] \, \mathrm{e}^{\mathrm{i}S[\Phi]}, \qquad (5.55)$$

where the Lebesgue integration is performed over all continuous $\Phi^i_{con} = paths + fields + geometries$, while summation is performed over all discrete processes and regional topologies Φ^j_{dis}. The symbolic differential $\mathcal{D}[w\Phi]$ in the general path integral (5.55), represents an adaptive path measure, defined as a weighted product (with $i = 1, \ldots, n = con + dis$)

$$\mathcal{D}[w\Phi] = \lim_{N \longrightarrow \infty} \prod_{s=1}^{N} w_s d\Phi^i_s. \qquad (5.56)$$

The adaptive path integral (5.55)–(5.56) represents an ∞-dimensional neural network, with weights w updating by the general rule [IA07]

$$new\ value(t+1) = old\ value(t) + innovation(t).$$

The adaptive path integral (5.55) incorporates the local Bernstein adaptation process [Ber67, Ber82] according to Bernstein's discriminator concept

$$desired\ state\ SW(t+1) = current\ state\ IW(t) + adjustment\ step\ \Delta W(t)$$

as well as the augmented finite state machine of Brooks' subsumption architecture [Bro86a, Bro89, Bro90], with a networked behavior function

$$final\ state\ w(t+1) = current\ state\ w(t) + adjustment\ behavior\ f(\Delta w(t)).$$

We remark here that the traditional neural networks approaches are known for their classes of functions they can represent. This limitation has been attributed to their low-dimensionality (the largest neural networks are limited to the order of 10^5 dimensions [IE08]). The proposed path integral approach represents a new family of function-representation methods, which potentially offers a basis for a fundamentally more expansive solution.

On the macro-level in *LSF* we have the (loco)*motion action principle*

$$\delta S[x] = 0,$$

with the *Newtonian-like action* $S[x]$ given by

$$S[x] = \int_{t_{ini}}^{t_{fin}} dt \left[\frac{1}{2} g_{ij} \dot{x}^i \dot{x}^j + \varphi^i(x^i) \right], \qquad (5.57)$$

where $\dot{x}^i$ represents motivational (loco)motion velocity vector with cognitive *processing speed*. The first bracket term in (5.57) represents the kinetic energy T,

$$T = \frac{1}{2} g_{ij} \dot{x}^i \dot{x}^j,$$

generated by the *Riemannian metric tensor* g_{ij}, while the second bracket term, $\varphi^i(x^i)$, denotes the family of potential force-fields, driving the (loco)motions $x^i = x^i(t)$ (the *strengths* of the fields $\varphi^i(x^i)$ depend on their positions x^i in LSF). The corresponding Euler–Lagrangian equation gives the Newtonian-like equation of motion

$$\frac{d}{dt} T_{\dot{x}^i} - T_{x^i} = -\varphi^i_{x^i}, \qquad (5.58)$$

(subscripts denote the partial derivatives), which can be put into the standard Lagrangian form

$$\frac{d}{dt} L_{\dot{x}^i} = L_{x^i}, \quad \text{with } L = T - \varphi^i(x^i).$$

Now, according to Lewin, the life space also has a sophisticated topological structure. As a Riemannian smooth n-manifold, the LSF-manifold Σ gives rise to its fundamental n-groupoid, or n-category $\Pi_n(\Sigma)$ (see [II06c, II07e]). In $\Pi_n(\Sigma)$, 0-cells are *points* in Σ; 1-cells are *paths* in Σ (i.e., parameterized smooth maps $f : [0,1] \rightarrow \Sigma$); 2-cells are *smooth homotopies* (denoted by $\simeq$) *of paths* relative to endpoints (i.e., parameterized smooth maps $h : [0,1] \times [0,1] \rightarrow \Sigma$); 3-cells are *smooth homotopies of homotopies* of paths in Σ (i.e., parameterized smooth maps $j : [0,1] \times [0,1] \times [0,1] \rightarrow \Sigma$). Categorical *composition* is defined by *pasting* paths and homotopies. In this way, the following *recursive homotopy dynamics* emerges on the LSF-manifold Σ (**):

0-cell: $x_0 \bullet$ $x_0 \in M$;   in the higher cells below: $t, s \in [0,1]$;

1-cell: $x_0 \bullet \xrightarrow{\quad f \quad} \bullet x_1$ $f : x_0 \simeq x_1 \in M$,

$f : [0,1] \rightarrow M$, $f : x_0 \mapsto x_1$, $x_1 = f(x_0)$, $f(0) = x_0$, $f(1) = x_1$;

e.g., linear path: $f(t) = (1-t)x_0 + t x_1$;   or

Euler–Lagrangian f-dynamics with endpoint conditions (x_0, x_1):

$$\frac{d}{dt}f_{\dot{x}^i} = f_{x^i}, \quad \text{with } x(0) = x_0, \; x(1) = x_1, \; (i = 1, \dots, n);$$

2-cell: $x_0 \bullet$ $\overset{f}{\underset{g}{\Downarrow h}}$ $\bullet x_1$ $h : f \simeq g \in M$,

$h : [0,1] \times [0,1] \rightarrow M$, $h : f \mapsto g$, $g = h(f(x_0))$,

$h(x_0, 0) = f(x_0)$, $h(x_0, 1) = g(x_0)$, $h(0,t) = x_0$, $h(1,t) = x_1$

e.g., linear homotopy: $h(x_0, t) = (1-t)f(x_0) + t g(x_0)$;   or

homotopy between two Euler–Lagrangian (f, g)-dynamics

with the same endpoint conditions (x_0, x_1):

$$\frac{d}{dt}f_{\dot{x}^i} = f_{x^i}, \quad \text{and} \quad \frac{d}{dt}g_{\dot{x}^i} = g_{x^i} \quad \text{with } x(0) = x_0, \; x(1) = x_1;$$

3-cell: $x_0 \bullet$ $h \left(\overset{f}{\underset{g}{\overset{j}{\Longrightarrow}}} \right) i$ $\bullet x_1$ $j : h \simeq i \in M$,

$j : [0,1] \times [0,1] \times [0,1] \rightarrow M$, $j : h \mapsto i$, $i = j(h(f(x_0)))$

$j(x_0, t, 0) = h(f(x_0))$, $j(x_0, t, 1) = i(f(x_0))$,

$j(x_0, 0, s) = f(x_0)$, $j(x_0, 1, s) = g(x_0)$,

$j(0, t, s) = x_0$, $j(1, t, s) = x_1$

e.g., linear composite homotopy: $j(x_0, t, s) = (1 - t) h(f(x_0)) + t\, i(f(x_0))$; or, homotopy between two homotopies between above two Euler–Lagrangian (f, g)-dynamics with the same endpoint conditions (x_0, x_1).

On the micro-LSF level, instead of a single path defined by the Newtonian-like equation of motion (5.58), we have an ensemble of fluctuating and crossing paths with weighted probabilities (of the unit total sum). This ensemble of micro-paths is defined by the simplest instance of our adaptive path integral (5.55), similar to the Feynman's original *sum over histories*,

$$\langle Action | Intention \rangle_{paths} = \oint \mathcal{D}[wx]\, \mathrm{e}^{\mathrm{i}S[x]}, \tag{5.59}$$

where $\mathcal{D}[wx]$ is a functional measure on the *space of all weighted paths*, and the exponential depends on the action $S[x]$ given by (5.57). This procedure can be redefined in a mathematically cleaner way if we Wick-rotate the time variable t to imaginary values, $t \mapsto \tau = \mathrm{i}t$, thereby making all integrals real:

$$\oint \mathcal{D}[wx]\, \mathrm{e}^{\mathrm{i}S[x]} \xrightarrow{\ Wick\ } \oint \mathcal{D}[wx]\, \mathrm{e}^{-S[x]}. \tag{5.60}$$

Discretization of (5.60) gives the standard *thermo-dynamic-like partition function*

$$Z = \sum_j \mathrm{e}^{-w_j E^j / T}, \tag{5.61}$$

where E^j is the motion energy eigenvalue (reflecting each possible motivational energetic state), T is the temperature-like environmental control parameter, and the sum runs over all motion energy eigenstates (labeled by the index j). From (5.61), we can further calculate all thermodynamic-like and statistical properties of MD and CD, as for example, *transition entropy*, $S = k_B \ln Z$, etc.

Noisy Decision Making in the LSF

From CD-perspective, our adaptive path integral (5.59) calculates all (alternative) pathways of information flow during the transition *Intention $\rightarrow$ Action*. In the connectionist language, (5.59) represents *activation dynamics*, to which our *Monitor* process gives a kind of *backpropagation* feedback, a common type of supervised learning

$$w_s(t + 1) = w_s(t) - \eta \nabla J(t), \tag{5.62}$$

where η is a small constant, called the *step size*, or the *learning rate*, and $\nabla J(n)$ denotes the gradient of the 'performance hyper-surface' at the t-th iteration.

Now, the basic question about our local decision making process, occurring under uncertainty at the intention formation faze $\mathcal{F}$, is: Which alternative to choose? In our path-integral language this reads: Which path (alternative) should be given the highest probability weight w? This problem can be either iteratively solved by the learning process (5.62), controlled by the *MONITOR* feedback, which we term *algorithmic approach*, or by the local decision making process under uncertainty, which we term *heuristic approach* [IA07]. This qualitative analysis is based on the micro-level interpretation of the Newtonian-like action $S[x]$, given by (5.57) and figuring both processing speed $\dot{x}$ and LTM (i.e., the force-field $\varphi(x)$, see next subsection). Here we consider three different cases:

1. If the potential $\varphi(x)$ is not very dependent upon position $x(t)$, then the more direct paths contribute the most, as longer paths, with higher mean square velocities $[\dot{x}(t)]^2$ make the exponent more negative (after Wick rotation (5.60)).

2. On the other hand, suppose that $\varphi(x)$ does indeed depend on position x. For simplicity, let the potential increase for the larger values of x. Then a direct path does not necessarily give the largest contribution to the overall transition probability, because the integrated value of the potential is higher than over another paths.

3. Finally, consider a path that deviates widely from the direct path. Then $\varphi(x)$ decreases over that path, but at the same time the velocity $\dot{x}$ increases. In this case, we expect that the increased velocity $\dot{x}$ would more than compensate for the decreased potential over the path.

Therefore, the most important path (i.e., the path with the highest weight w) would be the one for which any smaller integrated value of the surrounding field potential $\varphi(x)$ is more than compensated for by an increase in kinetic-like energy $\frac{m}{2}\dot{x}^2$. In principle, this is neither the most direct path, nor the longest path, but rather a middle way between the two. Formally, it is the path along which the average Lagrangian is minimal,

$$\left\langle \frac{m}{2}\dot{x}^2 + \varphi(x) \right\rangle \longrightarrow \min, \tag{5.63}$$

i.e., the *path that requires minimal memory* (both LTM and WM) and *processing speed*. This mechanical result is consistent with the 'cognitive filter theory' of *selective attention* [Bro58], which postulates a low level filter that allows only a limited number of percepts to reach the brain at any time. In this theory, the importance of conscious, directed attention is minimized. The type of attention involving low level filtering corresponds to the concept of *early selection*.

Although we termed this 'heuristic approach' in the sense that we can instantly feel both the processing speed $\dot{x}$ and the LTM field $\varphi(x)$ involved, there is clearly a psycho-physical rule in the background, namely the averaging minimum relation (5.63).

From the decision making point of view, all possible paths (alternatives) represent the *consequences* of decision making. They are, by default, *short-term consequences*, as they are modeled in the micro-time-level. However, the path integral formalism allows calculation of the *long-term consequences*, just by extending the integration time, $t_{fin} \to \infty$. Besides, this *averaging decision mechanics*—choosing the optimal path—actually performs the 'averaging lift' in the LSF: from the micro-level to the macro-level.

For example, one of the simplest types of performance-degrading disturbances in the LSF is what we term motivational fatigue—a motivational drag factor that slows the actors' progress towards their goal. There are two fundamentally different sources of this motivational drag, both leading to apparently the same reduction in performance: (a) tiredness/exhaustion and (b) satiation (e.g., boredom). Both involve the same underlying mechanism (the raising valence of the alternatives to continuing the action) but the alternatives will differ considerably, depending on the properties of the task, from self-preservation/recuperation in the exhaustion case through to competing goals in the satiation case.

The spatial representation of this motivational drag is relatively simple: uni-dimensional LSF-coordinates may be sufficient for most purposes, which makes it attractive for the initial validation of our predictive model. Similarly uncomplicated spatial representations can be achieved for what we term motivational boost derived from the proximity to the goal (including the well-known phenomenon of 'the home stretch'): the closer the goal (e.g., a finishing line) is perceived to be, the stronger its 'pulling power' [Lew51, Lew97]. Combinations of motivational drag and motivational boost effects may be of particular interest in a range of applications. These combinations can be modeled within relatively simple uni-dimensional LSF-coordinate systems.

5.8.3 Geometric Chaos and Topological Phase Transitions

In this section we extend the LSF-formalism to incorporate geometrical chaos [IJP08, II07a, II08a] and associated topological phase transitions.

It is well-known that on the basis of the ergodic hypothesis, statistical mechanics describes the physics of many-degrees of freedom systems by replacing time averages of the relevant observables with ensemble averages. Therefore, instead of using statistical ensembles, we can investigate the Hamiltonian (microscopic) dynamics of a system undergoing a phase transition. The reason for tackling dynamics is twofold. First, there are observables, like Lyapunov exponents, that are intrinsically dynamical. Second, the geometrization of Hamiltonian dynamics in terms of Riemannian geometry provides new observables and, in general, an interesting framework to investigate the phenomenon of phase transitions [CCC97, Pet07]. The geometrical formulation of the dynamics of conservative systems [II06c, II08a] was first used by [Kry79] in his studies on the dynamical foundations of statistical mechanics and subsequently became a standard tool to study abstract systems in ergodic theory.

The simplest, mechanical-like LSF-action in the individual's LSF-manifold Σ has a Riemannian locomotion form [IA07] (summation convention is always assumed)

$$S[q] = \frac{1}{2} \int_{t_{ini}}^{t_{fin}} [a_{ij}\dot{q}^i\dot{q}^j - V(q)]\,dt, \tag{5.64}$$

where a_{ij} is the 'material' metric tensor that generates the total 'kinetic energy' of cognitive (loco)motions defined by their configuration coordinates q^i and velocities $\dot{q}^i$, with the motivational potential energy $V(q)$ and the standard Hamiltonian

$$H(p,q) = \sum_{i=1}^{N} \frac{1}{2}p_i^2 + V(q), \tag{5.65}$$

where p_i are the canonical (loco)motion momenta.

Dynamics of N DOF mechanical-like systems with action (5.64) and Hamiltonian (5.65) are commonly given by the set of geodesic equations [II06c, II07e]

$$\frac{d^2q^i}{ds^2} + \Gamma^i_{jk}\frac{dq^j}{ds}\frac{dq^k}{ds} = 0, \tag{5.66}$$

where Γ^i_{jk} are the Christoffel symbols of the affine Levi-Civita connection of the Riemannian LSF-manifold Σ.

Alternatively, a description of the extrema of the Hamilton's action (5.64) can be obtained using the Eisenhart metric [Eis29] on an enlarged LSF space-time manifold (given by $\{q^0 \equiv t, q^1, \ldots, q^N\}$ plus one real coordinate q^{N+1}), whose arc-length is

$$ds^2 = -2V(q)(dq^0)^2 + a_{ij}dq^idq^j + 2dq^0dq^{N+1}. \tag{5.67}$$

The manifold has a Lorentzian structure [Pet07] and the dynamical trajectories are those geode-sics satisfying the condition $ds^2 = C dt^2$, where C is a positive constant. In this geometrical framework, the instability of the trajectories is the instability of the geodesics, and it is completely determined by the curvature properties of the LSF-manifold Σ according to the Jacobi equation of geodesic deviation [II06c, II07e]

$$\frac{D^2J^i}{ds^2} + R^i_{jkm}\frac{dq^j}{ds}J^k\frac{dq^m}{ds} = 0, \tag{5.68}$$

whose solution J, usually called Jacobi variation field, locally measures the distance between nearby geodesics; D/ds stands for the covariant derivative along a geodesic and R^i_{jkm} are the components of the Riemann curvature tensor of the LSF-manifold Σ.

Using the Eisenhart metric (5.67), the relevant part of the Jacobi equation (5.68) is given by the tangent dynamics equation [CCP96b, CCC97]

$$\frac{d^2J^i}{dt^2} + R^i_{0k0}J^k = 0, \quad (i = 1, \ldots, N), \tag{5.69}$$

where the only non-vanishing components of the curvature tensor of the LSF-manifold Σ are

$$R^i_{0k0} = \partial^2 V / \partial q^i \partial q^j.$$

The tangent dynamics equation (5.69) is commonly used to define Lyapunov exponents in dynamical systems given by the Riemannian action (5.64) and Hamiltonian (5.65), using the formula [CPC00]

$$\lambda_1 = \lim_{t \to \infty} 1/2t \log \left(\sum_{i=1}^N [J_i^2(t) + \dot{J}_i^2(t)] \bigg/ \sum_{i=1}^N [J_i^2(0) + \dot{J}_i^2(0)] \right). \quad (5.70)$$

Lyapunov exponents measure the strength of dynamical chaos.

Now, to relate these results to topological phase transitions within the LSF-manifold Σ, recall that any two high-dimensional manifolds Σ_v and $\Sigma_{v'}$ have the same topology if they can be continuously and differentiably deformed into one another, that is if they are diffeomorphic. Thus by topology change the loss of diffeomorphicity is meant [Pet07]. In this respect, the so-called topological theorem [FP04] says that non-analyticity is the 'shadow' of a more fundamental phenomenon occurring in the system's configuration manifold (in our case the LSF-manifold): a topology change within the family of equipotential hypersurfaces

$$\Sigma_v = \{(q^1, \ldots, q^N) \in \mathbb{R}^N \mid V(q^1, \ldots, q^N) = v\},$$

where V and q^i are the microscopic interaction potential and coordinates respectively. This topological approach to PTs stems from the numerical study of the dynamical counterpart of phase transitions, and precisely from the observation of discontinuous or cuspy patterns displayed by the largest Lyapunov exponent λ_1 at the transition energy [CPC00]. Lyapunov exponents cannot be measured in laboratory experiments, at variance with thermodynamic observables, thus, being genuine dynamical observables they are only be estimated in numerical simulations of the microscopic dynamics. If there are critical points of V in configuration space, that is points $q_c = [\bar{q}_1, \ldots, \bar{q}_N]$ such that $\nabla V(q)|_{q=q_c} = 0$, according to the Morse Lemma [Hir76], in the neighborhood of any critical point q_c there always exists a coordinate system $\tilde{q}(t) = [\tilde{q}^1(t), \ldots, \tilde{q}^N(t)]$ for which

$$V(\tilde{q}) = V(q_c) - \tilde{q}_1^2 - \cdots - \tilde{q}_k^2 + \tilde{q}_{k+1}^2 + \cdots + \tilde{q}_N^2, \quad (5.71)$$

where k is the index of the critical point, i.e., the number of negative eigenvalues of the Hessian of the potential energy V. In the neighborhood of a critical point of the LSF-manifold Σ, (5.71) yields

$$\partial^2 V / \partial q^i \partial q^j = \pm \delta_{ij},$$

which gives k unstable directions which contribute to the exponential growth of the norm of the tangent vector J [CPC00].

This means that the strength of dynamical chaos within the individual's LSF-manifold Σ, measured by the largest Lyapunov exponent λ_1 given by (5.70), is affected by the existence of critical points q_c of the potential energy $V(q)$. However, as $V(q)$ is bounded below, it is a good Morse function, with no vanishing eigenvalues of its Hessian matrix. According to Morse theory [Hir76], the existence of critical points of V is associated with topology changes of the hypersurfaces $\{\Sigma_v\}_{v \in \mathbb{R}}$.

More precisely, let $V_N(q_1, \ldots, q_N) : R^N \to R$, be a smooth, bounded from below, finite-range and confining potential.[8] Denote by $\Sigma_v = V^{-1}(v)$, $v \in R$, its level sets, or equipotential hypersurfaces, in the LSF-manifold Σ. Then let $\bar{v} = v/N$ be the potential energy per degree of freedom. If there exists N_0, and if for any pair of values $\bar{v}$ and $\bar{v}'$ belonging to a given interval $I_{\bar{v}} = [\bar{v}_0, \bar{v}_1]$ and for any $N > N_0$ then the sequence of the Helmoltz free energies $\{F_N(\beta)\}_{N \in \mathbb{N}}$— where $\beta = 1/T$ (T is the temperature) and $\beta \in I_\beta = (\beta(\bar{v}_0), \beta(\bar{v}_1))$—is uniformly convergent at least in $C^2(I_\beta)$ [the space of twice differentiable functions in the interval I_β], so that $\lim_{N \to \infty} F_N \in C^2(I_\beta)$ and neither first nor second order phase transitions can occur in the (inverse) temperature interval $(\beta(\bar{v}_0), \beta(\bar{v}_1))$, where the inverse temperature is defined as [FP04, Pet07]

$$\beta(\bar{v}) = \partial S_N^{(-)}(\bar{v})/\partial \bar{v}, \quad \text{while } S_N^{(-)}(\bar{v}) = N^{-1} \log \int_{V(q) \leq \bar{v}N} d^N q$$

is one of the possible definitions of the micro-canonical configurational entropy. The intensive variable $\bar{v}$ has been introduced to ease the comparison between quantities computed at different N-values.

This theorem means that a topology change of the $\{\Sigma_v\}_{v \in \mathbb{R}}$ at some v_c is a necessary condition for a phase transition to take place at the corresponding energy value. The topology changes implied here are those described within the framework of Morse theory through 'attachment of handles' [Hir76] to the LSF-manifold Σ.

In the LSF path-integral language [IA07], we can say that suitable topology changes of equipotential submanifolds of the individual's LSF-manifold Σ can entail thermodynamic-like phase transitions [Hak83, Hak93, Hak96], according to the general formula:

$$\langle \text{phase out} \mid \text{phase in} \rangle := \oint_{\text{topology-change}} \mathcal{D}[w\Phi]\, e^{iS[\Phi]}.$$

The statistical behavior of the LSF-(loco)motion system (5.64) with the standard Hamiltonian (5.65) is encompassed, in the canonical ensemble, by its partition function, given by the phase-space path integral [II07e]

$$Z_N = \oint_{\text{top-ch}} \mathcal{D}[p]\mathcal{D}[q] \exp\left\{ i \int_t^{t'} [p\dot{q} - H(p,q)]\, d\tau \right\}, \qquad (5.72)$$

---

[8] These requirements for V are fulfilled by standard interatomic and intermolecular interaction potentials, as well as by classical spin potentials.

where we have used the shorthand notation

$$\oint_{\text{top-ch}} \mathcal{D}[p]\mathcal{D}[q] \equiv \int \prod_{\tau} \frac{dq(\tau)dp(\tau)}{2\pi}.$$

The phase-space path integral (5.72) can be calculated as the partition function [FPS00a],

$$Z_N(\beta) = \int \prod_{i=1}^{N} dp_i dq^i e^{-\beta H(p,q)} = \left(\frac{\pi}{\beta}\right)^{\frac{N}{2}} \int \prod_{i=1}^{N} dq^i e^{-\beta V(q)}$$

$$= \left(\frac{\pi}{\beta}\right)^{\frac{N}{2}} \int_0^\infty dv\, e^{-\beta v} \int_{\Sigma_v} \frac{d\sigma}{\|\nabla V\|}, \qquad (5.73)$$

where the last term is written using the so-called co-area formula [Fed69], and v labels the equipotential hypersurfaces Σ_v of the LSF-manifold Σ,

$$\Sigma_v = \{(q^1,\dots,q^N) \in \mathbb{R}^N | V(q^1,\dots,q^N) = v\}.$$

Equation (5.73) shows that the relevant statistical information is contained in the canonical configurational partition function

$$Z_N^C = \int \prod dq^i V(q) e^{-\beta V(q)}.$$

Note that Z_N^C is decomposed, in the last term of (5.73), into an infinite summation of geometric integrals,

$$\int_{\Sigma_v} d\sigma/\|\nabla V\|,$$

defined on the $\{\Sigma_v\}_{v \in \mathbb{R}}$. Once the microscopic interaction potential $V(q)$ is given, the configuration space of the system is automatically foliated into the family $\{\Sigma_v\}_{v \in \mathbb{R}}$ of these equipotential hypersurfaces. Now, from standard statistical mechanical arguments we know that, at any given value of the inverse temperature β, the larger the number N, the closer to $\Sigma_v \equiv \Sigma_{u_\beta}$ are the microstates that significantly contribute to the averages, computed through $Z_N(\beta)$, of thermodynamic observables. The hypersurface Σ_{u_β} is the one associated with

$$u_\beta = (Z_N^C)^{-1} \int \prod dq^i V(q) e^{-\beta V(q)},$$

the average potential energy computed at a given β. Thus, at any β, if N is very large the effective support of the canonical measure shrinks very close to a single $\Sigma_v = \Sigma_{u_\beta}$. Hence, the basic origin of a phase transition lies in a suitable topology change of the $\{\Sigma_v\}$, occurring at some v_c [FPS00a]. This topology change induces the singular behavior of the thermodynamic observables at a phase transition. It is conjectured that the counterpart of a phase transition is a breaking of diffeomorphicity among the surfaces Σ_v, it is appropriate to choose a diffeomorphism invariant to probe if and how the topology of the Σ_v changes as a function of v. Fortunately, such a topological invariant exists, the Euler characteristic of the LSF-manifold Σ, defined by [II06c, II07e]

$$\chi(\Sigma) = \sum_{k=0}^{N} (-1)^k b_k(\Sigma), \tag{5.74}$$

where the Betti numbers $b_k(\Sigma)$ are diffeomorphism invariants.[9] This homological formula can be simplified by the use of the Gauss–Bonnet–Hopf theorem, that relates $\chi(\Sigma)$ with the total Gauss–Kronecker curvature K_G of the LSF-manifold Σ

$$\chi(\Sigma) = \int_{\Sigma} K_G \, d\sigma,$$
$$\text{where } d\sigma = \sqrt{\det(a)} \, dx^1 dx^2 \cdots dx^n \tag{5.75}$$

is the invariant volume measure of the LSF-manifold Σ and a is the determinant of the LSF metric tensor a_{ij}. For technical details of this topological approach, see [II07e, II08a].

The domain of validity of the 'quantum' is not restricted to the microscopic world [Ume93]. There are macroscopic features of classically behaving systems, which cannot be explained without recourse to the quantum dynamics. This field theoretic model leads to the view of the phase transition as a condensation that is comparable to the formation of fog and rain drops from water vapor, and that might serve to model both the gamma and beta phase transitions. According to such a model, the production of activity with long-range correlation in the brain takes place through the mechanism of spontaneous breakdown of symmetry (SBS), which has for decades been shown to describe long-range correlation in condensed matter physics. The adoption of such a field theoretic approach enables modeling of the whole cerebral hemisphere and its hierarchy of components down to the atomic level as a fully integrated macroscopic quantum system, namely as a macroscopic system which is a quantum system not in the trivial sense that it is made, like all existing matter, by quantum components such as atoms and molecules, but in the sense that some of its macroscopic properties can best be described with recourse to quantum dynamics (see [FV06] and references therein).

Phase transitions can also be associated with autonomous robot competence levels, as informal specifications of desired classes of behaviors for robots over all environments they will encounter, as described by Brooks' subsumption architecture approach [Bro86a, Bro89, Bro90]. The distributed network

---

[9] The Betti numbers b_k are the dimensions of the de Rham's cohomology vector spaces $H^k(\Sigma; \mathbb{R})$ (therefore the b_k are integers).

of augmented finite-state machines can exist in different phases or modalities of their state-space variables, which determine the systems intrinsic behavior. The phase transition represented by this approach is triggered by either internal (a set-point) or external (a command) control stimuli, such as a command to transition from a sleep mode to awake mode, or walking to running.

5.8.4 Joint Action of Several Agents

In this section we propose an LSF-based model of the joint action of two or more actors, where actors can be both humans and robots. This joint action takes place in the joint LSF manifold Σ_J, composed of individual LSF manifolds $\Sigma_\alpha, \Sigma_\beta, \ldots$. It has a sophisticated geometrical and dynamical structure as follows.

To model the dynamics of the two-actor co-action, we propose to associate each of the actors with a set of their own time dependent trajectories, which constitutes an n-dimensional Riemannian LSF-manifold, $\Sigma_\alpha = \{x^i(t_i)\}$ and $\Sigma_\beta = \{y^j(t_j)\}$, respectively. Their associated tangent bundles contain their individual nD (loco)motion velocities, $T\Sigma_\alpha = \{\dot{x}^i(t_i) = dx^i/dt_i\}$ and $T\Sigma_\beta = \{\dot{y}^j(t_j) = dy^j/dt_j\}$. Further, following the general formalism of [IA07], outlined in the introduction, we use the modeling machinery consisting of: (i) Adaptive joint action at the top-master level, describing the externally-appearing deterministic, continuous and smooth dynamics, and (ii) Corresponding adaptive path integral (5.55) at the bottom-slave level, describing a wildly fluctuating dynamics including both continuous trajectories and Markov chains. This lower-level joint dynamics can be further discretized into a partition function of the corresponding statistical dynamics.

The smooth joint action with two terms, representing cognitive/motivational potential energy and physical kinetic energy, is formally given by:

$$A[x, y; t_i, t_j] = \frac{1}{2} \int_{t_i} \int_{t_j} \alpha_i \beta_j \delta(I_{ij}^2) \dot{x}^i(t_i) \dot{y}^j(t_j) \, dt_i dt_j + \frac{1}{2} \int_t g_{ij} \dot{x}^i(t) \dot{x}^j(t) \, dt,$$

with $I_{ij}^2 = \left[x^i(t_i) - y^j(t_j)\right]^2$, where $IN \leq t_i, t_j, t \leq OUT$.   (5.76)

The first term in (5.76) represents potential energy of the cognitive/motivational interaction between the two agents α_i and β_j.[10] It is a double integral over a delta function of the square of interval I^2 between two points on the paths in their Life-Spaces; thus, interaction occurs only when this interval, representing the motivational cognitive distance between the two agents, vanishes. Note that the cognitive (loco) motions of the two agents $\alpha_i[x^i(t_i)]$ and $\beta_j[y^j(t_j)]$, generally occur at different times t_i and t_j unless $t_i = t_j$, when cognitive synchronization occurs.

The second term in (5.76) represents kinetic energy of the physical interaction. Namely, when the cognitive synchronization in the first term takes

---

[10] Although, formally, this term contains cognitive velocities, it still represents 'potential energy' from the physical point of view.

place, the second term of physical kinetic energy is activated in the common manifold, which is one of the agents' Life Spaces, say $\Sigma_\alpha = \{x^i(t_i)\}$.

The reason why we have chosen the action (5.76) as a macroscopic model for human joint action is that (5.76) naturally represents the transition map,

$$A[x,y;t_i,t_j] : \text{MENTAL INTENTION} \overset{Synch}{\Longrightarrow} \text{PHYSICAL ACTION},$$

from mutual cognitive intention to joint physical action, in which the joint action starts after the mutual cognitive intention is synchronized. In simple words, "we can efficiently act together only after we have tuned-up our intentions".

Conversely, if we have a need to represent coaction of three actors, say α_i, β_j and γ_k (e.g., α_i in charge of acceleration, β_j—deceleration and γ_k-steering), we can associate each of them with an nD Riemannian Life–Space manifold, $\Sigma_\alpha = \{x^i(t_i)\}$, $\Sigma_\beta = \{y^j(t_j)\}$, and $\Sigma_\gamma = \{z^k(t_k)\}$, respectively, with the corresponding tangent bundles containing their individual (loco) motion velocities, $T\Sigma_\alpha = \{\dot{x}^i(t_i) = dx^i/dt_i\}$, $T\Sigma_\beta = \{\dot{y}^j(t_j) = dy^j/dt_j\}$ and $T\Sigma_\gamma = \{\dot{z}^k(t_k) = dz^k/dt_k\}$. Then, instead of (5.99) we have

$$A[t_i,t_j,t_k;t] = \frac{1}{2}\int_{t_i}\int_{t_j}\int_{t_k} \alpha_i(t_i)\beta_j(t_j)\gamma_k(t_k)$$
$$\times \delta(I_{ijk}^2)\dot{x}^i(t_i)\dot{y}^j(t_j)\dot{z}^k(t_k)\, dt_i dt_j dt_k$$
$$+ \frac{1}{2}\int_t W_{rs}^M(t,q,\dot{q})\dot{q}^r\dot{q}^s\, dt, \qquad (5.77)$$

where $IN \le t_i, t_j, t_k, t \le OUT$, with

$$I_{ijk}^2 = [x^i(t_i) - y^j(t_j)]^2 + [y^j(t_j) - z^k(t_k)]^2 + [z^k(t_k) - x^i(t_i)]^2.$$

Due to an intrinsic chaotic coupling, the three-actor (or, n-actor, $n > 3$) joint action (5.77) has a considerably more complicated geometrical structure then the bilateral co-action (5.76).[11] It actually happens in the common $3n$D Finsler manifold $\Sigma_J = \Sigma_\alpha \cup \Sigma_\beta \cup \Sigma_\gamma$, parameterized by the local joint coordinates dependent on the common time t. That is, $\Sigma_J = \{q^r(t), r = 1, \ldots, 3n\}$. Geometry of the joint manifold Σ_J is defined by the Finsler metric function $ds = F(q^r, dq^r)$, defined by

$$F^2(q,\dot{q}) = g_{rs}(q,\dot{q})\dot{q}^r\dot{q}^s, \qquad (5.78)$$

and the Finsler tensor $C_{rst}(q,\dot{q})$, defined by (see [II06c, II07e])

$$C_{rst}(q,\dot{q}) = \frac{1}{4}\frac{\partial^3 F^2(q,\dot{q})}{\partial \dot{q}^r \partial \dot{q}^s \partial \dot{q}^t} = \frac{1}{2}\frac{\partial g_{rs}}{\partial \dot{q}^r \partial \dot{q}^s}. \qquad (5.79)$$

---

[11] Recall that the necessary condition for chaos in continuous temporal or spatio-temporal systems is to have three variables with nonlinear couplings between them.

From the Finsler definitions (5.78)–(5.79), it follows that the partial interaction manifolds, $\Sigma_\alpha \cup \Sigma_\beta$, $\Sigma_\beta \cup \Sigma_\gamma$ and $\Sigma_\alpha \cup \Sigma_\gamma$, have Riemannian structures with the corresponding interaction kinetic energies,

$$T_{\alpha\beta} = \frac{1}{2}g_{ij}\dot{x}^i\dot{y}^j, \qquad T_{\alpha\gamma} = \frac{1}{2}g_{ik}\dot{x}^i\dot{z}^k, \qquad T_{\beta\gamma} = \frac{1}{2}g_{jk}\dot{y}^j\dot{z}^k.$$

At the slave level, the adaptive path integral (see [IA07]), representing an ∞-dimensional neural network, corresponding to the adaptive bilateral joint action (5.76), reads

$$\langle OUT|IN \rangle := \oint \mathcal{D}[w,x,y]\, e^{iA[x,y;t_i,t_j]}, \tag{5.80}$$

where the Lebesgue integration is performed over all continuous paths $x^i = x^i(t_i)$ and $y^j = y^j(t_j)$, while summation is performed over all associated discrete Markov fluctuations and jumps. The symbolic differential in the path integral (5.80) represents an adaptive path measure, defined as a weighted product

$$\mathcal{D}[w,x,y] = \lim_{N \to \infty} \prod_{s=1}^{N} w_{ij}^s dx^i dy^j, \quad (i,j = 1,\dots,n). \tag{5.81}$$

Similarly, in case of the triple joint action, the adaptive path integral reads,

$$\langle OUT|IN \rangle := \oint \mathcal{D}[w;x,y,z;q]\, e^{iA[t_i,t_j,t_k;t]}, \tag{5.82}$$

with the adaptive path measure defined by

$$\mathcal{D}[w;x,y,z;q] = \lim_{N \to \infty} \prod_{S=1}^{N} w_{ijkr}^S dx^i dy^j dz^k dq^r,$$
$$(i,j,k = 1,\dots,n; r = 1,\dots,3n). \tag{5.83}$$

The proposed path integral approach represents a new family of more expansive function-representation methods, which is now capable of representing input/output behavior of more than one actor. However, as we add the second and subsequent actors to the model, the requirements for the rigorous geometrical representations of their respective LSFs become nontrivial. For a single actor or a two-actor co-action the Riemannian geometry was sufficient, but it becomes insufficient for modeling the n-actor (with $n \geq 3$) joint action, due to an intrinsic chaotic coupling between the individual actors' LSFs. To model an n-actor joint LSF, we have to use the Finsler geometry, which is a generalization of the Riemannian one. This progression may seem trivial, both from standard psychological point of view, and from computational point of view, but it is not trivial from the geometrical perspective.

5.8.5 Chaos and Bernstein–Brooks Adaptation

From previous sections, we can see that for modeling a two-actor co-action the Riemannian geometry is sufficient. However, it becomes insufficient for modeling the joint action of 3 or more actors, due to an *intrinsic chaotic coupling* between the individual actors. In this case we have to use the Finsler geometry, which is a generalization of the Riemannian one. This corresponds to the well-known fact in chaos theory that in continuous-time systems chaos cannot exist in the phase plane—the third dimension of the system phase-space is necessary for its existence. This also corresponds to the well-known fact of life that a trilateral (or, multilateral) relation is many times more complex then a bilateral relation. (It is so in politics, in business, in marriage, in romantic relationships, in friendship, everywhere... Physicists would say that any bilateral relation(ship) between Alice and Bob is very likely to crash if Chris comes in between, or at least it becomes much more complicated.)

The adaptive path integrals (5.80) and (5.82) incorporate the *local Bernstein adaptation process* [Ber67, Ber82] according to Bernstein's discriminator concept

$$desired\ state\ SW(t+1) = current\ state\ IW(t) + adjustment\ step\ \Delta W(t).$$

The robustness of biological motor control systems in handling excess degrees of freedom has been attributed to a combination of tight hierarchical central planning and multiple levels of sensory feedback-based self-regulation that are relatively autonomous in their operation [BLT96]. These two processes are connected through a top-down process of action script delegation and bottom-up emergency escalation mechanisms. There is a complex interplay between the continuous sensory feedback and motion/action planning to achieve effective operation in uncertain environments (such as movement on uneven terrain cluttered with obstacles). In case of three or more actors, the multilateral feedback/planning loop has the purpose of *chaos control* [OGY90, II07a].

Complementing Bernstein's motor/chaos control principles is Brooks' concept of computational *subsumption architectures* [Bro86a, Bro90], which provides a method for structuring reactive systems from the bottom up using layered sets of behaviors. Each layer implements a particular goal of the agent, which subsumes that of the underlying layers.

For example, a robot's lowest layer could be "avoid an object", on top of it would be the layer "wander around", which in turn lies under "explore the world". The top layer in such a case could represent the ultimate goal of "creating a map". In this configuration, the lowest layers can work as fast-responding mechanisms (i.e., reflexes), while the higher layers can control the main direction to be taken in order to achieve a more abstract goal.

The substrate for this architecture comprises a network of finite state machines augmented with timing elements. A subsumption compiler compiles *augmented finite state machine* descriptions into a special-purpose scheduler to simulate parallelism and a set of finite state machine simulation routines. The resulting networked behavior function can be described conceptually as:

final state $w(t + 1) =$ *current state* $w(t) +$ *adjustment behavior* $f(\Delta w(t))$.

The Bernstein *weights*, or *Brooks nodes*, $w_{ij}^s = w_{ij}^s(t)$ in (5.81) are updated by the *Bernstein loop* during the joint transition process, according to one of the two standard neural learning schemes, in which the micro-time level is traversed in discrete steps, i.e., if $t = t_0, t_1, \ldots, t_s$ then $t + 1 = t_1, t_2, \ldots, t_{s+1}$:

1. A *self-organized, unsupervised* (e.g., Hebbian-like [Heb49]) learning rule:

$$w_{ij}^s(t + 1) = w_{ij}^s(t) + \frac{\sigma}{\eta}(w_{ij}^{s,d}(t) - w_{ij}^{s,a}(t)), \qquad (5.84)$$

where $\sigma = \sigma(t)$, $\eta = \eta(t)$ denote *signal* and *noise*, respectively, while new superscripts d and a denote   *desired* and *achieved* micro-states, respectively; or

2. A certain form of a *supervised gradient descent learning*:

$$w_{ij}^s(t + 1) = w_{ij}^s(t) - \eta \nabla J(t), \qquad (5.85)$$

where η is a small constant, called the *step size*, or the *learning rate*, and $\nabla J(n)$ denotes the gradient of the 'performance hyper-surface' at the t-th iteration.

Both Hebbian and supervised learning[12] are used in local decision making processes, e.g., at the intention formation phase (see [IA07]). Overall, the model presents a set of formalisms to represent time-critical aspects of collective performance in tactical teams. Its applications include hypotheses generation for real and virtual experiments on team performance, both in human teams (e.g., emergency crews) and hybrid human-machine teams (e.g., human-robotic crews). It is of particular value to the latter, as the increasing autonomy of robotic platforms poses non-trivial challenges, not only for the design of their operator interfaces, but also for the design of the teams themselves and their concept of operations.

Our extended LSF formalism is closely related to the Haken–Kelso–Bunz (HKB) model of self-organization in the motor system [HKB85, Kel95], including: multi-stability, phase transitions and hysteresis effects, presenting a contrary view to the purely feedback driven systems. HKB uses the concepts of synergetics (order parameters, control parameters, instability, etc.) and

---

[12] Note that we could also use a reward-based, *reinforcement learning* rule [SB98], in which system learns its *optimal policy*:

$$innovation(t) = |reward(t) - penalty(t)|.$$

the mathematical tools of nonlinearly coupled (nonlinear) dynamical systems to account for self-organized behavior both at the cooperative, coordinative level and at the level of the individual coordinating elements. The HKB model stands as a building block upon which numerous extensions and elaborations have been constructed. In particular, it has been possible to derive it from a realistic model of the cortical sheet in which neural areas undergo a re-organization that is mediated by intra- and inter-cortical connections. Also, the HKB model describes phase transitions ('switches') in coordinated human movement as follows: (i) when the agent begins in the anti-phase mode and speed of movement is increased, a spontaneous switch to symmetrical, in-phase movement occurs; (ii) this transition happens swiftly at a certain critical frequency; (iii) after the switch has occurred and the movement rate is now decreased the subject remains in the symmetrical mode, i.e. she does not switch back; and (iv) no such transitions occur if the subject begins with symmetrical, in-phase movements. The HKB dynamics of the order parameter relative phase as is given by a nonlinear first-order ODE:

$$\dot{\phi} = (\alpha + 2\beta r^2)\sin\phi - \beta r^2 \sin 2\phi,$$

where ϕ is the phase relation (that characterizes the observed patterns of behavior, changes abruptly at the transition and is only weakly dependent on parameters outside the phase transition), r is the oscillator amplitude, while α, β are coupling parameters (from which the critical frequency where the phase transition occurs can be calculated).

From a quantum perspective, closely related to the LSF model are the recent developments in motor control that deal with feedback information and environmental uncertainty [HN08a, HN08b]. The probabilistic nature of human action can be characterized by entropies at the level of the organism, task, and environment. Systematic changes in motor adaptation are characterized as task-organism and environment-organism tradeoffs in entropy. Such compensatory adaptations lead to a view of goal-directed motor control as the product of an underlying conservation of entropy across the task-organism-environment system. The conservation of entropy supports the view that context dependent adaptations in human goal-directed action are guided fundamentally by natural law and provides a novel means of examining human motor behavior. This is fundamentally related to the *Heisenberg uncertainty principle* and further support the argument for the primacy of a probabilistic approach toward the study of bio-psychological systems.

5.9 Quantum Cognition in Crowd Dynamics

In this section, following [IR08] we present an entropic geometrical model of *psycho-physical crowd dynamics* (with dissipative crowd kinematics), using Feynman action-amplitude formalism that operates on three synergetic levels: macro, meso and micro. The intent is to explain the dynamics of crowds simultaneously and consistently across these three levels, in order to characterize their geometrical properties particularly with respect to behavior regimes and the state changes between them. Its most natural statistical descriptor is crowd entropy S that satisfies the Prigogine's extended second law of thermodynamics, $\partial_t S \geq 0$ (for any non-isolated multi-component system). Qualitative similarities and superpositions between individual and crowd configuration manifolds motivate our claim that goal-directed crowd movement operates under entropy conservation, $\partial_t S = 0$, while natural crowd dynamics operates under (monotonically) increasing entropy function, $\partial_t S > 0$. Between these two distinct topological phases lies a phase transition with a chaotic inter-phase. Both inertial crowd dynamics and its dissipative kinematics represent diffusion processes on the crowd manifold governed by the Ricci flow, with the associated Perelman entropy-action.

5.9.1 Cognition and Crowd Behavior

Recall that the term *cognition*[13] is used in several loosely related ways to refer to a faculty for the human-like processing of information, applying knowledge and changing preferences. In psychology, cognition refers to an information processing view of an individual psychological functions (see [Mat08, Ree06, Wil06, Ash05, Ash94]). More generally, cognitive processes can be natural and artificial, conscious and not conscious; therefore, they are analyzed from different perspectives and in different contexts, e.g., anesthesia, neurology, psychology, philosophy, logic (both Aristotelian and mathematical), systemics, computer science, artificial intelligence (AI) and computational intelligence (CI). Both in psychology and in AI/CI, cognition refers to the mental functions, mental processes and states of intelligent entities (humans, human organizations, highly autonomous robots), with a particular focus toward the study of comprehension, inferencing, decision-making, planning and learning (see, e.g. [BD02]). The recently developed Scholarpedia, the free peer reviewed web encyclopedia of computational neuroscience is largely based on cognitive neuroscience (see, e.g. [Pes08]). The concept of cognition is closely related to such abstract concepts as mind, reasoning, perception, intelligence, learning, and many others that describe numerous capabilities of the human mind and expected properties of AI/CI (see [II07b, II07d] and references therein).

Yet disembodied cognition is a myth, albeit one that has had profound influence in Western science since Rene Descartes and others gave it credence during the Scientific Revolution. In fact, the mind-body separation had much more to do with explanation of method than with explanation of the mind and cognition, yet it is with respect to the latter that its impact is most widely felt. We find it to be an unsustainable assumption in the realm of crowd behavior. Mental intention is (almost immediately) followed by a physical action, that is, a human or animal movement [Sch07]. In animals, this physical action

---

[13] Latin: "cognoscere = to know".

would be jumping, running, flying, swimming, biting or grabbing. In humans, it can be talking, walking, driving, or shooting, etc. Mathematical description of human/animal movement in terms of the corresponding neuro-musculo-skeletal equations of motion, for the purpose of prediction and control, is formulated within the realm of biodynamics (see [IS01, Iva02, Iva04, IB05, Iva06, II06a, II06b, II06c]).

The crowd (or, collective) psycho-physical behavior is clearly formed by some kind of *superposition, contagion, emergence*, or *social convergence* from the individual agents' behavior. Le Bon's 1895 contagion theory, presented in "The Crowd: A Study of the Popular Mind" influenced many 20th century figures. Sigmund Freud criticized Le Bon's concept of "collective soul", asserting that crowds do not have a soul of their own. The main idea of Freudian crowd behavior theory was that people who were in a crowd acted differently towards people than those who were thinking individually: the minds of the group would merge together to form a collective way of thinking. This idea was further developed in Jungian famous "collective unconscious" [Jun70]. The term "collective behavior" [Blu51] refers to social processes and events which do not reflect existing social structure (laws, conventions, and institutions), but which emerge in a "spontaneous" way. Collective behavior might also be defined as action which is neither conforming (in which actors follow prevailing norms) nor deviant (in which actors violate those norms). According to the emergence theory [TK93], crowds begin as collectivities composed of people with mixed interests and motives; especially in the case of less stable crowds (expressive, acting and protest crowds) norms may be vague and changing; people in crowds make their own rules as they go along. According to currently popular convergence theory, crowd behavior is not a product of the crowd itself, but is carried into the crowd by particular individuals, thus crowds amount to a convergence of like-minded individuals.

We propose that the contagion and convergence theories may be unified by acknowledging that both factors may coexist, even within a single scenario: we propose to refer to this third approach as *behavioral composition*. It represents a substantial philosophical shift from traditional analytical approaches, which have assumed either reduction of a whole into parts or the emergence of the whole from the parts. In particular, both contagion and convergence are related to social entropy, which is the natural decay of structure (such as law, organization, and convention) in a social system [Dow07]. Thus, social entropy provides an entry point into realizing a behavioral-compositional theory of crowd dynamics.

Thus, while all mentioned psycho-social theories of crowd behavior are explanatory only, in this section we attempt to formulate a geometrically predictive model-theory of crowd psycho-physical behavior.

We propose the entropy formulation of crowd dynamics as a three-step process involving individual psycho-physical dynamics and collective psycho-physical dynamics.

5.9.2 Generic Three-Step Crowd Behavioral Dynamics

In this subsection we propose a generic crowd psycho-physical dynamics as a three-step process based on a general partition function formalism. Following the framework of the Prigogine's Extended Second Law of Thermodynamics (for any non-isolated multi-component system), $\partial_t S \geq 0$, for entropy S in any complex system described by its partition function, we formulate a generic crowd psycho-physical dynamics, based on above partition functions, as the following three-step process:

1. Individual psycho-physical dynamics ($\mathcal{ID}$) is a transition process from an entropy-growing "loading" phase of mental preparation, to the entropy-conserving "hitting/shooting" phase of physical action. Formally, $\mathcal{ID}$ is given by the phase-transition map:

$$\mathcal{ID} : \overbrace{\text{MENTAL PREPARATION}}^{\text{"LOADING"}:\partial_t S>0} \Longrightarrow \overbrace{\text{PHYSICAL ACTION}}^{\text{"HITTING"}:\partial_t S=0} \tag{5.86}$$

defined by the individual (chaotic) phase-transition amplitude

$$\left\langle \overset{\partial_t S=0}{\text{PHYS. ACTION}} \middle| CHAOS \middle| \overset{\partial_t S>0}{\text{MENTAL PREP.}} \right\rangle_{\text{ID}} := \int \mathcal{D}[\Phi] \, e^{iS_{\text{ID}}[\Phi]},$$

where the right-hand-side is the Lorentzian path-integral (or complex path-integral in real time), with the individual psycho-physical action

$$S_{\text{ID}}[\Phi] = \int_{t_{ini}}^{t_{fin}} L_{\text{ID}}[\Phi] \, dt,$$

where $L_{\text{ID}}[\Phi]$ is the psycho-physical Lagrangian, consisting of mental cognitive potential and physical kinetic energy.

2. Aggregate psycho-physical dynamics ($\mathcal{AD}$) represents the behavioral composition-transition map:

$$\mathcal{AD} : \sum_{i \in \text{AD}} \overbrace{\text{MENTAL PREPARATION}}^{\text{"LOADING"}:\partial_t S>0} \Longrightarrow \sum_{i \in \text{AD}} \overbrace{\text{PHYSICAL ACTION}_i}^{\text{"HITTING"}:\partial_t S=0}$$

$$\tag{5.87}$$

where the (weighted) aggregate sum is taken over all individual agents, assuming equipartition of the total psycho-physical energy. It is defined by the aggregate (chaotic) phase-transition amplitude

$$\left\langle \overset{\partial_t S=0}{\text{PHYS. ACTION}} \middle| CHAOS \middle| \overset{\partial_t S>0}{\text{MENTAL PREP.}} \right\rangle_{\text{AD}} := \int \mathcal{D}[\Phi] \, e^{-S_{\text{AD}}[\Phi]},$$

with the Euclidean path-integral in real time, that is the SFT-partition function, based on the aggregate psycho-physical action

$$S_{\text{AD}}[\Phi] = \int_{t_{ini}}^{t_{fin}} L_{\text{AD}}[\Phi]\, dt, \quad \text{with } L_{\text{AD}}[\Phi] = \sum_{i \in \text{AD}} L_{\text{ID}}^i[\Phi].$$

3. Crowd psycho-physical dynamics ($\mathcal{CD}$) represents the cumulative transition map:

$$\mathcal{CD}: \sum_{i \in \text{CD}} \overbrace{\text{MENTAL PREPARATION}}^{\text{``LOADING'':}\partial_t S > 0} \Longrightarrow \sum_{i \in \text{CD}} \overbrace{\text{PHYSICAL ACTION}_i}^{\text{``HITTING'':}\partial_t S = 0}$$

(5.88)

where the (weighted) cumulative sum is taken over all individual agents, assuming equipartition of the total psycho-physical energy. It is defined by the crowd (chaotic) phase-transition amplitude

$$\left\langle \overset{\partial_t S=0}{\text{PHYS. ACTION}} \middle| CHAOS \middle| \overset{\partial_t S>0}{\text{MENTAL PREP.}} \right\rangle_{\text{CD}} := \int \mathcal{D}[\Phi]\, e^{iS_{\text{CD}}[\Phi]},$$

with the general Lorentzian path-integral, that is, the QFT-partition function), based on the crowd psycho-physical action

$$S_{\text{CD}}[\Phi] = \int_{t_{ini}}^{t_{fin}} L_{\text{CD}}[\Phi]\, dt, \quad \text{with}$$

$$L_{\text{CD}}[\Phi] = \sum_{i \in \text{CD}} L_{\text{ID}}^i[\Phi] = \sum_{k=\#\text{ of ADs in CD}} L_{\text{AD}}^k[\Phi].$$

All three entropic phase-transition maps, $\mathcal{ID}$, $\mathcal{AD}$ and $\mathcal{CD}$, are spatio-temporal biodynamic cognition systems, evolving within their respective configuration manifolds (i.e., sets of their respective degrees-of-freedom with equipartition of energy), according to biphasic action-functional formalisms with psycho-physical Lagrangian functions L_{ID}, L_{AD} and L_{CD}, each consisting of:

1. Cognitive mental potential (which is a mental preparation for the physical action), and
2. Physical kinetic energy (which describes the physical action itself).

To develop $\mathcal{ID}$, $\mathcal{AD}$ and $\mathcal{CD}$ formalisms, we extend into a physical (or, more precisely, biodynamic) crowd domain a purely-mental individual Life-Space Foam (LSF) framework for motivational cognition [IA07], based on the *quantum-probability* concept.

5.9.3 Formal Individual, Aggregate and Crowd dynamics

In this section we formally develop a three-step crowd psycho-physical dynamics, conceptualized by transition maps (5.86)–(5.87)–(5.88), in agreement with Haken's synergetics [Hak83, Hak93]. We first develop a macro-level individual psycho-physical dynamics $\mathcal{ID}$. Then we generalize $\mathcal{ID}$ into an 'orchestrated' behavioral-compositional crowd dynamics $\mathcal{CD}$, using a quantum-like micro-level formalism with individual agents representing 'crowd quanta'. Finally we develop a meso-level aggregate statistical-field dynamics $\mathcal{AD}$, such that composition of the aggregates $\mathcal{AD}$ makes-up the crowd.

Individual Behavioral Dynamics ($\mathcal{ID}$)

$\mathcal{ID}$ transition map (5.86) is developed using the following action-amplitude formalism (see [IA07, IAY08]):

1. Macroscopically, as a smooth Riemannian n-manifold M_{ID} with steady force-fields and behavioral paths, modeled by a real-valued classical action functional $S_{\mathrm{ID}}[\Phi]$, of the form

$$S_{\mathrm{ID}}[\Phi] = \int_{t_{ini}}^{t_{fin}} L_{\mathrm{ID}}[\Phi]\, dt,$$

   (where macroscopic paths, fields and geometries are commonly denoted by an abstract field symbol Φ^i) with the potential-energy based Lagrangian L given by

$$L_{\mathrm{ID}}[\Phi] = \int d^n x \mathcal{L}_{\mathrm{ID}}(\Phi_i, \partial_{x^j}\Phi^i),$$

   where $\mathcal{L}$ is Lagrangian density, the integral is taken over all n local coordinates $x^j = x^j(t)$ of the ID, and $\partial_{x^j}\Phi^i$ are time and space partial derivatives of the Φ^i-variables over coordinates. The standard least action principle

$$\delta S_{\mathrm{ID}}[\Phi] = 0,$$

   gives, in the form of the Euler–Lagrangian equations, a shortest path, an extreme force-field, with a geometry of minimal curvature and topology without holes. We will see below that high Riemannian curvature generates chaotic behavior, while holes in the manifold produce topologically induced phase transitions.

2. Microscopically, as a collection of wildly fluctuating and jumping paths (histories), force-fields and geometries/topologies, modeled by a complex-valued adaptive path integral, formulated by defining a multi-phase and multi-path (multi-field and multi-geometry) transition amplitude from the entropy-growing state of Mental Preparation to the entropy-conserving state of Physical Action,

$$\langle \text{Physical Action} \mid \text{Mental Preparation}\rangle_{\mathrm{ID}} := \int_{\mathrm{ID}} \mathcal{D}[\Phi]\, e^{iS_{\mathrm{ID}}[\Phi]} \quad (5.89)$$

   where the functional ID-measure $\mathcal{D}[w\Phi]$ is defined as a weighted product

$$\mathcal{D}[w\Phi] = \lim_{N \to \infty} \prod_{s=1}^{N} w_s d\Phi_s^i, \quad (i = 1, \dots, n = con + dis), \qquad (5.90)$$

representing an ∞-dimensional neural network [IA07], with weights w_s updating by updating by one of the standard ANN-learning rules (5.84) or (1.16).[14]

In this way, we effectively derive a unique and globally smooth, causal and entropic phase-transition map (5.86), performed at a macroscopic (global) time-level from some initial time t_{ini} to the final time t_{fin}. Thus, we have obtained macro-objects in the ID: a single path described by Newtonian-like equation of motion, a single force-field described by Maxwellian-like field equations, and a single obstacle-free Riemannian geometry (with global topology without holes).

In particular, on the macro-level, we have the ID-paths, that is biodynamical trajectories generated by the Hamilton action principle

$$\delta S_{ID}[x] = 0,$$

with the Newtonian action $S_{ID}[x]$ given by (summation over repeated indices)

$$S_{ID}[x] = \int_{t_{ini}}^{t_{fin}} \left[\varphi + \frac{1}{2} g_{ij} \dot{x}^i \dot{x}^j \right] dt, \qquad (5.91)$$

where $\varphi = \varphi(t, x^i)$ denotes the mental LSF-potential field, while the second term,

$$T = \frac{1}{2} g_{ij} \dot{x}^i \dot{x}^j,$$

represents the physical (biodynamic) kinetic energy generated by the Riemannian inertial metric tensor g_{ij} of the configuration biodynamic manifold M_{ID} (see Fig. 5.1). The corresponding Euler–Lagrangian equations give the Newtonian equations of human movement

$$\frac{d}{dt} T_{\dot{x}^i} - T_{x^i} = F_i, \qquad (5.92)$$

where subscripts denote the partial derivatives and we have defined the covariant muscular forces $F_i = F_i(t, x^i, \dot{x}^i)$ as negative gradients of the mental potential $\varphi(x^i)$,

---

[14] The traditional neural networks approaches are known for their classes of functions they can represent. Here we are talking about functions in an *extensional* rather than merely *intensional* sense; that is, function can be read as input/output behavior [Bar84, Ben91, For03, Han04]. This limitation has been attributed to their low-dimensionality (the largest neural networks are limited to the order of 10^5 dimensions [IE08]). The proposed path integral approach represents a new family of function-representation methods, which potentially offers a basis for a fundamentally more expansive solution.

$$F_i = -\varphi_{x^i}. \tag{5.93}$$

Equation (5.92) can be put into the standard Lagrangian form as

$$\frac{d}{dt}L_{\dot{x}^i} = L_{x^i}, \quad \text{with } L = T - \varphi(x^i), \tag{5.94}$$

or (using the Legendre transform) into the forced, dissipative Hamiltonian form [Iva04, II06a]

$$\dot{x}^i = \partial_{p_i} H + \partial_{p_i} R, \qquad \dot{p}_i = F_i - \partial_{x^i} H + \partial_{x^i} R, \tag{5.95}$$

where p_i are the generalized momenta (canonically-conjugate to the coordinates x^i), $H = H(p, x)$ is the Hamiltonian (total energy function) and $R = R(p, x)$ is the general dissipative function.

The human motor system possesses many independently controllable components that often allow for more than a single movement pattern to be performed in order to achieve a goal. Hence, the motor system is endowed with a high level of adaptability to different tasks and also environmental contexts [HN08b]. The multiple $SE(3)$-dynamics applied to human musculo-skeletal system gives the fundamental law of biodynamics, which is the *covariant force law*:

Force co-vector field = Mass distribution × Acceleration vector-field, (5.96)

which is formally written:

$$F_i = g_{ij}a^j, \quad (i, j = 1, \ldots, n = \dim(M))$$

where F_i are the covariant force/torque components, g_{ij} is the inertial metric tensor of the configuration Riemannian manifold $M = \prod_i SE(3)^i$ (g_{ij} defines the mass-distribution of the human body), while a^j are the contravariant components of the linear and angular acceleration vector-field. (This fundamental biodynamic law states that contrary to common perception, acceleration and force are not quantities of the same nature: while acceleration is a non-inertial vector-field, force is an inertial co-vector-field. This apparently insignificant difference becomes crucial in injury prediction/prevention, especially in its derivative form in which the 'massless jerk' ($= \dot{a}$) is relatively benign, while the 'massive jolt' ($= \dot{F}$) is deadly.) Both Lagrangian and (topologically equivalent) Hamiltonian development of the covariant force law is fully elaborated in [II06a, II06b, II06c, II07e]. This is consistent with the postulation that human action is guided primarily by natural law [KT87].

On the micro-ID level, instead of each single trajectory defined by the Newtonian equation of motion (5.92), we have an ensemble of fluctuating and crossing paths on the configuration manifold M with weighted probabilities (of the unit total sum). This ensemble of micro-paths is defined by the simplest instance of our adaptive path integral (5.89), similar to the Feynman's original sum over histories,

$$\langle \text{Physical Action} \mid \text{Mental Preparation} \rangle_M = \int_{\text{ID}} \mathcal{D}[wx]\, e^{iS[x]}, \qquad (5.97)$$

where $\mathcal{D}[wx]$ is the functional ID-measure on the space of all weighted paths, and the exponential depends on the action $S_{\text{ID}}[x]$ given by (5.91).

Crowd Behavioral-Compositional Dynamics ($\mathcal{CD}$)

In this subsection we develop a generic crowd $\mathcal{CD}$, as a unique and globally smooth, causal and entropic phase-transition map (5.88), in which agents (or, crowd's individual entities) can be both humans and robots. This crowd psycho-physical action takes place in a crowd smooth Riemannian $3n$-manifold M. Recall from Fig. 5.1 that each individual segment of a human body moves in the Euclidean 3-space $\mathbb{R}^3$ according to its own constrained $SE(3)$-group. Similarly, each individual agent's trajectory, $x^i = x^i(t)$, $i = 1, \ldots, n$, is governed by the Euclidean $SE(2)$-group of rigid body motions in the plane. (Recall that a Lie group $SE(2) \equiv SO(2) \times \mathbb{R}$ is a set of all 3×3-matrices of the form:

$$\begin{bmatrix} \cos\theta & \sin\theta & x \\ -\sin\theta & \cos\theta & y \\ 0 & 0 & 1 \end{bmatrix},$$

including both rigid translations (i.e., Cartesian x, y-coordinates) and rotation matrix $\begin{bmatrix} \cos\theta & \sin\theta \\ -\sin\theta & \cos\theta \end{bmatrix}$ in Euclidean plane $\mathbb{R}^2$ (see [II06c, II07e]).) The crowd configuration manifold M is defined as a union of Euclidean $SE(2)$-groups for all n individual agents in the crowd, that is crowd's configuration $3n$-manifold is defined as a set

$$M = \sum_{k=1}^{n} SE(2)^k \equiv \sum_{k=1}^{n} SO(2)^k \times \mathbb{R}^k,$$

$$\text{coordinated by } \mathbf{x}^k = \{x^k, y^k, \theta^k\}, \text{ (for } k = 1, 2, \ldots, n). \qquad (5.98)$$

In other words, the crowd configuration manifold M is a *dynamical planar graph* with individual agents' $SE(2)$-groups of motion in the vertices and time-dependent inter-agent distances $I_{ij} = [x^i(t_i) - x^j(t_j)]$ as edges.

Similarly to the individual case, the crowd action functional includes mental cognitive potential and physical kinetic energy, formally given by (with $i, j = 1, \ldots, 3n$):

$$A[x^i, x^j; t_i, t_j] = \frac{1}{2} \int_{t_i} \int_{t_j} \delta(I_{ij}^2)\dot{x}^i(t_i)\dot{x}^j(t_j)\, dt_i dt_j + \frac{1}{2} \int_t g_{ij}\dot{x}^i(t)\dot{x}^j(t)\, dt,$$

$$\text{with } I_{ij}^2 = \left[x^i(t_i) - x^j(t_j)\right]^2, \text{ where } IN \leq t_i, t_j, t \leq OUT. \qquad (5.99)$$

The first term in (5.99) represents the mental potential for the interaction between any two agents x^i and x^i within the total crowd matrix x^{ij}. (Although, formally, this term contains cognitive velocities, it still represents 'potential energy' from the physical point of view.) It is defined as a double integral over a delta function of the square of interval I^2 between two points on the paths in their individual cognitive LSFs. Interaction occurs only when this LSF-distance between the two agents x^i and x^j vanishes. Note that the cognitive intentions of any two agents generally occur at different times t_i and t_j unless $t_i = t_j$, when cognitive synchronization occurs. This term effectively represents the *crowd cognitive controller* (see [IAY08]).

The second term in (5.99) represents kinetic energy of the physical interaction of agents. Namely, after the above cognitive synchronization is completed, the second term of physical kinetic energy is activated in the common CD manifold, reducing it to just one of the agents' individual manifolds, which is equivalent to the center-of-mass segment in the human musculo-skeletal system. Therefore, from (5.99) we can derive a generic Euler–Lagrangian dynamics that is a composition of (5.94), which also means that we have in place a generic Hamiltonian dynamics that is a amalgamate of (5.95), as well as the crowd covariant force law (5.96), the governing law of crowd biodynamics:

Crowd force co-vector field = Crowd mass distribution

$$\times \text{ Crowd acceleration vector-field,}$$

$$\text{formally: } F_i = g_{ij}a^j, \tag{5.100}$$

where g_{ij} is the inertial metric tensor of crowd manifold M. The left-hand side of this equation defines forces acting on the crowd, while right-hand defines its mass distribution coupled to the crowd kinematics ($\mathcal{CK}$, described in the next subsection).

At the slave level, the adaptive path integral, representing an ∞-dimensional neural network, corresponding to the psycho-physical crowd action (5.99), reads

$$\langle \text{Physical Action} \mid \text{Mental Preparation} \rangle_{\text{CD}} = \int_{\text{CD}} \mathcal{D}[w, x, y] \, e^{iA[x, y; t_i, t_j]}, \tag{5.101}$$

where the Lebesgue-type integration is performed over all continuous paths $x^i = x^i(t_i)$ and $y^j = y^j(t_j)$, while summation is performed over all associated discrete Markov fluctuations and jumps. The symbolic differential in the path integral (5.101) represents an adaptive path measure, defined as the weighted product

$$\mathcal{D}[w, x, y] = \lim_{N \to \infty} \prod_{s=1}^{N} w_{ij}^s dx^i dy^j, \quad (i, j = 1, \dots, n). \tag{5.102}$$

The quantum-field path integral (5.101)–(5.102) defines the microstate $\mathcal{CD}$-level, an ensemble of fluctuating and crossing paths on the crowd $3n$-manifold M.

Dissipative Crowd Kinematics ($\mathcal{CK}$)

The crowd action (5.99) with its amalgamate Lagrangian dynamics (5.94) and amalgamate Hamiltonian dynamics (5.95), as well as the crowd force law (5.100) define the macroscopic crowd dynamics, $\mathcal{CD}$. Suppose, for a moment, that $\mathcal{CD}$ is force-free and dissipation free, therefore conservative. Now, the basic characteristic of the conservative Lagrangian/Hamiltonian systems evolving in the phase space spanned by the system coordinates and their velocities/momenta, is that their *flow* φ_t^L preserves the phase-space volume, as proposed by the Liouville theorem, which is the well-known fact in statistical mechanics. However, the preservation of the phase volume causes structural instability of the conservative system, i.e., the phase-space spreading effect by which small phase regions R_t will tend to get distorted from the initial one R_o during the conservative system evolution. This problem, governed by entropy growth ($\partial_t S > 0$), is much more serious in higher dimensions than in lower dimensions, since there are so many 'directions' in which the region can locally spread (see [Pen89, II06c]). This phenomenon is related to *conservative Hamiltonian chaos* (see Sect. 5.9.4 below).

However, this situation is not very frequent in case of 'organized' human crowd. Its self-organization mechanisms are clearly much stronger than the conservative statistical mechanics effects, which we interpret in terms of Prigogine's dissipative structures. Formally, if dissipation of energy in a system is much stronger then its inertial characteristics, then instead of the second-order Newton–Lagrangian dynamic equations of motion, we are actually dealing with the first-order driftless (non-acceleration, non-inertial) kinematic equations of motion, which is related to *dissipative chaos* [Nic86]. Briefly, the dissipative crowd flow can be depicted like this: from the set of initial conditions for individual agents, the crowd evolves in time towards the set of the corresponding *entangled attractors*, which are mutually separated by fractal (non-integer dimension) separatrices.

In this subsection we elaborate on the dissipative crowd kinematics ($\mathcal{CK}$), which is self-controlled and dominates the $\mathcal{CD}$ if the crowd's inertial forces are much weaker then the the crowd's dissipation of energy, presented here in the form of nonlinear velocity controllers.

Recall that the essential concept in dynamical systems theory is the notion of a *vector-field* (that we will denote by a boldface symbol), which assigns a tangent vector to each point p in the manifold in case. In particular, $\mathbf{v}$ is a gradient vector-field if it equals the gradient of some scalar function. A *flow-line* of a vector-field $\mathbf{v}$ is a path $\boldsymbol{\gamma}(t)$ satisfying the vector ODE, $\dot{\boldsymbol{\gamma}}(t) = \mathbf{v}(\boldsymbol{\gamma}(t))$, that is, $\mathbf{v}$ yields the velocity field of the path $\boldsymbol{\gamma}(t)$. The set of all flow lines of a vector-field $\mathbf{v}$ comprises its flow φ_t that is (technically, see e.g., [II06c, II07e]) a one-parameter Lie group of diffeomorphisms (smooth bijective functions) generated by a vector-field $\mathbf{v}$ on M, such that

$$\varphi_t \circ \varphi_s = \varphi_{t+s}, \qquad \varphi_0 = \text{identity}, \quad \text{which gives: } \gamma(t) = \varphi_t(\gamma(0)).$$

Analytically, a vector-field $\mathbf{v}$ is defined as a set of autonomous ODEs. Its solution gives the flow φ_t, consisting of integral curves (or, flow lines) $\gamma(t)$ of the vector-field, such that all the vectors from the vector-field are tangent to integral curves at different representative points $p \in M$. In this way, through every representative point $p \in M$ passes both a curve from the flow and its tangent vector from the vector-field. Geometrically, vector-field is defined as a cross-section of the tangent bundle TM of the manifold M.

In general, given an nD frame $\{\partial_i\} \equiv \{\partial/\partial x^i\}$ on a smooth n-manifold M (that is, a basis of tangent vectors in a local coordinate chart $x^i = (x^1, \ldots, x^n) \subset M$), we can define any vector-field $\mathbf{v}$ on M by its components $v^i = v^i(t)$ as

$$\mathbf{v} = v^i \partial_i = v^i \frac{\partial}{\partial x^i} = v^1 \frac{\partial}{\partial x^1} + \cdots + v^n \frac{\partial}{\partial x^n}.$$

Thus, a vector-field $\mathbf{v} \in \mathcal{X}(M)$ (where $\mathcal{X}(M)$ is the set of all smooth vector-fields on M) is actually a differential operator that can be used to differentiate any smooth scalar function $f = f(x^1, \ldots, x^n)$ on M, as a *directional derivative* of f in the direction of $\mathbf{v}$. This is denoted simply $\mathbf{v}f$, such that

$$\mathbf{v}f = v^i \partial_i f = v^i \frac{\partial f}{\partial x^i} = v^1 \frac{\partial f}{\partial x^1} + \cdots + v^n \frac{\partial f}{\partial x^n}.$$

In particular, if $\mathbf{v} = \dot{\gamma}(t)$ is a velocity vector-field of a space curve $\gamma(t) = (x^1(t), \ldots, x^n(t))$, defined by its components $v^i = \dot{x}^i(t)$, directional derivative of $f(x^i)$ in the direction of $\mathbf{v}$ becomes

$$\mathbf{v}f = \dot{x}^i \partial_i f = \frac{dx^i}{dt} \frac{\partial f}{\partial x^i} = \frac{df}{dt} = \dot{f},$$

which is a rate-of-change of f along the curve $\gamma(t)$ at a point $x^i(t)$.

Given two vector-fields, $\mathbf{u} = u^i \partial_i, \mathbf{v} = v^i \partial_i \in \mathcal{X}(M)$, their Lie bracket (or, commutator) is another vector-field $[\mathbf{u}, \mathbf{v}] \in \mathcal{X}(M)$, defined by

$$[\mathbf{u}, \mathbf{v}] = \mathbf{uv} - \mathbf{vu} = u^i \partial_i v^j \partial_j - v^j \partial_j u^i \partial_i,$$

which, applied to any smooth function f on M, gives

$$[\mathbf{u}, \mathbf{v}](f) = \mathbf{u}\,(\mathbf{v}(f)) - \mathbf{v}\,(\mathbf{u}(f)).$$

The Lie bracket measures the failure of 'mixed directional derivatives' to commute. Clearly, mixed partial derivatives *do* commute, $[\partial_i, \partial_j] = 0$, while in general it is *not* the case, $[\mathbf{u}, \mathbf{v}] \neq 0$. In addition, suppose that $\mathbf{u}$ generates the flow φ_t and $\mathbf{v}$ generates the flow φ_s. Then, for any smooth function f on M, we have at any point p on M,

$$[\mathbf{u}, \mathbf{v}](f)(p) = \frac{\partial^2}{\partial t \partial s} \left(f(\varphi_s(\varphi_t(p))) - f(\varphi_t(\varphi_s(p))) \right),$$

which means that in $f(\varphi_s(\varphi_t(p)))$ we are starting at p, flowing along $\mathbf{v}$ a little bit, then along $\mathbf{u}$ a little bit, and then evaluating f, while in $f(\varphi_t(\varphi_s(p)))$ we are flowing first along $\mathbf{u}$ and then $\mathbf{v}$. Therefore, the Lie bracket infinitesimally measures how these flows fail to commute.

The Lie bracket satisfies the following three properties (for any three vector-fields $\mathbf{u}, \mathbf{v}, \mathbf{w} \in M$ and two constants a, b—thus forming a Lie algebra on the crowd manifold M):

(i) $[\mathbf{u}, \mathbf{v}] = -[\mathbf{v}, \mathbf{u}]$—skew-symmetry;
(ii) $[\mathbf{u}, a\mathbf{v} + b\mathbf{w}] = a[\mathbf{u}, \mathbf{v}] + b[\mathbf{u}, \mathbf{w}]$—bilinearity; and
(iii) $[\mathbf{u}, [\mathbf{v}, \mathbf{w}]] + [\mathbf{v}, [\mathbf{w}, \mathbf{u}]] + [\mathbf{w}, [\mathbf{u}, \mathbf{v}]]$—Jacobi identity.
   A new set of vector-fields on M can be generated by repeated Lie brackets of $\mathbf{u}, \mathbf{v}, \mathbf{w} \in M$.

The Lie bracket is a standard tool in geometric nonlinear control theory (see, e.g. [II06c, II07e]). Its action on vector-fields can be best visualized using the popular car parking example, in which the driver has two different vector-field transformations at his disposal. They can turn the steering wheel, or they can drive the car forward or backward. Here, we specify the state of a car by four coordinates: the (x, y) coordinates of the center of the rear axle, the direction θ of the car, and the angle ϕ between the front wheels and the direction of the car. l is the constant length of the car. Therefore, the 4D configuration manifold of a car is a set $M \equiv SO(2) \times \mathbb{R}^2$, coordinated by $\mathbf{x} = \{x, y, \theta, \phi\}$, which is slightly more complicated than the individual crowd agent's 3D configuration manifold $SE(2) \equiv SO(2) \times \mathbb{R}$, coordinated by $\mathbf{x} = \{x, y, \theta\}$. The driftless car kinematics can be defined as a vector ODE:

$$\dot{\mathbf{x}} = \mathbf{u}(\mathbf{x})c_1 + \mathbf{v}(\mathbf{x})c_2, \tag{5.103}$$

with two vector-fields, $\mathbf{u}, \mathbf{v} \in \mathcal{X}(M)$, and two scalar control inputs, c_1 and c_2. The infinitesimal car-parking transformations will be the following vector-fields

$$\mathbf{u}(\mathbf{x}) \equiv \text{DRIVE} = \cos\theta \frac{\partial}{\partial x} + \sin\theta \frac{\partial}{\partial y} + \frac{\tan\phi}{l} \frac{\partial}{\partial \theta} \equiv \begin{pmatrix} \cos\theta \\ \sin\theta \\ \frac{1}{l}\tan\phi \\ 0 \end{pmatrix},$$

and

$$\mathbf{v}(\mathbf{x}) \equiv \text{STEER} = \frac{\partial}{\partial \phi} \equiv \begin{pmatrix} 0 \\ 0 \\ 0 \\ 1 \end{pmatrix}.$$

The car kinematics (5.103) therefore expands into a matrix ODE:

$$\begin{pmatrix} \dot{x} \\ \dot{y} \\ \dot{\theta} \\ \dot{\phi} \end{pmatrix} = \text{DRIVE} \cdot c_1 + \text{STEER} \cdot c_2 \equiv \begin{pmatrix} \cos\theta \\ \sin\theta \\ \frac{1}{l}\tan\phi \\ 0 \end{pmatrix} \cdot c_1 + \begin{pmatrix} 0 \\ 0 \\ 0 \\ 1 \end{pmatrix} \cdot c_2.$$

However, STEER and DRIVE do not commute (otherwise we could do all your steering at home before driving of on a trip). Their combination is given by the Lie bracket

$$[\mathbf{v}, \mathbf{u}] \equiv [\text{STEER}, \text{DRIVE}] = \frac{1}{l \cos^2 \phi} \frac{\partial}{\partial \theta} \equiv \text{WRIGGLE}.$$

The operation $[\mathbf{v}, \mathbf{u}] \equiv$ WRIGGLE $\equiv$ [STEER,DRIVE] is the infinitesimal version of the sequence of transformations: steer, drive, steer back, and drive back, i.e.,

$$\{\text{STEER}, \text{DRIVE}, \text{STEER}^{-1}, \text{DRIVE}^{-1}\}.$$

Now, WRIGGLE can get us out of some parking spaces, but not tight ones: we may not have enough room to WRIGGLE out. The usual tight parking space restricts the DRIVE transformation, but not STEER. A truly tight parking space restricts STEER as well by putting your front wheels against the curb.

Fortunately, there is still another commutator available:

$$[\mathbf{u}, [\mathbf{v}, \mathbf{u}]] \equiv [\text{DRIVE}, [\text{STEER}, \text{DRIVE}]] = [[\mathbf{u}, \mathbf{v}], \mathbf{u}]$$
$$\equiv [\text{DRIVE}, \text{WRIGGLE}] = \frac{1}{l \cos^2 \phi} \left(\sin \theta \frac{\partial}{\partial x} - \cos \theta \frac{\partial}{\partial y} \right) \equiv \text{SLIDE}.$$

The operation $[[\mathbf{u}, \mathbf{v}], \mathbf{u}] \equiv$ SLIDE $\equiv$ [DRIVE, WRIGGLE] is a displacement at right angles to the car, and can get us out of any parking place. We just need to remember to steer, drive, steer back, drive some more, steer, drive back, steer back, and drive back:

$$\{\text{STEER}, \text{DRIVE}, \text{STEER}^{-1}, \text{DRIVE}, \text{STEER}, \text{DRIVE}^{-1}, \text{STEER}^{-1}, \text{DRIVE}^{-1}\}.$$

We have to reverse steer in the middle of the parking place. This is not intuitive, and no doubt is part of a common problem with parallel parking.

Thus, from only two controls, c_1 and c_2, we can form the vector-fields DRIVE $\equiv \mathbf{u}$, STEER $\equiv \mathbf{v}$, WRIGGLE $\equiv [\mathbf{v}, \mathbf{u}]$, and SLIDE $\equiv [[\mathbf{u}, \mathbf{v}], \mathbf{u}]$, allowing us to move anywhere in the car configuration manifold $M \equiv SO(2) \times \mathbb{R}^2$. All above computations are straightforward in $Mathematica^{TM\,15}$ if we define the following three symbolic functions:

1. Jacobian matrix: JacMat[v_List, x_List] := Outer[D, v, x];
2. Lie bracket: LieBrc[u_List, v_List, x_List] := JacMat[v, x] . u—JacMat[u, x] . v;
3. Repeated Lie bracket: Adj[u_List, v_List, x_List, k_] := If[k == 0, v, LieBrc[u, Adj[u, v, x, k - 1], x]];

In case of the human crowd, we have a slightly simpler, but multiplied problem, i.e., superposition of n individual agents' motions. So, we can define the dissipative crowd kinematics as a system of n vector ODEs:

---

[15] The above computations could instead be done in other available packages, such as Maple, by suitably translating the provided example code.

$$\dot{\mathbf{x}}^k = \mathbf{u}^k(\mathbf{x})c_1^k + \mathbf{v}^k(\mathbf{x})c_2^k, \tag{5.104}$$

where

$$\mathbf{u}^k(\mathbf{x}) \equiv \text{DRIVE}^k = \cos^k\theta\frac{\partial}{\partial x^k} + \sin^k\theta\frac{\partial}{\partial y^k} \equiv \begin{pmatrix} \cos^k\theta \\ \sin^k\theta \\ 0 \end{pmatrix},$$

and

$$\mathbf{v}^k(\mathbf{x}) \equiv \text{STEER}^k = \frac{\partial}{\partial\theta^k} \equiv \begin{pmatrix} 0 \\ 0 \\ 1 \end{pmatrix},$$

while c_1^k and c_2^k are crowd controls.

Thus, the crowd kinematics (5.104) expands into the matrix ODE:

$$\begin{pmatrix} \dot{x} \\ \dot{y} \\ \dot{\theta} \end{pmatrix} = \text{DRIVE}^k \cdot c_1^k + \text{STEER}^k \cdot c_2^k \equiv \begin{pmatrix} \cos^k\theta \\ \sin^k\theta \\ 0 \end{pmatrix} \cdot c_1^k + \begin{pmatrix} 0 \\ 0 \\ 1 \end{pmatrix} \cdot c_2^k. \tag{5.105}$$

The dissipative crowd kinematics (5.104)–(5.105) obeys the set of n-tuple integral rules of motion that are similar (though slightly simpler) to the above rules of the car kinematics, including the following derived vector-fields:

$$\text{WRIGGLE}^k \equiv [\text{STEER}^k, \text{DRIVE}^k] \equiv [\mathbf{v}^k, \mathbf{u}^k] \quad \text{and}$$
$$\text{SLIDE}^k \equiv [\text{DRIVE}^k, \text{WRIGGLE}^k] \equiv [[\mathbf{u}^k, \mathbf{v}^k], \mathbf{u}^k].$$

Thus, controlled by the two vector controls c_1^k and c_2^k, the crowd can form the vector-fields: DRIVE $\equiv \mathbf{u}^k$, STEER $\equiv \mathbf{v}^k$, WRIGGLE $\equiv [\mathbf{v}^k, \mathbf{u}^k]$, and SLIDE $\equiv [[\mathbf{u}^k, \mathbf{v}^k], \mathbf{u}^k]$, allowing it to move anywhere within its configuration manifold M given by (5.98). Solution of the dissipative crowd kinematics (5.104)–(5.105) defines the dissipative crowd flow, ϕ_t^K.

Now, the general $\mathcal{CD}$–$\mathcal{CK}$ crowd behavior can be defined as a amalgamate flow (psycho-physical Lagrangian flow, ϕ_t^L, plus dissipative kinematic flow, ϕ_t^K) on the crowd manifold M defined by (5.98),

$$C_t = \phi_t^L + \phi_t^K : t \mapsto (M(t), g(t)),$$

which is a one-parameter family of homeomorphic (topologically equivalent) Riemannian manifolds[16] $(M, g = g_{ij})$, parameterized by a 'time' parameter t.

---

[16] Proper differentiation of vector and tensor fields on a smooth Riemannian manifold (like the crowd $3n$-manifold M) is performed using the *Levi-Civita covariant derivative* (see, e.g., [II06c, II07e]). Formally, let M be a Riemannian N-manifold with the tangent bundle TM and a local coordinate system $\{x^i\}_{i=1}^N$ defined in an open set $U \subset M$. The covariant derivative operator, $\nabla_X : C^\infty(TM) \to C^\infty(TM)$, is the unique linear map such that for any vector-fields X, Y, Z, constant c, and scalar function f the following properties are valid:

That is, C_t can be used for describing smooth deformations of the crowd manifold M over time. The manifold family $(M(t), g(t))$ at time t determines the manifold family $(M(t + dt), g(t + dt))$ at an infinitesimal time $t + dt$ into the future, according to some prescribed geometric flow, like the celebrated *Ricci flow* [Ham82, Ham86, Ham93, Ham88] (that was an instrument for a proof of a 100-year old Poincaré conjecture),

$$\partial_t g_{ij}(t) = -2R_{ij}(t), \qquad (5.106)$$

where R_{ij} is the Ricci curvature tensor (see [II07e]) of the crowd manifold M and $\partial_t g(t)$ is defined as

$$\nabla_{X+cY} = \nabla_X + c\nabla_Y, \qquad \nabla_X(Y + fZ) = \nabla_X Y + (Xf)Z + f\nabla_X Z,$$
$$\nabla_X Y - \nabla_Y X = [X, Y],$$

where $[X, Y]$ is the Lie bracket of X and Y. In local coordinates, the metric g is defined for any orthonormal basis $(\partial_i = \partial/\partial x^i)$ in $U \subset M$ by $g_{ij} = g(\partial_i, \partial_j) = \delta_{ij}$, $\partial_k g_{ij} = 0$. Then the affine *Levi-Civita connection* is defined on M by

$$\nabla_{\partial_i} \partial_j = \Gamma_{ij}^k \partial_k, \quad \text{where } \Gamma_{ij}^k = \frac{1}{2} g^{kl} \left(\partial_i g_{jl} + \partial_j g_{il} - \partial_l g_{ij} \right)$$

are the Christoffel symbols.

Now, using the covariant derivative operator ∇_X we can define the *Riemann curvature* $(3, 1)$-tensor $\mathfrak{Rm}$ by

$$\mathfrak{Rm}(X, Y)Z = \nabla_X \nabla_Y Z - \nabla_Y \nabla_X Z - \nabla_{[X,Y]} Z,$$

which measures the curvature of the manifold by expressing how noncommutative covariant differentiation is. The $(3, 1)$-components R_{ijk}^l of $\mathfrak{Rm}$ are defined in $U \subset M$ by

$$\mathfrak{Rm}(\partial_i, \partial_j) \partial_k = R_{ijk}^l \partial_l, \quad \text{or} \quad R_{ijk}^l = \partial_i \Gamma_{jk}^l - \partial_j \Gamma_{ik}^l + \Gamma_{jk}^m \Gamma_{im}^l - \Gamma_{ik}^m \Gamma_{jm}^l.$$

Also, the Riemann $(4, 0)$-tensor $R_{ijkl} = g_{lm} R_{ijk}^m$ is defined as the g-based inner product on M,

$$R_{ijkl} = \langle \mathfrak{Rm}(\partial_i, \partial_j) \partial_k, \partial_l \rangle.$$

The first and second Bianchi identities for the Riemann $(4, 0)$-tensor R_{ijkl} hold,

$$R_{ijkl} + R_{jkil} + R_{kijl} = 0, \qquad \nabla_i R_{jklm} + \nabla_j R_{kilm} + \nabla_k R_{ijlm} = 0,$$

while the twice contracted second Bianchi identity reads: $2\nabla_j R_{ij} = \nabla_i R$.

The $(0, 2)$ *Ricci tensor* $\mathfrak{Rc}$ is the trace of the Riemann $(3, 1)$-tensor $\mathfrak{Rm}$,

$$\mathfrak{Rc}(Y, Z) + \text{tr}(X \to \mathfrak{Rm}(X, Y)Z), \quad \text{so that } \mathfrak{Rc}(X, Y) = g(\mathfrak{Rm}(\partial_i, X)\partial_i, Y).$$

Its components $R_{jk} = \mathfrak{Rc}(\partial_j, \partial_k)$ are given in $U \subset M$ by the contraction

$$R_{jk} = R_{ijk}^i, \quad \text{or} \quad R_{jk} = \partial_i \Gamma_{jk}^i - \partial_k \Gamma_{ji}^i + \Gamma_{mi}^i \Gamma_{jk}^m - \Gamma_{mk}^i \Gamma_{ji}^m.$$

Finally, the scalar curvature R is the trace of the Ricci tensor $\mathfrak{Rc}$, given in $U \subset M$ by: $R = g^{ij} R_{ij}$.

$$\partial_t g(t) \equiv \frac{d}{dt}g(t) := \lim_{dt\to 0}\frac{g(t+dt)-g(t)}{dt}. \qquad (5.107)$$

Aggregate Behavioral-Compositional Dynamics ($\mathcal{AD}$)

To formally develop the meso-level aggregate behavioral-compositional dynamics ($\mathcal{AD}$), we start with the crowd path integral (5.101), which can be redefined if we Wick-rotate the time variable t to imaginary values, $t \mapsto \tau = it$, thereby transforming the Lorentzian path integral in real time into the Euclidean path integral in imaginary time. Furthermore, if we rectify the time axis back to the real line, we get the adaptive SFT-partition function as our proposed $\mathcal{AD}$-model:

$$\langle \text{Physical Action} \mid \text{Mental Preparation}\rangle_{\text{AD}} = \int_{\text{CD}} \mathcal{D}[w,x,y]\,e^{-A[x,y;t_i,t_j]}.$$
$$(5.108)$$

The adaptive $\mathcal{AD}$-transition amplitude: $\langle \text{Physical Action} \mid \text{Mental Preparation}\rangle_{\text{AD}}$, as defined by the SFT-partition function (5.108), is a general model of the *Markov stochastic process*. Recall from Sect. 5.8.1 above that the *Markov assumption* can be formulated in terms of the conditional probabilities $P(x^i, t_i)$: if the times t_i increase from right to left, the conditional probability is determined entirely by the knowledge of the most recent condition. Markov process is generated by a set of conditional probabilities whose probability-density $P = P(x', t'|x'', t'')$ evolution obeys the general *Chapman–Kolmogorov integro-differential equation*

$$\partial_t P = -\sum_i \frac{\partial}{\partial x^i}\{A_i[x(t),t]P\} + \frac{1}{2}\sum_{ij}\frac{\partial^2}{\partial x^i \partial x^j}\{B_{ij}[x(t),t]P\}$$

$$+ \int dx\,\{W(x'|x'',t)P - W(x''|x',t)P\}$$

including *deterministic drift*, *diffusion fluctuations* and *discontinuous jumps* (given respectively in the first, second and third terms on the r.h.s.). This general Chapman–Kolmogorov integro-differential equation (5.109), with its conditional probability density evolution, $P = P(x', t'|x'', t'')$, is represented by our SFT-partition function (5.108).

Furthermore, discretization of the adaptive SFT-partition function (5.108) gives the standard thermodynamic partition function

$$Z = \sum_j e^{-w_j E^j/T}, \qquad (5.109)$$

where E^j is the motion energy eigenvalue (reflecting each possible motivational energetic state), T is the temperature-like environmental control parameter, and the sum runs over all ID energy eigenstates (labeled by the index j). From (5.109), we can calculate the *transition entropy*, as $S = k_B \ln Z$ (see the next section).

5.9.4 Crowd Entropy, Chaos and Phase Transitions

Recall that non-equilibrium phase transitions [Hak83, Hak93, Hak96, Hak00, Hak02] are phenomena which bring about qualitative physical changes at the macroscopic level in presence of the same microscopic forces acting among the constituents of a system. In this section we extend the $\mathcal{CD}$ formalism to incorporate both algorithmic and geometrical entropy as well as dynamical chaos [IJP08, II07a, II08a] between the entropy-growing phase of Mental Preparation and the entropy-conserving phase of Physical Action, together with the associated topological phase transitions.

5.9.5 Crowd Ricci Flow and Perelman Entropy

Recall that the inertial metric crowd flow, $C_t : t \mapsto (M(t), g(t))$ on the crowd $3n$-manifold (5.98) is a one-parameter family of homeomorphic Riemannian manifolds (M, g), evolving by the Ricci flow (5.106)–(5.107).

Now, given a smooth scalar function $u : M \to \mathbb{R}$ on the Riemannian crowd $3n$-manifold M, its Laplacian operator Δ is locally defined as

$$\Delta u = g^{ij} \nabla_i \nabla_j u,$$

where ∇_i is the covariant derivative (or, Levi-Civita connection, see [II07e]). We say that a smooth function $u : M \times [0, T) \to \mathbb{R}$, where $T \in (0, \infty]$, is a solution to the heat equation on M if

$$\partial_t u = \Delta u. \tag{5.110}$$

One of the most important properties satisfied by the heat equation is the maximum principle, which says that for any smooth solution to the heat equation, whatever point-wise bounds hold at $t = 0$ also hold for $t > 0$ [CC99]. This property exhibits the smoothing behavior of the heat diffusion (5.110) on M.

Closely related to the heat diffusion (5.110) is the (the Fields medal winning) Perelman entropy-action functional, which is on a $3n$-manifold M with a Riemannian metric g_{ij} and a (temperature-like) scalar function f given by [Per02]

$$\mathcal{E} = \int_M (R + |\nabla f|^2) e^{-f} d\mu \tag{5.111}$$

where R is the scalar Riemann curvature on M, while $d\mu$ is the volume $3n$-form on M, defined as

$$d\mu = \sqrt{\det(g_{ij})} \, dx^1 \wedge dx^2 \wedge \cdots \wedge dx^{3n}. \tag{5.112}$$

During the Ricci flow (5.106)–(5.107) on the crowd manifold (5.98), that is, during the inertial metric crowd flow, $C_t : t \mapsto (M(t), g(t))$, the Perelman entropy functional (5.111) evolves as

$$\partial_t \mathcal{E} = 2 \int |R_{ij} + \nabla_i \nabla_j f|^2 e^{-f} d\mu. \tag{5.113}$$

Now, the *crowd breathers* are solitonic crowd behaviors, which could be given by localized periodic solutions of some nonlinear soliton PDEs, including the exactly solvable sine-Gordon equation and the focusing nonlinear Schrödinger equation. In particular, the time-dependent crowd inertial metric $g_{ij}(t)$, evolving by the Ricci flow $g(t)$ given by (5.106)–(5.107) on the crowd $3n$-manifold M is the *Ricci crowd breather*, if for some $t_1 < t_2$ and $\alpha > 0$ the metrics $\alpha g_{ij}(t_1)$ and $g_{ij}(t_2)$ differ only by a diffeomorphism; the cases $\alpha = 1, \alpha < 1, \alpha > 1$ correspond to steady, shrinking and expanding crowd breathers, respectively. Trivial crowd breathers, for which the metrics $g_{ij}(t_1)$ and $g_{ij}(t_2)$ on M differ only by diffeomorphism and scaling for each pair of t_1 and t_2, are the *crowd Ricci solitons*. Thus, if we consider the Ricci flow (5.106)–(5.107) as a biodynamical system on the space of Riemannian metrics modulo diffeomorphism and scaling, then crowd breathers and solitons correspond to periodic orbits and fixed points respectively. At each time the Ricci soliton metric satisfies on M an equation of the form [Per02]

$$R_{ij} + cg_{ij} + \nabla_i b_j + \nabla_j b_i = 0,$$

where c is a number and b_i is a 1-form; in particular, when $b_i = \frac{1}{2}\nabla_i a$ for some function a on M, we get a gradient Ricci soliton.

Define $\lambda(g_{ij}) = \inf \mathcal{E}(g_{ij}, f)$, where infimum is taken over all smooth f, satisfying

$$\int_M e^{-f} d\mu = 1. \tag{5.114}$$

$\lambda(g_{ij})$ is the lowest eigenvalue of the operator $-4\Delta + R$. Then the entropy evolution formula (5.113) implies that $\lambda(g_{ij}(t))$ is non-decreasing in t, and moreover, if $\lambda(t_1) = \lambda(t_2)$, then for $t \in [t_1, t_2]$ we have $R_{ij} + \nabla_i \nabla_j f = 0$ for f which minimizes $\mathcal{E}$ on M [Per02]. Therefore, a steady breather on M is necessarily a steady soliton.

If we define the conjugate heat operator on M as

$$\square^* = -\partial/\partial t - \Delta + R$$

then we have the conjugate heat equation: $\square^* u = 0$.

The entropy functional (5.111) is nondecreasing under the coupled Ricci-diffusion flow on M [II08d]

$$\partial_t g_{ij} = -2R_{ij}, \qquad \partial_t u = -\Delta u + \frac{R}{2}u - \frac{|\nabla u|^2}{u}, \tag{5.115}$$

where the second equation ensures $\int_M u^2 d\mu = 1$, to be preserved by the Ricci flow $g(t)$ on M. If we define $u = e^{-\frac{f}{2}}$, then (5.115) is equivalent to f-evolution equation on M (the nonlinear backward heat equation),

$$\partial_t f = -\Delta f + |\nabla f|^2 - R,$$

which instead preserves (5.114). The coupled Ricci-diffusion flow (5.115) is the most general biodynamic model of the crowd reaction-diffusion processes on M. In a recent study [AIJ08] this general model has been implemented for modeling a generic perception-action cycle with applications to robot navigation in the form of a dynamical grid.

Perelman's functional $\mathcal{E}$ is analogous to negative thermodynamic entropy [Per02]. Recall that thermodynamic partition function for a generic canonical ensemble at temperature β^{-1} is given by

$$Z = \int e^{-\beta E} d\omega(E), \tag{5.116}$$

where $\omega(E)$ is a 'density measure', which does not depend on β. From it, the average energy is given by $\langle E \rangle = -\partial_\beta \ln Z$, the entropy is $S = \beta \langle E \rangle + \ln Z$, and the fluctuation is $\sigma = \langle (E - \langle E \rangle)^2 \rangle = \partial_{\beta^2} \ln Z$.

If we now fix a closed $3n$-manifold M with a probability measure m and a metric $g_{ij}(\tau)$ that depends on the temperature τ, then according to equation

$$\partial_\tau g_{ij} = 2(R_{ij} + \nabla_i \nabla_j f),$$

the partition function (5.116) is given by

$$\ln Z = \int \left(-f + \frac{n}{2} \right) dm. \tag{5.117}$$

From (5.117) we get (see [Per02])

$$\langle E \rangle = -\tau^2 \int_M \left(R + |\nabla f|^2 - \frac{n}{2\tau} \right) dm,$$

$$S = -\int_M (\tau(R + |\nabla f|^2) + f - n) \, dm,$$

$$\sigma = 2\tau^4 \int_M \left| R_{ij} + \nabla_i \nabla_j f - \frac{1}{2\tau} g_{ij} \right|^2 dm,$$

where $dm = u \, dV$, $u = (4\pi\tau)^{-\frac{n}{2}} e^{-f}$.

From the above formulas, we see that the fluctuation σ is nonnegative; it vanishes only on a gradient shrinking soliton. $\langle E \rangle$ is nonnegative as well, whenever the flow exists for all sufficiently small $\tau > 0$. Furthermore, if the heat function u: (a) tends to a δ-function as $\tau \to 0$, or (b) is a limit of a sequence of partial heat functions u_i, such that each u_i tends to a δ-function as $\tau \to \tau_i > 0$, and $\tau_i \to 0$, then the entropy S is also nonnegative. In case (a), all the quantities $\langle E \rangle, S, \sigma$ tend to zero as $\tau \to 0$, while in case (b), which may be interesting if $g_{ij}(\tau)$ becomes singular at $\tau = 0$, the entropy S may tend to a positive limit.

5.9.6 Chaotic Inter-Phase in Crowd Dynamics

Recall that $\mathcal{CD}$ transition map (5.88) is defined by the chaotic crowd phase-transition amplitude

$$\left\langle \overset{\partial_t S=0}{\text{PHYS. ACTION}} \middle| CHAOS \middle| \overset{\partial_t S>0}{\text{MENTAL PREP.}} \right\rangle := \int_M \mathcal{D}[x]\, e^{iA[x]},$$

where we expect the inter-phase chaotic behavior (see [IAY08]). To show that this chaotic inter-phase is caused by the change in Riemannian geometry of the crowd $3n$-manifold M, we will first simplify the $\mathcal{CD}$ action functional (5.99) as

$$A[x] = \frac{1}{2} \int_{t_{ini}}^{t_{fin}} [g_{ij}\dot{x}^i \dot{x}^j - V(x, \dot{x})]\, dt, \tag{5.118}$$

with the associated standard Hamiltonian, corresponding to the amalgamate version of (5.95),

$$H(p, x) = \sum_{i=1}^{N} \frac{1}{2}p_i^2 + V(x, \dot{x}), \tag{5.119}$$

where p_i are the $SE(2)$-momenta, canonically conjugate to the individual agents' $SE(2)$-coordinates x^i, $(i = 1, \ldots, 3n)$. Biodynamics of systems with action (5.118) and Hamiltonian (5.119) are given by the set of *geodesic equations* [II06c, II07e]

$$\frac{d^2 x^i}{ds^2} + \Gamma^i_{jk} \frac{dx^j}{ds} \frac{dx^k}{ds} = 0, \tag{5.120}$$

where Γ^i_{jk} are the Christoffel symbols of the affine Levi-Civita connection of the Riemannian $\mathcal{CD}$ manifold M. In this geometrical framework, the instability of the trajectories is the instability of the geodesics, and it is completely determined by the curvature properties of the $\mathcal{CD}$ manifold M according to the Jacobi equation of geodesic deviation [II06c, II07e]

$$\frac{D^2 J^i}{ds^2} + R^i_{jkm} \frac{dx^j}{ds} J^k \frac{dx^m}{ds} = 0, \tag{5.121}$$

whose solution J, usually called Jacobi variation field, locally measures the distance between nearby geodesics; D/ds stands for the covariant derivative along a geodesic and R^i_{jkm} are the components of the Riemann curvature tensor of the $\mathcal{CD}$ manifold M.

The relevant part of the Jacobi equation (5.121) is given by the tangent dynamics equation [CCP96b, CCC97]

$$\ddot{J}^i + R^i_{0k0} J^k = 0, \quad (i, k = 1, \ldots, 3n), \tag{5.122}$$

where the only non-vanishing components of the curvature tensor of the $\mathcal{CD}$ manifold M are

$$R^i_{0k0} = \partial^2 V / \partial x^i \partial x^k. \tag{5.123}$$

The tangent dynamics equation (5.122) can be used to define Lyapunov exponents in dynamical systems given by the Riemannian action (5.118) and Hamiltonian (5.119), using the formula [CPC00]

$$\lambda_1 = \lim_{t\to\infty} 1/2t \log(M_{i=1}^N[J_i^2(t) + J_i^2(t)]/M_{i=1}^N[J_i^2(0) + J_i^2(0)]). \qquad (5.124)$$

Lyapunov exponents measure the strength of dynamical chaos in the crowd psycho-physical behavior. The sum of positive Lyapunov exponents defines the *Kolmogorov–Sinai entropy* (see [II07a]).

5.9.7 Crowd Phase Transitions

Now, to relate these results to topological phase transitions within the $\mathcal{CD}$ manifold M given by (5.98), recall that any two high-dimensional manifolds M_v and $M_{v'}$ have the same topology if they can be continuously and differentiably deformed into one another, that is if they are diffeomorphic. Thus by topology change the 'loss of diffeomorphicity' is meant [Pet07]. In this respect, the so-called topological theorem [FP04] says that non-analyticity is the 'shadow' of a more fundamental phenomenon occurring in the system's configuration manifold (in our case the $\mathcal{CD}$ manifold): a topology change within the family of equipotential hypersurfaces

$$M_v = \{(x^1, \ldots, x^{3n}) \in \mathbb{R}^{3n} \mid V(x^1, \ldots, x^{3n}) = v\},$$

where V and x^i are the microscopic interaction potential and coordinates respectively. This topological approach to PTs stems from the numerical study of the dynamical counterpart of phase transitions, and precisely from the observation of discontinuous or cuspy patterns displayed by the largest Lyapunov exponent λ_1 at the transition energy [CPC00]. Lyapunov exponents cannot be measured in laboratory experiments, at variance with thermodynamic observables, thus, being genuine dynamical observables they are only be estimated in numerical simulations of the microscopic dynamics. If there are critical points of V in configuration space, that is points $x_c = [\bar{x}_1, \ldots, \bar{x}_{3n}]$ such that $\nabla V(x)|_{x=x_c} = 0$, according to the Morse Lemma [Hir76], in the neighborhood of any critical point x_c there always exists a coordinate system $x(t) = [x^1(t), \ldots, x^{3n}(t)]$ for which [CPC00]

$$V(x) = V(x_c) - x_1^2 - \cdots - x_k^2 + x_{k+1}^2 + \cdots + x_{3n}^2, \qquad (5.125)$$

where k is the index of the critical point, i.e., the number of negative eigenvalues of the Hessian of the potential energy V. In the neighborhood of a critical point of the $\mathcal{CD}$-manifold M, equation (5.125) yields the simplified form of (5.123), $\partial^2 V/\partial x^i \partial x^j = \pm \delta_{ij}$, giving j unstable directions that contribute to the exponential growth of the norm of the tangent vector J.

This means that the strength of dynamical chaos within the $\mathcal{CD}$-manifold M, measured by the largest Lyapunov exponent λ_1 given by (5.124), is affected by the existence of critical points x_c of the potential energy $V(x)$. However, as $V(x)$ is bounded below, it is a good Morse function, with no vanishing eigenvalues of its Hessian matrix. According to Morse theory [Hir76], the existence of critical points of V is associated with topology changes of the hypersurfaces $\{M_v\}_{v \in \mathbb{R}}$. The topology change of the $\{M_v\}_{v \in \mathbb{R}}$ at some v_c is a necessary condition for a phase transition to take place at the corresponding energy value [FP04]. The topology changes implied here are those described within the framework of Morse theory through 'attachment of handles' [Hir76] to the $\mathcal{CD}$-manifold M.

In our path-integral language this means that suitable topology changes of equipotential submanifolds of the $\mathcal{CD}$-manifold M can entail thermodynamic-like phase transitions [Hak83, Hak93, Hak96], according to the general formula:

$$\langle \text{phase out} \mid \text{phase in} \rangle := \int_{\text{top-ch}} \mathcal{D}[w\Phi]\, e^{iS[\Phi]}.$$

The statistical behavior of the crowd biodynamics system with the action functional (5.118) and the Hamiltonian (5.119) is encompassed, in the canonical ensemble, by its partition function, given by the Hamiltonian path integral [II07e]

$$Z_{3n} = \int_{\text{top-ch}} \mathcal{D}[p]\mathcal{D}[x]\, \exp\left\{ i \int_t^{t'} [p_i \dot{x}^i - H(p,x)]\, d\tau \right\}, \qquad (5.126)$$

where we have used the shorthand notation

$$\int_{\text{top-ch}} \mathcal{D}[p]\mathcal{D}[x] \equiv \int \prod_\tau \frac{dx(\tau)dp(\tau)}{2\pi}.$$

For more technical details, see [IR08].

6

Quantum Information, Games and Computation

In this Chapter we give a review of quantum information, quantum game theory, quantum computation and classical electronic for quantum computing.

6.1 Quantum Information and Computing

6.1.1 Entanglement, Teleportation and Information

Any storage, transmission, and processing of information relies on a physical carrier [Lan91]. In a handwritten note the sheet of paper serves as the carrier of information, in a desk top computer it is the random access memory and the hard drive on which the relevant data are stored. Communication makes use of sound waves, radio waves, or light pulses. The new field of *quantum information* is based on the idea that single quantum systems can be used as the elementary carriers of information, such as single photons, atoms, and ions. Quantum theory opens up new possibilities for information processing and communication [BEZ00, EJ96, NC00, Ste98, PV98]. In principle, quantum systems allow for information processing tasks which are very difficult or impossible to be performed when using classical systems. Envisioned applications range from the factorization of large numbers on a quantum computer to communication protocols, and key distribution in *quantum cryptography*.

Quantum theory may become relevant to technical development in information processing mainly for two reasons [Eis01]. On the one hand, the information processing and storage units in ordinary, "classical" computers are becoming smaller and smaller. The dimensions of transistor elements in silicon-based microchips are decreasing to the extent that they will be approaching scales in which quantum effects become relevant in the near future. On the other hand, it has become technologically possible to store and manipulate single quantum systems, e.g., with sophisticated methods from *quantum optics* and *solid state physics* [BEZ00, NC00].

V.G. Ivancevic, T.T. Ivancevic, *Quantum Neural Computation*,
Intelligent Systems, Control and Automation: Science and Engineering 40,
DOI 10.1007/978-90-481-3350-5_6, © Springer Science+Business Media B.V. 2010

The superior "performance" of quantum systems in computation and communication applications is predominantly rooted in a property of quantum mechanical states called *entanglement*. Essentially, entanglement comes along with new kinds of correlations. Even in classical composite systems measurement outcomes may be perfectly correlated. However, *entangled quantum states* may show stronger statistical correlations than those attainable in a classical composite system, where the correlation is produced by a classical random generator.[1]

On a purely theoretical level it is important to understand what kinds of tasks may be achieved with entangled quantum systems. For example, it is impossible to transmit the particular "quantum information" of a quantum system through a classical channel. This means that the statistical predictions of quantum mechanics cannot fully be reproduced when trying to extract information about the preparation procedure from a state of one quantum system, transmitting the classical information through the channel, and preparing another quantum system in a certain state on the basis of this information. It is nevertheless possible to transfer a quantum state to a different quantum system at a distant location without physically transmitting the actual quantum system in the particular state—provided that the parties initially share a pair of two-level systems in a maximally entangled state. This transfer of a quantum state was named *quantum teleportation*, a term borrowed from science-fiction literature [Eis01].

The teleportation protocol (see Fig. 6.1) was proposed in 1993 by C. Bennett et al. [BBS93]. It represented a major breakthrough in the field. Assume that one experimenter, from now on referred to as Alice, would like to send the unknown state of a given quantum system to Bob, another experimenter at a distant location. Bob prepares a bi-partite quantum system in a particular entangled state and gives one part of the system to Alice. In the next step

---

[1] In 1935 A. Einstein, B. Podolsky, and N. Rosen (*EPR*) published a seminal paper entitled "Can Quantum-Mechanical Description of Physical Reality Be Considered Complete?", which started a long lasting debate about the status of quantum theory [EPR35b]. On the basis of the predicted outcomes of measurements on two space-like separated quantum particles in an entangled state, EPR came to the conclusion that quantum mechanics could not be a complete theory, suggesting the view that additional hidden variables should be appended to a quantum state in order to restore causality and locality. N. Bohr, one of the spokesmen of the so-called Copenhagen school of the interpretation of quantum mechanics, argued against the assumption of a more complete underlying local deterministic level [Boh35]. It was not until 1964 that J. Bell presented a way to empirically test the two competing hypotheses [Bel64, Bel66, Bel87]. *Bell's theorem* is not strictly about quantum mechanics. It is a statement concerning correlations of measurement outcomes at distant events that any physical theory may predict under the assumption of an underlying local classical model [Bal87, Mer93a]. Starting from the 1980s many experiments were performed, each in favor of quantum mechanics and against local hidden variable models [AGR81, CS90, TBG98].

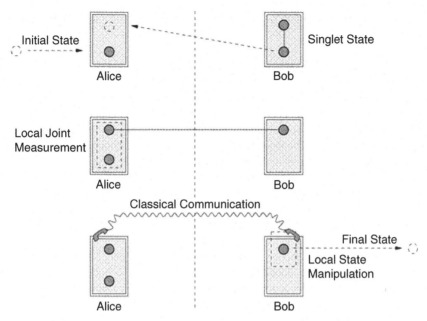

Fig. 6.1. The teleportation protocol as in [BBS93] (modified and adapted from [Eis01]).

Alice performs a local joint quantum measurement on both her quantum system and on the one she has received from Bob. Then she phones Bob and tells him about the outcome. Depending on the outcome of Alice's measurement, Bob can finally transform his part of the maximally entangled system by use of a simple manipulation of the state. The state of his system is eventually identical to the state of Alice's original system: the state has been "sent" from Alice to Bob.

Another important proposal of this type, also by C. Bennett and S. Wiesner, is the dense coding protocol [BW92a] concerning the transmission of classical information. A single quantum two-level system sent from Alice to Bob can carry a single bit of classical information. Surprisingly, if the two parties initially share a maximally entangled state, two bits of classical information can be transmitted with the same two-level system. Successful experimental quantum optical implementations of dense coding and teleportation were performed by [BPM97] and [BBD98]. Following the publication of these proposals, many other applications of entanglement were suggested,[2] the spectrum ranging from *quantum cryptography* using entangled quantum systems

---

[2] However, one has to bear in mind that notwithstanding the possible technical implications the research in quantum entanglement is to a large extent motivated by the wish to better understand one of the most important properties that distinguishes quantum mechanics from classical mechanics. Recent research in the theory of en-

by [Eke91] to improved frequency standards [CEH98] and clock synchronization [JAD00].

6.1.2 The Circuit Model for Quantum Computers

Recall that a classical digital computer operates on a string of input bits and returns a string of output bits. The function in between can be described as a logical circuit build up out of many elementary logic operations. That is, the whole computation can be decomposed into an array of smaller operations, or *gates*, acting only on one or two bits like the AND, OR and NOT operation. In fact, these three gates together with the COPY (or FANOUT) operation form a *universal* set of gates into which every *well-defined input–output function* can be decomposed. The *complexity of an algorithm* is then essentially the number of required elementary gates, resp. its asymptotic growth with the size of the input [EW06].

The circuit model for the quantum computer [Deu89, NC00] is actually very reminiscent of the classical circuit model: we only have to replace the input–output function by a quantum operation mapping quantum states *onto* quantum states. It is sufficient to consider operations only that have the property to be unitary, which means that the computation is taken to be logically reversible. In turn, any unitary operation can be decomposed into elementary gates acting only on one or two qubits. A set of elementary gates that allows for a realization of any unitary to arbitrary approximation is again referred to as being *universal* [BDE96, Deu89]. An important example of a set of universal gates is in this case any randomly chosen one-qubit rotation together with the *CNOT (Controlled NOT)* operation, which acts as

$$|x, y\rangle \mapsto |x, y \oplus x\rangle, \qquad (6.1)$$

where $\oplus$ means addition modulo 2 [Pre98a]. Like in the classical case there are infinitely many sets of universal gates. Notably, also any generic (i.e., randomly chosen) two-qubit gate (together with the possibility of switching the leads in order to swap qubits) is itself a universal set, very much like the NAND gate is for classical computing [BDE96].[3] Notably, any quantum circuit that makes use of a certain universal set of quantum gates can be simulated by a different quantum circuit based on another universal set of gates with only poly-logarithmic overhead [Kit97, NC00]. A particularly useful single-qubit gate is the *Hadamard gate*, acting as

---

tanglement shows that there are still fundamental open questions in non-relativistic quantum mechanics waiting to be solved.

[3] Any such generic quantum gate has so-called *entangling power* [Col06], in that it may transform a product state vector into one that can no longer be written as a tensor product. Such quantum mechanical pure states are called *entangled*. In intermediate steps of a quantum algorithm the physical state of the system is in general highly multi-particle entangled. In turn, the implementation of quantum gates in distributed quantum computation requires entanglement as a resource [EJP00].

$$|0\rangle \mapsto H|0\rangle = (|0\rangle + |1\rangle)/\sqrt{2}, \qquad |1\rangle \mapsto H|1\rangle = (|0\rangle - |1\rangle)/\sqrt{2}. \quad (6.2)$$

A *phase gate* does nothing but multiplying one of the basis vectors with a phase,

$$|0\rangle \mapsto |0\rangle, \qquad |1\rangle \mapsto i|1\rangle, \qquad\qquad\qquad (6.3)$$

and a *Pauli gate* corresponds to one of the three unitary Pauli matrices. The CNOT, the Hadamard, the phase gate, and the Pauli gate are quantum gates of utmost importance. Given their key status in many quantum algorithms, one might be tempted to think that with these ingredients alone (together with measurements of Pauli operators, see below), powerful quantum algorithms may be constructed that outperform the best known classical algorithm to a problem. This intuition is yet not correct: it is the content of the *Gottesman–Knill theorem* that any quantum circuit consisting of only these ingredients can be simulated efficiently on a classical computer [NC00, Got97].

One of the crucial differences between classical and quantum circuits is, that in the quantum case the COPY operation is not possible. In fact, the linearity of quantum mechanics forbids a device which copies an unknown quantum state—this is known as the *no-cloning theorem*.[4] The latter has far-reaching consequences, of which the most prominent one is the possibility of quantum cryptography coining this "no-go theorem" into an application [GRT02].

6.1.3 Elementary Quantum Algorithms

In the same scientific paper in which D. Deutsch introduced the notion of the universal quantum computer, he also presented the first *quantum algorithm* [Deu85].[5] The problem that this algorithm addresses, later referred to as *Deutsch's problem*, is a very simple one. Yet the *Deutsch algorithm* already exemplifies the advantages of a quantum computer through skillfully exploiting *quantum parallelism*. Like the Deutsch algorithm, all other elementary quantum algorithms amount to deciding which *black box* out of finitely many alternatives one has at hand. Such a black box is often also referred to as *oracle*. An input may be given to the oracle, one may read out or use the outcome in later steps of the quantum algorithm, and the objective is to find out the functioning of the black box. It is assumed that this oracle operation can be implemented with some sequence of quantum logic gates. The complexity of the quantum algorithm is then quantified in terms of the number of queries to the oracle [EW06].

---

[4] If one knows that the state vector is either $|0\rangle$ or $|1\rangle$ then a cloning machine is perfectly consistent with rules of quantum mechanics. However, producing perfect *clones* of an arbitrary quantum state is prohibited [WZ82].

[5] Quantum Turing machines were first considered by [Ben80] and developed by Deutsch in [Deu85].

With the help of the Deutsch algorithm, it is possible to decide whether a function has a certain property with a single call of the function, instead of two calls that are necessary classically.

Let $f : \{0,1\} \longrightarrow \{0,1\}$ be a function that has both a one-bit domain and range. This function can be either *constant* or *balanced*, which means that either $f(0) \oplus f(1) = 0$ or $f(0) \oplus f(1) = 1$ holds. The problem is to find out with the minimal number of function calls whether this function f is constant or balanced.

Classically, it is obvious that two function calls are required to decide which of the two allowed cases is realized, or, equivalently, what the value of $f(0) \oplus f(1)$ is. A way to compute the function f on a quantum computer is to transform the state vector of two qubits according to

$$|x, y\rangle \mapsto U_f|x, y\rangle = |x, f(x) \oplus y\rangle. \qquad (6.4)$$

In this manner, the evaluation can be realized unitarily. The above map is what is called a standard quantum oracle. The claim now is that using such an oracle, a single function call is sufficient for the evaluation of $f(0) \oplus f(1)$. In order to show this, let us assume that we have prepared two qubits in the state with state vector $|\Psi\rangle = (H \otimes H)|0,1\rangle$, where H denotes the Hadamard gate. We now apply the unitary U_f once to this state, and finally apply another Hadamard gate to the first qubit. The resulting state vector hence reads as

$$|\Psi'\rangle = (H \otimes 1)U_f(H \otimes H)|0,1\rangle. \qquad (6.5)$$

A short calculation shows that $|\Psi'\rangle$ can be evaluated to

$$|\Psi'\rangle = \pm|f(0) \oplus f(1)\rangle H|1\rangle. \qquad (6.6)$$

The second qubit is in the state corresponding to the vector $H|1\rangle$, which is of no relevance to our problem. The state of the first qubit, however, is quite remarkable: encoded is $|f(0) \oplus f(1)\rangle$, and both alternatives are decidable with unit probability in a measurement in the computational basis, as the two state vectors are orthogonal. That is, with a single measurement of the state, and notably, with a single call of the function f, of the first qubit we can decide whether f was constant or balanced.

For more details on quantum algorithms, see [EW06].

6.2 Quantum Games

Recall from the previous section on classical game theory that the important two player game is Tucker's *Prisoner's Dilemma* game [Tuc50].[6] In this game two players suspected of having committed a minor crime (in the following

---

[6] Recall that the story goes as follows: Alice and Bob are arrested for a minor crime. Despite lacking evidence, the prosecutor is convinced that they are guilty of a more

referred to as Alice and Bob) can independently decide whether they intend to "cooperate" or "defect". Being well aware of the consequences of their decisions the players obtain a certain pay-off according to their respective choices. This pay-off provides a quantitative characterization of their personal preferences. Both players are assumed to want to maximize their individual pay-off, yet they must pick their choice while they have to take their choices without the other player's decision. The players face a dilemma since dilemma lies in the fact that rational reasoning in such a situation dictates the players to defect, although they would be much happier if they both cooperated.

Formally, Alice has two basic choices, meaning that she can select from two possible *strategies* $s_A = C$ (cooperation) and $s_A = D$ (defection). Bob may also take $s_B = C$ or $s_B = D$. The game is defined by these possible strategies on the one hand, and on the other hand by a specification of how to evaluate the pay-off once the combination of chosen strategies (s_A, s_B) is known, i.e., the utility functions mapping (s_A, s_B) on a number [Mye91]. The expected pay-off quantifies the preference of the players.

A (strategic-form) *two player game* $\Gamma = (\{A, B\}, S, U)$ is fully specified by the set S of pairs of *strategies* (s_A, s_B), the *utility functions* $U = (u_A, u_B)$, and additional *rules of the game* consistent with the set of strategies.[7]

In this subsection, following [EW00], we further develop the idea of identifying strategic moves with quantum operations as introduced by [EWL99, MB98]. This approach appears to be fruitful in at least two ways [EWL99, MB98, GVW99]. On the one hand several recently proposed applications of quantum information theory can already be conceived as competitive situations where several parties with more or less opposed motives interact. These parties may, for example, apply quantum operations on a *bi-partite quantum system* [PW98]. In the same context, quantum cloning has been formulated as a game between two players [Wer98]. Similarly, eavesdropping in quantum cryptography [BB84] can be regarded as a game between the eavesdropper and the sender, and there are similarities of the extended form of quantum versions of games and quantum algorithms [Sho97, Eke91]. On the other hand a generalization of the theory of decisions into the domain of quantum probabilities seems interesting, as the roots of game theory are partly in probability theory. In this context it is of interest to investigate what solutions are attainable if superpositions of strategies are allowed for [EWL99, MB98].

Game theory does not explicitly concern itself with how the information is transmitted once a decision is taken. Yet, it should be clear that the practical implementation of any (classical) game inevitably makes use of the exchange of

---

severe crime and reveals the following three sets of strategies: If both defect, i.e. refuse to confess, they will equally be convicted for minor crimes. Alternatively, if both confess, i.e. cooperate, they will be convicted, albeit not charged with maximum penalty. The third option is for one of them to cooperate and for the other one to defect. If, e.g., Alice cooperates and Bob defects, Alice will be released shortly whereas Bob will have to face maximum penalty, and vice versa.

[7] The set $\{A, B\}$ is the set of players.

voting papers, faxes, emails, ballots, and the like. In the Prisoners' Dilemma, e.g., the two parties have to communicate with an advocate by talking to her or by writing a short letter on which the decision is indicated. The process of decision making in a game clearly requires the transmission of information. However, game theory does not concern itself with how information is transmitted once a decision is taken.

In the Prisoners' Dilemma the two parties clearly have to communicate with an advocate by talking to her or by writing a short letter on which the decision is indicated. While game theory (as an abstract theory which is broad in its scope) rarely bothers to deal with the transmission of information explicitly, it should be clear that the practical implementation of any (classical) game inevitably makes use of the exchange of voting papers, faxes, emails, ballots, and so on. That is, bearing in mind that a game is also about the transfer of information, it becomes legitimate to ask what happens if these carriers of information are taken to be quantum systems, quantum information being a fundamental notion of information.

By classical means a two player binary choice game may be played as follows: An arbiter takes two coins and forwards one coin each to the players. The players then receive their coin with head up and may keep it as it is ("cooperate") or turn it upside down so that tails is up ("defection"). Both players then return the coins to the arbiter who calculates the players' final pay-off corresponding to the combination of strategies he obtains from the players. Here, the coins serve as the physical carrier of information in the game. In a quantum version of such a game quantum systems would be used as such carriers of information. For a binary choice two player game an implementation making use of minimal resources involves two qubits as physical carriers.

6.2.1 Quantum Strategies

In this subsection, following [MB98], we analyze the effectiveness of quantum strategies, exemplified in the simple game $PQ - penny - flip$: The starship *Enterprise* is facing some immanent—and apparently inescapable—calamity when Q appears on the bridge and offers to help, provided Captain Picard[8] can beat him at penny flipping: Picard is to place a penny head up in a box, whereupon they will take turns (Q, then Picard, then Q) flipping the penny (or not), without being able to see it. Q wins if the penny is head up when they open the box. This is a *two-person zero-sum strategic game* which might be analyzed traditionally using the payoff matrix:

---

[8] Captain Picard and Q are characters in the popular American television (and movie) series *Star Trek: The Next Generation* whose initials and abilities are ideal for this illustration.

	NN	NF	FN	FF
N	-1	1	1	-1
F	1	-1	-1	1

where the rows and columns are labeled by Picard's and Q's *pure strategies*, respectively; F denotes a flip and N denotes no flip; and the numbers in the matrix are Picard's payoffs: 1 indicating a win and -1 a loss. For example, consider the top entry in the second column: Q's strategy is to flip the penny on his first turn and then not flip it on his second, while Picard's strategy is to not flip the penny on his turn. The result is that the state of the penny is, successively: H, T, T, T, so Picard wins.

Now it is natural to define a two dimensional vector space V with basis $\{H, T\}$ and to represent player strategies by sequences of 2×2 matrices. That is, the matrices

$$F = \begin{array}{c} H \\ T \end{array}\begin{pmatrix} \overset{H}{0} & \overset{T}{1} \\ 1 & 0 \end{pmatrix} \quad \text{and} \quad N = \begin{array}{c} H \\ T \end{array}\begin{pmatrix} \overset{H}{1} & \overset{T}{0} \\ 0 & 1 \end{pmatrix}$$

correspond to flipping and not flipping the penny, respectively, since we define them to act by left multiplication on the vector representing the state of the penny [MB98]. A *mixed* action is a convex linear combination of F and N, which acts as a 2×2 (doubly) stochastic matrix:

$$\begin{array}{c} H \\ T \end{array}\begin{pmatrix} \overset{H}{1-p} & \overset{T}{p} \\ p & 1-p \end{pmatrix}$$

if the player flips the penny with probability $p \in [0, 1]$. A sequence of mixed actions puts the state of the penny into a convex linear combination $aH + (1-a)T$, $0 \le a \le 1$, which means that if the box is opened the penny will be head up with probability a.

However, Q is utilizing a *quantum* strategy, namely a sequence of unitary, rather than stochastic, matrices. In standard Dirac notation the basis of V is written $\{|H\rangle, |T\rangle\}$. A *pure* quantum state for the penny is a linear combination $a|H\rangle + b|T\rangle$, $a, b \in \mathbb{C}$, $a\bar{a} + b\bar{b} = 1$, which means that if the box is opened, the penny will be head up with probability $a\bar{a}$. Since the penny starts in state $|H\rangle$, this is the state of the penny if Q's first action is the unitary operation

$$U_1 = U(a, b) = \begin{array}{c} H \\ T \end{array}\begin{pmatrix} \overset{H}{a} & \overset{T}{b} \\ \bar{b} & -\bar{a} \end{pmatrix}.$$

On the other hand, Captain Picard is utilizing a *classical* probabilistic strategy in which he flips the penny with probability p. After his action the penny is in a *mixed* quantum state, i.e., it is in the pure state $b|H\rangle + a|T\rangle$ with probability p and in the pure state $a|H\rangle + b|T\rangle$ with probability $1 - p$. Mixed states are conveniently represented as *density matrices*, elements of $V \otimes V^\dagger$ with trace 1; the diagonal entry (i, i) is the probability that the system is observed to be in state $|i\rangle$. The density matrix for a pure state $|\psi\rangle \in V$ is the projection matrix $|\psi\rangle\langle\psi|$ and the density matrix for a mixed state is the corresponding convex linear combination of pure density matrices. Unitary transformations act on density matrices by conjugation: The penny starts in the pure state $\rho_0 = |H\rangle\langle H|$ and Q's first action puts it into the pure state:

$$\rho_1 = U_1 \rho_0 U_1^\dagger.$$

Picard's mixed action acts on this density matrix, not as a stochastic matrix on a probabilistic state, but as a convex linear combination of unitary (deterministic) transformations [MB98]:

$$\rho_2 = pF\rho_1 F^\dagger + (1 - p)N\rho_1 N^\dagger.$$

For $p = \frac{1}{2}$ the diagonal elements of ρ_2 are each $\frac{1}{2}$. If the game were to end here, Picard's strategy would ensure him an expected payoff of 0, independently of Q's strategy. In fact, if Q were to employ any strategy for which $a\bar{a} \neq b\bar{b}$, Picard could obtain an expected payoff of $|a\bar{a} - b\bar{b}| > 0$ by setting $p = 0, 1$ according to whether $b\bar{b} > a\bar{a}$, or the reverse. Similarly, if Picard were to choose $p \neq \frac{1}{2}$, Q could obtain an expected payoff of $|2p - 1|$ by setting $a = 1$ or $b = 1$ according to whether $p < \frac{1}{2}$, or the reverse. Thus the mixed/quantum equilibria for the two-move game are pairs $([\frac{1}{2}F + \frac{1}{2}N], [U(a, b)])$ for which $a\bar{a} = \frac{1}{2} = b\bar{b}$ and the outcome is the same as if both players utilize optimal mixed strategies.

But Q has another move U_3 which again transforms the state of the penny by conjugation to $\rho_3 = U_3 \rho_2 U_3^\dagger$. If Q's strategy consists of $U_1 = U(1/\sqrt{2}, 1/\sqrt{2}) = U_3$, his first action puts the penny into a simultaneous eigenvalue 1 eigenstate of both F and N, which is therefore invariant under any mixed strategy $pF + (1 - p)N$ of Picard; and his second action inverts his first to give $\rho_3 = |H\rangle\langle H|$. That is, with probability 1 the penny is head up! Since Q can do no better than to win with probability 1, this is an optimal quantum strategy for him. All the pairs $([pF + (1-p)N], [U(1/\sqrt{2}, 1/\sqrt{2}), U(1/\sqrt{2}, 1/\sqrt{2})])$ are mixed/quantum equilibria for $PQ - penny - flip$, with value -1 to Picard; this is why he loses every game.

Therefore, $PQ - penny - flip$ is a very simple game, but it is structurally similar to the oracle problems for which efficient quantum algorithms are known—with Picard playing the role of the oracle. In Simon's problem the functions $f : \{0, 1\}^n \longrightarrow \{0, 1\}^n$ which satisfy $f(x) = f(y)$ if and only if $y = x \oplus s$ for some $s \in \{0, 1\}^n$ ($\oplus$ denotes componentwise addition, mod 2), correspond to Picard's pure strategies; we may imagine the oracle choosing a mixed strategy intended to minimize our chances of efficiently determining s probabilistically. Simon's algorithm is a quantum strategy which is more successful than any mixed, i.e., probabilistic, one. Similarly, in the problem of searching a database of size N, the locations in the database correspond to pure strategies; again we may imagine the oracle choosing a mixed strategy designed to frustrate our search for an item at some specified location. Grover's algorithm is a quantum strategy for a game of $2m$ moves alternating between us and the oracle, where $m = O(\sqrt{N})$, which out performs any mixed strategy. These three examples suggest the following theorem: *There is always a mixed/quantum equilibrium for a two-person zero-sum game, at which the expected payoff for the player utilizing a quantum strategy is at least as great as his expected payoff with an optimal mixed strategy.* For the proof, see [MB98].

Another natural question to ask is what happens if both players utilize quantum strategies. Meyer proposes the following statement: *A two-person zero-sum game need not have a quantum/quantum equilibrium.* That is, the situation when both players utilize quantum strategies is the same as when they both utilize pure (classical) strategies: there need not be any equilibrium solution. This suggests looking for the analogue of von Neumann's result on the existence of mixed strategy equilibria. So we should consider strategies which are convex linear combinations of unitary actions—*mixed quantum* strategies: *A two-person zero-sum game always has a mixed quantum/mixed quantum equilibrium.* For the proofs, see [MB98].

6.2.2 Quantum Games

A quantum game could be any quantum system where more than one party can implement quantum operations and where the utility of the moves can be quantified in an appropriate manner.

A *two-player quantum game* $\Gamma = (\mathcal{H}, \rho, S_A, S_B, P_A, P_B)$ is completely specified by the underlying Hilbert space $\mathcal{H}$ of the physical system, the initial state $\rho \in \mathcal{S}(\mathcal{H})$, where $\mathcal{S}(\mathcal{H})$ is the associated state space, the sets S_A and S_B of permissible quantum operations of the two players, and the *utility functionals* P_A and P_B, which specify the expected pay-off utility for each player. A *quantum strategy* $s_A \in S_A$, $s_B \in S_B$ is a quantum operation, that is, a completely positive trace-preserving map mapping the state space on itself.[9] The quantum game's definition also includes certain implicit rules, such as the order of the implementation of the respective quantum strategies. Rules also exclude certain actions, as the alteration of the pay-off during the game.

The quantum games proposed in [EWL99, MB98, GVW99] can be cast into this form. Also, the quantum cloning device as described in [Wer98] can be said to be a quantum game in this sense and additional rules of the game consistent with the set of quantum strategies. The utility functionals $P_A, P_B \to \mathbb{R}$

---

[9] E.g., Alice may allow a coupling of the original quantum system to an auxiliary quantum system and let the two unitarily interact. After performing a projective measurement on the composite system she could eventually consider the original system again by taking a partial trace with respect to the auxiliary part.

map the final state after applying the respective strategies on the set of real numbers; they quantify the individual preference of each player. The rules of the game include the order in which the two players may implement their respective quantum strategy. Rules also exclude certain actions, as the alteration of the pay-off during the game. A quantum game is called *zero-sum game*, if the expected pay-offs sum up to zero for all pairs of strategies, that is, if $P_A(s_A, s_B) = -P_B(s_A, s_B)$ for all $s_A \in S_A$, $s_B \in S_B$. Otherwise, it is called a *non-zero sum game*.

Note that we do not require a set of allowed strategies for a player to be a closed set. The identity operation is included, as "not to do anything" always seems to be a natural option, but in the following it is not assumed that these sets are closed sets. The allowed set of operations does not necessarily include only unitary operations, but any trace-preserving completely positive map corresponding with a generalized measurement.

It is natural to call two quantum strategies of Alice s_A and s'_A equivalent, if $P_A(s_A, s_B) = P_A(s'_A, s_B)$ and $P_B(s_A, s_B) = P_A(s'_A, s_B)$ for *all* possible s_B. That is, if s_A and s'_A yield the same expected pay-off for both players for all allowed strategies of Bob. In the same way strategies s_B and s'_B of Bob will be identified.

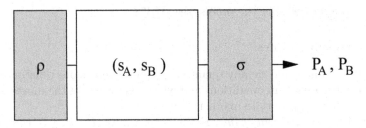

Fig. 6.2. The general setup of a quantum game (adapted from [EW00]).

A solution concept provides advice to the players with respect to the action they should take. Such a solution must be based on the particular utility functionals which define the preference of each player. Indeed, in the Prisoners' Dilemma game it was obvious what to accept as a solution, as to "defect" implied an advantage, regardless of the strategy of the other player. These definitions are fully analogous to the corresponding definitions in standard game theory [Mye91].

A quantum strategy s_A (implemented by Alice) is called *dominant strategy* of Alice if

$$P_A(s_A, s'_B) \geq P_A(s'_A, s'_B)$$

for all $s'_A \in S_A$, $s'_B \in S_B$. Analogously we can define a dominant strategy for Bob. A pair (s_A, s_B) is said to be an *equilibrium in dominant strategies* if s_A and s_B are the players' respective dominant strategies. A combination of strategies (s_A, s_B) is called a *Nash equilibrium* if

$$P_A(s_A, s_B) \geq P_A(s'_A, s_B),$$
$$P_B(s_A, s_B) \geq P_B(s_A, s'_B)$$

for all $s'_A \in S_A$, $s'_B \in S_B$. (s_A, s_B) is a *solution in maxi-min-strategies*, if

$$\min_{s'_B \in S_B} P_A(s_A, s'_B) \geq \min_{s'_B \in S_B} P_A(s'_A, s'_B) \quad \text{for all } s'_A \in S_A,$$

$$\min_{s'_A \in S_A} P_B(s'_A, s_B) \geq \min_{s'_A \in s_A} P_B(s'_A, s'_B) \quad \text{for all } s'_B \in S_B.$$

A pair of strategies (s_A, s_B) is called *Pareto optimal*, if it is not possible to increase one player's pay-off without lessening the pay-off of the other player. A solution in dominant strategies is the strongest solution concept for a non-zero sum game. In the Prisoner's Dilemma defection is the dominant strategy, as it is favorable regardless what strategy the other party picks. A dominant strategy exists if a particular strategy provides the maximum pay-off regardless of the strategy selected by the other player.

Typically, however, the optimal strategy depends on the strategy chosen by the other party. A Nash equilibrium implies that neither player has a motivation to unilaterally alter his or her strategy from this equilibrium solution, as this action will lessen his or her pay-off. Given that the other player will stick to the strategy corresponding to the equilibrium, the best result is achieved by also playing the equilibrium solution. The concept of Nash equilibria is of paramount importance to studies of non-zero-sum games. It is, however, only an acceptable solution concept if the Nash equilibrium is unique. For games with multiple equilibria the application of a hierarchy of natural refinement concepts may finally eliminate all but one of the Nash equilibria. Note that a Nash equilibrium is not necessarily efficient. In the Prisoners' Dilemma for example there is a unique equilibrium, but it is not Pareto optimal, meaning that there is another outcome which would make both players better off.

6.2.3 Two-Qubit Quantum Games

In the subsequent investigation we turn to specific games where the classical version of the game is faithfully entailed in the quantum game. In a quantum version of a binary choice game two qubits are prepared by a arbiter in a particular initial state, the qubits are sent to the two players who have physical instruments at hand to manipulate their qubits in an appropriate manner. Finally, the qubits are sent back to the arbiter who performs a measurement to evaluate the pay-off. The source, the devices to manipulate the state, and the measurement apparatus are known to the players when they are obliged to take their decision.

For such a bi-partite quantum game the system of interest is a quantum system with underlying Hilbert space

$$\mathcal{H} = \mathcal{H}_A \otimes \mathcal{H}_B$$

and associated state space $\mathcal{S}(\mathcal{H})$. Quantum strategies s_A and s_B of Alice and Bob are local quantum operations acting in $\mathcal{H}_A$ and $\mathcal{H}_B$ respectively.[10] That is, they are completely positive maps of the form

$$s_A \otimes s_B : \mathcal{S}(\mathcal{H}) \to \mathcal{S}(\mathcal{H}).$$

That is, Alice and Bob are restricted to implement their respective quantum strategy s_A and s_B on their qubit only. In this step they may choose any quantum strategy that is included in the set of strategies S. They are both well aware of the set S, but they do not know which particular quantum strategy the other party would actually implement. As the application of both quantum strategies amounts to a map $s_A \otimes s_B : \mathcal{S}(\mathcal{H}) \to \mathcal{S}(\mathcal{H})$, after execution of the moves the system is in the state

$$\sigma = (s_A \otimes s_B)(\rho).$$

Particularly important will be unitary operations s_A and s_B. They are associated with unitary operators U_A and U_B, written as $s_A \sim U_A$ and $s_B \sim U_B$. In this case the final state σ is given by

$$\sigma = (U_A \otimes U_B)\rho(U_A \otimes U_B)^\dagger.$$

From now on both the sets of strategies of Alice and Bob and the pay-off functionals are taken to be identical, that is,

$$S_A = S_B = S \quad \text{and} \quad P_A = P_B = P,$$

such that both parties face the same situation.

The quantum game $\Gamma = (\mathbb{C}^2 \otimes \mathbb{C}^2, \rho, S, S, P, P)$ can be played in the following way: The initial state ρ is taken to be a maximally entangled state in the respective state space. In order to be consistent with [EWL99] let $\rho = |\psi\rangle\langle\psi|$ with

$$|\psi\rangle = (|00\rangle + i|11\rangle)/\sqrt{2}, \tag{6.7}$$

where the first entry refers to $\mathcal{H}_A$ and the second to $\mathcal{H}_B$. The expected pay-off is determined from this state σ. The two qubits are forwarded to the arbiter who performs a projective selective measurement on the final state σ with the so-called *Kraus operators*, π_{CC}, π_{CD}, π_{DC}, and π_{DD}, where

$$
\begin{aligned}
\pi_{CC} &= |\psi_{CC}\rangle\langle\psi_{CC}|, & |\psi_{CC}\rangle &= (|00\rangle + i|11\rangle)/\sqrt{2}, \\
\pi_{CD} &= |\psi_{CD}\rangle\langle\psi_{CD}|, & |\psi_{CD}\rangle &= (|01\rangle - i|10\rangle)/\sqrt{2}, \\
\pi_{DC} &= |\psi_{DC}\rangle\langle\psi_{DC}|, & |\psi_{DC}\rangle &= (|10\rangle - i|01\rangle)/\sqrt{2}, \\
\pi_{DD} &= |\psi_{DD}\rangle\langle\psi_{DD}|, & |\psi_{DD}\rangle &= (|11\rangle + i|00\rangle)/\sqrt{2}.
\end{aligned}
\tag{6.8}
$$

---

[10] The quantum strategies $s_A \otimes 1$ and $1 \otimes s_B$ are in the following identified with s_A and s_B, respectively.

According to the outcome of the measurement, a pay-off of A_{CC}, A_{CD}, A_{DC}, or A_{DD} is given to Alice, Bob receives B_{CC}, B_{CD}, B_{DC}, or B_{DD}. The utility functionals, also referred to as expected pay-off of Alice and Bob, read

$$P_A(s_A, s_B) = A_{CC} \operatorname{tr}[\pi_{CC}\sigma] + A_{CD} \operatorname{tr}[\pi_{CD}\sigma]$$
$$+ A_{DC} \operatorname{tr}[\pi_{DC}\sigma] + A_{DD} \operatorname{tr}[\pi_{DD}\sigma], \tag{6.9}$$
$$P_B(s_A, s_B) = B_{CC} \operatorname{tr}[\pi_{CC}\sigma] + B_{CD} \operatorname{tr}[\pi_{CD}\sigma]$$
$$+ B_{DC} \operatorname{tr}[\pi_{DC}\sigma] + B_{DD} \operatorname{tr}[\pi_{DD}\sigma]. \tag{6.10}$$

It is important to note that the Kraus operators are chosen in such a way that the classical game is fully entailed in the quantum game: The *classical strategies* C and D are associated with particular unitary operators with a matrix representation

$$C \sim \begin{pmatrix} 1 & 0 \\ 0 & 1 \end{pmatrix}, \qquad D \sim \begin{pmatrix} 0 & 1 \\ -1 & 0 \end{pmatrix}.$$

C does not change the state at all, D implements a "spin-flip". If both parties stick to these classical strategies, (6.9) and (6.10) guarantee that the expected pay-off is exactly the pay-off of the corresponding classical game defined by the numbers A_{CC}, A_{CD}, A_{DC}, A_{DD}, B_{CC}, B_{CD}, B_{DC}, and B_{DD}. E.g., if Alice plays C and Bob chooses D, the state σ after implementation of the strategies is given by

$$\sigma = (C \otimes D)(\rho) = |\psi_{CD}\rangle\langle\psi_{CD}|,$$

such that Alice obtains A_{CD} units and Bob B_{CD} units pay-off. In this way the peculiarities of strategic moves in the quantum domain can be adequately studied. The players may make use of additional degrees of freedom which are not available with randomization of the classical strategies, but they can also stick to mere classical strategies. This scheme can be applied to any two player binary choice game and is to a high extent canonical.

Prisoners' Dilemma

We now investigate the solution concepts for the quantum analogue of the Prisoners' Dilemma as mentioned above,[11]

$$A_{CC} = B_{CC} = 3, \qquad A_{DD} = B_{DD} = 1,$$
$$A_{CD} = B_{DC} = 0, \qquad A_{DC} = B_{CD} = 5.$$

---

[11] From a game theoretical viewpoint, any positive numbers satisfying the symmetry conditions $A_{CC} = B_{CC}$, $A_{DD} = B_{DD}$, $A_{CD} = B_{DC}$, $A_{DC} = B_{CD}$ and the inequalities $A_{DC} > A_{CC} > A_{DD} > A_{CD}$ and $A_{CC} \geq (A_{CD} + A_{DC})/2$ define a (strong) Prisoners' Dilemma.

The solution provided by the introduced concepts clearly depends on the rules of the game, and in particular, on the sets of strategies. In all of the following sets of allowed strategies S the classical options (to defect and to cooperate) are included. Several interesting sets of strategies and concomitant solution concepts will at this point be studied. The first three subsections involve local unitary operations only, while in the last subsection other quantum operations are considered as well.

One-Parameter Set of Strategies

The first set of strategies $S^{(CL)}$ involves quantum operations s_A and s_B which are local rotations with one parameter. The matrix representation of the corresponding unitary operators is taken to be

$$U(\theta) = \begin{pmatrix} \cos(\theta/2) & \sin(\theta/2) \\ -\sin(\theta/2) & \cos(\theta/2) \end{pmatrix},$$

with $\theta \in [0, \pi]$. Hence, in this simple case, selecting strategies s_A and s_B amounts to choosing two angles θ_A and θ_B. The classical strategies of defection and cooperation are also included in the set of possible strategies. We associate cooperation with the operation C and defection with D, defined as

$$C \sim U(0) = \begin{pmatrix} 1 & 0 \\ 0 & 1 \end{pmatrix}, \qquad D \sim U(\pi) = \begin{pmatrix} 0 & 1 \\ -1 & 0 \end{pmatrix}.$$

An analysis of the expected pay-offs P_A and P_B,

$$P_A(\theta_A, \theta_B) = 3|\cos(\theta_A/2)\cos(\theta_B/2)|^2 + 5|\cos(\theta_B/2)\sin(\theta_A/2)|^2$$
$$+ |\sin(\theta_A/2)\sin(\theta_B/2)|^2,$$
$$P_B(\theta_A, \theta_B) = 3|\cos(\theta_A/2)\cos(\theta_B/2)|^2 + 5|\sin(\theta_B/2)\cos(\theta_A/2)|^2$$
$$+ |\sin(\theta_A/2)\sin(\theta_B/2)|^2,$$

shows that this game is the classical Prisoners' Dilemma game [EWL99]. The pay-off functions are actually identical to the analogous functions in the ordinary Prisoners' Dilemma with mixed (randomized) strategies, where cooperation is chosen with the classical probability $p = \cos^2(\theta/2)$. The inequalities

$$P_A(D, s_B) \geq P_A(s_A, s_B), \qquad P_B(s_A, D) \geq P_B(s_A, s_B),$$

hold for all $s_A, s_B \in S^{(CL)}$, therefore, (D, D) is an equilibrium in dominant strategies and thus the unique Nash equilibrium. As explained in the introduction this equilibrium is far from being efficient, because $P_A(D, D) = P_B(D, D) = 1$ instead of the Pareto optimal pay-off which would be 3.

Two-Parameter Set of Strategies

A more general set of strategies is the following two-parameter set $S^{(TP)}$. The matrix representation of operators corresponding to quantum strategies from this set is given by An even more efficient solution is achieved when one considers the following two-parameter set $S^{(TP)}$ of local unitary operations. The matrix representation of the operators corresponding to quantum strategies s_A and s_B is given by Indeed, in this case the Pareto optimum can be realized. Both s_A and s_B are taken from a certain subset of local unitary operations. The matrix representation of the operators forming a subset of $SU(2)$ is given by

$$U(\theta, \phi) = \begin{pmatrix} e^{i\phi}\cos(\theta/2) & \sin(\theta/2) \\ -\sin(\theta/2) & e^{-i\phi}\cos(\theta/2) \end{pmatrix},$$

with $\theta \in [0, \pi]$ and $\phi \in [0, \pi/2]$. Selecting a strategy s_A, s_B then means choosing appropriate angles θ_A, ϕ_A and θ_B, ϕ_B. The classical pure strategies can be realized as

$$C \sim U(0,0) \quad \text{and} \quad D \sim U(\pi, 0).$$

This case has also been considered in [EWL99]. The expected pay-off for Alice, e.g., explicitly reads

$$
\begin{aligned}
P_A(\theta_A, & \phi_A, \theta_B, \theta_B) \\
&= 3 \left| \cos(\phi_A + \phi_B)\cos(\theta_A/2)\cos(\theta_B/2) \right|^2 \\
&\quad + 5 \left| \sin(\phi_A)\cos(\theta_A/2)\sin(\theta_B/2) - \cos(\phi_B)\cos(\theta_B/2)\sin(\theta_A/2) \right|^2 \\
&\quad + \left| \sin(\phi_A + \phi_B)\cos(\theta_A/2)\cos(\theta_B/2) + \sin(\theta_A/2)\sin(\theta_B/2) \right|^2 . \quad (6.11)
\end{aligned}
$$

Due to entanglement of the initial state as defined in the general setup (6.7) Alices's expected pay-off depends also on Bob's choice. It turns out that the previous Nash equilibrium (D, D) of $S^{(CL)}$ is no longer an equilibrium solution, as both players can benefit from deviating from D. However, concomitant with the disappearance of this solution another Nash equilibrium has emerged, given by (Q, Q). The strategy Q is associated with a matrix

$$Q \sim U(0, \pi/2) = \begin{pmatrix} i & 0 \\ 0 & -i \end{pmatrix}.$$

This Nash equilibrium is unique [EWL99] and serves as the only acceptable solution of the game. The astonishing fact is that $P_A(Q,Q) = P_B(Q,Q) = 3$ (instead of 1) so that the Pareto optimum is realized. No player could gain without lessening the other player's expected pay-off. In this sense one can say that the dilemma of the original game has fully disappeared. In the classical game only mutual cooperation is Pareto optimal, but this pair of strategies does not correspond to a Nash equilibrium.

General Unitary Operations

One can generalize the previous setting to the case where Alice and Bob can implement operations s_A and s_B taken from $S^{(GU)}$, where $S^{(GU)}$ is the set of general local unitary operations, $\mathcal{H}_A$ and $\mathcal{H}_B$, respectively. Here, it could be suspected that the solution becomes more efficient the larger the sets of allowed operations are. But this is not the case. The previous Pareto optimal unique Nash equilibrium (Q, Q) ceases to be an equilibrium solution if the set is enlarged: For any strategy $s_B \in S^{(GU)}$ there exists an *optimal answer* local unitary operation $s_A \in S^{(GU)}$ of Alice resulting in

$$(s_A \otimes s_B)(\rho) = |\psi_{DC}\rangle\langle\psi_{DC}|, \tag{6.12}$$

with ρ given in (6.7). This quantum strategy is given by and $|\psi_{CD}\rangle$ in (6.8): For s_A That is, for any strategy of Bob s_B there is a strategy s_A of Alice such that

$$P_A(s_A, s_B) = 5 \quad \text{and} \quad P_B(s_A, s_B) = 0.$$

Take

$$s_A \sim \begin{pmatrix} a & b \\ c & d \end{pmatrix}, \quad s_B \sim \begin{pmatrix} -ib & a \\ -d & -ic \end{pmatrix},$$

where a, b, c, d are appropriate complex numbers. Given that Bob plays the strategy s_B associated with a particular Nash equilibrium (s_A, s_B), Alice can always apply the optimal answer s_A to achieve the maximal possible pay-off. However, the resulting pair of quantum strategies cannot be an equilibrium since again, the game being symmetric, Bob can improve his pay-off by changing his strategy to his optimal answer s'_B. Hence, there is no pair (s_A, s_B) of pure strategies with the property that the players can only lose from unilaterally deviating from this pair of strategies. In particular, one can always gain from choosing a different strategy than the pure strategy of other party.

Yet, there remain to be Nash equilibria in *mixed strategies* which are much more efficient than the classical outcome of the equilibrium in dominant strategies $P_A(D, D) = P_B(D, D) = 1$. In a mixed strategy of Alice, say, she selects a particular quantum strategy s_A (which is also called *pure strategy*) from the set of strategies S_A with a certain classical probability. She selects the strategy $s_A^{(i)}$ (which is also called *pure strategy*) out of a set of quantum strategies $s_A^{(1)}, \ldots, s_A^{(N)}$, $(i = 1, \ldots, N)$, with probability $p_A^{(i)} \in [0, 1]$, where $p_A^{(i)} \in [0, 1]$ and $\sum_{i=1}^N p_A^{(i)} = 1$.

One can also allow for *mixed strategies* by specifying a set $s_A^{(1)}, \ldots, s_A^{(N)}$ of non-randomized quantum strategies and probabilities $p_A^{(1)}, \ldots, p_A^{(N)}$ with $p_A^{(i)} \in [0, 1]$, $i = 1, \ldots, N$, and $\sum_{i=1}^N p_A^{(i)} = 1$. Alice then selects the strategy (which is also called *pure strategy*) $s_A^{(i)}$ with the classical probability $p_A^{(i)}$. That is, mixed strategies of Alice and Bob are associated with maps of the form

$$\rho \longmapsto \sigma = \sum_{i,j} p_A^{(i)} p_B^{(j)} (U_A^{(i)} \otimes U_B^{(j)}) \rho (U_A^{(i)} \otimes U_B^{(j)})^\dagger, \tag{6.13}$$

$p_A^{(i)}, p_B^{(i)} \in [0,1]$, $i, j = 1, 2, \ldots, N$, with $\sum_i p_A^{(i)} = \sum_j p_B^{(j)} = 1$. $U_A^{(i)}$ and $U_B^{(j)}$ are local unitary operators acting in $\mathcal{H}_A$ only corresponding to pure strategies $s_A^{(i)}$ and $s_B^{(j)}$.

The map given by (6.13) acts in $\mathcal{H}_A$ and $\mathcal{H}_A$ as a doubly stochastic map, that is, as a completely positive unital map. As a result, by Uhlmann's theorem [MO79] the final reduced states $\text{tr}_B[\sigma]$ and $\text{tr}_A[\sigma]$ must be more mixed than the reduced initial states $\text{tr}_B[\rho]$ and $\text{tr}_A[\rho]$ in the sense of memorization theory [MO79]. As the initial state ρ is a maximally entangled state, all accessible states after application of a mixed strategy of Alice and Bob are locally identical to the maximally mixed state $1/\dim(\mathcal{H}_A) = 1/\dim(\mathcal{H}_B)$, which is a multiple of 1.

A particular Nash equilibrium in mixed strategies can easily be identified, more than one Nash equilibrium can be found in this quantum game. The following construction, e.g., yields an equilibrium in mixed quantum strategies: Allow Alice to choose from two strategies $s_A^{(1)}$ and $s_A^{(2)}$ with probabilities $p_A^{(1)} = 1/2$ and $p_A^{(2)} = 1/2$, while Bob may take $s_B^{(1)}$ or $s_B^{(2)}$, with

$$s_A^{(1)} \sim \begin{pmatrix} 1 & 0 \\ 0 & 1 \end{pmatrix}, \qquad s_A^{(2)} \sim \begin{pmatrix} -i & 0 \\ 0 & i \end{pmatrix}, \tag{6.14}$$

$$s_B^{(1)} \sim \begin{pmatrix} 0 & 1 \\ -1 & 0 \end{pmatrix}, \qquad s_B^{(2)} \sim \begin{pmatrix} 0 & -i \\ -i & 0 \end{pmatrix}. \tag{6.15}$$

His probabilities are also given by $p_B^{(1)} = 1/2$ and $p_B^{(2)} = 1/2$. The quantum strategies of (6.14) and (6.15) are mutually optimal answers and have the property that

$$P_A(s_A^{(i)}, s_B^{(i)}) = 0, \qquad P_B(s_A^{(i)}, s_B^{(i)}) = 5, \tag{6.16}$$

$$P_A(s_A^{(i)}, s_B^{(3-i)}) = 5, \qquad P_B(s_A^{(i)}, s_B^{(3-i)}) = 0, \tag{6.17}$$

for $i = 1, 2$.

Such quantum strategies always exist (as implied by (6.12)). Take, e.g., the continuous set can be achieved by choosing an arbitrary local operation $s_A^{(1)}$ and by designing $s_A^{(2)}$, $s_B^{(1)}$, and $s_B^{(2)}$ according to (6.16) and (6.17).

Due to the particular constraints of (6.16) and (6.17) there exists no other mixed strategy for Bob yielding a better pay-off than the above mixed strategy, given that Alice sticks to the equilibrium strategy. This can be seen as follows. Let Alice use this particular mixed quantum strategy as above and let Bob use any mixed quantum strategy

$$s_B^{(1)}, \ldots, s_B^{(N)} \tag{6.18}$$

together with $p_A^{(1)}, \ldots, p_A^{(N)}$. The final state σ after application of the strategies is given by the convex combination

$$\sigma = \sum_{i=1,2} \sum_j p_A^{(i)} p_B^{(j)} (s_A^{(i)} \otimes s_B^{(j)})(\rho). \tag{6.19}$$

This convex combination can only lead to a smaller expected pay-off for Bob than the optimal pure strategy $s_B^{(k)}$ in (6.18), $k \in \{1, \ldots, N\}$. Such optimal pure strategies are given by $s_B^{(1)}$ and $s_B^{(2)}$ as in (6.15) leading to an expected pay-off for Bob of $P_B(s_A, s_B) = 2.5$; there are no pure strategies which achieve a larger expected pay-off. While both pure strategies $s_B^{(1)}$ and $s_B^{(2)}$ do not correspond to an equilibrium, the mixed strategy where $s_B^{(1)}$ and $s_B^{(2)}$ are chosen with $p_B^{(1)} = 1/2$ and $p_B^{(2)} = 1/2$ actually does. Nash equilibria consist of pairs of mutually optimal answers, and only for this choice of Bob the original mixed quantum strategy of Alice is *her* optimal choice, as the same argument applies also to her, the game being symmetric.

This Nash equilibrium is however not the only one. There exist also other four-tuples of matrices than the ones presented in (6.14) and (6.15) that satisfy (6.16) and (6.17). Such matrices can be made out by appropriately rotating the matrices of (6.14) and (6.15). In the light of the fact that there is more than one equilibrium it is not obvious which Nash equilibrium the players will realize. It is at first not even evident whether a Nash equilibrium will be played at all. But the game theoretical concept of the *focal point effect* [Sch60, Mye91] helps to resolve this issue.

To explore the general structure of any Nash equilibrium in mixed strategies we continue as follows: let

$$U_A^{(1)}, \ldots, U_A^{(N)} \tag{6.20}$$

together with $p_A^{(1)}, \ldots, p_A^{(N)}$ specify the mixed strategy pertinent to a Nash equilibrium of Alice. Then there exists a mixed strategy $U_B^{(1)}, \ldots, U_B^{(N)}, p_B^{(1)},$ $\ldots, p_B^{(N)}$ of Bob which rewards Bob with the best achievable pay-off, given that Alice plays this mixed strategy. Yet, the pair of mixed strategies associated with

$$QU_A^{(1)}Q^\dagger, \ldots, QU_A^{(N)}Q^\dagger, \qquad QU_B^{(1)}Q^\dagger, \ldots, QU_B^{(N)}Q^\dagger \tag{6.21}$$

with $p_A^{(1)}, \ldots, p_A^{(N)}, p_B^{(1)}, \ldots, p_B^{(N)}$ is another Nash equilibrium. This equilibrium leads to the same expected pay-off for both players, and is fully symmetric to the previous one. Doubly applying Q as $QQU_A^{(1)}Q^\dagger Q^\dagger, \ldots, QQU_A^{(N)}Q^\dagger Q^\dagger$ results again into a situation with equivalent strategies as the original ones. For a given Nash equilibrium as above the one specified by (6.21) will be called dual equilibrium.

However, there is a single Nash equilibrium (R, R) which is the only one which gives an expected pay-off of $P_A(R, R) = P_B(R, R) = 2.25$ and which is identical to its dual equilibrium: it is the simple map

$$\rho \longmapsto \sigma = 1/\dim(\mathcal{H}). \tag{6.22}$$

Indeed, there exist probabilities $p_A^{(1)}, \ldots, p_A^{(N)}$ and unitary operators $U_A^{(1)}, \ldots,$ $U_A^{(N)}$ such that [MO79]

$$\sum_i p_A^{(i)} (U_A^{(i)} \otimes 1) \rho (U_A^{(i)} \otimes 1)^\dagger = 1/\dim(\mathcal{H}).$$

If Alice has already selected $s_A = R$, the application of $s_B = R$ will not change the state of the quantum system any more.

Assume that (6.20) and (6.21) are associated with equivalent quantum strategies. This means that they have to produce the same expected pay-off for all quantum strategies s_B of Bob. If Alice and Bob apply $s_A \otimes s_B$ they get an expected pay-off according to (6.9) and (6.10); if Alice after implementation of s_A manipulates the quantum system by applying the local unitary operator $Q \otimes 1$, they obtain

$$
\begin{aligned}
P_A'(s_A, s_B) &= A_{DD} \text{tr}[\pi_{CC}\sigma] + A_{DC} \text{tr}[\pi_{CD}\sigma] \\
&\quad + A_{CD} \text{tr}[\pi_{DC}\sigma] + A_{CC} \text{tr}[\pi_{DD}\sigma], \\
P_B'(s_A, s_B) &= B_{DD} \text{tr}[\pi_{CC}\sigma] + B_{DC} \text{tr}[\pi_{CD}\sigma] \\
&\quad + B_{CD} \text{tr}[\pi_{DC}\sigma] + B_{CC} \text{tr}[\pi_{DD}\sigma].
\end{aligned}
$$

The only s_A with the property that $P_A'(s_A, s_B) = P_A(s_A, s_B)$ and $P_B'(s_A, s_B) = P_B(s_A, s_B)$ for all s_B is the map given by (6.22).

In principle, any Nash equilibrium may become a self-fulfilling prophecy if the particular Nash equilibrium is expected by both players. It has been pointed out that in a game with more than one equilibrium, anything that attracts the players' attention towards one of the equilibria may make them expect and therefore realize it [Sch60]. The corresponding *focal equilibrium* [Mye91] is the one which is conspicuously distinguished from the other Nash equilibria. In this particular situation there is indeed one Nash equilibrium different from all the others: it is the one which is equivalent to its dual equilibrium, the map which simply maps the initial state on the maximally mixed state. For all other expected pay-offs both players are ambivalent between (at least) two symmetric equilibria. The expected pay-off the players will receive in this focal equilibrium, $P_A(R, R) = P_B(R, R) = 2.25$, is not fully Pareto optimal, but it is again much more efficient than the classically achievable outcome of 1.[12]

Completely Positive Trace-Preserving Maps Corresponding to Local Operations

In this scenario both Alice and Bob may perform any operation that is allowed by quantum mechanics. That is, the set of strategies $S^{(CP)}$ is made up of (s_A, s_B), where both s_A and s_B correspond to a completely positive trace-preserving map

---

[12] In the classical game both players could also play C and D with probabilities $p = 1/2$ yielding the same expected pay-off of 2.25 for both players, but this pair of mixed strategies would be no equilibrium solution, as any player could benefit from simply choosing the dominant strategy D.

$$(s_A \otimes s_B)(\rho) = \sum_i \sum_j (A_i \otimes B_j)\rho(A_i \otimes B_j)^\dagger,$$

corresponding to a local operation, associated with Kraus operators A_i and B_j with $i, j = 1, 2, \ldots$. The trace-preserving property requires that

$$\sum_i A_i^\dagger A_i = 1 \quad \text{and} \quad \sum_i B_i^\dagger B_i = 1.$$

This case has already been mentioned in [EWL99]. The quantum strategies s_A and s_B do no longer inevitably act as unital maps in the respective Hilbert spaces as before. In other words, the reduced states of Alice and Bob after application of the quantum strategy are not necessarily identical to the maximally mixed state $1/\dim(\mathcal{H}_A)$.

As already pointed out in [EWL99], the pair of strategies (Q, Q) of the two-parameter set of strategies $S^{(TP)}$ is again no equilibrium solution. It is straightforward to prove that the Nash equilibria of the type of (6.14) and (6.15) of mixed strategies with general local unitary operations are, however, still present, and each of these equilibria yields an expected pay-off of 2.5. In the language of this subsection these equilibria are equilibria in pure strategies, as mixing is already included in this most general case.

In addition, as strategies do no longer have to be locally unital maps, it is not surprising that new Nash equilibria emerge: Alice and Bob may, e.g., perform a measurement associated with Kraus operators

$$A_1 = |0\rangle\langle 0|, \qquad A_2 = |1\rangle\langle 1|, \qquad B_1 = D|0\rangle\langle 0|, \qquad B_2 = D|1\rangle\langle 1|.$$

This operation yields a final state

$$\sigma = (s_A \otimes s_B)(\rho) = (|01\rangle\langle 01| + |10\rangle\langle 10|)/2.$$

Clearly neither Alice nor Bob can gain from unilaterally deviating from their strategy. Another Nash equilibrium is associated with Kraus operators

$$A_1 = |1\rangle\langle 1|, \qquad A_2 = D|0\rangle\langle 0|, \qquad B_1 = |1\rangle\langle 1|, \qquad B_2 = D|0\rangle\langle 0|.$$

One can nevertheless argue as in the previous case. Again, all Nash equilibria occur at least in pairs. First, there are again the dual equilibria from $S^{(GU)}$. Second, there are Nash equilibria (s_A, s_B), $s_A \neq s_B$, with the property that (s_B, s_A) is also a Nash equilibrium yielding the same expected pay-off. The only Nash equilibrium invariant under application of Q and exchange of the strategies of the players is again (R, R) defined in the previous subsection, which yields a pay-off $P_A(R, R) = P_B(R, R) = 2.25$. (R, R) is still a Nash equilibrium, because—given that one player picks R—the application of any strategy taken from $S^{(CP)}$ of the other player results in a final state which yields a pay-off smaller or equal to 2.25. This is the solution of the game is the most general case. While both players could in principle do better (as the solution lacks Pareto optimality), the efficiency of this focal equilibrium is much higher than the equilibrium in dominant strategies of the classical game. Hence, also in this most general case both players gain from using quantum strategies.

This study shows that the efficiency of the equilibrium the players can reach in this game depends on the actions the players may take. One feature, however is present in each of the considered sets: both players can increase their expected pay-offs drastically by resorting to quantum strategies.

Chicken Game

In the previous classical game, the Prisoners' Dilemma, an unambiguous solution can be specified consisting of a unique Nash equilibrium. However, this solution was not efficient, thus giving rise to the dilemma. In the so-called Chicken game [Mye91, Pou92],

$$A_{CC} = B_{CC} = 6, \qquad A_{CD} = B_{DC} = 8,$$
$$A_{DC} = B_{CD} = 2, \qquad A_{DD} = B_{DD} = 0.$$

This game has two Nash equilibria (C, D) and (D, C): it is not clear how to anticipate what the players' decision would be. In addition to the two Nash equilibria in pure strategies there is an equilibrium in mixed strategies, yielding an expected pay-off 4 [Mye91].

In order to investigate the new features of the game if superpositions of classical strategies are allowed for, three set of strategies are briefly discussed:

1. One-Parameter Set of Strategies

Again, we consider the set of strategies $S^{(CL)}$ of one-dimensional rotations. The strategies s_A and s_B are associated with local unitary operators

$$U(\theta) = \begin{pmatrix} \cos(\theta/2) & \sin(\theta/2) \\ -\sin(\theta/2) & \cos(\theta/2) \end{pmatrix},$$

with $\theta \in [0, \pi]$,

$$C \sim U(0) = \begin{pmatrix} 1 & 0 \\ 0 & 1 \end{pmatrix}, \qquad D \sim U(\pi) = \begin{pmatrix} 0 & 1 \\ -1 & 0 \end{pmatrix}.$$

Then as before, the quantum game yields the same expected pay-off as the classical game in randomized strategies. This means that still two Nash equilibria in pure strategies are present.

2. Two-Parameter Set of Strategies

The players can actually take advantage of an additional degree of freedom which is not accessible in the classical game. If they may apply unitary operations from $S^{(TP)}$ of the type

$$U(\theta, \phi) = \begin{pmatrix} e^{i\phi} \cos(\theta/2) & \sin(\theta/2) \\ -\sin(\theta/2) & e^{-i\phi} \cos(\theta/2) \end{pmatrix},$$

with $\theta \in [0, \pi]$ and $\phi \in [0, \pi/2]$ the situation is quite different than with $S^{(CL)}$. (C, D) and (C, D) with $C \sim U(0, 0)$ and $D \sim U(\pi, 0)$ are no longer equilibrium solutions. E.g., given that $s_A = D$ the pair of strategies (D, Q) with $Q \sim U(0, \pi/2)$ yields a better expected pay-off for Bob than (D, C), that is to say $P_B(D, Q) = 8$, $P_B(D, C) = 2$. In fact (Q, Q) is now the unique Nash equilibrium with $P_A(Q, Q) = P_B(Q, Q) = 6$, which follows from an investigation of the actual expected pay-offs of Alice and Bob analogous to (6.11). This solution is not only the unique acceptable solution of the game, but it is also an equilibrium that is Pareto optimal. This contrasts very much with the situation in the classical game, where the two equilibria were not that efficient.

3. Completely Positive Trace-Preserving Maps

As in the considerations concerning the Prisoner's Dilemma game, more than one Nash equilibrium is present, if both players can take quantum strategies from the set $S^{(CP)}$, and all Nash equilibria emerge at least in pairs as above. The focal equilibrium is given by (R, R), resulting in a pay-off of $P_A(R, R) = P_B(R, R) = 4$, which is the same as the mixed strategy of the classical game.

6.2.4 Quantum Cryptography and Quantum Gambling

Recall that *quantum cryptography* is a field which combines quantum theory with information theory. The goal of this field is to use the laws of physics to provide secure information exchange, in contrast to classical methods based on (unproven) complexity assumptions. In particular, quantum key distribution protocols [Eke91] became especially important due to technological advances which allow their implementation in the laboratory. However, the last important theoretical result in the field was of a negative character: [May97] and [LC97] showed that quantum bit commitment is not secure. Their work also raised serious doubts on the possibility of obtaining any secure two-party protocol, such as oblivious transfer and coin tossing [Lo97]. In this subsection, following [GVW99] we present a secure two-party quantum cryptographic task—quantum gambling, which has no classical counterpart.

Coin tossing is defined as a method of generating a random bit over a communication channel between two distant parties. The parties, traditionally named Alice and Bob, do not trust each other, or a third party. They create the random bit by exchanging quantum and classical information. At the end of the protocol the generated bit is known to both of them. If a party cheats, i.e. changes the occurrence probability of an outcome, the other party should be able to detect the cheating. We would consider a coin tossing protocol to be secure if it defines a parameter such that when it goes to infinity the probability to detect any finite change of probabilities goes to 1. Using a secure protocol the parties can make certain decisions depending on the value of the random bit, without being afraid that the opponent may have some advantage. For instance, Alice and Bob can play a game in which Alice wins if the outcome is '0' and Bob wins if it is '1'. Note that if bit commitment were secure, it could be used to implement coin tossing trivially: Alice commits a bit a to Bob; Bob tells Alice the value of a bit b; the random bit is the parity bit of a and b.

It is not known today if a secure quantum coin tossing protocol can be found.[13] It is only known that *ideal* coin tossing, i.e., in which no party can change the expected distribution of the outcomes, is impossible [LC97]. Based on our efforts in this direction, we are skeptical about the possibility to have secure (non-ideal) coin tossing. Nevertheless, we were able to construct a protocol which gives a solution to a closely related task. "Quantum gambling" is very close to playing in a casino located in a remote site, such as gambling over the Internet. As in a real casino, for instance when playing Roulette, the player's possible choices give him some probability to win twice the amount of his bet, or a smaller probability to win a bigger sum. However, in our protocol the player has only a partial control over these choices. In spite of its limitations our protocol provides a quantum solution to a useful task, which cannot be performed securely today in the classical framework. Assuming ideal apparatus and communication channels, the protocol is unconditionally secure, depending solely on the laws of physics.

Let us start by defining exactly the gambling task considered here. The casino (Alice) and the player (Bob) are physically separated, communicating via quantum and classical channels. The bet of Bob in a single game is taken for simplicity to be 1 coin. At the end of a game the player wins 1 or R coins, or loses 1 coin (his bet), depending on the result of the game. We have found a protocol which implement this game while respecting two requirements: First, the player can ensure that, irrespective of what the casino does, his expected gain is not less than δ coins, where δ is a negative function of R which goes to zero when R goes to infinity. The exact form of $\delta(R)$ will be specified below. Second, the casino can ensure that, irrespective of what the player does, its expected gain is not less than 0 coins.

---

[13] If we limit ourselves to spatially extended secure sites located one near the other, then secure coin tossing can be realized classically, by simultaneous exchange of information at the opposite sides of the sites. The security of this method relies on relativistic causality.

In order to define the protocol rigorously, we will first present the rules of the game, then the strategies of the players which ensure the outcomes quoted above and finally we will prove the security of the method.

The Rules of the Game: Alice has two boxes, A and B, which can store a particle [GVW99]. The quantum states of the particle in the boxes are denoted by $|a\rangle$ and $|b\rangle$, respectively. Alice prepares the particle in some state and sends box B to Bob.

Bob wins in one of the two cases:

1. If he finds the particle in box B, then Alice pays him 1 coin (after checking that box A is empty).
2. If he asks Alice to send him box A for verification and he finds that she initially prepared a state *different* from

$$|\psi_0\rangle = \frac{1}{\sqrt{2}}(|a\rangle + |b\rangle), \qquad (6.23)$$

then Alice pays him R coins.

In any other case Alice wins, and Bob pays her 1 coin.

The players' strategies which ensure (independently) an expectation value of Alice's gain $G_A \geq 0$ (irrespective of Bob's actions) and an expectation value of Bob's gain $G_B \geq \delta$ (irrespective of Alice's actions) are as follows:

Alice's Strategy: Alice prepares the equally distributed state $|\psi_0\rangle$ (given in (6.23)).

Bob's Strategy: After receiving box B, Bob splits the particle in two parts; specifically, he performs the following unitary operation:

$$|b\rangle \rightarrow \sqrt{1-\eta}|b\rangle + \sqrt{\eta}|b'\rangle, \qquad (6.24)$$

where $\langle b'|b\rangle = 0$. The particular splitting parameter η he uses is $\eta = \tilde{\eta}(R)$ (to be specified below). After the splitting Bob measures the projection operator on the state $|b\rangle$, and then

I. If the measurement yields a positive result, i.e. he finds the particle, he announces Alice that he won.
II. If the measurement yields a negative result, he asks Alice for box A and verifies the preparation.

This completes the formal definition of our protocol.

In order to prove the security of the scheme, we will analyze the average gain of each party as a result of her/his specific strategy. It is straightforward to see that Alice's strategy ensures $G_A \geq 0$. If Alice prepares the state $|\psi_0\rangle$, Bob has no meaningful way of increasing his odds beyond 50%: if he decides to open box B he has a probability of 0.5 to win 1 coin and a probability of 0.5 to lose 1 coin. He cannot cheat by claiming that he found the particle when he did not, since Alice learns the result by opening box A. If, instead, he decides to verify the preparation he will find the expected state, so he will lose 1 coin. Therefore $G_B \leq 0$, and since this is a zero-sum game, Alice's gain is $G_A \geq 0$, whatever Bob does.

Now we will prove that Bob, using the splitting parameter $\eta = \tilde{\eta}$, can ensure $G_B \geq \delta$. The values of $\tilde{\eta}$ and δ are determined by the calculation of Bob's expected gain, G_B. Bob tries to maximize G_B under the assumption that Alice uses the worse strategy for him, namely the one which minimizes G_B for Bob's particular strategy. Therefore, we will first minimize the function G_B for any η, and then we will find the maximum of the obtained function, with that computing δ. We will also compute the value of η at the peak, $\tilde{\eta}$, which will be the chosen splitting parameter of Bob.

Let us first write down the expression for G_B. Bob gets 1 coin if he detects the state $|b\rangle$; denote the probability for this event to occur by P_b. He gets R coins if he detects a different preparation than $|\psi_0\rangle$ (after failing to find the state $|b\rangle$, an event with a related probability of $1 - P_b$); denote the probability to detect a different preparation by P_D. He loses 1 coin if he does not detect a different preparation than $|\psi_0\rangle$ (after failing to find $|b\rangle$); the probability for this event is $(1 - P_D)$. Thus, the expectation value of Bob's gain is

$$G_B = P_b + (1 - P_b)[P_D R - (1 - P_D)]. \tag{6.25}$$

For the calculations of P_b and P_D we will consider the most general state Alice can prepare. In this case the particle may be located not only in boxes A and B, but also in other boxes C_i. The states $|a\rangle$, $|b\rangle$ and $|c_i\rangle$ are mutually orthogonal. She can also correlate the particle to an ancilla $|\Phi\rangle$, such that the most general preparation is

$$|\Psi_0\rangle = \alpha|a\rangle|\Phi_a\rangle + \beta|b\rangle|\Phi_b\rangle + \sum_i \gamma_i|c_i\rangle|\Phi_{c_i}\rangle, \tag{6.26}$$

where $|\Phi_a\rangle, |\Phi_b\rangle, |\Phi_{c_i}\rangle$ are the states of the ancilla and $|\alpha|^2 + |\beta|^2 + \sum_i |\gamma_i|^2 = 1$. After Bob splits $|b\rangle$, as described by (6.24), the state changes to

$$|\Psi_1\rangle = \alpha|a\rangle|\Phi_a\rangle + \beta\left(\sqrt{1-\eta}|b\rangle + \sqrt{\eta}|b'\rangle\right)|\Phi_b\rangle + \sum_i \gamma_i|c_i\rangle|\Phi_{c_i}\rangle. \tag{6.27}$$

The probability to find the state $|b\rangle$ (in step I. of Bob's strategy) is

$$P_b = \|\langle b|\Psi_1\rangle\|^2 = |\beta|^2(1 - \eta). \tag{6.28}$$

If Bob does not find $|b\rangle$, then the state reduces to

$$|\Psi_2\rangle = \mathcal{N}\left(\alpha|a\rangle|\Phi_a\rangle + \beta\sqrt{\eta}|b'\rangle|\Phi_b\rangle + \sum_i \gamma_i|c_i\rangle|\Phi_{c_i}\rangle\right), \tag{6.29}$$

where $\mathcal{N}$ is the normalization factor given by $\mathcal{N} = (1 - (1 - \eta)|\beta|^2)^{-1/2}$. On the other hand, if Alice prepares the state $|\psi_0\rangle$ instead of $|\Psi_0\rangle$, then at this stage the particle is in the state

$$|\psi_2\rangle = \sqrt{\frac{1}{1+\eta}}|a\rangle + \sqrt{\frac{\eta}{1+\eta}}|b'\rangle. \tag{6.30}$$

Thus, the best verification measurement of Bob is to make a projection measurement on this state. If the outcome is negative, Bob knows with certainty that Alice did not prepared the state $|\psi_0\rangle$. The probability of detecting such a different preparation is given by

$$P_D = 1 - \|\langle\psi_2|\Psi_2\rangle\|^2$$

$$= 1 - \mathcal{N}^2 \left\| \frac{\alpha}{\sqrt{1+\eta}}|\Phi_a\rangle + \frac{\beta\eta}{\sqrt{1+\eta}}|\Phi_b\rangle \right\|^2. \tag{6.31}$$

Since Alice wants to minimize G_B, she tries to minimize both P_b and P_D. From (6.31) we see that in order to minimize P_D, the states of the ancilla $|\Phi_a\rangle$ and $|\Phi_b\rangle$ have to be identical (up to some arbitrary phase), i.e., $|\langle\Phi_a|\Phi_b\rangle| = 1$. That is, Alice gets no advantage using an ancilla, so it can be eliminated. Then, in order to maximize $\mathcal{N}|\alpha + \beta\eta|$, Alice should set all γ_i to zero, as it is clear from the normalization constraint

$$|\alpha|^2 + |\beta|^2 = 1 - \sum_i |\gamma_i|^2.$$

This operation has no conflict with the minimization of P_b, since (6.28) contains only $|\beta|$. Also, the maximization is possible if the coefficients α and β, if seen as vectors in the complex space, point in the same direction. Therefore, Alice gains nothing by taking α and β to be complex numbers; it is sufficient to use real positive coefficients. Taking all these considerations into account, the state prepared by Alice can be simplified to

$$|\psi_0'\rangle = \sqrt{\frac{1}{2} + \epsilon}|a\rangle + \sqrt{\frac{1}{2} - \epsilon}|b\rangle. \tag{6.32}$$

Now, the state after Bob splits $|b\rangle$ reads

$$|\psi_1'\rangle = \sqrt{\frac{1}{2} + \epsilon}|a\rangle + \sqrt{\frac{1}{2} - \epsilon}\left(\sqrt{1-\eta}|b\rangle + \sqrt{\eta}|b'\rangle\right), \tag{6.33}$$

and so the probability to find $|b\rangle$ becomes

$$P_b = \|\langle b|\psi_1'\rangle\|^2 = \left(\frac{1}{2} - \epsilon\right)(1 - \eta). \tag{6.34}$$

When Bob does not find the state $|b\rangle$, $|\psi_1'\rangle$ reduces to

$$|\psi_2'\rangle = \frac{\sqrt{1+2\epsilon}|a\rangle + \sqrt{\eta(1-2\epsilon)}|b'\rangle}{\sqrt{1+2\epsilon + \eta(1-2\epsilon)}}, \tag{6.35}$$

which in turn leads to

$$P_D = 1 - \|\langle \psi_2 | \psi_2' \rangle\|^2 = \frac{2\eta \left(1 - \sqrt{1 - 4\epsilon^2}\right)}{(1 + \eta)^2 + 2\epsilon(1 - \eta^2)}. \tag{6.36}$$

Substituting (6.34) and (6.36) in (6.25), we find G_B in terms of the splitting parameter η, the preparation parameter ϵ and R:

$$G_B = -\frac{1}{1 + \eta} \Big[2\epsilon(1 - \eta^2) + \eta(\eta + \sqrt{1 - 4\epsilon^2})$$
$$- \eta(1 - \sqrt{1 - 4\epsilon^2})R \Big]. \tag{6.37}$$

In order to calculate the minimal gain of Bob, δ, irrespective of the particular strategy of Alice, we will first minimize G_B for ϵ and then maximize the result for η:

$$\delta(R) = Max_\eta [Min_\epsilon G_B(R, \eta, \epsilon)]. \tag{6.38}$$

The calculations yield

$$\delta = -\frac{1}{1 + \sqrt{R + 2 - \sqrt{(R + 2)^2 - 1}}}$$
$$\times \left\{ 2 + \left[R - \sqrt{(R + 2)^2 - 1} \right] \right.$$
$$\times \left. \left[1 - \sqrt{R + 2 - \sqrt{(R + 2)^2 - 1}} \right] \right\}, \tag{6.39}$$

obtained for Bob's splitting parameter

$$\tilde{\eta} = \sqrt{R + 2 - \sqrt{(R + 2)^2 - 1}}. \tag{6.40}$$

In the range of $R \gg 1$, these results can be simplified to

$$\delta \approx -\sqrt{\frac{2}{R}}, \tag{6.41}$$

$$\tilde{\eta} \approx \sqrt{\frac{1}{2R}}. \tag{6.42}$$

We have shown that if Bob follows the proposed strategy with $\eta = \tilde{\eta}$, then his average gain is not less than δ; this bound converges to 0, i.e. to the limit of a fair game, for $R \to \infty$. This is true for any possible strategy of Alice, therefore, the security of the protocol is established [GVW99].

To compare our scheme to a real gambling situation, let us consider the well-known Roulette game. A bet of 1 coin on the red/black numbers, i.e. half of the 36 numbers on the table, rewards the gambler with 1 coin once in 18/38 turns (on average, for a spinning wheel with 38 slots); this gives an expected gain of about -0.053 coins. To assure the same gain in our scheme, $R = 700$ is required. Note that extremely large values of R are practically meaningless, one reason being the limited total amount of money in use. Nevertheless, the bound on δ is not too restrictive when looking at the first prizes of some lottery games: a typical value of $R = 10^6$ gives a reasonably small δ of about -0.0014.

It is also interesting to consider the case of $R = 1$. This case corresponds to coin tossing, since it has only two outcomes: Bob's gain is either -1 coin (stands for bit '0') or 1 coin (stands for bit '1'). The minimal average gain of Bob is about -0.657, which translates to an occurrence probability of bit '1' of at least 0.172 (instead of 0.5 ideally), whatever Alice does. This is certainly not a good coin tossing scheme, however, no classical or quantum method is known to assure (unconditionally) *any* bound for the occurrence probability of both outcomes.

Our analysis so far was restricted to a single instance of the game, but the protocol may be repeated several times. After N games Bob's expected gain is $G_B \geq N\delta$ and Alice's expected gain is $G_A \geq 0$. Of course, Alice may choose now a complex strategy using ancillas and correlations between particles/ancillas from different runs. In this way she may change the probability distribution of her winnings, but she cannot reduce the minimal expected gain of Bob. Indeed, our proof considers the most general actions of Alice, so the average gain of Bob in each game is not less than δ, and consequently, it is not less then $N\delta$ after N games. A similar argument is valid for Bob's actions, so the average gain of Alice remains non-negative even after N games. In gambling games, in addition to the average gain, it is important to analyze the standard deviation of the gain, ΔG. Bob will normally accept to play a game with a negative gain only if $\Delta G_B \gg |G_B|$ (unless he has some specific target in mind). In a single application of our protocol, $\Delta G_B \geq 1$, so the condition is attained for big enough values of R (see (6.41)). However, increasing the number of games makes the gambling less attractive to Bob: if Alice follows the proposed strategy, $|G_B|$ grows as N while ΔG_B grows only as $\sqrt{N}$. Therefore, Bob should accept to play N times only if $N \ll 1/\delta^2 \sim R$.

Another important point to consider is the possible "cheating" of the parties. Alice has no meaningful way to cheat, since she is allowed to prepare any quantum state and she sends no classical information to Bob. Any operation other than preparing $|\psi_0\rangle$, as adding ancillas or putting more/less than one particle in the boxes, just decreases her minimal gain. Bob, however, may try to cheat. He may claim that he detected a different preparation than $|\psi_0\rangle$, even when his verification measurement does not show that. If Alice prepares the initial state $|\psi_0'\rangle$ (with $\epsilon > 0$), she is vulnerable to this cheating attempt: she has no way to know if Bob is lying or not. For this reason Alice's proposed strategy is to prepare $|\psi_0\rangle$ every time, such that any cheating of Bob could be invariably detected. When both parties follow the proposed strategies, i.e. $\epsilon = 0$ and $\eta = \tilde{\eta}$, the game is more fair for Bob than assumed in the proof [GVW99]:

$$G_{B_{prot}} = -G_{A_{prot}} = -\sqrt{R + 2 - \sqrt{(R+2)^2 - 1}}. \qquad (6.43)$$

For $R \gg 1$ we get $G_{B_{prot}} \approx -1/\sqrt{2R}$, which is approximately half of the value of δ calculated in (6.41).

The discussion up to this point assumed an ideal experimental setup. In practice errors are unavoidable, of course, and our protocol is very sensitive to the errors caused by the devices used in its implementation (communication channels, detectors, etc.). In the presence of errors, if the parties disagree about the result of a particular run it should be canceled. If such conflicts occur more than expected based on the experimental error rate, it means that (at least) one party is cheating, and the game should be stopped. The most sensitive part to errors is the verification measurement of Bob, i.e. the detection of the possible deviation of the initial state from $|\psi_0\rangle$. In the ideal case, using $\tilde{\eta}$ and the corresponding ϵ (the worst for honest Bob), the detection probability is very small: $P_D \approx \sqrt{2/R^3}$, for $R \gg 1$. Clearly, for a successful realization of the protocol, the error rate has to be lower than this number. Thus, in practice, the experimental error rate will constrain the maximal possible value of R.[14]

6.2.5 Formal Quantum Games

Recall that Meyer [MB98] and Eisert et al. [EWL99] brought the game theory into the physics community and created a new field, quantum game theory. They both quantized a classical game and found interesting new properties which the original classical game does not possess. Nevertheless, their quantized games seem quite different. *PQ* penny flipover studied by Meyer is a quantum sequential game, in which players take turns in performing some operations on a quantum system. On the other hand, quantum Prisoners' Dilemma studied by Eisert et al. is a quantum simultaneous game, in which there are n players and a quantum system which consists of n subsystems, and player i performs an operation only on the ith subsystem.

Since the seminal works of Meyer and Eisert et al., many studies have been made to quantize classical games and find interesting phenomena [EW00]. Most of the quantum games ever studied are classified into either quantum simultaneous games or quantum sequential games, although not much has been done on the latter.

Now that we see that game theory is combined with quantum theory and there are two types of quantum games, several questions naturally arise: (a) Are quantum games truly different from classical games? (b) If so, in what sense are they different? (c) What is the relationship between quantum simultaneous games and quantum sequential games? To answer these questions, it

---

[14] In the presence of errors an alternative strategy of Bob, without splitting state $|b\rangle$, is more efficient: Bob requests box A for verification randomly (at a rate optimized for the value of R) and measures the projection on $|\psi_0\rangle$ (not on $|\psi_2\rangle$). Although for a given R this strategy offers him a lower gain ($\delta \approx -2/\sqrt{R}$, for $R \gg 1$), it allows using a higher R for a given fidelity of the setup, since now $P_D \approx 1/R$.

is necessary to examine the whole structure of game theory including classical games and quantum games, not a particular phenomenon of a particular game.

A work by [LJ03] is a study along this line. They developed a formalism of games including classical games and quantum games. With the formalism they addressed the questions (a) and (b), concluding that "playing games quantum mechanically can be more efficient" and that "finite classical games consist of a strict subset of finite quantum games". However, they did not give a precise definition of the phrase "consist of a strict subset".

In this section, following [Kob07], we present various types of quantum games and try to answer the above questions.

Formal Framework

In order to discuss relationships between different types of games, we need a common framework in which various types of games are described. As the first step in our analysis, we will construct such a framework for our theory.

For the construction, a good place to start is to consider what is game theory. Game theory is the mathematical study of *game situations* which is characterized by the following three features: (i) There are two or more decision-makers, or players; (ii) Each player develops his/her strategy for pursuing his/her objectives. On the basis of the strategy, he/she chooses his/her action from possible alternatives; and (iii) As a result of all players' actions, some situation is realized. Whether the situation is preferable or not for one player depends not only on his/her action, but also on the other players' actions.

How much the realized situation is preferable for a player is quantified by a real number called a pay-off. Using this term, we can rephrase the second feature as "each player develops his/her strategy to maximize the expectation value of his/her pay-off". The reason why the expectation value is used to evaluate strategies is that we can determine the resulting situation only probabilistically in general, even when all players' strategies are known.

As a mathematical representation of the three features of game situations, we define a normal form of a game.

A normal form of a game is a triplet (N, Ω, f) whose components satisfy the following conditions: $N = \{1, 2, \ldots, n\}$ is a finite set; $\Omega = \Omega_1 \times \cdots \times \Omega_n$, where Ω_i is a nonempty set; and f is a function from Ω to $\mathbb{R}^n$.

Here, N denotes a set of players. Ω_i is a set of player i's strategies, which prescribes how he/she acts. The ith element of $f(\omega_1, \ldots, \omega_n)$ is the expectation value of the pay-off for player i, when player j adopts a strategy ω_j.

Next, we propose a general definition of games, which works as a framework for discussing relationship between various kinds of games. We can regard a game as consisting of some 'entities' (like players, cards, coins, etc.) and a set of rules under which a game situation occurs. We model the 'entities' in the form of a tuple T. Furthermore, we represent the game situation caused by the 'entities' T under a rule R as a normal form of a game, and write it as $R(T)$. Using these formulations, we define a game as follows.

We define a game as a pair (T, R), where T is a tuple, and R is a rule which determines uniquely a normal form from T. When $G = (T, R)$ is a game, we refer to $R(T)$ as the normal form of the game G.

The conception of the above definition will be clearer if we describe various kinds of games in the form of the pair defined above. Thus far, we have implicitly regarded strategies and actions of *individual* players as elementary components of a game. In classical game theory, such modeling of games is referred to as a noncooperative game, in contrast to a cooperative game in which strategies and actions of *groups* of players are elementary.

Various Types of Formal Games

In this subsection, following [Kob07], we introduce various types of classical and quantum games. First, we confirm that strategic games, which is a well-established representation of games in classical game theory (see, e.g., [OR94], as well as Appendix), can be described in our framework. Then, we define two quantum games, namely, quantum simultaneous games and quantum sequential games.

Strategic Games

We can redefine strategic games using our formal framework as follows.

A strategic game is a game (T, R) which has the following form: (i) $T = (N, S, f)$, and each component satisfies the following conditions: $N = \{1, 2, \ldots, n\}$ is a finite set; $S = S_1 \times \cdots \times S_n$, where S_i is a nonempty set; and $f : S \mapsto \mathbb{R}^n$ is a function from S to $\mathbb{R}^n$. (ii) $R(T) = T = (N, S, f)$.

If the set S_i is finite for all i, then we call the game (T, R) a finite strategic game. We denote the set of all strategic games by SG, and the set of all finite strategic games by FSG.

Let $G = ((N, S, f), R)$ be a finite strategic game. Then the mixed extension of G is a game $G^* = ((N, S, f), R^*)$, where the rule R^* is described as follows: $R^*(N, S, f) = (N, Q, F)$, where Q and F are of the following forms: $Q = Q_1 \times \cdots \times Q_n$, where Q_i is the set of all probability distribution over S_i; $F : Q \mapsto \mathbb{R}^n$ assigns to each $(q_1, \ldots, q_n) \in Q$ the expected value of f. That is, the value of F is given by

$$F(q_1, \ldots, q_n) = \sum_{s_1 \in S_1} \cdots \sum_{s_n \in S_n} \left\{ \prod_{i=1}^{n} q_i(s_i) \right\} f(s_1, \ldots, s_n),$$

where $q_i(s_i)$ is the probability attached to s_i.

We denote the set of all mixed extensions of finite strategic games by $MEFSG$.

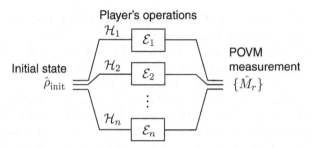

Fig. 6.3. The setup of a quantum simultaneous game (see text for explanation).

Quantum Simultaneous Games

Quantum simultaneous games are quantum games in which a quantum system is used according to a protocol depicted in Fig. 6.3. In quantum simultaneous games, there are n players who can not communicate with each other, and a referee. The referee prepares a quantum system in the initial state $\hat{\rho}_{\mathrm{init}}$. The quantum system is composed of n subsystems, where the Hilbert space for the ith subsystem is $\mathcal{H}_i$. The referee provides player i with the ith subsystem. Each player performs some quantum operation on the provided subsystem. It is determined in advance which operations are available for each player. After all players finish their operations, they return the subsystems to the referee. Then the referee performs a POVM measurement $\{\hat{M}_r\}$ on the total system. If the rth measurement outcome is obtained, player i receives a pay-off a_r^i.

Many studies on quantum simultaneous games have been carried out, starting with [EWL99, EW00].

The protocol of the quantum simultaneous games is formulated as follows.

A quantum simultaneous game is a game (T, R) which has the following form.

1. $T = (N, \mathcal{H}, \hat{\rho}_{\mathrm{init}}, \Omega, \{\hat{M}_r\}, \{\mathbf{a}_r\})$, and each component satisfies the following conditions: (i) $N = \{1, 2, \ldots, n\}$ is a finite set; (ii) $\mathcal{H} = \mathcal{H}_1 \otimes \mathcal{H}_2 \otimes \cdots \otimes \mathcal{H}_n$, where $\mathcal{H}_i$ is a Hilbert space; (iii) $\hat{\rho}_{\mathrm{init}}$ is a density operator on $\mathcal{H}$; (iv) $\Omega = \Omega_1 \times \Omega_2 \times \cdots \times \Omega_n$, where Ω_i is a subset of the set of all CPTP (completely positive trace preserving) maps on the set of density operators on $\mathcal{H}_i$. In other words, Ω_i is a set of quantum operations available for player i; (v) $\{\hat{M}_r\}$ is a POVM on $\mathcal{H}$; and (vi) $\mathbf{a}_r = (a_r^1, a_r^2, \ldots, a_r^n) \in \mathbb{R}^n$. The index r of $\mathbf{a}_r$ runs over the same domain as that of $\hat{M}_r$.
2. $R(T) = (N, \Omega, f)$. The value of f is given by

$$f(\mathcal{E}_1, \ldots, \mathcal{E}_n) = \sum_r \mathbf{a}_r \operatorname{Tr}\left[\hat{M}_r (\mathcal{E}_1 \otimes \cdots \otimes \mathcal{E}_n)(\hat{\rho}_{\mathrm{init}}) \right]$$

for all $(\mathcal{E}_1, \ldots, \mathcal{E}_n) \in \Omega$.

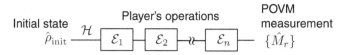

Fig. 6.4. The setup of a quantum sequential game (see text for explanation).

If $\mathcal{H}_i$ is finite dimensional for all i, then we refer to the game (T, R) as a finite quantum simultaneous game. We denote the set of all quantum simultaneous games by $QSim$, and the set of all finite quantum simultaneous games by $FQSim$.

Quantum Sequential Games

Quantum sequential games are another type of quantum games, in which a quantum system is used according to a protocol depicted in Fig. 6.4. In quantum sequential games, there are n players who can not communicate each other and a referee. The referee prepares a quantum system in the initial state $\hat{\rho}_{init}$. The players performs quantum operations on the quantum system in turn. The order of the turn may be regular like $1 \rightarrow 2 \rightarrow 3 \rightarrow 1 \rightarrow 2 \rightarrow 3 \rightarrow \cdots$, or may be irregular like $1 \rightarrow 3 \rightarrow 2 \rightarrow 3 \rightarrow 1 \rightarrow 2 \rightarrow \cdots$, yet it is determined in advance. After all the m operations are finished, the referee performs a POVM measurement $\{\hat{M}_r\}$. If the rth measurement outcome is obtained, then player i receives a pay-off a_r^i.

Games which belong to quantum sequential games include *PQ penny flipover* [MB98], *quantum Monty Hall* problem [FA02], and *quantum truel* [FA04].

The protocol of the quantum sequential games is formulated as follows. A quantum sequential game is a game (T, R) which has the following form.

1. $T = (N, \mathcal{H}, \hat{\rho}_{init}, Q, \mu, \{\hat{M}_r\}, \{\mathbf{a}_r\})$, and each component satisfies the following conditions: (i) $N = \{1, 2, \ldots, n\}$ is a finite set; (ii) $\mathcal{H}$ is a Hilbert space; (iii) $\hat{\rho}_{init}$ is a density operator on $\mathcal{H}$; (iv) $Q = Q_1 \times Q_2 \times \cdots \times Q_m$, where Q_k is a subset of the set of all CPTP maps on the set of density operators on $\mathcal{H}$; the total number of operations is denoted by m; (v) μ is a bijection from $\bigcup_{i=1}^{n}\{(i,j)|1 \leq j \leq m_i\}$ to $\{1, \ldots, m\}$, where m_i's are natural numbers satisfying $m_1 + \cdots + m_n = m$. The meaning of μ is that the jth operation for player i is the $\mu(i,j)$th operation in total; (vi) $\{\hat{M}_r\}$ is a POVM on $\mathcal{H}$; (vii) $\mathbf{a}_r = (a_r^1, a_r^2, \ldots, a_r^n) \in \mathbb{R}^n$; the index r of $\mathbf{a}_r$ runs over the same domain as that of $\hat{M}_r$.
2. $R(T) = (N, \Omega, f)$. The strategy space $\Omega = \Omega_1 \times \cdots \times \Omega_n$ is constructed as
$$\Omega_i = Q_{\mu(i,1)} \times Q_{\mu(i,2)} \times \cdots \times Q_{\mu(i,m_i)}.$$
The value of f is given by

$$f\left(\left(\mathcal{E}_{\mu(1,1)},\ldots,\mathcal{E}_{\mu(1,m_1)}\right),\ldots,\left(\mathcal{E}_{\mu(n,1)},\ldots,\mathcal{E}_{\mu(n,m_n)}\right)\right)$$
$$= \sum_r \mathbf{a}_r \operatorname{Tr}\left[\hat{M}_r \mathcal{E}_m \circ \mathcal{E}_{m-1} \circ \cdots \circ \mathcal{E}_1(\hat{\rho}_{\text{init}})\right]$$

for all $\left(\left(\mathcal{E}_{\mu(1,1)},\ldots,\mathcal{E}_{\mu(1,m_1)}\right),\ldots,\left(\mathcal{E}_{\mu(n,1)},\ldots,\mathcal{E}_{\mu(n,m_n)}\right)\right) \in \Omega$.

If $\mathcal{H}$ is finite dimensional, then we refer to the game (T, R) as a finite quantum sequential game. We denote the set of all quantum sequential games by $QSeq$, and the set of all finite quantum sequential games by $FQSeq$.

Equivalence of Formal Games

Now we define equivalence between two games. The basic idea is that two games are equivalent if their normal forms have the same structure, for the essence of a game is a game situation which is modeled by a normal form. The difficulty of this idea is that a strategy set Ω_i may have some redundancy; that is, two or more elements in Ω_i may represent essentially the same strategy. If this is the case, it does not work well to compare the strategy sets directly to judge whether two games are equivalent or not. Instead, we should define a new normal form in which the redundancy in the strategy set is excluded from the original normal form, and then compare the new normal forms of the two games.

As the first step to define equivalence between games, we clarify what it means by "two elements in Ω_i represent essentially the same strategy".

Let (N, Ω, f) be a normal form of a game. Two strategies $\omega_i, \omega_i' \in \Omega_i$ for player i are said to be redundant if

$$f(\omega_1 \ldots \omega_{i-1}, \omega_i, \omega_{i+1} \ldots \omega_n) = f(\omega_1 \ldots \omega_{i-1}, \omega_i', \omega_{i+1} \ldots \omega_n)$$

for all $\omega_1 \in \Omega_1, \ldots, \omega_{i-1} \in \Omega_{i-1}, \omega_{i+1} \in \Omega_{i+1}, \ldots, \omega_n \in \Omega_n$. If two strategies $\omega_i, \omega_i' \in \Omega_i$ are redundant, we write $\omega_i \sim \omega_i'$.

We can show that the binary relation $\sim$ is an *equivalence relation*. Namely, for all elements ω, ω', and ω'' of Ω_i, the following holds: (i) $\omega \sim \omega$; (ii) If $\omega \sim \omega'$ then $\omega' \sim \omega$; and (iii) If $\omega \sim \omega'$ and $\omega' \sim \omega''$ then $\omega \sim \omega''$.

Since $\sim$ is an equivalence relation, we can define the quotient set $\tilde{\Omega}_i$ of a strategy set Ω_i by $\sim$. The quotient set $\tilde{\Omega}_i$ is the set of all equivalence classes in Ω_i. An equivalence class in Ω_i is a subset of Ω_i which has the form of $\{\omega \mid \omega \in \Omega_i, a \sim \omega\}$, where a is an element of Ω_i. We denote by $[\omega]$ an equivalence class in which ω is included, and we define $\tilde{\Omega}$ as $\tilde{\Omega} \equiv \tilde{\Omega}_1 \times \cdots \times \tilde{\Omega}_n$. This $\tilde{\Omega}$ is a new strategy set which has no redundancy.

Next, we define a new expected pay-off function $\tilde{f}$ which maps $\tilde{\Omega}$ to $\mathbb{R}^n$ by

$$\tilde{f}([\omega_1], \ldots, [\omega_n]) = f(\omega_1, \ldots, \omega_n).$$

This definition says that for $(C_1, \ldots, C_n) \in \tilde{\Omega}$, the value of $\tilde{f}(C_1, \ldots, C_n)$ is determined by taking one element ω_i from each C_i and evaluating $f(\omega_1, \ldots, \omega_n)$.

$\tilde{f}$ is well-defined. That is to say, the value of $\tilde{f}(C_1, \ldots, C_n)$ is independent of which element in C_i one would choose. To show this, suppose $(C_1, \ldots, C_n) \in \tilde{\Omega}$ and $\alpha_i, \beta_i \in C_i$. Then $\alpha_i \sim \beta_i$ for every i, so that

$$f(\alpha_1, \alpha_2, \alpha_3, \ldots, \alpha_n) = f(\beta_1, \alpha_2, \alpha_3, \ldots, \alpha_n)$$
$$= f(\beta_1, \beta_2, \alpha_3, \ldots, \alpha_n)$$
$$\vdots$$
$$= f(\beta_1, \beta_2, \beta_3, \ldots, \beta_n).$$

Thus the value of $\tilde{f}(C_1, \ldots, C_n)$ is determined uniquely.

Using $\tilde{\Omega}$ and $\tilde{f}$ constructed from the original normal form (N, Ω, f), we define the new normal form as follows.

Let (N, Ω, f) be the normal form of a game G. We refer to $(N, \tilde{\Omega}, \tilde{f})$ as the reduced normal form of G.

Whether two games are equivalent or not is judged by comparing the reduced normal forms of these games, as we mentioned earlier.

Let $(N^{(1)}, \tilde{\Omega}^{(1)}, \tilde{f}^{(1)})$ be the reduced normal form of a game G_1, and let $(N^{(2)}, \tilde{\Omega}^{(2)}, \tilde{f}^{(2)})$ be the reduced normal form of a game G_2. Then, G_1 is said to be equivalent to G_2 if the following holds: (i) $N^{(1)} = N^{(2)} = \{1, \ldots, n\}$; and (ii) There exists a sequence $(\phi_1, \ldots, \phi_n)$ of bijection $\phi_k : \tilde{\Omega}_k^{(1)} \mapsto \tilde{\Omega}_k^{(2)}$, such that for all $(C_1, \ldots, C_n) \in \tilde{\Omega}^{(1)}$

$$\tilde{f}^{(1)}(C_1, \ldots, C_n) = \tilde{f}^{(2)}(\phi_1(C_1), \ldots, \phi_n(C_n)). \tag{6.44}$$

If G_1 is equivalent to G_2, we write $G_1 \parallel G_2$.

To give an example of equivalent games, let us consider classical PQ penny flipover [MB98], in which both player P and player Q are classical players. In this game, a penny is placed initially heads up in a box. Players take turns $(Q \to P \to Q)$ flipping the penny over or not. Each player can not know what the opponent did, nor see inside the box. Finally the box is opened, and Q wins if the penny is heads up. This game can be formulated as a finite strategic game whose pay-off matrix is given in Table 6.1.

	Q: NN	Q: NF	Q: FN	Q: FF
P: N	$(-1, 1)$	$(1, -1)$	$(1, -1)$	$(-1, 1)$
P: F	$(1, -1)$	$(-1, 1)$	$(-1, 1)$	$(1, -1)$

Table 6.1. Pay-off matrix for PQ penny flipover. F denotes a flipover and N denotes no flipover. The first entry in the parenthesis denotes P's pay-off and the second one denotes Q's pay-off.

Intuitively, Q does not benefit from the second move, so that it does not matter whether Q can do the second move or not. The notion of equivalence captures this intuition; the above penny flipover game is equivalent to a finite strategic game whose pay-off matrix is given in Table 6.2. It represents another penny flipover game in which both players act only once. Proof of the equivalence is easy and we omit it.

	Q: N	Q: F
P: N	$(-1, 1)$	$(1, -1)$
P: F	$(1, -1)$	$(-1, 1)$

Table 6.2. Pay-off matrix for another PQ penny flipover in which both players act only once.

We now return to the general discussion on the notion of equivalence. The following is a basic property of the equivalence between two games.

The binary relation $\|$ is an equivalence relation; namely, for any games G_1, G_2, and G_3, the following holds: reflexivity: $G_1 \| G_1$; symmetry: If $G_1 \| G_2$, then $G_2 \| G$; and transitivity: If $G_1 \| G_2$ and $G_2 \| G_3$, then $G_1 \| G_3$. For the proof, see [Kob07].

In some cases, we can find that two games are equivalent by comparing the normal forms of the games, not the reduced normal forms. In the following lemma, sufficient conditions for such cases are presented.

Let $(N^{(1)}, \Omega^{(1)}, f^{(1)})$ be the normal form of a game G_1, and let $(N^{(2)}, \Omega^{(2)}, f^{(2)})$ be the normal form of a game G_2. If the following conditions are satisfied, G_1 is equivalent to G_2: (i) $N^{(1)} = N^{(2)} = \{1, \ldots, n\}$; and (ii) There exists a sequence $(\psi_1, \ldots, \psi_n)$ of bijection $\psi_k : \Omega_k^{(1)} \mapsto \Omega_k^{(2)}$, such that for all $(\omega_1, \ldots, \omega_n) \in \Omega^{(1)}$,

$$f^{(1)}(\omega_1, \ldots, \omega_n) = f^{(2)}(\psi_1(\omega_1), \ldots, \psi_n(\omega_n)). \tag{6.45}$$

For the proof, see [Kob07].

We define a map ϕ_i from $\tilde{\Omega}_i^{(1)}$ to the set of all subsets of $\Omega_i^{(2)}$ as

$$\phi_i(C_i) = \{\psi_i(\omega') \mid \omega' \in C_i\}.$$

We show that the range of ϕ_i is a subset of $\tilde{\Omega}_i^{(2)}$; that is, for any $[\omega_i] \in \tilde{\Omega}_i^{(1)}$ there exists $\xi_i \in \Omega_i^{(2)}$ such that $\phi_i([\omega_i]) = [\xi_i]$. In fact, $\psi_i(\omega_i)$ is such a ξ_i:

$$\phi_i([\omega_i]) = [\psi_i(\omega_i)]. \tag{6.46}$$

Below, we will prove $\phi_i([\omega_i]) \subset [\psi_i(\omega_i)]$ first, and then prove $[\psi_i(\omega_i)] \subset \phi_i([\omega_i])$.

To prove $\phi_i([\omega_i]) \subset [\psi_i(\omega_i)]$, we will show that an arbitrary element $\sigma_i \in \phi_i([\omega_i])$ satisfies $\sigma_i \in [\psi_i(\omega_i)]$. For this purpose, it is sufficient to show that $\sigma_i \sim \psi_i(\omega_i)$; that is, for an arbitrary $\sigma_k \in \Omega_k^{(2)}$ $(k \neq i)$

$$f^{(2)}(\sigma_1, \ldots, \sigma_{i-1}, \sigma_i, \sigma_{i+1}, \ldots, \sigma_n)$$
$$= f^{(2)}(\sigma_1, \ldots, \sigma_{i-1}, \psi_i(\omega_i), \sigma_{i+1}, \ldots, \sigma_n). \tag{6.47}$$

Since ψ_k is a bijection, there exists $\omega_k \in \Omega_k^{(1)}$ such that $\psi_k(\omega_k) = \sigma_k$. In addition, because $\sigma_i \in \phi_i([\omega_i])$, there exists $\omega_i' \in [\omega_i]$ such that $\sigma_i = \psi_i(\omega_i')$. Thus,

$$f^{(2)}(\sigma_1, \ldots, \sigma_{i-1}, \sigma_i, \sigma_{i+1}, \ldots, \sigma_n)$$
$$= f^{(2)}(\psi_1(\omega_1), \ldots, \psi_{i-1}(\omega_{i-1}), \psi_i(\omega_i'), \psi_{i+1}(\omega_{i+1}), \ldots, \psi_n(\omega_n))$$
$$= f^{(1)}(\omega_1, \ldots, \omega_{i-1}, \omega_i', \omega_{i+1}, \ldots, \omega_n). \tag{6.48}$$

Because $\omega_i' \in [\omega_i]$ and $\omega_i \in [\omega_i]$, it follows that $\omega_i' \sim \omega_i$. Hence,

$$(6.48) = f^{(1)}(\omega_1, \ldots, \omega_{i-1}, \omega_i, \omega_{i+1}, \ldots, \omega_n)$$
$$= f^{(2)}(\psi_1(\omega_1), \ldots, \psi_{i-1}(\omega_{i-1}), \psi_i(\omega_i), \psi_{i+1}(\omega_{i+1}), \ldots, \psi_n(\omega_n))$$
$$= f^{(2)}(\sigma_1, \ldots, \sigma_{i-1}, \psi_i(\omega_i), \sigma_{i+1}, \ldots, \sigma_n), \tag{6.49}$$

which leads to the conclusion that (6.47) holds for any $\sigma_i \in \phi_i([\omega_i])$.

Conversely, we can show that $[\psi_i(\omega_i)] \subset \phi_i([\omega_i])$. Let σ_i be an arbitrary element of $[\psi_i(\omega_i)]$. Since ψ_i is a bijection, there exists $\omega_i' \in \Omega_i^{(1)}$ such that $\psi_i(\omega_i') = \sigma_i$. For such ω_i', it holds that $\psi_i(\omega_i') \sim \psi_i(\omega_i)$, because $\psi_i(\omega_i') \in [\psi_i(\omega_i)]$. Hence,

$$f^{(1)}(\omega_1, \ldots, \omega_{i-1}, \omega_i', \omega_{i+1}, \ldots, \omega_n)$$
$$= f^{(2)}(\psi_1(\omega_1), \ldots, \psi_{i-1}(\omega_{i-1}), \psi_i(\omega_i'), \psi_{i+1}(\omega_{i+1}), \ldots, \psi_n(\omega_n))$$
$$= f^{(2)}(\psi_1(\omega_1), \ldots, \psi_{i-1}(\omega_{i-1}), \psi_i(\omega_i), \psi_{i+1}(\omega_{i+1}), \ldots, \psi_n(\omega_n))$$
$$= f^{(1)}(\omega_1, \ldots, \omega_{i-1}, \omega_i, \omega_{i+1}, \ldots, \omega_n),$$

which indicates that $\omega_i' \sim \omega_i$. Thus, $\omega_i' \in [\omega_i]$. Therefore, we conclude that if $\sigma_i \in [\psi_i(\omega_i)]$, then $\sigma_i = \psi_i(\omega_i') \in \phi_i([\omega_i])$; that is, $[\psi_i(\omega_i)] \subset \phi_i([\omega_i])$.

We have shown above that ϕ_i is a map from $\tilde{\Omega}_i^{(1)}$ to $\tilde{\Omega}_i^{(2)}$. The next thing we have to show is that ϕ_i is a bijection from $\tilde{\Omega}_i^{(1)}$ to $\tilde{\Omega}_i^{(2)}$. We will show the bijectivity of ϕ_i by proving injectivity and surjectivity separately.

First, we show that ϕ_i is injective. Suppose $[\omega_i], [\omega_i'] \in \tilde{\Omega}_i^{(1)}$ and $[\omega_i] \neq [\omega_i']$. Because $[\omega_i] \neq [\omega_i']$, it follows that $\omega_i \nsim \omega_i'$, so that there exists $(\omega_1, \ldots, \omega_{i-1}, \omega_{i+1}, \ldots, \omega_n) \in \Omega_1^{(1)} \times \cdots \times \Omega_{i-1}^{(1)} \times \Omega_{i+1}^{(1)} \times \cdots \times \Omega_n^{(1)}$ such that

$$f^{(1)}(\omega_1, \ldots, \omega_{i-1}, \omega_i, \omega_{i+1}, \ldots, \omega_n) \neq f^{(1)}(\omega_1, \ldots, \omega_{i-1}, \omega_i', \omega_{i+1}, \ldots, \omega_n).$$

For such $(\omega_1, \ldots, \omega_{i-1}, \omega_{i+1}, \ldots, \omega_n)$,

$$f^{(2)}(\psi_1(\omega_1), \ldots, \psi_{i-1}(\omega_{i-1}), \psi_i(\omega_i), \psi_{i+1}(\omega_{i+1}), \ldots, \psi_n(\omega_n))$$
$$= f^{(1)}(\omega_1, \ldots, \omega_{i-1}, \omega_i, \omega_{i+1}, \ldots, \omega_n)$$
$$\neq f^{(1)}(\omega_1, \ldots, \omega_{i-1}, \omega_i', \omega_{i+1}, \ldots, \omega_n)$$
$$= f^{(2)}(\psi_1(\omega_1), \ldots, \psi_{i-1}(\omega_{i-1}), \psi_i(\omega_i'), \psi_{i+1}(\omega_{i+1}), \ldots, \psi_n(\omega_n)).$$

This indicates that $\psi_i(\omega_i) \not\sim \psi_i(\omega_i')$. Hence, $[\psi_i(\omega_i)] \neq [\psi_i(\omega_i')]$. Thus, using (6.46), we conclude that $\phi_i([\omega_i]) \neq \phi_i([\omega_i'])$.

Next, we show that ϕ_i is surjective. Let $[\sigma]$ be an arbitrary element of $\tilde{\Omega}_i^{(2)}$. Define $\omega \in \Omega_i^{(1)}$ as $\omega \equiv \psi_i^{-1}(\sigma)$. Then,

$$\phi_i([\omega]) = [\psi_i(\omega)] = [\sigma].$$

The first equation follows from (6.46). Thus, for an arbitrary $[\sigma] \in \tilde{\Omega}_i^{(2)}$, there exists $[\omega] \in \tilde{\Omega}_i^{(1)}$ such that $\phi_i([\omega]) = [\sigma]$.

Lastly, we show that $(\phi_1, \ldots, \phi_n)$ satisfies (6.44). For an arbitrary $([\omega_1], \ldots, [\omega_n]) \in \tilde{\Omega}^{(1)}$,

$$\tilde{f}^{(1)}([\omega_1], \ldots, [\omega_n]) = f^{(1)}(\omega_1, \ldots, \omega_n) \tag{6.50}$$

$$= f^{(2)}(\psi_1(\omega_1), \ldots, \psi_n(\omega_n)) \tag{6.51}$$

$$= \tilde{f}^{(2)}([\psi_1(\omega_1)], \ldots, [\psi_n(\omega_n)]) \tag{6.52}$$

$$= \tilde{f}^{(2)}(\phi_1([\omega_1]), \ldots, \phi_n([\omega_n])). \tag{6.53}$$

Equations (6.50) and (6.52) follow from the definition of $\tilde{f}^{(1)}$ and $\tilde{f}^{(2)}$. Equation (6.51) follows from (6.45). The last equation follows from (6.46).

Game Classes

This short subsection is devoted to explaining game classes and some binary relations between game classes. These notions simplify the statements of our main theorems.

First, we define a game class as a subset of G. We defined previously G, SG, FSG, $MEFSG$, $QSim$, $FQSim$, $QSeq$, and $FQSeq$. All of these are game classes. Note that G is itself a game class.

Next, we introduce some symbols. Let A and B be game classes. If for any game $G \in A$ there exists a game $G' \in B$ such that $G \parallel G'$, then we write $A \trianglelefteq B$. If $A \trianglelefteq B$ and $B \trianglelefteq A$, we write $A \bowtie B$. If $A \trianglelefteq B$ but $B \ntrianglelefteq A$, we write $A \triangleleft B$.

The binary relation $\trianglelefteq$ is a pre-order. Namely, for any game classes A, B, and C, the following holds: reflexivity: $A \trianglelefteq A$; and transitivity: If $A \trianglelefteq B$ and $B \trianglelefteq C$, then $A \trianglelefteq C$. For the proof, see [Kob07].

Summary on Formal Quantum Games

The following relationships exist between game classes $QSim$, $QSeq$, $FQSim$, and $FQSeq$ [Kob07]: $QSim \trianglelefteq QSeq$. Also, we have: $FQSim \trianglelefteq FQSeq$ and the converse $FQSeq \trianglelefteq FQSim$. For the proof, see [Kob07].

When a statement "if $G \in FQSim$ then G has a property P" is true, another statement "if $G \in FQSeq$ then G has a property P" is also true, and

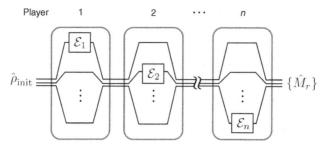

Fig. 6.5. A quantum sequential game G^{seq} which is equivalent to a quantum simultaneous game G depicted in Fig. 6.3 (see text for explanation).

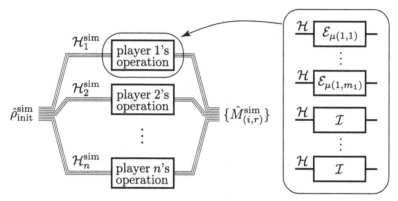

Fig. 6.6. A finite quantum simultaneous game G^{sim} which is equivalent to a given finite quantum sequential game G (see text for explanation).

vice versa. Here, P must be such a property that if a game G has the property P and $G \parallel G'$, then G' also has the property P. We call such P a property preserved under $\parallel$. For example, "a Nash equilibrium exists" is a property preserved under $\parallel$.

Unfortunately, no results are known which have the form "if $G \in FQSim$ ($FQSeq$) then G has a property P, but otherwise G does not necessarily have the property P". Consequently, we cannot reap the benefits of the above-mentioned deduction. However, numerous results exist which have the form "for a certain subset S of $FQSim$ ($FQSeq$), if $G \in S$ then G has a property Q preserved under $\parallel$, but otherwise G does not necessarily have the property Q". For such S and Q, the above theorems guarantee that there exists a subset $\mathbf{S'}$ of $FQSeq$ ($FQSim$) which satisfies the following: "If $G \in \mathbf{S'}$ then G has the property Q, but otherwise G does not necessarily have the property Q". In this sense, many of the results so far on $FQSim$ ($FQSeq$) can be translated into statements on $FQSeq$ ($FQSim$) [Kob07].

It is worth noting that efficiency of a game[15] is not a property preserved under $\|$. In an original quantum sequential game G, it is necessary to transmit a qudit $m+1$ times, while $4m$ times are needed in the constructed game G^{sim}. Thus, G^{sim} is far more inefficient than G, despite G^{sim} and G are equivalent games.

Relevant to the present subsection is the study by Lee and Johnson [LJ03]. To describe their argument, we have to introduce a new game class.

A finite quantum simultaneous game with all CPTP maps available is a subclass of finite quantum simultaneous games, in which a strategy set Ω_i is the set of all CPTP maps on the set of density operators on $\mathcal{H}_i$ for every i. We denote the set of all finite quantum simultaneous game with all CPTP maps available by $FQSimAll$.

We can easily prove that $FQSimAll \lhd FQSim$, by showing that the range of expected pay-off functions for a game in $FQSimAll$ must be connected, while the one for a game in $FQSim$ can be disconnected.

Lee and Johnson claimed that (i) "any game could be played classically" and (ii) "finite classical games consist of a strict subset of finite quantum games". Using the terms of this subsection, we may interpret these claims as follows.

We have: $SG \bowtie G$, $MEFSG \unlhd FQSimAll$ and $FQSimAll \ntrianglelefteq MEFSG$. For the proof, see [Kob07].

The relationships between various game classes can be summarized as:

$$MEFSG \lhd FQSimAll \lhd FQSim \bowtie FQSeq$$
$$\unlhd QSim \unlhd QSeq \unlhd SG \bowtie G. \tag{6.54}$$

Replacing $\unlhd$ in (6.54) with either $\lhd$ or $\bowtie$ will be a possible extension of this subsection.

Besides that, it remains to be investigated what the characterizing features of each game class in (6.54) are. Especially, further research on games which is in $FQSim$ (or equivalently $FQSeq$) but not equivalent to any games in $MEFSG$ would clarify the truly quantum mechanical nature of quantum games.

6.3 Hardware for Quantum Computers

This section addresses modern electronic devices called *Josephson junctions*, which promise to be a basic building blocks of the future quantum computers. Apparently, they can exhibit chaotic behavior, both as single junctions (which have macroscopic dynamics analogous to those of the forced nonlinear oscillators), and as arrays (or ladders) of junctions, which can show high-dimensional chaos.

---

[15] Amount of information exchange between players and a referee, required to play a game. See [LJ03] for more details.

A *Josephson junction* is a type of electronic circuit capable of switching at very high speeds, i.e., frequency of typically 10^{10}–10^{11} Hz, when operated at temperatures approaching absolute zero. It is an insulating barrier separating two superconducting materials and producing the *Josephson effect*. The terms are named eponymously after British physicist Brian David Josephson, who predicted the existence of the Josephson effect in 1962 [Jos74]. Josephson junction exploits the phenomenon of *superconductivity*, the ability of certain materials to conduct electric current with practically zero resistance. Josephson junctions have important applications in quantum-mechanical circuits. They have great technological promises as amplifiers, voltage standards, detectors, mixers, and fast switching devices for digital circuits. They are used in certain specialized instruments such as highly-sensitive microwave detectors, magnetometers, and QUIDs. Finally, Josephson junctions allow the realization of *qubits*, the key elements of *quantum computers*.

Josephson junctions have been particularly useful for experimental studies of nonlinear dynamics as the equation governing a single junction dynamics is the same as that for a pendulum [Str94]. Their dynamics can be analyzed both in a simple overdamped limit and in the more complex underdamped one, either for single junctions and for arrays of large numbers of coupled junctions.

A Josephson junction is made up of two superconductors, separated by a weak coupling non-superconducting layer, so thin that electrons can cross through the insulating barrier. It can be conceptually represented as:

$$Superconductor\ 1\ :\ \psi_1 e^{i\phi_1}$$
$$Weak\ Coupling\ \ \Updownarrow$$
$$Superconductor\ 2\ :\ \psi_2 e^{i\phi_2}$$

where the two superconducting regions are characterized by simple *quantum-mechanical wave functions*, $\psi_1 e^{i\phi_1}$ and $\psi_2 e^{i\phi_2}$, respectively. Normally, a much more complicated description would be necessary, as there are $\sim 10^{23}$ electrons to deal with, but in the superconducting ground state, these electrons form the so-called *Cooper pairs* that can be described by a single macroscopic wave function $\psi e^{i\phi}$. The flow of current between the superconductors in the absence of an applied voltage is called a *Josephson current*, and the movement of electrons across the barrier is known as *Josephson tunneling* (see Fig. 6.7). Two or more junctions joined by superconducting paths form what is called a *Josephson interferometer*.

One of the characteristics of a Josephson junction is that as the temperature is lowered, superconducting current flows through it even in the absence of voltage between the electrodes, part of the *Josephson effect*. The Josephson effect in particular results from two superconductors acting to preserve their long-range order across an insulating barrier. With a thin enough barrier, the phase of the electron wave-function in one superconductor maintains a fixed relationship with the phase of the wave-function in another superconductor.

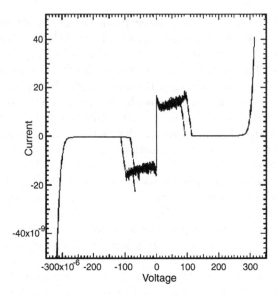

Fig. 6.7. *Josephson junction:* the current–voltage curve obtained at low temperature. The vertical portions (zero voltage) of the curve represent Cooper pair tunneling. There is a small magnetic field applied, so that the maximum Josephson current is severely reduced. Hysteresis is clearly visible around 100 microvolts. The portion of the curve between 100 and 300 microvolts is current independent, and is the regime where the device can be used as a detector.

This linking up of phase is called phase coherence. It occurs throughout a single superconductor, and it occurs between the superconductors in a Josephson junction. The *phase coherence*, or *long-range order*, is the essence of the Josephson effect.

While researching superconductivity, B.D. Josephson studied the properties of a junction between two superconductors. Following up on earlier work by L. Esaki and I. Giaever, he demonstrated that in a situation when there is electron flow between two superconductors through an insulating layer (in the absence of an applied voltage), and a voltage is applied, the current stops flowing and oscillates at a high frequency. The Josephson effect is influenced by magnetic fields in the vicinity, a capacity that enables the Josephson junction to be used in devices that measure extremely weak magnetic fields, such as superconducting quantum interference devices (SQUIDs). For their efforts, Josephson, Esaki, and Giaever shared the Nobel Prize for Physics in 1973.

The *Josephson-junction quantum computer* was demonstrated in April 1999 by Nakamura, Pashkin and Tsai of NEC Fundamental Research Laboratories in Tsukuba, Japan [NPT99]. In the same month, only about one week earlier, Ioffe, Geshkenbein, Feigel'man, Fauchère and Blatter, independently, described just such a computer in Nature [IGF99].

Nakamura, Pashkin and Tsai's computer is built around a *Cooper pair box*, which is a small superconducting island electrode weakly coupled to a bulk superconductor. Weak coupling between the superconductors creates a Josephson junction between them. Like most other junctions, the Josephson junction is also a capacitor, which is charged by the current that flows through it. A gate voltage is applied between the two superconducting electrodes. If the Cooper box is sufficiently small, e.g., as small as a quantum dot, the charging current breaks into discrete transfer of individual Cooper pairs, so that ultimately it is possible to just transfer a single Cooper pair across the junction. The effectiveness of the Cooper pair transfer depends on the energy difference between the box and the bulk and a maximum is reached when a voltage is applied, which equalizes this energy difference. This leads to *resonance* and observable *coherent quantum oscillations* [Ave99].

This contraption, like the Loss–Vincenzo quantum dot computer [LD98], has the advantage that it is controlled electrically. Unlike Loss–Vincenzo computer, this one actually exists in the laboratory. Nakamura, Pashkin and Tsai did not perform any computations with it though. At this stage it was enough of an art to observe the coherence for about 6 cycles of the Cooper pair oscillations, while the chip was cooled to about and carefully shielded from external electromagnetic radiation.

There are two general types of Josephson junctions: overdamped and underdamped. In overdamped junctions, the barrier is conducting (i.e., it is a normal metal or superconductor bridge). The effects of the junction's internal electrical resistance will be large compared to its small capacitance. An overdamped junction will quickly reach a unique equilibrium state for any given set of conditions.

The barrier of an underdamped junction is an insulator. The effects of the junction's internal resistance will be minimal. Underdamped junctions do not have unique equilibrium states, but are hysteretic.

A Josephson junction can be transformed into the so-called *Giaever tunneling junction* by the application of a small, well defined magnetic field. In such a situation, the new device is called a superconducting tunneling junction (STJ) and is used as a very sensitive photon detector throughout a wide range of the spectrum, from infrared to hard X-ray. Each photon breaks up a number of Cooper pairs. This number depends on the ratio of the photon energy to approximately twice the value of the gap parameter of the material of the junction. The detector can be operated as a photon-counting spectrometer, with a spectral resolution limited by the statistical fluctuations in the number of released charges. The detector has to be cooled to extremely low temperature, typically below 1 kelvin, to distinguish the signals generated by the detector from the thermal noise. Small arrays of STJs have demonstrated their potential as spectro-photometers and could further be used in astronomy [ESA05]. They are also used to perform energy dispersive X-ray spectroscopy and in principle they could be used as elements in infrared imaging devices as well [Ens05].

6.3.1 Josephson Effect and Pendulum Analog

Josephson Effect

The basic equations governing the dynamics of the Josephson effect are (see, e.g., [BP82]):

$$U(t) = \frac{\hbar}{2e}\frac{\partial \phi}{\partial t}, \qquad I(t) = I_c \sin\phi(t),$$

where $U(t)$ and $I(t)$ are the voltage and current across the Josephson junction, $\phi(t)$ is the phase difference between the wave functions in the two superconductors comprising the junction, and I_c is a constant, called the *critical current* of the junction. The critical current is an important phenomenological parameter of the device that can be affected by temperature as well as by an applied magnetic field. The physical constant $\hbar/2e$ is the magnetic flux quantum, the inverse of which is the *Josephson constant*.

The three main effects predicted by Josephson follow from these relations:

1. The DC Josephson effect. This refers to the phenomenon of a direct current crossing the insulator in the absence of any external electromagnetic field, owing to *Josephson tunneling*. This DC Josephson current is proportional to the sine of the phase difference across the insulator, and may take values between $-I_c$ and I_c.
2. The AC Josephson effect. With a fixed voltage U_{DC} across the junctions, the phase will vary linearly with time and the current will be an AC current with amplitude I_c and frequency $2e/\hbar U_{DC}$. This means a Josephson junction can act as a perfect *voltage-to-frequency converter*.
3. The inverse AC Josephson effect. If the phase takes the form

$$\phi(t) = \phi_0 + n\omega t + a\sin(\omega t),$$

the voltage and current will be

$$U(t) = \frac{\hbar}{2e}\omega[n + a\cos(\omega t)], \qquad I(t) = I_c \sum_{m=-\infty}^{\infty} J_n(a)\sin[\phi_0 + (n+m)\omega t].$$

The DC components will then be

$$U_{DC} = n\frac{\hbar}{2e}\omega, \qquad I(t) = I_c J_{-n}(a)\sin\phi_0.$$

Hence, for distinct DC voltages, the junction may carry a DC current and the junction acts like a perfect *frequency-to-voltage converter*.

Pendulum Analog

To show a driven pendulum analog of a microscopic description of a single Josephson junction, we start with:

1. The *Josephson current–phase relation*

$$I = I_c \sin \phi,$$

where I_c is the *critical current*, I is the bias current, and $\phi = \phi_2 - \phi_1$ is the constant *phase difference* between the phases of the two superconductors that are weakly coupled; and

2. The *Josephson voltage–phase relation*

$$V = \frac{\hbar}{2e} \dot{\phi},$$

where $V = V(t)$ is the instantaneous voltage across the junction, $\hbar$ is the Planck constant (divided by 2π), and e is the charge on the electron.

Now, if we apply Kirhoff's voltage and current laws for the parallel RC-circuit with resistance R and capacitance C, we come to the first-order ODE

$$C\dot{V} + \frac{V}{R} + I_c \sin \phi = I,$$

which can be recast solely in terms of the phase difference ϕ as the second-order pendulum-like ODE,

Josephson junction: $$\frac{\hbar C}{2e}\ddot{\phi} + \frac{\hbar}{2eR}\dot{\phi} + I_c \sin \phi = I,$$

$$(6.55)$$

Pendulum: $$ml^2\ddot{\theta} + b\dot{\theta} + mgl \sin \theta = \tau.$$

This mechanical analog has often proved useful in visualizing the dynamics of Josephson Junctions [Str94]. If we divide (6.55) by I_c and define a dimensionless time

$$\tau = \frac{2eI_c R}{\hbar}t,$$

we get the dimensionless oscillator equation for Josephson junction,

$$\beta\phi'' + \phi' + \sin \phi = \frac{I}{I_c},$$

$$(6.56)$$

where $\phi' = d\phi/d\tau$. The dimensionless group β, defined by

$$\beta = \frac{2eI_c R^2 C}{\hbar},$$

is called the McCumber parameter and represents a dimensionless capacitance.

In a simple *overdamped limit* $\beta \ll 1$ with *resistive loading*, the 'inertial term' $\beta\phi''$ may be neglected (as if oscillating in a highly-viscous medium), and so (6.56) reduces to a non-uniform oscillator

$$\phi' = \frac{I}{I_c} - \sin \phi, \tag{6.57}$$

with solutions approaching a stable fixed-point for $I < I_c$, and periodically varying for $I < I_c$. To find the current–voltage curve in the overdamped limit, we take the average voltage $\langle V \rangle$ as a function of the constant applied current I, assuming that all transients have decayed and the system has reached steady-state, and get

$$\langle V \rangle = I_c R \langle \phi' \rangle.$$

An overdamped *array of N Josephson Junctions* (6.57), parallel with a resistive load R, can be described by the system of first-order dimensionless ODEs [Str94]

$$\phi_k' = \Omega + a \sin \phi_k + \frac{1}{N} \sum_{j=1}^{N} \sin \phi_j, \quad k = 1, \ldots, N,$$

where

$$\Omega = I_b R_0 / I_c r, \qquad a = -(R_0 + r)/r, \qquad R_0 = R/N,$$

$$I_b = I_c \sin \phi_k + \frac{\hbar}{2eR} \dot{\phi}_k + \frac{\hbar}{2eR} \sum_{j=1}^{N} \dot{\phi}_j.$$

6.3.2 Dissipative Josephson Junction

The past decade has seen a considerable interest and remarkable activity in an area which presently is often referred to as macroscopic quantum mechanics. Specifically, one has been interested in quantum phenomena of macroscopic objects [Leg86].

In particular, macroscopic quantum tunneling [CL81] (quantum decay of a meta-stable state), and quantum coherence [LCD87] have been studied. Soon, it became clear that dissipation has a profound influence on these quantum phenomena. Phenomenologically, dissipation is the consequence of an interaction of the object with an environment which can be thought of as consisting of infinitely many degrees of freedom. Specifically, the environmental degrees of freedom may be chosen to be harmonic oscillators such that we may consider the dissipation as a process where excitations, that are phonons, are emitted and absorbed. This, Caldeira–Leggett model has been used in [CL81] where the influence of dissipation on tunneling has been explored.

As far as quantum coherence is concerned, the most simple system is an object with two different quantum states: it is thought to represent the limiting case of an object in a double-well potential where only the lowest energy states in each of the two wells is relevant and where the tunneling through the separating barrier allows for transitions that probe the coherence. Since a 2-state system is equivalent to a spin-one-half problem, this standard system is often referred to by this name. In particular, with the standard coupling to a dissipative environment made of harmonic oscillators, it is called the spin-boson problem which has been studied repeatedly in the past [LCD87, SW90].

Level quantization and resonant tunneling have been observed recently [Vaa95] in a double-well quantum-dot system. However, the influence of dissipation was not considered in this experiment. On the other hand, it seems that Josephson junctions are also suitable systems for obtaining experimental evidence pertaining to macroscopic quantum effects. In this context, evidence for level quantization and for quantum decay have been obtained [MDC85].

Recall that a Josephson junction may be characterized by a *current–phase relation*

$$I(\phi) = I_J \sin \phi, \tag{6.58}$$

where the phase ϕ is related to the voltage difference U by

$$\hbar \dot{\phi} = 2eU. \tag{6.59}$$

Therefore, the phase of a Josephson junction shunted by a capacitance C and biased by an external current I_x obeys a classical type of equation of motion

$$M\ddot{\phi} = -\frac{\partial V(\phi)}{\partial \phi}, \tag{6.60}$$

with the mass

$$M = \left(\frac{\hbar}{2e}\right)^2 C, \tag{6.61}$$

and the potential energy

$$V(\phi) = -\frac{\hbar}{2e}\left[I_J \cos \phi + I_x \phi\right]. \tag{6.62}$$

A widely discussed model of a dissipative object is the one where the Josephson junction is also shunted by an Ohmic resistor R. In this case, the classical equation of motion (6.60) has to be replaced by

$$M\ddot{\phi} = -\frac{\partial V(\phi)}{\partial \phi} - \eta \dot{\phi}, \quad \eta = \left(\frac{\hbar}{2e}\right)^2 \frac{1}{R}. \tag{6.63}$$

The model of a dissipative environment according to the above specification has been discussed by [CL81].

The potential energy $V(\phi)$ of (6.62) displays wells at $\phi \simeq 2n\pi$ with depth shifted by an amount $\Delta \simeq (2\pi\hbar/2e)I_x$. If the wells are sufficiently deep, one needs to concentrate only on transitions between pairs of adjacent wells. Thus, one arrives at the double well problem mentioned above.

The analysis in this section goes beyond the limiting situation where only the lowest level in each of the two wells is of importance. Roughly, this is realized when the level separation $\hbar(2E_J/M)^{1/2} \simeq (2e\hbar I_J/C)^{1/2}$ is smaller than or comparable with Δ. In particular, we will concentrate on resonance phenomena which are expected to show up whenever two levels in the adjacent wells happen to cross when the bias current I_x, that is Δ, is varied.

For such values of the bias current, there appear sharp asymmetric peaks in the current–voltage characteristic of the Josephson junction. This phenomenon has been studied by [LOS88] within the standard model in the one-phonon approximation. For bias currents that correspond to crossings of the next and next nearest levels (e.g., ground state in the left well and the first or second excited state at the right side), it is possible to neglect processes in the reverse direction provided that the temperature is sufficiently low. Thus, the restriction to a double well system receives additional support.

The transfer of the object from the left to the right potential well is accompanied by the emission of an infinite number of phonons. Therefore, in [OS94] the fact is taken into account that in the resonance region, the contribution of phonons of small energy is important as well as the contribution of resonance phonons with energy equal to the distance between levels in the wells.

Junction Hamiltonian and Its Eigenstates

The model of [MS95b] consists of a particle, called 'object' (coordinate R_1), which is coupled (in the sense of [CL81]) to a 'bath' of harmonic oscillators (coordinates R_j). We shall use the conventions $j \in \{2, \ldots, N\}$ for the bath oscillators and $k \in \{1, \ldots, N\}$ for the indices of all coordinates in the model. The double-well potential is approximated by two parabolas about the minima of the two wells.

The phase ϕ of the Josephson contact then corresponds to the object coordinate R_1 of the model, and the voltage U is related to the tunneling rate J by $2eU = \dot{\phi} = 2\pi J$. As it has already been remarked, the current I_x is proportional to the bias Δ of the two wells. Thus, calculating the transition rate for different values of the bias Δ is equivalent to the determination of the I–V characteristics.

Specifically, following [MS95b], we want to write the Hamiltonian of the model in the form

$$\hat{H} = \frac{1}{2m} \sum_k \hat{p}_k^2 + \hat{v}(\hat{R}_1) + \frac{m}{2} \sum_j \omega_j^2 (\hat{R}_j - \hat{R}_1)^2,$$

$$\hat{v}(\hat{R}_1) \approx \frac{m}{2} \sum_{\pm} \Omega^2 (\hat{R}_1 \pm a)^2 \pm \frac{\Delta}{2}.$$

(6.64)

The states for the two situations 'object in the left well' and 'object in the right well' will be denoted by $|\Lambda_L, L\rangle$ and $|\Lambda_R, R\rangle$, respectively. If one projects onto the eigenstates $|n\rangle$ of the 1D harmonic oscillator and takes into account the shift of the wells, one arrives at the following decomposition ($\phi_n(R) = \langle R|n\rangle$):

$$\langle n_L, \{R_j\}|\Lambda_L, L\rangle = \int dR_1 \phi_{n_L}(R_1 + a)\phi^L_{\Lambda_L}(\{R_k\}),$$

$$\langle n_R, \{R_j\}|\Lambda_R, R\rangle = \int dR_1 \phi_{n_R}(R_1 - a)\phi^R_{\Lambda_R}(\{R_k\}). \tag{6.65}$$

The situations 'object on the left' and 'object on the right' differ only by the shift and the bias of the wells. Therefore, one can find a unified representation by noting that $\phi^L_\Lambda(\{R_k\}) = \Phi_\Lambda(\{R_k + a\})$ and $\phi^R_\Lambda(\{R_k\}) = \Phi_\Lambda(\{R_k - a\})$. The eigenstates Φ_Λ are defined by the relations

$$\Phi_\Lambda(\{R_k\}) = \langle\{R_k\}|\Lambda\rangle, \qquad \hat{H}_0|\Lambda\rangle = E_\Lambda|\Lambda\rangle,$$

$$\hat{H}_0 = \frac{1}{2m}\sum_k \hat{p}_k^2 + \frac{m}{2}\sum_j \omega_j^2(\hat{R}_j - \hat{R}_1)^2 + \frac{m}{2}\Omega^2 \hat{R}_1^2.$$

Thus, it follows from (6.65) that

$$\langle n_L, \{R_j\}|\Lambda_L, L\rangle = \langle n_L, \{R_j\}|\exp\left(ia\sum_j \hat{p}_j\right)|\Lambda_L\rangle,$$

$$\langle n_R, \{R_j\}|\Lambda_R, R\rangle = \langle n_R, \{R_j\}|\exp\left(-ia\sum_j \hat{p}_j\right)|\Lambda_R\rangle, \tag{6.66}$$

where we have used the shift property of the momentum operator $\hat{p}$.

The coupling of the two wells is taken into account by means of a tunneling Hamiltonian $\hat{H}_T$ which we represent in the form

$$\langle\Lambda_L, L|\hat{H}_T|\Lambda_R, R\rangle = \int d\{R_j\} \sum_{n_L n_R} T_{n_L n_R}\langle\Lambda_L, L|n_L, \{R_j\}\rangle\langle n_R, \{R_j\}|\Lambda_R, R\rangle.$$

Using again the momentum operator, one can write

$$|x\rangle\langle x'| = e^{i\hat{p}(x'-x)}|x'\rangle\langle x'| = e^{i\hat{p}(x'-x)}\delta(x' - \hat{x}).$$

From this, we conclude that

$$\langle\Lambda_L, L|\hat{H}_T|\Lambda_R, R\rangle = \sum_{n_L n_R} T_{n_L n_R}\int dR_1 dR_1' \frac{dQ}{2\pi}\phi^*_{n_R}(R_1')\phi_{n_L}(R_1)$$

$$\times \langle\Lambda_L, L|e^{i\hat{p}_1(R_1'-R_1)}e^{iQ(R_1'-\hat{R}_1)}|\Lambda_R, R\rangle.$$

Transition Rate

The net transition rate from the left well to the right one is then in second order perturbation theory given by [MS95b]

$$J = 2\pi Z_0^{-1}\sum_{\Lambda_L, \Lambda_R}|\langle\Lambda_L, L|\hat{H}_T|\Lambda_R, R\rangle|^2\delta(E_{\Lambda_L} - E_{\Lambda_R} + \Delta)$$

$$\times [e^{-\beta E_{\Lambda_L}} - e^{\beta E_{\Lambda_R}}],$$

where $Z_0 = \operatorname{Tr} \exp(-\beta H_0)$. The δ-function may be written in Fourier representation, and the fact that the E_A are eigen-energies of $\hat{H}_0$ serves us to incorporate the energy conservation into Heisenberg's time-dependent operators $\hat{A}(t) = \exp(i\hat{H}_0 t)\hat{A}\exp(-i\hat{H}_0 t)$, i.e.,

$$\langle \Lambda_L, L|\hat{H}_T|\Lambda_R, R\rangle \delta(E_{\Lambda_L} - E_{\Lambda_R} + \Delta)$$
$$= \int dt e^{i\Delta t} \langle \Lambda_L|e^{-ia\sum_k \hat{p}_k(t)}\hat{H}_T(t)e^{-ia\sum_k \hat{p}_k(t)}|\Lambda_R\rangle.$$

Then, collecting our results from above we arrive at the expression

$$J = Z_0^{-1}(1 - e^{-\beta\Delta}) \int dt\, e^{i\Delta t} \sum_{n_L, n_R} \sum_{\bar{n}_L, \bar{n}_R} T_{n_L, n_R} T^*_{\bar{n}_L, \bar{n}_R}$$

$$\times \int \frac{dQ d\bar{Q}}{(2\pi)^2} \int dR_1 dR_1' d\bar{R}_1 d\bar{R}_1' \phi_{n_L}(R_1)\phi^*_{n_R}(R_1')\phi^*_{\bar{n}_L}(\bar{R}_1')\phi_{\bar{n}_R}(\bar{R}_1)$$

$$\times \operatorname{Tr}\{e^{-\beta\hat{H}_0}\, e^{-2ia\sum_k \hat{p}_k(t)}\, e^{i\hat{p}_1(t)(R_1'-R_1+2a)}\, e^{-iQ(\hat{R}_1(t)-R_1')}$$

$$\times e^{2ia\sum_k \hat{p}_k}\, e^{i\hat{p}_1(\bar{R}_1'-\bar{R}_1-2a)}\, e^{-i\bar{Q}(\hat{R}_1-\bar{R}_1')}\}.$$

Let us now use the relation

$$e^{-i(\hat{H}_0+\hat{W})t} = e^{-i\hat{H}_0 t}\hat{T} e^{-i\int_0^t dt'\hat{W}(t')}$$

which holds for $t > 0$ when $\hat{T}$ is the *time-ordering operator* and for $t < 0$ when the anti time-ordering is used. If we define $\langle\hat{A}\rangle = \operatorname{Tr}\exp(-\beta\hat{H}_0)\hat{A}/Z_0$, we can write the following result for the transition rate:

$$J = (1 - e^{-\beta\Delta}) \int dt\, e^{i(\Delta - 2m\Omega^2 a^2)t} \sum_{n_L, n_R} \sum_{\bar{n}_L, \bar{n}_R} T_{n_L, n_R} T^*_{\bar{n}_L, \bar{n}_R}$$

$$\times \int \frac{dQ d\bar{Q}}{(2\pi)^2} \int dR dR' d\bar{R} d\bar{R}'\, e^{iQ\frac{R+R'}{2}+i\bar{Q}\frac{\bar{R}+\bar{R}'}{2}}$$

$$\times \phi_{n_L}(R)\phi^*_{n_R}(R'-2a)\phi^*_{\bar{n}_L}(\bar{R}')\phi_{\bar{n}_R}(\bar{R}-2a)$$

$$\times \left\langle \hat{T}\exp\left[-iQ\hat{R}_1(t) + i\hat{p}_1(t)(R'-R) + 2im\Omega^2 a \int_0^t dt'\hat{R}_1(t')\right.\right.$$

$$\left.\left. - i\bar{Q}\hat{R}_1(0) + i\hat{p}_1(0)(\bar{R}'-\bar{R})\right]\right\rangle. \tag{6.67}$$

We are now in the position to make use of the fact that the Hamiltonian is quadratic in all coordinates so that we can evaluate exactly

$$\langle \hat{T}e^{i\int dt'\eta(t')\hat{R}_1(t')}\rangle = e^{-\frac{i}{2}\iint dt'dt''\eta(t')D(t',t'')\eta(t'')},$$
$$D(t',t'') = -i\langle \hat{T}\hat{R}_1(t')\hat{R}_1(t'')\rangle. \tag{6.68}$$

By comparison with the last two lines in (6.67), the function $\eta(t')$ is given by

$$\eta(t') = -Q\delta(t' - t) - \overline{Q}\delta(t') + 2m\Omega^2 a[\Theta(t') - \Theta(t' - t)]$$
$$+ m(R - R')\delta'(t' - t) + m(\overline{R} - \overline{R}')\delta'(t').$$

$\Theta(t)$ is meant to represent the step function. The derivatives of the δ-function arise from a partial integration of terms containing $\hat{p}(t) = md\hat{x}(t)/dt$. Note, that these act only on the coordinates but not on the step functions which arise due to the time ordering.

Moreover, the degrees of freedom of the bath can be integrated out in the usual way [CL81] leading to a dissipative influence on the object. One is then lead to the following form of the Fourier transform of $D(t, t') \equiv D(t - t')$:

$$D(\omega) = \frac{D^R(\omega)}{1 - \exp\left(-\hbar\omega/k_B T\right)} + \frac{D^R(-\omega)}{1 - \exp\left(\hbar\omega/k_B T\right)}, \quad (6.69)$$
$$(D^R)^{-1}(\omega) = m[(\omega + i0)^2 - \Omega^2] + i\eta\omega,$$

where we will use a spectral density $J(\omega) = \eta\omega$, $0 \leq \omega \leq \omega_c$ for the bath oscillators.

From (6.67) and (6.68) one can conclude that the integrations with respect to $Q, \overline{Q}, R, R', \overline{R}, \overline{R}'$ can be done exactly as only Gaussian integrals are involved (note, that the eigenstates of the harmonic oscillator are Gaussian functions and derivatives of these, respectively). Therefore, for given $n_L, n_R, \overline{n}_L, \overline{n}_R$, one has to perform a 6D Gaussian integral [MS95b].

6.3.3 Josephson Junction Ladder (JJL)

2D arrays of Josephson junctions have attracted much recent theoretical and experimental attention. Interesting physics arises as a result of competing vortex–vortex and vortex–lattice interactions. It is also considered to be a convenient experimental realization of the so-called *frustrated XY models*. Here, mainly following [DT95], we discuss the simplest such system, namely the *Josephson junction ladder* (JJL, see Fig. 6.8) [Kar84, Gra90].

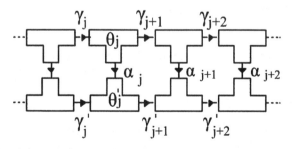

Fig. 6.8. Josephson junction Ladder (JJL).

To construct the system, superconducting elements are placed at the ladder sites. Below the bulk *superconducting-normal transition* temperature, the

state of each element is described by its charge and the phase of the superconducting wave ψ-function [And64]. In this section we neglect charging effects,
which corresponds to the condition that $4e^2/C \ll J$, with C being the capacitance of the element and J the Josephson coupling. Let θ_j (θ'_j) denote
the phase on the upper (lower) branch of the ladder at the jth rung. The
Hamiltonian for the array [Tin75] can be written in terms the gauge invariant
phase differences [DT95],

$$\gamma_j = \theta_j - \theta_{j-1} - (2\pi/\phi_0) \int_{j-1}^{j} A_x dx, \gamma'_j = \theta'_j - \theta'_{j-1} - (2\pi/\phi_0) \int_{j'-1}^{j'} A_x dx,$$

$$\text{and} \quad \alpha_j = \theta'_j - \theta_j - (2\pi/\phi_0) \int_{j}^{j'} A_y dx \quad \text{as}$$

$$H = -\sum_j (J_x \cos \gamma_j + J_x \cos \gamma'_j + J_y \cos \alpha_j), \tag{6.70}$$

where A_x and A_y are the components of the magnetic vector potential along
and transverse to the ladder, respectively, and ϕ_0 the flux quantum. The sum
of the phase differences around a plaquette is constrained by

$$\gamma_j - \gamma'_j + \alpha_j - \alpha_{j-1} = 2\pi(f - n_j),$$

where $n_j = 0, \pm 1, \pm 2, \ldots$ is the vortex occupancy number and $f = \phi/\phi_0$ with
ϕ being the magnetic flux through a plaquette. With this constraint, it is
convenient to write (6.70) in the form

$$H = -J\sum_j \{2\cos \eta_j \cos[(\alpha_{j-1} - \alpha_j)/2 + \pi(f - n_j)] + J_t \cos \alpha_j\},$$

$$\text{where } \eta_j = (\gamma_j + \gamma'_j)/2, J = J_x \text{ and } J_t = J_y/J_x. \tag{6.71}$$

The Hamiltonian is symmetric under $f \to f+1$ with $n_j \to n_j+1$, and $f \to -f$
with $n_j \to -n_j$, thus it is sufficient to study only the region $0 \leq f \leq 0.5$.
Since in one dimension ordered phases occur only at zero temperature, the
main interest is in the ground states of the ladder and the low temperature
excitations. Note that in (6.71) η_j decouples from α_j and n_j, so that all the
ground states have $\eta_j = 0$ to minimize H. The ground states will be among
the solutions to the current conservation equations: $\partial_{\alpha_j} H = 0$, i.e., [DT95]

$$J_t \sin \alpha_j = \sin[(\alpha_{j-1} - \alpha_j)/2 + \pi(f - n_j)] - \sin[(\alpha_j - \alpha_{j+1})/2 + \pi(f - n_{j+1})]. \tag{6.72}$$

For any given f there are a host of solutions to (6.72). The solution that
minimizes the energy must be selected to get the ground state.

If one expands the inter-plaquette cosine coupling term in (6.71) about
it's maximum, the discrete sine-Gordon model is obtained. A vortex $(n_j = 1)$
in the JJL corresponds to a kink in the sine-Gordon model. This analogy was
used by [Kar84] as an argument that this system should show similar behavior

to the discrete sine-Gordon model which has been studied by several authors [AA80, CS83, PTB86]. This analogy is only valid for J_t very small so that the inter-plaquette term dominates the behavior of the system making the expansion about its maximum a reasonable assumption. However, much of the interesting behavior of the discrete sine-Gordon model occurs in regions of large J_t ($J_t \sim 1$). Furthermore, much of the work by Aubry [AA80] on the sine-Gordon model relies on the convexity of the coupling potential which we do not have in the JJL.

Following [DT95], here we formulate the problem in terms of a transfer matrix obtained from the full partition function of the ladder. The eigenvalues and eigenfunctions of the transfer matrix are found numerically to determine the phases of the ladder as functions of f and J_t. We study the properties of various ground states and the low temperature excitations. As J_t is varied, all incommensurate ground states undergo a superconducting–normal transition at certain J_t which depends on f. One such transition will be analyzed. Finally we discuss the critical current.

The partition function for the ladder, with periodic boundary conditions and $K = J/k_B T$, is

$$Z = \prod_i^N \int_{-\pi}^{\pi} \sum_{\{n_i\}} d\alpha_i d\eta_i \exp\{K(2\cos\eta_i \cos[(\alpha_{i-1} - \alpha_i)/2$$
$$+ \pi(f - n_i)] + J_t \cos\alpha_i)\}.$$

The η_i can be integrated out resulting in a simple transfer matrix formalism for the partition function involving only the transverse phase differences:

$$Z = \prod_i^N \int_{-\pi}^{\pi} d\alpha_i P(\alpha_{i-1}, \alpha_i) = \text{Tr } \hat{P}^N.$$

The transfer matrix elements $P(\alpha, \alpha')$ are

$$P(\alpha, \alpha') = 4\pi \exp[K J_t (\cos\alpha + \cos\alpha')/2] I_0(2K \cos[(\alpha - \alpha')/2 + \pi f]), \quad (6.73)$$

where I_0 is the zeroth order modified Bessel function. Note that the elements of $\hat{P}$ are real and positive, so that its largest eigenvalue λ_0 is real, positive and nondegenerate. However, since $\hat{P}$ is not symmetric (except for $f = 0$ and $f = 1/2$) other eigenvalues can form complex conjugate pairs. As we will see from the correlation function, these complex eigenvalues determine the spatial periodicity of the ground states.

The two point correlation function of α_j's is [DT95]

$$\langle e^{i(\alpha_0 - \alpha_l)} \rangle = \lim_{N \to \infty} \frac{(\prod_i^N \int_{-\pi}^{\pi} d\alpha_i P(\alpha_{i-1}, \alpha_i)) e^{i(\alpha_0 - \alpha_l)}}{Z} = \sum_n c_n \left(\frac{\lambda_n}{\lambda_0}\right)^l,$$
$$(6.74)$$

where we have made use of the completeness of the left and right eigen-functions. (Note that since $\hat{P}$ is not symmetric both right ψ_n^R and left ψ_n^L eigenfunctions are need for the evaluation of correlation functions.) The λ_n in (6.74) are the eigenvalues ($|\lambda_n| \geq |\lambda_{n+1}|$ and $n = 0, 1, 2, \ldots$), and the constants

$$c_n = \int_{-\pi}^{\pi} da' \psi_0^L(\alpha') e^{i\alpha'} \psi_n^R(\alpha') \int_{-\pi}^{\pi} d\alpha \psi_n^L(\alpha) e^{-i\alpha} \psi_0^R(\alpha).$$

In the case where λ_1 is real and $|\lambda_1| > |\lambda_2|$, (6.74) simplifies for large l to

$$\langle e^{i(\alpha_0 - \alpha_l)} \rangle = c_0 + c_1 \left(\frac{\lambda_1}{\lambda_0} \right)^l, \quad |\lambda_1| > |\lambda_2|.$$

In the case where $\lambda_1 = \lambda_2^* = |\lambda_1| e^{i2\pi\Xi}$, (6.74) for large l is [16]

$$\langle e^{i(\alpha_0 - \alpha_l)} \rangle = c_0 + \left(c_1 e^{i2\pi\Xi l} + c_2 e^{-i2\pi\Xi l} \right) \left| \frac{\lambda_1}{\lambda_0} \right|^l, \quad \lambda_1 = \lambda_2^*.$$

There is no phase coherence between upper and lower branches of the ladder and hence no superconductivity in the transverse direction. In this case, we say that the α's are unpinned. If there exist finite intervals of α on which $\rho(\alpha) = 0$, there will be phase coherence between the upper and lower branches and we say that the α's are pinned. In term of the transfer matrix, the phase density is the product of the left and right eigenfunctions of λ_0 [GM79],

$$\rho(\alpha) = \psi_0^L(\alpha) \psi_0^R(\alpha).$$

We first discuss the case where $f < f_{c1}$. These are the *Meissner-states* in the sense that there are no vortices ($n_i = 0$) in the ladder. The ground state is simply $\alpha_i = 0$, $\gamma_j = \pi f$ and $\gamma_j' = -\pi f$, so that there is a global screening current $\pm J_x \sin \pi f$ in the upper and lower branches of the ladder [Kar84]. The phase density $\rho(\alpha) = \delta(\alpha)$. The properties of the Meissner state can be studied by expanding (6.71) around $\alpha_i = 0$,

$$H_M = (J/4) \sum_j [\cos(\pi f)(\alpha_{j-1} - \alpha_j)^2 + 2J_t \alpha_i^2].$$

The current conservation (6.72) becomes

$$\alpha_{j+1} = 2(1 + J_t / \cos \pi f) \alpha_j - \alpha_{j-1}. \tag{6.75}$$

---

[16] While the correlation length is given by $\xi = [\ln|\lambda_0/\lambda_1|]^{-1}$ the quantity $\Xi = Arg(\lambda_1)/2\pi$ determines the spatial periodicity of the state. By numerical calculation of λ_n, it is found that for f smaller than a critical value f_{c1} which depends on J_t, both λ_1 and λ_2 are real. These two eigenvalues become degenerate at f_{c1}, and then bifurcate into a complex conjugate pair [DT95].

Besides the ground state $\alpha_j = 0$, there are other two linearly independent solutions $\alpha_j = e^{\pm j/\xi_M}$ of (6.75) which describe collective fluctuations about the ground state, where

$$\frac{1}{\xi_M} = \ln\left[1 + \frac{J_t}{\cos \pi f} + \sqrt{\frac{2J_t}{\cos \pi f} + \left(\frac{J_t}{\cos \pi f}\right)^2}\right]. \qquad (6.76)$$

ξ_M is the low temperature correlation length for the Meissner state.[17] As f increases, the Meissner state becomes unstable to the formation of vortices. A vortex is constructed by patching the two solutions of (6.75) together using a matching condition. The energy ϵ_v of a single vortex is found to be [DT95]

$$\epsilon_v \approx [2 + (\pi^2/8)\tanh(1/2\xi_M)]\cos \pi f - (\pi + 1)\sin \pi f + 2J_t, \qquad (6.77)$$

for J_t close to one. The zero of ϵ_v determines f_{c1} which is in good agreement with the numerical result from the transfer matrix. For $f > f_{c1}$, ϵ_v is negative and vortices are spontaneously created. When vortices are far apart their interaction is caused only by the exponentially small overlap. The corresponding repulsion energy is of the order $J\exp(-l/\xi_M)$, where l is the distance between vortices. This leads to a free energy per plaquette of $F = \epsilon_v/l + J\exp(-l/\xi_M)/l$ [PTB86]. Minimizing this free energy as a function of l gives the vortex density for $f > f_{c1}$: $\langle n_j \rangle = l^{-1} = [\xi_M \ln|f_{c1} - f|]^{-1}$ where a linear approximation is used for f close to f_{c1}.

We now discuss the commensurate vortex states, taking the one with $\Xi = 1/2$ as an example. This state has many similarities to the Meissner state but some important differences. The ground state is

$$\alpha_0 = \arctan\left[\frac{2}{J_t}\sin(\pi f)\right], \qquad \alpha_1 = -\alpha_0, \qquad \alpha_{i\pm 2} = \alpha_i;$$

$$n_0 = 0, \qquad n_1 = 1, \qquad n_{i\pm 2} = n_i, \qquad (6.78)$$

so that there is a global screening current in the upper and lower branches of the ladder of $\pm 2\pi J(f - 1/2)/\sqrt{4 + J_t^2}$. The existence of the global screening, which is absent in an infinite 2D array, is the key reason for the existence of the steps at $\Xi = p/q$. It is easy to see that the symmetry of this vortex state is that of the (anti-ferromagnetic) Ising model. The ground state is two-fold degenerate. The low temperature excitations are domain boundaries between the two degenerate ground states. The energy of the domain boundary $J\epsilon_b$ can be estimated using similar methods to those used to derive (6.77) for the Meissner state. We found that $\epsilon_b = \epsilon_b^0 - (\pi^2/\sqrt{4 + J_t^2})|f - 1/2|$, where ϵ_b^0 depends only on

$$J_t = c\arctan^2(2/J_t)J_t^2\coth(1/\xi_b)/\sqrt{4 + J_t^2},$$

---

[17] Here, $\xi_M < 1$ for $J_t \sim 1$ making a continuum approximation invalid.

with c being a constant of order one and

$$\xi_b^{-1} = \ln(1 + J_t^2/2 + J_t\sqrt{1 + J_t^2/4}).$$

Thus the correlation length diverges with temperature as $\xi \sim \exp(2J\epsilon_b/k_B T)$. The transition from the $\varXi = 1/2$ state to nearby vortex states happens when f is such that $\epsilon_b = 0$; it is similar to the transition from the Meissner state to its nearby vortex states. All other steps $\varXi = p/q$ can be analyzed similarly. For comparison, we have evaluated ξ for various values of f and T from the transfer matrix and found that ξ fits $\xi \sim \exp(2J\epsilon_b/k_B T)$ (typically over several decades) at low temperature.

We now discuss the superconducting-normal transition in the transverse direction. For $J_t = 0$, the ground state has $\gamma_i = \gamma_i' = 0$ and

$$\alpha_j = 2\pi f j + \alpha_0 - 2\pi \sum_{i=0}^{i=j} n_i. \tag{6.79}$$

The average vortex density $\langle n_j \rangle$ is f; there is no screening of the magnetic field. α_0 in (6.79) is arbitrary; the α's are unpinned for all f. The system is simply two un-coupled 1D XY chains, so that the correlation length $\xi = 1/k_B T$. The system is superconducting at zero temperature along the ladder, but not in the transverse direction. As J_t rises above zero we observe a distinct difference between the system at rational and irrational values of f. For f rational, the α's become pinned for $J_t > 0$ ($\rho(\alpha)$ is a finite sum of delta functions) and the ladder is superconducting in *both* the longitudinal and transverse directions at zero temperature. The behavior for irrational f is illustrated in the following for the state with $\varXi = a_g$, where $a_g \approx 0.381966\ldots$ is one minus the inverse of the golden mean.

Finally, we consider critical currents along the ladder. One can get an estimate for the critical current by performing a perturbation expansion around the ground state (i.e., $\{n_j\}$ remain fixed) and imposing the current constraint of $\sin \gamma_j + \sin \gamma_j' = I$. Let $\delta\gamma_j$, $\delta\gamma_j'$ and $\delta\alpha_j$ be the change of γ_j, γ_j' and α_j in the current carrying state. One finds that stability of the ground state requires that $\delta\alpha_j = 0$, and consequently $\delta\gamma_j = \delta\gamma_j' = I/2\cos\gamma_j$. The critical current can be estimated by the requirement that the γ_j do not pass through $\pi/2$, which gives $I_c = 2(\pi/2 - \gamma_{\max})\cos\gamma_{\max}$, where $\gamma_{\max} = \max_j(\gamma_j)$. In all ground states we examined, commensurate and incommensurate, we found that $\gamma_{\max} < \pi/2$, implying a finite critical current for all f. See [DT95] for more details.

Underdamped JJL

Recall that the *discrete sine-Gordon equation* has been used by several groups to model so-called hybrid Josephson ladder arrays [UMM93, WSZ95]. Such an array consists of a ladder of parallel Josephson junctions which are inductively coupled together (e.g., by superconducting wires). The sine-Gordon equation then describes the phase differences across the junctions. In an applied magnetic field, this equation predicts remarkably complex behavior, including flux flow resistance below a certain critical current, and a field-independent resistance above that current arising from so-called *whirling modes* [WSZ95]. In the flux flow regime, the fluxons in this ladder propagate as localized solitons, and the IV characteristics exhibit voltage plateaus arising from the locking of solitons to linear *spin-wave modes*. At sufficiently large values of the anisotropy parameter η_J defined later, the solitons may propagate 'ballistically' on the plateaus, i.e., may travel a considerable distance even after the driving current is turned off.

Here, mainly following [RYD96], we show that this behavior is all found in a model in which the ladder is treated as a network of coupled small junctions arranged along both the edges and the rungs of the ladder. This model is often used to treat 2D Josephson networks, and includes *no* inductive coupling between junctions, other than that produced by the other junctions. To confirm our numerical results, we derive a discrete sine-Gordon equation from our coupled-network model. Thus, these seemingly quite different models produce nearly identical behavior for ladders. By extension, they suggest that some properties of 2D arrays might conceivably be treated by a similar simplification. In simulations [Bob92, GLW93, SIT95], underdamped arrays of this type show some similarities to ladder arrays, exhibiting the analogs of both voltage steps and whirling modes.

We consider a ladder consisting of coupled superconducting grains, the ith of which has order parameter

$$\Phi_i = \Phi_0 e^{i\theta_i}.$$

Grains i and j are coupled by *resistively-shunted Josephson junctions* (RSJ's) with current I_{ij}, shunt resistance R_{ij} and shunt capacitance C_{ij}, with periodic boundary conditions.

The phases θ_i evolve according to the coupled RSJ equations

$$\hbar\dot\theta_i/(2e) = V_i,$$
$$M_{ij}\dot V_j = I_i^{ext}/I_c - (R/R_{ij})(V_i - V_j) - (I_{ij}/I_c)\sin(\theta_{ij} - A_{ij}).$$

Here the time unit is $t_0 = \hbar/(2eRI_c)$, where R and I_c are the shunt resistance and critical current across a junction in the x-direction; I_i^{ext} is the external current fed into the ith node; the spatial distances are given in units of the lattice spacing a, and the voltage V_i in units of $I_c R$.

$$M_{ij} = -4\pi eCI_c R^2/h \quad \text{for } i \neq j, \quad \text{and}$$
$$M_{ii} = -\sum_{j\neq i} M_{ij},$$

where C is the intergrain capacitance. Finally,

$$A_{ij} = (2\pi/\Phi_0) \int_i^j A \cdot dl,$$

where A is the vector potential. Following [RYD96], we assume N plaquettes in the array, and postulate a current I uniformly injected into each node on the outer edge and extracted from each node on the inner edge of the ring. We also assume a uniform transverse magnetic field $B \equiv f\phi_0/a^2$, and use the *Landau gauge* $A = -Bx\hat{y}$.

We now show that this model reduces approximately to a discrete sine-Gordon equation for the *phase differences*. Label each grain by (x, y) where $x/a = 0, \ldots, N - 1$ and $y/a = 0, 1$. Subtracting the equation of motion for $\theta(x, 1)$ from that for $\theta(x, 2)$, and defining

$$\Psi(x) = \frac{1}{2}[\theta(x,1) + \theta(x,2)], \qquad \chi(x) = [\theta(x,2) - \theta(x,1)],$$

we get a differential equation for $\chi(x)$ which is second-order in time. This equation may be further simplified using the facts that $A_{x,y;x\pm1,y} = 0$ in the Landau gauge, and that $A_{x,1;x,2} = -A_{x,2;x,1}$, and by defining the *discrete Laplacian*

$$\chi(x + 1) - 2\chi(x) + \chi(x - 1) = \nabla^2 \chi(x).$$

Finally, using the boundary conditions,

$$I^{ext}(x, 2) = -I^{ext}(x, 1) \equiv I,$$

and introducing $\phi(x) = \chi(x) - A_{x,2;x,1}$, we get

$$[1 - \eta_c^2 \nabla^2]\beta\ddot{\phi} = i - [1 - \eta_r^2 \nabla^2]\dot{\phi} - \sin(\phi) + 2\eta_J^2$$
$$\times \sum_{i=\pm1} \cos\{\Psi(x) - \Psi(x+i)\}$$
$$\times \sin\{[\phi(x) - \phi(x+i)]/2\}, \tag{6.80}$$

where we have defined a dimensionless current $i = I/I_{cy}$, and anisotropy factors

$$2\eta_r^2 = R_y/R_x, \qquad 2\eta_c^2 = C_x/C_y, \qquad 2\eta_J^2 = I_{cx}/I_{cy}.$$

We now neglect all combined space and time derivatives of order three or higher. Similarly, we set the cosine factor equal to unity (this is also checked numerically to be valid *a posteriori*) and linearize the sine factor in the last term, so that the final summation can be expressed simply as $\nabla^2\phi$. With these approximations, (6.80) reduces to *discrete driven sine-Gordon equation with dissipation*:

$$\beta\ddot{\phi} + \dot{\phi} + \sin(\phi) - \eta_J^2 \nabla^2 \phi = i, \quad \text{where } \beta = 4\pi e I_{cy} R_y^2 C_y/h. \tag{6.81}$$

Soliton Behavior

In the absence of damping and driving, the continuum version of (6.81) has, among other solutions, the sine-Gordon soliton [Raj82], given by

$$\phi_s(x,t) \sim 4\tan^{-1}\left[\exp\left\{(x-v_vt)/\sqrt{\eta_J^2-\beta v_v^2,}\right\}\right]$$

where v_v is the velocity. The phase in this soliton rises from ~ 0 to $\sim 2\pi$ in a width $d_k \sim \sqrt{\eta_J^2 - \beta v_v^2}$.

The transition to the resistive state occurs at $n_{min} = 4, 2, 2, 1$ for $\eta_J^2 = 0.5, 1.25, 2.5, 5$. This can also be understood from the *kink-phason resonance* picture. To a phason mode, the passage of a kink of width d_k will appear like the switching on of a step-like driving current over a time of order d_k/v_v. The kink will couple to the phasons only if $d_k/v_v \geq \pi/\omega_1$, the half-period of the phason, or equivalently

$$\frac{1}{\sqrt{\beta}v_v} \geq \frac{\sqrt{1+\pi^2}}{\eta_J} = \frac{3.3}{\eta_J}.$$

This condition agrees very well with our numerical observations, even though it was obtained by considering soliton solutions from the continuum sine-Gordon equation.

The fact that the voltage in regime I is approximately linear in f can be qualitatively understood from the following argument. Suppose that ϕ for Nf fluxons can be approximated as a sum of well-separated solitons, each moving with the same velocity and described by

$$\phi(x,t) = \sum_{j=1}^{Nf} \phi_j, \quad \text{where } \phi_j = \phi_s(x-x_j,t).$$

Since the solitons are well separated, we can use following properties:

$$\sin\left[\sum_j \phi_j\right] = \sum_j \sin\phi_j \quad \text{and} \quad \int \dot{\phi}_j\dot{\phi}_i dx \propto \delta_{ij}.$$

By demanding that the energy dissipated by the damping of the moving soliton be balanced by that the driving current provides ($\propto \int dx i\dot{\phi}(x)$), one can show that the Nf fluxons should move with the same velocity v as that for a single fluxon driven by the same current. In the *whirling regime*, the *f-independence* of the voltage can be understood from a somewhat different argument. Here, we assume a periodic solution of the form

$$\phi = \sum_j^{Nf} \phi_w(x - \tilde{v}t - j/f),$$

moving with an unknown velocity $\tilde{v}$ where $\phi_w(\xi)$ describes a whirling solution containing one fluxon. Then using the property $\phi(x + m/f) = \phi(x) + 2\pi m$, one can show that [RYD96]

$$\sin\left[\sum_j^{Nf} \phi_w(x - \tilde{v}t - j/f)\right] = \sin[Nf\phi_w(x - \tilde{v}t)].$$

Finally, using the approximate property $\phi_w(\xi) \sim \xi$ of the whirling state, one finds $\tilde{v} = v/(Nf)$, leading to an f-independent voltage.

Ballistic Soliton Motion and Soliton Mass

A common feature of massive particles is their 'ballistic motion', defined as inertial propagation after the driving force has been turned off. Such propagation has been reported experimentally but as yet has not been observed numerically in either square or triangular lattices [GLW93]. In the so-called *flux-flow regime* at $\eta_J = 0.71$, we also find no ballistic propagation, presumably because of the large pinning energies produced by the periodic lattice.

We can define the fluxon mass in our ladder by equating the *charging energy* $E_c = C/2 \sum_{ij} V_{ij}^2$ to the kinetic energy of a soliton of mass M_v: $E_{kin} = \frac{1}{2}M_v v_v^2$ [GLW93]. Since E_c can be directly calculated in our simulation, while v_v can be calculated from $\langle V \rangle$, this gives an unambiguous way to determine M_v. For $\eta_J^2 = 0.5$, we find $E_c/C \sim 110(\langle V \rangle/I_c R)^2$, in the flux-flow regime. This gives $M_v^I \sim 3.4C\phi_0^2/a^2$, more than six times the usual estimate for the vortex mass in a 2D square lattice. Similarly, the vortex friction coefficient γ can be estimated by equating the rate of energy dissipation,

$$E_{dis} = 1/2 \sum_{ij} V_{ij}^2/R_{ij}, \quad \text{to} \quad \frac{1}{2}\gamma v_v^2.$$

This estimate yields $\gamma^I \sim 3.4\phi_0^2/(Ra^2)$, once again more than six times the value predicted for 2D arrays [GLW93]. This large dissipation explains the absence of ballistic motion for this anisotropy [GLW93]. At larger values $\eta_J^2 = 5$ and 2.5, a similar calculation gives $M_v^I \sim 0.28$ and $0.34\phi_0^2/(Ra^2)$, $\gamma^I \sim 0.28$ and $0.34\phi_0^2/(Ra^2)$. These lower values of γ^I, but especially the low pinning energies, may explain why ballistic motion is possible at these values of η_J. See [RYD96] for more details.

6.3.4 Synchronization in Arrays of Josephson Junctions

The *synchronization of coupled nonlinear oscillators* has been a fertile area of research for decades [PRK01]. In particular, *Winfree-type phase models* [Win67] have been extensively studied. In 1D, a generic version of this model for N oscillators reads

$$\dot{\theta}_j = \Omega_j + \sum_{k=1}^{N} \sigma_{j,k} \Gamma \left(\theta_k - \theta_j \right), \tag{6.82}$$

where θ_j is the phase of oscillator j, which can be envisioned as a point moving around the unit circle with angular velocity $\dot{\theta}_j = d\theta_j/dt$. In the absence of coupling, this overdamped oscillator has an angular velocity Ω_j. $\Gamma(\theta_k - \theta_j)$ is the coupling function, and $\sigma_{j,k}$ describes the range and nature (e.g., attractive or repulsive) of the coupling. The special case

$$\Gamma(\theta_k - \theta_j) = \sin(\theta_k - \theta_j), \qquad \sigma_{j,k} = \alpha/N, \qquad \alpha = \text{const},$$

corresponds to the uniform, sinusoidal coupling of each oscillator to the remaining $N-1$ oscillators. This mean-field system is usually called the *globally-coupled Kuramoto model* (GKM). Kuramoto was the first to show that for this particular form of coupling and in the $N \rightarrow \infty$ limit, there is a continuous dynamical phase transition at a critical value of the coupling strength α_c and that for $\alpha > \alpha_c$ both phase and frequency synchronization appear in the system [Kur84, Str00]. If $\sigma_{j,k} = \alpha\delta_{j,k\pm1}$ while the coupling function retains the form $\Gamma(\theta_j - \theta_k) = \sin(\theta_k - \theta_j)$, then we have the so-called *locally-coupled Kuramoto model* (LKM), in which each oscillator is coupled only to its nearest neighbors. Studies of synchronization in the LKM [SSK87], including extensions to more than one spatial dimension, have shown that α_c grows without bound in the $N \rightarrow \infty$ limit [Sm88].

Watts and Strogatz introduced a simple model for tuning collections of coupled dynamical systems between the two extremes of random and regular networks [WS98]. In this model, connections between nodes in a regular array are randomly rewired with a probability p, such that $p = 0$ means the network is regularly connected, while $p = 1$ results in a random connection of nodes. For a range of intermediate values of p between these two extremes, the network retains a property of regular networks (a large clustering coefficient) and also acquires a property of random networks (a short characteristic path length between nodes). Networks in this intermediate configuration are termed *small-world networks*. Many examples of such small worlds, both natural and human-made, have been discussed [Str]. Not surprisingly, there has been much interest in the synchronization of dynamical systems connected in a small-world geometry [BP02, NML03]. Generically, such studies have shown that the presence of small-world connections make it easier for a network to synchronize, an effect generally attributed to the reduced path length between the linked systems. This has also been found to be true for the special case in which the dynamics of each oscillator is described by a Kuramoto model [HCK02a, HCK02b].

As an example of *physically-controllable systems of nonlinear oscillators*, which can be studied both theoretically and experimentally, Josephson junction (JJ) arrays are almost without peer. Through modern fabrication techniques and careful experimental methods one can attain a high degree of control over the dynamics of a JJ array, and many detailed aspects of array behavior have been studied [NLG00]. Among the many different geometries of JJ arrays, *ladder* arrays deserve special attention. For example, they have been observed to support stable time-dependent, spatially-localized states known as *discrete breathers* [TMO00]. In addition, the ladder geometry is more complex than that of better understood serial arrays but less so than fully two-dimensional (2D) arrays. In fact, a ladder can be considered as a special kind of 2D array, and so the study of ladders could throw some light on the behavior of such 2D arrays. Also, linearly-stable synchronization of the horizontal, or rung, junctions in a ladder is observed in the absence of a load over a wide range of dc bias currents and junction parameters (such as junction capacitance), so that synchronization in this geometry appears to be robust [TSS05].

In the mid 1990's it was shown that a serial array of zero-capacitance, i.e., overdamped, junctions coupled to a load could be mapped onto the GKM [WCS96, WCS98]. The load in this case was essential in providing an all-to-all coupling among the junctions. The result was based on an averaging process, in which (at least) two distinct time scales were identified: the 'short' time scale set by the rapid voltage oscillations of the junctions (the array was current biased above its critical current) and 'long' time scale over which the junctions synchronize their voltages. If the *resistively-shunted junction* (RSJ) equations describing the dynamics of the junctions are integrated over one cycle of the 'short' time scale, what remains is the 'slow' dynamics, describing the synchronization of the array. This mapping is useful because it allows knowledge about the GKM to be applied to understanding the dynamics of the serial JJ array. For example, the authors of [WCS96] were able, based on the GKM, to predict the level of critical current disorder the array could tolerate before frequency synchronization would be lost. Frequency synchronization, also described as entrainment, refers to the state of the array in which all junctions not in the zero-voltage state have equal (to within some numerical precision) time-averaged voltages: $(\hbar/2e)\langle\dot{\theta}_j\rangle_t$, where θ_j is the gauge-invariant phase difference across junction j. More recently, the 'slow' synchronization dynamics of finite-capacitance serial arrays of JJ's has also been studied [CS95, WS97]. Perhaps surprisingly, however, no experimental work on JJ arrays has verified the accuracy of this GKM mapping. Instead, the first detailed experimental verification of Kuramoto's theory was recently performed on systems of coupled electrochemical oscillators [KZH02].

Recently, [DDT03] showed, with an eye toward a better understanding of synchronization in 2D JJ arrays, that a ladder array of *overdamped junctions* could be mapped onto the LKM. This work was based on an averaging process, as in [WCS96], and was valid in the limits of weak critical current disorder (less than about 10%) and large dc bias currents, I_B, along the rung junctions $(I_B/\langle I_c\rangle \gtrsim 3$, where $\langle I_c\rangle$ is the arithmetic average of the critical currents of the rung junctions. The result demonstrated, for both open and periodic boundary conditions, that synchronization of the current-biased rung junctions in the ladder is well described by (6.82).

In this subsection, following [TSS05], we demonstrate that a ladder array of *underdamped junctions* can be mapped onto a second-order Winfree-type oscillator model of the form

$$a\ddot{\theta}_j + \dot{\theta}_j = \Omega_j + \sum_{k=1}^{N} \sigma_{j,k} \Gamma(\theta_k - \theta_j), \tag{6.83}$$

where a is a constant related to the average capacitance of the rung junctions. This result is based on the *resistively & capacitively-shunted junction* (RCSJ) model and a multiple time scale analysis of the classical equations for the array. Secondly, we study the effects of *small world* (SW) connections on the synchronization of both overdamped and underdamped ladder arrays. It appears that SW connections make it easier for the ladder to synchronize, and that a Kuramoto or Winfree type model (6.82) and (6.83), suitably generalized to include the new connections, accurately describes the synchronization of this ladder.

Phase Model for Underdamped JJL

Following [TSS05] we analyze synchronization in disordered Josephson junction arrays. The ladder geometry used consists of an array with $N = 8$ plaquettes, periodic boundary conditions, and uniform dc bias currents, I_B, along the rung junctions (see Fig. 6.9). The *gauge-invariant phase difference* across rung junction j is γ_j, while the phase difference across the off-rung junctions along the outer(inner) edge of plaquette j is $\psi_{1,j}(\psi_{2,j})$. The critical current, resistance, and capacitance of rung junction j are denoted I_{cj}, R_j, and C_j, respectively. For simplicity, we assume all off-rung junctions are identical, with critical current I_{co}, resistance R_o, and capacitance C_o. We also assume that the product of the junction critical current and resistance is the same for all junctions in the array [Ben95], with a similar assumption about the ratio of each junction's critical current with its capacitance:

$$I_{cj} R_j = I_{co} R_o = \frac{\langle I_c \rangle}{\langle R^{-1} \rangle}, \tag{6.84}$$

$$\frac{I_{cj}}{C_j} = \frac{I_{co}}{C_o} = \frac{\langle I_c \rangle}{\langle C \rangle}, \tag{6.85}$$

where for any generic quantity X, the angular brackets with no subscript denote an arithmetic average over the set of rung junctions,

$$\langle X \rangle \equiv (1/N) \sum_{j=1}^{N} X_j.$$

For convenience, we work with dimensionless quantities. Our dimensionless time variable is

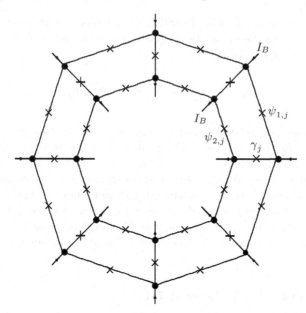

Fig. 6.9. A ladder array of Josephson junctions with periodic boundary conditions and $N = 8$ plaquettes. A uniform, dc bias current I_B is inserted into and extracted from each rung as shown. The gauge-invariant phase difference across the rung junctions is denoted by γ_j where $1 \le j \le N$, while the corresponding quantities for the off-rung junctions along the outer(inner) edge are $\psi_{1,j} (\psi_{2,j})$ (see text for explanation).

$$\tau \equiv \frac{t}{t_c} = \frac{2e\langle I_c \rangle t}{\hbar \langle R^{-1} \rangle}, \tag{6.86}$$

where t is the ordinary time. In the following, derivatives with respect to τ will be denoted by *prime* (e.g., $\psi' = d\psi/d\tau$). The dimensionless bias current is

$$i_B \equiv \frac{I_B}{\langle I_c \rangle}, \tag{6.87}$$

while the dimensionless critical current of rung junction j is $i_{cj} \equiv I_{cj}/\langle I_c \rangle$. The McCumber parameter in this case is

$$\beta_c \equiv \frac{2e\langle I_c \rangle \langle C \rangle}{\hbar \langle R^{-1} \rangle^2}. \tag{6.88}$$

Note that β_c is proportional to the mean capacitance of the rung junctions. An important dimensionless parameter is

$$\alpha \equiv \frac{I_{co}}{\langle I_c \rangle}, \tag{6.89}$$

which will effectively tune the nearest-neighbor interaction strength in our phase model for the ladder.

Conservation of charge applied to the superconducting islands on the outer and inner edge, respectively, of rung junction j yields the following equations in dimensionless variables [TSS05]:

$$i_B - i_{cj}\sin\gamma_j - i_{cj}\gamma_j' - i_{cj}\beta_c\gamma_j'' - \alpha\sin\psi_{1,j} - \alpha\psi_{1,j}'$$
$$- \alpha\beta_c\psi_{1,j}'' + \alpha\sin\psi_{1,j-1} + \alpha\psi_{1,j-1}' + \alpha\beta_c\psi_{1,j-1}'' = 0, \qquad (6.90)$$
$$-i_B + i_{cj}\sin\gamma_j + i_{cj}\gamma_j' + i_{cj}\beta_c\gamma_j'' - \alpha\sin\psi_{2,j} - \alpha\psi_{2,j}'$$
$$- \alpha\beta_c\psi_{2,j}'' + \alpha\sin\psi_{2,j-1} + \alpha\psi_{2,j-1}' + \alpha\beta_c\psi_{2,j-1}'' = 0, \qquad (6.91)$$

where $1 \leq j \leq N$. The result is a set of $2N$ equations in $3N$ unknowns: γ_j, $\psi_{1,j}$, and $\psi_{2,j}$. We supplement (6.91) by the constraint of fluxoid quantization in the absence of external or induced magnetic flux. For plaquette j this constraint yields the relationship

$$\gamma_j + \psi_{2,j} - \gamma_{j+1} - \psi_{1,j} = 0. \qquad (6.92)$$

Equations (6.91) and (6.92) can be solved numerically for the $3N$ phases γ_j, $\psi_{1,j}$ and $\psi_{2,j}$ [TSS05].

We assign the rung junction critical currents in one of two ways, randomly or nonrandomly. We generate random critical currents according to a parabolic *probability distribution function* (PDF) of the form

$$P(i_c) = \frac{3}{4\Delta^3}\left[\Delta^2 - (i_c - 1)^2\right], \qquad (6.93)$$

where $i_c = I_c/\langle I_c \rangle$ represents a scaled critical current, and Δ determines the spread of the critical currents. Equation (6.93) results in critical currents in the range $1 - \Delta \leq i_c \leq 1 + \Delta$. Note that this choice for the PDF (also used in [WCS96]) avoids extreme critical currents (relative to a mean value of unity) that are occasionally generated by PDF's with tails. The nonrandom method of assigning rung junction critical currents was based on the expression (with $1 \leq j \leq N$)

$$i_{cj} = 1 + \Delta - \frac{2\Delta}{(N-1)^2}\left[4j^2 - 4(N+1)j + (N+1)^2\right], \qquad (6.94)$$

which results in the i_{cj} values varying quadratically as a function of position along the ladder and falling within the range $1 - \Delta \leq i_{cj} \leq 1 + \Delta$. We usually use $\Delta = 0.05$.

Multiple Time-Scale Analysis

Now, our goal is to derive a Kuramoto-like model for the phase differences across the rung junctions, γ_j, starting with (6.91). We begin with two reasonable assumptions. First, we assume there is a simple phase relationship between the two off-rung junctions in the same plaquette [TSS05]:

$$\psi_{2,j} = -\psi_{1,j}, \tag{6.95}$$

the validity of which has been discussed in detail in [DDT03, FW95]. As a result, (6.92) reduces to

$$\psi_{1,j} = \frac{\gamma_j - \gamma_{j+1}}{2}, \tag{6.96}$$

which implies that (6.90) can be written as

$$i_{cj}\beta_c\gamma_j'' + i_{cj}\gamma_j' + \frac{\alpha\beta_c}{2}\left[\gamma_{j+1}'' - 2\gamma_j'' + \gamma_{j-1}''\right] + \frac{\alpha}{2}\left[\gamma_{j+1}' - 2\gamma_j' + \gamma_{j-1}'\right]$$
$$= i_B - i_{cj}\sin\gamma_j + \alpha\sum_{\delta=\pm1}\sin\left(\frac{\gamma_{j+\delta} - \gamma_j}{2}\right). \tag{6.97}$$

Our second assumption is that we can neglect the discrete Laplacian terms in (6.97), namely

$$\nabla^2\gamma_j' \equiv \gamma_{j+1}' - 2\gamma_j' + \gamma_{j-1}' \quad \text{and} \quad \nabla^2\gamma_j'' \equiv \gamma_{j+1}'' - 2\gamma_j'' + \gamma_{j-1}''.$$

We find numerically, over a wide range of bias currents i_B, *McCumber parameters* β_c, and coupling strengths α that $\nabla^2\gamma_j'$ and $\nabla^2\gamma_j''$ oscillate with a time-averaged value of approximately zero. Since the multiple time scale method is similar to averaging over a fast time scale, it seems reasonable to drop these terms. In light of this assumption, (6.97) becomes

$$i_{cj}\beta_c\gamma_j'' + i_{cj}\gamma_j' = i_B - i_{cj}\sin\gamma_j + \alpha\sum_{\delta=\pm1}\sin\left(\frac{\gamma_{j+\delta} - \gamma_j}{2}\right). \tag{6.98}$$

We can use (6.98) as the starting point for a multiple time scale analysis. Following [CS95] and [WS97], we divide (6.98) by i_B and define the following quantities:

$$\tilde{\tau} \equiv i_B\tau, \qquad \tilde{\beta}_c \equiv i_B\beta_c, \qquad \epsilon = 1/i_B. \tag{6.99}$$

In terms of these scaled quantities, (6.98) can be written as

$$i_{cj}\tilde{\beta}_c\frac{d^2\gamma_j}{d\tilde{\tau}^2} + i_{cj}\frac{d\gamma_j}{d\tilde{\tau}} + \epsilon i_{cj}\sin\gamma_j - \epsilon\alpha\sum_{\delta}\sin\left(\frac{\gamma_{j+\delta} - \gamma_j}{2}\right) = 1. \tag{6.100}$$

Next, we introduce a series of four (dimensionless) time scales,

$$T_n \equiv \epsilon^n\tilde{\tau}, \quad (n = 0, 1, 2, 3), \tag{6.101}$$

which are assumed to be independent of each other. Note that $0 < \epsilon < 1$ since $\epsilon = 1/i_B$. We can think of each successive time scale, T_n, as being 'slower' than the scale before it. For example, T_2 describes a slower time scale than T_1. The time derivatives in 6.100 can be written in terms of the new time scales, since we can think of $\tilde{\tau}$ as being a function of the four independent

T_n's, $\tilde{\tau} = \tilde{\tau}(T_0, T_1, T_2, T_3)$. Letting $\partial_n \equiv \partial/\partial T_n$, the first and second time derivatives can be written as [TSS05]

$$\frac{d}{d\tilde{\tau}} = \partial_0 + \epsilon\partial_1 + \epsilon^2\partial_2 + \epsilon^3\partial_3 \qquad (6.102)$$

$$\frac{d^2}{d\tilde{\tau}^2} = \partial_0^2 + 2\epsilon\partial_0\partial_1 + \epsilon^2\left(2\partial_0\partial_2 + \partial_1^2\right) + 2\epsilon^3\left(\partial_0\partial_3 + \partial_1\partial_2\right), \qquad (6.103)$$

where in (6.103) we have dropped terms of order ϵ^4 and higher.

Next, we expand the phase differences in an ϵ expansion

$$\gamma_j = \sum_{n=0}^{\infty} \epsilon^n \gamma_{n,j}(T_0, T_1, T_2, T_3). \qquad (6.104)$$

Substituting this expansion into (6.100) and collecting all terms of order ϵ^0 results in the expression

$$1 = i_{cj}\tilde{\beta}_c\partial_0^2\gamma_{0,j} + i_{cj}\partial_0\gamma_{0,j}, \qquad (6.105)$$

for which we find the solution

$$\gamma_{0,j} = \frac{T_0}{i_{cj}} + \phi_j(T_1, T_2, T_3), \qquad (6.106)$$

where we have ignored a transient term of the form $e^{-T_0/\tilde{\beta}_c}$, and where $\phi_j(T_i)$, $(i = 1, 2, 3)$ is assumed constant over the fastest time scale T_0. Note that the expression for $\gamma_{0,j}$ consists of a rapid phase rotation described by T_0/i_{cj} and slower-scale temporal variations, described by ϕ_j, on top of that overturning. In essence, the goal of this technique is to solve for the dynamical behavior of the slow phase variable, ϕ_j. The resulting differential equation for the ϕ_j is [TSS05]:

$$\beta_c\phi_j'' + \phi_j' = \Omega_j + K_j \sum_{\delta=\pm 1} \sin\left[\frac{\phi_{j+\delta} - \phi_j}{2}\right] + L_j \sum_{\delta=\pm 1} \sin\left[3\left(\frac{\phi_{j+\delta} - \phi_j}{2}\right)\right]$$

$$+ M_j \sum_{\delta=\pm 1} \left\{\cos\left[\frac{\phi_{j+\delta} - \phi_j}{2}\right] - \cos\left[3\left(\frac{\phi_{j+\delta} - \phi_j}{2}\right)\right]\right\}, \qquad (6.107)$$

where Ω_j is given by the expression (letting $x_j \equiv i_{cj}/i_B$ for convenience)

$$\Omega_j = \frac{1}{x_j}\left[1 - \frac{x_j^4}{\left(2\beta_c^2 + x_j^2\right)}\right], \qquad (6.108)$$

and the three coupling strengths are

$$K_j = \frac{\alpha}{i_{cj}}\left[1 + \frac{x_j^4\left(3x_j^2 + 23\beta_c^2\right)}{16\left(\beta_c^2 + x_j^2\right)^2}\right], \qquad (6.109)$$

$$L_j = \frac{\alpha}{i_{cj}} \frac{x_j^4 \left(3\beta_c^2 - x_j^2\right)}{16 \left(\beta_c^2 + x_j^2\right)^2}, \tag{6.110}$$

$$M_j = -\frac{\alpha}{i_{cj}} \frac{x_j^5 \beta_c}{4 \left(\beta_c^2 + x_j^2\right)^2}. \tag{6.111}$$

We emphasize that (6.107) is expressed in terms of the original, unscaled, time variable τ and McCumber parameter β_c.

We will generally consider bias current and junction capacitance values such that $x_j^2 \ll \beta_c^2$. In this limit, (6.109)–(6.111) can be approximated as follows [TSS05]:

$$K_j \to \frac{\alpha}{i_{cj}} \left[1 + \mathcal{O}\left(\frac{1}{i_B^4}\right)\right], \tag{6.112}$$

$$L_j \to \frac{\alpha}{i_{cj}} \left(\frac{3x_j^4}{16\beta_c^2}\right) \sim \mathcal{O}\left(\frac{1}{i_B^4}\right), \tag{6.113}$$

$$M_j \to -\frac{\alpha}{i_{cj}} \left(\frac{x_j^5}{4\beta_c^3}\right) \sim \mathcal{O}\left(\frac{1}{i_B^5}\right). \tag{6.114}$$

For large bias currents, it is reasonable to truncate (6.107) at $\mathcal{O}(1/i_B^3)$, which leaves

$$\beta_c \phi_j'' + \phi_j' = \Omega_j + \frac{\alpha}{i_{cj}} \sum_{\delta=\pm 1} \sin\left[\frac{\phi_{j+\delta} - \phi_j}{2}\right], \tag{6.115}$$

where all the cosine coupling terms and the third harmonic sine term have been dropped as a result of the truncation.

In the absence of any coupling between neighboring rung junctions ($\alpha = 0$) the solution to (6.115) is

$$\phi_j^{(\alpha=0)} = A + B e^{-\tau/\beta_c} + \Omega_j \tau,$$

where A and B are arbitrary constants. Ignoring the transient exponential term, we see that $d\phi_j^{(\alpha=0)}/d\tau = \Omega_j$, so we can think of Ω_j as the voltage across rung junction j in the un-coupled limit. Alternatively, Ω_j can be viewed as the angular velocity of the strongly-driven rotator in the un-coupled limit.

Equation (6.115) is our desired phase model for the rung junctions of the underdamped ladder [TSS05]. The result can be described as a locally-coupled Kuramoto model with a second-order time derivative (LKM2) and with junction coupling determined by α. In the context of systems of coupled rotators, the second derivative term is due to the non-negligible rotator inertia, whereas in the case of Josephson junctions the second derivative arises because of the junction capacitance. The *globally-coupled* version of the second-order Kuramoto model (GKM2) has been well studied; in this case the oscillator inertia leads to a first-order synchronization phase transition as well as to hysteresis between a weakly and a strongly coherent synchronized state [TLO97b, ABS00].

Comparison of LKM2 and RCSJ Models

We now compare the synchronization behavior of the RCSJ ladder array with the LKM2. We consider frequency and phase synchronization separately. For the rung junctions of the ladder, frequency synchronization occurs when the time average voltages, $\langle v_j \rangle_\tau = \langle \phi'_j \rangle_\tau$ are equal for all N junctions, within some specified precision. In the language of coupled rotators, this corresponds to phase points moving around the unit circle with the same average angular velocity. We quantify the degree of frequency synchronization via an 'order parameter' [TSS05]

$$f = 1 - \frac{s_v(\alpha)}{s_v(0)}, \tag{6.116}$$

where $s_v(\alpha)$ is the standard deviation of the N time-average voltages, $\langle v_j \rangle_\tau$:

$$s_v(\alpha) = \sqrt{\frac{\sum_{j=1}^{N}(\langle v_j \rangle_\tau - \frac{1}{N}\sum_{k=1}^{N}\langle v_k \rangle_\tau)^2}{N-1}} \tag{6.117}$$

In general, this standard deviation will be a function of the coupling strength α, so $s_v(0)$ is a measure of the spread of the $\langle v_j \rangle_\tau$ values for N independent junctions. Frequency synchronization of all N junctions is signaled by $f = 1$, while $f = 0$ means all N average voltages have their un-coupled values.

Phase synchronization of the rung junctions is measured by the usual *Kuramoto order parameter*

$$r \equiv \frac{1}{N}\sum_{j=1}^{N}e^{i\phi_j}. \tag{6.118}$$

Lastly in this subsection, we address the issue of the linear stability of the frequency synchronized states ($\alpha > \alpha_c$) by calculating their *Floquet exponents* numerically for the RCSJ model as well as analytically based on the LKM2, (6.115). The analytic technique used has been described in [TM01], giving as a result for the real part of the Floquet exponents:

$$\mathrm{Re}(\lambda_m t_c) = -\frac{1}{2\beta_c}\left[1 \pm \mathrm{Re}\sqrt{1 - 4\beta_c\left(\bar{K} + 3\bar{L}\right)\omega_m^2}\right], \tag{6.119}$$

where stable solutions correspond to exponents, λ_m, with a negative real part. One can think of the ω_m as the normal mode frequencies of the ladder. We find that for a ladder with periodic boundary conditions and N plaquettes

$$\omega_m^2 = \frac{4\sin^2(\frac{m\pi}{N})}{1 + 2\sin^2(\frac{m\pi}{N})}, \quad 0 \le m \le N-1. \tag{6.120}$$

To arrive at (6.119) we have ignored the effects of disorder so that $\bar{K}$ and $\bar{L}$ are obtained from (6.109) and (6.110) with the substitution $i_{cj} \to 1$ throughout.

This should be reasonable for the levels of disorder we have considered (5%). Substituting the expressions for $\bar{K}$ and $\bar{L}$ into 6.119 results in [TSS05]

$$\mathrm{Re}(\lambda_m t_c) = -\frac{1}{2\beta_c}\left[1 \pm \mathrm{Re}\sqrt{1 - 2\beta_c\alpha\left\{1 + \frac{2\beta_c^2}{(i_B^2\beta_c^2 + 1)^2}\right\}\omega_m^2}\right]. \quad (6.121)$$

We are most interested in the Floquet exponent of minimum magnitude, $\mathrm{Re}(\lambda_{\min}t_c)$, which essentially gives the lifetime of the longest-lived perturbations to the synchronized state.

'Small-World' Connections in JJL Arrays

Many properties of small world networks have been studied in the last several years, including not only the effects of network topology but also the dynamics of the node elements comprising the network [New00, Str]. Of particular interest has been the ability of oscillators to synchronize when configured in a small-world manner. Such synchronization studies can be broadly sorted into several categories [TSS05]:

(1) Work on coupled lattice maps has demonstrated that synchronization is made easier by the presence of random, *long-range connections* [GH00, BPV03].

(2) Much attention has been given to the synchronization of continuous time dynamical systems, including the first order *locally-coupled Kuramoto model* (LKM), in the presence of small-world connections [HCK02a, HCK02b, Wat99]. For example, Hong and coworkers [HCK02a, HCK02b] have shown that the LKM, which does not exhibit a true dynamical phase transition in the thermodynamic limit ($N \to \infty$) in the *pristine* case, does exhibit such a phase synchronization transition for even a small number of shortcuts. But the assertion [WC02] that any small world network can synchronize for a given coupling strength and large enough number of nodes, even when the pristine network would not synchronize under the same conditions, is not fully accepted [BP02].

(3) More general studies of synchronization in small world and scale-free networks [BP02, NML03] have shown that the small world topology does not guarantee that a network can synchronize. In [BP02] it was shown that one could calculate the average number of shortcuts per node, s_{sync}, required for a given dynamical system to synchronize. This study found no clear relation between this synchronization threshold and the onset of the small world region, i.e., the value of s such that the average path length between all pairs of nodes in the array is less than some threshold value. [NML03] studied arrays with a power-law distribution of node connectivities (scale-free networks) and found that a broader distribution of connectivities makes a network *less* synchronizable even though the average path length is smaller. It was argued that this behavior was caused

by an increased number of connections on the hubs of the scale-free network. Clearly it is dangerous to assume that merely reducing the average path length between nodes of an array will make such an array easier to synchronize.

Now, regarding Josephson-junction arrays, if we have a disordered array biased such that some subset of the junctions are in the voltage state, i.e., undergoing limit cycle oscillations, the question is will the addition of random, long-range connections between junctions aid the array in attaining frequency and/or phase synchronization? Can we address this question by using the mapping discussed above between the RCSJ model for the *underdamped ladder array* and the second-order, locally-coupled Kuramoto model (LKM2). Based on the results of [DDT03], we also know that the RSJ model for an *overdamped ladder* can be mapped onto a first-order, locally-coupled Kuramoto model (LKM). Because of this mapping, the ladder array falls into category (2) of the previous paragraph. In other words, we should expect the existence of shortcuts to drastically improve the ability of ladder arrays to synchronize [TSS05].

We add connections between pairs of rung junctions that will result in interactions that are longer than nearest neighbor in range. We do so by adding two, nondisordered, off-rung junctions for each such connection. We argue that the RCSJ equations for the underdamped junctions in the ladder array can be mapped onto a straightforward variation of (6.115), in which the sinusoidal coupling term for rung junction j also includes the longer-range couplings due to the added shortcuts. Imagine a ladder with a shortcut between junctions j and l, where $l \neq j, j \pm 1$. Conservation of charge applied to the two superconducting islands that comprise rung junction j will lead to equations very similar to (6.91). For example, the analog to (6.90) will be

$$
\begin{aligned}
&i_B - i_{cj}\sin\gamma_j - i_{cj}\gamma_j' - \beta_c i_{cj}\gamma_j'' - \alpha\sin\psi_{1,j} - \alpha\psi_{1,j}' \\
&- \beta_c\alpha\psi_{1,j}'' + \alpha\sin\psi_{1,j-1} + \alpha\psi_{1,j-1}' + \beta_c\alpha\psi_{1,j-1}'' \\
&+ \sum_l \left[\alpha\sin\psi_{1;jl} + \alpha\psi_{1;jl}' + \beta_c\alpha\psi_{1;jl}''\right] = 0,
\end{aligned}
$$

with an analogous equation corresponding to the inner superconducting island that can be generalized from (6.91). The sum over the index l accounts for all junctions connected to junction j via an added shortcut. Fluxoid quantization still holds, which means that we can augment (6.92) with

$$
\gamma_j + \psi_{2;jl} - \gamma_l - \psi_{1;jl} = 0. \tag{6.122}
$$

We also assume the analog of (6.95) holds:

$$
\psi_{2;jl} = -\psi_{1;jl}. \tag{6.123}
$$

Equations (6.122) and (6.123) allow us to write the analog to (6.96) for the case of shortcut junctions:

$$\psi_{1;jl} = \frac{\gamma_j - \gamma_l}{2} \tag{6.124}$$

Equation (6.122), in light of (6.124), can be written as

$$i_B - i_{cj}\sin\gamma_j - i_{cj}\gamma_j' - \beta_c i_{cj}\gamma_j'' + \alpha \sum_{\delta=\pm1}\sin\left(\frac{\gamma_{j+\delta} - \gamma_j}{2}\right)$$

$$+ \alpha \sum_l \sin\left(\frac{\gamma_j - \gamma_l}{2}\right) + \frac{\alpha}{2}\nabla^2\gamma_j' + \frac{\alpha}{2}\nabla^2\gamma_j''$$

$$+ \frac{\alpha}{2}\sum_l(\gamma_j' - \gamma_l') + \frac{\alpha}{2}\sum_l(\gamma_j'' - \gamma_l'') = 0,$$

where the sums $\sum_l$ are over all rung junctions connected to j via an added shortcut. As we did with the pristine ladder, we will drop the two discrete Laplacians, since they have a very small time average compared to the terms $i_{cj}\gamma_j' + i_{cj}\beta_c\gamma_j''$. The same is also true, however, of the terms $\alpha/2\sum_l(\gamma_j' - \gamma_l')$ and $\alpha/2\sum_l(\gamma_j'' - \gamma_l'')$, as direct numerical solution of the full RCSJ equations in the presence of shortcuts demonstrates. So we shall drop these terms as well. Then we have

$$i_B - i_{cj}\sin\gamma_j - i_{cj}\gamma_j' - \beta_c i_{cj}\gamma_j'' + \frac{\alpha}{2}\sum_{k\in\Lambda_j}\sin\left(\frac{\gamma_k - \gamma_j}{2}\right), \tag{6.125}$$

where the sum is over all junctions in Λ_j, which is the set of all junctions connected to junction j. From above results we can predict that a multiple time scale analysis of (6.125) results in a phase model of the form

$$\beta_c\frac{d^2\phi_j}{d\tau^2} + \frac{d\phi_j}{d\tau} = \Omega_j + \frac{\alpha}{2}\sum_{k\in\Lambda_j}\sin\left(\frac{\phi_k - \phi_j}{2}\right), \tag{6.126}$$

where Ω_j is give by (6.108). A similar analysis for the *overdamped ladder* leads to the result

$$\phi_j' = \Omega_j^{(1)} + \frac{\alpha}{2}\sum_{k\in\Lambda_j}\sin\left(\frac{\phi_k - \phi_j}{2}\right), \tag{6.127}$$

where the time-averaged voltage across each overdamped rung junction in the un-coupled limit is

$$\Omega_j^{(1)} = \sqrt{\left(\frac{i_B}{i_{cj}}\right)^2 - 1}. \tag{6.128}$$

Although the addition of shortcuts makes it easier for the array to synchronize, we should also consider the effects of such random connections on the stability of the synchronized state. The Floquet exponents for the synchronized state allow us to quantify this stability. Using a general technique discussed in [PC98], we can calculate the Floquet exponents λ_m for the LKM based on the expression

$$\lambda_m t_c = \alpha E_m^G, \tag{6.129}$$

where E_m^G are the eigenvalues of $\mathbf{G}$, the matrix of coupling coefficients for the array. A specific element, G_{ij}, of this matrix is unity if there is a connection between rung junctions i and j. The diagonal terms, G_{ii}, is merely the negative of the number of junctions connected to junction i. This gives the matrix the property $\sum_j G_{ij} = 0$. In the case of the pristine ladder, the eigenvalues of $\mathbf{G}$ can be calculated analytically, which yields Floquet exponents of the form

$$\lambda_m^{(p=0)} t_c = -4\alpha \sin^2 \left(\frac{m\pi}{N} \right). \tag{6.130}$$

See [TSS05] for more details.

6.4 Topological Machinery for Quantum Computation

In this section, following [NSS08], we review recent advances in topological quantum computation, which has emerged as one of the most exciting approaches to constructing a fault-tolerant quantum computer. For 'softer' review, see Introduction, as well as Sect. 3.2.6 above. The proposal relies on the existence of topological states of matter whose quasi-particle excitations are neither bosons nor fermions, but are particles known as *non-Abelian anyons*, meaning that they obey *non-Abelian braiding statistics*. Quantum information is stored in states with multiple quasi-particles, which have a topological degeneracy. The unitary gate operations which are necessary for quantum computation are carried out by braiding quasi-particles, and then measuring the multi-quasi-particle states. The fault-tolerance of a topological quantum computer arises from the non-local encoding of the states of the quasi-particles, which makes them immune to errors caused by local perturbations. To date, the only such topological states thought to have been found in nature are fractional quantum Hall states, most prominently the $\nu = 5/2$ state, although several other prospective candidates have been proposed in systems as disparate as ultra-cold atoms in optical lattices and thin film superconductors. In this review article, we describe current research in this field, focusing on the general theoretical concepts of *non-Abelian statistics* as it relates to topological quantum computation, on understanding *non-Abelian quantum Hall states*, on proposed experiments to detect non-Abelian anyons, and on proposed architectures for a topological quantum computer. We address both the mathematical underpinnings of topological quantum computation and the physics of the subject using the $\nu = 5/2$ fractional quantum Hall state as the archetype of a non-Abelian topological state enabling *fault-tolerant quantum computation*.

In recent years, physicists' understanding of the quantum properties of matter has undergone a major revolution precipitated by surprising experimental discoveries and profound theoretical revelations. Landmarks include the discoveries of the *fractional quantum Hall effect* and *high-temperature superconductivity* and the advent of *topological quantum field theories*. At the same time, new potential applications for quantum matter burst on the scene, punctuated by the discoveries of *Shor's factorization algorithm* and *quantum error correction* protocols. There has been a convergence between these developments. Nowhere is this more dramatic than in topological quantum computation, which seeks to exploit the emergent properties of many-particle systems to encode and manipulate quantum information in a manner which is resistant to error [NSS08].

It is rare for a new scientific paradigm, with its attendant concepts and mathematical formalism, to develop in parallel with potential applications, with all of their detailed technical issues. However, the physics of topological phases of matter is not only evolving alongside topological quantum computation but is even informed by it. Therefore, this review must necessarily be rather sweeping in scope, simply to introduce the concepts of non-Abelian anyons and topological quantum computation, their inter-connections, and how they may be realized in physical systems, particularly in several fractional quantum Hall states. (For a popular account, see [Col06]; for a more technical one, see [DFN06].) This exposition will take us on a tour extending from knot theory and topological quantum field theory to conformal field theory and the quantum Hall effect to quantum computation and all the way to the physics of gallium-arsenide devices.

6.4.1 Non-Abelian Anyons

Non-Abelian Braiding Statistics

Recall that *quantum statistics* is one of the basic pillars of the quantum-mechanical view of the world. It is the property which distinguishes *fermions* from *bosons*: the wave function that describes a system of many identical particles should satisfy the proper symmetry under the interchange of any two particles. In 3 spatial dimension and one time dimension, or $(3+1)$D, there are only two possible symmetries: the wave function of bosons is symmetric under exchange while that of fermions is anti-symmetric. One cannot overemphasize, the importance of the symmetry of the wave-function, which is the root of the *Pauli principle*, super-fluidity, the metallic state, *Bose–Einstein condensation*, and a long list of other phenomena [NSS08].

The limitation to one of two possible types of quantum symmetry originates from the observation that a process in which two particles are adiabatically interchanged twice is equivalent to a process in which one of the particles is adiabatically taken around the other. Since, in three dimensions, wrapping one particle all the way around another is topologically equivalent to a process in which none of the particles move at all, the wave function should be left unchanged by two such interchanges of particles. The only two possibilities are for the wave-function to change by a $\pm$ sign under a single interchange, corresponding to the cases of bosons and fermions, respectively.

We can recast this in the *path-integral* language (see Sect. 3.3 above, as well as [II08a, II08b]). Suppose we consider all possible trajectories in $(3+1)$D which take N particles from initial positions $R_1, R_2, \ldots, R_N$ at time t_i to final positions $R_1, R_2, \ldots, R_N$ at time t_f. If the particles are distinguishable, then there are no topologically non-trivial trajectories, i.e., all trajectories can be continuously deformed into the trajectory in which the particles do not move at all (straight lines in the time direction). If the particles are indistinguishable, then the different trajectories fall into topological classes corresponding to the elements of the *permutation group* S_N, with each element of the group specifying how the initial positions are permuted to obtain the final positions. To define the quantum evolution of such a system, we must specify how the permutation group acts on the states of the system. Fermions and bosons correspond to the only two 1D *irreducible representations* of the permutation group of N identical particles.

Two and one spatial dimensions are qualitatively different from three in that respect. In 1D systems an *adiabatic exchange*[18] of two particles is impossible without the particles passing through one another. Thus, when particles interact through a hard core interaction, bosons become equivalent to fermions.

2D systems are qualitatively different from 3D (and higher dimensions) in this respect. A particle loop that encircles another particle in two dimensions cannot be deformed to a point without cutting through the other particle. Consequently, the notion of a winding of one particle around another in two dimensions is well-defined. Then, when two particles are interchanged twice in a clockwise manner, their trajectory involves a non-trivial winding, and the system does not necessarily come back to the same state. This topological difference between two and three dimensions, first realized by [LM77, Wil82a], leads to a profound difference in the possible quantum-mechanical properties,

---

[18] An *adiabatic invariant* is a property of a physical system which stays constant when changes are made slowly. In particular, in thermodynamics, an *adiabatic process* is a change that occurs without *heat flow* and slowly compared to the time to reach *equilibrium*. In an adiabatic process, the system is in equilibrium at all stages. Under these conditions the *entropy* is constant. On the other hand, in mechanics, an adiabatic change is a slow deformation of the Hamiltonian where the fractional rate of change of the energy is much slower than the orbital frequency. The area enclosed by the different motions in phase space are then the adiabatic invariants. Finally, in quantum mechanics, an adiabatic change is one that occurs at a rate much slower than the difference in frequency between energy eigenstates. In this case, the energy states of the system do not make transitions, so that the quantum number is an adiabatic invariant. (Recall that the old quantum theory was formulated by equating the quantum number of a system with its classical adiabatic invariant. This determined the form of the *Bohr–Sommerfeld quantization* rule: the quantum number is the area in phase space of the classical orbit.)

at least as a matter of principle, for quantum systems when particles are confined to $(2+1)D$[19] (see also [GMS81, Wu84]).

Suppose that we have two identical particles in two dimensions. Then when one particle is exchanged in a counter-clockwise manner with the other, the wave-function can change by an arbitrary phase,

$$\psi(\mathbf{r_1}, \mathbf{r_2}) \rightarrow e^{i\theta} \psi(\mathbf{r_1}, \mathbf{r_2}). \tag{6.131}$$

The phase need not be merely a $\pm$ sign because a second counter-clockwise exchange need not lead back to the initial state but can result in a non-trivial phase:

$$\psi(\mathbf{r_1}, \mathbf{r_2}) \rightarrow e^{2i\theta} \psi(\mathbf{r_1}, \mathbf{r_2}). \tag{6.132}$$

The special cases $\theta = 0, \pi$ correspond to bosons and fermions, respectively. Recall from Introduction that particles with other values of the 'statistical angle' θ are called *anyons* (see e.g., [Wil90]). We will often refer to such particles as anyons with statistics θ.

Let us now consider the general case of N particles, where a more complex structure arises. The topological classes of trajectories which take these particles from initial positions $R_1, R_2, \ldots, R_N$ at time t_i to final positions $R_1, R_2, \ldots, R_N$ at time t_f are in one-to-one correspondence with the elements of the braid group $\mathcal{B}_N$. An element of the braid group can be visualized by thinking of trajectories of particles as world-lines (or strands) in $(2+1)D$ space-time originating at initial positions and terminating at final positions, as shown in Fig. 1.4. The time direction will be represented vertically on the page, with the initial time at the bottom and the final time at the top. An element of the N-particle braid group is an equivalence class of such trajectories up to smooth deformations. To represent an element of a class, we will draw the trajectories on paper with the initial and final points ordered along lines at the initial and final times. When drawing the trajectories, we must be careful to distinguish when one strand passes over or under another, corresponding to a clockwise or counter-clockwise exchange. We also require that any intermediate time slice must intersect N strands. Strands cannot 'double back', which would amount to particle creation/annihilation at intermediate stages. We do not allow this because we assume that the particle number is known. Then, the multiplication of two elements of the braid group is simply the successive execution of the corresponding trajectories.

The *braid group* can be represented algebraically in terms of generators σ_i, with $1 \leq i \leq N-1$. We choose an arbitrary ordering of the particles $1, 2, \ldots, N$.[20] σ_i is a counter-clockwise exchange of the ith and $(i+1)$th par-

---

[19] Note that in $(1+1)D$, quantum statistics is not well-defined since particle interchange is impossible without one particle going through another, and bosons with hard-core repulsion are equivalent to fermions.

[20] Choosing a different ordering would amount to a relabeling of the elements of the braid group, as given by conjugation by the braid which transforms one ordering into the other.

ticles. σ_i^{-1} is, therefore, a clockwise exchange of the ith and $(i+1)^{\text{th}}$ particles. The σ_is satisfy the defining relations (see Fig. 1.4),

$$\sigma_i \sigma_j = \sigma_j \sigma_i, \quad \text{for } |i - j| \geq 2,$$
$$\sigma_{i+1} \sigma_i = \sigma_{i+1} \sigma_i \sigma_{i+1}, \quad \text{for } 1 \leq i \leq n - 1.$$

The only difference from the permutation group S_N is that $\sigma_i^2 \neq 1$, but this makes an enormous difference. While the permutation group is finite, the number of elements in the group $|S_N| = N!$ the braid group is infinite, even for just two particles. Furthermore, there are non-trivial topological classes of trajectories even when the particles are distinguishable, e.g., in the two-particle case those trajectories in which one particle winds around the other an integer number of times. These topological classes correspond to the elements of the 'pure' braid group, which is the subgroup of the braid group containing only elements which bring each particle back to its own initial position, not the initial position of one of the other particles. The richness of the braid group is the key fact enabling quantum computation through quasi-particle braiding.

To define the quantum evolution of a system, we must now specify how the braid group acts on the states of the system. The simplest possibilities are 1D representations of the braid group. In these cases, the wave-function acquires a phase θ when one particle is taken around another, analogous to (6.131), (6.132). The special cases $\theta = 0, \pi$ are bosons and fermions, respectively, while particles with other values of θ are *anyons* [Wil90]. These are straightforward many-particle generalizations of the two-particle case considered above. An arbitrary element of the braid group is represented by the factor $e^{im\theta}$ where m is the total number of times that one particle winds around another in a counter-clockwise manner (minus the number of times that a particle winds around another in a clockwise manner). These representations are Abelian since the order of braiding operations in unimportant. However, they can still have a quite rich structure since there can be n_s different particle species with parameters θ_{ab}, where $a, b = 1, 2, \ldots, n_s$, specifying the phases resulting from braiding a particle of type a around a particle of type b. Since distinguishable particles can braid non-trivially, i.e., θ_{ab} can be non-zero for $a \neq b$ as well as for $a = b$, anyonic 'statistics' is, perhaps, better understood as a kind of topological interaction between particles.

We now turn to non-Abelian braiding statistics, which are associated with higher-dimensional representations of the braid group. Higher-dimensional representations can occur when there is a degenerate set of g states with particles at fixed positions $R_1, R_2, \ldots, R_n$. Let us define an orthonormal basis ψ_α, ($\alpha = 1, 2, \ldots, g$) of these degenerate states. Then an element of the braid group—say σ_1, which exchanges particles 1 and 2, is represented by a $g \times g$ unitary matrix $\rho(\sigma_1)$ acting on these states [NSS08]

$$\psi_\alpha \rightarrow [\rho(\sigma_1)]_{\alpha\beta} \psi_\beta.$$

On the other hand, exchanging particles 2 and 3 leads to:

$$\psi_\alpha \to [\rho(\sigma_2)]_{\alpha\beta}\psi_\beta.$$

Both $\rho(\sigma_1)$ and $\rho(\sigma_2)$ are $g \times g$ dimensional unitary matrices, which define unitary transformation within the subspace of degenerate ground states. If $\rho(\sigma_1)$ and $\rho(\sigma_1)$ do not commute,

$$[\rho(\sigma_1)]_{\alpha\beta}[\rho(\sigma_2)]_{\beta\gamma} \neq [\rho(\sigma_2)]_{\alpha\beta}[\rho(\sigma_1)]_{\beta\gamma},$$

the particles obey *non-Abelian braiding statistics*. Unless they commute for any interchange of particles, in which case the particles' braiding statistics is Abelian, braiding quasi-particles will cause non-trivial rotations within the degenerate many-quasi-particle Hilbert space. Furthermore, it will essentially be true at low energies that the *only* way to make non-trivial unitary operations on this degenerate space is by braiding quasi-particles around each other. This statement is equivalent to a statement that no local perturbation can have nonzero matrix elements within this degenerate space.

A system with anyonic particles must generally have multiple types of anyons. For instance, in a system with Abelian anyons with statistics θ, a bound state of two such particles has statistics 4θ. Even if no such stable bound state exists, we may wish to bring two anyons close together while all other particles are much further away. Then the two anyons can be approximated as a single particle whose quantum numbers are obtained by combining the quantum numbers, including the topological quantum numbers, of the two particles. As a result, a complete description of the system must also include these 'higher' particle species. For instance, if there are $\theta = \pi/m$ anyons in system, then there are also $\theta = 4\pi/m, 9\pi/m, \ldots, (m-1)^2\pi/m$. Since the statistics parameter is only well-defined up to 2π, $\theta = (m-1)^2\pi/m = -\pi/m$ for m even and $\pi - \pi/m$ for m odd. The formation of a different type of anyon by bringing together two anyons is called *fusion*. When a statistics π/m particle is fused with a statistics $-\pi/m$ particle, the result has statistics $\theta = 0$. It is convenient to call this the 'trivial' particle. As far as topological properties are concerned, such a boson is just as good as the absence of any particle, so the 'trivial' particle is also sometimes simply called the 'vacuum'. We will often denote the trivial particle by **1**.

With Abelian anyons which are made by forming successively larger composites of π/m particles, the *fusion rule* is [NSS08]

$$\frac{n^2\pi}{m} \times \frac{k^2\pi}{m} = \frac{(n+k)^2\pi}{m},$$

where we use $a \times b$ to denote a *fused with* b. However, for non-Abelian anyons, the situation is more complicated. As with ordinary quantum numbers, there might not be a unique way of combining topological quantum numbers (e.g., two spin-1/2 particles could combine to form either a spin-0 or a spin-1 particle). The different possibilities are called the different *fusion channels*. This is usually denoted by

$$\phi_a \times \phi_b = \sum_c N_{ab}^c \phi_c,$$

which represents the fact that when a particle of species a fuses with one of species b, the result can be a particle of species c if $N_{ab}^c \neq 0$. For Abelian anyons, the fusion multiplicities $N_{ab}^c = 1$ for only one value of c and $N_{ab}^{c'} = 0$ for all $c' \neq c$. For particles of type k with statistics $\theta_k = \pi k^2/m$, i.e., $N_{kk'}^{k''} = \delta_{k+k',k''}$. For non-Abelian anyons, there is at least one a, b such that there are multiple fusion channels c with $N_{ab}^c \neq 0$. In the examples which we will be considering in this section, $N_{ab}^c = 0$ or 1, but there are theories for which $N_{ab}^c > 1$ for some a, b, c. In this case, a and b can fuse to form c in $N_{ab}^c > 1$ different distinct ways. We will use $\bar{a}$ to denote the antiparticle of particle species a. When a and $\bar{a}$ fuse, they can always fuse to 1 in precisely one way, i.e., $N_{a\bar{a}}^1 = 1$; in the non-Abelian case, they may or may not be able to fuse to other particle types as well.

The different fusion channels are one way of accounting for the different degenerate multi-particle states. Let us see how this works in one simple model of non-Abelian anyons. This model is associated with *Ising anyons*, $SU(2)_2$, and chiral p-superconductors. There are slight differences between these three theories, relating to Abelian phases, but these are unimportant for the present discussion. This model has three different types of anyons, which can be variously called $1, \sigma, \psi$ or $0, \frac{1}{2}, 1$.[21] The fusion rules for such anyons are[22]

$$\sigma \times \sigma = 1 + \psi, \qquad \text{::::} \; \sigma \times \psi = \sigma, \qquad \text{::::} \; \psi \times \psi = 1,$$
$$\times x = x \quad (\text{for } x = 1, \sigma, \psi).$$

Note that there are two different fusion channels for two σs. As a result, if there are four σs which fuse together to give 1, there is a 2D space of such states. If we divided the four σs into two pairs, by grouping particles $1, 2$ and $3, 4$, then a basis for the 2D space is given by the state in which $1, 3$ fuse to 1 or $1, 3$ fuse to ψ ($2, 4$ must fuse to the same particle type as $1, 3$ do in order that all four particles fuse to 1). We can call these states Ψ_1 and Ψ_ψ; they are a basis for the four-quasi-particle Hilbert space with total topological charge 1.[23]

Of course, our division of the four σs into two pairs was arbitrary. We could have divided them differently, say, into the pairs $1, 3$ and $2, 4$. We would

---

[21] Unfortunately, the notation is a little confusing because the trivial particle is called '1' in the first model but '0' in the second, however, we will avoid confusion by using bold-faced 1 to denote the trivial particle.

[22] Translating these rules into the notation of $SU(2)_2$, we see that these fusion rules are very similar to the decomposition rules for tensor products of irreducible $SU(2)$ representations, but differ in the important respect that 1 is the maximum spin so that $\frac{1}{2} \times \frac{1}{2} = 0 + 1$, as in the $SU(2)$ case, but $\frac{1}{2} \times 1 = \frac{1}{2}$ and $1 \times 1 = 0$.

[23] Similarly, if they all fused to give ψ, there would be another 2D degenerate space; one basis is given by the state in which the first pair fuses to 1 while the second fuses to ψ and the state in which the opposite occurs.

thereby obtain two different basis states, $\tilde{\Psi}_1$ and $\tilde{\Psi}_\psi$, in which both pairs fuse to
1 or to ψ, respectively. This is just a different basis in the same 2D space. The
matrix parametrizing this basis change is called the F-matrix: $\tilde{\Psi}_a = F_{ab}\Psi_b$,
where $a, b = 1, \psi$. There should really be 6 indices on F if we include indices to
specify the 4 particle types: $[F_l^{ijk}]_{ab}$, but we have dropped these other indices
since $i = j = k = l = \sigma$ in our case. The F-matrices are sometimes called $6j$
symbols since they are analogous to the corresponding quantities for $SU(2)$
representations. Recall that in $SU(2)$, there are multiple states in which spins
$\mathbf{j}_1, \mathbf{j}_2, \mathbf{j}_3$ couple to form a total spin $\mathbf{J}$. For instance, $\mathbf{j}_1$ and $\mathbf{j}_2$ can add to
form $\mathbf{j}_{12}$, which can then add with $\mathbf{j}_3$ to give $\mathbf{J}$. The eigenstates of $(\mathbf{j}_{12})^2$
form a basis of the different states with fixed $\mathbf{j}_1, \mathbf{j}_2, \mathbf{j}_3$, and $\mathbf{J}$. Alternatively,
$\mathbf{j}_2$ and $\mathbf{j}_3$ can add to form $\mathbf{j}_{23}$, which can then add with $\mathbf{j}_1$ to give $\mathbf{J}$. The
eigenstates of $(\mathbf{j}_{23})^2$ form a different basis. The $6j$ symbol gives the basis
change between the two. The F-matrix of a system of anyons plays the same
role when particles of topological charges i, j, k fuse to total topological charge
l. If i and j fuse to a, which then fuses with k to give topological charge l,
the different allowed a define a basis. If j and k fuse to b and then fuse with
i to give topological charge l, this defines another basis, and the F-matrix is
the unitary transformation between the two bases. States with more than 4
quasi-particles can be understood by successively fusing additional particles.
The F-matrix can be applied to any set of 4 consecutively fused particles.

The different states in this degenerate multi-anyon state space transform
into each other under braiding. However, two particles cannot change their
fusion channel simply by braiding with each other since their total topological
charge can be measured along a far distant loop enclosing the two particles.
They must braid with a third particle in order to change their fusion channel.
Consequently, when two particles fuse in a particular channel (rather than a
linear superposition of channels), the effect of taking one particle around the
other is just multiplication by a phase. This phase resulting from a counter-
clockwise exchange of particles of types a and b which fuse to a particle of
type c is called R_c^{ab}. In the Ising anyon case we have [NSS08]

$$R_1^{\sigma\sigma} = e^{-\pi i/8}, \qquad R_\psi^{\sigma\sigma} = e^{3\pi i/8}, \qquad R_1^{\psi\psi} = -1, \qquad R_\sigma^{\sigma\psi} = i.$$

For an example of how this works, suppose that we create a pair of σ quasi-
particles out of the vacuum. They will necessarily fuse to 1. If we take one
around another, the state will change by a phase $e^{-\pi i/8}$. If we take a third σ
quasi-particle and take it around one, but not both, of the first two, then the
first two will now fuse to ψ. If we now take one of the first two around the
other, the state will change by a phase $e^{3\pi i/8}$.

Quasi-particles obeying non-Abelian braiding statistics or, simply non-
Abelian anyons, were first considered in the context of conformal field theory
by [MS88, MS89] and in the context of *Chern–Simons theory* by Ed Witten
[Wit89] (see Sect. 3.4.3 above). They were discussed in the context of discrete
gauge theories and linked to the representation theory of *quantum groups* by
[Bai80, BDW92, BDW93, BMW93]. They were discussed in a more general

context by [FRS89] and [FG90]. The properties of non-Abelian quasi-particles make them appealing for use in a quantum computer. But before discussing this, we will briefly review how they could occur in nature and then the basic ideas behind quantum computation.

6.4.2 Emergent Anyons

The preceding considerations show that exotic braiding statistics is a theoretical possibility in $(2 + 1)$D, but they do not tell us when and where they might occur in nature. Electrons, protons, atoms, and photons, are all either fermions or bosons even when they are confined to move in a 2D plane. However, if a system of many electrons (or bosons, atoms, etc.) confined to a 2D plane has excitations which are localized disturbances of its quantum-mechanical ground state, known as *quasi-particles*, then these quasi-particles can be anyons. When a system has anyonic quasi-particle excitations above its ground state, it is in a *topological phase of matter*.

Let us see how anyons might arise as an emergent property of a many-particle system. For the sake of concreteness, consider the ground state of a $(2 + 1)$D system of electrons, whose coordinates are $(r_1, \ldots, r_n)$. We assume that the ground state is separated from the excited states by an energy gap (i.e, it is incompressible), as is the situation in fractional quantum Hall states in 2D electron systems. The lowest energy electrically-charged excitations are known as quasi-particles or quasi-holes, depending on the sign of their electric charge.[24] These quasi-particles are local disturbances to the wave-function of the electrons corresponding to a quantized amount of total charge.

We now introduce into the system's Hamiltonian a scalar potential composed of many local 'traps', each sufficient to capture exactly one quasi-particle [NSS08]. These traps may be created by impurities, by very small gates, or by the potential created by tips of scanning microscopes. The quasi-particle's charge screens the potential introduced by the trap and the 'quasi-particle-tip' combination cannot be observed by local measurements from far away. Let us denote the positions of these traps to be $(R_1, \ldots, R_k)$, and assume that these positions are well spaced from each other compared to the microscopic length scales. A state with quasi-particles at these positions can be viewed as an excited state of the Hamiltonian of the system without the trap potential or, alternatively, as the ground state in the presence of the trap potential. When we refer to the ground state(s) of the system, we will often be referring to multi-quasi-particle states in the latter context. The quasi-particles' coordinates $(R_1, \ldots, R_k)$ are parameters both in the Hamiltonian and in the resulting ground state wave-function for the electrons.

We are concerned here with the effect of taking these quasi-particles around each other. We imagine making the quasi-particles coordinates:

---

[24] The term 'quasi-particle' is also sometimes used in a generic sense to mean both quasi-particle and quasi-hole as in the previous paragraph.

$\mathbf{R} = (R_1, \ldots, R_k)$ adiabatically time-dependent. In particular, we consider a trajectory in which the final configuration of quasi-particles is just a permutation of the initial configuration (i.e., at the end, the positions of the quasi-particles are identical to the initial positions, but some quasi-particles may have interchanged positions with others.) If the ground state wave function is single-valued with respect to $(R_1, \ldots, R_k)$, and if there is only one ground state for any given set of R_i's, then the final ground state to which the system returns to after the winding is identical to the initial one, up to a phase. Part of this phase is simply the dynamical phase which depends on the energy of the quasi-particle state and the length of time for the process. In the adiabatic limit, it is $\int dt E(\mathbf{R}(t))$. There is also a geometric phase which does not depend on how long the process takes. This *Berry phase* is [Ber84],

$$\alpha = \mathrm{i} \oint d\mathbf{R} \cdot \langle \psi(\mathbf{R}) | \nabla_{\mathbf{R}} | \psi(\mathbf{R}) \rangle, \qquad (6.133)$$

where $|\psi(\mathbf{R})\rangle$ is the ground state with the quasi-particles at positions $\mathbf{R}$, and where the integral is taken along the trajectory $\mathbf{R}(t)$. It is manifestly dependent only on the trajectory taken by the particles and not on how long it takes to move along this trajectory.

The phase α has a piece that depends on the geometry of the path traversed (typically proportional to the area enclosed by all of the loops), and a piece θ that depends only on the topology of the loops created. If $\theta \neq 0$, then the quasi-particles excitations of the system are anyons. In particular, if we consider the case where only two quasi-particles are interchanged clockwise (without wrapping around any other quasi-particles), θ is the statistical angle of the quasi-particles.

There were two key conditions to our above discussion of the Berry phase. The single valuedness of the wave function is a technical issue. The non-degeneracy of the ground state, however, is an important physical condition. In fact, most of this section deals with the situation in which this condition does not hold. We will generally be considering systems in which, once the positions $(R_1, \ldots, R_k)$ of the quasi-particles are fixed, there remain multiple degenerate ground states (i.e., ground states in the presence of a potential which captures quasi-particles at positions $(R_1, \ldots, R_k)$), which are distinguished by a set of internal quantum numbers. For reasons that will become clear later, we will refer to these quantum numbers as 'topological'.

When the ground state is degenerate, the effect of a closed trajectory of the R_i's is not necessarily *just* a phase factor. The system starts and ends in ground states, but the initial and final ground states may be different members of this degenerate space. The constraint imposed by adiabaticity in this case is that the adiabatic evolution of the state of the system is confined to the subspace of ground states. Thus, it may be expressed as a unitary transformation within this subspace. The inner product in (6.133) must be generalized to a matrix of such inner products:

$$\mathbf{m}_{ab} = \langle \psi_a(\mathbf{R}) | \boldsymbol{\nabla}_{\mathbf{R}} | \psi_b(\mathbf{R}) \rangle, \qquad (6.134)$$

where $|\psi_a(\mathbf{R})\rangle$, $(a = 1, 2, \ldots, g)$ are the g degenerate ground states. Since these matrices at different points $\mathbf{R}$ do not commute, we must path-order the integral in order to compute the transformation rule for the state, $\psi_a \to M_{ab}\psi_b$ where

$$
\begin{aligned}
M_{ab} &= \mathcal{P} \exp\left(i \oint d\mathbf{R} \cdot \mathbf{m}\right) \\
&= \sum_{n=0}^{\infty} i^n \int_0^{2\pi} ds_1 \int_0^{s_1} ds_2 \cdots \int_0^{s_{n-1}} ds_n \\
&\quad \times \left[\dot{\mathbf{R}}(s_1) \cdot \mathbf{m}_{aa_1}\left(\mathbf{R}(s_1)\right) \cdots \dot{\mathbf{R}}(s_n) \cdot \mathbf{m}_{a_nb}\left(\mathbf{R}(s_n)\right) \right],
\end{aligned}
$$

where $\mathbf{R}(s)$, $s \in [0, 2\pi]$ is the closed trajectory of the particles and the path-ordering symbol $\mathcal{P}$ is defined by the second equality. Again, the matrix M_{ab} may be the product of topological and non-topological parts. In a system in which quasi-particles obey non-Abelian braiding statistics, the non-topological part will be Abelian, that is, proportional to the unit matrix. Only the topological part will be non-Abelian.

The requirements for quasi-particles to follow non-Abelian statistics are then, first, that the N-quasi-particle ground state is degenerate. In general, the degeneracy will not be exact, but it should vanish exponentially as the quasi-particle separations are increased. Second, that adiabatic interchange of quasi-particles applies a unitary transformation on the ground state, whose non-Abelian part is determined only by the topology of the braid, while its non-topological part is Abelian. If the particles are not infinitely far apart, and the degeneracy is only approximate, then the adiabatic interchange must be done faster than the inverse of the energy splitting [TG91] between states in the nearly-degenerate subspace. Third, the only way to make unitary operations on the degenerate ground state space, so long as the particles are kept far apart, is by braiding. The simplest (albeit uninteresting) example of degenerate ground states may arise if each of the quasi-particles carried a spin $1/2$ with a vanishing g-factor. If that were the case, the system would satisfy the first requirement. Spin orbit coupling may conceivably lead to the second requirement being satisfied. Satisfying the third one, however, is much harder, and requires the subtle structure that we describe below.

The degeneracy of N-quasi-particle ground states is conditioned on the quasi-particles being well separated from one another. When quasi-particles are allowed to approach one another too closely, the degeneracy is lifted. In other words, when non-Abelian anyonic quasi-particles are close together, their different fusion channels are split in energy. This dependence is analogous to the way the energy of a system of spins depends on their internal quantum numbers when the spins are close together and their coupling becomes significant. The splitting between different fusion channels is a means for a measurement of the internal quantum state, a measurement that is of importance in the context of quantum computation.

Topological Quantum Computation

Basics of Quantum Computation

As the components of computers become smaller and smaller, we are approaching the limit in which quantum effects become important. One might ask whether this is a problem or an opportunity. The founders of the field of quantum computation ([Man80, Fey82, Fey86, Deu85], and most dramatically, [Sho94]) answered in favor of the latter. They showed that a computer which operates coherently on quantum states has potentially much greater power than a classical computer [NC00].

The problem which Feynman had in mind for a quantum computer was the simulation of a quantum system [Fey82]. He showed that certain many-body quantum Hamiltonians could be simulated *exponentially faster* on a quantum computer than they could be on a classical computer. This is an extremely important potential application of a quantum computer since it would enable us to understand the properties of complex materials, e.g., solve high-temperature superconductivity. Digital simulations of large scale quantum many-body Hamiltonians are essentially hopeless on classical computers because of the exponentially-large size of the Hilbert space. A quantum computer, using the physical resource of an exponentially-large Hilbert space, may also enable progress in the solution of lattice gauge theory and quantum chromodynamics, thus shedding light on strongly-interacting nuclear forces.

In 1994 Peter Shor found an application of a quantum computer which generated widespread interest not just inside but also outside of the physics community [Sho94]. He invented an algorithm by which a quantum computer could find the prime factors of an m digit number in a length of time $\sim m^2 \log m \log \log m$. This is much faster than the fastest known algorithm for a classical computer, which takes $\sim \exp(m^{1/3})$ time. Since many encryption schemes depend on the difficulty of finding the solution to problems similar to finding the prime factors of a large number, there is an obvious application of a quantum computer which is of great basic and applied interest.

The computation model set forth by these pioneers of quantum computing (and refined in [DiV00]), is based on three steps: initialization, unitary evolution and measurement. We assume that we have a system at our disposal with Hilbert space $\mathcal{H}$. We further assume that we can initialize the system in some known state $|\psi_0\rangle$. We unitarily evolve the system until it is in some final state $U(t)|\psi_0\rangle$. This evolution will occur according to some Hamiltonian $H(t)$ such that

$$\dot{U} = iH(t)U(t)/\hbar.$$

We require that we have enough control over this Hamiltonian so that $U(t)$ can be made to be any unitary transformation that we desire. Finally, we need to measure the state of the system at the end of this evolution. Such a process is called *quantum computation* [NC00]. The Hamiltonian $H(t)$ is the software program to be run. The initial state is the input to the calculation, and the final measurement is the output.

The need for versatility, i.e., for one computer to efficiently solve many different problems, requires the construction of the computer out of smaller pieces that can be manipulated and reconfigured individually. Typically the fundamental piece is taken to be a quantum two state system known as a 'qubit' which is the quantum analog of a bit. While a classical bit, i.e., a classical 2-state system, can be either 'zero' or 'one' at any given time, a qubit can be in one of the infinitely many superpositions $a|0\rangle + b|1\rangle$. For n qubits, the state becomes a vector in a 2^n-dimensional Hilbert space, in which the different qubits are generally entangled with one another.

The quantum phenomenon of superposition allows a system to traverse many trajectories in parallel, and determine its state by their coherent sum. In some sense this coherent sum amounts to a massive quantum parallelism. It should not, however, be confused with classical parallel computing, where many computers are run in parallel, and no coherent sum takes place [NSS08].

The biggest obstacle to building a practical quantum computer is posed by errors, which would invariably happen during any computation, quantum or classical. For any computation to be successful one must devise practical schemes for error correction which can be effectively implemented (and which must be sufficiently fault-tolerant). Errors are typically corrected in classical computers through redundancies, i.e., by keeping multiple copies of information and checking against these copies.

With a quantum computer, however, the situation is more complex. If we measure a quantum state during an intermediate stage of a calculation to see if an error has occurred, we collapse the wave function and thus destroy quantum superpositions and ruin the calculation. Furthermore, errors need not be merely a discrete flip of $|0\rangle$ to $|1\rangle$, but can be continuous: the state $a|0\rangle + b|1\rangle$ may drift, due to an error, to the state $\rightarrow a|0\rangle + be^{i\theta}|1\rangle$ with arbitrary θ.

Remarkably, in spite of these difficulties, error correction is possible for quantum computers (see, e.g. [CS96, Got98]). One can represent information redundantly so that errors can be identified without measuring the information. For instance, if we use three spins to represent each qubit, $|0\rangle \rightarrow |000\rangle$, $|1\rangle \rightarrow |111\rangle$, and the spin-flip rate is low, then we can identify errors by checking whether all three spins are the same (here, we represent an up spin by 0 and a down spin by 1). Suppose that our spins are in the state $\alpha|000\rangle + \beta|111\rangle$. If the first spin has flipped erroneously, then our spins are in the state $\alpha|100\rangle + \beta|011\rangle$. We can detect this error by checking whether the first spin is the same as the other two; this does not require us to measure the state of the qubit. If the first spin is different from the other two, then we just need to flip it. We repeat this process with the second and third spins. So long as we can be sure that two spins have not erroneously flipped (i.e., so long as the basic spin-flip rate is low), this procedure will correct spin-flip errors. A more elaborate encoding is necessary in order to correct phase errors, but the key observation is that a phase error in the σ_z basis is a bit flip error in the σ_x basis.

However, the error correction process may itself be a little noisy. More errors could then occur during error correction, and the whole procedure will fail unless the basic error rate is very small. Estimates of the threshold error rate above which error correction is impossible depend on the particular error correction scheme, but fall in the range 10^{-4}–10^{-6} (see, e.g., [AB97]). This means that we must be able to perform 10^4–10^6 operations perfectly before an error occurs. This is an extremely stringent constraint and it is presently unclear if local qubit-based quantum computation can ever be made fault-tolerant through quantum error correction protocols.

Random errors are caused by the interaction between the quantum computer and the environment. As a result of this interaction, the quantum computer, which is initially in a pure superposition state, becomes entangled with its environment. This can cause errors as follows. Suppose that the quantum computer is in the state $|0\rangle$ and the environment is in the state $|E_0\rangle$ so that their combined state is $|0\rangle|E_0\rangle$. The interaction between the computer and the environment could cause this state to evolve to $\alpha|0\rangle|E_0\rangle + \beta|1\rangle|E_1\rangle$, where $|E_1\rangle$ is another state of the environment (not necessarily orthogonal to $|E_0\rangle$). The computer undergoes a transition to the state $|1\rangle$ with probability $|\beta|^2$. Furthermore, the computer and the environment are now entangled, so the reduced density matrix for the computer alone describes a mixed state, e.g.,

$$\rho = \mathrm{diag}(|\alpha|^2, |\beta|^2) \quad \text{if } \langle E_0|E_1\rangle = 0.$$

Since we cannot measure the state of the environment accurately, information is lost, as reflected in the evolution of the density matrix of the computer from a pure state to a mixed one. In other words, the environment has caused *decoherence*. Decoherence can destroy quantum information even if the state of the computer does not undergo a transition. Although whether or not a transition occurs is basis-dependent (a bit flip in the σ_z basis is a phase flip in the σ_x basis), it is a useful distinction because many systems have a preferred basis, for instance the ground state $|0\rangle$ and excited state $|1\rangle$ of an ion in a trap. Suppose the state $|0\rangle$ evolves as above, but with $\alpha = 1$, $\beta = 0$ so that no transition occurs, while the state $|1\rangle|E_0\rangle$ evolves to $|1\rangle|E_1'\rangle$ with $\langle E_1'|E_1\rangle = 0$. Then an initial pure state $(a|0\rangle + b|1\rangle)|E_0\rangle$ evolves to a mixed state with density matrix: $\rho = \mathrm{diag}(|a|^2, |b|^2)$. The correlations in which our quantum information resides is now transferred to correlation between the quantum computer and the environment. The quantum state of a system invariably loses coherence in this way over a characteristic time scale T_{coh}. It was universally assumed until the advent of quantum error correction [Sho95] that quantum computation is intrinsically impossible since decoherence-induced quantum errors simply cannot be corrected in any real physical system. However, when error-correcting codes are used, the entanglement is transferred from the quantum computer to ancillary qubits so that the quantum information remains pure while the entropy is in the ancillary qubits.

Of course, even if the coupling to the environment were completely eliminated, so that there were no random errors, there could still be systematic errors. These are unitary errors which occur while we process quantum information. For instance, we may wish to rotate a qubit by 90 degrees but might inadvertently rotate it by 90.01 degrees.

From a practical standpoint, it is often useful to divide errors into two categories: (i) errors that occur when a qubit is being processed (i.e., when computations are being performed on that qubit) and (ii) errors that occur when a qubit is simply storing quantum information and is not being processed (i.e., when it is acting as a quantum memory). From a fundamental standpoint, this is a bit of a false dichotomy, since one can think of quantum information storage (or quantum memory) as being a computer that applies the identity operation over and over to the qubit (i.e., leaves it unchanged). Nonetheless, the problems faced in the two categories might be quite different. For quantum information processing, unitary errors, such as rotating a qubit by 90.01 degrees instead of 90, are an issue of how precisely one can manipulate the system. On the other hand, when a qubit is simply storing information, one is likely to be more concerned about errors caused by interactions with the environment. This is instead an issue of how well isolated one can make the system. As we will see below, a topological quantum computer is protected from problems in both of these categories.

Fault-Tolerance from Non-Abelian Anyons

Topological quantum computation is a scheme for using a system whose excitations satisfy non-Abelian braiding statistics to perform quantum computation in a way that is naturally immune to errors. The Hilbert space $\mathcal{H}$ used for quantum computation is the subspace of the total Hilbert space of the system comprised of the degenerate ground states with a fixed number of quasi-particles at fixed positions. Operations within this subspace are carried out by braiding quasi-particles. As we discussed above, the subspace of degenerate ground states is separated from the rest of the spectrum by an energy gap. Hence, if the temperature is much lower than the gap and the system is weakly-perturbed using frequencies much smaller than the gap, the system evolves only within the ground state subspace. Furthermore, that evolution is severely constrained, since it is essentially the case (with exceptions which we will discuss) that the only way the system can undergo a non-trivial unitary evolution, that is, an evolution that takes it from one ground state to another—is by having its quasi-particles braided. The reason for this exceptional stability is that any local perturbation[25] has no nontrivial matrix elements within the ground state subspace. Thus, the system is rather immune from decoherence [NSS08]. Unitary errors are also unlikely since the unitary

---

[25] Such as the electron–phonon interaction and the hyperfine electron-nuclear interaction, two major causes for decoherence in non-topological solid state spin-based quantum computers [WD06].

transformations associated with braiding quasi-particles are sensitive only to the topology of the quasi-particle trajectories, and not to their geometry or dynamics.

A model in which non-Abelian quasi-particles are utilized for quantum computation starts with the construction of qubits. In sharp contrast to most realizations of a quantum computer, a qubit here is a non-local entity, being comprised of several well-separated quasi-particles, with the two states of the qubit being two different values for the internal quantum numbers of this set of quasi-particles. In the simplest non-Abelian quantum Hall state, which has *Landau-level* filling factor $\nu = 5/2$, two quasi-particles can be put together to form a qubit. Unfortunately, this system turns out to be incapable of universal topological quantum computation using only braiding operations; some unprotected operations are necessary in order to perform universal quantum computation. The simplest system that is capable of universal topological quantum computation utilizes three quasi-particles to form one qubit.

As mentioned above, to perform a quantum computation, one must be able to initialize the state of qubits at the beginning, perform arbitrary controlled unitary operations on the state, and then measure the state of qubits at the end. We now address each of these in turn.

Initialization may be performed by preparing the quasi-particles in a specific way. For example, if a quasi-particle–anti-quasi-particle pair is created by 'pulling' it apart from the vacuum (e.g., pair creation from the vacuum by an electric field), the pair will begin in an initial state with the pair necessarily having conjugate quantum numbers (i.e., the 'total' quantum number of the pair remains the same as that of the vacuum). This gives us a known initial state to start with. It is also possible to use measurement and unitary evolution (both to be discussed below) as an initialization scheme, if one can measure the quantum numbers of some quasi-particles, one can then perform a controlled unitary operation to put them into any desired initial state.

Once the system is initialized, controlled unitary operations are then performed by physically dragging quasi-particles around one another in some specified way. When quasi-particles belonging to different qubits braid, the state of the qubits changes. Since the resulting unitary evolution depends only on the topology of the braid that is formed and not on the details of how it is done, it is insensitive to wiggles in the path, resulting, e.g., from the quasi-particles being scattered by phonons or photons. Determining which braid corresponds to which computation is a complicated but eminently solvable task.

Once the unitary evolution is completed, there are two ways to measure the state of the qubits. The first relies on the fact that the degeneracy of multi-quasi-particle states is split when quasi-particles are brought close together (within some microscopic length scale). When two quasi-particles are brought close together, for instance, a measurement of this energy (or a measurement of the force between two quasi-particles) measures the topological charge of the pair. A second way to measure the topological charge of a group of quasi-particles is by carrying out an Aharanov–Bohm type interference experiment. We take a 'beam' of test quasi-particles, send it through a beamsplitter, send one partial wave to the right of the group to be measured and another partial wave to the left of the group and then re-interfere the two waves. Since the two different beams make different braids around the test group, they will experience different unitary evolution depending on the topological quantum numbers of the test group. Thus, the re-interference of these two beams will reflect the topological quantum number of the group of quasi-particles enclosed.

This concludes a rough description of the way a topological quantum computation is to be performed. While the unitary transformation associated with a braid depends only on the topology of the braid, one may be concerned that errors could occur if one does not return the quasi-particles to precisely the correct position at the end of the braiding. This apparent problem, however, is evaded by the nature of the computations, which correspond to closed world lines that have no loose ends: when the computation involves creation and annihilation of a quasi-particle quasi-hole pair, the world-line is a closed curve in space-time. If the measurement occurs by bringing two particles together to measure their quantum charge, it does not matter where precisely they are brought together. Alternatively, when the measurement involves an interference experiment, the interfering particle must close a loop. In other words, a computation corresponds to a set of *links* rather than open braids, and the initialization and measurement techniques *necessarily* involve bringing quasi-particles together in some way, closing up the trajectories and making the full process from initialization to measurement completely topological.

Due to its special characteristics, then, topological quantum computation intrinsically guarantees fault-tolerance, at the level of 'hardware', without 'software'-based error correction schemes that are so essential for non-topological quantum computers. This immunity to errors results from the stability of the ground state subspace with respect to external local perturbations. In non-topological quantum computers, the qubits are local, and the operations on them are local, leading to a sensitivity to errors induced by local perturbations. In a topological quantum computer the qubits are non-local, and the operations, quasi-particle braiding, are non-local, leading to an immunity to local perturbations.

Such immunity to local perturbation gives topological quantum memories exceptional protection from errors due to the interaction with the environment. However, it is crucial to note that topological quantum computers are also exceptionally immune to unitary errors due to imprecise gate operation. Unlike other types of quantum computers, the operations that can be performed on a topological quantum computer (braids) naturally take a discrete set of values. As discussed above, when one makes a 90 degree rotation of a spin-based qubit, for example, it is possible that one will mistakenly rotate by 90.01 degrees thus introducing a small error. In contrast, braids are discrete: either a particle is taken around another, or it is not. There is no way to make a small error by having slight imprecision in the way the quasi-particles are moved [NSS08].

Given the exceptional stability of the ground states, and their insensitivity to local perturbations that do not involve excitations to excited states, one may ask then which physical processes do cause errors in such a topological quantum computer. Due to the topological stability of the unitary transformations associated with braids, the only error processes that we must be concerned about are processes that might cause us to form the wrong link, and hence the wrong computation. Certainly, one must keep careful track of the positions of *all* of the quasi-particles in the system during the computation and assure that one makes the correct braid to do the correct computation. This includes not just the 'intended' quasi-particles which we need to manipulate for our quantum computation, but also any 'unintended' quasi-particle which might be lurking in our system without our knowledge. Two possible sources of these unintended quasi-particles are thermally excited quasi-particle–quasi-hole pairs, and randomly localized quasi-particles trapped by disorder (e.g., impurities, surface roughness, etc.). In a typical thermal fluctuation, for example, a quasi-particle–quasi-hole pair is thermally created from the vacuum, braids with existing intended quasi-particles, and then gets annihilated. Typically, such a pair has opposite electrical charges, so its constituents will be attracted back to each other and annihilate. However, entropy or temperature may lead the quasi-particle and quasi-hole to split fully apart and wander relatively freely through part of the system before coming back together and annihilating. This type of process may change the state of the qubits encoded in the intended quasi-particles, and hence disrupt the computation. Fortunately, there is a whole class of such processes that do not in fact cause error. This class includes all of the most likely such thermal processes to occur: including when a pair is created, encircles a single already existing quasi-particle and then re-annihilates, or when a pair is created and one of the pair annihilates an already existing quasi-particle. For errors to be caused, the excited pair must braid at least two intended quasi-particles. Nonetheless, the possibility of thermally-excited quasi-particles wandering through the system creating unintended braids and thereby causing error is a serious one. For this reason, topological quantum computation must be performed at temperatures well below the energy gap for quasi-particle–quasi-hole creation so that these errors will be exponentially suppressed.

Similarly, localized quasi-particles that are induced by disorder (e.g., randomly-distributed impurities, surface roughness, etc.) are another serious obstacle to overcome, since they enlarge the dimension of the subspace of degenerate ground states in a way that is hard to control. In particular, these unaccounted-for quasi-particles may couple by tunneling to their intended counterparts, thereby introducing dynamics to what is supposed to be a topology-controlled system, and possibly ruining the quantum computation. We further note that, in quantum Hall systems, slight deviations in density or magnetic field will also create unintended quasi-particles that must be carefully avoided.

Finally, we also note that while non-Abelian quasi-particles are natural candidates for the realization of topological qubits, not every system where quasi-particles satisfy non-Abelian statistics is suitable for quantum computation. For this suitability it is essential that the set of unitary transformations induced by braiding quasi-particles is rich enough to allow for all operations needed for computation.

Non-Abelian Quantum Hall States

A necessary condition for topological quantum computation using non-Abelian anyons is the existence of a physical system where non-Abelian anyons can be found, manipulated (e.g., braided), and conveniently read out. Several theoretical models and proposals for systems having these properties have been introduced in recent years [FNS05a, FF05, Kit06]. Despite the theoretical work in these directions, the only real physical system where there is even indirect experimental evidence that non-Abelian anyons exist are quantum Hall systems in 2D electron gases (2DEGs) in high magnetic fields. Consequently, we will devote a considerable part of our discussion to putative non-Abelian quantum Hall systems which are also of great interest in their own right.

Quick Review of Quantum Hall Physics

A comprehensive review of the quantum Hall effect is well beyond the scope of this article and can be found in the literature (see, e.g. [DP97]). This effect, realized for two dimensional electronic systems in a strong magnetic field, is characterized by a gap between the ground state and the excited states (incompressibility); a vanishing longitudinal resistivity $\rho_{xx} = 0$, which implies a dissipationless flow of current; and the quantization of the Hall resistivity precisely to values of $\rho_{xy} = \frac{1}{\nu}\frac{h}{e^2}$, with ν being an integer (the integer quantum Hall effect), or a fraction (the fractional quantum Hall effect). These values of the two resistivities imply a vanishing longitudinal conductivity $\sigma_{xx} = 0$ and a quantized Hall conductivity $\sigma_{xy} = \nu \frac{e^2}{h}$.

To understand the quantized Hall effect, we begin by ignoring electron-electron Coulomb interactions, then the energy eigenstates of the single-electron Hamiltonian in a magnetic field,

$$H_0 = \frac{1}{2m}\left(p_i - \frac{e}{c}A(x_i)\right)^2,$$

break up into an equally-spaced set of degenerate levels called Landau levels. In symmetric gauge,

$$\mathbf{A}(\mathbf{x}) = \frac{1}{2}\mathbf{B} \times \mathbf{x},$$

a basis of single particle wave-functions in the lowest Landau level (LLL) is given by [NSS08]

$$\varphi_m(z) = z^m \exp(-|z|^2/(4\ell_0{}^2)), \quad \text{where } z = x + iy.$$

If the electrons are confined to a disk of area A pierced by magnetic flux $B \cdot A$, then there are $N_\Phi = BA/\Phi_0 = BAe/hc$ states in the lowest Landau level (and in each higher Landau level), where B is the magnetic field; h, c, and e are, respectively, Planck's constant, the speed of light, and the electron charge; and $\Phi_0 = hc/e$ is the flux quantum. In the absence of disorder, these single-particle states are all precisely degenerate. When the chemical potential lies between the νth and $(\nu + 1)$th Landau levels, the Hall conductance takes the quantized value

$$4\sigma_{xy} = \nu \frac{e^2}{h}, \quad \text{while } \sigma_{xx} = 0.$$

The 2D electron density, n, is related to ν via the formula $n = \nu eB/(hc)$. In the presence of a periodic potential and/or disorder (e.g., impurities), the Landau levels broaden into bands. However, except at the center of a band, all states are localized when disorder is present (see [DP97] and references therein). When the chemical potential lies in the region of localized states between the centers of the νth and $(\nu + 1)$th Landau bands, the Hall conductance again takes the quantized value $\sigma_{xy} = \nu \frac{e^2}{h}$ while $\sigma_{xx} = 0$. The density will be near but not necessarily equal to $\nu eB/(hc)$. This is known as the Integer quantum Hall effect (since ν is an integer).

The neglect of Coulomb interactions is justified when an integer number of Landau levels is filled, so long as the energy splitting between Landau levels, $\hbar\omega_c = \frac{\hbar eB}{mc}$ is much larger than the scale of the Coulomb energy, $\frac{e^2}{\ell_0}$, where $\ell_0 = \sqrt{hc/eB}$ is the magnetic length. When the electron density is such that a Landau level is only partially filled, Coulomb interactions may be important.

In the absence of disorder, a partially-filled Landau level has a very highly degenerate set of multi-particle states. This degeneracy is broken by electron–electron interactions. For instance, when the number of electrons is $N = N_\Phi/3$, i.e., $\nu = 1/3$, the ground state is non-degenerate and there is a gap to all excitations. When the electrons interact through Coulomb repulsion, the *Laughlin state*,

$$\Psi = \prod_{i>j} (z_i - z_j)^3 \, e^{-\sum_i |z_i|^2/4\ell_0{}^2}, \tag{6.135}$$

is an approximation to the ground state. Such ground states survive even in the presence of disorder if it is sufficiently weak compared to the gap to excited states. More delicate states with smaller excitation gaps are, therefore, only seen in extremely clean devices. However, some disorder is necessary to pin the charged quasi-particle excitations which are created if the density or magnetic field are slightly varied. When these excitations are localized, they do not contribute to the Hall conductance and a plateau is observed.

Quasi-particle excitations above fractional quantum Hall ground states, such as the $\nu = 1/3$ Laughlin state (6.135), are emergent anyons in the sense described above. An explicit calculation of the Berry phase, along the lines of (6.133) shows that quasi-particle excitations above the $\nu = 1/k$ Laughlin states have charge e/k and statistical angle $\theta = \pi/k$ [ASW84]. The charge is obtained from the non-topological part of the Berry phase which is proportional to the flux enclosed by a particle's trajectory times the quasi-particle charge. This is in agreement with a general argument that such quasi-particles must have fractional charge [Lau83]. The result for the statistics of the quasi-particles follows from the topological part of the Berry phase; it is in agreement with strong theoretical arguments which suggest that fractionally charged excitations are necessarily Abelian anyons (see [Wil90] and references therein). Definitive experimental evidence for the existence of fractionally charged excitations at $\nu = 1/3$ has been accumulating in [GS95, DRH97]. The observation of fractional statistics is much more subtle. First steps in that direction have been recently reported [CZG05] but are still debated (see [God07]).

The Laughlin states, with $\nu = 1/k$, are the best understood fractional quantum Hall states, both theoretically and experimentally. To explain more complicated observed fractions, with ν not of the form $\nu = 1/k$, Haldane and Halperin [Hal83a, Hal84] used a hierarchical construction in which quasi-particles of a principle $\nu = 1/k$ state can then themselves condense into a quantized state. In this way, quantized Hall states can be constructed for any odd-denominator fraction ν—but only for odd-denominator fractions. These states all have quasi-particles with fractional charge and Abelian fractional statistics. Later, it was noticed by Jain [Jai89] that the most prominent fractional quantum Hall states are of the form $\nu = p/(2p+1)$, which can be explained by noting that a system of electrons in a high magnetic field can be approximated by a system of auxiliary fermions, called 'composite fermions', in a lower magnetic field. If the electrons are at $\nu = p/(2p+1)$, then the lower magnetic field seen by the 'composite fermions' is such that they fill an integer number of Landau levels $\nu' = p$. Since the latter state has a gap, one can hope that the approximation is valid. The composite fermion picture of fractional quantum Hall states has proven to be qualitatively and semi-quantitatively correct in the LLL [MS03].

Systems with filling fraction $\nu > 1$, can be mapped to $\nu' \leq 1$ by keeping the fractional part of ν and using an appropriately modified Coulomb interaction to account for the difference between cyclotron orbits in the LLL and those in higher Landau levels [NSS08]. This involves the assumption that the inter-Landau level coupling is negligibly small. We note that this may not be a particularly good assumption for higher Landau levels, where the composite fermion picture less successful.

Our confidence in the picture described above for the $\nu = 1/k$ Laughlin states and the hierarchy of odd-denominator states which descend from them derives largely from numerical studies. Experimentally, most of what is known about quantum Hall states comes from transport experiments, measurements of the conductance (or resistance) tensor. While such measurements make it reasonably clear when a quantum Hall plateau exists at a given filling fraction, the nature of the plateau (i.e., the details of the low-energy theory) is extremely hard to discern. Because of this difficulty, numerical studies of small systems (exact diagonalizations and Monte Carlo) have played a very prominent role in providing further insight. Indeed, even Laughlin's original work [Lau83] on the $\nu = 1/3$ state relied heavily on accompanying numerical work. The approach taken was the following. One assumed that the splitting between Landau levels is the largest energy in the problem. The Hamiltonian is projected into the lowest Landau level, where, for a finite number of electrons and a fixed magnetic flux, the Hilbert space is finite-dimensional. Typically, the system is given periodic boundary conditions (i.e., is on a torus) or else is placed on a sphere; occasionally, one works on the disk, e.g., to study edge excitations. The Hamiltonian is then a finite-sized matrix which can be diagonalized by a computer so long as the number of electrons is not too large. Originally, Laughlin examined only 3 electrons, but modern computers can handle sometimes as many as 18 electrons. The resulting ground state wave-function can be compared to a proposed trial wave-function. Throughout the history of the field, this approach has proven to be extremely powerful in identifying the nature of experimentally-observed quantum Hall states when the system in question is deep within a quantum Hall phase, so that the associated correlation length is short and the basic physics is already apparent in small systems.

There are several serious challenges in using such numerical work to interpret experiments. First of all, there is always the challenge of extrapolating finite-size results to the thermodynamic limit. Secondly, simple overlaps between a proposed trial state and an exact ground state may not be sufficiently informative. For example, it is possible that an exact ground state will be adiabatically connected to a particular trial state, i.e., the two wave-functions represent the same phase of matter, but the overlaps may not be very high. For this reason, it is necessary to also examine quantum numbers and symmetries of the ground state, as well as the response of the ground state to various perturbations, particularly the response to changes in boundary conditions and in the flux.

Another difficulty is the choice of Hamiltonian to diagonalize. One may think that the Hamiltonian for a quantum Hall system is just that of 2D electrons in a magnetic field interacting via Coulomb forces. However, the small but finite width (perpendicular to the plane of the system) of the quantum well slightly alters the effective interaction between electrons. Similarly, screening (from any nearby conductors, or from inter-Landau-level virtual excitations), in-plane magnetic fields, and even various types of disorder may alter the Hamiltonian in subtle ways. To make matters worse, one may not even know all the physical parameters (dimensions, doping levels, detailed chemical composition, etc.) of any particular experimental system very accurately. Finally, Landau-level mixing is not small because the energy splitting between Landau levels is not much larger than the other energies in the problem. Thus, it is not even clear that it is correct to truncate the Hilbert space to the finite-dimensional Hilbert space of a single Landau level.

In the case of very robust states, such as the $\nu = 1/3$ state, these subtle effects are unimportant; the ground state is essentially the same irrespective of these small deviations from the idealized Hamiltonian. However, in the case of weaker states, such as those observed between $\nu = 2$ and $\nu = 4$ (some of which we will discuss below), it appears that very small changes in the Hamiltonian can indeed greatly affect the resulting ground state. Therefore, a very valuable approach has been to guess a likely Hamiltonian, and search a space of 'nearby' Hamiltonians, slightly varying the parameters of the Hamiltonian, to map out the phase diagram of the system. These phase diagrams suggest the exciting technological possibility that detailed numerics will allow us to engineer samples with just the right small perturbations so as display certain quantum Hall states more clearly [NSS08].

Possible Non-Abelian States

The observation of a quantum Hall state with an even denominator filling fraction [WES87], the $\nu = 5/2$ state, was the first indication that not all fractional quantum Hall states fit the above hierarchy (or equivalently composite fermion) picture. Independently, it was recognized [FL91, Fub91, MR91] that conformal field theory gives a way to write a variety of trial wave-functions for quantum Hall states, as we describe below. Using this approach, the so-called *Moore–Read Pfaffian wave-function* was constructed [MR91]

$$\Psi_{\mathrm{Pf}} = \mathrm{Pf}\left(\frac{1}{z_i - z_j}\right)\prod_{i<j}(z_i - z_j)^m e^{-\sum_i |z_i|^2/4\ell_0^2}. \tag{6.136}$$

The *Pfaffian* is the square root of the determinant of an anti-symmetric matrix or, equivalently, the anti-symmetrized sum over pairs:

$$\mathrm{Pf}\left(\frac{1}{z_j - z_k}\right) = \mathcal{A}\left(\frac{1}{z_1 - z_2}\frac{1}{z_3 - z_4}\cdots\right).$$

For m even, this is an even-denominator quantum Hall state in the lowest Landau level. Moore and Read suggested that its quasi-particle excitations would exhibit non-Abelian statistics [MR91]. This wave-function is the exact ground state of a 3-body repulsive interaction; as we discuss below, it is also an approximate ground state for more realistic interactions. This wave-function is a representative of a universality class which has remarkable properties which we discuss in detail in this section. There are 2^{n-1} states with $2n$ quasi-holes at fixed positions, thereby establishing the degeneracy of multi-quasi-particle states which is required for non-Abelian statistics [NSS08]. Furthermore, these quasi-hole wave-functions can also be related to conformal field theory, from which it can be deduced that the 2^{n-1}-dimensional vector space of states can be understood as the spinor representation of $SO(2n)$; braiding particles i and j has the action of a $\pi/2$ rotation in the $i - j$ plane in $\mathbb{R}^{2n}$. In short, these quasi-particles are essentially Ising anyons (with the difference being an additional Abelian component to their statistics). Although these properties were uncovered using specific wave-functions which are eigenstates of the 3-body interaction for which the *Pfaffian wave-function* is the exact ground state, they are representative of an entire universality class. The effective field theory for this universality class is $SU(2)$ Chern–Simons theory at level $k = 2$ together with an additional Abelian Chern–Simons term [FNT98, FHZ01]. Chern–Simons theory is the archetypal topological quantum field theory (TQFT). As we describe, Chern–Simons theory is related to the *Jones polynomial* of knot theory [Wit89]; consequently, the current through an interferometer in such a non-Abelian quantum Hall state would give a direct measure of the Jones polynomial for the link produced by the quasi-particle trajectories [FNT98].

One interesting feature of the Pfaffian wave-function is that it is the quantum Hall analog of a $p + ip$ superconductor: the anti-symmetrized product over pairs is the real-space form of the BCS wave-function [GWW92]. Read and Green [RG00] showed that the same topological properties mentioned above are realized by a $p + ip$-wave superconductor, thereby cementing the identification between such a paired state and the *Moore–Read state*. Ivanov [Iva01] computed the braiding matrices by this approach (see also [SOM04, SC06]). Consequently, we will often be able to discuss $p + ip$-wave superconductors and super-fluids in parallel with the $\nu = 5/2$ quantum Hall state, although the experimental probes are significantly different.

As we discuss below, all of these theoretical developments garnered greater interest when numerical work [Mor98] showed that the ground state of systems of up to 18 electrons in the $N = 1$ Landau level at filling fraction $1/2$ is in the universality class of the Moore–Read state. These results revived the conjecture that the lowest Landau level ($N = 0$) of both spins is filled and inert and the electrons in the $N = 1$ Landau level form the analog of the *Pfaffian state* [GWW92]. Consequently, it is the leading candidate for the experimentally-observed $\nu = 5/2$ state.

Read and Rezayi [RR99] constructed a series of non-Abelian quantum Hall states at filling fraction $\nu = N + k/(Mk + 2)$ with M odd, which generalize the Moore–Read state. These states are referred to as the Read–Rezayi $\mathbb{Z}_k$ parafermion states. Recently, a quantum Hall state was observed experimentally with $\nu = 12/5$ [XPV04]. It is suspected (see below) that the $\nu = 12/5$ state may be (the particle hole conjugate of) the $\mathbb{Z}_3$ Read–Rezayi state, although it is also possible that $12/5$ belongs to the conventional Abelian hierarchy as the $2/5$ state does. Such an option is not possible at $\nu = 5/2$ as a result of the even denominator.

In summary, it is well-established that if the observed $\nu = 5/2$ state is in the same universality class as the Moore–Read Pfaffian state, then its quasi-particle excitations are non-Abelian anyons. Similarly, if the $\nu = 12/5$ state is in the universality class of the $\mathbb{Z}_3$ Read–Rezayi state, its quasi-particles are non-Abelian anyons. There is no direct experimental evidence that the $\nu = 5/2$ is in this particular universality class, but there is evidence from numerics, as we further discuss below. There is even less evidence in the case of the $\nu = 12/5$ state. Below we will discuss proposed experiments which could directly verify the non-Abelian character of the $\nu = 5/2$ state and will briefly mention their extension to the $\nu = 12/5$ case. Both of these states, as well as others (e.g., [S99]), were constructed on the basis of very deep connections between conformal field theory, knot theory, and low-dimensional topology [Wit89]. Using methods from these different branches of theoretical physics and mathematics, we will explain the structure of the non-Abelian statistics of the $\nu = 5/2$ and $12/5$ states within the context of a large class of non-Abelian topological states.

The $\nu = 5/2$ fractional quantum Hall state is a useful case history for how numerics can elucidate experiments. This incompressible state is easily destroyed by the application of an in-plane magnetic field [EWS90]. At first it was assumed that this implied that the $5/2$ state is spin-unpolarized or partially polarized since the in-plane magnetic field presumably couples only to the electron spin. Careful finite-size numerical work changed this perception, leading to our current belief that the $5/2$ FQH state is actually in the universality class of the spin-polarized Moore–Read Pfaffian state.

In rather pivotal work [Mor98], it was shown that spin-polarized states at $\nu = 5/2$ have lower energy than spin-unpolarized states. Furthermore, it was shown that varying the Hamiltonian slightly caused a phase transition between a gapped phase that has high overlap with the Moore–Read wave-function and a compressible phase. The proposal put forth was that the most important effect of the in-plane field was not on the electron spins, but rather was to slightly alter the shape of the electron wave-function perpendicular to the sample which, in turn, slightly alters the effective electron–electron interaction, pushing the system over a phase boundary and destroying the gapped state. Further experimental work showed that the effect of the in-plane magnetic field is to drive the system across a phase transition from a gapped quantum Hall phase into an anisotropic compressible phase. Further numerical work [RH00] then mapped out a full phase diagram showing the transition between gapped and compressible phases and showing further that the experimental systems lie exceedingly close to the phase boundary. The correspondence between numerics and experiment has been made more quantitative by comparisons between the energy gap obtained from numerics and the one measured in experiments [MD03]. Very recently, this case has been further strengthened by the application of the density-matrix renormalization group method (DMRG) to this problem [FRD07].

One issue worth considering is possible competitors to the Moore–Read Pfaffian state. Experiments have already told us that there is a fractional quantum Hall state at $\nu = 5/2$. Therefore, our job is to determine which of the possible states is realized there. Serious alternatives to the Moore–Read Pfaffian state fall into two categories. On the one hand, there is the possibility that the ground state at $\nu = 5/2$ is not fully spin-polarized. If it were completely unpolarized, the so-called $(3, 3, 1)$ state [Hal83b, DP97] would be a possibility. However, Morf's numerics [Mor98] and a recent variational Monte Carlo study [DHN07] indicate that an unpolarized state is higher in energy than a fully-polarized state. This can be understood as a consequence of a tendency towards spontaneous ferromagnetism; however, a partially-polarized alternative (which may be either Abelian or non-Abelian) to the Pfaffian is not ruled out [DHN07]. Secondly, even if the ground state at $\nu = 5/2$ is fully spin-polarized, the Pfaffian is not the only possibility. It was very recently noticed that the Pfaffian state is not symmetric under a particle–hole transformation of a single Landau level (which, in this case, is the $N = 1$ Landau level, with the $N = 0$ Landau level filled and assumed inert), even though this is an exact symmetry of the Hamiltonian in the limit that the energy splitting between Landau levels is infinity. Therefore, there is a distinct state, dubbed the anti-Pfaffian [LHR07], which is an equally good state in this limit. Quasi-particles in this state are also essentially Ising anyons, but they differ from Pfaffian quasi-particles by Abelian statistical phases. In experiments, Landau-level mixing is not small, so one or the other state is lower in energy. On a finite torus, the symmetric combination of the Pfaffian and the anti-Pfaffian will be lower in energy, but as the thermodynamic limit is approached, the anti-symmetric combination will become equal in energy. This is a possible factor which complicates the extrapolation of numerics to the thermodynamic limit. On a finite sphere, particle–hole symmetry is not exact; it relates a system with $2N - 3$ flux quanta with a system with $2N + 1$ flux quanta. Thus, the anti-Pfaffian would not be apparent unless one looked at a different value of the flux. To summarize, the only known alternatives to the Pfaffian state, partially-polarized states and the anti-Pfaffian, have not really been tested by numerics, either because the spin-polarization was assumed to be 0% or 100% [Mor98] or because Landau-level mixing was neglected.

With this caveat in mind, it is instructive to compare the evidence placing the $\nu = 5/2$ FQH state in the Moore–Read Pfaffian universality class with the evidence placing the $\nu = 1/3$ FQH state in the corresponding Laughlin universality class. In the latter case, there have been several spectacular experiments [GS95, DRH97] which have observed quasi-particles with electrical charge $e/3$, in agreement with the prediction of the Laughlin universality class. In the case of the $\nu = 5/2$ FQH state, we do not yet have the corresponding measurements of the quasi-particle charge, which should be $e/4$. However, the observation of charge $e/3$, while consistent with the Laughlin universality class, does not uniquely fix the observed state in this class (see, e.g. [Woj01]). Thus, much of our confidence derives from the amazing (99% or better) overlap between the ground state obtained from exact diagonalization for a finite size 2D system with up to 14 electrons and the Laughlin wave-function. In the case of the $\nu = 5/2$ FQH state, the corresponding overlap (for 18 electrons on the sphere) between the $\nu = 5/2$ ground state and the Moore–Read Pfaffian state is reasonably impressive ($\sim 80\%$). This can be improved by modifying the wave-function at short distances without leaving the Pfaffian phase [MS08]. However, on the torus, as we mentioned above, the symmetric combination of the Pfaffian and the anti-Pfaffian is a better candidate wave-function in a finite-size system than the Pfaffian itself (or the anti-Pfaffian). Indeed, the symmetric combination of the Pfaffian and the anti-Pfaffian has an overlap of 97% for 14 electrons [RH00].

To summarize, the overlap is somewhat smaller in the 5/2 case than in the 1/3 case when particle-hole symmetry is not accounted for, but only slightly smaller when it is. This is an indication that Landau-level mixing—which will favor either the Pfaffian or the anti-Pfaffian—is an important effect at $\nu = 5/2$, unlike at $\nu = 1/3$. Moreover, Landau-level mixing is likely to be large because the 5/2 FQH state is typically realized at relatively low magnetic fields, making the Landau level separation energy relatively small.

Given that potentially large effects have been neglected, it is not too surprising that the gap obtained by extrapolating numerical results for finite-size systems [MAD02, MD03] is substantially larger than the experimentally-measured activation gap. Also, the corresponding excitation gap obtained from numerics for the $\nu = 1/3$ state is much larger than the measured activation gap. The discrepancy between the theoretical excitation gap and the measured activation gap is a generic problem of all FQH states, and may be related to poorly understood disorder effects and Landau-level mixing.

Finally, it is important to mention that several recent (2006–07) numerical works in the literature have raised some questions about the identification of the observed 5/2 FQH state with the Moore–Read Pfaffian [TJ06, Tok07, WQ06]. Considering the absence of a viable alternative (apart from the anti-Pfaffian and partially-polarized states, which were not considered by these authors) it seems unlikely that these doubts will continue to persist, as more thorough numerical work indicates (see, e.g. [MS08]).

While our current understanding of the 5/2 state is relatively good, the situation for the experimentally observed 12/5 state is more murky, although the possibilities are even more exciting, at least from the perspective of topological quantum computation. One (relatively dull) possibility is that the 12/5 state is essentially the same as the observed $\nu = 2/5$ state, which is Abelian. However, Read and Rezayi, in their initial work on non-Abelian generalizations of the Moore–Read state [RR99] proposed that the 12/5 state might be (the particle-hole conjugate of) their $\mathbb{Z}_3$ para-fermion (or $SU(2)$ level 3) state. This is quite an exciting possibility because, unlike the non-Abelian Moore–Read state at 5/2, the $\mathbb{Z}_3$ para-fermion state would have braiding statistics that allow universal topological quantum computation.

The initial numerics by Read and Rezayi [RR99] indicated that the 12/5 state is very close to a phase transition between the Abelian hierarchy state and the non-Abelian para-fermion state. Some more recent work by the same authors has mapped out a detailed phase diagram showing precisely for what range of parameters a system should be in the non-Abelian phase. It was found that the non-Abelian phase is not very 'far' from the results that would be expected from most real experimental systems. This again suggests that (if the system is not already in the non-Abelian phase), we may be able to engineer slight changes in an experimental sample that would push the system over the phase boundary into the non-Abelian phase.

Experimentally, very little is actually known about the 12/5 state. Indeed, a well quantized plateau has only ever been seen in a single published [XPV04] experiment. Furthermore, there is no experimental information about spin polarization (the non-Abelian phase should be polarized whereas the Abelian phase could be either polarized or unpolarized), and it is not at all clear why the 12/5 state has been seen, but its particle-hole conjugate, the 13/5 state, has not (in the limit of infinite Landau level separation, these two states will be identical in energy). Nonetheless, despite the substantial uncertainties, there is a great deal of excitement about the possibility that this state will provide a route to topological quantum computation.

The most strongly observed fractional quantum Hall states are the composite fermion states $\nu = p/(2p + 1)$, or are simple generalizations of them. There is little debate that these states are likely to be Abelian. However, there are a number of observed exotic states whose origin is not currently agreed upon. An optimist may look at any state of unknown origin and suggest that it is a non-Abelian state. Indeed, non-Abelian proposals (published and unpublished) have been made for a great variety states of uncertain origin (see, e.g. [WWQ06]): 3/8, 4/11, 8/3, and 7/3. More conventional Abelian proposals have been made for each of these states too [CJ04, GLS04, WQ02, WYQ04]. For each of these states, there is a great deal of research left to be done, both theoretical and experimental, before any sort of definitive conclusion is reached.

In this context, it is worthwhile to mention another class of quantum Hall systems where non-Abelian anyons could exist, namely bilayer or multilayer $2D$ systems [GWW91, DP97]. More work is necessary in investigating the possibility of non-Abelian multilayer quantum Hall states.

Interference Experiments

While numerics give useful insight about the topological nature of observed quantum Hall states, experimental measurements will ultimately play the decisive role. So far, rather little has been directly measured experimentally about the topological nature of the $\nu = 5/2$ state and even less is known about other putative non-Abelian quantum Hall states such as $\nu = 12/5$. In particular, there is no direct experimental evidence for the non-Abelian nature of the quasi-particles. The existence of a degenerate, or almost degenerate, subspace of ground states leads to a zero-temperature entropy and heat capacity, but those are very hard to measure experimentally. Furthermore, this degeneracy is just one requirement for non-Abelian statistics to take place. How then does one demonstrate experimentally that fractional quantum Hall states, particularly the $\nu = 5/2$ state, are indeed non-Abelian?

The fundamental quasi-particles (i.e., the ones with the smallest electrical charge) of the Moore–Read Pfaffian state have charge $e/4$ [MR91,

GWW92]. The fractional charge does not uniquely identify the state—the Abelian $(3, 3, 1)$ state has the same quasi-particle charge—but a different value of the minimal quasi-particle charge at $\nu = 5/2$ would certainly rule out the Pfaffian state. Hence, the first important measurement is the quasi-particle charge, which was done more than 10 years ago in the case of the $\nu = 1/3$ state [GS95, DRH97].

If the quasi-particle charge is shown to be $e/4$, then further experiments which probe the braiding statistics of the charge $e/4$ quasi-particles will be necessary to pin down the topological structure of the state. One way to do this is to use a mesoscopic interference device. Consider a *Fabry–Perot interferometer*, as depicted in Fig. 6.10. A Hall bar lying parallel to the x-axis is put in a field such that it is at filling fraction $\nu = 5/2$. It is perturbed by two constrictions, as shown in the figure. The two constrictions introduce two amplitudes for inter-edge tunneling, $t_{1,2}$. To lowest order in $t_{1,2}$, the four-terminal longitudinal conductance of the Hall bar, is [NSS08]

$$G_L \propto |t_1|^2 + |t_2|^2 + 2\operatorname{Re}\left\{t_1^* t_2 e^{i\phi}\right\}. \tag{6.137}$$

For an integer Landau filling, the relative phase ϕ may be varied either by a variation of the magnetic field or by a variation of the area of the 'cell' defined by the two edges and the two constrictions, since that phase is $2\pi\Phi/\Phi_0$, with $\Phi = BA$ being the flux enclosed in the cell, A the area of the cell, and Φ_0 the flux quantum. Thus, when the area of the cell is varied by means of a side gate, the back-scattered current should oscillate.

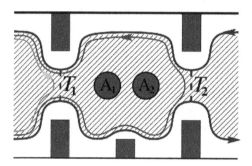

Fig. 6.10. A quantum Hall analog of a Fabry–Perot interferometer. Quasi-particles can tunnel from one edge to the other at either of two point contacts. To lowest order in the tunneling amplitudes, the backscattering probability, and hence the conductance, is determined by the interference between these two processes. The area in the cell can be varied by means of a side gate in order to observe an interference pattern. It is asymptotically characterized as projection of the target.s anyonic charge AND decoherence of anyonic charge entanglement between the interior and exterior of the target region (modified and adapted from [NSS08]).

For fractional quantum Hall states, the situation is different [CFK97]. In an approximation in which the electronic density is determined by the requirement of charge neutrality, a variation of the area of the cell varies the flux it encloses and keeps its bulk Landau filling unaltered. In contrast, a variation of the magnetic field changes the filling fraction in the bulk, and consequently introduces quasi-particles in the bulk. Since the statistics of the quasi-particles is fractional, they contribute to the phase ϕ. The back-scattering probability is then determined not only by the two constrictions and the area of the cell they define, but also by the number of localized quasi-particles that the cell encloses. By varying the voltage applied to an anti-dot in the cell, we can independently vary the number of quasi-particles in the cell. Again, however, as the area of the cell is varied, the back-scattered current oscillates.

For non-Abelian quantum Hall states, the situation is more interesting (see [FNT98, DFN05, SH06, BKS06, BSS06, CS06]). Consider the case of the Moore–Read Pfaffian state. For clarity, we assume that there are localized $e/4$ quasi-particles only within the cell (either at the anti-dot or elsewhere in the cell). If the current comes from the left, the portion of the current that is back-reflected from the left constriction does not encircle any of these quasi-particles, and thus does not interact with them. The part of the current that is back-scattered from the right constriction, on the other hand, does encircle the cell, and therefore applies a unitary transformation on the subspace of degenerate ground states. The final state of the ground state subspace that is coupled to the left back-scattered wave, $|\xi_0\rangle$, is then different from the state coupled to the right partial wave, $\hat{U}|\xi_0\rangle$. Here $\hat{U}$ is the unitary transformation that results from the encircling of the cell by the wave scattered from the right constriction. The interference term in the four-terminal longitudinal conductance, the final term in (6.137), is then multiplied by the matrix element $\langle\xi_0|\hat{U}|\xi_0\rangle$:

$$G_L \propto |t_1|^2 + |t_1|^2 + 2\,\mathrm{Re}\left\{t_1^* t_2 e^{i\phi}\left\langle\xi_0|\hat{U}|\xi_0\right\rangle\right\}. \tag{6.138}$$

For the Moore–Read Pfaffian state, which is believed to be realized at $\nu = 5/2$, the expectation value $\langle\xi_0|\hat{U}|\xi_0\rangle$ depends first and foremost on the parity of the number of $e/4$ quasi-particles localized in the cell. When that number is odd, the resulting expectation value is zero. When that number is even, the expectation value is non-zero and may assume one of two possible values, that differ by a minus sign. As a consequence, when the number of localized quasi-particles is odd, *no interference pattern is seen*, and the back-scattered current does not oscillate with small variations of the area of the cell. When that number is even, the back-scattered current oscillates as a function of the area of the cell.

A way to understand this striking result is to observe that the localized quasi-particles in the cell can be viewed as being created in pairs from the vacuum. Let us suppose that we want to have N quasi-particles in the cell. If N is odd, then we can create $(N + 1)/2$ pairs and take one of the resulting quasi-particles outside of the cell, where it is localized. Fusing all $N+1$ of these particles gives the trivial particle since they were created from the vacuum.

Now consider what happens when a current-carrying quasi-particle tunnels at one of the two point contacts. If it tunnels at the second one, it braids around the N quasi-particles in the cell (but not the $(N + 1)$th, which is outside the cell). This changes the fusion channel of the $N + 1$ localized quasi-particles. In the language introduced above, each $e/4$ quasi-particle is a σ particle. An odd number N of them can only fuse to σ; fused now with the $(N+1)$th, they can either give $\mathbf{1}$ or ψ. Current-carrying quasi-particles, when they braid with the N in the cell, toggle the system between these two possibilities. Since the state of the localized quasi-particles has been changed, such a process cannot interfere with a process in which the current-carrying quasi-particle tunnels at the first junction and does not encircle any of the localized quasi-particles. Therefore, the localized quasi-particles 'measure' which trajectory the current-carrying quasi-particles take [BSS07]. If N is even, then we can create $(N + 2)/2$ pairs and take two of the resulting quasi-particles outside of the cell. If the N quasi-particles in the cell all fuse to the trivial particle, then this is not necessary, we can just create $N/2$ pairs. However, if they fuse to a neutral fermion ψ, then we will need a pair outside the cell which also fuses to ψ so that the total fuses to $\mathbf{1}$, as it must for pair creation from the vacuum. A current-carrying quasi-particle picks up a phase depending on whether the N quasi-particles in the cell fuse to $\mathbf{1}$ or ψ.

The Fabry–Perot interferometer allows also for the interference of waves that are back-reflected several times. For an integer filling factor, in the limit of strong back-scattering at the constrictions, the sinusoidal dependence of the Hall bar's conductance on the area of the cell gives way to a resonance-like dependence: the conductance is zero unless a Coulomb peak develops. For the $\nu = 5/2$ state, again, the parity of the number of localized quasi-particles matters: when it is odd, the Coulomb blockade peaks are equally spaced. When it is even, the spacing between the peaks alternate between two values [SH06].

The Moore–Read Pfaffian state, which is possibly realized at $\nu = 5/2$, is the simplest of the non-Abelian states. The other states are more complex, but also richer. The geometry of the Fabry–Perot interferometer may be analyzed for these states as well. In general, for all non-Abelian states the conductance of the Hall bar depends on the internal state of the quasi-particles localized between the constrictions—i.e., the quasi-particle to which they fuse. However, only for the Moore–Read Pfaffian state is the effect quite so dramatic. For example, for the $\mathbb{Z}_3$ para-fermion state which may be realized at $\nu = 12/5$, when the number of localized quasi-particles is larger than three, the fusion channel of the quasi-particles determines whether the interference is fully visible or suppressed by a factor of $-\varphi^{-2}$ (with φ being the golden ratio $(\sqrt{5} + 1)/2$) [BSS06, CS06]. The number of quasi-particles, on the other hand, affects only the phase of the interference pattern. Similar to the case of $\nu = 5/2$ here too the position of Coulomb blockade peaks on the two parameter plane of area and magnetic field reflects the non-Abelian nature of the quasi-particles [NSS08].

A Fractional Quantum Hall Quantum Computer

We now describe how the constricted Hall bar may be utilized as a quantum bit [DFN05]. To that end, an even number of $e/4$ quasi-particles should be trapped in the cell between the constrictions, and a new, tunable, constriction should be added between the other two so that the cell is broken into two cells with an odd number of quasi-particles in each. One way to tune the number of quasi-particles in each half is to have two anti-dots in the Hall bar. By tuning the voltage on the anti-dots, we can change the number of quasi-holes on each. Let us assume that we thereby fix the number of quasi-particles in each half of the cell to be odd. For concreteness, let us take this odd number to be one (i.e., let us assume that we are in the idealized situation in which there are no quasi-particles in the bulk, and one quasi-hole on each anti-dot). These two quasi-holes then form a two-level system, i.e., a qubit. This two-level system can be understood in several ways, which we discuss in detail below. In brief, the two states correspond to whether the two σs fuse to 1 or ψ or, in the language of chiral p-wave superconductivity, the presence or absence of a neutral *Majorana fermion*; or, equivalently, as the fusion of two quasi-particles carrying the spin-$1/2$ representation of an $SU(2)$ gauge symmetry in the spin-0 or spin-1 channels.

The interference between the t_1 and t_2 processes depends on the state of the two-level system, so the qubit can be read by a measurement of the 4-terminal longitudinal conductance [NSS08]

$$G_L \propto |t_1|^2 + |t_2|^2 \pm 2 \, \mathrm{Re} \left\{ t_1^* t_2 e^{i\phi} \right\}, \qquad (6.139)$$

where the $\pm$ comes from the dependence of $\langle \xi_0 | \hat{U} | \xi_0 \rangle$ on the state of the qubit.

The purpose of the middle constriction is to allow us to manipulate the qubit. The state may be flipped, i.e., a σ_x or NOT gate can be applied, by the passage of a single quasi-particle from one edge to the other, provided that its trajectory passes in between the two localized quasi-particles. This is a simple example of how braiding causes non-trivial transformations of multi-quasi-particle states of non-Abelian quasi-particles. If we measure the 4-terminal longitudinal conductance G_L before and after applying this NOT gate, we will observe different values according to (6.139).

For this operation to be a NOT gate, it is important that just a single quasi-particle (or any odd number) tunnel from one edge to the other across the middle constriction. In order to regulate the number of quasi-particles which pass across the constriction, it may be useful to have a small anti-dot in the middle of the constriction with a large charging energy so that only a single quasi-particle can pass through at a time. If we do not have good control over how many quasi-particles tunnel, then it will be essentially random whether an even or odd number of quasi-particles tunnel across; half of the time, a NOT gate will be applied and the backscattering probability (hence the conductance) will change while the other half of the time, the backscattering probability is unchanged. If the constriction is pinched down to such an extreme that the 5/2 state is disrupted between the quasi-particles, then when it is restored, there will be an equal probability for the qubit to be in either state.

This qubit is topologically protected because its state can only be affected by a charge $e/4$ quasi-particle braiding with it. If a charge $e/4$ quasi-particle winds around one of the anti-dots, it effects a NOT gate on the qubit. The probability for such an event can be very small because the density of thermally-excited charge $e/4$ quasi-particles is exponentially suppressed at low temperatures, $n_{qp} \sim e^{-\Delta/(2T)}$. The simplest estimate of the error rate Γ (in units of the gap) is then of activated form:

$$\Gamma/\Delta \sim (T/\Delta) \, e^{-\Delta/(2T)}.$$

The most favorable experimental situation [XPV04] considered in [DFN05] has $\Delta \approx 500$ mK and $T \sim 5$ mK, producing an astronomically low error rate $\sim 10^{-15}$. This should be taken as an overly optimistic estimate. A more definitive answer is surely more complicated since there are multiple gaps which can be relevant in a disordered system. Furthermore, at very low temperatures, we would expect quasi-particle transport to be dominated by variable-range hopping of localized quasi-particles rather than thermal activation. Indeed, the crossover to this behavior may already be apparent [PXS99], in which case, the error suppression will be considerably weaker at the lowest temperatures. Although the error rate, which is determined by the probability for a quasi-particle to wind around the anti-dot, is not the same as the longitudinal resistance, which is the probability for it to go from one edge of the system to the other, the two are controlled by similar physical processes. A more sophisticated estimate would require a detailed analysis of the quasi-particle transport properties which contribute to the error rate. In addition, this error estimate assumes that all of the trapped (unintended) quasi-particles are kept very far from the quasi-particles which we use for our qubit so that they cannot exchange topological quantum numbers with our qubit via tunneling.

The device envisioned above can be generalized to one with many anti-dots and, therefore, many qubits. More complicated gates, such as a CNOT gate can be applied by braiding quasi-particles. It is not clear how to braid quasi-particles localized in the bulk, perhaps by transferring them from one anti-dot to another in a kind of 'bucket brigade'. This is an important problem for any realization of topological quantum computing. However, even if this were solved, there would still be the problem that braiding alone is not sufficient for universal quantum computation in the $\nu = 5/2$ state (assuming that it is the Moore–Read Pfaffian state). One must either use some unprotected operations (just two, in fact) or else use the $\nu = 12/5$, if it turns out to be the $\mathbb{Z}_3$ para-fermion non-Abelian state.

Physical Systems and Materials Considerations

As seen in the device described in the previous subsection, topological protection in non-Abelian fractional quantum Hall states hinges on the energy gap (Δ) separating the many-body degenerate ground states from the low-lying excited states. This excitation gap also leads to the incompressibility of the quantum Hall state and the quantization of the Hall resistance. Generally speaking, the larger the size of this excitation gap compared to the temperature, the more robust the topological protection, since thermal excitation of stray quasi-particles, which goes as $\exp(-\Delta/(2T))$, would potentially lead to errors [NSS08].

It must be emphasized that the relevant T here is the temperature of the electrons (or more precisely, the quasi-particles) and not that of the GaAs-AlGaAs lattice surrounding the 2D electron layer. Although the surrounding bath temperature could be lowered to 1 mK or below by using adiabatic de-magnetization in dilution refrigerators, the 2D electrons themselves thermally decouple from the bath at low temperatures and it is very difficult to cool the 2D electrons below $T \approx 20$ mK. It will be a great boost to hopes for topological quantum computation using non-Abelian fractional quantum Hall states if the electron temperature can be lowered to 1 mK or even below, and serious efforts are currently underway in several laboratories with this goal.

Unfortunately, the excitation gaps for the expected non-Abelian fractional quantum Hall states are typically very small (compared, for example, with the $\nu = 1/3$ fractional quantum Hall state). The early measured gap for the 5/2 state was around $\Delta \sim 25$ mK (in 1987) [WES87], but steady improvement in materials quality, as measured by the sample mobility, has considerably enhanced this gap. In the highest mobility samples available in 2007, $\Delta \approx 600$ mK [CKD07]. Indeed, there appears to be a close connection between the excitation gap Δ and the mobility (or the sample quality). Although the details of this connection are not well-understood, it is empirically well-established that enhancing the 2D mobility invariably leads to larger measured excitation gaps. In particular, an empirical relation, $\Delta = \Delta_0 - \Gamma$, where Δ is the measured activation gap and Δ_0 the ideal excitation gap with Γ being the level broadening arising from impurity and disorder scattering, has often been discussed in the literature (see, e.g., [DST93]). Writing the mobility $\mu = e\tau/m$, with τ the zero field Drude scattering time, we can write (an approximation of) the level broadening as $\Gamma = \hbar/(2\tau)$, indicating $\Gamma \sim \mu^{-1}$ in this simple picture, and therefore increasing the mobility should steadily enhance the operational excitation gap, as is found experimentally. It has recently been pointed out [MAD02] that by reducing Γ, an FQH gap of 2–3 K may be achievable in the 5/2 FQH state. Much less is currently known about the 12/5 state, but recent numeric al studies suggest that the maximal gap in typical samples will be quite a bit lower than for 5/2 [NSS08].

It is also possible to consider designing samples that would inherently have particularly large gaps. First of all, the interaction energy (which sets the overall scale of the gap) is roughly of the $1/r$ Coulomb form, so it scales as the inverse of the interparticle spacing. Doubling the density should therefore increase the gaps by roughly 40%. Although there are efforts underway to increase the density of samples [WMP07], there are practical limitations to how high a density one can obtain since, if one tries to over-fill a quantum well with electrons, the electrons will no longer remain strictly two dimensional (i.e., they will start filling higher subbands, or they will not remain in the well at all). Secondly, as discussed above, since the non-Abelian states appear generally to be very sensitive to the precise parameters of the Hamiltonian, another possible route to increased excitation gap would be to design the precise form of the inter-electron interaction, so that the Hamiltonian is at a point in the phase diagram with maximal gap. With all approaches for re-designing samples, however, it is crucial to keep the disorder level low, which is an exceedingly difficult challenge.

Note that a large excitation gap (and correspondingly low temperature) suppresses thermally excited quasi-particles but does not preclude stray localized quasi-particles which could be present even at $T = 0$. As long as their positions are known and fixed, and as long as they are few enough in number to be sufficiently well separated, these quasi-particles would not present a problem, as one could avoid moving other quasi-particles near their positions and one could then tailor algorithms to account for their presence. If the density of stray localized quasi-particles is sufficiently high, however, this would no longer be possible. Fortunately, these stray particles can be minimized in the same way as one of the above discussed solutions to keeping the energy gap large—improve the mobility of the 2D electron sample on which the measurements (i.e., the computation operations) are being carried out. Improvement in the mobility leads to both the enhancement of the excitation gap and the suppression of unwanted quasi-particle localization by disorder.

We should emphasize, however, how extremely high quality the current samples already are. Current 'good' sample mobilities are in the range of $10\text{--}30 \times 10^6$ cm^2/(V s). To give the reader an idea of how impressive this is, we note that under such conditions, at low temperatures, the mean free path for an electron may be a macroscopic length of a tenth of a millimeter or more.[26]

Nonetheless, further MBE technique and design improvement may be needed to push low-temperature 2D electron mobilities to 100×10^6 cm^2/[V s] or above for topological quantum computation to be feasible. At lower temperatures, $T < 100$ mK, the phonon scattering is very strongly suppressed [SPB90], and therefore, there is essentially no intrinsic limit to how high the 2D electron mobility can be since the extrinsic scattering associated with impurities and disorder can, in principle, be eliminated through materials improvement. In fact, steady materials improvement in modulation-doped 2D GaAs–AlGaAs heterostructures grown by the MBE technique has enhanced the 2D electron mobility from 10^4 cm^2/[V s] in the early 1980's to 30×10^6 cm^2/[V s] in 2004, a three orders of magnitude improvement in ma-

---

[26] Compare this to, say, copper at room temperature, which has a mean free path of tens of nanometers or less.

terials quality in roughly twenty years. Indeed, the vitality of the entire field of quantum Hall physics is a result of these amazing advances.

Other Proposed Non-Abelian Systems

This review devotes a great deal of attention to the non-Abelian anyonic properties of certain fractional quantum Hall states (e.g., $\nu = 5/2, 12/5$, etc. states) in 2D semiconductor structures, mainly because theoretical and experimental studies of such (possibly) non-Abelian fractional quantized Hall states is a mature subject, dating back to 1986, with many concrete results and ideas, including a recent proposal [DFN05] for the construction of qubits and a NOT gate for topological quantum computation. But there are several other systems which are potential candidates for topological quantum computation, and we briefly discuss these systems in this subsection. Indeed, the earliest proposals for fault-tolerant quantum computation with anyons were based on spin systems, not the quantum Hall effect [NSS08].

First, we emphasize that the most crucial necessary condition for carrying out topological quantum computation is the existence of appropriate 'topological matter', i.e., a physical system in a topological phase. Such a phase of matter has suitable ground state properties and quasi-particle excitations manifesting non-Abelian statistics. Unfortunately, the necessary and sufficient conditions for the existence of topological ground states are not known even in theoretical models. We note that the topological symmetry of the ground state is an emergent symmetry at low energy, which is not present in the microscopic Hamiltonian of the system. Consequently, given a Hamiltonian, it is very difficult to determine if its ground state is in a topological phase. It is certainly no easier than showing that any other low-energy emergent phenomenon occurs in a particular model. Except for rare exactly solvable models (see, e.g. [Kit06]), topological ground states are inferred on the basis of approximations and inspired guesswork. On the other hand, if topological states exist at all, they will be robust.[27] For this reason, we believe that if it can be shown that some model Hamiltonian has a topological ground state, then a real material which is described approximately by that model is likely to have a topological ground state as well.

One theoretical model which is known to have a non-Abelian topological ground state is a $p + ip$ wave superconductor (i.e., a superconductor where the order parameter is of $p_x + ip_y$ symmetry). Vortices in a superconductor of $p + ip$ pairing symmetry exhibit non-Abelian braiding statistics. This is really just a reincarnation of the physics of the Pfaffian state (believed to be realized at the $\nu = 5/2$ quantum Hall plateau) in zero magnetic field. Chiral p-wave superconductivity/super-fluidity is currently the most transparent route to non-Abelian anyons. As we discuss below, there are multiple physical systems

---

[27] In other words, their topological nature should be fairly insensitive to local perturbations, e.g., electron–phonon interaction or charge fluctuations between traps.

which may host such a reincarnation. The Kitaev honeycomb model [Kit06] is
a seemingly different model which gives rise to the same physics. In it, spins
interact anisotropically in such a way that their Hilbert space can be mapped
onto that of a system of Majorana fermions. In various parameter ranges, the
ground state is in either an Abelian topological phase, or a non-Abelian one
in the same universality class as a $p + ip$ superconductor.

Chiral p-wave superconductors, like quantum Hall states, break parity and
time-reversal symmetries, although they do so spontaneously, rather than as a
result of a large magnetic field. However, it is also possible to have a topological
phase which does not break these symmetries. Soluble theoretical models of
spins on a lattice have been constructed which have P, T-invariant topological
ground states. A very simple model of this type with an *Abelian topological
ground state*, called the *toric code*, was proposed in [Kit03]. Even though it is
not sufficient for topological quantum computation because it is Abelian, it is
instructive to consider this model because non-Abelian models can be viewed
as more complex versions of this model. It describes $s = 1/2$ spins on a lattice
interacting through the following Hamiltonian:

$$H = -J_1 \sum_i A_i - J_2 \sum_p F_p. \tag{6.140}$$

This model can be defined on an arbitrary lattice. The spins are assumed
to be on the links of the lattice. $A_i \equiv \prod_{\alpha \in \mathcal{N}(i)} \sigma_z^\alpha$, where $\mathcal{N}(i)$ is the set of
spins on links α which touch the vertex i, and $F_p \equiv \prod_{\alpha \in p} \sigma_x^\alpha$, where p is
a plaquette and $\alpha \in p$ are the spins on the links comprising the plaquette.
This model is exactly soluble because the A_is and F_ps all commute with each
other. For any $J_1, J_2 > 0$, the ground state $|0\rangle$ is given by $A_i|0\rangle = F_p|0\rangle = |0\rangle$
for all i, p. Quasi-particle excitations are sites i at which $A_i|0\rangle = -|0\rangle$ or
plaquettes p at which $F_p|0\rangle = -|0\rangle$. A pair of excited sites can be created
at i and i' by acting on the ground state with $\prod_{\alpha \in \mathcal{C}} \sigma_x^\alpha$, where the product
is over the links in a chain $\mathcal{C}$ on the lattice connecting i and i'. Similarly, a
pair of excited plaquettes can be created by acting on the ground state with
connected $\prod_{\alpha \in \tilde{\mathcal{C}}} \sigma_z^\alpha$ where the product is over the links crossed by a chain $\tilde{\mathcal{C}}$
on the dual lattice connecting the centers of plaquettes p and p'. Both types
of excitations are bosons, but when an excited site is taken around an excited
plaquette, the wave-function acquires a minus sign. Thus, these two types of
bosons are *relative semions*.

The toric code model is not very realistic, but it is closely related to some
more realistic models such as the *quantum dimer model* (see, e.g. [CCK89]).
The degrees of freedom in this model are dimers on the links of a lattice, which
represent a spin singlet bond between the two spins on either end of a link.
The quantum dimer model was proposed as an effective model for frustrated
anti-ferromagnets, in which the spins do not order, but instead form singlet
bonds which resonate among the links of the lattice—the resonating valence
bond (RVB) state [And73, And87, BZA87] which, in modern language, we
would describe as a specific realization of a simple Abelian topological state

[BFN99, BFN00]. While the quantum dimer model on the square lattice does not have a topological phase for any range of parameter values (the RVB state is only the ground state at a critical point), the model on a triangular lattice does have a topological phase.

Levin [LW05a, LW05b] constructed a model which is, in a sense, a non-Abelian generalization of Kitaev's toric code model. It is an exactly soluble model of spins on the links (two on each link) of the honeycomb lattice with three-spin interactions at each vertex and twelve-spin interactions around each plaquette. This model realizes a non-Abelian phase which supports Fibonacci anyons, which permits universal topological quantum computation (and generalizes straightforwardly to other non-Abelian topological phases). Other models have been constructed [FNS05a, FF05] which are not exactly soluble but have only two-body interactions and can be argued to support topological phases in some parameter regime. However, there is still a considerable gulf between models which are soluble or quasi-soluble and models which might be considered realistic for some material.

Models such as the *Kitaev model* and *Levin–Wen model* are deep within topological phases; there are no other competing states nearby in their phase diagram. However, simple models such as the *Heisenberg model*,[28] or extensions of the *Hubbard model*,[29] are not of this form. The implication is that

---

[28] The Heisenberg model is a statistical-mechanical model used in the study of critical points and phase transitions of magnetic systems, in which the spin of the magnetic systems are treated quantum mechanically. In the prototypical *Ising model*, defined on a nD lattice, at each lattice site, a spin $\sigma_i \in \{\pm\}$ represents a microscopic magnetic dipole to which the magnetic moment is either up or down. For quantum-mechanical reasons, the dominant coupling between two dipoles may cause nearest-neighbors to have lowest energy when they are aligned. Under this assumption (so that magnetic interactions only occur between adjacent dipoles) the Hamiltonian can be written in the form:

$$\hat{H} = -J \sum_{j=1}^{N} \sigma_j \sigma_{j+1} - h \sum_{j=1}^{N} \sigma_j,$$

for a 1D model consisting of N dipoles, subject to the periodic boundary condition $\sigma_{N+1} = \sigma_1$. The Heisenberg model is a more realistic model in that it treats the spins quantum-mechanically, by replacing the spin by a quantum operator (Pauli spin-1/2 matrices at spin 1/2), and the coupling constants J_x, J_y, and J_z.

[29] The Hubbard model, named after John Hubbard, is an approximate model used, especially in solid state physics, to describe the transition between conducting and insulating systems. It is the simplest model of interacting particles in a lattice, with only two terms in the Hamiltonian: a kinetic term allowing for tunneling ('hopping') of particles between sites of the lattice and a potential term consisting of an on-site interaction. The particles can either be fermions, as in Hubbard's original work (in 1963), or bosons, when the model is referred to as either the *Bose–Hubbard model* or the 'boson Hubbard model'. The Hubbard model is a good approximation for particles in a periodic potential at sufficiently low temperatures that all the particles

such models are not deep within a topological phase, and topological phases must compete with other phases, such as broken symmetry phases. In the quantum dimer model (see, e.g. [CCK89]), for instance, an Abelian topological phase must compete with various crystalline phases which occupy most of the phase diagram. This is presumably one obstacle to finding topological phases in more realistic models, i.e., models which would give an approximate description of some concrete physical system.

There are several physical systems—apart from fractional quantum Hall states, which might be promising hunting grounds for topological phases, including transition metal oxides and ultra-cold atoms in optical traps. The transition metal oxides have the advantage that we already know that they give rise to striking collective phenomena such as high-T_c superconductivity, colossal magnetoresistance, stripes, and thermoelectricity. Unfortunately, their physics is very difficult to unravel both theoretically and experimentally for this very reason: there are often many different competing phenomena in these materials. This is reflected in the models which describe transition metal oxides. They tend to have many closely competing phases, so that different approximate treatments find rather different phase diagrams. There is a second advantage to the transition metal oxides, namely that many sophisticated experimental techniques have been developed to study them, including transport, thermodynamic measurements, photoemission, neutron scattering, X-ray scattering, and NMR. Unfortunately, however, these methods are tailored for detecting broken-symmetry states or for giving a detailed understanding of metallic behavior, not for uncovering a topological phase. Nevertheless, this is such a rich family of materials that it would be surprising if there weren't a topological phase hiding there. There is one particular material in this family, Sr_2RuO_4, for which there is striking evidence that it is a chiral p-wave superconductor at low temperatures, $T_c \approx 1.5$ K [XMB06]. Half-quantum vortices in a thin film of such a superconductor would exhibit non-Abelian braiding statistics (since Sr_2RuO_4 is not spin-polarized, one must use half quantum vortices, not ordinary vortices). However, half quantum vortices are usually not the lowest energy vortices in a chiral p-wave superconductor, and a direct experimental observation of the half vortices themselves would be a substantial milestone on the way to topological quantum computation [DNT06].

The current status of research is as follows. 3D single-crystals and thin films of Sr_2RuO_4 have been fabricated and studied. The nature of the superconductivity of these samples has been studied by many experimental probes,

---

are in the lowest Bloch band, as long as any long-range interactions between the particles can be ignored. If interactions between particles on different sites of the lattice are included, the model is often referred to as the 'extended Hubbard model'. For electrons in a solid, the Hubbard model can be considered as an improvement on the *tight-binding model*, which includes only the hopping term. For strong interactions, it can give qualitatively different behavior from the tight-binding model, and correctly predicts the existence of so called *Mott insulators*, which are prevented from becoming conducting by the strong repulsion between the particles.

with the goal of identifying the symmetry of the Cooper-pair. There are many indications that support the identification of the Sr_2RuO_4 as a $p_x + ip_y$ super-conductor. First, experiments that probe the spins of the Cooper pair strongly indicate triplet pairing [MM03]. Such experiments probe the spin susceptibility through measurements of the NMR Knight shift and of neutron scattering. For singlet spin pairing the susceptibility vanishes at zero temperature, since the spins keep a zero polarization state in order to form Cooper pairs. In contrast, the susceptibility remains finite for triplet pairing, and this is indeed the observed behavior. Second, several experiments that probe time reversal symmetry have indicated that it is broken, as expected from a $p \pm ip$-super-conductor. These experiments include muon spin relaxation [MM03] and the polar Kerr effect [XMB06]. In contrast, magnetic imaging experiments designed to probe the edge currents that are associated with a super-conductor that breaks time reversal symmetry did not find the expected signal. The absence of this signal may be attributed to the existence of domains of $p + ip$ interleaved with those of $p - ip$. Altogether, then, Sr_2RuO_4 is likely to be a three dimensional $p + ip$ super-conductor, that may open the way for a realization of a 2D super-conductor that breaks time reversal symmetry.

The other very promising direction to look for topological phases, ultracold atoms in traps, also has several advantages. The Hamiltonian can often be tuned by, for instance, tuning the lasers which define an optical lattice or by tuning through a Feshbach resonance. For instance, there is a specific scheme for realizing the *Hubbard model* in this way. At present there are relatively few experimental probes of these systems, as compared with transition metal oxides or even semiconductor devices. However, to look on the bright side, some of the available probes give information that cannot be measured in electronic systems. Furthermore, new probes for cold atoms systems are being developed at a remarkable rate.

There are two different schemes for generating topological phases in ultracold atomic gases that seem particularly promising at the current time. The first is the approach of using fast rotating dilute Bose gases [WGS98] to make quantum Hall systems of bosons [CWK01]. Here, the rotation plays the role of an effective magnetic field, and the filling fraction is given by the ratio of the number of bosons to the number of vortices caused by rotation. Experimental techniques (see [BSS04, ARV01]) have been developed that can give very large rotation rates and filling fractions can be generated which are as low as $\nu = 500$. While this is sufficiently low that all of the bosons are in a single Landau level, it is still predicted to be several orders of magnitude too high to see interesting topological states. Theoretically, the interesting topological states occur for $\nu < 10$ [CWK01]. In particular, evidence is very strong that $\nu = 1$, should it be achieved, would be the bosonic analogue of the Moore–Read state, and (slightly less strong) $\nu = 3/2$ and $\nu = 2$ would be the Read–Rezayi states, if the inter-boson interaction is appropriately adjusted [RRC05, CR07]. In order to access this regime, either rotation rates will need to be increased substantially, or densities will have to be decreased substantially. While the latter sounds easier, it then results in all of the interaction scales being correspondingly lower, and hence implies that temperature would have to be lower also, which again becomes a challenge. Several other works have proposed using atomic lattice systems where manipulation of parameters of the Hamiltonian induces effective magnetic fields and should also result in quantum hall physics [NSS08].

The second route to generating topological phases in cold atoms is the idea of using a gas of ultra-cold fermions with a p-wave *Feschbach resonance*, which could form a spin-polarized chiral p-wave super-fluid [GRA05]. Preliminary studies of such p-wave systems have been made experimentally [GSB07] and unfortunately, it appears that the decay time of the Feshbach bound states may be so short that thermalization is impossible.

We note that both the $\nu = 1$ rotating boson system and the chiral p-wave super-fluid would be quite closely related to the putative non-Abelian quantum Hall state at $\nu = 5/2$ (as is Sr_2RuO_4). However, there is an important difference between a p-wave super-fluid of cold fermions and the $\nu = 5/2$ state. 2D superconductors, as well as super-fluids in any dimension, have a gapless Goldstone mode. Therefore, there is the danger that the motion of vortices may cause the excitation of low-energy modes. Superfluids of cold atoms may, however, be good test grounds for the detection of localized Majorana modes associated with localized vortices, as those are expected to have a clear signature in the absorption spectrum of RF radiation [TDN07], in the form of a discrete absorption peak whose density and weight are determined by the density of the vortices [GCS07]. One can also realize, using suitable laser configurations, Kitaev's honeycomb lattice model with cold atoms on an optical lattice [DDL03]. It has recently been shown how to braid anyons in such a model [ZST06].

A major difficulty in finding a topological phase in either a transition metal oxide or an ultra-cold atomic system is that topological phases are hard to detect directly. If the phase breaks parity and time-reversal symmetries, either spontaneously or as a result of an external magnetic field, then there is usually an experimental handle through transport, as in the fractional quantum Hall states or chiral p-wave superconductors. If the state does not break parity and time-reversal, however, there is no 'smoking gun' experiment, short of creating quasi-particles, braiding them, and measuring the outcome.

For more technical details, see [NSS08].

6.5 Option Price Modeling Using Quantum Neural Computation

The celebrated Black–Scholes partial differential equation (PDE) describes the time-evolution of the market value of a *stock option* [BS73, Mer73]. Formally, for a function $u = u(t, s)$ defined on the domain $0 \leq s < \infty$, $0 \leq t \leq T$ and describing the market value of a stock option with the stock (asset) price s, the *Black–Scholes PDE* can be written (using the physicist notation: $\partial_z u = \partial u/\partial z$) as a diffusion-type equation,[30] which also resembles the backward Fokker–Planck equation[31]

$$\partial_t u = -\frac{1}{2}(\sigma s)^2 \partial_{ss} u - r s \partial_s u + r u, \qquad (6.141)$$

where $\sigma > 0$ is the standard deviation, or *volatility* of s, r is the short-term prevailing continuously-compounded risk-free interest rate, and $T > 0$ is the time to maturity of the stock option. In this formulation it is assumed that the *underlying* (typically the stock) follows a *geometric Brownian motion* with 'drift' μ and volatility σ, given by the stochastic differential equation (SDE) [Osb59]

$$ds(t) = \mu s(t)dt + \sigma s(t)dW(t), \qquad (6.142)$$

where W is the standard Wiener process.[32]

---

[30] Recall that this similarity with diffusion equation led Black and Scholes to obtain their option price formula as the solution of the diffusion equation with the initial and boundary conditions given by the option contract terms.

[31] Recall that the forward Fokker–Planck equation (also known as the Kolmogorov forward equation, in which the probabilities diffuse outwards as time moves forwards) describes the time evolution of the probability density function $p = p(t, x)$ for the position x of a particle, and can be generalized to other observables as well [Kad00]. Its first use was statistical description of Brownian motion of a particle in a fluid. Applied to the option-pricing process $p = p(t, s)$ with *drift* $D_1 = D_1(t, s)$, *diffusion* $D_2 = D_2(t, s)$ and volatility σ^2, the forward Fokker–Planck equation reads:

$$\partial_t p = \frac{1}{2}\partial_{ss}\left(D_2\sigma^2 p\right) - \partial_s\left(D_1 p\right).$$

The corresponding backward Fokker–Planck equation (which is probabilistic diffusion in reverse, i.e., starting at the final forecasts, the probabilities diffuse outwards as time moves backwards) reads:

$$\partial_t p = -\frac{1}{2}\sigma^2\partial_{ss}\left(D_2 p\right) - \partial_s\left(D_1 p\right).$$

[32] The economic ideas behind the Black–Scholes option pricing theory translated to the stochastic methods and concepts are as follows (see [PPM00]). First, the option price depends on the stock price and this is a random variable evolving with time. Second, the efficient market hypothesis [Fam65, Jen78], i.e., the market incorporates instantaneously any information concerning future market evolution, implies that the random term in the stochastic equation must be delta-correlated. That is: speculative prices are driven by white noise. It is known that any white noise can be written as a combination of the derivative of the Wiener process [Wie61] and white shot noise (see [Gar85]). In this framework, the Black–Scholes option pricing method was first based on the geometric Brownian motion [BS73, Mer73], and it was lately extended to include white shot noise.

The solution of the PDE (6.141) depends on boundary conditions, subject to a number of interpretations, some requiring minor transformations of the basic BS equation or its solution.[33] In practice, the volatility is the least known parameter in (6.141), and its estimation is generally the most important part of pricing options. Usually, the volatility is given in a yearly basis, baselined to some standard, e.g., 252 trading days per year, or 360 or 365 calendar days. However, and especially after the 1987 crash, the geometric Brownian motion model and the BS formula were unable to reproduce the option price data of real markets.[34] As an alternative, models of financial dynamics based on two-dimensional diffusion processes, known as stochastic volatility (SV) models [FPS00b], are being widely accepted as a reasonable explanation for many empirical observations collected under the name of 'stylized facts' [Con01]. In such models the volatility, that is, the standard deviation of returns, originally thought to be a constant, is a random process coupled with the return in a

---

The PDE (6.141) is usually derived from SDEs describing the geometric Brownian motion (6.142), with the solution given by:

$$s(t) = s(0)e^{(\mu - \frac{1}{2}\sigma^2)t + \sigma W(t)}.$$

In mathematical finance, derivation is usually performed using Itô lemma [Ito51] (assuming that the underlying asset obeys the Itô SDE), while in econophysics it is performed using Stratonovich interpretation (assuming that the underlying asset obeys the Stratonovich SDE [Str66]) [Gar85, PPM00].

[33] The basic equation (6.141) can be applied to a number of one-dimensional models of interpretations of prices given to u, e.g., puts or calls, and to s, e.g., stocks or futures, dividends, etc. In the first (and most important) example, $u(t, s) = c(t, s)$ is a call on a *European vanilla option* with exercise price X and maturity at T; then the solution to (6.141) is given by (see, e.g. [Ing00])

$$c(s, t) = sN(d_1) - Xe^{-r(T-t)}N(d_2),$$

$$d_1 = \frac{\ln(s/X) + (r + \frac{1}{2}\sigma^2)(T - t)}{\sigma(T - t)^{1/2}},$$

$$d_2 = \frac{\ln(s/X) + (r - \frac{1}{2}\sigma^2)(T - t)}{\sigma(T - t)^{1/2}}, \quad \text{where}$$

$$N(d) = \frac{1}{2}[1 + \text{Erf}(d/\sqrt{2})].$$

[34] Recall that Black–Scholes model assumes that the underlying volatility is constant over the life of the derivative, and unaffected by the changes in the price level of the underlying. However, this model cannot explain long-observed features of the implied volatility surface such as *volatility smile* and skew, which indicate that implied volatility does tend to vary with respect to strike price and expiration. By assuming that the volatility of the underlying price is a stochastic process itself, rather than a constant, it becomes possible to model derivatives more accurately.

SDE of the form similar to (6.142), so that they both form a two-dimensional diffusion process governed by a pair of Langevin equations [FPS00b, PSM08, MP08].

Using the standard *Kolmogorov probability* approach, instead of the market value of an option given by the Black–Scholes equation (6.141), we could consider the corresponding probability density function (PDF) given by the backward Fokker–Planck equation (see [Gar85]). Alternatively, we can obtain the same PDF (for the market value of a stock option), using the *quantum-probability* formalism [II08a, II08b], as a solution to a time-dependent linear *Schrödinger equation* for the evolution of the complex-valued wave ψ-function for which the absolute square, $|\psi|^2$, is the PDF (see [Voi]).

In this paper we go a step further and propose a novel general quantum-probability based, option-pricing model, which is both *nonlinear* (see [Tri95, Rot99, AR01]) and *adaptive* (see [Tse96, Ing97, II08a]). In other words, we propose a quantum neural computation approach to option price modeling and simulation.

6.5.1 Bidirectional, Spatio-Temporal, Complex-Valued Associative Memory Machine

The new model is defined as a self-organized system of two coupled nonlinear Schrödinger (NLS) equations: one defining the *option-price wave function* $\psi = \psi(t, s)$, with the corresponding *option-price PDF* defined by $|\psi(t, s)|^2$, and the other defining the *volatility wave function* $\sigma = \sigma(t, \sigma)$, with the corresponding *volatility PDF* defined by $|\sigma(t, \sigma)|^2$. The two focusing NLS equations are coupled so that the volatility PDF is a parameter in the option-price NLS, while the option-price PDF is a parameter in the volatility NLS. In addition, both processes evolve in a common self-organizing *market heat potential*.

Formally, we propose an adaptive, semi-symmetrically coupled, volatility + option-pricing model (with interest rate r, imaginary unit $i = \sqrt{-1}$ and Hebbian learning rate c), which represents a bidirectional NLS-based spatio-temporal associative memory. The model is defined (in natural quantum units) by the following stiff NLS-system:

$$\text{Volatility NLS:} \quad i\partial_t \sigma = -\frac{1}{2}s^2|\psi|^2\partial_{ss}\sigma + V(w)|\sigma|^2\sigma, \qquad (6.143)$$

$$\text{Option price NLS:} \quad i\partial_t \psi = -\frac{1}{2}s^2|\sigma|^2\partial_{ss}\psi + |\psi|^2\psi + r\psi, \quad \text{with}$$

$$V(w) = \sum_{i=1}^{N} w_i g_i, \qquad (6.144)$$

$$\text{Adaptation ODE:} \quad \dot{w}_i = -w_i + c|\sigma|g_i|\psi|. \qquad (6.145)$$

In the proposed model, the σ-NLS (6.143) governs the short-range PDF-evolution for stochastic volatility, which plays the role of a nonlinear (variable) coefficient in (6.144); the ψ-NLS (6.144) defines the long-range PDF-evolution for stock price, which plays the role of a nonlinear coefficient in (6.143). The purpose of this coupling is to generate a *leverage effect*, i.e. stock volatility is (negatively) correlated to stock returns[35] (see, e.g. [RPD08]). The w-ODE (6.145) defines the (σ, ψ)-based continuous Hebbian learning [II07b, II07d]. The adaptive volatility potential $V(w)$ is defined as a scalar product of the weight vector w_i and the Gaussian kernel vector g_i and can be related to the market *temperature* (which obeys Boltzmann distribution [Kle02]). The Gaussian vector g_i is defined as:

$$g_i = \exp[-(d - m_i d)^2], \qquad d = y_{target} - y, \qquad (6.146)$$

with

$$y = 2\sin(60t), \qquad y_{target} = s|\sigma|^2 ds, \quad (i = 1, \ldots, N), \qquad (6.147)$$

where $m_i = \text{random}(-1.0, 1.0)$, while $ds = (s_1 - s_0)/(N - 1)$ represents the stock-price increment defined using the *method of lines* (where each NLS was decomposed into N first-order ODEs; see Appendix).

In this way, the whole model effectively performs quantum neural computation, by giving a spatio-temporal and quantum generalization of Kosko's BAM family of neural networks [Kos93, Kos96]. In addition, the solitary nature of NLS equations may describe brain-like effects frequently occurring in financial markets: volatility/price soliton propagation, reflection and collision.

Simulation Results

The model (6.143)–(6.145) has been numerically solved for the following initial conditions (IC) and repeatable boundary conditions (BC):

$$\text{IC:} \quad \sigma_i(0, s) = 0.25, \qquad \psi_i(0, s) = 1, \qquad (6.148)$$

$$\text{BC:} \quad \partial_t \sigma(t, s_0) = \partial_t \sigma(t, s_1), \qquad \partial_t \psi(t, s_0) = \partial_t \psi(t, s_1), \qquad (6.149)$$

using the method of lines (with $N = 30$ lines per NLS discretization, see Appendix). The average simulation time (depending on the random initial

---

[35] The hypothesis that financial leverage can explain the leverage effect was first discussed by F. Black [Bl76].

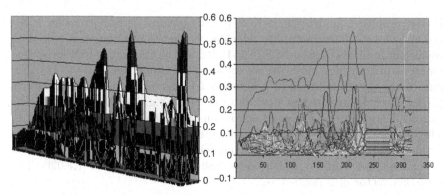

Fig. 6.11. A sample daily evolution of the short-range volatility PDF: global surface (*left*) and individual lines (*right*).

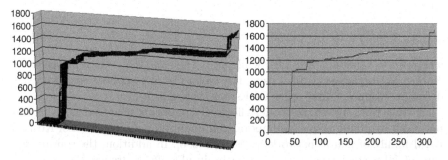

Fig. 6.12. A sample daily evolution of the long-range option-price PDF (corresponding to the volatility from Fig. 6.11): global surface (*left*) and individual lines (*right*).

weights and Gaussians) was 10–30 seconds on a standard Pentium 4 PC, using Visual C++ compiler[36]

A sample daily evolution of the volatility PDF is given in Fig. 6.11. The corresponding option-price PDF is given in Fig. 6.12. The corresponding Hebbian weights and Gaussian kernels are given in Fig. 6.13.

Complex-Valued Wave-Function Lines

Complex-valued quantum probability lines for option price evolution have some interesting properties of their own. Imaginary and real wave-functions

---

[36] In this respect it is important to note that in order to have a useful option pricing model, the speed of calculation is paramount. Especially for European options it is extremely important to have a fast calculation of the price and the hedge parameters, since these are the most liquidly traded financial options.

corresponding to the option-price PDF from Fig. 6.12 are given in Fig. 6.14. They can be combined in a phase-space fashion to give complex-plane plots, as in Fig. 6.15. These complex phase plots of individual wave-function lines depict small fluctuations around circular periodic motions.

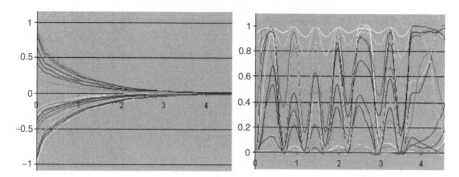

Fig. 6.13. A bimonthly evolution of Hebbian weights (*left*) and Gaussian kernels (*right*). We can see the stabilizing effect of the Hebbian self-organization (all weights tend to zero).

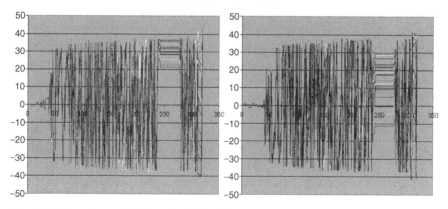

Fig. 6.14. Imaginary (*left*) and real (*right*) option-price wave-function, corresponding to the PDF from Fig. 6.12. Notice: (i) a slow start of the fluctuation, and (ii) the phase transition from 240–240 days from the beginning; this qualitative change of behavior *is not* visible in the PDF.

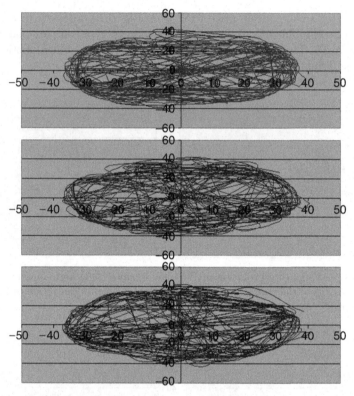

Fig. 6.15. Complex-plane phase-plots for three sample wave-function lines corresponding to real versus imaginary option-price lines from Fig. 6.14. Notice that all three lines fluctuate around a circular periodic motion.

7

Appendix: Mathematical and Computational Tools

In the Appendix we give a brief review of some mathematical [II06c, II07e] and physiological concepts [II06b]—necessary for comprehensive reading of the book, followed by classical computational tools used in quantum neural computation.

7.1 Meta-Language of Categories and Functors

In modern mathematical sciences whenever one defines a new class of mathematical objects, one proceeds almost in the next breath to say what kinds of maps between objects will be considered [Swi75, II06c]. A general framework for dealing with situations where we have some *objects* and *maps between objects*, like sets and functions, vector spaces and linear operators, points in a space and paths between points, etc.—gives the modern metalanguage of categories and functors. Categories are mathematical universes and functors are 'projectors' from one universe onto another. For this reason, in this book we extensively use this associative meta-language, mainly following its founder, S. MacLane [Mac71].

7.1.1 Maps

Notes from Set Theory

Given a map (or, a function) $f : A \rightarrow B$, the set A is called the *domain* of f, and denoted Dom f. The set B is called the *codomain* of f, and denoted Cod f. The codomain is not to be confused with the *range* of $f(A)$, which is in general only a subset of B.

A map $f : X \rightarrow Y$ is called *injective* or 1–1 or an *injection* if for every y in the codomain Y there is at most one x in the domain X with $f(x) = y$. Put another way, given x and x' in X, if $f(x) = f(x')$, then it follows that $x = x'$. A map $f : X \rightarrow Y$ is called *surjective* or *onto* or a *surjection* if for

V.G. Ivancevic, T.T. Ivancevic, *Quantum Neural Computation*,
Intelligent Systems, Control and Automation: Science and Engineering 40,
DOI 10.1007/978-90-481-3350-5_7, © Springer Science+Business Media B.V. 2010

every y in the codomain $\operatorname{Cod} f$ there is at least one x in the *domain* X with $f(x) = y$. Put another way, the *range* $f(X)$ is equal to the codomain Y. A map is *bijective* iff it is both injective and surjective. Injective functions are called the *monomorphisms*, and surjective functions are called the *epimorphisms* in the *category of sets* (see below).

Two main classes of maps (or, functions) that we will use int this book are: (i) continuous maps (denoted as C^0-class), and (ii) smooth or differentiable maps (denoted as C^∞-class). The former class is the core of topology, the letter of differential geometry. They are both used in the core concept of manifold.

A *relation* is any subset of a *Cartesian product* (see below). By definition, an *equivalence relation* α on a set X is a relation which is *reflexive, symmetrical* and *transitive*, i.e., relation that satisfies the following three conditions:

1. *Reflexivity*: each element $x \in X$ is equivalent to itself, i.e., $x\alpha x$,
2. *Symmetry*: for any two elements $a, b \in X$, $a\alpha b$ implies $b\alpha a$, and
3. *Transitivity*: $a\alpha b$ and $b\alpha c$ implies $a\alpha c$.

Similarly, a relation $\leq$ defines a *partial order* on a set S if it has the following properties:

1. *Reflexivity*: $a \leq a$ for all $a \in S$,
2. *Antisymmetry*: $a \leq b$ and $b \leq a$ implies $a = b$, and
3. *Transitivity*: $a \leq b$ and $b \leq c$ implies $a \leq c$.

A *partially ordered set* (or *poset*) is a set taken together with a partial order on it. Formally, a partially ordered set is defined as an ordered pair $P = (X, \leq)$, where X is called the *ground set* of P and $\leq$ is the partial order of P.

Notes From Calculus

Maps

Recall that a *map* (or, *function*) f is a *rule* that assigns to each element x in a set A exactly one element, called $f(x)$, in a set B. A map could be thought of as a *machine* $[[f]]$ with x-input (the *domain* of f is the set of all possible inputs) and $f(x)$-output (the *range* of f is the set of all possible outputs) [Stu99]

$$x \to [[f]] \to f(x)$$

There are four possible ways to represent a function (or map): (i) verbally (by a description in words); (ii) numerically (by a table of values); (iii) visually (by a graph); and (iv) algebraically (by an explicit formula). The most common method for visualizing a function is its *graph*. If f is a function with domain A, then its graph is the set of ordered input–output pairs

$$\{(x, f(x)) : x \in A\}.$$

A generalization of the graph concept is a concept of a *cross-section of a fibre bundle*, which is one of the core geometrical objects for dynamics of complex systems.

Algebra of Maps

Let f and g be maps with domains A and B. Then the maps $f + g$, $f - g$, fg, and f/g are defined as follows [Stu99]

$$(f + g)(x) = f(x) + g(x) \quad \text{domain} = A \cap B,$$
$$(f - g)(x) = f(x) - g(x) \quad \text{domain} = A \cap B,$$
$$(fg)(x) = f(x)g(x) \quad \text{domain} = A \cap B,$$
$$\left(\frac{f}{g}\right)(x) = \frac{f(x)}{g(x)} \quad \text{domain} = \{x \in A \cap B : g(x) \neq 0\}.$$

Compositions of Maps

Given two maps f and g, the composite map $f \circ g$ (also called the *composition* of f and g) is defined by

$$(f \circ g)(x) = f(g(x)).$$

The $(f \circ g)$-machine is composed of the g-machine (first) and then the f-machine [Stu99],

$$x \to [[g]] \to g(x) \to [[f]] \to f(g(x)).$$

For example, suppose that $y = f(u) = \sqrt{u}$ and $u = g(x) = x^2 + 1$. Since y is a function of u and u is a function of x, it follows that y is ultimately a function of x. We calculate this by substitution

$$y = f(u) = f \circ g = f(g(x)) = f(x^2 + 1) = \sqrt{x^2 + 1}.$$

The Chain Rule

If f and g are both differentiable (or smooth, i.e., C^∞) maps and $h = f \circ g$ is the composite map defined by $h(x) = f(g(x))$, then h is differentiable and h' is given by the product [Stu99]

$$h'(x) = f'(g(x))g'(x).$$

In Leibniz notation, if $y = f(u)$ and $u = g(x)$ are both differentiable maps, then

$$\frac{dy}{dx} = \frac{dy}{du}\frac{du}{dx}.$$

The reason for the name *chain rule* becomes clear if we add another link to the chain. Suppose that we have one more differentiable map $x = h(t)$. Then, to calculate the derivative of y with respect to t, we use the chain rule twice,

$$\frac{dy}{dt} = \frac{dy}{du}\frac{du}{dx}\frac{dx}{dt}.$$

Integration and Change of Variables

Given a 1–1 continuous (i.e., C^0) map F with a nonzero *Jacobian* $|\frac{\partial(x,\ldots)}{\partial(u,\ldots)}|$ that maps a region S onto a region R (see [Stu99]), we have the following substitution formulas:

1. for a single integral,

$$\int_R f(x)dx = \int_S f(x(u))\frac{\partial x}{\partial u}du,$$

2. for a double integral,

$$\iint_R f(x,y)dA = \iint_S f(x(u,v),y(u,v))\left|\frac{\partial(x,y)}{\partial(u,v)}\right|dudv,$$

3. for a triple integral,

$$\iiint_R f(x,y,z)dV$$
$$= \iiint_S f(x(u,v,w),y(u,v,w),z(u,v,w))\left|\frac{\partial(x,y,z)}{\partial(u,v,w)}\right|dudvdw,$$

4. similarly for n-tuple integrals.

Notes from General Topology

Topology is a kind of *abstraction* of Euclidean geometry, and also a natural framework for the study of *continuity*.[1] Euclidean geometry is abstracted by regarding triangles, circles, and squares as being the same basic object. Continuity enters because in saying this one has in mind a *continuous deformation* of a triangle into a square or a circle, or any arbitrary shape. On the other hand, a disk with a hole in the center is topologically different from a circle or a square because one cannot create or destroy holes by continuous deformations. Thus using topological methods one does not expect to be able to identify a geometrical figure as being a triangle or a square. However, one does expect to be able to detect the presence of gross features such as holes or the fact that the figure is made up of two disjoint pieces etc. In this way topology produces theorems that are usually qualitative in nature—they may assert, for example, the existence or non-existence of an object. They will not, in general, give the means for its construction [NS83].

---

[1] Intuitively speaking, a function $f : \mathbb{R} \longrightarrow \mathbb{R}$ is continuous near a point x in its domain if its value does not jump there. That is, if we just take δx to be small enough, the two function values $f(x)$ and $f(x + \delta x)$ should approach each other arbitrarily closely. In more rigorous terms, this leads to the following definition: A function $f : \mathbb{R} \longrightarrow \mathbb{R}$ is continuous at $x \in \mathbb{R}$ if for all $\epsilon > 0$, there exists a $\delta > 0$ such that for all $y \in \mathbb{R}$ with $|y - x| < \delta$, we have that $|f(y) - f(x)| < \epsilon$. The whole function is called continuous if it is continuous at every point x.

Topological Space

Study of topology starts with the fundamental notion of *topological space*. Let X be any *set* and $Y = \{X_\alpha\}$ denote a collection, finite or infinite of subsets of X. Then X and Y form a topological space provided the X_α and Y satisfy:

1. Any finite or infinite subcollection $\{Z_\alpha\} \subset X_\alpha$ has the property that $\bigcup Z_\alpha \in Y$,
2. Any *finite subcollection* $\{Z_{\alpha_1}, \ldots, Z_{\alpha_n}\} \subset X_\alpha$ has the property that $\bigcap Z_{\alpha_i} \in Y$, and
3. Both X and the empty set belong to Y.

The set X is then called a topological space and the X_α are called *open sets*. The choice of Y satisfying (2) is said to give a topology to X.

Given two topological spaces X and Y, a *function* (or, a *map*) $f : X \to Y$ is *continuous* if the inverse image of an open set in Y is an open set in X.

The main general idea in topology is to study spaces which can be continuously deformed into one another, namely the idea of *homeomorphism*. If we have two topological spaces X and Y, then a map $f : X \to Y$ is called a homeomorphism iff

1. f is continuous (C^0), and
2. There exists an inverse of f, denoted f^{-1}, which is also continuous.

Definition (2) implies that if f is a homeomorphism then so is f^{-1}. Homeomorphism is the main topological example of *reflexive, symmetrical* and *transitive relation*, i.e., *equivalence relation*. Homeomorphism divides all topological spaces up into *equivalence classes*. In other words, a pair of topological spaces, X and Y, belong to the same equivalence class if they are homeomorphic.

The second example of topological equivalence relation is *homotopy*. While homeomorphism generates equivalence classes whose members are topological spaces, homotopy generates equivalence classes whose members are continuous (C^0) maps. Consider two continuous maps $f, g : X \to Y$ between topological spaces X and Y. Then the map f is said to be *homotopic* to the map g if f can be continuously deformed into g (see below for the precise definition of homotopy). Homotopy is an equivalence relation which divides the space of continuous maps between two topological spaces into equivalence classes [NS83].

Another important notions in topology are *covering, compactness* and *connectedness*. Given a family of sets $\{X_\alpha\} = X$ say, then X is a *covering* of another set Y if $\bigcup X_\alpha$ contains Y. If all the X_α happen to be open sets the covering is called an *open covering*. Now consider the set Y and all its possible open coverings. The set Y is *compact* if for every open covering $\{X_\alpha\}$ with $\bigcup X_\alpha \supset Y$ there always exists a finite subcovering $\{X_1, \ldots, X_n\}$ of Y with $X_1 \cup \cdots \cup X_n \supset Y$. Again, we define a set Z to be *connected* if it cannot be written as $Z = Z_1 \cup Z_2$, where Z_1 and Z_2 are both open non-empty sets and $Z_1 \cap Z_2$ is an empty set.

Let $A_1, A_2, \ldots, A_n$ be closed subspaces of a topological space X such that $X = \bigcup_{i=1}^{n} A_i$. Suppose $f_i : A_i \to Y$ is a function, $1 \leq i \leq n$, such that

$$f_i | A_i \cap A_j = f_j | A_i \cap A_j, \quad 1 \leq i, \, j \leq n. \tag{7.1}$$

In this case f is continuous iff each f_i is. Using this procedure we can define a C^0-function $f : X \to Y$ by cutting up the space X into closed subsets A_i and defining f on each A_i separately in such a way that $f|A_i$ is obviously continuous; we then have only to check that the different definitions agree on the *overlaps* $A_i \cap A_j$.

The *universal property of the Cartesian product*: let $p_X : X \times Y \to X$, and $p_Y : X \times Y \to Y$ be the *projections* onto the first and second factors, respectively. Given any pair of functions $f : Z \to X$ and $g : Z \to Y$ there is a unique function $h : Z \to X \times Y$ such that $p_X \circ h = f$, and $p_Y \circ h = g$. Function h is continuous iff both f and g are. This property characterizes $X \times Y$ up to isomorphism. In particular, to check that a given function $h : Z \to X$ is continuous it will suffice to check that $p_X \circ h$ and $p_Y \circ h$ are continuous.

The *universal property of the quotient*: let α be an equivalence relation on a topological space X, let X/α denote the *space of equivalence classes* and $p_\alpha : X \to X/\alpha$ the *natural projection*. Given a function $f : X \to Y$, there is a function $f' : X/\alpha \to Y$ with $f' \circ p_\alpha = f$ iff $x\alpha x'$ implies $f(x) = f(x')$, for all $x \in X$. In this case f' is continuous iff f is. This property characterizes X/α up to homeomorphism.

Homotopy

Now we return to the fundamental notion of homotopy. Let I be a compact unit interval $I = [0, 1]$. A *homotopy* from X to Y is a continuous function $F : X \times I \to Y$. For each $t \in I$ one has $F_t : X \to Y$ defined by $F_t(x) = F(x, t)$ for all $x \in X$. The functions F_t are called the 'stages' of the homotopy. If $f, g : X \to Y$ are two continuous maps, we say f is homotopic to g, and write $f \simeq g$, if there is a homotopy $F : X \times I \to Y$ such that $F_0 = f$ and $F_1 = g$. In other words, f can be continuously deformed into g through the stages F_t. If $A \subset X$ is a subspace, then F is a homotopy relative to A if $F(a, t) = F(a, 0)$, for all $a \in A, t \in I$.

The homotopy relation $\simeq$ is an equivalence relation. To prove that we have $f \simeq f$ is obvious; take $F(x, t = f(x)$, for all $x \in X$, $t \in I$. If $f \simeq g$ and F is a homotopy from f to g, then $G : X \times I \to Y$ defined by $G(x, t) = F(x, 1 - t)$, is a homotopy from g to f, i.e., $g \simeq f$. If $f \simeq g$ with homotopy F and $g \simeq f$ with homotopy G, then $f \simeq h$ with homotopy H defined by

$$H(x, t) = \begin{cases} F(x, t), & 0 \leq t \leq 1/2, \\ G(x, 2t - 1), & 1/2 \leq t \leq 1. \end{cases}$$

To show that H is continuous we use the relation (7.1).

In this way, the set of all C^0-functions $f : X \to Y$ between two topological spaces X and Y, called the *function space* and denoted by Y^X, is partitioned into equivalence classes under the relation $\simeq$. The equivalence classes are called *homotopy classes*, the homotopy class of f is denoted by $[f]$, and the set of all homotopy classes is denoted by $[X;Y]$.

If α is an equivalence relation on a topological space X and $F : X \times I \to Y$ is a homotopy such that each stage F_t factors through X/α, i.e., $x\alpha x'$ implies $F_t(x) = F_t(x')$, then F induces a homotopy $F' : (X/\alpha) \times I \to Y$ such that $F' \circ (p_\alpha \times 1) = F$.

Homotopy theory has a range of applications of its own, outside topology and geometry, as for example in proving Cauchy theorem in complex variable theory, or in solving nonlinear equations of artificial neural networks.

A *pointed set* (S, s_0) is a set S together with a distinguished point $s_0 \in S$. Similarly, a *pointed topological space* (X, x_0) is a space X together with a distinguished point $x_0 \in X$. When we are concerned with pointed spaces $(X, x_0), (Y, y_0)$, etc., we always require that all functions $f : X \to Y$ shell preserve base points, i.e., $f(x_0) = y_0$, and that all homotopies $F : X \times I \to Y$ be relative to the base point, i.e., $F(x_0, t) = y_0$, for all $t \in I$. We denote the homotopy classes of base point-preserving functions by $[X, x_0; Y, y_0]$ (where homotopies are relative to x_0). $[X, x_0; Y, y_0]$ is a pointed set with base point f_0, the constant function: $f_0(x) = y_0$, for all $x \in X$.

A *path* $\gamma(t)$ from x_0 to x_1 in a topological space X is a continuous map $\gamma : I \to X$ with $\gamma(0) = x_0$ and $\gamma(1) = x_1$. Thus X^I is the space of all paths in X with the compact-open topology. We introduce a relation $\sim$ on X by saying $x_0 \sim x_1$ iff there is a path $\gamma : I \to X$ from x_0 to x_1. $\sim$ is clearly an equivalence relation, and the set of equivalence classes is denoted by $\pi_0(X)$. The elements of $\pi_0(X)$ are called the *path components*, or *0-components* of X. If $\pi_0(X)$ contains just one element, then X is called *path connected*, or *0-connected*. A *closed path*, or *loop* in X at the point x_0 is a path $\gamma(t)$ for which $\gamma(0) = \gamma(1) = x_0$. The *inverse loop* $\gamma^{-1}(t)$ based at $x_0 \in X$ is defined by $\gamma^{-1}(t) = \gamma(1 - t)$, for $0 \le t \le 1$. The *homotopy of loops* is the particular case of the above defined homotopy of continuous maps.

If (X, x_0) is a pointed space, then we may regard $\pi_0(X)$ as a pointed set with the 0-component of x_0 as a base point. We use the notation $\pi_0(X, x_0)$ to denote $p_0(X, x_0)$ thought of as a pointed set. If $f : X \to Y$ is a map then f sends 0-components of X into 0-components of Y and hence defines a function $\pi_0(f) : \pi_0(X) \to \pi_0(Y)$. Similarly, a base point-preserving map $f : (X, x_0) \to (Y, y_0)$ induces a map of pointed sets $\pi_0(f) : \pi_0(X, x_0) \to \pi_0(Y, y_0)$. In this way defined π_0 represents a 'functor' from the 'category' of topological (point) spaces to the underlying category of (point) sets (see the next subsection).

The *fundamental group* (introduced by Poincaré), denoted $\pi_1(X)$, of a pointed space (X, x_0) is the group (see Appendix) formed by the equivalence classes of the set of all *loops*, i.e., closed homotopies with initial and final points at a given base point x_0. The identity element of this group is the set of all paths homotopic to the degenerate path consisting of the point x_0.[2] The fundamental group $\pi_1(X)$ only depends on the homotopy type of the space X, that is, fundamental groups of homeomorphic spaces are isomorphic.

Combination of topology and calculus gives differential topology, or differential geometry.

Commutative Diagrams

The *category theory* (see below) was born with an observation that many properties of mathematical systems can be unified and simplified by a presentation with *commutative diagrams of arrows* [Mac71]. Each arrow $f : X \to Y$ represents a function (i.e., a map, transformation, operator); that is, a source (domain) set X, a target (codomain) set Y, and a rule $x \mapsto f(x)$ which assigns to each element $x \in X$ an element $f(x) \in Y$. A typical diagram of sets and functions is

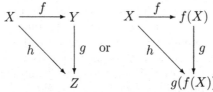

This diagram is *commutative* iff $h = g \circ f$, where $g \circ f$ is the usual composite function $g \circ f : X \to Z$, defined by $x \mapsto g(f(x))$.

Similar commutative diagrams apply in other mathematical, physical and computing contexts; e.g., in the 'category' of all topological spaces, the letters X, Y, and Z represent topological spaces while f, g, and h stand for continuous maps. Again, in the category of all groups, X, Y, and Z stand for groups, f, g, and h for homomorphisms.

Less formally, composing maps is like following directed paths from one object to another (e.g., from set to set). In general, a diagram is commutative iff any two paths along arrows that start at the same point and finish at the same point yield the same 'homomorphism' via compositions along successive arrows. Commutativity of the whole diagram follows from commutativity of its triangular components (depicting a 'commutative flow', see Fig. 7.1). Study of commutative diagrams is popularly called 'diagram chasing', and provides a powerful tool for mathematical thought.

Many properties of mathematical constructions may be represented by *universal properties* of diagrams [Mac71]. Consider the *Cartesian product* $X \times Y$ of two sets, consisting as usual of all ordered pairs $\langle x, y \rangle$ of elements $x \in X$ and $y \in Y$. The projections $\langle x, y \rangle \mapsto x$, $\langle x, y \rangle \mapsto y$ of the product on its

---

[2] The group product $f * g$ of loop f and loop g is given by the path of f followed by the path of g. The identity element is represented by the constant path, and the inverse f^{-1} of f is given by traversing f in the opposite direction. The fundamental group $\pi_1(X)$ is independent of the choice of base point x_0 because any loop through x_0 is homotopic to a loop through any other point x_1.

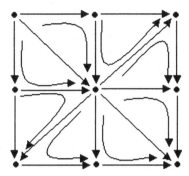

Fig. 7.1. A *commutative flow* (denoted by *curved arrows*) on a *triangulated digraph*. Commutativity of the whole diagram follows from commutativity of its triangular components.

'axes' X and Y are functions $p : X \times Y \to X$, $q : X \times Y \to Y$. Any function $h : W \to X \times Y$ from a third set W is uniquely determined by its composites $p \circ h$ and $q \circ h$. Conversely, given W and two functions f and g as in the diagram below, there is a unique function h which makes the following diagram commute:

$$
\begin{array}{ccc}
 & W & \\
f \swarrow & \downarrow h & \searrow g \\
X \xleftarrow{\ p\ } & X \times Y & \xrightarrow{\ q\ } Y
\end{array}
$$

This property describes the Cartesian product $X \times Y$ uniquely; the same diagram, read in the category of topological spaces or of groups, describes uniquely the Cartesian product of spaces or of the direct product of groups.

The construction 'Cartesian product' is technically called a 'functor' because it applies suitably both to the sets and to the functions between them; two functions $k : X \to X'$ and $l : Y \to Y'$ have a function $k \times l$ as their Cartesian product:

$$k \times l : X \times Y \to X' \times Y', \qquad \langle x, y \rangle \mapsto \langle kx, ly \rangle.$$

7.1.2 Categories

A category is a generic mathematical structure consisting of a collection of *objects* (sets with possibly additional structure), with a corresponding collection of *arrows*, or *morphisms*, between objects (agreeing with this additional structure). A category $\mathcal{K}$ is defined as a pair $(\mathrm{Ob}(\mathcal{K}), \mathrm{Mor}(\mathcal{K}))$ of generic objects $A, B, \ldots$ in $\mathrm{Ob}(\mathcal{K})$ and generic arrows $f : A \to B$, $g : B \to C, \ldots$ in $\mathrm{Mor}(\mathcal{K})$ between objects, with *associative composition*:

$$A \xrightarrow{\ f\ } B \xrightarrow{\ g\ } C = A \xrightarrow{\ f \circ g\ } C,$$

and *identity (loop)* arrow. (Note that in topological literature, $\text{Hom}(\mathcal{K})$ or $\text{hom}(\mathcal{K})$ is used instead of $\text{Mor}(\mathcal{K})$; see [Swi75]).

A category $\mathcal{K}$ is usually depicted as a *commutative diagram* (i.e., a diagram with a common *initial object A* and *final object D*):

$$
\begin{array}{ccc}
A & \xrightarrow{\ f\ } & B \\
h \downarrow & \mathcal{K} & \downarrow g \\
C & \xrightarrow[k]{} & D
\end{array}
$$

To make this more precise, we say that a *category* $\mathcal{K}$ is defined if we have:

1. A *class of objects* $\{A, B, C, \ldots\}$ of $\mathcal{K}$, denoted by $\text{Ob}(\mathcal{K})$;
2. A *set of morphisms*, or *arrows* $\text{Mor}_{\mathcal{K}}(A, B)$, with elements $f : A \to B$, defined for any *ordered pair* $(A, B) \in \mathcal{K}$, such that for two different pairs $(A, B) \neq (C, D)$ in $\mathcal{K}$, we have $\text{Mor}_{\mathcal{K}}(A, B) \cap \text{Mor}_{\mathcal{K}}(C, D) = \emptyset$;
3. For any *triplet* $(A, B, C) \in \mathcal{K}$ with $f : A \to B$ and $g : B \to C$, there is a *composition* of morphisms

$$\text{Mor}_{\mathcal{K}}(B, C) \times \text{Mor}_{\mathcal{K}}(A, B) \ni (g, f) \to g \circ f \in \text{Mor}_{\mathcal{K}}(A, C),$$

written schematically as

$$\frac{f : A \to B, \qquad g : B \to C}{g \circ f : A \to C}.$$

If we have a morphism $f \in \text{Mor}_{\mathcal{K}}(A, B)$, (otherwise written $f : A \to B$, or $A \xrightarrow{\ f\ } B$), then $A = \text{dom}(f)$ is a *domain* of f, and $B = \text{cod}(f)$ is a *codomain* of f (of which *range* of f is a subset, $B = \text{ran}(f)$).

To make $\mathcal{K}$ a category, it must also fulfill the following two properties:

1. *Associativity of morphisms*: for all $f \in \text{Mor}_{\mathcal{K}}(A, B)$, $g \in \text{Mor}_{\mathcal{K}}(B, C)$, and $h \in \text{Mor}_{\mathcal{K}}(C, D)$, we have $h \circ (g \circ f) = (h \circ g) \circ f$; in other words, the following diagram is commutative

$$
\begin{array}{ccc}
A & \xrightarrow{\ h \circ (g \circ f) = (h \circ g) \circ f\ } & D \\
f \downarrow & & \uparrow h \\
B & \xrightarrow[\quad g \quad]{} & C
\end{array}
$$

2. *Existence of identity morphism*: for every object $A \in \text{Ob}(\mathcal{K})$ exists a unique identity morphism $1_A \in \text{Mor}_{\mathcal{K}}(A, A)$; for any two morphisms

$f \in \mathrm{Mor}_{\mathcal{K}}(A, B)$, and $g \in \mathrm{Mor}_{\mathcal{K}}(B, C)$, compositions with identity morphism $1_B \in \mathrm{Mor}_{\mathcal{K}}(B, B)$ give $1_B \circ f = f$ and $g \circ 1_B = g$, i.e., the following diagram is commutative:

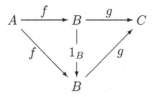

The set of all morphisms of the category $\mathcal{K}$ is denoted

$$\mathrm{Mor}(\mathcal{K}) = \bigcup_{A,B \in Ob(\mathcal{K})} \mathrm{Mor}_{\mathcal{K}}(A, B).$$

If for two morphisms $f \in \mathrm{Mor}_{\mathcal{K}}(A, B)$ and $g \in \mathrm{Mor}_{\mathcal{K}}(B, A)$ the equality $g \circ f = 1_A$ is valid, then the morphism g is said to be *left inverse* (or *retraction*), of f, and f *right inverse* (or *section*) of g. A morphism which is both right and left inverse of f is said to be *two-sided inverse* of f.

A morphism $m : A \rightarrow B$ is called *monomorphism* in $\mathcal{K}$ (i.e., *1-1*, or *injection* map), if for any two parallel morphisms $f_1, f_2 : C \rightarrow A$ in $\mathcal{K}$ the equality $m \circ f_1 = m \circ f_2$ implies $f_1 = f_2$; in other words, m is monomorphism if it is *left cancellable*. Any morphism with a left inverse is monomorphism.

A morphism $e : A \rightarrow B$ is called *epimorphism* in $\mathcal{K}$ (i.e., *onto*, or *surjection* map), if for any two morphisms $g_1, g_2 : B \rightarrow C$ in $\mathcal{K}$ the equality $g_1 \circ e = g_2 \circ e$ implies $g_1 = g_2$; in other words, e is epimorphism if it is *right cancellable*. Any morphism with a right inverse is epimorphism.

A morphism $f : A \rightarrow B$ is called *isomorphism* in $\mathcal{K}$ (denoted as $f : A \cong B$) if there exists a morphism $f^{-1} : B \rightarrow A$ which is a two-sided inverse of f in $\mathcal{K}$. The relation of isomorphism is reflexive, symmetric, and transitive, i.e., equivalence relation.

For example, an isomorphism in the category of sets is called a set-isomorphism, or a *bijection*, in the category of topological spaces is called a topological isomorphism, or a *homeomorphism*, in the category of differentiable manifolds is called a differentiable isomorphism, or a *diffeomorphism*.

A morphism $f \in \mathrm{Mor}_{\mathcal{K}}(A, B)$ is *regular* if there exists a morphism $g : B \rightarrow A$ in $\mathcal{K}$ such that $f \circ g \circ f = f$. Any morphism with either a left or a right inverse is regular.

An object T is a *terminal object* in $\mathcal{K}$ if to each object $A \in Ob(\mathcal{K})$ there is exactly one arrow $A \rightarrow T$. An object S is an *initial object* in $\mathcal{K}$ if to each object $A \in Ob(\mathcal{K})$ there is exactly one arrow $S \rightarrow A$. A *null object* $Z \in Ob(\mathcal{K})$ is an object which is both initial and terminal; it is unique up to isomorphism. For any two objects $A, B \in Ob(\mathcal{K})$ there is a unique morphism $A \rightarrow Z \rightarrow B$ (the composite through Z), called the *zero morphism* from A to B.

A notion of subcategory is analogous to the notion of subset. A subcategory $\mathcal{L}$ of a category $\mathcal{K}$ is said to be a *complete subcategory* iff for any objects $A, B \in \mathcal{L}$, every morphism $A \rightarrow B$ of $\mathcal{L}$ is in $\mathcal{K}$.

A *groupoid* is a category in which every morphism is invertible. A typical groupoid is the *fundamental groupoid* $\Pi_1(X)$ of a topological space X. An object of $\Pi_1(X)$ is a point $x \in X$, and a morphism $x \to x'$ of $\Pi_1(X)$ is a homotopy class of paths f from x to x'. The *composition* of paths $g : x' \to x''$ and $f : x \to x'$ is the path h which is 'f followed by g'. Composition applies also to homotopy classes, and makes $\Pi_1(X)$ a category and a groupoid (the inverse of any path is the same path traced in the opposite direction).

A *group* (see Appendix) is a groupoid with one object, i.e., a *category with one object* in which *all morphisms are isomorphisms*. Therefore, if we try to generalize the concept of a group, keeping associativity as an essential property, we get the notion of a category.

A category is *discrete* if every morphism is an identity. A *monoid* is a category with one object. A *group* is a category with one object in which every morphism has a two-sided inverse under composition.

Homological algebra was the progenitor of category theory (see e.g., [Die88]). Generalizing L. Euler's formula $f + v = e + 2$ for the faces, vertices and edges of a convex polyhedron, E. Betti defined *numerical invariants of spaces* by formal addition and subtraction of faces of various dimensions; H. Poincaré formalized these and introduced homology. E. Noether stressed the fact that these calculations go on in Abelian groups, and that the operation ∂_n taking a face of dimension n to the alternating sum of faces of dimension $n - 1$ which form its boundary is a homomorphism, and it also satisfies $\partial_n \circ \partial_{n+1} = 0$. There are many ways of approximating a given space by polyhedra, but the quotient $H_n = \operatorname{Ker} \partial_n / \operatorname{Im} \partial_{n+1}$ is an invariant, the *homology group*. Since Noether, the groups have been the object of study instead of their dimensions, which are the *Betti numbers*.

7.1.3 Functors

In algebraic topology, one attempts to assign to every topological space X some algebraic object $\mathcal{F}(X)$ in such a way that to every C^0-function $f : X \to Y$ there is assigned a homomorphism $\mathcal{F}(f) : \mathcal{F}(X) \to \mathcal{F}(Y)$ (see [Swi75, II06c]). One advantage of this procedure is, e.g., that if one is trying to prove the non-existence of a C^0-function $f : X \to Y$ with certain properties, one may find it relatively easy to prove the non-existence of the corresponding algebraic function $\mathcal{F}(f)$ and hence deduce that f could not exist. In other words, $\mathcal{F}$ is to be a 'homomorphism' from one category (e.g., $\mathcal{T}$) to another (e.g., $\mathcal{G}$ or $\mathcal{A}$). Formalization of this notion is a *functor*.

A functor is a generic *picture* projecting (all objects and morphisms of) a source category into a target category. Let $\mathcal{K} = (\operatorname{Ob}(\mathcal{K}), \operatorname{Mor}(\mathcal{K}))$ be a *source* (or domain) *category* and $\mathcal{L} = (\operatorname{Ob}(\mathcal{L}), \operatorname{Mor}(\mathcal{L}))$ be a *target* (or codomain) category. A functor $\mathcal{F} = (\mathcal{F}_O, \mathcal{F}_M)$ is defined as a pair of maps, $\mathcal{F}_O : \operatorname{Ob}(\mathcal{K}) \to \operatorname{Ob}(\mathcal{L})$ and $\mathcal{F}_M : \operatorname{Mor}(\mathcal{K}) \to \operatorname{Mor}(\mathcal{L})$, preserving categorical symmetry (i.e., commutativity of all diagrams) of $\mathcal{K}$ in $\mathcal{L}$.

More precisely, a *covariant functor*, or simply a *functor*, $\mathcal{F}_* : \mathcal{K} \to \mathcal{L}$ is a *picture* in the target category $\mathcal{L}$ of (all objects and morphisms of) the source category $\mathcal{K}$:

Similarly, a *contravariant functor*, or a *cofunctor*, $\mathcal{F}^* : \mathcal{K} \to \mathcal{L}$ is a *dual picture* with reversed arrows:

In other words, a *functor* $\mathcal{F} : \mathcal{K} \to \mathcal{L}$ from a *source* category $\mathcal{K}$ to a *target* category $\mathcal{L}$, is a pair $\mathcal{F} = (\mathcal{F}_O, \mathcal{F}_M)$ of maps $\mathcal{F}_O : \mathtt{Ob}(\mathcal{K}) \to \mathtt{Ob}(\mathcal{L})$, $\mathcal{F}_M : \mathtt{Mor}(\mathcal{K}) \to \mathtt{Mor}(\mathcal{L})$, such that

1. If $f \in \mathtt{Mor}_{\mathcal{K}}(A, B)$ then $\mathcal{F}_M(f) \in \mathtt{Mor}_{\mathcal{L}}(\mathcal{F}_O(A), \mathcal{F}_O(B))$ in case of the *covariant* functor $\mathcal{F}_*$, and $\mathcal{F}_M(f) \in \mathtt{Mor}_{\mathcal{L}}(\mathcal{F}_O(B), \mathcal{F}_O(A))$ in case of the *contravariant* functor $\mathcal{F}^*$;
2. For all $A \in \mathtt{Ob}(\mathcal{K}) : \mathcal{F}_M(1_A) = 1_{\mathcal{F}_O(A)}$;
3. For all $f, g \in \mathtt{Mor}(\mathcal{K})$: if $\mathrm{cod}(f) = \mathrm{dom}(g)$, then
   $\mathcal{F}_M(g \circ f) = \mathcal{F}_M(g) \circ \mathcal{F}_M(f)$ in case of the *covariant* functor $\mathcal{F}_*$, and
   $\mathcal{F}_M(g \circ f) = \mathcal{F}_M(f) \circ \mathcal{F}_M(g)$ in case of the *contravariant* functor $\mathcal{F}^*$.

Category theory originated in algebraic topology, which tried to assign algebraic invariants to topological structures. The golden rule of such *invariants* is that they should be *functors*. For example, the *fundamental group* π_1 is a functor. Algebraic topology constructs a group called the *fundamental group* $\pi_1(X)$ from any topological space X, which keeps track of how many holes the space X has. But also, any map between topological spaces determines a homomorphism $\phi : \pi_1(X) \to \pi_1(Y)$ of the fundamental groups. So the fundamental group is really a functor $\pi_1 : \mathcal{T} \to \mathcal{G}$. This allows us to completely transpose any situation involving *spaces* and *continuous maps* between them to a parallel situation involving *groups* and *homomorphisms* between them, and thus reduce some topology problems to algebra problems.

Also, singular homology in a given dimension n assigns to each topological space X an Abelian group $H_n(X)$, its nth *homology group* of X, and also to each continuous map $f : X \to Y$ of spaces a corresponding homomorphism

$H_n(f) : H_n(X) \to H_n(Y)$ of groups, and this in such a way that $H_n(X)$ becomes a functor $H_n : \mathcal{T} \to \mathcal{A}$.

The leading idea in the *use of functors in topology* is that H_n or π_n gives an algebraic picture or image not just of the topological spaces X, Y but also of all the continuous maps $f : X \to Y$ between them.

Similarly, there is a functor $\Pi_1 : \mathcal{T} \to \mathcal{G}$, called the 'fundamental groupoid functor', which plays a very basic role in algebraic topology. Here's how we get from any space X its 'fundamental groupoid' $\Pi_1(X)$. To say what the groupoid $\Pi_1(X)$ is, we need to say what its objects and morphisms are. The objects in $\Pi_1(X)$ are just the *points* of X and the morphisms are just certain equivalence classes of *paths* in X. More precisely, a morphism $f : x \to y$ in $\Pi_1(X)$ is just an equivalence class of continuous paths from x to y, where two paths from x to y are decreed equivalent if one can be continuously deformed to the other while not moving the endpoints. If this equivalence relation holds we say the two paths are 'homotopic', and we call the equivalence classes 'homotopy classes of paths' (see [Mac71, Swi75]).

Another examples are covariant *forgetful* functors:

- From the category of topological spaces to the category of sets; it 'forgets' the topology-structure.
- From the category of metric spaces to the category of topological spaces with the topology induced by the metrics; it 'forgets' the metric.

For each category $\mathcal{K}$, the *identity functor* $I_\mathcal{K}$ takes every $\mathcal{K}$-object and every $\mathcal{K}$-morphism to itself.

Given a category $\mathcal{K}$ and its subcategory $\mathcal{L}$, we have an *inclusion functor* $\text{In} : \mathcal{L} \to \mathcal{K}$.

Given a category $\mathcal{K}$, a *diagonal functor* $\Delta : \mathcal{K} \to \mathcal{K}$ takes each object $A \in \mathcal{K}$ to the object (A, A) in the product category $\mathcal{K} \times \mathcal{K}$.

Given a category $\mathcal{K}$ and a category of sets $\mathcal{S}$, each object $A \in \mathcal{K}$ determines a *covariant Hom-functor* $\mathcal{K}[A, \_] : \mathcal{K} \to \mathcal{S}$, a *contravariant Hom-functor* $\mathcal{K}[\_, A] : \mathcal{K} \to \mathcal{S}$, and a *Hom-bifunctor* $\mathcal{K}[\_, \_] : \mathcal{K}^{op} \times \mathcal{K} \to \mathcal{S}$.

A functor $\mathcal{F} : \mathcal{K} \to \mathcal{L}$ is a *faithful functor* if for all $A, B \in \text{Ob}(\mathcal{K})$ and for all $f, g \in \text{Mor}_\mathcal{K}(A, B)$, $\mathcal{F}(f) = \mathcal{F}(g)$ implies $f = g$; it is a *full functor* if for every $h \in \text{Mor}_\mathcal{L}(\mathcal{F}(A), \mathcal{F}(B))$, there is $g \in \text{Mor}_\mathcal{K}(A, B)$ such that $h = \mathcal{F}(g)$; it is a *full embedding* if it is both full and faithful.

A *representation of a group* is a functor $\mathcal{F} : \mathcal{G} \to \mathcal{V}$.

Similarly, we can define a *representation of a category* to be a functor $\mathcal{F} : \mathcal{K} \to \mathcal{V}$ from the 2-category $\mathcal{K}$ (a 'big' category including all ordinary, or 'small' categories, see Sect. 7.1.7 below) to the category of vector spaces $\mathcal{V}$. In this way, a category is a generalization of a group and group representations are a special case of category representations.

7.1.4 Natural Transformations

A *natural transformation* (i.e., a *functor morphism*) $\tau : \mathcal{F} \overset{\cdot}{\to} \mathcal{G}$ is a *map between two functors of the same variance*, $(\mathcal{F}, \mathcal{G}) : \mathcal{K} \rightrightarrows \mathcal{L}$, preserving categorical symmetry:

More precisely, all functors of the same variance from a source category $\mathcal{K}$ to a target category $\mathcal{L}$ form themselves objects of the *functor category* $\mathcal{L}^{\mathcal{K}}$. Morphisms of $\mathcal{L}^{\mathcal{K}}$, called *natural transformations*, are defined as follows.

Let $\mathcal{F} : \mathcal{K} \to \mathcal{L}$ and $\mathcal{G} : \mathcal{K} \to \mathcal{L}$ be two functors of the same variance from a category $\mathcal{K}$ to a category $\mathcal{L}$. Natural transformation $\mathcal{F} \overset{\tau}{\to} \mathcal{G}$ is a family of morphisms such that for all $f \in \mathrm{Mor}_{\mathcal{K}}(A, B)$ in the source category $\mathcal{K}$, we have $\mathcal{G}(f) \circ \tau_A = \tau_B \circ \mathcal{F}(f)$ in the target category $\mathcal{L}$. Then we say that the *component* $\tau_A : \mathcal{F}(A) \to \mathcal{G}(A)$ *is natural in A.*

If we think of a functor $\mathcal{F}$ as giving a *picture* in the target category $\mathcal{L}$ of (all the objects and morphisms of) the source category $\mathcal{K}$, then a natural transformation τ represents a set of morphisms mapping the picture $\mathcal{F}$ to another picture $\mathcal{G}$, preserving the commutativity of all diagrams.

An invertible natural transformation, such that all components τ_A are isomorphisms, is called a *natural equivalence* (or, *natural isomorphism*). In this case, the inverses $(\tau_A)^{-1}$ in $\mathcal{L}$ are the components of a natural isomorphism $(\tau)^{-1} : \mathcal{G} \overset{*}{\longrightarrow} \mathcal{F}$. Natural equivalences are among the most important *metamathematical constructions* in algebraic topology (see [Swi75]).

For example, let $\mathcal{B}$ be the category of Banach spaces over $\mathbb{R}$ and bounded linear maps. Define $D : \mathcal{B} \to \mathcal{B}$ by taking $D(X) = X^* =$ Banach space of bounded linear functionals on a space X and $D(f) = f^*$ for $f : X \to Y$ a bounded linear map. Then D is a cofunctor. $D^2 = D \circ D$ is also a functor. We also have the identity functor $1 : \mathcal{B} \to \mathcal{B}$. Define $T : 1 \to D \circ D$ as follows: for every $X \in \mathcal{B}$ let $T(X) : X \to D^2 X = X^{**}$ be the *natural inclusion*—that is, for $x \in X$ we have $[T(X)(x)](f) = f(x)$ for every $f \in X^*$. T is a natural transformation. On the subcategory of nD Banach spaces T is even a natural equivalence. The largest subcategory of $\mathcal{B}$ on which T is a natural equivalence is called the category of reflexive Banach spaces [Swi75].

As S. Eilenberg and S. MacLane first observed, 'category' has been defined in order to define 'functor' and 'functor' has been defined in order to define 'natural transformation' [Mac71]).

Compositions of Natural Transformations

Natural transformations can be *composed* in two different ways. First, we have an 'ordinary' composition: if $\mathcal{F}, \mathcal{G}$ and $\mathcal{H}$ are three functors from the source category $\mathcal{A}$ to the target category $\mathcal{B}$, and then $\alpha : \mathcal{F} \xrightarrow{\cdot} \mathcal{G}$, $\beta : \mathcal{G} \xrightarrow{\cdot} \mathcal{H}$ are two natural transformations, then the formula

$$(\beta \circ \alpha)_A = \beta_A \circ \alpha_A, \quad \text{for all } A \in \mathcal{A}, \tag{7.2}$$

defines a new natural transformation $\beta \circ \alpha : \mathcal{F} \xrightarrow{\cdot} \mathcal{H}$. This composition law is clearly associative and possesses a unit $1_{\mathcal{F}}$ at each functor $\mathcal{F}$, whose $\mathcal{A}$-component is $1_{\mathcal{F}A}$.

Second, we have the *Godement product* of natural transformations, usually denoted by $*$. Let $\mathcal{A}$, $\mathcal{B}$ and $\mathcal{C}$ be three categories, $\mathcal{F}, \mathcal{G}, \mathcal{H}$ and $\mathcal{K}$ be four functors such that $(\mathcal{F}, \mathcal{G}) : \mathcal{A} \rightrightarrows \mathcal{B}$ and $(\mathcal{H}, \mathcal{K}) : \mathcal{B} \rightrightarrows \mathcal{C}$, and $\alpha : \mathcal{F} \xrightarrow{\cdot} \mathcal{G}$, $\beta : \mathcal{H} \xrightarrow{\cdot} \mathcal{K}$ be two natural transformations. Now, instead of (7.2), the Godement composition is given by

$$(\beta * \alpha)_A = \beta_{GA} \circ H(\alpha_A) = K(\alpha_A) \circ \beta_{FA}, \quad \text{for all } A \in \mathcal{A}, \tag{7.3}$$

which defines a new natural transformation $\beta * \alpha : \mathcal{H} \circ \mathcal{F} \xrightarrow{\cdot} \mathcal{K} \circ \mathcal{G}$.

Finally, the two compositions (7.2) and (7.3) of natural transformations can be combined as

$$(\delta * \gamma) \circ (\beta * \alpha) = (\delta \circ \beta) * (\gamma \circ \alpha),$$

where $\mathcal{A}$, $\mathcal{B}$ and $\mathcal{C}$ are three categories, $\mathcal{F}, \mathcal{G}, \mathcal{H}, \mathcal{K}, \mathcal{L}, \mathcal{M}$ are six functors, and $\alpha : \mathcal{F} \xrightarrow{\cdot} \mathcal{H}$, $\beta : \mathcal{G} \xrightarrow{\cdot} \mathcal{K}$, $\gamma : \mathcal{H} \xrightarrow{\cdot} \mathcal{L}$, $\delta : \mathcal{K} \xrightarrow{\cdot} \mathcal{M}$ are four natural transformations.

Dinatural Transformations

Double natural transformations are called *dinatural transformations*. An *end of a functor* $S : C^{op} \times C \to X$ is a universal dinatural transformation from a constant e to S. In other words, an end of S is a pair $\langle e, \omega \rangle$, where e is an object of X and $\omega : e \xrightarrow{\cdot} S$ is a *wedge (dinatural) transformation* with the property that to every wedge $\beta : x \xrightarrow{\cdot} S$ there is a unique arrow $h : x \to e$ of B with $\beta_c = \omega_c h$ for all $a \in C$. We call ω the *ending wedge* with *components* ω_c, while the object e itself, by abuse of language, is called the end of S and written with integral notation as $\int_c S(c, c)$; thus

$$S(c, c) \xrightarrow{\omega_c} \int_c S(c, c) = e.$$

Note that the 'variable of integration' c appears twice under the integral sign (once contravariant, once covariant) and is 'bound' by the integral sign, in that the result no longer depends on c and so is unchanged if 'c' is replaced by any other letter standing for an object of the category C. These properties are like those of the letter x under the usual integral symbol $\int f(x)\, dx$ of calculus.

Every end is manifestly a limit (see below)—specifically, a limit of a suitable diagram in X made up of pieces like $S(b,b) \to S(b,c) \to S(c,c)$.

For each functor $T : C \to X$ there is an isomorphism

$$\int_c S(c,c) = \int_c Tc \cong \operatorname{Lim} T,$$

valid when either the end of the limit exists, carrying the ending wedge to the limiting cone; the indicated notation thus allows us to write any limit as an integral (an end) without explicitly mentioning the dummy variable (the first variable c of S).

A functor $H : X \to Y$ is said to *preserve the end* of a functor $S : C^{op} \times C \to X$ when $\omega : e \overset{..}{\to} S$ an end of S in X implies that $H\omega : He \overset{..}{\to} HS$ is an and for HS; in symbols

$$H \int_c S(c,c) = \int_c HS(c,c).$$

Similarly, H *creates* the end of S when to each end $v : y \overset{..}{\to} HS$ in Y there is a unique wedge $\omega : e \overset{..}{\to} S$ with $H\omega = v$, and this wedge ω is an end of S.

The definition of the coend of a functor $S : C^{op} \times C \to X$ is dual to that of an end. A *coend* of S is a pair $\langle d, \zeta \rangle$, consisting of an object $d \in X$ and a wedge $\zeta : S \overset{..}{\to} d$. The object d (when it exists, unique up to isomorphism) will usually be written with an integral sign and with the bound variable c as superscript; thus

$$S(c,c) \overset{\zeta_c}{\to} \int^c S(c,c) = d.$$

The formal properties of coends are dual to those of ends. Both are much like those for integrals in calculus (see [Mac71], for technical details).

7.1.5 Limits and Colimits

In abstract algebra constructions are often defined by an abstract property which requires the existence of unique morphisms under certain conditions. These properties are called *universal properties*. The *limit* of a functor generalizes the notions of inverse limit and product used in various parts of mathematics. The dual notion, *colimit*, generalizes direct limits and direct sums. Limits and colimits are defined via universal properties and provide many examples of *adjoint functors*.

A *limit* of a covariant functor $\mathcal{F} : \mathcal{J} \to \mathcal{C}$ is an object L of $\mathcal{C}$, together with morphisms $\phi_X : L \to \mathcal{F}(X)$ for every object X of $\mathcal{J}$, such that for every morphism $f : X \to Y$ in $\mathcal{J}$, we have $\mathcal{F}(f)\phi_X = \phi_Y$, and such that the following *universal property* is satisfied: for any object N of $\mathcal{C}$ and any set of morphisms $\psi_X : N \to \mathcal{F}(X)$ such that for every morphism $f : X \to Y$ in $\mathcal{J}$, we have $\mathcal{F}(f)\psi_X = \psi_Y$, there exists precisely one morphism $u : N \to L$ such that $\phi_X u = \psi_X$ for all X. If $\mathcal{F}$ has a limit (which it need not), then the limit is defined up to a unique isomorphism, and is denoted by $\lim \mathcal{F}$.

Analogously, a *colimit* of the functor $\mathcal{F} : \mathcal{J} \to \mathcal{C}$ is an object L of $\mathcal{C}$, together with morphisms $\phi_X : \mathcal{F}(X) \to L$ for every object X of $\mathcal{J}$, such that for every morphism $f : X \to Y$ in $\mathcal{J}$, we have $\phi_Y \mathcal{F}(X) = \phi_X$, and such that the following universal property is satisfied: for any object N of $\mathcal{C}$ and any set of morphisms $\psi_X : \mathcal{F}(X) \to N$ such that for every morphism $f : X \to Y$ in $\mathcal{J}$, we have $\psi_Y \mathcal{F}(X) = \psi_X$, there exists precisely one morphism $u : L \to N$ such that $u\phi_X = \psi_X$ for all X. The colimit of $\mathcal{F}$, unique up to unique isomorphism if it exists, is denoted by $\operatorname{colim} \mathcal{F}$.

Limits and colimits are related as follows: A functor $\mathcal{F} : \mathcal{J} \to \mathcal{C}$ has a colimit iff for every object N of $\mathcal{C}$, the functor $X \longmapsto Mor_{\mathcal{C}}(\mathcal{F}(X), N)$ (which is a covariant functor on the dual category $\mathcal{J}^{op}$) has a limit. If that is the case, then $Mor_{\mathcal{C}}(\operatorname{colim} \mathcal{F}, N) = \lim Mor_{\mathcal{C}}(\mathcal{F}(-), N)$ for every object N of $\mathcal{C}$.

7.1.6 Adjunction

The most important functorial operation is adjunction; as S. MacLane once said, "Adjoint functors arise everywhere" [Mac71].

The *adjunction* $\varphi : \mathcal{F} \dashv \mathcal{G}$ between two functors $(\mathcal{F}, \mathcal{G}) : \mathcal{K} \leftrightarrows \mathcal{L}$ of *opposite variance* [Kan58], represents a *weak functorial inverse*

$$\frac{f : \mathcal{F}(A) \to B}{\varphi(f) : A \to \mathcal{G}(B)}$$

forming a *natural equivalence* $\varphi : Mor_{\mathcal{K}}(\mathcal{F}(A), B) \xrightarrow{\varphi} Mor_{\mathcal{L}}(A, \mathcal{G}(B))$. The adjunction isomorphism is given by a *bijective correspondence* (a 1–1 and onto map on objects) $\varphi : Mor(\mathcal{K}) \ni f \to \varphi(f) \in Mor(\mathcal{L})$ of isomorphisms in the two categories, $\mathcal{K}$ (with a representative object A), and $\mathcal{L}$ (with a representative object B). It can be depicted as a (non-commutative) diagram

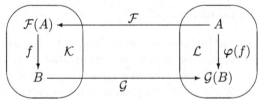

In this case $\mathcal{F}$ is called *left adjoint*, while $\mathcal{G}$ is called *right adjoint*.

In other words, an *adjunction* $F \dashv G$ between two functors $(\mathcal{F}, \mathcal{G})$ of opposite variance, from a source category $\mathcal{K}$ to a target category $\mathcal{L}$, is denoted by $(\mathcal{F}, \mathcal{G}, \eta, \varepsilon) : \mathcal{K} \leftrightarrows \mathcal{L}$. Here, $\mathcal{F} : \mathcal{L} \to \mathcal{K}$ is the *left (upper) adjoint functor*, $\mathcal{G} : \mathcal{L} \leftarrow \mathcal{K}$ is the *right (lower) adjoint functor*, $\eta : 1_{\mathcal{L}} \to \mathcal{G} \circ \mathcal{F}$ is the *unit natural transformation* (or, *front adjunction*), and $\varepsilon : \mathcal{F} \circ \mathcal{G} \to 1_{\mathcal{K}}$ is the *counit natural transformation* (or, *back adjunction*).

For example, $\mathcal{K} = \mathcal{S}$ is the category of sets and $\mathcal{L} = \mathcal{G}$ is the category of groups. Then $\mathcal{F}$ turns any set into the *free group* on that set, while the 'forgetful' functor $\mathcal{F}^*$ turns any group into the *underlying set* of that group. Similarly, all sorts of other 'free' and 'underlying' constructions are also left and right adjoints, respectively.

Right adjoints preserve *limits*, and left adjoints preserve *colimits*.

The category $\mathcal{C}$ is called a *cocomplete category* if every functor $\mathcal{F} : \mathcal{J} \to \mathcal{C}$ has a colimit. The following categories are cocomplete: $\mathcal{S}, \mathcal{G}, \mathcal{A}, \mathcal{T}$, and $\mathcal{PT}$.

The importance of adjoint functors lies in the fact that every functor which has a left adjoint (and therefore is a right adjoint) is continuous. In the category $\mathcal{A}$ of Abelian groups, this e.g., shows that the kernel of a product of homomorphisms is naturally identified with the product of the kernels. Also, limit functors themselves are continuous. A covariant functor $\mathcal{F} : \mathcal{J} \to \mathcal{C}$ is *cocontinuous* if it transforms colimits into colimits. Every functor which has a right adjoint (and is a left adjoint) is cocontinuous.

The *analogy* between *adjoint functors* and *adjoint linear operators* relies upon a deeper analogy: just as in quantum theory the inner product $\langle \phi, \psi \rangle$ represents the *amplitude* to pass from ϕ to ψ, in category theory $Mor(A, B)$ represents the *set of ways* to go from A to B. These are to Hilbert spaces as categories are to sets. The analogues of adjoint linear operators between Hilbert spaces are certain adjoint functors between 2-Hilbert spaces [Bae97, Bae98]. Similarly, the *adjoint representation* of a Lie group G is the linearized version of the action of G on itself by conjugation, i.e., for each $g \in G$, the inner automorphism $x \mapsto gxg^{-1}$ gives a linear transformation $Ad(g) : \mathfrak{g} \to \mathfrak{g}$, from the Lie algebra $\mathfrak{g}$ of G to itself.

Neurophysiological Sensory-Motor Adjunction

Recall that sensations from the skin, muscles, and internal organs of the body, are transmitted to the central nervous system via axons that enter via spinal nerves. They are called *sensory pathways*. On the other hand, the motor system executes control over the skeletal muscles of the body via several major tracts (including pyramidal and extrapyramidal). They are called *motor pathways*. Sensory-motor (or, sensorimotor) control/coordination concerns relationships between sensation and movement or, more broadly, between perception and action. The interplay of sensory and motor processes provides the basis of observable human behavior. Anatomically, its top-level, association link can be visualized as a talk between sensory and motor Penfield's homunculi. This sensory-motor control system can be modeled as an adjunction between the afferent sensory functor $\mathcal{S} : \mathcal{BODY} \to \mathcal{BRAIN}$ and the efferent motor functor $\mathcal{M} : \mathcal{BRAIN} \to \mathcal{BODY}$. Thus, we have $\mathcal{SMC} : \mathcal{S} \dashv \mathcal{M}$, with $(\mathcal{S}, \mathcal{M}) : \mathcal{BRAIN} \leftrightarrows \mathcal{BODY}$ and depicted as

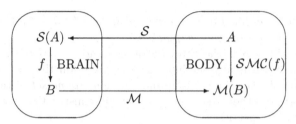

This adjunction offers a mathematical answer to the fundamental question: How would *Nature* solve a general biodynamics control/coordination problem? *By using a weak functorial inverse of sensory neural pathways and motor neural pathways, Nature controls human behavior in general, and human motion in particular.*

More generally, normal functioning of human body is achieved through interplay of a number of physiological systems—Objects of the category BODY: musculoskeletal system, circulatory system, gastrointestinal system, integumentary system, urinary system, reproductive system, immune system and endocrine system. These systems are all interrelated, so one can say that the Morphisms between them make the proper functioning of the BODY as a whole. On the other hand, BRAIN contains the images of all above functional systems (Brain objects) and their interrelations (Brain morphisms), for the purpose of body control. This body-control performed by the brain is partly unconscious, through neuro-endocrine complex, and partly conscious, through neuro-muscular complex. A generalized sensory functor SS sends the information about the state of all Body objects (at any time instant) to their images in the Brain. A generalized motor functor MM responds to these upward sensory signals by sending downward corrective action-commands from the Brain's objects and morphisms to the Body's objects and morphisms.

For physiological details, see [II06b]. For other bio-physical applications of categorical meta-language, see [II06a, II06c, II07e].

7.1.7 n-Categories and n-Functors

Generalization from 'Small' to 'Big' Categories

If we think of a point in geometrical space (either natural, or abstract) as an *object* (or, a 0-cell), and a path between two points as an *arrow* (or, a 1-*morphism*, or a 1-cell), we could think of a 'path of paths' as a 2-arrow (or, a 2-morphism, or a 2-cell), and a 'path of paths of paths' (or, a 3-morphism, or a 3-cell), etc. Here a 'path of paths' is just a continuous 1-parameter family of paths from between source and target points, which we can think of as tracing out a 2D surface, etc. In this way we get a 'skeleton' of an n-category, where a 1-category operates with 0-cells (objects) and 1-cells (arrows, causally connecting *source* objects with *target* ones), a 2-category operates with all the cells up to 2-cells [Bén67], a 3-category operates with all the cells up to 3-cells, etc. This skeleton clearly demonstrates the *hierarchical self-similarity* of *n-categories*:

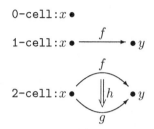

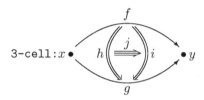

where triple arrow goes in the third direction, perpendicular to both single and double arrows. Categorical composition is defined by pasting arrows.

Thus, a 1-category can be depicted as a commutative triangle:

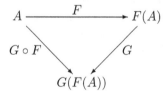

a 2-category is a commutative triangle:

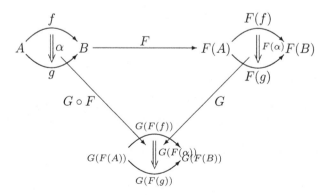

a 3-category is a commutative triangle:

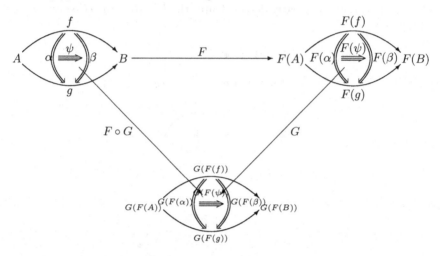

etc., up to n-categories.

For example, a *braided monoidal category* is a tricategory with one 0-cell and one 1-cell.

Many deep-sounding results in mathematical sciences are get by the process of *categorification*[3] of the high school mathematics [CF94, Bae98].

An n-category is a generic mathematical structure consisting of a collection of objects, a collection of arrows between objects, a collection of 2-arrows between arrows [Bén67], a collection of 3-arrows between 2-arrows, and so on up to n [Bae97, Bae98, Lei02, Lei03, Lei04].

More precisely, an n-category (for $n \geq 0$) consists of:

- 0-cells, or objects, $A, B, \ldots$

- 1-cells, or arrows, $A \xrightarrow{f} B$, with a composition

$$A \xrightarrow{f} B \xrightarrow{g} C = A \xrightarrow{g \circ f} C$$

- 2-cells, 'arrows between arrows', $A \underset{g}{\overset{f}{\Rightarrow}} \alpha \, B$, with vertical compositions

(denoted by $\circ$) and horizontal compositions (denoted by $*$), respectively given by

---

[3] Categorification means replacing sets with categories, functions with functors, and equations between functions by natural equivalences between functors. Iterating this process requires a theory of n-categories.

$$A \xrightarrow[h]{\overset{f}{\underset{\Downarrow\beta}{\overset{\Downarrow\alpha}{\underset{g}{\longrightarrow}}}}} B = A \xrightarrow[h]{\overset{f}{\underset{\Downarrow\beta\circ\alpha}{}}} B \quad \text{and}$$

$$A \xrightarrow[g]{\overset{f}{\Downarrow\alpha}} A' \xrightarrow[g']{\overset{f'}{\Downarrow\alpha'}} A'' = A \xrightarrow[g'\circ g]{\overset{f'\circ f}{\Downarrow\alpha'*\alpha}} A''$$

- 3-cells, 'arrows between arrows between arrows', $A \; \alpha \left(\overset{\Gamma}{\underset{}{\Longrightarrow}} \right) \beta \; B$ with f above and g below

(where the Γ-arrow goes in a direction perpendicular to f and α), with various kinds of vertical, horizontal and mixed compositions,
- etc., up to n-cells.

Calculus of n-categories has been developed as follows. First, there is $\mathcal{K}_2$, the 2-category of all ordinary (or small) categories. $\mathcal{K}_2$ has categories $\mathcal{K}, \mathcal{L}, \dots$ as objects, functors $\mathcal{F}, \mathcal{G} : \mathcal{K} \rightrightarrows \mathcal{L}$ as arrows, and natural transformations, like $\tau : \mathcal{F} \overset{.}{\to} \mathcal{G}$ as 2-arrows.

In a similar way, the arrows in a 3-category $\mathcal{K}_3$ are 2-functors $\mathcal{F}_2, \mathcal{G}_2, \dots$ sending objects in $\mathcal{K}_2$ to objects in $\mathcal{L}_2$, arrows to arrows, and 2-arrows to 2-arrows, strictly preserving all the structure of $\mathcal{K}_2$

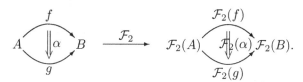

$$A \xrightarrow[g]{\overset{f}{\Downarrow\alpha}} B \quad \xrightarrow{\mathcal{F}_2} \quad \mathcal{F}_2(A) \xrightarrow[\mathcal{F}_2(g)]{\overset{\mathcal{F}_2(f)}{\Downarrow\mathcal{F}_2(\alpha)}} \mathcal{F}_2(B).$$

The 2-arrows in $\mathcal{K}_3$ are 2-natural transformations, like $\tau_2 : \mathcal{F}_2 \overset{2:}{\Rightarrow} \mathcal{G}_2$ between 2-functors $\mathcal{F}_2, \mathcal{G}_2 : \mathcal{K}_2 \to \mathcal{L}_2$ that sends each object in $\mathcal{K}_2$ to an arrow in $\mathcal{L}_2$ and each arrow in $\mathcal{K}_2$ to a 2-arrow in $\mathcal{L}_2$, and satisfies natural transformation-like conditions. We can visualize τ_2 as a prism going from one functorial picture of $\mathcal{K}_2$ in $\mathcal{L}_2$ to another, built using commutative squares:

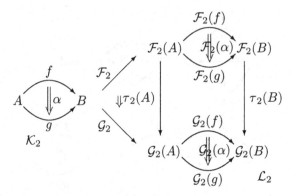

Similarly, the arrows in a 4-category $\mathcal{K}_4$ are 3-functors $\mathcal{F}_3, \mathcal{G}_3, \dots$ sending objects in $\mathcal{K}_3$ to objects in $\mathcal{L}_3$, arrows to arrows, and 2-arrows to 2-arrows, strictly preserving all the structure of $\mathcal{K}_3$

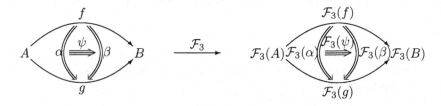

The 2-arrows in $\mathcal{K}_4$ are 3-natural transformations, like $\tau_3 : \mathcal{F} \overset{3}{\Rightarrow} \mathcal{G}$ between 3-functors $\mathcal{F}_3, \mathcal{G}_3 : \mathcal{K}_3 \to \mathcal{L}_3$ that sends each object in $\mathcal{K}_3$ to a arrow in $\mathcal{L}_3$ and each arrow in $\mathcal{K}_3$ to a 2-arrow in $\mathcal{L}_3$, and satisfies natural transformation-like conditions. We can visualize τ_3 as a prism going from one picture of $\mathcal{K}_3$ in $\mathcal{L}_3$ to another, built using commutative squares:

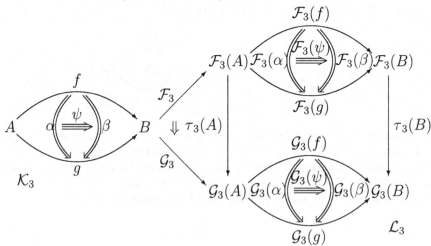

7.1.8 Topological Structure of n-Categories

We already emphasized the topological nature of ordinary category theory. This fact is even more obvious in the general case of n-categories (see [Lei02, Lei03, Lei04]).

Homotopy Theory

Any topological manifold M induces an *n-category* $\Pi_n(M)$ (its *fundamental n-groupoid*), in which 0-cells are *points* in M; 1-cells are *paths* in M (i.e., parameterized continuous maps $f : [0,1] \to M$); 2-cells are *homotopies* (denoted by $\simeq$) *of paths* relative to endpoints (i.e., parameterized continuous maps $h : [0,1] \times [0,1] \to M$); 3-cells are *homotopies of homotopies* of paths in M (i.e., parameterized continuous maps $j : [0,1] \times [0,1] \times [0,1] \to M$); categorical *composition* is defined by *pasting* paths and homotopies. In this way the following 'homotopy skeleton' emerges:

0-cell: $x \bullet$ $x \in M$;

1-cell: $x \bullet \xrightarrow{\ f\ } \bullet y$ $f : x \simeq y \in M$,
$f : [0,1] \to M$, $f : x \mapsto y$, $y = f(x)$, $f(0) = x$, $f(1) = y$;
e.g., linear path: $f(t) = (1-t)x + ty$;

2-cell: $x \bullet \Downarrow h \bullet y$ $h : f \simeq g \in M$,

$h : [0,1] \times [0,1] \to M$, $h : f \mapsto g$, $g = h(f(x))$,
$h(x,0) = f(x)$, $h(x,1) = g(x)$, $h(0,t) = x$, $h(1,t) = y$
e.g., linear homotopy: $h(x,t) = (1-t)f(x) + tg(x)$;

3-cell: $x \bullet\ h \left(\overset{j}{\Longrightarrow}\right) i\ \bullet y$ $j : h \simeq i \in M$,

$j : [0,1] \times [0,1] \times [0,1] \to M$, $j : h \mapsto i$, $i = j(h(f(x)))$
$j(x,t,0) = h(f(x))$, $j(x,t,1) = i(f(x))$,
$j(x,0,s) = f(x)$, $j(x,1,s) = g(x)$,
$j(0,t,s) = x$, $j(1,t,s) = y$
e.g., linear composite homotopy: $j(x,t,s) = (1-t)\,h(f(x)) + t\,i(f(x))$.

If M is a *smooth* manifold, then all included paths and homotopies need to be *smooth*. Recall that a *groupoid* is a category in which every morphism is invertible; its special case with only one object is a *group*.

Category $\mathcal{TT}$

Topological n-category $\mathcal{TT}$ has:

- 0-cells: topological spaces X
- 1-cells: continuous maps $X \xrightarrow{\;f\;} Y$
- 2-cells: homotopies h between f and g : X

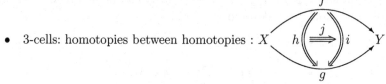

  i.e., continuous maps $h : X \times [0,1] \to Y$, such that $\forall x \in X$, $h(x,0) = f(x)$ and $h(x,1) = g(x)$

- 3-cells: homotopies between homotopies : X

  i.e., continuous maps $j : X \times [0,1] \times [0,1] \to Y$.

Category $\mathcal{CK}$

Consider an n-category $\mathcal{CK}$, which has:

- 0-cells: chain complexes A (of Abelian groups, say)
- 1-cells: chain maps $A \xrightarrow{\;f\;} B$
- 2-cells: chain homotopies A

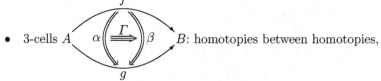

  B,

  i.e., maps $\alpha : A \to B$ of degree 1

- 3-cells A $\quad \alpha \Longrightarrow \beta \quad$ B: homotopies between homotopies,

  i.e., maps $\Gamma : A \to B$ of degree 2 such that $d\Gamma - \Gamma d = \beta - \alpha$.

There ought to be some kind of map $\mathcal{CC} : \mathcal{TT} \Rightarrow \mathcal{CK}$ (see [Lei02, Lei03, Lei04]).

Emerging Categories: *MATTER* $\Rightarrow$ *LIFE* $\Rightarrow$ *MIND*

The solitary thought nets effectively simulate the following 3-categorical structure of MIND, emerging from the 2-categorical structure of LIFE, which is itself emerging from the 1-categorical structure of MATTER:

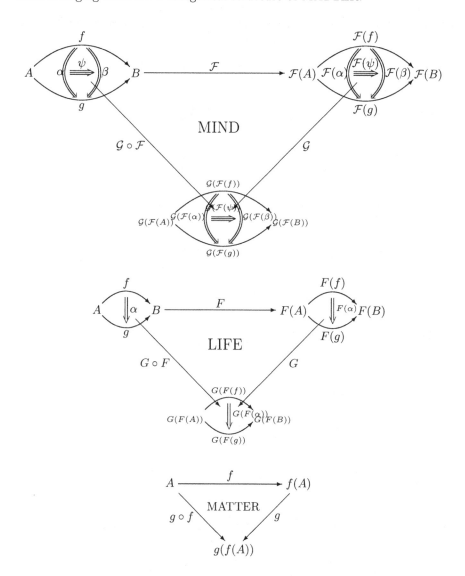

7.2 Frequently Used Mathematical Concepts

7.2.1 Groups and Related Algebraic Structures

As already stated, the basic functional unit of lower biomechanics is the special Euclidean group $SE(3)$ of rigid body motions. In general, a *group* is a pointed set (G, e) with a *multiplication* $\mu : G \times G \to G$ and an *inverse* $\nu : G \to G$ such that the following diagrams commute [Swi75]:

1.

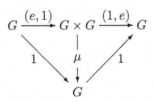

   (e is a two-sided identity)

2.

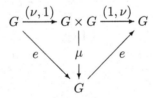

   (associativity)

3.

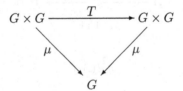

   (inverse).

Here $e : G \to G$ is the constant map $e(g) = e$ for all $g \in G$. $(e, 1)$ means the map such that $(e, 1)(g) = (e, g)$, etc. A group G is called *commutative* or *Abelian group* if in addition the following diagram commutes

$$G \times G \xrightarrow{\quad T \quad} G \times G$$

with μ pointing down to G from both sides.

where $T : G \times G \to G \times G$ is the switch map $T(g_1, g_2) = (g_1, g_2)$, for all $(g_1, g_2) \in G \times G$.

A group G *acts* (on the left) on a set A if there is a function $\alpha : G \times A \to A$ such that the following diagrams commute [Swi75]:

1.

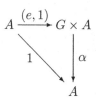

2.

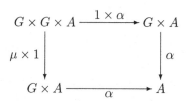

where $(e, 1)(x) = (e, x)$ for all $x \in A$. The *orbits* of the action are the sets $Gx = \{gx : g \in G\}$ for all $x \in A$.

Given two groups $(G, *)$ and $(H, \cdot)$, a *group homomorphism* from $(G, *)$ to $(H, \cdot)$ is a function $h : G \to H$ such that for all x and y in G it holds that

$$h(x * y) = h(x) \cdot h(y).$$

From this property, one can deduce that h maps the identity element e_G of G to the identity element e_H of H, and it also maps inverses to inverses in the sense that $h(x^{-1}) = h(x)^{-1}$. Hence one can say that h is *compatible* with the *group structure.*

The *kernel* $\operatorname{Ker} h$ of a group homomorphism $h : G \to H$ consists of all those elements of G which are sent by h to the identity element e_H of H, i.e.,

$$\operatorname{Ker} h = \{x \in G : h(x) = e_H\}.$$

The *image* $\operatorname{Im} h$ of a group homomorphism $h : G \to H$ consists of all elements of G which are sent by h to H, i.e.,

$$\operatorname{Im} h = \{h(x) : x \in G\}.$$

The kernel is a *normal subgroup* of G and the image is a *subgroup* of H. The homomorphism h is *injective* (and called a *group monomorphism*) iff $\operatorname{Ker} h = e_G$, i.e., iff the kernel of h consists of the identity element of G only.

Similarly, a *ring* is a set S together with two binary operators $+$ and $*$ (commonly interpreted as addition and multiplication, respectively) satisfying the following conditions:

1. Additive associativity: For all $a, b, c \in S$, $(a + b) + c = a + (b + c)$,
2. Additive commutativity: For all $a, b \in S$, $a + b = b + a$,
3. Additive identity: There exists an element $0 \in S$ such that for all $a \in S$, $0 + a = a + 0 = a$,

4. Additive inverse: For every $a \in S$ there exists $-a \in S$ such that $a + (-a) = (-a) + a = 0$,
5. Multiplicative associativity: For all $a, b, c \in S$, $(a * b) * c = a * (b * c)$,
6. Left and right distributivity: For all $a, b, c \in S$, $a * (b + c) = (a * b) + (a * c)$ and $(b + c) * a = (b * a) + (c * a)$.

A ring (the term introduced by *David Hilbert*) is therefore an Abelian group under addition and a semigroup under multiplication. A ring that is commutative under multiplication, has a unit element, and has no divisors of zero is called an *integral domain*. A ring which is also a commutative multiplication group is called a *field*. The simplest rings are the integers $\mathbb{Z}$, polynomials $R[x]$ and $R[x, y]$ in one and two variables, and square $n \times n$ real matrices.

An *ideal* is a subset $\mathfrak{I}$ of elements in a ring R which forms an additive group and has the property that, whenever x belongs to R and y belongs to $\mathfrak{I}$, then xy and yx belong to $\mathfrak{I}$. For example, the set of even integers is an ideal in the ring of integers $\mathbb{Z}$. Given an ideal $\mathfrak{I}$, it is possible to define a factor ring $R/\mathfrak{I}$.

A ring is called *left* (respectively, *right*) *Noetherian* if it does not contain an infinite ascending chain of left (respectively, right) ideals. In this case, the ring in question is said to satisfy the ascending chain condition on left (respectively, right) ideals. A *ring* is said to be *Noetherian* if it is both left and right Noetherian. If a ring R is Noetherian, then the following are equivalent:

1. R satisfies the ascending chain condition on ideals.
2. Every ideal of R is finitely generated.
3. Every set of ideals contains a maximal element.

A *module* is a mathematical object in which things can be added together commutatively by multiplying coefficients and in which most of the rules of manipulating vectors hold. A module is abstractly very similar to a vector space, although in modules, coefficients are taken in rings which are much more general algebraic objects than the fields used in vector spaces. A module taking its coefficients in a ring R is called a module over R or R-module. Modules are the basic tool of homological algebra.

Examples of modules include the set of integers $\mathbb{Z}$, the cubic lattice in d dimensions $\mathbb{Z}^d$, and the group ring of a group. $\mathbb{Z}$ is a module over itself. It is closed under addition and subtraction. Numbers of the form $n\alpha$ for $n \in \mathbb{Z}$ and α a fixed integer form a submodule since, for $(n, m) \in \mathbb{Z}$, $n\alpha \pm m\alpha = (n \pm m)\alpha$ and $(n \pm m)$ is still in $\mathbb{Z}$. Also, given two integers a and b, the smallest module containing a and b is the module for their greatest common divisor, $\alpha = GCD(a, b)$.

A module M is a *Noetherian module* if it obeys the ascending chain condition with respect to inclusion, i.e., if every set of increasing sequences of submodules eventually becomes constant. If a module M is Noetherian, then the following are equivalent:

1. M satisfies the ascending chain condition on submodules.
2. Every submodule of M is finitely generated.
3. Every set of submodules of M contains a maximal element.

Let I be a partially ordered set. A *direct system* of R-*modules* over I is an ordered pair $\{M_i, \varphi^i_j\}$ consisting of an indexed family of modules $\{M_i : i \in I\}$ together with a family of homomorphisms $\{\varphi^i_j : M_i \to M_j\}$ for $i \le j$, such that $\varphi^i_i = 1_{M_i}$ for all i and such that the following diagram commutes whenever $i \le j \le k$

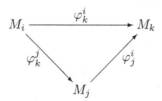

Similarly, an *inverse system* of R-*modules* over I is an ordered pair $\{M_i, \psi^j_i\}$ consisting of an indexed family of modules $\{M_i : i \in I\}$ together with a family of homomorphisms $\{\psi^j_i : M_j \to M_i\}$ for $i \le j$, such that $\psi^i_i = 1_{M_i}$ for all i and such that the following diagram commutes whenever $i \le j \le k$

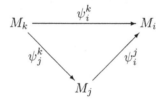

Braid Group

Let C_n be the space of unordered n-tuples of distinct points in the complex plane $\mathbb{C}$. The *braid group* B_n is the *fundamental group* (see Sect. 7.1.1) of C_n. A closed path γ on this space is a set of n paths $\gamma_i : [0,1] \to \mathbb{C}$ ($i = 1, \dots, n$) with $\gamma_i(t) \ne \gamma_j(t)$ and $\gamma_i(1) = \gamma_{\sigma(i)}(0)$, where σ is some permutation of $\{1, \dots, n\}$. Drawing the graphs of all these paths in 3 space, what we see is n strands between the $z = 0$ and $z = 1$ planes, possibly tangled, with composition given by stacking these braids on top of each other. Homotopy corresponds to isotopy of the braid, homotopies of the strands such that none of them cross. This is the origin of the name *braid group*. The braid group determines a *homomorphism* $\phi : B_n \to S_n$ where S_n is the symmetric group on n letters. For $\gamma \in B_n$, we get an element of S_n from the map $i \mapsto \gamma_i(1)$.

The braid group B_n can be given the following presentation (from Artin's first paper in 1925). If we label the braids as before, let σ_i be the braid that twists strands i and $i + 1$, with i passing beneath $i + 1$. Then the twists σ_i generate B_n, and the only relations needed are

$$\sigma_i \sigma_j = \sigma_j \sigma_i \quad \text{for } |i - j| \geq 2,\ 1 \leq i,\ j \leq n - 1,$$
$$\sigma_i \sigma_{i+1} \sigma_i = \sigma_{i+1} \sigma_i \sigma_{i+1} \quad \text{for } 1 \leq i \leq n - 2.$$

7.2.2 Manifolds, Bundles and Lie Groups

Geometrically, a manifold is a nonlinear (i.e., curved) space which is locally homeomorphic (i.e., topologically equivalent) to a linear (i.e., flat) Euclidean space $\mathbb{R}^n$; e.g., in a magnifying glass, each local patch of the apple surface looks like a plane, although globally (as a whole) the apple surface is totally different from the plane. Physically, a configuration manifold is a set of all degrees of freedom of a dynamical system.

More precisely, consider Consider a set M (see Fig. 7.2) which is a *candidate* for a manifold. Any point $x \in M^4$ has its *Euclidean chart*, given by a 1–1 and *onto* map $\varphi_i : M \to \mathbb{R}^n$, with its *Euclidean image* $V_i = \varphi_i(U_i)$. More precisely, a chart φ_i is defined by

$$\varphi_i : M \supset U_i \ni x \mapsto \varphi_i(x) \in V_i \subset \mathbb{R}^n,$$

where $U_i \subset M$ and $V_i \subset \mathbb{R}^n$ are open sets.

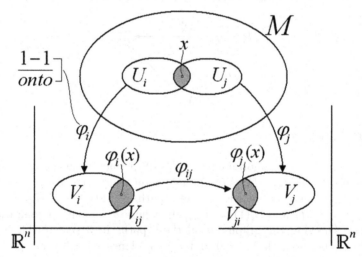

Fig. 7.2. Geometric picture of the manifold concept.

Clearly, any point $x \in M$ can have several different charts (see Fig. 7.2). Consider a case of two charts, $\varphi_i, \varphi_j : M \to \mathbb{R}^n$, having in their images two open sets, $V_{ij} = \varphi_i(U_i \cap U_j)$ and $V_{ji} = \varphi_j(U_i \cap U_j)$. Then we have *transition functions* φ_{ij} between them,

---

[4] Note that sometimes we will denote the point in a manifold M by m, and sometimes by x (thus implicitly assuming the existence of coordinates $x = (x^i)$).

$$\varphi_{ij} = \varphi_j \circ \varphi_i^{-1} : V_{ij} \to V_{ji}, \quad \text{locally given by } \varphi_{ij}(x) = \varphi_j(\varphi_i^{-1}(x)).$$

If transition functions φ_{ij} exist, then we say that two charts, φ_i and φ_j are *compatible*. Transition functions represent a general (nonlinear) *transformations of coordinates*, which are the core of classical *tensor calculus*.

A set of compatible charts $\varphi_i : M \to \mathbb{R}^n$, such that each point $x \in M$ has its Euclidean image in at least one chart, is called an *atlas*. Two atlases are *equivalent* iff all their charts are compatible (i.e., transition functions exist between them), so their union is also an atlas. A *manifold structure* is a class of equivalent atlases.

Finally, as charts $\varphi_i : M \to \mathbb{R}^n$ were supposed to be 1–1 and onto maps, they can be either *homeomorphisms*, in which case we have a *topological* (C^0) manifold, or *diffeomorphisms*, in which case we have a *smooth* (C^k) manifold.

On the other hand, tangent and cotangent bundles, TM and T^*M, respectively, of a smooth manifold M, are special cases of a more general geometrical object called *fibre bundle*, where the word *fiber* V of a map $\pi : Y \to X$ denotes the *preimage* $\pi^{-1}(x)$ of an element $x \in X$. It is a space which *locally* looks like a product of two spaces (similarly as a manifold locally looks like Euclidean space), but may possess a different *global* structure. To get a visual intuition behind this fundamental geometrical concept, we can say that a fibre bundle Y is a *homeomorphic generalization* of a *product space* $X \times V$ (see Fig. 7.3), where X and V are called the *base* and the *fibre*, respectively. $\pi : Y \to X$ is called the *projection*, $Y_x = \pi^{-1}(x)$ denotes a fibre over a point x of the base X, while the map $f = \pi^{-1} : X \to Y$ defines the *cross-section*, producing the *graph* $(x, f(x))$ in the bundle Y (e.g., in case of a tangent bundle, $f = \dot{x}$ represents a velocity vector-field, so that the graph in a the bundle Y reads $(x, \dot{x})$).

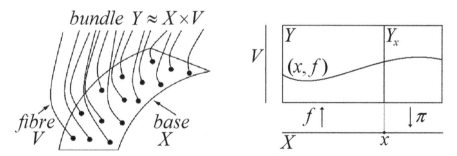

Fig. 7.3. A sketch of a fibre bundle $Y \approx X \times V$ as a generalization of a product space $X \times V$; *left*—main components; *right*—a few details (see text for explanation).

A principal G-bundle is a bundle $\pi : Y \to X$ generated by a Lie group G (see below) such that the group G preserves the fibers of the bundle Y.

The main reason why we need to study fibre bundles is that *all dynamical objects* (including vectors, tensors, differential forms and gauge potentials) are their *cross-sections*, representing *generalizations of graphs of continuous functions*. For more technical details, see [II06c, II07e].

Riemann and Ricci Curvatures on a Smooth Manifold

Recall that proper differentiation of vector and tensor fields on a smooth Riemannian n-manifold is performed using the *Levi-Civita covariant derivative* (see, e.g., [II06c, II07e]). Formally, let M be a Riemannian n-manifold with the tangent bundle TM and a local coordinate system $\{x^i\}_{i=1}^n$ defined in an open set $U \subset M$. The covariant derivative operator, $\nabla_X : C^\infty(TM) \to C^\infty(TM)$, is the unique linear map such that for any vector fields X, Y, Z, constant c, and function f the following properties are valid:

$$\nabla_{X+cY} = \nabla_X + c\nabla_Y,$$
$$\nabla_X(Y + fZ) = \nabla_X Y + (Xf)Z + f\nabla_X Z, \quad \text{with}$$
$$\nabla_X Y - \nabla_Y X = [X, Y], \quad \text{(torsion free property)}$$

where $[X, Y]$ is the Lie bracket of X and Y (see, e.g., [II06c, II07e]). In local coordinates, the metric g is defined for any orthonormal basis $(\partial_i = \partial_{x^i})$ in $U \subset M$ by

$$g_{ij} = g(\partial_i, \partial_j) = \delta_{ij}, \qquad \partial_k g_{ij} = 0.$$

Then the affine *Levi-Civita connection* is defined on M by

$$\nabla_{\partial_i}\partial_j = \Gamma_{ij}^k \partial_k, \quad \text{where } \Gamma_{ij}^k = \frac{1}{2}g^{kl}\left(\partial_i g_{jl} + \partial_j g_{il} - \partial_l g_{ij}\right)$$

are the (second-order) *Christoffel symbols*.

Now, using the covariant derivative operator ∇_X we can define the *Riemann curvature* $(3, 1)$-tensor $\mathfrak{Rm}$ by (see, e.g., [II06c, II07e])

$$\mathfrak{Rm}(X, Y)Z = \nabla_X \nabla_Y Z - \nabla_Y \nabla_X Z - \nabla_{[X,Y]}Z.$$

$\mathfrak{Rm}$ measures the curvature of the manifold by expressing how noncommutative covariant differentiation is. The $(3, 1)$-components R_{ijk}^l of $\mathfrak{Rm}$ are defined in $U \subset M$ by

$$\mathfrak{Rm}(\partial_i, \partial_j)\partial_k = R_{ijk}^l \partial_l, \quad \text{which expands (see [II06c, II07e]) as}$$
$$R_{ijk}^l = \partial_i \Gamma_{jk}^l - \partial_j \Gamma_{ik}^l + \Gamma_{jk}^m \Gamma_{im}^l - \Gamma_{ik}^m \Gamma_{jm}^l.$$

Also, the Riemann $(4, 0)$-tensor $R_{ijkl} = g_{lm}R_{ijk}^m$ is defined as the g-based inner product on M,

$$R_{ijkl} = \langle \mathfrak{Rm}(\partial_i, \partial_j)\partial_k, \partial_l \rangle.$$

The first and second Bianchi identities for the Riemann $(4, 0)$-tensor R_{ijkl} hold,

$$R_{ijkl} + R_{jkil} + R_{kijl} = 0, \qquad \nabla_i R_{jklm} + \nabla_j R_{kilm} + \nabla_k R_{ijlm} = 0,$$

while the twice contracted second Bianchi identity reads

$$2\nabla_j R_{ij} = \nabla_i R. \qquad (7.4)$$

The $(0,2)$ *Ricci tensor* $\mathfrak{Rc}$ is the trace of the Riemann $(3,1)$-tensor $\mathfrak{Rm}$,

$$\mathfrak{Rc}(Y,Z) + \mathrm{tr}(X \to \mathfrak{Rm}(X,Y)Z), \quad \text{so that} \quad \mathfrak{Rc}(X,Y) = g(\mathfrak{Rm}(\partial_i, X)\partial_i, Y).$$

Its components $R_{jk} = \mathfrak{Rc}(\partial_j, \partial_k)$ are given in $U \subset M$ by the contraction [II06c, II07e]

$$R_{jk} = R^i_{ijk}, \quad \text{or, in terms of Christoffel symbols,}$$
$$R_{jk} = \partial_i \Gamma^i_{jk} - \partial_k \Gamma^i_{ji} + \Gamma^i_{mi}\Gamma^m_{jk} - \Gamma^i_{mk}\Gamma^m_{ji}.$$

Being a symmetric second-order tensor, $\mathfrak{Rc}$ has $\left(\frac{n+1}{2}\right)$ independent components on an n-manifold M. In particular, on a 3-manifold, it has 6 components, and on a 2-surface it has only the following 3 components:

$$R_{11} = g^{22}R_{2112}, \qquad R_{12} = g^{12}R_{2121}, \qquad R_{22} = g^{11}R_{1221},$$

which are all proportional to the corresponding coordinates of the metric tensor,

$$\frac{R_{11}}{g_{11}} = \frac{R_{12}}{g_{12}} = \frac{R_{22}}{g_{22}} = -\frac{R_{1212}}{\det(g)}. \qquad (7.5)$$

Finally, the scalar curvature R is the trace of the Ricci tensor $\mathfrak{Rc}$, given in $U \subset M$ by: $R = g^{ij}R_{ij}$.

Lie Groups

A Lie group is a both a group and a manifold. More precisely, a *Lie group* is a smooth manifold M that has at the same time a group G-structure consistent with its manifold M-structure in the sense that *group multiplication* $\mu : G \times G \to G$, $(g,h) \mapsto gh$ and the *group inversion* $\nu : G \to G$, $g \mapsto g^{-1}$ are smooth functions. A point $e \in G$ is called the *group identity element*. For any Lie group G in a neighborhood of its identity element e it can be expressed in terms of a set of generators T^a $(a = 1, \ldots, \dim G)$ as

$$D(g) = \exp[-i\alpha_a T^a] \equiv \sum_{n=0}^{\infty} \frac{(-i)^n}{n!} \alpha_{a_1} \cdots \alpha_{a_n} T^{a_1} \cdots T^{a_n},$$

where $\alpha_a \in \mathbb{C}$ are a set of coordinates of M in a neighborhood of e. Because of the general *Baker–Campbell–Haussdorf formula*, the multiplication of two group elements is encoded in the value of the commutator of two generators, that in general has the form

$$[T^a, T^b] = \mathrm{i} f^{abc} T^c,$$

where $f^{abc} \in \mathbb{C}$ are called the structure constants. The set of generators with the commutator operation form the Lie algebra associated with the Lie group. Hence, given a representation of the Lie algebra of generators we can construct a representation of the group by exponentiation (at least locally near the identity).

In particular, for $SU(2)$-group, each group element is labeled by three real numbers α_k, $(k = 1, 2, 3)$. We have two basic representations: one is the fundamental representation (or spin $\frac{1}{2}$) defined by

$$D_{\frac{1}{2}}(\alpha_k) = e^{-\frac{i}{2} \alpha_k \sigma^k},$$

with σ^i the Pauli matrices. The second one is the adjoint (or spin 1) representation which can be written as

$$D_1(\alpha_k) = e^{-i \alpha_k J^k}, \quad \text{where}$$

$$J^1 = \begin{pmatrix} 0 & 0 & 0 \\ 0 & 0 & 1 \\ 0 & -1 & 0 \end{pmatrix}, \qquad J^2 = \begin{pmatrix} 0 & 0 & -1 \\ 0 & 0 & 0 \\ 1 & 0 & 0 \end{pmatrix}, \qquad J^3 = \begin{pmatrix} 0 & 1 & 0 \\ -1 & 0 & 0 \\ 0 & 0 & 0 \end{pmatrix}.$$

Actually, J^k generate rotations around the x, y and z axis respectively.

Let M be a smooth manifold. An *action of a Lie group* G (with the unit element e) on M is a smooth map $\phi : G \times M \to M$, such that for all $x \in M$ and $g, h \in G$, (i) $\phi(e, x) = x$ and (ii) $\phi(g, \phi(h, x)) = \phi(gh, x)$. For more technical details, see [II06c, II07e].

7.2.3 Unitary Matrix and Group

Unitary and Hermitian Matrices

A *unitary matrix* is an invertible $n \times n$ complex matrix U satisfying the condition

$$U^\dagger U = U U^\dagger = I_n \quad \text{or} \quad U^{-1} = U^\dagger,$$

where I_n is the identity n-matrix and $U^\dagger$ is the *conjugate transpose* (also called the *Hermitian transpose*, or *adjoint* matrix, which is obtained by first taking the transpose and then taking the complex conjugate of each entry[5]) of U.

---

[5] For example, if

$$U = \begin{bmatrix} 3+i & 5 \\ 2-2i & i \end{bmatrix}$$

then

$$U^\dagger = \begin{bmatrix} 3-i & 2+2i \\ 5 & -i \end{bmatrix}.$$

The following are all equivalent conditions: (i) U is unitary; (ii) $U^\dagger$ is unitary; (iii) the columns of U form an orthonormal basis of $\mathbb{C}^n$ with respect to $\langle \cdot, \cdot \rangle$; (iv) the rows of U form an orthonormal basis of $\mathbb{C}^n$ with respect to $\langle \cdot, \cdot \rangle$; (v) U is an *isometry* with respect to the norm from to $\langle \cdot, \cdot \rangle$. It follows from the isometry property that all eigenvalues of a unitary matrix are complex numbers of absolute value 1 (i.e., they lie on the unit circle centered at 0 in the complex plane). The same is true for the determinant, $|\det(U)| = 1$.

A unitary matrix in which all entries are real is an *orthogonal matrix*. Just as an orthogonal matrix G preserves the real inner product of two real vectors,

$$\langle Gx, Gy \rangle = \langle x, y \rangle,$$

so also a unitary matrix U preserves the complex inner product of two complex vectors,

$$\langle Ux, Uy \rangle = \langle x, y \rangle,$$

for all complex vectors x and y, where $\langle \cdot, \cdot \rangle$ denotes the standard inner product on $\mathbb{C}^n$.

All unitary matrices are normal, and the *spectral theorem* therefore applies to them: every unitary matrix U has a decomposition of the form

$$U = V \Sigma V^\dagger,$$

where V is unitary, and Σ is diagonal and unitary.

The set of all $n \times n$ unitary matrices with matrix multiplication forms a *unitary* group $U(n)$.

In particular, a *self-adjoint*, or *Hermitian matrix* is an $n \times n$ complex matrix H that is equal to its own conjugate transpose $H^\dagger$,[6]

$$H = H^\dagger.$$

The entries on the main diagonal of any Hermitian matrix are necessarily real. Every Hermitian matrix is normal, and the finite-dimensional spectral theorem applies: any Hermitian matrix can be diagonalized by a unitary matrix, and the resulting diagonal matrix has only real entries. This means that all eigenvalues of a Hermitian matrix are real, and, moreover, eigenvectors with distinct eigenvalues are orthogonal. It is possible to find an orthonormal basis of $\mathbb{C}^n$ consisting only of eigenvectors.

The sum of any two *Hermitian* matrices is Hermitian, and the inverse of an invertible Hermitian matrix is Hermitian as well. However, the product of two Hermitian matrices A and B will only be Hermitian if they commute, i.e., if $AB = BA$. Thus An is Hermitian if A is Hermitian and n is a positive

---

[6] For example,

$$H = \begin{bmatrix} 3 & 2+i \\ 2-i & 1 \end{bmatrix}$$

is a Hermitian matrix.

integer. If the eigenvalues of a Hermitian matrix are all positive, then the matrix is positive definite; if they are all non-negative, then the matrix is positive semidefinite. The Hermitian $n \times n$ matrices form a vector space of dimension n^2 (one DOF per main diagonal element, and two DOF per element above the main diagonal). over the real numbers (but not over the complex numbers).

The sum of a square matrix and its conjugate transpose is Hermitian, $A + A^\dagger = H$.

The difference of a square matrix and its conjugate transpose is *skew-Hermitian* (or, antihermitian), $A - A^\dagger = -B$, such that $B^\dagger = -B$.[7] The eigenvalues of a skew-Hermitian matrix are all purely imaginary. Furthermore, skew-Hermitian matrices are normal. Hence they are diagonalizable and their eigenvectors for distinct eigenvalues must be orthogonal. If B is skew-Hermitian, then $iB = H$ is Hermitian. If B is skew-Hermitian, then $e^B = U$ is unitary.

An arbitrary square matrix A can be written as the sum of a Hermitian matrix H and a skew-Hermitian matrix B:

$$A = H + B \ \text{ with } H = \frac{1}{2}(A + A^\dagger) \text{ and } B = \frac{1}{2}(A - A^\dagger).$$

The Special Unitary Group

The unitary group of degree n, denoted $U(n)$, is the group of $n \times n$ unitary matrices, with matrix multiplication as the group operation. The unitary group is a subgroup of the general linear group $GL(n, \mathbb{C})$.

In the simple case $n = 1$, the group $U(1)$ corresponds to the circle group, consisting of all complex numbers with absolute value 1 under multiplication. All the unitary groups contain copies of this group.

The unitary group $U(n)$ is a real Lie group of dimension n^2. The corresponding Lie algebra $\mathfrak{u}(n)$ consists of complex $n \times n$ skew-Hermitian matrices, with the Lie bracket $[B_i, B_j]$ given by the commutator.

Since the determinant of a unitary matrix is a complex number with norm 1, the determinant gives a *group homomorphism*:

$$\det : U(n) \to U(1).$$

The *kernel* of this homomorphism is the set of unitary matrices with unit determinant. This subgroup is called the *special unitary group*, denoted $SU(n)$. We then have a *short exact sequence* of Lie groups:

---

[7] For example, the following matrix is skew-Hermitian,

$$B = \begin{pmatrix} i & 2+i \\ -2+i & 3i \end{pmatrix}.$$

$$1 \to SU(n) \to U(n) \to U(1) \to 1.$$

This short exact sequence splits so that $U(n)$ may be written as a *semidirect product*

$$U(n) = SU(n) \rhd U(1).$$

Here the $U(1)$ subgroup of $U(n)$ consists of matrices of the form

$$\mathrm{diag}(e^{i\theta}, 1, 1, \ldots, 1).$$

The unitary group $U(n)$ is non-Abelian for $n > 1$.

The special unitary groups $SU(n)$ have wide applications in the *Standard model* of particle physics, especially $SU(2)$ in the *electro-weak interaction* and $SU(3)$ in QCD.

The simplest case, $SU(1)$, is a trivial group, having only a single element.

The group $SU(2)$ is isomorphic to the group of quaternions of absolute value 1, and is thus diffeomorphic to the 3-sphere S^3. Since unit quaternions can be used to represent rotations in 3D space (up to a sign), we have a surjective homomorphism

$$h : SU(2) \to SO(3),$$

whose kernel is $\{+I, -I\}$.

The special unitary group $SU(n)$ is a real matrix Lie group of dimension $n^2 - 1$. Topologically, it is compact and simply connected. Algebraically, it is a simple Lie group (meaning its Lie algebra is simple). The center of $SU(n)$ is isomorphic to the cyclic group $\mathbb{Z}_n$. Its outer automorphism group, for $n \geq 3$, is $\mathbb{Z}_2$, while the outer automorphism group of $SU(2)$ is the trivial group.

The $SU(n)$ algebra is generated by n^2 operators O_{ij}, which satisfy the commutator relationship (for $i, j, k, l = 1, 2, \ldots, n$),

$$[O_{ij}, O_{kl}] = \delta_{jk} O_{il} - \delta_{il} O_{kj}.$$

Additionally, the *trace operator* $N = \sum_{i=1}^{n} O_{ii}$ satisfies

$$[N, O_{ij}] = 0,$$

which implies that the number of independent generators of $SU(n)$ is $n^2 - 1$.

Generators of $SU(n)$

In general the generators T_a of $SU(n)$, are represented as traceless Hermitian matrices

$$\mathrm{tr}(T_a) = 0 \quad \text{and} \quad T_a = T_a^\dagger.$$

In the *fundamental representation* the generators are represented by

$$T_a T_b = \frac{1}{2n}\delta_{ab}I_n + \frac{1}{2}\sum_{c=1}^{n^2-1}(if_{abc} + d_{abc})T_c,$$

where f_{abc} are the structure constants, antisymmetric in all indices, while d_{abc} are constants symmetric in all indices, satisfying a normalization convention

$$\sum_{c,e=1}^{n^2-1} d_{ace}d_{bce} = \frac{n^2-4}{n}\delta_{ab}.$$

As a consequence we have

$$[T_a, T_b]_+ = \frac{1}{n}\delta_{ab} + \sum_{c=1}^{n^2-1} d_{abc}T_c, \quad \text{and}$$

$$[T_a, T_b]_- = i\sum_{c=1}^{n^2-1} f_{abc}T_c.$$

In the *adjoint representation* the generators are represented by $(n^2-1) \times (n^2-1)$-matrices whose elements are defined by the structure constants:

$$(T_a)_{jk} = -if_{ajk}.$$

$SU(2)$

For $SU(2)$, the generators T_a in the fundamental representation are proportional to the *Pauli matrices*,

$$\sigma_1 = \begin{pmatrix} 0 & 1 \\ 1 & 0 \end{pmatrix}, \qquad \sigma_2 = \begin{pmatrix} 0 & -i \\ i & 0 \end{pmatrix}, \qquad \sigma_3 = \begin{pmatrix} 1 & 0 \\ 0 & -1 \end{pmatrix},$$

such that

$$T_a = \frac{\sigma_a}{2},$$

so that T_a are traceless Hermitian matrices, as required. The structure constants f_{abc} for $SU(2)$ are the Levi-Civita symbols ϵ_{abc}, while the symmetric constants $d_{abc} = 0$.

$SU(3)$

For $SU(3)$,[8] the generators T_a in the fundamental representation are proportional to the *Gell-Mann matrices* λ_a, which are the $SU(3)$ analog of the Pauli matrices for $SU(2)$,

---

[8] The structure constants f_{abc} for $SU(3)$ have values given by

$$\lambda_1 = \begin{pmatrix} 0 & 1 & 0 \\ 1 & 0 & 0 \\ 0 & 0 & 0 \end{pmatrix}, \qquad \lambda_2 = \begin{pmatrix} 0 & -i & 0 \\ i & 0 & 0 \\ 0 & 0 & 0 \end{pmatrix}, \qquad \lambda_3 = \begin{pmatrix} 1 & 0 & 0 \\ 0 & -1 & 0 \\ 0 & 0 & 0 \end{pmatrix},$$

$$\lambda_4 = \begin{pmatrix} 0 & 0 & 1 \\ 0 & 0 & 0 \\ 1 & 0 & 0 \end{pmatrix}, \qquad \lambda_5 = \begin{pmatrix} 0 & 0 & -i \\ 0 & 0 & 0 \\ i & 0 & 0 \end{pmatrix}, \qquad \lambda_6 = \begin{pmatrix} 0 & 0 & 0 \\ 0 & 0 & 1 \\ 0 & 1 & 0 \end{pmatrix},$$

$$\lambda_7 = \begin{pmatrix} 0 & 0 & 0 \\ 0 & 0 & -i \\ 0 & i & 0 \end{pmatrix}, \qquad \lambda_8 = \frac{1}{\sqrt{3}} \begin{pmatrix} 1 & 0 & 0 \\ 0 & 1 & 0 \\ 0 & 0 & -2 \end{pmatrix},$$

such that

$$T_a = \frac{\lambda_a}{2},$$

are all traceless Hermitian matrices, as required. They satisfy the relations

$$[T_a, T_b] = i \sum_{c=1}^{8} f_{abc} T_c.$$

Lie Algebra of SU(n)

The Lie algebra $\mathfrak{su}(n)$ of $SU(n)$ consists of the traceless skew-Hermitian complex matrices, with the regular commutator as Lie bracket. A imaginary-unit factor i is often inserted by particle physicists, so that all matrices become Hermitian.

For example, the following skew-Hermitian matrices used in quantum mechanics form a basis for $\mathfrak{su}(2)$ over $\mathbb{R}$:

$$i\sigma_x = \begin{bmatrix} 0 & i \\ i & 0 \end{bmatrix}, \qquad i\sigma_y = \begin{bmatrix} 0 & 1 \\ -1 & 0 \end{bmatrix}, \qquad i\sigma_z = \begin{bmatrix} i & 0 \\ 0 & -i \end{bmatrix}.$$

This representation is often used in quantum mechanics, to represent the spin of fundamental particles such as electrons. Note that the product of any

$$f^{123} = 1,$$

$$f^{147} = -f^{156} = f^{246} = f^{257} = f^{345} = -f^{376} = \frac{1}{2},$$

$$f^{458} = f^{678} = \frac{\sqrt{3}}{2},$$

while the symmetric constants d_{abc} take the values

$$d^{118} = d^{228} = d^{338} = -d^{888} = \frac{1}{\sqrt{3}},$$

$$d^{448} = d^{558} = d^{668} = d^{778} = -\frac{1}{2\sqrt{3}},$$

$$d^{146} = d^{157} = -d^{247} = d^{256} = d^{344} = d^{355} = -d^{366} = -d^{377} = \frac{1}{2}.$$

two different generators is another generator, and that the generators anti-commute. Together with

$$iI_2 = \begin{bmatrix} i & 0 \\ 0 & i \end{bmatrix},$$

these are also generators of the Lie algebra $\mathfrak{u}(2)$ of the Lie group $U(2)$.

7.2.4 Differential Forms and Stokes Theorem

Given the space of exterior differential p-forms $\Omega^p(M)$ on a smooth manifold M, we have the *exterior derivative operator* $d : \Omega(M) \rightarrow \Omega^{p+1}(M)$ which generalizes ordinary vector differential operators (*grad, div* and *curl* see [Rha84, Fla63, II06c]) and transforms p-forms ω into $(p+1)$-forms $d\omega$, with the main property: $dd = d^2 = 0$. Given a p-form $\alpha \in \Omega^p(M)$ and a q-form $\beta \in \Omega^q(M)$, their exterior product is a $(p+q)$-form $\alpha \wedge \beta \in \Omega^{p+q}(M)$, where $\wedge$ is their anti-commutative exterior (or, 'wedge') product.

As differential forms are meant for integration, we have a generalization of all integral theorems from vector calculus in the form of the Stokes theorem: for the p-form ω, in an oriented nD domain C, which is a p-chain with a $(p-1)$-boundary ∂C,

$$\int_{\partial C} \omega = \int_C d\omega. \tag{7.6}$$

For any p-chain on a manifold M, *the boundary of a boundary is zero* [MTW73], that is, $\partial\partial C = \partial^2 = 0$.

A p-form β is called *closed* if its exterior derivative $d = \partial_i dx^i$ is equal to zero, $d\beta = 0$. From this condition one can see that the closed form (the *kernel* of the exterior derivative operator d) is conserved quantity. Therefore, closed p-forms possess certain invariant properties, physically corresponding to the conservation laws (see e.g., [AMR88, II07e]).

Also, a p-form β that is an exterior derivative of some $(p-1)$-form α, $\beta = d\alpha$, is called *exact* (the *image* of the exterior derivative operator d). By Poincaré lemma, exact forms prove to be closed automatically, $d\beta = d(d\alpha) = 0$.

Since $d^2 = 0$, *every exact form is closed.* The converse is only partially true, by Poincaré lemma: every closed form is *locally exact*. In particular, there is a Poincaré lemma for contractible manifolds: Any closed form on a smoothly contractible manifold is exact. The Poincaré lemma is a generalization and unification of two well-known facts in vector calculus:

(i) If $\text{curl}\, F = 0$, then locally $F = \text{grad}\, f$; and (ii) If $\text{div}\, F = 0$, then locally $F = \text{curl}\, G$.

A *cycle* is a p-chain, (or, an oriented p-domain) $C \in \mathcal{C}_p(M)$ such that $\partial C = 0$. A *boundary* is a chain C such that $C = \partial B$, for any other chain $B \in \mathcal{C}_p(M)$. Similarly, a *cocycle* (i.e., a *closed form*) is a cochain ω such that $d\omega = 0$. A *coboundary* (i.e., an *exact form*) is a cochain ω such that $\omega = d\theta$, for any other cochain θ. All exact forms are closed ($\omega = d\theta \Rightarrow d\omega = 0$) and all boundaries are cycles ($C = \partial B \Rightarrow \partial C = 0$). Converse is true only for smooth contractible manifolds, by Poincaré lemma.

Integration on a smooth manifold M should be thought of as a nondegenerate bilinear pairing $(\,,\,)$ between p-forms and p-chains (spanning a finite domain on M). Duality of p-forms and p-chains on M is based on the de Rham's 'period', defined as [Rha84, CD82]

$$\text{Period} := \int_C \omega := (C, \omega),$$

where C is a cycle, ω is a cocycle, while $\langle C, \omega \rangle = \omega(C)$ is their inner product $(C, \omega) : \Omega^p(M) \times C_p(M) \to \mathbb{R}$. From the Poincaré lemma, a closed p-form ω is exact iff $(C, \omega) = 0$.

The fundamental topological duality is based on the Stokes theorem (7.6), which can be re written as

$$(\partial C, \omega) = (C, d\omega),$$

where ∂C is the boundary of the p-chain C oriented coherently with C on M. While the *boundary operator* ∂ is a global operator, the coboundary operator d is local, and thus more suitable for applications. The main property of the exterior differential,

$$d \circ d \equiv d^2 = 0 \implies \partial \circ \partial \equiv \partial^2 = 0, \quad \text{(and converse)},$$

can be easily proved using the Stokes' theorem as

$$0 = (\partial^2 C, \omega) = (\partial C, d\omega) = (C, d^2\omega) = 0.$$

Now, in the Euclidean 3D space $\mathbb{R}^3$ we have the following de Rham *cochain complex*

$$0 \to \Omega^0(\mathbb{R}^3) \xrightarrow[\text{grad}]{d} \Omega^1(\mathbb{R}^3) \xrightarrow[\text{curl}]{d} \Omega^2(\mathbb{R}^3) \xrightarrow[\text{div}]{d} \Omega^3(\mathbb{R}^3) \to 0.$$

Using the *closure property* for the exterior differential in $\mathbb{R}^3$, $d \circ d \equiv d^2 = 0$, we get the standard identities from vector calculus

$$\text{curl} \cdot \text{grad} = 0 \quad \text{and} \quad \text{div} \cdot \text{curl} = 0.$$

As a duality, in $\mathbb{R}^3$ we have the following *chain complex*

$$0 \leftarrow C_0(\mathbb{R}^3) \xleftarrow{\partial} C_1(\mathbb{R}^3) \xleftarrow{\partial} C_2(\mathbb{R}^3) \xleftarrow{\partial} C_3(\mathbb{R}^3) \leftarrow 0,$$

(with the closure property $\partial \circ \partial \equiv \partial^2 = 0$) which implies the following three boundaries:

$$C_1 \xmapsto{\partial} C_0 = \partial(C_1), \qquad C_2 \xmapsto{\partial} C_1 = \partial(C_2), \qquad C_3 \xmapsto{\partial} C_2 = \partial(C_3),$$

where $C_0 \in \mathcal{C}_0$ is a 0-boundary (or, a point), $C_1 \in \mathcal{C}_1$ is a 1-boundary (or, a line), $C_2 \in \mathcal{C}_2$ is a 2-boundary (or, a surface), and $C_3 \in \mathcal{C}_3$ is a 3-boundary (or, a hypersurface). Similarly, the de Rham complex implies the following three coboundaries:

$$C^0 \overset{d}{\mapsto} C^1 = d(C^0), \qquad C^1 \overset{d}{\mapsto} C^2 = d(C^1), \qquad C^2 \overset{d}{\mapsto} C^3 = d(C^2),$$

where $C^0 \in \Omega^0$ is 0-form (or, a function), $C^1 \in \Omega^1$ is a 1-form, $C^2 \in \Omega^2$ is a 2-form, and $C^3 \in \Omega^3$ is a 3-form.

In general, on a smooth nD manifold M we have the following de Rham cochain complex [Rha84]

$$0 \to \Omega^0(M) \overset{d}{\longrightarrow} \Omega^1(M) \overset{d}{\longrightarrow} \Omega^2(M) \overset{d}{\longrightarrow} \Omega^3(M) \overset{d}{\longrightarrow} \cdots \overset{d}{\longrightarrow} \Omega^n(M) \to 0,$$

satisfying the closure property on M, $d \circ d \equiv d^2 = 0$.

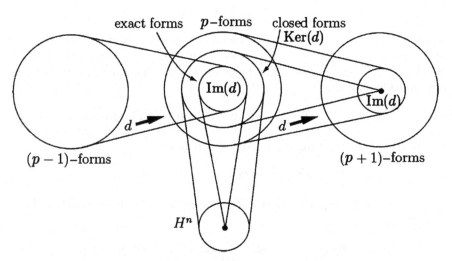

Fig. 7.4. A small portion of the de Rham cochain complex, showing a homomorphism of cohomology groups.

Informally, the de Rham cohomology is the (functional) space of closed differential p-forms modulo exact ones on a smooth manifold.

More precisely, the subspace of all closed p-forms (cocycles) on a smooth manifold M is the kernel $\mathrm{Ker}(d)$ of the de Rham d-homomorphism (see Fig. 7.4), denoted by $Z^p(M) \subset \Omega^p(M)$, and the sub-subspace of all exact p-forms (coboundaries) on M is the image $\mathrm{Im}(d)$ of the de Rham homomorphism denoted by $B^p(M) \subset Z^p(M)$. The *quotient space*

$$H^p_{DR}(M) := \frac{Z^p(M)}{B^p M} = \frac{\mathrm{Ker}(d : \Omega^p(M) \to \Omega^{p+1}(M))}{\mathrm{Im}(d : \Omega^{p-1}(M) \to \Omega^p(M))},$$

is called the pth de Rham *cohomology group* of a manifold M. It is a topological invariant of a manifold. Two p-cocycles $\alpha, \beta \in \Omega^p(M)$ are *cohomologous*, or belong to the same *cohomology class* $[\alpha] \in H^p(M)$, if they differ by a $(p-1)$-coboundary $\alpha - \beta = d\theta \in \Omega^{p-1}(M)$. The dimension $b_p = \dim H^p(M)$ of the de Rham cohomology group $H^p_{DR}(M)$ of the manifold M is called the Betti number b_p.

Similarly, the subspace of all p-cycles on a smooth manifold M is the kernel $\text{Ker}(\partial)$ of the ∂-homomorphism, denoted by $Z_p(M) \subset C_p(M)$, and the sub-subspace of all p-boundaries on M is the image $\text{Im}(\partial)$ of the ∂-homomorphism, denoted by $B_p(M) \subset C_p(M)$. Two p-cycles $C_1, C_2 \in C_p$ are *homologous*, if they differ by a $(p-1)$-boundary $C_1 - C_2 = \partial B \in C_{p-1}(M)$. Then C_1 and C_2 belong to the same *homology class* $[C] \in H_p(M)$, where $H_p(M)$ is the homology group of the manifold M, defined as

$$H_p(M) := \frac{Z_p(M)}{B_p(M)} = \frac{\text{Ker}(\partial : C_p(M) \to C_{p-1}(M))}{\text{Im}(\partial : C_{p+1}(M) \to C_p(M))},$$

where Z_p is the vector space of cycles and $B_p \subset Z_p$ is the vector space of boundaries on M. The dimension $b_p = \dim H_p(M)$ of the homology group $H_p(M)$ is, by the de Rham theorem, the same Betti number b_p.

If we know the Betti numbers for all (co)homology groups of the manifold M, we can calculate the *Euler–Poincaré characteristic* of M as

$$\chi(M) = \sum_{p=1}^{n} (-1)^p b_p.$$

For example, consider a small portion of the de Rham cochain complex of Fig. 7.4 spanning a space-time 4-manifold M,

$$\Omega^{p-1}(M) \xrightarrow{d_{p-1}} \Omega^p(M) \xrightarrow{d_p} \Omega^{p+1}(M).$$

As we have seen above, cohomology classifies topological spaces by comparing two subspaces of Ω^p: (i) the space of p-cocycles, $Z^p(M) = \text{Ker}\, d_p$, and (ii) the space of p-coboundaries, $B^p(M) = \text{Im}\, d_{p-1}$. Thus, for the cochain complex of any space-time 4-manifold we have,

$$d^2 = 0 \quad \Rightarrow \quad B^p(M) \subset Z^p(M),$$

that is, every p-coboundary is a p-cocycle. Whether the converse of this statement is true, according to Poincaré lemma, depends on the particular topology of a space-time 4-manifold. If every p-cocycle is a p-coboundary, so that B^p and Z^p are equal, then the cochain complex is exact at $\Omega^p(M)$. In topologically interesting regions of a space-time manifold M, exactness may fail [Wis06], and we measure the failure of exactness by taking the pth cohomology group

$$H^p(M) = Z^p(M)/B^p(M).$$

7.2.5 Symmetry Breaking and Partition Function

Big Bang, Symmetry Breaking and the Four Forces of Nature

One of the corner-stones of modern physics is the phenomenon of *symmetry breaking* and related *phase transitions*. Classical example is freezing of water below 0 degrees Celsius. Another example is splitting of the unified force of the Universe into the four forces observed today (see Fig. 7.5).

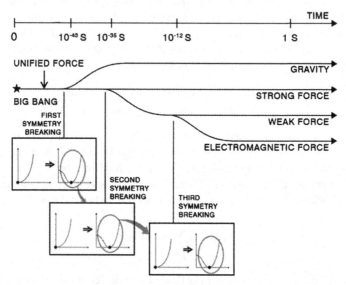

Fig. 7.5. From the Big Bang to the four forces of nature: time line for the splitting of the original Cosmic force into four distinct forces, along side the energy diagrams associated with symmetry breaking. The figure shows how the splitting of the forces is related to symmetry-breaking phase transitions. The symmetry-breaking curves show how the energy of the system varies with the phase of the system, as natural systems always choose states with the lowest energy (adapted from [Vaz03]).

Feynman's Partition Function

In statistical mechanics, the so-called *partition function* Z is a quantity that encodes the statistical properties of a system in thermodynamic equilibrium. It is a function of temperature and other parameters, such as the volume enclosing a gas. Other thermodynamic variables of the system, such as the total energy, free energy, entropy, and pressure, can be expressed in terms of the partition function or its derivatives.[9]

---

[9] There are actually several different types of partition functions, each corresponding to different types of statistical ensemble (or, equivalently, different types of free

The partition function of a *canonical ensemble*[10] is defined as a sum $Z(\beta) = \sum_j e^{-\beta E_j}$, where $\beta = 1/(k_B T)$ is the 'inverse temperature', where T is an ordinary temperature and k_B is the Boltzmann's constant. However, as the position x_i and momentum p_i variables of an ith particle in a system can vary continuously, the set of microstates is actually uncountable. In this case, some form of *coarse-graining* procedure must be carried out, which essentially amounts to treating two mechanical states as the same microstate if the differences in their position and momentum variables are 'small enough'. The partition function then takes the form of an integral. For instance, the partition function of a gas consisting of N molecules is proportional to the $6N$-dimensional phase-space integral,

$$Z(\beta) \sim \int_{\mathbb{R}^{6N}} d^3 p_i d^3 x_i \exp[-\beta H(p_i, x_i)],$$

where $H = H(p_i, x_i)$, $(i = 1, \ldots, N)$ is the classical Hamiltonian (total energy) function.

More generally, the so-called *configuration integral*, as used in probability theory, information science and dynamical systems, is an abstraction of the above definition of a partition function in statistical mechanics. It is a special case of a normalizing constant in probability theory, for the Boltzmann distribution. The partition function occurs in many problems of probability theory because, in situations where there is a natural symmetry, its associated probability measure, the *Gibbs measure* (see below), which generalizes the notion of the canonical ensemble, has the *Markov property*.

Given a set of random variables X_i taking on values x_i, and purely potential Hamiltonian function $H(x_i)$, $(i = 1, \ldots, N)$, the partition function is defined as

$$Z(\beta) = \sum_{x_i} \exp\left[-\beta H(x_i)\right].$$

---

energy.) The canonical partition function applies to a canonical ensemble, in which the system is allowed to exchange heat with the environment at fixed temperature, volume, and number of particles. The grand canonical partition function applies to a grand canonical ensemble, in which the system can exchange both heat and particles with the environment, at fixed temperature, volume, and chemical potential. Other types of partition functions can be defined for different circumstances.

[10] A canonical ensemble is a statistical ensemble representing a probability distribution of microscopic states of the system. Its probability distribution is characterized by the proportion p_i of members of the ensemble which exhibit a measurable macroscopic state i, where the proportion of microscopic states for each macroscopic state i is given by the Boltzmann distribution,

$$p_i = \frac{1}{Z} e^{-E_i/(kT)} = e^{-(E_i - A)/(kT)},$$

where E_i is the energy of state i. It can be shown that this is the distribution which is most likely, if each system in the ensemble can exchange energy with a heat bath, or alternatively with a large number of similar systems. In other words, it is the distribution which has *maximum entropy* for a given average energy $\langle E_i \rangle$.

The function H is understood to be a real-valued function on the space of states $\{X_1, X_2, \ldots\}$ while β is a real-valued free parameter (conventionally, the inverse temperature). The sum over the x_i is understood to be a sum over all possible values that the random variable X_i may take. Thus, the sum is to be replaced by an integral when the X_i are continuous, rather than discrete. Thus, one writes

$$Z(\beta) = \int dx_i \exp\left[-\beta H(x_i)\right],$$

for the case of continuously-varying random variables X_i.

The Gibbs measure of a random variable X_i having the value x_i is defined as the probability density function

$$P(X_i = x_i) = \frac{1}{Z(\beta)} \exp\left[-\beta E(x_i)\right] = \frac{\exp[-\beta H(x_i)]}{\sum_{x_i} \exp[-\beta H(x_i)]},$$

where $E(x_i) = H(x_i)$ is the energy of the configuration x_i. This probability, which is now properly normalized so that $0 \leq P(x_i) \leq 1$, can be interpreted as a likelihood that a specific configuration of values x_i, $(i = 1, 2, \ldots, N)$ occurs in the system.

As such, the partition function $Z(\beta)$ can be understood to provide the Gibbs measure on the space of states, which is the unique statistical distribution that maximizes the entropy for a fixed expectation value of the energy,

$$\langle H \rangle = -\frac{\partial \log(Z(\beta))}{\partial \beta}.$$

The associated entropy is given by

$$S = -\sum_{x_i} P(x_i) \ln P(x_i) = \beta \langle H \rangle + \log Z(\beta).$$

The principle of maximum entropy related to the expectation value of the energy $\langle H \rangle$, is a postulate about a universal feature of any probability assignment on a given set of propositions (events, hypotheses, indices, etc.). Let some testable information about a probability distribution function be given. Consider the set of all trial probability distributions which encode this information. Then the probability distribution which maximizes the information entropy is the true probability distribution, with respect to the testable information prescribed.

Now, the number of variables X_i need not be countable, in which case the set of coordinates $\{x_i\}$ becomes a field $\phi = \phi(x)$, so the sum is to be replaced by the *Euclidean path integral* (that is a Wick-rotated Feynman transition amplitude in imaginary time), as

$$Z(\phi) = \int \mathcal{D}[\phi] \exp\left[-H(\phi)\right].$$

More generally, in quantum field theory, instead of the field Hamiltonian $H(\phi)$ we have the action $S(\phi)$ of the theory. Both Euclidean path integral,

$$Z(\phi) = \int \mathcal{D}[\phi] \exp\left[-S(\phi)\right], \quad \text{real path integral in imaginary time} \quad (7.7)$$

and Lorentzian one,

$$Z(\phi) = \int \mathcal{D}[\phi] \exp\left[iS(\phi)\right], \quad \text{complex path integral in real time,} \quad (7.8)$$

are usually called 'partition functions'. While the Lorentzian path integral (7.8) represents a quantum-field theory-generalization of the Schrödinger equation, the Euclidean path integral (7.7) represents a statistical-field-theory generalization of the Fokker–Planck equation.

7.2.6 Basics of Kalman Filtering

In this subsection we give a brief description of the celebrated *Kalman filter*, a stochastic dynamical signal-processing system, rooted in the linear multiple-input multiple-output (MIMO) systems control theory.[11]

Radar-Tracking Problem and 1D Filters

As a soft introduction to Kalman filtering, consider a typical *fan-beam surveillance radar*[12] (see Fig. 7.6), in which the fan beam rotates continually through 360° azimuth angle θ, typically with a period of 10 seconds. Such a radar provides a 2D information about a target aircraft. The first dimension is the target range R (i.e., the time it takes for a transmitted pulse to go from the transmitter to the target and back); the second dimension is the azimuth of the target, which is determined from the azimuth angle θ the fan beam is pointing at when the target is detected [Bro88, Bro98]. Let us assume that at time $t = t_1$ the radar is pointing at scan angle and two targets are detected at ranges R_1 and R_2. Also, assume that on the next scan at time $t = t_1 + T$, again

---

[11] An extended, adaptive version of the Kalman filter provides also a solid engineering basis for the theory of artificial neural networks.

[12] For example, the ASR-11 Airport Surveillance Radar (ASR) is a commercial air traffic control radar. Another example is the fan-beam marine radar, used for tracking ships and for collision avoidance. They are called track-while-scan (TWS) radars; they do target tracking while the radar antenna rotates at a constant rate. The same tracking algorithms are also used for precision guidance of aircraft onto the runway during final approach (such guidance is especially needed during bad weather). An example is the GPS-22 High Performance Precision Approach Radar (HiPAR), which uses electronic scanning of the radar beam over a limited angle (20° in azimuth 8° in elevation) instead of mechanical scanning. All of the above radars do target search while doing target track [Bro88, Bro98].

two targets are detected. The question arises as to whether these two targets detected on the second scan are the same two targets or two new targets. The answer to this question is important both for civilian air traffic control radars and for military radars. In the case of the air traffic control radar, correct knowledge of the number of targets present is important in preventing target collisions. In the case of the military radar it is important for properly assessing the number of targets in a threat and for target interception.

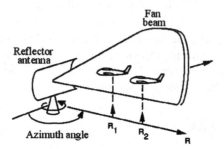

Fig. 7.6. Schematic representation of a fan-beam surveillance radar with an azimuth angle and range R (see text for explanation).

Now, let us assume two echoes are detected on the second scan and we correctly determine these two echoes are from the same two targets as observed on the first scan. The question then arises how to achieve the proper association of the echo from target 1 on the second scan with the echo from target 1 on the first scan and correspondingly the echo of target 2 on the second scan with that of target 2 on the first scan. If an incorrect association is made, then an incorrect velocity is attached to a given target. For the air traffic control radar this error in the target's speed could possibly lead to an aircraft collision; for a military radar, a missed target interception could occur.

The chances of incorrect association could be greatly reduced if we could accurately predict ahead of time where the echoes of targets 1 and 2 are to be expected on the second scan. Such a prediction is easily made if we had an estimate of the velocity and position of targets 1 and 2 at the time of the first scan. Then we could predict the distance target 1 would move during the scan-to-scan period and as a result have an estimate of the target's future position. Because the exact velocity and position of the target are not known at the time of the first scan, this prediction is not exact. If the inaccuracy of this prediction is known, we can set up a $\pm 3\sigma$ (or $\pm 2\sigma$) window about the expected value, where σ is the root-mean-square (RMS) error, or equivalently, the standard deviation of the sum of the prediction plus the RMS-error of the range measurement. If an echo is detected in this window for target 1 on the second scan, then with high probability it will be the echo from target 1. Similarly, a $\pm 3\sigma$ window is set for target 2 at the time of the second scan.

For simplicity assume we have a 1D world and a target moving radially away or toward the radar, with x_n representing the slant range to the target at a discrete time-scan n. In addition, for further simplicity we assume the target's velocity is constant; then the prediction of the target position (range) and velocity at the second scan can be made using the following simple *target equations of motion*:

$$x_{n+1} = x_n + T\dot{x}_n, \qquad \dot{x}_{n+1} = \dot{x}_n, \qquad (7.9)$$

where x_n is the target range at scan n, $\dot{x}_n$ is the target velocity at scan n, and T the scan-to-scan period.

Now, if we add the corresponding *position and velocity innovations* to each of the equations of motion, as well as the second subscript indicating the time at which the last measurement was made for use in estimating the target position and velocity, the above target equations of motion (7.9) become the so-called $g - h$ *tracking-filter prediction equations*, which are used extensively in radar systems [Mor69, Bla86],

$$x_{n+1,n} = x_{n,n-1} + T\dot{x}_{n+1,n} + g_n(z_n - x_{n,n-1}), \qquad (7.10)$$

$$\dot{x}_{n+1,n} = \dot{x}_{n,n-1} + \frac{h_n}{T}(z_n - x_{n,n-1}). \qquad (7.11)$$

In the tracking-filter equations (7.10)–(7.11), g_n and h_n are position and velocity step-sizes, both g- and h-innovations depend on the range measurement z_n, while all x_n and $\dot{x}_n$ values now become estimates rather than exact values (usually denoted by carets, which we here avoid for simplicity). These equations predict the target position and velocity at the next scan time. Their circuit form is given in Fig. 7.7.

An important class of $g - h$ filters are those for which g and h are fixed. For this case the computations required by the radar tracker are very simple. Specifically, for each target update only four adds and three multiplies are required for each target update. The memory requirements are very small. Specifically, for each target only two storage bins are required, one for the latest predicted target velocity and one for the latest predicted target position, past measurements and past predicted values not being needed for future predictions [Bro88, Bro98].

The tracking-filter equations (7.10)–(7.11) are actually 1D Kalman-filter prediction equations, which can be easily extended to an nD matrix form. The $g - h$ filter predicts position of a constant-velocity target perfectly in a steady state if there are no measurement errors. On the other hand, when tracking a target having a constant acceleration $\ddot{x}$ with a constant $g - h$ filter we get a constant *lag error* $e = -\ddot{x}T^2/h$, so for the filtered target position $x_{n,n}$ and velocity $x_{n,n}$ we have

$$e_{n,n} = -\ddot{x}T^2\frac{1-g}{h}, \qquad \dot{e}_{n,n} = -\ddot{x}T^2\frac{2g-h}{2h}.$$

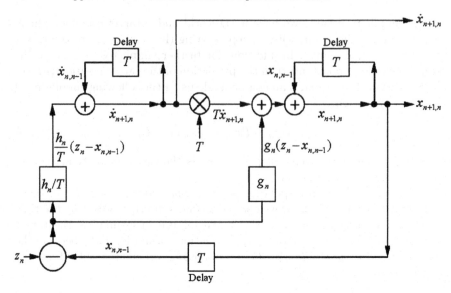

Fig. 7.7. Equivalent circuit diagram of a $g - h$ filter.

Finally, if we include the constant acceleration $\ddot{x}_n$ into the target equations of motion (7.9), we get

$$x_{n+1} = x_n + T\dot{x}_n + \frac{T^2}{2}\ddot{x}_n, \qquad \dot{x}_{n+1} = \dot{x}_n + T\ddot{x}_n, \qquad \ddot{x}_{n+1} = \ddot{x}_n. \quad (7.12)$$

The corresponding $g - h - k$ tracking-filter equations for updating the prediction estimates of position, velocity, and acceleration for the constant-accelerating target model become

$$x_{n+1,n} = x_{n,n-1} + T\dot{x}_{n+1,n} + g_n(z_n - x_{n,n-1}) + \frac{T^2}{2}\ddot{x}_{n+1,n},$$

$$\dot{x}_{n+1,n} = \dot{x}_{n,n-1} + \frac{h_n}{T}(z_n - x_{n,n-1}) + T\ddot{x}_{n+1,n},$$

$$\ddot{x}_{n+1,n} = \ddot{x}_{n,n-1} + \frac{2k_n}{T^2}(z_n - x_{n,n-1}),$$

where k is the acceleration step-size. This is the well-known $g - h - k$ filter, which can track a constant-accelerating target with zero-lag error in the steady state. It will have a constant-lag error for a target having a constantly changing acceleration with time, that is, for a target having a constant jerk. It is a three state tracking-filter for estimating the moving target's position, velocity, and acceleration. To get the corresponding nD Kalman filter prediction equations, we just replace the above scalar equations with the corresponding matrix equations.

Linear Kalman Filter

Classical linear Kalman filter is an effective procedure for *combining noisy sensor outputs* to *estimate* the *state* of a *system with uncertain dynamics*. The Kalman filter provides a *recursive solution* to the *linear optimal filtering problem*. It applies to stationary as well as non-stationary environments. The solution is recursive in that each updated estimate of the state is computed from the previous estimate and the new input data, so only the previous estimate requires storage. In addition to eliminating the need for storing the entire past observed data, the Kalman filter is computationally more efficient than computing the estimate directly from the entire past observed data at each step of the filtering process.

Recall that the *Kalman linear-quadratic regulator*, widely used in state-space control theory, represents a *linear state feedback control law*

$$u = -Kx,$$

for the linear MIMO system[13]

$$\dot{x} = Ax + Bu,$$

which minimizes a *quadratic cost function*

---

[13] It is well-known that *linear* multiple-input multiple-output (MIMO) control systems can always be put into Kalman's canonical (modular) state-space form of order n, with m inputs and k outputs [KFA69]. In the case of *continual-time systems* we have the state and output equation of the form

$$\begin{aligned} d\mathbf{x}/dt &= \mathbf{A}(t)\mathbf{x}(t) + \mathbf{B}(t)\mathbf{u}(t), \\ \mathbf{y}(t) &= \mathbf{C}(t)\mathbf{x}(t) + \mathbf{D}(t)\mathbf{u}(t), \end{aligned} \tag{7.13}$$

while in case of *discrete-time systems* we have the state and output equation of the form

$$\begin{aligned} \mathbf{x}(n+1) &= \mathbf{A}(n)\mathbf{x}(n) + \mathbf{B}(n)\mathbf{u}(n), \\ \mathbf{y}(n) &= \mathbf{C}(n)\mathbf{x}(n) + \mathbf{D}(n)\mathbf{u}(n). \end{aligned} \tag{7.14}$$

Both in (7.13) and in (7.14) the variables have the following meaning:
$\mathbf{x}(t) \in \mathbb{X}$ is an n-vector of *state variables* belonging to the *state space* $\mathbb{X} \subset \mathbb{R}^n$;
$\mathbf{u}(t) \in \mathbb{U}$ is an m-vector of *inputs* belonging to the *input space* $\mathbb{U} \subset \mathbb{R}^m$;
$\mathbf{y}(t) \in \mathbb{Y}$ is a k-vector of *outputs* belonging to the *output space* $\mathbb{Y} \subset \mathbb{R}^k$;
$\mathbf{A}(t) : \mathbb{X} \to \mathbb{X}$ is an $n \times n$ matrix of *state dynamics*;
$\mathbf{B}(t) : \mathbb{U} \to \mathbb{X}$ is an $n \times m$ matrix of *input map*;
$\mathbf{C}(t) : \mathbb{X} \to \mathbb{Y}$ is a $k \times n$ matrix of *output map*;
$\mathbf{D}(t) : \mathbb{U} \to \mathbb{Y}$ is a $k \times m$ matrix of *input–output transform*.

Input $\mathbf{u}(t) \in \mathbb{U}$ can be empirically determined by trial and error; it is properly defined by quadratic optimization process called *Kalman regulator*, or more generally (in the presence of noise), by (extended) *Kalman filter* [Kal60].

$$J = \int_0^\infty \left(x(t)^T Q x(t) + u(t)^T R u(t) \right) dt.$$

The control law is called *optimal with respect to the cost function J*.

Now, one might ask whether there is an optimal design technique for a *state estimator*. That is, is there an approach to *observer design* which is equivalent, in some sense, to the linear quadratic regulator?

Given the *observable system*

$$\dot{x} = Ax + Bu, \qquad y = Cx,$$

one may define the *dual system*

$$\dot{\theta} = A^T \theta + C^T \gamma,$$

and design an LQR controller to minimize the quadratic cost function

$$J = \int_0^\infty \left(\theta(t)^T Q \theta(t) + \gamma(t)^T R \gamma(t) \right) dt.$$

However, it is unclear how one should 'penalize' θ and γ in the cost function. Instead, consider the *extended observable system*

$$\dot{x} = Ax + Bu + w, \qquad y = Cx + v,$$

in which the dynamics are subject to random disturbances w and the measurements are subject to random noise v. In parallel with the development of the linear quadratic regulator, Rudolph Kalman examined the following *optimal estimator problem*: Construct a full state observer which minimizes the combined effect of the disturbances and the noise, thus providing a 'most likely' estimate of the system state. Solving this problem requires some information about the random processes. If the processes are zero-mean, Gaussian white noise processes (see Appendix), then the optimal estimator design problem becomes perfectly analogous to the LQR control design problem. In 1960, Kalman published his famous paper describing a recursive solution to the discrete-data linear filtering problem [Kal60]. Since that time, due in large part to advances in digital computing, the Kalman filter has been the subject of extensive particularly in the area of autonomous or assisted navigation (see, e.g., [Jaz70, AM79, Bh92, Hay01, GWA01]).

The Kalman filter is a *discrete-time, two-step process*, the steps of which are usually called *predictor* and *corrector*, thus resembling a popular *Adams–Bashforth–Moulton integrator* for ODEs (see, e.g., [WB95]). The *predictor*, or *time update*, projects the current system's state estimate ahead in time. The *corrector*, or *measurement update*, adjusts the projected state estimate by an actual system's measurement at that time. In this way, the correction step makes corrections to an estimate, based on new information obtained from sensor measurements. The continuous-time version is usually referred to as *Kalman–Bucy filter* or *smoother* [Sl03, Asi04].

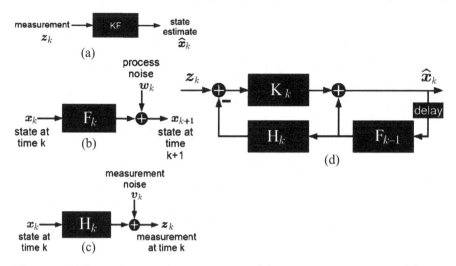

Fig. 7.8. Kalman filter predictor corrector: (**a**) input–output relation; (**b**) system model (time update); (**c**) measurement model (measurement update); and (**d**) closed-loop diagram.

Consider a generic linear, discrete-time dynamical system. The concept of *discrete state* is fundamental to this description. The *state vector*, denoted by x_k, is defined as the minimal set of data that is sufficient to uniquely describe the unforced dynamical behavior of the system; the subscript k denotes discrete time. In other words, the state is the least amount of data on the past behavior of the system that is needed to predict its future behavior. Typically, the state x_k is unknown. To estimate it, we use a set of observed data, denoted by the *observable* vector z_k.

The state-space model of a generic linear, discrete-time dynamical system includes the *process equation* (7.15) and the *measurement equation* (7.16)

$$x_{k+1} = F_{k+1,k}x_k + w_k, \tag{7.15}$$

$$z_k = H_k x_k + v_k, \tag{7.16}$$

where $F_{k+1,k}$ is the *transition matrix* taking the state x_k from time k to time $k+1$, H_k is the *measurement sensitivity matrix*, while w_k and v_k are independent, additive, zero-mean, white *Gaussian noise* processes, defined below.

The covariance matrix of the *process noise* w_k is defined by

$$E[w_n, w_k^T] = \begin{cases} Q_k, & \text{for } n = k, \\ 0, & \text{for } n = k. \end{cases}$$

Similarly, the covariance matrix of the *measurement noise* v_k is defined by

$$E[v_n, v_k^T] = \begin{cases} R_k, & \text{for } n = k, \\ 0, & \text{for } n = k. \end{cases}$$

The *Kalman filtering problem*, namely, the problem of jointly solving the process and measurement equations for the unknown state in an optimum manner may now be formally stated as follows: Use the entire observed data, consisting of the vectors $z_1, z_2, \ldots, z_k$, to find for each $k \geq 1$ the minimum mean-square error estimate of the state x_i. The problem is called *filtering* if $i = k$, *prediction* if $i > k$, and *smoothing* if $1 \leq i < k$.

The derivation of the Kalman filter is based on the following two theorems (see [Kal60, Hay01]):

- *Conditional mean estimator.* If the stochastic processes $\{x_k\}$ and $\{z_k\}$ are jointly Gaussian, then the optimum estimate $\hat{x}_k$ that minimizes the mean-square error J_k is the conditional mean estimator:

$$\hat{x}_k = E[x_k | z_1, z_2, \ldots, z_k].$$

- *Principle of orthogonality.* Let the stochastic processes $\{x_k\}$ and $\{z_k\}$ be of zero means; that is,

$$E[x_k] = E[z_k] = 0, \quad \text{for all } k.$$

Then:
(i) the stochastic process $\{x_k\}$ and $\{z_k\}$ are jointly Gaussian; or
(ii) if the optimal estimate $\hat{x}_k$ is restricted to be a linear function of the observables and the cost function is the mean-square error,
(iii) then the optimum estimate $\hat{x}_k$, given the observables $z_1, z_2, \ldots, z_k$, is the orthogonal projection of x_k on the space spanned by these observables.

The *Kalman filter design algorithm* consists of (see [Kal60, Hay01]):

1. *Initialization*: For $k = 0$, set

$$\hat{x}_0 = E[x_0], \qquad P_0 = E[(x_0 - E[x_0])(x_0 - E[x_0])^T]$$

and

2. *Computation*: For $k = 1, 2, \ldots$, compute:
(i) *State estimate propagation*

$$\hat{x}_{\bar{k}} = F_{k, k-1} \hat{x}_{\overline{k-1}};$$

(ii) *Error covariance propagation*

$$P_{\bar{k}} = F_{k, k-1} P_{k-1} F_{k, k-1}^T + Q_{k-1};$$

(iii) *Kalman gain matrix*

$$K_k = P_{\bar{k}} H_k^T [H_k P_{\bar{k}} H_k^T + R_k]^{-1};$$

3. (iv) *State estimate update*

$$\hat{x}_k = \hat{x}_{\bar{k}} + K_k(z_k - H_k\hat{x}_{\bar{k}});$$

(v) *Error covariance update*

$$P_k = (I - K_k H_k)P_{\bar{k}}.$$

Therefore, the basic Kalman filter is a linear, discrete-time, finite-dimensional system, which is endowed with a recursive structure that makes a digital computer well suited for its implementation. A key property of the Kalman filter is that it is the minimum mean-square (variance) estimator of the state of a linear dynamical system. The model is stochastic owing to the additive presence of process noise and measurement noise, which are assumed to be Gaussian with zero mean and known covariance matrices.

Extended Kalman Filter

The Kalman filtering problem considered so far has addressed the estimation of a state vector in a linear model of a dynamical system. If, however, the model is nonlinear, we may extend the use of Kalman filtering through a linearization procedure. The resulting filter is referred to as the *extended Kalman filter* (EKF) (see, e.g., [BLK01, Hay01]). Such an extension is feasible by virtue of the fact that the Kalman filter is described in terms of difference equations in the case of discrete-time systems. While the ordinary (i.e., linear) Kalman filter is defined in terms of the measurement sensitivity matrix H_k, the extended Kalman filter can be defined in terms of a suitably differentiable vector-valued measurement sensitivity function $h(k, x_k)$.

To set the stage for a development of the extended Kalman filter, consider a nonlinear dynamical system described by the state-space model

$$x_{k+1} = f(k, x_k) + w_k, \qquad z_k = h(k, x_k) + v_k, \qquad (7.17)$$

where, as before, w_k and v_k are independent zero-mean white Gaussian noise processes with covariance matrices R_k and Q_k, respectively. Here, however, the functional $f(k, x_k)$ denotes a nonlinear transition matrix function that is possibly time-variant. Likewise, the functional $h(k, x_k)$ denotes a vector-valued measurement sensitivity function, i.e., a nonlinear measurement matrix that may be time-variant, too [BLK01, Hay01].

The basic idea of the extended Kalman filter is to linearize the state-space model (7.17) at each time instant around the most recent state estimate, which is taken to be either $\hat{x}_k$ or $\hat{x}_{\bar{k}}$, depending on which particular functional is being considered. Once a linear model is obtained, the standard Kalman filter equations are applied.

The *EKF design algorithm* consists of [Hay01]:

1. The *discrete state-space model* (7.17).

2. *Definitions*

$$F_{k,k} = \left.\frac{\partial f(k,x)}{\partial x}\right|_{x=x_k}, \qquad H_k = \left.\frac{\partial h(k,x)}{\partial x}\right|_{x=x_{\bar{k}}}.$$

3. *Initialization*: For $k = 0$, set

$$\hat{x}_0 = E[x_0], \qquad P_0 = E[(x_0 - E[x_0])(x_0 - E[x_0])^T].$$

4. *Computation*: For $k = 1, 2, \ldots$, compute:
   (i) *State estimate propagation*

   $$\hat{x}_{\bar{k}} = F_{k,k-1}\,\hat{x}_{\bar{k}-1};$$

   (ii) *Error covariance propagation*

   $$P_{\bar{k}} = F_{k,k-1}\,P_{k-1}F_{\bar{k},k-1}^T + Q_{k-1};$$

   (iii) *Kalman gain matrix*

   $$K_k = P_{\bar{k}}H_k^T[H_k P_{\bar{k}}H_k^T + R_k]^{-1};$$

   (iv) *State estimate update*

   $$\hat{x}_k = \hat{x}_{\bar{k}} + K_k(z_k - H_k\hat{x}_{\bar{k}});$$

   (v) *Error covariance update*

   $$P_k = (I - K_k H_k)P_{\bar{k}}.$$

Sensor Fusion in Hybrid Systems

Kalman filter can be used to *combine* or *fuse* information from different sensors for hybrid systems (see Fig. 7.9), like accelerometers and gyroscopes (see text below). The basic idea is to use the Kalman filter to weight the different mediums most heavily in the circumstances where they each perform best, thus providing more accurate and stable estimates than a system based on any one medium alone (see [Lui02]). In particular, the *indirect feedback Kalman filter* (also called a *complementary* or *error-state Kalman filter*) is often used to combine the two mediums [May79]. In such a configuration, the Kalman filter is used to estimate the difference between the current inertial and optical (or acoustic) outputs, i.e. it continually estimates the error in the inertial estimates by using the optical system as a second (redundant) reference. This error estimate is then used to correct the inertial estimates. The *tuning* of the Kalman filter parameters then adjusts the weight of the correction as a function of frequency. By slightly modifying the Kalman filter, adaptive velocity response can be incorporated also. This can be accomplished by adjusting (in real time) the expected optical measurement error as a function of the magnitude of velocity [WB95]. Kalman filter has been used to investigate the human balancing system [KJH01, KD03].

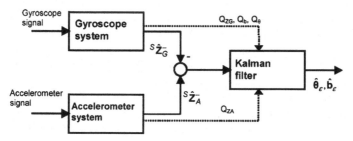

Fig. 7.9. Structure of *sensor fusion in Kalman filter estimation*. Both the accelerometer and the gyroscope system are used to make an estimate of the global vertical unit vector **Z**. The difference between the two estimates is written as a function of orientation error θ_ε and offset error $\mathbf{b}_\varepsilon$. Kalman filter estimates both θ_ε and $\mathbf{b}_\varepsilon$ using the error covariances of the orientation $\mathbf{Q}_\theta$, offset $\mathbf{Q}_\mathbf{b}$ and inclination estimation ($\mathbf{Q}_{\mathbf{ZG}}$ and $\mathbf{Q}_{\mathbf{ZA}}$). These estimated errors are used to correct the estimated.

Physiological Examples

Quaternion Filter for Attitude Estimation in Biomechanics

Recall that the so-called *inertial sensors* measure angular velocity (i.e., rotation rate) and translational (or, linear) acceleration, which are both vector-valued variables (see [GWA01]):

(i) *Gyroscopes* are sensors for measuring rotation: rate gyroscopes measure rotation rate, and displacement gyroscopes (also called whole-angle gyroscopes) measure rotation angle.

(ii) *Accelerometers* are sensors for measuring acceleration (however, they cannot measure gravitational acceleration; that is, an accelerometer in free fall (or in orbit) has no detectable input).

Mechanics of inertial sensors [Sod94] actually resembles mechanics of the *human vestibular system* in an attempt to capture the three translations and three rotations from the $SE(3)$ group of rigid body motions. The input axis of an inertial sensor defines which vector component it measures. Multiaxial sensors measure more than one component.[14]

---

[14] In *biomechanics*, inertial sensors (mainly accelerometers) have been used for *inverse dynamic analysis of human movement*. In [BRN96] a method was developed to calculate total resultant force and moment on a 3D body segment, from accelerometer data. The method was applied for an analysis of intersegmental loading at the hip joint during the single support phase of working and running, using four triaxial accelerometers mounted on the upper body. Results were compared to a conventional analysis using simultaneously recorded kinematics and ground reaction forces. The loading patterns obtained by both methods were similar, but the *accelerometry method* systematically underestimated the intersegmental force and moment at the hip by about 20%. This could be explained by the inertial and grav-

Now, although Euler angles are physically most plausible for representing rotations (as parameterizations of the $SO(3)$-group of rigid-body rotations), they 'flip' (i.e., have singularity) at the angle of $\pm\pi/2$. This is the reason for using Hamilton's quaternions instead (see Appendix). The *quaternion attitude estimation filter* was proposed at Monterey Naval Postgraduate School in a series of Master theses supervised by R. McGhee (see [Ghe96]) as an alternative representation and improvement to filters based on Euler angles. The quaternion attitude estimation filter is designed to track human limb segments through all orientations as part of an inertial tracking system. It uses a Kalman-fusion of three different types of sensors to obtain the information about the orientation of a tracked object. These sensors are a 3-axial accelerometer, a 3-axial gyroscope and a 3-axial *magnetometer* (digital compass).

An extended *Kalman-quaternion filter* for real-time estimation of rigid body motion altitude was proposed in [Joa00]. A process model for rigid body angular motions and angular rate measurements is defined. The process model converts angular rates into quaternion rates, which are in turn integrated to obtain quaternions. The outputs of the model are values of 3D angular rates, 3D linear accelerations, and 3D magnetic field vector. Gauss–Newton iteration is utilized to find the best quaternion that relates the measured linear accelerations and earth magnetic field in the body coordinate frame to calculated values in the earth coordinate frame. The quaternion obtained from the optimization algorithm is used as part of the observations for the Kalman filter. As a result, the measurement equations become linear.

Adaptive Estimation in Human Spatial Orientation

The extended Kalman filter is widely used in biomechanical experiments. For example, an adaptive Kalman estimator model of human spatial orientation is presented in [KJH01, KD03]. The adaptive Kalman filter dynamically weights sensory error signals. More specific, the model weights the difference between expected and actual sensory signals as a function of environmental conditions. The model does not require any changes in model parameters. Differences with existing models of spatial orientation are in the following: (i) Environmental conditions are not specified but estimated; (ii) The sensor noise characteristics are the only parameters supplied by the model designer; (iii) History-dependent effects and mental resources can be modeled; and (iv) Vestibular

---

itational forces originating from the swing leg which were neglected in the analysis. In addition, the accelerometry analysis was not reliable during the impact phase of running, when the upper body and accelerometers did not behave as a rigid body. For applications where these limitations are acceptable, the accelerometry method has the advantage that it does not require a gait laboratory environment and can be used for field studies with a completely body-mounted recording system. The method does not require differentiation or integration, and therefore provides the possibility of real-time inverse dynamics analysis.

thresholds are not included in the model; instead vestibular-related threshold effects are predicted by the model.

The model was applied to human stance control and evaluated with results of a visually induced sway experiment [KJH01, Pet02a]. From these experiments it is known that the amplitude of visually induced sway reaches a saturation level as the stimulus level increases. This saturation level is higher when the support base is sway referenced. For subjects experiencing vestibular loss, these saturation effects do not occur. Unknown sensory noise characteristics were found by matching model predictions with these experimental results. Using only five model parameters, far more than five data points were successfully predicted. Model predictions showed that both the saturation levels are vestibular related since removal of the vestibular organs in the model removed the saturation effects, as was also shown in the experiments. It seems that the nature of these vestibular-related threshold effects is not physical, since in the model no threshold is included. The model results suggest that vestibular-related thresholds are the result of the processing of noisy sensory and motor output signals. Model analysis suggests that, especially for slow and small movements, the environment postural orientation can not be estimated optimally, which causes sensory illusions. The model also confirms the experimental finding that postural orientation is history dependent and can be shaped by instruction or mental knowledge. In addition the model predicts the following: (i) Vestibular-loss patients cannot handle sensory conflicting situations and will fall down; (ii) During sinusoidal support-base translations vestibular function is needed to prevent falling; (iii) During sinusoidal support-base translations vestibular function is needed to prevent falling; (iv) During sinusoidal support-base translations vestibular function is needed to prevent falling; and (v) Loss of vestibular function results in falling for large support-base rotations with the eyes closed. These predictions agree with experiments [Pet02a].

To relate neuromuscular disorders (impairments) with balance disorders (disabilities) a well-defined method that identifies the different factors determining balance control is essential. An adequate approach is to isolate different sensory sources and to calculate transfer functions between the input (e.g., perturbation) and output (e.g., body sway) signals. Using this system identification approach the dynamical behavior of postural control can be obtained and is well defined. The adaptive Kalman model of balance control [KJH01, Pet02a] was successfully used to reproduce the experimental findings on *sensory-motor integration* in human postural control [Pet02a] (see Fig. 7.10).

7.2.7 Basics of Wavelet Transforms

Recall that classical *Fourier transform* uses, as its orthonormal basis functions, sinusoidal waves, which extend to infinity in both directions. The basis

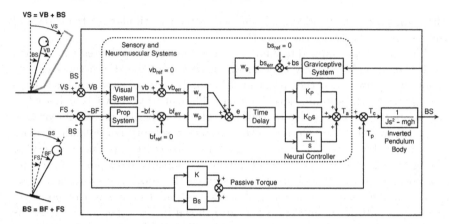

Fig. 7.10. 'Independent channel model' of *sensory-motor integration* in postural control showing a weighted addition of contributions from visual, proprioceptive, and graviceptive systems to the generation of an active corrective torque, T_a, as well as a passive torque contribution, T_p, related to body movement relative to the feet.

vectors of the *discrete Fourier transform* are also nonzero over their entire domain; technically speaking, they do not have *compact support*. On the other hand, *transient signal components* are nonzero only during a short interval. Such components do not resemble any of the *Fourier basis functions*, and they are not represented compactly in the transform coefficients (i.e., the *frequency spectrum*).

Wavelets are waves of limited duration that vary both in time-point (or position) and frequency, suitable as basis functions for some integral or discrete transforms. They form the basis of modern *time-frequency analysis*, resembling both recording and playing of musical notes.

Certain ideas of the wavelet theory appeared quite a long time ago. For example, early in 1910 A. Haar published full orthonormal system of basis functions with local definition domain (which are now known as *Haar wavelets*). The first record of wavelets was in the literature on digital processing and analysis of the seismic signals (the works by A. Grossman and J. Morlet). In the recent years there even appeared a separate scientific area to deal with wavelet analysis and the wavelet transformation theory. Wavelets are extensively used for the purposes of filtration and preprocessing data, analysis and prediction of stock markets situations, image recognition, as well as for processing and synthesizing various signals, like speech or medical signals, for images compressing and processing, training neural networks and so on (see, e.g., [Dau01, Mal89, Mal99]).

The term *wavelet analysis* and the *wavelet transform* refers to the representation of a signal in terms of scaled and translated copies (known as 'daughter', or 'baby wavelets') of a finite length or fast decaying oscillating

waveform (known as the 'mother wavelet'). In formal terms, this representation is a wavelet series[15] representation of a square-integrable function[16]
with respect to either a complete, orthonormal set of basis functions, or an
over-complete set of *frame functions* (also known as a *Riesz basis*), for the
Hilbert space of square-integrable functions. Note that the wavelets in the
JPEG2000 standard are bi-orthogonal, which means that although the frame
is over-complete, the frame is a tight frame, and the same frame functions
(except for conjugation in the case of complex wavelets) are used for both
analysis and synthesis, i.e., in both the forward and inverse transform.

Wavelet theory is applicable to several other subjects. All wavelet transforms may be considered to be forms of time-frequency representation and
are, therefore, related to the subject of harmonic analysis. Almost all practically useful discrete wavelet transforms make use of filter-banks containing finite impulse response filters. The wavelets forming a continuous wavelet
transform are subject to *Heisenberg uncertainty principle* and, equivalently,
discrete wavelet bases may be considered in the context of other forms of the
uncertainty principle.

---

[15] The word *wavelet* (plural *wavelets*) denotes a small wave, or a *ripple* (that is,
a moving disturbance or undulation in the surface of a liquid, or a sound similar
to that or undulating water). Derived terms include: wavelet compression, wavelet
matrix, wavelet modulation, wavelet packet decomposition and wavelet series.

[16] Recall that an *integrable function* is a function whose integral exists. Unless specifically stated, the integral in question is usually the *Lebesgue integral*. Otherwise, one
can say that the function is 'Riemann-integrable' (i.e., its Riemann integral exists).
Here we briefly examine the concept of *Lebesgue integrability*. Given a set X with
sigma-algebra σ defined on X and a measure μ defined on σ, a real- or complex-
valued function $f : X \to \mathbb{R}$ (or, $f : X \to \mathbb{C}$) is called *integrable* if both f^+ and f^-
are measurable functions with finite Lebesgue integral. Let

$$f^+ = \max(f, 0), \quad \text{and}$$
$$f^- = \max(-f, 0)$$

be the 'positive' and 'negative' part of f. If f is integrable, then its integral is defined
as

$$\int f = \mu(f^+) - \mu(f^-).$$

For a real number $p \geq 0$, the function f is called *p-integrable* if the function $|f|^p$
is integrable; for $p = 1$, one says *absolutely integrable*. The term *p-summable* is
sometimes used as well, especially if the function f is a sequence and μ is discrete.
The L^p spaces are one of the main objects of study of *functional analysis*.

In particular, a real- or complex-valued function of a real or complex variable is
square-integrable on an interval if the integral of the square of its absolute value, over
that interval, is finite. The set of all measurable functions that are square-integrable
forms a particular *Hilbert space*, the so-called L^2 space. This is especially useful in
quantum mechanics as wave functions must be square-integrable over all space if a
physically possible solution is to be obtained from the theory.

The wavelet transform is often compared with the *Fourier transform*, in which signals are represented as a sum of sinusoids.[17] The main difference is

---

[17] Recall that the Fourier transform is a certain linear operator that maps functions to other functions. Loosely speaking, the Fourier transform decomposes a function into a continuous spectrum of its frequency components, and the inverse transform synthesizes a function from its spectrum of frequency components. A useful analogy is the relationship between a series of pure notes (the frequency components) and a musical chord (the function itself). In mathematical physics, the Fourier transform of a signal $x(t)$ can be thought of as that signal in the *frequency domain*. This is similar to the basic idea of the various other Fourier transforms including the Fourier series of a periodic function (see, e.g., [Wil95]).

More precisely, suppose that x is a complex-valued *Lebesgue integrable function*. The Fourier transform to the frequency domain ω is given by the function

$$X(\omega) = \frac{1}{\sqrt{2\pi}} \int_{-\infty}^{\infty} x(t)e^{-i\omega t} \, dt, \quad \text{for every } \omega \in \mathbb{R}.$$

When the independent variable t represents time (with SI unit of seconds), the transform variable ω represents *angular frequency* (in radians per second). Other notations for this same function are: $\hat{x}(\omega)$ and $\mathcal{F}\{x\}(\omega)$. The function is complex-valued in general (with $i = \sqrt{-1}$).

If $X(\omega)$ is defined as above, and $x(t)$ is sufficiently smooth, then it can be reconstructed by the *inverse Fourier transform*,

$$x(t) = \frac{1}{\sqrt{2\pi}} \int_{-\infty}^{\infty} X(\omega)e^{i\omega t} \, d\omega, \quad \text{for every } t \in \mathbb{R}.$$

The interpretation of $X(\omega)$ is aided by expressing it in polar coordinate form, $X(\omega) = A(\omega) \cdot e^{i\phi(\omega)}$, where $A(\omega) = |X(\omega)|$ is the *amplitude* and $\phi(\omega) = \angle X(\omega)$ is the phase. Then the inverse transform can be written as

$$x(t) = \frac{1}{\sqrt{2\pi}} \int_{-\infty}^{\infty} A(\omega)e^{i(\omega t + \phi(\omega))} \, d\omega,$$

which is a recombination of all the frequency components of $x(t)$. Each component is a complex sinusoid of the form $e^{i\omega t}$ whose amplitude is proportional to $A(\omega)$ and whose initial phase angle (at $t = 0$) is $\phi(\omega)$.

Among a number of significant properties of the Fourier transform, probably the most important is the *convolution theorem*: If $f(t)$ and $h(t)$ are integrable functions with Fourier transforms $F(\omega)$ and $H(\omega)$ respectively, and if the convolution of f and h exists and is absolutely integrable, then the Fourier transform of the convolution is given by the product of the Fourier transforms $F(\omega)H(\omega)$, i.e., if

$$g(t) = \{f * h\}(t) = \int_{-\infty}^{\infty} f(s)h(t-s) \, ds$$

(where $*$ denotes the convolution operation), then

$$G(\omega) = \sqrt{2\pi} \cdot F(\omega)H(\omega).$$

Analogously, if $g(t)$ is the *cross-correlation* of $f(t)$ and $h(t)$, i.e., that wavelets are localized in both time and frequency whereas the standard Fourier transform is only localized in frequency. The *short-time Fourier transform* (STFT) is also time and frequency localized but there are issues with the frequency time resolution and wavelets often give a better signal representation using multiresolution analysis. The discrete wavelet transform is also less computationally complex, taking $O(N)$ time as compared to $O(N \log N)$ for the *fast Fourier transform* (where N is the data size).

Wavelet transforms are broadly divided into continuous, discrete and multiresolution wavelet transforms.

Continuous Wavelet Transforms

In the *continuous wavelet transform* (CWT), a given signal of finite energy is projected on a continuous family of frequency bands (or similar subspaces of the function space $L^2(\mathbb{R})$), for instance on every frequency band of the form $[f, 2f]$ for all positive frequencies $f > 0$. By a suitable integration over all the thus obtained frequency components one can reconstruct the original signal.

The frequency bands or subspaces are scaled versions of a subspace at scale 1. This subspace in turn is in most situations generated by the shifts of one generating function $\psi \in L^2(\mathbb{R})$, the *mother wavelet*.[18] For the example of the scale one frequency band $[1, 2]$, this function is given by

$$g(t) = (f \star h)(t) = \int_{-\infty}^{\infty} \bar{f}(s)h(t + s)\, ds,$$

then the Fourier transform of $g(t)$ is

$$G(\omega) = \sqrt{2\pi}\bar{F}(\omega)H(\omega),$$

where capital letters are again used to denote the Fourier transform.

[18] For practical applications one prefers for efficiency reasons continuously-differentiable functions with compact support as mother (prototype) wavelet functions. However, to satisfy analytical requirements (in the continuous WT) and in general for theoretical reasons one chooses the wavelet functions from a subspace of the space $L^1(\mathbb{R}) \cap L^2(\mathbb{R})$. This is the space of *measurable functions* that are both absolutely and square integrable:

$$\int_{-\infty}^{\infty} |\psi(t)|\, dt < \infty \quad \text{and} \quad \int_{-\infty}^{\infty} |\psi(t)|^2\, dt < \infty.$$

Being in this space ensures that one can formulate the conditions of zero-mean and square-norm one:

$$\text{condition for zero-mean:} \quad \int_{-\infty}^{\infty} \psi(t)\, dt = 0, \quad \text{and}$$

$$\text{condition for square-norm one:} \quad \int_{-\infty}^{\infty} |\psi(t)|^2\, dt = 1.$$

$$\psi(t) = 2\operatorname{sinc}(2t) - \operatorname{sinc}(t) = \frac{\sin(2\pi t) - \sin(\pi t)}{\pi t},$$

with the (normalized) *sinc function*.[19] Other examples of mother wavelets are given in Fig. 7.11.

For ψ to be a wavelet for the continuous wavelet transform (see there for exact statement), the mother wavelet must satisfy an admissibility criterion (loosely speaking, a kind of half-differentiability) in order to get a stably invertible transform.

For the discrete wavelet transform, one needs at least the condition that the wavelet series is a representation of the identity in the space $L^2(\mathbb{R})$. Most constructions of discrete WT make use of the multiresolution analysis (see below), which defines the wavelet by a scaling function. This *scaling function* itself is solution to a certain *functional equation*.

In most situations it is useful to restrict ψ to be a continuous function with a higher number M of vanishing moments, i.e., for all integer,

$$\int_{-\infty}^{\infty} t^m \psi(t)\, dt = 0, \quad \text{for all } m < M \in \mathbb{Z}.$$

[19] Recall that the sinc function, denoted by $\operatorname{sinc}(x)$ has two definitions, sometimes distinguished as the normalized sinc function and un-normalized sinc function: (i) in *digital signal processing* and *information theory*, the normalized sinc function is commonly defined by

$$\operatorname{sinc}(x) = \frac{\sin(\pi x)}{\pi x},$$

while (ii) in mathematics, the historical un-normalized sinc function (for *sinus cardinalis*), is defined by

$$\operatorname{sinc}(x) = \frac{\sin(x)}{x}.$$

In both cases, the value of the function at the *removable singularity* at zero is sometimes specified explicitly as the limit value 1. The sinc function is analytic everywhere. The normalized sinc function has properties that make it ideal in relationship to *interpolation* and *bandlimited functions*: (i) $\operatorname{sinc}(0) = 1$ and $\operatorname{sinc}(k) = 0$ for $k \neq 0$ and $k \in \mathbb{Z}$ (integers); that is, it is an interpolating function; (ii) the functions $x_k(t) = \operatorname{sinc}(t - k)$ form an orthonormal basis for bandlimited functions in the function space $L^2(\mathbb{R})$, with highest angular frequency $\omega_H = \pi$ (that is, highest cycle frequency $f_H = 1/2$); (iii) the *continuous Fourier transform* of the normalized sinc function is

$$\int_{-\infty}^{\infty} \operatorname{sinc}(t)\, e^{-2\pi i f t}\, dt = \operatorname{rect}(f),$$

where the rectangular function $\operatorname{rect}(f)$ is 1 for argument between $-1/2$ and $1/2$, and zero otherwise; (iv) the normalized sinc function is related to the *Dirac-delta distribution* $\delta(x)$ by

$$\lim_{a \to 0} \int_{-\infty}^{\infty} \frac{1}{a}\operatorname{sinc}(x/a)\varphi(x)\, dx = \int_{-\infty}^{\infty} \delta(x)\varphi(x)\, dx = \varphi(0),$$

for any smooth function $\varphi(x)$ with compact support.

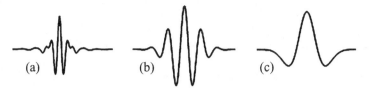

Fig. 7.11. Examples of mother wavelets: (**a**) Meyer wavelet, (**b**) Morlet wavelet, and (**c**) Mexican Hat wavelet.

The subspace of scale a or frequency band $[1/a, 2/a]$ is generated by the functions (sometimes called 'baby wavelets')

$$\psi_{a,b}(t) = \frac{1}{\sqrt{a}} \psi \left(\frac{t-b}{a} \right),$$

where a is positive and defines the scale and b is any real number and defines the shift. The pair (a, b) defines a point in the upper half-plane $\mathbb{R}_+ \times \mathbb{R}$.

The projection of a function x onto the subspace of scale a has then the form

$$x_a(t) = \int_{\mathbb{R}} WT_\phi\{x\}(a, b) \cdot \psi_{a,b}(t) \, db,$$

with wavelet coefficients

$$WT_\phi\{x\}(a, b) = \langle x, \psi_{a,b} \rangle = \int_{\mathbb{R}} x(t) \overline{\psi_{a,b}(t)} \, dt.$$

For the analysis of the signal x, one can assemble the wavelet coefficients into a scaleogram of the signal.

In other words, the CWT of a function f is a wavelet transform defined by

$$\gamma(\tau, s) = \int_{-\infty}^{+\infty} f(t) \frac{1}{\sqrt{|s|}} \overline{\psi \left(\frac{t-\tau}{s} \right)} \, dt, \quad \text{where}$$

$$C_\psi = \int_{-\infty}^{+\infty} \frac{|\hat{\psi}(\zeta)|^2}{|\zeta|} d\zeta$$

is called the *admissibility constant* and $\hat{\psi}$ is the *Fourier transform* of ψ. For a successful inverse transform, the admissibility constant has to satisfy the admissibility condition: $0 < C_\psi < +\infty$. It is possible to show that the *admissibility condition* implies that $\hat{\psi}(0) = 0$, so that a wavelet must integrate to zero. The function ψ serves as the prototype for the *daughter wavelets* the signal is convolved with. For this reason, it is called the *mother wavelet*. The daughter wavelets are scaled and shifted copies of the mother wavelet,

$$\psi_{s,\tau}(t) = \frac{1}{\sqrt{|s|}} \psi \left(\frac{t-\tau}{s} \right).$$

The CWT of a discretized signal is typically computed over the temporal domain (translation) of the signal and a range of scales equivalent to the *Nyquist range*. Computation can either be performed using direct inner products (possibly taking advantage of the sparseness of the wavelet) or via the Fast Fourier transform (FFT).[20] In the latter case, the continuous wavelet transform is noted to be a convolution at each scale, which can be performed efficiently via a discrete Fourier transform using the FFT.

The common CWT applications include:

(i) Determination of the *fractal dimension* of a signal, which looks at extrema of the CWT with respect to translation in order to quantify the fractal dimension of a signal; and

(ii) The *time-frequency analysis*, which relates extrema of the CWT with respect to scale to conventional Fourier components in order to decompose

---

[20] Recall that the *discrete Fourier transform* (DFT), occasionally called the *finite Fourier transform*, is a transform for Fourier analysis of finite-domain discrete-time signals. It is widely employed in signal processing and related fields to analyze the frequencies contained in a sampled signal, to solve partial differential equations, and to perform other operations such as convolutions. The DFT can be computed efficiently in practice using a fast Fourier transform (FFT) algorithm. The FFT is an efficient algorithm to compute the discrete Fourier transform (DFT) and its inverse. FFTs are of great importance to a wide variety of applications, from digital signal processing to solving partial differential equations to algorithms for quickly multiplying large integers (see, e.g., [Ora88]).

Let $x_0, \ldots, x_{N-1}$ be complex numbers. The DFT is defined by the formula

$$X_k = \sum_{n=0}^{N-1} x_n e^{-\frac{2\pi i}{N} nk}, \quad k = 0, \ldots, N-1,$$

while the *inverse discrete Fourier transform* (IDFT) is given by

$$x_n = \frac{1}{N} \sum_{k=0}^{N-1} X_k e^{\frac{2\pi i}{N} kn}, \quad n = 0, \ldots, N-1.$$

Evaluating these sums directly would take $O(N^2)$ arithmetical operations. An FFT is an algorithm to compute the same result in only $O(N \log N)$ operations. In general, such algorithms depend upon the factorization of N, but (contrary to popular misconception) there are $O(N \log N)$ FFTs for all N, even prime N. Many FFT algorithms only depend on the fact that $e^{-\frac{2\pi i}{N}}$ is a primitive root of unity, and thus can be applied to analogous transforms over any finite field, such as number-theoretic transforms. Since the inverse DFT is the same as the DFT, but with the opposite sign in the exponent and a $1/N$ factor, any FFT algorithm can easily be adapted for it as well.

By far the most common FFT is the *Cooley–Tukey algorithm*. This is a *divide and conquer algorithm* that recursively breaks down a DFT of any composite size $N = N_1 N_2$ into many smaller DFTs of sizes N_1 and N_2, along with $O(N)$ multiplications by complex roots of unity traditionally called *twiddle factors*.

a signal in terms of both time and frequency simultaneously. Continuous wavelets used for time-frequency analysis are designed to mimic the complex sinusoidal basis functions of the Fourier transform. CWT-based time-frequency analysis has many benefits over other time-frequency methods. Time-frequency analysis has applications in many subjects including physics (quantum mechanics, seismic geophysics, turbulence), chemistry (diffraction), biology (EEG, ECG, protein- and DNA-sequence analysis), engineering (electrical transient response, impulse-shock response for non-destructive testing, fatigue analysis), finance, climatology and speech recognition.

Discrete Wavelet Transforms

The *discrete wavelet transform* (DWT) refers to wavelet transforms for which the wavelets are discretely sampled. It is computationally impossible to analyze a signal using all wavelet coefficients. So one may wonder if it is sufficient to pick a discrete subset of the upper half-plane to be able to reconstruct a signal from the corresponding wavelet coefficients. One such system is the affine system for some real parameters $a > 1, b > 0$. The corresponding discrete subset of the half-plane consists of all the points $(a^m, na^m b)$ with integers $m, n \in \mathbb{Z}$. The corresponding baby wavelets are now given by

$$\psi_{m,n}(t) = a^{-m/2} \psi(a^{-m} t - nb).$$

A sufficient condition for the reconstruction of any signal x of finite energy by the formula

$$x(t) = \sum_{m \in \mathbb{Z}} \sum_{n \in \mathbb{Z}} \langle x, \psi_{m,n} \rangle \cdot \psi_{m,n}(t)$$

is that the functions $\{\psi_{m,n} : m, n \in \mathbb{Z}\}$ form a tight frame of $L^2(\mathbb{R})$.

The first DWT was invented by the Hungarian mathematician Alfréd Haar. For an input represented by a list of 2^n numbers, the *Haar wavelet transform* may be considered to simply pair-up input values, storing the difference and passing the sum. This process is repeated recursively, pairing-up the sums to provide the next scale: finally resulting in 2^{n-1} differences and one final sum.

This simple DWT illustrates the desirable properties of wavelets in general. Firstly, the discrete transform can be performed in $O(n)$ operations. Secondly, the transform captures not only some notion of the frequency content of the input, by examining it at different scales, but also captures temporal content, i.e., the times at which these frequencies occur. Combined, these two properties make the *Fast Wavelet Transform* (FWT), an alternative to the conventional *Fast Fourier Transform*.

The most common set of discrete wavelet transforms were formulated by the Belgian mathematician Ingrid Daubechies in 1988. This formulation is based upon the use of recurrence relations to generate progressively finer discrete samplings of an implicit mother wavelet function, each resolution being twice that of the previous scale. In her seminal paper, Daubechies derives a family of wavelets, the first of which is the Haar wavelet. Interest in this field has exploded since then, with the development of many descendants of Daubechies' original family of wavelets.

Other forms of DWT include the non- or *undecimated wavelet transform* (where down-sampling is omitted), the *Newland transform* (where an orthonormal basis of wavelets is formed from appropriately constructed top-hat filters in frequency space). Wavelet packet transforms are also related to the discrete wavelet transform. Complex wavelet transform is another form.

The DWT has a huge number of applications in science, engineering, mathematics and computer science. Most notably, the discrete wavelet transform is used for *signal coding*, where the properties of the transform are exploited to represent a discrete signal in a more redundant form, often as a preconditioning for *data compression*.

One DWT-Level

The DWT of a signal x is calculated by passing it through a series of filters. First the samples are passed through a *low-pass filter* with *impulse response* g resulting in a *convolution* of the two, defined by

$$y[n] = (x * g)[n] = \sum_{k=-\infty}^{\infty} x[k]g[n-k].$$

The signal is also decomposed simultaneously using a *high-pass filter* h. The outputs giving the detail coefficients (from the high-pass filter) and approximation coefficients (from the low-pass). It is important that the two filters are related to each other and they are known as a *quadrature mirror filter*. However, since half the frequencies of the signal have now been removed, half the samples can be discarded according to Nyquist's rule. The filter outputs are then down-sampled by 2,

$$y_{\text{low}}[n] = \sum_{k=-\infty}^{\infty} x[k]g[2n-k], \qquad y_{\text{high}}[n] = \sum_{k=-\infty}^{\infty} x[k]h[2n-k].$$

This decomposition has halved the time resolution since only half of each filter output characterizes the signal. However, each output has half the frequency band of the input so the frequency resolution has been doubled. This is in keeping with the *Heisenberg uncertainty principle*.

Using the *down-sampling operator* '$\downarrow$' defined as $(y \downarrow k)[n] = y[kn]$, the above summation can be written more concisely as

$$y_{\text{low}} = (x * g) \downarrow 2, \qquad y_{\text{high}} = (x * h) \downarrow 2.$$

However, computing a complete convolution $x * g$ with subsequent down-sampling would waste computation time. The so-called *lifting scheme* is an optimization where these two computations are interleaved.

DWT-Cascading and Filter Banks

This decomposition is repeated to further increase the frequency resolution and the approximation coefficients decomposed with high and low pass filters and then down-sampled. This is represented as a binary tree with nodes representing a sub-space with a different time-frequency localization. The tree is known as a *filter bank*, see Fig. 7.12. At each DWT-filter level the signal is decomposed into low and high frequencies. Due to the decomposition process the input signal must be a multiple of 2n where n is the number of levels. For

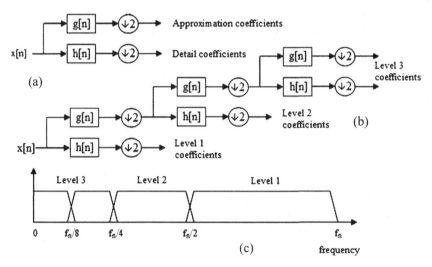

Fig. 7.12. DWT-filtering: (a) block-diagram of filter analysis; (b) a 3 level filter bank; (c) Frequency domain representation.

example a signal with 32 samples, frequency range 0 to f_n and 3 levels of decomposition, 4 output scales are produced as (corresponding to Fig. 7.12(c)):

Level	Frequencies	Samples
3	0 to $f_n/8$	4
	$f_n/8$ to $f_n/4$	4
2	$f_n/4$ to $f_n/2$	8
1	$f_n/2$ to f_n	16

Examples of DWT-Code

In its simplest form, the DWT is remarkably easy to compute. For example, the code for Haar wavelet in Java reads:

```
public static int[] invoke(int[] input) {
    //WARNING: This will destroy the contents of the input
```

```
//array
//This function assumes input.length = 2^{n},~n>1
int[] output = new int[input.length];
for(int length = input.length >> 1; ; length >>= 1) {
//length = 2^{n}, WITH DECREASING n
    for(int i = 0; i < length; i++) {
        int sum = input[i*2]+input[i*2+1];
        int difference = input[i*2]-input[i*2+1];
        output[i] = sum;
        output[length+i] = difference;
    }
    if (length ==1)
        return output;
    //Swap arrays to do next iteration
    System.arraycopy(output, 0, input, 0, length<<1);
}
}
```

The actual fast wavelet packet (see Fig. 7.13) analysis algorithms (wavelets being a special cases) permit us to perform an adapted *Fourier windowing* directly in time domain by successive filtering of a function into different regions in frequency. The dual version of the window selection provides an adapted *Subband coding algorithm*. The wavelet packet library is constructed by iterating the wavelet algorithm. This library contains the wavelet basis, the so-called *Walsh functions*, and smooth versions of Walsh functions called wavelet packets (see [CMQ91]).

Multiresolution Analysis and Wavelet Transforms

Suppose we have a signal (which can be whatever from sensor readings to digitized voice or image). The idea behind the multiresolution analysis (MRA) is that the signal is looked at very closely—first under a microscope, then with a magnifying lens, then we make a couple of steps aside, and finally take a look at it from afar. What do we get from all this? First, by consecutively roughening (or refining) the signal we can reveal its local characteristics (e.g., emphasis in speech, or distinctive details of an image) and range them according to their intensiveness. Second, this demonstrates how the dynamics of the signal changes depend on the zoom. While sudden changes (like emergency deviation of a sensor readings) are usually visible for 'unaided eye', interactions of events on a small scale that develop into large-scale events (like intensive traffic on the highway consists of movements of multiple separate cars) are pretty hard to make out. And vice versa, while concentrating on small details only, we can easily overlook some global level events.

The idea of using wavelets for the MRA is that the signal is expanded by the basis formed with the offsets and non-uniformly scaled copies of the

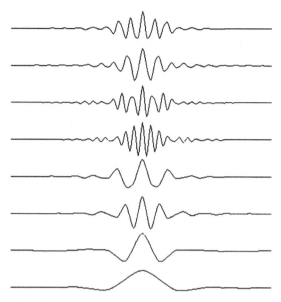

Fig. 7.13. Mutually-orthogonal wavelet packets.

function prototype (which means that wavelet transform is essentially *fractal*). Such basis functions are called *wavelets* if they are defined in the Hilbert space $L^2(\mathbb{R})$; they are oscillating about the abscissa axis and converge to zero as the absolute value of the argument increases (see Fig. 7.11). Hence, the signal convolution with a wavelet allows finding differential peculiarities of the signal in the localization area of the wavelet. Besides, the larger is the scale of the wavelet the wider portion of the signal influences the result of the convolution.

According to the uncertainty principle, the better the function is concentrated in time the more it is smeared in the frequency domain. When the function is re-scaled, the product of time and frequency ranges remains constant and represents the area of the cell in the time-and-frequency (phase) plane (see Fig. 7.14). Thanks to this, the signal's low-frequency components can be localized in the frequency domain (dominant harmonics), and high-frequency ones in the time domain (sudden changes, peaks etc.) Moreover, wavelet analysis allows investigating behavior of fractal functions, i.e. the ones that have no derivatives in any point. In each instance of the discretized wavelet transform, there are only a finite number of wavelet coefficients for each bounded rectangular region in the upper half-plane. Still, each coefficient requires the evaluation of an integral. To avoid this numerical complexity one needs one auxiliary function, the *father wavelet* $\phi \in L^2(\mathbb{R})$. Further, one has to restrict a to be an integer number. A typical choice is $a = 2$ and $b = 1$. The most famous pair of father and mother wavelets is the *Daubechies 4 tap wavelet*, see Fig. 7.15.

From the mother and father wavelets one constructs the subspaces

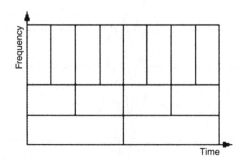

Fig. 7.14. Time-frequency phase-plane of wavelet transform.

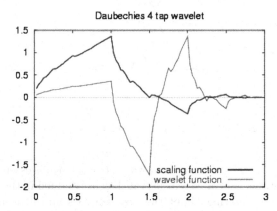

Fig. 7.15. Daubechies D4 tap wavelet.

$$V_m = \mathrm{span}(\phi_{m,n} : n \in \mathbb{Z}), \quad \text{where } \phi_{m,n}(t) = 2^{-m/2}\phi(2^{-m}t - n),$$
$$W_m = \mathrm{span}(\psi_{m,n} : n \in \mathbb{Z}), \quad \text{where } \psi_{m,n}(t) = 2^{-m/2}\psi(2^{-m}t - n).$$

From these one requires that the sequence

$$\{0\} \subset \cdots \subset V_1 \subset V_0 \subset V_{-1} \subset \cdots \subset L^2(\mathbb{R})$$

forms a multiresolution analysis of $L^2(\mathbb{R})$ and that the subspaces $\ldots, W_1, W_0,$ $W_{-1}, \ldots$ are the orthogonal 'differences' of the above sequence, that is, W_m is the orthogonal complement of V_m inside the subspace V_{m-1}. In analogy to the *sampling theorem* one may conclude that the space V_m with sampling distance 2^m more or less covers the frequency baseband from 0 to 2^{-m-1}. As an orthogonal complement, W_m roughly covers the band $[2^{-m-1}, 2^{-m}]$.

From those inclusions and orthogonality relations follows the existence of sequences $h = \{h_n\}_{n\in\mathbb{Z}}$ and $g = \{g_n\}_{n\in\mathbb{Z}}$ that satisfy the identities

$$h_n = \langle \phi_{0,0}, \phi_{1,n} \rangle, \quad \text{and} \quad \phi(t) = \sqrt{2}\sum_{n\in\mathbb{Z}} h_n \phi(2t - n), \quad \text{and}$$

$$g_n = \langle \psi_{0,0}, \phi_{1,n} \rangle, \quad \text{and} \quad \psi(t) = \sqrt{2}\sum_{n\in\mathbb{Z}} g_n \phi(2t - n).$$

The second identity of the first pair is a *refinement equation* for the father wavelet ϕ. Both pairs of identities form the basis for the algorithm of the *fast wavelet transform*.

Wavelet Series

Informally, a wavelet series is a representation of a square-integrable (real-, or complex-valued) function by a certain orthonormal series generated by a wavelet. Formally, a function $\psi \in L^2(\mathbb{R})$ is called an *orthonormal wavelet* if it can be used to define a *Hilbert basis*, that is a complete orthonormal system, for the Hilbert space $L^2(\mathbb{R})$ of square integrable functions. The Hilbert basis is constructed as the family of functions $\{\psi_{jk} : j, k \in \mathbb{Z}\}$ by means of *dyadic translations* and *dilations* of ψ,

$$\psi_{jk}(x) = 2^{j/2}\psi(2^j x - k), \quad \text{for integers } j, k \in \mathbb{Z}.$$

This family is an orthonormal system if it is orthonormal under the *inner product*

$$\langle \psi_{jk}, \psi_{lm} \rangle = \delta_{jl}\delta_{km},$$

where δ_{ik} is the *Kronecker delta* and $\langle f, g \rangle$ is the standard inner product on $L^2(\mathbb{R})$, given by

$$\langle f, g \rangle = \int_{-\infty}^{\infty} \overline{f(x)}g(x)\, dx.$$

The requirement of completeness is that every function may be expanded in the basis as

$$f(x) = \sum_{j,k=-\infty}^{\infty} c_{jk}\psi_{jk}(x),$$

with convergence of the series understood to be convergence in the norm. Such a representation of a function f is known as a wavelet series. This implies that an orthonormal wavelet is self-dual. If we define the wavelet transform by

$$[W_\psi f](a, b) = \frac{1}{\sqrt{|a|}} \int_{-\infty}^{\infty} \overline{\psi\left(\frac{x-b}{a}\right)} f(x)dx,$$

then the *wavelet coefficients* c_{jk} are given by

$$c_{jk} = [W_\psi f](2^{-j}, k2^{-j}).$$

Here, $a = 2^{-j}$ is called the *binary (dyadic) dilation*, and $b = k2^{-j}$ is the *binary (dyadic) position*.

Unlike the Fourier transform, which is an integral transform in both directions, the wavelet series is an integral transform in one direction, and a series in the other, much like the Fourier series. The canonical example of an orthonormal wavelet, that is, a wavelet that provides a complete set of basis elements for $L^2(\mathbb{R})$, is the *Haar wavelet*.

Signal De-Noising with Wavelet Transforms

Recall that signal de-noising represents one of the most important problems in *digital signal processing* (DSP). Any practical signal apart from useful information contains traces of irrelevant influence (interference, or noise). The model of such a signal can be written as

$$s(t) = f(t) + \sigma e(t),$$

where $f(t)$ is the useful signal, $e(t)$ is the noise, σ is the noise level and $s(t)$—the total (raw) signal under consideration. In most cases it can be suggested that the function $e(t)$ is described by the *Gaussian white noise* model, and information about the noise is contained in the high-frequency spectral region of the signal, while the useful information is contained in the low-frequency one.

For such a model, there exist four stages of de-noising with wavelet transform [Don95]:

1. Signal expansion by the wavelets basis.
2. Selection of the noise threshold value for each level of expansion.
3. Threshold filtering of detail coefficients.
4. Signal reconstruction.

From the statistical point of view, such a method represents a nonparametric estimation of the signal regression model using the orthogonal basis. The method works best with comparatively smooth signals, i.e., those in decomposition of which only small number of detail coefficients significantly differ from zero. In general case, the choice of a wavelet and a depth of decomposition depends on properties of a certain signal. Here we list a few recommendations:

1. The smoother wavelets create more smooth approximation of the signal. Vice versa, 'short' wavelets keep better track of peaks of the approximable function.
2. The depth of decomposition influences the scale of details to be discarded. In other words, with larger depth of decomposition the model subtracts noise of larger level until the scale of details gets 'over-enlarged' and the transform begins distorting the initial signal shape. It is noteworthy that with further increase of the decomposition depth the transform proceeds to forming a smoothed version of the initial signal, i.e., not only the noise gets filtered out, but also some local singularities (peaks) of the initial signal (see Fig. 7.16). As a rule, when selecting the noise threshold (stage 2) criterions are used that minimize the quadratic loss for the selected noise model. Let us take as an example an expression for the so-called 'omnibus test', a criterion which suits well for the Gaussian white noise model with zero average of distribution and dispersion of 1: $\theta = \sqrt{2\ln(n)}$, where n is length of a sample, and θ is the threshold value. If noise level σ (for the Gaussian distribution it is the mean-square deviation) differs from 1, then the threshold value must be scaled to the value.

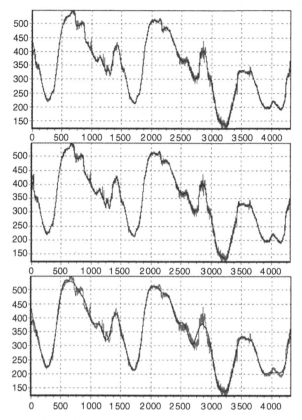

Fig. 7.16. Signal de-noising using wavelet transforms. *Top*: using the smooth wavelet (Daubechies, 7th order, 5 levels of decomposition). *Middle*: using a shorter wavelet (Daubechies, 2nd order, signal peaks are traced better). *Down*: signal 'over-enlargement' (7 levels of decomposition, local peculiarities of the signal are smoothed).

Example: Analysis of Heartbeat Intervals

Multiresolution wavelet analysis [Dau92, Mal89, Mey86, AU96, Aka97] has proved to be a useful technique for analyzing signals at multiple scales, even in the presence of non-stationarities which often obscure such signals [AGH88, THL96]. The sequence of times between human heartbeats (R–R intervals) is a prototype of a non-stationary time series that carries information about the state of cardiovascular health of the patient [KLS82, BLW94].

By projecting this sequence into a wavelet space, a new set of variables is obtained, whose statistics allow us, for the first time, to correctly classify every patient in a standard data set as either heart-failure or normal, with 100% accuracy. It is clear from our results that the R–R intervals alone suffice as a measure for the presence of heart-failure; the full electrocardiogram (ECG) in not required. This remarkable result arises from the ability of multi-resolution analysis to simultaneously and compactly monitor multiple time scales and thereby to expose a hitherto unknown scale window (between 16 and 32 heartbeats) over which the widths of the R–R wavelet coefficients fall into disjoint sets for normal and heart-failure patients. The emergence of this particular scale window should help shed light on the underlying dynamics of cardiovascular function [BLW94]. Previous approaches [PMH93, PHS95, VPS97], even those that have made use of wavelets [IRP96], have been successful only in providing a *statistically* significant measure, rather than the *clinically significant* one we have developed. The analysis method presented here is applicable to a wide variety of non-stationary physical and biological signals, regardless of whether the underlying fluctuations have stochastic origins or arise from nonlinear dynamical processes.

The series of intervals between adjacent heartbeats τ_i (known as R–R or inter-beat intervals in cardiology; see Fig. 7.17) is thought to result from a complex superposition of multiple physiological processes at their respective characteristic time scales [BLW94]. In this subsection, following [TFT98], we demonstrate that is is possible, without any *a priori* knowledge of the physiological time scales or underlying heart dynamics, to determine a range of scales over which a statistic of the wavelet coefficients permits each heart-failure and normal patient to be correctly categorized.

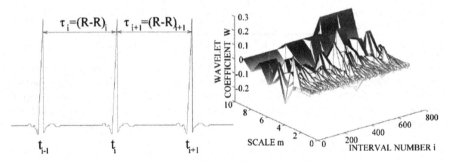

Fig. 7.17. A diagram of an ECG segment, showing the beat occurrence times t_i and the inter-beat (R–R) intervals τ_i (*left*). 3D representation of the wavelet coefficient W as a function of scale ($1 \le m \le 10$) and interval number i, over a portion of the data set, using a Daubechies 10-tap wavelet (*right*, see text for explanation).

Scale-dependent statistics are constructed by transforming the discrete-time sequence of R–R intervals $s = \{\tau_i\}$ into a space of wavelet coefficients. One can think of the transformed signal in terms of a landscape over a 2D plane whose axes are inter-beat-interval number i and scale m (see Fig. 7.17, right). Smaller scales correspond to more rapid variations and therefore to higher frequencies. The height is the value of the corresponding wavelet coefficient. With such a three-dimensional construct it is possible to trace the relative importance of different scales as the heartbeat sequence proceeds. Technically, the coefficients are obtained by carrying out a discrete wavelet transform (DWT) [Dau92, AU96],

$$W_{m,n}^{\text{wav}}(s) = 2^{-m/2} \sum_{i=1}^{M} \tau_i \psi(2^{-m}i - n),$$

where the scale variable m and the translation variable n are integers, and M represents the total number of R–R intervals analyzed. The discrete wavelet transform is evaluated at the points (m, n) in the scale–interval–number plane.

This transform was performed in [TFT98] using a broad range of orthonormal, compactly supported analyzing wavelets (including Daubechies 10-tap and Haar wavelets). Orthogonality in the DWT provides that the information represented at a certain scale m is disjoint from the information at other scales. Because certain wavelets ψ have vanishing moments, polynomial trends in the signal are automatically eliminated in the process of wavelet transform [AGH88, THL96]. This is salutatory in the case of the heartbeat time series, which are eliminated by the wavelet transform.

Since the signal s fluctuates in time, so too does the sequence of wavelet coefficients at any given scale, though its mean is zero [Dau92]. The wavelet coefficients for the heart-failure patient evidently exhibit substantially reduced variability, particularly at intermediate scales. A natural measure for this variability is the wavelet-coefficient standard deviation, as a function of scale [TFT98]

$$\sigma_{\text{wav}}(m) = \left[\frac{1}{N-1} \sum_{n=1}^{N} (W_{m,n}^{\text{wav}}(s) - \langle W_{m,n}^{\text{wav}}(s) \rangle)^2 \right]^{\frac{1}{2}},$$

where N is the number of wavelet coefficients at a given scale m ($N = M/2^m$).

The principal results of this subsection were obtained using the Haar wavelet and the Daubechies 10-tap wavelet. At scales 4 and 5, corresponding to $2^4 - 2^5 = 16 - 32$ heartbeats, σ_{wav} serves to completely separate the two classes of patients for both types of wavelets (white regions), thereby providing a clinically significant measure [Swe88, TT96] of the presence of heart failure with 100% sensitivity at 100% specificity (such that all normals are so identified). One can do no better. Though it has been previously shown that complete separation can be achieved using heart *rate* analysis [TT96, TT93], this is the first instance that we know of, in which the R–R intervals can be used as a definitive determinant of the presence of a heart disorder in an individual patient. Both at smaller, and at larger scales, there are multiple overlaps of the heart-failures and the normals, though the measure certainly remains statistically significant over all scales shown. Similar results are obtained for other analyzing wavelets.

The results indicate that healthy patients exhibit greater fluctuations than those afflicted with heart failure over a time scale of 16–32 heartbeats (roughly

0.2 to 0.5 minutes). This also appears to apply for sudden cardiac death (SCD). It is tempting to ascribe the physiological origin of this window to baroreflex modulations of the sympathetic or parasympathetic tone, which lie in the range 0.04 to 0.09 Hz (0.2 to 0.5 minutes), but we do not believe that this is correct. Rather, we expect that this window likely has its origin in the intrinsic behavior of the heart itself. It will be important to carry out a thorough study in which our multiresolution wavelet-analysis technique is applied to the R–R intervals from transplanted hearts to assess the role that the autonomic system might play in heart-rate variability.

It is useful to tease apart the roles played by the *magnitudes* τ_i of the inter-beat intervals and their *ordering* in achieving this complete separation. The effects of the former continue to reside in the randomly reordered (shuffled) sequence of R–R intervals; however information about the ordering is removed in this surrogate data set [TT96]. We therefore calculate the standard deviation $\sigma_{\text{wav}}^{\text{shuf}}$ for all 27 heartbeat time series after shuffling the R–R intervals. It is clear that the two classes of patients are no longer completely separated; three heart-failure patients fall among the normals at all scales, yielding a sensitivity of 80% at a specificity of 100%. The shuffled-wavelet result is essentially identical to that obtained by using the standard deviation σ_{int} of the inter-beat-interval histogram (IIH) obtained from these data sets, a measure that has long been used in cardiology [WVH78, KMB87]. The shuffled inter-beat intervals essentially comprise a renewal process so that the IIH contains all of the available information. All dependencies among intervals, and therefore long-term correlations, are removed from the shuffled surrogate data [TT96], leaving behind only short-term information. Indeed, for the Haar wavelet (in the absence or in the presence of shuffling), $\sigma_{\text{wav}}(m = 0)$ is analytically identical to σ_{int}.

The ordering of the inter-beat intervals gives rise to scaling behavior. For normal patients approximate scaling is maintained across all scales whereas for heart-failure patients (filled circles, dashed lines) the relatively flat nature of the curves in the region $m \leq 3$ indicates that σ_{wav} is essentially scale independent in this region. The distinction can be examined quantitatively by calculating the average scaling exponents α[21] in the two ranges ($1 \leq m \leq 3$ and $3 \leq m \leq 10$), for both classes of data. These observations lead us to consider a heart-failure index determined by the difference of these scaling exponents: $\Delta = \alpha(3 \leq m \leq 10) - \alpha(1 \leq m \leq 3)$. Evaluating Δ we find that two heart-failures fall among the normals, corresponding to a sensitivity of 87% at a specificity of 100%. Thus considering only the scaling information in σ_{wav}, while ignoring the magnitude differences for normals and heart-failures

---

[21] The scaling exponent α is computed from an individual wavelet-coefficient standard-deviation curve by calculating the slope of its square (rendering it a variance so that it corresponds to other standard scaling-exponent measures) on a base-10 log–log plot. Thus $\alpha = [d(\log_{10} \sigma_{\text{wav}}^2(m))]/[d(\log_{10} 2^m)] = [2/\log_{10} 2][d(\log_{10} \sigma_{\text{wav}}(m))/dm]$.

associated with short-term information, fails to give rise to complete separation.

Over the years, using this same collection of data, a number of measures based on scaling have been evaluated for their accuracy in discriminating between normal and heart-failure patients. [PMH93] examined the correlation properties of the heartbeat-interval increments $I_i = \{\tau_{i+1} - \tau_i\}$, and obtained the exponent of the associated power-law spectrum, which they denoted as β. It was shown subsequently [TT96] that this measure is isomorphic to the exponent δ of the interval-based spectrum [AGU81] at low frequencies [KM82] and therefore reveals only long-term correlations.

As a conclusion, the presented multiresolution wavelet approach succeeds not only because it eliminates trends in a mathematically acceptable way, but also because it crisply reveals a range of scales over which heart-failure patients differ from normals, both in short- and long-term heartbeat behavior. In contrast, inter-beat-interval measures reflect only short-term behavior, whereas scaling measures reflect only long-term behavior. For more details, see [TFT98].

7.2.8 Basic of Nonlinear Dynamics and Chaos Theory

Basic Terms of Nonlinear Dynamics

Recall that nonlinear dynamics is a language to talk about dynamical systems. Here, brief definitions are given for the basic terms of this language. All these terms will be illustrated at the pendulum example.

- *Dynamical system:* A part of the world which can be seen as a self-contained entity with some temporal behavior. In nonlinear dynamics, speaking about a dynamical system usually means to speak about an abstract mathematical system which is a model for such an entity. Mathematically, a dynamical system is defined by its *state* and by its *dynamics*. A pendulum is an example for a dynamical system.

- *State of a system:* A number or a vector (i.e., a list of numbers) defining the state of the dynamical system uniquely. For the free (un-driven) pendulum, the state is uniquely defined by the angle θ and the angular velocity $\dot{\theta} = d\theta/dt$. In the case of driving, the driving phase ϕ is also needed because the pendulum becomes a non-autonomous system. In spatially extended systems, the state is often a *field* (a scalar-field or a vector-field). Mathematically spoken, fields are functions with space coordinates as independent variables. The velocity vector-field of a fluid is a well-known example.

- *Phase space:* All possible states of the system. Each point in the phase-space corresponds to a unique state (see Fig. 7.18). In the case of the free pendulum, the phase-space has 2D whereas for driven pendulum it has 3D. The dimension of the phase-space is infinite in cases where the system state is defined by a field.

- *Dynamics, or equation of motion:* The causal relation between the present state and the next state in the future. It is a deterministic rule which tells us what happens in the next time step. In the case of a continuous time, the time step is infinitesimally small. Thus, the equation of motion is an ordinary differential equation (ODE) (or a system of ODEs):

$$\dot{x} = f(x),$$

where x is the state and t is the time variable (overdot is the time derivative—as always). An example is the equation of motion of an un-driven and un-damped pendulum. In the case of a discrete time, the time steps are nonzero and the dynamics is a map:

$$x_{n+1} = f(x_n),$$

with the discrete time n. Note, that the corresponding physical time points t_n do not necessarily occur equidistantly. Only the order has to be the same. That is,

$$n < m \quad \Longrightarrow \quad t_n < t_m.$$

The dynamics is *linear* if the causal relation between the present state and the next state is linear. Otherwise it is *nonlinear*. If we have the case in which the next state is not uniquely defined by the present one, this is generally an indication that the *phase-space is not complete*. Thus, there are important variables determining the state which had been forgotten. This is a crucial point while modeling a real-life systems. Beside this, there are two important classes of systems where the phase-space is incomplete: the *non-autonomous and stochastic systems*. A non-autonomous system has an equation of motion which depends explicitly on time. Thus, the dynamical rule governing the next state not only depends on the present state but also at the time it applies. A driven pendulum is a classical example of a *non-autonomous system*. Fortunately, there is an easy way to make the phase-space complete: we simply include the time into the definition of the state. Mathematically, this is done by introducing a new state variable: t. Its dynamics reads

$$\dot{t} = 1, \quad \text{or} \quad t_{n+1} = t_n,$$

depending on whether time is continuous or discrete. For the periodically driven pendula, it is also natural to take the driving phase as the new state variable. Its equation of motion reads

$$\dot{\theta} = 2\pi w,$$

where w is the driving frequency (so that the angular driving frequency is $2\pi w$). On the other hand, in a *stochastic system*, the number and the nature of the variables necessary to complete the phase-space is usually unknown. Therefore, the next state can not be deduced from the present one. The deterministic rule is replaced by a stochastic one. Instead of the next state, it gives only the probabilities of all points in the phase-space to be the next state.

- *Orbit or trajectory:* A solution of the equation of motion. In the case of continuous time, it is a curve in phase-space parametrized by the time variable. For a discrete system it is an ordered set of points in the phase-space.

- *Phase Flow:* The mapping (or, map) of the whole phase-space of a continuous dynamical system onto itself for a given time step t. If t is an infinitesimal time step dt, the flow is just given by the right-hand side of the equation of motion (i.e., f). In general, the flow for a finite time step is not known analytically because this would be equivalent to have a solution of the equation of motion. For example, Fig. 7.18 shows the *phase-flow* of a *damped pendulum* in the $(\theta, \dot{\theta})$-phase-plane.

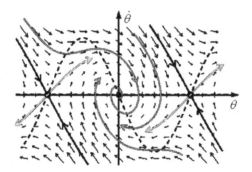

Fig. 7.18. Phase-portrait of a damped pendulum: *Arrows* denote the phase-flow, *dashed line* is a null-cline, *filled dot* is a stable fixed-point, *open dot* is an unstable fixed-point, *dark gray curves* are trajectories starting from sample initial points, *dark lines with arrows* are stable directions (manifolds), *light lines with arrows* are unstable directions (manifolds), the area between the stable manifolds is basin of attraction.

Phase Plane: Nonlinear Dynamics without Chaos

The general form of a 2D vector-field on the phase plane (similar to one in Fig. 7.18) is given by

$$\dot{x}_1 = f_1(x_1, x_2), \qquad \dot{x}_2 = f_2(x_1, x_2),$$

where f_i ($i = 1, 2$) are given function. By 'flowing along' the above vector-field, a *phase point* 'traces out' a solution $x_i(t)$, corresponding to a *trajectory* which is tangent to the vector-field. The entire phase plane is filled with trajectories (since each point can play the role of initial condition, depicting the so-called *phase portrait*. Every phase portrait has the following salient features (see [Str94]):

1. The fixed points, which satisfy: $f_i(x) = 0$, and correspond to the system's steady states or equilibria.
2. The closed orbits, corresponding to the *periodic solutions* (for which $x(t + T) = x(t)$, for all t, for some $T > 0$.
3. The specific *flow pattern*, i.e., the arrangement of trajectories near the fixed points and closed orbits.
4. The *stability* (attracting property) or *instability* (repelling property) of the fixed points and closed orbits.

Nothing more complicated than the fixed points and closed orbits can exist in the phase plane, according to the celebrated *Poincaré–Bendixson theorem*, which says that the dynamical possibilities in the phase plane are very limited. Specifically, *there cannot be chaotic behavior in the phase plane*. In other words, there is *no chaos in continuous 2D systems*.

However, there can exist chaotic behavior in *non-autonomous 2D continuous systems*, namely in the *forced nonlinear oscillators*, where explicit time-dependence actually represents the third dimension.

Free vs. Forced Nonlinear Oscillators

Here we give three examples of classical nonlinear oscillators, each in two modes: free (non-chaotic) and forced (possibly chaotic). For the simulation we use the technique called *time-phase plot*, combining an ordinary time plot with a phase-plane plot. We can see the considerable difference in complexity between unforced and forced oscillators (with all other parameters being the same). The reason for this is that *all forced 2D oscillators actually have dimension 3, although they are commonly written as a second-order ODE*. That is why for development of non-autonomous mechanics we use the *formalism of jet bundles*, see [II06c].

Spring

- Free (Rayleigh) spring (see Fig. 7.19):

$$\dot{x} = y,$$

$$\dot{y} = -\frac{1}{m}(ax^3 + bx + cy),$$

where x is displacement, y is velocity, $m > 0$ is mass, $ax^3 + bx + cy$ is the restoring force of the spring, with $b > 0$; we have three possible cases: hard spring $(a > 0)$, linear (Hooke) spring $(a = 0)$, or soft spring $(a < 0)$.[22]

---

[22] In his book *The Theory of Sound*, Lord Rayleigh introduced a series of methods that would prove quite general, such as the notion of a *limit cycle—a periodic motion a system goes to regardless of the initial conditions.*

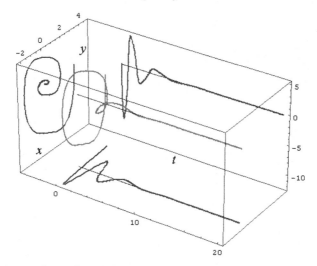

Fig. 7.19. Time-phase plot of the free hard spring with the following parameters: $m = 0.5$ kg, $a = 1.3$, $b = 0.7$, $c = 0.5$, $x_0 = 3$, $y_0 = 0$, $t_{max} = 20$ s. Simulated using $Mathematica^{TM}$.

- Forced (Duffing) spring (see Fig. 7.20):

$$\dot{x} = y,$$
$$\dot{y} = -\frac{1}{m}(ax^3 + bx + cy) + F\cos(wt),$$
$$\dot{\theta} = w,$$

where F is the force amplitude, θ is the driving phase and w is the driving frequency; the rest is the same as above.

Self-Sustained Oscillator

- Free (Rayleigh) self-sustained oscillator (see Fig. 7.21):

$$\dot{x} = y,$$
$$\dot{y} = -\frac{1}{CL}(x + By^3 - Ay),$$

where x is current, y is voltage, $C > 0$ is capacitance and $L > 0$ is inductance; $By^3 - Ay$ (with $A, B > 0$) is the characteristic function of vacuum tube.
- Forced (Rayleigh) self-sustained oscillator (see Fig. 7.22):

$$\dot{x} = y,$$
$$\dot{y} = -\frac{1}{CL}(x + By^3 - Ay) + F\cos(wt),$$
$$\dot{\theta} = w.$$

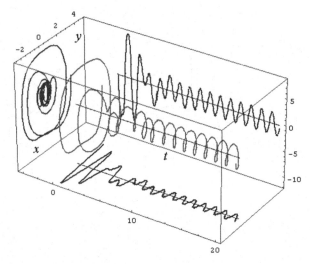

Fig. 7.20. Time-phase plot of the forced hard spring with the following parameters: $m = 0.5$ kg, $a = 1.3$, $b = 0.7$, $c = 0.5$, $x_0 = 3$, $y_0 = 0$, $t_{max} = 20$ s, $F = 10$, $w = 5$. Simulated using $Mathematica^{TM}$.

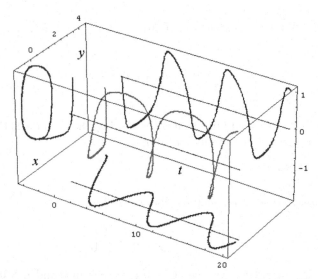

Fig. 7.21. Time-phase plot of the free Rayleigh's self-sustained oscillator with the following parameters: $A = 1.3$, $B = 1.5$, $C = 0.7$, $L = 1.5$, $x_0 = 3$, $y_0 = 0$, $t_{max} = 20$ s. Simulated using $Mathematica^{TM}$.

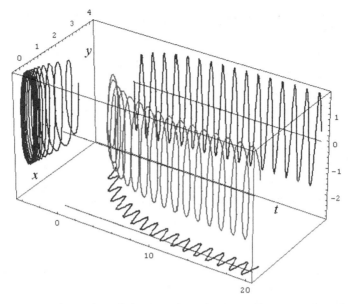

Fig. 7.22. Time-phase plot of the forced Rayleigh's self-sustained oscillator with the following parameters: $A = 1.3$, $B = 1.5$, $C = 0.7$, $L = 1.5$, $x_0 = 3$, $y_0 = 0$, $t_{max} = 20$ s, $F = 10$, $w = 5$. Simulated using $Mathematica^{TM}$.

Van der Pol Oscillator

- Free Van der Pol oscillator (see Fig. 7.23):

$$\dot{x} = y,$$
$$\dot{y} = -\frac{1}{CL}[x + (Bx^2 - A)y].$$
(7.18)

- Forced Van der Pol oscillator (see Fig. 7.24):

$$\dot{x} = y,$$
$$\dot{y} = -\frac{1}{CL}[x + (Bx^2 - A)y] + F\cos(wt),$$
$$\dot{\theta} = w.$$

A Brief History of Chaos Theory

Now, without pretending to give a complete history of chaos theory, in this section we present only its most prominent milestones (in our view). For a number of other important contributors, see [Gle87]). Before we embark on the quick historical journey of chaos theory, note that classical mechanics has not stood still since the foundational work of its father, *Sir Isaac Newton*. The mechanical formalism that we use today was developed mostly by the

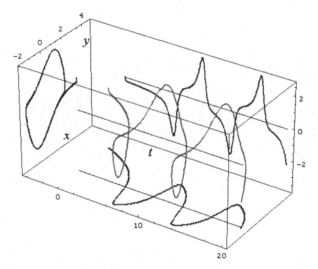

Fig. 7.23. Time-phase plot of the free Van der Pol oscillator with the following parameters: $A = 1.3$, $B = 1.5$, $C = 0.7$, $L = 1.5$, $x_0 = 3$, $y_0 = 0$, $t_{max} = 20$ s. Simulated using $Mathematica^{TM}$.

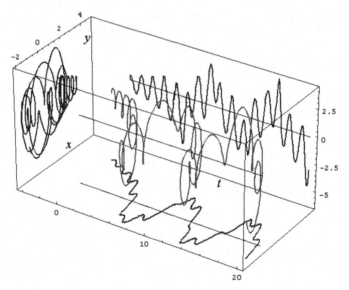

Fig. 7.24. Time-phase plot of the forced Van der Pol oscillator with the following parameters: $A = 1.3$, $B = 1.5$, $C = 0.7$, $L = 1.5$, $x_0 = 3$, $y_0 = 0$, $t_{max} = 20$ s, $F = 10$, $w = 5$. Simulated using $Mathematica^{TM}$.

three giants: *Leonhard Euler, Joseph Louis Lagrange* and *Sir William Rowan Hamilton*. By the end of the 1800's the three problems that would lead to the notion of chaotic dynamics were already known: the *3-body problem* (see

Fig. 7.25), the *ergodic hypothesis*,[23] and *nonlinear oscillators* (see Figs. 7.19–7.24).

Poincaré: *Qualitative Dynamics, Topology and Chaos*

Chaos theory really started with *Henry Jules Poincaré*, the last mathematical universalist, the father of both *dynamical systems* and *topology* (which he considered to be the two sides of the same coin). Together with the four dynamics giants mentioned above, Poincaré has been considered as one of the great scientific geniuses of all time.[24]

---

[23] The second problem that played a key role in development of chaotic dynamics was the *ergodic hypothesis of Boltzmann*. Recall that *James Clerk Maxwell* and *Ludwig Boltzmann* had combined the mechanics of Newton with notions of probability in order to create statistical mechanics, deriving thermodynamics from the equations of mechanics. To evaluate the heat capacity of even a simple system, Boltzmann had to make a great simplifying assumption of ergodicity: that the dynamical system would visit every part of the phase-space allowed by conservations law equally often. This hypothesis was extended to other averages used in statistical mechanics and was called the ergodic hypothesis. It was reformulated by Poincaré to say that a trajectory comes as close as desired to any phase-space point.

Proving the ergodic hypothesis turned out to be very difficult. By the end of our own century it has only been shown true for a few systems and wrong for quite a few others. Early on, as a mathematical necessity, the proof of the hypothesis was broken down into two parts. First one would show that the mechanical system was ergodic (it would go near any point) and then one would show that it would go near each point equally often and regularly so that the computed averages made mathematical sense. Koopman took the first step in proving the ergodic hypothesis when he noticed that it was possible to reformulate it using the recently developed methods of *Hilbert spaces*. This was an important step that showed that it was possible to take a finite-dimensional nonlinear problem and reformulate it as a infinite-dimensional linear problem. This does not make the problem easier, but it does allow one to use a different set of mathematical tools on the problem. Shortly after Koopman started lecturing on his method, *John von Neumann* proved a version of the ergodic hypothesis, giving it the status of a theorem. He proved that if the mechanical system was ergodic, then the computed averages would make sense. Soon afterwards *George Birkhoff* published a much stronger version of the theorem (see [CAM05]).

[24] Recall that Henri Poincaré (April 29, 1854–July 17, 1912), was one of France's greatest mathematicians and theoretical physicists, and a philosopher of science. Poincaré is often described as the last 'universalist' (after Gauss), capable of understanding and contributing in virtually all parts of mathematics. As a mathematician and physicist, he made many original fundamental contributions to pure and applied mathematics, mathematical physics, and celestial mechanics. He was responsible for formulating the Poincaré conjecture, one of the most famous problems in mathematics. In his research on the three-body problem, Poincaré became the first person to discover a *deterministic chaotic system*. Besides, Poincaré introduced the modern principle of relativity and was the first to present the Lorentz transformations in their modern symmetrical form (Poincaré group). Poincaré discovered the remaining relativistic velocity transformations and recorded them in a letter to Lorentz in

Poincaré conjectured and proved a number of theorems. Two of them related to chaotic dynamics are:

1. The *Poincaré–Bendixson theorem* says: Let F be a dynamical system on the real plane defined by

$$(\dot{x}, \dot{y}) = (f(x, y), g(x, y)),$$

where f and g are continuous differentiable functions of x and y. Let S be a closed bounded subset of the 2D phase-space of F that does not contain a stationary point of F and let C be a trajectory of F that never leaves S. Then C is either a limit-cycle or C converges to a limit-cycle. The Poincaré–Bendixson theorem limits the types of long term behavior that can be exhibited by continuous planar dynamical systems. One important implication is that a 2D continuous dynamical system cannot give rise to a *strange attractor*. If a strange attractor C did exist in such a system, then it could be enclosed in a closed and bounded subset of the phase-space. By making this subset small enough, any nearby stationary points could be excluded. But then the Poincaré–Bendixson theorem says that C is not a strange attractor at all—it is either a limit-cycle or it converges to a limit-cycle. The Poincaré–Bendixson theorem says that chaotic behavior can only arise in continuous dynamical systems whose phase-space has 3 or more dimensions. However, this restriction does not apply to discrete dynamical systems, where chaotic behavior can arise in two or even one-dimensional.

2. The *Poincaré–Hopf index theorem* says: Let M be a compact differentiable manifold and v be a vector-field on M with isolated zeroes. If M

1905. Thus he got perfect invariance of all of Maxwell's equations, the final step in the discovery of the theory of special relativity. As a mathematician and physicist, he made many original fundamental contributions to pure and applied mathematics, mathematical physics, and celestial mechanics. He was responsible for formulating the Poincaré conjecture, one of the most famous problems in mathematics. In his research on the three-body problem, Poincaré became the first person to discover a *deterministic chaotic system*. Besides, Poincaré introduced the modern principle of relativity and was the first to present the Lorentz transformations in their modern symmetrical form (Poincaré group). Poincaré discovered the remaining relativistic velocity transformations and recorded them in a letter to Lorentz in 1905. Thus he got perfect invariance of all of Maxwell's equations, the final step in the discovery of the theory of special relativity.

Poincaré had the opposite philosophical views of Bertrand Russell and Gottlob Frege, who believed that mathematics were a branch of logic. Poincaré strongly disagreed, claiming that *intuition* was the *life of mathematics*. Poincaré gives an interesting point of view in his book 'Science and Hypothesis': "For a superficial observer, scientific truth is beyond the possibility of doubt; the logic of science is infallible, and if the scientists are sometimes mistaken, this is only from their mistaking its rule."

has boundary, then we insist that v be pointing in the outward normal direction along the boundary. Then we have the formula

$$\sum_i index_v = \chi(M),$$

where the sum is over all the isolated zeroes of v and $\chi(M)$ is the *Euler characteristic* of M. systems.

In 1887, in honor of his 60th birthday, Oscar II, King of Sweden offered a prize to the person who could answer the question "Is the Solar system stable?" Poincaré won the prize with his famous work on the *3-body problem*. He considered the Sun, Earth and Moon orbiting in a plane under their mutual gravitational attractions (see Fig. 7.25). Like the pendulum, this system has some unstable solutions. Introducing a *Poincaré section*, he saw that *homoclinic tangles* must occur. These would then give rise to *chaos* and *unpredictability*.

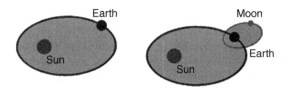

Fig. 7.25. The 2-body, problem solved by Newton (*left*), and the 3-body problem, first attacked by Poincaré, and still the point of active research (*right*).

Recall that trying to predict the motion of the Moon has preoccupied astronomers since antiquity. Accurate understanding of its motion was important for determining the longitude of ships while traversing open seas. The *Rudolphine Tables* of *Johannes Kepler* had been a great improvement over previous tables, and Kepler was justly proud of his achievements. Bernoulli used Newton's work on mechanics to derive the elliptic orbits of Kepler and set an example of how equations of motion could be solved by integrating. But the motion of the Moon is not well approximated by an ellipse with the Earth at a focus; at least the effects of the Sun have to be taken into account if one wants to reproduce the data the classical Greeks already possessed. To do that one has to consider the motion of three bodies: the Moon, the Earth, and the Sun. When the planets are replaced by point particles of arbitrary masses, the problem to be solved is known as the 3-body problem. The 3-body problem was also a model to another concern in astronomy. In the Newtonian model of the Solar system it is possible for one of the planets to go from an elliptic orbit around the Sun to an orbit that escaped its domain or that plunged right into it. Knowing if any of the planets would do so became the problem of the stability of the Solar system. A planet would not meet this terrible end if Solar system consisted of two celestial bodies, but whether such fate could befall in the 3-body case remained unclear.

After many failed attempts to solve the 3-body problem, natural philosophers started to suspect that it was impossible to integrate. The usual technique for integrating problems was to find the conserved quantities, quantities that do not change with time and allow one to relate the momenta and positions different times. The first sign on the impossibility of integrating the 3-body problem came from a result of Burns that showed that there were no conserved quantities that were polynomial in the momenta and positions. Burns' result did not preclude the possibility of more complicated conserved quantities. This problem was settled by Poincaré and Sundman in two very different ways.

In an attempt to promote the journal *Acta Mathematica*, *Gustaf Mittag-Leffler* got the permission of the King Oscar II of Sweden and Norway to establish a mathematical competition. Several questions were posed (although the king would have preferred only one), and the prize of 2500 kroner would go to the best submission. One of the questions was formulated by the 'father of modern analysis', *Karl Weierstrass*:

> "Given a system of arbitrary mass points that attract each other according to Newton's laws, under the assumption that no two points ever collide, try to find a representation of the coordinates of each point as a series in a variable that is some known function of time and for all of whose values the series converges uniformly.
> This problem, whose solution would *considerably extend our understanding of the Solar system...*"

Poincaré's submission won the prize. He showed that *conserved quantities that were analytic in the momenta and positions could not exist*. To show that he introduced methods that were very geometrical in spirit: the importance of phase flow, the role of *periodic orbits* and their cross sections, the *homoclinic points* (see [CAM05]).[25]

---

[25] The interesting thing about Poincaré's work was that it did not solve the problem posed. He did not find a function that would give the coordinates as a function of time for all times. He did not show that it was impossible either, but rather that it could not be done with the Bernoulli technique of finding a conserved quantity and trying to integrate. Integration would seem unlikely from Poincaré's prize-winning memoir, but it was accomplished by the Finnish-born Swedish mathematician Sundman, who showed that to integrate the 3-body problem one had to confront the 2-body collisions. He did that by making them go away through a trick known as regularization of the collision manifold. The trick is not to expand the coordinates as a function of time t, but rather as a function of $\sqrt[3]{t}$. To solve the problem for all times he used a conformal map into a strip. This allowed Sundman to obtain a series expansion for the coordinates valid for all times, solving the problem that was proposed by Weirstrass in the King Oscar II's competition. Though Sundman's work deserves better credit than it gets, it did not live up to Weirstrass's expectations,

Poincaré pointed out that the problem was not correctly posed, and proved that a complete solution to it could not be found. His work was so impressive that in 1888 the jury recognized its value by awarding him the prize. He found that the evolution of such a system is often chaotic in the sense that a small perturbation in the initial state, such as a slight change in one body's initial position, might lead to a radically different later state. If the slight change is not detectable by our measuring instruments, then we will not be able to predict which final state will occur. One of the judges, the distinguished Karl Weierstrass, said, "This work cannot indeed be considered as furnishing the complete solution of the question proposed, but that it is nevertheless of such importance that its publication will inaugurate a new era in the history of celestial mechanics." Weierstrass did not know how accurate he was. In Poincaré's paper, he described new mathematical ideas such as homoclinic points. The memoir was about to be published in Acta Mathematica when an error was found by the editor. This error in fact led to further discoveries by Poincaré, which are now considered to be the beginning of *chaos theory*. The memoir was published later in 1890. Poincaré's research into orbits about Lagrange points and low-energy transfers was not utilized for more than a century afterwards.

In 1889 Poincaré proved that for the restricted three body problem no integrals exist apart from the Jacobian. In 1890 Poincaré proved his famous recurrence theorem, namely that in any small region of phase-space trajectories exist which pass through the region infinitely often. Poincaré published 3 volumes of 'Les méthods nouvelle de la mécanique celeste' between 1892 and 1899. He discussed convergence and uniform convergence of the series solutions discussed by earlier mathematicians and proved them not to be uniformly convergent. The stability proofs of Lagrange and Laplace became inconclusive after this result.

Poincaré introduced further topological methods in 1912 for the theory of stability of orbits in the 3-body problem. It fact Poincaré essentially invented topology in his attempt to answer stability questions in the three body problem. He conjectured that there are infinitely many periodic solutions of the restricted problem, the conjecture being later proved by George *Birkhoff*. The stability of the orbits in the three body problem was also investigated by Levi-Civita, Birkhoff and others (see [II06c] for technical details).

To examine chaos, Poincaré used the idea of a section, today called the *Poincaré section*, which cuts across the orbits in phase-space. While the original dynamical system always *flows in continuous time*, on the Poincaré section we can observe *discrete-time steps*. More precisely, the original *phase-space flow* (see [II06c]) is replaced by an *iterated map*, which reduces the dimension of the phase-space by one (see Fig. 7.26). Later, to show what a Poincaré section would look like, Hénon devised a simple 2D-map, which is today called

---

and the series solution did not 'considerably extend our understanding of the Solar system.' The work that followed from Poincaré did.

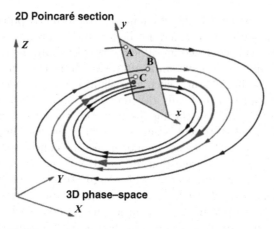

Fig. 7.26. The 2D Poincaré section, reducing the 3D phase-space, using the iterated map: $x_{new} = F(x, y)$, $y_{new} = G(x, y)$.

the *Hénon map*: $x_{new} = 1 - ax^2 + by$, $y_{new} = x$, with parameters $a = 1.4$, $b = 0.3$. Given any starting point, this map generates a sequence of points settling onto a chaotic attractor.

As an inheritance of Poincaré work, the chaos of the Solar system has been recently used for the SOHO project,[26] to minimize the fuel consumption need for the space flights. Namely, in a rotating frame, a spacecraft can remain stationary at 5 *Lagrange's points* (see Fig. 7.27).

Poincaré had two protegés in the development of chaos theory in the new world: *George D. Birkhoff* (see [Bir15, Bir27, Bir17] for the *Birkhoff curve shortening flow*), and *Stephen Smale* (see below).

In some detail, the theorems of John von Neumann and George Birkhoff on the *ergodic hypothesis* (see footnote 6 above) were published in 1912 and 1913. This line of inquiry developed in two directions. One direction took an abstract approach and considered *dynamical systems as transformations of*

---

[26] The SOHO project is being carried out jointly by ESA (European Space Agency) and NASA (US National Aeronautics and Space Administration), as a cooperative effort between the two agencies in the framework of the Solar Terrestrial Science Program (STSP) comprising SOHO and CLUSTER, and the International Solar–Terrestrial Physics Program (ISTP), with Geotail (ISAS–Japan), Wind, and Polar. SOHO was launched on December 2, 1995. The SOHO spacecraft was built in Europe by an industry team led by Matra, and instruments were provided by European and American scientists. There are nine European Principal Investigators (PI's) and three American ones. Large engineering teams and more than 200 co-investigators from many institutions support the PI's in the development of the instruments and in the preparation of their operations and data analysis. NASA is responsible for the launch and mission operations. Large radio dishes around the world which form NASA's Deep Space Network are used to track the spacecraft beyond the Earth's orbit. Mission control is based at Goddard Space Flight Center in Maryland.

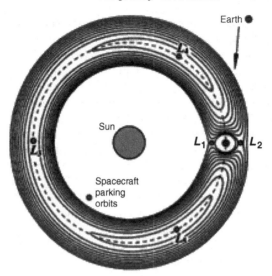

Fig. 7.27. The *Lagrange's points* $(L_1, \ldots, L_5)$ used for the space flights. Points L_1, L_2, L_3 on the Sun–Earth axis are unstable. The SOHO spacecraft used a *halo orbit* around L_1 to observe the Sun. The triangular points L_4 and L_5 are often stable. A Japanese rescue mission used a chaotic Earth–Moon trajectory.

measurable spaces into themselves. Could we classify these transformations in a meaningful way? This lead *Andrey N. Kolmogorov* to the introduction of the fundamental concept of *entropy* for dynamical systems. With entropy as a *dynamical invariant* it became possible to classify a set of abstract dynamical systems known as the *Bernoulli systems*.

The other line that developed from the ergodic hypothesis was in trying to find mechanical systems that are ergodic. *An ergodic system could not have stable orbits, as these would break ergodicity.* So, in 1898 *Jacques S. Hadamard* published a paper on *billiards*, where he showed that the *motion of balls on surfaces of constant negative curvature is everywhere unstable.* This dynamical system was to prove very useful and it was taken up by Birkhoff.

Marston Morse in 1923 showed that it was possible to enumerate the orbits of a ball on a surface of constant negative curvature.[27] He did this by introducing a symbolic code to each orbit and showed that the number of pos-

---

[27] Recall from [II06c] that in differential topology, the techniques of *Morse theory* give a very direct way of analyzing the topology of a manifold by studying differentiable functions on that manifold. According to the basic insights of Marston Morse, a differentiable function on a manifold will, in a typical case, reflect the topology quite directly. Morse theory allows one to find the so-called CW-structures and handle decompositions on manifolds and to obtain substantial information about their homology. Before Morse, Arthur Cayley and James Clerk Maxwell developed some of the ideas of Morse theory in the context of topography. Morse originally applied his theory to geodesics (critical points of the energy functional on paths).

sible codes grew exponentially with the length of the code. With contributions by E. Artin, G. Hedlund, and *Heinz Hopf* it was eventually proven that the motion of a ball on a surface of constant negative curvature was ergodic. The importance of this result escaped most physicists, one exception being N.M. *Krylov*, who understood that a physical billiard was a dynamical system on a surface of negative curvature, but with the curvature concentrated along the lines of collision. Sinai, who was the first to show that a physical billiard can be ergodic, knew Krylov's work well.

On the other hand, the work of Lord Rayleigh also received vigorous development. It prompted many experiments and some theoretical development by B. *Van der Pol*, G. *Duffing*, and D. *Hayashi*. They found other systems in which the nonlinear oscillator played a role and classified the possible motions of these systems. This concreteness of experiments, and the possibility of analysis was too much of temptation for M.L. *Cartwright* and J.E. *Littlewood*, who set out to prove that many of the structures conjectured by the experimentalists and theoretical physicists did indeed follow from the equations of motion.

Also, G. Birkhoff had found a 'remarkable curve' in a 2D map; it appeared to be non-differentiable and it would be nice to see if a smooth flow could generate such a curve. The work of Cartwright and Littlewood lead to the work of N. Levinson, which in turn provided the basis for the horseshoe construction of Steve Smale.

In Russia, *Aleksandr M. Lyapunov* paralleled the methods of Poincaré and initiated the strong Russian dynamical systems school. A. *Andronov*[28] carried on with the study of nonlinear oscillators and in 1937 introduced together with *Lev S. Pontryagin*[29] the notion of *coarse systems*. They were formalizing the understanding garnered from the study of nonlinear oscillators, the understanding that many of the details on how these oscillators work do not affect the overall picture of the phase-space: there will still be limit cycles if one changes the dissipation or spring force function by a little bit. And changing the system a little bit has the great advantage of eliminating exceptional cases in the mathematical analysis. Coarse systems were the concept that caught Smale's attention and enticed him to study dynamical systems (see [CAM05]).

The path traversed from ergodicity to entropy is a little more confusing. The general character of entropy was understood by *Norbert Wiener*,[30] who seemed to have spoken to *Claude E. Shannon*.[31] In 1948 Shannon published his results on *information theory*, where he discusses the entropy of the shift transformation.

These techniques were later used by *Raoul Bott* in his proof of the celebrated Bott periodicity theorem.

[28] Recall that both the *Andronov–Hopf bifurcation* and a crater on the Moon are named after Aleksandr Andronov.

[29] The father of modern optimal control theory (see [II06c]).

[30] The father of cybernetics.

[31] The father of information theory.

In Russia, *Andrey N. Kolmogorov* went far beyond and suggested a definition of the metric entropy of an area preserving transformation in order to classify Bernoulli shifts. The suggestion was taken by his student Ya.G. *Sinai* and the results published in 1959. In 1967 D.V. Anosov[32] and Sinai applied the notion of entropy to the study of dynamical systems. It was in the context of studying the entropy associated to a dynamical system that Sinai introduced *Markov partitions* (in 1968), which allow one *to relate dynamical systems and statistical mechanics*; this has been a very fruitful relationship. It adds measure notions to the topological framework laid down in Smale's dynamical systems paper. Markov partitions *divide the phase-space* of the dynamical system into nice little boxes that map into each other. Each box is labeled by a code and the dynamics on the phase-space maps the codes around, inducing a *symbolic dynamics*. From the number of boxes needed to cover all the space, Sinai was able to define the notion of entropy of a dynamical system. However, the relations with statistical mechanics became explicit in the work of *David Ruelle*.[33] Ruelle understood that the topology of the orbits could be specified by a symbolic code, and that one could associate an 'energy' to each orbit. The energies could be formally combined in a *partition function* (see [II06c]) to generate the invariant measure of the system.

Smale: *Topological Horseshoe and Chaos of Stretching and Folding*

The first deliberate, coordinated attempt to understand how global system's behavior might differ from its local behavior, came from topologist Steve Smale from the University of California at Berkeley. A young physicist, making a small talk, asked what Smale was working on. The answer stunned

---

[32] Recall that the *Anosov map* on a manifold M is a certain type of mapping, from M to itself, with rather clearly marked local directions of 'expansion' and 'contraction'. More precisely:

- If a differentiable map f on M has a hyperbolic structure on the tangent bundle, then it is called an *Anosov map*. Examples include the *Bernoulli map*, and *Arnold cat map*.
- If the Anosov map is a diffeomorphism, then it is called an *Anosov diffeomorphism*. Anosov proved that Anosov diffeomorphisms are *structurally stable*.
- If a flow on a manifold splits the tangent bundle into three invariant subbundles, with one subbundle that is exponentially contracting, and one that is exponentially expanding, and a third, non-expanding, non-contracting 1D sub-bundle, then the flow is called an *Anosov flow*.

[33] David Ruelle is a mathematical physicist working on statistical physics and dynamical systems. Together with *Floris Takens*, he coined the term *strange attractor*, and founded a modern *theory of turbulence*. Namely, in a seminal paper [RT71] they argued that, as a function of an external parameter, the *route to chaos in a fluid flow* is a transition sequence leading from stationary (S) to single periodic (P), double periodic (QP_2), triple periodic (QP_3) and, possibly, quadruply periodic (QP_4) motions, before the flow becomes chaotic (C).

him: "Oscillators." It was absurd. Oscillators (pendulums, springs, or electric circuits) where the sort of problem that a physicist finished off early in his training. They were easy. Why would a great mathematician be studying elementary physics? However, Smale was looking at nonlinear oscillators, chaotic oscillators—and seeing things that physicists had learned no to see [Gle87].

Smale's 1966 Fields Medal honored a famous piece of work in high-dimensional topology, proving *Poincaré conjecture* for all dimensions greater than 4; he later generalized the ideas in a 107 page paper that established the *H-cobordism theorem* (this seminal result provides algebraic topological criteria for establishing that higher-dimensional manifolds are diffeomorphic).

After having made great strides in topology, Smale then turned to the study of nonlinear dynamical systems, where he made significant advances as well.[34] His first contribution is the famous *horseshoe map* [Sma67] that started-off significant research in dynamical systems and chaos theory.[35] Smale also outlined a mathematical research program carried out by many others. Smale is also known for injecting *Morse theory* into mathematical economics, as well as recent explorations of various theories of computation. In 1998 he compiled a list of 18 problems in mathematics to be solved in the 21st century. This list was compiled in the spirit of Hilbert's famous list of problems produced in 1900. In fact, Smale's list includes some of the original Hilbert problems. Smale's problems include the Jacobian conjecture and the Riemann hypothesis, both of which are still unsolved.

The *Smale horseshoe map* (see Fig. 7.28) *is any member of a class of chaotic maps of the square into itself.* This topological transformation provided a basis for understanding the chaotic properties of dynamical systems. Its basis are simple: A space is stretched in one direction, squeezed in another,

---

[34] In the fall of 1961 Steven Smale was invited to Kiev where he met V.I. Arnol'd, (one of the fathers of modern geometrical mechanics [II06c]), D.V. Anosov, Sinai, and Novikov. He lectured there, and spent a lot of time with Anosov. He suggested a series of conjectures, most of which Anosov proved within a year. It was Anosov who showed that there are dynamical systems for which all points (as opposed to a non-wandering set) admit the hyperbolic structure, and it was in honor of this result that Smale named them *Axiom-A systems*. In Kiev Smale found a receptive audience that had been thinking about these problems. Smale's result catalyzed their thoughts and initiated a chain of developments that persisted into the 1970's.
[35] In his landmark 1967 Bulletin survey article entitled 'Differentiable dynamical systems' [Sma67], Smale presented his program for hyperbolic dynamical systems and stability, complete with a superb collection of problems. The major theorem of the paper was the Ω-Stability Theorem: the global foliation of invariant sets of the map into disjoint stable and unstable parts, whose proof was a tour de force in the new dynamical methods. Some other important ideas of this paper are the existence of a horseshoe and enumeration and ordering of all its orbits, as well as the use of zeta functions to study dynamical systems. The emphasis of the paper is on the global properties of the dynamical system, on how to understand the topology of the orbits. Smale's account takes us from a local differential equation (in the form of vector fields) to the global topological description in terms of horseshoes.

and then folded. When the process is repeated, it produces something like a many-layered pastry dough, in which a pair of points that end up close together may have begun far apart, while two initially nearby points can end completely far apart.[36]

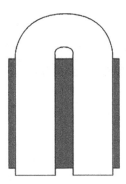

Fig. 7.28. The *Smale horseshoe map* consists of a sequence of operations on the unit square. First, stretch in the y-direction by more than a factor of two, then squeeze (compress) in the x-direction by more than a factor of two. Finally, fold the resulting rectangle and fit it back onto the square, overlapping at the top and bottom, and not quite reaching the ends to the left and right (and with a gap in the middle), as illustrated in the diagram. The shape of the stretched and folded map gives the horseshoe map its name. Note that it is vital to the construction process for the map to overlap and leave the middle and vertical edges of the initial unit square uncovered.

The horseshoe map was introduced by Smale while studying the behavior of the orbits of the *relaxation Van der Pol oscillator*. The action of the map is defined geometrically by squishing the square, then stretching the result into a long strip, and finally folding the strip into the shape of a horseshoe. Most points eventually leave the square under the action of the map f. They go to the side caps where they will, under iteration, converge to a *fixed-point* in one of the caps. The points that remain in the square under repeated iteration form a *fractal set* and are part of the *invariant set* of the map f (see Fig. 7.29).

The *stretching, folding* and *squeezing* of the horseshoe map are the essential elements that must be present in any chaotic system. In the horseshoe map the squeezing and stretching are uniform. They compensate each other so that the area of the square does not change. The folding is done neatly, so that the orbits that remain forever in the square can be simply described.

---

[36] Originally, Smale had hoped to explain all dynamical systems in terms of *stretching* and *squeezing*—with no folding, at least no folding that would drastically undermine a system's stability. But *folding* turned out to be necessary, and folding allowed sharp changes in dynamical behavior [Gle87].

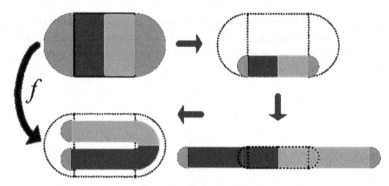

Fig. 7.29. The *Smale horseshoe map* f, defined by *stretching, folding* and *squeezing* of the system's phase-space.

Repeating this generates the horseshoe attractor. If one looks at a cross section of the final structure, it is seen to correspond to a *Cantor set*.

The Smale horseshoe map is the set of basic topological operations for constructing an attractor consist of stretching (which gives sensitivity to initial conditions) and folding (which gives the attraction). Since *trajectories in phase-space cannot cross*, the repeated stretching and folding operations result in an object of great topological complexity. For any horseshoe map we have:

- There is an infinite number of periodic orbits;
- Periodic orbits of arbitrarily long period exist;
- The number or periodic orbits grows exponentially with the period; and
- Close to any point of the fractal invariant set there is a point of a periodic orbit.

More precisely, the horseshoe map f is a *diffeomorphism* defined from a region S of the plane into itself. The region S is a square capped by two semi-disks. The action of f is defined through the composition of three geometrically defined transformations. First the square is contracted along the vertical direction by a factor $a < 1/2$. The caps are contracted so as to remain semi-disks attached to the resulting rectangle. Contracting by a factor smaller than one half assures that there will be a gap between the branches of the horseshoe. Next the rectangle is stretched by a factor of $1/a$; the caps remain unchanged. Finally the resulting strip is folded into a horseshoe-shape and placed back into S.

The interesting part of the dynamics is the image of the square into itself. Once that part is defined, the map can be extended to a diffeomorphism by defining its action on the caps. The caps are made to contract and eventually map inside one of the caps (the left one in the figure). The extension of f to the caps adds a fixed-point to the *non-wandering set* of the map. To keep the class of horseshoe maps simple, the curved region of the horseshoe should not map back into the square.

The horseshoe map is one-to-one (1–1, or injection): any point in the domain has a unique image, even though not all points of the domain are the image of a point. The inverse of the horseshoe map, denoted by f^{-1}, cannot have as its domain the entire region S, instead it must be restricted to the image of S under f, that is, the domain of f^{-1} is $f(S)$.

Fig. 7.30. Other types of horseshoe maps can be made by folding the contracted and stretched square in different ways.

By folding the contracted and stretched square in different ways, other types of horseshoe maps are possible (see Fig. 7.30). The contracted square cannot overlap itself to assure that it remains 1–1. When the action on the square is extended to a diffeomorphism, the extension cannot always be done on the plane. For example, the map on the right needs to be extended to a diffeomorphism of the sphere by using a 'cap' that wraps around the equator.

The horseshoe map is an Axiom A diffeomorphism that serves as a model for the general behavior at a transverse *homoclinic point*, where the *stable and unstable manifolds* of a periodic point intersect.

The horseshoe map was designed by Smale to reproduce the chaotic dynamics of a *flow* in the neighborhood of a given periodic *orbit*. The neighborhood is chosen to be a small disk perpendicular to the orbit. As the system evolves, points in this disk remain close to the given periodic orbit, tracing out orbits that eventually intersect the disk once again. Other orbits diverge.

The behavior of all the orbits in the disk can be determined by considering what happens to the disk. The intersection of the disk with the given periodic orbit comes back to itself every period of the orbit and so do points in its neighborhood. When this neighborhood returns, its shape is transformed. Among the points back inside the disk are some points that will leave the disk neighborhood and others that will continue to return. The set of points that never leaves the neighborhood of the given periodic orbit form a fractal.

A symbolic name can be given to all the orbits that remain in the neighborhood. The initial neighborhood disk can be divided into a small number of regions. Knowing the sequence in which the orbit visits these regions allows the orbit to be pinpointed exactly. The visitation sequence of the orbits provide the so-called *symbolic dynamics*[37]

---

[37] Symbolic dynamics is the practice of modeling a dynamical system by a space consisting of infinite sequences of abstract symbols, each sequence corresponding to

It is possible to describe the behavior of all initial conditions of the horse-shoe map. An initial point $u_0 = x, y$ gets mapped into the point $u_1 = f(u_0)$. Its iterate is the point $u_2 = f(u_1) = f^2(u_0)$, and repeated iteration generates the orbit $u_0, u_1, u_2, \ldots$ Under repeated iteration of the horseshoe map, most orbits end up at the fixed-point in the left cap. This is because the horseshoe maps the left cap into itself by an *affine transformation*, which has exactly one fixed-point. Any orbit that lands on the left cap never leaves it and converges to the fixed-point in the left cap under iteration. Points in the right cap get mapped into the left cap on the next iteration, and most points in the square get mapped into the caps. Under iteration, most points will be part of orbits that converge to the fixed-point in the left cap, but some points of the square never leave.

Under forward iterations of the horseshoe map, the original square gets mapped into a series of horizontal strips. The points in these horizontal strips come from vertical strips in the original square. Let S_0 be the original square, map it forward n times, and consider only the points that fall back into the square S_0, which is a set of horizontal stripes $H_n = f^n(S_0) \cap S_0$. The points in the horizontal stripes came from the vertical stripes $V_n = f^{-n}(H_n)$, which are the horizontal strips H_n mapped backwards n times. That is, a point in V_n will, under n iterations of the horseshoe map, end up in the set H_n of vertical strips (see Fig. 7.31).

Now, if a point is to remain indefinitely in the square, then it must belong to an *invariant set* Λ that maps to itself. Whether this set is empty or not has to be determined. The vertical strips V_1 map into the horizontal strips H_1, but not all points of V_1 map back into V_1. Only the points in the intersection of V_1 and H_1 may belong to Λ, as can be checked by following points outside the intersection for one more iteration. The intersection of the horizontal and vertical stripes, $H_n \cap V_n$, are squares that converge in the limit $n \to \infty$ to the invariant set Λ (see Fig. 7.32).

The structure of invariant set Λ can be better understood by introducing a system of labels for all the intersections, namely a *symbolic dynamics*. The intersection $H_n \cap V_n$ is contained in V_1. So any point that is in Λ under iteration must land in the left vertical strip A of V_1, or on the right vertical strip B. The lower horizontal strip of H_1 is the image of A and the upper horizontal strip is the image of B, so $H_1 = f(A) \cap f(B)$. The strips A and B can be used to label the four squares in the intersection of V_1 and H_1 (see Fig. 7.33) as:

---

a state of the system, and a shift operator corresponding to the dynamics. Symbolic dynamics originated as a method to study general dynamical systems, now though, its techniques and ideas have found significant applications in data storage and transmission, linear algebra, the motions of the planets and many other areas. The distinct feature in symbolic dynamics is that time is measured in discrete intervals. So at each time interval the system is in a particular state. Each state is associated with a symbol and the evolution of the system is described by an infinite sequence of symbols (see text below).

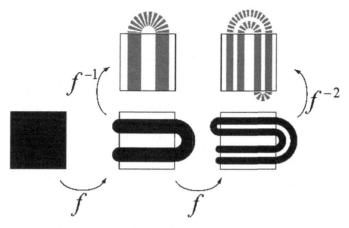

Fig. 7.31. Iterated horseshoe map: pre-images of the square region.

Fig. 7.32. Intersections that converge to the invariant set Λ.

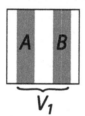

Fig. 7.33. The basic domains of the horseshoe map in symbolic dynamics.

$$\Lambda_{A\bullet A} = f(A) \cap A, \qquad \Lambda_{A\bullet B} = f(A) \cap B,$$
$$\Lambda_{B\bullet A} = f(B) \cap A, \qquad \Lambda_{B\bullet B} = f(B) \cap B.$$

The set $\Lambda_{B\bullet A}$ consist of points from strip A that were in strip B in the previous iteration. A dot is used to separate the region the point of an orbit is in from the region the point came from.

This notation can be extended to higher iterates of the horseshoe map. The vertical strips can be named according to the sequence of visits to strip A or strip B. For example, the set $ABB \subset V_3$ consists of the points from A that will all land in B in one iteration and remain in B in the iteration after that:

$$ABB = \{x \in A \mid f(x) \in B \text{ and } f^2(x) \in B\}.$$

Working backwards from that trajectory determines a small region, the set ABB, within V_3.

The horizontal strips are named from their vertical strip pre-images. In this notation, the intersection of V_2 and H_2 consists of 16 squares, one of which is

$$\Lambda_{AB \bullet BB} = f^2(AB) \cap BB.$$

All the points in $\Lambda_{AB \bullet BB}$ are in B and will continue to be in B for at least one more iteration. Their previous trajectory before landing in BB was A followed by B.

Any one of the intersections $\Lambda_{P \bullet F}$ of a horizontal strip with a vertical strip, where P and F are sequences of As and Bs, is an affine transformation of a small region in V_1. If P has k symbols in it, and if $f^{-k}(\Lambda_{P \bullet F})$ and $\Lambda_{P \bullet F}$ intersect, then the region $\Lambda_{P \bullet F}$ will have a *fixed-point*. This happens when the sequence P is the same as F. For example, $\Lambda_{ABAB \bullet ABAB} \subset V_4 \cap H_4$ has at least one fixed-point. This point is also the same as the fixed-point in $\Lambda_{AB \bullet AB}$. By including more and more ABs in the P and F part of the label of intersection, the area of the intersection can be made as small as needed. It converges to a point that is part of a *periodic orbit of the horseshoe map*. The periodic orbit can be labeled by the simplest sequence of As and Bs that labels one of the regions the periodic orbit visits. For every sequence of As and Bs there is a periodic orbit.

The Smale horseshoe map is the same topological structure as the *homoclinic tangle*. To dynamically introduce homoclinic tangles, let us consider a classical engineering problem of *escape from a potential well*. Namely, if we have a motion, $x = x(t)$, of a damped particle in a well with potential energy $V = x^2/2 - x^3/3$ (see Fig. 7.34) excited by a periodic driving force, $F \cos(wt)$ (with the period $T = 2\pi/w$), we are dealing with a nonlinear dynamical system given by [TS01]

$$\ddot{x} + a\dot{x} + x - x^2 = F \cos(wt). \tag{7.19}$$

Now, if the driving is switched off, i.e., $F = 0$, we have an autonomous 2D-system with the phase-portrait (and the safe basin of attraction) given in Fig. 7.34(below). The grey area of escape starts over the hilltop to infinity. Once we start driving, the system (7.19) becomes 3-dimensional, with its 3D phase-space. We need to see the basin in a *stroboscopic section* (see Fig. 7.35). The hill-top solution still has an inset and outset. As the driving increases, the inset and outset get tangled. They intersect one another an infinite number of times. The boundary of the safe basin becomes fractal. As the driving increases even more, the so-called fractal-fingers created by the homoclinic tangling, make a sudden incursion into the safe basin. At that point, the integrity of the in-well motions is lost [TS01].

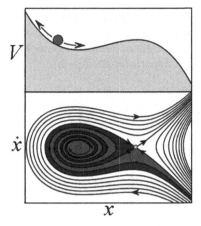

Fig. 7.34. Motion of a damped particle in a potential well, driven by a periodic force $F\cos(wt)$. *Up*: potential $(x-V)$-plot, with $V = x^2/2-x^3/3$; *down*: the corresponding phase $(x-\dot{x})$-portrait, showing the safe basin of attraction—if the driving is switched off ($F = 0$).

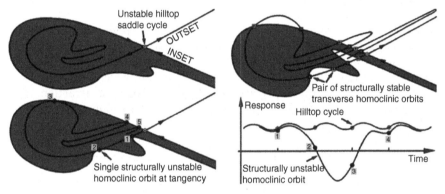

Fig. 7.35. Dynamics of a homoclinic tangle. The hill-top solution of a damped particle in a potential well driven by a periodic force. As the driving increases, the inset and outset get tangled.

Now, topologically speaking (referring to Fig. 7.36), let X be the point of intersection, with X' ahead of X on one manifold and ahead of X'' of the other. The map of each of these points TX' and TX'' must be ahead of the map of X, TX. The only way this can happen is if the manifold loops back and crosses itself at a new *homoclinic point*, i.e., a point where a stable and an unstable separatrix (invariant manifold) from the same fixed-point or same family intersect. Another loop must be formed, with T^2X another homoclinic point. Since T^2X is closer to the hyperbolic point than TX, the distance between T^2X and TX is less than that between X and TX. Area preservation requires the area to remain the same, so each new curve (which is closer than the previous one) must extend further. In effect, the loops become longer and thinner. The network of curves leading to a dense area of homoclinic points is known as a homoclinic tangle or tendril. Homoclinic points appear where chaotic regions touch in a *hyperbolic fixed-point*.

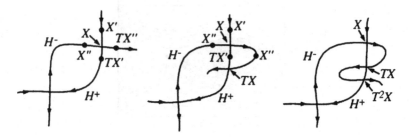

Fig. 7.36. More on homoclinic tangle (see text for explanation).

Lorenz: *Weather Prediction and Chaos*

Recall that an *attractor* is a set of system's states (i.e., points in the system's phase-space), invariant under the dynamics, towards which neighboring states in a given *basin of attraction* asymptotically approach in the course of dynamic evolution.[38] An attractor is defined as the smallest unit which cannot be itself decomposed into two or more attractors with distinct basins of attraction. This restriction is necessary since a dynamical system may have multiple attractors, each with its own basin of attraction.

Conservative systems do not have attractors, since the motion is periodic. For dissipative dynamical systems, however, volumes shrink exponentially, so attractors have 0 volume in nD phase-space.

In particular, a stable *fixed-point* surrounded by a dissipative region is an attractor known as a *map sink*.[39] Regular attractors (corresponding to 0 *Lyapunov exponents*) act as *limit cycles*, in which trajectories circle around a limiting trajectory which they asymptotically approach, but never reach. The so-called *strange attractors*[40] are bounded regions of phase-space (corresponding to positive Lyapunov characteristic exponents) having zero measure in the embedding phase-space and a *fractal dimension*. Trajectories within a strange attractor appear to skip around randomly.

---

[38] A *basin of attraction* is a set of points in the system's phase-space, such that initial conditions chosen in this set dynamically evolve to a particular attractor.

[39] A *map sink* is a stable fixed-point of a map which, in a dissipative dynamical system, is an attractor.

[40] A strange attractor is an attracting set that has zero measure in the embedding phase-space and has fractal dimension. Trajectories within a strange attractor appear to skip around randomly.

In 1963, Ed Lorenz from MIT was trying to improve weather forecasting. Using a primitive computer of those days, he discovered the first *chaotic attractor*. Lorenz used three Cartesian variables, (x, y, z), to define *atmospheric convection*. Changing in time, these variables gave him a trajectory in a (Euclidean) 3D-space. From all starts, trajectories settle onto a chaotic, or *strange attractor*.[41]

More precisely, Lorenz reduced the *Navier–Stokes equations* for *convective Bénard fluid flow* into three first order coupled nonlinear differential equations, already introduced above as (7.29) and demonstrated with these the idea of sensitive dependence upon initial conditions and chaos (see [Lor63, Spa82]).

---

[41] Edward Lorenz is a professor of meteorology at MIT who wrote the first clear paper on *deterministic chaos*. The paper was called 'Deterministic Nonperiodic Flow' and it was published in the Journal of Atmospheric Sciences in 1963. Before that, in 1960, Lorenz began a project to simulate weather patterns on a computer system called the Royal McBee. Lacking much memory, the computer was unable to create complex patterns, but it was able to show the interaction between major meteorological events such as tornados, hurricanes, easterlies and westerlies. A variety of factors was represented by a number, and Lorenz could use computer printouts to analyze the results. After watching his systems develop on the computer, Lorenz began to see patterns emerge, and was able to predict with some degree of accuracy what would happen next. While carrying out an experiment, Lorenz made an accidental discovery. He had completed a run, and wanted to recreate the pattern. Using a printout, Lorenz entered some variables into the computer and expected the simulation to proceed the same as it had before. To his surprise, the pattern began to diverge from the previous run, and after a few 'months' of simulated time, the pattern was completely different. Lorenz eventually discovered why seemingly identical variables could produce such different results. When Lorenz entered the numbers to recreate the scenario, the printout provided him with numbers to the thousandth position (such as 0.617). However, the computer's internal memory held numbers up to the millionth position (such as 0.617395); these numbers were used to create the scenario for the initial run. This small deviation resulted in a completely divergent weather pattern in just a few months. This discovery creates the groundwork of chaos theory: In a system, small deviations can result in large changes. This concept is now known as a *butterfly effect*.

Lorenz definition of chaos is: "The property that characterizes a dynamical system in which most orbits exhibit sensitive dependence." Dynamical systems (like the weather) are all around us. They have recurrent behavior (it is always hotter in summer than winter) but are very difficult to pin down and predict apart from the very short term. 'What will the weather be tomorrow?'—can be anticipated, but 'What will the weather be in a months time?' is an impossible question to answer.

Lorenz showed that with a set of simple differential equations seemingly very complex turbulent behavior could be created that would previously have been considered as random. He further showed that accurate longer range forecasts in any chaotic system were impossible, thereby overturning the previous orthodoxy. It had been believed that the more equations you add to describe a system, the more accurate will be the eventual forecast.

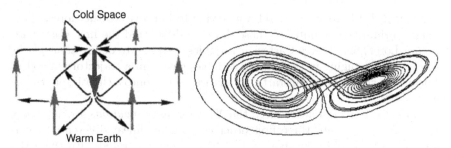

Fig. 7.37. Bénard cells, showing a typical vortex of a rolling air, with a warm air rising in a ring and a cool air descending in the center (*left*). A simple model of the Bénard cells provided by the celebrated 'Lorenz-butterfly' (or, 'Lorenz-mask') *strange attractor* (*right*).

We rewrite the celebrated Lorenz equations here as

$$\dot{x} = a(y - x), \qquad \dot{y} = bx - y - xz, \qquad \dot{z} = xy - cz, \qquad (7.20)$$

where x, y and z are dynamical variables, constituting the 3D *phase-space* of the *Lorenz system*; and a, b and c are the parameters of the system. Originally, Lorenz used this model to describe the unpredictable behavior of the weather, where x is the rate of convective overturning (convection is the process by which heat is transferred by a moving fluid), y is the horizontal temperature overturning, and z is the vertical temperature overturning; the parameters are: $a \equiv P$-proportional to the *Prandtl number* (ratio of the fluid viscosity of a substance to its thermal conductivity, usually set at 10), $b \equiv R$-proportional to the Rayleigh number (difference in temperature between the top and bottom of the system, usually set at 28), and $c \equiv K$-a number proportional to the physical proportions of the region under consideration (width to height ratio of the box which holds the system, usually set at 8/3). The Lorenz system (7.20) has the properties:

1. *Symmetry*: $(x, y, z) \rightarrow (-x, -y, z)$ for all values of the parameters, and
2. The z-axis ($x = y = 0$) is *invariant* (i.e., all trajectories that start on it also end on it).

Nowadays it is well-known that the Lorenz model is a paradigm for low-dimensional chaos in dynamical systems in synergetics and this model or its modifications are widely investigated in connection with modeling purposes in meteorology, hydrodynamics, laser physics, superconductivity, electronics, oil industry, chemical and biological kinetics, etc.

The 3D *phase-portrait* of the Lorenz system (7.37) shows the celebrated 'Lorenz mask', a special type of *fractal attractor* (see Fig. 7.37). It depicts the famous '*butterfly effect*', (i.e., sensitive dependence on initial conditions)— the popular idea in meteorology that 'the flapping of a butterfly's wings in Brazil can set off a tornado in Texas' (i.e., a tiny difference is amplified until two outcomes are totally different), so that the long term behavior becomes impossible to predict (e.g., long term weather forecasting). The Lorenz mask has the following characteristics:

1. Trajectory does not intersect itself in three dimensions;
2. Trajectory is not periodic or transient;
3. General form of the shape does not depend on initial conditions; and
4. Exact sequence of loops is very sensitive to the initial conditions.

Feigenbaum: *A Constant and Universality*

Mitchell Jay Feigenbaum (born December 19, 1944; Philadelphia, USA) is a mathematical physicist whose pioneering studies in chaos theory led to the discovery of the *Feigenbaum constant*.

In 1964 he began graduate studies at the MIT. Enrolling to study electrical engineering, he changed to physics and was awarded a doctorate in 1970 for a thesis on dispersion relations under Francis Low. After short positions at Cornell University and Virginia Polytechnic Institute, he was offered a longer-term post at Los Alamos National Laboratory to study turbulence. Although the group was ultimately unable to unravel the intractable theory of turbulent fluids, his research led him to study chaotic maps.

Many mathematical maps involving a single linear parameter exhibit apparently random behavior known as chaos when the parameter lies in a certain range. As the parameter is increased towards this region, the map undergoes bifurcations at precise values of the parameter. At first there is one stable point, then bifurcating to oscillate between two points, then bifurcating again to oscillate between four points and so on. In 1975 Feigenbaum, using the HP-65 computer he was given, discovered that the ratio of the difference between the values at which such successive *period-doubling bifurcations* (called the *Feigenbaum cascade*) occur tends to a constant of around 4.6692. He was then able to provide a mathematical proof of the fact, and showed that the same behavior and the same constant would occur in a wide class of mathematical functions prior to the onset of chaos. For the first time this universal result enabled mathematicians to take their first huge step to unraveling the apparently intractable 'random' behavior of chaotic systems. This 'ratio of convergence' is now known as the Feigenbaum constant.

More precisely, the Feigenbaum constant δ is a universal constant for functions approaching chaos via successive period doubling bifurcations. It was discovered by Feigenbaum in 1975, while studying the fixed-points of the iterated function $f(x) = 1 - \mu|x|^r$, and characterizes the geometric approach of the bifurcation parameter to its limiting value (see Fig. 7.38) as the parameter μ is increased for fixed x [Fei79].

The Logistic map is a well known example of the maps that Feigenbaum studied in his famous Universality paper [Fei78].

In 1986 Feigenbaum was awarded the Wolf Prize in Physics. He has been Toyota Professor at Rockefeller University since 1986.

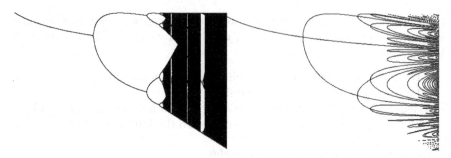

Fig. 7.38. Feigenbaum constant: approaching chaos via successive period doubling bifurcations. The *plot on the left* is made by iterating equation $f(x) = 1 - \mu|x|^r$ with $r = 2$ several hundred times for a series of discrete but closely spaced values of μ, discarding the first hundred or so points before the iteration has settled down to its fixed-points, and then plotting the points remaining. The *plot on the right* more directly shows the cycle may be constructed by plotting function $f^n(x) - x$ as a function of μ, showing the resulting curves for $n = 1, 2, 4$. Simulated in *Mathematica*^{TM}.

For details on Feigenbaum universality, see [Gle87].

May: *Population Modeling and Chaos*

Let $x(t)$ be the population of the species at time t; then the *conservation law* for the population is conceptually given by (see [Mur02])

$$\dot{x} = births - deaths + migration, \qquad (7.21)$$

where $\dot{x} = dx/dt$. The above conceptual equation gave rise to a series of *population models*. The simplest continuous-time model, due to Thomas Malthus from 1798 [Mal798],[42] has no migration, while the birth and death terms are proportional to x,

---

[42] The Rev. Thomas Robert Malthus, FRS (February, 1766–December 23, 1834), was an English demographer and political economist best known for his pessimistic but highly influential views. Malthus's views were largely developed in reaction to the optimistic views of his father, Daniel Malthus and his associates, notably Jean-Jacques Rousseau and William Godwin. Malthus's essay was also in response to the views of the Marquis de Condorcet. In An Essay on the Principle of Population, first published in 1798, Malthus made the famous prediction that population would outrun food supply, leading to a decrease in food per person: "The power of population is so superior to the power of the earth to produce subsistence for man, that premature death must in some shape or other visit the human race. The vices of mankind are active and able ministers of depopulation. They are the precursors in the great army of destruction; and often finish the dreadful work themselves. But should they fail in this war of extermination, sickly seasons, epidemics, pestilence, and plague, advance in terrific array, and sweep off their thousands and tens of thousands. Should success be still incomplete, gigantic inevitable famine stalks in the rear, and with one mighty blow levels the population with the food of the world."

$$\dot{x} = bx - dx \quad \Longrightarrow \quad x(t) = x_0 e^{(b-d)t}, \tag{7.22}$$

where b, d are positive constants and $x_0 = x(0)$ is the initial population. Thus, according to the *Malthus model* (7.22), if $b > d$, the population grows exponentially, while if $b < d$, it dies out. Clearly, this approach is fairly over-simplified and apparently fairly unrealistic. (However, if we consider the past and predicted growth estimates for the total world population from the 1900, we see that it has actually grown exponentially.)

This simple example shows that it is difficult to make long-term predictions (or, even relatively short-term ones), unless we know sufficient facts to incorporate in the model to make it a *reliable predictor*. In the long run, clearly, there must be some adjustment to such exponential growth. François Verhulst [Ver838, Ver845][43] proposed that a *self-limiting process* should operate when a population becomes too large. He proposed the so-called *logistic growth* population model,

$$\dot{x} = rx(1 - x/K), \tag{7.23}$$

where r, K are positive constants. In the Verhulst logistic model (7.23), the constant K is the *carrying capacity* of the environment (usually determined by the available sustaining resources), while the per capita birth rate $rx(1 -$

This Principle of Population was based on the idea that population if unchecked increases at a geometric rate, whereas the food supply grows at an arithmetic rate. Only natural causes (e.g. accidents and old age), misery (war, pestilence, and above all famine), moral restraint and vice (which for Malthus included infanticide, murder, contraception and homosexuality) could check excessive population growth. Thus, Malthus regarded his Principle of Population as an explanation of the past and the present situation of humanity, as well as a prediction of our future. The eight major points regarding evolution found in his 1798 *Essay* are: (i) Population level is severely limited by subsistence. (ii) When the means of subsistence increases, population increases. (iii) Population pressures stimulate increases in productivity. (iv) Increases in productivity stimulates further population growth. (v) Since this productivity can never keep up with the potential of population growth for long, there must be strong checks on population to keep it in line with carrying capacity. (vi) It is through individual cost/benefit decisions regarding sex, work, and children that population and production are expanded or contracted. (vii) Positive checks will come into operation as population exceeds subsistence level. (viii) The nature of these checks will have significant effect on the rest of the sociocultural system.

Evolutionists John Maynard Smith and Ronald Fisher were both critical of Malthus' theory, though it was Fisher who referred to the *growth rate* r (used in *logistic equation*) as the *Malthusian parameter*. Fisher referred to "...a relic of creationist philosophy..." in observing the fecundity of nature and deducing (as Darwin did) that this therefore drove natural selection. Smith doubted that famine was the great leveler that Malthus insisted it was.

[43] François Verhulst (October 28, 1804–February 15, 1849, Brussels, Belgium) was a mathematician and a doctor in number theory from the University of Ghent in 1825. Verhulst published in 1838 the logistic demographic model (7.23).

x/K) is dependent on x. There are two steady states (where $\dot{x} = 0$) for (7.23): (i) $x = 0$ (unstable, since linearization about it gives $\dot{x} \approx rx$); and (ii) $x = K$ (stable, since linearization about it gives $\frac{d}{dt}(x - K) \approx -r(x - K)$, so $\lim_{t \to \infty} x = K$). The carrying capacity K determines the size of the stable steady state population, while r is a measure of the rate at which it is reached (i.e., the measure of the dynamics)—thus $1/r$ is a representative timescale of the response of the model to any change in the population. The solution of (7.23) is

$$x(t) = \frac{x_0 K e^{rt}}{[K + x_0(e^{rt} - 1)]} \quad \Longrightarrow \quad \lim_{t \to \infty} x(t) = K.$$

In general, if we consider a population to be governed by

$$\dot{x} = f(x), \tag{7.24}$$

where typically $f(x)$ is a nonlinear function of x, then the equilibrium solutions x^* are solutions of $f(x) = 0$, and are linearly stable to small perturbations if $\dot{f}(x^*) < 0$, and unstable if $\dot{f}(x^*) > 0$ [Mur02].

In the mid 20th century, ecologists realized that many species had no overlap between successive generations and so population growth happens in discrete-time steps x_t, rather than in continuous-time $x(t)$ as suggested by the conservative law (7.21) and its Maltus–Verhulst derivations. This leads to study *discrete-time models* given by *difference equations*, or, *maps*, of the form

$$x_{t+1} = f(x_t), \tag{7.25}$$

where $f(x_t)$ is some generic nonlinear function of x_t. Clearly, (7.25) is a discrete-time version of (7.24). However, instead of solving differential equations, if we know the particular form of $f(x_t)$, it is a straightforward matter to evaluate x_{t+1} and subsequent generations by simple recursion of (7.25). The skill in modeling a specific population's growth dynamics lies in determining the appropriate form of $f(x_t)$ to reflect known observations or facts about the species in question.

In 1970s, Robert May, a physicist by training, won the Crafoord Prize for 'pioneering ecological research in theoretical analysis of the dynamics of populations, communities and ecosystems', by proposing a simple *logistic map* model for the generic population growth (7.25).[44] May's model of population growth is the celebrated *logistic map* [May76b, May73, May76a],

---

[44] Lord Robert May received his Ph.D. in theoretical physics from University of Sydney in 1959. He then worked at Harvard University and the University of Sydney before developing an interest in animal population dynamics and the relationship between complexity and stability in natural communities. He moved to Princeton University in 1973 and to Oxford and the Imperial College in 1988. May was able to make major advances in the field of population biology through the application of mathematics. His work played a key role in the development of *theoretical ecology* through the 1970s and 1980s. He also applied these tools to the study of disease and to the study of *bio-diversity*.

$$x_{t+1} = rx_t(1 - x_t), \tag{7.26}$$

where r is the *Malthusian parameter* that varies between 0 and 4, and the initial value of the population $x_0 = x(0)$ is restricted to be between 0 and 1. Therefore, in May's logistic map (7.26), the generic function $f(x_t)$ gets a specific quadratic form

$$f(x_t) = rx_t(1 - x_t).$$

For $r < 3$, the x_t have a single value. For $3 < r < 3.4$, the x_t oscillate between two values (see *bifurcation diagram*[45] on Fig. 7.39). As r increases, bifurcations occur where the number of iterates doubles. These *period doubling bifurcations* continue to a limit point at $r_{lim} = 3.569944$ at which the period is 2^∞ and the dynamics become chaotic. The r values for the first two bifurcations can be found analytically, they are $r_1 = 3$ and $r_2 = 1 + \sqrt{6}$. We can label the successive values of r at which bifurcations occur as r_1, r_2, ... The universal number associated with such period doubling sequences is

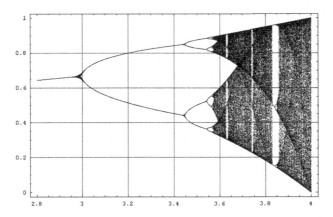

Fig. 7.39. Bifurcation diagram for the logistic map, simulated using *Mathematica*^{TM}.

called the *Feigenbaum number*,

$$\delta = \lim_{k \to \infty} \frac{r_k - r_{k-1}}{r_{k+1} - r_k} \approx 4.669.$$

This series of period-doubling bifurcations says that close enough to r_{lim} the distance between bifurcation points decreases by a factor of δ for each bifurcation. The complex *fractal pattern* got in this way shrinks indefinitely.

---

[45] A bifurcation diagram shows the possible long-term values a variable of a system can get in function of a parameter of the system.

Hénon: *A Special 2D Map and Its Strange Attractor*

Michel Hénon (born 1931 in Paris, France) is a mathematician and astronomer. He is currently at the Nice Observatory. In astronomy, Hénon is well known for his contributions to stellar dynamics, most notably the problem of *globular cluster* (see [Gle87]). In late 1960s and early 1970s he was involved in dynamical evolution of star clusters, in particular the globular clusters. He developed a numerical technique using *Monte Carlo methods*, to follow the dynamical evolution of a spherical star cluster much faster than the so-called n-body methods. In mathematics, he is well known for the Hénon map, a simple discrete dynamical system that exhibits chaotic behavior. Lately he has been involved in the restricted 3-body problem.

His celebrated *Hénon map* [Hen69] is a discrete-time dynamical system that is an extension of the *logistic map* (7.26) and exhibits a chaotic behavior. The map was introduced by Michel Hénon as a simplified model of the *Poincaré section* of the *Lorenz system* (7.20). This 2D-map takes a point (x, y) in the plane and maps it to a new point defined by equations

$$x_{n+1} = y_n + 1 - ax_n^2, \qquad y_{n+1} = bx_n.$$

The map depends on two parameters, a and b, which for the canonical Hénon map have values of $a = 1.4$ and $b = 0.3$ (see Fig. 7.40). For the canonical values the Hénon map is chaotic. For other values of a and b the map may be chaotic, intermittent, or converge to a periodic orbit. An overview of the

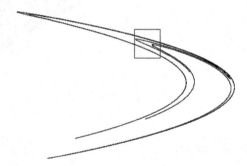

Fig. 7.40. *Hénon strange attractor* (see text for explanation), simulated using *Dynamics Solver*^{TM}.

type of behavior of the map at different parameter values may be obtained from its orbit (or, bifurcation) diagram (see Fig. 7.41). For the canonical map, an initial point of the plane will either approach a set of points known as the *Hénon strange attractor*, or diverge to infinity. The Hénon attractor is a fractal, smooth in one direction and a Cantor set in another. Numerical estimates yield a correlation dimension of 1.42 ± 0.02 (Grassberger, 1983) and a Hausdorff dimension of 1.261 ± 0.003 (Russel 1980) for the Hénon attractor.

As a dynamical system, the canonical Hénon map is interesting because, unlike the logistic map, its orbits defy a simple description. The Hénon map maps

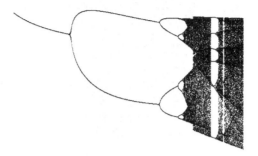

Fig. 7.41. Bifurcation diagram of the *Hénon strange attractor*, simulated using *Dynamics Solver*^{TM}.

two points into themselves: these are the invariant points. For the canonical values of a and b, one of these points is on the attractor: $x = 0.631354477\ldots$ and $y = 0.189406343\ldots$ This point is unstable. Points close to this fixed-point and along the slope 1.924 will approach the fixed-point and points along the slope -0.156 will move away from the fixed-point. These slopes arise from the linearizations of the *stable manifold* and *unstable manifold* of the fixed-point. The unstable manifold of the fixed-point in the attractor is contained in the strange attractor of the Hénon map. The Hénon map does not have a strange attractor for all values of the parameters a and b. For example, by keeping b fixed at 0.3 the bifurcation diagram shows that for a = 1.25 the Hénon map has a stable periodic orbit as an attractor. The authors of [CGP88] showed how the structure of the Hénon strange attractor could be understood in terms of unstable periodic orbits within the attractor.

For the (slightly modified) Hénon map: $x_{n+1} = ay_n + 1 - x_n^2$, $y_{n+1} = bx_n$, there are three *basins of attraction* (see Fig. 7.42).

The *generalized Hénon map* is a 3D-system (see Fig. 7.43)

$$x_{n+1} = ax_n - z(y_n - x_n^2)), \qquad y_{n+1} = zx_n + a(y_n - x_n^2)), \qquad z_{n+1} = z_n,$$

where $a = 0.24$ is a parameter. It is an *area-preserving map*, and simulates the *Poincaré map* of period orbits in *Hamiltonian systems*. Repeated random initial conditions are used in the simulation and their gray-scale color is selected at random.

Attractor vs. Chaotic Dynamics

Recall from [II06c] that the concept of *dynamical system* has its origins in *Newtonian mechanics*. There, as in other natural sciences and engineering

Fig. 7.42. Three basins of attraction for the Hénon map $x_{n+1} = ay_n + 1 - x_n^2$, $y_{n+1} = bx_n$, with $a = 0.475$.

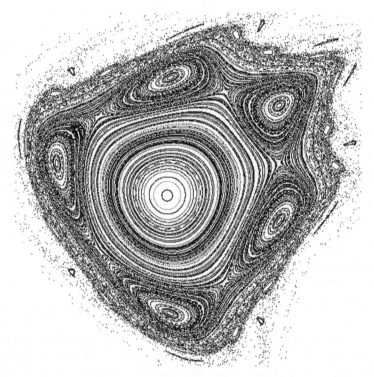

Fig. 7.43. Phase-plot of the *area-preserving generalized Hénon map*, simulated using *Dynamics Solver*[TM].

disciplines, the evolution rule of dynamical systems is given implicitly by a relation that gives the state of the system only a short time into the future. This relation is either a differential equation or difference equation. To determine the state for all future times requires iterating the relation many times—each advancing time a small step. The iteration procedure is referred to as solving the system or integrating the system. Once the system can be solved, given an initial point it is possible to determine all its future points, a collection known as a *trajectory* or *orbit*. All possible system trajectories comprise its *flow* in the phase-space.

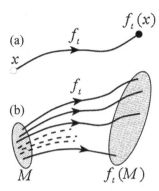

Fig. 7.44. Action of the *phase-flow* f_t in the phase-space manifold M: (**a**) Trajectory of a single initial point $x(t) \in M$, (**b**) Transporting the whole manifold M.

More precisely, recall from [II06c] that a *dynamical system* geometrically represents a *vector-field* (or, more generally, a *tensor-field*) in the system's phase-space manifold M, which upon *integration* (governed by the celebrated *existence & uniqueness theorems for ODEs*) defines a *phase-flow* in M (see Fig. 7.44). This phase-flow $f_t \in M$, describing the complete behavior of a dynamical system at every time instant, can be either linear, nonlinear or chaotic.

Before the advent of fast computers, solving a dynamical system required sophisticated mathematical techniques and could only be accomplished for a small class of linear dynamical systems. Numerical methods executed on computers have simplified the task of determining the orbits of a dynamical system.

For simple dynamical systems, knowing the trajectory is often sufficient, but most dynamical systems are too complicated to be understood in terms of individual trajectories. The difficulties arise because:

1. The systems studied may only be known approximately—the parameters of the system may not be known precisely or terms may be missing from the equations. The approximations used bring into question the validity or relevance of numerical solutions. To address these questions several notions of stability have been introduced in the study of dynamical systems, such as *Lyapunov stability* or *structural stability*. The stability of the dynamical system implies that there is a class of models or initial conditions

for which the trajectories would be equivalent. The operation for comparing orbits to establish their equivalence changes with the different notions of stability.

2. The type of trajectory may be more important than one particular trajectory. Some trajectories may be periodic, whereas others may wander through many different states of the system. Applications often require enumerating these classes or maintaining the system within one class. Classifying all possible trajectories has led to the qualitative study of dynamical systems, that is, properties that do not change under coordinate changes. Linear dynamical systems and systems that have two numbers describing a state are examples of dynamical systems where the possible classes of orbits are understood.

3. The behavior of trajectories as a function of a parameter may be what is needed for an application. As a parameter is varied, the dynamical systems may have *bifurcation points* where the qualitative behavior of the dynamical system changes. For example, it may go from having only periodic motions to apparently erratic behavior, as in the transition to *turbulence* of a fluid.

4. The trajectories of the system may appear erratic, as if random. In these cases it may be necessary to compute averages using one very long trajectory or many different trajectories. The averages are well defined for ergodic systems and a more detailed understanding has been worked out for *hyperbolic systems*. Understanding the probabilistic aspects of dynamical systems has helped establish the foundations of statistical mechanics and of chaos.

Now, let us start 'gently' with chaotic dynamics. Recall that a dynamical system may be defined as a deterministic rule for the time evolution of state observables. Well known examples are *ODEs* in which time is continuous,

$$\dot{\mathbf{x}}(t) = \mathbf{f}(\mathbf{x}(t)), \quad (\mathbf{x}, \mathbf{f} \in \mathbb{R}^n); \tag{7.27}$$

and *iterative maps* in which time is discrete:

$$\mathbf{x}(t+1) = \mathbf{g}(\mathbf{x}(t)), \quad (\mathbf{x}, \mathbf{g} \in \mathbb{R}^n). \tag{7.28}$$

In the case of maps, the evolution law is straightforward: from $\mathbf{x}(0)$ one computes $\mathbf{x}(1)$, and then $\mathbf{x}(2)$ and so on. For ODE's, under rather general assumptions on $\mathbf{f}$, from an initial condition $\mathbf{x}(0)$ one has a unique trajectory $\mathbf{x}(t)$ for $t > 0$ [Ott93]. Examples of regular behaviors (e.g., stable fixed-points, limit cycles) are well known, see Fig. 7.45.

A rather natural question is the possible existence of less regular behaviors i.e., different from stable fixed-points, periodic or quasi-periodic motion.

After the seminal works of Poincaré, Lorenz, Smale, May, and Hénon (to cite only the most eminent ones) it is now well established that the so called chaotic behavior is ubiquitous. As a relevant system, originated in the geophysical context, we mention the celebrated *Lorenz system* [Lor63, Spa82]

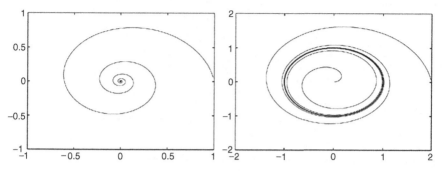

Fig. 7.45. Examples of regular attractors: fixed-point (*left*) and limit cycle (*right*). Note that limit cycles exist only in nonlinear dynamics.

$$\dot{x} = -\sigma(x - y),$$
$$\dot{y} = -xz + rx - y, \qquad (7.29)$$
$$\dot{z} = xy - bz.$$

This system is related to the *Rayleigh–Bénard convection* under very crude approximations. The quantity x is proportional the circulatory fluid particle velocity; the quantities y and z are related to the temperature profile; σ, b and r are dimensionless parameters. Lorenz studied the case with $\sigma = 10$ and $b = 8/3$ at varying r (which is proportional to the Rayleigh number). It is easy to see by linear analysis that the fixed-point $(0,0,0)$ is stable for $r < 1$. For $r > 1$ it becomes unstable and two new fixed-points appear

$$C_{+,-} = (\pm\sqrt{b(r - 1)}, \pm\sqrt{b(r - 1)}, r - 1), \qquad (7.30)$$

these are stable for $r < r_c = 24.74$. A nontrivial behavior, i.e., non periodic, is present for $r > r_c$, as is shown in Fig. 7.46.

In this 'strange', chaotic regime one has the so called sensitive dependence on initial conditions. Consider two trajectories, $\mathbf{x}(t)$ and $\mathbf{x}'(t)$, initially very close and denote with $\Delta(t) = \|\mathbf{x}'(t) - \mathbf{x}(t)\|$ their separation. Chaotic behavior means that if $\Delta(0) \to 0$, then as $t \to \infty$ one has $\Delta(t) \sim \Delta(0) \exp \lambda_1 t$, with $\lambda_1 > 0$ [BLV01].

Let us notice that, because of its chaotic behavior and its dissipative nature, i.e.,

$$\frac{\partial \dot{x}}{\partial x} + \frac{\partial \dot{y}}{\partial y} + \frac{\partial \dot{z}}{\partial z} < 0, \qquad (7.31)$$

the attractor of the Lorenz system cannot be a smooth surface. Indeed the attractor has a self-similar structure with a fractal dimension between 2 and 3. The Lorenz model (which had an important historical relevance in the development of chaos theory) is now considered a paradigmatic example of a chaotic system.

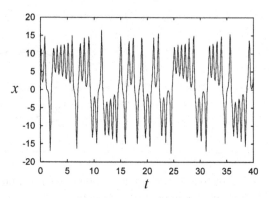

Fig. 7.46. Example of an aperiodic signal: the x variable of the Lorenz system (7.29) as function of time t, for $r = 28$.

Lyapunov Exponents

The sensitive dependence on the initial conditions can be formalized in order to give it a quantitative characterization. The main growth rate of trajectory separation is measured by the first (or maximum) *Lyapunov exponent*, defined as (see, e.g., [BLV01])

$$\lambda_1 = \lim_{t \to \infty} \lim_{\Delta(0) \to 0} \frac{1}{t} \ln \frac{\Delta(t)}{\Delta(0)}, \tag{7.32}$$

As long as $\Delta(t)$ remains sufficiently small (i.e., infinitesimal, strictly speaking), one can regard the separation as a tangent vector $\mathbf{z}(t)$ whose time evolution is

$$\dot{z}_i = \left. \frac{\partial f_i}{\partial x_j} \right|_{\mathbf{x}(t)} \cdot z_j, \tag{7.33}$$

and, therefore,

$$\lambda_1 = \lim_{t \to \infty} \frac{1}{t} \ln \frac{\|\mathbf{z}(t)\|}{\|\mathbf{z}(0)\|}. \tag{7.34}$$

In principle, λ_1 may depend on the initial condition $\mathbf{x}(0)$, but this dependence disappears for ergodic systems. In general there exist as many Lyapunov exponents, conventionally written in decreasing order $\lambda_1 \geq \lambda_2 \geq \lambda_3 \geq \cdots$, as the independent coordinates of the phase-space. Without entering the details, one can define the sum of the first k Lyapunov exponents as the growth rate of an infinitesimal kD volume in the phase-space. In particular, λ_1 is the growth rate of material lines, $\lambda_1 + \lambda_2$ is the growth rate of 2D surfaces, and so on. A numerical widely used efficient method is due to [BGG80].

It must be observed that, after a transient, the growth rate of any generic small perturbation (i.e., distance between two initially close trajectories) is measured by the first (maximum) Lyapunov exponent λ_1, and $\lambda_1 > 0$ means chaos. In such a case, the state of the system is unpredictable on long times.

Indeed, if we want to predict the state with a certain tolerance Δ then our forecast cannot be pushed over a certain time interval T_P, called *predictability time*, given by [BLV01]:

$$T_P \sim \frac{1}{\lambda_1} \ln \frac{\Delta}{\Delta(0)}. \tag{7.35}$$

The above relation shows that T_P is basically determined by $1/\lambda_1$, seen its weak dependence on the ratio $\Delta/\Delta(0)$. To be precise one must state that, for a series of reasons, relation (7.35) is too simple to be of actual relevance [BCF02].

Kolmogorov–Sinai Entropy

Deterministic chaotic systems, because of their irregular behavior, have many aspects in common with stochastic processes. The idea of using stochastic processes to mimic chaotic behavior, therefore, is rather natural [Chi79, Ben84]. One of the most relevant and successful approaches is symbolic dynamics [BS93]. For the sake of simplicity let us consider a discrete time dynamical system. One can introduce a partition $\mathcal{A}$ of the phase-space formed by N disjoint sets $A_1, \ldots, A_N$. From any initial condition one has a trajectory

$$\mathbf{x}(0) \rightarrow \mathbf{x}(1), \mathbf{x}(2), \ldots, \mathbf{x}(n), \ldots \tag{7.36}$$

dependently on the partition element visited, the trajectory (7.36), is associated to a symbolic sequence

$$\mathbf{x}(0) \rightarrow i_1, i_2, \ldots, i_n, \ldots \tag{7.37}$$

where i_n $(n = 1, 2, \ldots, N)$ means that $\mathbf{x}(n) \in A_{i_n}$ at the step n, for $n = 1, 2, \ldots$. The coarse-grained properties of chaotic trajectories are therefore studied through the discrete time process (7.37).

An important characterization of symbolic dynamics is given by the *Kolmogorov–Sinai entropy* (KS), defined as follows. Let $C_n = (i_1, i_2, \ldots, i_n)$ be a generic 'word' of size n and $P(C_n)$ its occurrence probability, the quantity [BLV01]

$$H_n = \sup_{A} \left[- \sum_{C_n} P(C_n) \ln P(C_n) \right], \tag{7.38}$$

is called *block entropy* of the n-sequences, and it is computed by taking the largest value over all possible partitions. In the limit of infinitely long sequences, the asymptotic entropy increment

$$h_{KS} = \lim_{n \to \infty} H_{n+1} - H_n, \tag{7.39}$$

is the Kolmogorov-Sinai entropy. The difference $H_{n+1} - H_n$ has the intuitive meaning of average information gain supplied by the $(n+1)$th symbol, provided that the previous n symbols are known. KS-entropy has an important connection with the positive Lyapunov exponents of the system [Ott93]:

$$h_{KS} = \sum_{\lambda_i > 0} \lambda_i. \tag{7.40}$$

In particular, for low-dimensional chaotic systems for which only one Lyapunov exponent is positive, one has $h_{KS} = \lambda_1$.

We observe that in (7.38) there is a technical difficulty, i.e., taking the sup over all the possible partitions. However, sometimes there exits a special partition, called generating partition, for which one finds that H_n coincides with its superior bound. Unfortunately the generating partition is often hard to find, even admitting that it exist. Nevertheless, given a certain partition, chosen by physical intuition, the statistical properties of the related symbol sequences can give information on the dynamical system beneath. For example, if the probability of observing a symbol (state) depends only by the knowledge of the immediately preceding symbol, the symbolic process becomes a *Markov chain* (see [II06c]) and all the statistical properties are determined by the transition matrix elements W_{ij} giving the probability of observing a transition $i \to j$ in one time step. If the memory of the system extends far beyond the time step between two consecutive symbols, and the occurrence probability of a symbol depends on k preceding steps, the process is called *Markov process* of order k and, in principle, a k rank tensor would be required to describe the dynamical system with good accuracy. It is possible to demonstrate that if $H_{n+1} - H_n = h_{KS}$ for $n \geq k + 1$, k is the (minimum) order of the required Markov process [Khi57]. It has to be pointed out, however, that to know the order of the suitable Markov process we need is of no practical utility if $k \gg 1$.

Periodic Orbit Theory

Confronted with a potentially chaotic dynamical system, we analyze it through a sequence of three distinct stages: (i) diagnose, (ii) count, (iii) measure. First we determine the intrinsic dimension of the system—the minimum number of coordinates necessary to capture its essential dynamics. If the system is very turbulent we are, at present, out of luck. We know only how to deal with the transitional regime between regular motions and chaotic dynamics in a few dimensions. That is still something; even an infinite-dimensional system such as a burning flame front can turn out to have a very few chaotic degrees of freedom. In this regime the chaotic dynamics is restricted to a space of low dimension, the number of relevant parameters is small, and we can proceed to step (ii); we count and classify all possible topologically distinct trajectories of the system into a hierarchy whose successive layers require increased precision and patience on the part of the observer. If successful, we can proceed with step (iii): investigate the weights of the different pieces of the system [CAM05].

With the game of pinball we are lucky: it is only a 2D system, free motion in a plane. The motion of a point particle is such that after a collision with one disk it either continues to another disk or it escapes. If we label the three disks by 1, 2 and 3, we can associate every trajectory with an itinerary, a sequence of labels indicating the order in which the disks are visited; for example, the two trajectories in Fig. 1.2 have itineraries 2313, 23132321 respectively. The itinerary is finite for a scattering trajectory, coming in from infinity and escaping after a finite number of collisions, infinite for a trapped trajectory, and infinitely repeating for a periodic orbit.[46] Such labeling is the simplest example of *symbolic dynamics*. As the particle cannot collide two times in succession with the same disk, any two consecutive symbols must differ. This is an example of *pruning*, a rule that forbids certain subsequences of symbols. Deriving pruning rules is in general a difficult problem, but with the game of pinball we are lucky, as there are no further pruning rules.[47]

Suppose you wanted to play a good game of pinball, that is, get the pinball to bounce as many times as you possibly can—what would be a winning strategy? The simplest thing would be to try to aim the pinball so it bounces many times between a pair of disks—if you managed to shoot it so it starts out in the periodic orbit bouncing along the line connecting two disk centers, it would stay there forever. Your game would be just as good if you managed to get it to keep bouncing between the three disks forever, or place it on any periodic orbit. The only rub is that any such orbit is unstable, so you have to aim very accurately in order to stay close to it for a while. So it is pretty clear that if one is interested in playing well, unstable periodic orbits are important—they form the skeleton onto which all trajectories trapped for long times cling.

Now, recall that a trajectory is *periodic* if it returns to its starting position and momentum. It is custom to refer to the set of periodic points that belong to a given periodic orbit as a *cycle*.

Short periodic orbits are easily drawn and enumerated, but it is rather hard to perceive the systematics of orbits from their shapes. In mechanics a trajectory is fully and uniquely specified by its position and momentum at a given instant, and no two distinct phase-space trajectories can intersect. Their projections on arbitrary subspaces, however, can and do intersect, in rather unilluminating ways. In the pinball example, the problem is that we are looking at the projections of a 4D phase-space trajectories onto its 2D subspace, the configuration space. A clearer picture of the dynamics is obtained by constructing a phase-space Poincaré section.

---

[46] The words *orbit* and *trajectory* here are synonymous.

[47] The choice of symbols is in no sense unique. For example, as at each bounce we can either proceed to the next disk or return to the previous disk, the above 3-letter alphabet can be replaced by a binary $\{0, 1\}$ alphabet. A clever choice of an alphabet will incorporate important features of the dynamics, such as its symmetries.

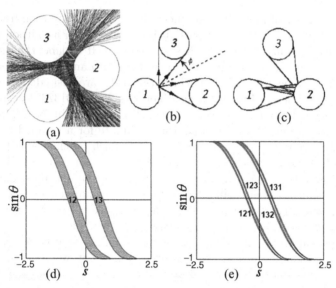

Fig. 7.47. A 3-disk pinball game. *Up*: (**a**) Elastic scattering around three hard disks (simulated in *Dynamics Solver^{TM}*); (**b**) A trajectory starting out from disk 1 can either hit another disk or escape; (**c**) Hitting two disks in a sequence requires a much sharper aim; the cones of initial conditions that hit more and more consecutive disks are nested within each other. *Down*: Poincaré section for the 3-disk pinball game, with trajectories emanating from the disk 1 with $x_0 = (arc\text{-}length, parallel\ momentum) = (s_0, p_0)$, disk radius: center separation ratio $a : R = 1 : 2.5$; (**d**) Strips of initial points M_{12}, M_{13} which reach disks 2, 3 in one bounce, respectively. (**e**) Strips of initial points M121, M131 M132 and M123 which reach disks 1, 2, 3 in two bounces, respectively; the Poincaré sections for trajectories originating on the other two disks are obtained by the appropriate relabeling of the strips (see text for explanation).

The position of the ball is described by a pair of numbers (the spatial coordinates on the plane), and the angle of its velocity vector. As far as a classical dynamist is concerned, this is a complete description. Now, suppose that the pinball has just bounced off disk 1. Depending on its position and outgoing angle, it could proceed to either disk 2 or 3. Not much happens in between the bounces—the ball just travels at constant velocity along a straight line—so we can reduce the 4D flow to a 2D map f that takes the coordinates of the pinball from one disk edge to another disk edge. Let us state this more precisely: the trajectory just after the moment of impact is defined by marking s_n, the arc-length position of the nth bounce along the billiard wall, and $p_n = p \sin \phi_n$ is the momentum component parallel to the billiard wall at the point of impact (see Fig. 7.47). Such a section of a flow is called a *Poincaré section*, and the particular choice of coordinates (due to Birkhoff) is particularly smart, as it conserves the phase-space volume. In terms of the Poincaré section, the dynamics is reduced to the *return map*

$$P : (s_n, p_n) \rightarrow (s_{n+1}, p_{n+1}),$$

from the boundary of a disk to the boundary of the next disk.

Next, we mark in the Poincaré section those initial conditions which do not escape in one bounce. There are two strips of survivors, as the trajectories originating from one disk can hit either of the other two disks, or escape without further ado. We label the two strips M_0, M_1. Embedded within them there are four strips, $M_{00}, M_{10}, M_{01}, M_{11}$ of initial conditions that survive for two bounces, and so forth (see Fig. 7.47). Provided that the disks are sufficiently separated, after n bounces the survivors are divided into 2^n distinct strips: the M_ith strip consists of all points with itinerary $i = s_1 s_2 s_3 \ldots s_n$, $s = \{0, 1\}$. The unstable cycles as a skeleton of chaos are almost visible here: each such patch contains a periodic point $\overline{s_1 s_2 s_3 \ldots s_n}$ with the basic block infinitely repeated. Periodic points are skeletal in the sense that as we look further and further, the strips shrink but the periodic points stay put forever.

We see now why it pays to utilize a symbolic dynamics; it provides a navigation chart through chaotic phase-space. There exists a unique trajectory for every admissible infinite length itinerary, and a unique itinerary labels every trapped trajectory. For example, the only trajectory labeled by 12 is the 2-cycle bouncing along the line connecting the centers of disks 1 and 2; any other trajectory starting out as 12... either eventually escapes or hits the 3rd disk [CAM05].

Now we can ask what is a good physical quantity to compute for the game of pinball? Such system, for which almost any trajectory eventually leaves a finite region (the pinball table) never to return, is said to be open, or a *repeller*. The repeller escape rate is an eminently measurable quantity. An example of such a measurement would be an unstable molecular or nuclear state which can be well approximated by a classical potential with the possibility of escape in certain directions. In an experiment many projectiles are injected into such a non-confining potential and their mean escape rate is measured. The numerical experiment might consist of injecting the pinball between the disks in some random direction and asking how many times the pinball bounces on the average before it escapes the region between the disks. On the other hand, for a theorist a good game of pinball consists in predicting accurately the asymptotic lifetime (or the escape rate) of the pinball.

Here we briefly show how Cvitanovic's *periodic orbit theory* [Cvi91] accomplishes this for us. Each step will be so simple that you can follow even at the cursory pace of this overview, and still the result is surprisingly elegant. Let us consider Fig. 7.47 again. In each bounce, the initial conditions get thinned out, yielding twice as many thin strips as at the previous bounce. The total area that remains at a given time is the sum of the areas of the strips, so that the fraction of survivors after n bounces, or the *survival probability* is given by

$$\hat{\Gamma}_1 = \frac{|M_0|}{|M|} + \frac{|M_1|}{|M|},$$

$$\hat{\Gamma}_2 = \frac{|M_{00}|}{|M|} + \frac{|M_{10}|}{|M|} + \frac{|M_{01}|}{|M|} + \frac{|M_{11}|}{|M|},$$

$$\vdots$$

$$\hat{\Gamma}_n = \frac{1}{|M|} \sum_{i=1}^{(n)} |M_i|,$$

(7.41)

where $i = 01, 10, 11, \ldots$ is a label of the ith strip (not a binary number), $|M|$ is the initial area, and $|M_i|$ is the area of the ith strip of survivors. Since at each bounce one routinely loses about the same fraction of trajectories, one expects the sum (7.41) to fall off exponentially with n and tend to the limit

$$\Gamma_{n+1}/\hat{\Gamma}_n = e^{-\gamma n} \to e^{-\gamma},$$

where the quantity γ is called the *escape rate* from the repeller. In [Cvi91] and subsequent papers, Cvitanovic has showed that the escape rate γ can be extracted from a highly convergent exact expansion by reformulating the sum (7.41) in terms of *unstable periodic orbits*.

Some Classical Attractor and Chaotic Systems

Here we present numerical simulations of several popular chaotic systems (see, e.g., [Wig90, BCB92, Ach97]). Generally, to observe chaos in continuous time system, it is known that the dimension of the equation must be three or higher. That is, *there is no chaos in any phase plane* (see [Str94]), we need the third dimension for chaos in continuous dynamics. However, note that *all forced oscillators have actually dimension 3, although they are commonly written as second-order ODEs.*[48] On the other hand, in discrete-time systems like logistic map or Hénon map, we can see chaos even if the dimension is one.

Simple Pendulum

Recall (see [II06a, II06b, II06c]) that a simple *un-damped pendulum* (see Fig. 7.48), given by equation

$$\ddot{\theta} + \frac{g}{l} \sin \theta = 0,$$

(7.42)

swings forever; it has closed orbits in a 2D phase-space (see Fig. 7.49).

---

[48] Both Newtonian equation of motion and RLC circuit can generate chaos, provided they have a forcing term. This forcing (driving) term in second-order ODEs is the motivational reason for development of the jet-bundle formalism for non-autonomous dynamics (see [II06c]).

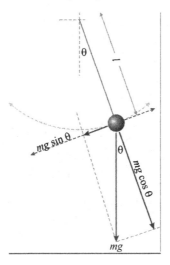

Fig. 7.48. Force diagram of a simple gravity pendulum.

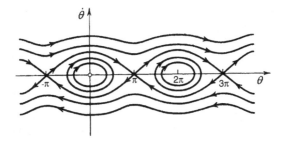

Fig. 7.49. Phase portrait of a simple gravity pendulum.

The conservative (un-damped) pendulum equation (7.42) does not take into account the effects of friction and dissipation. On the other hand, a simple *damped pendulum* (see Fig. 7.48) is given by modified equation, including a damping term proportional to the velocity,

$$\ddot{\theta} + \gamma\dot{\theta} + \frac{g}{l}\sin\theta = 0,$$

with the positive constant damping γ. This pendulum settles to rest (see Fig. 7.50). Its spiraling orbits lead to a point attractor (focus) in a 2D phase-space. All closed trajectories for periodic solutions are destroyed, and the trajectories spiral around one of the critical points, corresponding to the vertical equilibrium of the pendulum. On the phase plane, these critical points are stable spiral points for the underdamped pendulum, and they are stable nodes for the overdamped pendulum. The unstable equilibrium at the inverted vertical position remains an unstable saddle point. It is clear physically that damping means loss of energy. The dynamical motion of the pendulum decays due to the friction and the pendulum relaxes to the equilibrium state in the vertical position.

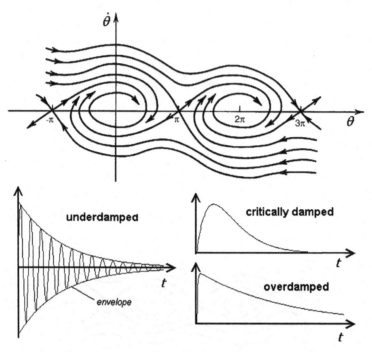

Fig. 7.50. A damped gravity pendulum settles to a rest: its phase portrait (*up*) shows spiraling orbits that lead to a focus attractor; its time plot (*down*) shows three common damping cases.

Finally, a *driven pendulum*, periodically forced by a force term $F\cos(w_Dt)$, is given by equation

$$\ddot{\theta} + \gamma\dot{\theta} + \frac{g}{l}\sin\theta = F\cos(w_Dt). \tag{7.43}$$

It has a 3D phase-space and can exhibit chaos (for certain values of its parameters, see Fig. 7.51).

Van der Pol Oscillator

The unforced Van der Pol oscillator has the form of a second order ODE (compare with (7.18) above)

$$\ddot{x} = \alpha(1 - x^2)\dot{x} - \omega^2 x. \tag{7.44}$$

Its celebrated *limit cycle* is given in Fig. 7.52. The simulation is performed with zero initial conditions and parameters $\alpha = \text{random}(0,3)$, and $\omega = 1$.

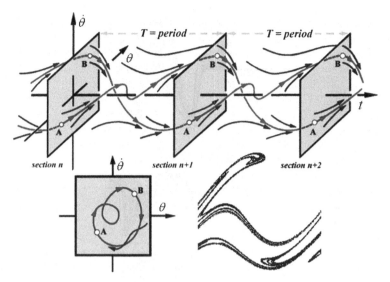

Fig. 7.51. A driven pendulum has a 3D phase-space with angle θ, angular velocity $\dot{\theta}$ and time t. *Dashed lines* denote steady states, while *solid lines* denote transients. *Right-down* we see a sample chaotic attractor (see text for explanation).

The Van der Pol oscillator was the first *relaxation oscillator*, used in 1928 as a model of human heartbeat (ω controls how much voltage is injected into the system, and α controls the way in which voltage flows through the system). The oscillator was also used as a model of an electronic circuit that appeared in very early radios in the days of vacuum tubes. The tube acts like a normal resistor when current is high, but acts like a negative resistor if the current is low. So this circuit pumps up small oscillations, but drags down large oscillations. α is a constant that affects how nonlinear the system is. For α equal to zero, the system is actually just a linear oscillator. As α grows the nonlinearity of the system becomes considerable.

The *sinusoidally-forced Van der Pol oscillator* is given by equation

$$\ddot{x} - \alpha(1 - x^2)\dot{x} + \omega^2 x = \gamma\cos(\phi t), \qquad (7.45)$$

where ϕ is the forcing frequency and γ is the amplitude of the forcing sinusoid.

Nerve Impulse Propagation

The nerve impulse propagation along the axon of a neuron can be studied by combining the equations for an excitable membrane with the differential equations for an electrical core conductor cable, assuming the axon to be an infinitely long cylinder. A well known approximation of FitzHugh [Fit61a] and Nagumo [NAY60] to describe the propagation of voltage pulses $V(x,t)$ along the membranes of nerve cells is the set of coupled PDEs

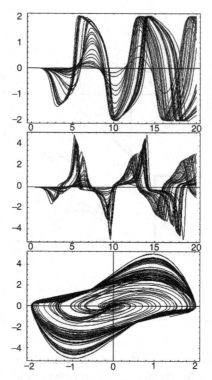

Fig. 7.52. Cascade of 30 unforced Van der Pol oscillators, simulated using $Mathematica^{TM}$; *top-down*: displacements, velocities and phase-plot (showing the celebrated limit cycle).

$$V_{xx} - V_t = F(V) + R - I, \qquad R_t = c(V + a - bR), \qquad (7.46)$$

where $R(x,t)$ is the recovery variable, I the external stimulus and a, b, c are related to the membrane radius, specific resistivity of the fluid inside the membrane and temperature factor respectively.

When the spatial variation of V, namely V_{xx}, is negligible, (7.46) reduces to the Van der Pol oscillator,

$$\dot{V} = V - \frac{V^3}{3} - R + I, \qquad \dot{R} = c(V + a - bR),$$

with $F(V) = -V + \frac{V^3}{3}$. Normally the constants in (7.46) satisfy the inequalities $b < 1$ and $3a + 2b > 3$, though from a purely mathematical point of view this need not be insisted upon. Then with a periodic (ac) applied membrane current $A_1 \cos \omega t$ and a (dc) bias A_0, the Van der Pol equation becomes

$$\dot{V} = V - \frac{V^3}{3} - R + A_0 + A_1 \cos \omega t, \qquad \dot{R} = c(V + a - bR). \qquad (7.47)$$

Further, (7.47) can be rewritten as a single second-order ODE by differentiating $\dot{V}$ with respect to time and using $\dot{R}$ for R,

$$\ddot{V} - (1-bc)\left\{1 - \frac{V^2}{1-bc}\right\}\dot{V} - c(b-1)V + \frac{bc}{3}V^3$$
$$= c(A_0 b - a) + A_1 \cos(\omega t + \phi), \qquad (7.48)$$

where $\phi = \tan^{-1}\frac{\omega}{bc}$. Using the transformation $x = (1-bc)^{-(1/2)}V$, $t \longrightarrow t' = t + \frac{\phi}{\omega}$, (7.48) can be rewritten as

$$\ddot{x} + p(x^2 - 1)\dot{x} + \omega_0^2 x + \beta x^3 = f_0 + f_1 \cos\omega t, \qquad (7.49)$$

where

$$p = (1-bc), \qquad \omega_0^2 = c(1-b), \qquad \beta = bc\frac{(1-bc)}{3},$$

$$f_0 = c\frac{(A_0 b - a)}{\sqrt{1-bc}}, \qquad f_1 = \frac{A_1}{\sqrt{1-bc}}.$$

Note that (7.49), or its rescaled form

$$\ddot{x} + p(kx^2 + g)\dot{x} + \omega_0^2 x + \beta x^3 = f_0 + f_1 \cos\omega t, \qquad (7.50)$$

is the *Duffing–Van der Pol equation*. In the limit $k = 0$, we have the Duffing equation discussed below (with $f_0 = 0$), and in the case $\beta = 0$ ($g = -1$, $k = 1$) we have the forced van der Pol equation. Equation (7.50) exhibits a very rich variety of bifurcations and chaos phenomena, including quasi-periodicity, phase lockings and so on, depending on whether the potential $V = \frac{1}{2}\omega_0^2 x^2 + \frac{\beta x^4}{4}$ is (i) a double well, (ii) a single well or (iii) a double hump [Lak97, Lak03].

Duffing Oscillator

The forced *Duffing oscillator* [Duf18] has the form similar to (7.45),

$$\ddot{x} + b\dot{x} - ax(1 - x^2) = \gamma\cos(\phi t). \qquad (7.51)$$

Stroboscopic *Poincaré sections* of a *strange attractor* can be seen (Fig. 7.53), with the *stretch-and-fold* action at work. The simulation is performed with parameters: $a = 1$, $b = 0.2$, and $\gamma = 0.3$, $\phi = 1$. The Duffing equation is used to model a double well oscillator such as the magneto-elastic mechanical system. This system consists of a beam positioned vertically between two magnets, with the top end fixed, and the bottom end free to swing. The beam will be attracted to one of the two magnets, and given some velocity will oscillate about that magnet until friction stops it. Each of the magnets creates a fixed-point where the beam may come to rest above that magnet and remain there in equilibrium. However, when this whole system is shaken by a periodic forcing term, the beam may jump back and forth from one magnet to the other in a seemingly random manner. Depending on how big the shaking term is, there may be no stable fixed-points and no stable fixed cycles in the system.

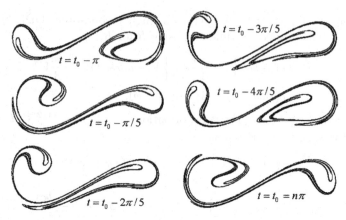

Fig. 7.53. Duffing strange attractor, showing stroboscopic Poincaré sections; simulated using *Dynamics Solver*[TM].

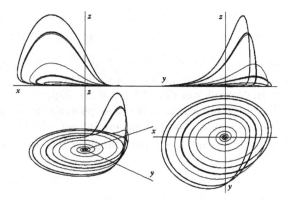

Fig. 7.54. The celebrated Rossler attractor, simulated using *Dynamics Solver*[TM].

Rossler System

Classical *Rossler system* is given by equations

$$\dot{x} = -y - z, \qquad \dot{y} = x + by, \qquad \dot{z} = b + z(x - a). \qquad (7.52)$$

Using the parameter values $a = 4$ and $b = 0.2$, the phase-portrait is produced (see Fig. 7.54), showing the celebrated attractor. The system is credited to O. *Rossler* and arose from work in chemical kinetics.

Chua's Circuit

Chua's circuit is a simple electronic circuit that exhibits classic chaotic behavior. First introduced in 1983 by Leon O. Chua, its ease of construction has made it an ubiquitous real-world example of a chaotic system, leading some to declare it 'a paradigm for chaos'. It has been the subject of much study; hundreds of papers have been published on this topic (see [Chu94]).

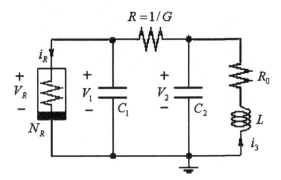

Fig. 7.55. Chua's circuit.

The *Chua's circuit* consists of two linear capacitors, two linear resistors, one linear inductor and a nonlinear resistor (see Fig. 7.55). By varying the various circuit parameters, we can get complicated nonlinear and chaotic phenomena. Let us consider the case where we vary the conductance G of the resistor R and keep the other components fixed. In particular, we choose $L = 18$ mH, $R_0 = 12.5$ Ω, $C_1 = 10$ nF, $C_2 = 100$ nF. The nonlinear resistor N_R (Chua's diode) is chosen to have a piecewise-linear V–I characteristic of the form:

$$i = - \begin{cases} G_b v + G_a - G_b & \text{if } v > 1, \\ G_a v & \text{if } |v| < 1, \\ G_b v + G_b - G_a & \text{if } v < -1, \end{cases}$$

with $G_a = -0.75757$ mS, and $G_b = -0.40909$ mS.

Starting from low G-values, the circuit is stable and all trajectories converge towards one of the two stable equilibrium points. As G is increased, a limit cycle appears due to a *Hopf-like bifurcation*. In order to observe the period-doubling route to chaos, we need to further increase G. At the end of the period-doubling bifurcations, we observe a chaotic attractor. Because of symmetry, there exists a twin attractor lying in symmetrical position with respect the origin. As G is further increased, these two chaotic attractors collide and form a 'double scroll' chaotic attractor.

After normalization, the state equations for the Chua's circuit read:

$$\dot{x} = a(y - x - f(x)), \qquad \dot{y} = x - y + z, \qquad \dot{z} = -by - cz, \qquad (7.53)$$

where $f(x)$ is a nonlinear function to be manipulated to give various chaotic behaviors.

By using a specific form of the nonlinearity $f(x)$, a family of *multi-spiral strange attractors* have been generated in [Ala99] (see Fig. 7.56).

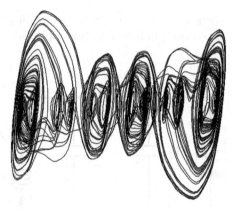

Fig. 7.56. A multi-spiral strange attractor of the Chua's circuit (see text for explanation).

Inverted Pendulum

Stability of the *inverted driven pendulum* given by equation

$$\ddot{\theta} + k\dot{\theta} + (1 + a\sqrt{\phi}\cos(\phi t))\sin\theta = 0,$$

where θ is the angle, is simulated in Fig. 7.57, using the parameter $a = 0.33$. It is possible to stabilize a mathematical pendulum around the upper vertical position by moving sinusoidally the suspension point in the vertical direction. Furthermore, the perturbed solution may be of two kinds: one goes to the vertical position while the other becomes periodic (see, e.g., [Ach97]).

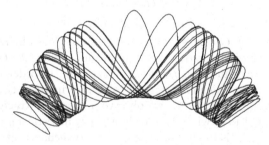

Fig. 7.57. Duffing strange attractor, showing stroboscopic Poincaré sections; simulated using *Dynamics Solver*^{TM}.

Elastic Pendulum

Elastic pendulum (Fig. 7.58) of proper length l, mass m and elastic constant k is given by equation

Fig. 7.58. Phase-portrait of an elastic pendulum showing *Lissajous curves*; simulated using *Dynamics Solver*^{TM}.

$$\ddot{x} = x\sqrt{\dot{y}} + \cos y - a(x-1), \qquad \ddot{y} = -(2\dot{x}\dot{y} + \sin y)/x,$$

where the parameter $a = kl/mg = 0.4$. High values of a give raise to a simple pendulum.

Lorenz–Maxwell–Haken System

In 1975, H. Haken showed [Hak83] that the *Lorenz equations* (7.37) were isomorphic to the *Maxwell–Haken laser equations*

$$\dot{E} = \sigma(P - E), \qquad \dot{P} = \beta(ED - P), \qquad \dot{D} = \gamma(\sigma - 1 - D - \sigma EP).$$

Here, the variables in the Lorenz equations, namely x,y and z correspond to the slowly varying amplitudes of the electric field E and polarization P and the inversion D respectively in the Maxwell–Haken equations. The parameters are related via $c = \frac{\gamma}{\beta}$, $a = \frac{\sigma}{\beta}$ and $b = \sigma + 1$, where γ is the relaxation rate of the inversion, β is the relaxation rate of the polarization, σ is the field relaxation rate, and σ represents the normalized pump power.

Autocatalator System

This 4D *autocatalator* system from *chemical kinetics* (see Fig. 7.59) is defined as (see, e.g., [BCB92])

$$\dot{x}_1 = -ax_1, \qquad \dot{x}_2 = ax_1 - bx_2 - x_2x_3^2,$$

$$\dot{x}_3 = bx_2 - x_3 + x_2x_3^2, \qquad \dot{x}_4 = x_3.$$

The simulation is performed with parameters: $a = 0.002$, and $b = 0.08$.

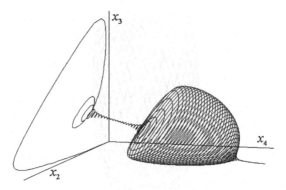

Fig. 7.59. 3D phase-portrait of the 4D autocatalator system, simulated using *Dynamics Solver*^{TM}.

Mandelbrot and Julia Sets

Recall that *Mandelbrot and Julia sets* (see Fig. 7.60) are celebrated *fractals*. Recall that fractals are sets with *fractional dimension* (see Fig. 7.61). The Mandelbrot and Julia sets are defined either by a quadratic *conformal z-map* [Man80a, Man80b]

$$z_{n+1} = z_n^2 + c,$$

or by a real (x, y)-map

$$x_{n+1} = \sqrt{x_n} - \sqrt{y_n} + c_1, \qquad y_{n+1} = 2x_n y_n + c_2,$$

where c, c_1 and c_2 are parameters. For almost every c, this conformal transformation generates a fractal (probably, only for $c = -2$ it is not a fractal). Julia set J_c with $c \ll 1$, the *capacity dimension* is

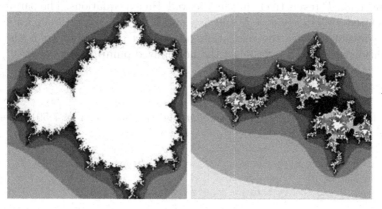

Fig. 7.60. The celebrated conformal Mandelbrot (*left*) and Julia (*right*) sets in the complex plane, simulated using *Dynamics Solver*^{TM}.

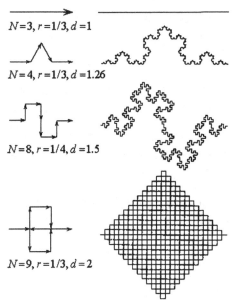

$N=3, r=1/3, d=1$

$N=4, r=1/3, d=1.26$

$N=8, r=1/4, d=1.5$

$N=9, r=1/3, d=2$

Fig. 7.61. Fractal dimension of curves in $\mathbb{R}^2$: $d = \frac{\log N}{\log 1/r}$.

$$d_{cap} = 1 + \frac{|c|^2}{4\ln 2} + O(|c|^3).$$

The set of all points for which J_c is connected is the Mandelbrot set.[49]

Biomorphic Systems

Closely related to the Mandelbrot and Julia sets are *biomorphic systems*, which look like one-celled organisms. The term '*biomorph*' was proposed by C. Pickover from IBM [Pic86, Pic87]. Pickover's biomorphs inhabit the complex plane like the Mandelbrot and Julia sets and exhibit a *protozoan morphology*. Biomorphs began for Pickover as a 'bug' in a program intended to probe the fractal properties of various formulas. He accidentally used an OR logical operator instead of an AND operator in the conditional test for the size of z's real and imaginary parts. The cilia that project from the biomorphs are a

---

[49] The Mandelbrot set has its place in complex-valued dynamics, a field first investigated by the French mathematicians Pierre Fatou [STZ93, SZT98] and Gaston Julia [MN95b] at the beginning of the 20th century. For general families of holomorphic functions, the boundary of the Mandelbrot set generalizes to the bifurcation locus, which is a natural object to study even when the connectedness locus is not useful. A related *Mandelbar set* was encountered by mathematician John Milnor in his study of parameter slices of real cubic polynomials; it is not locally connected; this property is inherited by the connectedness locus of real cubic polynomials.

consequence of this 'error'. Each biomorph is generated by multiple iterations of a particular conformal map,

$$z_{n+1} = f(z_n, c),$$

where c is a parameter. Each iteration takes the output of the previous operations as the input of the next iteration. To generate a biomorph, one first needs to lay out a grid of points on a rectangle in the complex plane [And01]. The coordinate of each point constitutes the real and imaginary parts of an initial value, z_0, for the iterative process. Each point is also assigned a pixel on the computer screen. Depending on the outcome of a simple test on the 'size' of the real and imaginary parts of the final value, the pixel is colored either black or white. The biomorphs presented in Fig. 7.62 are generated using the following conformal functions:

1. $f(z, c) = z^3$,
2. $f(z, c) = z^3 + c, c = 10$,
3. $f(z, c) = z^3 + c, c = 10 - 10i$,
4. $f(z, c) = z^5 + c, c = 0.77 - 0.77i$,
5. $f(z, c) = z^3 + \sin z + c, c = 1 - i$,
6. $f(z, c) = z^6 + \sin z + c, c = 0.5 - 0.5i$,
7. $f(z, c) = z^2 \sin z + c, c = 0.78 - 0.78i$,
8. $f(z, c) = z^c, c = 5 - i$,
9. $f(z, c) = |z|^c \sin z, c = 4$,
10. $f(z, c) = |z|^c \cos z + c, c = 3 + 3i$,
11. $f(z, c) = |z|^c(\cos z + z) + c, c = 3 + 2i$.

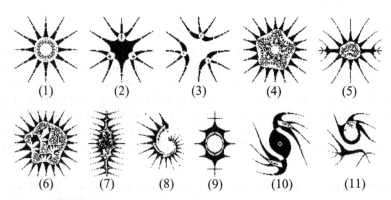

Fig. 7.62. *Pickover's biomorphs* (see text for details).

Basics of Chaos Control

Feedback and Non-Feedback Algorithms

Although the presence of chaotic behavior is generic and robust for suitable nonlinearities, ranges of parameters and external forces, there are practical situations where one wishes to avoid or control chaos so as to improve the performance of the dynamical system. Also, although chaos is sometimes useful as in a mixing process or in heat transfer, it is often unwanted or undesirable. For example, increased drag in flow systems, erratic fibrillations of heart beats, extreme weather patterns and complicated circuit oscillations are situations where chaos is harmful. Clearly, the ability to control chaos, that is to convert chaotic oscillations into desired regular ones with a periodic time dependence would be beneficial in working with a particular system. The possibility of purposeful selection and stabilization of particular orbits in a normally chaotic system, using minimal, predetermined efforts, provides a unique opportunity to maximize the output of a dynamical system. It is thus of great practical importance to develop suitable control methods and to analyze their efficacy.

Let us consider a general nD nonlinear dynamical system,

$$\dot{x} = F(x, p, t), \tag{7.54}$$

where $x = (x_1, x_2, x_3, \ldots, x_n)$ represents the n state variables and p is a control or external parameter. Let $x(t)$ be a chaotic solution of (7.54). Different control algorithms are essentially based on the fact that one would like to effect the most minimal changes to the original system so that it will not be grossly deformed. From this point of view, controlling methods or algorithms can be broadly classified into two categories:

(i) feedback methods, and
(ii) non-feedback algorithms.

Feedback methods essentially make use of the intrinsic properties of chaotic systems, including their sensitivity to initial conditions, to stabilize orbits already existing in the systems. Some of the prominent methods are the following (see, [Lak97, Lak03, Sch88, II06c]):

1. Adaptive control algorithm;
2. Nonlinear control algorithm;
3. Ott–Grebogi–Yorke (OGY) method of stabilizing unstable periodic orbits;
4. Singer's method of stabilizing unstable periodic orbits; and
5. Various control engineering approaches.

In contrast to feedback control techniques, non-feedback methods make use of a small perturbing external force such as a small driving force, a small noise term, a small constant bias or a weak modulation to some system parameter.

These methods modify the underlying chaotic dynamical system weakly so that stable solutions appear. Some of the important controlling methods of this type are the following.

1. Parametric perturbation method,
2. Addition of a weak periodic signal, constant bias or noise,
3. Entrainment-open loop control, and
4. Oscillator absorber method.

Here is a typical example of adaptive control algorithm. We can control the chaotic orbit $X_s = (x_s, y_s)$ of the *Van der Pol oscillator* (7.49) by introducing the following dynamics on the parameter A_1:

$$\dot{x} = x - \frac{x^3}{3} - y + A_0 + A_1 \cos \omega t, \qquad \dot{y} = c(x + a - by),$$
$$\dot{A}_1 = -\epsilon[(x - x_s) - (y - y_s)], \quad \epsilon \ll 1.$$

On the other hand, recall from [II06c] that a generic SISO nonlinear system

$$\dot{x} = f(x) + g(x)u, \qquad y = h(x) \tag{7.55}$$

is said to have *relative degree r* at a point x^o if

(i) $L_g L_f^k h(x) = 0$ for all x in a neighborhood of x^o and all $k < r - 1$
(ii) $L_g L_f^{r-1} h(x^o) \neq 0$, where L_g denotes the *Lie derivative* in the direction of the vector-field g.

Now, the Van der Pol oscillator (7.44) has the state space form

$$\dot{x} = f(x) + g(x)u = \begin{bmatrix} x_2 \\ 2\omega\zeta(1 - \mu x_1^2)x_2 - \omega^2 x_1 \end{bmatrix} + \begin{bmatrix} 0 \\ 1 \end{bmatrix} u. \tag{7.56}$$

Suppose the output function is chosen as

$$y = h(x) = x_1. \tag{7.57}$$

In this case we have

$$L_g h(x) = \frac{\partial h}{\partial x} g(x) = \begin{bmatrix} 1 & 0 \end{bmatrix} \begin{bmatrix} 0 \\ 1 \end{bmatrix} = 0, \tag{7.58}$$

and

$$L_f h(x) = \frac{\partial h}{\partial x} f(x) = \begin{bmatrix} 1 & 0 \end{bmatrix} \begin{bmatrix} x_2 \\ 2\omega\zeta(1 - \mu x_1^2)x_2 - \omega^2 x_1 \end{bmatrix} = x_2. \tag{7.59}$$

Moreover

$$L_g L_f h(x) = \frac{\partial(L_f h)}{\partial x} g(x) = \begin{bmatrix} 0 & 1 \end{bmatrix} \begin{bmatrix} 0 \\ 1 \end{bmatrix} = 1 \tag{7.60}$$

and thus we see that the Van der Pol oscillator system has relative degree 2 at any point x^o.

However, if the output function is, for instance

$$y = h(x) = \sin x_2 \tag{7.61}$$

then $L_g h(x) = \cos x_2$. The system has relative degree 1 at any point x^o, provided that $(x^o)_2 \neq (2k+1)\pi/2$. If the point x^o is such that this condition is violated, no relative degree can be defined.

Both adaptive and nonlinear control methods can be naturally extended to other chaotic systems, e.g., *Lorenz attractor* (see Fig. 7.63).

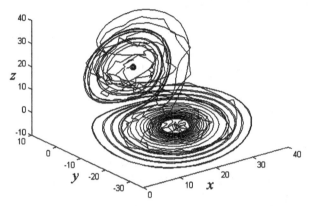

Fig. 7.63. Nonlinear control of the Lorenz system: targeting of unstable upper and lower states in the Lorenz attractor after applying random perturbations, using a MIMO nonlinear controller (see text for explanation).

Hybrid Systems and Homotopy ODEs

Consider a *hybrid dynamical system of variable structure*, given by an nD ODE-system (see [MWH01])

$$\dot{x} = f(t, x), \tag{7.62}$$

where $x = x(t) \in \mathbb{R}^n$ and $f = f(t, x) : \mathbb{R}^+ \times \mathbb{R}^n \to \mathbb{R}^n$. Let the domain $G \subset \mathbb{R}^+ \times \mathbb{R}^n$, on which the vector-field $f(t, x)$ is defined, be divided into two subdomains, G^+ and G^-, by means of a smooth $(n-1)$-manifold M. In $G^+ \cup M$, let there be given a vector-field $f^+(t, x)$, and in $G^- \cup M$, let there be given a vector-field $f^-(t, x)$. Assume that both $f^+ = f^+(t, x)$ and $f^- = f^-(t, x)$ are continuous in t and smooth in x. For the system (7.62), let

$$f = \begin{cases} f^+ & \text{when } x \in G^+, \\ f^- & \text{when } x \in G^-. \end{cases}$$

Under these conditions, a solution $x(t)$ of ODE (7.62) is well-defined while passing through G until the manifold M is reached.

Upon reaching the manifold M, in physical systems with inertia, the transition

$$\text{from } \dot{x} = f^-(t, x) \quad \text{to } \dot{x} = f^+(t, x)$$

does not take place instantly on reaching M, but after some delay. Due to this delay, the solution $x(t)$ oscillates about M, $x(t)$ being displaced along M with some mean velocity.

As the delay tends to zero, the limiting motion and velocity along M are determined by the *linear homotopy ODE*

$$\dot{x} = f^0(t, x) \equiv (1 - \alpha)f^-(t, x) + \alpha f^+(t, x), \tag{7.63}$$

where $x \in M$ and $\alpha \in [0, 1]$ is such that the *linear homotopy segment* $f^0(t, x)$ is tangential to M at the point x, i.e., $f^0(t, x) \in T_x M$, where $T_x M$ is the tangent space to the manifold M at the point x.

The vector-field $f^0(t, x)$ of the system (7.63) can be constructed as follows: at the point $x \in M$, $f^-(t, x)$ and $f^+(t, x)$ are given and their ends are joined by the linear homotopy segment. The point of intersection between this segment and $T_x M$ is the end of the required vector-field $f^0(t, x)$. The vector function $x(t)$ which satisfies (7.62) in G^- and G^+, and (7.63) when $x \in M$, can be considered as a *solution* of (7.62) in a *general sense*.

However, there are cases in which the solution $x(t)$ cannot consist of a finite or even countable number of arcs, each of which passes through G^- or G^+ satisfying (7.62), or moves along the manifold M and satisfies the homotopic ODE (7.63). To cover such cases, assume that the vector-field $f = f(t, x)$ in ODE (7.62) is a Lebesgue-measurable function in a domain $G \subset \mathbb{R}^+ \times \mathbb{R}^n$, and that for any closed bounded domain $D \subset G$ there exists a summable function $K(t)$ such that almost everywhere in D we have $|f(t, x)| \leq K(t)$. Then the absolutely continuous vector function $x(t)$ is called the *generalized solution* of the ODE (7.62) *in the sense of Filippov* (see [MWH01]) if for almost all t, the vector $\dot{x} = \dot{x}(t)$ belongs to the least convex closed set containing all the limiting values of the vector-field $f(t, x^*)$, where x^* tends towards x in an arbitrary manner, and the values of the function $f(t, x^*)$ on a set of measure zero in $\mathbb{R}^n$ are ignored.

Such *hybrid systems* of variable structure occur in the study of nonlinear electric networks (endowed with electronic switches, relays, diodes, rectifiers, etc.), in models of both natural and artificial neural networks, as well as in feedback control systems (usually with continuous-time plants and digital controllers/filters).

Exploiting Critical Sensitivity

The fact that some dynamical systems showing the necessary conditions for chaotic behavior possess such a critical dependence on the initial conditions was known since the end of the last century. However, only in the last thirty years, experimental observations have pointed out that, in fact, chaotic systems are common in nature. They can be found, e.g., in chemistry (*Belouzov–Zhabotinski reaction*), in nonlinear optics (lasers), in electronics (*Chua–Matsumoto circuit*), in fluid dynamics (*Rayleigh–Bénard convection*), etc. Many natural phenomena can also be characterized as being chaotic. They can be found in meteorology, solar system, heart and brain of living organisms and so on.

Due to their critical dependence on the initial conditions, and due to the fact that, in general, experimental initial conditions are never known perfectly, these systems are intrinsically unpredictable. Indeed, the prediction trajectory emerging from an initial condition and the real trajectory emerging from the real initial condition diverge exponentially in course of time, so that the error in the prediction (the distance between prediction and real trajectories) grows exponentially in time, until making the system's real trajectory completely different from the predicted one at long times.

For many years, this feature made chaos undesirable, and most experimentalists considered such characteristic as something to be strongly avoided. Besides their critical sensitivity to initial conditions, chaotic systems exhibit two other important properties. Firstly, there is an infinite number of unstable periodic orbits embedded in the underlying chaotic set. In other words, the skeleton of a chaotic attractor is a collection of an infinite number of periodic orbits, each one being unstable. Secondly, the dynamics in the chaotic attractor is *ergodic*, which implies that during its temporal evolution the system ergodically visits small neighborhood of every point in each one of the unstable periodic orbits embedded within the chaotic attractor.

A relevant consequence of these properties is that a chaotic dynamics can be seen as shadowing some periodic behavior at a given time, and erratically jumping from one to another periodic orbit. The idea of controlling chaos is then when a trajectory approaches ergodically a desired periodic orbit embedded in the attractor, one applies small perturbations to stabilize such an orbit. If one switches on the stabilizing perturbations, the trajectory moves to the neighborhood of the desired periodic orbit that can now be stabilized. This fact has suggested the idea that the critical sensitivity of a chaotic system to changes (perturbations) in its initial conditions may be, in fact, very desirable in practical experimental situations. Indeed, if it is true that a small perturbation can give rise to a very large response in the course of time, it is also true that a judicious choice of such a perturbation can direct the trajectory to wherever one wants in the attractor, and to produce a series of desired dynamical states. This is exactly the idea of *targeting* [BGL00].

The important point here is that, because of chaos, one is able to produce an infinite number of desired dynamical behaviors (either periodic and not periodic) using the same chaotic system, with the only help of tiny perturbations chosen properly. We stress that this is not the case for a non-chaotic dynamics, wherein the perturbations to be done for producing a desired behavior must, in general, be of the same order of magnitude as the un-perturbed evolution of the dynamical variables.

The *idea* of *chaos control* was enunciated in 1990 at the University of Maryland, by E. Ott, C. Grebogi and J.A. Yorke [OGY90], widely referred to as Ott–Grebogi–Yorke (OGY, for short). In OGY-paper [OGY90], the ideas for controlling chaos were outlined and a method for stabilizing an unstable periodic orbit was suggested, as a proof of principle. The main idea consisted in waiting for a natural passage of the chaotic orbit close to the desired periodic behavior, and then applying a small judiciously chosen perturbation, in order to stabilize such periodic dynamics (which would be, in fact, unstable for the un-perturbed system). Through this mechanism, one can use a given laboratory system for producing an infinite number of different periodic behavior (the infinite number of its unstable periodic orbits), with a great flexibility in switching from one to another behavior. Much more, by constructing appropriate goal dynamics, compatible with the chaotic attractor, an operator may apply small perturbations to produce any kind of desired dynamics, even not periodic, with practical application in the coding process of signals.

A branch of the theory of dynamical systems has been developed with the aim of formalizing and quantitatively characterizing the sensitivity to initial conditions. The *largest Lyapunov exponent* λ (together with the related *Kaplan–Yorke dimension* d_{KY}) and the *Kolmogorov–Sinai entropy* h_{KS} are the two indicators for measuring the *rate of error growth* and *information* produced by the dynamical system [ER85].

Lyapunov Exponents and Kaplan–Yorke Dimension

Recall that the characteristic Lyapunov exponents are somehow an extension of the linear stability analysis to the case of aperiodic motions. Roughly speaking, they measure the typical rate of exponential divergence of nearby trajectories. In this sense they give information on the rate of growth of a very small error on the initial state of a system [BCF02].

Consider an nD dynamical system given by the set of ODEs of the form

$$\dot{x} = f(x), \tag{7.64}$$

where $x = (x_1, \ldots, x_n) \in \mathbb{R}^n$ and $f : \mathbb{R}^n \to \mathbb{R}^n$. Recall that since the r.h.s of equation (7.64) does not depend on t explicitly, the system is called *autonomous*. We assume that f is smooth enough that the evolution is well-defined for time intervals of arbitrary extension, and that the motion occurs in a bounded region R of the system phase-space M. We intend to study the separation between two trajectories in M, $x(t)$ and $x'(t)$, starting from two close initial conditions, $x(0)$ and $x'(0) = x(0) + \delta x(0)$ in $R_0 \subset M$, respectively.

As long as the difference between the trajectories, $\delta x(t) = x'(t) - x(t)$, remains infinitesimal, it can be regarded as a vector, $z(t)$, in the tangent space $T_x M$ of M. The time evolution of $z(t)$ is given by the linearized differential equations:

$$\dot{z}_i(t) = \left.\frac{\partial f_i}{\partial x_j}\right|_{x(t)} z_j(t).$$

Under rather general hypothesis, [Ose68] proved that for almost all initial conditions $x(0) \in R$, there exists an orthonormal basis $\{e_i\}$ in the tangent space $T_x M$ such that, for large times,

$$z(t) = c_i e_i \exp(\lambda_i t), \tag{7.65}$$

where the coefficients $\{c_i\}$ depend on $z(0)$. The exponents $\lambda_1 \geq \lambda_2 \geq \cdots \geq \lambda_d$ are called *characteristic Lyapunov exponents*. If the dynamical system has an ergodic invariant measure on M, the spectrum of LEs $\{\lambda_i\}$ does not depend on the initial conditions, except for a set of measure zero with respect to the natural invariant measure.

Equation (7.65) describes how a dD spherical region $R = S^n \subset M$, with radius ϵ centered in $x(0)$, deforms, with time, into an ellipsoid of semi-axes $\epsilon_i(t) = \epsilon \exp(\lambda_i t)$, directed along the e_i vectors. Furthermore, for a generic small perturbation $\delta x(0)$, the distance between the reference and the perturbed trajectory behaves as

$$|\delta x(t)| \sim |\delta x(0)| \exp(\lambda_1 t) \left[1 + O\left(\exp -(\lambda_1 - \lambda_2)t\right)\right].$$

If $\lambda_1 > 0$ we have a rapid (exponential) amplification of an error on the initial condition. In such a case, the system is chaotic and, unpredictable on the long times. Indeed, if the initial error amounts to $\delta_0 = |\delta x(0)|$, and we purpose to predict the states of the system with a certain tolerance Δ, then the prediction is reliable just up to a *predictability time* given by

$$T_p \sim \frac{1}{\lambda_1} \ln\left(\frac{\Delta}{\delta_0}\right).$$

This equation shows that T_p is basically determined by the *positive leading Lyapunov exponent*, since its dependence on δ_0 and Δ is logarithmically weak. Because of its preeminent role, λ_1 is often referred as 'the leading positive Lyapunov exponent', and denoted by λ.

Therefore, Lyapunov exponents are average rates of expansion or contraction along the principal axes. For the ith principal axis, the corresponding Lyapunov exponent is defined as

$$\lambda_i = \lim_{t \to \infty} \{(1/t) \ln[L_i(t)/L_i(0)]\}, \tag{7.66}$$

where $L_i(t)$ is the radius of the ellipsoid along the ith principal axis at time t. For technical details on calculating Lyapunov exponents from any time series data, see [WSS85].

An initial volume V_0 of the phase-space region R_0 evolves on average as

$$V(t) = V_0 e^{(\lambda_1 + \lambda_2 + \cdots + \lambda_{2n})t}, \tag{7.67}$$

and therefore the rate of change of $V(t)$ is simply

$$\dot{V}(t) = \sum_{i=1}^{2n} \lambda_i V(t).$$

In the case of a 2D phase area A, evolving as $A(t) = A_0 e^{(\lambda_1 + \lambda_2)t}$, a *Lyapunov dimension* d_L is defined as

$$d_L = \lim_{\epsilon \to 0} \left[\frac{d(\ln(N(\epsilon)))}{d(\ln(1/\epsilon))} \right],$$

where $N(\epsilon)$ is the number of squares with sides of length ϵ required to cover $A(t)$, and d represents an ordinary *capacity dimension*,

$$d_c = \lim_{\epsilon \to 0} \left(\frac{\ln N}{\ln(1/\epsilon)} \right).$$

Lyapunov dimension can be extended to the case of nD phase-space by means of the *Kaplan–Yorke dimension* [KY79a, YAS96, OGY90]) as

$$d_{KY} = j + \frac{\lambda_1 + \lambda_2 + \cdots + \lambda_j}{|\lambda_{j+1}|},$$

where the λ_i are ordered (λ_1 being the largest) and j is the index of the smallest nonnegative Lyapunov exponent.

Kolmogorov–Sinai Entropy

Recall that the LE, λ, gives a first quantitative information on how rapidly we loose the ability of predicting the evolution of a system [BCF02]. A state, initially determined with an error $\delta x(0)$, after a time enough larger than $1/\lambda$, may be found almost everywhere in the region of motion $R \in M$. In this respect, the *Kolmogorov–Sinai* (KS) *entropy*, h_{KS}, supplies a more refined information. The error on the initial state is due to the maximal resolution we use for observing the system. For simplicity, let us assume the same resolution ϵ for each degree of freedom. We build a partition of the phase-space M with cells of volume ϵ^d, so that the state of the system at $t = t_0$ is found in a region R_0 of volume $V_0 = \epsilon^d$ around $x(t_0)$. Now we consider the trajectories starting from V_0 at t_0 and sampled at discrete times $t_j = j\tau$ ($j = 1, 2, 3, \ldots, t$). Since we are considering motions that evolve in a bounded region $R \subset M$, all the trajectories visit a finite number of different cells, each one identified by a symbol. In this way a unique sequence of symbols $\{s(0), s(1), s(2), \ldots\}$ is associated with a given trajectory $x(t)$. In a chaotic system, although each evolution $x(t)$ is univocally determined by $x(t_0)$, a great number of different symbolic sequences originates by the same initial cell, because of the divergence of nearby trajectories. The total number of the admissible symbolic

sequences, $\tilde{N}(\epsilon,t)$, increases exponentially with a rate given by the topological entropy

$$h_T = \lim_{\epsilon \to 0} \lim_{t \to \infty} \frac{1}{t} \ln \tilde{N}(\epsilon,t).$$

However, if we consider only the number of sequences $N_{eff}(\epsilon,t) \leq \tilde{N}(\epsilon,t)$
which appear with very high probability in the long time limit—those that
can be numerically or experimentally detected and that are associated with
the natural measure—we arrive at a more physical quantity, namely the
Kolmogorov–Sinai entropy [ER85]:

$$h_{KS} = \lim_{\epsilon \to 0} \lim_{t \to \infty} \frac{1}{t} \ln N_{eff}(\epsilon,t) \leq h_T. \tag{7.68}$$

h_{KS} quantifies the long time exponential rate of growth of the number of
the effective coarse-grained trajectories of a system. This suggests a link with
information theory where the Shannon entropy measures the mean asymptotic
growth of the number of the typical sequences—the ensemble of which has
probability almost one—emitted by a source.

We may wonder what is the number of cells where, at a time $t > t_0$, the
points that evolved from R_0 can be found, i.e., we wish to know how big is the
coarse-grained volume $V(\epsilon,t)$, occupied by the states evolved from the volume
V_0 of the region R_0, if the minimum volume we can observe is $V_{min} = \epsilon^d$. As
stated above (7.67), we have

$$V(t) \sim V_0 \exp\left(t \sum_{i=1}^{d} \lambda_i\right).$$

However, this is true only in the limit $\epsilon \to 0$. In this (unrealistic) limit,
$V(t) = V_0$ for a conservative system (where $\sum_{i=1}^{d} \lambda_i = 0$) and $V(t) < V_0$
for a dissipative system (where $\sum_{i=1}^{d} \lambda_i < 0$). As a consequence of limited
resolution power, in the evolution of the volume $V_0 = \epsilon^d$ the effect of the
contracting directions (associated with the negative Lyapunov exponents) is
completely lost. We can experience only the effect of the expanding directions,
associated with the positive Lyapunov exponents. As a consequence, in the
typical case, the coarse grained volume behaves as

$$V(\epsilon,t) \sim V_0 \, e^{\left(\sum_{\lambda_i > 0} \lambda_i\right)t},$$

when V_0 is small enough. Since $N_{eff}(\epsilon,t) \propto V(\epsilon,t)/V_0$, one has

$$h_{KS} = \sum_{\lambda_i > 0} \lambda_i.$$

This argument can be made more rigorous with a proper mathematical definition of the metric entropy. In this case one derives the Pesin relation [Pes77,
ER85]

$$h_{KS} \le \sum_{\lambda_i > 0} \lambda_i. \qquad (7.69)$$

Because of its relation with the Lyapunov exponents—or by the definition (7.68)—it is clear that also h_{KS} is a fine-grained and global characterization of a dynamical system.

The metric entropy is an invariant characteristic quantity of a dynamical system, i.e., given two systems with invariant measures, their KS-entropies exist and they are equal iff the systems are isomorphic [Bil65].

Chaos Control by Ott, Grebogi and Yorke (OGY)

Besides the occurrence of chaos in a large variety of natural processes, chaos may also occur because one may wish to design a physical, biological or chemical experiment, or to project an industrial plant to behave in a chaotic manner. The OGY-idea is that chaos may indeed be desirable since it can be controlled by using small perturbation to some accessible parameter.

The major key ingredient for the OGY-control of chaos is the observation that a chaotic set, on which the trajectory of the chaotic process lives, has embedded within it a large number of unstable low-period periodic orbits. In addition, because of ergodicity, the trajectory visits or accesses the neighborhood of each one of these periodic orbits. Some of these periodic orbits may correspond to a desired system's performance according to some criterion. The second ingredient is the realization that chaos, while signifying sensitive dependence on small changes to the current state and henceforth rendering unpredictable the system state in the long time, also implies that the system's behavior can be altered by using small perturbations. Then, the accessibility of the chaotic systems to many different periodic orbits combined with its sensitivity to small perturbations allows for the control and the manipulation of the chaotic process. Specifically, the OGY approach is then as follows. One first determines some of the unstable low-period periodic orbits that are embedded in the chaotic set. One then examines the location and the stability of these orbits and chooses one which yields the desired system performance. Finally, one applies small control to stabilize this desired periodic orbit. However, all this can be done from data by using nonlinear time series analysis for the observation, understanding and control of the system. This is particularly important since chaotic systems are rather complicated and the detailed knowledge of the equations of the process is often unknown [BGL00].

Simple Example of Chaos Control: a 1D Map

The basic idea of controlling chaos can be understood [Lai94] by considering May's classical *logistic map* [May76b] (7.26)

$$x_{n+1} = f(x_n, r) = r x_n (1 - x_n),$$

where x is restricted to the unit interval $[0, 1]$, and r is a control parameter. It is known that this map develops chaos via the *period-doubling bifurcation* route. For $0 < r < 1$, the asymptotic state of the map (or the attractor of the map) is $x = 0$; for $1 < r < 3$, the attractor is a nonzero fixed-point $x_F = 1 - 1/r$; for $3 < r < 1 + \sqrt{6}$, this fixed-point is unstable and the attractor is a stable period-2 orbit. As r is increased further, a sequence of period-doubling bifurcations occurs in which successive period-doubled orbits become stable. The period-doubling cascade accumulates at $r = r_\infty \approx 3.57$, after which chaos can arise.

Consider the case $r = 3.8$ for which the system is apparently chaotic. An important characteristic of a chaotic attractor is that there exists an infinite number of unstable periodic orbits embedded within it. For example, there are a fixed-point $x_F \approx 0.7368$ and a period-2 orbit with components $x(1) \approx 0.3737$ and $x(2) = 0.8894$, where $x(1) = f(x(2))$ and $x(2) = f(x(1))$.

Now suppose we want to avoid chaos at $r = 3.8$. In particular, we want trajectories resulting from a randomly chosen initial condition x_0 to be as close as possible to the period-2 orbit, assuming that this period-2 orbit gives the best system performance. Of course, we can choose the desired asymptotic state of the map to be any of the infinite number of unstable periodic orbits. Suppose that the parameter r can be finely tuned in a small range around the value $r_0 = 3.8$, i.e., r is allowed to vary in the range $[r_0 - \delta, r_0 + \delta]$, where $\delta \ll 1$. Due to the nature of the chaotic attractor, a trajectory that begins from an arbitrary value of x_0 will fall, with probability one, into the neighborhood of the desired period-2 orbit at some later time. The trajectory would diverge quickly from the period-2 orbit if we do not intervene. Our task is to program the variation of the control parameter so that the trajectory stays in the neighborhood of the period-2 orbit as long as the control is present. In general, the small parameter perturbations will be time dependent [BGL00].

The logistic map in the neighborhood of a periodic orbit can be approximated by a linear equation expanded around the periodic orbit. Denote the target period-m orbit to be controlled as $x(i)$, $i = 1, \ldots, m$, where $x(i + 1) = f(x(i))$ and $x(m + 1) = x(1)$. Assume that at time n, the trajectory falls into the neighborhood of component i of the period-m orbit. The linearized dynamics in the neighborhood of component $i + 1$ is then

$$x_{n+1} - x(i + 1) = \frac{\partial f}{\partial x}[x_n - x(i)] + \frac{\partial f}{\partial r}\Delta r_n$$
$$= r_0[1 - 2x(i)][x_n - x(i)] + x(i)[1 - x(i)]\Delta r_n,$$

where the partial derivatives are evaluated at $x = x(i)$ and $r = r_0$. We require x_{n+1} to stay in the neighborhood of m. Hence, we set $x_{n+1} - x(i + 1) = 0$, which gives

$$\Delta r_n = r_0 \frac{[2x(i) - 1][x_n - x(i)]}{x(i)[1 - x(i)]}. \tag{7.70}$$

Equation (7.70) holds only when the trajectory x_n enters a small neighborhood of the period-m orbit, i.e., when $|x_n - x(i)| \ll 1$, and hence the required parameter perturbation Δr_n is small. Let the length of a small interval defining the neighborhood around each component of the period-m orbit be 2ε. In general, the required maximum parameter perturbation δ is proportional to ε. Since ε can be chosen to be arbitrarily small, δ also can be made arbitrarily small. The average transient time before a trajectory enters the neighborhood of the target periodic orbit depends on ε (or δ). When the trajectory is outside the neighborhood of the target periodic orbit, we do not apply any parameter perturbation, so the system evolves at its nominal parameter value r_0. Hence we set $\Delta r_n = 0$ when $\Delta r_n > \delta$. The parameter perturbation Δr_n depends on x_n and is time-dependent.

The above strategy for controlling the orbit is very flexible for stabilizing different periodic orbits at different times. Suppose we first stabilize a chaotic trajectory around a period-2 orbit. Then we might wish to stabilize the fixed-point of the logistic map, assuming that the fixed-point would correspond to a better system performance at a later time. To achieve this change of control, we simply turn off the parameter control with respect to the period-2 orbit. Without control, the trajectory will diverge from the period-2 orbit exponentially. We let the system evolve at the parameter value r_0. Due to the nature of chaos, there comes a time when the chaotic trajectory enters a small neighborhood of the fixed-point. At this time we turn on a new set of parameter perturbations calculated with respect to the fixed-point. The trajectory can then be stabilized around the fixed-point [Lai94].

In the presence of external noise, a controlled trajectory will occasionally be 'kicked' out of the neighborhood of the periodic orbit. If this behavior occurs, we turn off the parameter perturbation and let the system evolve by itself. With probability one the chaotic trajectory will enter the neighborhood of the target periodic orbit and be controlled again. The effect of the noise is to turn a controlled periodic trajectory into an intermittent one in which chaotic phases (uncontrolled trajectories) are interspersed with laminar phases (controlled periodic trajectories). It is easy to verify that the averaged length of the laminar phase increases as the noise amplitude decreases [Lai94].

Blind Chaos Control

One of the most surprising successes of chaos theory has been in biology: the experimentally demonstrated ability to control the timing of spikes of electrical activity in complex and apparently chaotic systems such as heart tissue [GSD92] and brain tissue [SJD94]. In these experiments, PPF control, a modified formulation of OGY control [OGY90], was applied to set the timing of external stimuli; the controlled system showed stable periodic trajectories instead of the irregular inter-spike intervals seen in the uncontrolled system. The mechanism of control in these experiments was interpreted originally as analogous to that of OGY control: unstable periodic orbits riddle the chaotic attractor and the electrical stimuli place the system's state on the stable manifold of one of these periodic orbits [KY79a].

Alternative possible mechanisms for the experimental observations have been described by Zeng and Glass [GZ94] and Christini and Collins [CC95]. These authors point out that the controlling external stimuli serve to truncate the inter-spike interval to a maximum value. When applied, the control stimulus sets the next interval s_{n+1} to be on the line

$$s_{n+1} = As_n + C. \tag{7.71}$$

We will call this relationship the 'control line.' Zeng and Glass showed that if the uncontrolled relationship between inter-spike intervals is a chaotic 1D function, $s_{n+1} = f(s_n)$, then the control system effectively flattens the top of this map and the controlled dynamics may have fixed points or other periodic orbits [GZ94]. The authors of [CC95] showed that behavior analogous to the fixed-point control seen in the biological experiments can be accomplished even in completely random systems. Since neither chaotic 1D systems nor random systems have a stable manifold, the interval-truncation interpretation of the biological experiments is different than the OGY interpretation. The interval-truncation method differs also from OGY and related control methods in that the perturbing control input is a fixed-size stimulus whose timing can be treated as a continuous parameter. This type of input is conventional in cardiology (see e.g., [HCT97]).

Kaplan demonstrated in [KY79a] that the state-truncation interpretation was applicable in cases where there was a stable manifold of a periodic orbit as well as in cases where there were only unstable manifolds. He found that superior control could be achieved by intentionally placing the system's state off of any stable manifold. That suggested a powerful scheme for the rapid experimental identification of fixed points and other periodic orbits in systems where inter-spike intervals were of interest.

The chaos control in [GSD92] and [SJD94] was implemented in two stages. First, inter-spike intervals s_n from the uncontrolled, 'natural' system were observed. Modeling the system as a function of two variables

$$s_{n+1} = f(s_n, s_{n-1}),$$

the location $s^\star$ of a putative unstable flip-saddle type fixed-point and the corresponding stable eigenvalue λ_s were estimated from the data[50] (see [CK97]). The linear approximation to the stable manifold lies on a line given by (7.71) with

$$A = \lambda_s \quad \text{and} \quad C = (1 - \lambda_s)s^\star.$$

Second, using estimated values of A and C, the control system was turned on. Following each observed interval s_n, the maximum allowed value of the next inter-spike interval was computed as

$$S_{n+1} = As_n + C.$$

---

[50] Since the fixed-point is unstable, there is also an unstable eigenvalue λ_u.

If the next interval naturally was shorter than S_{n+1} no control stimulus was applied to the system. Otherwise, an external stimulus was provided to truncate the inter-spike interval at $s_{n+1} = S_{n+1}$.

In practice, the values of s^* and λ_s for a real fixed-point of the natural system are known only imperfectly from the data. Insofar as the estimates are inaccurate, the control system does not place the state on the true stable manifold. Therefore, we will analyze the controlled system without presuming that A and C in (7.71) correspond to the stable manifold.

If the natural dynamics of the system is modeled by

$$s_{n+1} = f(s_n, s_{n-1}),$$

then the dynamics of the controlled system is given by [KY79a]

$$s_{n+1} = \min \begin{cases} f(s_n, s_{n-1}): & \text{Natural Dynamics}, \\ As_n + C: & \text{Control Line}. \end{cases} \tag{7.72}$$

We can study the dynamics of the controlled system close to a natural fixed-point, s^*, by approximating the natural dynamics linearly as[51]

$$s_{n+1} = f(s_n, s_{n-1}) = (\lambda_s + \lambda_u)s_n - \lambda_s\lambda_u s_{n-1} + s^*(1 + \lambda_s\lambda_u - \lambda_s - \lambda_u).$$

Since the controlled system (7.72) is nonlinear even when $f()$ is linear, it is difficult to analyze its behavior by algebraic iteration. Nonetheless, the controlled system can be studied in terms of 1D maps.

Following any inter-spike interval when the controlling stimulus has been applied, the system's state (s_n, s_{n-1}) will lie somewhere on the control line. From this time onward the state will lie on an image of the control line even if additional stimuli are applied during future inter-spike intervals.

The stability of the controlled dynamics fixed-point and the size of its basin of attraction can be analyzed in terms of the control line and its image. When the previous inter-spike interval has been terminated by a control stimulus, the state lies somewhere on the control line. If the controlled dynamics are to have a stable fixed-point, this must be at the controller fixed-point x^* where the control line intersects the line of identity. However, the controller fixed-point need not be a fixed-point of the controlled dynamics. For example, if the image of the controller fixed-point is below the controller fixed-point, then the inter-spike interval following a stimulus will be terminated naturally.

For the controller fixed-point to be a fixed-point of the controlled dynamics, we require that the natural image of the controller fixed-point be at or above the controller fixed-point. Thus the dynamics of the controlled system, close to x^*, are given simply by

$$s_{n+1} = As_n + C. \tag{7.73}$$

---

[51] Equation (7.73) is simply the linear equation $s_{n+1} = as_n + bs_{n-1} + c$ with a, b, and c set to give eigenvalues λ_s and λ_u and fixed-point s^*.

Type of FP	λ_u	λ_s	$x^\star$ Locat.
Flip saddle	$\lambda_u < -1$	$-1 < \lambda_s < 1$	$x^\star < s^\star$
Saddle	$\lambda_u > 1$	$-1 < \lambda_s < 1$	$x^\star > s^\star$
Single-flip repeller	$\lambda_u > 1$	$\lambda_s < -1$	$x^\star > s^\star$
Double-flip repeller	$\lambda_u < -1$	$\lambda_s < -1$	$x^\star < s^\star$
Spiral (complex λ)	$|\lambda_u| > 1$	$|\lambda_s| > 1$	$x^\star < s^\star$

Table 7.1. Cases which lead to a stable fixed-point for the controlled dynamics. In all cases, it is assumed that $|\mathcal{A}| < 1$. (For the cases where $\lambda_s < -1$, the subscript s in λ_s is misleading in that the corresponding manifold is unstable. For the spiral, there is no stable manifold (adapted from [KY79a]).)

The fixed-point of these dynamics is stable so long as $-1 < A < 1$. In the case of a flip saddle, we therefore have a simple recipe for successful state-truncation control: position $x^\star$ below the natural fixed-point $s^\star$ and set $-1 < A < 1$.

Fixed points of the controlled dynamics can exist for natural dynamics other than flip saddles. This can be seen using the following reasoning: Let ξ be the difference between the controller fixed-point and the natural fixed-point: $s^\star = x^\star + \xi$. Then the natural image of the controller fixed-point can be found from (7.73) to be [KY79a]

$$s_{n+1} = (\lambda_s + \lambda_u)x^\star - \lambda_s\lambda_u x^\star + (1 + \lambda_s\lambda_u - \lambda_s - \lambda_u)(x^\star + \xi). \qquad (7.74)$$

The condition that

$$s_{n+1} \geq x^\star \qquad (7.75)$$

will be satisfied depending only on λ_s, λ_u, and $\xi = s^\star - x^\star$. This means that for any flip saddle, so long as $x^\star < s^\star$, the point $x^\star$ will be a fixed-point of the controlled dynamics and will be stable so long as $-1 < A < 1$.

Equations (7.74) and (7.75) imply that control can lead to a stable fixed-point for any type of fixed-point except those for which both λ_u and λ_s are greater than 1 (so long as $-1 < A < 1$). Since the required relationship between $x^\star$ and $s^\star$ for a stable fixed-point of the controlled dynamics depends on the eigenvalues, it is convenient to divide the fixed points into four classes, as given in Table 7.1.

Beyond the issue of the stability of the fixed-point of the controlled dynamics, there is the question of the size of the fixed-point's basin of attraction. Although the local stability of the fixed-point is guaranteed for the cases in Table 7.1 for $-1 < A < 1$, the basin of attraction of this fixed-point may be small or large depending on A, C, $s^\star$, λ_u and λ_s.

The endpoints of the basin of attraction can be derived analytically [KY79a]. The size of the basin of attraction will often be zero when A and C are chosen to match the stable manifold of the natural system. Therefore, in order to make the basin large, it is advantageous intentionally to misplace the control line and to put $x^\star$ in the direction indicated in Table 7.1. In addition, control may be enhanced by setting $A \neq \lambda_s$, for instance $A = 0$.

If the relationship between $x^\star$ and $s^\star$ is reversed from that given in Table 7.1, the controlled dynamics will not have a stable fixed points. To some extent, these can also be studied using 1D maps. The flip saddle and double-flip repeller can display stable period-2 orbits and chaos. For the non-flip saddle and single-flip repeller, control is unstable when $x^\star < s^\star$.

The fact that control may be successful or even enhanced when A and C are not matched to λ_s and $s^\star$ suggests that it may be useful to reverse the experimental procedure often followed in chaos control. Rather than first identifying the parameters of the natural unstable fixed points and then applying the control, one can blindly attempt control and then deduce the natural dynamics from the behavior of the controlled system. This use of PPF control is reminiscent of pioneering studies that used periodic stimulation to demonstrate the complex dynamics of biological preparations [GGS81].

As an example, consider the *Hénon map*:

$$s_{n+1} = 1.4 + 0.3s_{n-1} - s_n^2.$$

This system has two distinct fixed points. There is a flip-saddle at $s^\star = 0.884$ with $\lambda_u = -1.924$ and $\lambda_s = 0.156$ and a non-flip saddle at $s^\star = -1.584$ with $\lambda_u = 3.26$ and $\lambda_s = -0.092$. In addition, there is an unstable flip-saddle orbit of period 2 following the sequence $1.366 \to -0.666 \to 1.366$. There are no real orbits of period 3, but there is an unstable orbit of period 4 following the sequence $0.893 \to 0.305 \to 1.575 \to -0.989 \to 0.893$. These facts can be deduced by algebraic analysis of the equations.

In an experiment using the controlled system, the control parameter $x^\star = C/(1 - A)$ can be varied. The theory presented above indicates that the controlled system should undergo a bifurcation as $x^\star$ passes through $s^\star$. For each value of $x^\star$, the controlled system was iterated from a random initial condition and the values of s_n plotted after allowing a transient to decay. A bifurcation from a stable fixed-point to a stable period 2 as $x^\star$ passes through the flip-saddle value of $s^\star = 0.884$. A different type bifurcation occurs at the non-flip saddle fixed-point at $s^\star = -1.584$. To the left of the bifurcation point, the iterates are diverging to $-\infty$ and are not plotted.

Adding Gaussian dynamical noise (of standard deviation 0.05) does not substantially alter the bifurcation diagram, suggesting that examination of the truncation control bifurcation diagram may be a practical way to read off the location of the unstable fixed points in an experimental preparation.

Unstable periodic orbits can be difficult to find in uncontrolled dynamics because there is typically little data near such orbits. Application of PPF control, even blindly, can stabilize such orbits and dramatically improve the ability to locate them. This, and the robustness of the control, may prove particularly useful in biological experiments where orbits may drift in time as the properties of the system change [KY79a].

7.2.9 Basics of Nash's Game Theory

The 1994 Nobel Prize for Economic Sciences was jointly awarded to John F. Nash, Jr. from Princeton, Reinhard Selten from Bonn and John C. Harsanyi from Berkeley. This Prize recognized the central importance of *game theory* in modern economic theory. Also, the timing of these awards had a historical significance, since that year was the fiftieth anniversary of the publication of the classical book [NNM44] on *cooperative game theory* by John Von Neumann and Oskar Morgenstern.

The same situations that economists and mathematicians call *games*, psychologists usually call *social situations*. While game theory has applications to "games" such as poker and chess, the social situations are the core of modern research in game theory. Game theory has two main branches: Cooperative game theory focuses on the formation of coalitions and studies social situations axiomatically. Non-cooperative game theory models a social situation by specifying the options, incentives and information of the "players" and attempts to determine how they will play.

As the work of the other two 1994 Nobel Laureates was largely based on Nash's *non-cooperative game theory* in general, and his *governing equilibrium dynamics* in particular, in this section we will review Nash's work, discovering a way to predict the outcome of almost any type of *strategic interaction*.[52]

Since the work of John von Neumann, "games" have been a scientific metaphor for a much wider range of human interactions in which the outcomes depend on the interactive strategies of two or more persons, who have opposed or at best mixed motives. Among the issues discussed in game theory are the following:

(1) What does it mean to choose actions "rationally" when outcomes depend on the actions chosen by others and when information is incomplete?

(2) In "games" that allow mutual gain (or, mutual loss) is it "rational" to cooperate to realize the mutual gain (or avoid the mutual loss) or is it "rational" to act aggressively in seeking individual gain regardless of mutual gain or loss?

(3) If the answers to (2) are "sometimes," in what circumstances is aggression rational and in what circumstances is cooperation rational?

(4) In particular, do ongoing relationships differ from one-off encounters in this connection?

(5) Can moral rules of cooperation emerge spontaneously from the interactions of rational egoists?

(6) How does real human behavior correspond to "rational" behavior in these cases?

(7) If it differs, in what direction? Are people more cooperative than would be "rational?" More aggressive? Both?

---

[52] J. Nash's life motivated the Oscar-winning film "A Beautiful Mind".

Game theory starts from a description of the game. There are two distinct but related ways of describing a game mathematically. The *extensive form* is the most detailed way of describing a game. It describes play by means of a *game tree* that explicitly indicates *when players move*, which *moves* are available, and what they know about the moves of other players and *nature* when they move. Most important, it specifies the *pay-offs* that players receive at the end of the game [Lev01].

Fundamental to game theory is the notion of a *strategy*. A strategy a specification of how to play the game in every contingency, i.e., a *set of instructions* that a player could give to a friend or program on a computer so that the friend or computer could play the game on his behalf. Generally, strategies are contingent responses: in the game of chess, for example, a strategy should specify how to play for every possible arrangement of pieces on the board.

An alternative to the extensive form is the *normal* or *strategic form*— a description of a game by specifying the strategies and pay-offs. This is less detailed than the extensive form, specifying only the list of strategies available to each player. Since the strategies specify how each player is to play in each circumstance, we can work out from the *strategy profile* specifying each player's strategy what pay-off is received by each player. This *map from strategy profiles to pay-offs* is called the *normal* or *strategic* form. It is perhaps the most familiar form of a game, and is frequently given in the form of a *game matrix*.[53] For example, in the celebrated *Prisoner's Dilemma* game (Table 7.2) the two players are partners in a crime who have been captured by the police. Each suspect is placed in a separate cell, and offered the opportunity to confess to the crime. The rows of the matrix correspond to strategies of the first player. The columns are strategies of the second player. The numbers in the matrix are the pay-offs: the first number is the pay-off to the first player, the second the pay-off to the second player. Note that higher numbers are better (more utility); here, the total pay-off to both players is highest if neither confesses so each receives 5. However, game theory predicts that this will not be the outcome of the game (hence the dilemma). Each player reasons as follows: if the other player does not confess, it is best for me to confess (9 instead of 5). If the other player does confess, it is also best for me to confess (1 instead of 0). So no matter what I think the other player will do, it is best to confess. The theory predicts, therefore, that each player following his own self-interest will result in confessions by both players.

This game has fascinated game theorists for a variety of reasons [Lev01]. First, it is a simple representation of a variety of important situations. For

---

[53] This is the way of describing a game by listing the players (or individuals) participating in the game, and for each player, listing the alternative choices (strategies) available to that player. In the case of a two-player game, the actions of the first player form the rows, and the actions of the second player the columns, of a matrix. The entries in the matrix are two numbers representing the utility or pay-off to the first and second player respectively.

Player 1	Player 2	
	not confess	confess
not confess	5, 5	0, 9
confess	9, 0	1, 1

Table 7.2. Tucker's "Prisoner's Dilemma" game

example, instead of confess/not confess we could label the strategies "contribute to the common good" or "behave selfishly." This captures a variety of situations economists describe as public goods problems. An example is the construction of a bridge. It is best for everyone if the bridge is built, but best for each individual if someone else builds the bridge. This is sometimes refereed to in economics as an externality. Similarly this game could describe the alternative of two firms competing in the same market, and instead of confess/not confess we could label the strategies "set a high price" and "set a low price." Naturally, it is best for both firms if they both set high prices, but best for each individual firm to set a low price while the opposition sets a high price.

A second feature of this game, is that it is self-evident how an intelligent individual should behave. No matter what a suspect believes his partner is going to do, is is always best to confess. If the partner in the other cell is not confessing, it is possible to get 9 instead of 5. If the partner in the other cell is confessing, it is possible to get 1 instead of 0. Yet the pursuit of individually sensible behavior results in each player getting only 1 unit of utility, much less than the 5 units each that they would get if neither confessed. This conflict between the pursuit of individual goals and the common good is at the heart of many game theoretic problems.

A third feature of this game is that it changes in a very significant way if the game is repeated, or if the players will interact with each other again in the future. Suppose for example that after this game is over, and the suspects either are freed or are released from jail they will commit another crime and the game will be played again. In this case in the first period the suspects may reason that they should not confess because if they do not their partner will not confess in the second game. Strictly speaking, this conclusion is not valid, since in the second game both suspects will confess no matter what happened in the first game. However, repetition opens up the possibility of being rewarded or punished in the future for current behavior, and game theorists have provided a number of theories to explain the obvious intuition that if the game is repeated often enough, the suspects ought to cooperate.

The previous example illustrates the central concept in game theory, that of an *equilibrium*. This is an example of a *dominant strategy equilibrium*, a strategy profile in which each player plays best-response that does not depend on the strategies of other players. Here, the incentive of each player to confess does not depend on how the other player plays. Dominant strategy

Player 1	Player 2	
	opera	ball-game
opera	1, 2	0, 0
ball-game	0, 0	2, 1

Table 7.3. "Battle of the Sexes" game

is the most persuasive notion of equilibrium known to game theorists. In the experimental laboratory, however, players who play the prisoner's dilemma sometimes cooperate. The view of game theorists is that this does not contradict the theory, so much as reflect the fact that players in the laboratory have concerns besides monetary pay-offs. An important current topic of research in game theory is the study of the relationship between *monetary* pay-offs and the *utility* pay-offs that reflect players' real incentive for making decisions.

As a contrast to the prisoner's dilemma, consider the *Battle of the Sexes* game (Table 7.3). The story goes that a husband and wife must agree on how to spend the evening. The husband (player 1) prefers to go to the ball-game (2 instead of 1), and the wife (player 2) to the opera (also 2 instead of 1). However, they prefer agreement to disagreement, so if they disagree both get 0. This game does not admit a dominant strategy equilibrium. If the husband thinks the wife's strategy is to choose the opera, his *best response* is to choose opera rather than ball-game (1 instead of 0). Conversely, if he thinks the wife's strategy is to choose the ball-game, his best response is ball-game (2 instead of 0). While in the prisoner's dilemma, the best response does not depend on what the other player is thought to be doing, in the battle of the sexes, the best response depends entirely on what the other player is thought to be doing. This is sometime called a *coordination game* to reflect the fact that each player wants to coordinate with the other player.

For games without dominant strategies the equilibrium notion most widely used by game theorists is that of *Nash equilibrium*, a strategy profile in which *each player plays a best-response to the strategies of other players.*[54] The battle of the sexes game has two Nash equilibria: both go to the opera, or both go to the ball game: if each expects the other to go to the opera (ball-game) the best response is to go to the opera (ball-game). By way of contrast, one going to the opera and one to the ball-game is not a Nash equilibrium: since each correctly anticipates that the other is doing the opposite, neither one is playing a best response.

Games with more than one equilibrium pose a dilemma for game theory: how do we or the players know which equilibrium to choose? This question has been a focal point for research in game theory since its inception. Modern theorists incline to the view that *equilibrium is arrived at through learning*:

---

[54] If there is a set of strategies with the property that no player can benefit by changing her strategy while the other players keep their strategies unchanged, then that set of strategies and the corresponding payoffs constitute the Nash equilibrium.

Player 1	Player 2	
	Canterbury	Paris
Canterbury	−1, 1	1, −1
Paris	1, −1	−1, 1

Table 7.4. "Matching Pennies," or "Holmes–Moriarity" game

people have many opportunities to play various games, and through experience learn which is the "right" equilibrium.

Consider the well-known *Matching Pennies* game (Table 7.4). There is, however, a more colorful story from Conan Doyle's Sherlock Holmes story *The Last Problem*. Moriarity (player 2) is pursuing Holmes (player 1) by train in order to kill Holmes and save himself. The train stops at Canterbury on the way to Paris. If both stop at Canterbury, Moriarity catches Holmes and wins the game (−1 for Holmes, 1 for Moriarity). Similarly if both stop at Paris. Conversely, if they stop at different places, Holmes escapes (1 for Holmes and −1 for Moriarity). This is an example of a *zero-sum game: one player's loss is another player's gain.*[55] In the story, Holmes stops at Canterbury, while Moriarity continues on to Paris. But it is easy to see that this is not a Nash equilibrium: Moriarity should have anticipated that Holmes would get off at Canterbury, and so his best response was to get off also at Canterbury. As Holmes says "There are limits, you see, to our friend's intelligence. It would have been a coup-de-maître had he deduced what I would deduce and acted accordingly." However, this game does not have *any* Nash equilibrium: whichever player loses should anticipate losing, and so choose different strategy.

What do game theorists make of a game without a Nash equilibrium? The answer is that there are more ways to play the game than are represented in the matrix. Instead of simply choosing Canterbury or Paris, a player can flip a coin to decide what to do. This is an example of a *random* or *mixed strategy*, which simply means a particular way of choosing randomly among the different strategies. It is a mathematical fact, although not an easy one to prove, that *every game with a finite number of players and finite number of strategies has at least one mixed strategy Nash equilibrium*. The mixed strategy equilibrium of the matching pennies game is well known: each player should randomize 50–50 between the two alternatives. If Moriarity randomizes 50–50 between Canterbury and Paris, then Holmes has a 50% chance of winning and 50% chance of losing regardless of whether he choose to stop at Canterbury or Paris. Since he is indifferent between the two choices, he does not mind flipping a coin to decide between the two, and so there is no better choice than for him to randomize 50–50 himself. Similarly when Holmes is randomizing 50–50,

---

[55] If we add up the wins and losses in a game, treating losses as negatives, and we find that the sum is zero for each set of strategies chosen, then the game is a "zero-sum game."

there is no better choice for Moriarity to do the same. Each player, correctly anticipating that his opponent will randomize 50–50 can do no better than to do the same. So perhaps Holmes (or Conan Doyle) is not such a clever game theorist after all [Lev01].

Mixed strategy equilibrium points out an aspect of Nash equilibrium that is often confusing for beginners. Nash equilibrium does not require a positive reason for playing the equilibrium strategy. In matching pennies, Holmes and Moriarity are indifferent: they have no positive reason to randomize 50–50 rather than doing something else. However, it is only an equilibrium if they both happen to randomize 50–50. The central thing to keep in mind is that Nash equilibrium does not attempt to explain why players play the way they do. It merely proposes a way of playing so that no player would have an incentive to play differently. Like the issue of multiple equilibria, theories that provide a positive reason for players to be at equilibrium have been one of the staples of game theory research, and the notion of players learning over time has played a central role in this research.

Nash Equilibrium: The Focal Point of Game Theory

Background to Nash Equilibrium

To describe the spirit of the time, we might quote from R. Aumann [Aum87]: "The period of the late 40's and early 50's was a period of excitement in game theory. The discipline had broken out of its cocoon and was testing its wings. Giants walked the earth. At Princeton, John Nash laid the groundwork for the general non-cooperative theory and for cooperative bargaining theory. Lloyd Shapley defined the value for coalitional games, initiated the theory of stochastic games, co-invented the core with D.B. Gillies, and together with John Milnor developed the first game models with a continuum of players. Harold Kuhn reformulated the extensive form of a game, and worked on behavior strategies and perfect recall. Al Tucker discovered the Prisoner's Dilemma, and supported a number of young game theorists through the Office of Naval Research."[56]

The twenty-year old John Nash came to Princeton in September of 1948. He came to the Mathematics Department with a one sentence letter of recommendation from R.L. Duffin of Carnegie Institute of Technology.[57] This

---

[56] It is important to recognize that these results did not respond to some suggestion of von Neumann, nor did they follow work that he had outlined or proposed; rather they were revolutionary new ideas that ran counter to von Neumann's theory [NS94]. In almost every instance, it was a repair of some inadequacy of the game theory as outlined by von Neumann and Morgenstern. All of the results cited in [Aum87] were obtained by members of the Mathematics Department at Princeton University. At the same time, the RAND Corporation, funded by the US Air Force, which was to be for many years the other major center of game-theoretic research, had just opened its doors in Santa Monica.

[57] Today, this is the Carnegie Mellon University.

letter said, simply: "This man is a genius." The results for which he is being honored by Nobel Prize were obtained in his first fourteen months of graduate study [NS94].

Although the speed with which Nash obtained these results is surprising, equally surprising and certainly less widely known is that Nash had already completed an important piece of work on bargaining while still an undergraduate at the Carnegie Institute of Technology. This work, a subsection for an elective course in international economics, possibly the only formal course in economics he has ever had, was done in complete ignorance of the work of von Neumann and Morgenstern. In short, when he did this work he didn't know that game theory existed. This result, which is a model of theoretical elegance, posits *four reasonable requirements* or axioms on any game solution [Nas97]:

1. Any solution should be invariant under positive linear affine transformations[58] of the utility functions;
2. The solution should be efficient in the sense of *Pareto optimality*, i.e., the best that could be achieved without disadvantaging at least one group[59];
3. Irrelevant alternatives should not change the outcome of the solution; and
4. Bargaining problems with symmetric outcome sets should have symmetric solutions.

If these four reasonable conditions are satisfied then *there exists a unique solution* to the game, namely, *the outcome that maximizes the product of the players' utilities*.

The main result, the *definition of a Nash equilibrium*, and a proof of its existence had been completed prior to November 1949, the date of submission by Lefschetz to the National Academy of Sciences. The thesis [Nas50a] itself was completed and submitted after the persistent urging and counsel of Professor Al Tucker. The formal rules at Princeton require that the thesis must be read by two professors, who prepare a report evaluating the work. In this case, the readers were Tucker and the statistician, John Tukey; the evaluation was written by Tucker himself. He wrote, "This is a highly original and important contribution to the Theory of Games. It develops notions and properties of "non-cooperative games," finite n-person games which are very

---

[58] Aaffine transformation is any transformation preserving collinearity (i.e., all points lying on a line initially still lie on a line after transformation) and ratios of distances (e.g., the midpoint of a line segment remains the midpoint after transformation). An affine transformation may also be thought of as a *shearing transformation*.

[59] Pareto optimality is a measure of efficiency, named after Vilfredo Pareto. An outcome of a game is Pareto optimal if there is no other outcome that makes every player at least as well off and at least one player strictly better off. That is, a Pareto Optimal outcome cannot be improved upon without hurting at least one player. Often, a Nash Equilibrium is not Pareto Optimal implying that the players' payoffs can all be increased.

interesting in themselves and which may open up many hitherto untouched problems that lie beyond the zero-sum two-person case. Both in conception and in execution this thesis is entirely the author's own" [NS94].

The theory which fully occupies half of the von Neumann and Morgenstern book [NNM44] deals with cooperative game theory envisaging coalitions, side-payments, and binding agreements. In addition, they proposed as a solution concept a notion we now call a "stable set", which need not exist for every game. By contrast, Nash proved by page 6 of his thesis [Nas50a] that every n-person finite non-cooperative game has at least one (Nash) equilibrium point. This is a profile of mixed strategies, one for each player, which is such that no player can improve his pay-off by changing his mixed strategy unilaterally [NS94].

The Nash equilibrium is without doubt the single game theoretic solution concept that is most frequently applied in *economics*. Economic applications include oligopoly, entry and exit, market equilibrium, search, location, bargaining, product quality, auctions, insurance, principal-agent (problems), higher education, discrimination, public goods, what have you. On the political front, applications include voting, arms control and inspection, as well as most international political models (deterrence, etc.). Biological applications all deal with forms of strategic equilibrium; they suggest an interpretation of equilibrium quite different from the usual overt rationalism [Aum87].

In the short period of 1950–53, John Nash published four brilliant papers [Nas50b, Nas50c, Nas51, Nas53, Nas97], in which he made at least three fundamentally important contributions to game theory:

1. He introduced the distinction between cooperative and non-cooperative games. The former are games in which the players can make enforceable agreements and can also make irrevocable threats to other players. That is to say, they can fully commit themselves to specific strategies. In contrast, in noncooperative games, such self-commitment is not possible.
2. As a natural solution concept for non-cooperative games, he introduced the concept of equilibrium points ([Nas50b, Nas51]), now usually described as Nash equilibria. He also established their existence in all finite games.
3. As a solution concept for two-person cooperative games, he proposed the Nash bargaining solution, first for games with fixed threats [Nas50c], and later also for games with variable threats [Nas53]. He also showed that, in the latter case, the two players' optimal strategies will have maximin and minimax properties.

Von Neumann and Morgenstern's classical book contains an excellent mathematical analysis of one class of non-cooperative games, viz. of two-person zero-sum games and of the minimax solution for such games. It contains also an excellent mathematical discussion of one cooperative solution concept, that of stable sets, for many specific games. Yet, it so happens that the concept of two-person zero-sum games has very few real-life applications. The concept of stable sets has even fewer empirical applications.

Then appeared the so-called *Nash's program*, which proposes [Nas51]:

The writer has developed a "dynamical" approach to the study of co-operative games based on reduction to non-cooperative form. One pro-ceeds by constructing a model of the pre-play negotiation so that the steps of (this) negotiation become moves in a larger non-cooperative game... describing the total situation. This larger game is then treated in terms of the theory of this subsection... and if values are obtained (then) they are taken as the values of the cooperative game. Thus, the problem of analyzing a cooperative game becomes the problem of obtaining a suitable, and convincing, non-cooperative model for the negotiation.

When game theorists speak of 'Nash's program', it is this two-paragraph passage they have in mind. That is to say, they are talking about *the program of trying to reduce cooperative games to non-cooperative games by means of suitable non-cooperative models of the bargaining process among the players.*

It is an interesting fact of intellectual history that Nash's papers in the early 1950's at first encouraged game theorists to cultivate cooperative and non-cooperative game theory as largely independent disciplines, with a con-centration on cooperative theory. However, twenty-five years later they encour-aged a shift to non-cooperative game theory and to non-cooperative models of the negotiations among the players.

One of Reinhard Selten's important contributions was his distinction be-tween *perfect & imperfect Nash equilibria*. It was based on his realization that even strategy combinations fully satisfying Nash's definition of Nash equilibria might very well contain some irrational strategies. To exclude such imperfect Nash equilibria containing such irrational strategies, at first he proposed what now are called subgame-perfect equilibria.

A Nash equilibrium is defined as a strategy combination with the property that every player's strategy is a best reply to the other players' strategies. This of course is true also for Nash equilibria in mixed strategies. But in the latter case, besides his mixed equilibrium strategy, each player will also have infinitely many alternative strategies that are his best replies to the other players' strategies. This will make such equilibria potentially unstable.

In view of this fact, it was desirable to show, that "almost all" Nash equilibria can be interpreted as strict equilibria in pure strategies of a suitably chosen game with randomly fluctuating pay-off functions.

When John Nash published his basic papers on 'equilibrium points in n-person games', and 'non-cooperative games', nobody would have foretold the great impact of Nash equilibrium on economics and social science in general. It was even less expected that Nash's equilibrium point concept would ever have any significance for biological theory.

Originally, von Neumann and Morgenstern developed game theory as a mathematical method especially adapted to economics and social science in general. In the introduction of their book, they emphasized their view that methods taken over from the natural sciences are inadequate for their purpose. They succeeded in creating a new method of mathematical analysis not borrowed from physics. In the case of game theory the flow of methodological innovation did not go in the usual direction from the natural to the social sciences but rather in the opposite one. The basis for this extremely successful transfer is the concept of Nash equilibrium.

In his Ph.D. dissertation [Nas50a], John Nash provided *two interpretations* of his equilibrium concept for non-cooperative games, one *rationalistic* and one *population-statistic*:

- In the first, standard, rationalistic interpretation, one imagines that the game in question is played only once, that the participants are "rational," and that they know the full structure of the game. However, Nash comments: "It is quite strongly a rationalistic and idealizing interpretation."
- The second, population-statistic interpretation, which Nash calls the mass-action interpretation, was until recently largely unknown. Here Nash imagines that the game in question is played over and over again by participants who are not necessarily "rational" and who need not know the structure of the game:

  "It is unnecessary to assume that the participants have full knowledge of the total structure of the game, or the ability and inclination to go through any complex reasoning processes. But the participants are supposed to accumulate empirical information on the relative advantages of the various pure strategies at their disposal.

  To be more detailed, we assume that there is a population (in the sense of statistics) of participants for each position of the game. Let us also assume that the 'average playing' of the game involves n participants selected at random from the n populations, and that there is a stable average frequency with which each pure strategy is employed by the 'average member' of the appropriate population.

  Since there is to be no collaboration between individuals playing in different positions of the game, the probability that a particular n-tuple of pure strategies will be employed in a playing of the game should be the product of the probabilities indicating the chance of each of the n pure strategies to be employed in a random playing".

The Rationalistic Interpretation

A non-cooperative game is given by a set of players, each having a set of strategies and a pay-off function. A strategy vector is a Nash equilibrium if each player's strategy maximizes his pay-off if the strategies of the others are held fixed. In his Ph.D. thesis [Nas50a], Nash introduces this concept and he derives several properties of it, the most important one being existence of at least one equilibrium for every finite game. In published work [Nas50b, Nas51], Nash provides two alternative, elegant proofs, one based on Kakutani's fixed point theorem, the other based directly on Brouwer's fixed point theorem.

Player 1	Player 2	
	A	B
a	1, 2	−1, −4
b	−4, −1	2, 1

Table 7.5. Nash's game

These techniques have inspired many other existence proofs, for example, in the area of general equilibrium theory (see [Deb84]).

Nash's motivation for his standard "rationalistic and idealizing interpretation" which is applicable to a game played just once, but which requires that the players are rational and know the full structure of the game, runs as follows:

"We proceed by investigating the question: What would be a "rational" prediction of the behavior to be expected of rational playing the game in question? By using the principles that a rational prediction should be unique, that the players should be able to deduce and make use of it, and that such knowledge on the part of each player of what to expect the others to do should not lead him to act out of conformity with the prediction, one is led to the concept of a solution defined before" [Nas50a].

In other words, a theory of rational behavior has to prescribe the play of a Nash equilibrium since otherwise the theory is self-destroying. Note that the argument invokes three assumptions:

(i) players actively randomize in choosing their actions,
(ii) players know the game and the solution, and
(iii) the solution is unique.

Later work has scrutinized and clarified the role of each of these assumptions. Harsanyi [Har73] showed that a mixed strategy of one player can be interpreted as the beliefs (conjectures) of the other players concerning his behavior. This reinterpretation provides a "Bayesian" foundation for mixed strategy equilibria and eliminates the intuitive difficulties associated with them.

Since the rationalistic justification of equilibria relies on uniqueness, multiplicity of equilibria is problematic. Nash remarks that it sometimes happens that good heuristic reasons can be found for narrowing down the set of equilibria. One simple example that Nash provides [Nas51] is given in Table 7.5. This game has equilibria at (a, A) and (b, B), as well as a mixed equilibrium. Nash writes that "empirical tests show a tendency toward (a, A)," but he does not provide further details. One heuristic argument is that (a, A) is less risky than (b, B), an argument that is formalized by Harsanyi and Selten's concept of *risk dominance* [HS88]. It figures prominently both in the literature that builds on the "rationalistic interpretation" as well as in the literature that builds on the "mass-action" interpretation of Nash equilibrium.

The Mass-Action Interpretation

Notation and Preliminaries

Consider a finite n-player game G in normal (or strategic) form [Wei94]. Let A, be the pure-strategy set of player position $i \in I = \{1, \ldots, n\}$, S_i, its mixed-strategy simplex, and $S = \prod_{i \in I} S_i$, the polyhedron of mixed-strategy profiles. For any player position i, pure strategy $\alpha \in A_i$ and mixed strategy $s_i \in S_i$, let $s_{i\alpha}$ denote the probability assigned to α. A strategy profile s is called *interior* if *all* pure strategies are used with positive probability. The expected pay-off to player position i when a profile $s \in S$ is played is denoted $\pi_i(s)$, while $\pi_{i\alpha}(s)$ denotes the pay-off to player i when he uses pure strategy $\alpha \in A_i$ against the profile $s \in S$. A strategy profile $s \in S$ is a *Nash equilibrium* if and only if $s_{i\alpha} > 0$ implies $\pi_{i\alpha}(s) = \max_{\beta \in A_i} \pi_{i\beta}(s)$.

In the spirit of the *mass-action interpretation*, imagine that the game is played over and over again by individuals who are randomly drawn from (infinitely) large populations, one population for each player position i in the game. A population state is then formally identical with a mixed-strategy profile $s \in S$, but now each component $s_i \in S_i$ represents the distribution of pure strategies in player population i, i.e., $s_{i\alpha}$ is the probability that a randomly selected individual in population i will use pure strategy $\alpha \in A$ (the so-called α-strategist), when drawn to play the game. In this interpretation $\pi_{i\alpha}(s)$ is the (expected) pay-off to an individual in player population i who uses pure strategy α and $\pi_i(s) \sum_\beta s_{i\beta} \pi_{i\beta}(s)$ is the average (expected) pay-off in player population i, both quantities being evaluated in population state s.

Suppose that every now and then, say, according to a statistically independent Poisson process, each individual reviews her strategy choice. By the law of large numbers the aggregate process of strategy adaptation may then be approximated by deterministic flows, and these may be described in terms of ordinary differential equations.[60]

Innovative Adaptation

We first consider the case when strategy adaptation is memory-less in the sense that the time rate of strategy revision and the choice probabilities of strategy-reviewing individuals are functions of the current state s (only) [Wei94]:

$$\dot{s}_{i\alpha}(t) = f_{i\alpha}[s(t)], \tag{7.76}$$

(overdot denotes time derivative) for some functions $f_{i\alpha} : S \to \mathbb{R}$. The quantity $f_{i\alpha}(s)$ thus represents the net increase per time unit of the population share of α-strategists in player population i when the overall population state

---

[60] Note that these conditions can be relaxed to *stochastic flows*, commonly described either by Langevin stochastic ODEs, or by corresponding Fokker–Planck PDEs (defining time evolution of the solution's probability density).

is s. The (composite) function f is assumed to be Lipschitz continuous and such that all solution trajectories starting in S remain forever in S. Such a function f is usually called a vector-field for (7.76).

The class of population dynamics (7.76) clearly allows for an innovative element; some individuals may begin using earlier unused strategies, either intentionally, by calculation or experiment, or unintentionally, by mistakes or mutations. This apparently implies that only those population states that constitute Nash equilibria can be stationary. The requirement is simple: if there is some (used or unused) pure strategy that results in a pay-off above the current average pay-off in the player population in question, then some such pure strategy will grow in population share. Formally, for any population state $s \in S$ and player position $i \in I$, let $B_i(s)$ denote the (possibly empty) subset of better-then-average pure strategies, $B_i(s) = \{\alpha \in A_i : \pi_{i\alpha}(s) > \pi_i(s)\}$. Inventiveness can then be formalized as

[IN]:   If $B_i(s) \neq \emptyset$, then $f_{i\alpha}(s) > 0$ for some $\alpha \in B_i(s)$.

This condition is, for instance, met if reviewing individuals move toward the best replies to the current population state. Note that [IN] requires no knowledge about pay-offs to other player positions, nor is any detailed knowledge of the pay-offs to one's own strategy set necessary. It is sufficient that individuals on average tend to twitch toward some of the better-than-average performing strategies.

Proposition 1. *Suppose f meets [IN]. If a population state s is stationary under the associated dynamics (7.76) then s constitutes a Nash equilibrium of G.*

In order to incorporate memory in the dynamic process of strategy adaptation, one may introduce real variables $P_{i\alpha}$, one for each player position i and pure strategy $\alpha \in A$, that represent the ith population's recollection of earlier pay-offs to pure strategy α. Assume that the recalled pay-off to any pure strategy $\alpha \in A$ changes with time according to

$$P_{i\alpha}(t) = h_{i\alpha}(\pi_{i\alpha}[s(t)], p_{i\alpha}(t), t), \tag{7.77}$$

where $h_{i\alpha}$ is a Lipschitz continuous function such that the recalled pay-off changes only if the current pay-off differs from the recalled pay-off, i.e., if $h_{i\alpha}(\pi_{i\alpha}, p_{i\alpha}, t) = 0$ implies $\pi_{i\alpha} = p_{i\alpha}$.

The full adaptation dynamics with memory is then a system of differential equations in the state vector $x = (s, p)$, where p moves according to (7.77) and s moves according to

$$\dot{s}_{i\alpha}(t) = f_{i\alpha}[s(t), p(t)]. \tag{7.78}$$

A counterpart to the earlier requirement [IN] of inventiveness is: if there is some (used or unused) pure strategy which is *recalled* to result in a pay-off above the average of the currently recalled pay-offs in the player population in question, then some such pure strategy will increase its population share. Formally, for any state (s, p) and player position $i \in I$, let $B_i(s, p) = \{\alpha \in A_i : p_{i\alpha} > \sum_{\beta \in A_i} s_{i\beta} p_{i\beta}\}$. Inventiveness can then be formalized as

[IN']:   If $B_i(s, p) \neq \emptyset$, then $f_{i\alpha}(s, p) > 0$ for some $\alpha \in B_i(s, p)$.

The following extension of Proposition 1 is straightforward:

Proposition 2. *Suppose f meets [IN']. If a population state (s, p) is stationary under the associated dynamics (7.77)–(7.78), then s constitutes a Nash equilibrium of G.*

Imitative Adaptation

It may be argued that the above classes of population dynamics go somewhat beyond the spirit of the mass-action interpretation since they presume that individuals perform a certain amount of calculations. Therefore, now assume no memory and no inventiveness as defined above. Thus, individuals now switch only between strategies already in use, and they do so only on the basis of these strategies current performance. Technically, this means that the population dynamics (7.76) has a vector-field f of the form [Wei94]

$$f_{i\alpha}(s) = g_{i\alpha}(s) s_{i\alpha}. \tag{7.79}$$

The involved functions $g_{i\alpha}$ are usually called *growth-rate functions*, $g_{i\alpha}(s)$ being the growth rate of the population share of pure strategy α in player population i when the population state is s. No vector-field of the form (7.79) is innovative in the sense of condition [IN], because if all individuals in a player population initially use only one (or a few) pure strategy then they will continue doing so forever, irrespective of whether some unused strategy yields a high pay-off or not. Consequently, stationarity does not imply Nash equilibrium for the present class of dynamics, which is usually called *imitative*.

A prime example of such dynamics is the so-called *replicator dynamics* used in *evolutionary biology* [Tay79]. Here, pure strategies represent genetically programmed behaviors, reproduction is asexual, each offspring inherits its parents strategy, and pay-offs represent reproductive fitness. Thus $\pi_{i\alpha}(s)$ is the number of (surviving) offspring to an α-strategist in population i, and $\pi_i(s)$ is the average number of (surviving) offspring per individual in the same population. In the standard version of the population model, each pure strategy's growth rate is proportional to its current pay-off

$$g_{i\alpha}(s) = \pi_{i\alpha}(s) - \pi_i(s). \tag{7.80}$$

If there exists a pure strategy which results in a pay-off above average in its player population (whether this pure strategy be currently used or not), then some such pure strategy has a positive growth rate. Hence, if all such strategies are present in the population, then some such population share will grow. Formally:

[POS]: If $B_i(s) \neq \emptyset$, then $g_{i\alpha}(s) > 0$ for some $\alpha \in B_i(s)$.

The next proposition establishes the following implications under pay-off positive imitation:

(a) If all strategies are present in a stationary population state, then this constitutes a Nash equilibrium;
(b) A dynamically stable population state constitutes a Nash equilibrium,
(c) If a dynamic solution trajectory starts from a population state in which all pure strategies are present and the trajectory converges over time, then the limit state is a Nash equilibrium.

Proposition 3. *Suppose g meets [POS], and consider the associated population dynamics (7.76) where f is defined in (7.79).*

(a) If s is interior and stationary, then s is a Nash equilibrium.
(b) If s is dynamically stable, then s is a Nash equilibrium.
(c) If s is the limit of some interior solution trajectory, then s is a Nash equilibrium.

This demonstrates that the mass-action interpretation of Nash equilibria is in stark contrast with the usual rationalistic interpretation, but is closely related to ideas in evolutionary game theory. It opens new avenues for equilibrium and stability analysis of social and economic processes, and suggests new ways to combine insights in the social and behavior sciences with economic theory [Wei94].

Nash Equilibrium in Human Performance

We end this section on classical (Nash) game theory by giving an example of mixed strategies in the game of tennis.

Great sport players use a variety of moves to keep their opponents guessing. Game theory suggests that there is an optimal way to confuse one's opponent. This can be illustrated with the game of tennis [DN93].

Imagine that it is your serve. Your opponent's forehand is stronger than their backhand. However, if you continually serve to their backhand then they will anticipate correctly and gain an advantage. What should you do? It turns out that you can increase your performance by systematically favoring one side, although in an unpredictable way.

More specifically, consider the following opponent. Her forehand is her strong game. If she anticipates your serve correctly, her forehand return will be successful 90% of the time, while an anticipated backhand return will be successful only 60% of the time. Obviously, if she anticipates incorrectly she

Opponent Anticipates	Your Aim for Serve	
	forehand	backhand
forehand	90%	20%
backhand	30%	60%

Table 7.6. Probability that your opponent returns your serve

fares much worse. If she anticipates the backhand side and serve goes to the forehand side, she successfully returns 30% of the time, and if she anticipates the forehand side and serve is to the backhand side, she successfully returns 20% of the time.

Your opponent wants to maximize her success of return, while your goal is the exact opposite. Before the match, what strategy should you take? The chart (see Fig. 7.64) gives you your best strategy. To the left of where the two lines intersect, your opponent does better if she anticipates backhand, to the right she does better if she anticipates forehand. The 40%:60% mixture is the only strategy the receiver cannot exploit to her advantage. Using this strategy, her success of return will be only 48%.

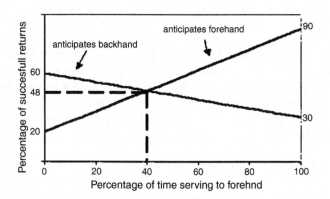

Fig. 7.64. Best serve strategy.

Both theoretical and empirical study was performed in [WW01], where a similar game was used as a theoretical model of the tennis serve and its relation to the winning of points in a tennis match. The assumption used was: Every point in a tennis match was played as a 2×2 constant-sum normal-form game with a unique equilibrium in strictly mixed strategies. The authors studied ten matches from the four "Grand Slam" tournaments and the ATP-Masters tournaments, using videotapes. They selected top matches, between players (Rosewall, Smith, Borg, McEnroe, Connors, Lendl, Edberg, Cash, Wilander, Becker, Sampras, Agassi, and Korda) who were well known to each other, which lasted long enough to have lots of points scored. They analyzed only first serves (if the first serve went out of bounds, there was a second serve, but it was not analyzed in this study).

The conclusions of the study are: The players did use the equilibrium mixing probabilities, but they tended to switch from one action to the other too often. Those two facts can be consistent because there are two ways to randomize with, for example, 50 percent probability between F and B: FBF-BFBFB and FFFBBB. Game theory predicts that each pattern would be equally likely, since if one were more likely, the other player could use that fact to better predict what would happen. The authors found that the players used the FBFBFB switching style more often than game theory would predict.

7.3 Frequently Used Computational Tools

Most of the working code in this section is written in Fortran 95, while the rest of the code is written in C/C++, $Delphi^{TM}$ and $Mathematica^{TM}$. The Fortran 95 code has been compiled by at least one of the three free Fortran 95 compilers: two versions of the GNU Fortran compiler: $gfortran$ (http://gcc.gnu.org/fortran/) and $g95$ (http://g95.org), as well as a personal edition of the Silverfrost $FTN95^{TM}$ compiler for MS Windows. The presented C/C++ code has been compiled by the free GNU compiler DEV-C++ (http://www.bloodshed.net/), which is compatible with MS $Visual$ C++. The presented Delphi code is extremely fast, several times faster then Fortran 95 or C/C++ and it compiles also under the free GNU compiler DEV-$Pascal$ (http://www.bloodshed.net/devpascal.html). The presented $Mathematica$ code shows very high-level computing, that is solving difficult symbolic and numerical problems without worrying about the underlying algorithms— $Mathematica$ chooses the most efficient algorithms for any particular problem, and in most cases it includes almost all know algorithms. The price for this huge source of embedded algorithms is two-fold: roughly a hundred times slower computation than $Delphi$ and almost impossible implementation in a stand-alone software packages.[61] Note that the Fortran 95 code can be automatically translated into C/C++ using specialized $For2C$ translators. Similarly, both Delphi and C/C++ code can be automatically translated into Java classes.

7.3.1 Basic Numerical Algorithms

Function Plot

The following Fortran 95 code plots a sample function (see Fig. 7.65) defined as a subroutine.

---

[61] Recall here that $Mathematica$ notebooks can be called from other programs through a protocol called MathLink.

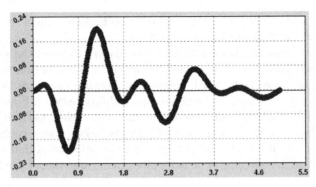

Fig. 7.65. Plot of a sample function: $f(t) = e^{-t}(\sin(2\pi t) - \sin(\pi t))/\pi t$.

```
Program IvFunPlot   ! V.G. Ivancevic, 2007
  ! plot a function

  implicit none
  real*8, parameter :: Tfin = 5.0
  real*8, parameter :: h = 0.01
  real*8 :: t, f, Pi
  Pi = 4.0 * atan(1.0)
  open (unit=7,file="Ivfunction.dat")     ! output file
  call define_function
  close(7)
contains

  subroutine define_function     ! define a function to plot
    t = 0.0
    do while (t < Tfin)
       f = exp(-t) * ( sin(2 * Pi * t) - sin(Pi * t) ) / Pi*t
       t = t + h
       write (7, *) t, f
    end do
  end subroutine define_function

end program IvFunPlot
```

The same program, with the same output, in C reads:

```
//Program IvFunPlot   ! V.G. Ivancevic, 2007
#include <stdio.h>

int main() {
    double t, Tfin, h, f, Pi;  // double times and step-size
```

```
    // initialize variables
    t = 0.0;   Tfin = 5.0;
    h = 0.01;  f = 0.0;
    Pi = 4.0 * atan(1.0);

    // output file
    FILE *fout;     // output file
    if ((fout = fopen("Ivfunction.dat", "w")) == NULL)
    { printf("Error opening file\n"); exit(1); }

    // time loop
    while (t < Tfin)         // function definition and output
    {
       f = exp(-t) * ( sin(2 * Pi * t) - sin(Pi * t) )
       / Pi*t;
       fprintf(fout,"%lf\t%lf\n",t,f);  // print t, f
       t += h;
    }
    fclose(fout);
    return(0);
}
```

The same program, with the same output, in Delphi reads:

```
Program IvFunPlot;  // V.G. Ivancevic, 2007 uses
  SysUtils;
const
  Tfin = 5.0;   h = 0.01;
var
  FF : text;      t, Fun: extended;
begin
assign(FF,'Ivfun.dat');{$I-}reset(FF);{$I+}{$I-}rewrite(FF);
{$I+}
  t := 0;
  while t <= Tfin do begin
    Fun := exp(-t) * ( sin(2 * Pi * t) - sin(Pi * t) )
    / Pi*t;
    WriteLn(FF, t,'   ',Fun);
    t := t + h;
  end; close(FF);
end.
```

Note that out of the three presented versions, by far the fastest (both in compiling and in execution) and most accurate (with 20 significant digits) is the Delphi version.

Logistic Map

The following Fortran 95 code iteratively solves May's logistic map

$$x :\longrightarrow rx(1-x), \quad r \text{ is a population parameter,}$$

and produces its bifurcation diagram (see Fig. 7.66).

```
Program IvLogistic  ! V.G. Ivancevic, 2007
    ! Bifurcation diagram of the logistic map:  r*x*(1-x)

    implicit none
    real*8 :: r_min, r_max, r, step, x
    integer :: t  ! discrete time
    r_min = 1.0    ! population-parameter range
    r_max = 4.0
    step = 0.01   ! iteration step

    open(unit=7,file='IvLogistic.out.dat')   ! the output file

    do r = r_min, (r_max-step), step   ! loop for r iterations
       x = 0.5      ! initial condition

       do t = 0, 200       ! wait until transients die out
          x = r * x * (1 - x)
       end do

       do t = 201, 401    ! output 200 points
          x = r * x * (1 - x)
          write (7, *) r, x
       end do
    end do
    close(7)
    stop 'data saved in IvLogistic.out.dat'
end program IvLogistic
```

Numerical Differentiation

The following Fortran 95 code performs numerical differentiation of analytically defined functions using forward, central and extrapolated differences.

```
Program Ivdiff  ! V.G. Ivancevic, 2007
    ! Numerical differentiation of a given function
    ! using forward, central and extrapolated differences

    implicit none
    real*8 :: f, h, deriv(3), x, xmin, xmax, xstep
```

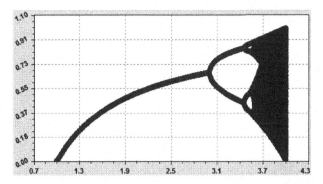

Fig. 7.66. Bifurcation diagram of the logistic map with the population parameter r ranging from 1.0 to 4.0.

```
open(unit=7,file='Ivdiff.out.dat')   ! the output file
h = 1.e-5     ! h step-size for approximation
xmin  = 0.0
xmax  = 10.0
xstep = 0.01   ! step for DO-loop iteration
do x = xmin, xmax, xstep
   deriv(1) = (f(x+h) - f(x)) / h
   ! forward difference
   deriv(2) = (f(x+h/2) - f(x-h/2)) / h
   ! central difference
   ! extrapolated difference
   deriv(3) = 8 * (f(x + h/4) - f(x - h/4)) / (3*h) &
            - (f(x + h/2) - f(x - h/2)) / (3*h)
   write (7, *) x, deriv(1), deriv(2), deriv(3)
end do
close(7)
stop 'data saved in Ivdiff.out.dat'
end program Ivdiff

function f(x)    ! sample function to differentiate
  implicit none
  real*8 :: f, x
  f = cos(x)
return; end
```

Random Walk

The following Fortran 95 code simulates random walk.

```
Program IvRandWalk  ! V.G. Ivancevic, 2007
  ! Random walk simulation
```

```fortran
implicit none
real*8 :: harv, root2, x, y, r(1:10000)
integer :: i, j, max, seed
max   = 10000      ! set parameters
seed  = 11168
root2 = 1.4142135623730950488E0

open(unit=7,file='IvRandWalk.out.dat')

call random_seed(seed)      ! random seed generator
do j = 1, max
    r(j) = 0       ! clear array
end do
call random_number(harv)    ! harv = random number

! average over 100 trials
do j = 1, 100
    x = 0
    y = 0
    ! take max steps
    do i = 1, max
        x = x + (harv - 0.5) * 2.0 * root2
        y = y + (harv - 0.5) * 2.0 * root2
        r(i) = r(i) + Sqrt(x * x + y * y)
    end do
end do

! output data for plot of r vs. sqrt(N)
do i = 1, max
    write (7, *) Sqrt(Real(i)), ' ', r(i) / 100
end do
close(7)
stop 'data saved in Ivdiff.out.dat'
end program IvRandWalk
```

7.3.2 Numerical Integration of Functions

Basic Monte Carlo Integration

The following Fortran 95 code presents the basic Monte Carlo integration of a given function $f(x) = 10\sin(x)$ on a given interval $[0, \pi]$.

```fortran
Program IvMCint ! V.G. Ivancevic, 2007
! Monte Carlo integration;
! computing the area under the curve y=10sin(x) on [0,pi]
```

```
! the exact solution is 20

  implicit none
  integer :: i,j,m,n
  real :: Area,R1,R2,x,y,A,B,y_max,err,Pi
  Pi = 4.0 * atan(1.0)
  y_max = 10.0
  B = Pi    ! limits
  A = 0.0
  call random_seed    ! initiate random number generator
  write (6, *) '        It.#       Area           err'
  do i = 1, 9
     n = 10**i
     m = 0
     do j = 1, n
        call random_number(R1)  ! call random number
        call random_number(R2)
        x  = A + (B - A) * R1
        y  = y_max * R2
        if ( y < F(x) ) m = m + 1
     end do
     Area  = (B - A) * y_max * m / n    ! calculate area
     err = abs(20.0 - Area)             ! error definition
     write (6, *) n, Area, err
  end do
contains

  real function F(x)         ! function defined
  real, intent(IN) :: x
     F = 10.0*sin(x)
  end function F

end program IvMCint
```

Trapezoid and Simpson's Integration

The following Fortran 95 code performs numerical integration of analytically
defined functions using trapezoid and Simpson's rules.

```
Program IvTrapSimp  ! V.G. Ivancevic, 2007
  ! Integrating a given function
  ! using trapezoid and Simpson's rules

  implicit none
  real*8 :: trapez, simpson, r1, r2, th_sol, vmin, vmax
```

```fortran
      integer :: i
      ! theoretical result (12 digits) and integration range
      th_sol = 0.580958373336
      vmin = 0.0
      vmax = 1.0

      open(unit=7,file='IvTrapSimp.dat')   ! the output file

      ! calculate the integral using both methods for
      steps = 3..501
      do i = 3, 501, 2
         r1 = trapez(i, vmin, vmax)
         r1 = abs(r1 - th_sol)
         ! computation error for trapezoid rule
         r2 = simpson(i, vmin, vmax)
         r2 = abs(r2 - th_sol)
         ! computation error for Simpson's rule
         write(7,*) i, r1, r2
      end do
      close(7)
      stop 'data saved in IvTrapSimp.dat'
   end program IvTrapSimp

   function f(x)          ! sample function used for integration
      implicit none
      real*8 :: f, x
      f = sin(exp(-x))
   return; end

   function trapez(i, min, max)     ! defining the trapezoid rule
      implicit none
      real*8 :: f, interval, min, max, trapez, x
      integer :: i, n
      trapez = 0
      interval = ((max - min) / (i-1))
      ! sum the midpoints
      do n = 2, (i-1)
         x = interval * (n-1)
         trapez = trapez + f(x)* interval
      end do
      ! add the endpoints
      trapez = trapez + 0.5 * (f(min) + f(max)) * interval
   return; end

   function simpson(i, min, max)   ! defining the Simpson rule
```

```
   implicit none
   real*8 :: f, interval, min, max, simpson, x
   integer :: i, n
   simpson = 0.0
   interval = ((max - min) / (i-1))
   ! loop for odd points
   do n = 2, (i-1), 2
      x = interval * (n-1)
      simpson = simpson + 4 * f(x)
   end do
   ! loop for even points
   do n = 3, (i-1), 2
      x = interval * (n-1)
      simpson = simpson + 2 * f(x)
   end do
   ! add the endpoints
   simpson = simpson + f(min) + f(max)
   simpson = simpson * interval / 3
return; end
```

Romberg Integration

The following Fortran 95 code presents the *Romberg integration* of functions,
combining the above trapezoid rule with the concept of *Richardson extrapolation*.

```
Program Ivromberg  ! V.G. Ivancevic, 2007
   ! Romberg integration of a sample function
   ! combining trapezoid rule with Richardson extrapolation

   real*8 :: r(0:20,0:6), a, b, x
   logical :: Precis

   pi = 4.0*atan(1.0)
   a = -2*pi
   b = 2*pi
   Precis = .false.
   nmax = 20
   n = 0
   nints = 1
   h = b - a
   r(0,0) = 0.5 * h * (f(a) + f(b))
   eps = 1.0e-8
   do while ((.not. Precis) .and. (n < nmax))
     n = n + 1
     nints = 2 * nints
```

```fortran
      twoh = h
      h = 0.5 * h
      r(n,0) = 0.0
      x = a + h
      ! Trapezoid evaluation
      do k = 1, nints, 2
        r(n,0) = r(n,0) + f(x)
        x = x + twoh
      end do
      r(n,0) = 0.5 * r(n-1,0) + h*r(n,0)
      m = 0
      power4 = 1
      ! Richardson extrapolation sequence
      do while ((.not. Precis) .and. m <= n)
        power4 = 4 * power4
        m = m + 1
        dif = (r(n,m-1) - r(n-1,m-1)) / (power4 - 1)
        r(n,m) = r(n,m-1) + dif
        Precis = (abs(r(n,m) - r(n,m-1)) <= &
                            eps * (1 + abs(r(n,m-1))) .and. &
                            n > 2)
      end do
      print 101,(r(n,m),m=0,n)
      101 format (6(1x,f14.11))
   end do
 end program Ivromberg

function f(x)   ! sample function to integrate
    real*8 :: x
    f = exp(-0.1*sin(x)**2)
end function f
```

7.3.3 Numerical Integration of ODEs

Symplectic Verlet Integrator for Simple ODEs

The following Fortran 95 code defines the symplectic Verlet integrator, commonly used in molecular dynamics and computer graphics. Here it is used to simulate a simple spring-mass oscillator (see Fig. 7.67).

```fortran
Program IvVerlet   ! V.G. Ivancevic, 2007
   ! Program for the symplectic Verlet integrator
   ! Simulating a spring-mass oscillator with energy
   ! conservation
```

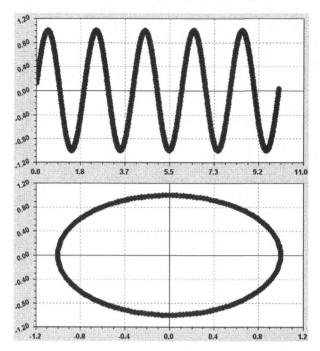

Fig. 7.67. Symplectic Verlet simulation (for 10 s) of a simple spring-mass oscillator: *up*—time plot, *down*—phase plot.

```
implicit none
! system parameters
integer, parameter :: dp = kind(1.0d0)
real(dp), parameter :: Tfin = 10.0d0
real(dp), parameter :: pi = 3.141592653589793d0
real(dp), parameter :: mass = 1.0d0 / pi
real(dp), parameter :: spring = pi
real(dp), parameter :: r_start = 0.0d0
real(dp), parameter :: v_start = 1.0d0 / mass
real(dp), parameter :: h = 0.01d0
real(dp) :: t, r_old, v_old, f_old, r_new, v_new, f_new,
energy

open(unit=7,file='IvVerlet.out.dat')  ! the output file

r_old = r_start
v_old = v_start
f_old = -spring * r_start   ! defining the spring force
write(7,*) 0.0d0, r_old, mass * v_old, 0.0d0
```

```
t = 0.0d0
do while (t < Tfin)    ! main time loop
   ! position = xo + vo*t + a*t^2/2
   r_new = r_old + v_old * h + 0.5d0 * f_new * h**2 / mass
   f_new = -spring * r_new
   ! velocity = vo + (a+a')*t/2
   v_new = v_old + 0.5d0 * h * (f_old + f_new) / mass
   r_old = r_new
   v_old = v_new
   f_old = f_new
   ! energy = m*v^2/2 + c*x^2/2
   energy = 0.5d0 * mass * v_new**2 + 0.5d0 * spring
   * r_new**2
   ! output writing to the file
   write(7,*) t, r_old, mass * v_old, energy - 0.5d0 * pi
   t = t + h        ! time increment
end do     ! end of time loop
close(7)
stop 'data saved in IvVerlet.out.dat'
end program IvVerlet
```

Runge–Kutta Integrator for ODEs

The following Fortran 95 code defines the 4th order Runge–Kutta (RK4) integrator for a system of first-order ODEs (see, e.g., [But64, DP80]). Here it is used to simulate a damped forced linear oscillator (see Fig. 7.68 below), defined by

$$\ddot{y} + 0.3\dot{y} + y = 5\sin(3t), \quad \text{with } y(0) = 1.0, \dot{y}(0) = 0.0.$$

```
Program Ivrk4   ! V.G. Ivancevic, 2007
   ! Runge-Kutta 4th order integrator
   ! Simulating a damped forced linear oscillator

   implicit none
   real*8 :: t, h, Tini, Tfin, y(2)
   integer :: n
   n = 2             ! number of first-order ODEs
   Tini = 0.0        ! initial time
   Tfin = 10.0       ! final time
   h = 0.01          ! step-size
   y(1) = 1.0        ! initial position
   y(2) = 0.0        ! initial velocity

   open(unit=7,file='Ivrk4.out.dat')   ! the output file
```

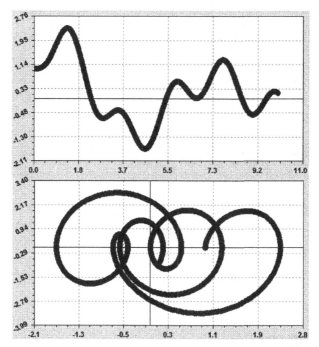

Fig. 7.68. RK4-simulation (for 10 s) of a damped forced linear oscillator, $\ddot{y} + 0.3\dot{y} + y = 5\sin(3t)$, with initial conditions $y(0) = 1.0$, $\dot{y}(0) = 0.0$: *up*—time plot, *down*—phase plot. Compare with Fig. 7.100 below.

```
   do t = Tini, Tfin, h
   ! do M=(Tfin-Tini)/h steps of Runga-Kutta algorithm
      call rk4(t, h, y, n)
      write (7,*) t, y(1), y(2)
      ! output: time, position, velocity
   end do
   close(7)
   stop 'data saved in Ivrk4.out.dat'
end program Ivrk4

subroutine rk4(t, xstep, y, n)
! defining Runge-Kutta 4th order
   implicit none
   real*8 :: deriv, h, t, xstep, y(5)
   real*8 :: k1(5), k2(5), k3(5), k4(5), t1(5), t2(5), t3(5)
   integer :: i, n
   h = xstep / 2.0
   do i = 1, n
```

```fortran
      k1(i) = xstep * deriv(t, y, i)
      t1(i) = y(i) + 0.5 * k1(i)
   end do
   do i = 1, n
      k2(i) = xstep * deriv(t + h, t1, i)
      t2(i) = y(i) + 0.5 * k2(i)
   end do
   do i = 1, n
      k3(i) = xstep * deriv(t + h, t2, i)
      t3(i) = y(i) + k3(i)
   end do
   do i = 1, n
      k4(i) = xstep * deriv(t + xstep, t3, i)
      y(i) = y(i) + (k1(i) + (2.0 * (k2(i) + k3(i))) + k4(i))
      / 6.0
   end do
return; end

function deriv(t, x, i)
! definition of equations (derivatives)
   implicit none
   real*8 :: deriv, t, x(2)
   integer :: i          ! damped forced linear oscillator:
   if (i == 1) deriv = x(2)
   if (i == 2) deriv = -x(1) -0.3 * x(2) + 5*sin(3*t)
return; end
```

Lorenz Butterfly Attractor in C

```c
// Lorenz Butterfly Attractor: Runge-Kutta 4 integrator
// Dr. Vladimir G. Ivancevic, 2008

#include <stdio.h>
#include <math.h>

// Lorenz equations
void ODEs(double t, double x_t , double y_t, double z_t,
  double* eq)
{
    eq[0] = -3 * (x_t - y_t);
    eq[1] = (-x_t * z_t) + (26.5 * x_t) - y_t;
    eq[2] = (x_t * y_t) - z_t;
}

// Runge-Kutta 4th order integrator
```

```
void RK4(double t, double* x, double* y, double* z, double h)
{
    double eq1[3];
    double eq2[3];
    double eq3[3];
    double eq4[3];
    double h2;

    ODEs(t, *x, *y, *z, eq1);
    h2=h/2.0;
    ODEs(t+h2, *x+h2*eq1[0], *y+h2*eq1[1], *z+h2*eq1[2],
    eq2);
    ODEs(t+h2, *x+h2*eq2[0], *y+h2*eq2[1], *z+h2*eq2[2],
    eq3);
    ODEs(t+h, *x+h*eq3[0],*y+h*eq3[1], *z+h*eq3[2], eq4);
    *x=*x+h*(eq1[0]+2.0*eq2[0]+2.0*eq3[0]+eq4[0])/6.0;
    *y=*y+h*(eq1[1]+2.0*eq2[1]+2.0*eq3[1]+eq4[1])/6.0;
    *z=*z+h*(eq1[2]+2.0*eq2[2]+2.0*eq3[2]+eq4[2])/6.0;
}

int main()
{
    FILE* pOut;              // file pointer to output file

    double t    = 0.0;
    double tl   = 200.0;  // ending t
    double t0   = 0.0;    // starting t
    double h    = 0.01;   // step size
    double fx   = 0.0;    // x
    double fy   = 1.0;    // y
    double fz   = 0.0;    // z

    pOut = fopen("out.csv", "w");  // open file
    fputs("t,x,y,z\n", pOut);   // write header

    while (t < tl)              // Main integration loop
    {
        RK4(t, &fx, &fy, &fz, h);    // Call for integrator
        fprintf(pOut, "%.4f,%.7f,%.7f,%.7f\n", t, fx, fy,
        fz);
        t+=h;
    }

    fclose(pOut);
    printf("finished...");
```

```
    getchar();

    return 0;
}
```

Runge–Kutta–Nystrom Integrator for 2nd Order ODEs

The following Fortran 95 code defines the Runge–Kutta–Nystrom mechanical integrator. The example ODE used is: $\ddot{y} = 0.5(t+y+\dot{y}+2)$, $y(0) = 0$, $\dot{y}(0) = 0$, with the analytical solution: $y(t) = e^t - t - 1$. Using the time step $h = 0.01$ s, the obtained numerical solution is correct up to 6 decimal places.

```
Program IvRKN  ! V.G. Ivancevic, 2007
  ! Runge-Kutta-Nystrom integrator for 2nd order ODEs

  real*8 :: h, t, y, ydot, fun, K, L, k1, k2, k3, k4, Tfin
  t = 0.d0     ! initial conditions
  y = 0.d0
  ydot = 0.d0
  Tfin = 1.d0   ! final time
  h = 0.01      ! time-step
  open(unit=7,file='IvRKN.out.dat')       ! output file
  write (7, *) 'RUNGE-KUTTA-NYSTROM METHOD for'
  write (7, *) 'ODE:  ydd = 0.5(t + y + yd + 2),   y(0) = 0,
  yd(0) = 0'
  write (7, *) 'Solution:'
  write (7, *) 't  y  y(t) = exp(t) - t - 1'
  write (7, *) t, y, exp(t)-t-1
  do while ( t < Tfin )
     K = 0.5d0 * h * ( ydot + 0.5d0 * k1 )
     L = h * ( ydot + k3 )
     k1 = 0.5d0 * h * fun( t                , y     , ydot )
     k2 = 0.5d0 * h * fun( t + 0.5d0 * h, y + K, ydot +  k1 )
     k3 = 0.5d0 * h * fun( t + 0.5d0 * h, y + K, ydot +  k2 )
     k4 = 0.5d0 * h * fun( t + h        , y + L,
     ydot +  2.d0 * k3 )
     t = t + h
     y = y + h * ( ydot + ( k1 + k2 + k3) / 3.d0 )
     write (7, *) t, y, exp(t) - t - 1
     ydot = ydot + ( k1 + 2.d0 * k2 + 2.d0 * k3 + k4) / 3.d0
  end do
  close(7)
  stop 'data saved in IvRKN.out.dat'
end program IvRKN
```

```fortran
      real*8 function fun(t, y, ydot)
        real*8 :: t, y, ydot              ! 2nd order ODE defined
        fun = 0.5d0 * ( t + y + ydot + 2 )
      end function fun
```

Runge–Kutta–Nystrom Integrator in C

```c
// Dr. Vladimir G. Ivancevic, 2008
// Runge-Kutta-Nystrom integrator
// for systems of second-order Newton-Lagrange ODEs

#include <math.h>
#include <stdio.h>
#include <stdlib.h>

// Define ODEs
static void ODE(int eqn,double t,double f[],double df[],
double ddf[]) {
        ddf[0] = 3*sin(5*t);   // X - acceleration
        ddf[1] = 5*sin(3*t);   // y - acceleration
        ddf[2] = 7*sin(2*t);   // W - acceleration
     return;
}

// RKN integrator
static void RKN(int eqn,int n,double h,double t,double f[],
double df[],double ddf[]){
     double k1,k2,k3,k4,K,L;
     double fmod[3];
     double dfmod[3];
     double ddfmod[3];

     fmod[eqn] = f[eqn];
     dfmod[eqn] = df[eqn];
     ddfmod[eqn] = ddf[eqn];
     ODE(eqn,t,fmod,dfmod,ddfmod);
     k1 = 0.5*h*ddfmod[eqn];
     K = 0.5*h*(df[eqn] + 0.5*k1);

     fmod[eqn] = f[eqn]+ K;
     dfmod[eqn] = df[eqn] + k1;
     ODE(eqn,t+0.5*h,fmod,dfmod,ddfmod);
     k2 = 0.5*h*ddfmod[eqn];

     fmod[eqn] = f[eqn]+ K;
```

```
        dfmod[eqn] = df[eqn] + k2;
        ODE(eqn,t+0.5*h,fmod,dfmod,ddfmod);
        k3 = 0.5*h*ddfmod[eqn];
        L = h*(df[eqn] + k3);

        fmod[eqn] = f[eqn]+ L;
        dfmod[eqn] = df[eqn] + 2*k3;
        ODE(eqn,t+h,fmod,dfmod,ddfmod);
        k4 = 0.5*h*ddfmod[eqn];

        df[eqn] = df[eqn]+(k1+2.0*k2+2.0*k3+k4)/3.0;
        f[eqn] = f[eqn]+h*(df[eqn]+(k1+k2+k3)/3.0);
        return;
}

int main() {

    /* declare variables */
    int n = 3;
    double t, Tend, h;
    double f0[3];              /* Displacement */
    double df0[3];             /* Velocity */
    double ddf0[3];            /* Acceleration */

    /* initialize variables */
    t = 0.00;
    Tend = 10.00;             /* 10 Sec Max for Method */
    h = 0.01;                  /* 1/100 Sec step interval */

    f0[0] = 0.00;                 /* Y Displacement */
    f0[1] = 0.00;                 /* X Displacement */
    f0[2] = 0.00;                 /* W Displacement */

    df0[0] = 2.196;               /* Y Velocity */
    df0[1] = 0.00;                /* X Velocity */
    df0[2] = 0.05;                /* W Velocity */

    ddf0[0] = 0.00;               /* Y Acceleration */
    ddf0[1] = 0.00;               /* X Acceleration */
    ddf0[2] = 0.00;               /* W Acceleration */

    FILE *fp1;
    if ((fp1 = fopen("X Displacement.txt", "w")) == NULL)
    {
     printf("Error opening file\n");   exit(1);
```

```
    }

FILE *fp2;
if ((fp2 = fopen("X Velocity.txt", "w")) == NULL)
{
 printf("Error opening file\n");    exit(1);
 }

FILE *fp3;
if ((fp3 = fopen("Y Displacement.txt", "w")) == NULL)
{
 printf("Error opening file\n");    exit(1);
 }

FILE *fp4;
if ((fp4 = fopen("Y Velocity.txt", "w")) == NULL)
{
 printf("Error opening file\n");    exit(1);
 }

FILE *fp5;
if ((fp5 = fopen("W Displacement.txt", "w")) == NULL)
{
 printf("Error opening file\n");    exit(1);
 }

FILE *fp6;
if ((fp6 = fopen("W Velocity.txt", "w")) == NULL)
{
 printf("Error opening file\n");    exit(1);
 }

while (t < Tend)                // Main time loop
{
      RKN(0,3,h,t,f0,df0,ddf0);
      RKN(1,3,h,t,f0,df0,ddf0);
      RKN(2,3,h,t,f0,df0,ddf0);
        fprintf(fp1,"%lf\t%lf\n",t+h,f0[0]);
        fprintf(fp2,"%lf\t%lf\n",t+h,df0[0]);
        fprintf(fp3,"%lf\t%lf\n",t+h,f0[1]);
        fprintf(fp4,"%lf\t%lf\n",t+h,df0[1]);
        fprintf(fp5,"%lf\t%lf\n",t+h,f0[2]);
        fprintf(fp6,"%lf\t%lf\n",t+h,df0[2]);
      t=t+h;
}
```

```
    // Close Output files
    fclose(fp1);
    fclose(fp2);
    fclose(fp3);
    fclose(fp4);
    fclose(fp5);
    fclose(fp6);
    return(0);
}
```

Complex-Valued Runge–Kutta–Nystrom Integrator in C

```
// Dr. Vladimir G. Ivancevic, 2008
// Complex Runge-Kutta-Nystrom integrator
// for systems of second-order complex-valued ODEs

#include <math.h>
#include <stdio.h>
#include <stdlib.h>
#include <complex.h>

// Define complex-valued ODEs
static void ODE(int eq, double t, double _Complex z[],
 double _Complex dz[], double _Complex ddz[]) {
// 2DOF Eq. of motion in the Complex-plane for one robot
    ddz[0] = 0.2*ccos(dz[0]) - 0.5*csin(z[0])
        + 0.3*csin(2.*t) - 0.2*I*ccos(5.*t);
/*  ddz[1] = ... for the second robot, etc */
    return;
}

// Complex-valued RKN integrator
static void RKN(int eq, int n, double h, double t,
 double _Complex z[], double _Complex dz[],
   double _Complex ddz[]) {
    double _Complex k1,k2,k3,k4,K,L;
    double _Complex zmod[3];
    double _Complex dzmod[3];
    double _Complex ddzmod[3];

    zmod[eq] = z[eq];
    dzmod[eq] = dz[eq];
    ddzmod[eq] = ddz[eq];
    ODE(eq,t,zmod,dzmod,ddzmod);
```

```
        k1 = 0.5*h*ddzmod[eq];
        K = 0.5*h*(dz[eq] + 0.5*k1);

        zmod[eq] = z[eq]+ K;
        dzmod[eq] = dz[eq] + k1;
        ODE(eq,t+0.5*h,zmod,dzmod,ddzmod);
        k2 = 0.5*h*ddzmod[eq];

        zmod[eq] = z[eq]+ K;
        dzmod[eq] = dz[eq] + k2;
        ODE(eq,t+0.5*h,zmod,dzmod,ddzmod);
        k3 = 0.5*h*ddzmod[eq];
        L = h*(dz[eq] + k3);

        zmod[eq] = z[eq]+ L;
        dzmod[eq] = dz[eq] + 2*k3;
        ODE(eq,t+h,zmod,dzmod,ddzmod);
        k4 = 0.5*h*ddzmod[eq];

        dz[eq] = dz[eq]+(k1+2.0*k2+2.0*k3+k4)/3.0;
        z[eq] = z[eq]+h*(dz[eq]+(k1+k2+k3)/3.0);
        return;
}


int main() {
    /* declare variables */
    // I is imaginary unit; do NOT use I as a loop variable!
    int n = 1;    // Number of complex-valued ODEs
    double t, Tfin, h;  // double times and step-size
    double _Complex z0[3];          /* Displacement */
    double _Complex dz0[3];         /* Velocity */
    double _Complex ddz0[3];        /* Acceleration */

    /* initialise variables */
    t = 0.0;
    Tfin = 10.0;
    // initial and final times and time-step
    h = 0.01;
    // Imaginary unit = I = (0.0F +  1.0iF);
    z0[0] = 0.1 + I*0.1;
    /* Initial 1.Displacement */
    dz0[0] = 0.1 + I*0.1;                /* Initial 1.Velocity */

    /* output files */
    FILE *fp1;
```

```
if ((fp1 = fopen("1.Compl.Displ.txt", "w")) == NULL)
{ printf("Error opening file\n"); exit(1); }

FILE *fp2;
if ((fp2 = fopen("1.Compl.Veloc.txt", "w")) == NULL)
{ printf("Error opening file\n"); exit(1); }

/* time loop */
while (t < Tfin)
{
   RKN(0,1,h,t,z0,dz0,ddz0);        // Integrate 1.ODE
/* RKN(1,1,h,t,z0,dz0,ddz0);    - Integrate 2.ODE, etc */
   fprintf(fp1,"%lf\t%lf\t%lf\n",t,creal(z0[0]),
     cimag(z0[0]));  // print t, real(z), imag(z)
   fprintf(fp2,"%lf\t%lf\t%lf\n",t,creal(dz0[0]),
     cimag(dz0[0])); // print t, real(dz), imag(dz)
   t += h;
}

// close Output files
fclose(fp1);
fclose(fp2);
return(0);
}
```

Predictor–Corrector Integrator for ODEs

The following Fortran 95 code presents the Adams predictor–corrector integrator for ODEs. It includes the above Runge–Kutta 4 routine as a starter.

```
Program IvAdams46   ! V.G. Ivancevic, 2007
   ! Adams 4th/6th order Predictor-Corrector integrator
   for ODEs

   real*8 :: h, x(100), y(100), f1
   integer :: n

   x(1) = 0.d0     ! initial conditions
   y(1) = 5.d0

   ! exact solution  x(t) = 7exp(t) - t^2 - 2t - 2,
   x(1) = f(1):
     f1 = 7. * exp(1.) - 5.

   write(6,*) ' '
```

```fortran
      write(6,*) '                ', '                  x(1)', '
     error'

      h = 0.1d0
      n = 10

      call Adams4(n,h,x,y)      ! Adams 4 output and error
      write(6,*)  'Adams4:  ', y(n+1),  abs(y(n+1)-f1)

      call Adams6(n,h,x,y)      ! Adams 6 output and error
      write(6,*)  'Adams6:  ', y(n+1),  abs(y(n+1)-f1)

end program IvAdams46

real*8 function f(t,v)
  real*8 t, v      ! differential equation defined
  f = t*t + v
end function f

subroutine RK4(h,x,y)     ! Runge-Kutta 4 procedure
  real*8 :: h, x(100), y(100), f, k1, k2, k3, k4
  integer :: i
  do i = 1,5
     k1 = h * f(x(i), y(i))
     k2 = h * f(x(i) + 0.5d0 * h, y(i) + 0.5d0 * k1)
     k3 = h * f(x(i) + 0.5d0 * h, y(i) + 0.5d0 * k2)
     k4 = h * f(x(i) + h, y(i) + k3)
     x(i+1) = x(i) + h
     y(i+1) = y(i) + (k1 + 2.d0 * k2 + 2.d0 * k3 + k4) / 6.d0
  end do
end subroutine RK4

subroutine Adams4(n,h,x,y)      ! Adams 4 predictor-corrector
procedure
  real*8 :: h, x(100), y(100), py(100), f
  integer :: n, i

  call RK4(h,x,y)   ! Runge-Kutta 4 starter

  do i = 4, n + 1
     x(i+1) = x(i) + h

     ! Adams-Bashford 4th order predictor :
     py(i+1) = y(i) + h * ( 55.d0 * f(x(i), y(i)) &
             - 59.d0 * f(x(i-1), y(i-1)) &
```

```
              + 37.d0 * f(x(i-2), y(i-2)) &
              - 9.d0 * f(x(i-3), y(i-3))  ) / 24.d0

      ! Adams-Moulton 4th order corrector:
      y(i+1) = y(i) + h * ( 9.d0 * f(x(i+1), py(i+1)) &
              + 19.d0 * f(x(i), y(i)) &
              - 5.d0 * f(x(i-1), y(i-1)) &
              + f(x(i-2), y(i-2))  ) / 24.d0
   end do
end subroutine Adams4

subroutine Adams6(n,h,x,y)      ! Adams 6 predictor-corrector
procedure
      real*8 h, x(100), y(100), py(100), f
      integer n, i

   call RK4(h,x,y)     ! Runge-Kutta 4 starter

   do i = 6, n + 1
      x(i+1) = x(i) + h

      ! Adams-Bashford 6th order predictor:
      py(i+1) = y(i) + h * ( 4277.d0 * f(x(i), y(i)) &
              - 7923.d0 * f(x(i-1), y(i-1)) &
              + 9982.d0 * f(x(i-2), y(i-2)) &
              - 7298.d0 * f(x(i-3), y(i-3)) &
              + 2877.d0 * f(x(i-4), y(i-4)) &
              -  475.d0 * f(x(i-5), y(i-5)) ) / 1440.d0

      ! Adams-Moulton 6th order corrector:
      y(i+1) = y(i) + h * ( 475.d0 * f(x(i+1), py(i+1)) &
              + 1427.d0 * f(x(i), y(i)) &
              - 798.d0 * f(x(i-1), y(i-1)) &
              + 482.d0 * f(x(i-2), y(i-2)) &
              - 173.d0 * f(x(i-3), y(i-3)) &
              +  27.d0 * f(x(i-4), y(i-4)) ) / 1440.d0
   end do
end subroutine Adams6
```

Damped, Random, Sinusoidally Driven Pendulum

Consider the damped, random, sinusoidally driven pendulum:

$$\dot{\omega} = -\omega/q - \sin\theta + g\cos\theta + rnd, \qquad \dot{\theta} = \omega, \qquad \dot{\phi} = w_d$$

Equations and initial conditions

$Eqns = \{\omega'[t] == -\omega[t]/q - \sin[\theta[t]] + g * \cos[\phi[t]] + Random[],$
$\theta'[t] == \omega[t], \phi'[t] == wd,$
$\omega[0] == Random[], \theta[0] == Random[], \phi[0] == Random[]\};$

Parameters

$Eqns = Eqns/.\{g \rightarrow 0.9, q \rightarrow 2., wd \rightarrow 2/3\}; Tfin = 50;$

Numerical solution and plots

$Sol = NDSolve[Eqns, \{\omega, \theta, \phi\}, \{t, Tfin\}];$
$\quad ParametricPlot[Evaluate[\{\theta[t], \omega[t]\}/.Sol], \{t, 0, Tfin\},$
$\quad PlotRange \rightarrow All,$
$Frame \rightarrow True, AxesLabel \rightarrow \{\theta, \omega\}];$

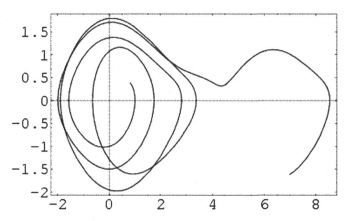

Fig. 7.69. Phase plot of the equation: $\dot{\omega} = -\omega/q - \sin\theta + g\cos\theta + rnd$, $\dot{\theta} = \omega$, $\dot{\phi} = w_d$.

$\quad Plot[Evaluate[\omega[t]/.Sol], \{t, 0, Tfin\}, PlotRange \rightarrow All,$
$Frame \rightarrow True, PlotLabel \rightarrow \omega];$
$\quad << Graphics`Graphics`;$
$Om = Table[\omega[s * Tfin/100]/.Sol, \{s, 1, 100\}];$
$S = Fourier[Om];$

Stochastic Dynamics: Langevin and Fokker–Planck Equations

Parameter Bifurcation

Potential function $V(q)$ & Probability-density function (or stationary distribution function) $P(q)$

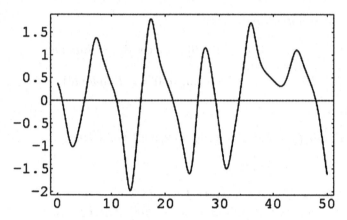

Fig. 7.70. Time evolution of angular velocity $\omega(t)$.

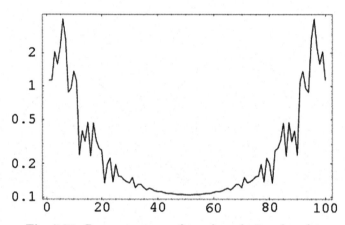

Fig. 7.71. Power spectrum of angular velocity—log plot.

$V = kq + \frac{q^2}{2} + \frac{q^3}{3} + \frac{q^4}{4} + \frac{q^5}{5}; \quad P = e^{-V};$
$Print["Potential : V = ", V]; Tfin = 3;$
$Do[Plot[\{P/. \ \{k \ \rightarrow \ i\}, V/. \ \ \{k \ \rightarrow \ i\}\}, \{q, -Tfin, Tfin\}, ImageSize \ \rightarrow$
$150, PlotLabel \rightarrow i], \{i, -2, -5, -0.5\}];$
    Potential: $V = k \, q + \frac{q^2}{2} + \frac{q^3}{3} + \frac{q^4}{4} + \frac{q^5}{5}$

Overdamped Langevin Equation

$F = -\partial_q V; Print["Equation : q' = ", F]; F = F/. \{q \rightarrow q[t]\};$
$Sol[K\_] \ := \ NDSolve[\{q'[t] \ == \ F + \sin[t]Random[], q[0] \ == \ 1\}/. \ k \ \rightarrow$
$K, q, \{t, \frac{Tfin}{3}\}];$
$[Plot[Evaluate[q[t]/. \ \ Sol[k]], \{t, 0, \frac{Tfin}{3}\}, PlotRange \ \rightarrow \ All, PlotLabel \ \rightarrow$
$k, ImageSize \rightarrow 150], \{k, -6, 1\}];$
Equation: $q' = -k - q - q^2 - q^3 - q^4$

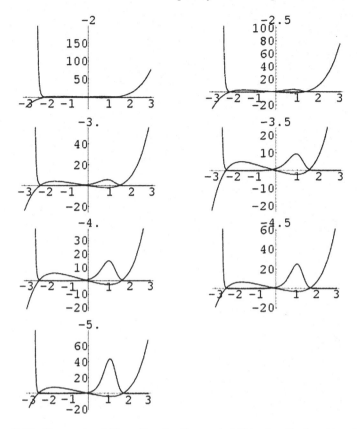

Fig. 7.72. Parameter bifurcation for the probability-density function $P(q)$.

7.3.4 Vector-Field and Lyapunov Function

In this subsection we give $Mathematica^{TM}$ code for a vector-field and Lyapunov function for a nonlinear differential equation.

```
<< Graphics`PlotField`
```
The system's differential equation is given as: $\ddot{x} = -\frac{6}{10}\dot{x} - 3x - x^2$
The equivalent Hamiltonian system is: $q' = p, \quad p' = -\frac{6}{10}p - 3q - q^2$
Hence, the associated vector field is: $F[q,p] = \{p, -\frac{6}{10}p - 3q - q^2\}$
```
F[q_, p_] := {p, -6/10 p - 3q - q^2}
```
Display the vector-field.
```
VField = PlotVectorField[F[q,p], {q, -4, 4}, {p, -4, 4}, PlotPoints ->
20,
ScaleFunction -> (0.2#&), ScaleFactor -> None, AspectRatio -> 1,
PlotRange -> {{-4, 2}, {-3, 3}}, Frame -> True];
```

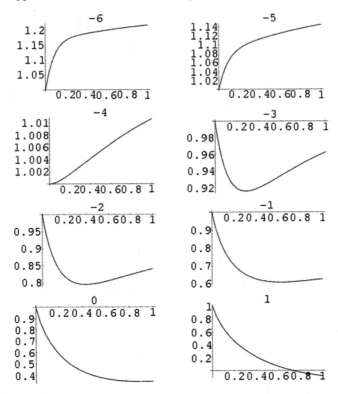

Fig. 7.73. Simulation of the overdamped Langevin equation.

Define the candidate Lyapunov function.
$$F = p^2 + \frac{pq}{5} + \frac{153q^2}{50} + \frac{2q^3}{3}$$
Plot contours of this function to visualize the regions defined by the candidate Lyapunov function (blue is used to indicate positive function values).

$CPlt1 = ContourPlot[F, \{q, -4, 2\}, \{p, -3, 3\}, PlotPoints \rightarrow 49,$
$Contours \rightarrow \{-0.1, 0., 0.5, 1., 5., 9.45, 10.\},$
$ContourStyle \rightarrow \{\{RGBColor[1, 0, 0]\}, \{RGBColor[0, 0, 0]\},$
$\{RGBColor[0, 0, 1]\},$
$\{RGBColor[0, 0, 1]\}, \{RGBColor[0, 0, 1]\}, \{RGBColor[0, 0, 1]\},$
$\{RGBColor[0, 0, 1]\}\}, ContourShading \rightarrow False, AspectRatio \rightarrow 1];$

Define the derivative of the Lyapunov function with respect to time.
$$DF = -p^2 - \frac{3q^2}{5} - \frac{q^3}{5}.$$
Plot contours of this function to determine regions where it is negative (black indicates zero contour, red indicates a negative contour and blue indicates a positive contour).

$CPlt2 = ContourPlot[DF, \{q, -4, 2\}, \{p, -3, 3\}, PlotPoints \rightarrow 49,$
$Contours \rightarrow \{-0.1, 0., 0.1\},$

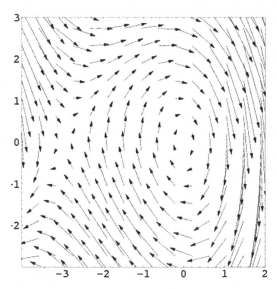

Fig. 7.74. Hamiltonian vector-field for the equation: $\ddot{x} = -\frac{6}{10}\dot{x} - 3x - x^2$.

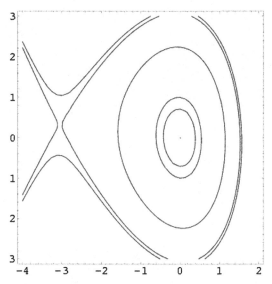

Fig. 7.75. Contour plot of the candidate Lyapunov function: $F = p^2 + \frac{pq}{5} + \frac{153q^2}{50} + \frac{2q^3}{3}$.

$ContourStyle \rightarrow \{\{RGBColor[1,0,0]\}, \{RGBColor[0,0,0]\},$
$\{RGBColor[0,0,1]\}\},$
$ContourShading \rightarrow False, AspectRatio \rightarrow 1];$

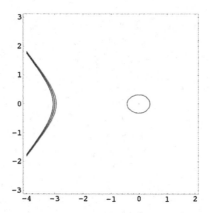

Fig. 7.76. Contour plot of the derivative $\dot{F}$ of the Lyapunov function F.

Numerical simulation.
$FF[q\_, p\_] = \{p, -\frac{6}{10}p - 3q - q^2\}/.\{q \to q[t], p \to p[t]\}; Tfin = 10;$

$Qeq = q'[t] == FF[q, p][[1]]; Peq = p'[t] == FF[q, p][[2]];$
$sys = \{Qeq, Peq\}; Inic = \{q[0] == 0, p[0] == 0.1\}; vars = \{q[t], p[t]\}; eqs = Join[sys, Inic]$
$\{q'[t] == p[t], p'[t] == -\frac{3p[t]}{5} - 3q[t] - q[t]^2, q[0] == 0, p[0] == 0.1\}$
$sol = NDSolve[eqs, vars, \{t, Tfin\}];$
$nPl = ParametricPlot[Evaluate[vars/.sol], \{t, 0, Tfin\}, Frame \to True];$

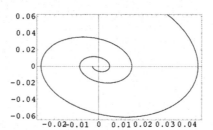

Fig. 7.77. Numerical simulation—phase plot of the original differential equation.

Put all the images together on one plot. Regions outlined by the candidate Lyapunov function contours where the derivative is zero define the system's stable trajectories.
$Show[\{VField, CPlt1, CPlt2, nPl\}, AspectRatio \to 1, Frame \to True];$

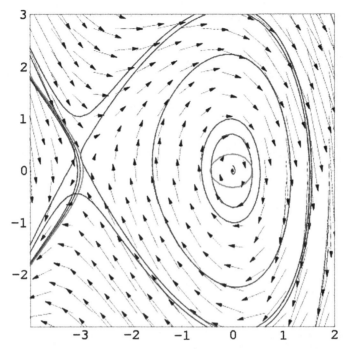

Fig. 7.78. Showing all the figures simultaneously: regions outlined by the candidate Lyapunov function contours where the derivative is zero define the system's stable trajectories.

7.3.5 Basics of Qualitative Dynamics

Dissipation

Divergence defined:

$Div[f\_List, x\_List] := Inner[D, f, x, Plus]$

Given ODEs $x' = F$ and $y' = G$, where the vector-fields F and G are defined as:

$$F := \begin{pmatrix} -\omega/q - \sin[\theta] + g\cos[\phi] \\ \omega \\ a \end{pmatrix};$$

$$G := \begin{pmatrix} -\sin[\theta] + g\cos[\phi] \\ \omega \\ a \end{pmatrix};$$

If the divergence of the vector-field is negative, the system is dissipative.

$D_F = Div[Flatten[F], \{\omega, \theta, \phi\}]$

$-\frac{1}{q}$

If the divergence of the vector-field is 0, the system is conservative.

$D_G = Div[Flatten[G], \{\omega, \theta, \phi\}]$

0

Linearization and Critical Points

Jacobian matrix defined:

$JacMat[f\_List, x\_List] := Outer[D, f, x]$

Given ODE $x\prime = F(x)$, where the vector-field F is defined as:

$$F = \begin{pmatrix} (1 - x_1)x_1 \\ -2x_2 \end{pmatrix}$$

$\{\{(1 - x_1)x_1\}, \{-2x_2\}\}$

Several ways to find zero-points of F:

$Solve[F == \{0, 0\}]$

$\{\{x_1 \to 0, x_2 \to 0\}, \{x_1 \to 1, x_2 \to 0\}\}$

$FindRoot[Flatten[F] == \{0, 0\}, \{x_1, 0\}, \{x_2, 0\}]$

$\{x_1 \to 0., x_2 \to 0.\}$

Jacobian matrix of F:

$J = JacMat[F, \{x_1, x_2\}]//Transpose//MatrixForm$

$$\begin{pmatrix} 1 - 2x_1 \\ 0 \end{pmatrix}$$

Jacobian matrix at the zero point $(0, 0)$:

$J/.\{x_1 \to 0, x_2 \to 0\}$

$$\begin{pmatrix} 1 \\ 0 \end{pmatrix}$$

gives one eigenvalue negative and one positive $\Rightarrow$ a saddle.

Jacobian matrix at the zero point $(1, 0)$:

$J/.\{x_1 \to 1, x_2 \to 0\}$

$$\begin{pmatrix} -1 \\ 0 \end{pmatrix}$$

gives both eigenvalues negative $\Rightarrow$ an attracting node.

Lyapunov Stability

Definiton of the Lie derivative $L_f h$ of function $h(x)$ along direction defined by the vector-field $f(x)$,

$L_f h = Grad(h) \cdot f$:

$LieDer[f\_List, h\_, x\_List] := \partial_\# h \& /@x.f$

Given ODE $x' = F$, where the vector-field F is defined as:

$$F = \begin{pmatrix} x_2 \\ -\alpha x_2 - x_1{}^3 \end{pmatrix};$$

When $\alpha = 0$ the ODE is Hamiltonian with Hamiltonian function H given as:

$H = \frac{1}{2}x_2{}^2 + \frac{1}{4}x_1{}^4;$

$x_1' = \partial_{x_2} H$

x_2

$x_2' = -\partial_{x_1} H$

$-x_1^3$

So we choose a positive-definite Lyapunov function $V(x) = H > 0$ for all x on R^2,

$V = H$;

Its orbital time derivative $V'(x)$ represents the Lie derivative along direction defined by the vector-field F:

$V' = Simplify[LieDer[F, V, \{x_1, x_2\}]]$

$\{-\alpha x_2^2\}$

Therefore, $V'(x) = -\alpha x_2^2 \leq 0$ on R^2 and the system is Lyapunov stable for all x on R^2 (according to the Lyapunov stability theorem, which says:

1. If $V'(x) < 0$ the system is asymptotically stable;
2. If $V'(x) \leq 0$ the system is stable;
3. If $V'(x) > 0$ the system is unstable).

7.3.6 Kuramoto's Neural Model

In this subsection we give $Mathematica^{TM}$ code for the Kuramoto's neural model.

$n = 100; \epsilon = 0.01; Table[\Omega_i = Random[], \{i, n\}];$
$a[t\_] = 10 \sin[5t]; X = Table[\theta_i[t], \{i, n\}];$
$Sys = Table\left[\theta_i'[t] == \Omega_i + \epsilon a[t] \sum_{j=1}^n \sin[\theta_j[t] - \theta_i[t]], \{i, n\}\right];$
$Eqns = Join[Sys, Table[\theta_i[0] == 0, \{i, n\}]];$
$Tfin = 10; Sol = NDSolve[Eqns, X, \{t, Tfin\}];$
$Plot[Evaluate[X/.Sol], \{t, 0, Tfin\}, PlotRange \to All, Frame \to True];$

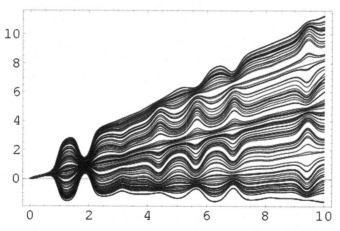

Fig. 7.79. Simulation of the 100D Kuramoto model: θ_i vs. time (10 s).

Cascade of Van der Pol Oscillators

In this subsection we give $Mathematica^{TM}$ code for a cascade of Van der Pol Oscillators.

Set up:
$n = 30; Tfin = 20;$
Input:
$x_0[t] = \sin[t]; x_0\prime[t] = \exp[-t];$
Random parameters:
$Table[\{\epsilon_i = Random[Real, \{0, 3\}], \alpha_i = Random[Real, \{-1, 1\}], \beta_i = Random[Real, \{0, 1\}]\}, \{i, n\}];$
$Eqn = Table[\{$
$x_i''[t] - \epsilon_i(1 - x_i[t]^2)x_i'[t] + x_i[t] == \alpha_i x_{i-1}[t] + \beta_i x'_{i-1}[t],$
$x_i[0] == 0.01, x'_i[0] == 0\}, \{i, n\}];$
$Sol = NDSolve[Flatten[Eqn], Table[x_i, \{i, n\}], \{t, Tfin\}];$
$Do[g_i = ParametricPlot[Evaluate[\{x_i[t], x_i'[t]\}/.Sol], \{t, 0, Tfin\},$
$PlotRange \rightarrow All, Frame \rightarrow True, DisplayFunction \rightarrow Identity], \{i, n\}];$
$Plot[Evaluate[Table[x_i[t], \{i, n\}]/.Sol], \{t, 0, Tfin\},$
$PlotRange \rightarrow All, Frame \rightarrow True, PlotLabel \rightarrow "DISPLACEMENTS in-TIME"];$

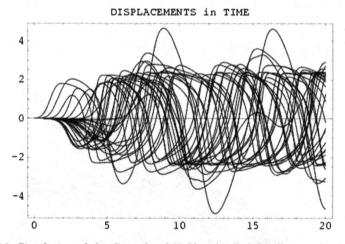

DISPLACEMENTS in TIME

Fig. 7.80. Simulation of the Cascade of 30 Van der Pol Oscillators: displacements vs. time (20 s).

$Plot[Evaluate[Table[x_i'[t], \{i, n\}]/.Sol], \{t, 0, Tfin\},$
$PlotRange \rightarrow All, Frame \rightarrow True, PlotLabel \rightarrow "VELOCITIES in - TIME"];$

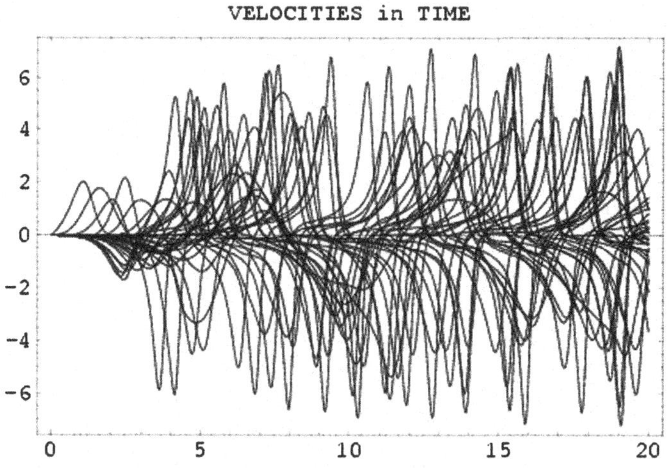

VELOCITIES in TIME

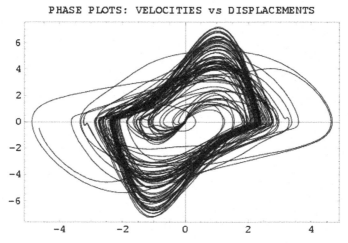

Fig. 7.81. Simulation of the Cascade of 30 Van der Pol Oscillators: velocities vs. time (20 s).

PHASE PLOTS: VELOCITIES vs DISPLACEMENTS

Fig. 7.82. Simulation of the Cascade of 30 Van der Pol Oscillators: phase portrait (20 s).

$Show[Table[g_i, \{i, n\}], DisplayFunction \rightarrow \$DisplayFunction,$
$Frame \rightarrow True, PlotLabel \rightarrow$
$"PHASEPLOTS : VELOCITIESvsDISPLACEMENTS"];$

Fuzzy Dynamical Oscillators

In this subsection we give $Mathematica^{TM}$ code for several fuzzified dynamical oscillators.

Fuzzy Linear Oscillator (Fig. 7.83)

$k = 5.; w = 2/3; v = 0.9; e = 1; Tfin = 20; n = 10;$
$For[i = 1, i <= n, i + +, \{A[i] = 3 * i + 4;$
    (* fuzzy parameter *)
$F[i] := A[i] * \exp[-k * t] * \cos[w * t];$
    (* equation *)
$Eqns[i] := x[i]"[t] + v * x[i]'[t] + e * x[i][t] == F[i],$
$x[i][0] == 0, x[i]'[0] == 0\};$
$Sol[i] = NDSolve[Eqns[i], x[i], \{t, Tfin\}];$
$g[i] = Plot[Evaluate[x[i][t]/.Sol[i]], \{t, 0, Tfin\},$
$PlotRange \to All, DisplayFunction \to Identity];$
$f[i] = ParametricPlot[Evaluate[\{x[i][t], x[i]'[t]\}/.Sol[i]], \{t, 0, Tfin\},$
$PlotRange \to All, Frame \to True, DisplayFunction \to Identity]\}];$
$Show[Table[g[i], \{i, n\}], DisplayFunction \to \$DisplayFunction, Frame \to$
$True];$
$Show[Table[f[i], \{i, n\}], DisplayFunction \to \$DisplayFunction, Frame \to$
$True];$

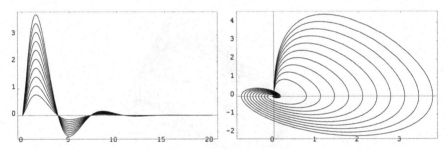

Fig. 7.83. Fuzzy linear oscillator: time evolution (*left*); phase-plot (*right*).

Fuzzy Hamilton Oscillator (Fig. 7.84)

$k = 5.; w = 2/3; v = 0.9; e = 1; Tfin = 20; n = 10;$
$For[i = 1, i <= n, i + +, \{A[i] = 3 * i + 4;$
    (* fuzzy parameter *)
$F[i] := A[i] * \exp[-k * t] * \cos[w * t];$
    (* equation *)
$Eqns[i] := x[i]"[t] + v * x[i]'[t] + e * \sin[x[i][t]] == F[i],$
$x[i][0] == 0, x[i]'[0] == 0$
$Sol[i] = NDSolve[Eqns[i], x[i], \{t, Tfin\}];$
$g[i] = Plot[Evaluate[x[i][t]/.Sol[i]], \{t, 0, Tfin\},$
    $PlotRange \to All, DisplayFunction \to Identity];$
$f[i] = ParametricPlot[Evaluate[\{x[i][t], x[i]'[t]\}/.Sol[i]], \{t, 0, Tfin\},$

$PlotRange \rightarrow All, Frame \rightarrow True, DisplayFunction \rightarrow Identity]\}]$;
$Show[Table[g[i], \{i, n\}], DisplayFunction \rightarrow \$DisplayFunction, Frame \rightarrow$
$True]$;
$Show[Table[f[i], \{i, n\}], DisplayFunction \rightarrow \$DisplayFunction, Frame \rightarrow$
$True]$;

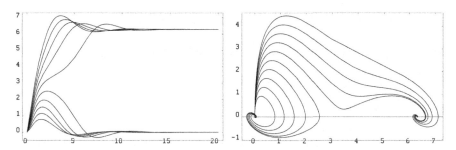

Fig. 7.84. Fuzzy Hamilton oscillator: time evolution (*left*); phase-plot (*right*).

Fuzzy Duffing Oscillator (Fig. 7.85)

$B = 0.15; Tfin = 10; n = 10$;
$For[i = 1, i <= n, i + +, \{A[i] = 0.2 * i$;
    (* fuzzy parameter *)
$F[i] := A[i] * \cos[w * t]$;
    (* equation*)
$Eqns[i] := \{x[i]"[t] + B * x[i]'[t] - x[i][t] + x[i][t]^3 == F[i]$,
$x[i][0] == -1, x[i]'[0] == 1\}$;
$Sol[i] = NDSolve[Eqns[i], x[i], \{t, Tfin\}]$;
$g[i] = Plot[Evaluate[x[i][t]/.Sol[i]], \{t, 0, Tfin\}$,
$PlotRange \rightarrow All, DisplayFunction \rightarrow Identity]$;
$f[i] = ParametricPlot[Evaluate[\{x[i][t], x[i]'[t]\}/.Sol[i]], \{t, 0, Tfin\}$,
$PlotRange \rightarrow All, Frame \rightarrow True, DisplayFunction \rightarrow Identity]\}]$;
$Show[Table[g[i], \{i, n\}], DisplayFunction \rightarrow \$DisplayFunction, Frame \rightarrow$
$True]$;
$Show[Table[f[i], \{i, n\}], DisplayFunction \rightarrow \$DisplayFunction, Frame \rightarrow$
$True]$;

Fuzzy (Forced) Van der Pol Oscillator (Fig. 7.86)

$b = 5.; eps = 1; wo = 2.; fr = 4.; Tfin = 5; n = 10$;
    (* parameters *)
$For[i = 1, i <= n, i + +, \{A[i] = 3 * i + 4$;
    (* interval amplitude *)
$F[i] := A[i] * \sin[fr * t]$;

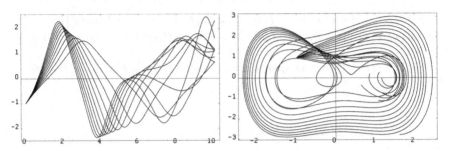

Fig. 7.85. Fuzzy Duffing oscillator: time evolution (*left*); phase-plot (*right*).

```
(* periodic forcing *)
Eqn[i] := {
    (* Van der Pol equation *)
x[i]"[t] − eps ∗ (1 − 4 ∗ b ∗ x[i][t]²) ∗ x[i]′[t] + wo² ∗ x[i][t] == F[i],
    x[i][0] == 0, x[i]′[0] == 0};
    Sol[i] = NDSolve[Eqn[i], x[i], {t, Tfin}];
    (* solution *)
g[i] = Plot[Evaluate[x[i][t]/.Sol[i]], {t, 0, Tfin},
    (* time plot *)
PlotRange → All, DisplayFunction → Identity];
    (* phase plot *)
f[i] = ParametricPlot[Evaluate[{x[i][t], x[i]′[t]}/.Sol[i]], {t, 0, Tfin},
PlotRange → All, Frame → True, DisplayFunction → Identity]};
Show[Table[g[i], {i, n}], DisplayFunction → $DisplayFunction, Frame →
True];
Show[Table[f[i], {i, n}], DisplayFunction → $DisplayFunction, Frame →
True];
```

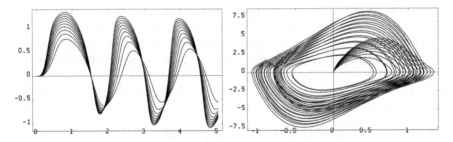

Fig. 7.86. Fuzzy Duffing oscillator: time evolution (*left*); phase-plot (*right*).

7.3.7 Boundary Value Problems and PDEs

Two-Point Boundary Value Problem for ODEs

The following Fortran 95 code defines the Shooting algorithm for the two-point boundary value problem (BVP) for ODEs. The example ODE used is:

$$\ddot{y} = -\pi^2(y+1)/4, \qquad y(0) = 0, \qquad y(1) = 1,$$

with the solution given at Fig. 7.87 below.

```fortran
Program IvShooting  ! V.G. Ivancevic, 2007

! Shooting algorithm for the boundary value problem (BVP)

! using RK4 and secant methods

  implicit none
  integer, parameter :: n = 101, m = 2
  integer :: i
  real*8 :: k11,k21,k12,k22,k13,k23,k14,k24
  real*8 :: dl,xl,xu,h,d,yl,yu,x0,dx,x1,x2,f0,f1
  real*8 :: x,y1,y2,g1,g1fun,g2,g2fun
  real*8, dimension (2,n) :: y
  dl = 1.0e-06     ! initialization
  xl = 0.0
  xu = 1.0
  h  = (xu - xl) / (n - 1)
  d  = 0.1
  yl = 0.0
  yu = 1.0
  x0 = (yu - yl) / (xu - xl)
  dx = 0.01
  x1 = x0 + dx
  open(unit=7,file='IvShoot.out.dat')
  y(1,1) = yl
  do while ( abs(d) > dl )     ! secant search for the root
     y(2,1) = x0
     do i = 1, n-1
     ! RK4 calculation of the first trial solution
        x  = xl + h*i
        y1 = y(1,i)
        y2 = y(2,i)
        k11 = h * g1fun(y1,y2,x)
        k21 = h * g2fun(y1,y2,x)
        k12 = h * g1fun((y1+k11/2.0),(y2+k21/2.0),(x+h/2.0))
```

```
      k22 = h * g2fun((y1+k11/2.0),(y2+k21/2.0),(x+h/2.0))
      k13 = h * g1fun((y1+k12/2.0),(y2+k22/2.0),(x+h/2.0))
      k23 = h * g2fun((y1+k12/2.0),(y2+k22/2.0),(x+h/2.0))
      k14 = h * g1fun((y1+k13),(y2+k23),(x+h))
      k24 = h * g2fun((y1+k13),(y2+k23),(x+h))
      y(1,i+1) = y(1,i) + (k11 + 2.0 * (k12 + k13) + k14)
      / 6.0
      y(2,i+1) = y(2,i) + (k21 + 2.0 * (k22 + k23) + k24)
      / 6.0
    end do
    f0 = y(1,n) - 1.0
    y(2,1) = x1
    do i = 1, n-1
    ! RK4 calculation of the second trial solution
      x  = x1 + h*i
      y1 = y(1,i)
      y2 = y(2,i)
      k11 = h * g1fun(y1,y2,x)
      k21 = h * g2fun(y1,y2,x)
      k12 = h * g1fun((y1+k11/2.0),(y2+k21/2.0),(x+h/2.0))
      k22 = h * g2fun((y1+k11/2.0),(y2+k21/2.0),(x+h/2.0))
      k13 = h * g1fun((y1+k12/2.0),(y2+k22/2.0),(x+h/2.0))
      k23 = h * g2fun((y1+k12/2.0),(y2+k22/2.0),(x+h/2.0))
      k14 = h * g1fun((y1+k13),(y2+k23),(x+h))
      k24 = h * g2fun((y1+k13),(y2+k23),(x+h))
      y(1,i+1) = y(1,i) + (k11 + 2.0 * (k12 + k13) + k14)
      / 6.0
      y(2,i+1) = y(2,i) + (k21 + 2.0 * (k22 + k23) + k24)
      / 6.0
    end do
    f1 = y(1,n) - 1.0
    d  = f1 - f0
    x2 = x1 - f1 * (x1 - x0) / d
    x0 = x1
    x1 = x2
  end do
  write (7,"(2F16.8)") (h*(i-1), y(1,i),i=1,n,m)
  close(7)
  stop 'data saved in IvShoot.out.dat'
end program IvShooting

real*8 function g1fun (y1,y2,t) result (g1)
  implicit none
  real*8 :: y1,y2,t,g1
  g1 = y2
```

```
    end function g1fun

    real*8 function g2fun (y1,y2,t) result (g2)
      implicit none
      real*8 :: pi,y1,y2,t,g2
      pi = 4.0*atan(1.0)
      g2 = - pi*pi * (y1 + 1.0) / 4.0
    end function g2fun
```

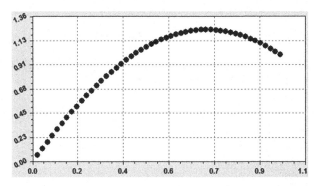

Fig. 7.87. Solution of the BVP with the ODE: $\ddot{y} = -\pi^2(y+1)/4,\, y(0) = 0, y(1) = 1$.

Wave Equation: Simple Time-Stepping Method

The following Fortran 95 code solves *hyperbolic* $(1+1)$D wave PDE (see [II06c])

$$\frac{\partial^2 U}{\partial t^2} = \frac{\partial^2 U}{\partial x^2}.$$

It has been used for simulating string vibrations (see Fig. 7.88).

```
    Program IvWave  ! V.G. Ivancevic, 2007
      ! Solution of the wave equation by simple time-stepping
      ! number of space points is 100

      implicit none
      real*8 :: u(101,3)
      integer :: i, k, max
      max = 1000    ! number of time-steps

      open(unit=7,file='IvWave.out.dat')  ! the output file

      do i = 1, 80
      ! initial conditions for the first 80 points
```

```
      u(i,1) = 0.00125*i
   end do
   do i = 81, 101     ! initial conditions for the rest points
      u(i,1) = 0.1 - 0.005 * (i - 81)
   end do

   do i = 2,100       ! the first step ahead in time
      u(i,2) = u(i,1) + 0.5 * (u(i+1, 1) + u(i-1, 1)
      - 2.0 * u(i,1))
   end do

   do k = 1, max          ! main time loop
      do i = 2, 100
         u(i,3) = 2.0 * u(i,2) - u(i,1) + (u(i+1,2) &
               + u(i-1,2) - 2.0 * u(i,2))
      end do
      do i = 1, 101
         u(i,1) = u(i,2)
         ! new iteration in time now becomes old
         u(i,2) = u(i,3)
      end do

      if ((mod(k,10) == 0)) then
         do i = 1, 101
            write (7, *) k, u(i,3)
            ! plot data every 10 time steps
         end do
         write (7, *)
      end if
   end do
   close(7)
   stop 'data saved in IvWave.out.dat'
end program IvWave
```

Heat Equation: Finite Differences Method

The following Fortran 95 code solves *parabolic* (1+1)D heat PDE (see [II06c]), with thermal conductivity k, specific heat σ and iron density ρ,

$$\frac{\partial U}{\partial t} = c\frac{\partial^2 U}{\partial x^2}, \quad \text{with } c = \frac{k}{\sigma\rho},$$

for constant initial conditions: $u(x,0) = 100°$ Celsius everywhere along an iron wire except for the endpoints which are always at $0°$C (zero boundary conditions: $u(0,t) = u(l,t) = 0$.). The plot of the simulated thermal flow along an iron wire is given in Fig. 7.89.

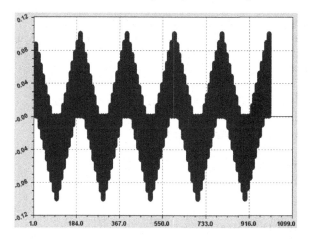

Fig. 7.88. String vibrations as a solution of the $(1+1)$D wave equation.

```
Program IvHeat   ! V.G. Ivancevic, 2007
  ! Solution of the heat equation using finite differences
  ! Simulating heat conduction along an iron wire

  implicit none
  real*8 :: dif_coef, rho, spec_heat, th_con, u(101,2)
  integer :: i, k, max

  open(unit=7,file='IvHeat.out.dat')  ! the output file

  spec_heat = 0.113        ! specific heat
  th_con = 0.12         ! thermal conductivity
  rho = 7.8          ! iron density
  dif_coef = th_con / (spec_heat * rho)
  max = 50000         ! number of iterations

  do i = 1, 100         ! initial condition
     u(i,1) = 100.0
  end do

  do i = 1, 2          ! boundary condition
     u(1,i) = 0.0
     u(101,i) = 0.0
  end do

  do k = 1, max
  ! main time loop - finite differences method
     do i = 2, 100
```

```fortran
      ! loop over space, endpoints stay fixed
         u(i,2) = u(i,1) + dif_coef * (u(i+1,1) + u(i-1,1) &
            - 2.0 * u(i,1))
      end do

      if ((mod(k,1000) == 0) .or. (k == 1)) then
         do i = 1, 101, 2
         ! output temperature every 1000 steps
            write (7, *) k, u(i,2)
         end do
         write (7, *)
      end if
      do i = 2, 100  ! new values now become old
         u(i,1) = u(i,2)
      end do
   end do
   close(7)
   stop 'data saved in IvHeat.out.dat'
end program IvHeat
```

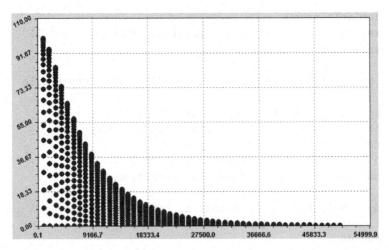

Fig. 7.89. Solution of a $(1+1)$D heat equation for constant initial conditions: $u(x,0) = 100$ and zero boundary conditions: $u(0,t) = u(l,t) = 0$.

Laplace Equation: Finite Differences Method

The following Fortran 95 code solves *elliptic* 2D Laplace PDE (see [II06c]),

$$\frac{\partial^2 U}{\partial x^2} + \frac{\partial^2 U}{\partial y^2} = 0$$

on a 50×50 spatial grid, using finite differences method. The plot of the solution is given in Fig. 7.90.

```
Program IvLaplace  ! V.G. Ivancevic, 2007
   ! Solution of the Laplace equation using finite differences
   ! on the grid of 50 x 50 space points

   implicit none
   integer :: max, i, j, iter
   parameter (max = 50)
   real*8 ::  u(max,max)

   open(unit=7,file='IvLaplace.out.dat')

   do i = 1, max            ! boundary condition:
      u(i,1) = 100.0        ! left side with constant potential
   end do

   do iter = 1, 1000         ! iteration algorithm
      do i = 2, (max-1)
         do j = 2, (max-1)
            u(i,j) = 0.25 * (u(i+1,j) + u(i-1,j) + u(i,j+1)
            + u(i,j-1))
         end do
      end do
   end do

   do i = 1, max
      do j = 1, max
         write (7, *) u(i,j)       ! 2D data output
      end do
      write (7, *)
   end do
   close(7)
   stop 'data saved in IvLaplace.out.dat'
end program IvLaplace
```

Method of Lines: Heat Equation

The following C/C++ code solves the $(1 + 1)$ *heat equation*

$$\partial_t u(t, x) = \frac{1}{2} \partial_{xx} u(t, x),$$

with initial condition

$$f(0, x) = 2 \cos(3x)$$

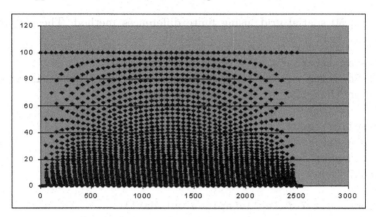

Fig. 7.90. Solution of the 2D Laplace equation with the boundary condition: $u(x,0) = 100$ on a 50×50 spatial grid.

and boundary conditions

$$f(t,0) = 2\sin(2t), \qquad f(t,5) = \cos(3t),$$

—using the *method of lines* (see Sect. 4.6.6 above). Discretization of the heat PDE, using 21 ODEs, was solved by the *Cash–Karp modification* of the *Runge–Kutta–Fehlberg integrator*.

```
// MOL: Heat Eq using adaptive Runge-Kutta-Fehlberg
// integrator
// V.G. Ivancevic, 2008

#include <stdio.h>
#include <math.h>
#include <time.h>

#define max(a, b) (((a) > (b)) ? (a) : (b)) // max macro

#define h0     0.01
#define MAX_h h0*20
#define MIN_h h0/20

#define abstol 0.00001
#define reltol 0.001

#define N 21

double h    = h0;      // Time step
double t    = 0.0;     // Initial time
double t1   = 10.0;    // Final time
```

```
double fx[N+1];


// Heat Eq MoL
void Eqs(double t, double fx[], double* eq)
{
    double xl=0.0;
    double xu=1.0;
    double dx2=pow(((xu-xl)/(N-1)),2);

    int i;

    eq[N]=cos(3.0*t);      //BC f(t,5)=cos(3t)
    eq[1]=2.0*sin(2.0*t); //BC f(t,0)=2sin(2t)

    for(i=2; i<N; i++)
    {
        eq[i]=0.5*(((fx[i+1]-2.0*fx[i]+fx[i-1])/dx2)); // PDE
    }
}

// Runge-Kutta-Fehlberg integrator (Cash-Karp modification)
void RKF45(double fx[])
{
    double k1[N+1], k2[N+1], k3[N+1], k4[N+1],
    k5[N+1], k6[N+1], err[N+1], x[N+1], tol[N+1], delta,
    maxerr;
    int i=0;

    //k1
    Eqs(t,fx,k1);
    for(i=1; i<=N; i++)
    {
        k1[i]*=h;
    }

    //k2
    for(i=1; i<=N; i++) x[i] = fx[i]+(k1[i]/4);
    Eqs(t+(h/4),x,k2);
    for(i=1; i<=N; i++) k2[i]*=h;

    //k3
    for(i=1; i<=N; i++) x[i] = fx[i]+(3*k1[i]/32)
    +(9*k2[i]/32);
```

```
Eqs(t+(3*h/8),x,k3);
for(i=1; i<=N; i++) k3[i]*=h;

//k4
for(i=1; i<=N; i++)
x[i] = fx[i]+(1932*k1[i]/2197)-(7200*k2[i]/2197)
+(7296*k3[i]/2197);
Eqs(t+(12*h/13),x,k4);
for(i=1; i<=N; i++) k4[i]*=h;

//k5
for(i=1; i<=N; i++)
x[i] = fx[i]+(439*k1[i]/216)-(8*k2[i])+(3680*k3[i]/513)
-(845*k4[i]/4104);
Eqs(t+h,x,k5);
for(i=1; i<=N; i++) k5[i]*=h;

//k6
for(i=1; i<=N; i++)
x[i] = fx[i]-(8*k1[i]/27)+(2*k2[i])-(3544*k3[i]/2565)
+(1859*k4[i]/4104)-(11*k5[i]/40);
Eqs(t+(h/2),x,k6);
for(i=1; i<=N; i++) k6[i]*=h;

maxerr = 0;
for(i=1; i<=N; i++)
{
    x[i] = fx[i]+(25*k1[i]/216)+(1408*k3[i]/2565)
    +(2197*k4[i]/4104)-(k5[i]/5);
    err[i] = fx[i]+(16*k1[i]/135)+(6656*k3[i]/12825)
    +(28561*k4[i]/56430)-(9*k5[i]/50)+(2*k6[i]/55);
    tol[i] = x[i]*reltol + abstol;
    maxerr = max(maxerr, fabs(err[i] - x[i])/tol[i]);
}

if(maxerr <= 1.0)
{
    t += h;
    delta = 0.84*pow(maxerr, -0.25);

    // Increase step if required
    if(delta > 1.5 && h < MAX_h) h*= 1.5;
    else if(delta > 1.0 && h < MAX_h) h*=delta;
    if(t+h > t1) h = t1-t;
```

```
        // Save updated values
        for(i=1; i<=N; i++)  fx[i] = x[i];
    }
    else
    {
        // Reduce step
        delta = 0.84*pow(maxerr,-0.25);
        if(delta < 0.1) h *= 0.1;
        else  h *= delta;
    }
}

void header(FILE* pOut)
{
    int i;
    fputs(",\t\n u,", pOut);
    for(i=1; i<=N; i++)
    {
        fprintf(pOut, ",x%i", i);
    }
    fputs("\n", pOut);
}

void print(FILE* pOut)
{
    int i;
    fprintf(pOut, ",%.4f", t);
    for(i=1; i<=N; i++) // start from 2 removes zeros.
        fprintf(pOut, ",%.7f", fx[i]);
    fprintf(pOut, "\n");
}

int main()
{
    int ticks;
    double ms;
    clock_t start, stop;

    double prev_t = -1;  // Previous t
    int i;
    double x;

    FILE* pOut;
    // File pointer to output file
    pOut = fopen("out.csv", "w");      // Write header
```

```
header(pOut); // print header

start = clock() * CLK_TCK;                    // Clock setup

for(i = 1; i <= N; i++)
{
    x = (double)(i-1)/(N-1) * 5;  // x value
    fx[i]=2*cos(3*x);             //IC f(0,x)=2cos(3x)
}

while (t < t1)
// Main time integration loop
{
    RKF45(fx);                    // Call RKF45

    if(prev_t != t)
        print(pOut);

    prev_t = t;
}

fclose(pOut);

stop  = clock() * CLK_TCK;         // Timing
ticks = stop - start;
ms = (float)ticks / 1000.0f;
printf("finished... (ms = %f)\n", ms);
getchar();

    return 0;
}
```

Time-Dependent Schrödinger Equation

The following Fortran 95 code solves the $(1 + 1)$ *Schrödinger equation* (see [II08b]),

$$i\partial_t\psi(x,t) = -\frac{1}{2}\partial_{x^2}\psi(x,t) + V(x)\psi(x,t)$$

for the *Gaussian wave-packet* in a quadratic potential $V(x) = \frac{1}{2}x^2$, i.e., *harmonic oscillator* (see Fig. 7.91).

```
Program IvSchrod  ! V.G. Ivancevic, 2007
    ! Time-dependent Schroedinger equation
    ! Gaussian wave-packet in a quadratic potential
```

```
implicit none
real*8 :: pi, dx, p0, dt, x, ps_re(751,2), ps_im(751,2)
real*8 :: v(751), p2(751)
complex :: exc, z_i
integer :: max,i,n

open(unit=7,file='IvSchrod.out.dat')

pi = 3.14159265358979323385
z_i = cmplx(0.0,1.0)
dx = 0.02
p0 = 3 * pi    ! initial momentum given to the wave packet
dt = dx * dx / 4.0
max = 750
x = -7.5     ! initial conditions
do i = 1, max+1
   exc = exp(z_i * p0 * x)
   ! real part of initial wave function
   ps_re(i,1) = real(exc * exp(-0.5 * (x/0.5)**2.0))
   ! imaginary part of initial wave function
   ps_im(i,1) = aimag(exc * exp(-0.5 * (x/0.5)**2.0))
   ! quadratic potential
   v(i) = 5.0 * x * x
   x = x + dx
end do


do n = 1, 20000        ! main time loop

   do i = 2, max       ! real part of the wave packet
      ps_re(i,2) = ps_re(i,1) - dt * (ps_im(i+1,1)
         + ps_im(i-1,1) &
         - 2.D0 * ps_im(i,1))
         / (dx*dx) + dt * v(i) * ps_im(i,1)
                     ! and probability:
      p2(i) = ps_re(i,1) * ps_re(i,2)
         + ps_im(i,1) * ps_im(i,1)
   end do

   do i = 2, max       ! imaginary part of the wave packet
      ps_im(i,2) = ps_im(i,1) + dt * (ps_re(i+1,2)
         + ps_re(i-1,2) &
         - 2.D0 * ps_re(i,2))
         / (dx*dx) - dt * v(i) * ps_re(i,2)
   end do
```

```
      if ((n.eq.1) .or. (MOD(n,2000).eq.0)) then
      ! outputting probability density every 2000 time-steps
        do i = 1, max+1, 10
           write (7, *) i, p2(i) + 0.0015 * v(i)
        end do
        write (7, *)
      end if

      do i = 1, max+1     ! new iterations now become old
         ps_im(i,1) = ps_im(i,2)
         ps_re(i,1) = ps_re(i,2)
      end do
   end do
   close(7)
   stop 'data saved in IvSchrod.out.dat'
end program IvSchrod
```

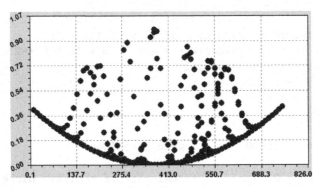

Fig. 7.91. Solution of the (1+1) Schrödinger equation for the Gaussian wave-packet in a quadratic potential.

Korteveg–de Vries Soliton

The following Fortran 95 code solves the nonlinear Korteveg–de Vries PDE (see [II06c]),

$$u_t + \mu u_{xxx} + \epsilon u u_x = 0, \quad \text{(subscripts here denote partial derivatives)},$$

with parameters μ, ϵ and sample initial and boundary conditions (given in the code below), using the finite difference method. Its solitary solution is given in Fig. 7.92).

```
Program IvKdV  ! V.G. Ivancevic, 2007
   ! Korteveg-de Vries soliton
```

```
! finite difference method

implicit none
real*8 :: dt, dx, max, mu, eps, a1, a2, a3, cc, t, u(131,3)
parameter (dt = 0.1, dx = 0.4, max = 2000, mu = 0.1,
eps = 0.2)
integer :: i, j, k

open(unit=7,file='IvKdV.out.dat')

do i = 1, 131        ! initial condition
   u(i,1) = 0.5 * (1.0 - tanh(0.2 * dx * (i-1) - 5.0))
end do

u(1,2) = 1.0       ! boundary conditions
u(1,3) = 1.0
u(131,2) = 0.0
u(131,3) = 0.0
cc  = mu * dt / (dx**3.0)
t = dt

do i = 2, 130    ! the first t step
   a1 = eps * dt * (u(i+1,1) + u(i,1) + u(i-1,1))
   / (dx * 6.0d0)
   if ((i > 2) .and. (i < 129)) then
      a2 = u(i+2,1) + 2.0 * u(i-1,1) - 2.0 * u(i+1,1)
      - u(i-2,1)
   end if
   if ((i == 2) .or. (i == 130)) then
      a2 = u(i-1,1) - u(i+1,1)
   end if
   a3 = u(i+1,1) - u(i-1,1)
   u(i,2) = u(i,1) - a1 * a3 - cc * a2 / 3.d0
end do

do j = 1, max    ! all other t steps
   do i = 2, 130
      a1 = eps * dt * (u(i+1,2) + u(i,2) + u(i-1,2))
      / (3.0d0 * dx)
      if ((i > 2) .and. (i < 129)) then
         a2 = u(i+2,2) + 2.0d0*u(i-1,2) - 2.0d0*u(i+1,2)
         - u(i-2,2)
      end if
      if ((i == 2) .or. (i == 130)) then
         a2 = u(i-1,2) - u(i+1,2)
```

```
      end if
      a3 = u(i+1,2) - u(i-1,2)
      u(i,3) = u(i,1) - a1 * a3 - 2.d0 * cc * a2 / 3.d0
      u(1,3) = 1.0D0
    end do

    do k = 1, 131        ! new iterations become old now
      u(k,1) = u(k,2)
      u(k,2) = u(k,3)
    end do

    if ((mod(j,200) == 0)) then
      do k = 1, 131    ! output the result every 200 t steps
        write (7, *) k, u(k,3)
      end do
      write (7, *)
    end if
    t = t + dt
  end do
  close(7)
  stop 'data saved in IvKdV.out.dat'
end program IvKdV
```

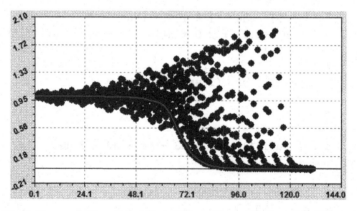

Fig. 7.92. Solitary solution of the $(1 + 1)$D nonlinear Korteveg–de Vries equation.

NLS Soliton: Method of Lines

The following C/C++ code solves the $(1 + 1)$ NLS

$$\mathrm{i}\frac{\partial \psi}{\partial t} = -\frac{1}{2}\frac{\partial^2 \psi}{\partial x^2} + V(x)\psi|\psi|^2,$$

with potential field $V(x) = \sin(x)$, final time $t_{fin} = 10$, final space $x_{fin} = 50$, initial condition $\psi(0, x) = V(t_{fin})$, and boundary conditions $\psi(t, 0) = \psi(t, x_{fin})$—using the *method of lines* (see Sect. 4.6.6 above). Discretization of the NLS, using the set of 31 approximating complex-valued ODEs:

$$i\frac{d\psi_k}{dt} = -\frac{1}{2}\frac{\psi_{k+1} - 2\psi_k + \psi_{k-1}}{\Delta x^2} + V(x_k)\psi_k|\psi_k|^2, \quad (k = 1, 2, \ldots 31),$$

was solved (see Fig. 7.93) by the *Cash–Karp modification* of the *Runge–Kutta–Fehlberg integrator*.

```
// MOL: NLS using adaptive Runge-Kutta-Fehlberg integrator
// V.G. Ivancevic, 2009


#include <stdio.h>
#include <math.h>
#include <time.h>
#include <stdlib.h>
#include <complex.h>

#define max(a, b) (((a) > (b)) ? (a) : (b)) // max macro

#define h0    0.01
#define MAX_h h0*20
#define MIN_h h0/20

#define abstol 0.00001
#define reltol 0.001

#define N 31

double h   = h0;      // Time step
double t   = 0.0;     // Initial time
double t1  = 10.0;    // Final time
double _Complex x0  =  0.0;    // x start.
double _Complex x1  = 50.0;    // x finish.

double _Complex fx[N+1];
double _Complex V[N+1];


// NLS PDE
void Eqs(double t, double _Complex fx[], double _Complex* eq)
{
    double _Complex dx2=pow((((x1-x0)/(N-1)),2), pdf;
```

```
    double coeff = -0.5;
    int i;

    for(i=2; i<=N; i++)
    {
        pdf = cabs(fx[i])*cabs(fx[i]);

        if(i==N)   //u(t,x1)
        {
            eq[i] = ((2.0*fx[N-1] - fx[i]) / dx2 * coeff) / I
            + ((V[i] * fx[i] * pdf / I));
        }
        else
        {
            eq[i] = ((fx[i+1] - 2.0*fx[i] + fx[i-1]) /
            dx2 * coeff) / I + ((V[i] * fx[i] * pdf / I));
        }
    }

    // BC: u(t,0)=u(t,x1);
    eq[1] = eq[N];
}


// Set the IC's and V(x)
void Initialize()
{
    int i;
    double x;

    // IC u(0,x) = sin(t1)
    for(i = 1; i <= N; i++)
        fx[i] = sin(t1) + 0*I;

    // Potential.
    for(i = 1; i <= N; i++)
    {
        x = (double)(i-1)/(N-1) * x1;
        V[i] = sin(x);
    }
}


// Runge-Kutta-Fehlberg integrator
void RKF45(double _Complex fx[])
{
```

```
double _Complex k1[N+1], k2[N+1], k3[N+1], k4[N+1],
k5[N+1], k6[N+1], err[N+1], x[N+1];
double  delta, maxerr, tol[N+1];

int i=0;

//k1
Eqs(t,fx,k1);
for(i=1; i<=N; i++) k1[i]*=h;

//k2
for(i=1; i<=N; i++) x[i] = fx[i]+(k1[i]/4);
Eqs(t+(h/4),x,k2);
for(i=1; i<=N; i++) k2[i]*=h;

//k3
for(i=1; i<=N; i++)
x[i] = fx[i]+(3*k1[i]/32)+(9*k2[i]/32);
Eqs(t+(3*h/8),x,k3);
for(i=1; i<=N; i++) k3[i]*=h;

//k4
for(i=1; i<=N; i++) x[i] = fx[i]+(1932*k1[i]/2197)
-(7200*k2[i]/2197)+(7296*k3[i]/2197);
Eqs(t+(12*h/13),x,k4);
for(i=1; i<=N; i++) k4[i]*=h;

//k5
for(i=1; i<=N; i++) x[i] = fx[i]+(439*k1[i]/216)
-(8*k2[i])+(3680*k3[i]/513)
-(845*k4[i]/4104);
Eqs(t+h,x,k5);
for(i=1; i<=N; i++) k5[i]*=h;

//k6
for(i=1; i<=N; i++) x[i] = fx[i]-(8*k1[i]/27)+(2*k2[i])
-(3544*k3[i]/2565)
+(1859*k4[i]/4104)-(11*k5[i]/40);
Eqs(t+(h/2),x,k6);
for(i=1; i<=N; i++) k6[i]*=h;

maxerr = 0;
for(i=1; i<=N; i++)
{
    x[i]   = fx[i]+(25*k1[i]/216)+(1408*k3[i]/2565)
```

```
              +(2197*k4[i]/4104)-(k5[i]/5);
        err[i] = fx[i]+(16*k1[i]/135)+(6656*k3[i]/12825)+
        (28561*k4[i]/56430)-(9*k5[i]/50)+(2*k6[i]/55);
        tol[i] = x[i]*reltol + abstol;
        maxerr = max(maxerr, fabs(err[i] - x[i])/tol[i]);
    }

    if(maxerr <= 1.0)
    {
        t += h;
        delta = 0.84*pow(maxerr, -0.25);

        // Increase step if required
        if(delta > 1.5 && h < MAX_h) h*= 1.5;
        else if(delta > 1.0 && h < MAX_h) h*=delta;
        if(t+h > t1) h = t1-t;

        // Save updated values
        for(i=1; i<=N; i++)  fx[i] = x[i];
    }
    else
    {
        // Reduce step
        delta = 0.84*pow(maxerr,-0.25);
        if(delta < 0.1) h *= 0.1;
        else  h *= delta;
    }
}

void header(FILE* pOut_r, FILE* pOut_i, FILE* pOut_pdf)
{
    int i;
    fputs(",t\n u,", pOut_r);
    fputs(",t\n u,", pOut_i);
    fputs(",t\n u,", pOut_pdf);
    for(i=1; i<=N; i++)
    {
        fprintf(pOut_r, ",x%i", i);
        fprintf(pOut_i, ",x%i", i);
        fprintf(pOut_pdf, ",x%i", i);
    }
    fputs("\n", pOut_r);
    fputs("\n", pOut_i);
    fputs("\n", pOut_pdf);
}
```

```c
void print(FILE* pOut_r, FILE* pOut_i, FILE* pOut_pdf)
{
    int i;
    fprintf(pOut_r, ",%.4f", t);
    fprintf(pOut_i, ",%.4f", t);
    fprintf(pOut_pdf, ",%.4f", t);
    for(i=1; i<=N; i++)
    {
        fprintf(pOut_r, ",%.4f",  creal(fx[i]));
        fprintf(pOut_i, ",%.4f",  cimag(fx[i]));
        fprintf(pOut_pdf, ",%.4f", cabs(fx[i])*cabs(fx[i]));
    }
    fprintf(pOut_r, "\n");
    fprintf(pOut_i, "\n");
    fprintf(pOut_pdf, "\n");
}

int main()
{
    int ticks;
    double ms, prev_t = -1;  // Previous t;
    clock_t start, stop;
    double _Complex d;

    FILE* pOut_r  = fopen("out_r.csv",  "w");
    FILE* pOut_i  = fopen("out_i.csv",  "w");
    FILE* pOut_pdf = fopen("out_pdf.csv", "w");

    header(pOut_r, pOut_i, pOut_pdf); // print header

    start = clock() * CLK_TCK;                 // Clock setup

    Initialize(); // Initialize the NLS/Integrator.

    while (t < t1)              // Main time integration loop
    {
        RKF45(fx);                            // Call RKF45

        if(prev_t != t)
            print(pOut_r, pOut_i, pOut_pdf);

        prev_t = t;
    }
```

```
fclose(pOut_r);
fclose(pOut_i);

stop  = clock() * CLK_TCK;              // Timing
ticks = stop - start;
ms = (float)ticks / 1000.0f;
printf("finished... (ms = %f)\n", ms);
getchar();

return 0;
}
```

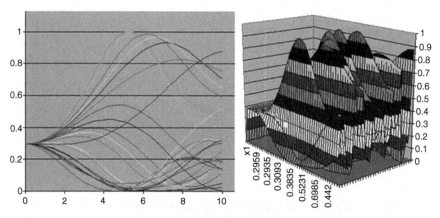

Fig. 7.93. Solitary solution of the $(1 + 1)$D NLS equation using *method of lines*.

7.3.8 Evolution PDEs in *Mathematica*^{TM}

In this subsection we give numerical solution and a 3D plot of several common PDEs in *Mathematica*^{TM}, using the built-in *method of lines* (see e.g., [HSG07]).

Heat Equation

Consider 1D heat equation. With fixed boundary conditions this is a model for diffusion of heat in an insulated rod with the temperatures at the endpoints held fixed.

This command solves the heat equation with the left end $(x = 0)$ held at fixed temperature 0, the right end $(x = 1)$ held at fixed temperature 1, and an initial heat profile given by the quadratic in x.

$sol = NDSolve[\{\partial_t u[x,t] == \partial_{x,x}u[x,t], u[x,0] == 6x - 6x^4 + x^2,$
$u[0,t] == 0, u[1,t] == 1\}, u, \{x,0,1\}, \{t,0,0.5\}]$

$\{\{u \rightarrow InterpolatingFunction[\{\{0.,1.\},\{0.,0.5\}\}, <>]\}\}$

The boundary conditions can be a linear combination of Dirichlet and Neumann type conditions and can be time dependent. For example, entering these command produces a plot of a solution with the temperature at the left edge varying sinusoidally.

$sol = u/.First[NDSolve[\{\partial_t u[x,t] == \frac{2\partial_{\{x,2\}}u[x,t]}{9\pi^2},$
$u[x,0] == 0, u[0,t] == \sin[t], u^{(1,0)}[1,t] == 0\}, u, \{x,0,1\}, \{t,0,6\pi\}]];$
$Plot3D[sol[x,t], \{x,0,1\}, \{t,0,6\pi\}, PlotRange \rightarrow All, PlotPoints \rightarrow 25];$

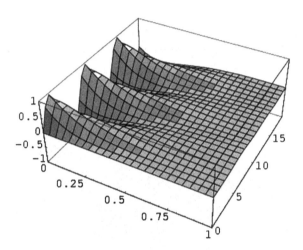

Fig. 7.94. Numerical solution of a 1D heat equation.

Linear and Nonlinear Wave Equation

Linear case

$NDSolve[\{\partial_{t,t}u[t,x] == \partial_{x,x}u[t,x], u[0,x] == e^{-x^2},$
$u^{(1,0)}[0,x] == 0, u[t,-6] == u[t,6]\}, u, \{t,0,6\}, \{x,-6,6\}];$
$Plot3D[Evaluate[u[t,x]/. First[\%]], \{t,0,6\}, \{x,-6,6\}, PlotPoints \rightarrow 50];$

Nonlinear case

$NDSolve[\{\partial_{t,t}u[t,x] == \partial_{x,x}u[t,x] + (1 - u[t,x]^2)(1 + 2u[t,x]), u[0,x] == e^{-x^2},$
$u^{(1,0)}[0,x] == 0, u[t,-10] == u[t,10]\}, u, \{t,0,10\}, \{x,-10,10\}];$
$Plot3D[Evaluate[u[t,x]/. First[\%]], \{t,0,10\}, \{x,-10,10\},$
$PlotPoints \rightarrow 80];$

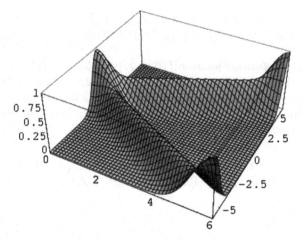

Fig. 7.95. Numerical solution of a 1D linear wave equation.

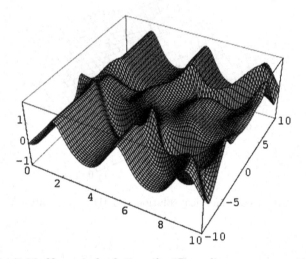

Fig. 7.96. Numerical solution of a 1D nonlinear wave equation.

Soliton: Nonlinear Schrodinger Equation

These commands set up a periodic initial condition (it happens to be a soliton), and computes the solution, and then produces 17 plots showing the modulus and real and imaginary parts of the solution.

$Soliton[a\_, b\_, m\_, x0\_][x\_, t\_] := Module[\{\omega, ek, xx = x - x0\}, \omega = (2m - 1)a^2 - b^2$;

$\sqrt{m}a\,JacobiCN[a(xx - 2bt), m]e^{I(bxx + \omega t)}]; Clear[per]$;

$per[m\_] := 4EllipticK[m]; m = mm/. FindRoot[per[mm] == 8$,
$\{mm, 0.5\}]$;

$solution = u/. First[NDSolve[\{Iu[x, t] + \partial_{\{x,2\}}u[x, t] + 2Abs[u[x, t]]^2u[x, t]$

$$== 0,$$
$$u[x, 0] == Soliton[1, \tfrac{\pi}{4}, m, 0][x, 0], u[0, t] == u[8, t]\}, u, \{x, 0, 8\}, \{t, 0, 4\}]];$$

$$Plot[\{Abs[solution[x, 2]], Re[solution[x, 2]], Im[solution[x, 2]]\}, \{x, 0, 8\},$$
$$PlotRange \to \{-0.8, 0.8\}, Frame \to True];$$

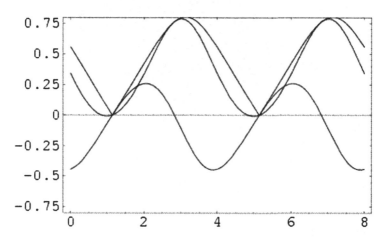

Fig. 7.97. Numerical solution of a 1D nonlinear Schrödinger equation.

Nonlinear Sine-Gordon Equation

$$sol = NDSolve[\{\partial_{t,t}y[x, t] == \partial_{x,x}y[x, t] + \sin[y[x, t]], y[x, 0] == e^{-x^2},$$
$$y^{(0,1)}[x, 0] == 0, y[-5, t] == y[5, t]\}, y, \{x, -5, 5\}, \{t, 0, 5\}];$$

$$Plot3D[Evaluate[y[x, t]/.First[sol]], \{x, -5, 5\}, \{t, 0, 5\},$$
$$PlotPoints \to 30];$$

7.3.9 Fourier Transforms

Simple Discrete Fourier Transform

The following Fortran 95 code presents the basic discrete Fourier transformation. It outputs real and imaginary frequencies of the input signal. In contrast, in the next paragraph we give the famous FFT algorithm.

```
Program IvFourier   ! V.G. Ivancevic, 2007
   ! Simple discrete Fourier transform

   implicit none
   parameter (max = 1000, pi = 3.1415926535897932385E0)
```

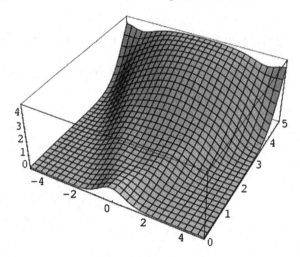

Fig. 7.98. Numerical solution of a 1D Sine-Gordon equation.

```
integer :: i, j, k, max
real*8 :: pi, input(max), real, imag

open(8, File='fft_in.dat', Status='OLD')  ! input file
open(9, File='Fourier.dat')  ! output file

do i = 1, max      ! read the input signal data
        read (8, *) input(i)
end do

do j = 1, i
   real = 0.0
   imag = 0.0

   do k = 1, i   ! loop for sums
      real = real + input(k) * cos((2 * pi * k * j) / i)
      imag = imag + input(k) * sin((2 * pi * k * j) / i)
   end do

   write (9,*) j, real/i, imag/i      ! data output
end do
close(8); close(9)
stop 'data saved in Fourier.dat'
end
```

Fast Fourier Transform

The following Fortran 95 code presents the Cooley–Tukey FFT algorithm
[CT65] for the discrete Fourier transform

$$X_k = \sum_{n=0}^{N-1} x_n e^{-\frac{2\pi i}{N}nk}, \quad (k = 0, \ldots, N-1).$$

```
Program IvFFT  ! V.G. Ivancevic, 2007
  ! Fast Fourier Transform
  ! frequency analysis of a discreet signal

  integer :: i,n1,ndata,ip
  real*8 :: df,dt,f,temp,tbegin,tend, signal(1024),A(1024,2)

  open (UNIT=1,FILE='fft-in.dat',STATUS='OLD')  ! input file
  open (UNIT=2,FILE='fft-out.dat')              ! output file

  read(1,*)  ndata      ! number of input signal points
  if  (ndata > 2048) then
     ndata = 2048;  ip = 11
     ! take the nearest power of two
  end if
  if (ndata < 2048 .and. ndata > 1023) then
     ndata = 1024;  ip = 10
  end if
  if (ndata < 1024 .and. ndata > 511) then
     ndata = 512;  ip = 9
  end if
  if (ndata < 512 .and. ndata > 255) then
     ndata = 256;  ip = 8
  end if
  if (ndata < 256 .and. ndata > 127) then
     ndata = 128;  ip = 7
  end if
  if (ndata < 128 .and. ndata > 63) then
     ndata = 64;  ip = 6
  end if
  if (ndata < 64 .and. ndata > 31) then
     ndata = 32;  ip = 5
  end if
  if  (ndata < 32) then
     write (2, *)
     ' Error: number of points too small (<32) !'
     stop
```

```
      end if

      do i = 1, ndata          ! read ndata couples of T(i), Y(i)
         read(1, *)  temp, signal(i)
         if (i. eq. 1)      tbegin = temp
         if (i. eq. ndata) tend = temp
      end do

      ! put the input signal into the real part of a complex
      ! vector A
      do i = 1, ndata
         A(i,1) = signal(i)
         A(i,2) = 0.d0
      end do

      call FFT(A,ip)     ! call FFT subroutine
      n1 = ndata / 2
      ! get frequencies from the signal (real vector)

      do i = 1, n1
         ! calculate modulus of a complex number for each i value
         temp = dsqrt(A(i,1) * A(i,1) + A(i,2) * A(i,2))
         if (i > 1)  temp = 2.d0 * temp
         signal(i) = temp
      end do

      dt = (tend - tbegin) / (ndata - 1)
      ! calculate sampling time range dt
      df = 1.d0 / (ndata * dt)        ! calculate frequency step df
      f = 0.d0
      write (2, *) '     Frequency          Value      '
      write (2, *) '----------------------------------'
      do i = 1, n1
         write (2, 100)  f, signal(i)
         ! output the frequency spectrum
         f = f + df
      end do
100   format('    ',F8.2,'          ',F10.6)
      stop 'data saved in fft-out.dat'
      end program IvFFT

      subroutine FFT(A,m)
         ! This procedure calculates the fast Fourier transform of a
         ! real function sampled at N points 0,1,....,N-1 (N must be
         ! a power of two). T being the sampling duration,
```

```
! the maximum frequency in the signal cannot be greater
! than fc = 1/(2T). The resulting spectrum H(k) is discreet
! and contains the frequencies:
!          fn = k/(NT)   with   k = -N/2,.. -1,0,1,2,..,N/2,
! where
! H(0) corresponds to the null frequency,
! H(N/2) corresponds to the fc frequency.

  real*8 :: Phi, Pi, A(1024,2), U(2), W(2), T(2)
  integer n, nv2, nm1, j, i, iip, k, l, le, le1
    Pi = 4.0 * atan(1.0)
    n = 2**m
    nv2 = n / 2
    nm1 = n - 1
    j = 1
    do i=1, nm1
       if (i < j) then
          T(1) = A(j,1)
          T(2) = A(j,2)
          A(j,1) = A(i,1)
          A(j,2) = A(i,2)
          A(i,1) = T(1)
          A(i,2) = T(2)
       end if
       k = nv2
100    if (k >= j) goto 200
       j = j - k
       k = k/2
       goto 100
200    j = j + k
    end do
    le = 1
    do l = 1, m
      le1 = le
      le = le * 2
      U(1) = 1.d0
      U(2) = 0.d0
      Phi = Pi / le1
      W(1) = dcos(Phi)
      W(2) = dsin(Phi)
      do j = 1, le1
         i = j - le
300      if (i >= n - le) goto 400
         i = i + le
         iip = i + le1
```

```
              T(1) = A(iip,1) * U(1) - A(iip,2) * U(2)
              T(2) = A(iip,1) * U(2) + A(iip,2) *U (1)
              A(iip,1) = A(i,1) - T(1)
              A(iip,2) = A(i,2) - T(2)
              A(i,1) = A(i,1) + T(1)
              A(i,2) = A(i,2) + T(2)
              goto 300
400           temp = U(1)
              U(1) = W(1) * U(1) - W(2) * U(2)
              U(2) = W(1) * U(2) + W(2) * temp
          end do
        end do
        do i = 1, n
           A(i,1) = A(i,1) / n
           A(i,2) = A(i,2) / n
        end do
      return
    end
```

The samples from the input and output files are given here:

```
1024
  0.00000000000000E+0000     0.00000000000000E+0000
  1.95503421309917E-0004     3.53914399999776E+0001
  3.91006842619834E-0004     5.95684899999760E+0001
  5.86510263929974E-0004     6.54621699999552E+0001
  7.82013685239669E-0004     5.24038399999845E+0001
  9.77517106550252E-0004     2.62296199999982E+0001
  1.17302052785995E-0003    -2.94931200000065E+0000
  1.36852394916964E-0003    -2.44801499999885E+0001
  1.56402737047934E-0003    -3.10349799999967E+0001
       Frequency          Value
    ------------------------------

         0.00           0.271690
         5.00           0.546336
         9.99           0.555560
        14.99           0.572242
        19.98           0.598828
        24.98           0.640061
        29.97           0.705645
```

Pendulum Power Sprectrum

The following Fortran 95 code presents the FFT power spectrum of a damped
forced pendulum solved by RK4.

```
! FFT power spectrum of a damped forced pendulum solved
```

```
! by RK4

module CB
  real*8 :: q,b,w
end module CB

Program IvFFTpend  ! V.G. Ivancevic, 2007
  use CB
  implicit none
  integer, parameter :: n = 65536, l = 128, m = 16, md = 16
  integer :: i, j
  real*8 :: pi, f1, h, od, t, y1, y2, g1, gx1, g2, gx2
  real*8 :: dk11, dk21, dk12, dk22, dk13, dk23, dk14, dk24
  real, dimension (n) :: AR, AI, WR, WI, O
  real, dimension (2,n) :: Y

  open(9, File='FFTpendul.dat')  ! output file

  pi = 4.0 * atan(1.0)
  f1 = 1.0 / sqrt(float(n))
  w  = 2.0 / 3.0
  h  = 2.0 * pi / (l * w)
  od = 2.0 * pi / (n * h * w * w)
  q  = 0.5
  b  = 1.15
  Y(1,1) = 0.0
  Y(2,1) = 2.0

  do i = 1, n - 1
  ! RK4 algorithm to integrate the equation
     t  = h * i
     y1 = Y(1,i)
     y2 = Y(2,i)
     dk11 = h * gx1(y1, y2, t)
     dk21 = h * gx2(y1, y2, t)
     dk12 = h * gx1((y1 + dk11/2.0),(y2 + dk21/2.0),
     (t + h/2.0))
     dk22 = h * gx2((y1 + dk11/2.0),(y2 + dk21/2.0),
     (t + h/2.0))
     dk13 = h * gx1((y1 + dk12/2.0),(y2 + dk22/2.0),
     (t + h/2.0))
     dk23 = h * gx2((y1 + dk12/2.0),(y2 + dk22/2.0),
     (t + h/2.0))
     dk14 = h * gx1((y1 + dk13),(y2 + dk23),(t + h))
     dk24 = h * gx2((y1 + dk13),(y2 + dk23),(t + h))
```

```fortran
      Y(1,i+1) = Y(1,i) + (dk11 + 2.0 * (dk12 + dk13) + dk14)
      / 6.0
      Y(2,i+1) = Y(2,i) + (dk21 + 2.0 * (dk22 + dk23) + dk24)
      / 6.0

      if (abs(Y(1,i+1)) > pi) then
      ! bring theta back to region [-pi,pi]
         Y(1,i+1) = Y(1,i+1) - 2.0 * pi * abs(Y(1,i+1))
         / Y(1,i+1)
      end if
   end do
!
   do i = 1, n
      AR(i) = Y(1,i)
      WR(i) = Y(2,i)
      AI(i) = 0.0
      WI(i) = 0.0
   end do
   call FFT (AR,AI,n,m)        ! call FFT routine
   call FFT (WR,WI,n,m)
!
   do i = 1, n
      O(i)  = (i - 1) * od
      AR(i) = (f1 * AR(i))**2 + (f1 * AI(i))**2
      WR(i) = (f1 * WR(i))**2 + (f1 * WI(i))**2
      AR(i) = ALOG10(AR(i))
      WR(i) = ALOG10(WR(i))
   end do
   write (9,"(3F16.10)") (O(i), AR(i), WR(i), i=1, (l*md), 4)
end program IvFFTpend


subroutine FFT (AR,AI,n,m)
! Fast Fourier transform subroutine
with n = 2**m

   implicit none
   integer, intent (IN) :: n,m
   integer :: n1,n2,i,j,k,l,l1,l2
   real*8 :: pi,a1,a2,q,u,v
   real, intent (INOUT), dimension (n) :: AR,AI

   pi = 4.0 * atan(1.0)
   n2 = n / 2
   n1 = 2**m
```

```
if(n1 .ne. n) stop 'Indices do not match'
l = 1     ! rearrange the data to the bit reversed order
do k = 1, n - 1
   if (k < l) then
      a1    = AR(l)
      a2    = AI(l)
      AR(l) = AR(k)
      AR(k) = a1
      AI(l) = AI(k)
      AI(k) = a2
   end if
   j = n2
   do while (j < l)
      l = l - j
      j = j / 2
   end do
   l = l + j
end do

l2 = 1
! perform additions at all levels with reordered data
do l = 1, m
   q  =  0.0
   l1 =  l2
   l2 =  2 * l1
   do k = 1, l1
      u  =  cos(q)
      v  = -sin(q)
      q  =  q + pi/l1
      do j = k, n, l2
         i     =  j + l1
         a1    =  AR(i) * u - AI(i) * v
         a2    =  AR(i) * v + AI(i) * u
         AR(i) =  AR(j) - a1
         AR(j) =  AR(j) + a1
         AI(i) =  AI(j) - a2
         AI(j) =  AI(j) + a2
      end do
   end do
end do
end subroutine FFT

! functions defining two first-order pendulum ODEs
! function gx1
(y1,y2,t) result (g1)
```

```
   g1 = y2
end function gx1

function gx2 (y1,y2,t) result (g2)
   use CB
   g2 = -q * y2 - sin(y1) + b * cos(w*t)
end function gx2
```

7.3.10 Sophisticated Random Algorithms

Ising Magnetic Dipole Model

Recall that in statistical physics, the Ising magnetic dipole model (or, the *Ising system*) serves as one of the simplest models of interacting bodies. The spin-flip dynamics version of the model has been used to model ferromagnets, anti-ferromagnetism and phase separation in binary alloys. It has also been applied to spin glasses and neural networks. It has been suggested (see, e.g. [Cal85]) that the Ising model is relevant to imitative behavior in general, including such disparate systems as flying birds, swimming fish, flashing fireflies, beating heart cells, spreading diseases and even fashion fads.

The following Fortran 95 code solves the Ising system using random walk techniques from the previous code.

```
Program IvIsing  ! V.G. Ivancevic, 2007
   ! Ising spin-up, spin-down model of magnetic dipole

   implicit none
   integer :: max; parameter(max=100)
   integer :: element, i, spins(max), seed, t
   real*8 :: harv, energy, kt, new, j, old
   ! define number temperature, exchange energy, random seed
   parameter(kt=100, j=-1, seed=68111)
   call random_seed(seed)

   open(unit=7,File='Ising.spin-up.dat')       ! output files
   open(unit=8,File='Ising.spin-down.dat')

   do i = 1, max
   ! generate a uniform configuration of spins
      spins(i) = 1
   end do

   call random_number(harv)   ! harv = pseudo-random number

   do t = 1, 500       ! main time loop
```

```fortran
      old = energy(spins, j, max)
      ! calculate system's energy
      element = harv * max + 1       ! pick one element
      spins(element) = spins(element)*(-1)     ! change spin
      new = energy(spins, j, max)      ! calculate new energy
      ! reject change if new energy is greater and the
      ! Boltzmann
      ! factor is less than another random number
      if ((new > old) .and. (exp((old - new)/kt) < harv)) then
         spins(element) = spins(element)*(-1)
      end if

      do i = 1, max      ! output a map of spins
         if (spins(i) == 1) then
            write (7, *) t, i
          end if
          if (spins(i) == (-1)) then
            write (8, *) t, i
          end if
      end do
   end do
   close(7);  close(8)
   stop 'data saved in Ising.spin-up.out.dat'
end program IvIsing

function energy(array, j, max)
! system's energy function
  implicit none
  integer :: array(max), i, max
  real*8 :: energy, j
    energy = 0.0
  do i = 1, (max-1)
     energy = energy + array(i) * array(i+1)
  end do
return; end
```

Markov Chain

The following Fortran 95 code Simulates an n-state Markov chain and gives the probability vector for each state.

```fortran
Program IvMarkovChain  ! V.G. Ivancevic, 2007
  ! Simulates an n-state Markov chain
  ! giving the probability vector for each state

  implicit none
```

```
integer :: state, n, i, j, Nsteps
real(kind(1.d0)) :: r
real(kind(1.d0)), allocatable :: T(:,:), Tc(:,:), pp(:)
integer, allocatable :: p(:)

call random_seed()

n = 4 ! assume a 4-state chain

allocate(T(n,n))
T = transpose(reshape((/&           ! make the T-matrix
    & 12.d0, 12.d0, 12.d0, 12.d0,&
    &  6.d0, 18.d0, 12.d0, 12.d0,&
    &  3.d0,  6.d0, 30.d0,  9.d0,&
    &  4.d0,  8.d0, 12.d0, 24.d0 &
    &/)/48.d0,(/4,4/)))

write(*,*) 'The T matrix is:'
call write_mat(T,6) ! output the matrix T for check
call write_mat(T*48.,6)

! cumulative probability Tc(i,j) is a probability to jump
! from state i to any state smaller or equal to j.
! Tc(:,0)=0 and Tc(:,n)=1; for example: r=0.3; if Tc
! for current state is: 0 .25 .5 .75 1.0 then we would
! move to state 2.

allocate(Tc(n,0:n))
Tc(:,0) = 0.d0 ! You can not jump to state 0.
do j = 1, n
  Tc(:,j) = Tc(:,j-1)+T(:,j)!
  ! add T(i -> j) to previous Tc(i,j-1)
end do

if (maxval(abs(Tc(:,n) - 1.d0)) > 0.d0) stop 'Error in Tc!'

allocate(p(n))
p = 0
write(*,*) 'Tc matrix is:'
call write_mat(Tc,6)

write(*,*) 'Give number of steps to make'
read(*,*) Nsteps
```

```fortran
call random_number(r)
state = min(int(r*n) + 1, n) ! start from random state
write(*,*) ' starts at state', state

big: do i = 1, Nsteps
  call random_number(r)    ! find a new state
  small: do j = 1, n
    if (Tc(state,j) >= r) exit small
  end do small

  state = j     ! move to a new state j
  p(state) = p(state) + 1
end do big

write(*,*) ''
write(*,*) ' The approximation for distribution is '
write(*,*) real(p)/real(Nsteps)
write(*,*) ''
write(*,*) ' and error estimate'
write(*,*) SQRT((real(p)/real(Nsteps) &
      -(real(p)/real(Nsteps))**2)/real(Nsteps))
write(*,*) ''
write(*,*) ' Actual error:'
write(*,*) real(p)/real(Nsteps)-(/.1,.2,.4,.3/)
write(*,*) ''

write(*,*) ' Let''s refine:'
allocate(pp(n))
write(*,*) ' how many steps?'
read(*,*) j
pp = real(p)/real(Nsteps)
do i = 1, j
  pp = matmul(pp,T)
  write(*,*) sum(abs(pp - (/.1d0,.2d0,.4d0,.3d0/))),&
      ' is sum abs of error vector at', i
end do
write(*,*) 'pp is '
write(*,*) real(pp)

write(*,*) 'Let''s print the n-step transition matrices'
write(*,*) ' how many steps?'
read(*,*) j
do i = 1, j
  T = matmul(T,T)
  write(*,'(a,i0,a)') ' at step ', i
```

```
      call write_mat(T,6)
   end do        ,
contains

   subroutine write_mat(A,to)
   implicit none
     real(kind=kind(0.d0)) :: A(:,:)
     integer :: i, j, to

     write(to,'(a)') 'matrix:------'
     do i = 1, size(A(:,1))
       write(to,'(a)', advance='no') '| '
       do j = 1, size(A(1,:))
         write(to,'(ES14.5E3)', advance='no') A(i,j)
       end do
       write(to,'(a)') ''
     end do
     write(to,'(a)') '----------end'
   end subroutine write_mat
end Program IvMarkovChain
```

Simulated Annealing and Traveling Salesman Problem

The following Fortran 95 code solves the traveling salesman problem (TSP),
using the simulated annealing (SA) method (see Fig. 7.99 below).

```
Program Iv_TSP_SA   ! V.G. Ivancevic, 2007
   ! Solves the traveling salesman problem (TSP)
   ! using simulated annealing (SA) algorithm

   implicit none
   integer :: n, i, j, ii(2), rejected, accepted, k
   real(kind(1.d0)), allocatable :: r(:,:), d(:,:)
   ! coordinates and distances
   integer, allocatable :: s(:) ! the path of visiting cities
   real(kind(1.d0)) :: l_current, T, rand, dl, l_best
   logical :: in_coords = .false.

   call random_seed()

   write(*,*) 'Give number of cities:
   (negative number to give coordinates)'
   read(*,*) n
   if (n < 0) then
      in_coords = .true.
      n = abs(n)
```

```fortran
      write(*,*) 'OK, you must give the coordinates!'
end if
allocate(s(0:n+1),r(n,2),d(n,n))

call random_number(r)
if (in_coords) then
   do i = 1, n
      read(*,*) r(i,1:2)
   end do
else
   do i = 1, n
      write(*,*) r(i,1:2)
   end do
end if

write(*,*) 'First city ', r(1,:)
write(*,*) 'Last city  ', r(n,:)

call do_d() ! This does the distances between cities

do i = 1, n
   S(i) = i
   ! In initial path, we move in order 1->2->3...
end do
S(0) = S(n)
S(n+1) = S(1)

call ps_out_path('initial_path.ps')

l_current = l()
write(*,*) 'Initial path has length ', l_current
l_best = l_current
k = 0
T = 0.5d0
big: do i = 1, huge(i)
   rejected = 0
   accepted = 0
   do j = 1, n*100
      ii(1) = random_site()          ! exchanging two roads
      ii(2) = ii(1)
      do while(abs(ii(2) - ii(1)) <= 1 .or.
      abs(ii(2) - ii(1)) > n/2)
         ii(2) = random_site() ! random two cities
      end do
      dl = d_l(ii(1), ii(2))          ! change in path length
```

```
       call random_number(rand)     ! ask Boltzmann factor...
       if (rand >= exp(-dl/T)) then
          rejected = rejected + 1
          ! ... if change is accepted
       else
          call exc2(ii(1),ii(2))    ! change is accepted
          accepted = accepted + 1
          l_current = l_current + dl
          if (l_current<l_best) l_best = l_current
       end if
    end do

    T = 0.999*T     ! lower the temperature
    if (MOD(i,100) == 1) write(*,*)
    ' current length ', l_current, &
      ' ac ', real(accepted) / real(accepted + rejected)
    if (accepted == 0) then
       if (l_current - l_best > 1.d - 10) then
          k = k + 1
          t = 0.25
       if (k > 3) then
          write(*,*) 'Giving up...'
       exit
       else
          write(*,*) 'Heat up'
       end if
      else
       exit
      end if
    end if
end do big
write(*,*) ' '
write(*,*) 'Rounding error due to dl above:',
abs(l() - l_current)
write(*,*) ' '
write(*,*) 'Final length ', l()
if (l_current - l_best > 1.d - 10) then
   write(*,*) ' '
   write(*,*) 'However, there was even shorter one:'
   write(*,*) 'Best  length ', l_best, ' diff ',
   l_current-l_best
else
   write(*,*) 'This looks fine.'
end if
```

```
  call ps_out_path('final_path.ps')

  write(*,*) ''
  write(*,*) 'See postscript figures *.ps for paths'
contains

  subroutine ps_out_path(name)
    ! draws the current path 'S' into a postscript file
    integer :: i
    character(len=*) :: name
    open(112,file=name)
    write(112,'(a)') '%!PS-Adobe-3.0 EPSF-3.0'
    write(112,'(a)') '%%BoundingBox: 0 0 500 500'
    write(112,'(a)') '50 50 translate 400 400 scale'
    write(112,'(a)') '0.005 setlinewidth 1 setlinejoin'
    do i = 1, n
       write(112,*) r(S(i),:), '0.01 0 360 arc stroke'
    end do
    write(112,*) 'newpath',r(S(1),:), 'moveto'
    do i = 2, n
       write(112,*) r(S(i),:), 'lineto'
    end do
    write(112,*) 'closepath stroke showpage'
    write(112,*) '%%EOF'
    close(112)
  end subroutine ps_out_path

  function d_1(i,j)
    real(kind(1.d0)) :: d_1
    integer :: i, j
    d_1 = -(d(S(i), S(i+1)) + d(S(j), S(j+1))) &
        + (d(S(i),S(j))+d(S(j+1),S(i+1)))
  end function d_1

  subroutine exc2(i,j)
    ! an exchange of two roads - reversing the path i->j
    integer :: i, j, ii, jj
    ii = min(i,j)
    jj = max(i,j)
    S(jj:ii+1:-1) = S(ii+1:jj)
    if (ii == 1) S(n+1) = S(1)
    if (jj == n) S(0) = S(n)
  end subroutine exc2

  function l()
```

```fortran
! calculates the length of the current path 'S'
  real(kind(1.d0)) :: l
  integer :: i
  l = 0.d0
  do i = 1, n
     l = l+d(S(i),S(i+1))
  end do
end function l

subroutine do_d()
  integer :: i, j
  d = 0.d0
  do i = 1, n
     do j = i+1, n
        d(i,j) = sqrt(sum((r(i,:) - r(j,:))**2))
        d(j,i) = d(i,j)
     end do
  end do
end subroutine do_d

function random_site()
  integer :: random_site
  real(kind(1.d0)) :: r
  call random_number(r)
  random_site = int(r*n) + 1
end function random_site
end Program Iv_TSP_SA
```

Feynman Path Integral Monte Carlo

The following Fortran 95 code solves the Feynman path integral for ground state wave ψ-function, using random Monte Carlo method (a more sophisticated version of the Ising spin and Markov chain models above).

```fortran
Program IvFeynMC  ! V.G. Ivancevic, 2007
  ! Feynman path integral Monte Carlo
  ! for ground state wave function

  implicit none
  integer :: i, j, max, element, prop(100), seed
  real*8 :: change, harv, energy, newE, oldE, out, path(100)
  parameter (max=250000, seed=68111)

  open(unit=7,File='IvFeyn.out.dat')       ! output file
```

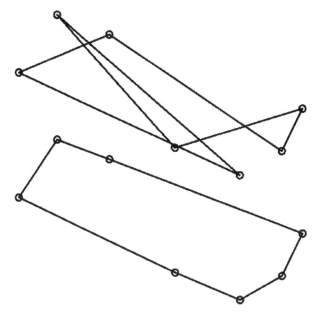

Fig. 7.99. Initial (*up*) and final (*down*) paths for the traveling salesman problem (TSP) with 7 cities, solved using simulated annealing (SA) algorithm.

```
call random_seed(seed)
do j = 1, 100     ! initial path and probability
   path(j) = 0.0
   prop(j) = 0
end do
call random_number(harv)   ! harv = pseudo-random number
oldE = energy(path, 100)   ! find energy of initial path

do i = 1, max     ! main time loop
   element = harv * 100 + 1    ! pick one random element
   change = 2 * (harv - 0.5)
   ! change it by a random value
   path(element) = path(element) + change
   newE = energy(path, 100)
   ! find the new energy
   ! reject change if new energy is greater and the
   ! Boltzmann factor is less than another random number
   if ((newE > oldE) .and. (exp(oldE - newE) < harv)) then
      path(element) = path(element) - change
   end if

   do j = 1, 100      ! add-up probabilities
```

```
            element = path(j) * 10 + 50
            prop(element) = prop(element) + 1
        end do
        oldE = newE
    end do

    do j = 1, 100          ! output data
        out = prop(j)
        write (7, *) j-50, out/max
    end do
    close(7);
    stop 'data saved in IvFeyn.out.dat'
end program IvFeynMC

function energy(array, max)        ! calculate system's energy
    implicit none
    integer :: i, max
    real*8 :: energy, array(max)
    energy = 0.0
    do i = 1,(max-1)
        energy = energy + (array(i+1) - array(i))**2 &
                + array(i)**2
    end do
return; end
```

7.3.11 Sophisticated Integration Algorithms

N-Dimensional Gauss–Legendre Integration

The following short Fortran 95 presents the N-dimensional Gauss–Legendre integration, using the standard 6-point formula. It has been used for calculation of the triple integral of the function $f = x^2 y^3 z^5$.

```
Program IvNdGauss
! N-dimensional Gauss--Legendre integration !
6-point formula, example of a triple integral of f(x,y,z)
! with x
in [A1,B1], y in [A2,B2], z in [A3,B3]

    dimension G(6),W(6)
    data A1,B1,A2,B2,A3,B3/-1.0,2.0,4.0,8.0,3.0,6.0/
    data G/-0.93246951,-0.66120939,-0.23861919,0.23861919, &
          0.66120939,0.93246951/
    data W/0.17132449,0.36076157,0.46791393,0.46791393, &
          0.36076157,0.17132449/
    open (UNIT=7,FILE="IvNGauss.out.dat")        ! output file
```

```
      write ( 7, * )
      'Triple Integration by Gauss-Legendre Quadrature'
      write ( 7, 101) A1,B1,A2,B2,A3,B3
101   format (/,' Limits of Integration:',/,'
for x:',2E12.4,/, &
        ' for y:',2E12.4,/,' for z:',2E12.4)
      sum = 0.0
   do i = 1, 6
      xx = ((B1 - A1) * G(i) + A1 + B1) / 2.0
      do j = 1, 6
         yy = ((B2 - A2) * G(j) + A2 + B2) / 2.0
         do k = 1, 6
            zz = ((B3 - A3) * G(k) + A3 + B3) / 2.0
            sum = sum + W(i) * W(j) * W(k) * f(xx,yy,zz)
         end do
      end do
   end do
   sum = sum * (B1 - A1) * (B2 - A2) * (B3 - A3) / 8.0
   write ( 7, '(a)' ) ' '
   write ( 7, * ) 'Triple Integral =', sum
   close(7)
   stop 'data saved in IvNGauss.out.dat'
end program IvNdGauss

   function f(x,y,z)      ! function to integrate
     f = (x**2)*(y**3)*(z**5)
   return
   end function f
```

N-Dimensional Romberg Integration

The following Fortran 95 code presents ND-Romberg integration of several functions, using a difference table for Richardson extrapolation.

```
Program IvNdRomberg  ! V.G. Ivancevic, 2007
  ! N-dimensional Romberg integration
  implicit none
  real*8 :: a, b
  real, external :: f1, f6, f5, f2, f3, f4
  a = 0.0E+00
  b = 1.0E+00
    open (UNIT=7,FILE="IvNromb.out.dat")      ! output file
  write ( 7, * ) '  ND-Romberg quadrature procedure'
  write ( 7, * ) '  for integral of F(X) on [A,B]**N'
  write ( 7, * ) ' '
  write ( 7, * ) ' A =', a
```

```
   write ( 7, * ) '  B =', b
   write ( 7, * ) ' '
   write ( 7, * ) ' '
   write ( 7, * ) '  F(X) = 1'
   write ( 7, * ) ' '
     call testRom ( f1 )
   write ( 7, * ) ' '
   write ( 7, * ) ' '
   write ( 7, * ) '  F(X) = X = SUM ( X(1:N) )'
   write ( 7, * ) ' '
     call testRom ( f2 )
   write ( 7, * ) ' '
   write ( 7, * ) ' '
   write ( 7, * ) '  F(X) = X**2 = SUM ( X(1:N)**2 )'
   write ( 7, * ) ' '
     call testRom ( f3 )
   write ( 7, * ) ' '
   write ( 7, * ) ' '
   write ( 7, * ) '  F(X) = X**3 = SUM ( X(1:N)**3 )'
   write ( 7, * ) ' '
     call testRom ( f4 )
   write ( 7, * ) ' '
   write ( 7, * ) ' '
   write ( 7, * ) '  F(X) = EXP(X) = EXP ( SUM ( X(1:N) ) )'
   write ( 7, * ) ' '
     call testRom ( f5 )
   write ( 7, * ) ' '
   write ( 7, * ) ' '
   write ( 7, * ) '  F(X) = 1 / (1 + X**2)  = 1
   / ( 1 + SUM ( X(1:N)**2 ) )'
   write ( 7, * ) ' '
     call testRom ( f6 )
   close(7)
   stop 'data saved in IvNromb.out.dat'
end program IvNdRomberg

function f1 ( n, x )
   ! F1DN(X(1:N)) = 1.
   real f1
      f1 = 1.0E+00
   return
end function f1

function f2 ( n, x )
   ! FXDN(X(1:N)) = SUM ( X(1:N) )
```

```
  integer :: n
  real*8 :: f2, x(n)
    f2 = sum ( x(1:n) )
  return
end function f2

function f3 ( n, x )
  ! FX2DN(X(1:N)) = SUM ( X(1:N)**2 )
  integer :: n
  real*8 :: f3, x(n)
    f3 = sum ( x(1:n)**2 )
  return
end function f3

function f4 ( n, x )
  ! FX3DN(X(1:N)) = SUM ( X(1:N)**3 )
  integer :: n
  real*8 :: f4, x(n)
    f4 = sum ( x(1:n)**3 )
  return
end function f4

function f5 ( n, x )
  ! FEDN(X(1:N)) = EXP ( SUM ( X(1:N) ) )
  integer :: n
  real*8 :: f5, x(n)
    f5 = exp ( sum ( x(1:n) ) )
  return
end function f5

function f6 ( n, x )
  ! FBDN(X(1:N)) = 1 / ( 1 + SUM ( X(1:N)**2 ) )
  integer :: n
  real*8 :: f6, x(n)
    f6 = 1.0E+00 / ( 1.0E+00 + sum ( x(1:n)**2 ) )
  return
end function f6

subroutine testRom ( func )
  ! testing NDROMB-procedure
  implicit none
  integer, parameter :: maxit = 3, ndim = 3,
  nwork = maxit + ndim
  real*8 :: eps, aval(ndim), bval(ndim)
  real, external :: func
```

```
      integer :: ind, iwork(nwork), nsub(ndim)
      real*8 :: work(nwork), result
      aval(1:ndim) = 0.0E+00          ! integration limits
      bval(1:ndim) = 1.0E+00
      eps = 0.001E+00                 ! error
      nsub(1:2) = (/ 10, 10 /)        ! number of subintervals
        call ndromb ( func, aval, bval, ndim, nsub, maxit, eps,
        iwork, &
          work, nwork, result, ind )
      write ( 7, '(a6,g13.6)' ) 'NDROMB ', result
      return
   end subroutine testRom

   subroutine ndromb ( func, a, b, ndim, nsub, maxit, eps,
   iwork, work, &
      nwork, result, ind )
      ! NDROMB-procedure approximates the integral
      ! of F(X) over an ND product region
      implicit none
      integer :: ndim, nwork, i, istep, ll, ind, kdim, maxit, &
         iwork(nwork), nsub(ndim)
      real*8 :: a(ndim), b(ndim), en, eps, factor, resold,
      result, &
         rnderr, submid, sum1, weight, work(nwork)
      real, external :: func

      if( ndim < 1 ) then
         write ( 7, '(a)' ) ' '
         write ( 7, '(a)' ) 'NDROMB - Fatal error!'
         write ( 7, '(a,i6)' ) '  NDIM is less than 1.
      NDIM = ',ndim
         stop
      end if

      if ( maxit < 1 ) then
         write ( 7, '(a)' ) ' '
         write ( 7, '(a)' ) 'NDROMB - Fatal error!'
         write ( 7, '(a,i6)' ) '  MAXIT is less than 1.
      MAXIT = ',maxit
         stop
      end if

      do i = 1, ndim
         if ( nsub(i) <= 0 ) then
            write ( 7, '(a)' ) ' '
```

```
      write ( 7, '(a)' ) 'NDROMB - Fatal error!'
      write ( 7, '(a)' ) '  NSUB(I) is less than 1.'
      write ( 7, '(a,i6)' ) '  for I = ',i
      write ( 7, '(a,i6)' ) '  NSUB(I)=',nsub(i)
      stop
   end if
end do

ind = 0
rnderr = epsilon ( 1.0E+00 )
iwork(ndim+1) = 1

if ( maxit > 1 ) then
   iwork(ndim+2) = 2
end if

istep = 1

  10 continue

sum1 = 0.0E+00

weight = 1.0E+00
do i = 1, ndim
   weight = (b(i) - a(i)) * weight / real(nsub(i))
   iwork(i) = 1
end do

  50 continue

kdim = 1
do i = 1, ndim
   submid = real(iwork(i)) - 0.5E+00
   work(i) = a(i) + (b(i) - a(i)) * submid / real(nsub(i))
end do
sum1 = sum1 + func(ndim,work)

  70 continue

if ( iwork(kdim) < nsub(kdim) ) then
   iwork(kdim) = iwork(kdim) + 1
   go to 50
end if

iwork(kdim) = 1
```

```
kdim = kdim+1

if ( kdim <= ndim ) then
   go to 70
end if

work(istep+ndim) = weight * sum1

if ( istep <= 1 ) then
   result = work(ndim+1)
   resold = result
   ind = 1
   if( istep >= maxit) return
   istep = istep+1
   go to 110
end if

en = real(iwork(ndim+istep))

do ll = 2, istep
! difference table for Richardson extrapolation
   i = istep+1-ll
   factor = real(iwork(ndim+i)**2 - 1) / en
   work(ndim+i) = work(ndim+i+1) + (work(ndim+i+1) &
                - work(ndim+i)) * factor
end do

result = work(ndim+1)

ind = 1

if ( abs(result - resold) < abs(result * (eps + rnderr)) )
then
   return
end if

ind = -1

if ( istep >= maxit ) then
   return
end if

resold = result
istep = istep+1
```

```
      iwork(ndim+istep) = int(1.5 * real(iwork(ndim+istep-1)))

      110 continue

      do i = 1, ndim
         nsub(i) = iwork(ndim+istep) * nsub(i)
      end do

      go to 10
   end
```

Here is the output of the above code:

```
   ND-Romberg quadrature procedure
   for integral of F(X) on [A,B]**N

   A =      0.00000
   B =      1.00000


   F(X) = 1

NDROMB  1.00000


   F(X) = X = SUM ( X(1:N) )

NDROMB  1.50000


   F(X) = X**2 = SUM ( X(1:N)**2 )

NDROMB  1.01653


   F(X) = X**3 = SUM ( X(1:N)**3 )

NDROMB 0.774794


   F(X) = EXP(X) = EXP ( SUM ( X(1:N) ) )

NDROMB  5.11486


   F(X) = 1 / (1 + X**2)  = 1 / ( 1 + SUM ( X(1:N)**2 ) )
```

```
NDROMB 0.533346
```

Runge–Kutta–Fehlberg Integrator for ODEs

The following Fortran 95 code presents a more sophisticated version of the Runge–Kutta integrator for ODEs. It is the 4/5th order Runge–Kutta–Fehlberg (RKF) ODE-integrator with adaptive steps-size control [Feh68b, Feh68a]. For comparison, it has been applied to the same damped forced linear oscillator as the above Runge–Kutta 4 program (see Fig. 7.100).

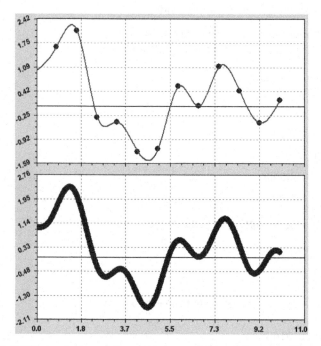

Fig. 7.100. RKF-simulation of the same damped forced linear oscillator, $\ddot{y} + 0.3\dot{y} + y = 5\sin(3t)$, as with the ordinary Runge–Kutta integrator (*above*): *up*—RKF time plot, *down*—RK4 time plot given here for visual comparison. We can see significant difference between the numbers of output data points.

```
Program IvRKF45   ! V.G. Ivancevic, 2007
! RKF45 adaptive step-size ODE integrator - double precision

   implicit none
   integer, parameter :: neqn = 2
   double precision :: t, t_out, t_start, t_stop, abserr,
   relerr, &
```

```fortran
  work(6*neqn+3), x(neqn)
integer :: i_step, iflag, iwork(5), n_step
external f

open ( unit=7, file='IvRKF.out.dat' )

abserr = 0.000000001d+00   ! absolute and relative errors
relerr = 0.000000001d+00

iflag = 1

t_start = 0.0d+00 ! initial time
t_stop = 10.0d+00 ! final time

n_step = 12
t_out = 0.0d+00

x(1) = 1.0d+00     ! initial conditions
x(2) = 0.0d+00

write (7, *) t_out, x(1), x(2)

do i_step = 1, n_step
! defining internal and output time flow

   t = ( ( n_step - i_step + 1 ) * t_start &
     + ( i_step - 1 ) * t_stop ) / dble ( n_step )
   t_out = ( ( n_step - i_step ) * t_start &
        + ( i_step ) * t_stop ) / dble ( n_step )

   call rkf45 ( f, neqn, x, t, t_out, relerr, abserr, iflag, &
   work, iwork )
   write (7, *) t_out, x(1), x(2)
   ! output: time, position, velocity

end do
close(7)
stop 'data saved in IvRKF.out.dat'
end program IvRKF45

!************************************************************
subroutine f (t, x, xp)  ! definition of OdEs (derivatives)
  implicit none
  double precision :: t, x(2), xp(2)
  xp(1) = x(2)
```

```fortran
      xp(2) = -x(1) - 0.3d+00 * x(2) + 5*sin(3.0d+00 * t)
      return
      end


!*************************************************************
      subroutine rkf45 ( f, neqn, x, t, tout, relerr, abserr,
        iflag, & work, iwork )

      ! Runge-Kutta-Fehlberg 4-5 order method
      ! interfacing routine to relieve the user of a long calling
      ! list via the splitting apart of two working storage
      ! arrays

      implicit none
      integer :: neqn, iflag, iwork(5), k1, k1m, k2, k3, k4, k5,
      k6
      double precision :: abserr, relerr, t, tout,
      work(6*neqn+3), x(neqn)
      external :: f

      k1m = neqn + 1
      ! compute indices for the splitting of the work array
      k1 = k1m + 1
      k2 = k1 + neqn
      k3 = k2 + neqn
      k4 = k3 + neqn
      k5 = k4 + neqn
      k6 = k5 + neqn

      call rkfs ( f, neqn, x, t, tout, relerr, abserr, iflag,
        work(1), & work(k1m), work(k1), work(k2), work(k3),
        work(k4), work(k5), work(k6), &
        work(k6+1), iwork(1), iwork(2),
        iwork(3), iwork(4), iwork(5) )

      return
      end


!*************************************************************
      subroutine rkfs ( f, neqn, x, t, tout, relerr, abserr, iflag,
      xp, h, &
        f1, f2, f3, f4, f5, savre, savae, nfe, kop, init, jflag,
        kflag )
      ! Runge-Kutta-Fehlberg method implemented
```

```fortran
implicit none
integer :: neqn, iflag, init, jflag, k, kflag, kop, mflag,
nfe
integer, parameter :: maxnfe = 3000
external f
logical :: output, hfaild
double precision :: a, abserr, relerr, ae, t, dt, ee,
eeoet, eps, esttol, &
  et, h, hmin, ypk, s, savae, savre, scale, rer, tol,
  toln, tout, f1(neqn), &
  f2(neqn), f3(neqn), f4(neqn), f5(neqn), x(neqn), xp(neqn)
double precision, parameter :: remin = 10d-12

! check the input parameters
eps = epsilon ( eps )

if ( neqn < 1 ) then
  iflag = 8
  return
end if

if ( relerr < 0.0d+00 ) then
  iflag = 8
  return
end if

if ( abserr < 0.0d+00 ) then
  iflag = 8
  return
end if

mflag = abs ( iflag )

if ( abs ( iflag ) < 1 .or. abs ( iflag ) > 8 ) then
  iflag = 8
  return
end if

if ( mflag == 1 ) then
  go to 50    ! is this the first call?
end if

! check continuation possibilities
if ( t == tout .and. kflag /= 3 ) then
  iflag = 8
```

```
    return
end if

if ( mflag /= 2 ) then
  go to 25
end if

if ( kflag == 3 ) go to 45      ! iflag = +2 or -2
if ( init == 0 ) go to 45
if ( kflag == 4 ) go to 40

if ( kflag == 5 .and. abserr == 0.0d+00 ) then
  stop
end if

if ( kflag == 6 .and. relerr <= savre .and.
abserr <= savae ) then
  stop
end if

go to 50

! iflag = 3,4,5,6,7 or 8
 25 continue

if ( iflag == 3 ) go to 45
if ( iflag == 4 ) go to 40
if ( iflag == 5 .and. abserr > 0.0d+00 ) go to 45

! integration cannot be continued since user did not
! respond to the instructions pertaining to
! iflag=5,6,7 or 8

stop

! reset function evaluation counter

40 continue

nfe = 0
if ( mflag == 2 ) then
  go to 50
end if

! reset flag value from previous call
```

```
45 continue

iflag = jflag

if ( kflag == 3 ) then
  mflag = abs ( iflag )
end if

! save input iflag and set continuation flag

50 continue

jflag = iflag
kflag = 0

! save relerr and abserr for checking input on
! subsequent calls

savre = relerr
savae = abserr

! the relative error tolerance

rer = 2.0d+00 * epsilon ( rer ) + remin

! the relative error tolerance is too small

if ( relerr < rer ) then
  relerr = rer
  iflag = 3
  kflag = 3
  return
end if

dt = tout - t

if ( mflag == 1 ) go to 60
if ( init == 0 ) go to 65
go to 80

60 continue

init = 0
kop = 0
```

```
a = t
call f ( a, x, xp )
nfe = 1

if ( t == tout ) then
  iflag = 2
  return
end if

 65 continue

init = 1
h = abs ( dt )
toln = 0.0d+00
do k = 1, neqn
  tol = relerr * abs ( x(k) ) + abserr
  if ( tol > 0.0d+00 ) then
    toln = tol
    ypk = abs ( xp(k) )
    if ( ypk * h**5 > tol) then
      h = ( tol / ypk )**0.2d+00
    end if
  end if
end do

if ( toln <= 0.0d+00 ) then
  h = 0.0d+00
end if

h = max ( h, 26.0d+00 * eps * max ( abs ( t ),
abs ( dt ) ) )
jflag =  sign ( 2, iflag )

! set stepsize for integration in the direction from T to
! TOUT

 80 continue

h = sign ( h, dt )

! test to see if RKF45 is being severely impacted by too
! many output points

if ( abs ( h ) >= 2.0d+00 * abs ( dt ) ) then
  kop = kop + 1
```

```
end if

! unnecessary frequency of output

if ( kop == 100 ) then
  kop = 0
  iflag = 7
  return
end if

! if too close to output point, extrapolate and return

if ( abs ( dt ) <= 26.0d+00 * eps * abs ( t ) ) then
  x(1:neqn) = x(1:neqn) + dt * xp(1:neqn)
  a = tout
  call f ( a, x, xp )
  nfe = nfe + 1
  t = tout
  iflag = 2
  return
end if

! initialize output point indicator

output = .false.

! scale the error tolerances to avoid premature underflow

scale = 2.0d+00 / relerr
ae = scale * abserr

! step by step integration

100 continue

hfaild = .false.

! at smallest allowable step-size

hmin = 26.0d+00 * eps * abs ( t )

! adjust step-size if necessary to hit the output point

dt = tout - t
if ( abs ( dt ) >= 2.0d+00 * abs ( h ) ) go to 200
```

```
! the next successful step will complete the integration
! to the output point

if ( abs ( dt ) <= abs ( h ) ) then
  output = .true.
  h = dt
  go to 200
end if

h = 0.5d+00 * dt

! core integrator for taking a single step

200 continue

if ( nfe > maxnfe ) then
  iflag = 4
  kflag = 4
  return
end if

! advance an approximate solution over one step of length H

call fehl ( f, neqn, x, t, h, xp, f1, f2, f3, f4, f5, f1 )
nfe = nfe + 5

! compute and test allowable tolerances versus local error
! estimates

eeoet = 0.0d+00

do k = 1, neqn

  et = abs ( x(k) ) + abs ( f1(k) ) + ae

  if ( et <= 0.0d+00 ) then
    iflag = 5
    return
  end if

  ee = abs ( ( ( -2090.0d+00 * xp(k)
    + ( 21970.0d+00 * f3(k) &
    - 15048.0d+00 * f4(k) ) ) &
    + ( 22528.0d+00 * f2(k) - 27360.0d+00 * f5(k) ) )
```

```
    eeoet = max ( eeoet, ee / et )

end do

esttol = abs ( h ) * eeoet * scale / 752400.0d+00

if ( esttol <= 10d+00 ) then
  go to 260
end if

! unsuccessful step - reduce the stepsize and try again

hfaild = .true.
output = .false.

if ( esttol < 59049.0d+00 ) then
  s = 0.9d+00 / esttol**0.2d+00
else
  s = 0.1d+00
end if

h = s * h

if ( abs ( h ) < hmin ) then
  iflag = 6
  kflag = 6
  return
else
  go to 200
end if

! successful step; store solution at T+H and evaluate
! derivatives there

260 continue

t = t + h
x(1:neqn) = f1(1:neqn)
a = t
call f ( a, x, xp )
nfe = nfe + 1

! choose next stepsize; the increase is limited to
! a factor of 5
```

```fortran
    if ( esttol > 0.0001889568d+00 ) then
      s = 0.9d+00 / esttol**0.2d+00
    else
      s = 5.0d+00
    end if

    if ( hfaild ) then
      s = min ( s, 1.0d+00 )
    end if

    h = sign ( max ( s * abs ( h ), hmin ), h )

    ! end of core integrator

    ! should we take another step?

    if ( output ) then
      t = tout
      iflag = 2
    end if

    if ( iflag > 0 ) go to 100

    ! integration successfully completed

    ! one-step mode

    iflag = - 2

    return
end



subroutine fehl (f, neqn, x, t, h, xp, f1, f2, f3, f4, f5, s)
    ! takes one Fehlberg 4-5 order step
    implicit none
    integer :: neqn
    double precision :: ch, h, t, f1(neqn), f2(neqn), f3(neqn), &
      f4(neqn), & f5(neqn), s(neqn), x(neqn), xp(neqn)
    external f

    ch = h / 4.0d+00

    f5(1:neqn) = x(1:neqn) + ch * xp(1:neqn)
```

```
call f ( t + ch, f5, f1 )

ch = 3.0d+00 * h / 32.0d+00

f5(1:neqn) = x(1:neqn)
+ ch * ( xp(1:neqn) + 3.0d+00 * f1(1:neqn) )

call f ( t + 3.0d+00 * h / 8.0d+00, f5, f2 )

ch = h / 2197.0d+00

f5(1:neqn) = x(1:neqn) + ch * ( 1932.0d+00 * xp(1:neqn) &
  + ( 7296.0d+00 * f2(1:neqn) - 7200.0d+00 * f1(1:neqn) ) )

call f ( t + 12.0d+00 * h / 13.0d+00, f5, f3 )

ch = h / 4104.0d+00

f5(1:neqn) = x(1:neqn) + ch * ( ( 8341.0d+00 * xp(1:neqn) &
  - 845.0d+00 * f3(1:neqn) ) + ( 29440.0d+00 * f2(1:neqn) &
  - 32832.0d+00 * f1(1:neqn) ) )

call f ( t + h, f5, f4 )

ch = h / 20520.0d+00

f1(1:neqn) = x(1:neqn)
  + ch * ( ( -6080.0d+00 * xp(1:neqn) &
  + ( 9295.0d+00 * f3(1:neqn)
  - 5643.0d+00 * f4(1:neqn) ) ) &
  + ( 41040.0d+00 * f1(1:neqn)
  - 28352.0d+00 * f2(1:neqn) ) )

call f ( t + h / 2.0d+00, f1, f5 )
! ! Ready to compute the approximate solution at T+H. !
ch = h / 7618050.0d+00

s(1:neqn) = x(1:neqn)
  + ch * ( ( 902880.0d+00 * xp(1:neqn) &
  + ( 3855735.0d+00 * f3(1:neqn)
  - 1371249.0d+00 * f4(1:neqn) ) ) &
  + ( 3953664.0d+00 * f2(1:neqn)
  + 277020.0d+00 * f5(1:neqn) ) )
```

```
    return
end
```

Fast General ODE Solver

The following Delphi code defines the fast 6th-order Runge–Kutta–Gill integrator for 2nd-order ODEs. Ten example ODEs are solved here: Bessel, Legendre, Laguerre, Chebyshev, Mathieu, Duffing, (modified) Van Der Pol, driven (damped) pendulum, FitzHugh–Nagumo neural (action) potential and Lotka–Volterra predator–prey equations. Three sample outputs are given in Figs. 7.101, 7.102 and 7.103 below.

```
Program IvODEsAll;
// V.G. Ivancevic, 2007 {$APPTYPE CONSOLE}
// Runge-Kutta-Gill 6th order ODE solver

uses SysUtils; const
  n = 10;
  // number of degrees of freedom, i.e., 2nd order ODEs
  Tfin = 10.0;      // final time
  h = 0.01;         // time-step
  c = 10.0;    wo = 3.0;  // input amplitude and frequency
  t1 = 1.0;    t2 : Single = 2.0;
  // stimulation-time intervals
type
  VektE = Array[1..n] of Extended;
var
  i, mm, Stim : Integer;      q, qd, p, pd, QQ, PP : VektE;
  t, ee, u : Extended;
  D1,D2,D3,D4,D5,D6,D7,D8,D9,D10 : Text;
  Label
  One, Four, Five, Three, Seven, Cont;

procedure Bessel;        // Bessel equation
                         // tx" + x' + x(t^2 - k^2)/t = 0
const  k = 0; begin
  qd[1] := p[1];
  pd[1] := -p[1] /t - q[1]*(t*t - k*k) /t/t;
  WriteLn(D1, t, #9, q[1], #9, p[1]);
end;

procedure Legendre;      // Legendre equation
                         // (1 - t^2)x" - 2tx' + k(k + 1)x = 0
const  k = 2; begin
  qd[2] := p[2];
  pd[2] := (2*t*p[2] - k*q[2]*(k + 1)) / (1 - t*t);
```

```
  WriteLn(D2, t, #9, q[2], #9, p[2]);
end;

procedure Laguerre;        // Laguerre equation
                           // tx" + (1 - t)x' + kx = 0
const  k = 3; begin
  qd[3] := p[3];
  pd[3] := -((1 - t) * p[3] + n * q[3]) / t;
  WriteLn(D3, t, #9, q[3], #9, p[3]);
end;

procedure Mathieu;         // Mathieu equation
                           // x" + (a - 2b cos(2t)) x = 0
const  a = 1.0;  b = 5.0; begin
  qd[4] := p[4];
  pd[4] := -(a - 2*b*cos(2*t)) * q[4];
  WriteLn(D4, t, #9, q[4], #9, p[4]);
end;

procedure Chebyshev;       // Chebyshev equation
                           // (1 - t^2)x" - tx' + k^2x = 0
const  k = 3; begin
  qd[5] := p[5];
  pd[5] := (t*p[5] - k*k*q[5]) / (1 - t*t);
  WriteLn(D5, t, #9, q[5], #9, p[5]);
end;

procedure Duffing;     // Duffing equation
                       // x" + cx + wo^2x^3 = u(t)
var  u : single; begin
  qd[6] := p[6];
  pd[6] := u*Stim - c*q[6] - wo*wo*q[6]*q[6]*q[6];
  WriteLn(D6, t, #9, q[6], #9, p[6]);
end;

procedure VDPmod;      // Modified Van der Pol equations
     // q' = u(t) + q - q^3/3 - p,    p' = e(q - a - b*p)
const  a = 0.7;    b = 0.8;    e = 0.01; var u : single; begin
  qd[7] := u*Stim + q[7] - q[7]*q[7]*q[7]/3 - p[7];
  pd[7] := e*(q[7] - a - b*p[7]);
  WriteLn(D7, t, #9, q[7], #9, p[7]);
end;

procedure Pendulum;        // driven damped pendulum equation
                           // x" + x' + (G/L)Sin(x) = u(t)
```

```
const    G = 9.81;    L = 0.5; var u : single; begin
  qd[8] := p[8];
  pd[8] := u*Stim - p[8] - (G / L) * Sin(q[8]);
  WriteLn(D8, t, #9, q[8], #9, p[8]);
end;

procedure FitzHugh;
// FitzHugh-Nagumo neural action potential
// q' = I - q(q-a)(q-1) - p,      p' = e(q-gp)
const
  Amp = 0.1;      a = 0.139;
  Eps = 0.008;    Gam = 2.54;
begin
  qd[9] := Amp*Stim - q[9]*(q[9] - a)*(q[9] - 1) - p[9];
  pd[9] := Eps*(q[9] - Gam*p[9]);
  WriteLn(D9, t, #9, q[9], #9, p[9]);
end;

procedure Lotka;    // Lotka-Volterra predator-prey equations
                    // q' = (A - Bp)q,     p' = (Cq - D)p
const
  A : Single = 0.4;    B : Single = 0.3;
  C : Single = 0.5;    D : Single = 0.2;
begin
  qd[10] := (A - B*p[10])*q[10];
  pd[10] := (C*q[10] - D)*p[10];
  WriteLn(D10, t, #9, q[10], #9, p[10]);
end;

Begin    { main }
  mm := 0;   t := 0.00000001;   u := 0.0;
  // initial time and input
  for i := 1 to n do  // initial conditions
  begin
    q[i] := 1.0;   p[i] := 0.0;    ee :=  0.0;
  end;
Assign(D1,'Bessel.dat');{$i-}Reset(D1);
{$i+}{$i-}Rewrite(D1);{$i+}
Assign(D2,'Legendre.dat');{$i-}Reset(D2);
{$i+}{$i-}Rewrite(D2);{$i+}
Assign(D3,'Laguerre.dat');{$i-}Reset(D3);
{$i+}{$i-}Rewrite(D3);{$i+}
Assign(D4,'Mathieu.dat');{$i-}Reset(D4);
{$i+}{$i-}Rewrite(D4);{$i+}
Assign(D5,'Chebyshev.dat');{$i-}Reset(D5);
```

```
{$i+}{$i-}Rewrite(D5);{$i+}
Assign(D6,'Duffing.dat');{$i-}Reset(D6);
{$i+}{$i-}Rewrite(D6);{$i+}
Assign(D7,'VDPmod.dat');{$i-}Reset(D7);
{$i+}{$i-}Rewrite(D7);{$i+}
Assign(D8,'Pendulum.dat');{$i-}Reset(D8);
{$i+}{$i-}Rewrite(D8);{$i+}
Assign(D9,'FitzHugh.dat');{$i-}Reset(D9);
{$i+}{$i-}Rewrite(D9);{$i+}
Assign(D10,'Lotka.dat');{$i-}Reset(D10);
{$i+}{$i-}Rewrite(D10);{$i+}


// Runge--Kutta--Gill integrator starts here
while ( t <= Tfin ) do begin      // main time loop

if t < t1 then Stim := 0 else if t < t2 then Stim := 1

else Stim := 0;

u := c * Sin(wo * t);              // input forcing

mm:=mm+1; Case mm of
  1:Goto One; 2:Goto Four; 3:Goto Five; 4:Goto Three;
  5:Goto Seven;
end; One:
   begin
     for i := 1 to n do begin
       QQ[i]:=0.0;  PP[i]:=0.0;
     end;
     ee:=0.5;
     Goto Cont;
   end;
Three: ee:=1.7071067811865475244; Four: t:=t+0.5*h; Five:
   begin
     for i := 1 to n do begin
       q[i]:=q[i]+ee*(h*qd[i]-QQ[i]);
       QQ[i]:=2.0*ee*h*qd[i]+(1.0-3.0*ee)*QQ[i];
       p[i]:=p[i]+ee*(h*pd[i]-PP[i]);
       PP[i]:=2.0*ee*h*pd[i]+(1.0-3.0*ee)*PP[i];
     end;
     ee:=0.2928932188134524756;
     Goto Cont;
   end;
Seven:
   begin
```

```
      for i := 1 to n do begin
        q[i]:=q[i]+h*qd[i]/6.0-QQ[i]/3.0;
        p[i]:=p[i]+h*pd[i]/6.0-PP[i]/3.0;
      end;
      mm:=0;
    end;
Cont:
  Bessel;    // call the Bessel equation
  Legendre;  // call the Legendre equation
  Laguerre;  // call the Laguerre equation
  Chebyshev; // call the Chebyshev equation
  Mathieu;   // call the Mathieu equation
  Duffing;   // call the forced Duffing equation
  VDPmod;    // call the modified Van Der Pol equation
  Pendulum;  // call the forced pendulum equation
  FitzHugh;  // call the FitzHugh-Nagumo neural equations
  Lotka;  // call the
  // Lotka-Volterra predator-prey equations
  end;      { end of time loop }
close(D1);close(D2);close(D3);close(D4);close(D5);
close(D6);close(D7);close(D8);close(D9);close(D10);
WriteLn('***DONE! ***'); End.
```

Linear BVP: a Finite Element Solution

The following Fortran 95 code presents the finite element solution of a 1D linear boundary value problem. The output of the program reads:

```
FE Method for a 1D BVP
data:
a = 2.00 b =-3.00 c = 4.00 d = 2.50
xl = 1.00 ne =    5

Boundary Conditions: phi(1) = 0.00 phi(6) = 0.00

System Stiffness Matrix:
  -8.23333    8.63333    0.00000    0.00000    0.00000    0.00000
  11.6333    -19.4667    8.63333    0.00000    0.00000    0.00000
  0.00000    11.6333   -19.4667    8.63333    0.00000    0.00000
  0.00000    0.00000    11.6333   -19.4667    8.63333    0.00000
  0.00000    0.00000    0.00000    11.6333   -19.4667    8.63333
  0.00000    0.00000    0.00000    0.00000    11.6333   -11.2333
System Load Vector:
  0.250000   0.500000   0.500000   0.500000   0.500000   0.250000
Solution Vector:
  0.00000   -0.100800   -0.169372   -0.188163   -0.138131   0.00000
```

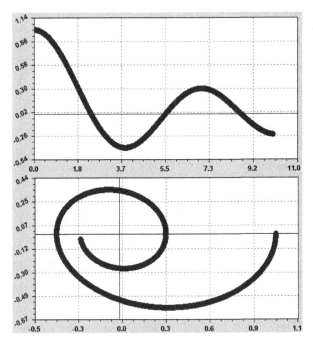

Fig. 7.101. RKG-simulation of the Bessel equation, $t\ddot{x} + \dot{x} + x(t^2 - k^2)/t = 0$: displacement (*up*) and phase-plot (*down*).

```
Program IvFE ! FE solution of 1D BVPs

    dimension cx(6),phi(6),xke(2,2),xk(6,6),xk2(4,4),pe(2), &
      p(6),p2(4),xkn(4,5),el(5)
    data cx/0.0,0.2,0.4,0.6,0.8,1.0/
    data a,b,c,d/2.0,-3.0,4.0,2.5/
    ! coefficients of a 2nd-order ODE
    data eps,xl,ne/1.0E-6,1.0,5/
    ! converg. crit., x-range, no. elements
    open(unit=7,file='IvFE.out.dat')
    phi(1) = 0.0        ! boundary conditions
    phi(6) = 0.0
    write (7, *) 'FE Method for 1D BVPs'
    write (7, 42) a,b,c,d,xl,ne
 42 format (' data:',/,' a =',F5.2,' b =',F5.2,' c =',F5.2, &
      ' d =',F5.2,/,' xl =',F5.2,' ne =',I4,/)
    write (7, 43) phi(1),phi(6)
 43 format (' Boundary Conditions:',' phi(1) =',F5.2,'
phi(6) =',F5.2,/)
    nn = ne + 1    ! number of nodes
```

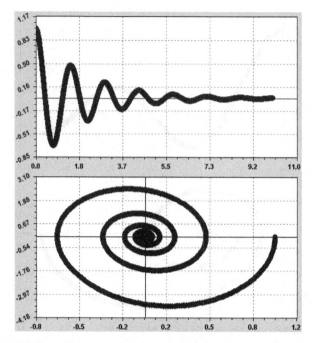

Fig. 7.102. RKG-simulation of the forced damped pendulum equation, $\ddot{x} + \dot{x} + (G/L)\operatorname{Sin}(x) = u(t)$: displacement (*up*) and phase-plot (*down*).

```
nn2 = nn - 2
nnp1 = nn2 + 1
do i = 1, ne
! length of element i = difference of x-coordinates
   el(i) = cx(i+1) - cx(i)
end do
do i = 1, nn
   p(i) = 0.0        ! system charact. vector
   do j = 1, nn
   xk(i,j) = 0.0     ! system charact. matrix
   end do
end do
do i = 1, ne
   xke(1,1) = -(a/el(i)) - (b/2.0) + (c*el(i)/3.0)
   xke(1,2) = (a/el(i)) + (b/2.0) + (c*el(i)/6.0)
   xke(2,1) = (a/el(i)) - (b/2.0) + (c*el(i)/6.0)
   xke(2,2) = -(a/el(i)) + (b/2.0) + (c*el(i)/3.0)
   pe(1) = d*el(i)/2.0     ! element charact. vector
   pe(2) = pe(1)
   xk(i,i) = xk(i,i) + xke(1,1)
```

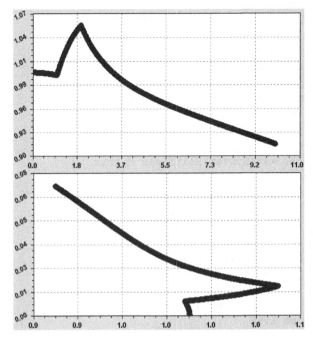

Fig. 7.103. RKG-simulation of the FitzHugh–Nagumo equations for the neural action potential, $\dot{q} = I - q(q - a)(q - 1) - p$, $\dot{p} = e(q - gp)$: displacement (*up*) and phase-plot (*down*).

```
      xk(i,i+1) = xk(i,i+1) + xke(1,2)
      xk(i+1,i) = xk(i+1,i) + xke(2,1)
      xk(i+1,i+1) = xk(i+1,i+1) + xke(2,2)
      p(i) = p(i) + pe(1)
      p(i+1) = p(i+1) + pe(2)
   end do
   do i = 1, nn2
      do j = 1, nn2
      xk2(i,j) = xk(i+1,j+1)
      p2(i) = p(i+1)
      end do
   end do
   do i = 1, nn2
      p2(i) = p2(i) - xk(i+1,1) * phi(1)
      - xk(i+1,nn) * phi(nn)
   end do
   do i = 1, nn2
      do j = 1, nn2
      xkn(i,j) = xk2(i,j)
```

```
      end do
   end do
   do i = 1, nn2
      xkn(i,nnp1) = p2(i)
   end do
   call GaussEl (xkn,p2,nn2,nnp1,eps)
   do i = 1, nn2
      phi(i+1) = p2(i)
   end do
   write (7, *) 'System Stiffness Matrix:'
   do i = 1, nn
      write (7, *) (xk(i,j), j=1,nn)
   end do
   write (7, *) 'System Load Vector:'
   write (7, *) (p(i), i=1,nn)
   write (7, *) 'Solution Vector:'
   write (7, *) (phi(i), i=1,nn)
   close(7)
   stop 'data saved in IvFE.out.dat'
end program IvFE

subroutine GaussEl (a,b,n,np1,eps)
   dimension a(n,np1),b(n)
   do i = 1, n
      a(i,np1) = b(i)
   end do
   do ii = 2, n
      i = ii - 1
      npivot = i
   do jj = ii, n
      if (abs(a(npivot,i)) < abs(a(jj,i))) npivot = jj
   end do
      if (abs(a(npivot,i)) < eps) goto 100
      if (npivot == i) goto 40
   do jj = 1, np1
      Y = a(i,jj)
      a(i,jj) = a(npivot,jj)
      a(npivot,jj) = Y
   end do
40 do ji = ii, n
      if (abs(a(ji,i)) < eps) goto 60
      con = a(ji,i) / a(i,i)
      a(ji,i) = con
   do kj = ii, np1
      a(ji,kj) = a(ji,kj) - con * a(i,kj)
```

```
        end do
      end do
60 end do
          if (abs(a(n,n)) < eps) goto 100
          a(n,np1) = a(n,np1) / a(n,n)
     do j = 2, n
          nn = np1 - j
          ll = nn + 1
          con = a(nn,np1)
     do k = ll, n
        con = con - a(nn,k) * a(k,np1)
     end do
        a(nn,np1) = con / a(nn,nn)
     end do
          goto 120
 100    write (7, 110)
 110    format (/,'Zero pivot element,
        matrix may be singular',/)
120 do i = 1, n
        b(i) = a(i,np1)
    end do
    return
end subroutine GaussEl
```

References

AA68.   Arnold, V.I., Avez, A.: Ergodic Problems of Classical Mechanics. Benjamin, New York (1968)

AA80.   Aubry, S., ré, G.: Colloquium on computational methods in theoretical physics. In: Horowitz, N. (ed.) Group Theoretical Methods in Physics. Ann. Israel Phys. Soc., vol. 3, pp. 133–164 (1980)

AB97.   Aharonov, D., Ben-Or, M.: In: Proceedings of the 29th Annual ACM Symposium on Theory of Computing, pp. 176–188. ACM Press, New York (1997)

ABB04.   Anderson, J.R., Bothell, D., Byrne, M.D., Douglass, S., Lebiere, C., Qin, Y.: An integrated theory of the mind. Psychol. Rev. **111**(4), 1036–1060 (2004)

ABB89.   Aneziris, C., Balachandran, A.P., Bourdeau, M., et al.: Statistics and general relativity. Mod. Phys. Lett. A **4**, 331 (1989)

ABL00.   Arhem, P., Blomberg, C., Liljenstrom, H. (eds.): Disorder Versus Order in Brain Function. Progress in Neural Processing, vol. 12. World Scientific, Singapore (2000)

ABM93.   Abeles, M., Bergman, H., Margalit, E., Vaadia, E.: Spatiotemporal firing patterns in frontal cortex of behaving monkeys. J. Neurophys. **70**, 1629–1638 (1993)

ABS00.   Acebrón, J.A., Bonilla, L.L., Spigler, R.: Synchronization in populations of globally coupled oscillators with inertial effects. Phys. Rev. E **62**, 3437–3454 (2000)

ABS01.   Ardonne, E., Bouwknegt, P., Schoutens, K.: Non-Abelian quantum Hall states—exclusion statistics, k-matrices and duality. J. Stat. Phys. **102**, 421 (2001)

AE87.   Allen, L., Eberly, J.H.: Optical Resonance and Two-level Atoms. Dover, New York (1987)

AEK87.   Akhmediev, N.N., Eleonskii, V.M., Kulagin, N.E.: First-order exact solutions of the nonlinear Schrödinger equation. Theor. Math. Phys. **72**, 809–818 (1987)

AF96.   Abry, P., Flandrin, P.: In: Wavelets in Medicine and Biology, pp. 413–437. CRC Press, Boca Raton (1996)

AG05.   Apps, R., Garwicz, M.: Anatomical and physiological foundations of cerebellar information processing. Nat. Rev. Neurosci. **6**, 297–311 (2005)

V.G. Ivancevic, T.T. Ivancevic, *Quantum Neural Computation*,
Intelligent Systems, Control and Automation: Science and Engineering 40,
DOI 10.1007/978-90-481-3350-5, ⓒ Springer Science+Business Media B.V. 2010

AGH88.   Arneodo, A., Grasseau, G., Holschneider, M.: Wavelet transform of mul-
          tifractals. Phys. Rev. Lett. **61**, 2281–2284 (1988)
AGR81.   Aspect, A., Grangier, P., Roger, G.: Experimental test of realistic local
          theories via Bell's theorem. Phys. Rev. Lett. **47**, 460 (1981)
AGU81.   Akselrod, S., Gordon, D., Ubel, F.A., et al.: Power spectrum analysis of
          heart rate fluctuation: a quantitative probe of beat-to-beat cardiovascu-
          lar control. Science **213**, 220–222 (1981)
AH94.   Allender, E., Hertrampf, U.: Depth reductions for circuits of unbounded
          fan-in. SIAM J. Comput. **112**, 217–238 (1994); An earlier version appears
          as: Allender, E., A note on the power of threshold circuits, Foundations
          of Computer Science, pp. 580–584 (1989)
AIJ08.   Aidman, E., Ivancevic, V., Jennings, A.: A coupled reaction–diffusion
          field model for perception—action cycle with applications to robot nav-
          igation. Int. J. Intel. Def. Sup. Sys. **1**(2), 93–116 (2008)
AJB99.   Albert, R., Jeong, H., Barabasi, A.-L.: Diameter of the world wide web.
          Nature **401**, 130–131 (1999)
AK07.   Ardonne, E., Kim, E.-A.: Hearing non-Abelian statistics from a Moore–
          Read double point contact interferometer. arXiv:0705.2902 (2007)
AK73.   Ablowitz, M.J., Kaup, D.J., Newell, A.C., Segur, H.: Method for solving
          the sine-Gordon equation. Phys. Rev. Let. **30**, 1262–1264 (1973)
AKN74.   Ablowitz, M., Kaup, D., Newell, A., Segur, H.: The inverse scattering
          transform-Fourier analysis for nonlinear problems. Stud. Appl. Math.
          53(4), 249–315 (1974)
AL98.   Abrams, D.S., Lloyd, S.: Nonlinear quantum mechanics implies
          polynomial-time solution for NP-complete and #P problems. Phys. Rev.
          Lett. **81**, 3992–3995 (1998)
ALW08.   Ambjorn, J., Loll, R., Watabiki, Y., Westra, W., Zohren, S.: A matrix
          model for 2D quantum gravity defined by causal dynamical triangula-
          tions. Phys. Lett. B **665**, 252–256 (2008)
AM79.   Anderson, B.D.O., Moore, J.B.: Optimal Filtering. Prentice Hall, New
          York (1979)
AMR88.   Abraham, R., Marsden, J., Ratiu, T.: Manifolds, Tensor Analysis and
          Applications. Springer, New York (1988)
AMS88.   Albano, A.M., Muench, J., Schwartz, C., Mees, A.I., Rapp, P.E.:
          Singular-value decomposition and the Grassberger–Procaccia algorithm.
          Phys. Rev. A **38**, 3017–3026 (1988)
AMV00.   Alfinito, E., Manka, R., Vitiello, G.: Vacuum structure for expanding
          geometry. Class. Quantum Gravity **17**, 93 (2000)
AN99.   Aoyagi, T., Nomura, M.: Oscillator neural network retrieving sparsely
          coded phase patterns. Phys. Rev. Lett. **83**, 1062–1065 (1999)
ARV01.   Abo-Shaeer, J.R., Raman, C., Vogels, J., Ketterle, W.: Observation of
          vortex lattices in Bose–Einstein condensate. Science **292**, 476 (2001)
AR01.   Ammann, M., Reich, C.: VaR for nonlinear financial instruments—Linear
          approximation or full Monte-Carlo? Financ. Mark. Portf. Manag. **15**(3),
          (2001)
ARV02.   Alfinito, E., Romei, O., Vitiello, G.: On topological defect formation in
          the process of symmetry breaking phase transitions. Mod. Phys. Lett. B
          16, 93 (2002)
AS07.   Ardonne, E., Schoutens, K.: Wavefunctions for topological quantum reg-
          isters. Ann. Phys. (N.Y.) **322**, 221 (2007)

AS72.   Abramowitz, M., Stegun, I.A.: Handbook of Mathematical Functions. Dover, New York (1972)

ASW84.   Arovas, D., Schrieffer, J.R., Wilczek, F.: Fractional statistics and the quantum Hall effect. Phys. Rev. Lett. **53**(7), 722 (1984)

AT06.   Anishchenko, A., Treves, A.: Autoassociative memory retrieval and spontaneous activity bumps in small-world networks of integrate-and-fire neurons. J. Physiol. (Paris) **100**(4), 225–236 (2006)

AU96.   Aldroubi, A., Unser, M. (eds.): Wavelets in Medicine and Biology. CRC Press, Boca Raton (1996)

AV00.   Alfinito, E., Vitiello, G.: Formation and life-time of memory domains in the dissipative quantum model of brain. Int. J. Mod. Phys. B **14**, 853–868 (2000)

Ach97.   Acheson, D.: From Calculus to Chaos. Oxford University Press, Oxford (1997)

Adl94.   Adleman, L.M.: Molecular computation of solutions to combinatorial problems. Science **266**(11), 1021–1024 (1994)

Aha98.   Aharonov, D.: Quantum computation. arXiv:quant-ph/9812037 (1998)

Aka97.   Akay, M. (ed.): Time Frequency and Wavelets in Biomedical Signal Processing. IEEE Press, Piscataway (1997)

Ala99.   Alaoui, A.M.: Differential equations with multispiral attractors. Int. J. Bifurc. Chaos **9**(6), 1009–1039 (1999)

Alb71.   Albus, J.S.: A theory of cerebellar function. Math. Biosci. **10**, 25–61 (1971)

Alt01.   Altaisky, M.V.: Quantum neural network. arXiv:quant-ph/0107012 (2001)

Alt07.   Altaisky, M.V.: Wavelet-based quantum field theory. SIGMA **3**, 105–117 (2007)

Ama77.   Amari, S.: Dynamics of pattern formation in lateral-inhibition type neural fields. Biol. Cybern. **27**, 77–87 (1977)

Ama83.   Amari, S.: Field theory of self-organizing neural nets. IEEE Trans. Syst. Man Cybern. **13**, 741–748 (1983)

Ami89a.   Amit, D.: Modeling Brain Function. Springer, Berlin (1989)

Ami89b.   Amit, D.J.: Modeling Brain Function: the World of Attractor Neural Networks. Cambridge University Press, Cambridge (1989)

And01.   Andrecut, M.: Biomorphs, Program for Mathcad^{TM}, Mathcad Application Files, Mathsoft (2001)

And04.   Anderson, M.T.: Geometrization of 3-manifolds via the Ricci flow. Not. Am. Math. Soc. **51**(2), 184–193 (2004)

And26.   Andrian, A.D.: The impulses produced by sensory nerve endings. J. Physiol. (London) **61**, 49–72 (1926)

And64.   Anderson, P.W.: In: Caianiello, E.R. (ed.) Lectures on the Many Body Problem. Academic Press, New York (1964)

And73.   Anderson, P.W.: Resonating valence bonds: a new kind of insulator? Mater. Res. Bull. **8**, 153 (1973)

And87.   Anderson, P.W.: The resonating valence bond state in La_2CuO_4 and superconductivity. Science **235**, 1196 (1987)

Ang92.   Angluin, D.: Computational learning theory: Survey and selected bibliography. In: Proceedings of the 24th Annual ACM Symposium on Theory of Computing, pp. 351–369 (1992)

Arb98.   Arbib, M. (ed.): Handbook of Brain Theory and Neural Networks, 2nd edn. MIT Press, Cambridge (1998)

Ard02.   Ardonne, E.: Parafermion statistics and the application to non-Abelian quantum Hall states. J. Phys. A **35**, 447 (2002)

Arn78.   Arnold, V.I.: Ordinary Differential Equations. MIT Press, Cambridge (1978)

Arn88.   Arnold, V.I.: Geometrical Methods in the Theory of Ordinary Differential Equations. Springer, New York (1988)

Arn93.   Arnold, V.I.: Dynamical Systems. Encyclopaedia of Mathematical Sciences. Springer, Berlin (1993)

Ash05.   Ashcraft, M.H.: Cognition, 4th edn. Prentice Hall, New York (2005)

Ash94.   Ashcraft, M.H.: Human Memory and Cognition, 2nd edn. HarperCollins, New York (1994)

Asi04.   Asif, A.: Fast implementations of the Kalman–Bucy filter for satellite data assimilation. IEEE Signal Process. Lett. **11**(2), 235–238 (2004)

Ati88.   Atiyah, M.: Topological quantum field theories. Pub. Math. IHÉS **68**, 175–186 (1988)

Atm04.   Atmanspacher, H.: Quantum theory and consciousness: an overview with selected examples. Discrete Dyn. **8**, 51–73 (2004)

Aum87.   Aumann, R.J.: Game theory. In: Milgate, M., Newman P. (eds.) The New Palgrave Dictionary of Economics, pp. 460–482 (1987)

Ave99.   Averin, D.V.: Solid-state qubits under control. Nature **398**, 748–749 (1999)

BAI96.   Blanco, S., D'Attellis, C.E., Isaacson, S.I., Rosso, O.A., Sirne, R.O.: Time-frequency analysis of electroencephalogram series, II: gabor and wavelet transforms. Phys. Rev. E **54**, 6661–6672 (1996)

BB84.   Bennett, C.H., Brassard, G.: Proceedings of the IEEE Int. Conf. Comp. Sys. Sig. Proc. Bangalore, India. IEEE, New York (1984); Also, Bennett, C.H., Bessette, F., Brassard, G., Salvail, L., Smolin, J.: J. Cryptol. **5**, 3 (1992)

BB95.   Berry, D.C., Broadbent, D.E.: Implicit learning in the control of complex systems: a reconsideration of some of the earlier claims. In: Frensch, P.A., Funke, J. (eds.) Complex Problem Solving: The European Perspective, pp. 131–150. Lawrence Erlbaum Associates, Hillsdale (1995)

BBC95.   Barenco, A., Bennett, C., Cleve, R., et al.: Elementary gates for quantum computation. Phys. Rev. A **52**, 34–57 (1995)

BBD92.   Bartnik, E.A., Blinowska, K.J., Durks, P.J.: Single evoked potential reconstruction by means of wavelet transform. Biol. Cybern. **67**, 175–181 (1992)

BBD98.   Boschi, D., Branca, S., DeMartini, F., Hardy, L., Popescu, S.: Experimental realization of teleporting an unknown pure quantum state via dual classical and Einstein–Podolski–Rosen channels. Phys. Rev. Lett. **80**, 1121 (1998)

BBP96.   Bennett, C.H., Brassard, G., Popescu, S., Smolin, J.A., Wootters, W.K.: Purification of noisy entanglement and faithful teleportation via noisy channels. Phys. Rev. Lett. **76**, 722–725 (1996)

BBP96c.   Bennett, C.H., Bernstein, H.J., Popescu, S., Schumacher, B.: Concentrating partial entanglement by local operations. Phys. Rev. A **53**, 2046 (1996)

BBS93.   Bennett, C.H., Brassard, G., Crépeau, C., Jozsa, R., Peres, A., Wootters,
            W.K.: Teleporting an unknown quantum state via dual classical and
            Einstein–Podolsky–Rosen channels. Phys. Rev. Lett. **70**, 1895 (1993)
BC81.   Bloedel, J.R., Courville, J.: Cerebellar afferent systems. In: Brookhart,
            J., Mountcastle, V., Brooks, V., Geiger, S. (eds.) Handbook of Physiol-
            ogy, Sect. 1. The Nervous System. Motor Control. American Physiolog-
            ical Society, Bethesda (1981)
BCB92.   Borelli, R.L., Coleman, C., Boyce, W.E.: Differential Equations Labora-
            tory Workbook. Wiley, New York (1992)
BCF02.   Boffetta, G., Cencini, M., Falcioni, M., Vulpiani, A.: Predictability: a way
            to characterize complexity. Phys. Rep. **356**, 367–474 (2002)
BCG98.   Behera, L., Chaudhury, S., Gopal, M.: Applications of self-organizing
            neural networks in robot tracking control. IEEE Proc. Control Theory
            Appl. **145**, 134 (1998)
BCI93.   Balmfort, N.J., Cvitanovic, P., Ierley, G.R., Spiegel, E.A., Vattay, G.:
            Advection of vector-fields by chaotic flows. Ann. N. Y. Acad. Sci. **706**,
            148 (1993)
BCJ99.   Braunstein, S.L., Caves, C.M., Jozsa, R., Linden, N., Popescu, S., Schack,
            R.: Separability of very noisy mixed states and implications for NMR
            quantum computing. Phys. Rev. Lett. **83**, 1054–1057 (1999)
BD02.   Busemeyer, J.R., Diederich, A.: Survey of decision field theory. Math.
            Soc. Sci. **43**, 345–370 (2002)
BD65.   Bjorken, J.D., Drell, S.D.: Relativistic Quantum Fields. McGraw-Hill,
            New York (1965)
BDD02.   Bremner, M.J., Dawson, C.M., Dodd, J.L., et al.: Practical scheme for
            quantum computation with any 2-qubit entangling gate. Phys. Rev. Lett.
            89(24), 247902 (2002)
BDE96.   Barenco, A., Deutsch, D., Ekert, A., Jozsa, R.: Conditional quantum
            dynamics and logic gates. Phys. Rev. Lett. **74**, 4083–4086 (1995)
BDG93.   Bielawski, S., Derozier, D., Glorieux, P.: Experimental characterization
            of unstable periodic orbits by controlling chaos. Phys. Rev. A **47**, 2492
            (1993)
BDI96.   Blanco, S., D'Attellis, C.E., Isaacson, S.I., Rosso, O.A., Sirne, R.O.:
            Time-frequency analysis of electroencephalogram series, II: gabor and
            wavelet transforms. Phys. Rev. E **54**, 6661–6672 (1996)
BDW92.   Bais, F., van Driel, A., de Wild Propitius, M.: Quantum symmetries in
            discrete gauge theories. Phys. Lett. B **280**, 63 (1992)
BDW93.   Bais, F., van Driel, A., de Wild Propitius, M.: Anyons in discrete gauge
            theories with Chern–Simons terms. Nucl. Phys. B **393**, 547 (1993)
BED96.   Bulsara, A.R., Elston, T.C., Doering, C.R., Lindenberg, K.: Cooperative
            behavior in periodically driven noisy integrate-fire models of neuronal
            dynamics. Phys. Rev. E **53**, 3958–3969 (1996)
BEZ00.   Bouwmeester, D., Ekert, A., Zeilinger, A. (eds.): The Physics of Quantum
            Information. Springer, Heidelberg (2000)
BF91.   Balatsky, A., Fradkin, E.: Singlet quantum Hall effect Chern–Simons
            theories. Phys. Rev. B **43**(13), 10622 (1991)
BFF81.   Brody, T.A., Flores, J., French, J.B., et al.: Random-matrix physics:
            spectrum and strength fluctuations. Rev. Mod. Phys. **53**, 385 (1981)
BFG02.   Balents, L., Fisher, M.P.A., Girvin, S.M.: Fractionalization in an easy-
            axis Kagome anti-ferromagnet. Phys. Rev. B **65**(9), 224412 (2002)

846   References

BFM97.   Back, T., Fogel, D., Michalewicz, Z.: Handbook of Evolutionary Computation. Oxford University Press, London (1997)
BFN00.   Balents, L., Fisher, M.P.A., Nayak, C.: Dual vortex theory of strongly interacting electrons: a non-Fermi liquid with a twist. Phys. Rev. B **61**(9), 6307 (2000)
BFN98.   Balents, L., Fisher, M.P.A., Nayak, C.: Nodal liquid theory of the pseudogap phase of high-T_c superconductors. Int. J. Mod. Phys. B **12**(10), 1033 (1998)
BFN99.   Balents, L., Fisher, M.P.A., Nayak, C.: Dual order parameter for the nodal liquid. Phys. Rev. B **60**, 1654 (1999)
BG96.   Baker, G.L., Gollub, J.P.: Chaotic Dynamics: An Introduction, 2nd edn. Cambridge University Press, Cambridge (1996)
BGC96.   Behera, L., Gopal, M., Chaudhury, S.: On adaptive trajectory tracking of … its neural emulator. IEEE Trans. Neural Netw. **7**, 1401–1414 (1996)
BGG80.   Benettin, G., Giorgilli, A., Galgani, L., Strelcyn, J.M.: Lyapunov exponents for smooth dynamical systems and for Hamiltonian systems; a method for computing all of them, part 1: theory, part 2: numerical applications. Meccanica **15**, 9–20 & 21–30 (1980)
BGL00.   Boccaletti, S., Grebogi, C., Lai, Y.-C., Mancini, H., Maza, D.: The control of chaos: theory and applications. Phys. Rep. **329**, 103–197 (2000)
BGS76.   Benettin, G., Galgani, L., Strelcyn, J.M.: Kolmogorov entropy and numerical experiments. Phys. Rev. A **14**, 2338 (1976)
BGS84.   Bohigas, O., Giannoni, M.J., Schmit, C.: Characterization of chaotic quantum spectra and universality of level fluctuation laws. Phys. Rev. Lett. **52**, 1 (1984)
BHM95.   Blanchet, C., Habegger, N., Masbaum, G., Vogel, P.: Topological quantum field theories derived from the Kauffman bracket. Topology **34**, 883 (1995)
BHZ05.   Bonesteel, N.E., Hormozi, L., Zikos, G., Simon, S.H.: Braid topologies for quantum computation. Phys. Rev. Lett. **95**(14), 140503 (2005)
BIT80.   Bahill, A.T., Iandolo, M.J., Troost, B.T.: Vis. Res. **20**, 923 (1980)
BJH95.   Le Bihan, D., Jezzard, P., Haxby, J., Sadato, N., Rueckert, L., Mattay, V.: Functional magnetic resonance imaging of the brain. Ann. Int. Med. **122**(4), 296–303 (1995)
BK01.   Bakalov, B., Kirillov, A.: Lectures on Tensor Categories and Modular Functors. University Lecture Series, vol. 21. American Mathematical Society, Providence (2001)
BK04.   Boyden, E.S., Katoh, A., Raymond, J.L.: Cerebellum-dependent learning: the role of multiple plasticity mechanisms. Annu. Rev. Neurosci. **27**, 581–609 (2004)
BK05.   Behera, L., Kar, I.: Quantum stochastic filtering. Proc. IEEE Int. Conf. SMC **3**, 2161–2167 (2005)
BKE05.   Behera, L., Kar, I., Elitzur, A.C.: Recurrent quantum neural network model to describe eye tracking of moving target. Found. Phys. Lett. **18**(4), 357–370 (2005)
BKE06.   Behera, L., Kar, I., Elitzur, A.C.: Recurrent quantum neural network and its applications. In: Tuszynski, J.A. (ed.) The Emerging Physics of Consciousness. Springer, Berlin (2006)

BKS06.   Bonderson, P., Kitaev, A., Shtengel, K.: Detecting non-Abelian statistics in the $\nu = 5/2$ fractional quantum Hall state. Phys. Rev. Lett. **96**(1), 016803 (2006)

BKW06.   Bergholtz, E.J., Kailasvuori, J., Wikberg, E., Hansson, T.H., Karlhede, A.: Pfaffian quantum Hall state made simple: multiple vacua and domain walls on a thin torus. Phys. Rev. B **74**(8), 081308 (2006)

BL79.   Berestycki, H., Lions, P.L.: Existence d'ondes solitaires dans des problèmes nonlinéaires du type Klein–Gordon. C. R. Acad. Sci. Paris Sér. A-B **288**(7), A395–A398 (1979)

BLK01.   Bar-Shalom, Y., Li, X.R., Kirubarajan, T.: Estimation with Applications to Tracking and Navigation. Wiley, New York (2001)

BLT96.   Bernstein, N.A., Latash, M.L., Turvey, M.T. (eds.): Dexterity and Its Development. Lawrence Erlbaum Associates, Hillsdale (1996)

BLV01.   Boffetta, G., Lacorata, G., Vulpiani, A.: Introduction to chaos and diffusion. Chaos in geophysical flows. In: ISSAOS (2001)

BLW94.   Bassingthwaighte, J.B., Liebovitch, L.S., West, B.J.: Fractal Physiology. Oxford University Press, New York (1994)

BM93.   Bal, T., McCormick, D.A.: Cellular mechanisms of a synchronized oscillation in the thalamus. Science **261**, 361–364 (1993)

BM94.   Baez, J.C., Muniain, J.P.: Gauge Fields, Knots and Gravity, Series on Knots and Everything, vol. 4. World Scientific, Singapore (1994)

BMB94.   Basser, P.J., Mattiello, J., Le Bihan, D.: MR diffusion tensor spectroscopy and imaging. Biophys. J. **66**, 259–267 (1994)

BMN96.   Bonesteel, N.E., McDonald, I.A., Nayak, C.: Gauge fields and pairing in double-layer composite fermion metals. Phys. Rev. Lett. **77**(14), 3009 (1996)

BMP01.   Le Bihan, D., Mangin, J.F., Poupon, C., Clark, C.A., Pappata, S., Molko, N., Chabriat, H.: Diffusion tensor imaging: concepts and applications. J. Magn. Reson. Imaging **13**(4), 534–546 (2001)

BMW93.   Bais, F.A., Morozov, A., de Wild Propitius, M.: Charge screening in the Higgs phase of Chern–Simons electrodynamics. Phys. Rev. Lett. **71**, 2383 (1993)

BN01.   Bastian, J., Nguyenkim, J.: Dendritic modulation of burst-like firing in sensory neurons. J. Neurophysiol. **85**, 10–22 (2001)

BN04.   Bertschinger, N., Natschläger, T.: Real-time computation neural networks. Neural. Comput. **16**, 1413–1436 (2004)

BN06.   Bena, C., Nayak, C.: Effects of non-Abelian statistics on two-terminal shot noise in a quantum Hall liquid in the Pfaffian state. Phys. Rev. B **73**(15), 155335 (2006)

BN07.   Bishara, W., Nayak, C.: Non-Abelian anyon superconductivity. Phys. Rev. Lett. **99**(6), 066401 (2007)

BNO03.   Breban, R., Nusse, H.E., Ott, E.: Scaling properties of saddle-node bifurcations on fractal basin boundaries. Phys. Rev. E **68**(6), 066213 (2003)

BP02.   Barahona, M., Pecora, L.M.: Synchronization in small-world systems. Phys. Rev. Lett. **89**, 054101 (2002)

BP82.   Barone, A., Paterno, G.: Physics and Applications of the Josephson Effect. Wiley, New York (1982)

BP92.   Benvenuto, N., Piazza, F.: On the complex backpropagation algorithm. IEEE Trans. Signal Process. **40**(4), 967–969 (1992)

BP96.   Basser, P.J., Pierpaoli, C.: Microstructural and physiological features of tissues elucidated by quantitative-diffusion-tensor MRI. J. Magn. Reson. B **111**, 209–219 (1996)

BP97.   Badii, R., Politi, A.: Complexity: Hierarchical Structures and Scaling in Physics. Cambridge University Press, Cambridge (1997)

BPM97.   Bouwmeester, D., Pan, J.-W., Mattle, K., Eibl, M., Weinfurter, H., Zeilinger, A.: Experimental quantum teleportation. Nature **390**, 575 (1997)

BPS70.   Baum, L.E., Petrie, T., Soules, G., Weiss, N.: A maximization technique occurring in the statistical analysis of probabilistic functions of Markov chains. Ann. Math. Stat. **41**(1), 164–171 (1970)

BPS75.   Belavin, A.A., Polyakov, A.M., Swartz, A.S., Tyupkin, Yu.S.: SU(2) instantpons discovered. Phys. Lett. B **59**, 85 (1975)

BPS98.   Blackmore, D.L., Prykarpatsky, Y.A., Samulyak, R.V.: The integrability of Lie-invariant geometric objects generated by ideals in the Grassmann algebra. J. Nonlinear Math. Phys. **5**(1), 54–67 (1998)

BPV03.   Batista, A.M., de Pinto, S.E., Viana, R.L., Lopes, S.R.: Mode locking in small-world networks of coupled circle maps. Physica A **322**, 118 (2003)

BPZ84.   Belavin, A.A., Polyakov, A.M., Zamolodchikov, A.B.: Infinite conformal symmetry in 2D quantum field theory. Nucl. Phys. B **241**, 333 (1984)

BRN96.   Van den Bogert, A.J., Read, L., Nigg, B.M.: A method for inverse dynamic analysis using accelerometry. J. Biomech. **29**(7), 949–954 (1996)

BS73.   Black, F., Scholes, M.: The pricing of options and corporate liabilities. J. Pol. Econ. **81**, 637–659 (1973)

BS79.   Bahill, A.T., Stark, L.: The trajectories of saccadic eye movements. Sci. Am. **240**, 108–117 (1979)

BS91.   Braitenberg, V., Shulz, A.: Anatomy of the Cortex. Springer, Berlin (1991)

BS93.   Beck, C., Schlogl, F.: Thermodynamics of Chaotic Systems. Cambridge University Press, Cambridge (1993)

BS99.   Bouwknegt, P., Schoutens, K.: Exclusion statistics in conformal field theory—generalized fermions and spinons for level-1 WZW theories. Nucl. Phys. B **547**, 501 (1999)

BSN85.   Babloyantz, A., Salazar, J.M., Nicolis, C.: Evidence for chaotic dynamics during the sleep cycle. Phys. Lett. A **111**, 152 (1985)

BSS04.   Bretin, V., Stock, S., Seurin, Y., Dalibard, J.: Fast rotation of a Bose–Einstein condensate. Phys. Rev. Lett. **92**(5), 050403 (2004)

BSS06.   Bonderson, P., Shtengel, K., Slingerland, J.K.: Probing non-Abelian statistics with quasiparticle interferometry. Phys. Rev. Lett. **97**(1), 016401 (2006)

BSS07.   Bonderson, P., Shtengel, K., Slingerland, J.K.: Decoherence of anyonic charge in interferometry measurements. Phys. Rev. Lett. **98**(7), 070401 (2007)

BST87.   Boebinger, G., Stomer, H., Tsui, D., et al.: Activation energies and localization in the fractional quantum Hall effect. Phys. Rev. B **36**, 7919 (1987)

BTS98.   Bergmann, K., Theuer, H., Shore, B.W.: Coherent population transfer among quantum states of atoms and molecules. Rev. Mod. Phys. **70**, 1003 (1998)

BTU93.   Bohigas, O., Tomsovic, S., Ullmo, D.: Strongly chaotic and mixed systems: some classical and quantum properties. Phys. Rep. **223**, 43–133 (1993)

BV93.   Bernstein, E., Vazirani, U.: Quantum complexity theory. SIAM J. Comput. **26**(5), 1411–1473 (1997)

BW90.   Blok, B., Wen, X.G.: Effective theories of the fractional quantum Hall effect: hierarchy construction. Phys. Rev. B **42**(13), 8145 (1990)

BW92a.   Bennett, C.H., Wiesner, S.J.: Communication via 1- and 2-particle operators on Einstein–Podolsky–Rosen states. Phys. Rev. Lett. **69**, 2881 (1992)

BW92b.   Blok, B., Wen, X.-G.: Many-body systems with non-Abelian statistics. Nucl. Phys. B **374**, 615 (1992)

BWT06.   Busemeyer, J.R., Wang, Z., Townsend, J.T.: Quantum dynamics of human decision-making. J. Math. Psychol. **50**, 220–241 (2006)

BY97.   Bar-Yam, Y.: Dynamics of Complex Systems. Perseus Books, Reading (1997)

BY00.   Bar-Yam, Y. (ed.): Unifying Themes in Complex Systems: Proc. Int. Conf. Complex Syst. Perseus, Reading (2000)

BY04.   Bar-Yam, Y.: Unifying principles in complex systems. In: Roco, M.C., Bainbridge, W.S. (eds.) Converging Technology (NBIC) for Improving Human Performance (2004)

BY07.   Bonesteel, N.E., Yang, K.: Infinite-randomness fixed points for chains of non-Abelian quasiparticles. Phys. Rev. Lett. **99**(14), 140405 (2007)

BZA87.   Baskaran, G., Zou, Z., Anderson, P.W.: The resonating valence bond state and high-T_c superconductivity—a mean field theory. Solid State Commun. **63**(11), 973 (1987)

Bén67.   Bénabou, J.: Introduction to bicategories. In: Lecture Notes in Mathematics. Springer, New York (1967)

Bac96.   Back, T.: Evolutionary Algorithms in Theory and Practice: Evolution Strategies, Evolutionary Programming, Genetic Algorithms. Oxford University Press, London (1996)

Bae97.   Baez, J.: An introduction to n-categories. In: Moggi, E., Rosolini, G. (eds.) 7th Conference on Category Theory and Computer Science. Lecture Notes in Computer Science. Springer, Berlin (1997)

Bae98.   Baez, J., Dolan, J.: Higher-dimensional algebra, III: n-categories and the algebra of opetopes. Adv. Math. **135**(2), 145–206 (1998)

Bai80.   Bais, F.A.: Flux metamorphosis. Nucl. Phys. B **170**, 32 (1980)

Bai89.   Bai-Iin, H.: Elementary Symbolic Dynamics and Chaos in Dissipative Systems. World Scientific, Singapore (1989)

Bal06.   Balakrishnan, J.: A geometric framework for phase synchronization in coupled noisy nonlinear systems. Phys. Rev. E **73**, 036206 (2006)

Bal87.   Ballentine, L.E.: Foundations of quantum mechanics since the Bell inequalities. Am. J. Phys. **55**, 785 (1987)

Ban97.   Bantay, P.: The Frobenius–Schur indicator in conformal field theory. Phys. Lett. B **394**, 87 (1997)

Bar08.   Barbiellini, B.: Quantum computing lectures. http://stardec.hpcc.neu.edu/~bba/RES/QCOMP/ (2008)

Bar84.   Barendregt, H.: The Lambda Calculus: Its Syntax and Semantics. Studies in Logic and the Foundations of Mathematics. North-Holland, Amsterdam (1984)

Bar91.   Barkley, D.: A model for fast computer-simulation of waves in excitable
            media. Physica D **49**, 61–70 (1991)
Bat89.   Battle, G.: Wavelets and Renormalization. World Scientific, Singapore
            (1989)
Bea95.   Bear, M.F., et al. (eds.): Neuroscience: Exploring The Brain. Williams
            and Wilkins, Baltimore (1995)
Bec08.   Beck, F.: Synaptic quantum tunnelling. Neuroquantology **6**(2), 140–151
            (2008)
Bel64.   Bell, J.S.: On the Einstein–Podolsky–Rosen paradox. Physics **1**, 195
            (1964)
Bel66.   Bell, J.S.: On the problem of hidden variables in quantum mechanics.
            Rev. Mod. Phys. **38**, 447 (1966)
Bel87.   Bell, J.S.: Speakable and Unspeakable in Quantum Mechanics. Cam-
            bridge University Press, Cambridge (1987)
Ben73.   Bennett, C.H.: Logical reversibility of computation. IBM J. Res. Dev.
            17, 525–532 (1973)
Ben80.   Benioff, P.: Quantum mechanical Hamiltonian models of discrete pro-
            cesses. J. Stat. Phys. **22**, 563–591 (1980)
Ben82.   Benioff, P.A.: Quantum mechanical Hamiltonian models of Turing ma-
            chines. J. Stat. Phys. **29**(3), 515–546 (1982)
Ben84.   Benettin, G.: Power law behaviour of Lyapunov exponents in some con-
            servative dynamical systems. Physica D **13**, 211–213 (1984)
Ben89.   Bennett, C.H.: Time/space trade-offs for reversible computation. SIAM
            J. Comput. **18**, 766–776 (1989)
Ben91.   Van Benthem, J.: Reflections on epistemic logic. Log. Anal. **133/134**,
            5–14 (1991)
Ben95.   Benz, S.P.: Superconductor-normal-superconductor junctions for pro-
            grammable voltage standards. Appl. Phys. Lett. **67**, 2714–2716 (1995)
Ben96.   Bennett, C., et al.: 4-year COBE DMR cosmic microwave background
            observations: maps and basic results. Astrophys. J. **464**, L1 (1996)
Ber67.   Bernstein, N.A.: The Coordination and Regulation of Movements. Perg-
            amon, London (1967)
Ber82.   Bernstein, N.A.: Some emergent problems of the regulation of motor
            acts. In: Whiting, H. (ed.) Human Motor Actions: Bernstein Reassessed,
            pp. 343–358. North-Holland, Amsterdam (1982)
Ber84.   Berry, M.V.: Quantal phase factors accompanying adiabatic changes.
            Proc. R. Soc. Lond. A **392**, 45 (1984)
Ber97.   Berthiaume, A.: Quantum computation. In: Hemaspaandra, L.A., Sel-
            man, A.L. (eds.) Complexity Theory Retrospective II, pp. 23–51.
            Springer, Berlin (1997)
Bey01.   Beyer, H.-G.: The Theory of Evolution Strategies. Springer, Berlin
            (2001)
Bh92.   Brown, R.G., Hwang, P.Y.C.: Introduction to Random Signals and Ap-
            plied Kalman Filtering. Wiley, New York (1992)
Bih00.   Le Bihan, D.: What to expect from MRI in the investigation of the
            central nervous system? C. R. Acad. Sci. III **323**(4), 341–350 (2000)
Bih03.   Le Bihan, D.: Looking into the functional architecture of the brain with
            diffusion MRI. Nat. Rev. Neurosci. **4**(6), 469–480 (2003)
Bih96.   Le Bihan, D.: Functional MRI of the brain: principles, applications and
            limitations. J. Neuroradiol. **23**(1), 1–5 (1996)

Bil65.   Billingsley, P.: Ergodic Theory and Information. Wiley, New York (1965)
Bir15.   Birkhoff, G.D.: The restricted problem of three-bodies. Rend. Circ. Mat. Palermo **39**, 255–334 (1915)
Bir17.   Birkhoff, G.D.: Dynamical systems with two degrees of freedom. Trans. Am. Math. Soc. **18**, 199–300 (1917)
Bir27.   Birkhoff, G.D.: Dynamical Systems. American Mathematical Society, Providence (1927)
Bl76.   Black, F.: Studies of stock price volatility changes. In: Proc. 1976 Meet. Ame. Stat. Assoc. Bus. Econ. Stat., pp. 177–181 (1976)
Bla86.   Blackman, S.S.: Multiple-Target Tracking with Radar Applications. Artech House, Norwood (1986)
Blu51.   Blumer, H.: Collective Behavior. In: Lee, A.M. (ed.) Principles of Sociology, pp. 67–121. Barnes & Noble, New York (1951)
Bob92.   Bobbert, P.A.: Simulation of vortex motion in underdamped two-dimensional arrays of Josephson junctions. Phys. Rev. B **45**, 7540–7543 (1992)
Bod97.   Bode, M.: Front-bifurcations in reaction–diffusion systems with inhomogeneous parameter distributions. Physica D **106**, 270–286 (1997)
Boh35.   Bohr, N.: Can quantum mechanical description of physical reality be considered complete? Phys. Rev. **48**, 696 (1935)
Boh51.   Bohm, D.: Quantum Theory. Prentice Hall, Englewood Cliffs (1951)
Boh92.   Bohm, D.: Thought as a System. Routledge, London (1992)
Bon00.   Bonesteel, N.E.: Chiral spin liquids and quantum error-correcting codes. Phys. Rev. A **62**(6), 062310 (2000)
Bon89.   Bonesteel, N.E.: Valence bonds and the Lieb–Schultz–Mattis theorem. Phys. Rev. B **40**(14), 8954 (1989)
Bou98.   Bourgain, J.: Scattering in the energy space and below for 3D NLS. J. Anal. Math. **75**, 267–297 (1998)
Bou99.   Bourgain, J.: New Global Well-Posedness Results for Non-Linear Schrödinger Equations. American Mathematical Society, Providence (1999)
Boy91.   Boyd, R.W.: Nonlinear Optics. Academic Press, New York (1991)
Bra01.   Branke, J.: Evolutionary Optimization in Dynamic Environments. Kluwer, Dordrecht (2001)
Bra06.   Bravyi, S.: Universal quantum computation with the $\nu = 5/2$ fractional quantum Hall state. Phys. Rev. A **73**(4), 042313 (2006)
Bro58.   Broadbent, D.E.: Perception and Communications. Pergamon, London (1958)
Bro86a.   Brooks, R.A.: A robust layered control system for a mobile robot. IEEE Trans. Robot. Autom. **2**(1), 14–23 (1986)
Bro86b.   Brooks, R.A.: A robust layered control system for a mobile robot. IEEE Trans. Robot. Autom. **2**(1), 14–23 (1986)
Bro89.   Brooks, R.A.: A robot that walks: emergent behavior form a carefully evolved network. Neural. Comput. **12**, 253–262 (1989)
Bro90.   Brooks, R.A.: Elephants don't play chess. Robot. Autom. Syst. **6**, 3–15 (1990)
Bro88.   Brookner, E. (ed.): Aspects of Modern Radar. LexBook, Lexington (1988)
Bro98.   Brookner, E.: Tracking and Kalman Filtering Made Easy. Wiley, New York (1998)

Buc70.   Bucy, R.S.: Linear and nonlinear filtering. IEEE Proc. **58**(6), (1970)

But64.   Butcher, J.C.: On Runge–Kutta processes of high order. J. Aust. Math. Soc. **4**, 179–194 (1964)

CAM05.   Cvitanovic, P., Artuso, R., Mainieri, R., Tanner, G., Vattay, G.: Chaos: Classical and Quantum. Niels Bohr Institute, Copenhagen (2005)

CBE95.   Clark, I., Biscay, R., Echeverria, M., Virues, T.: Multiresolution decomposition of non-stationary EEG signals: a preliminary study. Comput. Biol. Med. **25**, 373 (1995)

CC95.   Christini, D.J., Collins, J.J.: Controlling nonchaotic neuronal noise using chaos control techniques. Phys. Rev. Lett. **75**, 2782–2785 (1995)

CC99.   Cao, H.D., Chow, B.: Recent developments on the Ricci flow. Bull. Am. Math. Soc. **36**, 59–74 (1999)

CCC97.   Caiani, L., Casetti, L., Clementi, C., Pettini, M.: Geometry of dynamics Lyapunov exponents and phase transitions. Phys. Rev. Lett. **79**, 4361–4364 (1997)

CCF95.   Cattaneo, A.S., Cotta-Ramusino, P., Froehlich, J., Martellini, M.: Topological bf theories in 3 and 4 dimensions. J. Math. Phys. **36**, 6137 (1995)

CCI95a.   Collins, J.J., Chow, C.C., Imhoff, T.T.: Stochastic resonance without tuning. Nature **376**, 236–238 (1995)

CCI95b.   Collins, J.J., Chow, C.C., Imhoff, T.T.: Aperiodic stochastic resonance in excitable systems. Phys. Rev. E **52**, R3321–R3324 (1995)

CCK89.   Chayes, J.T., Chayes, L., Kivelson, S.A.: Valence bond ground states in a frustrated 2D spin-1/2 Heisenberg antiferromagnet. Commun. Math. Phys. **123**, 53 (1989)

CCP96a.   Christiansen, F., Cvitanovic, P., Putkaradze, V.: Hopf's last hope: spatio-temporal chaos in terms of unstable recurrent patterns. Nonlinearity **10**, 1 (1997)

CCP96b.   Casetti, L., Clementi, C., Pettini, M.: Riemannian theory of Hamiltonian chaos and Lyapunov exponents. Phys. Rev. E **54**, 5969 (1996)

CD82.   Choquet-Bruhat, Y., DeWitt-Morete, C.: Analysis, Manifolds and Physics, 2nd edn. North-Holland, Amsterdam (1982)

CEH98.   Cirac, J.I., Ekert, A., Huelga, S.F., Macchiavello, C.: On the improvement of frequency stardards with quantum entanglement. Phys. Rev. A **59**, 4249 (1999)

CER00.   Calvagno, M., Ermani, M., Rinaldo, R., Sartoretto, F.: In: Proceedings of the 2000 IEEE International Conference on Acoustics, Speech, and Signal Processing, vol. 6 (2000)

CF94.   Crane, L., Frenkel, I.: Four dimensional topological quantum field theory, Hopf categories, and the canonical bases. J. Math. Phys. **35**, 5136–5154 (1994)

CFK97.   De, C., Chamon, C., Freed, D.E., Kivelson, S.A., Sondhi, S.L., Wen, X.G.: Two point-contact interferometer for quantum Hall systems. Phys. Rev. B **55**(4), 2331 (1997)

CG03.   Carpenter, G.A., Grossberg, S.: Adaptive resonance theory. In: Arbib, M.A. (ed.) The Handbook of Brain Theory, Neural Networks, 2nd edn., pp. 87–90. MIT Press, Cambridge (2003)

CG83a.   Cohen, M.A., Grossberg, S.: Absolute stability of global pattern formation and parallel memory storage by competitive neural networks. IEEE Trans. Syst. Man Cybern. **13**, 815–826 (1983)

CG83b.   Cohen, M.A., Grossberg, S.: Absolute stability of global pattern for-
            mation, parallel memory storage by competitive neural networks. IEEE
            Trans. Syst. Man Cybern. **13**(5), 815–826 (1983)
CGC96.   Chapeau-Blondeau, F., Godivier, X., Chambet, N.: Stochastic resonance
            in a neuron model that transmits spike trains. Phys. Rev. E **53**, 1273–
            1275 (1996)
CGM98.   Casetti, L., Gatto, R., Modugno, M.: Chaos in effective classical and
            quantum dynamics. Phys. Rev. E **57**, 1223 (1998)
CGP88.   Cvitanovic, P., Gunaratne, G., Procaccia, I.: Topological, metric prop-
            erties of Hénon-type strange attractors. Phys. Rev. A **38**, 1503–1520
            (1988)
CGT01.   Cappelli, A., Georgiev, L.S., Todorov, I.T.: Parafermion Hall states from
            coset projections of Abelian conformal theories. Nucl. Phys. B **599**(3),
            499 (2001)
CGT99.   Cappelli, A., Georgiev, L.S., Todorov, I.T.: A unified conformal field
            theory description of paired quantum Hall states. Commun. Math. Phys.
            205, 657 (1999)
CH07.   Chen, H.-D., Hu, J.: Majorana fermions, exact mappings between clas-
            sical and topological orders. arXiv:cond-mat/0702366 (2007)
CH86.   Carr, C.E., Heiligenberg, W., Rose, G.J.: A time-comparison circuit in
            the electric fish midbrain, I: behavior and physiology. J. Neurosci. **6**,
            107–119 (1986)
CHK04.   Caron, L.A., Huard, D., Kröger, H., et al.: Comparison of classical chaos
            with quantum chaos. J. Phys. A, Math. Gen. **37**, 6251–6265 (2004)
CIS97.   Casdagli, M.C., Iasemidis, L.D., Savit, R.S., Gilmore, R.L., Roper, S.N.,
            Sackellares, J.C.: Non-linearity in invasive EEG recordings from patients
            with temporal lobe epilepsy. Electroencephalogr. Clin. Neurophysiol.
            102, 98 (1997)
CJ04.   Chang, C.-C., Jain, J.K.: Microscopic origin of the next-generation frac-
            tional quantum Hall effect. Phys. Rev. Lett. **92**(19), 196806 (2004)
CJP93.   Crisanti, A., Jensen, M.H., Paladin, G., Vulpiani, A.: Intermittency, pre-
            dictability in turbulence. Phys. Rev. Lett. **70**, 166 (1993)
CK04.   Chow, B., Knopf, D.: The Ricci Flow: An Introduction, Mathematical
            Surveys and Monographs. American Mathematical Society, Providence
            (2004)
CK90.   Crick, F., Koch, C.: Towards a neurobiological theory of consciousness.
            Sem. Neurosci. **2**, 263–275 (1990)
CK97.   Cutler, C.D., Kaplan, D.T. (eds.): Nonlinear Dynamics and Time Series.
            Fields Inst. Comm., vol. 11. American Mathematical Society, Providence
            (1997)
CKD07.   Choi, H., Kang, W., Das Sarma, S., Pfeiffer, L., West, K.: Density matrix
            renormalization group study of incompressible fractional quantum hall
            states. arXiv:0706.4469 (2007)
CKS03.   Colliander, J., Keel, M., Staffilani, G., Takaoka, H., Tao, T.: Polynomial
            growth and orbital instability bounds for L^2-subcritical NLS below the
            energy norm. Commun. Pure Appl. Anal. **2**, 33–50 (2003)
CKS04.   Colliander, J., Keel, M., Staffilani, G., Takaoka, H., Tao, T.: Global ex-
            istence and scattering for rough solutions of a nonlinear Schrödinger
            equation in $BbbR^3$. Commun. Pure Appl. Math. **57**, 987–1014 (2004)

CL81.   Caldeira, A.O., Leggett, A.J.: Influence of dissipation on quantum tun-
           neling in macroscopic systems. Phys. Rev. Lett. **46**, 211 (1981)
CL83.   Caldiera, A.O., Leggett, A.J.: Quantum tunneling in a dissipative sys-
           tem. Ann. Phys. (N.Y.) **149**, 347–456 (1983)
CL84.   Cheng, T.-P., Li, L.-F.: Gauge Theory of Elementary Particle Physics.
           Clarendon, Oxford (1984)
CMP98b.   Claussen, J.C., Mausbach, T., Piel, A., Schuster, H.G.: Memory differ-
           ence control of unknown unstable fixed-points: drifting parameter con-
           ditions, delayed measurement. Phys. Rev. E **58**(6), 7260–7273 (1998)
CMQ91.   Coifman, R., Meyer, Y., Quake, S., Wickerhauser, M.V.: Wavelet analysis
           and signal processing. In: Proceedings, Conference on Wavelets, Lowell,
           MA (1991)
CPC00.   Casetti, L., Pettini, M., Cohen, E.G.D.: Geometric approach to Hamil-
           tonian dynamics and statistical mechanics. Phys. Rep. **337**, 237–341
           (2000)
CQD05.   Chen, Y., Qiu, X.-J., Dong, X.-L.: Pseudo-spin model for the microtubule
           wall in external field. Biosystems **82**, 127–136 (2005)
CR07.   Cooper, N.R., Rezayi, E.H.: Competing compressible and incompressible
           phases in rotating atomic Bose gases at filling factor $\nu = 2$. Phys. Rev. A
           75(1), 013627 (2007)
CR08.   Conway, J.M., Riecke, H.: Superlattice patterns in the complex
           Ginzburg–Landau equation with multi-resonant forcing. arXiv:0803.
           0346 [nlin.PS] (2008)
CRV92.   Celeghini, E., Rasetti, M., Vitiello, G.: Quantum dissipation. Ann. Phys.
           215, 156 (1992)
CS00.   Cristianini, N., Shawe-Taylor, J.: Support Vector Machines. Cambridge
           University Press, Cambridge (2000)
CS06.   Chung, S.B., Stone, M.: Proposal for reading out anyon qubits in non-
           Abelian $\nu = 12/5$ quantum Hall state. Phys. Rev. B **73**(24), 245311
           (2006)
CS07.   Chung, S.B., Stone, M.: Explicit monodromy of Moore–Read wavefunc-
           tions on a torus. J. Phys. A, Math. Theor. **40**(19), 4923 (2007)
CS83.   Coppersmith, S.N., Fisher, D.S.: Pinning transition of the discrete sine-
           Gordon equation. Phys. Rev. B **28**, 2566–2581 (1983)
CS86.   Chang, S.-J., Shi, K.-J.: Evolution and exact eigenstates of a resonant
           quantum system. Phys. Rev. A **34**, 7 (1986)
CS90.   Clauser, J.F., Shimony, A.: Bell's theorem: experimental tests and im-
           plications. Rep. Prog. Phys. **41**, 1131 (1990)
CS94.   Coolen, A.C.C., Sherrington, D.: Order-parameter flow in the fully con-
           nected Hopfield model near saturation. Phys. Rev. E **49**, 1921–1934
           (1994); Erratum: Phys. Rev. E **49**, 5906 (1994)
CS95.   Chernikov, A.A., Schmidt, G.: Conditions for synchronization in
           Josephson-junction arrays. Phys. Rev. E **52**, 3415–3419 (1995)
CS96.   Calderbank, A.R., Shor, P.W.: Good quantum error-correcting codes ex-
           ist. Phys. Rev. A **54**, 1098 (1996)
CS98.   Claussen, J.C., Schuster, H.G.: Stability borders of delayed measurement
           from time-discrete systems. arXiv:nlin.CD/0204031 (1998)
CT07.   Craddock, T.J.A., Tuszynski, J.A.: On the role of the microtubules in
           cognitive brain functions. Neuroquantology **5**(1), 32–57 (2007)

CT65.   Cooley, J.W., Tukey, J.W.: An algorithm for the machine calculation of
             complex Fourier series. Math. Comput. **19**, 297–301 (1965)
CTT03.   Coldea, R., Tennant, D., Tylczynski, Z.: Observation of extended scatter-
             ing continua characteristic of spin fractionalization in the 2d frustrated
             quantum magnet Cs_2CuCl_4 by neutron scattering. Phys. Rev. B **68**,
             134424 (2003)
CW00.   Cleve, R., Watrous, J.: Fast parallel circuits for the quantum Fourier
             transform. arXiv:quant-ph/0006004 (2000)
CW90.   Cazenave, T., Weissler, F.B.: Critical nonlinear Schrödinger equation.
             Nonlinear Anal. TMA **14**, 807–836 (1990)
CWK01.   Cooper, N.R., Wilkin, N.K., Gunn, J.M.F.: Quantum phases of vor-
             tices in rotating Bose–Einstein condensates. Phys. Rev. Lett. **87**, 120405
             (2001)
CWM00.   Clark, C.A., Werring, D.J., Miller, D.H.: Diffusion imaging of the spinal
             cord in vivo: estimation of the principal diffusivities and application to
             multiple sclerosis. Magn. Reson. Med. **43**, 133–138 (2000)
CWW89.   Chen, Y., Wilczek, F., Witten, E., Halperin, B.: On anyon superconduc-
             tivity. Int. J. Mod. Phys. B **3**, 1001 (1989)
CY89.   Crutchfield, J.P., Young, K.: Computation at the onset of chaos. In:
             Zurek, W.H. (ed.) Complexity, Entropy and the Physics of Informa-
             tion. SFI Studies in the Sciences Complexity, vol. VIII, p. 223. Addison-
             Wesley, Reading (1989)
CZG05.   Camino, F.E., Zhou, W., Goldman, V.J.: Realization of a Laughlin quasi-
             particle interferometer: observation of fractional statistics. Phys. Rev. B
             72(7), 075342 (2005)
Cal85.   Callen, H.B.: Thermodynamics and an Introduction to Thermostatistics.
             Wiley, New York (1985)
Car86.   Cardy, J.L.: Operator content of 2D conformally invariant theories. Nucl.
             Phys. B **270**, 186 (1986)
Car76.   Carey, A.L.: Square-integrable representations of non-unimodular
             groups. Bull. Austr. Math. Soc. **15**, 1–12 (1976)
Cas89.   Casdagli, M.: Nonlinear prediction of chaotic time series. Physica D **35**,
             335–356 (1989)
Cas92.   Casdagli, M.: Chaos and deterministic versus stochastic nonlinear mod-
             eling. J. R. Stat. Soc. B **54**, 303–328 (1992)
Caz03.   Cazenave, T.: Semilinear Schrödinger Equations. Courant Lecture Notes
             in Mathematics, vol. 10. American Mathematical Society, Providence
             (2003)
Cha97.   Chalmers, D.: The Conscious Mind. Oxford University Press, Oxford
             (1997)
Cha08.   Chang, J.-J.: Studies and discussion of properties of biophotons and their
             functions. Neuroquantology **6**(4), 420–430 (2008)
Chi79.   Chirikov, B.V.: A universal instability of many-dimensional oscillator
             systems. Phys. Rep. **52**, 264–379 (1979)
Chu94.   Chua, L.O.: Chua's circuit: an overview ten years later. J. Circuits Syst.
             Comput. **4**(2), 117–159 (1994)
Chu92.   Chui, C.K.: An Introduction to Wavelets. Academic Press, New York
             (1992)
Cla02a.   Claussen, J.C.: Generalized winner relaxing Kohonen feature maps.
             arXiv:cond-mat/0208414 (2002)

Cla02b.   Claussen, J.C.: Floquet stability analysis of Ott–Grebogi–Yorke, differ-
            ence control. arXiv:nlin.CD/0204060 (2002)
Col06.   Collins, G.P.: Computing with Quantum Knots. Scientific American,
            April (2006)
Con01.   Cont, R.: Empirical properties of asset returns: stylized facts and statis-
            tical issues. Quant. Finance 1, 223–236 (2001)
Con08.   Conte, E.: Testing quantum consciousness. Neuroquantology 6(2), 126–
            139 (2008)
Cox92.   Cox, E.: Fuzzy fundamentals. IEEE Spectr. 58–61 (1992)
Cox94.   Cox, E.: The Fuzzy Systems Handbook. AP Professional (1994)
Cri94.   Crick, F.: The Astonishing Hypothesis. Charles Scribner's Sons, New
            York (1994)
Cvi00.   Cvitanovic, P.: Chaotic field theory: a sketch. Physica A 288, 61–80
            (2000)
Cvi91.   Cvitanovic, P.: Periodic orbits as the skeleton of classical and quantum
            chaos. Physica D 51, 138 (1991)
DBL02.   Dafilis, M.P., Bourke, P.D., Liley, D.T.J., Cadusch, P.J.: Visualising
            chaos in a model of brain electrical activity. Comput. Graph. 26(6),
            971–976 (2002)
DBO01.   DeShazer, D.J., Breban, R., Ott, E., Roy, R.: Detecting phase synchro-
            nization in a chaotic laser array. Phys. Rev. Lett. 87, 044101 (2001)
DCM03.   Doiron, B., Chacron, M.J., Maler, L., Longtin, A., Bastian, J.: Inhibitory
            feedback required for network oscillatory responses to communication
            but not prey stimuli. Nature 421, 539–543 (2003)
DDL03.   Duan, L.-M., Demler, E., Lukin, M.D.: Controlling spin exchange inter-
            actions of ultracold atoms in optical lattices. Phys. Rev. Lett. 91(9),
            090402 (2003)
DDT03.   Daniels, B.C., Dissanayake, S.T.M., Trees, B.R.: Synchronization of cou-
            pled rotators: Josephson junction ladders and the locally coupled Ku-
            ramoto model. Phys. Rev. E 67, 026216 (2003)
DF85.   Deutsch, D.: Quantum theory, the Church-Turing principle and the uni-
            versal quantum computer. Proc. R. Soc. (London) A 400, 97–117 (1985);
            Also, Feynman, R.P.: Quantum mechanical computers. Found. Phys.
            16(6), 507–531 (1985)
DFI05.   Doucot, B., Feigel'man, M.V., Ioffe, L.B., Ioselevich, A.S.: Protected
            qubits and Chern–Simons theories in Josephson junction arrays. Phys.
            Rev. B 71(2), 024505 (2005)
DFN05.   Das Sarma, S., Freedman, M., Nayak, C.: Topologically protected qubits
            from a possible non-Abelian fractional quantum Hall state. Phys. Rev.
            Lett. 94(16), 166802 (2005)
DFN06.   Das Sarma, S., Freedman, M., Nayak, C.: Topological quantum compu-
            tation. Phys. Today 59, 32 (2006)
DGA99.   Diesmann, M., Gewaltig, M., Aertsen, A.: Stable propagation of syn-
            chronous spiking in cortical neural networks. Nature 402, 529 (1999)
DGY97.   Ding, M., Grebogi, C., Yorke, J.A.: Chaotic dynamics. In: Grebogi, C.,
            Yorke, J.A. (eds.) The Impact of Chaos on Science and Society, pp. 1–17.
            United Nations University Press, Tokyo (1997)
DHN07.   Dimov, I., Halperin, B.I., Nayak, C.: Spin order in paired quantum Hall
            states. arXiv:0710.1921 (2007)

DHR71.   Doplicher, S., Haag, R., Roberts, J.E.: Local observables and particle statistics, I. Commun. Math. Phys. **23**, 199 (1971)

DHR74.   Doplicher, S., Haag, R., Roberts, J.E.: Local observables and particle statistics, II. Commun. Math. Phys. **35**, 49 (1974)

DHS91.   Domany, E., van Hemmen, J.L., Schulten, K. (eds.): Models of Neural Networks. Springer, Berlin (1991)

DHT05.   Droll, J.A., Hayhoe, M.M., Triesch, J., Sullivan, B.T.: Task demands control acquisition and storage of visual information. J. Exp. Psych. Hum. Perc. Perf. **31**, 1416–1438 (2005)

DIS97.   D'Attellis, Isaacson, S.I., Sirne, R.O.: Detection of epileptic events using wavelets. Ann. Biomed. Eng. **25**, 286–293 (1997)

DIV04.   Douçot, B., Ioffe, L.B., Vidal, J.: Discrete non-Abelian gauge theories in Josephson-junction arrays and quantum computation. Phys. Rev. B **69**(21), 214501 (2004)

DJT82.   Deser, S., Jackiw, R., Templeton, S.: Three-dimensional massive gauge theories. Phys. Rev. Lett. **48**(15), 975 (1982)

DKL02.   Dennis, E., Kitaev, A., Landahl, A., Preskill, J.: Topological quantum memory. J. Math. Phys. **43**, 4452 (2002)

DKM95.   Douglas, R.J., Koch, C., Mahowald, M., Martin, K., Suarez, H.: Recurrent excitation in neocortical circuits. Science **69**, 981–985 (1995)

DLC01.   Dafilis, M.P., Liley, D.T.J., Cadusch, P.J.: Robust chaos in a model of the electroencephalogram: implications for brain dynamics. Chaos **11**(3), 474–4748 (2001)

DLL02.   Doiron, B., Laing, C., Longtin, A., Maler, L.: Ghostbursting: a novel neuronal burst mechanism. J. Comput. Neurosci. **12**, 5–25 (2002)

DM76.   Duflo, M., Moore, C.C.: On regular representations of nonunimodular locally compact group. J. Funct. Anal. **21**, 209–243 (1976)

DM04.   Destexhe, A., Marder, E.: Plasticity in single neuron and circuit computations. Nature **431**, 789–795 (2004)

DM96.   De'Charmes, R.C., Merzenich, M.M.: Nature **381**, 610 (1996)

DM99.   Diambra, L., Malta, C.P.: Nonlinear models for detecting epileptic spikes. Phys. Rev. E **59**, 929–937 (1999)

DMS97.   Di Francesco, P., Mathieu, P., Sénéchal, D.: Conformal Field Theory. Springer, New York (1997)

DMS.   Destexhe, A., Mainen, Z.F., Sejnowski, T.J.: In: Arbib, M.A. (ed.) The Handbook of Brain Theory and Neural Networks, p. 956. MIT Press, Cambridge (1998)

DN93.   Dixit, A.K., Nalebuff, B.: Thinking Strategically: The Competitive Edge in Business, Politics, and Everyday Life. Norton, New York (1993)

DNT06.   Das Sarma, S., Nayak, C., Tewari, S.: Proposal to stabilize and detect half-quantum vortices in strontium ruthenate thin films: non-Abelian braiding statistics of vortices in a $p_x + ip_y$ superconductor. Phys. Rev. B **73**(22), 220502 (2006)

DP80.   Dormand, J.R., Prince, P.J.: A family of embedded Runge–Kutta formulae. J. Comput. Appl. Math. **6**, 19–26 (1980)

DP97.   Das Sarma, S., Pinczuk, A. (eds.): Perspectives in Quantum Hall Effects: Novel Quantum Liquids in Low-Dimensional Semiconductor Structures. Wiley, New York (1997)

DR08.   Dauxois, T., Ruffo, S.: Fermi–Pasta–Ulam nonlinear lattice oscillations. Scholarpedia **3**(8), 5538 (2008)

DRH97.   De Picciotto, R., Reznikov, M., Heiblum, M., et al.: Direct observation of a fractional charge. Nature **389**, 162 (1997)

DSI00.   Ditto, W.L., Spano, M.L., In, V., et al.: Control of human atrial fibrillation. Int. J. Bifurc. Chaos **10**, 593–601 (2000)

DSS96.   Dote, Y., Strefezza, M., Suitno, A.: Neuro fuzzy robust controllers for drive systems. In: Simpson, P.K. (ed.) Neural Networks Applications. IEEE, New York (1996)

DST93.   Du, R.R., Stormer, H.L., Tsui, D.C., Pfeiffer, L.N., West, K.W.: Experimental evidence for new particles in the fractional quantum Hall effect. Phys. Rev. Lett. **70**(19), 2944 (1993)

DT95.   Denniston, C., Tang, C.: Phases of Josephson junction ladders. Phys. Rev. Lett. **75**, 3930 (1995)

DTP91.   Douek, P., Turner, R., Pekar, J., Patronas, N.J., Le Bihan, D.: MR color mapping of myelin fiber orientation. J. Comput. Assist. Tomogr. **15**, 923–929 (1991)

DW90.   Dijkgraaf, R., Witten, E.: Topological gauge theories and group cohomology. Commun. Math. Phys. **129**, 393 (1990)

DWP93.   Douglass, J.K., Wilkens, L., Pantazelou, E., Moss, F.: Noise enhancement of information transfer in crayfish mechanoreceptors by stochastic resonance. Nature **365**, 337–340 (1993)

Das99.   Dasgupta, D. (ed.): Artificial Immune Systems and Their Applications. Springer, Berlin (1999)

Dau92.   Daubechies, I.: Ten Lectures on Wavelets. SIAM, Philadelphia (1992)

Dau01.   Daubechies, I.: Ten Lectures on Wavelets. R.Kh.D., Moscow (2001)

Dav82.   Davydov, A.S.: Biology and Quantum Mechanics. Pergamon, Oxford (1982)

Daw92.   Dawes, R.L.: Quantum neurodynamics: neural stochastic filtering with the Schroedinger equation. In: Proc. IJCNN (1992)

Deb84.   Debreu, G.: Economic theory in the mathematical mode. In: Les Prix Nobel (1983). Reprinted in Am. Ec. Rev. **74**, 267–278 (1984)

Del99.   Deligne, P., et al. (ed.): Quantum Fields and Strings: a Course for Mathematicians. American Mathematical Society, Providence (1999)

Deu85.   Deutsch, D.: Quantum theory, the Church-Turing principle and the universal quantum computer. Proc. R. Soc. Lond. A **400**, 97–117 (1985)

Deu89.   Deutsch, D.: Quantum Computational Networks. Proc. R. Soc. Lond. A **425**, 73–90 (1989)

Deu92.   Deutsch, D., Jozsa, R.: Rapid solution of problems by quantum computation. Proc. R. Soc. Lond. A **439**, 553–558 (1992)

DiV00.   DiVincenzo, D.P.: The physical implementation of quantum computation. Fortschr. Phys. Prog. Phys. **48**(9–11), 771 (2000)

Die88.   Dieudonne, J.A.: A History of Algebraic and Differential Topology 1900–1960. Birkhäuser, Basel (1988)

Dir25.   Dirac, P.A.M.: The fundamental equations of quantum mechanics. Proc. R. Soc. Lond. A **109**(752), 642–653 (1925)

Dir26a.   Dirac, P.A.M.: Quantum mechanics, a preliminary investigation of the hydrogen atom. Proc. R. Soc. Lond. A **110**(755), 561–579 (1926)

Dir26b.   Dirac, P.A.M.: The elimination of the nodes in quantum mechanics. Proc. R. Soc. Lond. A **111**(757), 281–305 (1926)

Dir26c.   Dirac, P.A.M.: Relativity quantum mechanics with an application to Compton scattering. Proc. R. Soc. Lond. A **111**(758), 281–305 (1926)

Dir26d.   Dirac, P.A.M.: On the theory of quantum mechanics. Proc. R. Soc.
            Lond. A **112**(762), 661–677 (1926)
Dir26e.   Dirac, P.A.M.: The physical interpretation of the quantum dynamics.
            Proc. R. Soc. Lond. A **113**(765), 1–40 (1927)
Dir28a.   Dirac, P.A.M.: The quantum theory of the electron. Proc. R. Soc.
            Lond. A **117**(778), 610–624 (1928)
Dir28b.   Dirac, P.A.M.: The quantum theory of the electron, part II. Proc. R.
            Soc. Lond. A **118**(779), 351–361 (1928)
Dir29.   Dirac, P.A.M.: Quantum mechanics of many-electron systems. Proc. R.
            Soc. Lond. A **123**(792), 714–733 (1929)
Dir32.   Dirac, P.A.M.: Relativistic quantum mechanics. Proc. R. Soc. Lond. A
            136(829), 453–464 (1932)
Dir36.   Dirac, P.A.M.: Relativistic wave equations. Proc. R. Soc. Lond. A
            155(886), 447–459 (1936)
Dir58.   Dirac, P.A.M.: Generalized Hamiltonian dynamics. Proc. R. Soc. Lond. A
            246(1246), 326–332 (1958)
Dir82.   Dirac, P.A.M.: Principles of Quantum Mechanics, 4th edn. Oxford Uni-
            versity Press, London (1982)
Do95.   Ding, M., Ott, E.: Chaotic scattering in systems with more than two
            degrees of freedom. Ann. N. Y. Acad. Sci. **751**, 182 (1995)
Don95.   Donoho, D.L.: De-Noising by soft-thresholding. IEEE Trans. Inf. Theory
            41(3), 613–627 (1995)
Dow07.   Downarowicz, T.: Entropy. Scholarpedia **2**(11), 3901 (2007)
Duf18.   Duffing, G.: Erzwungene Schwingungen bei veränderlicher Eigenfre-
            quenz. Vieweg, Braunschweig (1918)
Dus84.   Dustin, P.: Microtubules. Springer, Berlin (1984)
Dya06.   Dyakonov, M.I.: Is fault–tolerant quantum computation really possible?
            arXiv:quant-ph/0610117 (2006)
ECP02.   Eisenstein, J.P., Cooper, K.B., Pfeiffer, L.N., West, K.W.: Insulating and
            fractional quantum Hall states in the first excited Landau level. Phys.
            Rev. Lett. **88**(7), 076801 (2002)
EHM99.   Eiben, A.E., Hinterding, R., Michalewicz, Z.: Parameter control in evo-
            lutionary algorithms. IEEE Trans. Evol. Comput. **3**(2), 124–141 (1999)
EIS67.   Eccles, J.C., Ito, M., Szentagothai, J.: The Cerebellum as a Neuronal
            Machine. Springer, Berlin (1967)
EJ96.   Ekert, A., Jozsa, R.: Quantum computation and Shor's factoring algo-
            rithm. Rev. Mod. Phys. **68**, 733 (1996)
EJP00.   Eisert, J., Jacobs, K., Papadopoulos, P., Plenio, M.B.: Nonlocal content
            of quantum operations. Phys. Rev. A **62**, 052317 (2000)
EL78.   Eells, J., Lemaire, L.: A report on harmonic maps. Bull. Lond. Math.
            Soc. **10**, 1–68 (1978)
EMN92.   Ellis, J., Mavromatos, N., Nanopoulos, D.V.: String theory modifies
            quantum mechanics. CERN-TH/6595 (1992)
EMN99a.   Ellis, J., Mavromatos, N., Nanopoulos, D.V.: A microscopic Liouville
            arrow of time. Chaos Solitons Fractals **10**(2–3), 345–363 (1999)
EMN99b.   Ellis, J., Mavromatos, N., Nanopoulos, D.V.: A microscopic Liouville
            arrow of time. Chaos Solitons Fractals **10**(2–3), 345–363 (1999)
EMS89.   Elitzur, S., Moore, G., Schwimmer, A., Seiberg, N.: Remarks on
            the canonical quantization of the Chern–Simons–Witten theory. Nucl.
            Phys. B **326**(1), 108 (1989)

860   References

EP98.   Ershov, S.V., Potapov, A.B.: On the concept of stationary Lyapunov basis. Physica D **118**, 167 (1998)

EPG95.   Ernst, U., Pawelzik, K., Geisel, T.: Synchronization induced by temporal delays in pulse-coupled oscillators. Phys. Rev. Lett. **74**, 1570 (1995)

EPR35a.   Einstein, A., Podolsky, B., Rosen, N.: In: Zurek, W.H., Wheeler, J.A. (eds.) Quantum Theory and Measurement (1935)

EPR35b.   Einstein, A., Podolsky, B., Rosen, N.: Can quantum mechanical description of physical reality be considered complete? Phys. Rev. **47**, 777 (1935)

EPR99.   Eckmann, J.P., Pillet, C.A., Rey-Bellet, L.: Non-equilibrium statistical mechanics of anharmonic chains coupled to two heat baths at different temperatures. Commun. Math. Phys. **201**, 657–697 (1999)

ER85.   Eckmann, J.P., Ruelle, D.: Ergodic theory of chaos, strange attractors. Rev. Mod. Phys. **57**, 617–630 (1985)

ES03.   Eiben, A.E., Smith, J.E.: Introduction to Evolutionary Computing. Springer, New York (2003)

ES46.   Einstein, A., Strauss, E.: Ann. Math. **47**, 731 (1946)

ESA05.   European Space Agency: Payload and Advanced Concepts: Superconducting Tunnel Junction (STJ), February 17 (2005)

ESH98.   Elson, R.C., Selverston, A.I., Huerta, R., et al.: Synchronous behavior of two coupled biological neurons. Phys. Rev. Lett. **81**, 5692–5695 (1998)

EW00.   Eisert, J., Wilkens, M.: Quantum games. J. Mod. Opt. **47**, 2543–2556 (2000)

EW06.   Eisert, J., Wolf, M.M.: Quantum computing. In: Handbook of Nature-Inspired and Innovative Computing. Springer, Berlin (2006)

EWL99.   Eisert, J., Wilkens, M., Lewenstein, M.: Quantum games and quantum strategies. Phys. Rev. Lett. **83**, 3077–3080 (1999)

EWS90.   Eisenstein, J.P., Willett, R.L., Stormer, H.L., Pfeiffer, L.N., West, K.W.: Activation energies for the even-denominator fractional quantum Hall effect. Surf. Sci. **229**, 31 (1990)

Ebe02.   Ebersole, J.S.: Current Practice of Clinical Electroencephalography. Williams, Wilkins, Lippincott (2002)

Ecc64.   Eccles, J.C.: The Physiology of Synapses. Springer, Berlin (1964)

Ecc86.   Eccles, J.C.: Do mental events cause neural events analogously to the probability fields of quantum mechanics? Proc. R. Soc. B **227**, 411–428 (1986)

Ecc90.   Eccles, J.C.: A unitary hypothesis of mind-brain interaction in the cerebral cortex. Proc. R. Soc. Lond. B **240**, 433–451 (1990)

Ecc94.   Eccles, J.C.: How the Self Controls Its Brain. Springer, Berlin (1994)

Ein16.   Einstein, A.: The foundation of the general theory of relativity. Ann. Phys. **49**, 769–822 (1916)

Ein48.   Einstein, A.: Quantum mechanics and reality (Quanten-Mechanik und Wirklichkeit). Dialectica **2**, 320–324 (1948)

Eis01.   Eisert, J.: Entanglement in quantum information theory. PhD thesis, Univ. Potsdam (2001)

Eis29.   Eisenhart, L.P.: Dynamical trajectories and geodesics. Math. Ann. **30**, 591–606 (1929)

Eke91.   Ekert, A.K.: Quantum cryptography based on Bell's theorem. Phys. Rev. Lett. **67**, 661–663 (1991)

Elm90.   Elman, J.: Finding structure in time. Cogn. Sci. **14**, 179–211 (1990)

Eng01.   Engel, J.: Classification of epileptic disorders. Epilepsia **42**, 796–803 (2001)

Eng06.   Engelbrecht, A.: Fundamentals of Computational Swarm Intelligence. Wiley, New York (2006)

Ens05.   Enss, C. (ed.): Cryogenic Particle Detection. Topics in Applied Physics, vol. 99. Springer, New York (2005)

Erm81.   Ermentrout, G.B.: The behavior of rings of coupled oscillators. J. Math. Biol. **12**, 327 (1981)

FA02.   Flitney, A.P., Abbott, D.: Quantum version of the Monty Hall problem. Phys. Rev. A **65**, 062318 (2002)

FA04.   Flitney, A.P., Abbott, D.: Quantum two and three person duels. J. Opt. B **6**, S860–S866 (2004)

FB96.   Freud, T.F., Buzsaki, G.: Hippocampus **6**, 345 (1996)

FF05.   Fendley, P., Fradkin, E.: Realizing non-Abelian statistics in time-reversal-invariant systems. Phys. Rev. B **72**(2), 024412 (2005)

FFG05.   Fogassi, L., Ferrari, P.F., Gesierich, B., Rozzi, S., Chersi, F., Rizzolatti, G.: Parietal lobe: from action organization to intention understanding. Science **29**, 662–667 (2005)

FFN06.   Fendley, P., Fisher, M.P.A., Nayak, C.: Dynamical disentanglement across a point contact in a non-Abelian quantum Hall state. Phys. Rev. Lett. **97**(3), 036801 (2006)

FFN07a.   Fendley, P., Fisher, M.P.A., Nayak, C.: Edge states and tunneling of non-Abelian quasiparticles in the $\nu = 5/2$ quantum Hall state and p + ip superconductors. Phys. Rev. B **75**(4), 045317 (2007)

FFN07b.   Fendley, P., Fisher, M.P.A., Nayak, C.: Topological entanglement entropy from the holographic partition function. J. Stat. Phys. **126**, 1111 (2007)

FG90.   Fröhlich, J., Gabbiani, F.: Braid statistics in local quantum theory. Rev. Math. Phys. **2**, 251 (1990)

FGG00.   Farhi, E., Goldstone, J., Gutmann, S., Sipser, M.: Quantum computation by adiabatic evolution. arXiv:quant-ph/0001106 (2000)

FGK06.   Feldman, D.E., Gefen, Y., Kitaev, A., Law, K.T., Stern, A.: Shot noise in anyonic Mach–Zehnder interferometer. arXiv:cond-mat/0612608 (2006)

FH65.   Feynman, R.P., Hibbs, A.R.: Quantum Mechanics and Path Integrals. McGraw–Hill, New York (1965)

FHL89.   Fetter, A., Hanna, C., Laughlin, R.: Random-phase approximation in the fractional-statistics gas. Phys. Rev. B **39**, 9679 (1989)

FHZ01.   Fradkin, E., Huerta, M., Zemba, G.: Effective Chern–Simons theories of Pfaffian and parafermionic quantum Hall states, and orbifold conformal field theories. Nucl. Phys. B **601**, 591 (2001)

FK06.   Feldman, D.E., Kitaev, A.: Detecting non-Abelian statistics with an electronic Mach–Zehnder interferometer. Phys. Rev. Lett. **97**(18), 186803 (2006)

FK39.   Frenkel, J., Kontorova, T.: On the theory of plastic deformation and twinning. Phys. Z. Sov. Union **1**, 137–149 (1939)

FK80.   Fradkin, E., Kadanoff, L.P.: Disorder variables and para-fermions in 2D statistical mechanics. Nucl. Phys. B **170**(1), 1 (1980)

FK83.   Frölich, H., Kremer, F.: Coherent Excitations in Biological Systems. Springer, New York (1983)

FKL03a.   Freedman, M., Kitaev, A., Larsen, M., Wang, Z.: Topological quantum computation. Bull. Am. Math. Soc. **40**, 31 (2003)

FKL03b.   Freedman, M., Kitaev, A., Larsen, M., Wang, Z.: Topological quantum computation. Bull. Am. Math. Soc. **40**, 31–38 (2003)

FKN72.   Field, R.J., Kórós, E., Noyes, R.M.: Oscillations in chemical systems. J. Am. Chem. Soc. **94**, 8649–8664 (1972)

FL91.   Fubini, S., Lutken, C.A.: Vertex operators in the fractional quantum Hall effect. Mod. Phys. Lett. **A6**, 487 (1991)

FLW02a.   Freedman, M.H., Larsen, M.J., Wang, Z.: A modular functor which is universal for quantum computation. Commun. Math. Phys. **227**, 605 (2002)

FLW02b.   Freedman, M.H., Larsen, M.J., Wang, Z.: The two-eigenvalue problem and density of Jones representation of braid groups. Commun. Math. Phys. **228**, 177 (2002)

FM88.   Fröhlich, J., Marchetti, P.A.: Quantum field theory of anyons. Lett. Math. Phys. **16**, 347 (1988)

FM89.   Fröhlich, J., Marchetti, P.A.: Quantum field theories of vortices and anyons. Commun. Math. Phys. **121**, 177 (1989)

FM91.   Fröhlich, J., Marchetti, P.A.: Spin-statistics theorem and scattering in planar quantum field theories with braid statistics. Nucl. Phys. B **356**, 533 (1991)

FMS02.   Fendley, P., Moessner, R., Sondhi, S.L.: Classical dimers on the triangular lattice. Phys. Rev. B **66**, 214513 (2002)

FNS03.   Freedman, M., Nayak, C., Shtengel, K.: Non-Abelian topological phases in an extended Hubbard model (2003)

FNS04.   Freedman, M., Nayak, C., Shtengel, K., Walker, K., Wang, Z.: A class of P,T-invariant topological phases of interacting electrons. Ann. Phys. (N.Y.) **310**, 428 (2004)

FNS05a.   Freedman, M., Nayak, C., Shtengel, K.: Extended Hubbard model with ring exchange: a route to a non-Abelian topological phase. Phys. Rev. Lett. **94**(6), 066401 (2005)

FNS05b.   Freedman, M., Nayak, C., Shtengel, K.: Line of critical points in $2+1$ dimensions: quantum critical loop gases and non-Abelian gauge theory. Phys. Rev. Lett. **94**(14), 147205 (2005)

FNS99.   Fradkin, E., Nayak, C., Schoutens, K.: Landau–Ginzburg theories for non-Abelian quantum Hall states. Nucl. Phys. B **546**, 711 (1999)

FNT98.   Fradkin, E., Nayak, C., Tsvelik, A., Wilczek, F.: A Chern–Simons effective field theory for the Pfaffian quantum Hall state. Nucl. Phys. B **516**(3), 704 (1998)

FNW06.   Freedman, M., Nayak, C., Walker, K.: Towards universal topological quantum computation in the $\nu = 5/2$ fractional quantum Hall state. Phys. Rev. B **73**(24), 245307 (2006)

FOW66.   Fogel, L.J., Owens, A.J., Walsh, M.J.: Artificial Intelligence through Simulated Evolution. Wiley, New York (1966)

FP04.   Franzosi, R., Pettini, M.: Theorem on the origin of phase transitions. Phys. Rev. Lett. **92**, 060601 (2004)

FP67.   Faddeev, L.D., Popov, V.N.: Feynman diagrams for the Yang–Mills field. Phys. Lett. B **25**, 29 (1967)

FPP86.   Farmer, J.D., Packard, N., Perelson, A.: The immune system, adaptation, machine learning. Physica D **22**, 187–204 (1986)

FPR06.   Faber, J., Portugal, R., Rosa, L.P.: Information processing in brain mi-
            crotubules. Biosystems **83**, 1–9 (2006)
FPS00a.   Franzosi, R., Pettini, M., Spinelli, L.: Topology and phase transitions:
            a paradigmatic evidence. Phys. Rev. Lett. **84**, 2774–2777 (2000)
FPS00b.   Fouque, J.-P., Papanicolau, G., Sircar, K.R.: Derivatives in Financial
            Markets with Stochastic Volatility. Cambridge University Press, Cam-
            bridge (2000)
FPU55.   Fermi, E., Pasta, J., Ulam, S.: Studies of nonlinear problems, I. Los
            Alamos report LA-1940 (1955); Published later in E. Segré (ed.) Col-
            lected Papers of Enrico Fermi, Univ. Chicago Press (1965)
FPU74.   Fermi, E., Pasta, J., Ulam, S.: Studies of nonlinear problems, I. Los
            Alamos Report LA1940 (1955); reproduced in A.C. Newell (ed.) Nonlin-
            ear Wave Motion, American Mathematical Society, Providence, pp. 143–
            156 (1974)
FRD07.   Feiguin, A.E., Rezayi, C., Nayak, C., Das Sarma, S.: Density matrix
            renormalization group study of incompressible fractional quantum hall
            states. arXiv:0706.4469 (2007)
FRS89.   Fredenhagen, K., Rehren, K.H., Schroer, B.: Superselection sectors with
            braid group statistics and exchange algebras. Commun. Math. Phys. **125**,
            201 (1989)
FS75.   Fradkin, E., Shenker, S.H.: Phase diagrams of lattice gauge theories with
            Higgs fields. Phys. Rev. D **19**(12), 3682 (1975)
FS86.   Fraser, A.M., Swinney, H.L.: Independent coordinates for strange attrac-
            tors from mutual information. Phys. Rev. A **33**, 1134–1140 (1986)
FS92.   Freeman, J.A., Skapura, D.M.: Neural Networks: Algorithms, Applica-
            tions and Programming Techniques. Addison-Wesley, Reading (1992)
FSB97.   Faag, A.H., Sitkoff, N., Barto, A.G., Houk, J.C.: Cerebellar learning for
            control of a two-link arm in muscle space. In: Proc. IEEE Int. Conf. Rob.
            Aut. (1997)
FTL07.   Feiguin, A., Trebst, S., Ludwig, A.W.W., et al.: Interacting anyons in
            topological quantum liquids: the golden chain. Phys. Rev. Lett. **98**(16),
            160409 (2007)
FV06.   Freeman, W.J., Vitiello, G.: Nonlinear brain dynamics as macroscopic
            manifestation of underlying many-body field dynamics. Phys. Life Rev.
            3(2), 93–118 (2006)
FW89.   Friedrich, H., Wintgen, D.: The hydrogen atom in a uniform magnetic
            field—an example of chaos. Phys. Rep. **183**, 37 (1989)
FW95.   Filatrella, G., Wiesenfeld, K.: Magnetic-field effect in a two-dimensional
            array of short Josephson junctions. J. Appl. Phys. **78**, 1878–1883 (1995)
FY83.   Fujisaka, H., Yamada, T.: Amplitude equation of higher-dimensional
            Nikolaevskii turbulence. Prog. Theor. Phys. **69**, 32 (1983)
FZ85.   Fateev, V.A., Zamolodchikov, A.B.: Parafermionic currents in the 2D
            conformal quantum field theory and selfdual critical points in $z(n)$ in-
            variant statistical systems. Sov. Phys. JETP **62**, 215 (1985)
Fak98.   Fakir, R.: Nonstationary stochastic resonance. Phys. Rev. E **57**, 6996–
            7001 (1998)
Fam65.   Fama, E.: The behavior of stock market prices. J. Bus. **38**, 34–105 (1965)
Fed69.   Federer, H.: Geometric Measure Theory. Springer, New York (1969)
Fed81.   Federbush, P.G.: A mass zero cluster expansion. Commun. Math. Phys.
            81, 327–340 (1981)

Fed95.   Federbush, P.: A new formulation and regularization of gauge theories using a non-linear wavelet expansion. Prog. Theor. Phys. **94**, 1135–1146 (1995)

Feh68a.   Fehlberg, E.: Low order classical Runge–Kutta formulas with stepsize control and their application to some heat-transfer problems. TR R-315, NASA (1969)

Feh68b.   Fehlberg, E.: Classical fifth, sixth, seventh, and eighth order Runge–Kutta formulas with stepsize control. TR R-287, NASA (1968)

Fei78.   Feigenbaum, M.J.: Quantitative universality for a class of nonlinear transformations. J. Stat. Phys. **19**, 25–52 (1978)

Fei79.   Feigenbaum, M.J.: The universal metric properties of nonlinear transformations. J. Stat. Phys. **21**, 669–706 (1979)

Fen07.   Fendley, P.: Quantum loop models and the non-Abelian toric code. arXiv:0711.0014 (2007)

Fer23.   Fermi, E.: Beweis dass ein mechanisches Normalsysteme im Allgemeinen quasi-ergodisch ist. Phys. Z. **24**, 261–265 (1923)

Fer99.   Ferber, J.: Multi-Agent Systems. An Introduction to Distributed Artificial Intelligence. Addison-Wesley, Reading (1999)

Fey48.   Feynman, R.P.: Space-time approach to nonrelativistic quantum mechanics. Rev. Mod. Phys. **20**, 367–387 (1948)

Fey49.   Feynman, R.P.: Space-time approach to quantum electrodynamics. Phys. Rev. **76**, 769–789 (1949)

Fey50.   Feynman, R.P.: Mathematical formulation of the quantum theory of electromagnetic interaction. Phys. Rev. **80**, 440–457 (1950)

Fey51.   Feynman, R.P.: An operator calculus having applications in quantum electrodynamics. Phys. Rev. **84**, 108–128 (1951)

Fey72.   Feynman, R.P.: Statistical Mechanics, a Set of Lectures. Benjamin, Reading (1972)

Fey82.   Feynman, R.P.: Simulating physics with computers. Int. J. Theor. Phys. **21**(6–7), 467–488 (1982)

Fey86.   Feynman, R.P.: Quantum mechanical computers. Found. Phys. **16**(6), 507 (1986)

Fey98.   Feynman, R.P.: Quantum Electrodynamics. Advanced Book Classics. Perseus, Reading (1998)

Fid06.   Fidkowski, L., Freedman, M., Nayak, C., Walker, K., Wang, Z.: From string nets to nonabelions. cond-mat/0610583 (2006)

Fid07.   Fidkowski, L.: Double point contact in the $k = 3$ Read–Rezayi state. arXiv:0704.3291 (2007)

Fie07.   Field, R.J.: Oregonator. Scholarpedia **2**(5), 1386 (2007)

Fit61a.   FitzHugh, R.A.: Impulses, physiological states in theoretical models of nerve membrane. Biophys. J. **1**, 445–466 (1961)

Fit61b.   FitzHugh, R.: Impulses and physiological states in theoretical models of nerve membrane. Biophys. J. **1**, 445–466 (1961)

Fla63.   Flanders, H.: Differential Forms: with Applications to the Physical Sciences. Academic Press, New York (1963)

Fog98.   Fogel, D. (ed.): Evolutionary Computation: The Fossil Record. IEEE Press, New York (1998)

For03.   Forster, T.: Logic, Induction and the Theory of Sets. London Mathematical Society Student Texts, vol. 56. Cambridge University Press, Cambridge (2003)

Fra86.   Frampton, P.H.: Gauge Field Theories, Frontiers in Physics. Addison-Wesley, Reading (1986)

Fre00.   Freeman, W.J.: Neurodynamics. An Exploration of Mesoscopic Brain Dynamics. Springer, London (2000)

Fre01.   Freedman, M.H.: Quantum computation and the localization of modular functors. Found. Comput. Math. **1**, 183 (2001)

Fre03.   Freedman, M.H.: A magnetic model with a possible Chern–Simons phase. Commun. Math. Phys. **234**, 129 (2003)

Fre75.   Freeman, W.J.: Mass Action in the Nervous System. Academic Press, New York (1975/2004)

Fre90.   Freeman, W.J.: On the problem of anomalous dispersion in chaotic phase transitions of neural masses, its significance for the management of perceptual information in brains. In: Haken, H., Stadler, M. (eds.) Synergetics of Cognition, vol. 45, pp. 126–143. Springer, Berlin (1990)

Fre91.   Freeman, W.J.: The physiology of perception. Sci. Am. **264**(2), 78–85 (1991)

Fre92.   Freeman, W.J.: Tutorial on neurobiology: from single neurons to brain chaos. Int. J. Bifurc. Chaos **2**(3), 451–482 (1992)

Fre96.   Freeman, W.J.: Random activity at the microscopic neural level in cortex sustains is regulated by low dimensional dynamics of macroscopic cortical activity. Int. J. Neural Syst. **7**, 473 (1996)

Fri04.   Friedrich, R.: Group theoretic methods in the theory of pattern formation. In: Collective Dynamics of Nonlinear and Disordered Systems. Springer, Berlin (2004)

Fri94.   Fritzke, B.: Fast learning with incremental RBF networks. Neural Process. Lett. **1**, 1–5 (1994)

Fri99.   Frieden, R.B.: Physics from Fisher Information: A Unification. Cambridge University Press, Cambridge (1999)

Fub91.   Fubini, S.: Vertex operators and quantum Hall effect. Mod. Phys. Lett. A **6**, 347 (1991)

Fuc92.   Fuchs, J.: Affine Lie Algebras and Quantum Groups. Cambridge University Press, Cambridge (1992)

Fuk75.   Fukushima, K.: Cognitron: a self-organizing multilayered neural network. Biol. Cybern. **20**, 121–136 (1975)

Fun95.   Funke, J.: Solving complex problems: human identification, control of complex systems. In: Sternberg, R.J., Frensch, P.A. (eds.) Complex Problem Solving: Principles, Mechanisms, pp. 185–222. Lawrence Erlbaum Associates, Hillsdale (1995)

Fun99.   Funar, L.: On TQFT representations. Pac. J. Math. **188**, 251 (1999)

GA01.   Grewal, M.S., Andrews, A.P.: Kalman Filtering: Theory and Practice Using MATLAB. Wiley-Interscience, New York (2001)

GA99.   Gallagher, R., Appenzeller, T.: Beyond reductionism. Science **284**, 79 (1999)

GAM01.   Gutiérrez, J., Alcántara, R., Medina, V.: Analysis and localization of epileptic events using wavelet packets. Med. Eng. Phys. **23**, 623–631 (2001)

GAP06.   Gurevich, S.V., Amiranashvili, S., Purwins, H.G.: Breathing dissipative solitons in three-component reaction–diffusion system. Phys. Rev. E **74**, 066201 (2006)

GC98.   Gershenfeld, N., Chuang, I.L.: Quantum computing with molecules. Sci. Am. **June** (1998)

GCS07.   Grosfeld, E., Cooper, N.R., Stern, A., Ilan, R.: Predicted signatures of p-wave superfluid phases Majorana zero modes of fermionic atoms in rf absorption. Phys. Rev. B **76**(10), 104516 (2007)

GEG72.   Georgi, H., Glashow, S.: Unified weak and electromagnetic interactions without neutral currents. Phys. Rev. Lett. **28**, 1494 (1972)

GG06.   Georgiev, L.S., Geller, M.R.: Aharonov–Bohm effect in the non-Abelian quantum Hall fluid. Phys. Rev. B **73**(20), 205310 (2006)

GG76.   Gotman, J., Gloor, P.: Automatic recognition and qualification of interictal epileptic activity in the human scalp EEG. Electroencephalogr. Clin. Neurophysiol. **41**, 513–529 (1976)

GGK67.   Gardner, C.S., Greene, C.S., Kruskal, M.D., Miura, R.M.: Method for solving the Korteweg–de Vries equation. Phys. Rev. Lett. **19**, 1095–1097 (1967)

GGS81.   Guevara, M.R., Glass, L., Shrier, A.: Phase locking, period-doubling bifurcations and irregular dynamics in periodically stimulated cardiac cells. Science **214**, 1350–1353 (1981)

GH00.   Gade, P.M., Hu, C.K.: Synchronous chaos in coupled map lattices with small-world interactions. Phys. Rev. E **62**, 6409–6413 (2000)

GH83.   Guckenheimer, J., Holmes, P.: Nonlinear Oscillations, Dynamical Systems and Bifurcations of Vector Fields. Springer, Berlin (1983)

GHJ98.   Gammaitoni, L., Hännggi, P., Jung, P., Marchesoni, F.: Stochastic resonance. Rev. Mod. Phys. **70**, 223–287 (1998)

GHP00.   Green, F., Homer, S., Pollett, C.: On the complexity of quantum ACC. arXiv:quant-ph/0002057 (2000)

GK88.   Gawędzki, K., Kupiainen, A.: G/H conformal field theory from gauged WZW model. Phys. Lett. B **215**, 119 (1988)

GK92a.   Georgiou, G.M., Koutsougeras, C.: Complex domain backpropagation. IEEE Trans. Circuits Syst. **39**(5), 330–334 (1992)

GK92b.   Georgiou, G.M., Koutsougeras, C.: Complex domain backpropagation. Anal. Dig. Signal Process. **39**(5), 330–334 (1992)

GK94.   Gegenberg, J., Kunstatter, G.: The partition function for topological field theories. Ann. Phys. **231**, 270–289 (1994)

GK96.   Gomi, H., Kawato, M.: Equilibrium-point control hypothesis examined by measured arm-stiffness during multi-joint movement. Science **272**, 117–120 (1996)

GLS04.   Goerbig, M.O., Lederer, P., Smith, C.M.: Second generation of composite fermions and the self-similarity of the fractional quantum Hall effect. Int. J. Mod. Phys. B **18**, 3549 (2004)

GLW93.   Geigenmuller, U., Lobb, C.J., Whan, C.B.: Friction and inertia of a vortex in an underdamped Josephson array. Phys. Rev. B **47**, 348–358 (1993)

GM06.   Meinhardt, H.: Gierer–Meinhardt model. Scholarpedia **1**(12), 1418 (2006)

GM72.   Gierer, A., Meinhardt, H.: A theory of biological pattern formation. Kybernetik **12**, 30–39 (1972)

GM79.   Guyer, R.A., Miller, M.D.: Commensurability in one dimension at $T \neq 0$. Phys. Rev. Lett. **42**, 718–722 (1979)

GM87.   Girvin, S.M., MacDonald, A.H.: Off-diagonal long-range order, oblique confinement and the fractional quantum Hall effect. Phys. Rev. Lett. **58**(12), 1252 (1987)

GM88.   Glass, L., Mackey, M.C.: From Clocks to Chaos, The Rhythms of Life. Princeton University Press, Princeton (1988)

GMK94.   Gabbiani, F., Midtgaard, J., Knoepfl, T.: Synaptic integration in a model of cerebellar granule cells. J. Neurophysiol. **72**, 999–1009 (1994)

GMS81.   Goldin, G.A., Menikoff, R., Sharp, D.H.: Representations of a local current algebra in nonsimply connected space and the Aharonov–Bohm effect. J. Math. Phys. **22**(8), 1664 (1981)

GM.   Gray, C.M., McCormick, D.A.: Chattering cells: superficial pyramidal neurons contributing to the generation of synchronous oscillations in the visual cortex. Science **274**(5284), 109–113 (1996)

GN97.   Gurarie, V., Nayak, C.: A plasma analogy Berry matrices for non-Abelian quantum Hall states. Nucl. Phys. B **506**(3), 685 (1997)

GNN96.   Gluckman, B.J., Netoff, J.I., Neel, E.J., et al.: Stochastic resonance in a neuronal network from mammalian brain. Phys. Rev. Lett. **77**, 4098–4101 (1996)

GOP84.   Grebogi, C., Ott, E., Pelikan, S., Yorke, J.A.: Strange attractors that are not chaotic. Physica D **3**, 261–268 (1984)

GOY87.   Grebogi, C., Ott, E., Yorke, J.A.: Chaos, strange attractors, and fractal basin boundaries in nonlinear dynamics. Science **238**, 632–637 (1987)

GP83.   Grassberger, P., Procaccia, I.: Measuring the strangeness of strange attractors. Physica D **9**, 189–208 (1983)

GQ87.   Gepner, D., Qiu, Z.A.: Modular invariant partition functions for parafermionic field theories. Nucl. Phys. B **285**, 423 (1987)

GRA05.   Gurarie, V., Radzihovsky, L., Andreev, A.V.: Quantum phase transitions across a p-wave Feshbach resonance. Phys. Rev. Lett. **94**(23), 230403 (2005)

GRM90.   Gaspard, P., Rice, S.A., Mikeska, H.J., Nakamura, K.: Parametric motion of energy levels: curvature distribution. Phys. Rev. A **42**, 4015 (1990)

GRS96.   Goméz, C., Ruiz-Altaba, M., Sierra, G.: Quantum Groups in Two-Dimensional Physics. Cambridge Monographs on Mathematical Physics. Cambridge University Press, Cambridge (1996)

GRT02.   Gisin, N., Ribordy, G., Tittel, W., Zbinden, H.: Quantum cryptography. Rev. Mod. Phys. **74**, 145–195 (2002)

GRW86.   Ghirardi, G.C., Rimini, A., Weber, T.: Unified dynamics for microscopic and macroscopic systems. Phys. Rev. D **34**(2), 470–491 (1986)

GS06.   Grosfeld, E., Stern, A.: Electronic transport in an array of quasiparticles in the $\nu = 5/2$ non-Abelian quantum Hall state. Phys. Rev. B **73**(20), 201303 (2006)

GS89.   Gray, C.M., Singer, W.: Stimulus-specific neuronal oscillations in orientation columns of cat visual cortex. Proc. Natl. Acad. Sci. (USA) **86**, 1698–1702 (1989)

GS95.   Goldman, V.J., Su, B.: Resonant tunneling in the quantum Hall regime: measurement of fractional charge. Science **267**(5200), 1010 (1995)

GS98.   Grosche, C., Steiner, F.: Handbook of Feynman Path Integrals. Springer Tracts in Modern Physics, vol. 145. Springer, Berlin (1998)

GSB07.   Gaebler, J.P., Stewart, J.T., Bohn, J.L., Jin, D.S.: p-wave Feshbach molecules. Phys. Rev. Lett. **98**(20), 200403 (2007)

GSD92. Garfinkel, A., Spano, M.L., Ditto, W.L., Weiss, J.N.: Controlling cardiac chaos. Science **257**, 1230–1235 (1992)

GSS06. Grosfeld, E., Simon, S.H., Stern, A.: Switching noise as a probe of statistics in the fractional quantum Hall effect. Phys. Rev. Lett. **96**(22), 226803 (2006)

GV79. Ginibre, J., Velo, G.: On a class of nonlinear Schrödinger equations, I: the Cauchy problem, general case. J. Funct. Anal. **32**, 1–32 (1979)

GV85a. Ginibre, J., Velo, G.: The global Cauchy problem for the nonlinear Schrödinger equation revisited. Ann. Inst. H. Poincaré Anal. Non Linéaire **2**, 309–327 (1985)

GV85b. Ginibre, J., Velo, G.: Scattering theory in the energy space for a class of nonlinear Schrödinger equations. J. Math. Pure Appl. **64**, 363–401 (1985)

GVP94. Van Gelderen, P., de Vleeschouwer, M.H.M., des Pres, D., et al.: Water diffusion and acute stroke. Magn. Reson. Med. **31**, 154–163 (1994)

GVW99. Goldenberg, L., Vaidman, L., Wiesner, S.: Quantum gambling. Phys. Rev. Lett. **82**, 3356–3359 (1999)

GW81. Goldstone, J., Wilczek, F.: Fractional quantum numbers of solitons. Phys. Rev. Lett. **47**, 986 (1981)

GW82. Gotman, J., Wang, L.Y.: Automatic recognition of epileptic seizures in the EEG. Electroencephalogr. Clin. Neurophysiol. **54**, 530–540 (1982)

GW86. Gepner, D., Witten, E.: String theory on group manifolds. Nucl. Phys. B **278**, 493 (1986)

GW91. Gotman, J., Wang, L.Y.: State-dependent spike detection: concepts and preliminary results. Electroencephalogr. Clin. Neurophysiol. **79**, 11–19 (1991)

GWA01. Grewal, M.S., Weill, L.R., Andrews, A.P.: Global Positioning Systems, Inertial Navigation, and Integration. Wiley, New York (2001)

GWW00. Guhr, T., Wilke, T., Weidenmüller, H.A.: Stochastic field theory for a Dirac particle propagating in gauge field disorder. Phys. Rev. Lett. **85**, 2252–2255 (2000)

GWW91. Greiter, M., Wen, X.G., Wilczek, F.: Paired Hall state at half filling. Phys. Rev. Lett. **66**(24), 3205 (1991)

GWW92. Greiter, M., Wen, X.G., Wilczek, F.: Paired Hall states. Nucl. Phys. B **374**(3), 567 (1992)

GZ01. Gupta, S., Zia, R.: Quantum neural networks. J. Comput. Syst. Sci. **63**, 355 (2001)

GZ86. Gielen, C.C., van Zuylen, E.J.: Coordination of arm muscles during flexion and supination: application of the tensor analysis approach. Neurosci. **17**, 527–539 (1986)

GZ94. Glass, L., Zeng, W.: Bifurcations in flat-topped maps and the control of cardiac chaos. Int. J. Bifurc. Chaos **4**, 1061–1067 (1994)

Gam95. Gammaitoni, L.: Stochastic resonance and the dithering effect in threshold physical systems. Phys. Rev. E **52**, 4691–4698 (1995)

Gar85. Gardiner, C.W.: Handbook of Stochastic Methods for Physics, Chemistry, Natural Sciences, 2nd edn. Springer, New York (1985)

Geo06. Georgiev, L.S.: Topologically protected gates for quantum computation with non-Abelian anyons in the Pfaffian quantum Hall state. Phys. Rev. B **74**(23), 235112 (2006)

Ghe90.   Ghez, C.: Introduction to motor system. In: Kandel, E.K., Schwarz, J.H. (eds.) Principles of Neural Science, 2nd edn., pp. 429–442. Elsevier, Amsterdam (1990)

Ghe91.   Ghez, C.: Muscles: Effectors of the Motor Systems. In: Kandel, E.R., Schwartz, J.H., Jessell, T.M. (eds.) Principles of Neural Science, 3rd edn., pp. 548–563. Elsevier, Amsterdam (1991)

Ghe96.   McGhee, R.B.: Research Notes: A Quaternion Attitude Filter Using Angular Rate Sensors, Accelerometers, and a 3-Axis Magnetometer. Computer Science Department, Naval Postgraduate School, Monterey (1996)

Gla63a.   Glauber, R.J.: The Quantum Theory of Optical Coherence. Phys. Rev. **130**, 2529–2539 (1963)

Gla63b.   Glauber, R.J.: Coherent and incoherent states of the radiation field. Phys. Rev. **131**, 2766–2788 (1963)

Gla77.   Glassey, R.T.: On the blowing up of solutions to the Cauchy problem for nonlinear Schrödinger operators. J. Math. Phys. **8**, 1794–1797 (1977)

Gle87.   Gleick, J.: Chaos: Making a New Science. Penguin–Viking, New York (1987)

Gli05.   Glimcher, P.W.: Indeterminacy in brain and behaviour. Annu. Rev. Psychol. **56**, 25–56 (2005)

God07.   Godfrey, M.D., Jiang, P., Kang, W., et al.: Aharonov–Bohm-like oscillations in quantum Hall corrals. arXiv:0708.2448 (2007)

Gol89.   Goldberg, D.E.: Genetic Algorithms in Search, Optimization and Machine Learning. Kluwer Academic, Boston (1989)

Gol92.   Goldenfeld, N.: Lectures on Phase Transitions and the Renormalization Group. Addison-Wesley, Reading (1992)

Gol99a.   Goldberger, A.L.: Nonlinear dynamics, fractals, and chaos theory: implications for neuroautonomic heart rate control in health, disease. In: Bolis, C.L., Licinio, J. (eds.) The Autonomic Nervous System. World Health Organization, Geneva (1999)

Gol99b.   Gold, M.: A Kurt Lewin Reader the Complete Social Scientist. Am. Psychol. Assoc., Washington (1999)

Got97.   Gottesman, D.: Stabilizer codes and quantum error correction. PhD thesis, CalTech, Pasadena (1997)

Got98.   Gottesman, D.: Theory of fault–tolerant quantum computation. Phys. Rev. A **57**(1), 127 (1998)

Goz83.   Gozzi, E.: Functional-integral approach to Parisi–Wu stochastic quantization: scalar theory. Phys. Rev. D **28**, 1922 (1983)

Gra82.   Gray, G.: Rehabilitating the dendritic spine. Trends Neurosci. **5**, 5–6 (1982)

Gra90.   Granato, E.: Phase transitions in Josephson-junction ladders in a magnetic field. Phys. Rev. B **42**, 4797–4799 (1990)

Gro03.   Grossberg, S.: How does the cerebral cortex work? Development, learning, attention, and 3D vision by laminar circuits of visual cortex. Behav. Cogn. Neurosci. Rev. **2**, 47–76 (2003)

Gro69.   Grossberg, S.: Embedding fields: A theory of learning with physiological implications. J. Math. Psychol. **6**, 209–239 (1969)

Gro82.   Grossberg, S.: Studies of Mind and Brain. Kluwer, Dordrecht (1982)

Gro87.   Grossberg, S.: Competitive learning: from interactive activation to adaptive resonance. Cogn. Sci. **11**, 23–63 (1987)

870   References

Gro88.   Grossberg, S.: Neural Networks and Natural Intelligence. MIT Press,
          Cambridge (1988)
Gro99.   Grossberg, S.: How does the cerebral cortex work? Learning, attention
          and grouping by the laminar circuits of visual cortex. Spat. Vis. **12**,
          163–186 (1999)
Gru99.   Gruska, J.: Quantum Computing. McGraw-Hill, New York (1999)
Gun03.   Gunion, J.F.: Class Notes on Path-Integral Methods. Davis, Philadelphia
          (2003)
Gut90.   Gutzwiller, M.C.: Chaos in Classical and Quantum Mechanics. Springer,
          New York (1990)
Gut92.   Gutzwiller, M.: Quantum chaos. Sci. Am. **January** (1992)
Gut98.   Gutkin, B.S., Ermentrout, B.: Dynamics of membrane excitability de-
          termine interspike interval variability: a link between spike generation
          mechanisms, cortical spike train statistics. Neural Comput. **10**(5), 1047–
          1065 (1998)
HBB96.   Houk, J.C., Buckingham, J.T., Barto, A.G.: Models of the cerebellum,
          motor learning. Behav. Brain Sci. **19**(3), 368–383 (1996)
HCK02a.   Hong, H., Choi, M.Y., Kim, B.J.: Synchronization on small-world net-
          works. Phys. Rev. E **65**, 26139 (2002)
HCK02b.   Hong, H., Choi, M.Y., Kim, B.J.: Phase ordering on small-world networks
          with nearest-neighbor edges. Phys. Rev. E **65**, 047104 (2002)
HCT97.   Hall, K., Christini, D.J., Tremblay, M., Collins, J.J., Glass, L., Billette,
          J.: Dynamic control of cardiac alternans. Phys. Rev. Lett. **78**, 4518–4521
          (1997)
HH01.   Horodecki, P., Horodecki, R.: Distillation and bound entanglement.
          Quantum Inf. Comput. **1**(1), 45 (2001)
HH52a.   Hodgkin, A.L., Huxley, A.F.: A quantitative description of membrane
          current and application to conduction and excitation in nerve. J. Physiol.
          117, 500–544 (1952)
HH52b.   Hodgkin, A.L., Huxley, A.F.: A qualitative description of membrane cur-
          rent and its application to conduction and excitation in nerve. J. Physiol.
          117, 500 (1952)
HHB01a.   Hasson, U., Hendler, T., Bashat, D.B., Malach, R.: Vase or face? A neu-
          ral correlates of shape-selective grouping processes in the human brain.
          J. Cogn. Neurosci. **13**(6), 744–753 (2001)
HHB01b.   Hensinger, W.K., Häffner, H., Browaeys, A., et al.: Dynamical tunnelling
          of ultracold atoms. Nature **412**, 52 (2001)
HHH99.   Hirai, K., Hirose, M., Haikawa, Takenaka, T.: The development of Honda
          humanoid robot. In: Proc. of the IEEE Int. Conf. on Robotics and Au-
          tomation, Leuven, Belgium, pp. 1321–1326 (1999)
HHS91.   Nielsen, K., Hynne, F., Sorensen, P.G.: Hopf bifurcation in chemical
          kinetics. J. Chem. Phys. **94**, 1020–1029 (1991)
HHT02.   Hagan, S., Hameroff, S.R., Tuszynski, J.A.: Quantum computation in
          brain microtubules: decoherence and biological feasibility. Phys. Rev. D
          65, 061901 (2002)
HI01.   Hoppensteadt, F.C., Izhikevich, E.M.: Canonical neural models. In: Ar-
          bib, M.A. (ed.) Brain Theory, Neural Networks, 2nd edn. MIT Press,
          Cambridge (2001)
HI97.   Hoppensteadt, F.C., Izhikevich, E.M.: Weakly Connected Neural Net-
          works. Springer, New York (1997)

HI99.   Hoppensteadt, F.C., Izhikevich, E.M.: Oscillatory neurocomputers with dynamic connectivity. Phys. Rev. Let. **82**(14), 2983–2986 (1999)

HKB85.   Haken, H., Kelso, J.A.S., Bunz, H.: A theoretical model of phase transitions in human hand movements. Biol. Cybern. **51**, 347–356 (1985)

HKM03.   Huard, D., Kröger, H., Melkonyan, G., Moriarty, K.J.M., Nadeau, L.P.: Test of quantum action for inverse square potential. Phys. Rev. A **68**, 034101 (2003)

HKS99.   Hegger, R., Kantz, H., Schreiber, T.: Practical implementation of nonlinear time series methods: the TISEAN package. Chaos **9**, 413 (1999)

HL84.   Horsthemke, W., Lefever, R.: Noise-Induced Transitions. Springer, Berlin (1984)

HL86a.   Hale, J.K., Lin, X.B.: Symbolic dynamics and nonlinear semiflows. Ann. Mat. Pure Appl. **144**(4), 229–259 (1986)

HL86b.   Hale, J.K., Lin, X.B.: Examples of transverse homoclinic orbits in delay equations. Nonlinear Anal. **10**, 693–709 (1986)

HL93.   Hale, J.K., Lunel, S.M.V.: Introduction to Functional Differential Equations. Springer, New York (1993)

HM96.   Handy, C.M., Murenzi, R.: Continuous wavelet transform analysis of one-dimensional quantum bound states from first principles. Phys. Rev. A **54**, 3754–3763 (1996)

HMP97.   Huelga, S.F., Macchiavello, C., Pellizzari, T., Ekert, A.K., Plenio, M.B., Cirac, J.I.: Improvement of frequency standards with quantum entanglement. Phys. Rev. Lett. **79**, 3865–3868 (1997)

HN08a.   Hong, S.L., Newell, K.M.: Entropy conservation in the control of human action. Nonlinear Dyn. Psychol. Life Sci. **12**(2), 163–190 (2008)

HN08b.   Hong, S.L., Newell, K.M.: Entropy compensation in human motor adaptation. Chaos **18**(1), 013108 (2008)

HO96.   Hunt, B.R., Ott, E.: Optimal periodic orbits of chaotic systems occur at low period. Phys. Rev. E **54**, 328–337 (1996)

HOG88.   Hsu, G.H., Ott, E., Grebogi, C.: Strange saddles and dimensions of their invariant manifolds. Phys. Lett. A **127**, 199–204 (1988)

HOP98.   Hatsopoulas, N., Ojakangas, C.L., Paninski, L., Donohue, J.P.: Information about movement direction obtained from synchronous activity ofmotor cortical neurons. Proc. Natl. Acad. Sci. USA **95**, 15706 (1998)

HOS04.   Hansson, T.H., Oganesyan, V., Sondhi, S.L.: Superconductors are topologically ordered. Ann. Phys. (N.Y.) **313**, 497 (2004)

HOY96.   Hunt, B.R., Ott, E., Yorke, J.A.: Fractal dimensions of chaotic saddles of dynamical systems. Phys. Rev. E **54**, 4819–4823 (1996)

HP29a.   Heisenberg, W., Pauli, W.: Z. Phys. **56**, 1–61 (1929)

HP29b.   Heisenberg, W., Pauli, W.: Z. Phys. **59**, 168–190 (1929)

HP93.   Hameroff, S.R., Penrose, R.: Conscious events as orchestrated spacetime selections. J. Conscious. Stud. **3**(1), 36–53 (1996)

HP96a.   Hameroff, S.R., Penrose, R.: Orchestrated reduction of quantum coherence in brain microtubules: a model for consciousness. In: Hameroff, S.R., Kaszniak, A.W., Scott, A.C. (eds.) Toward a Science of Consciousness: the First Tucson Discussion and Debates, pp. 507–539. MIT Press, Cambridge (1996)

HP96b.   Hawking, S., Penrose, R.: The Nature of Space and Time. Princeton University Press, Princeton (1996)

HS74.   Hirsch, M.W., Smale, S.: Differential Equations, Dynamical Systems and
            Linear Algebra. Academic Press, New York (1974)
HS88.   Harsanyi, J.C., Selten, R.: A general theory of equilibrium selection in
            games. MIT Press, Cambridge (1988)
HSG07.   Hamdi, S., Schiesser, W.E., Griffiths, G.W.: Method of lines. Scholarpe-
            dia $2(7)$, 2859 (2007)
HSS00.   Hulata, E., Segev, R., Shapir, Y., Benveniste, M., Ben-Jacob, E.: Detec-
            tion and sorting of neural spikes using wavelet packets. Phys. Rev. Lett.
            85, 4637–4640 (2000)
HSW99.   Hitchin, N.J., Segal, G.B., Ward, R.S.: Integrable systems. Twistors,
            loop groups, and Riemann surfaces. In: Lectures from the Instructional
            Conference Held at the University of Oxford, Oxford, September 1997.
            Oxford Graduate Texts in Mathematics, vol. 4. Clarendon/Oxford Uni-
            versity Press, New York (1999)
HT05.   He, B., Teixeira, F.L.: On the degrees of freedom of lattice electrody-
            namics. Phys. Lett. A **336**, 1–7 (2005)
HT85.   Hopfield, J.J., Tank, D.W.: Neural computation of decisions in optimi-
            sation problems. Biol. Cybern. **52**, 114–152 (1985)
HTC83.   Huang, G.M., Tarn, T.J., Clark, J.W.: On the controllability of
            quantum–mechanical systems. J. Math. Phys. **24**, 2608 (1983)
HU79.   Hopcroft, J.E., Ullman, J.D.: Introduction to Automata Theory, Lan-
            guages and Computation. Addison-Wesley, New York (1979)
HW82.   Hameroff, S.R., Watt, R.C.: Information processing in microtubules.
            J. Theor. Biol. **98**, 549–561 (1982)
HW83.   Hameroff, S.R., Watt, R.C.: Do anesthetics act by altering electron mo-
            bility? Anesth. Analg. **62**, 936–940 (1983)
Haa01.   Haake, F.: Quantum Signatures of Chaos. Springer, Berlin (2001)
Hak00.   Haken, H.: Information, Self-Organization: A Macroscopic Approach to
            Complex Systems. Springer, Berlin (2000)
Hak02.   Haken, H.: Brain Dynamics, Synchronization and Activity Patterns in
            Pulse-Codupled Neural Nets with Delays and Noise. Springer, Berlin
            (2002)
Hak83.   Haken, H.: Synergetics: An Introduction, 3rd edn. Springer, Berlin
            (1983)
Hak91.   Haken, H.: Synergetic Computers and Cognition. Springer, Berlin (1991)
Hak93.   Haken, H.: Advanced Synergetics: Instability Hierarchies of Self-
            Organizing Systems and Devices, 3nd edn. Springer, Berlin (1993)
Hak96.   Haken, H.: Principles of Brain Functioning: A Synergetic Approach to
            Brain Activity, Behavior and Cognition. Springer, Berlin (1996)
Hal82.   Halperin, B.I.: Quantized Hall conductance, current-carrying edge states,
            and the existence of extended states in a 2D disordered potential. Phys.
            Rev. B **25**(4), 2185 (1982)
Hal83a.   Haldane, F.D.M.: Fractional quantization of the Hall effect: a hierarchy of
            incompressible quantum fluid states. Phys. Rev. Lett. **51**(7), 605 (1983)
Hal83b.   Halperin, B.I.: Theory of the quantized Hall conductance. Helv. Phys.
            Acta **56**, 75 (1983)
Hal84.   Halperin, B.I.: Statistics of quasiparticles and the hierarchy of fractional
            quantized Hall states. Phys. Rev. Lett. **52**(18), 1583 (1984)
Ham82.   Hamilton, R.S.: Three-manifolds with positive Ricci curvature. J. Differ.
            Geom. **17**, 255–306 (1982)

Ham86.   Hamilton, R.S.: Four-manifolds with positive curvature operator. J. Differ. Geom. **24**, 153–179 (1986)

Ham87.   Hameroff, S.R.: Ultimate Computing: Biomolecular Consciousness and Nanotechnology. North-Holland, Amsterdam (1987)

Ham88.   Hamilton, R.S.: The Ricci flow on surfaces. Contemp. Math. **71**, 237–261 (1988)

Ham93.   Hamilton, R.S.: The Harnack estimate for the Ricci flow. J. Differ. Geom. **37**, 225–243 (1993)

Ham94.   Hameroff, S.R.: Quantum coherence in microtubules: a neural basis for emergent consciousness. J. Conscious. Stud. **1**(1), 91–118 (1994)

Ham98.   Hameroff, S.R.: Quantum computation in brain microtubules? The Penrose–Hameroff "Orch OR" model of consciousness. Philos. Trans. R. Soc. Lond. A **356**, 1869–1896 (1998)

Ham99.   Hamilton, R.S.: Non-singular solutions of the Ricci flow on three-manifolds. Commun. Anal. Geom. **7**(4), 695–729 (1999)

Han04.   Hankin, C.: An introduction to Lambda Calculi for Computer Scientists. College Sci. Pub., State College (2004)

Har73.   Harsanyi, J.C.: Games with randomly disturbed pay-offs: a new rationale for mixedstrategy equilibrium points. Int. J. Game Theory **2**, 1–23 (1973)

Har92.   Harris-Warrick, R.M. (ed.): The Stomatogastric Nervous System. MIT Press, Cambridge (1992)

Har97.   Harting, J.K.: The Global Anatomy. Medical School, University Wisconsin (1997)

Has00.   Hasegawa, H.: Responses of a Hodgkin–Huxley neuron to various types of spike-train inputs. Phys. Rev. E **61**, 718–726 (2000)

Has01.   Hasegawa, H.: A wavelet analysis of transient spike trains of Hodgkin–Huxley neurons. arXiv:cond-mat/0109444 (2001)

Has02.   Hasegawa, H.: Stochastic resonance of ensemble neurons for transient spike trains: a wavelet analysis. Phys. Rev. E **66**, 21902 (2002)

Hat77a.   Hatze, H.: A myocybernetic control model of skeletal muscle. Biol. Cybern. **25**, 103–119 (1977)

Hat77b.   Hatze, H.: A complete set of control equations for the human musculoskeletal system. J. Biomech. **10**, 799–805 (1977b)

Hat78.   Hatze, H.: A general myocybernetic control model of skeletal muscle. Biol. Cybern. **28**, 143–157 (1978)

Hay01.   Haykin, S. (ed.): Kalman Filtering and Neural Networks. Wiley, New York (2001)

Hay91.   Haykin, S.: Adaptive Filter Theory. Prentice Hall, Englewood Cliffs (1991)

Hay98.   Haykin, S.S.: Neural Networks: A Comprehensive Foundation, 2nd edn. Prentice Hall, New York (1998)

Heb49.   Hebb, D.O.: The Organization of Behavior. Wiley, New York (1949)

Hec87.   Hecht-Nielsen, R.: Counterpropagation networks. Appl. Opt. **26**(23), 4979–4984 (1987)

Hec90.   Hecht-Nielsen, R.: NeuroComputing. Addison-Wesley, Reading (1990)

Hee90.   Heermann, D.W.: Computer Simulation Methods in Theoretical Physics, 2nd edn. Springer, Berlin (1990)

Hen66.   Hénon, M.: Sur la topologie des lignes de courant dans un cas particulier. C. R. Acad. Sci. Paris A **262**, 312–314 (1966)

Hen69. Hénon, M.: Numerical study of quadratic area preserving mappings. Q. Appl. Math. **27** (1969)

Hen76. Hénon, M.: A two-dimensional mapping with a strange attractor. Commun. Math. Phys. **50**, 69–77 (1976)

Hew69. Hewitt, C.: PLANNER: a language for proving theorems in robots. In: IJCAI (1969)

Hil00. Hilfer, R. (ed.): Applications of Fractional Calculus in Physics. World Scientific, Singapore (2000)

Hil38. Hill, A.V.: The heat of shortening and the dynamic constants of muscle. Proc. R. Soc. B **76**, 136–195 (1938)

Hil94. Hilborn, R.C.: Chaos and Nonlinear Dynamics: An Introduction for Scientists, Engineers. Oxford University Press, Oxford (1994)

Hir71. Hirota, R.: Exact solution of the Korteweg–de Vries equation for multiple collisions of solitons. Phys. Rev. Lett. **27**, 1192–1194 (1971)

Hir76. Hirsch, M.W.: Differential Topology. Springer, New York (1976)

Hod64. Hodgkin, A.L.: The Conduction of the Nervous Impulse. Liverpool University Press, Liverpool (1964)

Hol92. Holland, J.H.: Adaptation in Natural, Artificial Systems, 2nd edn. MIT Press, Cambridge (1992)

Hol95. Holland, J.H.: Hidden Order: How Adaptation Builds Complexity. Addison-Wesley, New York (1995)

Hop82. Hopfield, J.J.: Neural networks, physical systems with emergent collective computational abilities. Proc. Natl. Acad. Sci. USA **79**, 2554 (1982)

Hop84. Hopfield, J.J.: Neurons with graded response have collective computational properties like those of two-state neurons. Proc. Natl. Acad. Sci. USA **81**, 3088–3092 (1984)

Hop95. Hopfield, J.J.: Pattern recognition computation using action potential timing for stimulus representation. Nature **376**, 33–36 (1995)

Hou67. Houk, J.C.: Feedback control of skeletal muscles. Brain Res. **5**, 433–451 (1967)

Hou78. Houk, J.C.: Participation of reflex mechanisms and reaction-time processes in the compensatory adjustments to mechanical disturbances. Prog. Clin. Neurophysiol. **4**, 193–215 (1978)

Hou79. Houk, J.C.: Regulation of stiffness by skeletomotor reflexes. Annu. Rev. Physiol. **41**, 99–123 (1979)

Hsu08. Hsu, S.Y.: Some results for the Perelman LYH-type inequality. arXiv: 0801.3506 [math.DG] (2008)

Hun01. Hunt, E.L.: Multiple views of multiple intelligence. (Review of Intelligence Reframed: Multiple Intelligences for the 21st Century.) Contemp. Psychol. **46**, 5–7 (2001)

IA07. Ivancevic, V., Aidman, E.: Life-space foam: a medium for motivational and cognitive dynamics. Physica A **382**, 616–630 (2007)

IAY08. Ivancevic, V., Aidman, E., Yen, L.: Extending Feynman's formalisms for modelling human joint action coordination. Int. J. Biomath. (to appear)

IB05. Ivancevic, V., Beagley, N.: Brain-like functor control-machine for general humanoid biodynamics. Int. J. Math. Math. Sci. **11**, 1759–1779 (2005)

IDW03. Izhikevich, E.M., Desai, N.S., Walcott, E.C., Hoppensteadt, F.C.: Bursts as a unit of neural information: selective communication via resonance. Trends Neurosci. **26**, 161–167 (2003)

IE08.   Izhikevich, E.M., Edelman, G.M.: Large-scale model of mammalian tha-
            lamocortical systems. Proc. Natl. Acad. Sci. **105**, 3593–3598 (2008)
IGF99.   Ioffe, L.B., Geshkenbein, V.B., Feigel'man, M.V., Fauchère, A.L., Blat-
            ter, G.: Environmentally decoupled sds-wave Josephson junctions for
            quantum computing. Nature **398**, 679–681 (1999)
II06a.   Ivancevic, V., Ivancevic, T.: Human-Like Biomechanics. Springer, Dor-
            drecht (2005)
II06b.   Ivancevic, V., Ivancevic, T.: Natural Biodynamics. World Scientific, Sin-
            gapore (2006)
II06c.   Ivancevic, V., Ivancevic, T.: Geometrical Dynamics of Complex Systems.
            Springer, Dordrecht (2006)
II07a.   Ivancevic, V., Ivancevic, T.: High-Dimensional Chaotic and Attractor
            Systems. Springer, Dordrecht (2007)
II07b.   Ivancevic, V., Ivancevic, T.: Neuro-Fuzzy Associative Machinery for
            Comprehensive Brain and Cognition Modelling. Springer, Berlin (2007)
II07c.   Ivancevic, V., Ivancevic, T.: Complex Dynamics: Advanced System Dy-
            namics in Complex Variables. Springer, Dordrecht (2007)
II07d.   Ivancevic, V., Ivancevic, T.: Computational Mind: A Complex Dynamics
            Perspective. Springer, Berlin (2007)
II07e.   Ivancevic, V., Ivancevic, T.: Applied Differential Geometry: A Modern
            Introduction. World Scientific, Singapore (2007)
II08a.   Ivancevic, V., Ivancevic, T.: Complex Nonlinearity: Chaos, Phase Tran-
            sitions, Topology Change and Path Integrals. Springer, Berlin (2008)
II08b.   Ivancevic, V., Ivancevic, T.: Quantum Leap: From Dirac and Feynman,
            Across the Universe, to Human Body and Mind. World Scientific, Sin-
            gapore (2008)
II08c.   Ivancevic, V., Ivancevic, T.: Nonlinear quantum psychodynamics with
            topological phase transitions. Neuroquantology (2008, to appear)
II08d.   Ivancevic, V., Ivancevic, T.: Ricci flow and bio-reaction–diffusion sys-
            tems. Math. Mod. Methods Appl. Sci. (2009, to appear)
IJB99a.   Ivancevic, T., Jain, L.C., Bottema, M.: A new two-feature GBAM-
            neurodynamical classifier for breast cancer diagnosis. In: Proc. KES'99.
            IEEE Press, New York (1999)
IJB99b.   Ivancevic, T., Jain, L.C., Bottema, M.: A new two-feature FAM-matrix
            classifier for breast cancer diagnosis. In: Proc. KES'99. IEEE Press, New
            York (1999)
IJP08.   Ivancevic, T., Jain, L., Pattison, J., Hariz, A.: Nonlinear dynamics and
            chaos methods in neurodynamics and complex data analysis. Nonlinear
            Dyn. **56**(1–2), 23–44 (2009)
IKD88.   Ivry, R.B., Keele, S.W., Diener, H.C.: Dissociation of the lateral and
            medial cerebellum in movement timing and movement execution. Exp.
            Brain Res. **73**(1), 167–80 (1988)
IK91.   Isham, C.J., Klauder, J.R.: Coherent states for n-dimensional Euclidean
            groups $E(n)$ and their applications. J. Math. Phys. **32**, 607–620 (1991)
IKM03.   Imamizu, H., Kuroda, T., Miyauchi, S., Yoshioka, T., Kawato, M.: Mod-
            ular organization of internal models of tools in the human cerebellum.
            Proc. Natl. Acad. Sci. USA **100**, 5461–5466 (2003)
IMT00.   Imamizu, H., Miyauchi, S., Tamada, T., Sasaki, Y., Takino, R., Puetz,
            B., Yoshioka, T., Kawato, M.: Human cerebellar activity reflecting an
            acquired internal model of a novel tool. Nature **403**, 192–195 (2000)

IP01a.   Ivancevic, V., Pearce, C.E.M.: Topological duality in humanoid robot dynamics. ANZIAM J. **43**, 43183–43194 (2001)

IP01b.   Ivancevic, V., Pearce, C.E.M.: Poisson manifolds in generalized Hamiltonian biomechanics. Bull. Austral. Math. Soc. **64**, 515–526 (2001)

IP96.   Iacomelli, G., Pettini, M.: Regular and chaotic quantum motions. Phys. Lett. A **212**, 29 (1996)

IR08.   Ivancevic, V.G., Reid, D.J.: Entropic geometry of crowd dynamics. Entropy (2008, submitted)

IRP96.   Ivanov, P.C., Rosenblum, M.G., Peng, C.-K., et al.: Scaling behaviour of heartbeat intervals obtained by wavelet-based time-series analysis. Nature **383**, 323–327 (1996)

IS01.   Ivancevic, V., Snoswell, M.: Fuzzy-stochastic functor machine for general humanoid-robot dynamics. IEEE Trans. Syst. Man Cybern. B **31**(3), 319–330 (2001)

IZ80.   Itzykson, C., Zuber, J.: Quantum Field Theory. McGraw-Hill, New York (1980)

Ing82.   Ingber, L.: Statistical mechanics of neocortical interactions, I: basic formulation. Physica D **5**, 83–107 (1982)

Ing97.   Ingber, L.: Statistical mechanics of neocortical interactions: applications of canonical momenta indicators to electroencephalography. Phys. Rev. E **55**(4), 4578–4593 (1997)

Ing98.   Ingber, L.: Statistical mechanics of neocortical interactions: training, testing canonical momenta indicators of EEG. Math. Comput. Model. **27**(3), 33–64 (1998)

Ing00.   Ingber, L.: High-resolution path-integral development of financial options. Physica A **283**, 529–558 (2000)

Isi89.   Isidori, A.: Nonlinear Control Systems, An Introduction, 2nd edn. Springer, Berlin (1989)

Ito51.   Itô, K.: On stochastic differential equations. Mem. Am. Math. Soc. **4**, 1–51 (1951)

Ito60.   Ito, K.: Wiener integral, Feynman integral. Stat. Probab. **2**, 227–238 (1960)

Ito84.   Ito, M.: Cerebellum and Neural Control. Raven Press, New York (1984)

Ito90.   Ito, M.: A new physiological concept on cerebellum. Rev. Neurol. (1990)

Iva01.   Ivanov, D.A.: Non-Abelian statistics of half-quantum vortices in p-wave superconductors. Phys. Rev. Lett. **86**, 268 (2001)

Iva02.   Ivancevic, V.: Generalized Hamiltonian biodynamics and topology invariants of humanoid robots. Int. J. Math. Math. Sci. **31**(9), 555–565 (2002)

Iva04.   Ivancevic, V.: Symplectic rotational geometry in human biomechanics. SIAM Rev. **46**(3), 455–474 (2004)

Iva06.   Ivancevic, V.: Lie–Lagrangian model for realistic human bio-dynamics. Int. J. Hum. Robot. **3**(2), 205–218 (2006)

Iva06b.   Ivancevic, V.: Dynamics of humanoid robots: geometrical, topological duality. In: Misra, J.C. (ed.) Biomathematics: Modelling, Simulation. World Scientific, Singapore (2006)

Izh00.   Izhikevich, E.M.: Neural excitability, spiking and bursting. Int. J. Bifurc. Chaos **10**, 1171–1266 (2000)

Izh01a.   Izhikevich, E.M.: Synchronization of elliptic bursters. SIAM Rev. **43**(2), 315–344 (2001)

Izh01b.   Izhikevich, E.M.: Resonate-and-fire neurons. Neural Netw. **14**, 883–894 (2001)

Izh04.   Izhikevich, E.M.: Which model to use for cortical spiking neurons? IEEE Trans. Neural Netw. **15**, 1063–1070 (2004)

Izh07.   Izhikevich, E.M.: Dynamical Systems in Neuroscience: The Geometry of Excitability and Bursting. MIT Press, Cambridge (2007)

Izh99a.   Izhikevich, E.M.: Class 1 neural excitability, conventional synapses, weakly connected networks and mathematical foundations of pulse-coupled models. IEEE Trans. Neural Netw. **10**, 499–507 (1999)

Izh99b.   Izhikevich, E.M.: Weakly connected quasiperiodic oscillators, FM interactions and multiplexing in the brain. SIAM J. Appl. Math. **59**(6), 2193–2223 (1999)

Izr86.   Izrailev, F.M.: Limiting quasienergy statistics for simple quantum systems. Phys. Rev. Lett. **56**, 541 (1986)

JAD00.   Jozsa, R., Abrams, D.S., Dowling, J.P., Williams, C.P.: Quantum clock synchronization based on shared prior entanglement. Phys. Rev. Lett. **85**, 2010 (2000)

JKL01.   Jirari, H., Kröger, H., Luo, X.Q., Moriarty, K.J.M., Rubin, S.G.: Closed path integrals and the quantum action. Phys. Rev. Lett. **86**, 187 (2001)

JKL02.   Jirari, H., Kröger, H., Luo, X.Q., Melkonyan, G., Moriarty, K.J.M.: Renormalisation in quantum mechanics. Phys. Lett. A **303**, 299 (2002)

JPY96.   Jibu, M., Pribram, K.H., Yasue, K.: From conscious experience to memory storage and retrieval: the role of quantum brain dynamics, boson condensation of evanescent photons. Int. J. Mod. Phys. B **10**, 1735 (1996)

JR92.   Judson, R.S., Rabitz, H.: Teaching lasers to control molecules. Phys. Rev. Lett. **68**, 1500 (1992)

JY95.   Jibu, M., Yasue, K.: Quantum Brain Dynamics and Consciousness. John Benjamins, Amsterdam (1995)

Jac91.   Jackson, E.A.: Perspectives of Nonlinear Dynamics. Cambridge University Press, Cambridge (1991)

Jac98.   Jackson, J.D.: Classical Electrodynamics. Wiley, New York (1998)

Jai89.   Jain, J.K.: Composite-fermion approach for the fractional quantum Hall effect. Phys. Rev. Lett. **63**, 199 (1989)

Jal02.   Jallon, P.: Epilepsy and epileptic disorders, an epidemiological marker? Contribution of descriptive epidemiology. Epileptic Disord. **4**, 1 (2002)

Jaz70.   Jazwinski, A.H.: Stochastic Processes and Filtering Theory. Academic Press, New York (1970)

Jen78.   Jensen, M.C.: Some anomalous evidence regarding market efficiency, an editorial introduction. J. Financ. Econ. **6**, 95–101 (1978)

Joa00.   Joao, M.L.: An extended Kalman filter for quaternion-based attitude estimation. Master thesis, Computer Science Department, Naval Postgraduate School, Monterey (2000)

Jon85.   Jones, V.F.R.: A polynomial invariant for knots via von Neumann algebras. Bull. Am. Math. Soc. **12**, 103 (1985)

Jos74.   Josephson, B.D.: The discovery of tunnelling supercurrents. Rev. Mod. Phys. **46**(2), 251–254 (1974)

Jun70.   Jung, C.J.: Collected Works of C.G. Jung. Princeton University Press, Princeton (1970)

KBA99.   Koza, J.R., Bennett, F.H., Andre, D., Keane, M.A.: Genetic Program-
         ming III: Darwinian Invention, Problem Solving. Morgan Kaufmann, San
         Mateo (1999)
KBE93.   Kisvárday, Z.F., Beaulieu, C., Eysel, U.T.: Network of GABAergic large
         Basket cells in cat visual cortex (area 18). Implications for lateral disin-
         hibition. J. Comput. Neurol. **327**, 398–415 (1993)
KD03.   Van der Kooij, H., Donker, S.: Use of adaptive model of balance control
         in the identification of postural dynamics. In: Proc. ISB'03, Univ. Otago,
         Dunedin, NZ (2003)
KFA69.   Kalman, R.E., Falb, P., Arbib, M.A.: Topics in Mathematical System
         Theory. McGraw-Hill, New York (1969)
KG85.   Kantz, H., Grassberger, P.: Repellers, semi-attractors and long-lived
         chaotic transients. Physica D **17**, 75–86 (1985)
KHO00.   Kanamaru, T., Horita, T., Okabe, Y.: Theoretical analysis of array-
         enhanced stochastic resonance in the diffusively coupled FitzHugh–
         Nagumo equation. Phys. Rev. E **64**, 31908 (2000)
KHS93.   Koruga, D.L., Hameroff, S.I., Sundareshan, M.K., Withers, J., Loutfy,
         R.: Fullerence C60: History, Physics, Nanobiology and Nanotechnology.
         Elsevier Science, Amsterdam (1993)
KI92.   Kennel, M.B., Isabelle, S.: Method to distinguish possible chaos from
         colored noise and to determine embedding parameters. Phys. Rev. A **46**,
         3111–3118 (1992)
KJ03.   Knoblich, G., Jordan, S.: Action coordination in individuals and groups:
         learning anticipatory control. J. Exp. Psychol. Learn. Mem. Cogn. **29**,
         1006–1016 (2003)
KJH01.   Van der Kooij, H., Jacobs, R., van der Helm, F.: An adaptive model of
         sensory integration in a dynamic environment applied to human stance
         control. Biol. Cybern. **84**, 103–115 (2001)
KK00.   Kye, W.-H., Kim, C.-M.: Characteristic relations of type-I intermittency
         in the presence of noise. Phys. Rev. E **62**, 6304–6307 (2000)
KKS03.   Koza, J.R., Keane, M.A., Streeter, M.J., Mydlowec, W., Yu, J., Lanza,
         G.: Genetic Programming, IV: Routine Human-Competitive Machine In-
         telligence. Kluwer, Dordrecht (2003)
KLM01.   Knill, E., Laflamme, R., Milburn, G.J.: A scheme for efficient quantum
         computation with linear optics. Nature **409**, 46–57 (2001)
KL07.   Karafyllidis, I.G., Lagoudas, D.C.: Microtubules as mechanical force sen-
         sors. Biosystems **88**(1–2), 137–146 (2007)
KLM06.   Kröger, H., Laprise, J.F., Melkonyan, G., Zomorrodi, R.: Quantum chaos
         versus classical chaos: why is quantum chaos weaker? In: Ausloos, M.,
         Dirickx, M. (eds.) The Logistic Map and the Route to Chaos, pp. 355–
         367. Springer, Berlin (2006)
KLR03.   Kye, W.-H., Lee, D.-S., Rim, S., Kim, C.-M., Park, Y.-J.: Periodic
         phase synchronization in coupled chaotic oscillators. Phys. Rev. E **68**,
         025201(R) (2003)
KLS82.   Kitney, R.I., Linkens, D., Selman, A.C., McDonald, A.H.: The inter-
         action between heart rate and respiration. Nonlinear analysis based on
         computer modeling. Automedica **4**, 141–153 (1982)
KM75.   Kohler, G., Milstein, C.: Continuous cultures of fused cells secreting an-
         tibody of predefined specificity. Nature **256**, 495 (1975)

KM82.   Kobayashi, M., Musha, T.: 1/f fluctuation of heartbeat period. IEEE Trans. Biomed. Eng. **29**, 456–457 (1982)

KMB87.   Kleiger, R.E., Miller, J.P., Bigger, J.T., Moss, A.J.: Decreased heart rate variability and its association with increased mortality after acute myocardial infarction. Am. J. Cardiol. **59**, 256–262 (1987)

KMM94.   Konen, W., Maurer, T., von der Malsburg, C.: A fast dynamic link matching algorithm for invariant pattern recognition. Neural Netw. **7**, 1019–1030 (1994)

KMT07.   Knill, D.C., Maloney, L.T., Trommershauser, J.: Sensorimotor processing and goal-directed movement. J. Vis. **7**(5), 1–2 (2007)

KMY84.   Kaplan, J.L., Mallet-Paret, J., Yorke, J.A.: The Lyapunov dimension of a nowhere differentiable attracting torus. Ergod. Theory Dyn. Syst. **4**, 261 (1984)

KN00.   Kotz, S., Nadarajah, S.: Extreme Value Distributions. Imperial College Press, London (2000)

KP95.   Kocarev, L., Parlitz, U.: General approach for chaotic synchronization with applications to communication. Phys. Rev. Lett. **74**, 5028–5031 (1995)

KP96.   Kocarev, L., Parlitz, U.: Generalized synchronization, predictability and equivalence of unidirectionally coupled dynamical systems. Phys. Rev. Lett. **76**, 1816–1819 (1996)

KS02.   Kasabov, N., Song, Q.: Denfis: dynamic evolving neural fuzzy inference systems and its application for time series prediction. IEEE Trans. Fuzzy Syst. **10**(2), 144–154 (2002)

KS97.   Kantz, H., Schreiber, T.: Nonlinear Time Series Analysis. Cambridge Nonlinear Science Series. Cambridge University Press, Cambridge (1997)

KSJ91.   Kandel, E.R., Schwartz, J.H., Jessel, T.M.: Principles of Neural Sciences. Elsevier, Amsterdam (1991)

KT01.   Kye, W.-H., Topaj, D.: Attractor bifurcation and on-off intermittency. Phys. Rev. E **63**, 045202(R) (2001)

KT07.   Kolokolnikov, T., Tlidi, V.: Spot deformation and replication in the two-dimensional Belousov–Zhabotinsky reaction in water-in-oil microemulsion. Phys. Rev. Lett. **98**, 188303 (2007)

KT87.   Kugler, P.N., Turvey, M.T.: Information, Natural Law, and the Self-Assembly of Rhythmic Movement: Theoretical and Experimental Investigations. Erlbaum, Hillsdale (1987)

KT98.   Keel, M., Tao, T.: Endpoint Strichartz estimates. Am. Math. J. **120**, 955–980 (1998)

KVE05.   Kaminaga, A., Vanag, V.K., Epstein, I.R.: 'Black spots' in a surfactant-rich Belousov–Zhabotinsky reaction dispersed in a water-in-oil microemulsion system. J. Chem. Phys. **122**, 174706 (2005)

KVE06.   Kaminaga, A., Vanag, V.K., Epstein, I.R.: A reaction–diffusion memory device. Ang. Chem. **45**, 3087 (2006)

KY75.   Kaplan, J.L., Yorke, J.A.: On the stability of a periodic solution of a differential delay equation. SIAM J. Math. Anal. **6**, 268–282 (1975)

KY79a.   Kaplan, J.L., Yorke, J.A.: Numerical solution of a generalized eigenvalue problem for even mapping. In: Peitgen, H.O., Walther, H.O. (eds.) Functional Differential Equations, Approximations of Fixed Points. Lecture Notes in Mathematics, vol. 730, pp. 228–256. Springer, Berlin (1979)

KY79b.   Kaplan, J.L., Yorke, J.A.: Preturbulence: a regime observed in a fluid flow of Lorenz. Commun. Math. Phys. **67**, 93–108 (1979)

KY91.   Kennedy, J., Yorke, J.A.: Basins of Wada. Physica D **51**, 213–225 (1991)

KYR98.   Kim, C.M., Yim, G.S., Ryu, J.W., Park, Y.J.: Characteristic relations of type-III intermittency in an electronic circuit. Phys. Rev. Lett. **80**, 5317–5320 (1998)

KZH02.   Kiss, I.Z., Zhai, Y., Hudson, J.L.: Emerging coherence in a population of chemical oscillators. Science **296**, 1676–1678 (2002)

Kad00.   Kadanoff, L.P.: Statistical Physics: Statics, Dynamics and Renormalization. World Scientific, Singapore (2000)

Kad66.   Kadanof f, L.P.: Scaling laws for Ising models near Tc. Physics **2**, 263–272 (1966)

Kak93.   Kaku, M.: Quantum Field Theory: A Modern Introduction. Oxford University Press, London (1993)

Kal60.   Kalman, R.E.: A new approach to linear filtering and prediction problems. Trans. ASME, Ser. D, J. Bas. Eng. **82**, 34–45 (1960)

Kan58.   Kan, D.M.: Adjoint functors. Trans. Am. Math. Soc. **89**, 294–329 (1958)

Kan65.   Kant, I.: Critique of Pure Reason (trans. Norman Kemp Smith) Discussions of Transcendental Unity of Apperception, pp. 135–161; Transcendental Aesthetic, pp. 74–81; Phenomena and Noumena, pp. 257–275. St. Martin's Press, New York (1965)

Kar84.   Kardar, M.: Free energies for the discrete chain in a periodic potential and the dual Coulomb gas. Phys. Rev. B **30**, 6368–6378 (1984)

Kas02.   Kasabov, N.: Evolving Connectionist Systems: Methods and Applications in Bioinformatics, Brain Study and Intelligent Machines. Springer, London (2002)

Kat95.   Kato, T.: On nonlinear Schrödinger equations, II: H^s-solutions and unconditional well-posedness. J. Anal. Math. **67**, 281–306 (1995)

Kaw99.   Kawato, M.: Internal models for motor control and trajectory planning. Curr. Opin. Neurobiol. **9**, 718–727 (1999)

Kay91.   Kay, D.S.: Computer interaction: Debugging the problems. In: Sternberg, R.J., Frensch, P.A. (eds.) Complex Problem Solving: Principles, Mechanisms, Pp. 317–340. Lawrence Erlbaum Associates, Hillsdale (1991)

KdV895.   Korteweg, D.J., de Vries, G.: On the change of form of long waves advancing in a rectangular canal, and on a new type of long stationary waves. Philos. Mag. **539**, 422–443 (1895)

Kel95.   Kelso, J.A.S.: Dynamic Patterns: The Self Organization of Brain and Behavior. MIT Press, Cambridge (1995)

Khi57.   Khinchin, A.I.: Mathematical Foundations of Information Theory. Dover, New York (1957)

Kit06.   Kitaev, A.Y.: Anyons in an exactly solved model and beyond. Ann. Phys. N.Y. **321**, 2 (2006)

Kit03.   Kitaev, A.Y.: Fault-tolerant quantum computation by anyons. Ann. Phys. N.Y. **303**, 2 (2003)

Kit97.   Kitaev, A.Y.: Quantum computations: algorithms and error correction. Russ. Math. Surv. **52**, 1191–1249 (1997)

Kla00.   Klauder, J.R.: Beyond Conventional Quantization. Cambridge University Press, Cambridge (2000)

Kla97.   Klauder, J.R.: Understanding quantization. Found. Phys. **27**, 1467–1483 (1997)

Kle02.   Kleinert, H.: Path Integrals in Quantum Mechanics, Statistics, Polymer Physics, and Financial Markets, 3rd edn. World Scientific, Singapore (2002)

Kob07.   Kobayashi, N.: Equivalence between quantum simultaneous games and quantum sequential games. arXiv:0711.0630 [quant-ph] (2007)

Koh88.   Kohonen, T.: Self Organization, Associative Memory. Springer, Berlin (1988)

Koh91.   Kohonen, T.: Self-organizing maps: optimization approaches. In: Kohonen, T., et al. (eds.) Artificial Neural Networks. North-Holland, Amsterdam (1991)

Kos86.   Kosko, B.: Fuzzy cognitive maps. Int. J. Man-Mach. Stud. **24**, 65–75 (1986)

Kos88.   Kosko, B.: Bidirectional Associative Memory. IEEE Trans. Syst. Man Cybern. **18**, 49–60 (1988)

Kos92.   Kosko, B.: Neural Networks, Fuzzy Systems, A Dynamical Systems Approach to Machine Intelligence. Prentice Hall, New York (1992)

Kos93.   Kosko, B.: Fuzzy Thinking. Disney Books, Hyperion (1993)

Kos96.   Kosko, B.: Fuzzy Engineering. Prentice Hall, New York (1996)

Kos99.   Kosko, B.: The Fuzzy Future: From Society, Science to Heaven in a Chip. Random House, Harmony (1999)

Koz91.   Kozen, D.: The Design and Analysis of Algorithms. Texts and Monographs in Computer Science. Springer, Berlin (1991)

Koz92.   Koza, J.R.: Genetic Programming: On the Programming of Computers by Means of Natural Selection. MIT Press, Cambridge (1992)

Koz95.   Koza, J.R.: Genetic Programming II: Automatic Discovery of Reusable Programs. MIT Press, Cambridge (1995)

Krö02.   Kröger, H.: Existence of the quantum action. Phys. Rev. A **65**, 052118 (2002)

Kry79.   Krylov, N.S.: Works on the Foundations of Statistical Mechanics. Princeton University Press, Princeton (1979)

Kuh85.   Kuhl, J.: Volitional mediator of cognition-behaviour consistency: self-regulatory processes, action versus state orientation. In: Kuhl, J., Beckman, S. (eds.) Action Control: From Cognition to Behaviour. Springer, Berlin (1985)

Kuk00.   Kuksin, S.: Analysis of Hamiltonian PDEs. Oxford Lecture Series in Mathematics and Its Applications, vol. 19. Oxford University Press, Oxford (2000)

Kun99.   Kunz-Schughart, L.A.: Multicellular tumor spheroids: intermediates between monolayer culture and in vivo tumor. Cell Biol. Int. **23**(3), 157–161 (1999)

Kur05.   Kurita, Y.: Indispensable role of quantum theory in the brain dynamics. Biosystems **80**, 263–272 (2005)

Kur76.   Kuramoto, Y., Tsuzuki, T.: Persistent propagation of concentration waves in dissipative media far from thermal equilibrium. Prog. Theor. Phys. **55**, 365 (1976)

Kur84.   Kuramoto, Y.: Chemical Oscillations. Waves, Turbulence. Springer, New York (1984)

Kuz95.   Kuznetsov, Y.A.: Elements of Applied Bifurcation Theory. Applied Mathematical Sciences, vol. 112. Springer, Berlin (1995)

Kwo89.   Kwong, M.K.: Uniqueness of positive solutions of $\Delta\psi - \psi + \psi^p = 0$ in $\mathbb{R}^n$. Arch. Ration. Mech. Anal. **105**, 243–266 (1989)

LBM05a.   De Lucia, M., Bottaccio, M., Montuori, M., Pietronero, L.: A topological approach to neural complexity. Phys. Rev. E **71**, 016114 (2005)

LBM05b.   Tononi, G., Sporns, O., Edelman, G.M.: A measure for brain complexity: relating functional segregation and integration in the nervous system. Proc. Natl. Acad. Sci. USA **91**(11), 5033–5037 (1994)

LBY07.   Lu, C.-Y., Browne, D.E., Yang, T., Pan, J.-W.: Demonstration of Shor's quantum factoring algorithm using photonic qubits. Phys. Rev. Lett. **99**, 250504 (2007)

LC94.   Longtin, A., Chialvo, D.R.: Stochastic and deterministic resonances for excitable systems. Phys. Rev. Lett. **81**, 4012–4015 (1994)

LC97.   Lo, H.K., Chau, H.F.: Is quantum bit commitment really possible? Phys. Rev. Lett. **78**, 3410–3413 (1997)

LCD02.   Liley, D.T.J., Cadusch, P.J., Dafilis, M.P.: A spatially continuous mean field theory of electrocortical activity. Comput. Neural Syst. **13**(1), 67–113 (2002)

LCD87.   Leggett, A.J., Chakravarty, S., Dorsey, A.T., Fisher, M.P.A., Chang, A., Zwerger, W.: Dynamics of the dissipative two-state system. Rev. Mod. Phys. **59**, 1 (1987)

LCW99.   Liley, D.T.J., Cadusch, P.J., Wright, J.J.: A continuum theory of electrocortical activity. Neurocomputing **26**, 795–800 (1999)

LD98.   Loss, D., DiVincenzo, D.P.: Quantum computation with quantum dots. Phys. Rev. A **57**(1), 120–126 (1998)

LE98.   Lehnertz, K., Elger, C.E.: Can epileptic seizures be predicted? Evidence from nonlinear time series analysis of brain electrical activity. Phys. Rev. Lett. **80**, 5019–5022 (1998)

LEA00.   Lehnertz, K., Elger, C.E., Arnhold, J., Grassberger, P. (eds.): Chaos in Brain. World Scientific, Singapore (2000)

LGL03.   Latka, M., Glaubic-Latka, M., Latka, D., West, B.J.: Fractal rigidity in migraine. arXiv:physics/0301055 (2003)

LH01.   Land, M.F., Hayhoe, M.: In what ways do eye movements contribute to everyday activities? Vis. Res. **41**, 3559–3565 (2001)

LH91.   Lenz, G., Haake, F.: Reliability of small matrices for large spectra with nonuniversal fluctuations. Phys. Rev. Lett. **67**, 1–4 (1991)

LHR07.   Levin, M., Halperin, B.I., Rosenow, B.: Particle-hole symmetry and the Pfaffian state. Phys. Rev. Lett. **99**, 236806 (2007)

LHW00.   Liu, F., Hu, B., Wang, W.: Effects of correlated and independent noise on signal processing in neuronal systems. Phys. Rev. E **63**, 31907 (2000)

LJ03.   Lee, C.F., Johnson, N.F.: Efficiency and formalism of quantum games. Phys. Rev. A **67**, 022311 (2003)

LK99.   Lee, S., Kim, S.: Parameter dependence of stochastic resonance in the stochastic Hodgkin–Huxley neuron. Phys. Rev. E **60**, 826–830 (1999)

LL91.   Lesgold, A., Lajoie, S.: Complex problem solving in electronics. In: Sternberg, R.J., Frensch, P.A. (eds.) Complex Problem Solving: Principles, Mechanisms, pp. 287–316. Lawrence Erlbaum Associates, Hillsdale (1991)

LM04.   Laurent, C., Martel, Y.: Smoothness and exponential decay for L^2-compact solutions of the generalized Korteweg–de Vries equations. Commun. Partial Differ. Equ. **29**, 157–171 (2004)

LM77.   Leinaas, J.M., Myrheim, J.: On the theory of identical particles. Nuovo
            Cimento Soc. Ital. Fis., B **37**, 1 (1977)
LM96.   Levins, J.E., Miller, J.P.: Broadband neural encoding in the cricket cereal
            sensory system enhanced by stochastic resonance. Nature **380**, 165–168
            (1996)
LOS88.   Larkin, A.I., Ovchinnikov, Yu.N., Schmid, A.: Physica B **152**, 266 (1988)
LS01.   Lindner, B., Schimansky-Geier, L.: Transmission of noise coded versus
            additive signals through a neuronal ensemble. Phys. Rev. Lett. **86**, 2934–
            2937 (2001)
LW05a.   Levin, M.A., Wen, X.-G.: Colloquium: photons and electrons as emergent
            phenomena. Rev. Mod. Phys. **77**, 871 (2005)
LW05b.   Levin, M.A., Wen, X.-G.: String-net condensation: a physical mechanism
            for topological phases. Phys. Rev. B **71**, 045110 (2005)
LY85a.   Ledrappier, F., Young, L.-S.: The metric entropy of diffeomorphisms,
            I: characterization of measures satisfying Pesin's entropy formula. Ann.
            Math. **122**, 509–539 (1985)
LY85b.   Ledrappier, F., Young, L.-S.: The metric entropy of diffeomorphisms, II:
            relations between entropy, exponents and dimension. Ann. Math. **122**(3),
            540–574 (1985)
LY86.   Li, P., Yau, S.T.: On the parabolic kernel of the Schrödinger operator.
            Acta Math. **156**, 153–201 (1986)
LYG05.   Lu, T.T., Yu, F.T.S., Gregory, D.A.: Self-organizing optical neural net-
            work for unsupervised learning. Proc. SPIE **1296**, 378 (2005)
Lab97.   Labastida, J.M.F., Lozano, C.: Lectures on topological quantum field
            theory. CERN-TH/97-250; US-FT-30/97. arXiv:hep-th/9709192 (1997)
Lai94.   Lai, Y.-C.: Controlling chaos. Comput. Phys. **8**, 62–67 (1994)
Lak03.   Lakshmanan, M., Rajasekar, S.: Nonlinear Dynamics: Integrability,
            Chaos and Patterns. Springer, New York (2003)
Lak97.   Lakshmanan, M.: Bifurcations, chaos, controlling and synchronization of
            certain nonlinear oscillators. In: Kosmann-Schwarzbach, Y., Grammati-
            cos, B., Tamizhmani, K.M. (eds.) Lecture Notes in Physics, vol. 495,
            p. 206. Springer, Berlin (1997)
Lam02.   Lampinen, J.: A constraint handling method for the differential evolution
            algorithm. In: Sincak, P., Vascak, J., Kvasnicka, V., Pospichal, J. (eds.)
            Intelligent Technologies—Theory, Applications, pp. 152–158. IOS Press,
            Amsterdam (2002)
Lam76.   Lamb, G.L. Jr.: Bäcklund transforms at the turn of the century. In:
            Miura, R.M. (ed.) Bäcklund Transforms. Springer, Berlin (1976)
Lan91.   Landauer, R.: Information is physical. Phys. Today **5**, 23 (1991)
Lan95.   Landauer, R.: Is quantum mechanics useful? Philos. Trans. R. Soc.
            Lond. A **353**, 367–376 (1995)
Las42.   Lashley, K.S.: The problem of cerebral organization in vision. In: Bio-
            logical Symposia, VII: Visual Mechanisms, pp. 301–322. Jaques Cattell
            Press, Lancaster (1942)
Lau83.   Laughlin, R.B.: Anomalous quantum Hall effect: an incompressible quan-
            tum fluid with fractionally charged excitations. Phys. Rev. Lett. **50**, 1395
            (1983)
Lax68.   Lax, P.D.: Integrals of nonlinear equations of evolution and solitary
            waves. Commun. Pure Appl. Math. **21**, 467–490 (1968)

884   References

Lee90.   Lee, C.C.: Fuzzy logic in control systems. IEEE Trans. Syst. Man Cybern. **20**(2), 404–435 (1990)

Leg86.   Leggett, A.J.: In: de Boer, J., Dal, E., Ulfbeck, O. (eds.) The Lesson of Quantum Theory, Niels Bohr Centenary Symposium 1985. North-Holland, Amsterdam (1986)

Lei02.   Leinster, T.: A survey of definitions of n-category. Theor. Appl. Categ. **10**, 1–70 (2002)

Lei03.   Leinster, T.: Higher Operads, Higher Categories. London Mathematical Society Lecture Notes Series. Cambridge University Press, Cambridge (2003)

Lei04.   Leinster, T.: Operads in higher-dimensional category theory. Theor. Appl. Categ. **12**, 73–194 (2004)

Len04.   Lenz, F.: Topological concepts in gauge theories. FAU-TP3-04/3. arXiv: hep-th/0403286 (2004)

Lev01.   Levine, D.K.: Game Theory, Encyclopedia of Cognitive Science. Nature Pub. Group, Basingstoke (2001)

Lev92.   Levy, S.: Artificial Life: A Report from the Frontier where Computers Meet Biology. Vintage Books, Random House, New York (1992)

Lew51.   Lewin, K.: Field Theory in Social Science. University Chicago Press, Chicago (1951)

Lew97.   Lewin, K.: Resolving Social Conflicts and Field Theory in Social Science. Am. Psychol. Assoc., Washington (1997)

Li04.   Li, Y.: Chaos in Partial Differential Equations. Int. Press, Sommerville (2004)

Li04a.   Li, Y.: Persistent homoclinic orbits for nonlinear Schrödinger equation under singular perturbation. Dyn. Partial Differ. Equ. **1**(1), 87–123 (2004)

Li04b.   Li, Y.: Existence of chaos for nonlinear Schrödinger equation under singular perturbation. Dyn. Partial Differ. Equ. **1**(2), 225–237 (2004)

Li04c.   Li, Y.: Homoclinic tubes and chaos in perturbed sine-Gordon equation. Chaos Solitons Fractals **20**(4), 791–798 (2004)

Li04d.   Li, Y.: Chaos in Miles' equations. Chaos Solitons Fractals **22**(4), 965–974 (2004)

Li05.   Li, Y.: Invariant manifolds and their zero-viscosity limits for Navier-Stokes equations. Dyn. Partial Differ. Equ. **2**(2), 159–186 (2005)

Li07.   Li, J.: First variation of the log entropy functional along the Ricci flow. arXiv:0712.0832 [math.DG] (2007)

Lis97.   Lisman, J.: Bursts as a unit of neural information: making unreliable synapses reliable. Trends Neurosci. **20**, 38–43 (1997)

Lo97.   Lo, H.K.: Insecurity of quantum secure computations. Phys. Rev. A **56**, 1154–1162 (1997)

Lol08.   Loll, R.: The emergence of spacetime or quantum gravity on your desktop. Class. Quantum Gravity **25**, 114006 (2008)

Lon93.   Longtin, A.: Stochastic resonance in neuron models. J. Stat. Phys. **70**, 309 (1993)

Lor63.   Lorenz, E.N.: Deterministic nonperiodic flow. J. Atmos. Sci. **20**, 130–141 (1963)

Lug02.   Luger, G.F.: Artificial Intelligence: Structures and Strategies for Complex Problem Solving, 4th edn. Pearson Educ Ltd, Harlow (2002)

Lui02.   Luinge, H.J.: Inertial Sensing of Human Movement. PhD thesis, University of Twente. Twente University Press (2002)

Lwk03.   Latka, M., Was, Z., Kozik, A., West, B.J.: Wavelet analysis of epileptic spikes. Phys. Rev. E **67**, 052902 (2003)

MA02.   Mascalchi, M., et al.: Proton MR spectroscopy of the cerebellum and pons in patients with degenerative ataxia. Radiology **223**, 371 (2002)

MAD02.   Morf, R.H., d'Ambrumenil, N., Das Sarma, S.: Excitation gaps in fractional quantum Hall states: an exact diagonalization study. Phys. Rev. B **66**, 075408 (2002)

Mer73.   Merton, R.C.: Theory of rational option pricing. Bell J. Econ. Manag. Sci. **4**, 141–183 (1973)

MB98.   Meyer, D.A., Brown, T.A.: Statistical mechanics of voting. Phys. Rev. Lett. **81**, 1718–1721 (1998)

MB99.   Maass, W., Bishop, C.M.: Pulsed Neural Networks. MIT Press, Cambridge (1999)

MBB08.   Morgan, S.W., Biktasheva, I.V., Biktashev, V.N.: Control of scroll wave turbulence using resonant perturbations. Phys. Rev. E **78**, 046207 (2008)

MD03.   Morf, R., d'Ambrumenil, N.: Disorder in fractional quantum Hall states and the gap at $\nu = 5/2$. Phys. Rev. B **68**, 113309 (2003)

MD07.   Missel, A.R., Dahmen, K.A.: Hopping conduction and bacteria: transport in disordered reaction–diffusion systems. Phys. Rev. Let. **100**, 058301 (2007)

MDC85.   Martinis, J.M., Devoret, M.H., Clarke, J.: Energy-level quantization in the zero-voltage state of a current-biased Josephson junction. Phys. Rev. Lett. **55**, 1543–1546 (1985)

MG77.   Mackey, M.C., Glass, L.: Oscillation and chaos in physiological control systems. Science **197**, 287–289 (1977)

MGH94.   Melinger, J.S., Gandhi, S.R., Hariharan, A., Goswami, D., Warren, W.S.: Adiabatic population transfer with frequency-swept laser pulses. J. Chem. Phys. **101**, 6439 (1994)

MGO85.   McDonald, S.W., Grebogi, C., Ott, E., Yorke, J.A.: Fractal basin boundaries. Physica D **17**, 125–153 (1985)

MJS07.   Maass, W., Joshi, P., Sontag, E.D.: Computational aspects of feedback in neural circuits. PLoS Comput. Biol. **3**(1), e165 (2007)

MK05.   Moon, S.J., Kevrekidis, I.G.: An equation-free approach to coupled oscillator dynamics: the Kuramoto model example. Int. J. Bifurc. Chaos **16**(7), 2043–2052 (2006)

ML81.   Morris, C., Lecar, H.: Voltage oscillations in the barnacle giant muscle fiber. Biophys. J. **35**, 193–213 (1981)

MM03.   Mackenzie, A.P., Maeno, Y.: The superconductivity of Sr_2RuO_4 and the physics of spin-triplet pairing. Rev. Mod. Phys. **75**(2), 657 (2003)

MM97.   Matinyan, S.G., Müller, B.: Quantum fluctuations and dynamical chaos. Phys. Rev. Lett. **78**, 2515–2518 (1997)

MMT06.   Martel, Y., Merle, F., Tsai, T.P.: Stability in H^1 for the sum of K solitary waves to some nonlinear Schrödinger equations. Duke Math. J. **133**, 405–466 (2006)

MN95a.   Mavromatos, N.E., Nanopoulos, D.V.: A non-critical string (Liouville) approach to brain microtubules: state vector reduction and memory coding, capacity. ACT-19/95, CTP-TAMU-55/95, OUTP-95-52P (1995)

MN95b.   Mavromatos, N.E., Nanopoulos, D.V.: Non-critical string theory formulation of microtubule dynamics and quantum aspects of brain function. ENSLAPP-A-524/95 (1995)

MN98.   Moore, C., Nilsson, M.: Parallel quantum computation and quantum codes, arXiv:quant-ph/9808027 (1998)

MNL03.   Motter, A.E., Nishikawa, T., Lai, Y.-C.: Large-scale structural organization of social networks. Phys. Rev. E **68**, 036105 (2003)

MO79.   Marshall, A.W., Olkin, I.: Inequalities: Theory of Majorisation and Its Applications. Academic Press, New York (1979)

MP43.   McCulloch, W., Pitts, W.: A logical calculus of the ideas imminent in the nervous activity. Bull. Math. Biophys. **5**, 115–133 (1943)

MP47.   McCulloch, W.S., Pitts, W.: How we know universals. Bull. Math. Biophys. 127–147 (1947)

MP69.   Minsky, M., Papert, S.: Perceptrons. MIT Press, Cambridge (1969)

MR03.   Merle, F., Raphael, P.: Sharp upper bound on the blow-up rate for the critical nonlinear Schrödinger equation. Geom. Funct. Anal. **13**(3), 591–642 (2003)

MR04.   Merle, F., Raphael, P.: On universality of blow-up profile for L^2 critical nonlinear Schrödinger equation. Invent. Math. **156**(3), 565–672 (2004)

MR05.   Merle, F., Raphael, P.: The blow-up dynamic and upper bound on the blow-up rate for critical nonlinear Schrödinger equation. Ann. Math. (2) **161**(1), 157–222 (2005)

MR91.   Moore, G., Read, N.: Non-Abelians in the fractional quantum Hall effect. Nucl. Phys. B **360**, 362 (1991)

MRJ06.   Masgrau, L., Roujeinikova, A., Johannissen, L.O., et al.: Atomic description of an enzyme reaction dominated by proton tunneling. Science **312**, 237–241 (2006)

MS03.   Murthy, G., Shankar, R.: Hamiltonian theories of the fractional quantum Hall effect. Rev. Mod. Phys. **75**, 1101 (2003)

MS08.   Moller, G., Simon, S.H.: Paired composite fermion wavefunctions. Phys. Rev. B **77**, 075319 (2008)

MS88.   Moore, G., Seiberg, N.: Polynomial equations for rational conformal field theories. Phys. Lett. B **212**, 451 (1988)

MS89.   Moore, G., Seiberg, N.: Classical and quantum conformal field theory. Commun. Math. Phys. **123**, 177 (1989)

MS95a.   Mainen, Z.F., Sejnowsky, T.J.: Reliability of spike timing in neocortical neurons. Science **268**, 1503 (1995)

MS95b.   Müllers, J., Schmid, A.: Resonances in the current-voltage characteristics of a dissipative Josephson junction. arXiv:cond-mat/9508035 (1995)

MSM05.   Michalewicz, Z., Schmidt, M., Michalewicz, M., Chiriac, C.: A decision-support system based on computational intelligence: a case study. IEEE Intel. Syst. **20**(4), 44–49 (2005)

MTW04.   Markram, H., Toledo-Rodriguez, M., Wang, Y., Gupta, A., Silberberg, G., Wu, C.: Interneurons of the neocortical inhibitory system. Nat. Rev. Neurosci. **5**, 793–807 (2004)

MTW73.   Misner, C.W., Thorne, K.S., Wheeler, J.A.: Gravitation. Freeman, San Francisco (1973)

MWH01.   Michel, A.N., Wang, K., Hu, B.: Qualitative Theory of Dynamical Systems, 2nd edn. Dekker, New York (2001)

Mac06.   Mackenzie, D.: Perelman declines Math's top prize; three others honored in Madrid. Science **313**, 1027 (2006)

Mac71.   MacLane, S.: Categories for the Working Mathematician. Springer, New York (1971)

Mal798.   Malthus, T.R.: An Essay on the Principle of Population. Originally published in 1798. Penguin, Baltimore (1970)

Mal81.   Von der Mahlsburg, C.: The correlation theory of brain function. MPI Biophysical Chemistry, Internal Report, 81-82 (1981)

Mal85.   Von der Mahlsburg, C.: Nervous structures with dynamical links. Ber. Bunsenges. Phys. Chem. **89**, 703–710 (1985)

Mal88.   Von der Mahlsburg, C.: Pattern recognition by labelled graph matching. Neural Netw. **7**, 1019–1030 (1988)

Ma96.   Von der Mahlsburg, C.: The binding problem of neural networks. In: Llinas, R., Churchland, P.S. (eds.) The Mind-Brain Continuum. MIT Press, Cambridge (1996)

Mal89.   Mallat, S.: A theory for multiresolution signal decomposition: the wavelet representation. IEEE Trans. Pattern Anal. Mach. Intell. **11**, 674–693 (1989)

Mal99.   Mallat, S.: A Wavelet Tour of Signal Processing. Academic Press, New York (1999)

Man07.   Manousakis, E.: Quantum theory, consciousness and temporal perception: binocular rivalry. arXiv:0709.4516 (2007)

Man80.   Manin, Y.I.: Computable and Uncomputable, vol. 21. Sovetskoye Radio, Moscow (1980)

Man80a.   Mandelbrot, B.: Fractal aspects of the iteration of $z \mapsto \lambda z(1 - z)$ for complex λ, z. Ann. N.Y. Acad. Sci. **357**, 249–259 (1980)

Man80b.   Mandelbrot, B.: The Fractal Geometry of Nature. Freeman, New York (1980)

Mar98.   Marieb, E.N.: Human Anatomy and Physiology, 4th edn. Benjamin/Cummings, Menlo Park (1998)

Mar99.   Marsden, J.E.: Elementary Theory of Dynamical Systems. Lecture Notes. CDS, Caltech (1999)

Mas04.   Mashour, G.A.: The cognitive binding problem: from Kant to quantum neurodynamics. Neuroquantology **1**, 29–38 (2004)

MP08.   Masoliver, J., Perello, J.: The escape problem under stochastic volatility: the Heston model. Phys. Rev. E **78**, 056104 (2008)

Mat08.   Matlin, M.W.: Cognition, 7th edn. Wiley, New York (2008)

Mat69.   Matthews, P.B.C.: Evidence that the secondary as well as the primary endings of the muscle spindles may be responsible for the tonic stretch-reflex of the decerebrate eat. J. Physiol. Lond. **204**, 365–393 (1969)

Mat72.   Matthews, P.B.C.: Mammalian Muscle Receptors and Their Central Action. Williams & Wilkins, Baltimore (1972)

Mat98.   Mato, G.: Stochastic resonance in neural systems: effect of temporal correlation in the spike trains. Phys. Rev. E **58**, 876–880 (1998)

Mau07.   Mauroy, B.: Following red blood cells in a pulmonary capillary. arXiv: 0710.5399 [bio-ph] (2007)

May73.   May, R.M. (ed.): Stability and Complexity in Model Ecosystems. Princeton University Press, Princeton (1973)

May76a.   May, R.M. (ed.): Theoretical Ecology: Principles and Applications. Blackwell Sci., Oxford (1976)

May76b. May, R.: Simple mathematical models with very complicated dynamics. Nature **261**(5560), 459–467 (1976)

May79.   Maybeck, P.S.: Stochastic Models, Estimation, and Control, vol. 1. Academic Press, New York (1979)

May92.   Mayer, R.E.: Thinking, Problem Solving and Cognition, 2nd edn. Freeman, New York (1992)

May97.   Mayers, D.: Unconditionally secure quantum bit commitment is impossible. Phys. Rev. Lett. **78**, 3414–3417 (1997)

Men98.   Mendes, R.V.: Conditional exponents, entropies and a measure of dynamical self-organization. Phys. Lett. A **248**, 167–1973 (1998)

Mer01.   Merle, F.: Existence of blow-up solutions in the energy space for the critical generalized KdV equation. J. Am. Math. Soc. **14**, 555–578 (2001)

Mer93a. Mermin, D.: Hidden variables and the two theorems of John Bell. Rev. Mod. Phys. **65**, 803 (1993)

Mer93b. Merle, F.: Determination of blow-up solutions with minimal mass for nonlinear Schrödinger equation with critical power. Duke Math. J. **69**, 427–453 (1993)

Mes00.   Messiah, A.: Quantum Mechanics. Dover, New York (2000)

Met97.   Metzger, M.A.: Applications of nonlinear dynamical systems theory in developmental psychology: motor and cognitive development. Nonlinear Dyn. Psychol. Life Sci. **1**, 55–68 (1997)

Mey86.   Meyer, Y.: Ondelettes et Opérateurs. Hermann, Paris (1990)

Mic06.   Michalewicz, Z.: Adaptive business intelligence. Talk presented at DSTO-Adelaide (2006)

Mic99.   Michalewicz, Z.: Genetic Algorithms + Data Structures = Evolution Programs. Springer, Berlin (1999)

Mil03.   Milnor, J.: Towards the Poincaré conjecture and the classification of 3-manifolds. Not. Am. Math. Soc. **50**(10), 1226–1233 (2003)

Mil56.   Miller, G.A.: The magical number seven, plus or minus two: some limits on our capacity for processing information. Psychol. Rev. **63**, 81–97 (1956)

Mil99.   Milnor, J.: Periodic orbits, externals rays and the Mandelbrot set. Stony Brook IMS Preprint # 1999/3 (1999)

Mit96.   Mitchell, M.: An Introduction to Genetic Algorithms. MIT Press, Cambridge (1996)

Miu68.   Miura, R.: Korteweg–de Vries equation and generalizations, I: a remarkable explicit nonlinear transformation. J. Math. Phys. **9**, 1202–1204 (1968)

Miu76.   Miura, R.: The Korteweg–de Vries equation: a survey of results. SIAM Rev. **18**, 412–459 (1976)

Miz04.   Mizumachi, T.: Asymptotic stability of solitary wave solutions to the regularized long-wave equation. J. Differ. Equ. **200**(2), 312–341 (2004)

Mol97.   Molavi, D.W.: Neuroscience Tutorial. School of Medicine, Washington University, Washington (1997)

Moo89.   Moore, W.: Schrödinger: Life and Thought. Cambridge University Press, Cambridge (1989)

Moo99.   Moore, C.: Quantum circuits: fanout, parity, and counting. arXiv:quant-ph/9903046 (1999)

Mor69.   Morrison, N.: Introduction to Sequential Smoothing and Prediction. McGraw-Hill, New York (1969)

Mor98.   Morf, R.H.: Transition from quantum Hall to compressible states in the second Landau level: new light on the $\nu = 5/2$ enigma. Phys. Rev. Lett. **80**, 1505 (1998)

Mos73.   Moser, J.: Stable and Random Motions in Dynamical Systems. Princeton University Press, Princeton (1973)

Mos96.   Mosekilde, E.: Topics in Nonlinear Dynamics: Application to Physics, Biology and Economics. World Scientific, Singapore (1996)

Mou95.   Mould, R.: The inside observer in quantum mechanics. Found. Phys. **25**(11), 1621–1629 (1995)

Mou98.   Mould, R.: Consciousness and quantum mechanics. Found. Phys. **28**(11), 1703–1718 (1998)

Mou99.   Mould, R.: Quantum consciousness. Found. Phys. **29**(12), 1951–1961 (1999)

Mur02.   Murray, J.D.: Mathematical Biology, vol. I: An Introduction, 3rd edn. Springer, New York (2002)

Mye91.   Myerson, R.B.: Game Theory: An Analysis of Conflict. MIT Press, Cambridge (1991)

NAY60.   Nagumo, J., Arimoto, S., Yoshizawa, S.: An active pulse transmission line simulating 1214-nerve axons. Proc. IRL **50**, 2061–2070 (1960)

NC00.   Nielsen, M.A., Chuang, I.L.: Quantum Computation and Quantum Information. Cambridge University Press, Cambridge (2000)

NF91.   Nitta, T., Furuya, T.: A complex back-propagation learning. Trans. Inf. Proc. Soc. Jpn. **32**(10), 1319–1329 (1991)

NH76.   Nichols, T.R., Houk, J.C.: The improvement in linearity and the regulation of stiffness that results from the actions of the stretch-reflex. J. Neurophysiol. **39**, 119–142 (1976)

NLG00.   Newrock, R.S., Lobb, C.J., Geigenmüller, U., Octavio, M.: Solid State Physics, vol. 54. Academic Press, San Diego (2000)

NMG99.   Nozaki, D., Mar, D.J., Grigg, P., Collins, J.J.: Effects of colored noise on stochastic resonance in sensory neurons. Phys. Rev. Lett. **82**, 2402–2405 (1999)

NML02.   Nishikawa, T., Motter, A.E., Lai, Y.-C., Hoppensteadt, F.C.: Smallest small-world network. Phys. Rev. E **66**, 046139 (2002)

NML03.   Nishikawa, T., Motter, A.E., Lai, Y.C., Hoppensteadt, F.C.: Heterogeneity in oscillator networks: are smaller worlds easier to synchronize? Phys. Rev. Lett. **91**, 014101 (2003)

NNM00.   Nagao, N., Nishimura, H., Matsui, N.: A neural chaos model of multistable perception. Neural Process. Lett. **12**(3), 267–276 (2000)

NNM07.   Newman-Norlund, R.D., Noordzij, M.L., Meulenbroek, R.G.J., Bekkering, H.: Exploring the brain basis of joint action: co-ordination of actions goals and intentions. Soc. Neurosci. **2**(1), 48–65 (2007)

NNM44.   Von Neumann, J., Morgenstern, O.: Theory of Games and Economic Behavior. Princeton University Press, Princeton (1944)

NOY95.   Nusse, H.E., Ott, E., Yorke, J.A.: Saddle-node bifurcations on fractal basin boundaries. Phys. Rev. Lett. **75**(13), 2482 (1995)

NP77.   Nicolis, G., Prigogine, I.: Self-Organization in Nonequilibrium Systems: From Dissipative Structures to Order Through Fluctuations. Wiley, New York (1977)

NPT99.   Nakamura, Y., Pashkin, Yu.A., Tsai, J.S.: Coherent control of macro-
          scopic quantum states in a single-Cooper-pair box. Nature **398**, 786–788
          (1999)
NR02.   Neiman, A.B., Russell, D.F.: Synchronization of noise-induced bursts in
          noncoupled sensory neurons. Phys. Rev. Lett. **88**, 138103 (2002)
NS72.   Newell, A., Simon, H.A.: Human Problem Solving. Prentice Hall, Engle-
          wood Cliffs (1972)
NS83.   Nash, C., Sen, S.: Topology and Geometry for Physicists. Academic
          Press, London (1983)
NS90.   Nijmeijer, H., van der Schaft, A.J.: Nonlinear Dynamical Control Sys-
          tems. Springer, New York (1990)
NS94.   Nobel Seminar: The Work of John Nash in Game Theory, Chaired by
          H.W. Kuhn. Department of Mathematics, Princeton University, Decem-
          ber (1994)
NS98.   Nelson, D.R., Shnerb, N.M.: Non-hermitian localization and population
          biology. Phys. Rev. E **58**, 1383–1403 (1998)
NSS08.   Nayak, C., Simon, S.H., Stern, A., Freedman, M., Sarma, S.D.: Non-
          Abelian anyons and topological quantum computation. Rev. Mod. Phys.
          80, 1083 (2008)
NSS99.   Nunez, P.L., Silberstein, R.B., Shi, Z.P., et al.: EEG coherency, II: ex-
          perimental comparisons of multiple measures. Clin. Neurophysiol. **110**,
          469–486 (1999)
NSW97.   Nunez, P.L., Srinivasan, R., Westdorp, A.F., et al.: EEG coherency,
          I: statistics, reference electrode, volume conduction, Laplacians, corti-
          cal imaging, and interpretation at multiple scales. Electroencephalogr.
          Clin. Neurophysiol. **103**, 499–515 (1997)
NY92.   Nusse, H.E., Yorke, J.A.: The equality of fractal dimension and uncer-
          tainty dimension for certain dynamical systems. Commun. Math. Phys.
          150, 1 (1992)
Nai05.   Nair, V.P.: Quantum Field Theory. A Modern Perspective. Springer, New
          York (2005)
Nan95.   Nanopoulos, D.V.: Theory of brain function, quantum mechanics and
          superstrings. CERN-TH/95-128 (1995)
Nas50a.   Nash, J.F. Jr.: Non-cooperative games. PhD thesis, Math. Dep. Prince-
          ton Univ., Princeton (1950)
Nas50b.   Nash, J.F. Jr.: Equilibrium points in n-person games. Proc. Natl. Acad.
          Sci. **36**, 48–49 (1950)
Nas50c.   Nash, J.F. Jr.: The bargaining problem. Econometrica **18**, 155–162
          (1950)
Nas51.   Nash, J.F. Jr.: Non-cooperative games. Ann. Math. **54**, 286–295 (1951)
Nas53.   Nash, J.F. Jr.: Two-person cooperative games. Econometrica **21**, 128–
          140 (1953)
Nas97.   Nash, J.F. Jr.: Essays on Game Theory. Edward Elgar, Cheltenham Glos
          (1997)
Nay73.   Nayfeh, A.H.: Perturbation Methods. Wiley, New York (1973)
Neu58.   Von Neumann, J.: The Computer and the Brain. Yale University Press,
          Yale (1958)
New00.   Newman, M.E.J.: Models of the small world. J. Stat. Phys. **101**, 819
          (2000)

New74.   Newell, A.C. (ed.): Nonlinear Wave Motion. American Mathematical Society, Providence (1974)

Nic86.   Nicolis, J.S.: Dynamics of Hierarchical Systems: An Evolutionary Approach. Synergetics. Springer, Berlin (1986)

Nik95.   Nikitin, I.N.: Quantum string theory in the space of states in an indefinite metric. Theor. Math. Phys. **107**(2), 589–601 (1995)

Nit00.   Nitta, T.: An analysis on fundamental structure of complex-valued neuron. Neural Process. Lett. **12**(3), 239–246 (2000)

Nit04.   Nitta, T.: Reducibility of the complex-valued neural network. Neural Inf. Process. **2**(3), 53–56 (2004)

Nit97.   Nitta, T.: An extension of the back-propagation algorithm to complex numbers. Neural Netw. **10**(8), 1392–1415 (1997)

Nor99.   Normile, D.: Complex systems: building working cells 'in silico'. Science **284**, 80 (1999)

Nov80.   Novoksenov, V.J.: Asymptotic behavior as $t \to \infty$ of the solution to the Cauchy problem for a nonlinear Schrödinger equation. Dokl. Akad. Nauk SSSR **251**, 799–802 (1980) (in Russian)

Nun00.   Nunez, P.L.: Toward a quantitative description of largescale neocortical dynamic function. EEG. Behav. Brain Sci. **23**, 371–437 (2000)

Nun81.   Nunez, P.L.: Electric Fields of the Brain: The Neurophysics of EEG. Oxford University Press, New York (1981)

OCD04.   Oswald, A.M., Chacron, M.J., Doiron, B., Bastian, J., Maler, L.: Parallel processing of sensory input by bursts, isolated spikes. J. Neurosci. **24**(18), 4351–4362 (2004)

OGY90.   Ott, E., Grebogi, C., Yorke, J.A.: Controlling chaos. Phys. Rev. Lett. **64**, 1196–1199 (1990)

OR94.   Osborne, M., Rubinstein, A.: A Course in Game Theory. MIT Press, Cambridge (1994)

OS94.   Ovchinnikov, Yu.N., Schmid, A.: Resonance phenomena in the current-voltage characteristic of a Josephson junction. Phys. Rev. B **50**, 6332–6339 (1994)

OSB02.   Ott, E., So, P., Barreto, E., Antonsen, T.: The onset of synchronization in systems of globally coupled chaotic and periodic oscillators. Physica D **173**(1–2), 29–51 (2002)

OT91.   Ogawa, T., Tsutsumi, Y.: Blow-up of H^1 solution for the nonlinear Schrödinger equation. J. Differ. Equ. **92**(2), 317–330 (1991)

Oja82.   Oja, E.: A simplified neuron modeled as a principal component analyzer. J. Math. Biol. **15**, 267–273 (1982)

Oja98.   Oja, E.: In: Arbib, M.A. (ed.) The Handbook of Brain Theory and Neural Networks, p. 753. MIT Press, Cambridge (1998)

Ora88.   Oran, B.R.: The Fast Fourier Transform and Its Applications. Prentice Hall, Englewood Cliffs (1988)

Osb59.   Osborne, M.F.M.: Brownian motion in the stock market. Oper. Res. **7**, 145–173 (1959)

Ose68.   Oseledets, V.I.: A multiplicative ergodic theorem: characteristic Lyapunov exponents of dynamical systems. Trans. Mosc. Math. Soc., vol. 19, 197–231 (1968)

Ott89.   Ottino, J.M.: The Kinematics of Mixing: Stretching, Chaos and Transport. Cambridge University Press, Cambridge (1989)

Ott93.   Ott, E.: Chaos in Dynamical Systems. Cambridge University Press, Cambridge (1993)

PBL05.   Purwins, H.G., Bodeker, H.U., Liehr, A.W.: Dissipative solitons in reaction–diffusion systems. In: Akhmediev, N., Ankiewicz, A. (eds.) Dissipative Solitons. Lecture Notes in Physics. Springer, Berlin (2005)

PC05.   Park, J., Chung, W.-K.: Geometric integration on Euclidean group with application to articulated multibody systems. IEEE Trans. Robot. **21**(5), 850–863 (2005)

PC90.   Pecora, L.M., Carroll, T.L.: Synchronization in chaotic systems. Phys. Rev. Lett. **64**, 821–824 (1990)

PC91.   Pecora, L.M., Carroll, T.L.: Driving systems with chaotic signals. Phys. Rev. A **44**, 2374–2383 (1991)

PC98.   Pecora, L.M., Carroll, T.L.: Master stability functions for synchronized coupled systems. Phys. Rev. Lett. **80**, 2109–2112 (1998)

PCF80.   Packard, N.H., Crutchfield, J.P., Farmer, J.D., Shaw, R.S.: Geometry from a time series. Phys. Rev. Lett. **45**, 712–716 (1980)

PD92.   Pritchard, W.S., Duke, D.W.: Measuring chaos in the brain. Int. J. Neurosci. **67**, 31 (1992)

PD95.   Pritchard, W.S., Duke, D.W.: Measuring 'chaos' in the brain: a tutorial review of EEG dimension estimation. Brain Cogn. **27**, 353–397 (1995)

PDR88.   Peirce, A.P., Dahleh, M.A., Rabitz, H.: Optimal control of quantum-mechanical systems: existence, numerical approximation, and applications. Phys. Rev. A **37**, 4950 (1988)

PEL00.   Principe, J., Euliano, N., Lefebvre, C.: Neural and Adaptive Systems: Fundamentals Through Simulations. Wiley, New York (2000)

PG97.   Pakkenberg, B., Gundersen, H.J.: Neocortical neuron number in humans: effect of sex and age. J. Comput. Neurol. **384**(2), 312–320 (1997)

PGY06.   Politi, A., Ginelli, F., Yanchuk, S., Maistrenko, Y.: From synchronization to Lyapunov exponents, back. arXiv:nlin.CD/0605012 (2006)

PHS95.   Peng, C.-K., Havlin, S., Stanley, H.E., Goldberger, A.L.: Complexity measures for the analysis of heart rate variability. Chaos **5**, 82–87 (1995)

PHV86.   Posch, H.A., Hoover, W.G., Vesely, F.J.: Canonical dynamics of the Nosé oscillator: stability, order and chaos. Phys. Rev. A **33**(6), 4253–4265 (1986)

PJ01.   Pauli, W., Jung, C.G.: Atom and archetype. In: Meier, C.A. (ed.) The Pauli/Jung Letters, 1932–1958. Princeton University Press, Princeton (2001)

PJ55.   Pauli, W., Jung, C.G.: The Interpretation of Nature and the Psyche. Random House, New York (1955)

PK97.   Pikovsky, A., Kurth, J.: Coherence resonance in a noise-driven excitable systems. Phys. Rev. Lett. **78**, 775–778 (1997)

PL80.   Pellionisz, A., Llinas, R.: Tensorial approach to the geometry of brain function: cerebellar coordination via a metric tensor. Neuroscience **5**, 1125–1136 (1980)

PL82.   Pellionisz, A., Llinas, R.: Space-time representation in the brain. The cerebellum as a predictive space-time metric tensor. Neuroscience **7**(12), 2949–2970 (1982)

PL85.   Pellionisz, A., Llinas, R.: Tensor network theory of the meta-organization of functional geometries in the central nervous system. Neuroscience **16**(2), 245–273 (1985)

PM80.   Pomeau, Y., Manneville, P.: Intermittent transition to turbulence in
             dissipative dynamical systems. Commun. Math. Phys. **74**(2), 189–197
             (1980)

PMH93.   Peng, C.-K., Mietus, J., Hausdorff, J.M., et al.: Long-range anticorrela-
             tions and non-Gaussian behavior of the heartbeat. Phys. Rev. Lett. **70**,
             1343–1346 (1993)

PMW96.   Pei, X., Moss, F., Wilkens, L.A.: Light enhances hydrodynamic signaling
             in the caudal photoreceptor interneuron of the crayfish. J. Neurophysiol.
             76, 3002 (1996)

POR97.   Pikovsky, A., Osipov, G., Rosenblum, M., Zaks, M., Kurths, J.:
             Attractor–repeller collision and eyelet intermittency at the transition to
             phase synchronization. Phys. Rev. Lett. **79**, 47–50 (1997)

PRK01.   Pikovsky, A., Rosenblum, M., Kurths, J.: Synchronization: A Universal
             Concept in Nonlinear Sciences. Cambridge University Press, Cambridge
             (2001)

PPM00.   Perello, J., Porra, J.M., Montero, M., Masoliver, J.: Black-Scholes option
             pricing within Ito and Stratonovich conventions. Physica A **278**(1–2),
             260–274 (2000)

PSM08.   Perello, J., Sircar, R., Masoliver, J.: Option pricing under stochastic
             volatility: the exponential Ornstein–Uhlenbeck model. J. Stat. Mech.
             P06010 (2008)

PS01.   Van Putten, A.M., Stam, C.J.: Application of a neural complexity mea-
             sure to multi-channel EEG. Phys. Lett. A **281**, 131 (2001)

PS62.   Perring, J.K., Skyrme, T.R.H.: A model unified field equation. Nucl.
             Phys. **31**, 550–555 (1962)

PS95.   Peskin, M.E., Schroeder, D.V.: An Introduction to Quantum Field The-
             ory. Addison Wesley, Reading (1995)

PTB86.   Pokrovsky, V.L., Talapov, A.L., Bak, P.: In: Trullinger, Zakharov,
             Pokrovsky (eds.) Solitons, pp. 71–127. Elsevier Science, Amsterdam
             (1986)

PV03.   Pessa, E., Vitiello, G.: Quantum noise, entanglement and chaos in the
             quantum field theory of mind/brain states. Mind Matter **1**, 59–79 (2003)

PV04.   Pessa, E., Vitiello, G.: Quantum noise induced entanglement and chaos
             in the dissipative quantum model of brain. Int. J. Mod. Phys. B **18**,
             841–858 (2004)

PV98.   Plenio, M.B., Vedral, V.: Entanglement in quantum information theory.
             Contemp. Phys. **39**, 431 (1998)

PV99.   Pessa, E., Vitiello, G.: Quantum dissipation and neural net dynamics.
             Biolectrochem. Bioenergy **48**, 339–342 (1999)

PW94.   Pego, R., Weinstein, M.: Asymptotic stability of solitary waves. Com-
             mun. Math. Phys. **164**(2), 305–349 (1994)

PW98.   Plenio, M.B., Vedral, V.: Teleportation, entanglement and thermody-
             namics in the quantum world. Contemp. Phys. **39**, 431–446 (1998)

PWM96.   Pei, X., Wilkens, L., Moss, F.: Noise-mediated spike timing precision
             from aperiodic stimuli in an array of Hodgekin–Huxley-type neurons.
             Phys. Rev. Lett. **77**, 4679–4682 (1996)

PXS99.   Pan, W., Xia, J.-S., Shvarts, V., et al.: Exact quantization of the even-
             denominator fractional quantum Hall state at $\nu = 5/2$ Landau level
             filling factor. Phys. Rev. Lett. **83**, 3530 (1999)

PZD00.   Pouget, A., Zemel, R.S., Dayan, P.: Processing with population codes. Nat. Neurosci. **1**, 125–132 (2000)

PZR97.   Pikovsky, A., Zaks, M., Rosenblum, M., Osipov, G., Kurths, J.: Phase synchronization of chaotic oscillations in terms of periodic orbits. Chaos **7**, 680 (1997)

Pal59.   Palais, R.S.: Natural operations on differential forms. Trans. Am. Math. Soc. **92**, 125–141 (1959)

Pal97.   Palais, R.S.: The symmetries of solitons. Bull. Am. Math. Soc. **34**, 339–403 (1997)

Par02.   Partovi, M.H.: Hamilton–Jacobi formulation of Kolmogorov–Sinai entropy for classical and quantum dynamics. Phys. Rev. Lett. **89**, 144101 (2002)

Pen79.   Penrose, R.: Singularities and time-asymmetry. In: Hawking, S., Israel, W. (eds.) General Relativity: An Einstein Centenary Survey, pp. 581–638. Cambridge University Press, Cambridge (1979)

Pen89.   Penrose, R.: The Emperor's New Mind. Oxford University Press, Oxford (1989)

Pen94.   Penrose, R.: Shadows of the Mind. Oxford University Press, Oxford (1994)

Pen97.   Penrose, R.: The Large, the Small and the Human Mind. Cambridge University Press, Cambridge (1997)

Pen98.   Penrose, R.: Quantum computation, entanglement and state reduction. Philos. Trans. R. Soc. Lond. **356**, 1927–1939 (1998)

Per97a.   Perelman, G.: Some results on the scattering of weakly interacting solitons for nonlinear Schrödinger equations. In: Spectral Theory, Microlocal Analysis, and Singular Manifolds, pp. 78–137. Akademie Verlag, Berlin (1997)

Per02.   Perelman, G.: The entropy formula for the Ricci flow and its geometric applications. arXiv:math.DG/0211159 (2002)

Per03a.   Pereira, A.: The quantum mind/classical brain problem. Neuroquantology **1**, 94–118 (2003)

Per03b.   Perelman, G.: Ricci flow with surgery on three-manifolds. arXiv:math. DG/0303109 (2003)

Per04.   Perelman, G.S.: Asymptotic stability of multi-soliton solutions for nonlinear Schrödinger equations. Commun. Partial Differ. Equ. **29**, 1051–1095 (2004)

Per97b.   Pert, C.B.: Molecules of Emotion. Scribner, New York (1997)

Pes08.   Pessoa, L.: Cognition and Emotion. Scholarpedia, Cog. Neurosci. (2008)

Pes76.   Pesin, Ya.B.: Invariant manifold families which correspond to non-vanishing characteristic exponents. Izv. Akad. Nauk SSSR Ser. Mat. **40**(6), 1332–1379 (1976)

Pes77.   Pesin, Ya.B.: Lyapunov characteristic exponents, smooth ergodic theory. Russ. Math. Surv. **32**(4), 55–114 (1977)

Pet02a.   Peterka, R.J.: Sensorimotor integration in human postural control. J. Neurophysiol. **88**, 1097–1118 (2002)

Pet02b.   Petras, I.: Control of fractional-order Chua's system. J. Electr. Eng. **53**(7–8), 219–222 (2002)

Pet07.   Pettini, M.: Geometry and Topology in Hamiltonian Dynamics and Statistical Mechanics. Springer, New York (2007)

Pic86.   Pickover, C.A.: Computer displays of biological forms generated from mathematical feedback loops. Comput. Graph. Forum **5**, 313 (1986)

Pic87.   Pickover, C.A.: Mathematics and beauty: time-discrete phase planes associated with the cyclic system. Comput. Graph. Forum **11**, 217 (1987)

Pink97.   Pinker, S.: How the Mind Works. Norton, New York (1997)

Pop00.   Pope, S.B.: Turbulent Flows. Cambridge University Press, Cambridge (2000)

Pop08.   Popp, F.-A.: Consciousness as evolutionary process based on coherent states. Neuquantology **6**(4), 431–439 (2008)

Pou92.   Poundstone, W.: Prisoners' Dilemma. John von Neumann, Game Theory, and the Puzzle of the Bomb. Doubleday, New York (1992)

Pre03.   Preziosi, L.: Cancer Modeling and Simulation. CRC Press, Boca Raton (2003)

Pre97.   Preskill, J.: Fault-tolerant quantum computation. arXiv:quant-ph/9712048 (1997)

Pre98a.   Preskill, J.: Quantum Information and Computation. Lecture Notes for Physics, vol. 229. CalTech, Pasadena (1998)

Pre98b.   Preskill, J.: Physics 229. http://www.theory.caltech.edu/people/preskill/ph229 (1998)

Pri71.   Pribram, K.H.: Languages of the Brain. Prentice Hall, Englewood Cliffs (1971)

Pri80.   Prigogine, I.: From Being to Becoming: Time and Complexity in the Physical Sciences. Freeman, San Francisco (1980)

Pri91.   Pribram, K.H.: Brain and Perception. Lawrence Erlbaum, Hillsdale (1991)

Pul05.   Pulvermüller, F.: Brain mechanicsms, kinking language and action. Nat. Rev. Neurosci. **6**, 576–582 (2005)

Pyr92.   Pyragas, K.: Continuous control of chaos, by self-controlling feedback. Phys. Lett. A **170**, 421–428 (1992)

Pyr95.   Pyragas, K.: Control of chaos via extended delay feedback. Phys. Lett. A **206**, 323–330 (1995)

Qui00.   Quiroga, R.Q.: Obtaining single trial EPs with wavelet denoising. Physica D **145**, 278–292 (2000)

RBT01.   Roe, R.M., Busemeyer, J.R., Townsend, J.T.: Multi-alternative decision field theory: a dynamic connectionist model of decision making. Psychol. Rev. **108**, 370–392 (2001)

RBY01.   Rosso, O.A., Blanco, S., Yordanova, J., et al.: Wavelet entropy: a new tool for analysis of short-duration brain electrical signals. J. Neurosci. Methods **105**, 65 (2001)

RCL93.   Rosenstein, M.T., Collins, J.J., Luca, C.J.: A practical method for calculating largest Lyapunov exponents from small data sets. Physica D **65**, 117–134 (1993)

RCM07.   Roose, T., Chapman, S.J., Maini, P.K.: Mathematical models of avascular tumor growth. SIAM Rev. **49**(2), 179–208 (2007)

REW00.   Rabinovich, M.I., Ezersky, A.B., Weidman, P.D.: The Dynamics of Patterns. World Scientific, Singapore (2000)

RG00.   Read, N., Green, D.: Paired states of fermions in two dimensions with breaking of parity and time-reversal symmetries and the fractional quantum Hall effect. Phys. Rev. B **61**, 10267 (2000)

RG98.   Rao, A.S., Georgeff, M.P.: Decision procedures for BDI logics. J. Log.
            Comput. **8**(3), 292–343 (1998)
RGL99.   Rodriguez, E., George, N., Lachaux, J., Martinerie, J., Renault, B.,
            Varela, F.: Long-distance synchronization of human brain activity. Na-
            ture **397**, 430 (1999)
RGS99.   Reinagel, P., Godwin, D., Sherman, S.M., Koch, C.: Encoding of visual
            information by LGN bursts. J. Neurophysiol. **81**, 2558–2569 (1999)
RH00.   Rezayi, E.H., Haldane, F.D.M.: Incompressible paired Hall state, stripe
            order, and the composite fermion liquid phase in half-filled Landau levels.
            Phys. Rev. Lett. **84**, 4685 (2000)
RH85.   Rose, G., Heilingenberg, W.: Temporal hyperacuity in single neurons of
            electric fish. Nature **318**, 178 (1985)
RH89.   Rose, R.M., Hindmarsh, J.L.: The assembly of ionic currents in a thala-
            mic neuron, I: the three-dimensional model. Proc. R. Soc. Lond. B **237**,
            267–288 (1989)
RN03.   Russel, S., Norvig, P.: Artificial Intelligence: A Modern Approach. Pren-
            tice Hall, New Jersey (2003)
ROH98.   Rosa, E., Ott, E., Hess, M.H.: Transition to phase synchronization of
            chaos. Phys. Rev. Lett. **80**, 1642–1645 (1998)
RP98.   Rieffel, E.G., Polak, W.: An introduction to quantum computing for
            non-physicists. arXiv:quant-ph/9809016 (1998)
RPG94.   Rasmussen, J., Pejtersen, A.M., Goodstein, L.P.: Cognitive Systems En-
            gineering. Wiley, New York (1994)
RPK96.   Rosenblum, M., Pikovsky, A., Kurths, J.: Phase synchronization of
            chaotic oscillators. Phys. Rev. Lett. **76**, 1804 (1996)
RPK97.   Rosenblum, M., Pikovsky, A., Kurths, J.: From phase to lag synchro-
            nization in coupled chaotic oscillators. Phys. Rev. Lett. **78**, 4193–4196
            (1997)
RR96.   Ramakrishna, V., Rabitz, H.: Relation between quantum computing and
            quantum controllability. Phys. Rev. A **54**, 1715 (1996)
RR98.   Rezek, I.A., Roberts, S.J.: Stochastic complexity measures for physiolog-
            ical signal analysis. IEEE Trans. Biomed. Eng. **45**, 1186–1191 (1998)
RPD08.   Roman, H.E., Porto, M., Dose, C.: Skewness, long-time memory, and
            non-stationarity: application to leverage effect in financial time series.
            Eur. Phys. Lett. **84**, 28001 (2008)
RR99.   Read, N., Rezayi, E.: Beyond paired quantum Hall states: parafermions
            and incompressible states in the first excited Landau level. Phys. Rev. B
            59, 8084 (1999)
RRC05.   Rezayi, E.H., Read, N., Cooper, N.R.: Incompressible liquid state of
            rapidly rotating bosons at filling factor 3/2. Phys. Rev. Lett. **95**(16),
            160404 (2005)
RSE82.   Van Rotterdam, A., Lopes da Silva, F.H., van den Ende, J., Viergever,
            M.A., Hermans, A.J.: A model of the spatialtemporal characteristics of
            the alpha rhythm. Bull. Math. Biol. **44**(2), 283–305 (1982)
RSS05.   Rodnianski, I., Schlag, W., Soffer, A.D.: Asymptotic stability of N-
            soliton states of NLS. Commun. Pure Appl. Math. **58**(2), 149–216 (2005)
RT01.   Van Rullen, R., Thorpe, S.J.: Rate coding versus temporal order coding:
            what the retinal ganglion cells tell the visual cortex. Neural Comput. **13**,
            1255–1283 (2001)

RT71.   Ruelle, D., Takens, F.: On the nature of turbulence. Commun. Math. Phys. **20**, 167–192 (1971)

RT92.   Reif, J.H., Tate, S.R.: On threshold circuits and polynomial computation. SIAM J. Comput. **21**(5), 896–908 (1992)

RU67.   Ricciardi, L.M., Umezawa, H.: Brain physics and many-body problems. Kibernetik **4**, 44 (1967)

RWS96.   Rieke, F., Warland, D., Steveninck, R., Bialek, W.: Exploring the Neural Code. MIT Press, Cambridge (1996)

RYD96.   Ryu, S., Yu, W., Stroud, D.: Dynamics of an underdamped Josephson-junction ladder. Phys. Rev. E **53**, 2190–2195 (1996)

RZ00.   Rice, S.A., Zhao, M.: Optical Control of Molecular Dynamics. Wiley, New York (2000)

Rab89.   Rabiner, L.R.: A tutorial on hidden Markov models and selected applications in speech recognition. Proc. IEEE **77**(2), 257–286 (1989)

Raj82.   Rajaraman, R.: Solitons, Instantons. North-Holland, Amsterdam (1982)

Ram90.   Ramond, P.: Field Theory: A Modern Primer. Addison-Wesley, Reading (1990)

Rea06.   Read, N.: Wavefunctions and counting formulas for quasiholes of clustered quantum Hall states on a sphere. Phys. Rev. B **73**(24), 245334 (2006)

Ree06.   Reed, S.K.: Cognition: Theory and Applications, 7th edn. Wadsworth, Belmont (2006)

Rei93.   Reif, J.H.: Synthesis of Parallel Algorithms. Morgan Kaufmann, San Mateo (1993)

Rha84.   De Rham, G.: Differentiable Manifolds. Springer, Berlin (1984)

Rin85.   Rinzel, J.: Bursting oscillations in an excitable membrane model. In: Sleeman, B.D., Jarvis, R.J. (eds.) Ordinary, Partial Differential Equations. Proceedings of the 8th Dundee Conference. Lecture Notes in Mathematics, vol. 1151. Springer, Berlin (1985)

Rob95.   Robinson, C.: Dynamical Systems. CRC Press, Boca Raton (1995)

Ros58b.   Rosenblatt, F.: The perceptron: a probabilistic model for information storage and organization in the brain. Physiol. Rev. **65**, 386–408 (1958)

Rot01.   Roth, G.: The Brain and Its Reality. Cognitive Neurobiology and Its Philosophical Consequences. Suhrkamp, Frankfurt (2001)

Rot99.   Rothman, P.: Nonlinear Time Series Analysis of Economic and Financial Data. Springer, Berlin (1999)

Roy08.   Roy, A.: Connectionism, controllers, and a brain theory. IEEE Trans. Syst. Man Cybern. A **38**(6), 1434–1441 (2008)

Rue76.   Ruelle, D.: A measure associated with axiom A attractors. Am. J. Math. **98**, 619–654 (1976)

Rue78a.   Ruelle, D.: An inequality for the entropy of differentiable maps. Bol. Soc. Bras. Mat. **9**, 83–87 (1978)

Rue78b.   Ruelle, D.: Thermodynamic Formalism. Addison-Wesley, Reading (1978)

Rue78c.   Ruelle, D.: Thermodynamic formalism. In: Encyclopaedia of Mathematics and its Applications. Addison–Wesley, Reading (1978)

Rue79.   Ruelle, D.: Ergodic theory of differentiable dynamical systems. Publ. Math. IHES **50**, 27–58 (1979)

Rue89.   Ruelle, D.: The thermodynamical formalism for expanding maps. Commun. Math. Phys. **125**, 239–262 (1989)

Rue98.   Ruelle, D.: Nonequillibrium statistical mechanics near equilibrium: computing higher order terms. Nonlinearity **11**, 5–18 (1998)

Rue99.   Ruelle, D.: Smooth dynamics and new theoretical ideas in nonequilibrium statistical mechanics. J. Stat. Phys. **95**, 393–468 (1999)

Rul01.   Rulkov, N.F.: Regularization of synchronized chaotic bursts. Phys. Rev. Lett. **86**, 183–186 (2001)

Rus844.   Russell, J.S.: Report on waves. In: 14th Meeting of the British Association for the Advancement of Science. BAAS, London (1844)

Rus885.   Russell, J.S.: The Wave of Translation in the Oceans of Water, Air and Ether. Trübner, London (1885)

Ruz81.   Borodin, A.: On relating time and space to size and depth. SIAM J. Comput. **6**(4), 733–744 (1977)

Ryd96.   Ryder, L.: Quantum Field Theory. Cambridge University Press, Cambridge (1996)

S99.   Ardonne, E., Schoutens, K.: New class of non-Abelian spin-singlet quantum Hall states. Phys. Rev. Lett. **82**(25), 5096 (1999)

SA76.   Segur, H., Ablowitz, M.: Asymptotic solutions and conservation laws for the nonlinear Schrödinger equation, I. J. Math. Phys. **17**, 710–713 (1976)

SA98.   Schaal, S., Atkeson, C.G.: Constructive incremental learning from only local information. Neural Comput. **10**, 2047–2084 (1998)

SB01.   Slingerland, J.K., Bais, F.A.: Quantum groups and non-Abelian braiding in quantum Hall systems. Nucl. Phys. B **612**, 229 (2001)

SB03.   Shapiro, M., Brumer, P.: Laser control of product quantum state populations in unimolecular reactions. J. Chem. Phys. **84**, 4103 (1986); Principles of the Quantum Control of Molecular Processes. Wiley, New York (2003)

SB74.   Sayers, B.McA., Beagley, H.A.: Objective evaluation of auditory evoked EEG responses. Nature **251**, 608–609 (1974)

SB98.   Sutton, R.S., Barto, A.G.: Reinforcement Learning: An Introduction. MIT Press, Cambridge (1998)

SBK06.   Sebanz, N., Bekkering, H., Knoblich, G.: Joint action: bodies and minds moving together. Trends Cogn. Sci. **10**(2), 70–76 (2006)

SBR99.   Samar, V.J., Bopardikar, A., Rao, R., Swartz, K.: Brain Lang. **66**, 7 (1999)

SC06.   Stone, M., Chung, S.-B.: Fusion rules and vortices in $p_x + ip_y$ superconductors. Phys. Rev. B **73**(1), 014505 (2006)

SDK53.   Seeger, A., Donth, H., Kochendörfer, A.: Theorie der Versetzungen in eindimensionalen Atomreihen. Z. Phys. **134**, 173–193 (1953)

SE99.   Sartoretto, F., Ermani, M.: Automatic detection of epileptiform activity by single-level wavelet analysis. Clin. Neurophysiol. **110**, 239–249 (1999)

SF95.   Schwengelbeck, U., Faisal, F.H.M.: Definition of Lyapunov exponents and KS-entropy in quantum mechanics. Phys. Lett. A **199**, 281 (1995)

SG95.   Singer, W., Gray, C.M.: Visual feature integration and temporal correlation hypothesis. Annu. Rev. Neurosci. **18**, 555–586 (1995)

SH06.   Stern, A., Halperin, B.I.: Proposed experiments to probe the non-Abelian $\nu = 5/2$ quantum Hall state. Phys. Rev. Lett. **96**(1), 016802 (2006)

SIT95.   Shea, H.R., Itzler, M.A., Tinkham, M.: Inductance effects and dimensionality crossover in hybrid superconducting arrays. Phys. Rev. B **51**, 12690–12697 (1995)

SJD94.   Schiff, S.J., Jerger, K., Duong, D.H., Chang, T., Spano, M.L., Ditto, W.L.: Controlling chaos in the brain. Nature **370**, 615–620 (1994)

SK95.   Schreiber, T., Kaplan, D.T.: Signal separation by nonlinear projections: the fetal electrocardiogram. Phys. Rev. E **53**, R4326–R4329 (1995)

SKG93.   Shidara, M., Kawano, K., Gomi, H., Kawato, M.: Inverse-dynamics model eye movement control by Purkinje cells in the cerebellum. Nature **365**, 50–52 (1993)

SLK99.   Stam, C.J., van der Leij, E.M.H., Keunen, R.W.M., Tavy, D.L.J.: Nonlinear EEG changes in postanoxic encephalopathy. Theory Biosci. **118**, 209 (1999)

SM01.   Stocks, N.G., Mannella, R.: Generic noise-enhanced coding in neuronal arrays. Phys. Rev. E **64**, 30902 (2001)

SM90.   Sugihara, G., May, R.M.: Nonlinear forecasting as a way of distinguishing chaos from measurement error in time series. Nature **344**, 734–741 (1990)

SN94.   Shadlen, M.N., Newsome, W.T.: Noise, neural codes and cortical organization. Curr. Opin. Neurobiol. **4**, 569–579 (1994)

SOM04.   Stern, A., von Oppen, F., Mariani, E.: Geometric phases and quantum entanglement as building blocks for non-Abelian quasiparticle statistics. Phys. Rev. B **70**(20), 205338 (2004)

SOR01.   Steck, D.A., Oskay, W.H., Raizen, M.G.: Observation of chaos–assisted tunneling between islands of stability. Science **293**, 274 (2001)

SPB90.   Stormer, H.L., Pfeiffer, L.N., Baldwin, K.W., West, K.W.: Observation of a Bloch–Grüneisen regime in 2D electron transport. Phys. Rev. B **41**(2), 1278 (1990)

SPS99.   Shimokawa, T., Pakdaman, K., Sato, S.: Time-scale matching in the response of a leaky integrate-and-fire neuron model to periodic stimulus with additive noise. Phys. Rev. E **59**, 3427–3443 (1999)

SRK98.   Schäfer, C., Rosenblum, M.G., Kurths, J., Abel, H.-H.: Heartbeat synchronized with ventilation. Nature **392**, 239–240 (1998)

SRP99.   Shimokawa, T., Rogel, A., Pakdaman, K., Sato, S.: Time-scale matching in the response of a leaky integrate-and-fire neuron model to periodic stimulus with additive noise. Phys. Rev. E **59**, 3461–3443 (1999)

SS00.   Schreiber, T., Schmitz, A.: Surrogate time series. Physica D **142**, 346–382 (2000)

SS01.   Scholkopf, B., Smola, A.: Leaning with Kernels. MIT Press, Cambridge (2001)

SS81.   Su, W.P., Schrieffer, J.R.: Fractionally charged excitations in charge-density-wave systems with commensurability 3. Phys. Rev. Lett. **46**(11), 738 (1981)

SS85.   Shatah, J., Strauss, W.: Instability of nonlinear bound states. Commun. Math. Phys. **100**, 173–190 (1985)

SS99.   Sulem, C., Sulem, P.: The Nonlinear Schrödinger Equation: Self-Focusing and Wave Collapse. Appl. Math. Sci., vol. 139. Springer, New York (1999)

SSH80.   Su, W.P., Schrieffer, J.R., Heeger, A.J.: Soliton excitations in polyacetylene. Phys. Rev. B **22**(4), 2099 (1980)

SSK87.   Sakaguchi, H., Shinomoto, S., Kuramoto, Y.: Local and global self-entrainments in oscillator-lattices. Prog. Theor. Phys. **77**, 1005–1010 (1987)

STU78. Stuart, C.I.J., Takahashi, Y., Umezawa, H.: On the stability, non-local properties of memory. J. Theor. Biol. **71**, 605–618 (1978)

STU79. Stuart, C.I.J., Takahashi, Y., Umezawa, H.: Mixed system brain dynamics: neural memory as a macroscopic ordered state. Found. Phys. **9**, 301 (1979)

STZ93. Satarić, M.V., Tuszynski, J.A., Zakula, R.B.: Kinklike excitations as an energy-transfer mechanism in microtubules. Phys. Rev. E **48**, 589–597 (1993)

SUK01. Su, H., Alroy, G., Kirson, E.D., Yaari, Y.: Extracellular calcium modulates persistent sodium current-dependent burst-firing in hippocampal pyramidal neurons. J. Neurosci. **21**, 4173–4182 (2001)

SVW95. Srivastava, Y.N., Vitiello, G., Widom, A.: Quantum dissipation and quantum noise. Ann. Phys. **238**, 200 (1995)

SW02.   Senhadji, L., Wendling, F.: Epileptic transient detection: wavelets and time-frequency approaches. Neurophysiol. Clin. **32**, 175–192 (2002)

SW90.   Sassetti, M., Weiss, U.: Universality in the dissipative two-state system. Phys. Rev. Lett. **65**, 2262–2265 (1990)

SWV99. Srivastava, Y.N., Widom, A., Vitiello, G.: Quantum measurements, information and entropy production. Int. J. Mod. Phys. B **13**, 3369–3382 (1999)

SYC91. Sauer, T., Yorke, J.A., Casdagli, M.: Embedology. Stat. Phys. **65**, 579 (1991)

SZT98. Satarić, M.V., Zeković, S., Tuszyński, Pokorny, J.: Mössbauer effect as a possible tool in detecting nonlinear excitations in microtubules. Phys. Rev. E **58**, 6333–6339 (1998)

Sam01. Samal, M.K.: Speculations on a unified theory of matter and mind. In: Proc. Int. Conf. Science, Metaphysics: A Discussion on Consciousness, Genetics. NIAS, Bangalore (2001)

Sam89. Davies, P.: The New Physics. Cambridge University Press, Cambridge (1989)

Sam95. Samuels, A.: Jung and the Post-Jungians. Routledge, London (1985)

Sam99. Samal, M.K.: Can science 'explain' consciousness? In: Sreekantan, B.V., et al. (eds.) Proc. Nat. Conf. Scientific, Philosophical Studies on Consciousness. NIAS, Bangalore (1999)

Sch07. Schöner, G.: Dynamical systems approaches to cognition. In: Cambridge Handbook of Computational Cognitive Modeling. Cambridge University Press, Cambridge (2007)

Sch60. Schelling, T.C.: The Strategy of Conflict. Harvard University Press, Cambridge (1960)

Sch68. Schiff, L.I.: Quantum Mechanics. McGraw-Hill, New York (1968)

Sch88. Schuster, H.G. (ed.): Handbook of Chaos Control. Wiley-VCH, New York (1999)

Sch94. Schiff, S.J., et al.: Controlling chaos in the brain. Nature **370**, 615–620 (1994)

Sch95. Schumacher, B.: Quantum coding. Phys. Rev. A **51**, 2738–2747 (1995)

Sch98. Schaal, S.: Robot learning. In: Arbib, M. (ed.) Handbook of Brain Theory and Neural Networks. MIT Press, Cambridge (1998)

Sch99. Schaal, S.: Is imitation learning the route to humanoid robots? Trends Cogn. Sci. **3**, 233–242 (1999)

Sco04.   Scott, A. (ed.): Encyclopedia of Nonlinear Science. Routledge, New York (2004)

Sco99.   Scott, A.C.: Nonlinear Science: Emergence and Dynamics of Coherent Structures. Oxford University Press, Oxford (1999)

Ser99.   Service, R.F.: Complex systems: exploring the systems of life. Science **284**, 80 (1999)

Sha06.   Sharma, S.: An exploratory study of chaos in human-machine system dynamics. IEEE Trans. SMC B. **36**(2), 319–326 (2006)

Sha94.   Shankar, R.: Principles of Quantum Mechanics. Plenum, New York (1994)

She01.   Sherman, S.M.: Tonic, burst firing: dual modes of thalamocortical relay. Trends in Neurosci. **24**, 122–126 (2001)

Sho94.   Shor, P.W.: In: Goldwasser, S. (ed.) Proc. 35th Ann. Symp. Foundations of Computer Science, pp. 124–134. IEEE Comput. Soc. Press, Los Alamitos (1994)

Sho95.   Shor, P.W.: Scheme for reducing decoherence in quantum computer memory. Phys. Rev. A **52**, R2493 (1995)

Sho97.   Shor, P.W.: Polynomial-time algorithms for prime factorization and discrete logarithms on a quantum computer. SIAM J. Comput. **26**, 1484–1509 (1997)

Si89.   Sastri, S.S., Isidori, A.: Adaptive control of linearizable systems. IEEE Trans. Autom. Control **34**(1), 1123–1131 (1989)

Siv77.   Sivashinsky, G.I.: Nonlinear analysis of hydrodynamical instability in laminar flames, I: derivation of basic equations. Acta Astron. **4**, 1177 (1977)

Sl03.   Stevens, B.L., Lewis, F.L.: Aircraft Control and Simulation, 2nd edn. Wiley, Hoboken (2003)

Sm88.   Strogatz, S.H., Mirollo, R.E.: Phase-locking and critical phenomena in lattices of coupled nonlinear oscillators with random intrinsic frequencies. Physica D **31**, 143 (1988)

Sma67.   Smale, S.: Differentiable dynamical systems. Bull. Am. Math. Soc. **73**, 747–817 (1967)

Sma98.   Van der Smagt, P.: Cerebellar control of robot arms. Connect. Sci. **10**, 301–320 (1998)

Smi98.   Smith, R.G.: In: Arbib, M.A. (ed.) The Handbook of Brain Theory and Neural Networks, p. 816. MIT Press, Cambridge (1998)

Sod94.   Soderkvist, J.: Micromachined gyroscopes. Sens. Actuators A **43**, 65–71 (1994)

Sof06.   Soffer, A.: Soliton dynamics and scattering. In: Proc. Internat. Congress of Math., Madrid, vol. III, pp. 459–472 (2006)

Spa82.   Sparrow, C.: The Lorenz Equations: Bifurcations, Chaos and Strange Attractors. Springer, New York (1982)

Spo80.   Spohn, H.: Kinetic equations from Hamiltonian dynamics: Markovian limits. Rev. Mod. Phys. **52**, 569–615 (1980)

Spr93a.   Sprott, J.C.: Automatic generation of strange attractors. Comput. Graph. **17**(3), 325–332 (1993)

Spr93b.   Sprott, J.C.: Strange Attractors: Creating Patterns in Chaos. M&T Books, New York (1993)

Sre07.   Srednicki, M.A.: Quantum Field Theory. Cambridge University Press, Cambridge (2007)

Sta83.   Stapp, H.P.: Exact solution of the infrared problem. Phys. Phys. Rev. D **28**, 1386–1418 (1983)

Sta93.   Stapp, H.P.: Mind, Matter and Quantum Mechanics. Springer, New York (1993)

Sta95.   Stapp, H.P.: Chance, choice and consciousness: the role of mind in the quantum brain. arXiv:quant-ph/9511029 (1995)

Ste36.   Steuerwald, R.: Über Enneper'sche Flächen und Bäcklund'sche Transformation. In: Abhandlungen der Bayerischen Akademie der Wissenschaften München, pp. 1–105 (1936)

Ste92.   Stein, D.L.: Spin Glasses in Biology. World Scientific, Singapore (1992)

Ste93.   Stein, E.M.: Harmonic Analysis. Princeton University Press, Princeton (1993)

Ste96b.   Steane, A.M.: Error correcting codes in quantum theory. Phys. Rev. Lett. **77**(5), 793 (1996)

Ste96c.   Steane, A.M.: Simple quantum error-correcting codes. Phys. Rev. A **54**(6), 4741 (1996)

Ste98.   Steane, A.: Quantum computing. Rep. Prog. Phys. **61**, 117 (1998)

Sto00.   Stocks, N.G.: Suprathreshold stochastic resonance in multilevel threshold systems. Phys. Rev. Lett. **84**, 2310–2313 (2000)

Str00.   Strogatz, S.H.: From Kuramoto to Crawford: exploring the onset of synchronization in populations of coupled oscillators. Physica D **143**, 1–20 (2000)

Str68.   Strang, G.: On the construction and comparison of difference schemes. SIAM J. Numer. Anal. **5**, 506–517 (1968)

Str89.   Strauss, W.: Nonlinear Wave Equations. Regional Conf. Series in Math. (1989)

Str94.   Strogatz, S.: Nonlinear Dynamics and Chaos. Addison-Wesley, Reading (1994)

Str.   Strogatz, S.H.: Exploring complex networks. Nature **410**, 268 (2001)

Str66.   Stratonovich, R.L.: A new representation for stochastic integrals and equations. SIAM J. Control **4**, 362–371 (1966)

Stu99.   Stuart, J.: Calculus, 5th edn. Brooks/Cole Publ., Pacific Grove (2003)

Sut88a.   Sutherland, B.: Systems with resonating-valence-bond ground states: correlations and excitations. Phys. Rev. B **37**, 3786 (1988)

Sut88b.   Sutherland, R.M.: Cell and environment interactions in tumor microregions: the multicell spheroid model. Science **240**, 177–184 (1988)

Swe88.   Swets, J.A.: Visual sustained attention: image degradation produces rapid sensitivity decrement over time. Science **240**, 1285–1293 (1988)

Swi75.   Switzer, R.K.: Algebraic topology—homology and homotopy. In: Classics in Mathematics. Springer, New York (1975)

Sze78.   Szentagothai, J.: The neuron network of the cerebral cortex: a functional interpretation. Proc. R. Soc. Lond. B **201**, 219–248 (1978)

TBG98.   Tittel, W., Brendel, J., Gisin, B., Herzog, T., Zbinden, H., Gisin, N.: Experimental demonstration of quantum correlations over more than 10 km. Phys. Rev. A **57**, 3229 (1998)

TC98.   Torrence, C., Compo, G.P.: A practical guide to wavelet analysis. Bull. Am. Meteorol. Soc. **79**, 61–78 (1998)

TC99.   Teixeira, F.L., Chew, W.C.: Lattice electromagnetic theory from a topological viewpoint. J. Math. Phys. **40**, 169–187 (1999)

TDL07.   Tewari, S., Das Sarma, S., Lee, D.H.: An index theorem for the Majorana zero modes in chiral p-wave superconductors. Phys. Rev. Lett. **99**, 037001 (2007). arXiv:cond-mat/0609556

TDN07.   Tewari, S., Das Sarma, S., Nayak, C., Zhang, C., Zoller, P.: Quantum computation using vortices Majorana zero modes of a $p_x + \mathrm{i}p_y$ superfluid of fermionic cold atoms. Phys. Rev. Lett. **98**(1), 010506 (2007)

TEP07.   Tracy, L.A., Eisenstein, J.P., Pfeiffer, L.N., West, K.W.: Spin transition in the half-filled Landau level. Phys. Rev. Lett. **98**(8), 086801 (2007)

TF91.   Tsue, Y., Fujiwara, Y.: Time-dependent variational approach to (1 + 1)-dimensional scalar-field solitons. Prog. Theor. Phys. **86**(2), 469–489 (1991)

TFL98.   Thurner, S., Feurstein, M.C., Lowen, S.B., Teich, M.C.: Receiver-operating-characteristic analysis reveals superiority of scale-dependent wavelet and spectral measures for assessing cardiac dysfunction. Phys. Rev. Lett. **81**, 5688–5691 (1998)

TFM05.   Takami, T., Fujisaki, H., Miyadera, T.: Coarse-grained picture for controlling quantum chaos. Adv. Chem. Phys. A **130**, 435 (2005)

TFM96.   Thorpe, S., Fize, D., Marlot, C.: Speed of processing in the human visual system. Nature (London) **381**, 520 (1996)

TFT98.   Thurner, S., Feurstein, M.C., Teich, M.C.: Multiresolution wavelet analysis of heartbeat intervals discriminates healthy patients from those with cardiac pathology. Phys. Rev. Lett. **80**, 1544–1547 (1998)

TG91.   Thouless, D.J., Gefen, Y.: Fractional quantum Hall effect and multiple Aharonov–Bohm periods. Phys. Rev. Lett. **66**(6), 806 (1991)

TGK92.   Thach, W.T., Goodkin, H., Keating, J.: The cerebellum and the adaptive coordination of movement. Annu. Rev. Neurosci. 403–442 (1992)

TH92.   Takami, T., Hasegawa, H.: Curvature distribution of chaotic quantum systems: Universality and nonuniversality. Phys. Rev. Lett. **68**, 419 (1992)

THL96.   Teich, M.C., Heneghan, C., Lowen, S.B., Turcott, R.G.: In: Wavelets in Medicine and Biology, pp. 383–412. CRC Press, Boca Raton (1996)

TJ02.   Todorov, E., Jordan, M.I.: Optimal feedback control as a theory of motor coordination. Nat. Neurosci. **5**(11), 1226–1235 (2002)

TJ06.   Toke, C., Jain, J.K.: Understanding the 5/2 fractional quantum Hall effect without the Pfaffian wave function. Phys. Rev. Lett. **96**, 246805/14 (2006)

TJW99.   Traub, R.D., Jefferys, J.G.R., Whittington, M.A.: Fast Oscillations in Cortical Circuits. MIT Press, Cambridge (1999)

TK93.   Turner, R.H., Killian, L.M.: Collective Behavior, 4th edn. Prentice Hall, Englewood Cliffs (1993)

TLD07.   Tognoli, E., Lagarde, J., DeGuzman, G.C., Kelso, J.A.S.: The phi complex as a neuromarker of human social coordination. Proc. Natl. Acad. Sci. **104**(19), 8190–8195 (2007)

TLO97a.   Tanaka, H.A., Lichtenberg, A.J., Oishi, S.: Self-synchronization of coupled oscillators. with hysteretic response. Physica D **100**, 279 (1997)

TLO97b.   Tanaka, H.A., Lichtenberg, A.J., Oishi, S.: First order phase transition resulting from finite inertia in coupled oscillator systems. Phys. Rev. Lett. **78**, 2104–2107 (1997)

TLW03.   Telenczuk, B., Latka, M., West, B.J.: Deterministic uncertainty. arXiv: nlin/0306057 (2003)

TM01.   Trees, B.R., Murgescu, R.A.: Phase locking in Josephson ladders and the discrete sine-Gordon equation: the effects of boundary conditions, current-induced magnetic fields. Phys. Rev. E **64**, 046205 (2001)

TMO00.   Trías, E., Mazo, J.J., Orlando, T.P.: Discrete breathers in nonlinear lattices: experimental detection in a Josephson array. Phys. Rev. Lett. **84**, 741 (2000)

TN98.   Teranishi, Y., Nakamura, H.: Control of time-dependent nonadiabatic processes by an external field. Phys. Rev. Lett. **81**, 2032 (1998)

TR85.   Tannor, D.J., Rice, S.A.: Control of selectivity of chemical reaction via control of wave packet evolution. J. Chem. Phys. **83**, 5013 (1985)

TRW98.   Tass, P., Rosenblum, M.G., Weule, J., Kurths, J., et al.: Detection of $n : m$ phase locking from noisy data: application to magnetoencephalography. Phys. Rev. Lett. **81**, 3291–3294 (1998)

TS01.   Thompson, J.M.T., Stewart, H.B.: Nonlinear Dynamics, Chaos: Geometrical Methods for Engineers, Scientists. Wiley, New York (2001)

TS03.   Tserkovnyak, Y., Simon, S.H.: Monte Carlo evaluation of non-Abelian statistics. Phys. Rev. Lett. **90**(1), 016802 (2003)

TS91.   Thompson, J.M.T., Soliman, M.S.: Indeterminate jumps to resonance from a tangled saddle-node bifurcation. Proc. R. Soc. Lond. A **432**, 101–111 (1991)

TSG82.   Tsui, D.C., Stormer, H.L., Gossard, A.C.: Two-dimensional magnetotransport in the extreme quantum limit. Phys. Rev. Lett. **48**(22), 1559 (1982)

TSP99.   Tanabe, S., Sato, S., Pakdaman, K.: Response of an ensemble of noisy neuron models to a single input. Phys. Rev. E **60**, 7235–7238 (1999)

TSS05.   Trees, B.R., Saranathan, V., Stroud, D.: Synchronization in disordered Josephson junction arrays: small-world connections and the Kuramoto model. Phys. Rev. E **71**, 016215 (2005)

TT93.   Turcott, R.G., Teich, M.C.: Fractal character of the electrocardiogram: distinguishing heart-failure and normal patients. Proc. SPIE 2036, 22–39 (1993)

TT96.   Turcott, R.G., Teich, M.C.: Fractal character of the electrocardiogram: distinguishing heart-failure and normal patients. Ann. Biomed. Eng. **24**, 269–293 (1996)

TV02.   Tesch, C.M., de Vivie-Riedle, R.: Quantum computation with vibrationally excited molecules. Phys. Rev. Lett. **89**, 157901 (2002)

TV05.   Tao, T., Visan, M.: Stability of energy-critical nonlinear Schrödinger equations in high dimensions. Electron. J. Differ. Equ. **118**, 1–28 (2005)

TV92.   Turaev, V.G., Viro, O.Y.: State sum invariants of 3-manifolds and quantum 6j-symbols. Topology **31**(4), 865–902 (1992)

TVP99.   Tabony, J., Vuillard, L., Papaseit, C.: Biological self-organisation, pattern formation by way of microtubule reaction–diffusion processes. Adv. Complex Syst. **2**(3), 221–276 (1999)

TVZ08.   Tao, T., Visan, M., Zhang, X.: Global well-posedness and scattering for the mass-critical nonlinear Schrödinger equation for radial data in high dimensions. Duke Math. J. (2008, to appear)

TW01.   Tegmark, M., Wheeler, J.A.: 100 years of the quantum. Sci. Am. **February**, 68–75 (2001)

TW84.   Tao, R., Wu, Y.-S.: Gauge invariance and fractional quantum Hall effect. Phys. Rev. B **30**(2), 1097 (1984)

TWT07. Trebst, S., Werner, P., Troyer, M., Shtengel, K., Nayak, C.: Breakdown of a topological phase: quantum phase transition in a loop gas model with tension. Phys. Rev. Lett. **98**(7), 070602 (2007)

TZD07. Tewari, S., Zhang, C., Das Sarma, S., Nayak, C., Lee, D.-H.: Non-local quantum phenomena associated with the Majorana fermion bound states in a $p_x + ip_y$ superconductor. arXiv:cond-mat/0703717 (2007)

Tak81. Takens, F.: Dynamical systems and turbulence. In: Lecture Notes in Mathematics, vol. 898, pp. 366–381. Springer, Berlin (1981)

Tak89. Takahashi, K.: Distribution functions in classical and quantum mechanics. Prog. Theor. Phys. Suppl. **98**, 109 (1989)

Tak92. Takami, T.: Semiclassical interpretation of avoided crossings for classically nonintegrable systems. Phys. Rev. Lett. **68**, 3371 (1992); Also, Semiclassical study of avoided crossings. Phys. Rev. E **52**, 2434 (1995)

Tan93. Tanaka, K.: Neuronal mechanisms of object recognition. Science **262**, 685–688 (1993)

Tao04. Tao, T.: On the asymptotic behavior of large radial data for a focusing non-linear Schrödinger equation. Dyn. Partial Differ. Equ. **1**, 1–48 (2004)

Tao06. Tao, T.: Nonlinear dispersive equations: local and global analysis. In: CBMS Regional Series in Mathematics (2006)

Tao07a. Tao, T.: Scattering for the quartic generalised Korteweg–de Vries equation. J. Differ. Equ. **232**, 623–651 (2007)

Tao07b. Tao, T.: A (concentration-)compact attractor for high-dimensional non-linear Schrödinger equations. Dyn. Partial Differ. Equ. **4**, 1–53 (2007)

Tao08a. Tao, T.: Global behavior of nonlinear dispersive and wave equations. Curr. Dev. Math. **2006**, 255–340 (2008)

Tao08b. Tao, T.: Why are solitons stable? Bull. AMS, S 0273-0979(08)01228-7 (2008)

Tap74. Tappert, F.: Numerical solutions of the Korteweg–de Vries equations and its generalizations by the split-step Fourier method. In: Nonlinear Wave Motion. Lectures in Applied Mathematics, vol. 15, pp. 215–216. American Mathematical Society, Providence (1974)

Tay79. Taylor, P.: Evolutionarily stable strategies with two types of players. J. Appl. Probab. **16**, 76–83 (1979)

Teg00. Tegmark, M.: The importance of quantum decoherence in brain processes. Phys. Rev. E **61**, 4194–4206 (2000)

Tei98. Teich, M.C.: Multiresolution wavelet analysis of heart rate variability for heart–failure and heart-transplant patients. Proc. Int. Conf. IEEE Eng. Med. Biol. Soc. **20**, 1136–1141 (1998)

Tha99. Thaller, B.: Visual Quantum Mechanics. Springer, New York (1999)

Tha05. Thaller, B.: Advanced Visual Quantum Mechanics. Springer, New York (2005)

Thu82. Thurston, W.: Three-dimensional manifolds, Kleinian groups and hyperbolic geometry. Bull. Am. Math. Soc. **6**, 357–381 (1982)

Tin75. Tinkham, M.: Introduction to Superconductivity. McGraw-Hill, New York (1975)

Tod67. Toda, M.: Vibration of a chain with nonlinear interactions. J. Phys. Soc. Jpn. **22**, 431–36; also, Wave propagation in anharmonic lattices. J. Phys. Soc. Jpn. **23**, 501–506 (1967)

Tok07. Toke, C., Regnault, N., Jain, J.K.: Nature of excitations of the 5/2 fractional quantum Hall effect. Phys. Rev. Lett. **98**, 036806 (2007)

Tri95.   Trippi, R.R.: Chaos & Nonlinear Dynamics in the Financial Markets. Irwin Prof. Pub. (1995)

Tse96.   Tse, W.M.: Policy implications in an adaptive financial economy. J. Econ. Dyn. Control **20**(8), 1339–1366 (1996)

Tuc50.   Tucker, A.W.: unpublished material (1950)

Tur52.   Turing, A.M.: The Chemical Basis of Morphogenesis. Philos. Trans. R. Soc. Lond. B **237**, 37–72 (1952)

Tur94.   Turaev, V.G.: Quantum Invariants of Knots and 3-Manifolds. Walter de Gruyter, Berlin (1994)

Tys85.   Tyson, J.J.: A quantitative account of oscillations, bistability and travelling waves in the Belousov–Zhabotinskii reaction. In: Field, R.J., Burger, M. (eds.) Oscillation and Travelling Waves in Chemical Systems. Wiley-Interscience, New York (1985)

Tzv05.   Tzvetkov, N.: On the long time behavior of KdV type equations [after Martel-Merle], Séminaire Bourbaki, vol. 2003/2004. Astérisque No. 299, Exp. No. 933, viii, 219–248 (2005)

UA96.   Unser, M., Aldroubi, A.: A review of wavelets in biomedical applications. Proc. IEEE **84**, 626–638 (1996)

UMM93.   Ustinov, A.V., Cirillo, M., Malomed, B.A.: Fluxon dynamics in one-dimensional Josephson-junction arrays. Phys. Rev. B **47**, 8357–8360 (1993)

Ula91.   Ulam, S.M.: Adventures of a Mathematician. University of California Press, Berkeley (1991)

Ume93.   Umezawa, H.: Advanced Field Theory: Micro Macro and Thermal Concepts. Am. Inst. Phys., New York (1993)

Unr95.   Unruh, W.G.: Maintaining coherence in quantum computers. Phys. Rev. A **51**, 992–997 (1995)

VPS97.   Viswanathan, G.M., Peng, C.-K., Stanley, H.E., Goldberger, A.L.: Deviations from uniform power law scaling in nonstationary time series. Phys. Rev. E **55**, 845–849 (1997)

VSB01.   Vandersypen, L.M.K., Steffen, M., Breyta, G.: Experimental realization of Shor's quantum factoring algorithm using nuclear magnetic resonance. Nature **414**, 883–887 (2001)

Vaa95.   Van der Vaart, N.C., et al.: Resonant tunneling through two discrete energy states. Phys. Rev. Lett. **74**, 4702–4705 (1995)

Vap95.   Vapnik, V.: The Nature of Statistical Learning Theory. Springer, New York (1995)

Vap98.   Vapnik, V.: Statistical Learning Theory. Wiley, New York (1998)

Vas04.   Vasiliev, A.N.: The Field Theoretic Renormalization Group in Critical Behavior Theory and Stochastic Dynamics. Chapman and Hall/CRC, Boca Raton (2004)

Ven07.   Vazirani, U.: Quantum computing. http://www.cs.berkeley.edu/~vazirani/qc.html (2003)

Vaz03.   Venkataraman, G.: In: Quest of Infinity. Radio Sai (2007)

Ver838.   Verhulst, P.F.: Notice sur la loi que la population pursuit dans son accroissement. Corresp. Math. Phys. **10**, 113–121 (1838)

Ver845.   Verhulst, P.F.: Recherches mathematiques sur la loi d'accroissement de la population (Mathematical researches into the law of population growth increase). Nouv. Mem. Acad. R. Sci. Belles-Lettres Bruxelles **18**(1), 1–45 (1845)

Ver88.   Verlinde, E.: Fusion rules and modular transformations in 2d conformal field theory. Nucl. Phys. B **300**, 360 (1988)

Vit01.   Vitiello, G.: My Double Unveiled. John Benjamins, Amsterdam (2001)

Vit95.   Vitiello, G.: Dissipation and memory capacity in the quantum brain model. Int. J. Mod. Phys. B **9**, 973–989 (1995)

Voi02.   Voisin, C.: Hodge Theory and Complex Algebraic Geometry I. Cambridge University Press, Cambridge (2002)

Vol31.   Volterra, V.: Variations and fluctuations of the number of individuals in animal species living together. In: Animal Ecology. McGraw-Hill, New York (1931)

Voi.   Voit, J.: The Statistical Mechanics of Financial Markets. Springer, Berlin (2005)

Vol94.   Volovik, G.E.: The Universe in a Helium Droplet. Oxford University Press, Oxford (1994)

WB95.   Welch, G., Bishop, G.: An Introduction to the Kalman Filter. University North Carolina, Chapel Hill (1995)

WBG01.   Wynn, J.C., Bonn, D.A., Gardner, B.W., et al.: Limits on spin-charge separation from $h/2e$ fluxoids in very underdoped $YBa_2Cu_3O_{6+x}$. Phys. Rev. Lett. **87**, 197002 (2001)

WBI92.   Wineland, D.J., Bollinger, J.J., Itano, W.M., Moore, F.L.: Spin squeezing and reduced quantum noise in spectroscopy. Phys. Rev. A **46**, R6797–R6800 (1992)

WBI.   Weng, G., Bhalla, U.S., Iyengar, R.: Complexity in biological signaling systems. Science **284**, 92 (1999)

WC02.   Wang, X.F., Chen, G.: Synchronization in small-world dynamical networks. Int. J. Bifurc. Chaos **12**, 187 (2002)

WC98.   Williams, C.P., Clearwater, S.H.: Explorations in Quantum Computing. Springer, New York (1998)

WCS96.   Wiesenfeld, K., Colet, P., Strogatz, S.H.: Synchronization transitions in a disordered Josephson series array. Phys. Rev. Lett. **76**, 404–407 (1996)

WCS98.   Wiesenfeld, K., Colet, P., Strogatz, S.H.: Frequency locking in Josephson arrays: connection with the Kuramoto model. Phys. Rev. E **57**, 1563–1569 (1998)

WCW00.   Wang, Y., Chik, D.T.W., Wang, Z.D.: Coherence resonance and noise-induced synchronization in globally coupled Hodgkin–Huxley neurons. Phys. Rev. E **61**, 740–746 (2000)

WD06.   Witzel, W.M., Das Sarma, S.: Quantum theory for electron spin decoherence induced by nuclear spin dynamics in semiconductor quantum computer architectures: spectral diffusion of localized electron spins in the nuclear solid-state environment. Phys. Rev. B **74**(3), 035322 (2006)

WES87.   Willett, R., Eisenstein, J.P., Stormer, H.L., et al.: Observation of an even-denominator quantum number in the fractional quantum Hall effect. Phys. Rev. Lett. **59**(15), 1776 (1987)

WF49.   Wheeler, J.A., Feynman, R.P.: Classical electrodynamics in terms of direct interparticle action. Rev. Mod. Phys. **21**, 425–433 (1949)

WGS98.   Wilkin, N.K., Gunn, J.M.F., Smith, R.A.: Do attractive bosons condense? Phys. Rev. Lett. **80**(11), 2265 (1998)

WK98.   Wolpert, D., Kawato, M.: Multiple paired forward, inverse models for motor control. Neural Netw. **11**, 1317–1329 (1998)

908   References

WLG03.   West, B.J., Latka, M., Glaubic-Latka, M., Latka, D.: Multifractality of cerebral blood flow. Physica A **318**(3), 453–460 (2003)
WMH08.   Wasserman, A., Maitra, N.T., Heller, E.J.: Investigating interaction-induced chaos using time-dependent density functional theory. Phys. Rev. A **77**, 042503 (2008)
WMP07.   Willett, R.L., Manfra, M.J., Pfeiffer, L.N., West, K.W.: Confinement of fractional quantum Hall states in narrow conducting channels. Appl. Phys. Lett. **91** (2007)
WN90.   Wen, X.G., Niu, Q.: Ground-state degeneracy of the fractional quantum Hall states in the presence of a random potential and on high-genus Riemann surfaces. Phys. Rev. B **41**(13), 9377 (1990)
WNK95.   West, B.J., Novaes, M.N., Kovcic, V.: In: Iannoccone, P.M., Khokha, M. (eds.) Fractal Geometry in Biological Systems, pp. 267–316. CRC Press, Boca Raton (1995)
WPP94.   Wiesenfeld, K., Pierson, D., Pantazelou, E., Dames, C., Moss, F.: Stochastic resonance on a circle. Phys. Rev. Lett. **72**, 2125–2129 (1994)
WQ02.   Wojs, A., Quinn, J.J.: Electron correlations in a partially filled first excited Landau level. Physica E **12**, 63 (2002)
WQ06.   Wojs, A., Quinn, J.J.: Landau level mixing in the $\nu = 5/2$ fractional quantum Hall state. Phys. Rev. B **74**(23), 235319 (2006)
WS97.   Watanabe, S., Swift, J.W.: Stability of periodic solutions in series arrays of Josephson junctions with internal capacitance. J. Nonlinear Sci. **7**, 503 (1997)
WS98.   Watts, D.J., Strogatz, S.H.: Collective dynamics of 'small-world' networks. Nature **393**, 440–442 (1998)
WSR00.   Widman, G., Schreiber, T., Rehberg, B., Hoeft, A., Elger, C.E.: Quantification of depth of anesthesia by nonlinear time series analysis of brain electrical activity. Phys. Rev. E **62**, 4898–4903 (2000)
WSS85.   Wolf, A., Swift, J.B., Swinney, H.L., Vastano, J.A.: Determining Lyapunov exponents from a time series. Physica D **16**(3), 285–317 (1985)
WST88.   Willett, R., Stormer, H., Tsui, D., Gossard, A., English, J.: Quantitative experimental test for the theoretical gap energies in the fractional quantum Hall effect. Phys. Rev. B **37**, 8476 (1988)
WSZ95.   Watanabe, S., Strogatz, S.H., van der Zant, H.S.J., Orlando, T.P.: Whirling modes, parametric instabilities in the discrete sine-Gordon equation: experimental tests in Josephson rings. Phys. Rev. Lett. **74**, 379–382 (1995)
WVH78.   Wolf, M.M., Varigos, G.A., Hunt, D., Sloman, J.G.: Sinus arrhythmia in acute myocardial infarction. Med. J. Austr. **2**, 52–53 (1978)
WW01.   Walker, M., Wooders, J.: Minimax play at Wimbledon. Am. Econ. Rev. **91**(5), 1521–1538 (2001)
WW83a.   Wehner, M.F., Wolfer, W.G.: Numerical evaluation of path-integral solutions to Fokker–Planck equations, I. Phys. Rev. A **27**, 2663–2670 (1983)
WW83b.   Wehner, M.F., Wolfer, W.G.: Numerical evaluation of path-integral solutions to Fokker–Planck equations, II: restricted stochastic processes. Phys. Rev. A **28**, 3003–3011 (1983)
WW94.   Wen, X.G., Wu, Y.S.: Chiral operator product algebra hidden in certain fractional quantum Hall wave functions. Nucl. Phys. B **419**, 455 (1994)

WWQ06. Wojs, A., Wodzinski, D., Quinn, J.J.: Second generation of Moore–Read quasiholes in a composite-fermion liquid. Phys. Rev. B **74**(3), 035315 (2006)

WYQ04. Wojs, A., Yi, K.-S., Quinn, J.J.: Fractional quantum Hall states of clustered composite fermions. Phys. Rev. B **69**(20), 205322 (2004)

WYR06. Wan, X., Yang, K., Rezayi, E.H.: Edge excitations and non-Abelian statistics in the Moore–Read state: a numerical study in the presence of Coulomb interaction and edge confinement. Phys. Rev. Lett. **97**(25), 256804 (2006)

WZ71.   Wess, J., Zumino, B.: Consequences of anomalous ward identities. Phys. Lett. B **37**, 95 (1971)

WZ82.   Wootters, W.K., Zurek, W.H.: A single quantum cannot be cloned. Nature **299**, 802 (1982)

WZ92.   Wen, X.G., Zee, A.: Classification of Abelian quantum Hall states and matrix formulation of topological fluids. Phys. Rev. B **46**(4), 2290 (1992)

WZ98.   Wen, X.G., Zee, A.: Topological degeneracy of quantum Hall fluids. Phys. Rev. B **58**(23), 15717 (1998)

WZS99. West, B.J., Zhang, R., Sanders, A.W., et al.: Fractal fluctuations in transcranial Doppler signals. Phys. Rev. E **59**, 3492–3498 (1999)

Wal91. Walker, K.: On Witten's 3-manifold invariants. Available at http://canyon23.net/math/1991TQFTNotes.pdf (1991)

Wat90. Watson, L.T.: Globally convergent homotopy algorithms for nonlinear systems of equations. Nonlinear Dyn. **1**, 143–191 (1990)

Wat99. Watts, D.J.: Small Worlds. Princeton University Press, Princeton (1999)

Wei83. Weinstein, M.I.: Nonlinear Schrödinger equations and sharp interpolation estimates. Commun. Math. Phys. **87**, 567–576 (1983)

Wei85. Weinstein, M.: Modulational stability of ground states of nonlinear Schrödinger equations. SIAM J. Math. Anal. **16**, 472–491 (1985)

Wei86. Weinstein, M.: Lyapunov stability of ground states of nonlinear dispersive equations. Commun. Pure Appl. Math. **39**, 51–68 (1986)

Wei94. Weibull, J.: Evolutionary Game Theory. MIT Press, Cambridge (1994)

Wei95. Weinberg, S.: In: The Quantum Theory of Fields, vols. 1–3. Cambridge University Press, Cambridge (1995)

Wen04. Wen, X.G.: Quantum Field Theory of Many-Body Systems. Oxford University Press, Oxford (2004)

Wen90. Wen, X.G.: Topological orders in rigid states. Int. J. Mod. Phys. B **4**(2), 239 (1990)

Wen91a. Wen, X.G.: Mean-field theory of spin-liquid states with finite energy gap and topological orders. Phys. Rev. B **44**(6), 2664 (1991)

Wen91b. Wen, X.G.: Non-Abelian statistics in the fractional quantum Hall states. Phys. Rev. Lett. **66**(6), 802 (1991)

Wen92. Wen, X.G.: Theory of the edge states in fractional quantum Hall effects. Int. J. Mod. Phys. B **6**(10), 1711 (1992)

Wen99. Wen, X.G.: Projective construction of non-Abelian quantum Hall liquids. Phys. Rev. B **60**(12), 8827 (1999)

Wer89. Werbos, P.J.: Backpropagation, neurocontrol: a review and prospectus. In: IEEE/INNS Int. Joint Conf. Neu. Net., Washington, D.C., vol. 1, pp. 209–216 (1989)

Wer90. Werbos, P.: Backpropagation through time: what it does, how to do it. Proc. IEEE **78**(10) (1990)

Wer98.   Werner, R.F.: Optimal cloning of pure states. Phys. Rev. A **58**, 1827–1832 (1998)

Whe89.   Wheeler, J.A.: Information, physics and quantum: the search for the links. In: Proc. 3rd Int. Symp. Foundations of Quantum Mechanics, Tokyo, pp. 354–368 (1989)

Wie61.   Wiener, N.: Cybernetics. Wiley, New York (1961)

Wig90.   Wiggins, S.: Introduction to Applied Dynamical Systems and Chaos. Springer, New York (1990)

Wil06.   Willingham, D.T.: Cognition: The Thinking Animal, 3rd edn. Prentice Hall, New York (2006)

Wil56.   Wilkie, D.R.: The mechanical properties of muscle. Br. Med. Bull. **12**, 177–182 (1956)

Wil82a.   Wilczek, F.: Magnetic flux, angular momentum, and statistics. Phys. Rev. Lett. **48**(17), 1144 (1982)

Wil82b.   Wilczek, F.: Quantum mechanics of fractional-spin particles. Phys. Rev. Lett. **49**(14), 957 (1982)

Wil90.   Wilczek, F.: Fractional Statistics and Anyon Superconductivity. World Scientific, Singapore (1990)

Wil95.   Wilson, R.G.: Fourier Series and Optical Transform Techniques in Contemporary Optics. Wiley, New York (1995)

Win67.   Winfree, A.T.: Biological rhythms and the behavior of populations of coupled oscillators. J. Theor. Biol. **16**, 15 (1967)

Win80.   Winfree, A.T.: The Geometry of Biological Time. Springer, New York (1980)

Wis06.   Wise, D.K.: p-form electrodynamics on discrete spacetimes. Class. Quantum Gravity **23**, 5129–5176 (2006)

Wit03.   DeWitt, B.S.: The Global Approach to Quantum Field Theory, vols. 1 & 2. Oxford University Press, Oxford (2003)

Wit83.   Witten, E.: Global aspects of current algebra. Nucl. Phys. B **223**, 422 (1983)

Wit88.   Witten, E.: Topological quantum field theory. Commun. Math. Phys. **117**(3), 353–386 (1988)

Wit89.   Witten, E.: Quantum field theory and the Jones polynomial. Commun. Math. Phys. **121**, 351 (1989)

Woj01.   Wojs, A.: Electron correlations in partially filled lowest and excited Landau levels. Phys. Rev. B **63**(12), 125312 (2001)

Wol02.   Wolfram, S.: A New Kind of Science. Wolfram Media (2002)

Wol84.   Wolfram, S.: Cellular automata as models of complexity. Nature **311**, 419–424 (1984)

Woo00.   Wooldridge, M.: Reasoning about Rational Agents. MIT Press, Boston (2000)

Wu84.   Wu, Y.-S.: General theory for quantum statistics in two dimensions. Phys. Rev. Lett. **52**(24), 2103 (1984)

XMB06.   Xia, J., Maeno, Y., Beyersdorf, P.T., et al.: High resolution polar Kerr effect measurements of Sr_2RuO_4: evidence for broken time-reversal symmetry in the superconducting state. Phys. Rev. Lett. **97**(16), 167002 (2006)

XPV04.   Xia, J.S., Pan, W., Vicente, C.L., et al.: Electron correlation in the second Landau level: a competition between many nearly degenerate quantum phases. Phys. Rev. Lett. **93**(17), 176809 (2004)

YAS96.   Yorke, J.A., Alligood, K., Sauer, T.: Chaos: An Introduction to Dynamical Systems. Springer, New York (1996)

YL97.   Yalcinkaya, T., Lai, Y.-C.: Phase characterization of chaos. Phys. Rev. Lett. **79**, 3885–3888 (1997)

YLS04.   Yeh, S.L., Lo, R.C., Shi, C.Y.: Optical implementation of the Hopfield neural network with matrix gratings. Appl. Opt. **43**, 858–865 (2004)

YML00.   Yanchuk, S., Maistrenko, Yu., Lading, B., Mosekilde, E.: Effects of a parameter mismatch on the synchronization of two coupled chaotic oscillators. Int. J. Bifurc. Chaos **10**, 2629–2648 (2000)

Yag87.   Yager, R.R.: Fuzzy Sets and Applications: Selected Papers by L.A. Zadeh. Wiley, New York (1987)

Yao93.   Yao, A.: Quantum circuit complexity. In: Proc 34th IEEE Symposium on Foundations of Computer Science, pp. 352–361 (1993)

Yau06.   Yau, S.T.: Structure of three-manifolds—Poincaré and geometrization conjectures. Talk given at the Morningside Center of Mathematics, June (2006)

Ye07.   Ye, R.: The log entropy functional along the Ricci flow. arXiv:0708. 2008v3 [math.DG] (2007)

Yeo92.   Yeomans, J.M.: Statistical Mechanics of Phase Transitions. Oxford University Press, Oxford (1992)

ZBR98.   Zhu, W., Botina, J., Rabitz, H.: Rapidly convergent iteration methods for quantum optimal control of population. J. Chem. Phys. **108**, 1953 (1998)

ZD86.   Zhang, F., Das Sarma, S.: Excitation gap in the fractional quantum Hall effect: finite layer thickness corrections. Phys.Rev. B **33**, 2903 (1986)

ZD93.   Zakrzewski, J., Delande, D.: Parametric motion of energy levels in quantum chaotic systems, I: curvature distributions. Phys. Rev. E **47**, 1650 (1993)

ZHK89.   Zhang, S.C., Hansson, T.H., Kivelson, S.: Effective-field-theory model for the fractional quantum Hall effect. Phys. Rev. Lett. **62**(1), 82 (1989)

ZJ89.   Zinn-Justin, J.: Quantum field theory and critical phenomena. Oxford University Press, Oxford (1989)

ZK65a.   Zabusky, N.J., Kruskal, M.D.: Interactions of solitons in a collisionless plasma and the recurrence of initial states. Phys. Rev. Let. **15**, 240–243 (1965)

ZK65b.   Zabusky, N.J., Kruskal, M.D.: Interaction of 'solitons' in a collisionless plasma and the recurrence of initial states. Phys. Rev. Lett. **15**, 240 (1965)

ZM76.   Zakharov, V.E., Manakov, S.V.: Asymptotic behavior of non-linear wave systems integrated by the inverse scattering method. Sov. Phys. JETP **44**, 106–112 (1976)

ZR99.   Zhu, W., Rabitz, H.: Noniterative algorithms for finding quantum optimal controls. J. Chem. Phys. **110**, 7142 (1999)

ZS72.   Zakharov, V.E., Shabat, A.B.: Exact theory of two-dimensional self-focusing and one-dimensional self-modulation of waves in nonlinear media. Sov. Phys. JETP **34**, 62–69 (1972)

ZST06.   Zhang, C., Scarola, V., Tewari, S., Das Sarma, S.: Anyonic braiding in optical lattices. arXiv:quant-ph/0609101 (2006)

ZT97.   Zouridakis, G., Tam, D.C.: Comput. Biol. Methods **27**, 9 (1997)

ZTD07.   Zhang, C., Tewari, S., Das Sarma, S.: Bell's inequality and universal quantum gates in a cold atom chiral fermionic p-wave superfluid. arXiv: 0705.4647 (2007)

Zad65.   Zadeh, L.A.: Fuzzy sets. Inf. Control **8**, 338–353 (1965)

Zad78.   Zadeh, L.A.: Fuzzy sets as a basis for a theory of possibility. Fuzzy Sets Syst. **1**(1), 3–28 (1978)

Zee03.   Zee, A.: Quantum Field Theory in a Nutshell. Princeton University Press, Princeton (2003)

Zha07.   Zhabotinsky, A.M.: Belousov–Zhabotinsky reaction. Scholarpedia **2**(9), 1435 (2007)

Index

3-body problem, 646, 649

Abelian topological ground state, 547
accelerometer, 617
achieved, 25, 378, 412
across, 133
action initiation, 395
action of a Lie group, 594
action potential, 45
activation derivative, 99, 237
activation dynamics, 100, 237, 400
activator, 231
Adaline, 80
Adams–Bashforth–Moulton integrator, 612
adaptive control, 47
adaptive filter, 100, 237
adaptive functional measure, 25
adaptive Kalman filtering, 47
adaptive NLS, 20
adaptive quantum computation, 183
adaptive quantum systems, 19
adaptive resonance theory, 104
adaptive sensory-motor control, 105
adiabatic evolution, 330
adiabatic exchange, 513
adiabatic invariant, 513
adiabatic process, 513
adiabatic transport, 185
adjoint, 594
adjoint functors, 575
adjoint representation, 598
admissibility condition, 625
admissibility constant, 625

admissible vector, 196
afferent nerves, 57
affine transformation, 660
Airy equation, 314
algorithmic approach, 401
Ambrose–Singer theorem, 265
amplitude, 171, 622
analogous ART2 system, 2
anatomical MRI, 71
Andronov, 654
Andronov–Hopf bifurcation, 654
angular frequency, 622
ANN dynamorphism, 80
ANN evolution, 79
Anosov diffeomorphism, 655
Anosov flow, 655
Anosov map, 655
anticommutation relations, 177
antikink, 309
anyonic excitations, 185
anyonic statistics, 16
anyons, 14, 514, 515
area-preserving map, 673
Arnold cat map, 655
array of N Josephson junctions, 484
arrow of time, 361
artificial neural network, 1
associative composition, 567
associative memory, 48
associativity of morphisms, 568
asymptotic stability, 322
atlas, 591
atmospheric convection, 665

V.G. Ivancevic, T.T. Ivancevic, *Quantum Neural Computation*,
Intelligent Systems, Control and Automation: Science and Engineering 40,
DOI 10.1007/978-90-481-3350-5, © Springer Science+Business Media B.V. 2010

attack function, 105
attractor, 664
auditory cortex, 44
auto- and cross-correlations, 123
auto-overlap, 240
autocatalator, 693
autonomic nervous system, 54
average energy, 257
averaging decision mechanics, 402
Axiom-A systems, 656
axon, 51

Bäcklund transform, 290
backpropagation, 86, 400
Baker–Campbell–Haussdorf formula,
    593
bandlimited functions, 624
basic biomechanical unit, 134
basin of attraction, 664, 673
basket cell, 147
battle of the sexes, 716
behavioral composition, 415
Bell state, 5
Bell's inequality, 438
Bell's theorem, 438
Belouzov–Zhabotinski reaction, 701
Bernoulli map, 655
Bernoulli shift dynamics, 259, 261
Bernoulli systems, 653
Berry phase, 520
best response, 716
Betti numbers, 570
bi-partite quantum system, 443
bi-stable perception, 37
bidirectional associative memory, 244
bifurcation diagram, 671
bifurcation point, 676
billiard, 653
binary (dyadic) dilation, 633
binary (dyadic) position, 633
binary signals, 99
binding by assembly, 36
binding by convergence, 36
binding by synchrony, 36
binding problem, 36
binding unit, 36
binocular rivalry model, 39
bio-diversity, 670
biomechanics, 617

biomorph, 695
biomorphic systems, 695
biophotons, 39
bipolar neurons, 52
bipolar signals, 99
Birkhoff, 651, 652
Birkhoff curve shortening flow, 652
Bloch, 70
Bloch sphere, 5
block entropy, 679
Bohr–Sommerfeld quantization, 513
Bohr's correspondence principle, 268
Boltzmann, 647
Boltzmann's constant, 22
Bose–Einstein condensate, 18, 36, 313
Bose–Einstein condensation, 512
Bose–Hubbard model, 548
bosons, 512
bound states, 319
boundary, 600
boundary operator, 601
bounded-energy solutions, 339
bounded-energy strong solution, 339
Boussinesq equation, 301
BQP, 181
braid group, 16, 185, 514, 589
braid relation, 17
braid theory, 14
braided monoidal category, 580
braids, 14
brain plasticity, 46, 49
Bratteli diagram, 216
breather solution, 326, 341, 343
Brownian dynamics, 392
Burgers dynamical system, 264
butterfly effect, 666

cache, 4
canonical ensemble, 22
canonical micro-circuits, 32
Cantor set, 658
capacity dimension, 694, 704
carrying capacity, 669
Cartwright, 654
Cash–Karp modification, 774, 783
Cauchy problem for KdV, 307
cavity resonators, 39
central biodynamic adjunction, 49
central nervous system, 53

central processing unit, 3
cerebellar cortex, 43
cerebellum, 69
cerebral cortex, 31, 43
cerebral hemispheres, 68
chaos, 649
chaos control, 702
chaos theory, 258, 651
chaotic behavior, 258
Chapman–Kolmogorov equation, 394
Chapman–Kolmogorov integro-
    differential equation, 395,
    429
characteristic Lyapunov exponents, 703
charging energy, 498
chemical kinetics, 693
Chern–Simons theory, 518
Christoffel symbols, 592
Chua–Matsumoto circuit, 701
Chua's circuit, 691
classical brain, 20
classical perceptron, 21
classical solutions, 315
classical strategies, 451
climbing fibre, 146
closed, 600
closed form, 600
closure property, 601
CNOT (Controlled NOT), 440
CNOT-gate, 17
co-active neuro-fuzzy inference system,
    94
co-existence of alternatives, 157
coarse system, 654
coarse-graining, 22, 605
coboundary, 600
cocycle, 600
code, 361
code subspace, 185
cognition, 414
cognitive binding problem, 35
cognitive information processing, 104
Cohen–Grossberg activation equations,
    2, 105
Cohen–Grossberg theorem, 3, 105
coherent quantum oscillations, 481
cohomological, 201
cohomology class, 203, 603
cohomology group, 603

Coiflet wavelet, 121, 122
collapse of the tunneling state, 35
combining noisy sensor outputs, 611
commutator, 172
compact attractor, 337
compact support, 620
complete synchronization, 106
complex vector space, 157
complex-valued ANNs, 97
complex-valued Gaussian, 111
complex-valued order parameter, 112
complexity class BQP, 6
complexity measure, 26
complexity of an algorithm, 440
computational basis states, 7
computational summand, 186
condensed, 360
conditional Lyapunov exponents, 113
configuration integral, 23, 605
conformal power, 333
conformal transformations, 214
conformal z-map, 694
conjugate gradient method, 89
conjugate transpose, 594
connection, 174
connectionism, 1, 30
connectionist, 81
connectionist brain theory, 30
consciousness, 39
consequences, 402
conservation law, 668
conservative Hamiltonian chaos, 423
conserved energy, 331, 333
conserved momentum, 332
continual-time systems, 611
continuity equation, 176, 204
continuous Fourier transform, 624
continuous Hopfield circuit, 2
continuous Hopfield network, 239
continuous limit, 308
continuous model for neural activity in
    cortical structures, 228
continuous wavelet transform, 195, 196,
    623
controlled phase shifter gates, 7
convective Bénard fluid flow, 665
convex optimization problem, 95
convolution, 628
convolution theorem, 622

Cooley–Tukey algorithm, 626
Cooper pair box, 481
Cooper pairs, 479
cooperative game theory, 713
coordination game, 716
corrector, 612
cost function, 612
counter-propagation network, 3
covariant derivative, 207
covariant Dirac equation, 178
covariant force law, 420
Crick–Koch binding hypothesis, 49
critical, 327
critical current, 482, 483
critical dynamics, 19
critical regularity, 329
critical slowing down, 244
critique of pure reason, 36
cross-correlation, 622
cross-overlap, 240
crowd cognitive controller, 422
cubic NLS, 327
cumulative distribution function, 391
current state, 398, 411, 412
current–phase relation, 485
curvature, 174
curved arrows, 567
cycle, 600, 681
cyto-sceletal networks, 350

damped pendulum, 641, 685
data compression, 628
Daubechies 4 tap wavelet, 631
Daubechies wavelet, 121
de-noising filter, 124
decision problem, 182
decoherence, 159, 524
deep nuclei, 147
definition of a Nash equilibrium, 719
defocussing NLS, 327
degree of anisotropy, 76
delta-rule, 84
dendrites, 50
dendron, 33
depolarization, 59
desired, 25, 378, 412
desired state, 398, 411
deterministic chaos, 665
deterministic chaotic system, 647, 648

deterministic diffusion, 283
Deutsch algorithm, 441
diencephalon, 69
diffeomorphism, 658
difference equation, 670
different, 462
differential Hebbian law, 105
diffusion anisotropy, 73
diffusion coefficient, 71
diffusion equation, 223
diffusion flux, 223
diffusion magnetic resonance imaging,
    71
diffusion MRI, 71
diffusion tensor, 71
diffusion-weighted images, 74
digital signal processing, 624, 634
dilations, 633
Dirac, 177
Dirac bra-ket notation, 7
Dirac equation, 177
Dirac interaction picture, 163
Dirac-delta distribution, 624
Dirac's perturbation method, 165
directional derivative, 424
discrete, 120
discrete Amari equation, 228
discrete breathers, 500
discrete Fourier transform, 170, 620,
    626
discrete Laplacian, 496
discrete sine-Gordon equation, 494
discrete wavelet transform, 627
discrete-time models, 670
discrete-time steps, 651
discrete-time systems, 611
dispersion, 316
dispersion relation, 317
dispersive wave equations, 313
dissipation, 366
dissipative chaos, 423
dissipative solitons, 219
distorted plane waves, 319
distribution function, 391
divide and conquer algorithm, 626
DNA computers, 8
dominant strategy, 448
dominant strategy equilibrium, 715
Donaldson–Witten theory, 201

down-sampling operator, 628
driven pendulum, 686
dual series, 330
dual system, 612
Duffing, 654
Duffing oscillator, 689
Duffing–Van der Pol equation, 689
dyadic translations, 633
dynamical homeostasis, 19
dynamical invariant, 653
dynamical planar graph, 421
dynamical system, 647, 675
dynamics, 639
dynamics and topology of brain
    networks, 26

early selection, 401
echo attenuation, 75
echoplanar imaging, 75
economics, 720
eddy, 261
edge of chaos, 19, 219
eigenfunction, 320
Einstein–Hilbert functional, 251
elastic pendulum, 692
electric current density, 175
electro-weak interaction, 597
embedding methods, 89
encoding, 139
energy, 339
energy space, 339
energy surfaces, 316
ensemble code, 126
entangled attractors, 423
entangled quantum states, 438
entangled state, 365
entanglement, 5, 9, 438
entangling power, 440
enteric nervous system, 55
entropic motor control, 386
entropy, 257, 513, 653
EPR, 438
equilibrium, 513, 715
equilibrium in dominant strategies, 448
equipartition of energy, 292
equivalence relation, 472
ergodic hypothesis, 647, 652
ergodic hypothesis of Boltzmann, 647
error-correcting codes, 12

error-correction procedures, 14
escape rate, 684
Euclidean chart, 590
Euclidean image, 590
Euclidean path integral, 24
Euler, 646
Euler characteristic, 649
Euler–Lagrange equation, 274, 331
Euler–Poincaré characteristic, 603
every game with a finite number of
    players and finite number of
    strategies has at least one mixed
    strategy Nash equilibrium, 717
evolution PDE, 329
evolutionary biology, 726
evolutionary computation, 95
exact, 600
exact form, 600
excitation contraction coupling, 377
excitatory, 62
existence & uniqueness theorems for
    ODEs, 675
exocytosis, 33
exponentially faster, 522
extended Kalman filter, 89, 615
extended observable system, 612
extensive form, 714

Fabry–Perot interferometer, 539
fan-beam surveillance radar, 607
Faraday, 175
Faraday tensor, 174
fast Fourier transform, 623, 627
fast Hadamard transform, 7
fast wavelet transform, 627, 633
father wavelet, 631
fault-tolerant computation, 12
fault-tolerant quantum computation,
    511
feedforward neural network, 80
Feigenbaum cascade, 667
Feigenbaum constant, 667
Feigenbaum number, 671
fermions, 512
Feschbach resonance, 551
Feynman diagrams, 168
Feynman path integral, 168
Feynman–Vernon formalism, 363
Fibonacci anyons, 17

Fibonacci(2n), 186
field, 173, 639
field operator, 173
figure-ground vase, 37
filter bank, 629
filtering, 614
finite Fourier transform, 626
first quantization, 160
first-order-search methods, 88
Fisher's equation, 225
FitzHough–Nagumo model, 114
FitzHugh–Nagumo neuron, 233
fixed-point, 657, 662, 664
fixed-point attractors, 228
flame front, 264
Floquet exponents, 507
flow, 262, 659, 675
flow pattern, 642
flow-line, 423
fluctuation, 258
flux-flow regime, 498
focal equilibrium, 457
focal point effect, 456
focusing NLS, 327
Fokker–Planck equation, 224, 244, 394
folding, 657
for games without dominant strategies, 716
forced nonlinear oscillators, 642
forced Van der Pol oscillator, 687
forgetful memories, 241
formalism of jet bundles, 642
four reasonable requirements, 719
Fourier analysis, 271
Fourier basis functions, 620
Fourier transform, 194, 319, 325, 619, 622, 625
Fourier windowing, 630
FPU-paradox, 292
FPU-recurrence, 291
fractal, 631, 694
fractal attractor, 666
fractal dimension, 626, 664
fractal pattern, 671
fractal set, 657
fractional dimension, 694
fractional quantum Hall effect, 16, 18, 184, 512

fractional quantum Hall ground states, 185
fractionally charged quasi-particles, 16
frame functions, 621
free wave equation, 313
frequency domain, 622
frequency spectrum, 620
frequency-to-voltage converter, 482
Frölich waves, 37
frontal lobe, 43
frustrated XY models, 489
fully recurrent networks, 96
functional analysis, 621
functional equation, 624
functional manifold, 264
functional MRI, 71
fundamental group, 589
fundamental representation, 597
fusion, 516
fusion channels, 516
fusion rule, 516

Galilean invariance, 334, 338
game matrix, 714
game situations, 468
game theory, 713
game tree, 714
games, 713
gamma oscillations, 115
gates, 440
gauge field, 173, 174
gauge theory, 173
gauge transformation, 173
gauge-invariant phase difference, 501
Gauss–Bonnet formula, 246
Gaussian distribution, 283
Gaussian functional integral, 204
Gaussian noise, 613
Gaussian random vectors, 283
Gaussian soliton, 255
Gaussian wave-packet, 778
Gaussian white noise, 118, 634
Gell-Mann matrices, 598
general sense, 700
generalized feedforward network, 92
generalized Gaussian, 100
generalized Hebbian rule, 110
generalized Hénon map, 673

generalized Korteveg–de Vries (gKdV)
     equation, 314
generalized solution, 700
generalized synchronization, 106
generic system evolution, 40
genetic algorithm, 87, 95
genetic control, 95
geometrization conjecture, 246
George Birkhoff, 647
Giaever tunneling junction, 481
Gibbs ensemble, 166
Gibbs measure, 23
Gibbs statistical density function, 166
Ginzburg–Landau equation, 263
Glauber dynamics, 241
glia, 53
globally-coupled Kuramoto model, 499
globular cluster, 672
Goldstone theorem, 361
Golgi cells, 147
Golgi tendon organs, 144
Gottesman–Knill theorem, 441
governing equilibrium dynamics, 713
gradient descent method, 87
gradient information, 82
gradient of the performance surface, 82
grand conjecture, 341
graph, 591
Green functions, 194
grey matter, 43
ground state, 329
group action, 586
group homomorphism, 596
group identity element, 593
group inversion, 593
group multiplication, 593
group velocity, 317
groups, 469
growth rate, 669
growth-rate functions, 726
gyroscope, 617

H-cobordism theorem, 656
Haar wavelet, 620, 633
Haar wavelet transform, 627
Hadamard, 653
Hadamard gate, 440
Hadamard transform, 7
Hadamard transform gates, 7

halo orbit, 653
Hamilton, 646
Hamiltonian, 315, 338
Hamiltonian path problem, 8
Hamiltonian system, 673
Hamming distance, 18
Hamming hypercube, 239
harmonic, 202
harmonic oscillator, 778
Harr wavelet, 121
Hartman–Grobman theorem, 243
Hayashi, 654
heat equation, 773
heat flow, 513
Hebb rules, 47
Hebbian innovation, 242
Hebbian learning, 374
Hebbian theory, 63
Hecht–Nielsen counterpropagation
     network, 102
Heisenberg model, 548
Heisenberg picture, 163
Heisenberg representation, 165
Heisenberg uncertainty principle, 387,
     621, 628
Heisenberg uncertainty relation, 162
Hénon map, 652, 672, 712
Hénon strange attractor, 672
her, 456
Hermitian, 595
Hermitian inner (scalar) product, 158
Hermitian matrix, 595
Hermitian transpose, 594
heuristic approach, 401
high-pass filter, 628
high-temperature superconductivity,
     512
Hilbert basis, 633
Hilbert space, 95, 157, 339, 621, 647
Hodge star, 201
Hodgkin–Huxley, 60
Hodgkin–Huxley equation, 2
Hodgkin–Huxley model, 114
Hodgkin–Huxley neural model, 33
Hölder regularity index, 338
holographic hypothesis, 360
holonomy, 184
homeostasis, 67
homoclinic point, 650, 659, 663

homoclinic tangle, 649, 662
homological algebra, 570
homology class, 603
homology group, 570
homomorphism, 589
homotopy methods, 89
Hooke's law, 291
Hopf, 654
Hopf bifurcation, 232
Hopf-like bifurcation, 691
Hopfield model, 47
Hopfield synaptic matrix, 240
Hubbard model, 548, 550
human biodynamics engine, 381, 382
human brain, 43
human heart beat and respiration, 106
human memory, 138
human vestibular system, 617
hurricane, 260
hybrid dynamical system of variable
    structure, 699
hybrid systems, 700
hyperbolic fixed-point, 664
hyperbolic system, 676
hyperbolic tangent threshold activation
    functions, 81
hyperpolarization, 59

ideal, 125, 461
image, 600
imitative, 726
impulse response, 628
in the sense of Filippov, 700
independent component analysis
    networks, 94
individual, 469
inertial sensors, 617
inferior cerebellar peduncle, 147
information, 702
information theory, 624, 654
inhibitor, 231
inhibitory, 62
inner product, 633
inner-product space, 7
innovation, 25
instability, 642
integrable function, 621
integrate-and-fire model, 114
integration, 675

intention, 397
intention formation, 395
inter-spike-interval, 115
interior, 724
interpolation, 624
intuition, 648
invariant set, 657, 660
inverse, 119
inverse discrete Fourier transform, 626
inverse DWT, 120
inverse dynamic analysis of human
    movement, 617
inverse Fourier transform, 622
inverse scattering, 323
inverse scattering problem, 326
inverted driven pendulum, 692
inviscid Burgers' equation, 305
ion trap, 6
irreducible representations, 513
Ising anyons, 517
Ising Hamiltonian, 238, 240
Ising model, 548
Ising system, 800
Ising-spin Hopfield network, 102
isometry, 595
iterated map, 651
iterative maps, 676
Ito stochastic integral, 393

Jacobi identity, 180, 210
Jones polynomial, 17, 186, 534
Jordan and Elman networks, 93
Josephson constant, 482
Josephson current, 479
Josephson current–phase relation, 483
Josephson effect, 479
Josephson interferometer, 479
Josephson junction, 478, 479
Josephson junction ladder, 489
Josephson tunneling, 479, 482
Josephson voltage–phase relation, 483
Josephson-junction quantum computer,
    480

Kadanoff spin-blocking procedure, 195
Kalman filter, 607, 611
Kalman filtering problem, 614
Kalman linear-quadratic regulator, 611
Kalman regulator, 611

Kalman–Bucy filter, 612
Kalman-quaternion filter, 618
Kaplan–Yorke dimension, 704
Karhunen–Loeve covariance matrix, 101, 240
KdV–Burgers' equation, 307
Kelvin inversion, 333
Kepler, 649
kernel, 596, 600
kink, 309
kink-phason resonance, 497
Kitaev model, 548
Klein–Gordon equation, 293
knot theory, 14
knowledge of the manifold, 36
Kohonen continuous self organizing map, 103
Kohonen self-organizing map, 93
Kolmogorov, 653, 655
Kolmogorov–Arnold–Moser theorem, 268
Kolmogorov–Sinai, 704
Kolmogorov–Sinai entropy, 114, 271, 679, 702, 705
Korteveg–de Vries (KdV) equation, 290, 311, 314
Kraus operators, 450
Kronecker delta, 633
Krylov, 654
Kuramoto model, 28, 109
Kuramoto order parameter, 507
Kuramoto–Sivashinsky equation, 307

ladder, 500
lag error, 609
lag synchronization, 106
Lagrange, 646
Lagrange's points, 652
Lagrangian, 176
laminar flow, 259
Landau gauge, 496
Landau-level, 526
Laplace–Beltrami operator, 221
Laplace's equation, 224
Laplacian operator, 19
largest eigen-diffusivity, 76
largest Lyapunov exponent, 702
Laughlin state, 530
Lauterbur, 70

Lax pair equation, 324
Lax pairs, 324
learning dynamics, 100, 101, 237, 238
learning rate, 25, 83, 378, 400, 412
learning rate scheduling, 83
least means square algorithm, 83
Lebesgue integrability, 621
Lebesgue integrable function, 622
Lebesgue integral, 621
Left, 78
Levenberg–Marquardt algorithm, 89
Levi-Civita connection, 428, 592
Levi-Civita covariant derivative, 427, 592
Levin–Wen model, 548
Lie derivative, 698
Lie group, 593
Lie structure equations, 265
Lie–Poisson bracket, 309
life space foam, 396
lifting scheme, 628
light, 152
limit cycle, 642, 664, 686
linear, 611
linear homotopy, 90
linear homotopy ODE, 700
linear homotopy segment, 700
linear optimal filtering problem, 611
linear Schrödinger equation, 313
linear state feedback control law, 611
linear superposition, 152
links, 527
Liouville equation, 394
Liouville theorem, 355
Lissajous curves, 693
Littlewood, 654
local Bernstein adaptation process, 411
locally exact, 600
locally-coupled Kuramoto model, 499, 508
locally-optimal solution, 369
logarithmically divergent solitons, 326
logic gates, 14
logical qutrit, 184
logistic equation, 669
logistic growth, 669
logistic map, 670, 672, 706
long-range connections, 508
long-range correlation, 360

long-range order, 480
long-term consequences, 402
long-term memory, 100, 138, 237
Lorenz attractor, 699
Lorenz equations, 693
Lorenz mask, 666
Lorenz system, 666, 672, 676
Lotka–Volterra ensemble dynamics, 244
low-pass filter, 628
low-regularity well-posedness theory,
    315
Lyapunov, 654
Lyapunov dimension, 704
Lyapunov exponent, 271, 664, 678
Lyapunov function, 102
Lyapunov stability, 675

Mach–Zehnder interferometer, 156
machine learning, 1
magnetic resonance imaging, 70
magnetometer, 618
main direction of diffusivities, 76
maintaining the action, 395
Majorana fermion, 542
Malthus model, 669
Malthusian parameter, 669, 671
Mandelbrot and Julia sets, 694
manifold structure, 591
Mansfield, 70
map, 670
map from strategy profiles to pay-offs,
    714
map sink, 664
Markov assumption, 395, 429
Markov chain, 392, 680
Markov partitions, 655
Markov process, 680
Markov property, 23
Markov stochastic process, 392, 429
mass-action interpretation, 724
master equation, 394
match-based learning, 104
matching pennies, 717
matrices, 177
matrix cost function, 88
maximum entropy, 22, 99, 237
maximum principle, 248, 255
Maxwell, 175, 647
Maxwell–Haken laser equations, 693

McCulloch–Pitts neurons, 80
McCumber parameter, 504
mean diffusivity, 75
mean square error, 81
mean-field theory, 18
measurable functions, 623
measurement equation, 613
measurement noise, 613
measurement sensitivity matrix, 613
medulla oblongata, 69
Meissner-states, 492
method of lines, 345, 774, 783, 788
methods of regularization, 194
microtubules, 37, 349
midbrain, 69
middle cerebellar peduncle, 148
millimetric image resolution, 71
minimizing the error, 81
Mittag–Leffler, 650
mixed strategy, 454, 717
modified Korteveg–de Vries (mKdV)
    equation, 314
modular feedforward networks, 93
modular functors, 181
momentum learning, 86
momentum operator, 173
monetary, 716
Monte Carlo method, 672
Moore–Penrose pseudoinverse, 240
Moore–Read Pfaffian wave-function,
    533
Moore–Read state, 534
Moore's law, 181
Morse, 653
Morse theory, 653, 656
mossy fibers, 148
mother wavelet, 116, 119, 623, 625
motion action principle, 398
motor cortex, 44
motor dynamorphism, 134
motor nerve fibers, 54
motor servo, 144
motor-learning, 134
Mott insulators, 549
moves, 714
multi-soliton, 341
multi-soliton solutions, 344
multi-spiral strange attractor, 691
multilayer perceptron, 80, 90

multipolar neurons, 52
multisoliton states, 323
mutual overlap, 240

Nash equilibrium, 448, 716, 724
Nash's program, 721
nature, 714
Navier–Stokes equations, 258, 263, 665
neocortex, 31
neocortical biophysics, 34
neocortical dynamics, 33
network of networks, 347
neural action potential, 377
neural attractor dynamics, 228
neural communication, 45
neural network, 47
neural path integral, 377
neural state-space, 244
neural-networks complexity, 26
neuro-computing, 1
neurons, 50
NeuroQuantology, 30
new value, 25, 378, 398
Newland transform, 628
Newton, 645
Newton–Raphson method, 89
Newtonian mechanics, 673
Newtonian method, 87
Newtonian-like action, 398
Newton's Principia, 268
NLS-flow, 337
no-cloning theorem, 441
Noetherian ring, 588
Noether's theorem, 316
noise, 25, 378, 412
non-Abelian anyons, 18, 511
non-Abelian braiding statistics, 516
non-Abelian quantum Hall states, 511
non-Abelian statistics, 511
non-autonomous 2D continuous
    systems, 642
non-autonomous system, 640
non-compact Lie group, 344
non-cooperative game theory, 713
non-equilibrium phase transitions, 244
non-wandering set, 658
non-zero sum game, 448
nonlinear, 127
nonlinear Klein–Gordon equation, 335

nonlinear oscillators, 647
nonlinear resonances, 342
nonlinear Schrödinger equation (NLS),
    20, 294, 311, 313, 327
nonlinear wave equation, 331
nonlinearity, 288
normal, 714
normal mode, 291
normalization condition, 159
normalized Ricci flow, 250
normalized state, 158
nuclear magnetic resonance, 6
Nyquist range, 626

observable system, 612
observables, 199
observer design, 612
occipital lobe, 43
ODEs, 676
Oja–Hebb learning rule, 100
olfactory cortex, 44
operator product expansion, 215
optical cavity, 6
optical neural networks, 8
optimal answer, 454
optimal estimator problem, 612
optimal policy, 412
orbit, 344, 659, 675, 681
orbit Hilbert space, 169
orbital stability, 322
order parameter, 329
order parameter equation, 243
order parameters, 288
orthogonal matrix, 595
orthogonality, 158
orthonormal wavelet, 633
oscillator neural networks, 109
overdamped junction, 500
overdamped ladder, 509, 510
overdamped limit, 483
overlap, 240

parallel distributed processing, 1
parameter space, 18
parasympathetic nervous system, 55
Pareto optimal, 449
Pareto optimality, 719
parietal lobe, 43
partition function, 22, 205, 257, 655

path that requires minimal memory,
    401
path-integral, 513
pattern matching process, 104
pattern vector, 21
pattern-recognition, 83
Pauli gate, 441
Pauli matrices, 598
Pauli principle, 512
Pauli sigma matrices, 178
Pavlov, 63
Pavlov's conditional-reflex learning, 63
pay-off, 714
pdf, 375
peduncles, 147
penny flipover, 471
perception energy, 39
perceptron, 83
perceptron learning algorithm, 21
perfect & imperfect Nash equilibria, 721
performance surface, 82
period-doubling bifurcation, 667, 671,
    707
periodic orbit, 270, 650, 662
periodic orbit theory, 264, 683
periodic phase synchronization, 106
periodic solutions, 642
peripheral nerve fibers, 54
peripheral nervous system, 53
permutation group, 513
perturbation problem, 329
perturbation theory, 267, 329
petite conjecture, 341
Pfaffian, 533
Pfaffian state, 534
Pfaffian wave-function, 534
phase coherence, 480
phase difference, 107, 483, 496
phase invariance, 338
phase point, 641
phase portrait, 642
phase space, 268, 271
phase synchronization, 106
phase transitions, 604
phase velocity, 317
phase-flow, 641, 675
phase-space flow, 651
phases, 395
photoelectric effect, 151

photons, 152
physically-controllable systems of
    nonlinear oscillators, 499
Pickover's biomorphs, 696
Planck's constant, 154
Planck's quantum hypothesis, 151
plane wave, 317
plat closure, 187
Poincaré, 647
Poincaré conjecture, 220, 656
Poincaré map, 673
Poincaré section, 271, 649, 651, 672,
    682, 689
Poincaré–Bendixson theorem, 642, 648
Poincaré–Hopf index theorem, 648
point at infinity, 153
Poisson equation, 166
Poisson evolution equation, 309
Poisson Lie groups, 308
Poisson manifold, 308, 309
poly-time algorithm, 186
polynomial-time algorithm, 182
pons, 69
Pontryagin, 654
population code, 126
population models, 668
population-statistic, 722
position and velocity innovations, 609
positive leading Lyapunov exponent,
    703
postsynaptic potential, 238
Prandtl number, 666
predictability time, 703
prediction, 614
predictor, 612
principal component analysis, 116
principal component analysis networks,
    94
principle of cognitive optical illusions,
    38
Prisoner's dilemma, 442, 714
pristine, 508
probability amplitude, 12, 159, 165, 166
probability density, 195
probability density function, 23, 373
probability distribution function, 503
process equation, 613
process noise, 613
processing speed, 398, 401

product topology, 262
protozoan morphology, 695
pruning, 681
pseudoconformal symmetry, 334
pseudolocality theorem, 256
psycho-physical crowd dynamics, 413
punctures, 185
Purcell, 70
pure strategy, 454

QNN of QNNs, 347
quadratic cost function, 611
quadrature mirror filter, 628
quantum action functional, 272
quantum algorithm, 441
quantum bits, 170
quantum brain, 20, 374
quantum chaos, 267, 268
quantum chaos systems, 281
quantum circuit, 7, 17, 171, 181
quantum computation, 4, 522
quantum computer, 5, 171, 479
quantum computer algorithms, 7
quantum computing, 4, 170
quantum cryptography, 437, 439, 460
quantum dimer model, 547
quantum dot, 6
quantum error correction, 512
quantum field theory, 168
quantum Fourier transform, 7, 170
quantum gates, 7, 12
quantum groups, 518
quantum Hamilton's equations, 162
quantum information, 437
quantum logic gate, 17
quantum loop gases, 18
quantum mechanics, 621
quantum media, 313
quantum Monty Hall, 471
quantum neural computation, 30
quantum neural network, 11, 19
quantum observable, 171
quantum optics, 437
quantum parallelism, 441
quantum phase space, 268
quantum state ket-vector, 153
quantum statistics, 512
quantum strategy, 447
quantum superposition, 6

quantum superposition equation, 19
quantum teleportation, 438
quantum trigger, 33
quantum truel, 471
quantum tunneling effect, 33
quantum Turing machine, 17, 181
quantum-mechanical wave function, 479
quantum-probability, 417
quasi-particle tunneling, 33
quasi-particles, 519
quaternion attitude estimation filter,
    618
qubit, 3, 5, 12, 170, 182, 479
quotient space, 602

Rabi frequency, 284
Rabi oscillation, 284
radial basis function network, 94
radiation, 342
radiation state, 342
random, 717
random matrix theory, 283
random partial recursive function
    halting, 23
random variable, 391
random walk, 392
rate, 637
rate code, 126
rate of error growth, 702
rate of relaxation, 228
rationalistic, 722
Rayleigh–Bénard convection, 677, 701
re-polarization, 59
reaction–diffusion, 219
reactive neurodynamics, 105
recalled, 725
reciprocal activation, 145
reciprocal inhibition, 144
recurrence time, 293
recurrent neural networks, 18
recurrent QNN, 374
recursive homotopy dynamics, 399
recursive solution, 611
recursively hierarchical structure, 43
reduced curvature 1-form, 266
refinement equation, 633
reflectance pattern, 103
reflex, 65, 135
regional energy minimization, 78

reinforcement learning, 26, 412
relative degree, 698
relative phase gate, 182
relative semions, 547
relaxation oscillator, 687
reliable predictor, 669
removable singularity, 624
repeller, 683
replicator dynamics, 726
represent quantum field theory, 24
resistive loading, 483
resistively & capacitively-shunted
    junction, 501
resistively-shunted Josephson junctions,
    495
resistively-shunted junction, 500
resonance, 481
return map, 683
Reynolds number, 259
Ricci flow, 428
Ricci flow equation, 220
Ricci solitons, 255
Ricci tensor, 428, 593
Richardson extrapolation, 737
Riemann curvature, 428, 592
Riemann sphere, 153
Riemannian metric tensor, 398
Riesz basis, 621
ripple, 621
risk dominance, 723
Romberg integration, 737
Rosenblatt, 83
Rossler, 690
Rossler system, 690
rotating-wave approximation, 284
route to chaos, 655
route to turbulence, 261
Rubin face, 37
Rudolphine tables, 649
Ruelle, 655
rules of the game, 443
Runge–Kutta–Fehlberg integrator, 774,
    783

saltatory conduction, 60
sampling theorem, 632
scalar curvature, 247
scaling function, 624
scattering data, 325

scattering space, 344
scattering state, 340
Schrödinger equation, 19, 39, 152, 154,
    224, 294, 318, 778
Schrödinger operator, 320
Schrödinger's picture, 152, 163, 165
Schwann cells, 53
Schwartz function, 321
Schwarz-type, 199
search, 81
selective attention, 401
selectivity, 133
self-adjoint, 595
self-limiting process, 669
self-organized, 25, 378, 412
self-similar phenomena, 194
semidirect product, 597
sensitivity, 132
sensor fusion in hybrid systems, 616
sensory, 377
sensory memory, 138
sensory-motor integration, 619
sequential (threshold) dynamics, 238
set of instructions, 714
Shannon, 654
Shannon entropy, 26
Shor's algorithm, 6, 170
Shor's factorization algorithm, 512
short exact sequence, 596
short-term consequences, 402
short-term memory, 100, 138, 237
short-time Fourier transform, 116, 623
signal, 25, 378, 412
signal coding, 628
signal velocity, 100, 237
signal-to-noise ratio, 115
signal-to-noise ratio, 123
simulated annealing, 87
Sinai, 655
sinc function, 624
sine-Gordon (SG) equation, 261, 293,
    308
Sinfire neural network, 117
sinus cardinalis, 624
skew-Hermitian, 596
slope parameter, 81
Smale, 652
Smale horseshoe map, 656
small world, 501

small-world geometry, 28
small-world networks, 26, 499
smoothing, 614
Sobolev inequality, 332
social convergence, 415
social hierarchy, 27
social network, 27
social situations, 713
solid state physics, 437
solitary wave, 312
soliton, 288, 312, 332, 334
soliton resolution conjecture, 323, 341
soliton solution, 329
solution, 700
solution in maxi-min-strategies, 449
somatic nervous system, 54
somatosensory cortex, 44
space of all weighted paths, 400
spatiotemporal networks, 105
special unitary group, 596
spectral theorem, 595
spin-wave modes, 495
spinal nerves, 59
spindle receptors, 144
spine synapse, 33
split-stepping, 306
squeezing, 657
stability, 642
stable and unstable manifold, 659
stable manifold, 673
Standard model, 597
state, 639
state estimator, 612
state vector, 613
static backpropagation, 80
stationary soliton solutions, 341
stationary solutions, 332
statistically, 636
steepest descent method, 82
step size, 25, 83, 378, 400, 412
stimulus, 378
stochastic filtering problem, 373
stochastic integral, 393
stochastic system, 640
stochastic-gradient order parameter
    equations, 243
stock option, 551
storage, 139

strange attractor, 648, 655, 664, 665,
    666, 689
strategic, 714
strategic form, 714
strategic interaction, 713
strategy, 443, 714
strategy profile, 714
stream of photons, 155
strengths, 398
stretch-and-fold, 689
stretching, 657
stroboscopic section, 662
strong coupling expansion, 330
structural stability, 675
structurally stable, 655
structure constants, 215
sub-critical, 327
Subband coding algorithm, 630
subsumption architectures, 411
sum over gauge orbits, 213
sum over histories, 400
super-critical, 327
supercell thunderstorms, 260
superconducting-normal transition, 489
superconductivity, 479
superior cerebellar peduncle, 148
superposition, 5
supervised gradient descent learning,
    25, 378, 412
supervised network, 80
support vector machine, 94
survival probability, 683
symbolic dynamics, 655, 659, 660, 681
symmetry breaking, 604
symmetry breaking instability, 244
symmetry group, 344
sympathetic nervous system, 55
symplectic form, 333
symplectic structure, 331
synaptic junctions, 46
synaptic potential, 377
synchronization in chaotic oscillators,
    106
synchronization of coupled nonlinear
    oscillators, 498
syncytium, 37
synergetics, 243
system dynamorphism, 41
system with uncertain dynamics, 611

Takens, 655
target equations of motion, 609
targeting, 701
temporal associative memories, 241
temporal code, 126
temporal dynamical systems, 101, 237
temporal lobe, 43
temporary phase-locking, 106
tensor product, 182
tensor-field, 675
termination, 395
The Last Problem, 717
the outcome that maximizes the
    product of the players' utilities,
    719
theorem on local existence and
    uniqueness, 339
theoretical ecology, 670
theory of turbulence, 655
there exists a unique solution, 719
thermo-dynamic-like partition function,
    400
tight-binding model, 549
time-dependent Schrödinger equation,
    157
time-frequency, 332
time-frequency analysis, 620, 626
time-independent Schrödinger equation,
    319
time-lagged recurrent networks, 96
time-ordering operator, 488
time-phase plot, 642
Toda lattice, 293
topological, 199
topological defects, 184
topological insulators, 18
topological phase of matter, 519
topological properties, 14
topological quantum computer, 14
topological quantum field theories, 512
topological qubit, 16
topology, 647
toric code, 547
tornado, 260
trace formula, 270
trace operator, 597
tracking-filter prediction equations, 609
trajectory, 641, 675, 681

transcendental unity of apperception,
    36
transient chaos, 261
transient signal components, 620
transient spike inputs, 116
transition amplitude, 165
transition entropy, 400
transition functions, 590
transition matrix, 613
transition propagator, 396, 397
translation group, 194
trapped ion quantum computer, 9
trapped particles, 14
trapped quantum particles, 14
traveling wave solution, 290
turbulence, 258, 676
turbulent flow, 259
Turing bifurcation, 232
twiddle factors, 626
twisted, 201
two interpretations, 722
two player game, 443
two-person zero-sum strategic game,
    444
two-player quantum game, 447

un-damped pendulum, 684
uncertainty principle, 116
undecimated wavelet transform, 628
underdamped junction, 501
underdamped ladder array, 509
unipolar neurons, 52
unitary, 595
unitary evolution, 152
unitary matrix, 7, 594
unitary quantum evolution, 154
unitary shift operator, 344
unitary topological modular functor,
    186
universal approximation theorem, 90
universal Turing machine, 3
unpredictability, 649
unstable manifold, 673
unstable periodic orbits, 684
unsupervised, 25, 378, 412
Upanishads, 25, 30, 188
utility, 716
utility functionals, 447
utility functions, 443

vacuum state, 365
Van der Pol, 654
Van der Pol oscillator, 657, 698
variation formulas, 249
vector momentum, 334
vector-field, 423, 675
vertebrate brain, 68
virial inequality, 330
visual cortex, 44
voltage-to-frequency converter, 482
volume, 249
von Neumann, 647
von Neumann architecture, 3
von Neumann bottleneck, 4
von Neumann computer, 3
von Neumann's quantum density
    function, 166
vortex, 259
vorticity dynamics, 260

Walsh functions, 7, 630
wave d'Lambertian operator, 331
wave psi-function, 157
wave-function collapse, 39
wave-particle duality of matter, 151
wavelet, 621, 631

wavelet analysis, 620
wavelet coefficients, 633
wavelet transform, 620
wavelet transformation, 122
wavelet-based QFT, 194
weakly-connected neural network, 49,
    242
Weierstrass, 650
weights, 412
well-defined input–output function, 440
when players move, 714
whirling modes, 495
whirling regime, 497
white matter, 43
Wiener, 654
Wiener process, 394
Wigner distribution, 269, 274
Wigner function, 362
Winfree-type phase models, 498
Witten–Chern–Simons theory, 185
Witten-type, 201
working memory, 138
world lines, 14

zero-sum game, 448
Zhu–Botina–Rabitz functional, 282

Printed in the United States
By Bookmasters